For Extra Practice

P9-AEV-417

The following sections feature extra practice problems available when you register for MyMathLab.com. For your convenience, these exercises are also printed in the *Instructor's Resource Guide*. Please contact your instructor for assistance.

Elementary and Intermediate Algebra

CONCEPTS AND APPLICATIONS

Elementary and Intermediate Algebra

FOURTH EDITION

CONCEPTS AND APPLICATIONS

MARVIN L. BITTINGER
Indiana University Purdue University Indianapolis

DAVID J. ELLENBOGEN
Community College of Vermont

BARBARA L. JOHNSON
Indiana University Purdue University Indianapolis

PEARSON

Addison
Wesley

Boston   San Francisco   New York
London   Toronto   Sydney   Tokyo   Singapore   Madrid
Mexico City   Munich   Paris   Cape Town   Hong Kong   Montreal

Publisher	Greg Tobin
Editor in Chief	Maureen O'Connor
Acquisitions Editor	Jennifer Crum
Associate Editor	Katie Nopper
Editorial Assistant	Elizabeth Bernardi
Managing Editor	Ron Hampton
Production Supervisor	Kathleen A. Manley
Media Producer	Ceci Fleming
Art Editor and Photo Researcher	Geri Davis/The Davis Group, Inc.
Compositor	Beacon Publishing Services
Software Development	David Malone and Rebecca Williams
Marketing Manager	Dona Kenly
Marketing Coordinator	Tracy Rabinowitz
Prepress Supervisor	Caroline Fell
Manufacturing Buyer	Evelyn Beaton
Design Supervisor/Cover Designer	Dennis Schaefer
Text Designer	Geri Davis/The Davis Group, Inc.
Cover Photograph	Two Crocuses; © M. Botzek/Masterfile

For permission to use copyrighted material, grateful acknowledgment is made to the copyright holders on page G-8, which is hereby made part of this copyright page.

Library of Congress Cataloging-in-Publication Data
Bittinger, Marvin L.
  Elementary and intermediate algebra: concepts and applications
    Marvin L. Bittinger, David J. Ellenbogen, Barbara L. Johnson.—4th ed.
      p. cm.
  Includes index.
  ISBN 0-321-23383-2 (student ed.)
  1. Algebra—Textbooks. I. Ellenbogen, David. II. Johnson,
    Barbara L. (Barbara Loreen), 1962—III. Title.
QA152.3.B546 2006
512.9—dc22
                                                          2005042975

1 2 3 4 5 6 7 8 9 10—VH—09 08 07 06 05

For Jeff

Contents

14

Sequences, Series, and the Binomial Theorem

R

Elementary Algebra Review

Preface

It is with great pleasure that we introduce you to the fourth edition of *Elementary and Intermediate Algebra: Concepts and Applications*. Every time we work on a new edition, it's a balancing act. On the one hand, we want to preserve the features, applications, and explanations that faculty have come to rely on and expect. On the other hand, we want to blend our own ideas for improvement with the many insights that we receive from faculty and students throughout North America. The result is a living document in which new features and applications are developed at the same time that successful features and popular applications from previous editions are updated and refined. Our goal, as always, is to present content that is easy to understand and has the depth required for success in this and future courses.

Appropriate for a course, or courses, combining the study of elementary and intermediate algebra, this text covers both elementary and intermediate algebra topics without the repetition of instruction necessary in two separate texts. It is the third of three texts in an algebra series that also includes *Elementary Algebra: Concepts and Applications*, Seventh Edition, by Bittinger/Ellenbogen and *Intermediate Algebra: Concepts and Applications*, Seventh Edition, by Bittinger/Ellenbogen.

Approach

Our goal, quite simply, is to help today's students both learn and retain mathematical concepts. To achieve this goal, we feel that we must prepare developmental-mathematics students for the transition from "skills-oriented" elementary and intermediate algebra courses to more "concept-oriented" college-level mathematics courses. This requires that we teach these same students critical thinking skills: to reason mathematically, to communicate mathematically, and to identify and solve mathematical problems. Following are three aspects of our approach that we use to help meet the challenges we all face when teaching developmental mathematics.

Problem Solving

One distinguishing feature of our approach is our treatment of and emphasis on problem solving. We use problem solving and applications to motivate the material wherever possible, and we include real-life applications and problem-solving techniques throughout the text. Problem solving not only encourages students to think about how mathematics can be used, it helps to prepare them for more advanced material in future courses.

In Chapter 2, we introduce our five-step process for solving problems: (1) Familiarize, (2) Translate, (3) Carry out, (4) Check, and (5) State the answer. These steps are then used consistently throughout the text when encountering a problem-solving situation. Repeated use of this problem-solving strategy helps provide students with a starting point for any type of problem they encounter, and frees them to focus on the unique aspects of the particular problem situation. We often use estimation and carefully checked guesses to help with the *Familiarize* and *Check* steps (see pp. 115 and 420–421). We also use dimensional analysis as a quick check of certain problems (see pp. 179 and 428).

Applications

Interesting applications of mathematics help motivate both students and instructors. Solving applied problems gives students the opportunity to see their conceptual understanding put to use in a real way. In the Fourth Edition of *Elementary and Intermediate Algebra: Concepts and Applications*, we have increased the number of applications, the number of real-data problems, and the number of reference lines that specify the sources of the real-world data. As in the past, art is integrated into the applications and exercises to aid the student in visualizing the mathematics. (See pp. 116, 193, 255, 358, and 427.)

Pedagogy

New!   **CONCEPT REINFORCEMENT EXERCISES.**   This feature is designed to help students build their confidence and comprehension through true/false, matching, and fill-in-the-blank exercises at the beginning of most exercise sets. When appropriate, special attention is devoted to increasing student understanding of the new vocabulary and notation developed in that section. (See pp. 109, 154, 194, and 244.)

New!   **STUDY SKILLS.**   These remarks, located in the margin near the beginning of each section, provide suggestions for successful study habits that can be applied to both this and other college courses. Ranging from ideas for better time management to suggestions for test preparation, these comments can be useful to even experienced college students. (See pp. 8, 90, and 202.)

New!   **STUDENT NOTES.**   These comments, strategically located in the margin within each section, are specific to the mathematics appearing on that page. Remarks are more casual in format than the typical exposition and range from suggestions on how to avoid common mistakes to how to best read new mathematical notation. (See pp. 106, 191, and 317.)

New!   **STUDY SUMMARY.**   Each chapter closes with a Study Summary in which the authors have written notes that highlight the most important concepts and terminology from that chapter. Page numbers are provided so students can reference where the terminology is first introduced, important properties and formulas are listed, and examples of specific types of problems are often listed.

In short, the Study Summary provides a terrific point from which to begin reviewing for a chapter test. (See pp. 143, 298, and 368.)

New!   **THE ALGEBRAIC–GRAPHICAL CONNECTION.**   Because many of us master concepts more easily when provided with visual information, we now include a feature that offers students the opportunity to visualize algebraic concepts that might otherwise prove elusive. (See pp. 353 and 417.)

CONNECTING THE CONCEPTS.   To help students understand the "big picture," Connecting the Concepts subsections within each chapter (and highlighted in the table of contents) relate the concept at hand to previously learned and upcoming concepts. Because students may occasionally "lose sight of the forest because of the trees," this feature helps them to better keep their bearings as they encounter new material. (See pp. 348, 412, and 472.)

AHA! EXERCISES.   Designated by Aha!, these exercises can be solved quickly if the student has the proper insight. The Aha! designation is used the first time a new insight can be used on a particular type of exercise and tells the student that there is a simpler way to complete the exercise than the more routine approach. It's then up to the student to find the simpler approach and, in subsequent exercises, to determine if and when that particular insight can be used again. The Aha! exercises are not more difficult than the neighboring exercises, but are designed to reward students who "look before they leap" into a problem. (See pp. 87, 237, 280, and 404.)

TECHNOLOGY CONNECTIONS.   Throughout each chapter, optional Technology Connection boxes help students use graphing calculator technology to better visualize a concept that they have just learned. To connect this feature to the exercise sets, certain exercises are marked with a graphing calculator icon and reinforce the use of this optional technology. (See pp. 92, 262, 310, and 410.)

SKILL MAINTENANCE EXERCISES.   Retention of skills is critical to a student's success in this and future courses. To this end, beginning in Section 1.2, every exercise set includes Skill Maintenance exercises that review skills and concepts from preceding sections of the text. Whenever possible, these six to eight exercises provide extra practice with specific skills that may be "rusty" but are needed for the next section of the text. (See pp. 185, 247, and 337.)

SYNTHESIS EXERCISES.   Following the Skill Maintenance section, every exercise set ends with a group of Synthesis exercises that offer opportunities for students to synthesize skills and concepts from earlier sections with the present material, and often provide students with deeper insights into the current topic. Synthesis exercises are generally more challenging than those in the main body of the exercise set and occasionally include Aha! exercises. (See pp. 142, 200, 238, and 281.)

WRITING EXERCISES.   Every set of exercises includes at least four writing exercises. Two of these are more basic and appear just before the Skill Maintenance exercises. The other writing exercises are more challenging and appear as Synthesis exercises. All writing exercises are marked with and require answers that are one or more complete sentences. Because some instructors may collect answers to writing exercises, and because more than one answer may be correct, answers to writing exercises are listed at the back of the text only for Review Exercises. Writing exercises have been found to aid in student comprehension, critical thinking, and conceptualization. (See pp. 157, 319, 370, and 396.)

COLLABORATIVE CORNERS.   Studies have shown that students who study together generally outperform those who do not. Throughout the text, we continue to provide optional Collaborative Corner activities that require students to work in groups to explore and solve mathematical problems. There are typically two to three Collaborative Corners per chapter, each one appearing after the appropriate exercise set. (See pp. 12, 114, 158, and 320.) Additional Collaborative Corner activities and suggestions for directing collaborative learning appear in the *Printed Test Bank/Instructor's Resource Guide.*

CUMULATIVE REVIEW.   After Chapters 3, 6, 9, 12, and 14, we have included a Cumulative Review, which reviews skills and concepts from all preceding chapters of the text. (See pp. 226, 439, 630, 852, and 941.)

What's New in the Fourth Edition?

We have rewritten many key topics in response to user and reviewer feedback and have made significant improvements in design, art, pedagogy, and an expanded supplements package. Detailed information about the content changes is available in the form of a conversion guide. Please ask your local Addison-Wesley sales consultant for more information. Following is a list of the major changes in this edition.

New Design

While incorporating a new layout, a fresh palette of colors, and new features, we have maintained the larger page dimension for an open look, and a typeface that is easy to read. As always, it is our goal to make the text look mature without being intimidating. In addition, we continue to pay close attention to the pedagogical use of color to make sure that it is used to present concepts in the clearest possible manner.

Content Changes

A variety of content changes have been made throughout the text. Some of the more significant changes are listed below.

- Chapter 1 now includes more emphasis on translating English to mathematics.
- Chapter 2 includes new material on how to choose between the multiplication or addition principles when solving an equation. There is also new material on how to decide which principle to use first when solving linear equations of a more general variety.
- Chapter 3 now includes new material on how to choose an appropriate range and scale for the axes when graphing real-world data. There is also increased emphasis on visualizing the connection between rate and slope.
- Chapter 4 now makes greater use of prime factorizations as a tool for finding the largest common factor.

- Chapter 7 now includes a brief review of graphing linear equations before introducing linear functions.
- Chapter 11 now makes greater use of graphs as a way to visualize quadratic equations and functions.
- Throughout the text, there is an increased emphasis on students learning how to distinguish between equivalent expressions and equivalent equations.

Ancillaries

The following ancillaries are available to help both instructors and students use this text more effectively:

STUDENT SUPPLEMENTS	INSTRUCTOR SUPPLEMENTS
Student's Solutions Manual • By Judith A. Penna, *Indiana University Purdue University Indianapolis* • Contains completely worked-out solutions with step-by-step annotations for all the odd-numbered exercises in the text, with the exception of the writing exercises. ISBN: 0-321-28678-2	**Instructor's Solutions Manual** • By Judith A. Penna, *Indiana University Purdue University Indianapolis* • Contains fully worked-out solutions to the odd-numbered exercises and brief solutions to the even-numbered exercises in the exercise sets. ISBN: 0-321-28677-4
Videotapes • Present a series of lectures correlated directly to the content of each section of the text. • Feature an engaging team of instructors, including one of the authors, who present material using examples and exercises often from the text in a format that stresses student interaction. • New! A video guide correlated directly to the text provides quick check exercises for each video segment so students can assess their understanding. ISBN: 0-321-28680-4	**Answer Book** • By Judith A. Penna, *Indiana University Purdue University Indianapolis* • Contains answers to all the section exercises in the text. ISBN: 0-321-33555-4 **Printed Test Bank/Instructor's Resource Guide** • By Patricia A. Slipher • Contains two multiple-choice tests per chapter; six free-response tests per chapter; eight final exams; extra practice exercises on selected topics from the text; extra Collaborative Corners; correlation guide; video index; and transparency masters. ISBN: 0-321-28674-X
Digital Video Tutor • Complete set of digitized videos on CD-ROMs for student use at home or on campus. • Ideal for distance learning or supplemental instruction. • New! Each section concludes with quick check exercises and their solutions for student self-assessment. • Video guide available (see above). ISBN: 0-321-28681-2	**New! Adjunct Support Manual** • Includes resources designed to help both new and adjunct faculty with course preparation and classroom management. • Offers helpful teaching tips. ISBN: 0-321-28679-0

(continued)

STUDENT SUPPLEMENTS	INSTRUCTOR SUPPLEMENTS
Addison-Wesley Math Tutor Center • The Addison-Wesley Math Tutor Center is staffed by qualified mathematics instructors who provide students with tutoring on examples and odd-numbered exercises from the textbook. Tutoring is available via toll-free telephone, toll-free fax, e-mail, or the Internet. White Board technology allows tutors and students to actually see problems worked while they "talk" in real time over the Internet during tutoring sessions. For more information, go to www.aw-bc.com/tutorcenter **MathXL® Tutorials on CD** • Provides algorithmically generated practice exercises that correlate to exercises at the end of sections. • Every exercise is accompanied by an example and a guided solution, and selected exercises may also include a video clip. • The software recognizes student errors and provides feedback. It can also generate printed summaries of students' progress. ISBN: 0-321-28676-6	**TestGen with Quizmaster** • Enables instructors to build, edit, print, and administer tests. • Features a computerized bank of questions developed to cover all text objectives. • Available on a dual-platform Windows/Macintosh CD-ROM. ISBN: 0-321-28675-8

MathXL® www.mathxl.com

MathXL is a powerful online homework, tutorial, and assessment system that accompanies your Addison-Wesley textbook. With MathXL, instructors can create, edit, and assign online homework and tests using algorithmically generated exercises correlated to your textbook. All student work is tracked in MathXL's online gradebook. Students can take chapter tests in MathXL and receive personalized study plans based on their test results. The study plan diagnoses weaknesses and links students directly to tutorial exercises for the objectives they need to study and retest. Students can also access supplemental video clips and animations directly from selected exercises. MathXL is available to qualified adopters. For more information, visit our website at www.mathxl.com, or contact your Addison-Wesley sales representative.

MyMathLab www.mymathlab.com

MyMathLab is a series of text-specific, easily customizable online courses for Addison-Wesley textbooks in mathematics and statistics. MyMathLab is powered by CourseCompass™—Pearson Education's online teaching and learning environment—and by MathXL®—our online homework, tutorial, and assessment system. MyMathLab gives instructors the tools they need to deliver all or a portion of their course online, whether students are in a lab setting or working from home. MyMathLab provides a rich and flexible set of course materials, featuring free-response exercises that are algorithmically generated for unlimited practice and mastery. Students can also use online tools, such as video lectures, animations, and a multimedia textbook, to independently improve their understanding and performance. Instructors can use MyMathLab's homework and test managers to select and assign online exercises correlated directly to the textbook, and they can import TestGen tests into MyMathLab for added flexibility. MyMathLab's online gradebook—designed specifically for mathematics and statistics—automatically tracks students' homework and test results and gives the instructor control over how to calculate final grades. Instructors can also add offline (paper-and-pencil) grades to the gradebook. MyMathLab is available to qualified adopters. For more information, visit our Web site at www.mymathlab.com or contact your Addison-Wesley sales representative.

InterAct Math® Tutorial Web site   www.interactmath.com

Get practice and tutorial help online! This interactive tutorial Web site provides algorithmically generated practice exercises that correlate directly to the exercises in your textbook. You can retry an exercise as many times as you like with new values each time for unlimited practice and mastery. Every exercise is accompanied by an interactive guided solution that gives you helpful feedback if you enter an incorrect answer, and you can also view a worked-out sample problem that steps you through an exercise similar to the one you're working on.

Acknowledgments

No book can be produced without a team of professionals who take pride in their work and are willing to put in long hours. Judy Penna's outstanding work in organizing and preparing the printed supplements amounts to an inspection of the text that goes far beyond the call of duty and for which we are always extremely grateful. Elina Niemelä, Dawn Mulheron, Vince Koehler, Linda Wagner, and Enoch Bentley also deserve special thanks for their careful accuracy checks, well-thought-out suggestions, and uncanny eye for detail. Thanks to Patty Slipher for authoring the *Printed Test Bank*.

We are also indebted to Chris Burditt and Jann MacInnes for their many fine ideas that appear in our Collaborative Corners and Vince McGarry and Janet Wyatt for their recommendations for Teaching Tips featured in the *Annotated Instructor's Edition*.

Geri Davis, of the Davis Group, Inc., performed superb work as designer, art editor, and photo researcher, and is always a pleasure to work with. Network Graphics generated the graphs, charts, and many of the illustrations. Not only are the people at Network reliable, but they clearly take pride in their work. The many illustrations appear thanks to Jim Bryant and Bill Melvin, both of whom are artists with insights and creativity.

Our team at Addison-Wesley deserves special thanks. Associate Editor Katie Nopper and Editorial Assistant Elizabeth Bernardi helped with a variety of jobs and always in a timely way. Senior Acquisitions Editor Jenny Crum provided many fine suggestions and is superb at remaining involved and helpful throughout the project. Senior Production Supervisor Kathy Manley exhibited patience when others would have shown frustration. Designer Dennis Schaefer's willingness to listen and then creatively respond resulted in a book that is beautiful to look at. Senior Marketing Manager Dona Kenly and Marketing Coordinator Tracy Rabinowitz skillfully kept us in touch with the needs of faculty. Our Editor in Chief, Maureen O'Connor, and our Publisher, Greg Tobin, deserve credit for assembling this fine team.

We also thank the students at the Community College of Vermont and the following professors for their thoughtful reviews and insightful comments.

Prerevision Diary Reviewers

Tim Chappell, *Penn Valley Community College*
Sharon Hamsa, *Longview Community College*
Sally Keely, *Clark College*

Manuscript Reviewers

Jan Archibald, *Ventura College*
Don Brown, *Macon State College*
Anissa Florence, *University of Louisville*
Pat Horacek, *Pensacola Junior College*
Ana Leon, *Louisville Community College*
Linda Lohman, *Jefferson Community College*
Amy Petty, *South Suburban College*
Thomas Pulver, *Waubonsee Community College*
Angela Redmon, *Wenatchee Valley College*

Finally, a special thank you to all those who so generously agreed to discuss their professional use of mathematics in our chapter openers. These dedicated people all share a desire to make math more meaningful to students. We cannot imagine a finer set of role models.

M.L.B.
D.J.E.
B.L.J.

Feature Walkthrough

1 Introduction to Algebraic Expressions

AN APPLICATION

During the Northeast's electrical blackout of August 14, 2003, residents of Bloomfield, New Jersey, lost power at 4:00 P.M. One resident returned from vacation at 3:00 P.M. the following day to find the clocks in her apartment reading 8:00 A.M. At what time, and on what day, was power restored?

This problem appears as Exercise 147 in Section 1.6.

Rodney E. Johnson
TECHNICIAN
Indianapolis, Indiana

Math is one of the key elements enabling me to perform my job well. Although I often use calculators and computer software, I need a good understanding of math in order to interpret and apply the results.

Chapter Openers

Each chapter opens with a list of sections covered and a real-life application that includes a testimonial from a person in that field to show how important mathematics is in everyday problem solving. Real data often appear in these applications, in many other exercises, and in "on the job" examples similar to what students might find in the workplace.

To clear an equation of decimals, we count the greatest number of decimal places in any one number. If the greatest number of decimal places is 1, we multiply both sides by 10; if it is 2, we multiply by 100; and so on. This procedure is the same as multiplying by the least common denominator after converting the decimals to fractions.

EXAMPLE 7 Solve: $16.3 - 7.2y = -8.18$.

Student Notes

Compare the steps of Examples 4 and 7. Note that although the two approaches differ, they yield the same solution. Whenever you can use two approaches to solve a problem, try to do so, both as a check and as a valuable learning experience.

Solution The greatest number of decimal places in any one number is *two*. Multiplying by 100 will clear all decimals.

$$100(16.3 - 7.2y) = 100(-8.18) \qquad \text{Multiplying both sides by 100}$$
$$100(16.3) - 100(7.2y) = 100(-8.18) \qquad \text{Using the distributive law}$$
$$1630 - 720y = -818 \qquad \text{Simplifying}$$
$$-720y = -818 - 1630 \qquad \text{Subtracting 1630 from both sides}$$
$$-720y = -2448 \qquad \text{Combining like terms}$$
$$y = \frac{-2448}{-720} \qquad \text{Dividing both sides by } -720$$
$$y = 3.4$$

In Example 4, the same solution was found without clearing decimals. Finding the same answer in two ways is a good check. The solution is 3.4.

Student Notes

These comments, strategically located in the margin within each section, are specific to the mathematics appearing on that page. Remarks are more casual in format than the typical exposition and range from suggestions on how to avoid common mistakes to how to properly read new notation.

When denominators are different, we use the identity property of 1 and multiply to find a common denominator. Then we add, as in Example 9.

EXAMPLE 10 Add or subtract as indicated: (a) $\frac{7}{8} + \frac{5}{12}$; (b) $\frac{9}{8} - \frac{4}{5}$.

Study Skills

Do the Exercises

• Usually an instructor assigns some odd-numbered exercises. When you complete these, you can check your answers at the back of the book. If you miss any, closely examine your work, and if necessary, consult the *Student's Solutions Manual* or your instructor for guidance.

• Whether or not your instructor assigns the even-numbered exercises, try to do some on your own. There are no answers given for them, so you will gain practice doing exercises that are similar to quiz or test problems. Check your answers later with a friend or your instructor.

Solution

a) The number 24 is ... $\frac{5}{12}$ by suitable form ... of 24:

$$\frac{7}{8} + \frac{5}{12} = \frac{7}{8} \cdot \frac{3}{3}$$
$$= \frac{21}{24} + \frac{10}{2\,}$$
$$= \frac{31}{24}.$$

b) $\frac{9}{8} - \frac{4}{5} = \frac{9}{8} \cdot \frac{5}{5} - \frac{\,}{\,}$
$$= \frac{45}{40} - \frac{32}{40} =$$

Study Skills

These remarks, located in the margin near the beginning of each section, provide suggestions for successful study habits and can be extended to courses other than algebra. Ranging from ideas for better time management to how to prepare for tests, these comments can be useful to even experienced college students.

technology connection

Most graphing calculators have a TABLE feature that enables the calculator to evaluate a variable expression for different choices of x. For example, to evaluate $6x + 5 - 7x$ for $x = 0, 1, 2, \ldots$, we first use [Y=] to enter $6x + 5 - 7x$ as y_1. We then use [2ND] [TBLSET] to specify which x-values will be used. Using TblStart = 0, ΔTbl = 1, and selecting AUTO twice, we can generate a table in which the value of $6x + 5 - 7x$ is listed for values of x starting at 0 and increasing by ones.

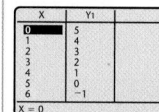

1. Create the above table on your graphing calculator. Scroll up and down to extend the table.
2. Enter $10 - 4x + 7$ as y_2. Your table should now have three columns.
3. For what x-value is y_1 the same as y_2? Compare this with the solution of Example 5(c). Is this a reliable way to solve equations? Why or why not?

$$2x - 4 = -3x + 1$$

Isolate variable terms on one side and constant terms on the other side.

$$2x + 3x - 4 = -3x + 3x + 1 \qquad \text{Adding } 3x \text{ to both sides}$$
$$5x - 4 = 1 \qquad \text{Simplifying}$$
$$5x - 4 + 4 = 1 + 4 \qquad \text{Adding 4 to both sides}$$
$$5x = 5 \qquad \text{Combining like terms}$$
$$\frac{5x}{5} = \frac{5}{5} \qquad \text{Dividing both sides by 5}$$
$$x = 1$$

Check:

$$\begin{array}{c|c} 2x - 4 = -3x + 1 \\ \hline 2 \cdot 1 - 4 & -3 \cdot 1 + 1 \\ 2 - 4 & -3 + 1 \\ -2 & \overset{?}{=} -2 \qquad \text{TRUE} \end{array}$$

The solution is 1.

c) $6x + 5 - 7x = 10 - 4x + 7$
$$-x + 5 = 17 - 4x \qquad \text{Com...}$$
$$-x + 5 + 4x = 17 - 4x + 4x \qquad \text{Add...}$$
$$5 + 3x = 17 \qquad \text{Sim...}$$
$$3x = 12 \qquad \text{Sub...}$$
$$\frac{3x}{3} = \frac{12}{3} \qquad \text{Divi...}$$
$$x = 4$$

Check:

$$\begin{array}{c|c} 6x + 5 - 7x = 10 - 4x \\ \hline 6 \cdot 4 + 5 - 7 \cdot 4 & 10 - 4 \\ 24 + 5 - 28 & 10 - 16 \\ 1 & \overset{?}{=} 1 \end{array}$$

The solution is 4.

d) $2 - 5(x + 5) = 3(x - 2) - 1$
$$2 - 5x - 25 = 3x - 6 - 1 \qquad \text{Using ... simila...}$$
$$-5x - 23 = 3x - 7 \qquad \text{Combining like terms on each side}$$
$$\left. \begin{array}{l} -5x - 23 + 7 = 3x \\ -23 + 7 = 3x + 5x \end{array} \right\} \quad \begin{array}{l} \text{Adding 7 and } 5x \text{ to both sides. This} \\ \text{isolates the } x\text{-terms on one side and the} \\ \text{constant terms on the other.} \end{array}$$
$$-16 = 8x \qquad \text{Simplifying}$$
$$\frac{-16}{8} = \frac{8x}{8} \qquad \text{Dividing both sides by 8}$$

Technology Connections

Optional Technology Connections appear throughout each chapter to help students visualize, through the use of technology, a concept that they have just learned. This feature is reinforced in many exercise sets through exercises marked with a graphing calculator icon.

$$\frac{x^2}{x-1} = \frac{1}{x-1}$$ We cannot have $x = 1$.

$$(x-1) \cdot \frac{x^2}{x-1} = (x-1) \cdot \frac{1}{x-1}$$ Multiplying both sides by $x-1$

$$x^2 = 1$$ Removing a factor eq $(x-1)/(x-1) = 1$

$$x^2 - 1 = 0$$ Subtracting 1 from b

$$(x-1)(x+1) = 0$$ Factoring

$$x - 1 = 0 \quad or \quad x + 1 = 0$$ Using the principle o products

$$x = 1 \quad or \quad x = -1.$$ Above, we stated that $x \neq 1$.

Because of the above restriction, 1 must be rejected as a soluti dent should check in the original equation that -1 *does* check. T is -1.

A Visual Interpretation

 ALGEBRAIC—GRAPHICAL CONNECTION

We can obtain a visual check of the solutions of a rational equation by graphing. For example, consider the equation

$$\frac{x}{4} + \frac{x}{2} = 6.$$

We can examine the solution by graphing the equations

$$y = \frac{x}{4} + \frac{x}{2} \quad and \quad y = 6$$

using the same set of axes, as shown at left.

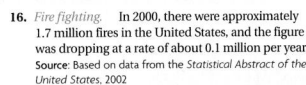

The y-values for each equation will be the same where the graphs intersect. The x-value of that point will yield that value, so it will be the solution of the equation. It appears from the graph that when $x = 8$, the value of $x/4 + x/2$ is 6. We can check by substitution:

$$\frac{x}{4} + \frac{x}{2} = \frac{8}{4} + \frac{8}{2} = 2 + 4 = 6.$$

Thus the solution is 8.

3.4 RATES **183**

16. *Fire fighting.* In 2000, there were approximately 1.7 million fires in the United States, and the figure was dropping at a rate of about 0.1 million per year.
Source: Based on data from the *Statistical Abstract of the United States*, 2002

17. *Train travel.* At 3:00 P.M., the Boston–Washington Metroliner had traveled 230 mi and was cruising at a rate of 90 miles per hour.

18. *Plane travel.* At 4:00 P.M., the Seattle–Los Angeles shuttle had traveled 400 mi and was cruising at a rate of 300 miles per hour.

19. *Wages.* By 2:00 P.M., Diane had earned $50. She continued earning money at a rate of $15 per hour.

20. *Wages.* By 3:00 P.M., Arnie had earned $70. He continued earning money at a rate of $12 per hour.

21. *Telephone bills.* Roberta's phone bill was already $7.50 when she made a call for which she was charged at a rate of $0.10 per minute.

22. *Telephone bills.* At 3:00 P.M., Larry's phone bill was $6.50 and increasing at a rate of 7¢ per minute.

In Exercises 23–32, use the graph provided to calculate a rate of change in which the units of the horizontal axis are used in the denominator.

23. *Hairdresser.* Eve's Custom Cuts has a graph displaying data from a recent day of work. At what rate does Eve work?

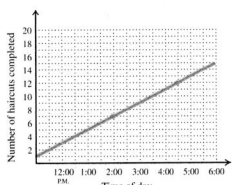

24. *Manicures.* The following graph shows data from a recent day's work at the O'Hara School of Cosmetology. At what rate do they work?

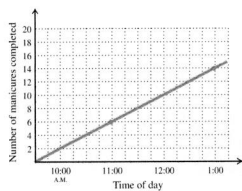

25. *Train travel.* The following graph shows data from a recent train ride from Chicago to St. Louis. At what rate did the train travel?

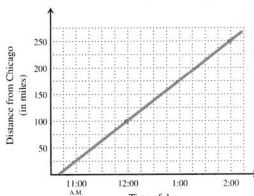

26. *Train travel.* The following graph shows data from a recent train ride from Denver to Kansas City. At what rate did the train travel?

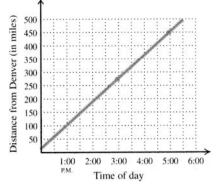

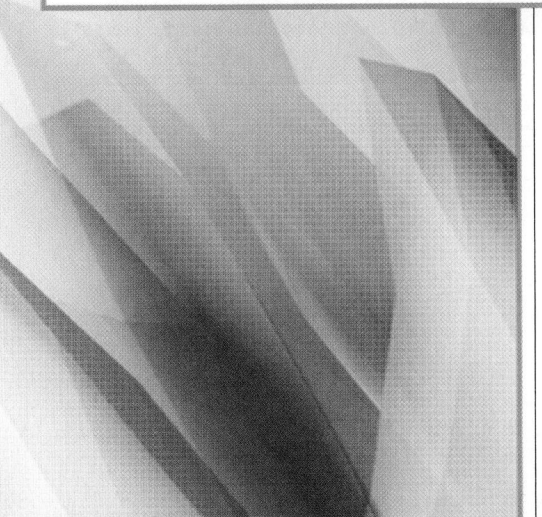

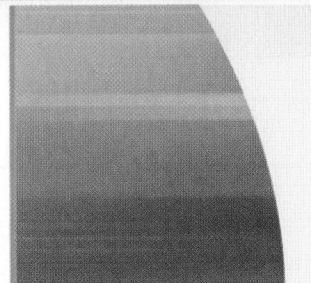

NEW!

Algebraic-Graphical Connection

To provide visual understanding of algebra, Algebraic-Graphical Connections are included as appropriate. This feature connects the algebra to a graphical interpretation.

Real-Data Applications

Real-data applications have always been a strength of this text, helping students to see realistic uses of data and math in the world around them.

Connecting the Concepts

This feature highlights the importance of connecting concepts and invites students to pause and check that they understand the "big picture." This helps assure that students understand how concepts work together in several sections at once.

NEW!

CONNECTING THE CONCEPTS

In this section, we have defined several types of functions and described some of their characteristics. They are listed together below for easy reference.

LINEAR FUNCTION

$f(x) = mx + b, m \neq 0$
Graph: straight line;
Domain: $\mathbb{R}$;
Range: $\mathbb{R}$

CONSTANT FUNCTION

$f(x) = b$
Graph: horizontal line;
Domain: $\mathbb{R}$;
Range: $\{b\}$

ABSOLUTE–VALUE FUNCTION

Described by an absolute-value equation
Domain: $\mathbb{R}$

QUADRATIC FUNCTION

$p(x) = ax^2 + bx + c, a \neq 0$
Domain: $\mathbb{R}$

POLYNOMIAL FUNCTION

Described by a polynomial equation
Domain: $\mathbb{R}$

RATIONAL FUNCTION

Described by a rational equation
Domain consists of all real numbers except those that make a denominator 0

Exercise Set

3.6

FOR EXTRA HELP

Student's Solutions Manual | Digital Video Tutor CD 2 Videotape 6 | AW Math Tutor Center | MathXL, Tutor on CD

Concept Reinforcement In each of Exercises 1–6, match the phrase with the most appropriate choice from the column on the right.

1. ___ The y-intercept of the graph of $y = 2x - 3$ a) $(0,$
2. ___ The y-intercept of the graph of $y = 3x - 2$ b) 2
3. ___ The slope of the graph of $y = 3x - 2$ c) $(0,$
4. ___ The slope of the graph of $y = 2x - 3$ d) $\frac{2}{3}$

Concept Reinforcement Exercises

This feature is designed to help students build their confidence and comprehension through true/false, matching, and fill-in-the-blank exercises at the start of nearly all exercise sets. Where appropriate, special attention is devoted to increasing student understanding of the new vocabulary and notation developed in that section.

Exercises

SKILL MAINTENANCE EXERCISES
Skill Maintenance exercises appear in all exercise sets as a means of keeping past concepts fresh and as a way of reviewing previously covered skills to prepare students for upcoming topics.

SYNTHESIS EXERCISES Synthesis exercises guarantee an extensive and wide-ranging variety of problems in every exercise set. The Synthesis exercises allow students to combine concepts from more than one section and provide challenge for even the strongest students.

WRITING EXERCISES Writing exercises, indicated by , provide opportunities for students to answer problems with one or more sentences. Often, these questions have more than one correct response and ask students to explain *why* a certain concept works as it does.

100. Jake keys $18/2 \cdot 3$ into his calculator and expects the result to be 3. What mistake is he probably making?

SKILL MAINTENANCE

Translate to an algebraic expression. [1.1]

101. Nine more than twice a number
102. Half of the sum of two numbers

SYNTHESIS

103. Write the sentence $(-x)^2 \neq -x^2$ in words. Explain why $(-x)^2$ and $-x^2$ are not equivalent.

104. Write the sentence $-|x| \neq -x$ in words. Explain why $-|x|$ and $-x$ are not equivalent.

Simplify.

105. $5t - \{7t - [4r - 3(t - 7)] + 6r\} - 4r$

106. $z - \{2z - [3z - (4z - 5z) - 6z] - 7z\} - 8z$

107. $\{x - [f - (f - x)] + [x - f]\} - 3x$

108. Is it true that for all real numbers a and b,
$ab = (-a)(-b)$?
Why or why not?

109. Is it true that for all real numbers a, b, and c,
$a|b - c| = ab - ac$?
Why or why not?

If $n > 0$, $m > 0$, and $n \neq m$, classify each of the following as either true or false.

110. $-n + m = -(n + m)$
111. $m - n = -(n - m)$
112. $n(-n - m) = -n^2 + nm$
113. $-m(n - m) = -(mn + m^2)$
114. $-n(-n - m) = n(n + m)$

Evaluate.

115. $[x + 3(2 - 5x) \div 7 + x](x - 3)$, for $x = 3$

116. $[x + 2 \div 3x] \div [x + 2 \div 3x]$, for $x = -7$

117. In Mexico, between 500 B.C. and 600 A.D., the Mayans represented numbers using powers of 20 and certain symbols. For example, the symbols

represent $4 \cdot 20^3 + 17 \cdot 20^2 + 10 \cdot 20^1 + 0 \cdot 20^0$. Evaluate this number.
Source: National Council of Teachers of Mathematics, 1906 Association Drive, Reston, VA 22091

118. Examine the Mayan symbols and the numbers in Exercise 117. What numbers do

●, ⬭, and ⬤

each represent?

119. Calculate the volume of the tower shown below.

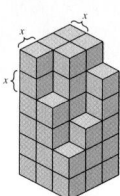

AHA! EXERCISES In many exercise sets, students will see the **Aha!** icon. This icon indicates that there is a simpler way to complete the exercise without going through a lengthy computation. It's then up to the student to discover that simpler approach. The **Aha!** icon appears the first time a new insight can be used on a particular type of exercise. After that, the student must determine if and when that particular insight can be reused.

3 Study Summary

Information can be communicated quickly and effectively in a visual form through the use of graphs. **Bar graphs, circle graphs,** and **line graphs** are three of the most commonly used types of graphs (pp. 148–150)

When information is given as **ordered pairs**, like $(-3, 2)$ or $(5, 17)$, the pairs can be **plotted** or **graphed** using a **coordinate system** that uses two **axes**, which are most often labeled x and y. The axes intersect at the **origin**, $(0, 0)$ (p. 151).

When an equation contains two variables, any solution is an ordered pair. When all ordered pairs that are solutions of an equation are graphed, we say that we have **graphed the equation** (p. 160). Some equations are said to be **linear**, like $y = 2x - 7$ or $2x + 3y = 12$, because their graphs are lines (p. 160). Other equations have **nonlinear graphs** (p. 165). When a graph crosses the y-axis, the point at which the axis is crossed is called a **y-intercept.** Similarly, an **x-intercept** is any point at which a graph crosses the x-axis (p. 169).

A line's slant is measured as its **slope.** Slope measures the **rate** at which the quantity measured on the vertical axis changes with respect to the quantity measured on the horizontal axis (p. 190).

$$\text{Slope} = m = \frac{\text{change in } y}{\text{change in } x} = \frac{\text{rise}}{\text{run}} = \frac{y_2 - y_1}{x_2 - x_1}$$

Any equation that can be written in **standard form**, $Ax + By = C$, is linear (p. 160). An equation in **slope–intercept form**, $y = mx + b$, represents a line with slope m and y-intercept $(0, b)$ (p. 203). An equation in **point–slope form**, $y - y_1 = m(x - x_1)$, represents a line with slope m passing through (x_1, y_1) (...)

Parallel lines have the same slope. **Perpendicular lines** ... reciprocals; the product of the slopes is -1 (pp. 206, 207...)

NEW!

Study Summary

Each chapter closes with a Study Summary in which the authors provide study notes highlighting the most important concepts and terminology from that chapter. Page numbers are provided to reference where the concepts and terminology first appear.

1–6 Cumulative Review

1. Use the commutative law of addition to write an expression equivalent to $a + 2b$. [1.2]

2. Write a true sentence using either $<$ or $>$:
$-3.1 \;\square\; -3.15$. [1.4]

3. Evaluate $(y - 1)^2$ for $y = -4$. [1.8]

4. Simplify: $-4[2(x - 3) - 1]$. [1.8]

Simplify.

5. $-\frac{1}{2} + \frac{3}{8} + (-6) + \frac{3}{4}$ [1.5]

6. $-\frac{72}{108} \div \left(-\frac{2}{3}\right)$ [1.7]

7. $-6.262 \div 1.01$ [1.7]

8. $4 \div (-2) \cdot 2 + 3 \cdot 4$ [1.8]

Solve.

9. $5(x - 2) = 40$ [2.2]

10. $49 = x^2$ [5.6]

11. $-6t = 20$ [2.1]

12. $3x - 4 = 7x + 8$ [2.2]

13. $4(y - 5) = -2(y + 2)$ [2.2]

14. $x^2 + 11x + 10 = 0$ [5.7]

15. $\frac{4}{9}t + \frac{2}{3} = \frac{1}{3}t - \frac{2}{9}$ [2.2]

16. $\frac{4}{x} + x = 5$ [6.6]

17. $6 - y \geq 2y + 8$ [2.6]

18. $\frac{2}{x - 3} = \frac{5}{3x + 1}$ [6.6]

19. $2x^2 + 7x = 4$ [5.7]

20. $4(x + 7) < 5(x - 3)$ [2.6]

21. $\frac{t^2}{t + 5} = \frac{25}{t + 5}$ [6.6]

22. $(2x + 7)(x - 5) = 0$ [5.7]

23. $\frac{2}{x^2 - 9} + \frac{5}{x - 3} = \frac{3}{x + 3}$ [6.6]

Solve each formula. [2.3]

24. $3a - b + 9 = c$, for b

25. $\frac{3}{4}(x + 2y) = z$, for y

Combine like terms. [1.6]

26. $x + 2y - 2z + \frac{1}{2}x - z$

27. $2x^3 - 7 + \frac{3}{7}x^2 - 6x^3 - \frac{4}{7}x^2 + 5$

Graph. [3.2], [3.3]

28. $y = \frac{3}{4}x + 5$

29. $x = -3$

30. $4x + 5y = 20$

31. $y = 6$

32. Find the slope of the line containing the points $(1, 5)$ and $(2, 3)$. [3.5]

33. Find the slope and the y-intercept of the line given by $3x - 6y = 12$. [3.6]

34. Write the slope–intercept equation for the line with slope -3 that contains the point $(2, -5)$. [3.7]

35. A large truck traveling 55 mph gets 7.6 miles per gallon of fuel. The same truck traveling 70 mph gets 6.1 miles per gallon.
Source: Kenworth Truck Co.
a) Graph the data and determine an equation that describes the miles per gallon y in terms of the speed x. [3.7]
b) Use the equation to estimate the miles per gallon for a truck traveling 60 mph. [3.7]

Simplify.

36. $\frac{x^{-5}}{x^{-3}}$ [4.8]

37. $y^2 \cdot y^{-8}$ [4.8]

38. $-(2a^2b^7)^2$ [4.1]

39. Subtract: [4.3]
$(-8y^2 - y + 2) - (y^3 - 6y^2 + y - 5)$.

Multiply.

40. $5(3a - 2b + c)$ [1.2]

41. $(2x^2 - 1)(x^3 + x - 3)$ [4.4]

42. $(6x - 5y)^2$ [4.5]

43. $(x + 8)(2x - 5)$ [4.5]

44. $(2x^3 + 1)(2x^3 - 1)$ [4.5]

Factor.

45. $6x - 2x^2 - 24x^4$ [5.1]

46. $16x^2 - 81$ [5.4]

Cumulative Review

A Cumulative Review is found after Chapters 3, 6, 9, 12, and 14 to help students review skills and concepts from all preceding chapters of the text. The final Cumulative Review is an excellent guide in preparing for a final exam.

1

Introduction to Algebraic Expressions

AN APPLICATION

During the Northeast's electrical blackout of August 14, 2003, residents of Bloomfield, New Jersey, lost power at 4:00 P.M. One resident returned from vacation at 3:00 P.M. the following day to find the clocks in her apartment reading 8:00 A.M. At what time, and on what day, was power restored?

This problem appears as Exercise 147 in Section 1.6.

Rodney E. Johnson
TECHNICIAN
Indianapolis, Indiana

Math is one of the key elements enabling me to perform my job well. Although I often use calculators and computer software, I need a good understanding of math in order to interpret and apply the results.

*P**roblem solving is the focus of this text. Chapter 1 presents important preliminaries that are needed for the problem-solving approach that is developed in Chapter 2 and used throughout the rest of the book. These preliminaries include a review of arithmetic, a discussion of real numbers and their properties, and an examination of how real numbers are added, subtracted, multiplied, divided, and raised to powers.*

1.1 Introduction to Algebra

Algebraic Expressions • Translating to Algebraic Expressions •
Translating to Equations

This section introduces some basic concepts and expressions used in algebra. Solving real-world problems is an important part of algebra, so we will focus on the wordings and mathematical expressions that often arise in applications.

Algebraic Expressions

Probably the greatest difference between arithmetic and algebra is the use of *variables* in algebra. When a letter can be any one of a set of numbers, that letter is a **variable**. For example, if n represents the number of tickets purchased for a Radiohead concert, then n will vary, depending on factors like price and day of the week. This makes n a variable. If each ticket costs \$25, then 3 tickets cost $25 \cdot 3$ dollars, 4 tickets cost $25 \cdot 4$ dollars, and n tickets cost $25 \cdot n$ dollars. Note that $25 \cdot n$ means 25 *times n*. The number 25 is an example of a **constant** because it does not change.

Price per Ticket (in dollars)	Number of Tickets Purchased	Total Paid (in dollars)
25	n	$25 \cdot n$

The expression $25 \cdot n$ is a **variable expression** because its value varies with the replacement for n. In this case, the total amount paid, $25 \cdot n$, will change with the number of tickets purchased. In the following chart, we replace n with a variety of values and compute the total amount paid. In doing so, we are **evaluating the expression** $25 \cdot n$.

Price per Ticket (in dollars), 25	Number of Tickets Purchased, *n*	Total Paid (in dollars), 25 · *n*
25	400	$10,000
25	500	12,500
25	600	15,000

Variable expressions are examples of *algebraic expressions*. An **algebraic expression** consists of variables and/or numerals, often with operation signs and grouping symbols. Examples are

$$t + 97, \quad 5 \cdot x, \quad 3a - b, \quad 18 \div y, \quad \frac{9}{7}, \quad \text{and} \quad 4r(s + t).$$

Recall that a fraction bar is a division symbol: $\frac{9}{7}$, or 9/7, means $9 \div 7$. Similarly, multiplication can be written in several ways. For example, "5 times x" can be written as $5 \cdot x, 5 \times x, 5(x)$, or simply $5x$. On many calculators, this appears as $5 * x$.

To **evaluate** an algebraic expression, we substitute a number for each variable in the expression. This replaces each variable with a number. We then calculate the result.

EXAMPLE 1   Evaluate each expression for the given values.

a) $x + y$ for $x = 37$ and $y = 28$

b) $5ab$ for $a = 2$ and $b = 3$

Solution

a) We substitute 37 for x and 28 for y and carry out the addition:

$$x + y = 37 + 28 = 65.$$

The number 65 is called the **value** of the expression.

b) We substitute 2 for a and 3 for b and multiply:

$$5ab = 5 \cdot 2 \cdot 3 = 10 \cdot 3 = 30. \qquad 5ab \text{ means 5 times } a \text{ times } b.$$

EXAMPLE 2   The area A of a rectangle of length l and width w is given by the formula $A = lw$. Find the area when l is 17 in. and w is 10 in.

Solution   We evaluate, using 17 in. for l and 10 in. for w, and carry out the multiplication:

$$A = lw = (17 \text{ in.})(10 \text{ in.})$$
$$= (17)(10)(\text{in.})(\text{in.})$$
$$= 170 \text{ in}^2, \text{ or } 170 \text{ square inches.}$$

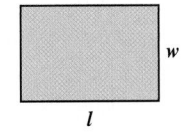

Note that we always use square units for area and $(\text{in.})(\text{in.}) = \text{in}^2$. Exponents like the 2 within the expression in^2 are discussed further in Section 1.8.

EXAMPLE 3   The area of a triangle with a base of length b and a height of length h is given by the formula $A = \frac{1}{2}bh$. Find the area when b is 8 m and h is 6.4 m.

Solution   We substitute 8 m for b and 6.4 m for h and then multiply:

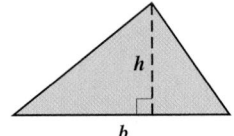

$$A = \tfrac{1}{2}bh = \tfrac{1}{2}(8\text{ m})(6.4\text{ m})$$
$$= \tfrac{1}{2}(8)(6.4)(\text{m})(\text{m})$$
$$= 4(6.4)\text{ m}^2$$
$$= 25.6\text{ m}^2, \text{ or } 25.6 \text{ square meters.}$$

Translating to Algebraic Expressions

Before attempting to translate problems to equations, we need to be able to translate certain phrases to algebraic expressions.

Important Words	Sample Phrase or Sentence	Translation
Addition (+)		
added to	700 lb was added to the car's weight.	$w + 700$
sum of	The sum of a number and 12	$n + 12$
plus	53 plus some number	$53 + x$
more than	8 more than Biloxi's population	$p + 8$
increased by	Jake's original guess, increased by 4	$n + 4$
Subtraction (−)		
subtracted from	2 oz was subtracted from the bag's weight.	$w - 2$
difference of	The difference of two scores	$m - n$
minus	A team of size s, minus 2 injured players	$s - 2$
less than	9 less than the given population	$p - 9$
decreased by	The car's speed, decreased by 8 mph	$s - 8$
Multiplication (·)		
multiplied by	The number of reservations, multiplied by 3	$r \cdot 3$
product of	The product of two numbers	$m \cdot n$
times	5 times the dog's weight	$5w$
twice	Twice Jackie's age	$2a$
of	$\frac{1}{2}$ of Amelia's salary	$\frac{1}{2}s$
Division (÷)		
divided by	A 2-lb coffee cake, divided by 3	$2 \div 3$
quotient of	The quotient of 14 and 7	$14 \div 7$
divided into	4 divided into the delivery fee	$f \div 4$
ratio of	The ratio of $500 to the cost of a new car	$500/n$
per	There were 18 models per class of size s.	$18/s$

EXAMPLE 4   Translate each phrase to an algebraic expression.

   a) Four less than Jean's height, in inches

   b) Eighteen more than a number

   c) A day's pay, in dollars, divided by eight

Solution   To help think through a translation, we sometimes begin with a specific number in place of a variable.

   a) If the height were 60, then 4 less than 60 would mean $60 - 4$. If the height were 70, the translation would be $70 - 4$. If we use h to represent "Jean's height, in inches," the translation of "Four less than Jean's height, in inches" is $h - 4$.

   b) If we knew the number to be 10, the translation would be $10 + 18$, or $18 + 10$. If we use t to represent "a number," the translation of "Eighteen more than a number" is

$$t + 18, \quad \text{or} \quad 18 + t.$$

   c) We let d represent "a day's pay, in dollars." If the pay were \$78, the translation would be $78 \div 8$, or $\frac{78}{8}$. Thus our translation of "a day's pay, in dollars, divided by eight" is

$$d \div 8, \quad \text{or} \quad \frac{d}{8}.$$

Caution!   The order in which we subtract and divide affects the answer! Answering $4 - h$ or $8 \div d$ in Examples 4(a) and 4(c) is incorrect.

EXAMPLE 5   Translate each of the following.

   a) Kate's age increased by five

   b) Half of some number

   c) Three more than twice a number

   d) Six less than the product of two numbers

   e) Forty-seven percent of the town's population

Solution

Phrase	*Algebraic Expression*
a) Kate's age increased by five	$a + 5$, or $5 + a$
b) Half of some number	$\frac{1}{2}t$, or $\frac{t}{2}$, or $t/2$, or $t \div 2$
c) Three more than twice a number	$2x + 3$, or $3 + 2x$
d) Six less than the product of two numbers	$mn - 6$
e) Forty-seven percent of the town's population	47% of p, or $0.47p$

Translating to Equations

The symbol = ("equals") indicates that the expressions on either side of the equals sign represent the same number. An **equation** is a number sentence with the verb =. Equations may be true, false, or neither true nor false.

EXAMPLE 6   Determine whether each equation is true, false, or neither.

a) $8 \cdot 4 = 32$ b) $7 - 2 = 4$ c) $x + 6 = 13$

Solution

a) $8 \cdot 4 = 32$ The equation is *true*.

b) $7 - 2 = 4$ The equation is *false*.

c) $x + 6 = 13$ The equation is *neither* true nor false, because we do not know what number x represents.

Solution

A replacement or substitution that makes an equation true is called a *solution*. Some equations have more than one solution, and some have no solution. When all solutions have been found, we have *solved* the equation.

To see if a number is a solution, we evaluate all expressions in the equation. If the values on both sides of the equation are the same, the number is a solution.

EXAMPLE 7   Determine whether 7 is a solution of $x + 6 = 13$.

Solution   We evaluate $x + 6$ and compare both sides of the equation:

$$\begin{array}{r|l}
x + 6 = 13 & \text{Writing the equation} \\
7 + 6 \mid 13 & \text{Substituting 7 for } x \\
13 \stackrel{?}{=} 13 & 13 = 13 \text{ is TRUE.}
\end{array}$$

Since the left-hand and the right-hand sides are the same, 7 is a solution.

Although we do not study solving equations until Chapter 2, we can translate certain problem situations to equations now. The words "is the same as," "equal," "is," and "are" translate to "=."

Words indicating equality, **=** : "is the same as," "equal," "is," "are"

EXAMPLE 8   Translate the following problem to an equation.

What number plus 478 is 1019?

Solution   We let *y* represent the unknown number. The translation then comes almost directly from the English sentence.

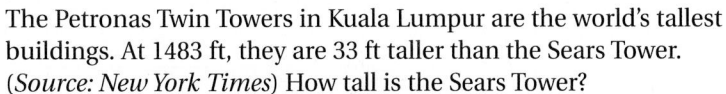

Note that "plus" translates to "+" and "is" translates to "=."

Sometimes it helps to reword a problem before translating.

EXAMPLE 9   Translate the following problem to an equation.

The Petronas Twin Towers in Kuala Lumpur are the world's tallest buildings. At 1483 ft, they are 33 ft taller than the Sears Tower. (*Source: New York Times*) How tall is the Sears Tower?

Solution   We let *h* represent the height, in feet, of the Sears Tower. A rewording and translation follow:

Rewording: The height of the   33 ft   the height of the
                    Twin Towers   is more than   Sears Tower.

Translating: 1483   = $h + 33$

---

technology connection

Technology Connections are activities that make use of the graphing calculator as a tool for learning algebra. These activities use only basic features that are common to most graphing calculators. (**Henceforth in this text we will refer to all graphing utilities as graphing calculators.**) In some cases, students may find the user's manual for their particular calculator helpful for exact keystrokes.

Although all graphing calculators are not the same, most share the following characteristics.

Screen.   The large screen can show graphs and tables as well as the keystrokes used. The screen has a different layout for different functions. Computations are performed in the **home screen**. On many calculators, the home screen is accessed by pressing **2ND** (QUIT). The **cursor** shows location on the screen, and the **contrast** determines how dark the characters appear.

Keypad.   There are options written above the keys as well as on them. To access those above the keys, we press **2ND** or **ALPHA** and then the key. Expressions are usually entered as they would appear in print. For example, to evaluate $3xy + x$ for $x = 65$ and $y = 92$, we press 3 (×) 65 (×) 92 (+) 65 and then **ENTER**. The value of the expression, 18005, will appear at the right of the screen.

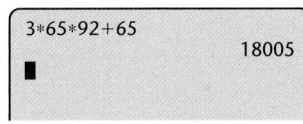

Evaluate each of the following.

1. $27a - 18b$, for $a = 136$ and $b = 13$
2. $19xy - 9x + 13y$, for $x = 87$ and $y = 29$

Study Skills _____

Get the Facts

Throughout this textbook, you will find a feature called Study Skills. These tips are intended to help improve your math study skills. On the first day of class, you should complete this chart.

Instructor: Name _____

Office hours and location _____

Phone number _____

Fax number _____

E-mail address _____

Find the names of two students whom you could contact for information or study questions:

1. Name _____

   Phone number _____

   E-mail address _____

2. Name _____

   Phone number _____

   E-mail address _____

Math lab on campus:

Location _____

Hours _____

Phone _____

Tutoring:

Campus location _____

Hours _____

Addison-Wesley Tutor Center _____

To order, call _____

(See the preface for important information concerning this tutoring.)

Important supplements:

(See the preface for a complete list of available supplements.)

Supplements recommended by the instructor.

_____

_____

Exercise Set

1.1

🔁 *Concept Reinforcement   Classify each of the following as either an expression or an equation.*

1. $4x + 7$

2. $3x = 21$

3. $2x - 5 = 9$

4. $8x - 3$

5. $38 = 2t$

6. $45 = a - 1$

7. $4a - 5b$

8. $3t + 4 = 19$

9. $2x - 3y = 8$

10. $12 - 4xy$

11. $7 - 4rt$

12. $9a + b$

Evaluate.

13. $4a$, for $a = 9$

14. $\dfrac{x - y}{6}$, for $x = 23$ and $y = 5$

15. $8x$, for $x = 7$

16. $t + 8$, for $t = 2$

17. $\dfrac{a}{b}$, for $a = 45$ and $b = 9$

18. $13 - r$, for $r = 9$

19. $\dfrac{x + y}{4}$, for $x = 2$ and $y = 14$

20. $\dfrac{m}{n}$, for $m = 54$ and $n = 9$

21. $\dfrac{p + q}{7}$, for $p = 15$ and $q = 20$

22. $\dfrac{9m}{q}$, for $m = 6$ and $q = 18$

23. $\dfrac{m - n}{2}$, for $m = 20$ and $n = 8$

24. $\dfrac{5z}{y}$, for $z = 9$ and $y = 15$

Substitute to find the value of each expression.

25. *Hockey.* The area of a rectangle with base b and height h is bh. A regulation hockey goal is 6 ft wide and 4 ft high. Find the area of the opening.

26. *Orbit time.* A communications satellite orbiting 300 mi above the earth travels about 27,000 mi in one orbit. The time, in hours, for an orbit is

$$\frac{27{,}000}{v},$$

where v is the velocity, in miles per hour. How long will an orbit take at a velocity of 1125 mph?

27. *Zoology.* A great white shark has triangular teeth. Each tooth measures about 5 cm across the base and has a height of 6 cm. Find the surface area of the front side of one such tooth. (See Example 3.)

28. *Work time.* Enrico takes five times as long to do a job as Rosa does. Suppose t represents the time it takes Rosa to do the job. Then $5t$ represents the time it takes Enrico. How long does it take Enrico if Rosa takes **(a)** 30 sec? **(b)** 90 sec? **(c)** 2 min?

29. *Olympic softball.* A softball player's batting average is h/a, where h is the number of hits and a is the number of "at bats." In the 2004 Summer Olympics, Laura Berg had 7 hits in 19 at bats. What was her batting average? Round to the nearest thousandth.

30. *Area of a parallelogram.* The area of a parallelogram with base b and height h is bh. Find the area of the parallelogram when the height is 6 cm (centimeters) and the base is 7.5 cm.

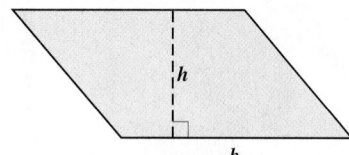

Translate to an algebraic expression.

31. 5 more than Ron's age

32. The product of 4 and a

33. 6 more than b

34. 7 more than Lou's weight

35. 9 less than c

36. 4 less than d

37. 6 increased by q

38. 11 increased by z

39. 9 times Phil's speed

40. c more than d

41. x less than y

42. 2 less than Lorrie's age

43. x divided by w

44. The quotient of two numbers

45. m subtracted from n

46. p subtracted from q

47. The sum of the box's length and height

48. The sum of d and f

49. The product of 9 and twice m

50. Penny's speed minus twice the wind speed

51. One quarter of some number

52. One third of the sum of two numbers

53. 64% of the women attending

54. 38% of a number

Determine whether the given number is a solution of the given equation.

55. 15; $x + 17 = 32$

56. 75; $y + 28 = 93$

57. 93; $a - 28 = 75$

58. 12; $8t = 96$

59. 63; $\dfrac{t}{7} = 9$

60. 52; $\dfrac{x}{8} = 6$

61. 3; $\dfrac{108}{x} = 36$

62. 7; $\dfrac{94}{y} = 12$

Translate each problem to an equation. Do not solve.

63. What number added to 73 is 201?

64. Seven times what number is 1596?

65. When 42 is multiplied by a number, the result is 2352. Find the number.

66. When 345 is added to a number, the result is 987. Find the number.

67. *Chess.* A chess board has 64 squares. If pieces occupy 19 squares, how many squares are unoccupied?

68. *Hours worked.* A carpenter charges $25 an hour. How many hours did she work if she billed a total of $53,400?

69. *Recycling.* Currently, Americans recycle or compost 27% of all municipal solid waste. This is the same as recycling or composting 56 million tons. What is the total amount of waste generated?

70. *Travel to work.* In the Northeast, the average commute to work is 24.5 min. The average commuting time in the West is 1.8 min less. How long is the average commute in the West?

In each of Exercises 71–78, match the phrase or sentence with the appropriate expression or equation from the column on the right.

71. _____ Twice the sum of two numbers

a) $\dfrac{x}{y} + 6$

72. _____ Five less than a number is nine.

b) $2(x + y) = 48$

73. _____ Two more than a number is five.

c) $\dfrac{1}{2} \cdot a \cdot b$

74. _____ Half of the product of two numbers

d) $t + 2 = 5$

75. _____ Three times the sum of a number and five

e) $ab - 1 = 49$

76. _____ Twice the sum of two numbers is 48.

f) $2(m + n)$

77. _____ One less than the product of two numbers is 49.

g) $3(t + 5)$

78. _____ Six more than the quotient of two numbers

h) $x - 5 = 9$

To the student and the instructor: Writing exercises, denoted by 📝, should be answered using one or more English sentences. Because answers to many writing exercises will vary, solutions are not listed in the answers at the back of the book.

📝 **79.** What is the difference between a variable, a variable expression, and an equation?

📝 **80.** What does it mean to evaluate an algebraic expression?

SYNTHESIS

To the student and the instructor: Synthesis exercises are designed to challenge students to extend the concepts or skills studied in each section. Many synthesis exercises will require the assimilation of skills and concepts from several sections.

📝 **81.** If the lengths of the sides of a square are doubled, is the area doubled? Why or why not?

📝 **82.** Write a problem that translates to $1998 + t = 2006$.

83. Signs of Distinction charges $90 per square foot for handpainted signs. The town of Belmar commissioned a triangular sign with a base of 3 ft and a height of 2.5 ft. How much will the sign cost?

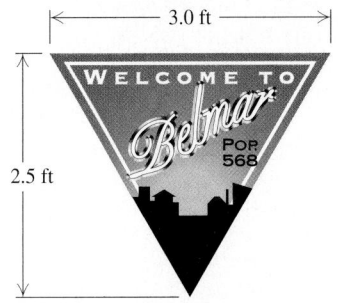

84. Find the area that is shaded.

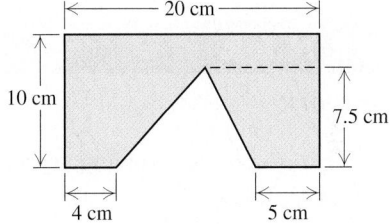

85. Evaluate $\dfrac{x - y}{3}$ when x is twice y and $x = 12$.

86. Evaluate $\dfrac{x + y}{2}$ when y is twice x and $x = 6$.

87. Evaluate $\dfrac{a + b}{4}$ when a is twice b and $a = 16$.

88. Evaluate $\dfrac{a - b}{3}$ when a is three times b and $a = 18$.

Answer each question with an algebraic expression.

89. If $w + 3$ is a whole number, what is the next whole number after it?

90. If $d + 2$ is an odd number, what is the preceding odd number?

Translate to an algebraic expression.

91. The perimeter of a rectangle with length l and width w (perimeter means distance around)

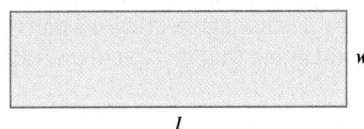

92. The perimeter of a square with side s (perimeter means distance around)

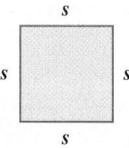

93. Ellie's race time, assuming she took 5 sec longer than Joe and Joe took 3 sec longer than Molly. Assume that Molly's time was t seconds.

94. Ray's age 7 yr from now if he is 2 yr older than Monique and Monique is a years old

95. If the length of the height of a triangle is doubled, is its area also doubled? Why or why not?

CORNER

Teamwork

Focus: Group problem solving; working collaboratively

Time: 15 minutes

Group size: 2

Working and studying as a team often enables students to solve problems that are difficult to solve alone.

ACTIVITY

1. The left-hand column below contains the names of 12 colleges. A scrambled list of the names of their sports teams is on the right. As a group, match the names of the colleges to the teams.

1. University of Texas	**a)** Antelopes
2. Western State College of Colorado	**b)** Fighting Banana Slugs
3. University of North Carolina	**c)** Sea Warriors
4. University of Massachusetts	**d)** Gators
5. Hawaii Pacific University	**e)** Mountaineers
6. University of Nebraska	**f)** Sailfish
7. University of California, Santa Cruz	**g)** Longhorns
8. University of Louisiana at Lafayette	**h)** Tar Heels
9. Grand Canyon University	**i)** Seawolves
10. Palm Beach Atlantic University	**j)** Ragin' Cajuns
11. University of Alaska, Anchorage	**k)** Cornhuskers
12. University of Florida	**l)** Minutemen

2. After working for 5 min, confer with another group and reach mutual agreement.
3. Does the class agree on all 12 pairs?
4. Do you agree that group collaboration enhances our ability to solve problems?

The Commutative, Associative, and Distributive Laws

Equivalent Expressions • The Commutative Laws •
The Associative Laws • The Distributive Law •
The Distributive Law and Factoring

In order to solve equations, we must be able to manipulate algebraic expressions. The commutative, associative, and distributive laws discussed in this section enable us to write *equivalent expressions* that will simplify our work. Indeed, much of this text is devoted to finding equivalent expressions.

Equivalent Expressions

The expressions $4 + 4 + 4, 3 \cdot 4$, and $4 \cdot 3$ all represent the same number, 12. Expressions that represent the same number are said to be **equivalent**. The equivalent expressions $t + 18$ and $18 + t$ were used on p. 5 when we translated "eighteen more than a number." To check that these expressions are equivalent, we make some choices for t:

$$\text{When } t = 3, \quad t + 18 = 3 + 18 \quad \text{and} \quad 18 + t = 18 + 3$$
$$= 21 \qquad\qquad\qquad = 21.$$

$$\text{When } t = 40, \quad t + 18 = 40 + 18 \quad \text{and} \quad 18 + t = 18 + 40$$
$$= 58 \qquad\qquad\qquad = 58.$$

The Commutative Laws

Recall that changing the order in addition or multiplication does not change the result. Equations like $3 + 78 = 78 + 3$ and $5 \cdot 14 = 14 \cdot 5$ illustrate this idea and show that addition and multiplication are **commutative**.

The Commutative Laws

For Addition. For any numbers a and b,

$$a + b = b + a.$$

(Changing the order of addition does not affect the answer.)

For Multiplication. For any numbers a and b,

$$ab = ba.$$

(Changing the order of multiplication does not affect the answer.)

EXAMPLE 1   Use the commutative laws to write an expression equivalent to each of the following: **(a)** $y + 5$; **(b)** $9x$; **(c)** $7 + ab$.

Solution

a) $y + 5$ is equivalent to $5 + y$ by the commutative law of addition.

b) $9x$ is equivalent to $x \cdot 9$ by the commutative law of multiplication.

c) $7 + ab$ is equivalent to $ab + 7$ by the commutative law of *addition*.

$7 + ab$ is also equivalent to $7 + ba$ by the commutative law of *multiplication*.

$7 + ab$ is also equivalent to $ba + 7$ by the two commutative laws, used together.

The Associative Laws

Parentheses are used to indicate groupings. We normally simplify within the parentheses first. For example,

$$3 + (8 + 4) = 3 + 12 \quad \text{and} \quad (3 + 8) + 4 = 11 + 4$$
$$= 15 \qquad\qquad\qquad = 15.$$

Similarly,

$$4 \cdot (2 \cdot 3) = 4 \cdot 6 \quad \text{and} \quad (4 \cdot 2) \cdot 3 = 8 \cdot 3$$
$$= 24 \qquad\qquad\qquad = 24.$$

Note that, so long as only addition or only multiplication appears in an expression, changing the grouping does not change the result. Equations such as $3 + (7 + 5) = (3 + 7) + 5$ and $4(5 \cdot 3) = (4 \cdot 5)3$ illustrate that addition and multiplication are **associative**.

The Associative Laws

For Addition. For any numbers a, b, and c,

$$a + (b + c) = (a + b) + c.$$

(Numbers can be grouped in any manner for addition.)

For Multiplication. For any numbers a, b, and c,

$$a \cdot (b \cdot c) = (a \cdot b) \cdot c.$$

(Numbers can be grouped in any manner for multiplication.)

EXAMPLE 2   Use an associative law to write an expression equivalent to each of the following: **(a)** $y + (z + 3)$; **(b)** $(8x)y$.

Solution

a) $y + (z + 3)$ is equivalent to $(y + z) + 3$ by the associative law of addition.

b) $(8x)y$ is equivalent to $8(xy)$ by the associative law of multiplication.

When only addition or only multiplication is involved, parentheses do not change the result. For that reason, we sometimes omit them altogether. Thus,

$$x + (y + 7) = x + y + 7, \quad \text{and} \quad l(wh) = lwh.$$

A sum such as $(5 + 1) + (3 + 5) + 9$ can be simplified by pairing numbers that add to 10. The associative and commutative laws allow us to do this:

$$(5 + 1) + (3 + 5) + 9 = 5 + 5 + 9 + 1 + 3$$
$$= 10 + 10 + 3 = 23.$$

EXAMPLE 3   Use the commutative and/or associative laws of addition to write two expressions equivalent to $(7 + x) + 3$. Then simplify.

Solution

$(7 + x) + 3 = (x + 7) + 3$	Using the commutative law; $(x + 7) + 3$ is one equivalent expression.
$= x + (7 + 3)$	Using the associative law; $x + (7 + 3)$ is another equivalent expression.
$= x + 10$	Simplifying

EXAMPLE 4   Use the commutative and/or associative laws of multiplication to write two expressions equivalent to $2(x \cdot 3)$.

Solution

$2(x \cdot 3) = 2(3x)$	Using the commutative law; $2(3x)$ is one equivalent expression.
$= (2 \cdot 3)x$	Using the associative law; $(2 \cdot 3)x$ is another equivalent expression.
$= 6x$	Simplifying

The Distributive Law

The *distributive law* is probably the single most important law for manipulating algebraic expressions. Unlike the commutative and associative laws, the distributive law uses multiplication together with addition.

You have already used the distributive law although you may not have realized it at the time. To illustrate, try to multiply $3 \cdot 21$ mentally. Many people find the product, 63, by thinking of 21 as $20 + 1$ and then multiplying 20 by 3 and 1 by 3. The sum of the two products, $60 + 3$, is 63. Note that if the 3 does not multiply both 20 and 1, the result will not be correct.

EXAMPLE 5

Compute in two ways: $4(7 + 2)$.

Solution

a) As in the discussion of $3(20 + 1)$ above, to compute $4(7 + 2)$, we can multiply both 7 and 2 by 4 and add the results:

$$4(7 + 2) = 4 \cdot 7 + 4 \cdot 2 \qquad \text{Multiplying both 7 and 2 by 4}$$
$$= 28 + 8 = 36. \qquad \text{Adding}$$

b) By first adding inside the parentheses, we get the same result in a different way:

$$4(7 + 2) = 4(9) \qquad \text{Adding; } 7 + 2 = 9$$
$$= 36. \qquad \text{Multiplying}$$

The Distributive Law

For any numbers a, b, and c,

$$a(b + c) = ab + ac.$$

(The product of a number and a sum can be written as the sum of two products.)

EXAMPLE 6

Multiply: $3(x + 2)$.

Solution   Since $x + 2$ cannot be simplified unless a value for x is given, we use the distributive law:

$$3(x + 2) = 3 \cdot x + 3 \cdot 2 \qquad \text{Using the distributive law}$$
$$= 3x + 6. \qquad \text{Note that } 3 \cdot x \text{ is the same as } 3x.$$

The expression $3x + 6$ has two *terms*, $3x$ and 6. In general, a **term** is a number, a variable, or a product or quotient of numbers and/or variables. Thus, t, 29, $5ab$, and $2x/y$ are terms in $t + 29 + 5ab + 2x/y$. Note that terms are separated by plus signs.

EXAMPLE 7

List the terms in the expression $7s + st + \dfrac{3}{t}$.

Solution   Terms are separated by plus signs, so the terms in $7s + st + 3/t$ are $7s$, st, and $3/t$.

The distributive law can also be used when more than two terms are inside the parentheses.

EXAMPLE 8   Multiply: $6(s + 2 + 5w)$.

Solution

$$6(s + 2 + 5w) = 6 \cdot s + 6 \cdot 2 + 6 \cdot 5w \qquad \text{Using the distributive law}$$
$$= 6s + 12 + (6 \cdot 5)w \qquad \text{Using the associative law}$$
$$\text{for multiplication}$$
$$= 6s + 12 + 30w$$

Because of the commutative law of multiplication, the distributive law can be used on the "right": $(b + c)a = ba + ca$.

EXAMPLE 9   Multiply: $(c + 4)5$.

Solution

$$(c + 4)5 = c \cdot 5 + 4 \cdot 5 \qquad \text{Using the distributive law on the right}$$
$$= 5c + 20$$

Caution!   To use the distributive law for removing parentheses, be sure to multiply *each* term inside the parentheses by the multiplier outside:

$$a(b + c) = ab + c.$$

The Distributive Law and Factoring

If we use the distributive law in reverse, we have the basis of a process called **factoring**: $ab + ac = a(b + c)$. To **factor** an expression means to write an equivalent expression that is a product. The parts of the product are called **factors**. Note that "factor" can be used as either a verb or a noun. Thus in the expression $5t$, the factors are 5 and t. In the expression $4(m + n)$, the factors are 4 and $(m + n)$.

EXAMPLE 10   Use the distributive law to factor each of the following.

a) $3x + 3y$ **b)** $7x + 21y + 7$

Solution

a) By the distributive law,

$$3x + 3y = 3(x + y). \qquad \text{The } common \ factor \text{ for } 3x \text{ and } 3y \text{ is } 3.$$

b) $7x + 21y + 7 = 7 \cdot x + 7 \cdot 3y + 7 \cdot 1 \qquad \text{The common factor is 7.}$
$$= 7(x + 3y + 1) \qquad \text{Using the distributive law}$$
$$\text{Be sure to include both the 1 and}$$
$$\text{the common factor, 7.}$$

To check our factoring, we multiply to see if the original expression is obtained. For example, to check the **factorization** in Example 10(b), note that

$$7(x + 3y + 1) = 7 \cdot x + 7 \cdot 3y + 7 \cdot 1$$
$$= 7x + 21y + 7.$$

Since $7x + 21y + 7$ is what we started with in Example 10(b), we have a check.

Exercise Set

1.2

FOR EXTRA HELP

 Student's Solutions Manual

 Digital Video Tutor CD 1 Videotape 1

 AW Math Tutor Center

 MathXL Tutorials on CD

 MathXL

 MyMathLab

↪ *Concept Reinforcement   Complete each sentence using one of these terms: commutative, associative, or distributive.*

1. $8 + t$ is equivalent to $t + 8$ by the _____ law for addition.

2. $3(xy)$ is equivalent to $(3x)y$ by the _____ law for multiplication.

3. $(5b)c$ is equivalent to $5(bc)$ by the _____ law for multiplication.

4. mn is equivalent to nm by the _____ law for multiplication.

5. $x(y + z)$ is equivalent to $xy + xz$ by the _____ law.

6. $(9 + a) + b$ is equivalent to $9 + (a + b)$ by the _____ law for addition.

7. $a + (6 + d)$ is equivalent to $(a + 6) + d$ by the _____ law for addition.

8. $t + 4$ is equivalent to $4 + t$ by the _____ law for addition.

9. $x \cdot 7$ is equivalent to $7 \cdot x$ by the _____ law for multiplication.

10. $2(a + b)$ is equivalent to $2 \cdot a + 2 \cdot b$ by the _____ law.

Use the commutative law of addition to write an equivalent expression.

11. $7 + x$

12. $a + 2$

13. $ab + c$

14. $x + 3y$

15. $9x + 3y$

16. $3a + 7b$

17. $5(a + 1)$

18. $9(x + 5)$

Use the commutative law of multiplication to write an equivalent expression.

19. $2 \cdot a$

20. xy

21. st

22. $4x$

23. $5 + ab$

24. $x + 3y$

25. $5(a + 1)$

26. $9(x + 5)$

Use the associative law of addition to write an equivalent expression.

27. $(a + 5) + b$

28. $(5 + m) + r$

29. $r + (t + 7)$

30. $x + (2 + y)$

31. $(ab + c) + d$

32. $(m + np) + r$

Use the associative law of multiplication to write an equivalent expression.

33. $(8x)y$

34. $(9a)b$

35. $2(ab)$

36. $9(rp)$

37. $3[2(a + b)]$

38. $5[x(2 + y)]$

Use the commutative and/or associative laws to write two equivalent expressions. Answers may vary.

39. $r + (t + 6)$

40. $5 + (v + w)$

41. $(17a)b$

42. $x(3y)$

Use the commutative and/or associative laws to show why the expression on the left is equivalent to the expression on the right. Write a series of steps with labels, as in Example 4.

43. $(5 + x) + 2$ is equivalent to $x + 7$

44. $(2a)4$ is equivalent to $8a$

45. $(m \cdot 3)7$ is equivalent to $21m$

46. $4 + (9 + x)$ is equivalent to $x + 13$

Multiply.

47. $4(a + 3)$

48. $3(x + 5)$

49. $6(1 + x)$

50. $6(v + 4)$

51. $3(x + 1)$

52. $9(x + 3)$

53. $8(3 + y)$

54. $7(s + 5)$

55. $9(2x + 6)$

56. $9(6m + 7)$

57. $5(r + 2 + 3t)$

58. $4(5x + 8 + 3p)$

59. $(a + b)2$

60. $(x + 2)7$

61. $(x + y + 2)5$

62. $(2 + a + b)6$

List the terms in each expression.

63. $x + xyz + 19$

64. $9 + 17a + abc$

65. $2a + \dfrac{a}{b} + 5b$

66. $3xy + 20 + \dfrac{4a}{b}$

Use the distributive law to factor each of the following. Check by multiplying.

67. $2a + 2b$

68. $5y + 5z$

69. $7 + 7y$

70. $13 + 13x$

71. $18x + 3$

72. $20a + 5$

73. $5x + 10 + 15y$

74. $3 + 27b + 6c$

75. $12x + 9$

76. $6x + 6$

77. $3a + 9b$

78. $5a + 15b$

79. $44x + 11y + 22z$

80. $14a + 56b + 7$

List the factors in each expression.

81. st

82. $5x$

83. $3(x + y)$

84. $(a + b)6$

85. $7 \cdot a$

86. $m \cdot 2$

87. $(a - b)(x - y)$

88. $(3 - a)(b + c)$

89. Is subtraction commutative? Why or why not?

90. Is division associative? Why or why not?

SKILL MAINTENANCE

To the student and the instructor: Exercises included for Skill Maintenance review skills previously studied in the text. Often these exercises provide preparation for the next section of the text. The numbers in brackets immediately following the directions or exercise indicate the section in which the skill was introduced. The answers to all Skill Maintenance exercises appear at the back of the book. If a Skill Maintenance exercise gives you difficulty, review the material in the indicated section of the text.

Translate to an algebraic expression. [1.1]

91. Twice Kara's salary

92. Half of m

SYNTHESIS

93. Are terms and factors the same thing? Why or why not?

94. Explain how the distributive, commutative, and associative laws can be used to show that $2(3x + 4y)$ is equivalent to $6x + 8y$.

Tell whether the expressions in each pairing are equivalent. Then explain why or why not.

95. $8 + 4(a + b)$ and $4(2 + a + b)$

96. $7 \div 3m$ and $m \cdot 3 \div 7$

97. $(rt + st)5$ and $5t(r + s)$

98. $yax + ax$ and $xa(1 + y)$

99. $30y + x15$ and $5[2(x + 3y)]$

100. $[c(2 + 3b)]5$ and $10c + 15bc$

101. Evaluate the expressions $3(2 + x)$ and $6 + x$ for $x = 0$. Do your results indicate that $3(2 + x)$ and $6 + x$ are equivalent? Why or why not?

102. Factor $15x + 40$. Then evaluate both $15x + 40$ and the factorization for $x = 4$. Do your results *guarantee* that the factorization is correct? Why or why not? (*Hint*: See Exercise 101.)

COLLABORATIVE

CORNER

Mental Addition

Focus: Application of commutative and associative laws

Time: 10 minutes

Group size: 2–3

Legend has it that while still in grade school, the mathematician Carl Friedrich Gauss (1777–1855) was able to add the numbers from 1 to 100 mentally. Gauss did not add them sequentially, but rather paired 1 with 99, 2 with 98, and so on.

ACTIVITY

1. Use a method similar to Gauss's to simplify the following:

$$1 + 2 + 3 + 4 + 5 + 6 + 7 + 8 + 9 + 10.$$

One group member should add from left to right as a check.

2. Use Gauss's method to find the sum of the first 25 counting numbers:

$$1 + 2 + 3 + \cdots + 23 + 24 + 25.$$

Again, one student should add from left to right as a check.

3. How were the associative and commutative laws applied in parts (1) and (2) above?

4. Now use a similar approach involving both addition and division to find the sum of the first 10 counting numbers:

$$\begin{aligned} 1 + 2 + 3 + \cdots + 10 \\ + \; 10 + 9 + 8 + \cdots + 1 \end{aligned}$$

5. Use the approach in step (4) to find the sum of the first 100 counting numbers. Are the associative and commutative laws applied in this method, too? How is the distributive law used in this approach?

1.3   Fraction Notation

Factors and Prime Factorizations • Fraction Notation •
Multiplication, Division, and Simplification • More
Simplifying • Addition and Subtraction

This section covers multiplication, addition, subtraction, and division with fractions. Although much of this may be review, note that fraction expressions that contain variables are also included.

Factors and Prime Factorizations

In preparation for work with fraction notation, we first review how *natural numbers* are factored. **Natural numbers** can be thought of as the counting numbers:

$$1, 2, 3, 4, 5, \ldots .^*$$

(The dots indicate that the established pattern continues without ending.)

---

*A similar collection of numbers, the **whole numbers,** includes 0: 0, 1, 2, 3, ...

To factor a number, we simply express it as a product of two or more numbers.

EXAMPLE 1   Write several factorizations of 12. Then list all factors of 12.

Solution   The number 12 can be factored in several ways:

$$1 \cdot 12, \qquad 2 \cdot 6, \qquad 3 \cdot 4, \qquad 2 \cdot 2 \cdot 3.$$

The factors of 12 are 1, 2, 3, 4, 6, and 12.

Some numbers have only two factors, the number itself and 1. Such numbers are called **prime**.

> **Prime Number**
>
> A *prime number* is a natural number that has exactly two different factors: the number itself and 1. The first several primes are 2, 3, 5, 7, 11, 13, 17, 19, and 23.

EXAMPLE 2   Which of these numbers are prime? 29, 4, 1

Solution

29 is prime. It has exactly two different factors, 29 and 1.

4 is not prime. It has three different factors, 1, 2, and 4.

1 is not prime. It does not have two *different* factors.

If a natural number, other than 1, is not prime, we call it **composite**. Every composite number can be factored into a product of prime numbers. Such a factorization is called the **prime factorization** of that composite number.

EXAMPLE 3   Find the prime factorization of 36.

Solution   We first factor 36 in any way that we can. One way is like this:

$$36 = 4 \cdot 9.$$

The factors 4 and 9 are not prime, so we factor them:

$$36 = 4 \cdot 9$$
$$= 2 \cdot 2 \cdot 3 \cdot 3. \qquad \text{2 and 3 are both prime.}$$

The prime factorization of 36 is $2 \cdot 2 \cdot 3 \cdot 3$.

Student Notes

When writing a factorization, you are writing an equivalent expression for the original number. Some students do this with a tree diagram:

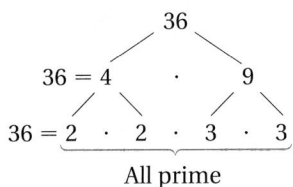

All prime

Fraction Notation

An example of **fraction notation** for a number is

$$\dfrac{2}{3}\begin{array}{l}\leftarrow \text{Numerator} \\ \leftarrow \text{Denominator}\end{array}$$

The top number is called the **numerator**, and the bottom number is called the **denominator**. When the numerator and the denominator are the same nonzero number, we have fraction notation for the number 1.

Fraction Notation for 1

For any number a, except 0,

$$\dfrac{a}{a} = 1.$$

(Any nonzero number divided by itself is 1.)

Multiplication, Division, and Simplification

Recall from arithmetic that fractions are multiplied as follows.

Multiplication of Fractions

For any two fractions a/b and c/d,

$$\dfrac{a}{b} \cdot \dfrac{c}{d} = \dfrac{ac}{bd}.$$

(The numerator of the product is the product of the two numerators. The denominator of the product is the product of the two denominators.)

EXAMPLE 4   Multiply: **(a)** $\dfrac{2}{3} \cdot \dfrac{7}{5}$;   **(b)** $\dfrac{4}{x} \cdot \dfrac{8}{y}$.

Solution   We multiply numerators as well as denominators.

a) $\dfrac{2}{3} \cdot \dfrac{7}{5} = \dfrac{2 \cdot 7}{3 \cdot 5} = \dfrac{14}{15}$

b) $\dfrac{4}{x} \cdot \dfrac{8}{y} = \dfrac{4 \cdot 8}{x \cdot y} = \dfrac{32}{xy}$

Two numbers whose product is 1 are **reciprocals**, or **multiplicative inverses**, of each other. All numbers, except zero, have reciprocals. For example,

the reciprocal of $\dfrac{2}{3}$ is $\dfrac{3}{2}$ because $\dfrac{2}{3} \cdot \dfrac{3}{2} = \dfrac{6}{6} = 1$;

the reciprocal of 9 is $\dfrac{1}{9}$ because $9 \cdot \dfrac{1}{9} = \dfrac{9}{9} = 1$; and

the reciprocal of $\dfrac{1}{4}$ is 4 because $\dfrac{1}{4} \cdot 4 = 1$.

Reciprocals are used to rewrite division in an equivalent form that uses multiplication.

Division of Fractions

To divide two fractions, multiply by the reciprocal of the divisor:

$$\frac{a}{b} \div \frac{c}{d} = \frac{a}{b} \cdot \frac{d}{c}.$$

EXAMPLE 5   Divide: $\dfrac{1}{2} \div \dfrac{3}{5}$.

Solution

$$\frac{1}{2} \div \frac{3}{5} = \frac{1}{2} \cdot \frac{5}{3} \qquad \frac{5}{3} \text{ is the reciprocal of } \frac{3}{5}.$$

$$= \frac{5}{6}$$

When one of the fractions being multiplied is 1, multiplying yields an equivalent expression because of the *identity property of* 1. A similar property could be stated for division, but there is no need to do so here.

The Identity Property of 1

For any number a,

$$a \cdot 1 = a.$$

(Multiplying a number by 1 gives that same number.)

EXAMPLE 6   Multiply $\dfrac{4}{5} \cdot \dfrac{6}{6}$ to find an expression equivalent to $\dfrac{4}{5}$.

Solution   Since $\frac{6}{6} = 1$, the expression $\frac{4}{5} \cdot \frac{6}{6}$ is equivalent to $\frac{4}{5} \cdot 1$, or simply $\frac{4}{5}$. We have

$$\frac{4}{5} \cdot \frac{6}{6} = \frac{4 \cdot 6}{5 \cdot 6} = \frac{24}{30}.$$

Thus, $\frac{24}{30}$ is equivalent to $\frac{4}{5}$.

The steps of Example 6 are reversed by "removing a factor equal to 1"—in this case, $\frac{6}{6}$. By removing a factor that equals 1, we can *simplify* an expression like $\frac{24}{30}$ to an equivalent expression like $\frac{4}{5}$.

To simplify, we factor the numerator and the denominator, looking for the largest factor common to both. This is sometimes made easier by writing prime factorizations. After identifying common factors, we can express the fraction as a product of two fractions, one of which is in the form a/a.

EXAMPLE 7   Simplify: **(a)** $\dfrac{15}{40}$; **(b)** $\dfrac{36}{24}$.

Solution

a) Note that 5 is a factor of both 15 and 40:

$$\frac{15}{40} = \frac{3 \cdot 5}{8 \cdot 5} \qquad \text{Factoring the numerator and the denominator, using the common factor, 5}$$

$$= \frac{3}{8} \cdot \frac{5}{5} \qquad \text{Rewriting as a product of two fractions; } \frac{5}{5} = 1$$

$$= \frac{3}{8} \cdot 1 = \frac{3}{8}. \qquad \text{Using the identity property of 1 (removing a factor equal to 1)}$$

b) $\dfrac{36}{24} = \dfrac{2 \cdot 2 \cdot 3 \cdot 3}{2 \cdot 2 \cdot 2 \cdot 3}$ Writing the prime factorizations and identifying common factors; 12/12 could also be used.

$$= \frac{3}{2} \cdot \frac{2 \cdot 2 \cdot 3}{2 \cdot 2 \cdot 3} \qquad \text{Rewriting as a product of two fractions; } \frac{2 \cdot 2 \cdot 3}{2 \cdot 2 \cdot 3} = 1$$

$$= \frac{3}{2} \cdot 1 = \frac{3}{2} \qquad \text{Using the identity property of 1}$$

It is always wise to check your result to see if any common factors of the numerator and the denominator remain. (This will never happen if prime factorizations are used correctly.) If common factors remain, repeat the process by removing another factor equal to 1 to simplify your result.

More Simplifying

"Canceling" is a shortcut that you may have used for removing a factor equal to 1 when working with fraction notation. With *great* concern, we mention it as a possible way to speed up your work. Canceling can be used only when removing common factors in numerators and denominators. Canceling *cannot* be used in sums or differences. Our concern is that "canceling" be used with understanding. Example 7(b) might have been done faster as follows:

$$\frac{36}{24} = \frac{2 \cdot 2 \cdot 3 \cdot \cancel{3}}{2 \cdot 2 \cdot 2 \cdot \cancel{3}} = \frac{3}{2}, \quad \text{or} \quad \frac{36}{24} = \frac{3 \cdot \cancel{12}}{2 \cdot \cancel{12}} = \frac{3}{2}, \quad \text{or} \quad \frac{\overset{3}{\underset{12}{\cancel{\overset{18}{\cancel{36}}}}}}{\underset{2}{\cancel{\underset{12}{\cancel{24}}}}} = \frac{3}{2}.$$

Caution!   Unfortunately, canceling is often performed incorrectly:

$$\frac{\cancel{2}+3}{\cancel{2}} = 3, \qquad \frac{\cancel{4}-1}{\cancel{4}-2} = \frac{1}{2}, \qquad \frac{\cancel{15}}{\cancel{54}} = \frac{1}{4},$$

The above cancellations are incorrect because the expressions canceled are *not* factors. For example, in $2 + 3$, the 2 and 3 are not factors. Correct simplifications are as follows:

$$\frac{2+3}{2} = \frac{5}{2}, \qquad \frac{4-1}{4-2} = \frac{3}{2}, \qquad \frac{15}{54} = \frac{5 \cdot \cancel{3}}{18 \cdot \cancel{3}} = \frac{5}{18}.$$

Remember: **If you can't factor, you can't cancel! If in doubt, don't cancel!**

Sometimes it is helpful to use 1 as a factor in the numerator or the denominator when simplifying.

EXAMPLE 8   Simplify: $\dfrac{9}{72}$.

Solution

$$\frac{9}{72} = \frac{1 \cdot 9}{8 \cdot 9} \qquad \text{Factoring and using the identity property of 1 to write 9 as } 1 \cdot 9$$

$$= \frac{1 \cdot \cancel{9}}{8 \cdot \cancel{9}} = \frac{1}{8} \qquad \text{Simplifying by removing a factor equal to 1: } \frac{9}{9} = 1$$

Addition and Subtraction

When denominators are the same, fractions are added or subtracted by adding or subtracting numerators and keeping the same denominator.

Addition and Subtraction of Fractions

For any two fractions a/d and b/d,

$$\frac{a}{d} + \frac{b}{d} = \frac{a+b}{d} \quad \text{and} \quad \frac{a}{d} - \frac{b}{d} = \frac{a-b}{d}.$$

EXAMPLE 9   Add and simplify: $\dfrac{4}{8} + \dfrac{5}{8}$.

Solution   The common denominator is 8. We add the numerators and keep the common denominator:

$$\frac{4}{8} + \frac{5}{8} = \frac{4+5}{8} = \frac{9}{8}.$$

You can think of this as
$$4 \cdot \frac{1}{8} + 5 \cdot \frac{1}{8} = 9 \cdot \frac{1}{8}, \text{ or } \frac{9}{8}.$$

In arithmetic, we often write $1\frac{1}{8}$ rather than the "improper" fraction $\frac{9}{8}$. In algebra, $\frac{9}{8}$ is generally more useful and is quite "proper" for our purposes.

When denominators are different, we use the identity property of 1 and multiply to find a common denominator. Then we add, as in Example 9.

EXAMPLE 10   Add or subtract as indicated: **(a)** $\dfrac{7}{8} + \dfrac{5}{12}$; **(b)** $\dfrac{9}{8} - \dfrac{4}{5}$.

Solution

a) The number 24 is divisible by both 8 and 12. We multiply both $\frac{7}{8}$ and $\frac{5}{12}$ by suitable forms of 1 to obtain two fractions with denominators of 24:

$$\frac{7}{8} + \frac{5}{12} = \frac{7}{8} \cdot \frac{3}{3} + \frac{5}{12} \cdot \frac{2}{2}$$

Multiplying by 1. Since $8 \cdot 3 = 24$, we multiply $\frac{7}{8}$ by $\frac{3}{3}$. Since $12 \cdot 2 = 24$, we multiply $\frac{5}{12}$ by $\frac{2}{2}$.

$$= \frac{21}{24} + \frac{10}{24}$$

Performing the multiplication

$$= \frac{31}{24}.$$

Adding fractions

b) $\dfrac{9}{8} - \dfrac{4}{5} = \dfrac{9}{8} \cdot \dfrac{5}{5} - \dfrac{4}{5} \cdot \dfrac{8}{8}$

Using 40 as a common denominator

$$= \frac{45}{40} - \frac{32}{40} = \frac{13}{40}$$

Subtracting fractions

After adding, subtracting, multiplying, or dividing, we may still need to simplify the answer.

EXAMPLE 11   Perform the indicated operation and, if possible, simplify.

a) $\dfrac{7}{10} - \dfrac{1}{5}$ b) $8 \cdot \dfrac{5}{12}$ c) $\dfrac{\frac{5}{6}}{\frac{25}{9}}$

Solution

a) $\dfrac{7}{10} - \dfrac{1}{5} = \dfrac{7}{10} - \dfrac{1}{5} \cdot \dfrac{2}{2}$ Using 10 as the common denominator

$= \dfrac{7}{10} - \dfrac{2}{10}$

$= \dfrac{5}{10} = \dfrac{1 \cdot \cancel{5}}{2 \cdot \cancel{5}} = \dfrac{1}{2}$ Removing a factor equal to 1: $\dfrac{5}{5} = 1$

b) $8 \cdot \dfrac{5}{12} = \dfrac{8 \cdot 5}{12}$ Multiplying numerators and denominators. Think of 8 as $\frac{8}{1}$.

$= \dfrac{2 \cdot 2 \cdot 2 \cdot 5}{2 \cdot 2 \cdot 3}$ Factoring; $\dfrac{4 \cdot 2 \cdot 5}{4 \cdot 3}$ can also be used.

$= \dfrac{\cancel{2} \cdot \cancel{2} \cdot 2 \cdot 5}{\cancel{2} \cdot \cancel{2} \cdot 3}$ Removing a factor equal to 1: $\dfrac{2 \cdot 2}{2 \cdot 2} = 1$

$= \dfrac{10}{3}$ Simplifying

c) $\dfrac{\frac{5}{6}}{\frac{25}{9}} = \dfrac{5}{6} \div \dfrac{25}{9}$ Rewriting horizontally. Remember that a fraction bar indicates division.

$= \dfrac{5}{6} \cdot \dfrac{9}{25}$ Multiplying by the reciprocal of $\frac{25}{9}$

$= \dfrac{5 \cdot 3 \cdot 3}{2 \cdot 3 \cdot 5 \cdot 5}$ Writing as one fraction and factoring

$= \dfrac{\cancel{5} \cdot \cancel{3} \cdot 3}{2 \cdot \cancel{3} \cdot \cancel{5} \cdot 5}$ Removing a factor equal to 1: $\dfrac{5 \cdot 3}{3 \cdot 5} = 1$

$= \dfrac{3}{10}$ Simplifying

---

technology connection

Some graphing calculators can perform operations using fraction notation. Others may be able to convert answers given in decimal notation to fraction notation. Often this conversion is done using a command found in a **menu**, or a list of options that appears when a key is pressed. To select an item from a menu, we highlight its number and press **ENTER** or simply press the number of the item.

For example, to find fraction notation for $\frac{2}{15} + \frac{7}{12}$, we enter the expression as $2/15 + 7/12$. The answer is given in decimal notation. To convert this to fraction notation, we press **MATH** and select the Frac option. In this case, the notation Ans▶ Frac shows that the graphing calculator will convert .7166666667 to fraction notation.

```
2/15+7/12
                .7166666667
Ans▶Frac
                     43/60
```

We see that $\frac{2}{15} + \frac{7}{12} = \frac{43}{60}$.

Exercise Set

1.3

To the student and the instructor: *Beginning in this section, selected exercises are marked with the icon* **Aha!** *. These "Aha!" exercises can be answered most easily if the student pauses to inspect the exercise rather than proceed mechanically. This is done to discourage rote memorization. Some later "Aha!" exercises in this exercise set are unmarked, to encourage students to always pause before working a problem.*

🖐 *Concept Reinforcement* *Label each of the following numbers as prime, composite, or neither.*

1. 9   **2.** 15   **3.** 31   **4.** 35

5. 25   **6.** 37   **7.** 2   **8.** 1

9. 0   **10.** 4

🖐 *Concept Reinforcement* *In each of Exercises 11–14, match the description with a number from the list on the right.*

11. ____ A factor of 35   **a)** 2

12. ____ A number that has 3 as a factor   **b)** 7

13. ____ An odd composite number   **c)** 60

14. ____ The only even prime number   **d)** 65

Write at least two factorizations of each number. Then list all the factors of the number.

15. 50   **16.** 70   **17.** 42   **18.** 60

Find the prime factorization of each number. If the number is prime, state this.

19. 26   **20.** 15   **21.** 30

22. 55   **23.** 27   **24.** 98

25. 18   **26.** 54   **27.** 40

28. 56   **29.** 43   **30.** 120

31. 210   **32.** 79   **33.** 115

34. 143

Simplify.

35. $\dfrac{14}{21}$ **36.** $\dfrac{10}{14}$ **37.** $\dfrac{16}{56}$

38. $\dfrac{72}{27}$ **39.** $\dfrac{6}{48}$ **40.** $\dfrac{12}{70}$

41. $\dfrac{49}{7}$ **42.** $\dfrac{132}{11}$ **43.** $\dfrac{19}{76}$

44. $\dfrac{17}{51}$ **45.** $\dfrac{150}{25}$ **46.** $\dfrac{170}{34}$

47. $\dfrac{42}{50}$ **48.** $\dfrac{75}{80}$ **49.** $\dfrac{120}{82}$

50. $\dfrac{75}{45}$ **51.** $\dfrac{210}{98}$ **52.** $\dfrac{140}{350}$

Perform the indicated operation and, if possible, simplify.

53. $\dfrac{1}{2} \cdot \dfrac{3}{7}$ **54.** $\dfrac{11}{10} \cdot \dfrac{8}{5}$ **55.** $\dfrac{9}{2} \cdot \dfrac{3}{4}$

Aha! **56.** $\dfrac{11}{12} \cdot \dfrac{12}{11}$ **57.** $\dfrac{1}{8} + \dfrac{3}{8}$ **58.** $\dfrac{1}{2} + \dfrac{1}{8}$

59. $\dfrac{4}{9} + \dfrac{13}{18}$ **60.** $\dfrac{4}{5} + \dfrac{8}{15}$ **61.** $\dfrac{3}{a} \cdot \dfrac{b}{7}$

62. $\dfrac{x}{5} \cdot \dfrac{y}{z}$ **63.** $\dfrac{4}{a} + \dfrac{3}{a}$ **64.** $\dfrac{7}{a} - \dfrac{5}{a}$

65. $\dfrac{3}{10} + \dfrac{8}{15}$ **66.** $\dfrac{7}{8} + \dfrac{5}{12}$ **67.** $\dfrac{9}{7} - \dfrac{2}{7}$

68. $\dfrac{12}{5} - \dfrac{2}{5}$ **69.** $\dfrac{13}{18} - \dfrac{4}{9}$ **70.** $\dfrac{13}{15} - \dfrac{8}{45}$

Aha! **71.** $\dfrac{20}{30} - \dfrac{2}{3}$ **72.** $\dfrac{5}{7} - \dfrac{5}{21}$ **73.** $\dfrac{7}{6} \div \dfrac{3}{5}$

74. $\dfrac{7}{5} \div \dfrac{3}{4}$ **75.** $\dfrac{8}{9} \div \dfrac{4}{15}$ **76.** $\dfrac{9}{4} \div 9$

77. $12 \div \dfrac{3}{7}$ **78.** $\dfrac{1}{10} \div \dfrac{1}{5}$ Aha! **79.** $\dfrac{7}{13} \div \dfrac{7}{13}$

80. $\dfrac{17}{8} \div \dfrac{5}{6}$ **81.** $\dfrac{\frac{2}{7}}{\frac{5}{3}}$ **82.** $\dfrac{\frac{3}{8}}{\frac{1}{5}}$

83. $\dfrac{9}{\frac{1}{2}}$ **84.** $\dfrac{\frac{7}{3}}{5}$

85. Under what circumstances would the sum of two fractions be easier to compute than the product of the same two fractions?

86. Under what circumstances would the product of two fractions be easier to compute than the sum of the same two fractions?

SKILL MAINTENANCE

Use a commutative law to write an equivalent expression. There can be more than one correct answer.
[1.2]

87. $5(x + 3)$ **88.** $7 + (a + b)$

SYNTHESIS

89. Bryce insists that $(2 + x)/8$ is equivalent to $(1 + x)/4$. What mistake do you think is being made and how could you demonstrate to Bryce that the two expressions are not equivalent?

90. Why are 0 and 1 considered neither prime nor composite?

91. In the following table, the top number can be factored in such a way that the sum of the factors is the bottom number. For example, in the first column, 56 is factored as $7 \cdot 8$, since $7 + 8 = 15$, the bottom number. Find the missing numbers in each column.

Product	56	63	36	72	140	96	168
Factor	7						
Factor	8						
Sum	15	16	20	38	24	20	29

92. *Packaging.* Tritan Candies uses two sizes of boxes, 6 in. long and 8 in. long. These are packed end to end in bigger cartons to be shipped. What is the shortest-length carton that will accommodate boxes of either size without any room left over? (Each carton must contain boxes of only one size; no mixing is allowed.)

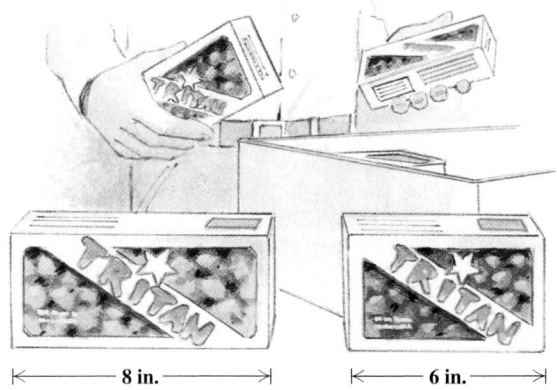

|← 8 in. →| |← 6 in. →|

Simplify.

93. $\dfrac{16 \cdot 9 \cdot 4}{15 \cdot 8 \cdot 12}$ **94.** $\dfrac{9 \cdot 8xy}{2xy \cdot 36}$

95. $\dfrac{27pqrs}{9prst}$ **96.** $\dfrac{512}{192}$

97. $\dfrac{15 \cdot 4xy \cdot 9}{6 \cdot 25x \cdot 15y}$ **98.** $\dfrac{10x \cdot 12 \cdot 25y}{2 \cdot 30x \cdot 20y}$

99. $\dfrac{\frac{27ab}{15mn}}{\frac{18bc}{25np}}$ **100.** $\dfrac{\frac{45xyz}{24ab}}{\frac{30xz}{32ac}}$

101. $\dfrac{5\frac{3}{4} rs}{4\frac{1}{2} st}$ **102.** $\dfrac{3\frac{5}{7} mn}{2\frac{4}{5} np}$

Find the area of each figure.

103.

$\frac{7}{9}$ m $\frac{7}{9}$ m

$\frac{4}{5}$ m

104.

$\frac{5}{4}$ m

$\frac{10}{7}$ m

> **Set of Integers**
>
> The set of integers $= \{\ldots, -4, -3, -2, -1, 0, 1, 2, 3, 4, \ldots\}$.

Integers are associated with many real-world problems and situations.

EXAMPLE 1   State which integer(s) corresponds to each situation.

a) In 2003, the Colonial Pipeline Company was fined a record $34 million for pollution crimes. (*Source*: GreenConsumerGuide.com)

b) The lowest point in New Orleans is 8 ft below sea level.

c) To lose one pound of fat, it is necessary for most people to create a 3500-calorie deficit. (*Source*: World Health Organization)

Solution

a) The integer $-34,000,000$ corresponds to a fine of $34 million.

b) The integer -8 corresponds to 8 ft below sea level.

New Orleans

Sea level

-8 ft

c) The integer -3500 corresponds to a deficit of 3500 calories.

The Rational Numbers

Although numbers like $\frac{5}{9}$ are built out of integers, these numbers are not themselves integers. Another set, the **rational numbers**, contains fractions and decimals, as well as the integers. Some examples of rational numbers are

$$\frac{5}{9}, \quad -\frac{4}{7}, \quad 95, \quad -16, \quad 0, \quad \frac{-35}{8}, \quad 2.4, \quad -0.31.$$

In Section 1.7, we show that $-\frac{4}{7}$ can be written as $\frac{-4}{7}$ or $\frac{4}{-7}$. Indeed, every number listed above can be written as an integer over an integer. For example, 95 can be written as $\frac{95}{1}$ and 2.4 can be written as $\frac{24}{10}$. In this manner, any *rational* number can be expressed as the *ratio* of two integers. Rather than attempt to list all rational numbers, we use this idea of ratio to describe the set as follows.

Set of Rational Numbers

$$\text{The set of rational numbers} = \left\{ \frac{a}{b} \,\middle|\, a \text{ and } b \text{ are integers and } b \neq 0 \right\}.$$

This is read "the set of all numbers $\frac{a}{b}$, where a and b are integers and b does not equal zero."

In Section 1.7, we explain why b cannot equal 0.

To *graph* a number is to mark its location on a number line.

EXAMPLE 2   Graph each of the following rational numbers: **(a)** $\frac{5}{2}$; **(b)** -3.2; **(c)** $\frac{11}{8}$.

Solution

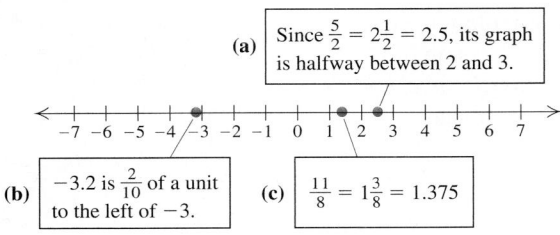

(a) Since $\frac{5}{2} = 2\frac{1}{2} = 2.5$, its graph is halfway between 2 and 3.

(b) -3.2 is $\frac{2}{10}$ of a unit to the left of -3.

(c) $\frac{11}{8} = 1\frac{3}{8} = 1.375$

It is important to remember that every rational number can be written as a fraction or a decimal.

EXAMPLE 3   Convert to decimal notation: $-\frac{5}{8}$.

Solution   We first find decimal notation for $\frac{5}{8}$. Since $\frac{5}{8}$ means $5 \div 8$, we divide.

$$
\begin{array}{r}
0.6\,2\,5 \\
8\overline{)5.0\,0\,0} \\
4\,8\,0\,0 \\
\hline
2\,0\,0 \\
1\,6\,0 \\
\hline
4\,0 \\
4\,0 \\
\hline
0
\end{array}
$$
$\leftarrow$ The remainder is 0.

Thus, $\frac{5}{8} = 0.625$, so $-\frac{5}{8} = -0.625$.

Because the division in Example 3 ends with the remainder 0, we consider -0.625 a **terminating decimal**. If we are "bringing down" zeros and a remainder reappears, we have a **repeating decimal**, as shown in the next example.

EXAMPLE 4   Convert to decimal notation: $\frac{7}{11}$.

Solution   We divide:

$$
\begin{array}{r}
0.6\,3\,6\,3\ldots \\
11\overline{)7.0\,0\,0\,0} \\
\underline{6\,6} \\
4\,0 \\
\underline{3\,3} \\
7\,0 \\
\underline{6\,6} \\
4\,0
\end{array}
$$

4 reappears as a remainder, so the pattern of 6's and 3's in the quotient will continue.

We abbreviate repeating decimals by writing a bar over the repeating part—in this case, $0.\overline{63}$. Thus, $\frac{7}{11} = 0.\overline{63}$.

Although we do not prove it here, every rational number can be expressed as either a terminating or repeating decimal, and every terminating or repeating decimal can be expressed as a ratio of two integers.

Real Numbers and Order

Some numbers, when written in decimal form, neither terminate nor repeat. Such numbers are called **irrational numbers**.

What sort of numbers are irrational? One example is π (the Greek letter *pi*, read "pie"), which is used to find the area and the circumference of a circle: $A = \pi r^2$ and $C = 2\pi r$.

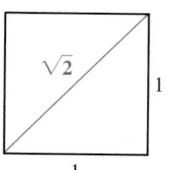

Another irrational number, $\sqrt{2}$ (read "the square root of 2"), is the length of the diagonal of a square with sides of length 1. It is also the number that, when multiplied by itself, gives 2. No rational number can be multiplied by itself to get 2, although some approximations come close:

1.4 is an *approximation* of $\sqrt{2}$ because $(1.4)(1.4) = 1.96$;

1.41 is a better approximation because $(1.41)(1.41) = 1.9881$;

1.4142 is an even better approximation because $(1.4142)(1.4142) = 1.99996164$.

To approximate $\sqrt{2}$ on some calculators, simply press ② and then √. With other calculators, press √ ②, and **ENTER**, or consult a manual.

EXAMPLE 5

Graph the real number $\sqrt{3}$ on a number line.

Solution   We use a calculator and approximate: $\sqrt{3} \approx 1.732$ (" $\approx$ " means "approximately equals"). Then we locate this number on a number line.

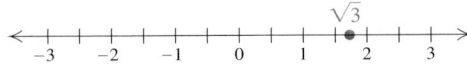

technology connection

To approximate $\sqrt{3}$ on most graphing calculators, we press ✓ and then enter 3 enclosed by parentheses. Some graphing calculators will supply the left parenthesis automatically when ✓ is pressed.

   Approximate each of the following to nine decimal places.

1. $\sqrt{5}$ **2.** $\sqrt{7}$

3. $\sqrt{13}$ **4.** $\sqrt{27}$

5. $\sqrt{38}$ **6.** $\sqrt{50}$

   The rational numbers and the irrational numbers together correspond to all the points on a number line and make up what is called the **real-number system**.

Set of Real Numbers

The set of real numbers = The set of all numbers corresponding to points on the number line.

   The following figure shows the relationships among various kinds of numbers.

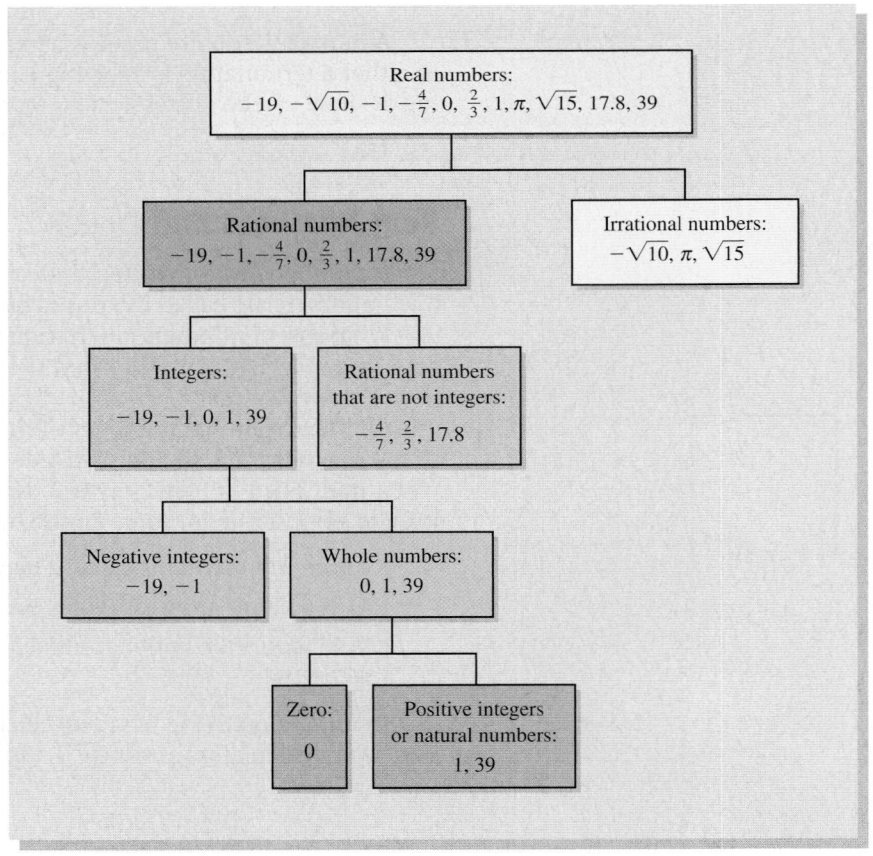

EXAMPLE 6   Which numbers in the following list are **(a)** whole numbers? **(b)** integers? **(c)** rational numbers? **(d)** irrational numbers? **(e)** real numbers?

$$-38, \quad -\frac{8}{5}, \quad 0, \quad 3, \quad 4.5, \quad \sqrt{30}, \quad 52$$

Solution

a) 0, 3, and 52 are whole numbers.

b) $-38, 0, 3,$ and 52 are integers.

c) $-38, -\frac{8}{5}, 0, 3, 4.5,$ and 52 are rational numbers.

d) $\sqrt{30}$ is an irrational number.

e) $-38, -\frac{8}{5}, 0, 3, 4.5, \sqrt{30},$ and 52 are real numbers.

Real numbers are named in order on the number line, with larger numbers further to the right. For any two numbers, the one to the left is less than the one to the right. We use the symbol **<** to mean "**is less than.**" The sentence $-8 < 6$ means "-8 is less than 6." The symbol **>** means "**is greater than.**" The sentence $-3 > -7$ means "-3 is greater than -7."

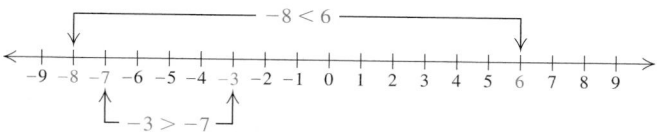

EXAMPLE 7   Use either $<$ or $>$ for ▨ to write a true sentence.

a) 2 ▨ 9

b) -3.45 ▨ 1.32

c) 6 ▨ -12

d) -18 ▨ -5

e) $\frac{7}{11}$ ▨ $\frac{5}{8}$

Solution

a) Since 2 is to the left of 9 on a number line, we know that 2 is less than 9, so $2 < 9$.

b) Since -3.45 is to the left of 1.32, we have $-3.45 < 1.32$.

c) Since 6 is to the right of -12, we have $6 > -12$.

d) Since -18 is to the left of -5, we have $-18 < -5$.

e) We convert to decimal notation: $\frac{7}{11} = 0.\overline{63}$ and $\frac{5}{8} = 0.625$. Thus, $\frac{7}{11} > \frac{5}{8}$.
    We also could have used a common denominator: $\frac{7}{11} = \frac{56}{88} > \frac{55}{88} = \frac{5}{8}$.

Sentences like "$a < -5$" and "$-3 > -8$" are **inequalities**. It is useful to remember that every inequality can be written in two ways. For example,

$$-3 > -8 \quad \text{has the same meaning as} \quad -8 < -3.$$

It may be helpful to think of an inequality sign as an "arrow" with the smaller side pointing to the smaller number.

Note that $a > 0$ means that a represents a positive real number and $a < 0$ means that a represents a negative real number.

Statements like $a \le b$ and $b \ge a$ are also inequalities. We read $a \le b$ as "a **is less than or equal to** b" and $a \ge b$ as "a **is greater than or equal to** b."

EXAMPLE 8   Classify each inequality as true or false.

a) $-3 \le 5$ **b)** $-3 \le -3$ **c)** $-5 \ge 4$

Solution

a) $-3 \le 5$ is *true* because $-3 < 5$ is true.

b) $-3 \le -3$ is *true* because $-3 = -3$ is true.

c) $-5 \ge 4$ is *false* since neither $-5 > 4$ nor $-5 = 4$ is true.

Student Notes————

It is important to remember that just because an equation or inequality is written or printed, it is not necessarily *true*. For instance, $6 = 7$ is an equation and $2 > 5$ is an inequality. Of course, both statements are *false*.

Absolute Value

There is a convenient terminology and notation for the distance a number is from 0 on a number line. It is called the **absolute value** of the number.

> **Absolute Value**
>
> We write $|a|$, read "the absolute value of a," to represent the number of units that a is from zero.

EXAMPLE 9   Find each absolute value: **(a)** $|-3|$; **(b)** $|7.2|$; **(c)** $|0|$.

Solution

a) $|-3| = 3$ since -3 is 3 units from 0.

b) $|7.2| = 7.2$ since 7.2 is 7.2 units from 0.

c) $|0| = 0$ since 0 is 0 units from itself.

Distance is never negative, so numbers that are opposites have the same absolute value. If a number is nonnegative, its absolute value is the number itself. If a number is negative, its absolute value is its opposite.

technology connection

Most graphing calculators use the notation abs (2) to indicate the absolute value of 2. This is often accessed using the NUM option of the MATH key. When using a graphing calculator for this, be sure to distinguish between the ⊝ and ⊝ keys (see p. 49 in Section 1.6).

Exercise Set

1.4

☞ *Concept Reinforcement*   *In each of Exercises 1–8, fill in the blank using one of the following words: natural number, whole number, integer, rational number, terminating, repeating, irrational number, absolute value.*

1. Division can be used to show that $\frac{4}{7}$ can be written as a(n) _____ decimal.

2. Division can be used to show that $\frac{3}{20}$ can be written as a(n) _____ decimal.

3. If a number is a(n) _____, it is either a whole number or the opposite of a whole number.

4. 0 is the only _____ that is not a natural number.

5. Any number of the form a/b, where a and b are integers, with $b \neq 0$, is an example of a(n) _____.

6. A number like $\sqrt{5}$, which cannot be written precisely in fraction or decimal notation, is an example of a(n) _____.

7. If a number is a(n) _____, then it can be thought of as a counting number.

8. When two numbers are opposites, they have the same _____.

State which real number(s) correspond to each situation.

9. The Dead Sea is 1349 feet below sea level, whereas Mt. Everest is 29,035 feet above sea level.

10. Using a NordicTrack exercise machine, Kit burned 150 calories. She then drank an isotonic drink containing 65 calories.

11. The highest temperature ever reached on the earth was 950 million degrees Fahrenheit (F). The lowest temperature ever reached on the earth was approximately 460°F below zero.
    Source: *Guinness World Records* 2001

12. In Burlington, Vermont, the record low temperature for Washington's Birthday is 19°F below zero. The record high for the date is 59°F above zero.

13. In the 2003 Masters Tournament, golfer Tiger Woods finished two over par. In the 2003 Nissan Open, he finished six under par.

14. Ignition occurs 10 seconds before liftoff. A spent fuel tank is detached 235 seconds after liftoff.

15. Janice deposited $750 in a savings account. Two weeks later, she withdrew $125.

16. *Birth and death rates.* Recently, the world birth rate was 2043 per hundred thousand. The death rate was 883 per hundred thousand.
    Source: *Central Intelligence Agency, 2003*

17. In bowling, the Jets are 34 pins behind the Strikers after one game. Describe the situation from the viewpoint of each team.

18. The halfback gained 8 yd on the first play. The quarterback was tackled for a 5-yd loss on the second play.

Graph each rational number on a number line.

19. $\frac{10}{3}$

20. $-\frac{17}{5}$

21. -4.3

22. 3.87

23. -2

24. 5

Write decimal notation for each number.

25. $\frac{7}{8}$ **26.** $-\frac{1}{8}$ **27.** $-\frac{3}{4}$

28. $\frac{5}{6}$ **29.** $\frac{7}{6}$ **30.** $\frac{5}{12}$

31. $\frac{2}{3}$ **32.** $\frac{1}{4}$ **33.** $-\frac{1}{2}$

34. $-\frac{3}{8}$ *Aha!* **35.** $\frac{13}{100}$ **36.** $-\frac{7}{20}$

Write a true sentence using either < or >.

37. $-9 \ \blacksquare \ 4$ **38.** $9 \ \blacksquare \ 0$

39. $7 \ \blacksquare \ 0$ **40.** $8 \ \blacksquare \ -8$

41. $-6 \ \blacksquare \ 6$ **42.** $0 \ \blacksquare \ -7$

43. $-8 \ \blacksquare \ -5$ **44.** $-4 \ \blacksquare \ -3$

45. $-5 \ \blacksquare \ -11$ **46.** $-3 \ \blacksquare \ -4$

47. $-12.5 \ \blacksquare \ -9.4$ **48.** $-10.3 \ \blacksquare \ -14.5$

49. $\frac{5}{12} \ \blacksquare \ \frac{11}{25}$ **50.** $-\frac{14}{17} \ \blacksquare \ -\frac{27}{35}$

For each of the following, write a second inequality with the same meaning.

51. $-7 > x$ **52.** $a > 9$

53. $-10 \leq y$ **54.** $12 \geq t$

Classify each inequality as either true or false.

55. $-3 \geq -11$ **56.** $5 \leq -5$

57. $0 \geq 8$ **58.** $-5 \leq 7$

59. $-8 \leq -8$ **60.** $8 \geq 8$

Find each absolute value.

61. $|-58|$ **62.** $|-47|$

63. $|17|$ **64.** $|3.1|$

65. $|5.6|$ **66.** $\left|-\frac{2}{5}\right|$

67. $|329|$ **68.** $|-456|$

69. $\left|-\frac{9}{7}\right|$ **70.** $|8.02|$

71. $|0|$ **72.** $|-1.07|$

73. $|x|$, for $x = -8$ **74.** $|a|$, for $a = -5$

For Exercises 75–80, consider the following list:

$$-83, \quad -4.7, \quad 0, \quad \tfrac{5}{9}, \quad \pi, \quad \sqrt{17}, \quad 8.31, \quad 62.$$

75. List all rational numbers.

76. List all natural numbers.

77. List all integers.

78. List all irrational numbers.

79. List all real numbers.

80. List all nonnegative integers.

81. Is every integer a rational number? Why or why not?

82. Is every integer a natural number? Why or why not?

SKILL MAINTENANCE

83. Evaluate $3xy$ for $x = 2$ and $y = 7$. [1.1]

84. Use a commutative law to write an expression equivalent to $ab + 5$. [1.2]

SYNTHESIS

85. Is the absolute value of a number always positive? Why or why not?

86. How many rational numbers are there between 0 and 1? Justify your answer.

87. Does "nonnegative" mean the same thing as "positive"? Why or why not?

List in order from least to greatest.

88. $13, -12, 5, -17$

89. $-23, 4, 0, -17$

90. $-\frac{2}{3}, \frac{1}{2}, -\frac{3}{4}, -\frac{5}{6}, \frac{3}{8}, \frac{1}{6}$

91. $\frac{4}{5}, \frac{4}{3}, \frac{4}{8}, \frac{4}{6}, \frac{4}{9}, \frac{4}{2}, -\frac{4}{3}$

Write a true sentence using either $<, >,$ *or* $=$.

92. $|-5| \ \blacksquare \ |-2|$

93. $|4| \ \blacksquare \ |-7|$

94. $|-8| \ \blacksquare \ |8|$

95. $|23| \ \blacksquare \ |-23|$

96. $|-3| \ \blacksquare \ |5|$

Solve. Consider only integer replacements.

Aha! **97.** $|x| = 7$

98. $|x| < 3$

99. $2 < |x| < 5$

Given that $0.3\overline{3} = \frac{1}{3}$ *and* $0.6\overline{6} = \frac{2}{3}$, *express each of the following as a ratio of two integers.*

100. $0.1\overline{1}$

101. $0.9\overline{9}$

102. $5.5\overline{5}$

103. $7.7\overline{7}$

To the student and instructor: *The calculator icon,* 🖩*, is used to indicate those exercises designed to be solved with a calculator.*

🖩 **104.** When Helga's calculator gives a decimal value for $\sqrt{2}$ and that value is promptly squared, the result is 2. Yet when that same decimal approximation is entered by hand and then squared, the result is not exactly 2. Why do you suppose this is?

🖩 **105.** Is the following statement true? Why or why not?
$$\sqrt{a^2} = |a| \quad \text{for any real number } a.$$

Addition of Real Numbers

1.5

Adding with a Number Line • Adding Without a Number Line •
Problem Solving • Combining Like Terms

We now consider addition of real numbers. To gain understanding, we will use a number line first. After observing the principles involved, we will develop rules that allow us to work more quickly without a number line.

Adding with a Number Line

To add $a + b$ on a number line, we start at a and move according to b.

a) If b is positive, we move to the right (the positive direction).

b) If b is negative, we move to the left (the negative direction).

c) If b is 0, we stay at a.

EXAMPLE 1 Add: $-4 + 9$.

Solution To add on a number line, we locate the first number, -4, and then move 9 units to the right. Note that it requires 4 units to reach 0. The difference between 9 and 4 is where we finish.

$$-4 + 9 = 5$$

Start at -4.

Move 9 units to the right.

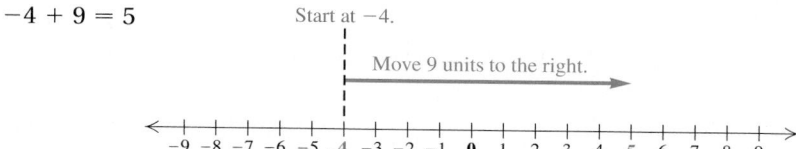

EXAMPLE 2   Add: $3 + (-5)$.

Student Notes _____

Parentheses are essential when a negative sign follows an operation. Just as we would never write $8 \div \times 2$, it is improper to write $3 + -5$.

Solution   We locate the first number, 3, and then move 5 units to the left. Note that it requires 3 units to reach 0. The difference between 5 and 3 is 2, so we finish 2 units to the left of 0.

$$3 + (-5) = -2$$

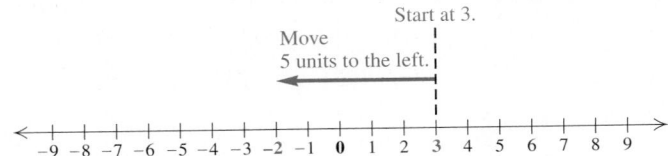

EXAMPLE 3   Add: $-4 + (-3)$.

Solution   After locating -4, we move 3 units to the left. We finish a total of 7 units to the left of 0.

$$-4 + (-3) = -7$$

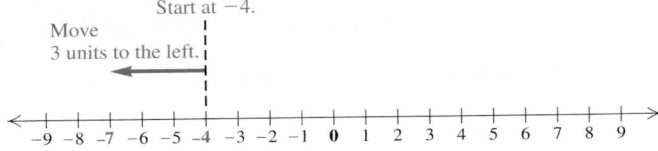

EXAMPLE 4   Add: $-5.2 + 0$.

Solution   We locate -5.2 and move 0 units. Thus we finish where we started, at -5.2.

$$-5.2 + 0 = -5.2$$

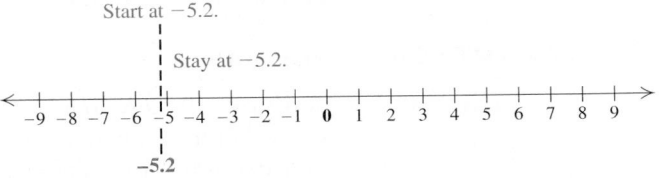

From Examples 1–4, the following rules emerge.

> **Rules for Addition of Real Numbers**
> **1.** *Positive numbers*: Add as usual. The answer is positive.
> **2.** *Negative numbers*: Add absolute values and make the answer negative (see Example 3).
> **3.** *A positive number and a negative number*: Subtract the smaller absolute value from the greater absolute value. Then:
>   **a)** If the positive number has the greater absolute value, the answer is positive (see Example 1).
>   **b)** If the negative number has the greater absolute value, the answer is negative (see Example 2).
>   **c)** If the numbers have the same absolute value, the answer is 0.
> **4.** *One number is zero*: The sum is the other number (see Example 4).

Rule 4 is known as the **identity property of 0**. It says that for any real number a, we have $a + 0 = a$.

Adding Without a Number Line

The rules listed above can be used without drawing a number line.

EXAMPLE 5   Add without using a number line.

a) $-12 + (-7)$ **b)** $-1.4 + 8.5$

c) $-36 + 21$ **d)** $1.5 + (-1.5)$

e) $-\frac{7}{8} + 0$ **f)** $\frac{2}{3} + \left(-\frac{5}{8}\right)$

Solution

a) $-12 + (-7) = -19$ Two negatives. *Think:* Add the absolute values, 12 and 7, to get 19. Make the answer *negative*, -19.

b) $-1.4 + 8.5 = 7.1$ A negative and a positive. *Think:* The difference of absolute values is $8.5 - 1.4$, or 7.1. The positive number has the larger absolute value, so the answer is *positive*, 7.1.

c) $-36 + 21 = -15$ A negative and a positive. *Think:* The difference of absolute values is $36 - 21$, or 15. The negative number has the larger absolute value, so the answer is *negative*, -15.

d) $1.5 + (-1.5) = 0$ A negative and a positive. *Think:* Since the numbers are opposites, they have the same absolute value and the answer is 0.

e) $-\dfrac{7}{8} + 0 = -\dfrac{7}{8}$ One number is zero. The sum is the other number, $-\frac{7}{8}$.

Study Skills _____

Two (or More) Heads Are Better Than One

Consider forming a study group with some of your fellow students. Exchange telephone numbers, schedules, and any e-mail addresses so that you can coordinate study time for homework and tests.

f) $\dfrac{2}{3} + \left(-\dfrac{5}{8}\right) = \dfrac{16}{24} + \left(-\dfrac{15}{24}\right)$ This is similar to part (b) above. We find a common denominator and then add.

$\qquad\qquad\qquad = \dfrac{1}{24}$

If we are adding several numbers, some positive and some negative, the commutative and associative laws allow us to add all the positives, then add all the negatives, and then add the results. Of course, we can also add from left to right, if we prefer.

EXAMPLE 6   Add: $15 + (-2) + 7 + 14 + (-5) + (-12)$.

Solution

$$15 + (-2) + 7 + 14 + (-5) + (-12)$$

$\qquad = 15 + 7 + 14 + (-2) + (-5) + (-12)$ Using the commutative law of addition

$\qquad = (15 + 7 + 14) + [(-2) + (-5) + (-12)]$ Using the associative law of addition

$\qquad = 36 + (-19)$ Adding the positives; adding the negatives

$\qquad = 17$ Adding a positive and a negative

Problem Solving

Addition of real numbers occurs in many real-world applications.

EXAMPLE 7   Lake level. Between early June 2001 through the first week of June 2003, the south end of the Great Salt Lake dropped $2\frac{1}{4}$ ft, rose $\frac{3}{4}$ ft, and dropped 2 ft (*Source*: U.S. Geological Survey). By how much did the level change over 2 yr?

Solution   The problem translates to a sum:

Rewording: The 1st   the 2nd   the 3rd   the total
         change   plus   change   plus   change   is   change.

Translating: $-2\dfrac{1}{4}$ $+$ $\dfrac{3}{4}$ $+$ (-2) $=$ Total change.

Adding from left to right, we have

$$-2\frac{1}{4} + \frac{3}{4} + (-2) = -\frac{9}{4} + \frac{3}{4} + (-2) = -\frac{6}{4} + (-2) = -1\frac{1}{2} + (-2) = -3\frac{1}{2}.$$

The lake level dropped $3\frac{1}{2}$ ft between early June 2001 and the first week of June 2003.

Combining Like Terms

When two terms have variable factors that are exactly the same, like $5ab$ and $7ab$, the terms are called **like**, or **similar**, **terms**.* The distributive law enables us to **combine**, or **collect**, **like terms**. The above rules for addition will again apply.

EXAMPLE 8   Combine like terms.

a) $-7x + 9x$ **b)** $2a + (-3b) + (-5a) + 9b$
c) $6 + y + (-3.5y) + 2$

Solution

a) $-7x + 9x = (-7 + 9)x$ Using the distributive law
          $= 2x$ Adding -7 and 9

b) $2a + (-3b) + (-5a) + 9b$
    $= 2a + (-5a) + (-3b) + 9b$ Using the commutative law of addition
    $= (2 + (-5))a + (-3 + 9)b$ Using the distributive law
    $= -3a + 6b$ Adding

c) $6 + y + (-3.5y) + 2 = y + (-3.5y) + 6 + 2$ Using the commutative law of addition
                 $= (1 + (-3.5))y + 6 + 2$ Using the distributive law
                 $= -2.5y + 8$ Adding

With practice we can leave out some steps, combining like terms mentally. Note that numbers like 6 and 2 in the expression $6 + y + (-3.5y) + 2$ are constants and are also considered to be like terms.

*Like terms are discussed in greater detail in Section 1.8.

Exercise Set

1.5

FOR EXTRA HELP

Student's
Solutions
Manual

Digital Video Tutor
CD 1
Videotape 1

AW Math
Tutor Center

MathXL Tutorials
on CD

Math⬩XL
MathXL

MyMathLab
MyMathLab

> *Concept Reinforcement*   *In each of Exercises 1–6, match the term with a like term from the column on the right.*

1. _____ $8n$

2. _____ $7m$

3. _____ 43

4. _____ $28z$

5. _____ $-2x$

6. _____ $-9t$

a) $-3z$

b) $5x$

c) $2t$

d) $-4m$

e) 9

f) $-3n$

Add using a number line.

7. $5 + (-8)$

8. $2 + (-5)$

9. $-5 + 9$

10. $-3 + 8$

11. $8 + (-8)$

12. $6 + (-6)$

13. $-3 + (-5)$

14. $-4 + (-6)$

Add. Do not use a number line except as a check.

15. $-27 + 0$

16. $-6 + 0$

17. $0 + (-8)$

18. $0 + (-2)$

19. $12 + (-12)$

20. $17 + (-17)$

21. $-24 + (-17)$

22. $-17 + (-25)$

23. $-13 + 13$

24. $-18 + 18$

25. $18 + (-11)$

26. $8 + (-5)$

27. $10 + (-12)$

28. $9 + (-13)$

29. $-3 + 14$

30. $13 + (-6)$

31. $-14 + (-19)$

32. $11 + (-9)$

33. $19 + (-19)$

34. $-20 + (-6)$

35. $23 + (-5)$

36. $-15 + (-7)$

37. $-31 + (-14)$

38. $40 + (-8)$

39. $40 + (-40)$

40. $-25 + 25$

41. $85 + (-65)$

42. $63 + (-18)$

43. $-3.6 + 1.9$

44. $-6.5 + 4.7$

45. $-5.4 + (-3.7)$

46. $-3.8 + (-9.4)$

47. $\frac{-3}{5} + \frac{4}{5}$

48. $\frac{-2}{7} + \frac{3}{7}$

49. $\frac{-4}{7} + \frac{-2}{7}$

50. $\frac{-5}{9} + \frac{-2}{9}$

51. $-\frac{2}{5} + \frac{1}{3}$

52. $-\frac{4}{13} + \frac{1}{2}$

53. $\frac{-4}{9} + \frac{2}{3}$

54. $\frac{-1}{6} + \frac{1}{3}$

55. $35 + (-14) + (-19) + (-5)$

56. $28 + (-44) + 17 + 31 + (-94)$

Aha! **57.** $-4.9 + 8.5 + 4.9 + (-8.5)$

58. $24 + 3.1 + (-44) + (-8.2) + 63$

Solve. Write your answer as a complete sentence.

59. *Gas prices.* In a recent year, the price of a gallon of 87-octane gasoline was $1.69. The price then rose 5¢, dropped 3¢, and then rose 7¢. By how much did the price change during that period?

60. *Oil prices.* In a recent year, the price of a gallon of home heating oil was $1.28. The price then rose 6¢, dropped 2¢, and then rose 5¢. By how much did the price change during that period?

61. *Telephone bills.* Maya's cell-phone bill for July was $82. She sent a check for $50 and then ran up $63 in charges for August. What was her new balance?

62. *Profits and losses.* The following table shows the profits and losses of Fitness City over a 3-yr period. Find the profit or loss after this period of time.

Year	Profit or loss
2002	−$26,500
2003	−$10,200
2004	+$32,400

63. *Yardage gained.* In an intramural football game, the quarterback attempted passes with the following results.

First try   13-yd gain
Second try   12-yd loss
Third try   21-yd gain

Find the total gain (or loss).

64. *Account balance.* Leah has $350 in a checking account. She writes a check for $530, makes a deposit of $75, and then writes a check for $90. What is the balance in the account?

65. *Credit-card bills.* Lyle's credit-card bill indicates that he owes $470. He sends a check to the credit-card company for $45, charges another $160 in merchandise, and then pays off another $500 of his bill. What is Lyle's new balance?

66. *Stock growth.* In the course of one day, the value of a share of America Online rose $0.82, fell $0.90, and then rose $0.25. How much had the stock's value risen or fallen at the end of the day?

67. *Peak elevation.* The tallest mountain in the world, as measured from base to peak, is Mauna Kea in Hawaii. From a base 19,684 ft below sea level, it rises 33,480 ft. What is the elevation of its peak?
Source: *The Guinness Book of Records*, 1999

68. *Class size.* During the first two weeks of the semester, 5 students withdrew from Elisa's algebra class, 8 students were added to the class, and 6 students were dropped as "no-shows." By how many students did the original class size change?

Combine like terms.

69. $8a + 6a$

70. $3x + 8x$

71. $-3x + 12x$

72. $2m + (-7m)$

73. $5t + 8t$

74. $5a + 8a$

75. $7m + (-9m)$

76. $-4x + 4x$

77. $-5a + (-2a)$

78. $10n + (-17n)$

79. $-3 + 8x + 4 + (-10x)$

80. $8a + 5 + (-a) + (-3)$

Find the perimeter of each figure.

81.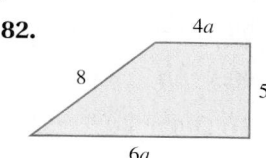

82.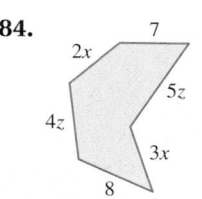

83.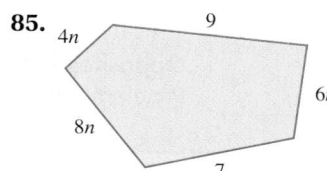

84.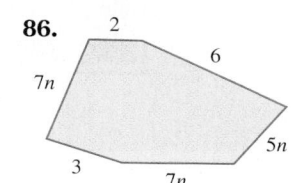

85.

86.

87. Explain in your own words why the sum of two negative numbers is negative.

88. Without performing the actual addition, explain why the sum of all integers from -10 to 10 is 0.

SKILL MAINTENANCE

89. Multiply: $7(3z + y + 2)$. [1.2]

90. Divide and simplify: $\frac{7}{2} \div \frac{3}{8}$. [1.3]

SYNTHESIS

91. Under what circumstances will the sum of one positive number and several negative numbers be positive?

92. Is it possible to add real numbers without knowing how to calculate $a - b$ with a and b both nonnegative and $a \geq b$? Why or why not?

93. *Stock prices.* The value of EKB stock rose $2.38 and then dropped $3.25 before finishing at $64.38. What was the stock's original value?

94. *Sports-card values.* The value of a sports card dropped $12 and then rose $17.50 before settling at $61. What was the original value of the card?

Find the missing term or terms.

95. $4x +$ ____ $+ (-9x) + (-2y) = -5x - 7y$

96. $-3a + 9b +$ ____ $+ 5a = 2a - 6b$

97. $3m + 2n +$ ____ $+ (-2m) = 2n + (-6m)$

98. ____ $+ 9x + (-4y) + x = 10x - 7y$

 99. $7t + 23 +$ ____ $+$ ____ $= 0$

100. *Geometry.* The perimeter of a rectangle is $7x + 10$. If the length of the rectangle is 5, express the width in terms of x.

101. *Golfing.* After five rounds of golf, a golf pro was 3 under par twice, 2 over par once, 2 under par once, and 1 over par once. On average, how far above or below par was the golfer?

1.6 Subtraction of Real Numbers

Opposites and Additive Inverses • Subtraction • Problem Solving

In arithmetic, when a number b is subtracted from another number a, the difference, $a - b$, is the number that when added to b gives a. For example, $45 - 17 = 28$ because $28 + 17 = 45$. We will use this approach to develop an efficient way of finding the value of $a - b$ for any real numbers a and b. Before doing so, however, we must develop some terminology.

Opposites and Additive Inverses

Numbers such as 6 and -6 are *opposites*, or *additive inverses*, of each other. Whenever opposites are added, the result is 0; and whenever two numbers add to 0, those numbers are opposites.

EXAMPLE 1 Find the opposite of each number: **(a)** 34; **(b)** -8.3; **(c)** 0.

Solution

a) The opposite of 34 is -34: $34 + (-34) = 0$.

b) The opposite of -8.3 is 8.3: $-8.3 + 8.3 = 0$.

c) The opposite of 0 is 0: $0 + 0 = 0$.

To write the opposite, we use the symbol $-$, as follows.

> **Opposite**
>
> The *opposite*, or *additive inverse*, of a number a is written $-a$ (read "the opposite of a" or "the additive inverse of a").

Note that if we take a number, say 8, and find its opposite, -8, and then find the opposite of the result, we will have the original number, 8, again.

> ### The Opposite of an Opposite
> For any real number a,
> $$-(-a) = a.$$
> (The opposite of the opposite of a is a.)

EXAMPLE 2   Find $-x$ and $-(-x)$ when $x = 16$.

Solution

   If $x = 16$, then $-x = -16$.   The opposite of 16 is -16.

   If $x = 16$, then $-(-x) = -(-16) = 16$.   The opposite of the opposite of 16 is 16.

EXAMPLE 3   Find $-x$ and $-(-x)$ when $x = -3$.

Solution

   If $x = -3$, then $-x = -(-3) = 3$.   The opposite of -3 is 3.

   If $x = -3$, then $-(-x) = -(-(-3)) = -(\ 3\ ) = -3$.

*Student Notes*_____

As you read mathematics, it is important to verbalize correctly the words and symbols to yourself. Consistently reading the expression $-x$ as "the opposite of x" is a good step in this direction.

Note in Example 3 that an extra set of parentheses is used to show that we are substituting the negative number -3 for x. The notation $-\ -x$ is not used.

A symbol such as -8 is usually read "negative 8." It could be read "the additive inverse of 8," because the additive inverse of 8 is negative 8. It could also be read "the opposite of 8," because the opposite of 8 is -8.

A symbol like $-x$, which has a variable, should be read "the opposite of x" or "the additive inverse of x" and *not* "negative x," since to do so suggests that $-x$ represents a negative number.

The symbol "$-$" is read differently depending on where it appears. For example, $-5 - (-x)$ is read "negative five minus the opposite of x."

EXAMPLE 4   Write each of the following in words.

a) $2 - 8$ **b)** $5 - (-4)$ **c)** $6 - (-x)$

Solution

a) $2 - 8$ is read "two minus eight."

b) $5 - (-4)$ is read "five minus negative four."

c) $6 - (-x)$ is read "six minus the opposite of x."

As we saw in Example 3, $-x$ can represent a positive number. This notation can be used to restate a result from Section 1.5 as *the law of opposites*:

The Law of Opposites

For any two numbers a and $-a$,

$$a + (-a) = 0.$$

(When opposites are added, their sum is 0.)

A negative number is said to have a "negative *sign*." A positive number is said to have a "positive *sign*." If we change a number to its opposite, or additive inverse, we say that we have "changed or reversed its sign."

EXAMPLE 5   Change the sign (find the opposite) of each number: **(a)** -3; **(b)** -10; **(c)** 14.

Solution

a) When we change the sign of -3, we obtain 3.

b) When we change the sign of -10, we obtain 10.

c) When we change the sign of 14, we obtain -14.

Subtraction

Opposites are helpful when subtraction involves negative numbers. To see why, look for a pattern in the following:

Subtracting		*Adding the Opposite*
$9 - 5 = 4$	since $4 + 5 = 9$	$9 + (-5) = 4$
$5 - 8 = -3$	since $-3 + 8 = 5$	$5 + (-8) = -3$
$-6 - 4 = -10$	since $-10 + 4 = -6$	$-6 + (-4) = -10$
$-7 - (-10) = 3$	since $3 + (-10) = -7$	$-7 + 10 = 3$
$-7 - (-2) = -5$	since $-5 + (-2) = -7$	$-7 + 2 = -5$

The matching results suggest that we can subtract by adding the opposite of the number being subtracted. This can always be done and often provides the easiest way to subtract real numbers.

Subtraction of Real Numbers

For any real numbers a and b,

$$a - b = a + (-b).$$

(To subtract, add the opposite, or additive inverse, of the number being subtracted.)

EXAMPLE 6   Subtract each of the following and then check with addition.

a) $2 - 6$ **b)** $4 - (-9)$ **c)** $-4.2 - (-3.6)$
d) $-1.8 - (-7.5)$ **e)** $\frac{1}{5} - \left(-\frac{3}{5}\right)$

Solution

a) $2 \overset{\frown}{-} 6 = 2 + (-6) = -4$ The opposite of 6 is -6. We change the
subtraction to addition and add the opposite.
Check: $-4 + 6 = 2$.

b) $4 - (-9) = 4 + 9 = 13$ The opposite of -9 is 9. We change the
subtraction to addition and add the opposite.
Check: $13 + (-9) = 4$.

c) $-4.2 - (-3.6) = -4.2 + 3.6$ Adding the opposite of -3.6.
Check: $-0.6 + (-3.6) = -4.2$.
$= -0.6$

d) $-1.8 - (-7.5) = -1.8 + 7.5$ Adding the opposite
$= 5.7$ *Check*: $5.7 + (-7.5) = -1.8$.

e) $\dfrac{1}{5} - \left(-\dfrac{3}{5}\right) = \dfrac{1}{5} + \dfrac{3}{5}$ Adding the opposite

$\phantom{\dfrac{1}{5} - \left(-\dfrac{3}{5}\right)} = \dfrac{1 + 3}{5}$ A common denominator exists so we
add in the numerator.

$\phantom{\dfrac{1}{5} - \left(-\dfrac{3}{5}\right)} = \dfrac{4}{5}$

Check: $\dfrac{4}{5} + \left(-\dfrac{3}{5}\right) = \dfrac{4}{5} + \dfrac{-3}{5} = \dfrac{4 + (-3)}{5} = \dfrac{1}{5}$.

---

technology connection

On nearly all graphing calculators, it is essential to distinguish between the key for negation and the key for subtraction. To enter a negative number, we use $\boxed{(-)}$ and to subtract, we use $\boxed{-}$. This said, be careful not to rely on a calculator for computations that you will be expected to do by hand.

EXAMPLE 7   Simplify: $8 - (-4) - 2 - (-5) + 3$.

Solution

$$8 - (-4) - 2 - (-5) + 3 = 8 + 4 + (-2) + 5 + 3$$ To subtract, we add the opposite.

$$= 18$$

Recall from Section 1.2 that the terms of an algebraic expression are separated by plus signs. This means that the terms of $5x - 7y - 9$ are $5x$, $-7y$, and -9, since $5x - 7y - 9 = 5x + (-7y) + (-9)$.

EXAMPLE 8   Identify the terms of $4 - 2ab + 7a - 9$.

Solution   We have

$$4 - 2ab + 7a - 9 = 4 + (-2ab) + 7a + (-9), \qquad \text{Rewriting as addition}$$

so the terms are 4, $-2ab$, $7a$, and -9.

EXAMPLE 9   Combine like terms.

a) $1 + 3x - 7x$ **b)** $-5a - 7b - 4a + 10b$

c) $4 - 3m - 9 + 7m$

Study Skills ———————

Indicate the Highlights

Most students find it helpful to draw a star or use a color, felt-tipped highlighter to indicate important concepts or trouble spots that require further study. Most campus bookstores carry a variety of highlighters that permit you to brightly color written material while keeping it easy to read.

Solution

a) $1 + 3x - 7x = 1 + 3x + (-7x)$ Adding the opposite

$$= 1 + (3 + (-7))x \left.\vphantom{\begin{matrix}a\\b\end{matrix}}\right\} \quad \begin{matrix}\text{Using the distributive law.}\\\text{Try to do this mentally.}\end{matrix}$$

$$= 1 + (-4)x$$

$$= 1 - 4x \qquad \begin{matrix}\text{Rewriting as subtraction to be more}\\\text{concise}\end{matrix}$$

b) $-5a - 7b - 4a + 10b = -5a + (-7b) + (-4a) + 10b$ Adding the opposite

$$= -5a + (-4a) + (-7b) + 10b \qquad \begin{matrix}\text{Using the}\\\text{commutative}\\\text{law of addition}\end{matrix}$$

$$= -9a + 3b \qquad \begin{matrix}\text{Combining like}\\\text{terms mentally}\end{matrix}$$

c) $4 - 3m - 9 + 7m = 4 + (-3m) + (-9) + 7m$ Rewriting as addition

$$= 4 + (-9) + (-3m) + 7m \qquad \begin{matrix}\text{Using the commutative}\\\text{law of addition}\end{matrix}$$

$$= -5 + 4m$$

Problem Solving

We use subtraction to solve problems involving differences. These include problems that ask "How much more?" or "How much higher?"

EXAMPLE 10   *Temperature extremes.* The highest temperature ever recorded in the United States is 134°F in Greenland Ranch, California, on July 10, 1913. The lowest temperature ever recorded is −80°F in Prospect Creek, Alaska, on January 23, 1971. (*Source*: National Oceanographic and Atmospheric Administration) How much higher was the temperature in Greenland Ranch than that in Prospect Creek?

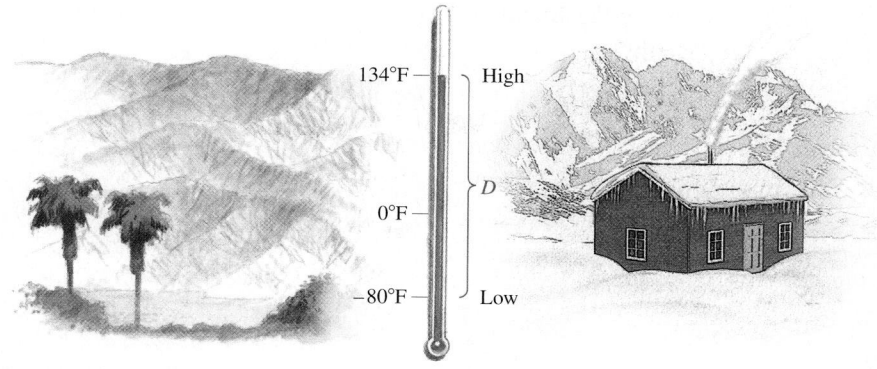

Solution   To find the difference between two temperatures, we always subtract the lower temperature from the higher temperature:

$$\underbrace{\text{Higher temperature}}_{134} \quad - \quad \underbrace{\text{Lower temperature}}_{(-80)}$$

$$= 134 + 80$$
$$= 214.$$

The temperature in Greenland Ranch was 214°F higher than the temperature in Prospect Creek.

Exercise Set

1.6

Concept Reinforcement   *In each of Exercises 1–8, match the expression with the appropriate wording from the column on the right.*

1. ____ $-x$

2. ____ $12 - x$

3. ____ $12 - (-x)$

4. ____ $x - 12$

5. ____ $x - (-12)$

6. ____ $-x - 12$

7. ____ $-x - x$

8. ____ $-x - (-12)$

a) x minus negative twelve

b) The opposite of x minus x

c) The opposite of x minus twelve

d) The opposite of x

e) The opposite of x minus negative twelve

f) Twelve minus the opposite of x

g) Twelve minus x

h) x minus twelve

Write each of the following in words.

9. $4 - 10$

10. $5 - 13$

11. $2 - (-9)$

12. $4 - (-1)$

13. $9 - (-t)$

14. $8 - (-m)$

15. $-x - y$

16. $-a - b$

17. $-3 - (-n)$

18. $-7 - (-m)$

Find the opposite, or additive inverse.

19. 39

20. -17

21. -9

22. $\frac{7}{2}$

23. -3.14

24. 48.2

Find −x when x is each of the following.

25. 23 **26.** −26 **27.** $-\frac{14}{3}$

28. $\frac{1}{328}$ **29.** 0.101 **30.** 0

Find −(−x) when x is each of the following.

31. 72 **32.** 29

33. $-\frac{2}{5}$ **34.** −9.1

Change the sign. (Find the opposite.)

35. −1 **36.** −7

37. 7 **38.** 10

Subtract.

39. 6 − 8 **40.** 4 − 13

41. 0 − 5 **42.** 0 − 8

43. 3 − 9 **44.** 3 − 13

45. 0 − 10 **46.** 0 − 7

47. −9 − (−3) **48.** −9 − (−5)

Aha! **49.** −8 − (−8) **50.** −10 − (−10)

51. 14 − 19 **52.** 12 − 16

53. 30 − 40 **54.** 20 − 27

55. −7 − (−9) **56.** −8 − (−3)

57. −9 − (−9) **58.** −40 − (−40)

59. 5 − 5 **60.** 7 − 7

61. 4 − (−4) **62.** 6 − (−6)

63. −7 − 4 **64.** −6 − 8

65. 6 − (−10) **66.** 3 − (−12)

67. −4 − 15 **68.** −14 − 2

69. −6 − (−5) **70.** −4 − (−3)

71. 5 − (−12) **72.** 5 − (−6)

73. 0 − 5 **74.** 0 − 6

75. −5 − (−2) **76.** −3 − (−1)

77. −7 − 14 **78.** −9 − 16

79. 0 − (−5) **80.** 0 − (−1)

81. −8 − 0 **82.** −9 − 0

83. 3 − (−7) **84.** 12 − (−5)

85. 2 − 25 **86.** 18 − 63

87. −42 − 26 **88.** −18 − 63

89. −51 − 7 **90.** −45 − 4

91. 3.2 − 8.7 **92.** 1.5 − 9.4

93. 0.072 − 1 **94.** 0.825 − 1

95. $\frac{2}{11} - \frac{9}{11}$ **96.** $\frac{3}{7} - \frac{5}{7}$

97. $\frac{-1}{5} - \frac{3}{5}$ **98.** $\frac{-2}{9} - \frac{5}{9}$

99. $-\frac{4}{17} - \left(-\frac{9}{17}\right)$ **100.** $-\frac{2}{13} - \left(-\frac{5}{13}\right)$

In each of Exercises 101–104, translate the phrase to mathematical language and simplify. See Example 10.

101. The difference between 3.8 and −5.2

102. The difference between −2.1 and −5.9

103. The difference between 114 and −79

104. The difference between 23 and −17

105. Subtract 37 from −21.

106. Subtract 19 from −7.

107. Subtract −25 from 9.

108. Subtract −31 from −5.

Simplify.

109. 25 − (−12) − 7 − (−2) + 9

110. 22 − (−18) + 7 + (−42) − 27

111. −31 + (−28) − (−14) − 17

112. −43 − (−19) − (−21) + 25

113. −34 − 28 + (−33) − 44

114. 39 + (−88) − 29 − (−83)

Aha! **115.** −93 + (−84) − (−93) − (−84)

116. 84 + (−99) + 44 − (−18) − 43

Identify the terms in each expression.

117. −7x − 4y

118. 7a − 9b

119. 9 − 5t − 3st

120. −4 − 3x + 2xy

Combine like terms.

121. 4x − 7x

122. 3a − 14a

123. $7a - 12a + 4$

124. $-9x - 13x + 7$

125. $-8n - 9 + n$

126. $-7 + 9n - 8$

127. $3x + 5 - 9x$

128. $2 + 3a - 7$

129. $2 - 6t - 9 - 2t$

130. $-5 + 3b - 7 - 5b$

131. $5y + (-3x) - 9x + 1 - 2y + 8$

132. $14 - (-5x) + 2z - (-32) + 4z - 2x$

133. $13x - (-2x) + 45 - (-21) - 7x$

134. $8x - (-2x) - 14 - (-5x) + 53 - 9x$

Solve.

135. *Changes in temperature.* The Viking 2 Lander spacecraft has determined that temperatures on Mars range from $-125°$ Celsius (C) to 25°C. Find the temperature range on Mars.
Source: The Lunar and Planetary Institute

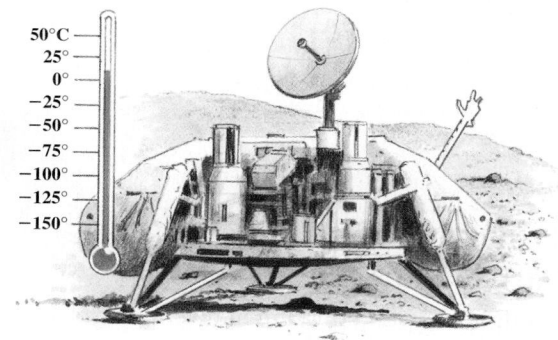

136. *Record temperature drop.* The greatest recorded temperature change in one day occurred in Browning, Montana, when the temperature fell from 44°F to $-56°$F. How much did the temperature drop?
Source: The Guinness Book of Records, 1999

137. *Elevation extremes.* The lowest elevation in Asia, the Dead Sea, is 1349 ft below sea level. The highest elevation in Asia, Mount Everest, is 29,035 ft. Find the difference in elevation.
Source: Time Almanac 2004

138. *Elevation extremes.* The elevation of Mount Whitney, the highest peak in California, is 14,776 ft more than the elevation of Death Valley, California. If Death Valley is 282 ft below sea level, find the elevation of Mount Whitney.
Source: 1999 Information Please Almanac

139. *Changes in elevation.* The lowest point in Africa is Lake Assal, which is 156 m below sea level. The lowest point in South America is the Valdes Peninsula, which is 40 m below sea level. How much lower is Lake Assal than the Valdes Peninsula?

140. *Underwater elevation.* The deepest point in the Pacific Ocean is the Marianas Trench, with a depth of 10,415 m. The deepest point in the Atlantic Ocean is the Puerto Rico Trench, with a depth of 8648 m. What is the difference in elevation of the two trenches?

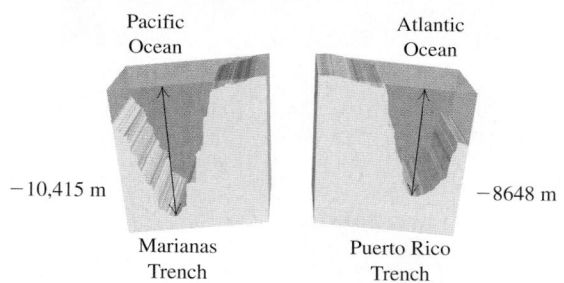

141. Jeremy insists that if you can *add* real numbers, then you can also *subtract* real numbers. Do you agree? Why or why not?

142. Are the expressions $-a + b$ and $a + (-b)$ opposites of each other? Why or why not?

SKILL MAINTENANCE

143. Find the area of a rectangle when the length is 36 ft and the width is 12 ft. [1.1]

144. Find the prime factorization of 864. [1.3]

SYNTHESIS

145. Why might it be advantageous to rewrite a long series of additions and subtractions as all additions?

146. If a and b are both negative, under what circumstances will $a - b$ be negative?

147. *Power outages.* During the Northeast's electrical blackout of August 14, 2003, residents of Bloomfield, New Jersey, lost power at 4:00 P.M. One resident returned from vacation at 3:00 P.M. the following day to find the clocks in her apartment reading 8:00 A.M. At what time, and on what day, was power restored?

Tell whether each statement is true or false for all real numbers m and n. Use various replacements for m and n to support your answer.

148. If $m > n$, then $m - n > 0$.

149. If $m > n$, then $m + n > 0$.

150. If m and n are opposites, then $m - n = 0$.

151. If $m = -n$, then $m + n = 0$.

152. A gambler loses a wager and then loses "double or nothing" (meaning the gambler owes twice as much) twice more. After the three losses, the gambler's assets are $-\$20$. Explain how much the gambler originally bet and how the $20 debt occurred.

153. List the keystrokes needed to compute $-9 - (-7)$.

154. If n is positive and m is negative, what is the sign of $n + (-m)$? Why?

1.7 Multiplication and Division of Real Numbers

Multiplication • Division

We now develop rules for multiplication and division of real numbers. Because multiplication and division are closely related, the rules are quite similar.

Multiplication

We already know how to multiply two nonnegative numbers. To see how to multiply a positive number and a negative number, consider the following pattern in which multiplication is regarded as repeated addition:

This number → $4(-5) = (-5) + (-5) + (-5) + (-5) = -20$ ← This number
decreases by $\quad$ $3(-5) =\qquad (-5) + (-5) + (-5) = -15$ $\quad$ increases by
1 each time. $\quad$ $2(-5) =\qquad\qquad (-5) + (-5) = -10$ $\quad$ 5 each time.
$\qquad\qquad\quad 1(-5) =\qquad\qquad\qquad\quad (-5) = \quad -5$
$\qquad\qquad\quad 0(-5) =\qquad\qquad\qquad\qquad\quad 0 = \quad\;\; 0$

This pattern illustrates that the product of a negative number and a positive number is negative.

The Product of a Negative Number and a Positive Number

To multiply a positive number and a negative number, multiply their absolute values. The answer is negative.

EXAMPLE 1

Multiply: **(a)** $8(-5)$; **(b)** $-\frac{1}{3} \cdot \frac{5}{7}$.

Solution

a) $8(-5) = -40$ *Think*: $8 \cdot 5 = 40$; make the answer negative.

b) $-\frac{1}{3} \cdot \frac{5}{7} = -\frac{5}{21}$ *Think*: $\frac{1}{3} \cdot \frac{5}{7} = \frac{5}{21}$; make the answer negative.

The pattern developed above includes not just products of positive and negative numbers, but a product involving zero as well.

The Multiplicative Property of Zero

For any real number a,

$$0 \cdot a = a \cdot 0 = 0.$$

(The product of 0 and any real number is 0.)

EXAMPLE 2

Multiply: $173(-452)0$.

Solution   We have

$$173(-452)0 = 173[(-452)0]$$ Using the associative law of multiplication

$$= 173[0]$$ Using the multiplicative property of zero

$$= 0.$$ Using the multiplicative property of zero again

Note that whenever 0 appears as a factor, the product is 0.

We can extend the above pattern still further to examine the product of two negative numbers.

This number → $2(-5) =$ $(-5) + (-5) = -10$ ← This number
decreases by $1(-5) =$ $(-5) = -5$ increases by
1 each time. $0(-5) =$ $0 = 0$ 5 each time.
                $-1(-5) =$ $-(-5) = 5$
                $-2(-5) = -(-5) - (-5) = 10$

According to the pattern, the product of two negative numbers is positive.

The Product of Two Negative Numbers

To multiply two negative numbers, multiply their absolute values. The answer is positive.

EXAMPLE 3   Multiply: **(a)** $(-6)(-8)$; **(b)** $(-1.2)(-3)$.

Solution

a) The absolute value of -6 is 6 and the absolute value of -8 is 8. Thus,

$$(-6)(-8) = 6 \cdot 8 \qquad \text{Multiplying absolute values. The answer is positive.}$$
$$= 48.$$

b) $(-1.2)(-3) = (1.2)(3) \qquad$ Multiplying absolute values. The answer is positive.
$$= 3.6 \qquad\qquad\quad \text{Try to go directly to this step.}$$

When three or more numbers are multiplied, we can order and group the numbers as we please, because of the commutative and associative laws.

EXAMPLE 4   Multiply: **(a)** $-3(-2)(-5)$; **(b)** $-4(-6)(-1)(-2)$.

Solution

a) $-3(-2)(-5) = 6(-5) \qquad$ Multiplying the first two numbers. The product of two negatives is positive.

$$= -30 \qquad\qquad \text{The product of a positive and a negative is negative.}$$

b) $-4(-6)(-1)(-2) = 24 \cdot 2 \qquad$ Multiplying the first two numbers and the last two numbers

$$= 48$$

We can see the following pattern in the results of Example 4.

The product of an even number of negative numbers is positive.

The product of an odd number of negative numbers is negative.

Division

Recall that $a \div b$, or $\frac{a}{b}$, is the number, if one exists, that when multiplied by b gives a. For example, to show that $10 \div 2$ is 5, we need only note that $5 \cdot 2 = 10$. Thus division can always be checked with multiplication.

EXAMPLE 5   Divide, if possible, and check your answer.

a) $14 \div (-7)$

b) $\dfrac{-32}{-4}$

c) $\dfrac{-10}{9}$

d) $\dfrac{-17}{0}$

Solution

a) $14 \div (-7) = -2$ We look for a number that when multiplied by -7 gives 14. That number is -2. *Check*: $(-2)(-7) = 14$.

b) $\dfrac{-32}{-4} = 8$ We look for a number that when multiplied by -4 gives -32. That number is 8. *Check*: $8(-4) = -32$.

c) $\dfrac{-10}{9} = -\dfrac{10}{9}$ We look for a number that when multiplied by 9 gives -10. That number is $-\frac{10}{9}$. *Check*: $-\frac{10}{9} \cdot 9 = -10$.

d) $\dfrac{-17}{0}$ is **undefined**.   We look for a number that when multiplied by 0 gives -17. There is no such number because if 0 is a factor, the product is 0, not -17.

Student Notes _____

Try to regard "undefined" as a mathematical way of saying "we do not give any meaning to this expression."

The sign rules for division are the same as those for multiplication: The quotient of a positive number and a negative number is negative; the quotient of two negative numbers is positive.

> **Rules for Multiplication and Division**
>
> To multiply or divide two nonzero real numbers:
>
> **1.** Using the absolute values, multiply or divide, as indicated.
> **2.** If the signs are the same, the answer is positive.
> **3.** If the signs are different, the answer is negative.

Had Example 5(a) been written as $-14 \div 7$ or $-\frac{14}{7}$, rather than $14 \div (-7)$, the result would still have been -2. Thus from Examples 5(a)–5(c), we have the following:

$$\frac{-a}{b} = \frac{a}{-b} = -\frac{a}{b} \qquad \text{and} \qquad \frac{-a}{-b} = \frac{a}{b}.$$

EXAMPLE 6   Rewrite each of the following in two equivalent forms: **(a)** $\frac{5}{-2}$; **(b)** $-\frac{3}{10}$.

Solution   We use one of the properties just listed.

a) $\dfrac{5}{-2} = \dfrac{-5}{2}$ and $\dfrac{5}{-2} = -\dfrac{5}{2}$

b) $-\dfrac{3}{10} = \dfrac{-3}{10}$ and $-\dfrac{3}{10} = \dfrac{3}{-10}$

Since $\dfrac{-a}{b} = \dfrac{a}{-b} = -\dfrac{a}{b}$

When a fraction contains a negative sign, it can be helpful to rewrite (or simply visualize) the fraction in an equivalent form.

EXAMPLE 7   Perform the indicated operation: **(a)** $\left(-\frac{4}{5}\right)\left(\frac{-7}{3}\right)$; **(b)** $-\frac{2}{7} + \frac{9}{-7}$.

Solution

a) $\left(-\dfrac{4}{5}\right)\left(\dfrac{-7}{3}\right) = \left(-\dfrac{4}{5}\right)\left(-\dfrac{7}{3}\right)$ Rewriting $\dfrac{-7}{3}$ as $-\dfrac{7}{3}$

$= \dfrac{28}{15}$ Try to go directly to this step.

b) Given a choice, we generally choose a positive denominator:

$-\dfrac{2}{7} + \dfrac{9}{-7} = \dfrac{-2}{7} + \dfrac{-9}{7}$ Rewriting both fractions with a common denominator of 7

$= \dfrac{-11}{7}$, or $-\dfrac{11}{7}$.

To divide with fraction notation, it is usually easiest to find a reciprocal and then multiply.

EXAMPLE 8   Find the reciprocal of each number, if it exists.

a) -27 **b)** $\frac{-3}{4}$ **c)** $-\frac{1}{5}$ **d)** 0

Solution   Recall from Section 1.3 that we can check that two numbers are reciprocals of each other by confirming that their product is 1.

a) The reciprocal of -27 is $\frac{1}{-27}$. More often, this number is written as $-\frac{1}{27}$.
   Check: $(-27)\left(-\frac{1}{27}\right) = \frac{27}{27} = 1$.

b) The reciprocal of $\frac{-3}{4}$ is $\frac{4}{-3}$, or, equivalently, $-\frac{4}{3}$.   *Check*: $\frac{-3}{4} \cdot \frac{4}{-3} = \frac{-12}{-12} = 1$.

c) The reciprocal of $-\frac{1}{5}$ is -5.   *Check*: $-\frac{1}{5}(-5) = \frac{5}{5} = 1$.

d) The reciprocal of 0 does not exist. To see this, recall that there is no number r for which $0 \cdot r = 1$.

EXAMPLE 9   Divide: **(a)** $-\frac{2}{3} \div \left(-\frac{5}{4}\right)$; **(b)** $-\frac{3}{4} \div \frac{3}{10}$.

Solution   We divide by multiplying by the reciprocal of the divisor.

a) $-\dfrac{2}{3} \div \left(-\dfrac{5}{4}\right) = -\dfrac{2}{3} \cdot \left(-\dfrac{4}{5}\right) = \dfrac{8}{15}$ Multiplying by the reciprocal

> Be careful not to change the sign when taking a reciprocal!

b) $-\dfrac{3}{4} \div \dfrac{3}{10} = -\dfrac{3}{4} \cdot \left(\dfrac{10}{3}\right) = -\dfrac{30}{12} = -\dfrac{5}{2} \cdot \dfrac{6}{6} = -\dfrac{5}{2}$ Removing a factor equal to 1: $\frac{6}{6} = 1$

To divide with decimal notation, it is usually easiest to carry out the division.

EXAMPLE 10   Divide: $27.9 \div (-3)$.

Solution

$$27.9 \div (-3) = \frac{27.9}{-3} = -9.3$$ Dividing: $3\overline{)27.9}$, quotient 9.3. The answer is negative.

In Example 5(d), we explained why we cannot divide -17 by 0. To see why *no* nonzero number b can be divided by 0, remember that $b \div 0$ would have to be the number that when multiplied by 0 gives b. But since the product of 0 and any number is 0, not b, we say that $b \div 0$ is **undefined** for $b \neq 0$. In the special case of $0 \div 0$, we look for a number r such that $0 \div 0 = r$ and $r \cdot 0 = 0$. But, $r \cdot 0 = 0$ for *any* number r. For this reason, we say that $b \div 0$ is undefined for any choice of b.*

Finally, note that $0 \div 7 = 0$ since $0 \cdot 7 = 0$. This can be written $0/7 = 0$. It is important not to confuse division *by* 0 with division *into* 0.

EXAMPLE 11   Divide, if possible: **(a)** $\frac{0}{-2}$; **(b)** $\frac{5}{0}$.

Solution

a) $\dfrac{0}{-2} = 0$ *Check*: $0(-2) = 0$.

b) $\dfrac{5}{0}$ is undefined.

---

*Sometimes $0 \div 0$ is said to be *indeterminate*.

Division Involving Zero

For any real number a,

$$\frac{a}{0} \text{ is undefined,}$$

and for $a \neq 0$,

$$\frac{0}{a} = 0.$$

Caution! It is important *not* to confuse *opposite* with *reciprocal*. Keep in mind that the opposite, or additive inverse, of a number is what we add to the number to get 0. The reciprocal, or multiplicative inverse, is what we multiply the number by to get 1.

Compare the following.

Number	Opposite (Change the sign.)	Reciprocal (Invert but do not change the sign.)
$-\dfrac{3}{8}$	$\dfrac{3}{8}$	$-\dfrac{8}{3}$ $\leftarrow \left(-\dfrac{3}{8}\right)\left(-\dfrac{8}{3}\right) = 1$
19	-19	$\dfrac{1}{19}$ $-\dfrac{3}{8} + \dfrac{3}{8} = 0$
$\dfrac{18}{7}$	$-\dfrac{18}{7}$	$\dfrac{7}{18}$
-7.9	7.9	$-\dfrac{1}{7.9}$, or $-\dfrac{10}{79}$
0	0	Not defined

Exercise Set

1.7

🔖 *Concept Reinforcement* In each of Exercises 1–10, replace the blank with either 0 or 1 to match the description given.

1. The product of two reciprocals ____

2. The sum of a pair of opposites ____

3. The sum of a pair of additive inverses ____

4. The product of two multiplicative inverses ____

5. This number has no reciprocal. ____

6. This number is its own reciprocal. ____

7. This number is the multiplicative identity. ____

8. This number is the additive identity. ____

9. A nonzero number divided by itself ____

10. Division by this number is undefined. ____

Multiply.

11. $-3 \cdot 8$

12. $-3 \cdot 7$

13. $-8 \cdot 7$

14. $-9 \cdot 2$

15. $8 \cdot (-3)$

16. $9 \cdot (-5)$

17. $-9 \cdot 8$

18. $-10 \cdot 3$

19. $-6 \cdot (-7)$

20. $-2 \cdot (-5)$

21. $-5 \cdot (-9)$

22. $-9 \cdot (-2)$

23. $-19 \cdot (-10)$

24. $-12 \cdot (-10)$

25. $-12 \cdot 12$

26. $-13 \cdot (-15)$

27. $-25 \cdot (-48)$

28. $39 \cdot (-43)$

29. $-3.5 \cdot (-28)$

30. $97 \cdot (-2.1)$

31. $6 \cdot (-13)$

32. $7 \cdot (-9)$

33. $-7 \cdot (-3.1)$

34. $-4 \cdot (-3.2)$

35. $\frac{2}{3} \cdot \left(-\frac{3}{5}\right)$

36. $\frac{5}{7} \cdot \left(-\frac{2}{3}\right)$

37. $-\frac{3}{8} \cdot \left(-\frac{2}{9}\right)$

38. $-\frac{5}{8} \cdot \left(-\frac{2}{5}\right)$

39. $(-5.3)(2.1)$

40. $(-4.3)(9.5)$

41. $-\frac{5}{9} \cdot \frac{3}{4}$

42. $-\frac{8}{3} \cdot \frac{9}{4}$

43. $3 \cdot (-7) \cdot (-2) \cdot 6$

44. $9 \cdot (-2) \cdot (-6) \cdot 7$

Aha! **45.** $-27 \cdot (-34) \cdot 0$

46. $-43 \cdot (-74) \cdot 0$

47. $-\frac{1}{3} \cdot \frac{1}{4} \cdot \left(-\frac{3}{7}\right)$

48. $-\frac{1}{2} \cdot \frac{3}{5} \cdot \left(-\frac{2}{7}\right)$

49. $-2 \cdot (-5) \cdot (-3) \cdot (-5)$

50. $-3 \cdot (-5) \cdot (-2) \cdot (-1)$

51. $(-31) \cdot (-27) \cdot 0$

52. $7 \cdot (-6) \cdot 5 \cdot (-4) \cdot 3 \cdot (-2) \cdot 1 \cdot 0$

53. $(-8)(-9)(-10)$

54. $(-7)(-8)(-9)(-10)$

55. $(-6)(-7)(-8)(-9)(-10)$

56. $(-5)(-6)(-7)(-8)(-9)(-10)$

Divide, if possible, and check. If a quotient is undefined, state this.

57. $14 \div (-2)$

58. $\frac{24}{-3}$

59. $\frac{36}{-9}$

60. $26 \div (-13)$

61. $\frac{-56}{8}$

62. $-32 \div (-4)$

63. $\frac{-48}{-12}$

64. $-63 \div (-9)$

65. $\frac{-72}{9}$

66. $\frac{-50}{25}$

67. $-100 \div (-50)$

68. $\frac{-200}{8}$

69. $-108 \div 9$

70. $\frac{-64}{-7}$

71. $\frac{400}{-50}$

72. $-300 \div (-13)$

73. $\frac{28}{0}$

74. $\frac{0}{-5}$

75. $-4.8 \div 1.2$

76. $-3.9 \div 1.3$

77. $\frac{0}{-9}$

78. $0 \div (-47)$

Aha! **79.** $\frac{9.7(-2.8)0}{4.3}$

80. $\frac{(-4.9)(7.2)}{0}$

Write each number in two equivalent forms, as in Example 6.

81. $\frac{-8}{3}$

82. $\frac{-12}{7}$

83. $\frac{29}{-35}$

84. $\frac{9}{-14}$

85. $-\frac{7}{3}$

86. $-\frac{4}{15}$

87. $\frac{-x}{2}$

88. $\frac{9}{-a}$

Find the reciprocal of each number, if it exists.

89. $\dfrac{4}{-5}$

90. $\dfrac{2}{-9}$

91. $-\dfrac{47}{13}$

92. $-\dfrac{31}{12}$

93. -10

94. 34

95. 4.3

96. -1.7

97. $\dfrac{-9}{4}$

98. $\dfrac{-6}{11}$

99. 0

100. -1

Perform the indicated operation and, if possible, simplify. If a quotient is undefined, state this.

101. $\left(\dfrac{-7}{4}\right)\left(-\dfrac{3}{5}\right)$

102. $\left(-\dfrac{5}{6}\right)\left(\dfrac{-1}{3}\right)$

103. $\left(\dfrac{-6}{5}\right)\left(\dfrac{2}{-11}\right)$

104. $\left(\dfrac{7}{-2}\right)\left(\dfrac{-5}{6}\right)$

105. $\dfrac{-3}{8} + \dfrac{-5}{8}$

106. $\dfrac{-4}{5} + \dfrac{7}{5}$

Aha! **107.** $\left(\dfrac{-9}{5}\right)\left(\dfrac{5}{-9}\right)$

108. $\left(-\dfrac{2}{7}\right)\left(\dfrac{5}{-8}\right)$

109. $\left(-\dfrac{3}{11}\right) + \left(-\dfrac{6}{11}\right)$

110. $\left(-\dfrac{4}{7}\right) + \left(-\dfrac{2}{7}\right)$

111. $\dfrac{7}{8} \div \left(-\dfrac{1}{2}\right)$

112. $\dfrac{3}{4} \div \left(-\dfrac{2}{3}\right)$

113. $\dfrac{9}{5} \cdot \dfrac{-20}{3}$

114. $\dfrac{-5}{12} \cdot \dfrac{7}{15}$

115. $\left(-\dfrac{18}{7}\right) + \left(-\dfrac{3}{7}\right)$

116. $\left(-\dfrac{12}{5}\right) + \left(-\dfrac{3}{5}\right)$

Aha! **117.** $-\dfrac{5}{9} \div \left(-\dfrac{5}{9}\right)$

118. $-\dfrac{5}{4} \div \left(-\dfrac{3}{4}\right)$

119. $-44.1 \div (-6.3)$

120. $-6.6 \div 3.3$

121. $\dfrac{5}{9} - \dfrac{7}{9}$

122. $\dfrac{2}{7} - \dfrac{6}{7}$

123. $\dfrac{-3}{10} + \dfrac{2}{5}$

124. $\dfrac{-5}{9} + \dfrac{2}{3}$

125. $\dfrac{7}{10} \div \left(\dfrac{-3}{5}\right)$

126. $\left(\dfrac{-3}{5}\right) \div \dfrac{6}{15}$

127. $\dfrac{5}{7} - \dfrac{1}{-7}$

128. $\dfrac{4}{9} - \dfrac{1}{-9}$

129. $\dfrac{-4}{15} + \dfrac{2}{-3}$

130. $\dfrac{3}{-10} + \dfrac{-1}{5}$

131. Most calculators have a key, often appearing as $\boxed{1/x}$, for finding reciprocals. To use this key, enter a number and then press $\boxed{1/x}$ to find its reciprocal. What should happen if you enter a number and then press the reciprocal key twice? Why?

132. Multiplication can be regarded as repeated addition. Using this idea and a number line, explain why $3 \cdot (-5) = -15$.

SKILL MAINTENANCE

133. Simplify: $\dfrac{264}{468}$. [1.3]

134. Combine like terms: $x + 12y + 11x - 14y - 9$. [1.5]

SYNTHESIS

135. If two nonzero numbers are opposites of each other, are their reciprocals opposites of each other? Why or why not?

136. If two numbers are reciprocals of each other, are their opposites reciprocals of each other? Why or why not?

137. Show that the reciprocal of a sum is *not* the sum of the two reciprocals.

138. Which real numbers are their own reciprocals?

Tell whether each expression represents a positive number or a negative number when m and n are negative.

139. $\dfrac{m}{-n}$

140. $\dfrac{-n}{-m}$

141. $-m \cdot \left(\dfrac{-n}{m}\right)$

142. $-\left(\dfrac{n}{-m}\right)$

143. $(m + n) \cdot \dfrac{m}{n}$

144. $(-n - m)\dfrac{n}{m}$

145. What must be true of m and n if $-mn$ is to be **(a)** positive? **(b)** zero? **(c)** negative?

146. The following is a proof that a positive number times a negative number is negative. Provide a reason for each step. Assume that $a > 0$ and $b > 0$.

$$a(-b) + ab = a[-b + b]$$
$$= a(0)$$
$$= 0$$

Therefore, $a(-b)$ is the opposite of ab.

147. Is it true that for any numbers a and b, if a is larger than b, then the reciprocal of a is smaller than the reciprocal of b? Why or why not?

1.8 Exponential Notation and Order of Operations

Exponential Notation • Order of Operations • Simplifying and the Distributive Law • The Opposite of a Sum

Algebraic expressions often contain *exponential notation*. In this section, we learn how to use exponential notation as well as rules for the *order of operations* in performing certain algebraic manipulations.

Exponential Notation

A product like $3 \cdot 3 \cdot 3 \cdot 3$, in which the factors are the same, is called a **power**. Powers occur often enough that a simpler notation called **exponential notation** is used. For

$\underbrace{3 \cdot 3 \cdot 3 \cdot 3}_{4 \text{ factors}},$ we write 3^4.   Because $3^4 = 81$, we sometimes say that 81 "is a power of 3."

This is read "three to the fourth power," or simply, "three to the fourth." The number 4 is called an **exponent** and the number 3 a **base**.

Expressions like s^2 and s^3 are usually read "*s* squared" and "*s* cubed," respectively. This comes from the fact that a square with sides of length *s* has an area *A* given by $A = s^2$ and a cube with sides of length *s* has a volume *V* given by $V = s^3$.

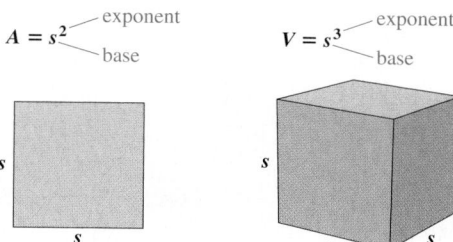

EXAMPLE 1 Write exponential notation for $10 \cdot 10 \cdot 10 \cdot 10 \cdot 10$.

Solution

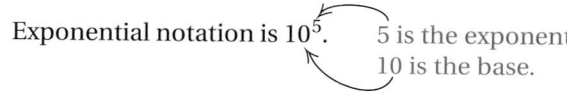

Exponential notation is 10^5.   5 is the exponent.
10 is the base.

EXAMPLE 2   Evaluate: **(a)** 5^2; **(b)** $(-5)^3$; **(c)** $(2n)^3$.

Solution

a) $5^2 = 5 \cdot 5 = 25$ The exponent 2 indicates two factors of 5.

b) $(-5)^3 = (-5)(-5)(-5)$ The exponent 3 indicates three factors of -5.

$\qquad = 25(-5)$ Using the associative law of multiplication

$\qquad = -125$

c) $(2n)^3 = (2n)(2n)(2n)$ The exponent 3 indicates three factors of $2n$.

$\qquad = 2 \cdot 2 \cdot 2 \cdot n \cdot n \cdot n$ Using the associative and commutative laws of multiplication

$\qquad = 8n^3$

To determine what the exponent 1 will mean, look for a pattern in the following:

$$7 \cdot 7 \cdot 7 \cdot 7 = 7^4$$
$$7 \cdot 7 \cdot 7 = 7^3$$
$$7 \cdot 7 = 7^2$$
$$7 = 7^?$$

We divide by 7 each time.

The exponents decrease by 1 each time. To extend the pattern, we say that

$$7 = 7^1.$$

Exponential Notation

For any natural number n,

$$b^n \quad \text{means} \quad \overbrace{b \cdot b \cdot b \cdot b \cdots b}^{n \text{ factors}}.$$

Order of Operations

How should $4 + 2 \times 5$ be computed? If we multiply 2 by 5 and then add 4, the result is 14. If we add 2 and 4 first and then multiply by 5, the result is 30. Since these results differ, the order in which we perform operations matters. If grouping symbols such as parentheses ( ), brackets [ ], braces { }, absolute-value symbols | |, or fraction bars appear, they tell us what to do first. For example,

$$(4 + 2) \times 5 \quad \text{indicates} \quad 6 \times 5, \quad \text{resulting in 30,}$$

and

$$4 + (2 \times 5) \quad \text{indicates} \quad 4 + 10, \quad \text{resulting in 14.}$$

technology connection

On most graphing calculators, grouping symbols, such as a fraction bar, must be replaced with parentheses. For example, to calculate

$$\frac{12(9 - 7) + 4 \cdot 5}{2^4 + 3^2},$$

we enter $(12(9 - 7) + 4 \cdot 5) \div (2^4 + 3^2)$. To enter an exponential expression, we enter the base, press ⌢ and then enter the exponent. The x² key can be used to enter an exponent of 2. We can also convert to fraction notation if we wish.

```
(12(9−7)+4∗5)/(2^4+3²)
                    1.76
Ans▶Frac
                   44/25
■
```

Besides grouping symbols, the following conventions exist for determining the order in which operations should be performed.

Rules for Order of Operations

1. Calculate within the innermost grouping symbols, ( ), [ ], { }, | |, and above or below fraction bars.
2. Simplify all exponential expressions.
3. Perform all multiplication and division, working from left to right.
4. Perform all addition and subtraction, working from left to right.

Thus the correct way to compute $4 + 2 \times 5$ is to first multiply 2 by 5 and then add 4. The result is 14.

EXAMPLE 3   Simplify: $15 - 2 \times 5 + 3$.

Solution   When no groupings or exponents appear, we *always* multiply or divide before adding or subtracting:

$$15 - 2 \times 5 + 3 = 15 - 10 + 3 \qquad \text{Multiplying}$$
$$= 5 + 3 \qquad \qquad \text{Subtracting and adding from left}$$
$$= 8. \qquad \qquad \text{to right}$$

Always calculate within parentheses first. When there are exponents and no parentheses, simplify powers before multiplying or dividing.

EXAMPLE 4   Simplify: **(a)** $(3 \cdot 4)^2$; **(b)** $3 \cdot 4^2$.

Solution

a) $(3 \cdot 4)^2 = (12)^2 \qquad$ Working within parentheses first
$$= 144$$

b) $3 \cdot 4^2 = 3 \cdot 16 \qquad$ Simplifying the power
$$= 48 \qquad \text{Multiplying}$$

Note that $(3 \cdot 4)^2 \neq 3 \cdot 4^2$.

Caution!   Example 4 illustrates that, in general, $(ab)^2 \neq ab^2$.

EXAMPLE 5   Evaluate for $x = 5$: **(a)** $(-x)^2$; **(b)** $-x^2$.

Solution

a) $(-x)^2 = (-5)^2 = (-5)(-5) = 25$ We square the opposite of 5.

b) $-x^2 = -5^2 = -25$ We square 5 and then find the opposite.

> *Caution!*   Example 5 illustrates that, in general, $(-x)^2 \neq -x^2$.

EXAMPLE 6   Evaluate $-15 \div 3(6 - a)^3$ for $a = 4$.

Solution

$$-15 \div 3(6 - a)^3 = -15 \div 3(6 - 4)^3$$ Substituting 4 for a

$$= -15 \div 3(2)^3$$ Working within parentheses first

$$= -15 \div 3 \cdot 8$$ Simplifying the exponential expression

$$\left. \begin{array}{l} = -5 \cdot 8 \\ = -40 \end{array} \right\}$$ Dividing and multiplying from left to right

Student Notes

The symbols ( ), [ ], and { } are all used in the same way. Used inside or next to each other, they make it easier to locate the left and right sides of a grouping. Try doing this in your own work to minimize mistakes.

When combinations of grouping symbols are used, the rules still apply within those grouping symbols. We begin with the innermost grouping symbols and work to the outside.

EXAMPLE 7   Simplify: $8 \div 4 + 3[9 + 2(3 - 5)^3]$.

Solution

$$8 \div 4 + 3[9 + 2(3 - 5)^3] = 8 \div 4 + 3[9 + 2(-2)^3]$$ Doing the calculations in the innermost parentheses first

$$= 8 \div 4 + 3[9 + 2(-8)]$$ $(-2)^3 = (-2)(-2)(-2)$ $= -8$

$$= 8 \div 4 + 3[9 + (-16)]$$

$$= 8 \div 4 + 3[-7]$$ Completing the calculations within the brackets

$$= 2 + (-21)$$ Multiplying and dividing from left to right

$$= -19$$

EXAMPLE 8   Calculate: $\dfrac{12(9-7)+4\cdot 5}{3^4+2^3}$.

Solution An equivalent expression with brackets is

$$[12(9-7)+4\cdot 5]\div[3^4+2^3].$$

In effect, we need to simplify the numerator, simplify the denominator, and then divide the results:

$$\frac{12(9-7)+4\cdot 5}{3^4+2^3}=\frac{12(2)+4\cdot 5}{81+8}$$

$$=\frac{24+20}{89}=\frac{44}{89}.$$

Simplifying and the Distributive Law

Sometimes we cannot simplify within grouping symbols. When a sum or difference is being grouped, the distributive law provides a method for removing the grouping symbols.

EXAMPLE 9   Simplify: $5x-9+2(4x+5)$.

Solution

$$5x-9+2(4x+5)=5x-9+8x+10 \qquad \text{Using the distributive law}$$

$$=13x+1 \qquad \text{Combining like terms}$$

Now that exponents have been introduced, we can make our definition of *like* or *similar terms* more precise. **Like**, or **similar**, **terms** are either constant terms or terms containing the same variable(s) raised to the same power(s). Thus, 5 and -7, $19xy$ and $2yx$, and $4a^3b$ and a^3b are all pairs of like terms.

EXAMPLE 10   Simplify: $7x^2+3[x^2+2x]-5x$.

Solution

$$7x^2+3[x^2+2x]-5x=7x^2+3x^2+6x-5x \qquad \text{Using the distributive law}$$

$$=10x^2+x \qquad \text{Combining like terms}$$

The Opposite of a Sum

When a number is multiplied by -1, the result is the opposite of that number. For example, $-1(7)=-7$ and $-1(-5)=5$.

The Property of −1

For any real number a,

$$-1 \cdot a = -a.$$

(Negative one times a is the opposite of a.)

When grouping symbols are preceded by a "−" symbol, we can multiply the grouping by −1 and use the distributive law. In this manner, we can find the *opposite*, or *additive inverse*, of a sum.

EXAMPLE 11   Write an expression equivalent to $-(3x + 2y + 4)$ without using parentheses.

Solution

$$
\begin{aligned}
-(3x + 2y + 4) &= -1(3x + 2y + 4) && \text{Using the property of } -1 \\
&= -1(3x) + (-1)(2y) + (-1)4 && \text{Using the distributive law} \\
&= -3x - 2y - 4 && \text{Using the associative law and the property of } -1
\end{aligned}
$$

Example 11 illustrates an important property of real numbers.

The Opposite of a Sum

For any real numbers a and b,

$$-(a + b) = -a + (-b) = -a - b.$$

(The opposite of a sum is the sum of the opposites.)

To remove parentheses from an expression like $-(x - 7y + 5)$, we can first rewrite the subtraction as addition:

$$
\begin{aligned}
-(x - 7y + 5) &= -(x + (-7y) + 5) && \text{Rewriting as addition} \\
&= -x + 7y - 5. && \text{Taking the opposite of a sum}
\end{aligned}
$$

This procedure is normally streamlined to one step in which we find the opposite by "removing parentheses and changing the sign of every term":

$$-(x - 7y + 5) = -x + 7y - 5.$$

EXAMPLE 12   Simplify: $3x - (4x + 2)$.

Solution

$$
\begin{aligned}
3x - (4x + 2) &= 3x + [-(4x + 2)] && \text{Adding the opposite of } 4x + 2 \\
&= 3x + [-4x - 2] && \text{Taking the opposite of } 4x + 2 \\
&= 3x + (-4x) + (-2) \\
&= 3x - 4x - 2 && \text{Try to go directly to this step.} \\
&= -x - 2 && \text{Combining like terms}
\end{aligned}
$$

In practice, the first three steps of Example 12 are generally skipped.

EXAMPLE 13   Simplify: $5t^2 - 2t - (4t^2 - 9t)$.

Solution

$$
\begin{aligned}
5t^2 - 2t - (4t^2 - 9t) &= 5t^2 - 2t - 4t^2 + 9t && \text{Removing} \\
& && \text{parentheses and} \\
& && \text{changing the sign of} \\
& && \text{each term inside} \\
&= t^2 + 7t && \text{Combining like terms}
\end{aligned}
$$

Expressions such as $7 - 3(x + 2)$ can be simplified as follows:

$$
\begin{aligned}
7 - 3(x + 2) &= 7 + [-3(x + 2)] && \text{Adding the opposite of } 3(x + 2) \\
&= 7 + [-3x - 6] && \text{Multiplying } x + 2 \text{ by } -3 \\
&= 7 - 3x - 6 && \text{Try to go directly to this step.} \\
&= 1 - 3x. && \text{Combining like terms}
\end{aligned}
$$

EXAMPLE 14   Simplify: **(a)** $3n - 2(4n - 5)$; **(b)** $7x^3 + 2 - [5(x^3 - 1) + 8]$.

Solution

a)
$$
\begin{aligned}
3n - 2(4n - 5) &= 3n - 8n + 10 && \text{Multiplying each term inside the} \\
& && \text{parentheses by } -2 \\
&= -5n + 10 && \text{Combining like terms}
\end{aligned}
$$

b)
$$
\begin{aligned}
7x^3 + 2 - [5(x^3 - 1) + 8] &= 7x^3 + 2 - [5x^3 - 5 + 8] && \text{Removing} \\
& && \text{parentheses} \\
&= 7x^3 + 2 - [5x^3 + 3] \\
&= 7x^3 + 2 - 5x^3 - 3 && \text{Removing brackets} \\
&= 2x^3 - 1 && \text{Combining like terms}
\end{aligned}
$$

CONNECTING THE CONCEPTS

Algebra is a tool that is used to solve problems. We have seen in this chapter that certain problems can be translated to algebraic expressions that can in turn be simplified (see Section 1.6). In Chapter 2, we will solve problems that require us to solve an equation.

As we progress through our study of algebra, it is important that we be able to distinguish between the two tasks of **simplifying an expression** and **solving an equation**. In Chapter 1, we did not solve equations, but we did simplify expressions. This enabled us to write *equivalent expressions* that were simpler than the given expression. In Chapter 2, we will continue to simplify expressions, but we will also begin to solve equations.

1.8 Exercise Set

FOR EXTRA HELP

Student's Solutions Manual | Digital Video Tutor CD 1 Videotape 1 | AW Math Tutor Center | MathXL Tutorials on CD | MathXL | MyMathLab

Concept Reinforcement In each part of Exercises 1 and 2, name the operation that should be performed first. Do not perform the calculations.

1. **a)** $4 + 8 \div 2 \cdot 2$
b) $7 - 9 + 15$
c) $5 - 2(3 + 4)$
d) $6 + 7 \cdot 3$
e) $18 - 2[4 + (3 - 2)]$
f) $\dfrac{5 - 6 \cdot 7}{2}$

2. **a)** $9 - 3 \cdot 4 \div 2$
b) $8 + 7(6 - 5)$
c) $5 \cdot [2 - 3(4 + 1)]$
d) $8 - 7 + 2$
e) $4 + 6 \div 2 \cdot 3$
f) $\dfrac{37}{8 - 2 \cdot 2}$

Write exponential notation.

3. $2 \cdot 2 \cdot 2$
4. $6 \cdot 6 \cdot 6 \cdot 6$
5. $x \cdot x \cdot x \cdot x \cdot x \cdot x \cdot x$
6. $y \cdot y \cdot y \cdot y \cdot y \cdot y$
7. $3t \cdot 3t \cdot 3t \cdot 3t \cdot 3t$
8. $5m \cdot 5m \cdot 5m \cdot 5m \cdot 5m$

Simplify.

9. 3^2
10. 5^3
11. $(-4)^2$
12. $(-7)^2$
13. -4^2
14. -7^2
15. 4^3
16. 9^1
17. $(-5)^4$
18. 5^4
19. 7^1
20. $(-1)^7$
21. $(3t)^4$
22. $(5t)^2$
23. $(-7x)^3$
24. $(-5x)^4$
25. $5 + 3 \cdot 7$
26. $3 - 4 \cdot 2$
27. $8 \cdot 7 + 6 \cdot 5$
28. $10 \cdot 5 + 1 \cdot 1$
29. $19 - 5 \cdot 3 + 3$
30. $14 - 2 \cdot 6 + 7$
31. $9 \div 3 + 16 \div 8$
32. $32 - 8 \div 4 - 2$
Aha! **33.** $14 \cdot 19 \div (19 \cdot 14)$
34. $18 - 6 \div 3 \cdot 2 + 7$
35. $3(-10)^2 - 8 \div 2^2$
36. $9 - 3^2 \div 9(-1)$
37. $8 - (2 \cdot 3 - 9)$

38. $(8 - 2 \cdot 3) - 9$

39. $(8 - 2)(3 - 9)$

40. $32 \div (-2)^2 \cdot 4$

41. $13(-10)^2 + 45 \div (-5)$

42. $5 \cdot 3^2 - 4^2 \cdot 2$

43. $2^4 + 2^3 - 10 \div (-1)^4$

44. $112 \div 28 - 112 \div 28$

45. $5 + 3(2 - 9)^2$

46. $9 - (3 - 5)^3 - 4$

47. $[2 \cdot (5 - 8)]^2$

48. $3(5 - 7)^4 \div 4$

49. $\dfrac{7 + 2}{5^2 - 4^2}$

50. $\dfrac{5^2 - 3^2}{2 \cdot 6 - 4}$

51. $8(-7) + |6(-5)|$

52. $|10(-5)| + 1(-1)$

53. $\dfrac{(-2)^3 + 4^2}{3 - 5^2 + 3 \cdot 6}$

54. $\dfrac{7^2 - (-1)^5}{3 - 2 \cdot 3^2 + 5}$

55. $\dfrac{27 - 2 \cdot 3^2}{8 \div 2^2 - (-2)^2}$

56. $\dfrac{(-5)^2 - 4 \cdot 5}{3^2 + 4 \cdot 2(-1)^5}$

Evaluate.

57. $9 - 4x$, for $x = 5$

58. $1 + x^3$, for $x = -2$

59. $24 \div t^3$, for $t = -2$

60. $20 \div a \cdot 4$, for $a = 5$

61. $45 \div 3 \cdot a$, for $a = -1$

62. $50 \div 2 \cdot t$, for $t = -5$

63. $5x \div 15x^2$, for $x = 3$

64. $6a \div 12a^3$, for $a = 2$

65. $45 \div 3^2 x(x - 1)$, for $x = 3$

66. $-30 \div t(t + 4)^2$, for $t = -6$

67. $-x^2 - 5x$, for $x = -3$

68. $(-x)^2 - 5x$, for $x = -3$

69. $\dfrac{3a - 4a^2}{a^2 - 20}$, for $a = 5$

70. $\dfrac{a^3 - 4a}{a(a - 3)}$, for $a = -2$

Write an equivalent expression without using grouping symbols.

71. $-(9x + 1)$

72. $-(3x + 5)$

73. $-[5 - 6x]$

74. $-(6x - 7)$

75. $-(4a - 3b + 7c)$

76. $-[5x - 2y - 3z]$

77. $-(3x^2 + 5x - 1)$

78. $-(8x^3 - 6x + 5)$

Simplify.

79. $8x - (6x + 7)$

80. $7y - (2y + 9)$

81. $2a - (5a - 9)$

82. $11n - (3n - 7)$

83. $2x + 7x - (4x + 6)$

84. $2a + 5a - (6a + 8)$

85. $9t - 5r - 2(3r + 6t)$

86. $4m - 9n - 3(2m - n)$

87. $15x - y - 5(3x - 2y + 5z)$

88. $4a - b - 4(5a - 7b + 8c)$

89. $3x^2 + 7 - (2x^2 + 5)$

90. $5x^4 + 3x - (5x^4 + 3x)$

91. $5t^3 + t - 3(t + 2t^3)$

92. $8n^2 + n - 2(n + 3n^2)$

93. $12a^2 - 3ab + 5b^2 - 5(-5a^2 + 4ab - 6b^2)$

94. $-8a^2 + 5ab - 12b^2 - 6(2a^2 - 4ab - 10b^2)$

95. $-7t^3 - t^2 - 3(5t^3 - 3t)$

96. $9t^4 + 7t - 5(9t^3 - 2t)$

97. $5(2x - 7) - [4(2x - 3) + 2]$

98. $3(6x - 5) - [3(1 - 8x) + 5]$

99. Some students use the mnemonic device PEMDAS to help remember the rules for the order of operations. Explain how this can be done and whether it is a valid approach.

100. Jake keys $18/2 \cdot 3$ into his calculator and expects the result to be 3. What mistake is he probably making?

SKILL MAINTENANCE

Translate to an algebraic expression. [1.1]

101. Nine more than twice a number

102. Half of the sum of two numbers

SYNTHESIS

103. Write the sentence $(-x)^2 \neq -x^2$ in words. Explain why $(-x)^2$ and $-x^2$ are not equivalent.

104. Write the sentence $-|x| \neq -x$ in words. Explain why $-|x|$ and $-x$ are not equivalent.

Simplify.

105. $5t - \{7t - [4r - 3(t - 7)] + 6r\} - 4r$

106. $z - \{2z - [3z - (4z - 5z) - 6z] - 7z\} - 8z$

107. $\{x - [f - (f - x)] + [x - f]\} - 3x$

108. Is it true that for all real numbers a and b,
$$ab = (-a)(-b)?$$
Why or why not?

109. Is it true that for all real numbers a, b, and c,
$$a|b - c| = ab - ac?$$
Why or why not?

If $n > 0$, $m > 0$, and $n \neq m$, classify each of the following as either true or false.

110. $-n + m = -(n + m)$

111. $m - n = -(n - m)$

112. $n(-n - m) = -n^2 + nm$

113. $-m(n - m) = -(mn + m^2)$

114. $-n(-n - m) = n(n + m)$

Evaluate.

Aha! 115. $[x + 3(2 - 5x) \div 7 + x](x - 3)$, for $x = 3$

Aha! 116. $[x + 2 \div 3x] \div [x + 2 \div 3x]$, for $x = -7$

117. In Mexico, between 500 B.C. and 600 A.D., the Mayans represented numbers using powers of 20 and certain symbols. For example, the symbols

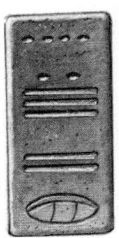

represent $4 \cdot 20^3 + 17 \cdot 20^2 + 10 \cdot 20^1 + 0 \cdot 20^0$. Evaluate this number.

Source: National Council of Teachers of Mathematics, 1906 Association Drive, Reston, VA 22091

118. Examine the Mayan symbols and the numbers in Exercise 117. What numbers do

•,   ⬭,   and   ⬭

each represent?

119. Calculate the volume of the tower shown below.

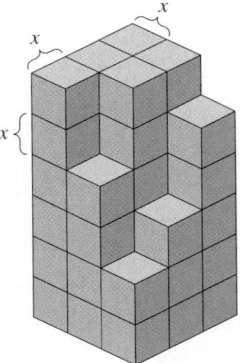

CORNER

Select the Symbols

COLLABORATIVE

Focus: Order of operations

Time: 15 minutes

Group size: 2

One way to master the rules for the order of operations is to insert symbols within a display of numbers in order to obtain a predetermined result. For example, the display

$$1 \quad 2 \quad 3 \quad 4 \quad 5$$

can be used to obtain the result 21 as follows:

$$(1 + 2) \div 3 + 4 \cdot 5.$$

Note that without an understanding of the rules for the order of operations, solving a problem of this sort is impossible.

ACTIVITY

1. Each group should prepare an exercise similar to the example shown above. (Exponents are not allowed.) To do so, first select five single-digit numbers for display. Then insert operations and grouping symbols and calculate the result.

2. Pair with another group. Each group should give the other its result along with its five-number display, and challenge the other group to insert symbols that will make the display equal the result given.

3. Share with the entire class the various mathematical statements developed by each group.

Study Summary

In a **variable expression** like $5ab^3$, the number 5 is a **constant** and a and b are **variables** (p. 2). When numbers are **substituted** for a and b, we say that we are **evaluating** the **algebraic expression** (p. 3). A substitution that makes an **equation** true is called a **solution** of the equation (p. 6).

Expressions like $5x + 2$ and $2 + x \cdot 5$ are **equivalent expressions** because, for any replacement of the variable, they will represent the same **value** (pp. 3, 13). The following laws are often used to create or **simplify** equivalent expressions (pp. 13–16, 23, 41):

Commutative laws:	$a + b = b + a;\quad ab = ba$
Associative laws:	$a + (b + c) = (a + b) + c;\quad a(bc) = (ab)c$
Distributive law:	$a(b + c) = ab + ac$
Identity property of 1:	$1 \cdot a = a \cdot 1 = a$
Identity property of 0:	$a + 0 = 0 + a = a$

Important **sets** of numbers are (p. 34)

Natural numbers:	$\{1, 2, 3, \ldots\}$
Whole numbers:	$\{0, 1, 2, 3, \ldots\}$
Integers:	$\{\ldots, -3, -2, -1, 0, 1, 2, 3, \ldots\}$
Rational numbers:	$\left\{\dfrac{a}{b} \,\middle|\, a \text{ and } b \text{ are integers and } b \neq 0\right\}$

Rational numbers can always be written as **terminating** or **repeating** decimals (p. 33). **Irrational numbers**, like $\sqrt{2}$ or π, can be thought of as nonterminating and nonrepeating decimals (p. 33). The set of **real numbers** consists of all rational and irrational numbers, taken together (p. 34). A natural number that has only itself and 1 as two different **factors**, like 2, 3, 5, 7, 11, 13, 17, 19, or 23, is said to be **prime** (p. 21). Natural numbers that have factors other than 1 and the number itself, like 4, 6, 8, and 9, are said to be **composite** (p. 21). The **prime factorization** of a composite number expresses that number as a product of prime numbers (p. 21).

In future chapters, the following properties and laws will be useful for solving equations (pp. 48, 55, 57, 60, 68):

Law of opposites:	$a + (-a) = 0$
Multiplicative property of 0:	$0 \cdot a = a \cdot 0 = 0$
Property of -1:	$-1 \cdot a = -a$
Opposite of a sum:	$-(a + b) = -a + (-b)$
Division involving 0:	$\dfrac{0}{a} = 0; \dfrac{a}{0}$ is undefined

$$\frac{-a}{b} = \frac{a}{-b} = -\frac{a}{b}, \qquad \frac{-a}{-b} = \frac{a}{b}$$

In geometric applications throughout the text, the following area formulas are used regularly (pp. 3, 4, 10):

Area of a rectangle:	$A = lw$
Area of a triangle:	$A = \frac{1}{2}bh$
Area of a parallelogram:	$A = bh$

The **terms** of a expression, like 4, $5x^2$, $-7y$, $2x$, and $3x^2$ in $4 + 5x^2 - 7y + 2x + 3x^2$, are separated by addition (p. 16). **Like terms** are either constants or have the same variable factors raised to the same power (p. 67).

When more than one operation appears in an algebraic expression or calculation, the following convention is used (p. 65).

Rules for Order of Operations

1. Calculate within the innermost grouping symbols, ( ), [ ], { }, | |, and above or below fraction bars.
2. Simplify all exponential expressions.
3. Perform all multiplication and division, working from left to right.
4. Perform all addition and subtraction, working from left to right.

1 Review Exercises

↪ *Concept Reinforcement*   *In each of Exercises 1–10, classify the statement as either true or false.*

1. $4x - 5y$ and $12 - 7a$ are both algebraic expressions containing two terms. [1.2]

2. $3t + 1 = 7$ and $8 - 2 = 9$ are both equations. [1.1]

3. The fact that $2 + x$ is equivalent to $x + 2$ is an illustration of the associative law for addition. [1.2]

4. The statement $4(a + 3) = 4 \cdot a + 4 \cdot 3$ illustrates the distributive law. [1.2]

5. The number 2 is neither prime nor composite. [1.3]

6. Every irrational number can be written as a repeating or terminating decimal. [1.4]

7. Every natural number is a whole number and every whole number is an integer. [1.4]

8. The expressions $9r^2s$ and $5rs^2$ are like terms. [1.5]

9. The opposite of x, written $-x$, never represents a positive number. [1.6]

10. The number 0 has no reciprocal. [1.3]

Evaluate.

11. $5t$, for $t = 3$ [1.1]

12. $\dfrac{x - y}{3}$, for $x = 17$ and $y = 5$ [1.1]

13. $9 - y^2$, for $y = 4$ [1.8]

14. $-10 + a^2 \div (b + 1)$, for $a = 5$ and $b = 4$ [1.8]

Translate to an algebraic expression. [1.1]

15. 7 less than z

16. The product of x and z

17. One more than the product of two numbers

18. Determine whether 35 is a solution of $x/5 = 8$.

19. Translate to an equation. Do not solve. [1.1]

According to Photo Marketing Association International, in 2003, 29.8 billion photos were taken using film. This number is 18.49 billion more than the number of digital photos taken. How many digital photos were taken in 2003?

20. Use the commutative law of multiplication to write an expression equivalent to $3t + 5$. [1.2]

21. Use the associative law of addition to write an expression equivalent to $(2x + y) + z$. [1.2]

22. Use the commutative and associative laws to write three expressions equivalent to $4(xy)$. [1.2]

Multiply. [1.2]

23. $6(3x + 5y)$

24. $8(5x + 3y + 2)$

Factor. [1.2]

25. $21x + 15y$

26. $35x + 14 + 7y$

27. Find the prime factorization of 52. [1.3]

Simplify. [1.3]

28. $\dfrac{20}{48}$

29. $\dfrac{18}{8}$

Perform the indicated operation and, if possible, simplify. [1.3]

30. $\dfrac{5}{12} + \dfrac{4}{9}$

31. $\dfrac{9}{16} \div 3$

32. $\dfrac{2}{3} - \dfrac{1}{15}$

33. $\dfrac{9}{10} \cdot \dfrac{16}{5}$

34. Tell which integers correspond to this situation: Renir has a debt of $45 and Raoul has $72 in his savings account. [1.4]

35. Graph on a number line: $\dfrac{-1}{3}$. [1.4]

36. Write an inequality with the same meaning as $-3 < x$. [1.4]

37. Classify as true or false: $2 \geq -8$. [1.4]

38. Classify as true or false: $0 \leq -1$. [1.4]

39. Find decimal notation: $-\dfrac{7}{8}$. [1.4]

40. Find the absolute value: $|-1|$. [1.4]

41. Find $-(-x)$ when x is -9. [1.6]

Simplify.

42. $4 + (-7)$ [1.5]

43. $-\dfrac{2}{3} + \dfrac{1}{12}$ [1.5]

44. $10 + (-9) + (-8) + 7$ [1.5]

45. $-3.8 + 5.1 + (-12) + (-4.3) + 10$ [1.5]

46. $-2 - (-7)$ [1.6]

47. $-\dfrac{9}{10} - \dfrac{1}{2}$ [1.6]

48. $-3.8 - 4.1$ [1.6]

49. $-9 \cdot (-6)$ [1.7]

50. $-2.7(3.4)$ [1.7]

51. $\dfrac{2}{3} \cdot \left(-\dfrac{3}{7}\right)$ [1.7]

52. $2 \cdot (-7) \cdot (-2) \cdot (-5)$ [1.7]

53. $35 \div (-5)$ [1.7]

54. $-5.1 \div 1.7$ [1.7]

55. $-\dfrac{3}{5} \div \left(-\dfrac{4}{5}\right)$ [1.7]

56. $|-3 \cdot 4 - 12 \cdot 2| - 8(-7)$ [1.8]

57. $|-12(-3) - 2^3 - (-9)(-10)|$ [1.8]

58. $120 - 6^2 \div 4 \cdot 8$ [1.8]

59. $(120 - 6^2) \div 4 \cdot 8$ [1.8]

60. $(120 - 6^2) \div (4 \cdot 8)$ [1.8]

61. $\dfrac{4(18 - 8) + 7 \cdot 9}{9^2 - 8^2}$ [1.8]

Combine like terms.

62. $11a + 2b + (-4a) + (-5b)$ [1.5]

63. $7x - 3y - 9x + 8y$ [1.6]

64. Find the opposite of -7. [1.6]

65. Find the reciprocal of -7. [1.7]

66. Write exponential notation for $2x \cdot 2x \cdot 2x \cdot 2x$. [1.8]

67. Simplify: $(-5x)^3$. [1.8]

Remove parentheses and simplify. [1.8]

68. $2a - (5a - 9)$

69. $3(b + 7) - 5b$

70. $3[11x - 3(4x - 1)]$

71. $2[6(y - 4) + 7]$

72. $[8(x + 4) - 10] - [3(x - 2) + 4]$

73. Explain the difference between a constant and a variable. [1.1]

74. Explain the difference between a term and a factor. [1.2]

SYNTHESIS

75. Describe at least three ways in which the distributive law was used in this chapter. [1.2]

76. Devise a rule for determining the sign of a negative number raised to a power. [1.8]

77. Evaluate $a^{50} - 20a^{25}b^4 + 100b^8$ for $a = 1$ and $b = 2$. [1.8]

78. If $0.090909\ldots = \frac{1}{11}$ and $0.181818\ldots = \frac{2}{11}$, what rational number is named by each of the following?
   a) $0.272727\ldots$ [1.4]   **b)** $0.909090\ldots$ [1.4]

Simplify. [1.8]

79. $-\left| \frac{7}{8} - \left(-\frac{1}{2}\right) - \frac{3}{4} \right|$

80. $(|2.7 - 3| + 3^2 - |-3|) \div (-3)$

Match the phrase in the left column with the most appropriate choice from the right column.

81. _____ A number is nonnegative. [1.4]

82. _____ The reciprocal of a sum [1.7]

83. _____ A number squared [1.8]

84. _____ The opposite of a sum [1.8]

85. _____ The opposite of an opposite is the original number. [1.6]

86. _____ The order in which numbers are added does not change the result. [1.2]

87. _____ A number is negative. [1.4]

88. _____ The absolute value of a product [1.4]

89. _____ A sum of a number and its reciprocal [1.7]

90. _____ The square of a sum [1.8]

91. _____ The absolute value of one number is less than the absolute value of another number. [1.4]

a) a^2

b) $a + b = b + a$

c) $a < 0$

d) $a + \dfrac{1}{a}$

e) $|ab|$

f) $(a + b)^2$

g) $|a| < |b|$

h) $-(a + b)$

i) $a \geq 0$

j) $\dfrac{1}{a + b}$

k) $-(-a) = a$

1 Chapter Test

1. Evaluate $\dfrac{2x}{y}$ for $x = 10$ and $y = 5$.

2. Write an algebraic expression: Nine less than some number.

3. Find the area of a triangle when the height h is 30 ft and the base b is 16 ft.

4. Use the commutative law of addition to write an expression equivalent to $3p + q$.

5. Use the associative law of multiplication to write an expression equivalent to $x \cdot (4 \cdot y)$.

6. Determine whether 7 is a solution of $65 - x = 69$.

7. Translate to an equation. Do not solve.

   On a hot summer day, Green River Electric met a demand of 2518 megawatts. This is only 282 megawatts less than its maximum production capability. What is the maximum capability of production?

Multiply.

8. $7(5 - x)$

9. $-5(y - 2)$

Factor.

10. $11 - 44x$

11. $7x + 21 + 14y$

12. Find the prime factorization of 300.

13. Simplify: $\frac{10}{35}$.

Write a true sentence using either $<$ or $>$.

14. $-4 \; \blacksquare \; 0$

15. $-3 \; \blacksquare \; -8$

Find the absolute value.

16. $\left| \frac{9}{4} \right|$

17. $|-2.7|$

18. Find the opposite of $\frac{2}{3}$.

19. Find the reciprocal of $-\frac{4}{7}$.

20. Find $-x$ when x is -8.

21. Write an inequality with the same meaning as $x \le -2$.

Perform the indicated operations and, if possible, simplify.

22. $3.1 - (-4.7)$

23. $-8 + 4 + (-7) + 3$

24. $3.2 - 5.7$

25. $\frac{1}{8} - \left(-\frac{3}{4}\right)$

26. $4 \cdot (-12)$

27. $-\frac{1}{2} \cdot \left(-\frac{3}{8}\right)$

28. $-54 \div 9$

29. $-\frac{3}{5} \div \left(-\frac{4}{5}\right)$

30. $4.864 \div (-0.5)$

31. $-2(16) - |2(-8) - 5^3|$

32. $9 + 7 - 4 - (-3)$

33. $256 \div (-16) \div 4$

34. $2^3 - 10[4 - (-2 + 18)3]$

35. Combine like terms: $18y + 30a - 9a + 4y$.

36. Simplify: $(-2x)^4$.

Remove parentheses and simplify.

37. $4x - (3x - 7)$

38. $4(2a - 3b) + a - 7$

39. $4\{3[5(y - 3) + 9] + 2(y + 8)\}$

SYNTHESIS

40. Evaluate $\dfrac{5y - x}{2}$ when $x = 20$ and y is 4 less than half of x.

41. Insert one pair of parentheses to make the following a true statement:
$$9 - 3 - 4 + 5 = 15.$$

Simplify.

42. $|-27 - 3(4)| - |-36| + |-12|$

43. $a - \{3a - [4a - (2a - 4a)]\}$

44. Classify the following as either true or false:
$$a|b - c| = |ab| - |ac|.$$

2

Equations, Inequalities, and Problem Solving

AN APPLICATION

Local light rail service in Denver, Colorado, costs $1.15 per trip (one way). A student monthly pass costs $21 (*Source*: rtd-denver.com). Gail is a student at Community College of Denver. Express as an inequality the number of trips per month that Gail should make if the pass is to save her money.

This problem appears as Exercise 23 in Section 2.7.

Lee Cryer
TRANSPORTATION PLANNER
Denver, Colorado

As a transportation planner, I use math to forecast the number of passengers we can anticipate and then take that information to determine how many buses, train cars, and park-n-ride spaces we need to serve our customers. I also use math to determine capital and operating costs and the revenues needed to pay those costs.

*S*olving equations and inequalities is a recurring theme in much of mathematics. In this chapter, we will study some of the principles used to solve equations and inequalities. We will then use equations and inequalities to solve applied problems.

2.1   Solving Equations

Equations and Solutions • The Addition Principle •
The Multiplication Principle • Selecting the Correct
Approach

Solving equations is essential for problem solving in algebra. In this section, we study two of the most important principles used for this task.

Equations and Solutions

We have already seen that an equation is a number sentence stating that the expressions on either side of the equals sign represent the same number. Some equations, like $3 + 2 = 5$ or $2x + 6 = 2(x + 3)$, are *always* true and some, like $3 + 2 = 6$ or $x + 2 = x + 3$, are *never* true. In this text, we will concentrate on equations like $x + 6 = 13$ or $7x = 141$ that are *sometimes* true, depending on the replacement value for the variable.

> **Solution of an Equation**
>
> Any replacement for the variable that makes an equation true is called a *solution* of the equation. To *solve* an equation means to find all of its solutions.

To determine whether a number is a solution, we substitute that number for the variable throughout the equation. If the values on both sides of the equals sign are the same, then the number that was substituted is a solution.

EXAMPLE 1   Determine whether 7 is a solution of $x + 6 = 13$.

Solution   We have

$$\frac{x + 6 = 13}{7 + 6 \mid 13} \qquad \text{Writing the equation}$$
$$\text{Substituting 7 for } x$$
$$13 \overset{?}{=} 13 \quad \text{TRUE} \qquad 13 = 13 \text{ is a true statement.}$$

Since the left-hand and the right-hand sides are the same, 7 is a solution.

> *Caution!* Note that in Example 1, the solution is 7, not 13.

EXAMPLE 2 Determine whether 19 is a solution of $7x = 141$.

Solution We have

$$7x = 141 \qquad \text{Writing the equation}$$
$$7(19) \mid 141 \qquad \text{Substituting 19 for } x$$
$$133 \overset{?}{=} 141 \quad \text{FALSE} \qquad \text{The statement } 133 = 141 \text{ is false.}$$

Since the left-hand and the right-hand sides differ, 19 is not a solution.

The Addition Principle

Consider the equation

$$x = 7.$$

We can easily see that the solution of this equation is 7. Replacing x with 7, we get

$$7 = 7, \quad \text{which is true.}$$

Now consider the equation

$$x + 6 = 13.$$

In Example 1, we found that the solution of $x + 6 = 13$ is also 7. Although the solution of $x = 7$ may seem more obvious, because $x + 6 = 13$ and $x = 7$ have identical solutions, the equations are said to be **equivalent**.

*Student Notes*_____

Be sure to remember that equivalent equations are *equations*, whereas equivalent expressions are *expressions* like $5a - 10$ or $5(a - 2)$.

> **Equivalent Equations**
>
> Equations with the same solutions are called *equivalent equations*.

There are principles that enable us to begin with one equation and end up with an equivalent equation, like $x = 7$, for which the solution is obvious. One such principle concerns addition. The equation $a = b$ says that a and b stand for the same number. Suppose this is true, and some number c is added to a. We get the same result if we add c to b, because a and b are the same number.

> **The Addition Principle**
>
> For any real numbers a, b, and c,
>
> $$a = b \quad \text{is equivalent to} \quad a + c = b + c.$$

To visualize the addition principle, consider a balance similar to one a jeweler might use. When the two sides of a balance hold equal weight, the balance is level. If weight is then added or removed, equally, on both sides, the balance will remain level.

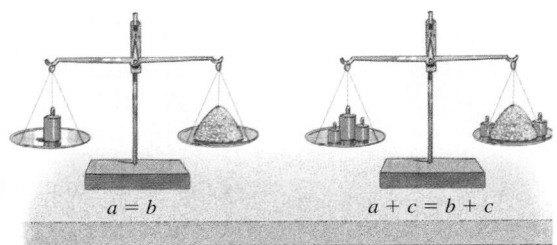

When using the addition principle, we often say that we "add the same number to both sides of an equation." We can also "subtract the same number from both sides," since subtraction can be regarded as the addition of an opposite.

EXAMPLE 3

Solve: $x + 5 = -7$.

Solution We can add any number we like to both sides. Since -5 is the opposite, or additive inverse, of 5, we add -5 to each side:

$$x + 5 = -7$$
$$x + 5 - 5 = -7 - 5 \qquad \text{Using the addition principle: adding } -5 \text{ to both sides or subtracting 5 from both sides}$$
$$x + 0 = -12 \qquad \text{Simplifying; } x + 5 - 5 = x + 5 + (-5) = x + 0$$
$$x = -12. \qquad \text{Using the identity property of 0}$$

It is obvious that the solution of $x = -12$ is the number -12. To check the answer in the original equation, we substitute.

Check:
$$\frac{x + 5 = -7}{-12 + 5 \mid -7}$$
$$-7 \overset{?}{=} -7 \quad \text{TRUE} \qquad -7 = -7 \text{ is true.}$$

The solution of the original equation is -12.

In Example 3, note that because we added the *opposite*, or *additive inverse*, of 5, the left side of the equation simplified to x plus the *additive identity*, 0, or simply x. These steps effectively replaced the 5 on the left with a 0. When solving $x + a = b$ for x, we simply add $-a$ to (or subtract a from) both sides.

EXAMPLE 4

Solve: $-6.5 = y - 8.4$.

Solution   The variable is on the right side this time. We can isolate y by adding 8.4 to each side:

$$-6.5 = y - 8.4$$ $y - 8.4$ can be regarded as $y + (-8.4)$.

$$-6.5 + 8.4 = y - 8.4 + 8.4$$ Using the addition principle: Adding 8.4 to both sides "eliminates" -8.4 on the right side.

$$1.9 = y.$$ $y - 8.4 + 8.4 = y + (-8.4) + 8.4 = y + 0 = y$

Check:
$$\begin{array}{c|c} -6.5 = y - 8.4 \\ \hline -6.5 & 1.9 - 8.4 \\ -6.5 \overset{?}{=} -6.5 \end{array}$$ TRUE $-6.5 = -6.5$ is true.

The solution is 1.9.

Note that the equations $a = b$ and $b = a$ have the same meaning. Thus, $-6.5 = y - 8.4$ could have been rewritten as $y - 8.4 = -6.5$.

The Multiplication Principle

A second principle for solving equations concerns multiplying. Suppose a and b are equal. If a and b are multiplied by some number c, then ac and bc will also be equal.

> **The Multiplication Principle**
>
> For any real numbers a, b, and c, with $c \neq 0$,
>
> $$a = b \quad \text{is equivalent to} \quad a \cdot c = b \cdot c.$$

EXAMPLE 5   Solve: $\frac{5}{4}x = 10$.

Solution   We can multiply both sides by any nonzero number we like. Since $\frac{4}{5}$ is the reciprocal of $\frac{5}{4}$, we multiply each side by $\frac{4}{5}$:

$$\frac{5}{4}x = 10$$

$$\frac{4}{5} \cdot \frac{5}{4}x = \frac{4}{5} \cdot 10$$ Using the multiplication principle: Multiplying both sides by $\frac{4}{5}$ "eliminates" the $\frac{5}{4}$ on the left.

$$1 \cdot x = 8$$ Simplifying

$$x = 8.$$ Using the identity property of 1

Check:
$$\begin{array}{c|c} \frac{5}{4}x = 10 \\ \hline \frac{5}{4} \cdot 8 & 10 \\ \frac{40}{4} & \\ 10 \overset{?}{=} 10 \end{array}$$ Think of 8 as $\frac{8}{1}$.   TRUE $10 = 10$ is true.

The solution is 8.

In Example 5, to get x alone, we multiplied by the *reciprocal*, or *multiplicative inverse* of $\frac{5}{4}$. We then simplified the left-hand side to x times the *multiplicative identity*, 1, or simply x. These steps effectively replaced the $\frac{5}{4}$ on the left with 1.

Because division is the same as multiplying by a reciprocal, the multiplication principle also tells us that we can "divide both sides by the same nonzero number." That is,

$$\text{if } a = b, \text{ then } \quad \frac{1}{c} \cdot a = \frac{1}{c} \cdot b \quad \text{and} \quad \frac{a}{c} = \frac{b}{c} \quad \text{(provided } c \neq 0).$$

In a product like $3x$, the multiplier 3 is called the **coefficient**. *When the coefficient of the variable is an integer or a decimal, it is usually easiest to solve an equation by dividing on both sides. When the coefficient is in fraction notation, it is usually easier to multiply by the reciprocal.*

EXAMPLE 6   Solve: **(a)** $-4x = 92$; **(b)** $-x = 9$; **(c)** $\dfrac{2y}{9} = \dfrac{8}{3}$.

Solution

a) In $-4x = 92$, the coefficient of x is an integer, so we *divide* on both sides:

$$\frac{-4x}{-4} = \frac{92}{-4}$$ Using the multiplication principle: Dividing both sides by -4 is the same as multiplying by $-\frac{1}{4}$.

$$1 \cdot x = -23$$ Simplifying

$$x = -23.$$ Using the identity property of 1

Check: $$\begin{array}{c|c} -4x = 92 \\ \hline -4(-23) & 92 \\ 92 \overset{?}{=} 92 & \text{TRUE} \end{array} \quad 92 = 92 \text{ is true.}$$

The solution is -23.

b) To solve an equation like $-x = 9$, remember that when an expression is multiplied or divided by -1, its sign is changed. Here we multiply both sides by -1 to change the sign of $-x$:

$$-x = 9$$

$$(-1)(-x) = (-1)9$$ Multiplying both sides by -1 (Dividing by -1 would also work). Note that the reciprocal of -1 is -1.

$$x = -9.$$ Note that $(-1)(-x)$ is the same as $(-1)(-1)x$.

Check: $$\begin{array}{c|c} -x = 9 \\ \hline -(-9) & 9 \\ 9 \overset{?}{=} 9 & \text{TRUE} \end{array} \quad 9 = 9 \text{ is true.}$$

The solution is -9.

c) To solve an equation like $\frac{2y}{9} = \frac{8}{3}$, we rewrite the left-hand side as $\frac{2}{9} \cdot y$ and then use the multiplication principle, multiplying by the reciprocal of $\frac{2}{9}$:

$$\frac{2y}{9} = \frac{8}{3}$$

$$\frac{2}{9} \cdot y = \frac{8}{3} \qquad \text{Rewriting } \frac{2y}{9} \text{ as } \frac{2}{9} \cdot y$$

$$\frac{9}{2} \cdot \frac{2}{9} \cdot y = \frac{9}{2} \cdot \frac{8}{3} \qquad \text{Multiplying both sides by } \frac{9}{2}$$

$$1y = \frac{3 \cdot 3 \cdot 2 \cdot 4}{2 \cdot 3} \qquad \text{Removing a factor equal to 1: } \frac{3 \cdot 2}{2 \cdot 3} = 1$$

$$y = 12.$$

Check:

$$\begin{array}{c|c} \frac{2y}{9} = \frac{8}{3} \\ \hline \frac{2 \cdot 12}{9} & \frac{8}{3} \\ \frac{24}{9} & \\ \frac{8}{3} \overset{?}{=} \frac{8}{3} \quad \text{TRUE} & \frac{8}{3} = \frac{8}{3} \text{ is true.} \end{array}$$

The solution is 12.

Selecting the Correct Approach

It is important that you be able to determine which principle should be used to solve a particular equation.

EXAMPLE 7   Solve: **(a)** $-\frac{2}{3} + x = \frac{5}{2}$; **(b)** $12.6 = 3t$.

Solution

a) To undo addition of $-\frac{2}{3}$, we subtract $-\frac{2}{3}$ or add $\frac{2}{3}$ to both sides:

$$-\frac{2}{3} + x = \frac{5}{2}$$

$$-\frac{2}{3} + x + \frac{2}{3} = \frac{5}{2} + \frac{2}{3} \qquad \text{Using the addition principle}$$

$$x = \frac{5}{2} + \frac{2}{3}$$

$$= \frac{5}{2} \cdot \frac{3}{3} + \frac{2}{3} \cdot \frac{2}{2} \qquad \text{Finding a common denominator}$$

$$= \frac{15}{6} + \frac{4}{6}$$

$$= \frac{19}{6}.$$

Check:

$$\begin{array}{c|c} -\frac{2}{3} + x = \frac{5}{2} \\ \hline -\frac{2}{3} + \frac{19}{6} & \frac{5}{2} \\ -\frac{4}{6} + \frac{19}{6} & \qquad -\frac{2}{3} \cdot \frac{2}{2} = -\frac{4}{6} \\ \frac{15}{6} & \\ \frac{5 \cdot 3}{2 \cdot 3} & \qquad \text{Removing a factor equal to 1: } \frac{3}{3} = 1 \\ \frac{5}{2} \overset{?}{=} \frac{5}{2} \quad \text{TRUE} & \frac{5}{2} = \frac{5}{2} \text{ is true.} \end{array}$$

The solution is $\frac{19}{6}$.

b) To undo multiplication by 3, we either divide both sides by 3 or multiply both sides by $\frac{1}{3}$:

$$12.6 = 3t$$

$$\frac{12.6}{3} = \frac{3t}{3} \qquad \text{Using the multiplication principle}$$

$$4.2 = 1t$$

$$4.2 = t \qquad \text{Simplifying}$$

Check:
$$\begin{array}{c|c} \multicolumn{2}{c}{12.6 = 3t} \\ \hline 12.6 & 3(4.2) \\ 12.6 & \stackrel{?}{=} 12.6 \end{array} \qquad \text{TRUE} \qquad 12.6 = 12.6 \text{ is true.}$$

The solution is 4.2.

Study Skills

Seeking Help?

Using some or all of these resources can make studying easier and more enjoyable.

A variety of resources are available to help you learn math.

- **Textbook supplements.** Are you aware of all the supplements that exist for this textbook? See the preface for a description of each supplement: the *Student's Solutions Manual*, a complete set of lessons in video or CD form, tutorial exercises on CD, and complete online courses in MathXL and MyMathLab.

- **Your college or university.** Your own college or university probably has resources to enhance your math learning.
    1. There may be a learning lab or tutoring center for drop-in tutoring.
    2. There may be study skills workshops or group tutoring sessions tailored for the specific course you are taking.
    3. Often, there is a bulletin board or network where you can locate the names of experienced private tutors.
    4. You might be able to find classmates interested in forming a study group.

- **Your instructor.** Although it may seem obvious, you should consider an often overlooked resource: your instructor. Find out your instructor's office hours and make it a point to visit when you need additional help. Many instructors welcome student e-mail. If you are hesitant to visit your instructor, e-mail may prove quite useful.

Exercise Set

2.1

FOR EXTRA HELP

| Student's Solutions Manual | Digital Video Tutor CD 1 Videotape 2 | Tutor Center AW Math Tutor Center | MathXL Tutorials on CD | Math XL MathXL | MyMathLab MyMathLab |

↪ *Concept Reinforcement   For each of Exercises 1–6, match the statement with the most appropriate choice from the column on the right.*

1. ____ A replacement that makes an equation true

a) Coefficient

2. ____ The equations $x + 3 = 7$ and $6x = 24$

b) Equivalent expressions

3. ____ The role of 9 in $9ab$

c) Equivalent equations

4. ____ The expressions $3(x - 2)$ and $3x - 6$

d) The multiplication principle

5. ____ The principle used to solve $\frac{2}{3} \cdot x = -4$

e) The addition principle

6. ____ The principle used to solve $\frac{2}{3} + x = -4$

f) Solution

Solve using the addition principle. Don't forget to check!

7. $x + 6 = 23$

8. $x + 5 = 8$

9. $y + 7 = -4$

10. $t + 6 = 43$

11. $t + 9 = -12$

12. $y + 7 = -3$

13. $-6 = y + 25$

14. $-5 = x + 8$

15. $x - 8 = 5$

16. $x - 9 = 6$

17. $12 = -7 + y$

18. $15 = -8 + z$

19. $-5 + t = -9$

20. $-6 + y = -21$

21. $r + \frac{1}{3} = \frac{8}{3}$

22. $t + \frac{3}{8} = \frac{5}{8}$

23. $x + \frac{3}{5} = -\frac{7}{10}$

24. $x + \frac{2}{3} = -\frac{5}{6}$

25. $x - \frac{5}{6} = \frac{7}{8}$

26. $y - \frac{3}{4} = \frac{5}{6}$

27. $-\frac{1}{5} + z = -\frac{1}{4}$

28. $-\frac{1}{8} + y = -\frac{3}{4}$

29. $m + 3.9 = 5.4$

30. $y + 5.3 = 8.7$

31. $-9.7 = -4.7 + y$

32. $-7.8 = 2.8 + x$

Solve using the multiplication principle. Don't forget to check!

33. $5x = 70$

34. $3x = 39$

35. $9t = 36$

36. $6x = 72$

37. $84 = 7x$

38. $56 = 7t$

39. $-x = 23$

40. $100 = -x$

Aha! **41.** $-t = -8$

42. $-68 = -r$

43. $7x = -49$

44. $9x = -36$

45. $-1.3a = -10.4$

46. $-3.4t = -20.4$

47. $\dfrac{y}{-8} = 11$

48. $\dfrac{a}{4} = 13$

49. $\dfrac{4}{5}x = 16$

50. $\dfrac{3}{4}x = 27$

51. $\dfrac{-x}{6} = 9$

52. $\dfrac{-t}{5} = 9$

53. $\dfrac{1}{9} = \dfrac{z}{5}$

54. $\dfrac{2}{7} = \dfrac{x}{3}$

Aha! **55.** $-\dfrac{3}{5}r = -\dfrac{3}{5}$

56. $-\dfrac{2}{5}y = -\dfrac{4}{15}$

57. $\dfrac{-3r}{2} = -\dfrac{27}{4}$

58. $\dfrac{5x}{7} = -\dfrac{10}{14}$

Solve. The icon ▤ *indicates an exercise designed to give practice using a calculator.*

59. $4.5 + t = -3.1$

60. $\dfrac{3}{4}x = 18$

61. $-8.2x = 20.5$

62. $t - 7.4 = -12.9$

63. $x - 4 = -19$

64. $y - 6 = -14$

65. $3 + t = 21$

66. $9 + t = 3$

67. $-12x = 72$

68. $-15x = 105$

69. $48 = -\dfrac{3}{8}y$

70. $14 = t + 27$

71. $a - \dfrac{1}{6} = -\dfrac{2}{3}$

72. $-\dfrac{x}{7} = \dfrac{2}{9}$

73. $-24 = \dfrac{8x}{5}$

74. $\dfrac{1}{5} + y = -\dfrac{3}{10}$

75. $-\dfrac{4}{3}t = -16$

76. $\dfrac{17}{35} = -x$

▤ **77.** $-483.297 = -794.053 + t$

▤ **78.** $-0.2344x = 2028.732$

▤ **79.** When solving an equation, how do you determine what number to add, subtract, multiply, or divide by on both sides of that equation?

▤ **80.** What is the difference between equivalent expressions and equivalent equations?

SKILL MAINTENANCE

Simplify. [1.8]

81. $9 - 2 \cdot 5^2 + 7$

82. $10 \div 2 \cdot 3^2 - 4$

83. $16 \div (2 - 3 \cdot 2) + 5$

84. $12 - 5 \cdot 2^3 + 4 \cdot 3$

SYNTHESIS

▤ **85.** To solve $-3.5 = 14t$, Anita adds 3.5 to both sides. Will this form an equivalent equation? Will it help solve the equation? Explain.

▤ **86.** Explain why it is not necessary to state a subtraction principle: For any real numbers a, b, and c, $a = b$ is equivalent to $a - c = b - c$.

Solve for x. Assume a, c, m ≠ 0.

87. $mx = 9.4m$

88. $x - 4 + a = a$

89. $cx + 5c = 7c$

90. $c \cdot \dfrac{21}{a} = \dfrac{7cx}{2a}$

91. $7 + |x| = 20$

92. $ax - 3a = 5a$

93. If $t - 3590 = 1820$, find $t + 3590$.

94. If $n + 268 = 124$, find $n - 268$.

▤ **95.** Lydia makes a calculation and gets an answer of 22.5. On the last step, she multiplies by 0.3 when she should have divided by 0.3. What should the correct answer be?

▤ **96.** Are the equations $x = 5$ and $x^2 = 25$ equivalent? Why or why not?

2.2   Using the Principles Together

Applying Both Principles • Combining Like Terms • Clearing Fractions and Decimals • Contradictions and Identities

CONNECTING THE CONCEPTS

We have stated that most of algebra involves either simplifying expressions (by writing equivalent expressions) or solving equations (by writing equivalent equations). In Section 2.1, we used the addition and multiplication principles to produce equivalent equations, like $x = 5$, from which the solution—in this case, 5—is obvious. Here in Section 2.2, we will find that more complicated equations can be solved by using both principles together and by using the commutative, associative,

and distributive laws to write equivalent expressions.

An important strategy for solving new problems is to find a way to make a new problem look like a problem that we already know how to solve. This is precisely the approach taken in this section. You will find that the last steps of the examples in this section are nearly identical to the steps used for solving the examples of Section 2.1. What is new in this section appears in the early steps of each example.

Applying Both Principles

The addition and multiplication principles, along with the laws discussed in Chapter 1, are our tools for solving equations. In this section, we will find that the sequence and manner in which these tools are used is especially important.

EXAMPLE 1   Solve: $5 + 3x = 17$.

Solution   Were we to evaluate $5 + 3x$, the rules for the order of operations direct us to *first* multiply by 3 and *then* add 5. Because of this, we can isolate $3x$ and then x by reversing these operations: We first subtract 5 from both sides and then divide both sides by 3. Our goal is an equivalent equation of the form $x = a$.

$$5 + 3x = 17$$

$$5 + 3x - 5 = 17 - 5$$ Using the addition principle: subtracting 5 from both sides (adding -5)

$$5 + (-5) + 3x = 12$$ Using a commutative law. Try to perform this step mentally.

First isolate the x-term.

$$3x = 12$$ Simplifying

$$\frac{3x}{3} = \frac{12}{3}$$ Using the multiplication principle: dividing both sides by 3 (multiplying by $\frac{1}{3}$)

Then isolate x.

$$x = 4.$$ Simplifying

Check:

$$\begin{array}{c|c} 5 + 3x = 17 \\ \hline 5 + 3 \cdot 4 & 17 \\ 5 + 12 & \\ 17 \overset{?}{=} 17 & \text{TRUE} \end{array}$$

We use the rules for order of operations: Find the product, $3 \cdot 4$, and then add.

The solution is 4.

EXAMPLE 2

Solve: $\frac{4}{3}x - 7 = 1$.

Solution In $\frac{4}{3}x - 7$, we multiply first and then subtract. To reverse these steps, we first add 7 and then either divide by $\frac{4}{3}$ or multiply by $\frac{3}{4}$.

$$\frac{4}{3}x - 7 = 1$$

$$\frac{4}{3}x - 7 + 7 = 1 + 7 \qquad \text{Adding 7 to both sides}$$

$$\frac{4}{3}x = 8$$

$$\frac{3}{4} \cdot \frac{4}{3}x = \frac{3}{4} \cdot 8 \qquad \text{Multiplying both sides by } \frac{3}{4}$$

$$\left.\begin{array}{c} 1 \cdot x = \dfrac{3 \cdot \cancel{4} \cdot 2}{\cancel{4}} \\ x = 6 \end{array}\right\} \quad \text{Simplifying}$$

Check:

$$\begin{array}{c|c} \frac{4}{3}x - 7 = 1 \\ \hline \frac{4}{3} \cdot 6 - 7 & 1 \\ 8 - 7 & \\ 1 \overset{?}{=} 1 & \text{TRUE} \end{array}$$

The solution is 6.

EXAMPLE 3

Solve: $45 - t = 13$.

Solution We have

$$45 - t = 13$$

$$45 - t - 45 = 13 - 45 \qquad \text{Subtracting 45 from both sides}$$

$$\left.\begin{array}{c} 45 + (-t) + (-45) = 13 - 45 \\ 45 + (-45) + (-t) = 13 - 45 \end{array}\right\} \quad \text{Try to do these steps mentally.}$$

$$-t = -32 \qquad \text{Try to go directly to this step.}$$

$$(-1)(-t) = (-1)(-32) \qquad \begin{array}{l}\text{Multiplying both sides by } -1 \\ \text{(Dividing by } -1 \text{ would also} \\ \text{work.)}\end{array}$$

$$t = 32.$$

Check: $\dfrac{45 - t = 13}{\begin{array}{c|c} 45 - 32 & 13 \end{array}}$

$$13 \overset{?}{=} 13 \quad \text{TRUE}$$

The solution is 32.

As our skills improve, certain steps can be streamlined.

EXAMPLE 4   Solve: $16.3 - 7.2y = -8.18$.

Solution   We have

$$16.3 - 7.2y = -8.18$$

$$16.3 - 7.2y - 16.3 = -8.18 - 16.3 \qquad \text{Subtracting 16.3 from both sides}$$

$$-7.2y = -24.48 \qquad \text{Simplifying}$$

$$\frac{-7.2y}{-7.2} = \frac{-24.48}{-7.2} \qquad \text{Dividing both sides by } -7.2$$

$$y = 3.4. \qquad \text{Simplifying}$$

Check: $\dfrac{16.3 - 7.2y = -8.18}{\begin{array}{c|c} 16.3 - 7.2(3.4) & -8.18 \\ 16.3 - 24.48 & \end{array}}$

$$-8.18 \overset{?}{=} -8.18 \quad \text{TRUE}$$

The solution is 3.4.

Combining Like Terms

If like terms appear on the same side of an equation, we combine them and then solve. Should like terms appear on both sides of an equation, we can use the addition principle to rewrite all like terms on one side.

EXAMPLE 5   Solve.

a) $3x + 4x = -14$ **b)** $2x - 4 = -3x + 1$

c) $6x + 5 - 7x = 10 - 4x + 7$ **d)** $2 - 5(x + 5) = 3(x - 2) - 1$

Solution

a) $3x + 4x = -14$

$$7x = -14 \qquad \text{Combining like terms}$$

$$\frac{7x}{7} = \frac{-14}{7} \qquad \text{Dividing both sides by 7}$$

$$x = -2 \qquad \text{Simplifying}$$

The check is left to the student. The solution is -2.

technology connection

Most graphing calculators have a TABLE feature that enables the calculator to evaluate a variable expression for different choices of x. For example, to evaluate $6x + 5 - 7x$ for $x = 0, 1, 2, \ldots$, we first use $\boxed{\text{Y=}}$ to enter $6x + 5 - 7x$ as y_1. We then use $\boxed{\text{2ND}}$ $\boxed{\text{TBLSET}}$ to specify which x-values will be used. Using TblStart $= 0$, ΔTbl $= 1$, and selecting AUTO twice, we can generate a table in which the value of $6x + 5 - 7x$ is listed for values of x starting at 0 and increasing by ones.

X	Y1	
0	5	
1	4	
2	3	
3	2	
4	1	
5	0	
6	−1	
X = 0		

1. Create the above table on your graphing calculator. Scroll up and down to extend the table.
2. Enter $10 - 4x + 7$ as y_2. Your table should now have three columns.
3. For what x-value is y_1 the same as y_2? Compare this with the solution of Example 5(c). Is this a reliable way to solve equations? Why or why not?

b) To solve $2x - 4 = -3x + 1$, we must first write only variable terms on one side and only constant terms on the other. This can be done by adding 4 to both sides, to get all constant terms on the right, and $3x$ to both sides, to get all variable terms on the left. We can add 4 first, or $3x$ first, or do both in one step.

> Isolate variable terms on one side and constant terms on the other side.

$$2x - 4 = -3x + 1$$
$$2x + 3x - 4 = -3x + 3x + 1 \qquad \text{Adding } 3x \text{ to both sides}$$
$$5x - 4 = 1 \qquad\qquad \text{Simplifying}$$
$$5x - 4 + 4 = 1 + 4 \qquad \text{Adding 4 to both sides}$$
$$5x = 5 \qquad\qquad \text{Combining like terms}$$
$$\tfrac{5x}{5} = \tfrac{5}{5} \qquad\qquad \text{Dividing both sides by 5}$$
$$x = 1$$

Check:
$$\begin{array}{c|c} \multicolumn{2}{c}{2x - 4 = -3x + 1} \\ \hline 2 \cdot 1 - 4 & -3 \cdot 1 + 1 \\ 2 - 4 & -3 + 1 \\ -2 \stackrel{?}{=} -2 & \quad \text{TRUE} \end{array}$$

The solution is 1.

c)
$$6x + 5 - 7x = 10 - 4x + 7$$
$$-x + 5 = 17 - 4x \qquad \text{Combining like terms within each side}$$
$$-x + 5 + 4x = 17 - 4x + 4x \qquad \text{Adding } 4x \text{ to both sides}$$
$$5 + 3x = 17 \qquad \text{Simplifying. This is identical to Example 1.}$$
$$3x = 12 \qquad \text{Subtracting 5 from both sides}$$
$$\tfrac{3x}{3} = \tfrac{12}{3} \qquad \text{Dividing both sides by 3}$$
$$x = 4$$

Check:
$$\begin{array}{c|c} \multicolumn{2}{c}{6x + 5 - 7x = 10 - 4x + 7} \\ \hline 6 \cdot 4 + 5 - 7 \cdot 4 & 10 - 4 \cdot 4 + 7 \\ 24 + 5 - 28 & 10 - 16 + 7 \\ 1 \stackrel{?}{=} 1 & \quad \text{TRUE} \end{array}$$

The solution is 4.

d)
$$2 - 5(x + 5) = 3(x - 2) - 1$$
$$2 - 5x - 25 = 3x - 6 - 1 \qquad \text{Using the distributive law. This is now similar to part (c) above.}$$
$$-5x - 23 = 3x - 7 \qquad \text{Combining like terms on each side}$$
$$\left.\begin{array}{l} -5x - 23 + 7 = 3x \\ -23 + 7 = 3x + 5x \end{array}\right\} \qquad \text{Adding 7 and } 5x \text{ to both sides. This isolates the } x\text{-terms on one side and the constant terms on the other.}$$
$$-16 = 8x \qquad \text{Simplifying}$$
$$\tfrac{-16}{8} = \tfrac{8x}{8} \qquad \text{Dividing both sides by 8}$$
$$-2 = x$$

The student can confirm that -2 checks and is the solution.

Clearing Fractions and Decimals

Equations are generally easier to solve when they do not contain fractions or decimals. The multiplication principle can be used to "clear" fractions or decimals, as shown here.

Clearing Fractions	Clearing Decimals
$\frac{1}{2}x + 5 = \frac{3}{4}$	$2.3x + 7 = 5.4$
$4(\frac{1}{2}x + 5) = 4 \cdot \frac{3}{4}$	$10(2.3x + 7) = 10 \cdot 5.4$
$2x + 20 = 3$	$23x + 70 = 54$

In each case, the resulting equation is equivalent to the original equation, but easier to solve.

The easiest way to clear an equation of fractions is to multiply *both sides* of the equation by the smallest, or *least*, common denominator.

EXAMPLE 6   Solve: **(a)** $\frac{2}{3}x - \frac{1}{6} = 2x$; **(b)** $\frac{2}{5}(3x + 2) = 8$.

Solution

a) We multiply both sides by 6, the least common denominator:

$$6\left(\frac{2}{3}x - \frac{1}{6}\right) = 6 \cdot 2x \quad \text{Multiplying both sides by 6}$$

$$6 \cdot \frac{2}{3}x - 6 \cdot \frac{1}{6} = 6 \cdot 2x \leftarrow \boxed{\text{Caution!   Be sure the distributive law is used to multiply \textit{all} the terms by 6.}}$$

$$4x - 1 = 12x \quad \text{Simplifying. Note that the fractions are cleared: } 6 \cdot \frac{2}{3} = 4, 6 \cdot \frac{1}{6} = 1, \text{ and } 6 \cdot 2 = 12.$$

$$-1 = 8x \quad \text{Subtracting } 4x \text{ from both sides}$$

$$\frac{-1}{8} = \frac{8x}{8} \quad \text{Dividing both sides by 8}$$

$$-\frac{1}{8} = x$$

The student can confirm that $-\frac{1}{8}$ checks and is the solution.

b) To solve $\frac{2}{5}(3x + 2) = 8$, we can multiply both sides by $\frac{5}{2}$ (or divide by $\frac{2}{5}$) to "undo" the multiplication by $\frac{2}{5}$ on the left side.

$$\frac{5}{2} \cdot \frac{2}{5}(3x + 2) = \frac{5}{2} \cdot 8 \quad \text{Multiplying both sides by } \frac{5}{2}$$

$$3x + 2 = 20 \quad \text{Simplifying; } \frac{5}{2} \cdot \frac{2}{5} = 1 \text{ and } \frac{5}{2} \cdot \frac{8}{1} = 20$$

$$3x = 18 \quad \text{Subtracting 2 from both sides}$$

$$x = 6 \quad \text{Dividing both sides by 3}$$

The student can confirm that 6 checks and is the solution.

To clear an equation of decimals, we count the greatest number of decimal places in any one number. If the greatest number of decimal places is 1, we multiply both sides by 10; if it is 2, we multiply by 100; and so on. This procedure is the same as multiplying by the least common denominator after converting the decimals to fractions.

EXAMPLE 7

Solve: $16.3 - 7.2y = -8.18$.

Solution The greatest number of decimal places in any one number is *two*. Multiplying by 100 will clear all decimals.

$$100(16.3 - 7.2y) = 100(-8.18)$$ Multiplying both sides by 100

$$100(16.3) - 100(7.2y) = 100(-8.18)$$ Using the distributive law

$$1630 - 720y = -818$$ Simplifying

$$-720y = -818 - 1630$$ Subtracting 1630 from both sides

$$-720y = -2448$$ Combining like terms

$$y = \frac{-2448}{-720}$$ Dividing both sides by -720

$$y = 3.4$$

In Example 4, the same solution was found without clearing decimals. Finding the same answer in two ways is a good check. The solution is 3.4.

Student Notes

Compare the steps of Examples 4 and 7. Note that although the two approaches differ, they yield the same solution. Whenever you can use two approaches to solve a problem, try to do so, both as a check and as a valuable learning experience.

An Equation-Solving Procedure

1. Use the multiplication principle to clear any fractions or decimals. (This is optional, but can ease computations. See Examples 6 and 7.)
2. If necessary, use the distributive law to remove parentheses. Then combine like terms on each side. (See Example 5.)
3. Use the addition principle, as needed, to isolate all variable terms on one side. Then combine like terms. (See Examples 1–7.)
4. Multiply or divide to solve for the variable, using the multiplication principle. (See Examples 1–7.)
5. Check all possible solutions in the original equation. (See Examples 1–5.)

Contradictions and Identities

All of the equations we have examined so far had a solution. Equations that are true for some values (solutions), but not for others, are called **conditional equations**. Equations that have no solution, such as $x + 1 = x + 2$, are called **contradictions**. If, when solving an equation, we obtain an equation that is false for any value of x, the equation has no solution.

EXAMPLE 8   Solve: $3x - 5 = 3(x - 2) + 4$.

Solution

$$3x - 5 = 3(x - 2) + 4$$

$$3x - 5 = 3x - 6 + 4 \qquad \text{Using the distributive law}$$

$$3x - 5 = 3x - 2 \qquad \text{Combining like terms}$$

$$-3x + 3x - 5 = -3x + 3x - 2 \qquad \text{Using the addition principle}$$

$$-5 = -2$$

Since the original equation is equivalent to $-5 = -2$, which is false, the original equation has no solution. There is no choice of x that will make $3x - 5 = 3(x - 2) + 4$ true. The equation is a contradiction. It is *never* true.

Some equations, like $x + 1 = x + 1$, are true for all replacements. Such an equation is called an **identity**.

EXAMPLE 9   Solve: $2x + 7 = 7(x + 1) - 5x$.

Solution $2x + 7 = 7(x + 1) - 5x$

$$2x + 7 = 7x + 7 - 5x \qquad \text{Using the distributive law}$$

$$2x + 7 = 2x + 7 \qquad \text{Combining like terms}$$

The equation $2x + 7 = 2x + 7$ is true regardless of the replacement for x, so all real numbers are solutions. Note that $2x + 7 = 2x + 7$ is equivalent to $2x = 2x$, $7 = 7$, or $0 = 0$. All real numbers are solutions and the equation is an identity.

Exercise Set

2.2

FOR EXTRA HELP

Student's Solutions Manual

Digital Video Tutor CD 1 Videotape 2

Tutor Center

AW Math Tutor Center

MathXL Tutorials on CD

Math XL

MathXL

MyMathLab

MyMathLab

↪ *Concept Reinforcement* In each of Exercises 1–6, match the equation with an equivalent equation from the column on the right that could be the next step in finding a solution.

1. ____ $3x - 1 = 7$

2. ____ $4x + 5x = 12$

3. ____ $6(x - 1) = 2$

4. ____ $7x = 9$

5. ____ $4x = 3 - 2x$

6. ____ $8x - 5 = 6 - 2x$

a) $6x - 6 = 2$

b) $4x + 2x = 3$

c) $3x = 7 + 1$

d) $8x + 2x = 6 + 5$

e) $9x = 12$

f) $x = \frac{9}{7}$

Solve and check. Label any contradictions or identities.

7. $2x + 9 = 25$

8. $3x + 6 = 30$

9. $6z + 4 = 46$

10. $6z + 3 = 57$

11. $7t - 8 = 27$

12. $6x - 3 = 15$

13. $3x - 9 = 33$

14. $5x - 9 = 41$

15. $8z + 2 = -54$

16. $4x + 3 = -21$

17. $-91 = 9t + 8$

18. $-39 = 1 + 8x$

19. $12 - 4x = 108$

20. $9 - 4x = 37$

21. $-6z - 18 = -132$

22. $-7x - 24 = -129$

23. $4x + 5x = 10$

24. $13 = 5x + 7x$

25. $32 - 7x = 11$

26. $27 - 6x = 99$

27. $\frac{3}{5}t - 1 = 8$

28. $\frac{2}{3}t - 1 = 5$

29. $4 + \frac{7}{2}x = -10$

30. $6 + \frac{5}{4}x = -4$

31. $-\dfrac{3a}{4} - 5 = 2$

32. $-\dfrac{7a}{8} - 2 = 1$

33. $2x = x + x$

34. $-3z + 8z = 45$

35. $4x - 6 = 6x$

36. $4x - x = 2x + x$

37. $5y - 2 = 28 - y$

38. $6x - 5 = 7 + 2x$

39. $7(2a - 1) = 21$

40. $5(2t - 2) = 30$

Aha! **41.** $8 = 8(x + 1)$

42. $9 = 3(5x - 2)$

43. $7r - (2r + 8) = 32$

44. $3(5 + 3m) - 8 = 88$

45. $6x + 3 = 2x + 3$

46. $6b - (3b + 8) = 16$

47. $5 - 2x = 3x - 7x + 25$

48. $10 - 3x = 2x - 8x + 40$

49. $7 + 3x - 6 = 3x + 5 - x$

50. $5 + 4x - 7 = 4x - 2 - x$

51. $4y - 4 + y + 24 = 6y + 20 - 4y$

52. $5y - 10 + y = 7y + 18 - 5y$

53. $13 - 3(2x - 1) = 4$

54. $5(d + 4) = 7(d - 2)$

55. $7(5x - 2) = 6(6x - 1)$

56. $5(t + 3) + 9 = 3(t - 2) + 6$

57. $19 - (2x + 3) = 2(x + 3) + x$

58. $13 - (2c + 2) = 2(c + 2) + 3c$

59. $3(x + 4) = 3(x - 1)$

60. $5(x - 7) = 3(x - 2) + 2x$

Clear fractions or decimals, solve, and check.

61. $\frac{5}{4}x + \frac{1}{4}x = 2x + \frac{1}{2} + \frac{3}{4}x$

62. $\frac{7}{8}x - \frac{1}{4} + \frac{3}{4}x = \frac{1}{16} + x$

63. $\frac{2}{3} + \frac{1}{4}t = 6$

64. $-\frac{1}{2} + x = -\frac{5}{6} - \frac{1}{3}$

65. $\frac{2}{3} + 4t = 6t - \frac{2}{15}$

66. $\frac{1}{2} + 4m = 3m - \frac{5}{2}$

67. $\frac{1}{3}x + \frac{2}{5} = \frac{4}{15} + \frac{3}{5}x - \frac{2}{3}$

68. $1 - \frac{2}{3}y = \frac{9}{5} - \frac{1}{5}y + \frac{3}{5}$

69. $2.1x + 45.2 = 3.2 - 8.4x$

70. $0.91 - 0.2z = 1.23 - 0.6z$

71. $0.76 + 0.21t = 0.96t - 0.49$

72. $1.7t + 8 - 1.62t = 0.4t - 0.32 + 8$

73. $\frac{2}{5}x - \frac{3}{2}x = \frac{3}{4}x + 2$

74. $\frac{5}{16}y + \frac{3}{8}y = 2 + \frac{1}{4}y$

75. $\frac{1}{3}(2x - 1) = 7$

76. $\frac{4}{3}(5x + 1) = 8$

77. $\frac{3}{4}(3t - 6) = 9$

78. $\frac{3}{2}(2x + 5) = -\frac{15}{2}$

79. $\frac{1}{6}\left(\frac{3}{4}x - 2\right) = -\frac{1}{5}$

80. $\frac{2}{3}\left(\frac{7}{8} - 4x\right) - \frac{5}{8} = \frac{3}{8}$

81. $0.7(3x + 6) = 1.1 - (x + 2)$

82. $0.9(2x + 8) = 20 - (x + 5)$

83. $a + (a - 3) = (a + 2) - (a + 1)$

84. $0.8 - 4(b - 1) = 0.2 + 3(4 - b)$

85. When an equation contains decimals, is it essential to clear the equation of decimals? Why or why not?

86. Why must the rules for the order of operations be understood before solving the equations in this section?

SKILL MAINTENANCE

Evaluate. [1.8]

87. $3 - 5a$, for $a = 2$

88. $12 \div 4 \cdot t$, for $t = 5$

89. $7x - 2x$, for $x = -3$

90. $t(8 - 3t)$, for $t = -2$

SYNTHESIS

91. What procedure would you follow to solve an equation like $0.23x + \frac{17}{3} = -0.8 + \frac{3}{4}x$? Could your procedure be streamlined? If so, how?

92. Dave is determined to solve the equation $3x + 4 = -11$ by first using the multiplication principle to "eliminate" the 3. How should he proceed and why?

Solve. Label any contradictions or identities.

93. $8.43x - 2.5(3.2 - 0.7x) = -3.455x + 9.04$

94. $0.008 + 9.62x - 42.8 = 0.944x + 0.0083 - x$

95. $-2[3(x - 2) + 4] = 4(5 - x) - 2x$

96. $0 = y - (-14) - (-3y)$

97. $2|x| = -14$

98. $|3x| = 6$

99. $2x(x + 5) - 3(x^2 + 2x - 1) = 9 - 5x - x^2$

100. $x(x - 4) = 3x(x + 1) - 2(x^2 + x - 5)$

101. $9 - 3x = 2(5 - 2x) - (1 - 5x)$

102. $2(7 - x) - 20 = 7x - 3(2 + 3x)$

Aha! **103.** $[7 - 2(8 \div (-2))]x = 0$

104. $\dfrac{x}{14} - \dfrac{5x + 2}{49} = \dfrac{3x - 4}{7}$

105. $\dfrac{5x + 3}{4} + \dfrac{25}{12} = \dfrac{5 + 2x}{3}$

COLLABORATIVE

CORNER

Step-by-Step Solutions

Focus: Solving linear equations

Time: 20 minutes

Group size: 3

In general, there is more than one correct sequence of steps for solving an equation. This makes it important that you write your steps clearly and logically so that others can follow your approach.

ACTIVITY

1. Each group member should select a different one of the following equations and, on a fresh sheet of paper, perform the first step of the solution.

$4 - 3(x - 3) = 7x + 6(2 - x)$
$5 - 7[x - 2(x - 6)] = 3x + 4(2x - 7) + 9$
$4x - 7[2 + 3(x - 5) + x] = 4 - 9(-3x - 19)$

2. Pass the papers around so that the second and third steps of each solution are performed by the other two group members. Before writing, make sure that the previous step is correct. If a mistake is discovered, return the problem to the person who made the mistake for repairs. Continue passing the problems around until all equations have been solved.

3. Each group should reach a consensus on what the three solutions are and then compare their answers to those of other groups.

2.3 Formulas

Evaluating Formulas • Solving for a Letter

Many applications of mathematics involve relationships among two or more quantities. An equation that represents such a relationship will use two or more letters and is known as a **formula**. Although most of the letters in this book represent variables, some—like c in $E = mc^2$ or π in $C = \pi d$—represent constants.

Evaluating Formulas

EXAMPLE 1   Outdoor concerts. The formula $d = 344t$ can be used to determine how far d, in meters, sound travels through room-temperature air in one second. In 2003,

technology connection

Sometimes we may wish to recall and modify a calculation. For example, suppose that after calculating $30 \cdot 1800$ we wish to find $30 \cdot 1870$. Pressing **2ND** (ENTRY) gives the following.

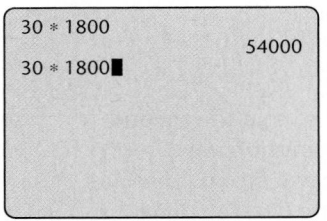

Moving the cursor left, we can change 1800 to 1870 and press **ENTER**.

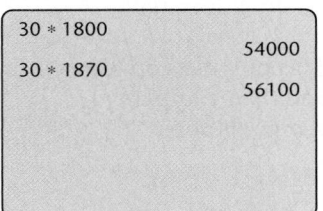

1. Verify the work above and then use **2ND** (ENTRY) to find $39 \cdot 1870$.

---

the Dave Matthews Band performed in New York City's Central Park. Fans near the back of the crowd experienced a 0.9-sec time lag between the time each word was pronounced on stage (as shown on large video monitors) and the time the sound reached their ears. How far were these fans from the stage?

Solution We substitute 0.9 for t in $d = 344t$ and calculate d:

$$d = 344(0.9) = 309.6.$$

The fans were about 309.6 m from the stage.

Solving for a Letter

In the Northeast, the formula $B = 30a$ is used to determine the minimum furnace output B, in British thermal units (Btu's), for a well-insulated home with a square feet of flooring. Suppose that a contractor has an extra furnace and wants to determine the size of the largest (well-insulated) house in which it can be used. The contractor can substitute the amount of the furnace's output in Btu's—say, 63,000—for B, and then solve for a:

$$63,000 = 30a \qquad \text{Replacing } B \text{ with } 63,000$$
$$2100 = a. \qquad \text{Dividing both sides by 30}$$

The home should have no more than 2100 ft^2 of flooring.

Were these calculations to be performed for a variety of furnaces, the contractor would find it easier to first solve $B = 30a$ for a, and *then* substitute values for B. This can be done in much the same way that we solved equations in Sections 2.1 and 2.2.

EXAMPLE 2   Solve for a: $B = 30a$.

Solution We have

$$B = 30a \qquad \text{We want this letter alone.}$$
$$\frac{B}{30} = a. \qquad \text{Dividing both sides by 30}$$

The equation $a = B/30$ gives a quick, easy way to determine the floor area of the largest (well-insulated) house that a furnace supplying B Btu's could heat.

To see how solving a formula is just like solving an equation, compare the following. In (A), we solve as usual; in (B), we show steps but do not simplify; and in (C), we *cannot* simplify since a, b, and c are unknown.

A. $5x + 2 = 12$

$5x = 12 - 2$

$5x = 10$

$x = \dfrac{10}{5} = 2$

B. $5x + 2 = 12$

$5x = 12 - 2$

$x = \dfrac{12 - 2}{5}$

C. $ax + b = c$

$ax = c - b$

$x = \dfrac{c - b}{a}$

EXAMPLE 3

Circumference of a circle. The formula $C = 2\pi r$ gives the *circumference* C of a circle with radius r. Solve for r.

Solution   The **circumference** is the distance around a circle.

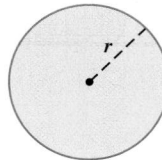

| Given a radius r, we can use this equation to find a circle's circumference C. |

$C = 2\pi r$ We want this letter alone.

$\dfrac{C}{2\pi} = \dfrac{2\pi r}{2\pi}$ Dividing both sides by 2π

| Given a circle's circumference C, we can use this equation to find the radius r. |

$\dfrac{C}{2\pi} = r$

EXAMPLE 4

Nutrition. The number of calories K needed each day by a moderately active woman who weighs w pounds, is h inches tall, and is a years old, can be estimated using the formula

$K = 917 + 6(w + h - a).$*

Solve for w.

Solution   We reverse the order in which the operations occur on the right side:

We want w alone.

$K = 917 + 6(w + h - a)$

$K - 917 = 6(w + h - a)$ Subtracting 917 from both sides

$\dfrac{K - 917}{6} = w + h - a$ Dividing both sides by 6

$\dfrac{K - 917}{6} + a - h = w.$ Adding a and subtracting h on both sides

This formula can be used to estimate a woman's weight, if we know her age, height, and caloric needs.

*Based on information from M. Parker (ed.), *She Does Math!* (Washington D.C.: Mathematical Association of America, 1995), p. 96.

The above steps are similar to those used in Section 2.2 to solve equations. We use the addition and multiplication principles just as before. An important difference that we will see in the next example is that we will sometimes need to factor.

> **To Solve a Formula for a Given Letter**
>
> 1. If the letter for which you are solving appears in a fraction, use the multiplication principle to clear fractions.
> 2. Isolate the term(s), with the letter you are solving for on one side of the equation.
> 3. If two or more terms contain the letter you are solving for, factor the letter out.
> 4. Multiply or divide to solve for the letter in question.

EXAMPLE 5

Solve for x: $y = ax + bx - 4$.

Solution   We solve as follows:

$$y = ax + bx - 4 \qquad \text{We want this letter alone.}$$

$$y + 4 = ax + bx \qquad \text{Adding 4 to both sides}$$

$$y + 4 = x(a + b) \qquad \text{Factoring}$$

$$\frac{y + 4}{a + b} = x. \qquad \begin{array}{l}\text{Dividing both sides by } a + b, \text{ or}\\ \text{multiplying both sides by } 1/(a + b)\end{array}$$

We can also write this as

$$x = \frac{y + 4}{a + b}.$$

> *Caution!*   Had we performed the following steps in Example 5, we would *not* have solved for x:
>
> $$y = ax + bx - 4$$
>
> $$y - ax + 4 = bx \qquad \text{Subtracting } ax \text{ and adding 4 to both sides}$$
>
> $$\text{Two occurrences of } x$$
>
> $$\frac{y - ax + 4}{b} = x. \qquad \text{Dividing both sides by } b$$
>
> The mathematics of each step is correct, but since x occurs on both sides of the formula, *we have not solved the formula for x.* Remember that the letter being solved for should be alone on one side of the equation, with no occurrence of that letter on the other side!

Exercise Set

2.3

FOR EXTRA HELP

Student's
Solutions
Manual

Digital Video Tutor
CD 1
Videotape 2

AW Math
Tutor Center

MathXL Tutorials
on CD

MathXL
MathXL

MyMathLab
MyMathLab

1. *Distance from a storm.* The formula $M = \frac{1}{5}t$ can be used to determine how far M, in miles, you are from lightning when its thunder takes t seconds to reach your ears. If it takes 10 sec for the sound of thunder to reach you after you have seen the lightning, how far away is the storm?

2. *Electrical power.* The power rating P, in watts, of an electrical appliance is determined by

$$P = I \cdot V,$$

where I is the current, in amperes, and V is the voltage, measured in volts. If a kitchen requires 30 amps of current and the voltage in the house is 115 volts, what is the wattage of the kitchen?

3. *College enrollment.* At many colleges, the number of "full-time-equivalent" students f is given by

$$f = \frac{n}{15},$$

where n is the total number of credits for which students have enrolled in a given semester. Determine the number of full-time-equivalent students on a campus in which students registered for a total of 21,345 credits.

4. *Wavelength of a musical note.* The wavelength w, in meters per cycle, of a musical note is given by

$$w = \frac{r}{f},$$

where r is the speed of the sound, in meters per second, and f is the frequency, in cycles per second. The speed of sound in air is 344 m/sec. What is the wavelength of a note whose frequency in air is 24 cycles per second?

5. *Furnace output.* Contractors in the Northeast use the formula $B = 30a$ to determine the minimum furnace output B, in British thermal units (Btu's),

for a well-insulated house with a square feet of flooring. Determine the minimum furnace output for an 1800-ft^2 house that is well insulated.

Source: U.S. Department of Energy

6. *Calorie density.* The calorie density D, in calories per ounce, of a food that contains c calories and weighs w ounces is given by

$$D = \frac{c}{w}.^*$$

Eight ounces of fat-free milk contains 84 calories. Find the calorie density of fat-free milk.

7. *Absorption of ibuprofen.* When 400 mg of the painkiller ibuprofen is swallowed, the number of milligrams n in the bloodstream t hours later (for $0 \le t \le 6$) is estimated by

$$n = 0.5t^4 + 3.45t^3 - 96.65t^2 + 347.7t.$$

How many milligrams of ibuprofen remain in the blood 1 hr after 400 mg has been swallowed?

8. *Size of a league schedule.* When all n teams in a league play every other team twice, a total of N games are played, where

$$N = n^2 - n.$$

If a soccer league has 7 teams and all teams play each other twice, how many games are played?

Solve each formula for the indicated letter.

9. $A = bh$, for b
(Area of parallelogram with base b and height h)

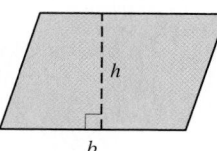

---

*Source: Nutrition Action Healthletter, March 2000, p. 9. Center for Science in the Public Interest, Suite 300; 1875 Connecticut Ave NW, Washington, D.C. 20008.

10. $A = bh$, for h

11. $d = rt$, for r
(A distance formula, where d is distance, r is speed, and t is time)

12. $d = rt$, for t

13. $I = Prt$, for P
(Simple-interest formula, where I is interest, P is principal, r is interest rate, and t is time)

14. $I = Prt$, for t

15. $H = 65 - m$, for m
(To determine the number of heating degree days H for a day with m degrees Fahrenheit as the average temperature)

16. $d = h - 64$, for h
(To determine how many inches d above average an h-inch-tall woman is)

17. $P = 2l + 2w$, for l
(Perimeter of a rectangle of length l and width w)

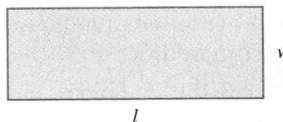

18. $P = 2l + 2w$, for w

19. $A = \pi r^2$, for π
(Area of a circle with radius r)

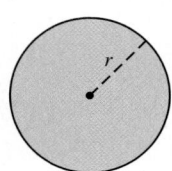

20. $A = \pi r^2$, for r^2

21. $A = \frac{1}{2}bh$, for h
(Area of a triangle with base b and height h)

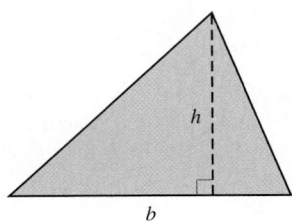

22. $A = \frac{1}{2}bh$, for b

23. $E = mc^2$, for m
(A relativity formula from physics)

24. $E = mc^2$, for c^2

25. $Q = \dfrac{c + d}{2}$, for d

26. $Q = \dfrac{p - q}{2}$, for p

27. $A = \dfrac{a + b + c}{3}$, for b

28. $A = \dfrac{a + b + c}{3}$, for c

29. $M = \dfrac{A}{s}$, for A
(To compute the Mach number M for speed A and speed of sound s)

30. $P = \dfrac{ab}{c}$, for b

31. $F = \dfrac{9}{5}C + 32$, for C
(To convert the Celsius temperature C to the Fahrenheit temperature F)

32. $M = \dfrac{3}{7}n + 29$, for n

33. $A = at + bt$, for t

34. $S = rx + sx$, for x

35. *Area of a trapezoid.* The formula
$$A = \tfrac{1}{2}ah + \tfrac{1}{2}bh$$
can be used to find the area A of a trapezoid with bases a and b and height h. Solve for h. (*Hint:* First clear fractions.)

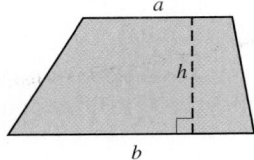

36. *Compounding interest.* The formula

$$A = P + Prt$$

is used to find the amount A in an account when simple interest is added to an investment of P dollars (see Exercise 13). Solve for P.

37. *Chess rating.* The formula

$$R = r + \frac{400(W - L)}{N}$$

is used to establish a chess player's rating R after that player has played N games, won W of them, and lost L of them. Here r is the average rating of the opponents. Solve for L.
Source: The U.S. Chess Federation

38. *Angle measure.* The angle measure S, of a sector of a circle, is given by

$$S = \frac{360A}{\pi r^2},$$

where r is the radius, A is the area of the sector, and S is in degrees. Solve for r^2.

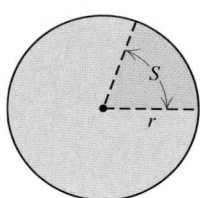

39. Naomi has a formula that allows her to convert Celsius temperatures to Fahrenheit temperatures. She needs a formula for converting Fahrenheit temperatures to Celsius temperatures. What advice can you give her?

40. Under what circumstances would it be useful to solve $d = rt$ for r? (See Exercise 11.)

SKILL MAINTENANCE

Multiply. [1.7]

Aha! **41.** $0.79(38.4)0$

42. $(0.085)(108)$

Simplify. [1.8]

43. $20 \div (-4) \cdot 2 - 3$

44. $5|8 - (2 - 7)|$

SYNTHESIS

45. The equations

$$P = 2l + 2w \quad \text{and} \quad w = \frac{P}{2} - l$$

are equivalent formulas involving the perimeter P, length l, and width w of a rectangle. Devise a problem for which the second of the two formulas would be more useful.

46. While solving $2A = ah + bh$ for h, Lea writes $\frac{2A - ah}{b} = h$. What is her mistake?

47. The number of calories K needed each day by a moderately active man who weighs w kilograms, is h centimeters tall, and is a years old, can be determined by

$$K = 19.18w + 7h - 9.52a + 92.4.*$$

If Janos is moderately active, weighs 82 kg, is 185 cm tall, and needs to consume 2627 calories a day, how old is he?

48. *Altitude and temperature.* Air temperature drops about 1° Celsius (C) for each 100-m rise above ground level, up to 12 km. If the ground level temperature is $t°$C, find a formula for the temperature T at an elevation of h meters.
Source: *A Sourcebook of School Mathematics,* Mathematical Association of America, 1980

49. *Surface area of a cube.* The surface area A of a cube with side s is given by

$$A = 6s^2.$$

If a cube's surface area is 54 in², find the volume of the cube.

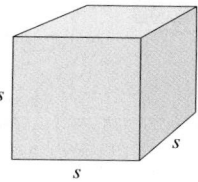

---

*Based on information from M. Parker (ed.), *She Does Math!* (Washington DC: Mathematical Association of America, 1995), p. 96.

50. *Weight of a fish.* An ancient fisherman's formula for estimating the weight of a fish is

$$w = \frac{lg^2}{800},$$

where w is the weight, in pounds, l is the length, in inches, and g is the girth (distance around the midsection), in inches. Estimate the girth of a 700-lb yellow tuna that is 8 ft long.

51. *Dosage size.* Clark's rule for determining the size of a particular child's medicine dosage c is

$$c = \frac{w}{a} \cdot d,$$

where w is the child's weight, in pounds, and d is the usual adult dosage for an adult weighing a pounds. Solve for a.
Source: Olsen, June Looby, et al., *Medical Dosage Calculations.* Redwood City, CA: Addison-Wesley, 1995

Solve each formula for the given letter.

52. $\dfrac{y}{z} \div \dfrac{z}{t} = 1$, for y

53. $ac = bc + d$, for c

54. $qt = r(s + t)$, for t

55. $3a = c - a(b + d)$, for a

56. *Furnace output.* The formula

$$B = 50a$$

is used in New England to estimate the minimum furnace output B, in Btu's, for an old, poorly insulated house with a square feet of flooring. Find an equation for determining the number of Btu's saved by insulating an old house. (*Hint*: See Exercise 5.)

57. Revise the formula in Example 4 so that a woman's weight in kilograms (2.2046 lb = 1 kg) and her height in centimeters (0.3937 in. = 1 cm) are used.

58. Revise the formula in Exercise 47 so that a man's weight in pounds (2.2046 lb = 1 kg) and his height in inches (0.3937 in. = 1 cm) are used.

2.4 Applications with Percent

Converting Between Percent Notation and Decimal Notation • Solving Percent Problems

Percent problems arise so frequently in everyday life that most often we are not even aware of them. In this section, we will solve some real-world percent problems. Before doing so, however, we need to review a few basics.

Converting Between Percent Notation and Decimal Notation

Nutritionists recommend that no more than 30% of the calories in a person's diet come from fat. This means that of every 100 calories consumed, no more than 30 should come from fat. Thus, 30% is a ratio of 30 to 100.

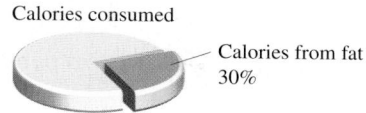

Calories consumed

Calories from fat
30%

The percent symbol % means "per hundred." We can regard the percent symbol as part of a name for a number. For example,

$$30\% \quad \text{is defined to mean} \quad \frac{30}{100}, \quad \text{or} \quad 30 \times \frac{1}{100}, \quad \text{or} \quad 30 \times 0.01.$$

Percent Notation

$$n\% \quad \text{means} \quad \frac{n}{100}, \quad \text{or} \quad n \times \frac{1}{100}, \quad \text{or} \quad n \times 0.01.$$

EXAMPLE 1   Convert to decimal notation: **(a)** 78%; **(b)** 1.3%.

Solution

a) $78\% = 78 \times 0.01$ Replacing % with ×0.01
$ = 0.78$

b) $1.3\% = 1.3 \times 0.01$ Replacing % with ×0.01
$ = 0.013$

As shown above, multiplication by 0.01 simply moves the decimal point two places to the left.

To convert from percent notation to decimal notation, move the decimal point two places to the left and drop the percent symbol.

EXAMPLE 2   Convert the percent notation in the following sentence to decimal notation: 60% of the pollution from a typical car trip is emitted in the first few minutes (*Source*: Chittenden County Transportation Authority).

Solution

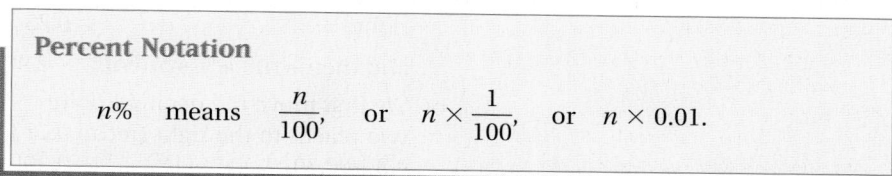

$$60\% = 60.0\% \qquad 0.60.0 \qquad 60\% = 0.60, \text{ or simply } 0.6$$

Move the decimal point two places to the left.

The procedure used in Examples 1 and 2 can be reversed:

$$0.38 = 38 \times 0.01$$
$$= 38\%. \qquad \text{Replacing} \times 0.01 \text{ with } \%$$

To convert from decimal notation to percent notation, move the decimal point two places to the right and write a percent symbol.

EXAMPLE 3   Convert to percent notation: **(a)** 1.27; **(b)** $\frac{1}{4}$; **(c)** 0.3.

Solution

a) We first move the decimal point
two places to the right: 1.27

and then write a % symbol: 127%   This is the same as multiplying
1.27 by 100 and writing %.

b) Note that $\frac{1}{4} = 0.25$. We move the
decimal point two places to the
right: 0.25

and then write a % symbol: 25%   Multiplying by 100 and writing %

c) We first move the decimal point
two places to the right (recall that
$0.3 = 0.30$):   0.30

and then write a % symbol: 30%   Multiplying by 100 and writing %

Solving Percent Problems

In solving percent problems, we first *translate* the problem to an equation. Then we *solve* the equation using the techniques discussed in Sections 2.1–2.3. The key words in the translation are as follows.

Key Words in Percent Translations

"**Of**" translates to " $\cdot$ " or " $\times$ ".   "**Is**" or "**Was**" translates to " $=$ ".

"**What**" translates to a variable.   **%** translates to "$\times \frac{1}{100}$" or "$\times 0.01$".

EXAMPLE 4   What is 11% of 49?

Solution

Translate: What   is   11%   of   49?
                 a $=$ 0.11 $\cdot$ 49   "of" means multiply; 11% = 0.11

$$a = 5.39$$

Thus, 5.39 is 11% of 49. The answer is 5.39.

Student Notes

A way of checking answers is by estimating as follows:

$$11\% \times 49 \approx 10\% \times 50$$
$$= 0.10 \times 50 = 5.$$

Since 5 is close to 5.39, our answer is reasonable.

EXAMPLE 5   3 is 16 percent of what?

Solution

Translate:

3	is	16 percent	of	what?
↓	↓	↓	↓	↓
3	=	0.16	·	y

$$\frac{3}{0.16} = y \qquad \text{Dividing both sides by 0.16}$$

$$18.75 = y$$

Thus, 3 is 16 percent of 18.75. The answer is 18.75.

EXAMPLE 6   What percent of $50 is $34?

Solution

Translate:

What percent	of	$50	is	$34?
↓	↓	↓	↓	↓
n	·	50	=	34

$$n = \frac{34}{50} \qquad \text{Dividing both sides by 50}$$

$$n = 0.68 = 68\% \qquad \text{Converting to per-cent notation}$$

Thus, 34 is 68% of 50. The answer is 68%.

Examples 4–6 represent the three basic types of percent problems.

EXAMPLE 7   *Coronary heart disease.* In 2003, there were 285 million people in the United States. About 2.5% of them had heart disease (*Source*: American Heart Association). How many had heart disease?

Solution   To solve the problem, we first reword and then translate. We let $a = $ the number of people in the United States with heart disease.

Rewording:	What	is	2.5%	of	285?
	↓	↓	↓	↓	↓
Translating:	a	=	0.025	×	285

The letter is by itself. To solve the equation, we need only multiply:

$$a = 0.025 \times 285 = 7.125.$$

Thus, 7.125 million is 2.5% of 285 million, so in 2003, about 7.125 million people in the United States had heart disease.

Student Notes _____

Always look for connections between examples. Here you should look for similarities between Examples 4 and 7 as well as between Examples 5 and 8 and between Examples 6 and 9.

EXAMPLE 8

Two-year college enrollments. In 2003, about 5.6 million students were enrolled in two-year institutions. This number was 38% of all enrollments in higher education. How many students in all were enrolled in higher education?

Solution Before translating the problem to mathematics, we reword and let S represent the total number of students, in millions, enrolled in higher education.

$$
\begin{array}{cccccc}
\textit{Rewording:} & 5.6 & \text{is} & 38\% & \text{of} & S. \\
& \downarrow & \downarrow & \downarrow & \downarrow & \downarrow \\
\textit{Translate:} & 5.6 & = & 0.38 & \cdot & S
\end{array}
$$

$$\frac{5.6}{0.38} = S$$

$$14.7 \approx S \qquad \text{The symbol } \approx \text{ means } \textit{is approximately equal to.}$$

There were about 14.7 million students enrolled in higher education in 2003.

EXAMPLE 9

Automobile prices. Recently, Harken Motors reduced the price of a 2003 Ford Focus from the manufacturer's suggested retail price (MSRP) of $17,500 to $15,925.

a) What percent of the MSRP does the sale price represent?

b) What is the percent of discount?

Solution

a) We reword and translate, using n for the unknown percent.

$$
\begin{array}{ccccc}
\textit{Rewording:} & \underline{\text{What percent}} & \text{of} & 17,500 & \text{is} & 15,925? \\
& \downarrow & \downarrow & \downarrow & \downarrow \\
\textit{Translating:} & n & \cdot & 17,500 & = & 15,925
\end{array}
$$

$$n = \frac{15,925}{17,500} \qquad \text{Dividing both sides by 17,500}$$

$$n = 0.91 = 91\% \qquad \text{Converting to percent notation}$$

The sale price is 91% of the MSRP.

b) Since the original price of $17,500 represents 100% of the MSRP, the sale price represents a discount of $(100 - 91)\%$, or 9%.

Exercise Set

2.4

Concept Reinforcement *In each of Exercises 1–10, match the question with the most appropriate translation from the column on the right. Some choices are used more than once.*

1. ____ What percent of 57 is 23?
2. ____ What percent of 23 is 57?
3. ____ 23 is 57% of what number?
4. ____ 57 is 23% of what number?
5. ____ 57 is what percent of 23?
6. ____ 23 is what percent of 57?
7. ____ What is 23% of 57?
8. ____ What is 57% of 23?
9. ____ 23% of what number is 57?
10. ____ 57% of what number is 23?

a) $a = (0.57)23$
b) $57 = 0.23y$
c) $n \cdot 23 = 57$
d) $n \cdot 57 = 23$
e) $23 = 0.57y$
f) $a = (0.23)57$

Convert the percent notation in each sentence to decimal notation.

11. *Pluto.* The planet Pluto is 30% ice.
    Source: www.childrensmuseum.org/cosmicquest/fieldguide/pluto.html

12. *Pluto.* The planet Pluto is 70% rock.
    Source: www.childrensmuseum.org/cosmicquest/fieldguide/pluto.html

13. *Dehydration.* A 2% drop in water content of the body can affect one's ability to study mathematics.
    Source: High Performance Nutrition

14. *Left-handed golfers.* Of those who golf, 7% are left-handed.
    Source: National Association of Left-Handed Golfers

15. *Wealth in the aged.* Those 60 and older own 77% of the nation's financial assets.

16. *Wealth in the aged.* Those 60 and older make up 66% of the nation's stockholders.

17. *Women in the workforce.* Women comprise 9% of all engineers.

18. *Women in the workforce.* Women comprise 30% of all chemists.

Convert to decimal notation.

19. 62.58%

20. 39.81%

21. 0.7%

22. 0.3%

23. 125%

24. 150%

Convert the decimal notation in each sentence to percent notation.

25. *NASCAR fans.* Of those who are fans of NASCAR racing, 0.64 of them have attended college or beyond.
    Source: NASCAR, Goodyear

26. *NASCAR fans.* Of those who are fans of NASCAR racing, 0.41 of them earn more than $50,000 per year.
    Source: NASCAR, Goodyear

27. *Foreign student enrollment.* Of all the foreign students studying in the United States, 0.106 are from China.
Source: Institute of International Education

28. *Heart attacks.* Of those suffering heart attacks, 0.67 will survive.
Source: American Heart Association

29. *Women in the workforce.* Women comprise 0.42 of all college faculty.

30. *Women in the workforce.* Women comprise 0.19 of all architects.

31. *Water in watermelon.* Watermelon is 0.9 water.

32. *Tax savings.* The tax cut enacted in 2003 will save 0.88 of all U.S. taxpayers less than $100.
Source: Citizens for Tax Justice

Convert to percent notation.

33. 0.0049 **34.** 0.0008

35. 1.08 **36.** 1.05

37. 2.3 **38.** 2.9

39. $\frac{4}{5}$ **40.** $\frac{3}{4}$ **41.** $\frac{8}{25}$ **42.** $\frac{3}{8}$

Solve.

43. What percent of 68 is 17?

44. What percent of 150 is 39?

45. What percent of 125 is 30?

46. What percent of 300 is 57?

47. 14 is 30% of what number?

48. 54 is 24% of what number?

49. 0.3 is 12% of what number?

50. 7 is 175% of what number?

51. What number is 35% of 240?

52. What number is 1% of one million?

53. What percent of 60 is 75?

Aha! **54.** What percent of 70 is 70?

55. What is 2% of 40?

56. What is 40% of 2?

Aha! **57.** 25 is what percent of 50?

58. 8 is 2% of what number?

Costs of owning a dog. The American Pet Products Manufacturers Association estimates that the total cost of owning a dog for its lifetime is $6600. The following circle graph shows the relative costs of raising a dog from birth to death.

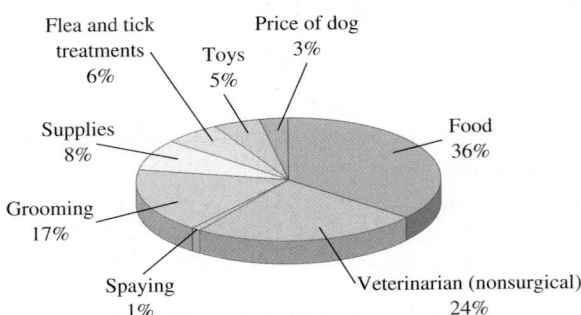

Costs of Owning a Dog

Source: The American Pet Products Manufacturers Association

In each of Exercises 59–64, determine the cost of owning a dog for its lifetime.

59. Price of dog

60. Food

61. Veterinarian

62. Grooming

63. Supplies

64. Flea and tick treatments

65. *College graduation.* To obtain his bachelor's degree in nursing, Frank must complete 125 credit hours of instruction. If he has completed 60% of his requirement, how many credits did Frank complete?

66. *College graduation.* To obtain her bachelor's degree in journalism, Beth must complete 125 credit hours of instruction. If 20% of Beth's credit hours remain to be completed, how many credits does she still need to take?

67. *Batting average.* At one point in a recent season, Ichiro Suzuki of the Seattle Mariners had 194 hits. His batting average was 0.310, or 31%. That is, of the total number of at-bats, 31% were hits. How many at-bats did he have?
Source: Major League Baseball

68. *Pass completions.* At one point in a recent season, Peyton Manning of the Indianapolis Colts had completed 357 passes. This was 62.5% of his attempts. How many attempts did he make?
Source: National Football League

69. *Tipping.* Leon left a $4 tip for a meal that cost $25.
a) What percent of the cost of the meal was the tip?
b) What was the total cost of the meal including the tip?

70. *Tipping.* Selena left a $12.76 tip for a meal that cost $58.
a) What percent of the cost of the meal was the tip?
b) What was the total cost of the meal including the tip?

71. *Auto sales.* In 2002, 8.3 million cars were sold in the United States. Of these, 6.0 million were manufactured in the United States. What percent were manufactured within the United States? outside the United States?
Source: The World Almanac and Book of Facts, 2004

72. *Auto sales.* Of the 2.3 million foreign cars sold in the United States in 2002, about 1.0 million were manufactured in Japan. What percent of foreign cars were manufactured in Japan? outside Japan?
Source: Autodata

73. *Student loans.* To finance her community college education, Sarah takes out a Stafford loan for $3500. After a year, Sarah decides to pay off the interest, which is 8% of $3500. How much will she pay?

74. *Student loans.* Paul takes out a subsidized federal Stafford loan for $2400. After a year, Paul decides to pay off the interest, which is 7% of $2400. How much will he pay?

75. *Infant health.* In a study of 300 pregnant women with "good-to-excellent" diets, 95% had babies in good or excellent health. How many women in this group had babies in good or excellent health?

76. *Infant health.* In a study of 300 pregnant women with "poor" diets, 8% had babies in good or excellent health. How many women in this group had babies in good or excellent health?

77. *Cost of self-employment.* Because of additional taxes and fewer benefits, it has been estimated that a self-employed person must earn 20% more than a non–self-employed person performing the same task(s). If Joy earns $15 an hour working for Village Copy, how much would she need to earn on her own for a comparable income?

78. Refer to Exercise 77. Ray earns $12 an hour working for Round Edge stairbuilders. How much would Ray need to earn on his own for a comparable income?

79. *Triathletes.* The number of USA Triathlon Members grew from 16,000 in 1993 to 40,000 in 2002. Calculate the percentage by which the number increased.
Source: USA Triathlon

80. *Fastest human.* In 2002, Tim Montgomery of the United States set a world record by sprinting 100 m in 9.78 sec. This broke the record of 9.79 sec established in 1999 by Maurice Green, also of the United States. Calculate the percentage by which the record decreased.
Source: www.runnersworld.com

81. A bill at Officeland totaled $37.80. How much did the merchandise cost if the sales tax is 5%?

82. Doreen's checkbook shows that she wrote a check for $987 for building materials. What was the price of the materials if the sales tax is 5%?

83. *Deducting sales tax.* A tax-exempt school group received a bill of $157.41 for educational software. The bill incorrectly included sales tax of 6%. How much should the school group pay?

84. *Deducting sales tax.* A tax-exempt charity received a bill of $145.90 for a sump pump. The bill incorrectly included sales tax of 5%. How much does the charity owe?

85. *Body fat.* One author of this text exercises regularly at a local YMCA that recently offered a body-fat percentage test to its members. The device used measures the passage of a very low voltage of electricity through the body. The author's body-fat percentage was found to be 16.5% and he weighs 191 lb. What part, in pounds, of his body weight is fat?

86. *Areas of Alaska and Arizona.* The area of Arizona is 19% of the area of Alaska. The area of Alaska is 586,400 mi^2. What is the area of Arizona?

87. *Junk mail.* The U.S. Postal Service reports that of the junk mail that is sent out, 78% is actually opened and read. A business sends out 9500 advertising brochures. How many of them can the business expect to be opened and read?

88. *Kissing and colds.* In a medical study, it was determined that if 800 people kiss someone else who has a cold, only 56 will actually catch the cold. What percent is this?

89. *Calorie content.* Pepperidge Farm Light Style 7 Grain Bread® has 140 calories in a 3-slice serving.

This is 15% less than the number of calories in a serving of regular bread. How many calories are in a serving of regular bread?

90. *Fat content.* Peek Freans Shortbread Reduced Fat Cookies® contain 35 calories of fat in each serving. This is 40% less than the fat content in the leading imported shortbread cookie. How many calories of fat are in a serving of the leading shortbread cookie?

91. Campus Bookbuyers pays $30 for a book and sells it for $60. Is this a 100% markup or a 50% markup? Explain.

92. If Julian leaves a $12 tip for a $90 dinner, is he being generous, stingy, or neither? Explain.

SKILL MAINTENANCE

Translate to an algebraic expression. [1.1]

93. 5 more than some number

94. 4 less than Tino's weight

95. The product of 8 and twice *a*

96. 1 more than the product of two numbers

SYNTHESIS

97. Does the following advertisement provide a convincing argument that summertime is when most burglaries occur? Why or why not?

98. Erin is returning a tent that she bought during a 25%-off storewide sale that has ended. She is offered store credit for 125% of what she paid (not to be used on sale items). Is this fair to Erin? Why or why not?

99. The community of Bardville has 1332 left-handed females. If 48% of the community is female and 15% of all females are left-handed, how many people are in the community?

100. It has been determined that at the age of 15, a boy has reached 96.1% of his final adult height. Jaraan is 6 ft 4 in. at the age of 15. What will his final adult height be?

101. It has been determined that at the age of 10, a girl has reached 84.4% of her final adult height. Dana is 4 ft 8 in. at the age of 10. What will her final adult height be?

102. *U.S. birth rate.*   Between 2001 and 2002, the U.S. birth rate dropped from 14.1 to 13.9 babies per 1000 women. Calculate the percentage by which the birth rate dropped and use that percentage to estimate birth rates for the United States in 2003 and 2004.
Source: www.socialsecurity.org

103. *Photography.*   A 6-in. by 8-in. photo is framed using a mat meant for a 5 in. by 7 in. photo. What percentage of the photo will be hidden by the mat?

104. Would it be better to receive a 5% raise and then, a year later, an 8% raise or the other way around? Why?

105. Herb is in the 30% tax bracket. This means that 30¢ of each dollar earned goes to taxes. Which would cost him the least: contributing $50 that is tax-deductible or contributing $40 that is not tax-deductible? Explain.

CORNER

Sales and Discounts

Focus: Applications and models using percent

Time: 15 minutes

Group size: 3

Materials: Calculators are optional.

Often a store will reduce the price of an item by a fixed percentage. When the sale ends, the items are returned to their original prices. Suppose a department store reduces all sporting goods 20%, all clothing 25%, and all electronics 10%.

ACTIVITY

1. Each group member should select one of the following items: a $50 basketball, an $80 jacket, or a $200 portable sound system. Fill in the first three columns of the first three rows of the chart below.
2. Apply the appropriate discount and determine the sale price of your item. Fill in the fourth column of the chart.

3. Next, find a multiplier that can be used to convert the sale price back to the original price and fill in the remaining column of the chart. Does this multiplier depend on the price of the item?
4. Working as a group, compare the results of part (3) for all three items. Then develop a formula for a multiplier that will restore a sale price to its original price, p, after a discount r has been applied. Complete the fourth row of the table and check that your formula will duplicate the results of part (3).
5. Use the formula from part (4) to find the multiplier that a store would use to return an item to its original price after a "30% off" sale expires. Fill in the last line on the chart.
6. Inspect the last column of your chart. How can these multipliers be used to determine the percentage by which a sale price is increased when a sale ends?

Original Price, p	Discount, r	$1 - r$	Sale Price	Multiplier to convert back to p
p	r	$1 - r$		
	0.30			

2.5 Problem Solving

Five Steps for Problem Solving • Applying the Five Steps

Probably the most important use of algebra is as a tool for problem solving. In this section, we develop a problem-solving approach that is used throughout the remainder of the text.

Five Steps for Problem Solving

In Section 2.4, we solved several real-world problems. To solve them, we first *familiarized* ourselves with percent notation. We then *translated* each problem into an equation, *solved* the equation, *checked* the solution, and *stated* the answer.

Five Steps for Problem Solving in Algebra
1. *Familiarize* yourself with the problem.
2. *Translate* to mathematical language. (This often means writing an equation.)
3. *Carry out* some mathematical manipulation. (This often means *solving* an equation.)
4. *Check* your possible answer in the original problem.
5. *State* the answer clearly, using a complete English sentence.

Of the five steps, the most important is probably the first one: becoming familiar with the problem. Here are some hints for familiarization.

To Become Familiar with a Problem
1. Read the problem carefully. Try to visualize the problem.
2. Reread the problem, perhaps aloud. Make sure you understand all important words.
3. List the information given and the question(s) to be answered. Choose a variable (or variables) to represent the unknown and specify what the variable represents. For example, let L = length in centimeters, d = distance in miles, and so on.
4. Look for similarities between the problem and other problems you have already solved.
5. Find more information. Look up a formula in a book, at a library, or online. Consult a reference librarian or an expert in the field.
6. Make a table that uses all the information you have available. Look for patterns that may help in the translation.
7. Make a drawing and label it with known and unknown information, using specific units if given.
8. Think of a possible answer and check the guess. Note the manner in which the guess is checked.

Applying the Five Steps

EXAMPLE 1   Hiking.   In 1948, Earl Shaffer became the first person to hike all 2100 miles of the Appalachian trail—from Springer Mountain, Georgia, to Mt. Katahdin, Maine. Shaffer repeated the feat 50 years later, and at age 79 became the oldest person to hike the entire trail. When Shaffer stood atop Big Walker Mountain, Virginia, he was three times as far from the northern end of the trail as from the southern end. At that point, how far was he from each end of the trail?

Earl Shaffer 1918–2002

Solution

1. **Familiarize.**   It may be helpful to make a drawing.

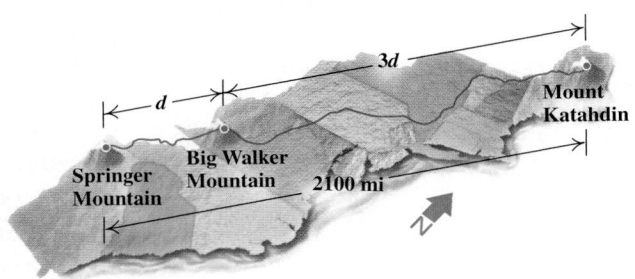

To gain some familiarity, let's suppose that Shaffer stood 600 mi from Springer Mountain. Three times 600 mi is 1800 mi. Since 600 mi + 1800 mi = 2400 mi and 2400 mi > 2100 mi, we see that our guess is too large. Rather than guess again, we let

   d = the distance, in miles, to the southern end

and

   $3d$ = the distance, in miles, to the northern end.

(We could also let x = the distance to the northern end and $\frac{1}{3}x$ = the distance to the southern end.)

2. **Translate.**   From the drawing, we see that the lengths of the two parts of the trail must add up to 2100 mi. This leads to our translation.

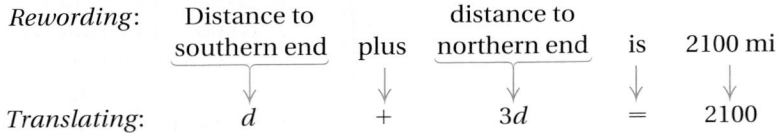

3. **Carry out.**   We solve the equation:

$$d + 3d = 2100$$
$$4d = 2100 \qquad \text{Combining like terms}$$
$$d = 525. \qquad \text{Dividing both sides by 4}$$

4. **Check.**   As predicted in the *Familiarize* step, d is less than 600 mi. If $d = 525$ mi, then $3d = 1575$ mi. Since 525 mi + 1575 mi = 2100 mi, we have a check.

5. **State.**   Atop Big Walker Mountain, Shaffer stood 525 mi from Springer Mountain and 1575 mi from Mount Katahdin.

   Before we solve the next problem, we need to learn some additional terminology regarding integers.
   The following are examples of **consecutive integers:** 16, 17, 18, 19, 20; and $-31, -30, -29, -28$. Note that consecutive integers can be represented in the form $x, x + 1, x + 2$, and so on.

The following are examples of **consecutive even integers:** 16, 18, 20, 22, 24; and $-52, -50, -48, -46$. Note that consecutive even integers can be represented in the form $x, x + 2, x + 4$, and so on.

The following are examples of **consecutive odd integers:** 21, 23, 25, 27, 29; and $-71, -69, -67, -65$. Note that consecutive odd integers can be represented in the form $x, x + 2, x + 4$, and so on.

EXAMPLE 2

Interstate mile markers. U.S. interstate highways post numbered markers at every mile to indicate location in case of an emergency (*Source*: Federal Highway Administration, Ed Rotalewski). The sum of two consecutive mile markers on I-70 in Kansas is 559. Find the numbers on the markers.

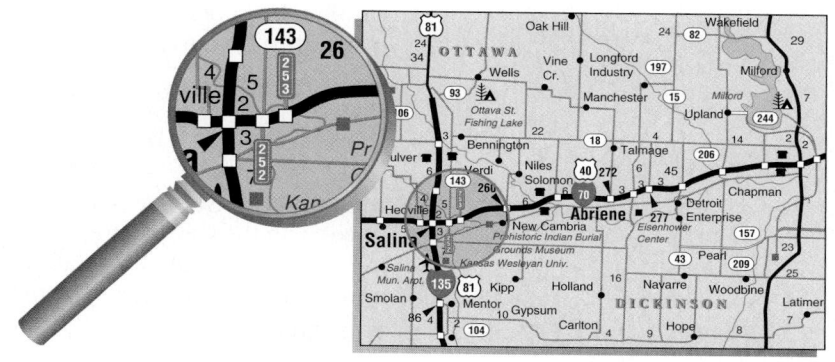

Study Skills _____

Set Reasonable Expectations

Do not be surprised if your success rate drops some as you work through the exercises in this section. *This is normal.* Your success rate will increase as you gain experience with these types of problems and use some of the study skills already listed.

x	$x + 1$	**Sum of x and $x + 1$**
114	115	229
252	253	505
302	303	605

Solution

1. **Familiarize.** The numbers on the mile markers are consecutive positive integers. Thus if we let $x =$ the smaller number, then $x + 1 =$ the larger number.

   To become familiar with the problem, we can make a table, as shown at left. First, we guess a value for x; then we find $x + 1$. Finally, we add the two numbers and check the sum.

   From the table, we see that the first marker will be between 252 and 302. We could continue guessing and solve the problem this way, but let's work on developing our algebra skills.

2. **Translate.** We reword the problem and translate as follows.

   Rewording: First integer   plus   second integer   is   559.

   Translating: x $+$ $(x + 1)$ $=$ 559

3. **Carry out.** We solve the equation:

$$x + (x + 1) = 559$$
$$2x + 1 = 559 \qquad \text{Using an associative law and combining like terms}$$
$$2x = 558 \qquad \text{Subtracting 1 from both sides}$$
$$x = 279. \qquad \text{Dividing both sides by 2}$$

If x is 279, then $x + 1$ is 280.

4. **Check.**   Our possible answers are 279 and 280. These are consecutive positive integers and $279 + 280 = 559$, so the answers check.

5. **State.**   The mile markers are 279 and 280.

EXAMPLE 3

IKON copiers.   IKON Office Solutions rents a Canon IR330 copier for $225 per month plus 1.2¢ per copy. A law firm needs to lease a copy machine for use during a special case that they anticipate will take 3 months. If they allot a budget of $1100, how many copies can they make?

Source: IKON Office Solutions, Nathan DuMond, Sales Manager

Solution

1. **Familiarize.**   Suppose that the law firm makes 20,000 copies. Then the cost is monthly charges plus copy charges, or

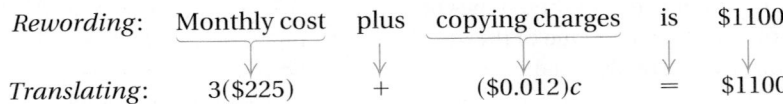

3($225)   plus   cost per copy   times   number of copies

$675   + $0.012   ·   20,000,

which is $915. Our guess of 20,000 is too small, but we have familiarized ourselves with the way in which a calculation is made. Note that we convert 1.2¢ to $0.012 so that all information is in the same unit, dollars. We let $c =$ the number of copies that can be made for $1100.

2. **Translate.**   We reword the problem and translate as follows.

Rewording: Monthly cost   plus   copying charges   is $1100.

Translating: 3($225)   +   ($0.012)c = $1100

3. **Carry out.**   We solve the equation:

$$3(225) + 0.012c = 1100$$
$$675 + 0.012c = 1100$$
$$0.012c = 425 \qquad \text{Subtracting 675 from both sides}$$
$$c = \frac{425}{0.012} \qquad \text{Dividing both sides by 0.012}$$
$$c \approx 35{,}417. \qquad \text{Rounding to the nearest one}$$

4. **Check.**   We check in the original problem. The cost for 35,417 pages is $35{,}417(\$0.012) = \$425.004$. The rental for 3 months is $3(\$225) = \$675$. The total cost is then $\$425.004 + \$675 \approx \$1100$, which is the $1100 that was allotted. Our answer exceeds 20,000, as we expected from the *Familiarize* step.

5. **State.**   The law firm can make 35,417 copies on the copy rental allotment of $1100.

EXAMPLE 4

Perimeter of NBA court.   The perimeter of an NBA basketball court is 288 ft. The length is 44 ft longer than the width. (*Source*: National Basketball Association) Find the dimensions of the court.

Student Notes

For most students, the most challenging step is step 2, "Translate." The table on p. 4 (Section 1.1) can be helpful in this regard.

Solution

1. **Familiarize.** Recall that the perimeter of a rectangle is twice the length plus twice the width. Suppose the court were 30 ft wide. The length would then be 30 + 44, or 74 ft, and the perimeter would be 2 · 30 ft + 2 · 74 ft, or 208 ft. This shows that in order for the perimeter to be 288 ft, the width must exceed 30 ft. Instead of guessing again, we let w = the width of the court, in feet. Since the court is "44 ft longer than it is wide," we let $w + 44$ = the length of the court, in feet.

2. **Translate.** To translate, we use $w + 44$ as the length and 288 as the perimeter. To double the length, $w + 44$, parentheses are essential.

   Rewording: Twice the length plus twice the width is 288 ft.

   Translating: $2(w + 44)$ + $2w$ = 288

3. **Carry out.** We solve the equation:

$$2(w + 44) + 2w = 288$$
$$2w + 88 + 2w = 288 \quad \text{Using the distributive law}$$
$$4w + 88 = 288 \quad \text{Combining like terms}$$
$$4w = 200$$
$$w = 50.$$

   The dimensions appear to be $w = 50$ ft, and $l = w + 44 = 94$ ft.

4. **Check.** If the width is 50 ft and the length is 94 ft, then the court is 44 ft longer than it is wide. The perimeter is $2(50 \text{ ft}) + 2(94 \text{ ft}) = 100 \text{ ft} + 188 \text{ ft}$, or 288 ft, as specified. We have a check.

5. **State.** An NBA court is 50 ft wide and 94 ft long.

Caution! Always be sure to answer the original problem completely. For instance, in Example 1 we needed to find *two* numbers: the distances from *each* end of the trail to the hiker. Similarly, in Example 4 we needed to find two dimensions, not just the width. Be sure to label each answer with the proper unit.

EXAMPLE 5

Selling a home.   The McCanns are planning to sell their home. If they want to be left with $117,500 after paying 6% of the selling price to a realtor as a commission, for how much must they sell the house?

Solution

1. **Familiarize.**   Suppose the McCanns sell the house for $120,000. A 6% commission can be determined by finding 6% of $120,000:

$$6\% \text{ of } \$120,000 = 0.06(\$120,000) = \$7200.$$

Subtracting this commission from $120,000 would leave the McCanns with

$$\$120,000 - \$7200 = \$112,800.$$

This shows that in order for the McCanns to clear $117,500, the house must sell for more than $120,000. To determine what the sale price must be, we could check more guesses. Instead, we let $x =$ the selling price, in dollars. With a 6% commission, the realtor would receive $0.06x$.

2. **Translate.**   We reword the problem and translate as follows.

Rewording:	Selling price	less	commission	is	amount remaining.
Translating:	x	$-$	$0.06x$	$=$	$117{,}500$

3. **Carry out.**   We solve the equation:

$$x - 0.06x = 117{,}500$$
$$1x - 0.06x = 117{,}500$$
$$0.94x = 117{,}500 \qquad \text{Combining like terms. Had we noted that after the commission has been paid, 94\% remains, we could have begun with this equation.}$$

$$x = \frac{117{,}500}{0.94} \qquad \text{Dividing both sides by 0.94}$$

$$x = 125{,}000.$$

4. **Check.**   To check, we first find 6% of $125,000:

$$6\% \text{ of } \$125,000 = 0.06(\$125,000) = \$7500. \qquad \text{This is the commission.}$$

Next, we subtract the commission to find the remaining amount:

$$\$125,000 - \$7500 = \$117,500.$$

Since, after the commission, the McCanns are left with $117,500, our answer checks. Note that the $125,000 sale price is greater than $120,000, as predicted in the *Familiarize* step.

5. **State.**   To be left with $117,500, the McCanns must sell the house for $125,000.

EXAMPLE 6   Cross section of a roof.   In a triangular gable end of a roof, the angle of the peak is twice as large as the angle on the back side of the house. The measure of the angle on the front side is 20° greater than the angle on the back side. How large are the angles?

Solution

1. **Familiarize.** We make a drawing. In this case, the measure of the back angle is x, the measure of the front angle is $x + 20$, and the measure of the peak angle is $2x$.

2. **Translate.** To translate, we need to recall that the sum of the measures of the angles in a triangle is 180°.

Rewording: Measure of   measure of   measure of
                back angle   +   front angle   +   peak angle   is   180°

Translating: x + $(x + 20)$ + $2x$ =   180

3. **Carry out.** We solve:

$$x + (x + 20) + 2x = 180$$
$$4x + 20 = 180$$
$$4x = 160$$
$$x = 40.$$

The measures for the angles appear to be:

Back angle: $x = 40°$,
Front angle: $x + 20 = 40 + 20 = 60°$,
Peak angle: $2x = 2(40) = 80°$.

4. **Check.** Consider 40°, 60°, and 80°. The measure of the front angle is 20° greater than the measure of the back angle, the measure of the peak angle is twice the measure of the back angle, and the sum is 180°. These numbers check.

5. **State.** The measures of the angles are 40°, 60°, and 80°.

We close this section with some tips to aid you in problem solving.

Problem-Solving Tips

1. The more problems you solve, the more your skills will improve.
2. Look for patterns when solving problems. Each time you study an example in a text, you may observe a pattern for problems found later in the exercise sets or in other practical situations.
3. When translating in mathematics, consider the dimensions of the variables and constants in the equation. The variables that represent length should all be in the same unit, those that represent money should all be in dollars or all in cents, and so on.
4. Make sure that units appear in the answer whenever appropriate and that you completely answer the original problem.

Exercise Set

2.5

FOR EXTRA HELP

Student's Solutions Manual | Digital Video Tutor CD 1 Videotape 2 | Tutor Center AW Math Tutor Center | MathXL Tutorials on CD | Math XL MathXL | MyMathLab MyMathLab

Solve. Even though you might find the answer quickly in some other way, practice using the five-step problem-solving process.

1. Two fewer than ten times a number is 78. What is the number?

2. Three less than twice a number is 19. What is the number?

3. Five times the sum of 3 and some number is 70. What is the number?

4. Twice the sum of 4 and some number is 34. What is the number?

5. *Price of sneakers.* Amy paid $72.25 for a pair of New Balance 765 running shoes during a 15%-off sale. What was the regular price?

6. *Price of a CD player.* Doug paid $72 for a shockproof portable CD player during a 20%-off sale. What was the regular price?

7. *Price of a calculator.* Evelyn paid $89.25, including 5% tax, for her graphing calculator. How much did the calculator itself cost?

8. *Price of a printer.* Jake paid $100.70, including 6% tax, for a color printer. How much did the printer itself cost?

9. *Running.* In 1997, Yiannis Kouros of Australia set the record for the greatest distance run in 24 hr by running 188 mi. After 8 hr, he was approximately twice as far from the finish line as he was from the start. How far had he run?
   Source: *Guinness World Records 2004 Edition*

10. *Sled-dog racing.*   The Iditarod sled-dog race extends for 1049 mi from Anchorage to Nome. If a musher is twice as far from Anchorage as from Nome, how many miles has the musher traveled?

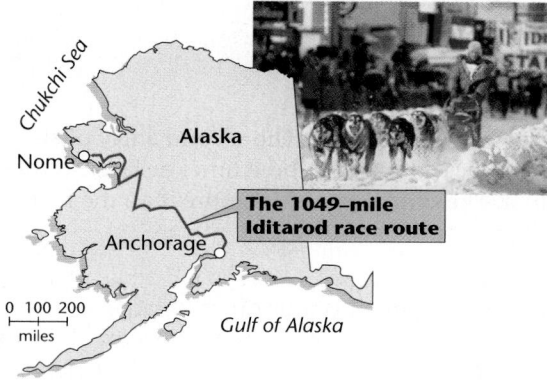

The 1049–mile Iditarod race route

11. *Car racing.*   In April 2004, Dan Wheldon won the Indy Japan 300 with a time of 1:49:48.2611 for the 300-mi race. At one point, Wheldon was 80 mi closer to the finish than the start. How far had Wheldon traveled at that point?

12. *Car racing.*   In August 2004, Jeff Gordon won his fourth Brickyard 400 with a time of 3:29:56 for the 400-mi race. At one point, Gordon was 70 mi closer to the finish than the start. How far had Gordon traveled at that point?

13. *Apartment numbers.*   The apartments in Joan's apartment house are consecutively numbered on each floor. The sum of her number and her next-door neighbor's number is 2409. What are the two numbers?

14. *Apartment numbers.*   The apartments in Vincent's apartment house are numbered consecutively on each floor. The sum of his number and his next-door neighbor's number is 1419. What are the two numbers?

15. *Street addresses.*   The houses on the south side of Elm Street are consecutive even numbers. Wanda and Larry are next-door neighbors and the sum of their house numbers is 794. Find their house numbers.

16. *Street addresses.*   The houses on the west side of Lincoln Avenue are consecutive odd numbers. Sam and Colleen are next-door neighbors and the sum of their house numbers is 572. Find their house numbers.

17. The sum of three consecutive page numbers is 60. Find the numbers.

18. The sum of three consecutive page numbers is 99. Find the numbers.

19. *Oldest bride.*   The world's oldest bride was 19 yr older than her groom. Together, their ages totaled 185 yr. How old were the bride and the groom?
Source: *Guinness World Records 2004 Edition*

20. *Oldest marrying couple.*   As half of the world's oldest marrying couple, the woman was 2 yr younger than the man. Together, their ages totaled 190 yr. How old were the man and the woman?
Source: *Guinness World Records 2004 Edition*

21. *Home remodeling.*   In a recent year, Americans spent a total of $35 billion to remodel bathrooms and kitchens. Twice as much was spent on kitchens as bathrooms. How much was spent on each?

22. *Women's clothing.*   Recently a total of $12.8 billion was spent on women's blouses and dresses. Of that total, $0.2 billion more was spent on blouses than on dresses. How much was spent on each type of clothing?

23. *Page numbers.*   The sum of the page numbers on the facing pages of a book is 281. What are the page numbers?

24. *Perimeter of a triangle.*   The perimeter of a triangle is 195 mm. If the lengths of the sides are consecutive odd integers, find the length of each side.

25. *Hancock Building dimensions.*   The top of the John Hancock Building in Chicago is a rectangle whose length is 60 ft more than the width. The perimeter is 520 ft. Find the width and the length of the rectangle. Find the area of the rectangle.

26. *Dimensions of a state.* The perimeter of the state of Wyoming is 1280 mi. The width is 90 mi less than the length. Find the width and the length.

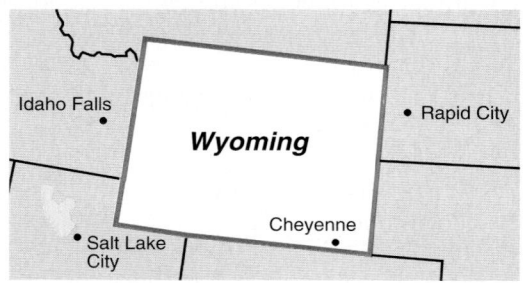

27. *Perimeter of a high school basketball court.* The perimeter of a standard high school basketball court is 268 ft. The length is 34 ft longer than the width. Find the dimensions of the court.
Source: Indiana High School Athletic Association

28. A rectangular community garden is to be enclosed with 92 m of fencing. In order to allow for compost storage, the garden must be 4 m longer than it is wide. Determine the dimensions of the garden.

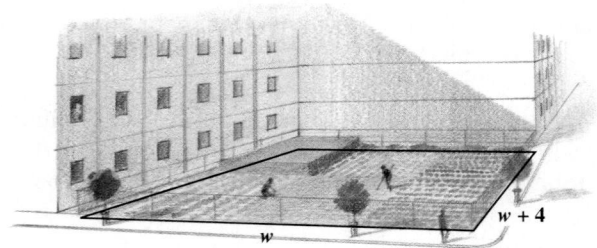

29. *Two-by-four.* The perimeter of a cross section of a "two-by-four" piece of lumber is $10\frac{1}{2}$ in. The length is twice the width. Find the actual dimensions of the cross section of a two-by-four.

Two-by-four $\quad P = 10\frac{1}{2}$ in.

30. *Standard billboard sign.* A standard rectangular highway billboard sign has a perimeter of 124 ft.

The length is 6 ft more than three times the width. Find the dimensions of the sign.

31. *Angles of a triangle.* The second angle of an architect's triangle is three times as large as the first. The third angle is 30° more than the first. Find the measure of each angle.

32. *Angles of a triangle.* The second angle of a triangular garden is four times as large as the first. The third angle is 45° less than the sum of the other two angles. Find the measure of each angle.

33. *Angles of a triangle.* The second angle of a triangular building lot is three times as large as the first. The third angle is 10° more than the sum of the other two angles. Find the measure of the third angle.

34. *Angles of a triangle.* The second angle of a triangular kite is four times as large as the first. The third angle is 5° more than the sum of the other two angles. Find the measure of the second angle.

35. *Rocket sections.* A rocket is divided into three sections: the payload and navigation section in the top, the fuel section in the middle, and the rocket engine section in the bottom. The top section is one-sixth the length of the bottom section. The middle section is one-half the length of the bottom section. The total length is 240 ft. Find the length of each section.

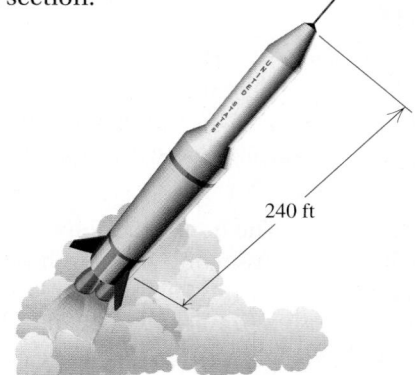

36. *Gourmet sandwiches.*   Jenny, Demi, and Sarah buy an 18-in. long gourmet sandwich and take it back to their apartment. Since they have different appetites, Jenny cuts the sandwich so that Demi gets half of what Jenny gets and Sarah gets three-fourths of what Jenny gets. Find the length of each person's sandwich.

37. *Taxi rates.*   In Chicago, a taxi ride costs $1.90 plus $1.60 for each mile traveled. Debbie and Alex have budgeted $18 for a taxi ride (excluding tip). How far can they travel on their $18 budget?

38. *Taxi fares.*   In New York City, taxis charge $2.50 plus $2.00 per mile for off-peak fares. How far can Ralph travel for $17.50 (assuming an off-peak fare)?
Source: *Burlington Free Press*, 3/31/04

39. *Truck rentals.*   Truck-Rite Rentals rents trucks at a daily rate of $49.95 plus 39¢ per mile. Concert Productions has budgeted $100 for renting a truck to haul equipment to an upcoming concert. How far can they travel in one day and stay within their budget?

40. *Truck rentals.*   Fine Line Trucks rents an 18-ft truck for $42 plus 35¢ per mile. Judy needs a truck for one day to deliver a shipment of plants. How far can she drive and stay within a budget of $70?

41. *Complementary angles.*   The sum of the measures of two *complementary* angles is 90°. If one angle measures 15° more than twice the measure of its complement, find the measure of each angle.

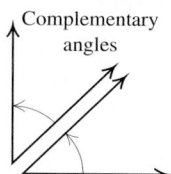

42. *Supplementary angles.*   The sum of the measures of two *supplementary* angles is 180°. If one angle measures 45° less than twice the measure of its supplement, find the measure of each angle.

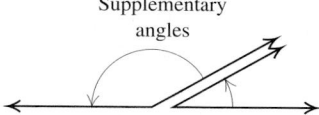

43. *Copier paper.*   The perimeter of standard-size copier paper is 99 cm. The width is 6.3 cm less than the length. Find the length and the width.

44. *Stock prices.*   Sarah's investment in Jet Blue stock grew 28% to $448. How much did she invest?

45. *Savings interest.*   Sharon invested money in a savings account at a rate of 6% simple interest. After 1 yr, she has $6996 in the account. How much did Sharon originally invest?

46. *Credit cards.*   The balance in Will's Mastercard® account grew 2%, to $870, in one month. What was his balance at the beginning of the month?

47. *Scrabble®.*   The highest one-game score for a Scrabble player occurred in the same game in which the record margin of victory, 796 points, was established. The winning and losing scores totaled 1302 points. What was the winning score?
Source: *Guinness World Records 2004 Edition*

48. *IKON copiers.*   The law firm in Example 3 decides to raise its budget to $1400 for the 3-month period. How many copies can they make for $1400?

49. *Cricket chirps and temperature.*   The equation $T = \frac{1}{4}N + 40$ can be used to determine the temperature T, in degrees Fahrenheit, given the number of times N a cricket chirps per minute. Determine the number of chirps per minute for a temperature of 80°F.

50. *Race time.* The equation $R = -0.028t + 20.8$ can be used to predict the world record in the 200-m dash, where R is the record in seconds and t is the number of years since 1920. In what year will the record be 18.0 sec?

51. Sean claims he can solve most of the problems in this section by guessing. Is there anything wrong with this approach? Why or why not?

52. When solving Exercise 19, Beth used a to represent the bride's age and Ben used a to represent the groom's age. Is one of these approaches preferable to the other? Why or why not?

SKILL MAINTENANCE

Write a true sentence using either $<$ *or* $>$. [1.4]

53. -9 ■ 5 **54.** 1 ■ 3

55. -4 ■ 7 **56.** -9 ■ -12

SYNTHESIS

57. Write a problem for a classmate to solve. Devise it so that the problem can be translated to the equation $x + (x + 2) + (x + 4) = 375$.

58. Write a problem for a classmate to solve. Devise it so that the solution is "Audrey can drive the rental truck for 50 mi without exceeding her budget."

59. *Discounted dinners.* Kate's "Dining Card" entitles her to $10 off the price of a meal after a 15% tip has been added to the cost of the meal. If, after the discount, the bill is $32.55, how much did the meal originally cost?

60. *Test scores.* Pam scored 78 on a test that had 4 fill-ins worth 7 points each and 24 multiple-choice questions worth 3 points each. She had one fill-in wrong. How many multiple-choice questions did Pam get right?

61. *Gettysburg Address.* Abraham Lincoln's 1863 Gettysburg Address refers to the year 1776 as "four *score* and seven years ago." Determine what a score is.

62. One number is 25% of another. The larger number is 12 more than the smaller. What are the numbers?

63. A storekeeper goes to the bank to get $10 worth of change. She requests twice as many quarters as half dollars, twice as many dimes as quarters, three times as many nickels as dimes, and no pennies or dollars. How many of each coin did the storekeeper get?

64. *Perimeter of a rectangle.* The width of a rectangle is three fourths of the length. The perimeter of the rectangle becomes 50 cm when the length and the width are each increased by 2 cm. Find the length and the width.

65. *Sharing fruit.* Apples are collected in a basket for six people. One third, one fourth, one eighth, and one fifth of the apples are given to four people, respectively. The fifth person gets ten apples, and one apple remains for the sixth person. Find the original number of apples in the basket.

66. *Discounts.* In exchange for opening a new credit account, Filene's Department Stores® subtracts 10% from all purchases made the day the account is established. Julio is opening an account and has a coupon for which he receives 10% off the first day's reduced price of a camera. If Julio's final price is $77.75, what was the price of the camera before the two discounts?

67. *Winning percentage.* In a basketball league, the Falcons won 15 of their first 20 games. In order to win 60% of the total number of games, how many more games will they have to play, assuming they win only half of the remaining games?

68. *Music-club purchases.* During a recent sale, BMG Music Service® charged $8.49 for the first CD ordered and $3.99 for all others. For shipping and handling, BMG charged $2.47 for the first CD, $2.28 for the second CD, and $1.99 for all others. The total cost of a shipment (excluding tax) was $65.07. How many CD's were in the shipment?

69. *Test scores.* Ella has an average score of 82 on three tests. Her average score on the first two tests is 85. What was the score on the third test?

70. *Taxi fares.* In New York City, a taxi ride costs $2.50 plus 40¢ per $\frac{1}{5}$ mile and 20¢ per minute stopped in traffic. Due to traffic, Glenda's taxi took 20 min to complete what is usually a 10-min drive. If she is charged $16.50 for the ride, how far did Glenda travel?

71. A school purchases a piano and must choose between paying $2000 at the time of purchase or $2150 at the end of one year. Which option should the school select and why?

Aha! **72.** Annette claims the following problem has no solution: "The sum of the page numbers on facing pages is 191. Find the page numbers." Is she correct? Why or why not?

73. The perimeter of a rectangle is 101.74 cm. If the length is 4.25 cm longer than the width, find the dimensions of the rectangle.

74. The second side of a triangle is 3.25 cm longer than the first side. The third side is 4.35 cm longer than the second side. If the perimeter of the triangle is 26.87 cm, find the length of each side.

2.6   Solving Inequalities

Solutions of Inequalities   •   Graphs of Inequalities   •   Solving Inequalities Using the Addition Principle   •   Solving Inequalities Using the Multiplication Principle   •   Using the Principles Together

Many real-world situations translate to *inequalities*. For example, a student might need to register for *at least* 12 credits; an elevator might be designed to hold *at most* 2000 pounds; a tax credit might be allowable for families with incomes of *less than* $25,000; and so on. Before solving applications of this type, we must adapt our equation-solving principles to the solving of inequalities.

Solutions of Inequalities

Recall from Section 1.4 that an inequality is a number sentence containing $>$ (is greater than), $<$ (is less than), $\geq$ (is greater than or equal to), or $\leq$ (is less than or equal to). Inequalities like

$$-7 > x, \qquad t < 5, \qquad 5x - 2 \geq 9, \quad \text{and} \quad -3y + 8 \leq -7$$

are true for some replacements of the variable and false for others.

EXAMPLE 1

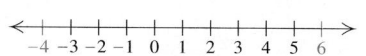

Determine whether the given number is a solution of $x < 2$: **(a)** -3; **(b)** 2.

Solution

a) Since $-3 < 2$ is true, -3 is a solution.

b) Since $2 < 2$ is false, 2 is not a solution.

EXAMPLE 2

Determine whether the given number is a solution of $y \geq 6$: **(a)** 6; **(b)** -4.

Solution

a) Since $6 \geq 6$ is true, 6 is a solution.

b) Since $-4 \geq 6$ is false, -4 is not a solution.

Graphs of Inequalities

Because the solutions of inequalities like $x < 2$ are too numerous to list, it is helpful to make a drawing that represents all the solutions. The **graph** of an inequality is such a drawing. Graphs of inequalities in one variable can be drawn on a number line by shading all points that are solutions. Open dots are used to indicate endpoints that are *not* solutions and closed dots indicate endpoints that *are* solutions.

EXAMPLE 3   Graph each inequality: **(a)** $x < 2$; **(b)** $y \geq -3$; **(c)** $-2 < x \leq 3$.

Solution

a) The solutions of $x < 2$ are those numbers less than 2. They are shown on the graph by shading all points to the left of 2. The open dot at 2 and the shading to its left indicate that 2 is *not* part of the graph, but numbers like 1.2 and 1.99 are.

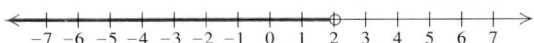

b) The solutions of $y \geq -3$ are shown on the number line by shading the point for -3 and all points to the right of -3. The closed dot at -3 indicates that -3 *is* part of the graph.

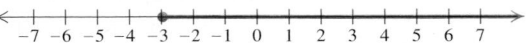

Student Notes _____

Note that $-2 < x < 3$ means $-2 < x$ *and* $x < 3$. Because of this, statements like $2 < x < 1$ make no sense — no number is both greater than 2 and less than 1.

c) The inequality $-2 < x \leq 3$ is read "-2 is less than x *and* x is less than or equal to 3," or "x is greater than -2 *and* less than or equal to 3." To be a solution of $-2 < x \leq 3$, a number must be a solution of both $-2 < x$ *and* $x \leq 3$. The number 1 is a solution, as are -0.5, 1.9, and 3. The open dot indicates that -2 is *not* a solution, whereas the closed dot indicates that 3 *is* a solution. The other solutions are shaded.

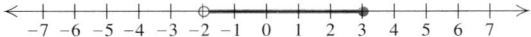

Solving Inequalities Using the Addition Principle

Consider a balance similar to one that appears in Section 2.1. When one side of the balance holds more weight than the other, the balance tips in that direction. If equal amounts of weight are then added to or subtracted from both sides of the balance, the balance remains tipped in the same direction.

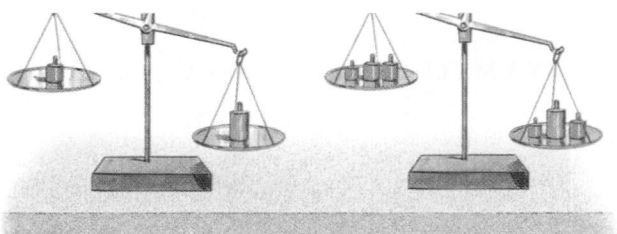

The balance illustrates the idea that when a number, such as 2, is added to (or subtracted from) both sides of a true inequality, such as $3 < 7$, we get another true inequality:

$$3 + 2 < 7 + 2, \quad \text{or} \quad 5 < 9.$$

Similarly, if we add -4 to both sides of $x + 4 < 10$, we get an *equivalent* inequality:

$$x + 4 + (-4) < 10 + (-4), \quad \text{or} \quad x < 6.$$

We say that $x + 4 < 10$ and $x < 6$ are **equivalent**, which means that both inequalities have the same solution set.

> **The Addition Principle for Inequalities**
> For any real numbers a, b, and c:
>
> $a < b$ is equivalent to $a + c < b + c$;
> $a \leq b$ is equivalent to $a + c \leq b + c$;
> $a > b$ is equivalent to $a + c > b + c$;
> $a \geq b$ is equivalent to $a + c \geq b + c$.

As with equations, our goal is to isolate the variable on one side.

EXAMPLE 4   Solve $x + 2 > 8$ and then graph the solution.

Solution   We use the addition principle, subtracting 2 from both sides:

$$x + 2 - 2 > 8 - 2 \qquad \text{Subtracting 2 from, or adding } -2 \text{ to, both sides}$$
$$x > 6.$$

From the inequality $x > 6$, we can determine the solutions easily. Any number greater than 6 makes $x > 6$ true and is a solution of that inequality as well as the inequality $x + 2 > 8$. The graph is as follows:

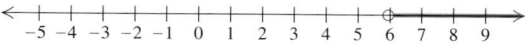

Because most inequalities have an infinite number of solutions, we cannot possibly check them all. A partial check can be made using one of the possible solutions. For this example, we can substitute any number greater than 6—say, 6.1—into the original inequality:

$$\frac{x + 2 > 8}{6.1 + 2 \mid 8}$$
$$8.1 \overset{?}{>} 8 \quad \text{TRUE} \qquad 8.1 > 8 \text{ is a true statement.}$$

Since $8.1 > 8$ is true, 6.1 is a solution. Any number greater than 6 is a solution.

Although the inequality $x > 6$ is easy to solve (we merely replace x with numbers greater than 6), it is worth noting that $x > 6$ is an *inequality*, not a *solution*. In fact, the solutions of $x > 6$ are numbers. To describe the set of all solutions, we will use **set-builder notation** to write the *solution set* of Example 4 as

$$\{x \mid x > 6\}.$$

This notation is read

"The set of all x such that x is greater than 6."

Thus a number is in $\{x \mid x > 6\}$ if that number is greater than 6. From now on, solutions of inequalities will be written using set-builder notation.

EXAMPLE 5

Solve $3x - 1 \le 2x - 5$ and then graph the solution.

Solution   We have

$$3x - 1 \le 2x - 5$$
$$3x - 1 + 1 \le 2x - 5 + 1 \qquad \text{Adding 1 to both sides}$$
$$3x \le 2x - 4 \qquad \text{Simplifying}$$
$$3x - 2x \le 2x - 4 - 2x \qquad \text{Subtracting } 2x \text{ from both sides}$$
$$x \le -4. \qquad \text{Simplifying}$$

The graph is as follows:

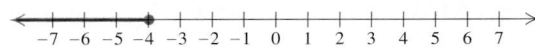

The student should check that any number less than or equal to -4 is a solution. The solution set is $\{x \mid x \le -4\}$.

technology connection

As a partial check of Example 5, we can let $y_1 = 3x - 1$ and $y_2 = 2x - 5$. By scrolling up or down, you can note that for $x \le -4$, we have $y_1 \le y_2$.

X	Y₁	Y₂
−5	−16	−15
−4	−13	−13
−3	−10	−11
−2	−7	−9
−1	−4	−7
0	−1	−5
1	2	−3

X = −5

Solving Inequalities Using the Multiplication Principle

There is a multiplication principle for inequalities similar to that for equations, but it must be modified when multiplying both sides by a negative number. Consider the true inequality

$$3 < 7.$$

If we multiply both sides by a *positive* number, say 2, we get another true inequality:

$$3 \cdot 2 < 7 \cdot 2, \quad \text{or} \quad 6 < 14. \qquad \text{TRUE}$$

If we multiply both sides by a negative number, say -2, we get a *false* inequality:

$$3 \cdot (-2) < 7 \cdot (-2), \quad \text{or} \quad -6 < -14. \qquad \text{FALSE}$$

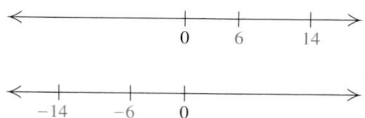

The fact that $6 < 14$ is true, but $-6 < -14$ is false, stems from the fact that the negative numbers, in a sense, *mirror* the positive numbers. Whereas 14 is to the *right* of 6, the number -14 is to the *left* of -6. Thus if we reverse the inequality symbol in $-6 < -14$, we get a true inequality:

$$-6 > -14. \quad \text{TRUE}$$

The Multiplication Principle for Inequalities

For any real numbers a and b, and for any *positive* number c:

$\quad a < b$ is equivalent to $ac < bc$, and

$\quad a > b$ is equivalent to $ac > bc$.

For any real numbers a and b, and for any *negative* number c:

$\quad a < b$ is equivalent to $ac > bc$, and

$\quad a > b$ is equivalent to $ac < bc$.

Similar statements hold for $\leq$ and $\geq$.

Caution!   When multiplying or dividing both sides of an inequality by a negative number, don't forget to reverse the inequality symbol!

EXAMPLE 6   Solve and graph each inequality: **(a)** $\frac{1}{4}x < 7$; **(b)** $-2y < 18$.

Solution

a) $\quad \frac{1}{4}x < 7$ $\qquad$ Multiplying both sides by 4, the reciprocal of $\frac{1}{4}$

$\quad 4 \cdot \frac{1}{4}x < 4 \cdot 7$

$\qquad\qquad$ The symbol stays the same, since 4 is positive.

$\qquad x < 28$ $\qquad$ Simplifying

The solution set is $\{x \mid x < 28\}$. The graph is as shown at left.

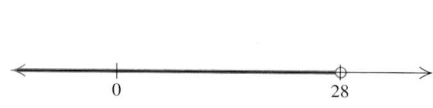

b) $-2y < 18$

$\quad \dfrac{-2y}{-2} > \dfrac{18}{-2}$ $\qquad$ Multiplying both sides by $-\frac{1}{2}$, or dividing both sides by -2

$\qquad\qquad$ *At this step*, we reverse the inequality, because $-\frac{1}{2}$ is negative.

$\quad y > -9$ $\qquad$ Simplifying

As a partial check, we substitute a number greater than -9, say -8, into the original inequality:

$$\frac{-2y < 18}{-2(-8) \mid 18}$$

$\qquad\qquad 16 \overset{?}{<} 18$ TRUE $\qquad$ $16 < 18$ is a true statement.

The solution set is $\{y \mid y > -9\}$. The graph is as shown at left.

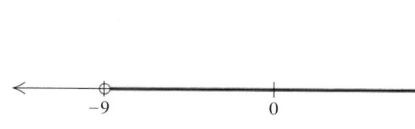

Using the Principles Together

We use the addition and multiplication principles together to solve inequalities much as we did when solving equations.

EXAMPLE 7   Solve: **(a)** $6 - 5y > 7$; **(b)** $2x - 9 \leq 7x + 1$.

Solution

a)
$$6 - 5y > 7$$

$-6 + 6 - 5y > -6 + 7$	Adding -6 to both sides
$-5y > 1$	Simplifying
$-\frac{1}{5} \cdot (-5y) < -\frac{1}{5} \cdot 1$	Multiplying both sides by $-\frac{1}{5}$, or dividing both sides by -5

Remember to reverse the inequality symbol!

$y < -\frac{1}{5}$	Simplifying

As a check, we substitute a number smaller than $-\frac{1}{5}$, say -1, into the original inequality:

$$\begin{array}{c|c} 6 - 5y > 7 \\ \hline 6 - 5(-1) & 7 \\ 6 - (-5) & \\ 11 \overset{?}{>} 7 & \text{TRUE} \end{array} \qquad 11 > 7 \text{ is a true statement.}$$

The solution set is $\left\{y \mid y < -\frac{1}{5}\right\}$. We show the graph in the margin for reference.

b)
$$2x - 9 \leq 7x + 1$$

$2x - 9 - 1 \leq 7x + 1 - 1$	Subtracting 1 from both sides
$2x - 10 \leq 7x$	Simplifying
$2x - 10 - 2x \leq 7x - 2x$	Subtracting $2x$ from both sides
$-10 \leq 5x$	Simplifying
$\dfrac{-10}{5} \leq \dfrac{5x}{5}$	Dividing both sides by 5
$-2 \leq x$	Simplifying

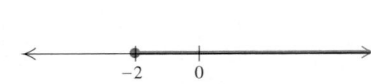

The solution set is $\{x \mid -2 \leq x\}$, or $\{x \mid x \geq -2\}$.

All of the equation-solving techniques used in Sections 2.1 and 2.2 can be used with inequalities provided we remember to reverse the inequality symbol when multiplying or dividing both sides by a negative number.

EXAMPLE 8   Solve: **(a)** $16.3 - 7.2p \le -8.18$; **(b)** $3(x - 9) - 1 < 2 - 5(x + 6)$.

Solution

a) The greatest number of decimal places in any one number is *two*. Multiplying both sides by 100 will clear decimals. Then we proceed as before.

$$16.3 - 7.2p \le -8.18$$

$$100(16.3 - 7.2p) \le 100(-8.18) \quad \text{Multiplying both sides by 100}$$

$$100(16.3) - 100(7.2p) \le 100(-8.18) \quad \text{Using the distributive law}$$

$$1630 - 720p \le -818 \quad \text{Simplifying}$$

$$-720p \le -818 - 1630 \quad \text{Subtracting 1630 from both sides}$$

$$-720p \le -2448 \quad \text{Simplifying}$$

$$p \ge \frac{-2448}{-720} \quad \text{Dividing both sides by } -720$$

Remember to reverse the symbol!

$$p \ge 3.4$$

The solution set is $\{p \mid p \ge 3.4\}$.

b) $3(x - 9) - 1 < 2 - 5(x + 6)$

$$3x - 27 - 1 < 2 - 5x - 30 \quad \text{Using the distributive law to remove parentheses}$$

$$3x - 28 < -5x - 28 \quad \text{Simplifying}$$

$$3x - 28 + 28 < -5x - 28 + 28 \quad \text{Adding 28 to both sides}$$

$$3x < -5x$$

$$3x + 5x < -5x + 5x \quad \text{Adding } 5x \text{ to both sides}$$

$$8x < 0$$

$$x < 0 \quad \text{Dividing both sides by 8}$$

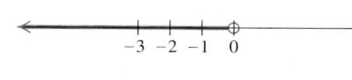

The solution set is $\{x \mid x < 0\}$.

Exercise Set

2.6

↶ *Concept Reinforcement*   *Insert the symbol* $<, >, \le$, *or* $\ge$ *to make each pair of inequalities equivalent.*

1. $-5x \le 30$; x ▨ -6

2. $-7t \ge 56$; t ▨ -8

3. $-2t > -14$; t ▨ 7

4. $-3x < -15$; x ▨ 5

↶ *Concept Reinforcement*   *Classify each pair of inequalities as "equivalent" or "not equivalent."*

5. $x < -2$; $-2 > x$

6. $t > -1$; $-1 < t$

7. $-4x - 1 \le 15$;
$-4x \le 16$

8. $-2t + 3 \ge 11$;
$-2t \ge 14$

Determine whether each number is a solution of the given inequality.

9. $x > -2$
  a) 5   b) 0   c) -1.9
  d) -7.3 e) 1.6

10. $y < 5$
  a) 0   b) 5   c) 4.99
  d) -13 e) $7\frac{1}{4}$

11. $x \geq 6$
  a) -6 b) 0   c) 6
  d) 6.01   e) $-3\frac{1}{2}$

12. $x \leq 10$
  a) 4   b) -10 c) 0
  d) 10.2   e) -4.7

Graph on a number line.

13. $y < 2$ **14.** $x \leq 7$

15. $y > 4$ **16.** $t > -2$

17. $0 \leq t$ **18.** $1 \leq m$

19. $-5 \leq x < 2$ **20.** $-3 < x \leq 5$

21. $-5 \leq x \leq 0$ **22.** $0 < x < 3$

Describe each graph using set-builder notation.

23.
24.
25.
26.
27.
28.
29.
30.

Solve using the addition principle. Graph and write set-builder notation for each answer.

31. $y + 6 > 9$ **32.** $y + 2 > 9$

33. $x + 9 \leq -12$ **34.** $x + 8 \leq -10$

35. $x - 3 < 14$ **36.** $x - 3 < 7$

37. $y - 10 > -16$ **38.** $y - 7 > -12$

39. $2x \leq x + 9$ **40.** $3x \leq 2x + 7$

41. $y + \frac{1}{3} \leq \frac{5}{6}$ **42.** $x + \frac{1}{4} \leq \frac{1}{2}$

43. $t - \frac{1}{8} > \frac{1}{2}$ **44.** $y - \frac{1}{3} > \frac{1}{4}$

45. $-9x + 17 > 17 - 8x$ **46.** $-8n + 12 > 12 - 7n$

Aha! **47.** $-23 < -t$ **48.** $19 < -x$

Solve using the multiplication principle. Graph and write set-builder notation for each answer.

49. $5x < 35$ **50.** $8x \geq 32$

51. $-7x < 13$ **52.** $8y < 17$

53. $-24 > 8t$ **54.** $-16x < -64$

55. $7y \geq -2$ **56.** $5x > -3$

57. $-2y \leq \frac{1}{5}$ **58.** $-2x \geq \frac{1}{5}$

59. $-\frac{8}{5} > -2x$ **60.** $-\frac{5}{8} < -10y$

Solve using the addition and multiplication principles.

61. $7 + 3x < 34$ **62.** $5 + 4y < 37$

63. $6 + 5y \geq 26$ **64.** $7 + 8x \geq 71$

65. $4t - 5 \leq 23$ **66.** $13x - 7 < -7$

67. $16 < 4 - 3y$ **68.** $22 < 6 - 8x$

69. $39 > 3 - 9x$ **70.** $5 > 5 - 7y$

71. $5 - 6y > 25$ **72.** $8 - 2y > 14$

73. $-3 < 8x + 7 - 7x$ **74.** $-5 < 9x + 8 - 8x$

75. $6 - 4y > 4 - 3y$ **76.** $7 - 8y > 5 - 7y$

77. $7 - 9y \leq 4 - 8y$ **78.** $6 - 13y \leq 4 - 12y$

79. $33 - 12x < 4x + 97$ **80.** $27 - 11x > 14x - 18$

81. $2.1x + 43.2 > 1.2 - 8.4x$

82. $0.96y - 0.79 \leq 0.21y + 0.46$

83. $0.7n - 15 + n \geq 2n - 8 - 0.4n$

84. $1.7t + 8 - 1.62t < 0.4t - 0.32 + 8$

85. $\dfrac{x}{3} - 4 \leq 1$

86. $\dfrac{2}{3} - \dfrac{x}{5} < \dfrac{4}{15}$

87. $3 < 5 - \dfrac{t}{7}$

88. $2 > 9 - \dfrac{x}{5}$

89. $4(2y - 3) < 36$

90. $3(2y - 3) > 21$

91. $3(t - 2) \geq 9(t + 2)$

92. $8(2t + 1) > 4(7t + 7)$

93. $3(r - 6) + 2 < 4(r + 2) - 21$

94. $5(t + 3) + 9 > 3(t - 2) + 6$

95. $\frac{2}{3}(2x - 1) \geq 10$

96. $\frac{4}{5}(3x + 4) \leq 20$

97. $\frac{3}{4}\left(3x - \frac{1}{2}\right) - \frac{2}{3} < \frac{1}{3}$

98. $\frac{2}{3}\left(\frac{7}{8} - 4x\right) - \frac{5}{8} < \frac{3}{8}$

99. Are the inequalities $x > -3$ and $x \geq -2$ equivalent? Why or why not?

100. Are the inequalities $t < -7$ and $t \leq -8$ equivalent? Why or why not?

SKILL MAINTENANCE

Translate to an algebraic expression. [1.1]

101. The sum of 3 and some number

102. Twice the sum of two numbers

103. Three less than twice a number

104. Five more than twice a number

SYNTHESIS

105. Explain how it is possible for the graph of an inequality to consist of just one number. (*Hint*: See Example 3c.)

106. Explain in your own words why it is necessary to reverse the inequality symbol when multiplying both sides of an inequality by a negative number.

Solve.

Aha! **107.** $x < x + 1$

108. $6[4 - 2(6 + 3t)] > 5[3(7 - t) - 4(8 + 2t)] - 20$

109. $27 - 4[2(4x - 3) + 7] \geq 2[4 - 2(3 - x)] - 3$

Solve for x.

110. $-(x + 5) \geq 4a - 5$

111. $\frac{1}{2}(2x + 2b) > \frac{1}{3}(21 + 3b)$

112. $y < ax + b$ (Assume $a > 0$.)

113. $y < ax + b$ (Assume $a < 0$.)

114. Graph the solutions of $|x| < 3$ on a number line.

Aha! **115.** Determine the solution set of $|x| > -3$.

116. Determine the solution set of $|x| < 0$.

2.7 Solving Applications with Inequalities

Translating to Inequalities • Solving Problems

The five steps for problem solving can be used for problems involving inequalities.

Translating to Inequalities

Before solving problems that involve inequalities, we list some important phrases to look for. Sample translations are listed as well.

Important Words	Sample Sentence	Translation
is at least	Bill is at least 21 years old.	$b \geq 21$
is at most	At most 5 students dropped the course.	$n \leq 5$
cannot exceed	To qualify, earnings cannot exceed $12,000.	$r \leq 12,000$
must exceed	The speed must exceed 15 mph.	$s > 15$
is less than	Tucker's weight is less than 50 lb.	$w < 50$
is more than	Boston is more than 200 miles away.	$d > 200$
is between	The film was between 90 and 100 minutes long.	$90 < t < 100$
no more than	Bing weighs no more than 90 lb.	$w \leq 90$
no less than	Valerie scored no less than 8.3.	$s \geq 8.3$

The following phrases deserve special attention.

Translating "at least" and "at most"

The quantity x is at least some amount q: $x \geq q$.

(If x is *at least* q, it cannot be less than q.)

The quantity x is at most some amount q: $x \leq q$.

(If x is *at most* q, it cannot be more than q.)

Solving Problems

EXAMPLE 1

Catering costs.   To cater a party, Curtis' Barbeque charges a $50 setup fee plus $15 per person. The cost of Hotel Pharmacy's end-of-season softball party cannot exceed $450. How many people can attend the party?

Solution

1. **Familiarize.** Suppose that 20 people were to attend the party. The cost would then be $50 + $15 · 20, or $350. This shows that more than 20 people could attend without exceeding $450. Instead of making another guess, we let n = the number of people in attendance.

2. **Translate.** The cost of the party will be $50 for the setup fee plus $15 times the number of people attending. We can reword as follows:

Rewording:	The setup fee	plus	the cost of the meals	cannot exceed	$450.
Translating:	50	+	$15 \cdot n$	$\leq$	450

3. **Carry out.**   We solve for n:

$$50 + 15n \leq 450$$
$$15n \leq 400 \qquad \text{Subtracting 50 from both sides}$$
$$n \leq \frac{400}{15} \qquad \text{Dividing both sides by 15}$$
$$n \leq 26\frac{2}{3}. \qquad \text{Simplifying}$$

4. **Check.**   Although the solution set of the inequality is all numbers less than or equal to $26\frac{2}{3}$, since n represents the number of people in attendance, we round *down* to 26. If 26 people attend, the cost will be $\$50 + \$15 \cdot 26$, or $\$440$, and if 27 attend, the cost will exceed $\$450$.

5. **State.**   At most 26 people can attend the party.

> *Caution!*   Solutions of problems should always be checked using the original wording of the problem. In some cases, answers might need to be whole numbers or integers or rounded off in a particular direction.

Some applications with inequalities involve *averages* or *means*. You are already familiar with the concept of averages from grades in courses that you have taken.

> *Average or mean*
>
> To find the **average** or **mean** of a set of numbers, add the numbers and then divide by the number of addends.

EXAMPLE 2

Nutrition.   The U.S. Department of Health and Human Services and the Department of Agriculture recommend that for a typical 2000-calorie daily diet, no more than 65 g of fat be consumed. In the first three days of a four-day vacation, Phil consumed 70 g, 62 g, and 80 g of fat. Determine (in terms of an inequality) how many grams of fat Phil can consume on the fourth day if he is to average no more than 65 g of fat per day.

Solution

1. **Familiarize.**   Suppose Phil consumed 64 g of fat on the fourth day. His daily average for the vacation would then be

$$\frac{70\,\text{g} + 62\,\text{g} + 80\,\text{g} + 64\,\text{g}}{4} = 69\,\text{g}. \qquad \text{There are 4 addends, so we divide by 4.}$$

This shows that Phil cannot consume 64 g of fat on the fourth day, if he is to average no more than 65 g of fat per day. Let's have x represent the number of grams of fat that Phil consumes on the fourth day.

2. **Translate.** We reword the problem and translate as follows:

Rewording: The average consumption of fat | should be no more than | 65 g.

Translating: $\dfrac{70 + 62 + 80 + x}{4}$ $\leq$ 65

3. **Carry out.** Because of the fraction, it is convenient to use the multiplication principle first:

$$\frac{70 + 62 + 80 + x}{4} \leq 65$$

$$4\left(\frac{70 + 62 + 80 + x}{4}\right) \leq 4 \cdot 65 \qquad \text{Multiplying both sides by 4}$$

$$70 + 62 + 80 + x \leq 260$$

$$212 + x \leq 260 \qquad \text{Simplifying}$$

$$x \leq 48. \qquad \text{Subtracting 212 from both sides}$$

4. **Check.** As a partial check, we show that Phil can consume 48 g of fat on the fourth day and not exceed a 65-g average for the four days:

$$\frac{70 + 62 + 80 + 48}{4} = \frac{260}{4} = 65. \qquad \text{Note that 65 does not, itself, exceed 65.}$$

5. **State.** Phil's average fat intake for the vacation will not exceed 65 g per day if he consumes no more than 48 g of fat on the fourth day.

Exercise Set

2.7

FOR EXTRA HELP

Student's Solutions Manual | Digital Video Tutor CD 1 Videotape 2 | AW Math Tutor Center | MathXL Tutorials on CD | MathXL | MyMathLab

↪ *Concept Reinforcement* In each of Exercises 1–8, match the sentence with one of the following:
$a < b;\ a \leq b;\ b < a;\ b \leq a.$

1. a is at least b.

2. a exceeds b.

3. a is at most b.

4. a is exceeded by b.

5. b is no more than a.

6. b is no less than a.

7. b is less than a.

8. b is more than a.

Translate to an inequality.

9. A number is at least 8.

10. A number is greater than or equal to 4.

11. The temperature is at most $-3°C$.

12. The average credit-card holder is more than $4000 in debt.

13. The price of Pat's PT Cruiser exceeded $21,900.

14. The time of the test was between 45 and 55 min.

15. Normandale Community College is no more than 15 mi away.

16. Tania's weight is less than 110 lb.

17. That number is greater than -2.

18. The costs of production of that CD-ROM cannot exceed $12,500.

19. Between 400,000 and 1,200,000 people attended the Million Man March.

20. The cost of gasoline was at most $2 per gallon.

Use an inequality and the five-step process to solve each problem.

21. *Furnace repairs.*   RJ's Plumbing and Heating charges $25 plus $30 per hour for emergency service. Gary remembers being billed over $100 for an emergency call. How long was RJ's there?

22. *College tuition.*   Karen's financial aid stipulates that her tuition not exceed $1000. If her local community college charges a $35 registration fee plus $375 per course, what is the greatest number of courses for which Karen can register?

23. *Mass transit.*   Local light rail service in Denver, Colorado, costs $1.15 per trip (one way). A student monthly pass costs $21. Gail is a student at Community College of Denver. Express as an inequality the number of trips per month that Gail should make if the pass is to save her money.
Source: rtd-denver.com

24. *Grade average.*   Nadia is taking a literature course in which four tests are given. To get a B, a student must average at least 80 on the four tests. Nadia scored 82, 76, and 78 on the first three tests. What scores on the last test will earn her at least a B?

25. *Quiz average.*   Rod's quiz grades are 73, 75, 89, and 91. What scores on a fifth quiz will make his average quiz grade at least 85?

26. *Nutrition.*   Following the guidelines of the Food and Drug Administration, Dale tries to eat at least 5 servings of fruits or vegetables each day. For the first six days of one week, she had 4, 6, 7, 4, 6, and 4 servings. How many servings of fruits or vegetables should Dale eat on Saturday, in order to average at least 5 servings per day for the week?

27. *College course load.*   To remain on financial aid, Millie needs to complete an average of at least 7 credits per quarter each year. In the first three quarters of 2004, Millie completed 5, 7, and 8 credits. How many credits of course work must Millie complete in the fourth quarter if she is to remain on financial aid?

28. *Music lessons.*   Band members at Colchester Middle School are expected to average at least 20 min of practice time per day. One week Monroe practiced 15 min, 28 min, 30 min, 0 min, 15 min, and 25 min. How long must he practice on the seventh day if he is to meet expectations?

29. *Electrician visits.* Dot's Electric made 17 customer calls last week and 22 calls this week. How many calls must be made next week in order to maintain an average of at least 20 for the three-week period?

30. *Perimeter of a rectangle.* The width of a rectangle is fixed at 8 ft. What lengths will make the perimeter at least 200 ft? at most 200 ft?

31. *Perimeter of a triangle.* One side of a triangle is 2 cm shorter than the base. The other side is 3 cm longer than the base. What lengths of the base will allow the perimeter to be greater than 19 cm?

32. *Perimeter of a pool.* The perimeter of a rectangular swimming pool is not to exceed 70 ft. The length is to be twice the width. What widths will meet these conditions?

33. *Well drilling.* All Seasons Well Drilling offers two plans. Under the "pay-as-you-go" plan, they charge $500 plus $8 a foot for a well of any depth. Under their "guaranteed-water" plan, they charge a flat fee of $4000 for a well that is guaranteed to provide adequate water for a household. For what depths would it save a customer money to use the pay-as-you-go plan?

34. *Cost of road service.* Rick's Automotive charges $50 plus $15 for each (15-min) unit of time when making a road call. Twin City Repair charges $70 plus $10 for each unit of time. Under what circumstances would it be more economical for a motorist to call Rick's?

35. *Insurance-covered repairs.* Most insurance companies will replace a vehicle if an estimated repair exceeds 80% of the "blue-book" value of the vehicle. Michelle's insurance company paid $8500 for repairs to her Subaru after an accident. What can be concluded about the blue-book value of the car?

36. *Insurance-covered repairs.* Following an accident, Jeff's Ford pickup was replaced by his insurance company because the damage was so extensive. Before the damage, the blue-book value of the truck was $21,000. How much would it have cost to repair the truck? (See Exercise 35.)

37. *Sizes of envelopes.* Rhetoric Advertising is a direct-mail company. It determines that for a particular campaign, it can use any envelope with a fixed width of $3\frac{1}{2}$ in. and an area of at least $17\frac{1}{2}$ in². Determine (in terms of an inequality) those lengths that will satisfy the company constraints.

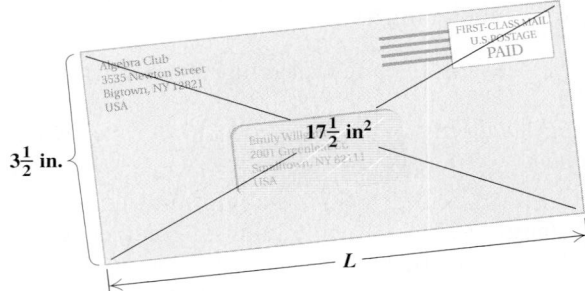

38. *Sizes of packages.* An overnight delivery service accepts packages of up to 165 in. in length and girth combined. (Girth is the distance around the package.) A package has a fixed girth of 53 in. Determine (in terms of an inequality) those lengths for which a package is acceptable.

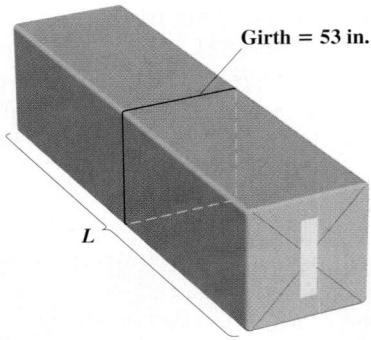

39. *Body temperature.* A person is considered to be feverish when his or her temperature is higher than 98.6°F. The formula $F = \frac{9}{5}C + 32$ can be used to convert Celsius temperatures C to Fahrenheit temperatures F. For which Celsius temperatures is a person considered feverish?

40. *Gold temperatures.* Gold stays solid at Fahrenheit temperatures below 1945.4°. Determine (in terms of an inequality) those Celsius temperatures for which gold stays solid. Use the formula given in Exercise 39.

41. *Area of a triangular flag.* As part of an outdoor education course, Wanda needs to make a bright-colored triangular flag with an area of at least 3 ft². What heights can the triangle be if the base is $1\frac{1}{2}$ ft?

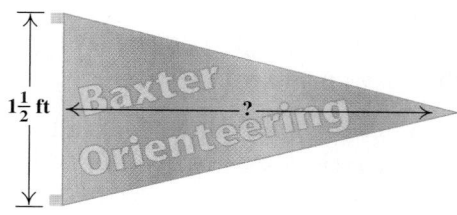

42. *Area of a triangular sign.* Zoning laws in Harrington prohibit displaying signs with areas exceeding 12 ft². If Flo's Marina is ordering a triangular sign with an 8-ft base, how tall can the sign be?

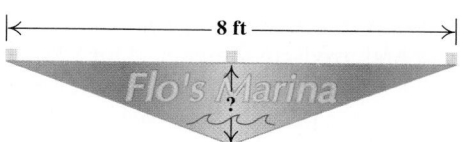

43. *Fat content in foods.* Reduced Fat Skippy® peanut butter contains 12 g of fat per serving. In order for a food to be labeled "reduced fat," it must have at least 25% less fat than the regular item. What can you conclude about the number of grams of fat in a serving of the regular Skippy peanut butter?
Source: Best Foods

44. *Fat content in foods.* Reduced Fat Chips Ahoy!® cookies contain 5 g of fat per serving. What can you conclude about the number of grams of fat in regular Chips Ahoy! cookies (see Exercise 43)?
Source: Nabisco Brands, Inc.

45. *Weight gain.* In the last weeks before the yearly Topsfield Weigh In, heavyweight pumpkins gain about 26 lb per day. Charlotte's heaviest pumpkin weighs 532 lb on September 5. For what dates will its weight exceed 818 lb?

46. *Pond depth.* On July 1, Garrett's Pond was 25 ft deep. Since that date, the water level has dropped $\frac{2}{3}$ ft per week. For what dates will the water level not exceed 21 ft?

47. *Blueprints.* To make copies of blueprints, Vantage Reprographics charges a $6 setup fee plus $4 per copy. Myra can spend no more than $65 for the copying. What numbers of copies will allow her to stay within budget?

48. *Banquet costs.* The women's volleyball team can spend at most $450 for its awards banquet at a local restaurant. If the restaurant charges a $40 setup fee plus $16 per person, at most how many can attend?

49. *World records in the mile run.* The formula
$$R = -0.0065t + 4.3222$$
can be used to predict the world record, in minutes, for the 1-mi run t years after 1900. Determine (in terms of an inequality) those years for which the world record will be less than 3.7 min.
Source: Based on information from Norris McWhirter, *Book of Historical Records.* New York: Sterling, 2000.

50. *World records in the women's 1500-m run.*   The formula

$$R = -0.0223t + 4.4078$$

can be used to predict the world record, in minutes, for the 1500-m run t years after 1930. Determine (in terms of an inequality) those years for which the world record will be less than 3.5 min.

51. *Toll charges.*   The equation $y = 0.03x + 0.21$ can be used to determine the approximate cost y, in dollars, of driving x miles on the Indiana toll road. For what mileages x will the cost be at most $6?

52. *Price of a movie ticket.*   The average price of a movie ticket can be estimated by the equation $P = 0.227Y - 448.71$, where Y is the year and P is the average price, in dollars. The price is lower than what might be expected due to senior-citizen discounts, children's prices, and special volume discounts. For what years will the average price of a movie ticket be at least $7? (Include the year in which the $7 ticket first occurs.)

53. If f represents Fran's age and t represents Todd's age, write a sentence that would translate to $t + 3 < f$.

54. Explain how the meanings of "Five more than a number" and "Five is more than a number" differ.

SKILL MAINTENANCE

Simplify. [1.8]

55. $\dfrac{9 - 5}{6 - 4}$

56. $\dfrac{8 - 5}{12 - 6}$

57. $\dfrac{8 - (-2)}{1 - 4}$

58. $\dfrac{7 - 9}{4 - (-6)}$

SYNTHESIS

59. Write a problem for a classmate to solve. Devise the problem so the answer is "At most 18 passengers can go on the boat." Design the problem so that at least one number in the solution must be rounded down.

60. Write a problem for a classmate to solve. Devise the problem so the answer is "The Rothmans can drive 90 mi without exceeding their truck rental budget."

61. *Ski wax.*   Green ski wax works best between 5° and 15° Fahrenheit. Determine those Celsius temperatures for which green ski wax works best. (See Exercise 39.)

62. *Parking fees.*   Mack's Parking Garage charges $4.00 for the first hour and $2.50 for each additional hour. For how long has a car been parked when the charge exceeds $16.50?

Aha! **63.** The area of a square can be no more than 64 cm². What lengths of a side will allow this?

Aha! **64.** The sum of two consecutive odd integers is less than 100. What is the largest pair of such integers?

65. *Nutritional standards.*   In order for a food to be labeled "lowfat," it must have fewer than 3 g of fat per serving. Reduced fat Tortilla Pops® contain 60% less fat than regular nacho cheese tortilla chips, but still cannot be labeled lowfat. What can you conclude about the fat content of a serving of nacho cheese tortilla chips?

66. *Parking fees.*   When asked how much the parking charge is for a certain car (see Exercise 62), Mack replies "between 14 and 24 dollars." For how long has the car been parked?

67. *Frequent buyer bonus.*   Alice's Books allows customers to select one free book for every 10 books purchased. The price of that book cannot exceed the average cost of the 10 books. Neoma has bought 9 books that average $12 per book. How much should her tenth book cost if she wants to select a $15 book for free?

68. *Grading.*   After 9 quizzes, Blythe's average is 84. Is it possible for Blythe to improve her average by two points with the next quiz? Why or why not?

69. *Discount card.*   Arnold and Diaz Booksellers offers a preferred-customer card for $25. The card entitles a customer to a 10% discount on all purchases for a period of one year. Under what circumstances would an individual save money by purchasing a card?

Study Summary

Equations like $3x - 1 = 10$, $3x = 11$, and $x = \frac{11}{3}$, which share the same solution, are said to be **equivalent equations** (p. 81). **The addition principle** and **the multiplication principle** are the methods most frequently used for producing equivalent equations that lead to a solution (pp. 81, 83).

Addition principle (for equations):

For any real numbers a, b, and c,

$$a = b \quad \text{is equivalent to} \quad a + c = b + c.$$

Multiplication principle (for equations):

For any real numbers a, b, and c, with $c \neq 0$,

$$a = b \quad \text{is equivalent to} \quad a \cdot c = b \cdot c.$$

When we are solving equations like $5t - 3 = 14$, the addition principle is generally used first. When the **coefficient** of the variable is a fraction or a decimal, as in $\frac{2}{3}x - 5 = 8$ or $7 + 0.2r = 3$, the multiplication principle is often used first to **clear fractions** or **decimals** (pp. 93–94). The same principles and strategies are used when **solving formulas** for a specified letter (p. 98).

Inequalities, like $2x + 5 > -1$ and $9 \leq -7t + 3$, are solved in much the same manner as equations, using **the addition** and **multiplication principles for inequalities** (pp. 129, 131). When using the multiplication principle for inequalities, it is important to remember to reverse the direction of the inequality symbol whenever both sides are multiplied or divided by a negative number.

Addition principle (for inequalities):

For any real numbers a, b, and c,

$$a < b \quad \text{is equivalent to} \quad a + c < b + c;$$
$$a > b \quad \text{is equivalent to} \quad a + c > b + c.$$

Multiplication principle (for inequalities):

For any real numbers a and b and any *positive* number c,

$$a < b \quad \text{is equivalent to} \quad ac < bc;$$
$$a > b \quad \text{is equivalent to} \quad ac > bc.$$

For any real numbers a and b and any *negative* number c,

$$a < b \quad \text{is equivalent to} \quad ac > bc;$$
$$a > b \quad \text{is equivalent to} \quad ac < bc.$$

Similar statements hold for $\leq$ and $\geq$.

Because the solution of an inequality usually consists of too many numbers to list, **set-builder notation** and **graphs** are used to represent the **solution set** (p. 130).

Many real-world problems translate to equations or inequalities, and some involve **percent notation** (p. 105). Such problems can be solved using five steps (p. 115):

Five Steps for Problem Solving in Algebra
1. *Familiarize* yourself with the problem.
2. *Translate* to mathematical language. (This often means writing an equation.)
3. *Carry out* some mathematical manipulation. (This often means *solving* an equation.)
4. *Check* your possible answer in the original problem.
5. *State* the answer clearly.

2 Review Exercises

↪ *Concept Reinforcement   Classify each of the following as either true or false.*

1. $5x - 4 = 2x$ and $3x = 4$ are equivalent equations. [2.1]
2. $5 - 2t < 9$ and $t > 6$ are equivalent inequalities. [2.6]
3. Some equations have no solution. [2.1]
4. Consecutive odd integers are 2 units apart. [2.5]
5. For any number a, $a \le a$. [2.6]
6. The addition principle is always used before the multiplication principle. [2.2]
7. A 10% discount results in a sale price that is 90% of the original price. [2.4]
8. Often it is impossible to list all solutions of an inequality number by number. [2.6]

Solve. Label any contradictions or identities.

9. $x + 9 = -16$ [2.1]
10. $-8x = -56$ [2.1]
11. $-\dfrac{x}{5} = 13$ [2.1]
12. $n - 7 = -6$ [2.1]
13. $12x = -60$ [2.1]
14. $x - 0.1 = 1.01$ [2.1]
15. $-\frac{2}{3} + x = -\frac{1}{6}$ [2.1]
16. $\frac{4}{5}y = -\frac{3}{16}$ [2.1]
17. $5z + 3 = 41$ [2.2]
18. $5 - x = 13$ [2.2]

19. $5t + 9 = 3t - 1$ [2.2]
20. $7x - 6 = 25x$ [2.2]
21. $\frac{1}{4}x - \frac{5}{8} = \frac{3}{8}$ [2.2]
22. $14y = 23y - 17 - 10$ [2.2]
23. $0.22y - 0.6 = 0.12y + 3 - 0.8y$ [2.2]
24. $\frac{1}{4}x - \frac{1}{8}x = 3 - \frac{1}{16}x$ [2.2]
25. $3(x + 5) = 36$ [2.2]
26. $4(5x - 7) = -56$ [2.2]
27. $8(x - 2) = 5(x + 4)$ [2.2]
28. $-5x + 3(x + 8) = 16 - 2x$ [2.2]

Solve each formula for the given letter. [2.3]

29. $C = \pi d$, for d
30. $V = \dfrac{1}{3}Bh$, for B
31. $A = \dfrac{a + b}{2}$, for b
32. Find decimal notation: 0.9%. [2.4]
33. Find percent notation: $\frac{11}{25}$. [2.4]
34. What percent of 60 is 42? [2.4]
35. 42 is 30% of what number? [2.4]

Determine whether each number is a solution of $x \leq 4$.
[2.6]

36. -3 **37.** 7 **38.** 4

Graph on a number line. [2.6]

39. $5x - 6 < 2x + 3$ **40.** $-2 < x \leq 5$

41. $t > 0$

Solve. Write the answers in set-builder notation. [2.6]

42. $t + \frac{2}{3} \geq \frac{1}{6}$ **43.** $9x \geq 63$

44. $2 + 6y > 20$ **45.** $7 - 3y \geq 27 + 2y$

46. $3x + 5 < 2x - 6$ **47.** $-4y < 28$

48. $3 - 4x < 27$ **49.** $4 - 8x < 13 + 3x$

50. $13 \leq -\frac{2}{3}t + 5$ **51.** $7 \leq 1 - \frac{3}{4}x$

Solve.

52. A can of powdered infant formula makes 120 oz of formula. How many 6-oz bottles of formula will the can make? [2.5]

53. A 32-ft beam is cut into two pieces. One piece is 2 ft longer than the other. How long are the pieces? [2.5]

54. About 55% of all charitable contributions are made to religious organizations. In 1999, $69.3 billion was given to religious organizations. How much was given to charities in general? [2.4]
Source: *Statistical Abstract of the United States*, 2002

55. The sum of two consecutive odd integers is 116. Find the integers. [2.5]

56. The perimeter of a rectangle is 56 cm. The width is 6 cm less than the length. Find the width and the length. [2.5]

57. After a 25% reduction, a picnic table is on sale for $120. What was the regular price? [2.4]

58. In 2002, the average male with a bachelor's degree or above earned about $78,000. This was 66% more than the average woman with a comparable background. How much did the average woman with a comparable background earn? [2.4]
Source: *The New York Times Almanac* 2003

59. The measure of the second angle of a triangle is 50° more than that of the first. The measure of the third angle is 10° less than twice the first. Find the measures of the angles. [2.5]

60. Lisa has budgeted an average of $95 a month for entertainment. For the first five months of the year, she has spent $98, $89, $110, $85, and $83. How much can Lisa spend in the sixth month without exceeding her average budget? [2.7]

61. The length of a rectangular frame is 43 cm. For what widths would the perimeter be greater than 120 cm? [2.7]

62. How does the multiplication principle for equations differ from the multiplication principle for inequalities? [2.1], [2.6]

SYNTHESIS

63. Explain how checking the solutions of an equation differs from checking the solutions of an inequality. [2.1], [2.6]

64. Children (aged 2–17) with access to computers, video games, and television spend an average of 33 hr 36 min per week viewing a screen of some type. This represents 31% more time than that spent by children who view only television. How much time each week does the average child spend in front of a television if he or she lives in a home without a computer or video games? [2.4]
Source: Stanger, J. D., and N. Gridina, *Media in the Home* 1999; the fourth annual survey of parents and children. Philadelphia: Annenberg Public Policy Center, University of Pennsylvania, 1999.

65. The combined length of the Nile and Amazon Rivers is 13,113 km. If the Amazon were 233 km longer, it would be as long as the Nile. Find the length of each river. [2.5]
Source: *The New York Times Almanac* 2003

66. Consumer experts advise us never to pay the sticker price for a car. A rule of thumb is to pay the sticker price minus 20% of the sticker price, plus $200. A car is purchased for $15,080 using the rule. What was the sticker price? [2.4], [2.5]

Solve.

67. $2|n| + 4 = 50$ [1.4], [2.2]

68. $|3n| = 60$ [1.4], [2.1]

69. $y = 2a - ab + 3$, for a [2.3]

2 Chapter Test

Solve. Label any contradictions or identities.

1. $t + 7 = 16$

2. $t - 3 = 12$

3. $6x = -18$

4. $-\frac{4}{7}x = -28$

5. $3t + 7 = 2t - 5$

6. $\frac{1}{2}x - \frac{3}{5} = \frac{2}{5}$

7. $8 - y = 16$

8. $-\frac{2}{5} + x = -\frac{3}{4}$

9. $4(x + 2) = 36$

10. $9 - 3x = 6(x + 4)$

11. $\frac{5}{6}(3x + 1) = 20$

12. $3(2x - 8) = 6(x - 4)$

Solve. Write the answers in set-builder notation.

13. $x + 6 > 1$

14. $14x + 9 > 13x - 4$

15. $\frac{1}{3}x < \frac{7}{8}$

16. $-2y \geq 26$

17. $4y \leq -32$

18. $-5x \geq \frac{1}{4}$

19. $4 - 6x > 40$

20. $5 - 9x \geq 19 + 5x$

Solve each formula for the given letter.

21. $A = 2\pi rh$, for r

22. $w = \dfrac{P + l}{2}$, for l

23. Find decimal notation: 230%.

24. Find percent notation: 0.054.

25. What number is 32% of 50?

26. What percent of 75 is 33?

Graph on a number line.

27. $y < 4$

28. $-2 \leq x \leq 2$

Solve.

29. The perimeter of a rectangular calculator is 36 cm. The length is 4 cm greater than the width. Find the width and the length.

30. Kari is taking a 240-mi bicycle trip through Vermont. She has three times as many miles to go as she has already ridden. How many miles has she biked so far?

31. The perimeter of a triangle is 249 mm. If the sides are consecutive odd integers, find the length of each side.

32. By lowering the temperature of their electric hot-water heater from 140°F to 120°F, the Kellys' average electric bill dropped by 7% to $60.45. What was their electric bill before they lowered the temperature of their hot water?

33. *Van rentals.*   Atlas rents a cargo van at a daily rate of $44.95 plus $0.39 per mile. A business has budgeted $250 for a one-day van rental. What mileages will allow the business to stay within budget? (Round to the nearest tenth of a mile.)

SYNTHESIS

Solve.

34. $c = \dfrac{2cd}{a - d}$, for d

35. $3|w| - 8 = 37$

36. A concert promoter had a certain number of tickets to give away. Five people got the tickets. The first got one third of the tickets, the second got one fourth of the tickets, and the third got one fifth of the tickets. The fourth person got eight tickets, and there were five tickets left for the fifth person. Find the total number of tickets given away.

3

Introduction to Graphing

AN APPLICATION

The plans for a skateboard "Fun Box" call for a ramp with a rise of 61 cm over a run of 167.6 cm (*Source*: www.heckler.com). Find the slope or grade of the ramp.

This problem appears as Exercise 83 in Section 3.5.

David M. Wood
CUSTOM SKATE-PARK DEVELOPER
Burlington, Vermont

I use geometry primarily when designing a skate park. Since all skate ramps are designed on an 8-ft radius, I use measurements that are multiples of 8. I make use of the concept of slope as well to design a bank ramp.

W*e now begin our study of graphing. First we will examine graphs as they commonly appear in newspapers or magazines and develop some terminology. Following that, we will graph certain equations and study the connection between rate and slope. We will also learn how graphs can be used as a problem-solving tool in many applications.*

Our work in this chapter centers on solving equations that contain two variables.

3.1 Reading Graphs, Plotting Points, and Scaling Graphs

Problem Solving with Bar, Circle, and Line Graphs • Points and Ordered Pairs • Numbering the Axes Appropriately

Today's print and electronic media make almost constant use of graphs. This can be attributed to the widespread availability of graphing software and the large quantity of information that a graph can display. In this section, we consider problem solving with bar graphs, line graphs, and circle graphs. Then we examine graphs that use a coordinate system.

Problem Solving with Bar, Circle, and Line Graphs

A *bar graph* is a convenient way of showing comparisons. In every bar graph, certain categories, such as body weight in the example below, are paired with certain numbers.

EXAMPLE 1

Driving under the influence.   A blood-alcohol level of 0.08% or higher makes driving illegal in the United States. This bar graph shows how many drinks a person of a certain weight would need to consume in 1 hr to achieve a blood-alcohol level of 0.08% (*Source*: Adapted from vsa.vassar.edu/~source/drugs/alcohol.html). Note that a 12-oz beer, a 5-oz glass of wine, or a cocktail containing $1\frac{1}{2}$ oz of distilled liquor all count as one drink.

Friends Don't Let Friends Drive Drunk

a) Approximately how many drinks would a 200-lb person have consumed if he or she had a blood-alcohol level of 0.08%?

b) What can be concluded about the weight of someone who can consume 3 drinks in an hour without reaching a blood-alcohol level of 0.08%?

Solution

a) We go to the top of the bar that is above the body weight 200 lb. Then we move horizontally from the top of the bar to the vertical scale listing numbers of drinks. It appears that approximately 4 drinks will give a 200-lb person a blood-alcohol level of 0.08%.

b) By moving up the vertical scale to the number 3, and then moving horizontally, we see that the first bar to reach a height of 3 corresponds to a weight of 140 lb. Thus an individual should weigh over 140 lb if he or she wishes to consume 3 drinks in an hour without exceeding a blood-alcohol level of 0.08%.

Circle graphs, or *pie charts*, are often used to show what percent of the whole each particular item in a group represents.

EXAMPLE 2   Color preference.   The circle graph below shows the favorite colors of Americans and the percentage that prefers each color (*Source*: Vitaly Komar and Alex Melamid: The Most Wanted Paintings on the Web). Because of rounding, the total is slightly less than 100%. There are approximately 292 million Americans, and three quarters of them live in cities. Assuming that color preference is unaffected by geographical location, how many urban Americans choose red as their favorite color?

What is Your Favorite Color?

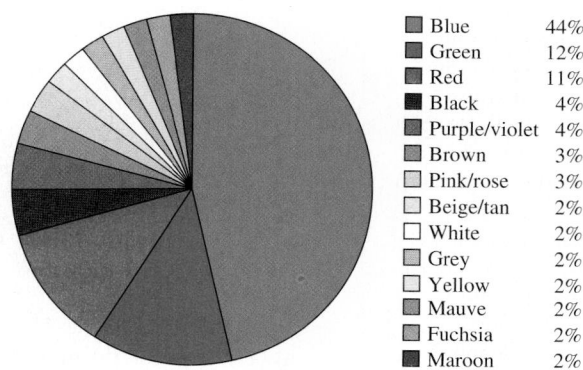

■ Blue	44%
■ Green	12%
■ Red	11%
■ Black	4%
■ Purple/violet	4%
■ Brown	3%
☐ Pink/rose	3%
☐ Beige/tan	2%
☐ White	2%
☐ Grey	2%
☐ Yellow	2%
■ Mauve	2%
■ Fuchsia	2%
■ Maroon	2%

Solution

1. **Familiarize.** The problem involves percents, so if we were unsure of how to solve percent problems, we might review Section 2.4. We are told that three quarters of all Americans live in cities. Since there are about 292 million Americans, this amounts to

$$\frac{3}{4} \cdot 292, \quad \text{or 219 million urban Americans.}$$

The chart indicates that 11% of the U.S. population prefers red. We let r = the number of urban Americans who select red as their favorite color. We assume that the color preferences of urban Americans are typical of all Americans.

2. **Translate.** We reword and translate the problem as follows:

Rewording: What   is   11%   of   219 million?

Translating: r =   11%   ·   219,000,000

3. **Carry out.** We solve the equation:

$$r = 0.11 \cdot 219{,}000{,}000 = 24{,}090{,}000.$$

4. **Check.** The check is left to the student.

5. **State.** About 24,090,000 urban Americans choose red as their favorite color.

EXAMPLE 3

Exercise and pulse rate.   The following line graph shows the relationship between a person's resting pulse rate and months of regular exercise.* Note that the symbol $\lessgtr$ is used to indicate that counting on the vertical axis begins at 50.

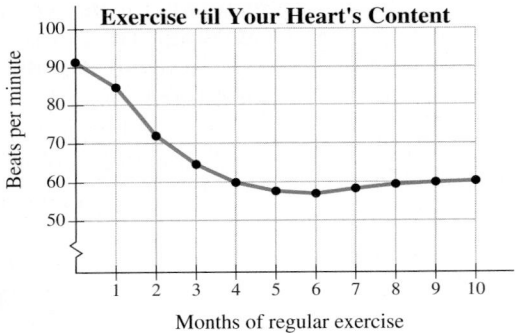

Exercise 'til Your Heart's Content

Beats per minute / Months of regular exercise

a) How many months of regular exercise are required to lower the pulse rate as much as possible?

b) How many months of regular exercise are needed to achieve a pulse rate of 65 beats per minute?

Solution

a) The lowest point on the graph occurs above the number 6. Thus, after 6 months of regular exercise, the pulse rate is lowered as much as possible.

b) To determine how many months of exercise are needed to lower a person's resting pulse rate to 65, we locate 65 midway between 60 and 70 on the vertical axis. From that location, we move right until the line is reached.

*Data from *Body Clock* by Dr. Martin Hughes (New York: Facts on File, Inc.), p. 60.

To set up a $[-100, 100, -5, 5]$ window, we press (WINDOW) and use the following settings. A scale of 10 for the x-axis is large enough for the marks on the x-axis to be distinct.

```
WINDOW
 Xmin=-100
 Xmax=100
 Xscl=10
 Ymin=-5
 Ymax=5
 Yscl=1
 Xres=1
```

When (GRAPH) is pressed, a graph extending from -100 to 100 along the x-axis (counted by 10's) and from -5 to 5 along the y-axis appears. When the arrow keys are pressed, a cursor can be moved, its coordinates appearing at the bottom of the window.

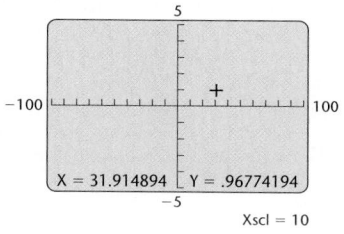

Set up the following viewing windows, choosing an appropriate scale for each axis. Then move the cursor and practice reading coordinates.

1. $[-10, 10, -10, 10]$
2. $[-5, 5, 0, 100]$
3. $[-1, 1, -0.1, 0.1]$

At that point, we move down to the horizontal scale and read the number of months required, as shown.

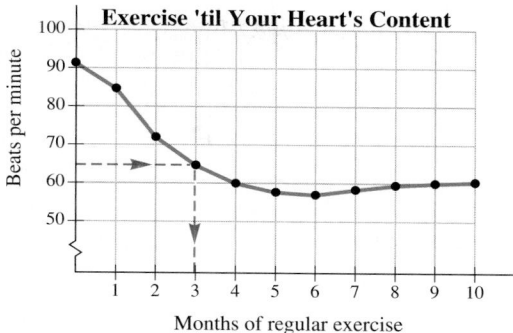

The pulse rate is 65 beats per minute after 3 months of regular exercise.

Points and Ordered Pairs

The line graph in Example 3 contains a collection of points. Each point pairs up a number of months of exercise with a pulse rate. To create such a graph, we **graph**, or **plot**, pairs of numbers on a plane. This is done using two perpendicular number lines called **axes** (pronounced "ak-sēz"; singular, **axis**). The point at which the axes cross is called the **origin**. Arrows on the axes indicate the positive directions.

Consider the pair $(3, 4)$. The numbers in such a pair are called **coordinates**. The **first coordinate** in this case is 3 and the **second coordinate** is 4. To plot $(3, 4)$, we start at the origin, move horizontally to the 3, move up vertically 4 units, and then make a "dot." Thus, $(3, 4)$ is located above 3 on the first axis and to the right of 4 on the second axis.

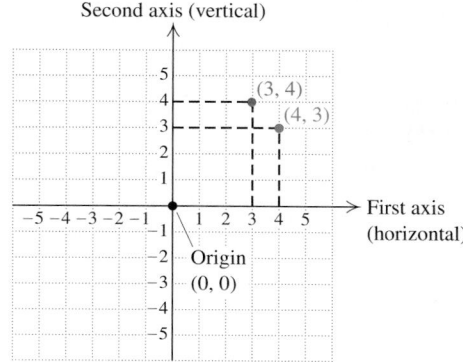

The point $(4, 3)$ is also plotted in the figure above. Note that $(3, 4)$ and $(4, 3)$ are different points. For this reason, coordinate pairs are called **ordered pairs**—the order in which the numbers appear is important.

EXAMPLE 4   Plot the point $(-3, 4)$.

Solution   The first number, -3, is negative. Starting at the origin, we move 3 units in the negative horizontal direction (3 units to the left). The second number, 4, is positive, so we move 4 units in the positive vertical direction (up). The point $(-3, 4)$ is above -3 on the first axis and to the left of 4 on the second axis.

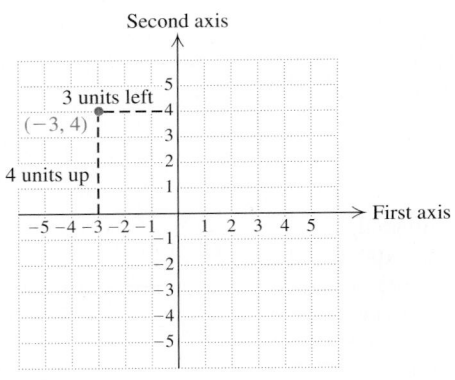

To find the coordinates of a point, we see how far to the right or left of the origin the point is and how far above or below the origin it is. Note that the coordinates of the origin itself are $(0, 0)$.

EXAMPLE 5   Find the coordinates of points *A, B, C, D, E, F,* and *G*.

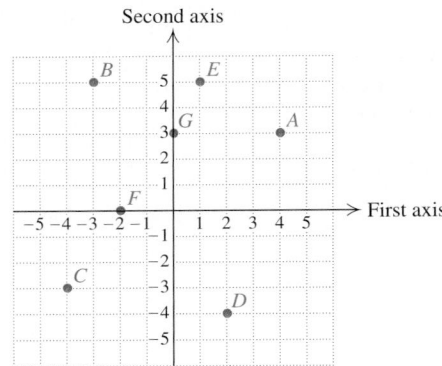

Solution   Point *A* is 4 units to the right of the origin and 3 units above the origin. Its coordinates are $(4, 3)$. The coordinates of the other points are as follows:

B: $(-3, 5)$;   *C*: $(-4, -3)$;   *D*: $(2, -4)$;

E: $(1, 5)$;   *F*: $(-2, 0)$;   *G*: $(0, 3)$.

Numbering the Axes Appropriately

Often it is necessary to graph a range of *x*-values and/or *y*-values that is too large to be displayed if each square of the grid is one unit wide and one unit high. For example, if we wish to plot the points $(-34, 450)$, $(48, 95)$, and $(10, -200)$ using a grid that is 10 squares wide and 10 squares high, we must make sure that *x*-values range, at least, from -34 to 48 and *y*-values extend, at least, from -200 to 450.

EXAMPLE 6 Use a grid 10 squares wide and 10 squares high to plot $(-34, 450)$, $(48, 95)$, and $(10, -200)$.

Solution   Since *x*-values vary from a low of -34 to a high of 48, the 10 horizontal squares must span $48 - (-34)$, or 82 units. Because 82 is not a multiple of 10, we round *up* to the next multiple of 10, which is 90. Dividing 90 by 10, we note that if each square is 9 units wide (has a *scale* of 9), we could count backward from 0 four squares and forward six squares to include *x*-values from -36 to 54. However, since it is more convenient to count by 10's, we will instead count backward to -40 and forward to 60.

This is how we will arrange the *y*-axis.

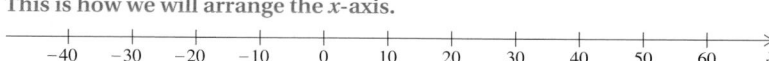

This is how we will arrange the *x*-axis.

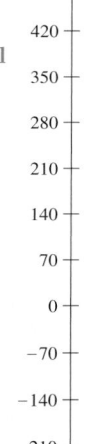

There is more than one correct way to cover the values from -34 to 48 using 10 increments. For instance, we could have counted from -60 to 90, using a scale of 15. In general, we try to use the smallest range and scale that will cover the given coordinates. Scales that are multiples of 2, 5, or 10 are especially convenient. It is essential that the numbering always starts from the origin.

Since we must be able to show *y*-values from -200 to 450, the 10 vertical squares must span $450 - (-200)$, or 650 units. For convenience, we round 650 *up* to 700 and then divide by 10: $700 \div 10 = 70$. Using 70 as the scale, we count *down* from 0 until we pass -200 and *up* from 0 until we pass 450, as shown at left.

Next, we combine our work with the *x*-values and the *y*-values to draw a graph in which the *x*-axis extends from -40 to 60 with a scale of 10 and the *y*-axis extends from -210 to 490 with a scale of 70. To correctly locate the axes on the grid, the two 0's must coincide where the axes cross. Finally, once the graph has been numbered, we plot the points as shown below.

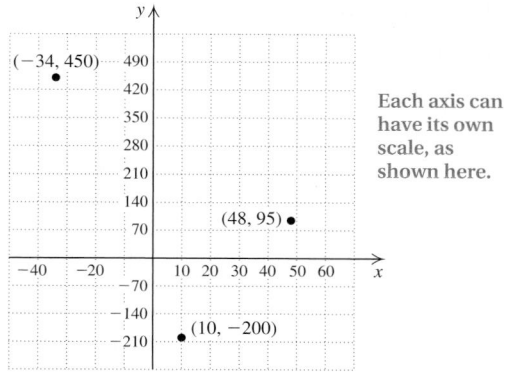

Each axis can have its own scale, as shown here.

The horizontal and vertical axes divide the plane into four regions, or **quadrants**, as indicated by Roman numerals in the following figure. In region I (the *first quadrant*), both coordinates of any point are positive. In region II (the *second quadrant*), the first coordinate is negative and the second is positive. In region III (the *third quadrant*), both coordinates are negative. In region IV (the *fourth quadrant*), the first coordinate is positive and the second is negative.

Note that the point $(-4, 5)$ is in the second quadrant and the point $(5, -5)$ is in the fourth quadrant. The points $(3, 0)$ and $(0, 1)$ are on the axes and are not considered to be in any quadrant.

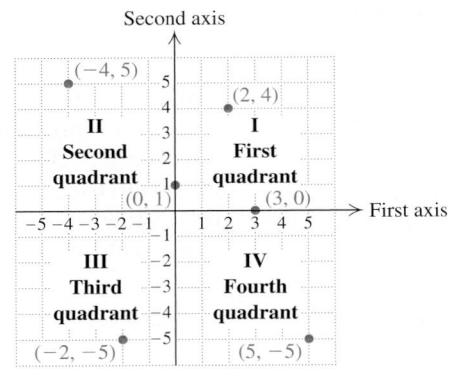

Exercise Set

3.1

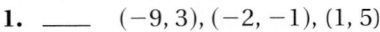 *Concept Reinforcement   In each of Exercises 1–4, match the set of coordinates with the graph from the column on the right that would be the best for plotting the points.*

1. _____ $(-9, 3), (-2, -1), (1, 5)$

2. _____ $(-2, -1), (1, 5), (7, 3)$

3. _____ $(-2, -9), (2, 1), (4, -6)$

4. _____ $(-2, -1), (-9, 3), (-4, -6)$

a)
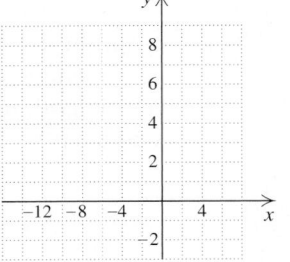

b)
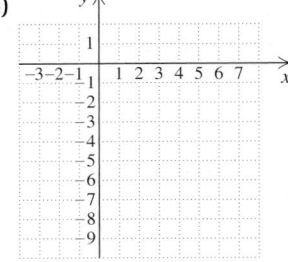

c)
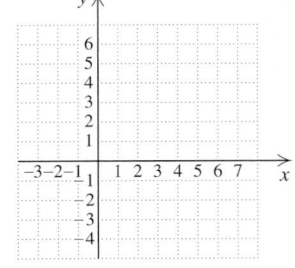

d)
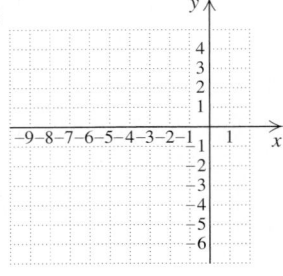

Blood alcohol level. Use the bar graph in Example 1 to answer Exercises 5–8.

5. Approximately how many drinks would a 100-lb person have consumed in 1 hr to reach a blood-alcohol level of 0.08%?

6. Approximately how many drinks would a 160-lb person have consumed in 1 hr to reach a blood-alcohol level of 0.08%?

7. What can you conclude about the weight of someone who has consumed 4 drinks in 1 hr without reaching a blood-alcohol level of 0.08%?

8. What can you conclude about the weight of someone who has consumed 5 drinks in 1 hr without reaching a blood-alcohol level of 0.08%?

Favorite color. Use the information in Example 2 to answer Exercises 9–12. Assume that color preference is unaffected by geographical location or age.

9. About one third of all Americans live in the South. How many Southerners choose brown as their favorite color?

10. About 50% of all Americans are at least 35 years old. How many Americans who are 35 or older choose purple/violet as their favorite color?

11. About one eighth of all Americans are senior citizens. How many senior citizens choose black as their favorite color?

12. About one fourth of all Americans are minors (under 18 years old). How many minors choose yellow as their favorite color?

Sorting solid waste. Use the following pie chart to answer Exercises 13–16.

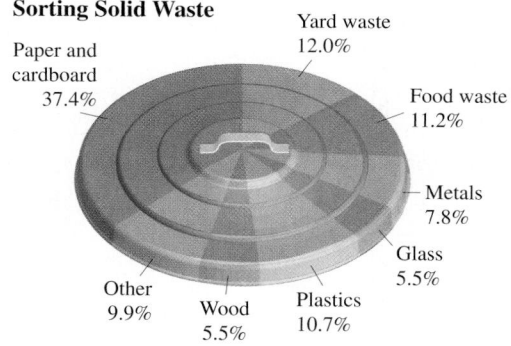

Sorting Solid Waste

Paper and cardboard 37.4%
Yard waste 12.0%
Food waste 11.2%
Metals 7.8%
Glass 5.5%
Plastics 10.7%
Wood 5.5%
Other 9.9%

Source: Statistical Abstract of the United States, 2002, based on data from 2000

13. In 2000, Americans generated 231.9 million tons of waste. How much of the waste was plastic?

14. In 2000, the average American generated 4.5 lb of waste per day. How much of that was paper and cardboard?

15. Americans are recycling about 22.7% of all glass that is in the waste stream. How much glass did Americans recycle in 2000? (See Exercise 13.)

16. Americans are recycling about 5.3% of all plastic waste. How much plastic did the average American recycle per day in 2000? (Use the information in Exercise 14.)

Cell-phone Internet access. The graph below shows estimates of the number of cell phones with access to the Internet.
Source: Forrester Research

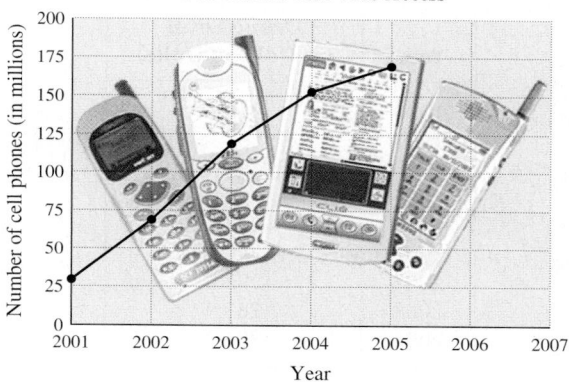

Cell Phones with Web Access

17. Approximately how many cell phones had Internet access in 2003?

18. Approximately how many cell phones had Internet access in 2002?

19. In what year did approximately 150 million cell phones have Internet access?

20. In what year did approximately 29 million cell phones have Internet access?

Plot each group of points.

21. $(1, 2), (-2, 3), (4, -1), (-5, -3), (4, 0), (0, -2)$

22. $(-2, -4), (4, -3), (5, 4), (-1, 0), (-4, 4), (0, 5)$

23. $(4, 4), (-2, 4), (5, -3), (-5, -5), (0, 4), (0, -4),$
$(3, 0), (-4, 0)$

24. $(2, 5), (-1, 3), (3, -2), (-2, -4), (0, 4), (0, -5),$
$(5, 0), (-5, 0)$

25. *Cell phones.* Listed below are estimates of the number of cell phones with or without access to the Internet. Make a line graph of the data.

Year	Number of Cell Phones (in millions)
2001	119.8
2002	135.3
2003	150.2
2004	163.8
2005	176.9

Source: Forrester Research

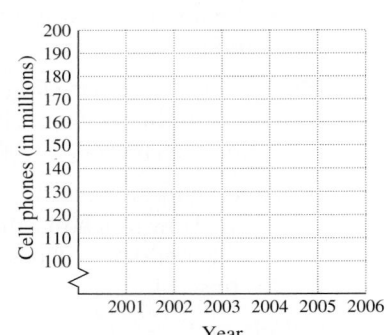

26. *Ozone layer.* Make a line graph of the data, listing years on the horizontal scale.

Year	Ozone Level (in Dobson Units)
1996	290
1997	293
1998	294
1999	293
2000	292.1
2001	290.4
2002	292.6
2003	287.9
2004	284.3

Source: johnstonsarchive.net

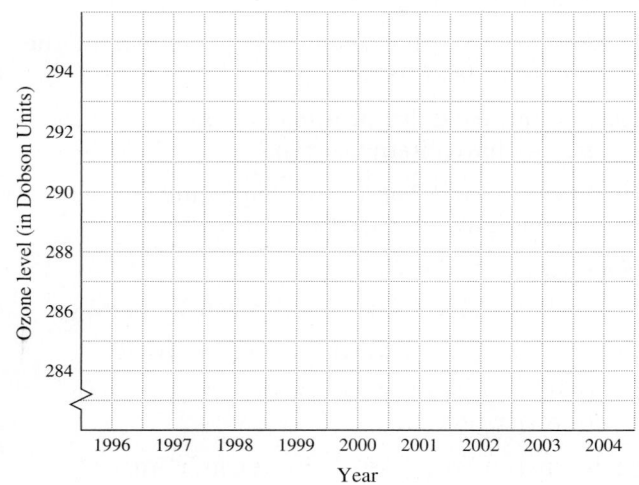

In Exercises 27–30, find the coordinates of points A, B, C, D, and E.

27.

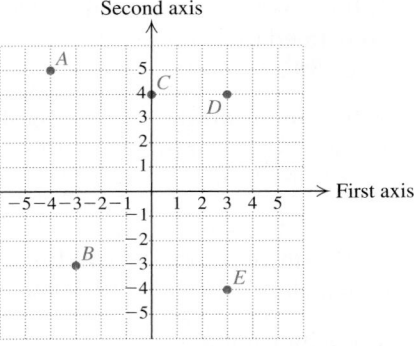

28.

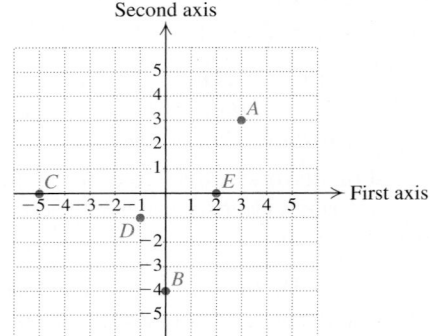

29.

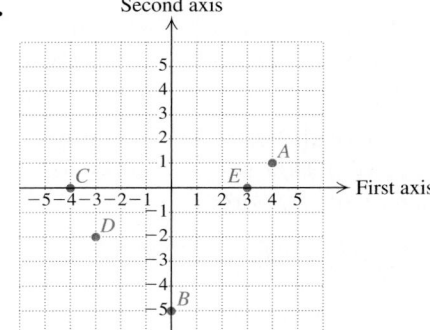

30.

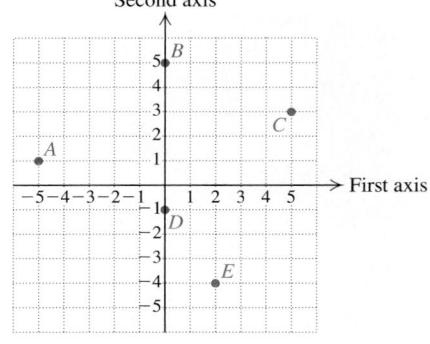

In Exercises 31–40, use a grid 10 squares wide and 10 squares high to plot the given coordinates. Choose your scale carefully. Scales may vary.

31. $(-75, 5), (-18, -2), (9, -4)$

32. $(-13, 3), (48, -1), (62, -4)$

33. $(-1, 83), (-5, -14), (5, 37)$

34. $(2, -79), (4, -25), (-4, 12)$

35. $(-10, -4), (-16, 7), (3, 15)$

36. $(5, -16), (-7, -4), (12, 3)$

37. $(-100, -5), (350, 20), (800, 37)$

38. $(750, -8), (-150, 17), (400, 32)$

39. $(-83, 491), (-124, -95), (54, -238)$

40. $(738, -89), (-49, -6), (-165, 53)$

In which quadrant is each point located?

41. $(7, -2)$ **42.** $(-1, -4)$ **43.** $(-4, -3)$

44. $(1, -5)$ **45.** $(2, 1)$ **46.** $(-4, 6)$

47. $(-4.9, 8.3)$ **48.** $(7.5, 2.9)$

49. In which quadrants are the first coordinates positive?

50. In which quadrants are the second coordinates negative?

51. In which quadrants do both coordinates have the same sign?

52. In which quadrants do the first and second coordinates have opposite signs?

53. The following graph was included in a mailing sent by Agway® to their oil customers in 2000. What information is missing from the graph and why is the graph misleading?

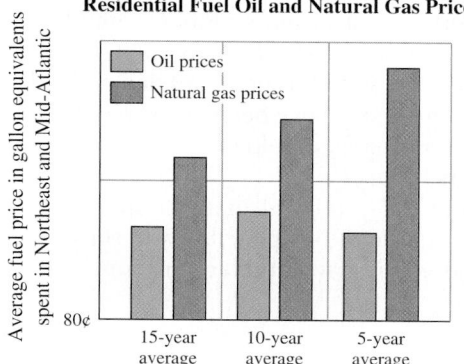

Source: Energy Research Center, Inc. *3/1/99–2/29/00

54. What do all of the points on the vertical axis of a graph have in common?

SKILL MAINTENANCE

Simplify. [1.8]

55. $4 \cdot 3 - 6 \cdot 5$ **56.** $5(-2) + 3(-7)$

57. $-\frac{1}{2}(-6) + 3$ **58.** $-\frac{2}{3}(-12) - 7$

Solve for y. [2.3]

59. $3x - 2y = 6$ **60.** $7x - 4y = 14$

SYNTHESIS

61. Describe what the result would be if the first and second coordinates of every point in the following graph of an arrow were interchanged.

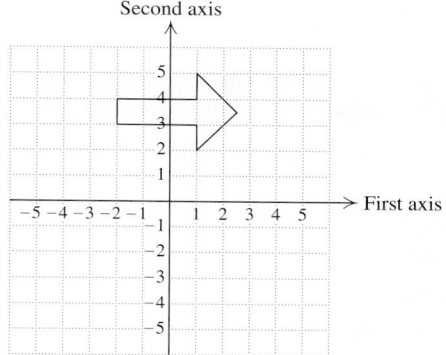

62. What advantage(s) does the use of a line graph have over that of a bar graph?

63. In which quadrant(s) could a point be located if its coordinates are opposites of each other?

64. In which quadrant(s) could a point be located if its coordinates are reciprocals of each other?

65. The points $(-1, 1), (4, 1),$ and $(4, -5)$ are three vertices of a rectangle. Find the coordinates of the fourth vertex.

66. The pairs $(-2, -3), (-1, 2),$ and $(4, -3)$ can serve as three (of four) vertices for three different parallelograms. Find the fourth vertex of each parallelogram.

67. Graph eight points such that the sum of the coordinates in each pair is 7. Answers may vary.

68. Find the perimeter of a rectangle if three of its vertices are $(5, -2)$, $(-3, -2)$, and $(-3, 3)$.

69. Find the area of a triangle whose vertices have coordinates $(0, 9)$, $(0, -4)$, and $(5, -4)$.

Coordinates on the globe. *Coordinates can also be used to describe the location on a sphere:* 0° *latitude is the equator and* 0° *longitude is a line from the North Pole to the South Pole through France and Algeria. In the figure*

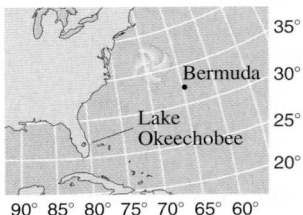

shown here, hurricane Clara is at a point about 260 mi northwest of Bermuda near latitude 36.0° *North, longitude* 69.0° *West.*

70. Approximate the latitude and the longitude of Bermuda.

71. Approximate the latitude and the longitude of Lake Okeechobee.

72. In the *Star Trek* science-fiction series, a three-dimensional coordinate system is used to locate objects in space. If the center of a planet is used as the origin, how many "quadrants" will exist? Why? If possible, sketch a three-dimensional coordinate system and label each "quadrant."

73. The graph accompanying Example 3 flattens out. Why do you think this occurs?

CORNER

COLLABORATIVE

You Sank My Battleship!

Focus: Graphing points
Time: 15–25 minutes
Group size: 3–5
Materials: Graph paper

In the game Battleship®, a player places a miniature ship on a grid that only that player can see. An opponent guesses at coordinates that might "hit" the "hidden" ship. The following activity is similar to this game.

ACTIVITY

1. Using only integers from -10 to 10 (inclusive), one group member should secretly record the coordinates of a point on a slip of paper. (This point is the hidden "battleship.")

2. The other group members can then ask up to 10 "yes/no" questions in an effort to determine the coordinates of the secret point. Be sure to phrase each question mathematically (for example, "Is the x-coordinate negative?")

3. The group member who selected the point should answer each question. On the basis of the answer given, another group member should cross out the points no longer under consideration. All group members should check that this is done correctly.

4. If the hidden point has not been determined after 10 questions have been answered, the secret coordinates should be revealed to all group members.

5. Repeat parts (1)–(4) until each group member has had the opportunity to select the hidden point and answer questions.

3.2 | Graphing Linear Equations

Solutions of Equations   •   Graphing Linear Equations   •
Applications

We have seen how bar, line, and circle graphs can represent information. Now we begin to learn how graphs can be used to represent solutions of equations.

Solutions of Equations

When an equation contains two variables, solutions must be ordered pairs in which each number in the pair replaces a letter in the equation. Unless stated otherwise, the first number in each pair replaces the variable that occurs first alphabetically.

EXAMPLE 1   Determine whether each of the following pairs is a solution of $4b - 3a = 22$: **(a)** $(2, 7)$; **(b)** $(1, 6)$.

Solution

a) We substitute 2 for a and 7 for b (alphabetical order of variables):

$$\frac{4b - 3a = 22}{\begin{array}{c|c} 4 \cdot 7 - 3 \cdot 2 & 22 \\ 28 - 6 & \end{array}}$$
$$22 \stackrel{?}{=} 22 \quad \text{TRUE}$$

Since $22 = 22$ is *true*, the pair $(2, 7)$ *is* a solution.

b) In this case, we replace a with 1 and b with 6:

$$\frac{4b - 3a = 22}{\begin{array}{c|c} 4 \cdot 6 - 3 \cdot 1 & 22 \\ 24 - 3 & \end{array}}$$
$$21 \stackrel{?}{=} 22 \quad \text{FALSE}$$

Since $21 = 22$ is *false*, the pair $(1, 6)$ is *not* a solution.

EXAMPLE 2   Show that the pairs $(3, 7)$, $(0, 1)$, and $(-3, -5)$ are solutions of $y = 2x + 1$. Then graph the three points to determine another pair that is a solution.

Solution   To show that a pair is a solution, we substitute, replacing x with the first coordinate and y with the second coordinate of each pair:

$$\frac{y = 2x + 1}{\begin{array}{c|c} 7 & 2 \cdot 3 + 1 \\ & 6 + 1 \end{array}}$$
$$7 \stackrel{?}{=} 7 \quad \text{TRUE}$$

$$\frac{y = 2x + 1}{\begin{array}{c|c} 1 & 2 \cdot 0 + 1 \\ & 0 + 1 \end{array}}$$
$$1 \stackrel{?}{=} 1 \quad \text{TRUE}$$

$$\frac{y = 2x + 1}{\begin{array}{c|c} -5 & 2(-3) + 1 \\ & -6 + 1 \end{array}}$$
$$-5 \stackrel{?}{=} -5 \quad \text{TRUE}$$

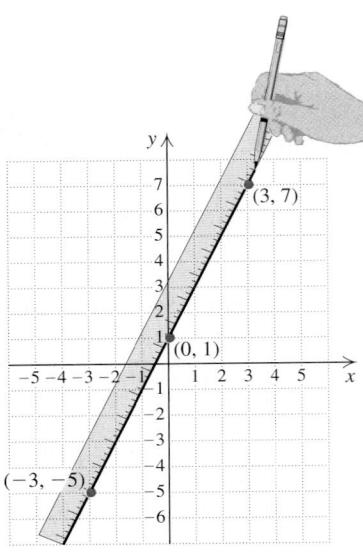

In each of the three cases, the substitution results in a true equation. Thus the pairs (3, 7), (0, 1), and (−3, −5) are all solutions. We graph them at left, labeling the "first" axis x and the "second" axis y. Note that the three points appear to "line up." Will other points that line up with these points also represent solutions of $y = 2x + 1$? To find out, we use a ruler and draw a line passing through (−3, −5), (0, 1), and (3, 7).

The line appears to pass through (2, 5). Let's check if this pair is a solution of $y = 2x + 1$:

$$
\begin{array}{c|c}
\multicolumn{2}{c}{y = 2x + 1} \\
\hline
5 & 2 \cdot 2 + 1 \\
 & 4 + 1 \\
5 & \overset{?}{=} 5 \qquad \text{TRUE}
\end{array}
$$

We see that (2, 5) *is* a solution. You should perform a similar check for at least one other point that appears to be on the line.

Example 2 leads us to suspect that *any* point on the line passing through (3, 7), (0, 1), and (−3, −5) represents a solution of $y = 2x + 1$. In fact, every solution of $y = 2x + 1$ is represented by a point on this line and every point on this line represents a solution. The line is called the **graph** of the equation.

Graphing Linear Equations

Equations like $y = 2x + 1$ or $4b - 3a = 22$ are said to be **linear** because the graph of each equation is a line. In general, any equation that can be written in the form $y = mx + b$ or $Ax + By = C$ (where m, b, A, B, and C are constants and A and B are not both 0) is linear. An equation of the form $Ax + By = C$ is said to be written in **standard form**.

To *graph* an equation is to make a drawing that represents its solutions. Linear equations can be graphed as follows.

To Graph a Linear Equation

1. Select a value for one coordinate and calculate the corresponding value of the other coordinate. Form an ordered pair. This pair is one solution of the equation.
2. Repeat step (1) to find a second ordered pair. A third ordered pair can be used as a check.
3. Plot the ordered pairs and draw a straight line passing through the points. The line represents all solutions of the equation.

EXAMPLE 3   Graph: $y = -3x + 1$.

Solution   We select a convenient value for x, compute y, and form an ordered pair. Then we repeat the process for other choices of x.

If $x = 2$,   then $y = -3 \cdot 2 + 1 = -5$,   and (2, −5) is a solution.

If $x = 0$,   then $y = -3 \cdot 0 + 1 = 1$,   and (0, 1) is a solution.

If $x = -1$,   then $y = -3(-1) + 1 = 4$,   and (−1, 4) is a solution.

Results are often listed in a table, as shown below. The points corresponding to each pair are then plotted.

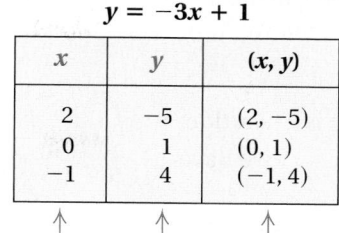

$$y = -3x + 1$$

x	y	(x, y)
2	−5	(2, −5)
0	1	(0, 1)
−1	4	(−1, 4)

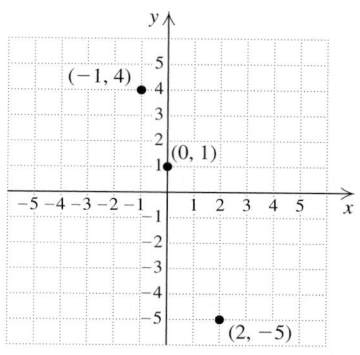

(1) Choose x.
(2) Compute y.
(3) Form the pair (x, y).
(4) Plot the points.

Note that all three points line up. If they didn't, we would know that we had made a mistake, because the equation is linear. When only two points are plotted, an error is more difficult to detect.

Finally, we use a ruler or other straightedge to draw a line. Every point on the line represents a solution of $y = -3x + 1$.

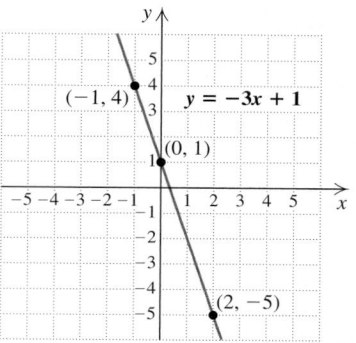

EXAMPLE 4   Graph: $y = 2x - 3$.

Solution   We select some convenient x-values and compute y-values.

If $x = 0$, then $y = 2 \cdot 0 - 3 = -3$, and $(0, -3)$ is a solution.
If $x = 1$, then $y = 2 \cdot 1 - 3 = -1$, and $(1, -1)$ is a solution.
If $x = 4$, then $y = 2 \cdot 4 - 3 = 5$, and $(4, 5)$ is a solution.

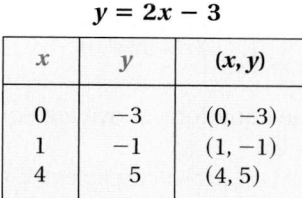

$$y = 2x - 3$$

x	y	(x, y)
0	−3	(0, −3)
1	−1	(1, −1)
4	5	(4, 5)

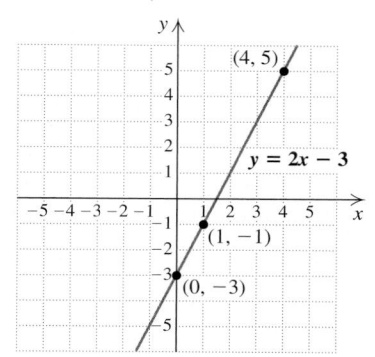

EXAMPLE 5   Graph: $4x + 2y = 12$.

Solution   To form ordered pairs, we can replace either variable with a number and then calculate the other coordinate:

If $y = 0$, we have $4x + 2 \cdot 0 = 12$
$$4x = 12$$
$$x = 3,$$

so $(3, 0)$ is a solution.

If $x = 0$, we have $4 \cdot 0 + 2y = 12$
$$2y = 12$$
$$y = 6,$$

so $(0, 6)$ is a solution.

If $y = 2$, we have $4x + 2 \cdot 2 = 12$
$$4x + 4 = 12$$
$$4x = 8$$
$$x = 2,$$

so $(2, 2)$ is a solution.

$4x + 2y = 12$

x	y	(x, y)
3	0	$(3, 0)$
0	6	$(0, 6)$
2	2	$(2, 2)$

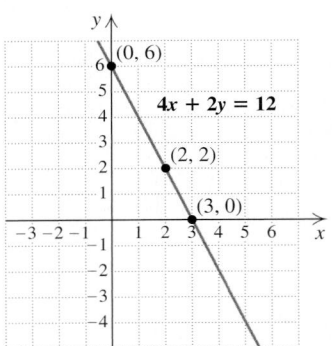

Note that in Examples 3 and 4 the variable y is isolated on one side of the equation. This generally simplifies calculations, so it is important to be able to solve for y before graphing.

EXAMPLE 6   Graph $3y = 2x$ by first solving for y.

Solution   To isolate y, we divide both sides by 3, or multiply both sides by $\frac{1}{3}$:

$$3y = 2x$$
$$\tfrac{1}{3} \cdot 3y = \tfrac{1}{3} \cdot 2x \qquad \text{Using the multiplication principle to multiply both sides by } \tfrac{1}{3}$$
$$\left. \begin{aligned} 1y &= \tfrac{2}{3} \cdot x \\ y &= \tfrac{2}{3}x. \end{aligned} \right\} \qquad \text{Simplifying}$$

Because all the equations above are equivalent, we can use $y = \frac{2}{3}x$ to draw the graph of $3y = 2x$.

To graph $y = \frac{2}{3}x$, we can select x-values that are multiples of 3. This will allow us to avoid fractions when the corresponding y-values are computed.

$$\left. \begin{aligned} \text{If } x = 3, \quad &\text{then } y = \tfrac{2}{3} \cdot 3 = 2. \\ \text{If } x = -3, \quad &\text{then } y = \tfrac{2}{3}(-3) = -2. \\ \text{If } x = 6, \quad &\text{then } y = \tfrac{2}{3} \cdot 6 = 4. \end{aligned} \right\}$$
Note that when multiples of 3 are substituted for x, the y-coordinates are not fractions.

The following table lists these solutions. Next, we plot the points and see that they form a line. Finally, we draw and label the line.

$$3y = 2x, \text{ or } y = \frac{2}{3}x$$

x	y	(x, y)
3	2	(3, 2)
−3	−2	(−3, −2)
6	4	(6, 4)

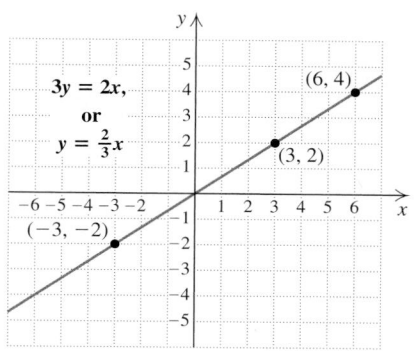

EXAMPLE 7

Graph $x + 5y = -10$ by first solving for y.

Solution   We have

$$x + 5y = -10$$
$$5y = -x - 10 \qquad \text{Adding } -x \text{ to both sides}$$
$$y = \tfrac{1}{5}(-x - 10) \qquad \text{Multiplying both sides by } \tfrac{1}{5}$$
$$y = -\tfrac{1}{5}x - 2. \qquad \text{Using the distributive law}$$

> **Caution!**   It is very important to multiply *both* $-x$ and -10 by $\frac{1}{5}$.

Student Notes _____

If you are troubled by calculations involving fractions, please review the manner in which fractions are added and subtracted. If necessary, ask your instructor for additional practice in this area.

Thus, $x + 5y = -10$ is equivalent to $y = -\frac{1}{5}x - 2$. If we choose x-values that are multiples of 5, we can avoid fractions when calculating the corresponding y-values.

If $x = 5,$ then $y = -\frac{1}{5} \cdot 5 - 2 = -1 - 2 = -3.$

If $x = 0,$ then $y = -\frac{1}{5} \cdot 0 - 2 = 0 - 2 = -2.$

If $x = -5,$ then $y = -\frac{1}{5}(-5) - 2 = 1 - 2 = -1.$

$$x + 5y = -10, \text{ or } y = -\frac{1}{5}x - 2$$

x	y	(x, y)
5	−3	(5, −3)
0	−2	(0, −2)
−5	−1	(−5, −1)

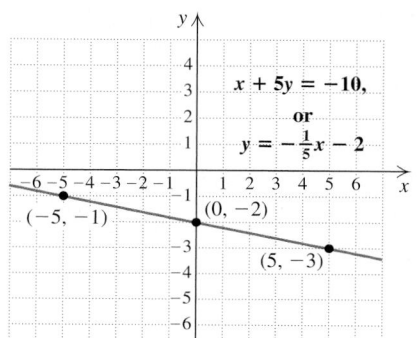

Applications

Linear equations appear in many real-life situations.

EXAMPLE 8   FedEx mailing costs.   The cost c, in dollars, of shipping a FedEx Priority Overnight package weighing 1 lb or more a distance of 1001 to 1400 mi is given by

$$c = 2.8w + 21.05,$$

where w is the package's weight in pounds (*Source*: Federal Express Corporation). Graph the equation and then use the graph to estimate the cost of shipping a $6\frac{1}{2}$-lb package.

Solution   We graph $c = 2.8w + 21.05$ by selecting values for w and then calculating the associated value c. In selecting values for w, note that, according to the problem, we must have $w \geq 1$.

If $w = 1$,   then $c = 2.8(1) + 21.05 = \$23.85$.

If $w = 4$,   then $c = 2.8(4) + 21.05 = \$32.25$.

If $w = 7$,   then $c = 2.8(7) + 21.05 = \$40.65$.

w	c
1	23.85
4	32.25
7	40.65

Because we are *selecting* values for w and *calculating* values for c, we represent w on the horizontal axis and c on the vertical axis. Counting by 2's horizontally and by 5's vertically will allow us to plot all three pairs, as shown on the left below.

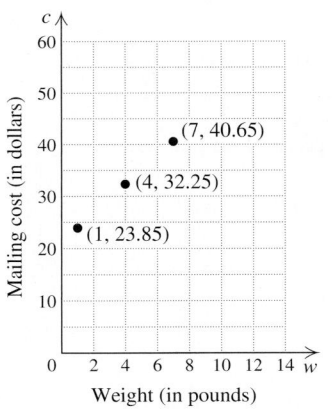

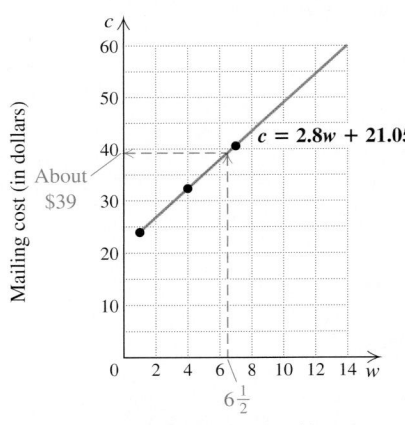

Since the three points line up, our calculations are probably correct. We draw a line, beginning at (1, 23.85). To estimate the cost of shipping a $6\frac{1}{2}$-lb package, we locate the point on the line that is above $6\frac{1}{2}$ and then find the value on the c-axis that corresponds to that point, as shown on the right above. The cost of shipping a $6\frac{1}{2}$-lb package is about \$39.

> **Caution!**   When the coordinates of a point are read from a graph, as in Example 8, values should not be considered exact.

Many equations in two variables have graphs that are not straight lines. Three such graphs are shown below. As before, each graph represents the solutions of the given equation. Graphing calculators are especially helpful when drawing these *nonlinear* graphs. Nonlinear graphs are studied in Chapter 11 and in more advanced courses.

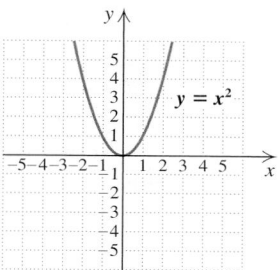
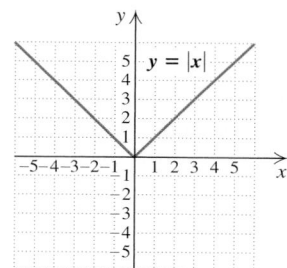
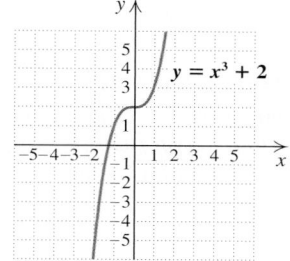

---

technology connection

Most graphing calculators require that y be alone on one side before the equation is entered. For example, to graph $5y + 4x = 13$, we would first solve for y. The student can check that solving for y yields the equation $y = -\frac{4}{5}x + \frac{13}{5}$.

We enter $-\frac{4}{5}x + \frac{13}{5}$ as Y1 and press ⟨GRAPH⟩. The standard viewing window $[-10, 10, -10, 10]$ results in the graph shown here.

Using a graphing calculator, graph each of the following. Select the "standard" $[-10, 10, -10, 10]$ window.

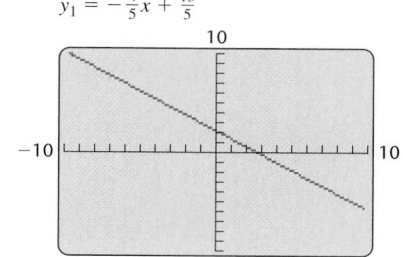

1. $y = -5x + 6.5$
2. $y = 3x + 4.5$
3. $7y - 4x = 22$
4. $5y + 11x = -20$
5. $2y - x^2 = 0$
6. $y + x^2 = 8$

Exercise Set

3.2

FOR EXTRA HELP

Student's
Solutions
Manual

Digital Video Tutor
CD 2
Videotape 3

AW Math
Tutor Center

MathXL Tutorials
on CD

MathXL

MyMathLab
MyMathLab

Determine whether each equation has the given ordered pair as a solution.

1. $y = 5x + 1$; $(0, 2)$

2. $y = 2x + 3$; $(0, 3)$

3. $3y + 2x = 12$; $(4, 2)$

4. $5x - 3y = 15$; $(0, 5)$

5. $4a - 3b = 11$; $(2, -1)$

6. $3q - 2p = -8$; $(1, -2)$

In Exercises 7–14, an equation and two ordered pairs are given. Show that each pair is a solution of the equation. Then graph the two pairs to determine another solution. Answers may vary.

7. $y = x + 3$; $(-1, 2)$, $(4, 7)$

8. $y = x - 2$; $(3, 1)$, $(-2, -4)$

9. $y = \frac{1}{2}x + 3$; $(4, 5)$, $(-2, 2)$

10. $y = \frac{1}{2}x - 1$; $(6, 2)$, $(0, -1)$

11. $y + 3x = 7$; $(2, 1)$, $(4, -5)$

12. $2y + x = 5$; $(-1, 3)$, $(7, -1)$

13. $4x - 2y = 10$; $(0, -5)$, $(4, 3)$

14. $6x - 3y = 3$; $(1, 1)$, $(-1, -3)$

Graph each equation.

15. $y = x + 1$

16. $y = x - 1$

17. $y = -x$

18. $y = x$

19. $y = \frac{1}{3}x$

20. $y = \frac{1}{2}x$

21. $y = x + 3$

22. $y = x + 2$

23. $y = 2x + 2$

24. $y = 3x - 2$

25. $y = \frac{1}{3}x - 4$

26. $y = \frac{1}{2}x + 1$

27. $x + y = 4$

28. $x + y = -5$

29. $y = \frac{5}{2}x + 3$

30. $y = \frac{5}{3}x - 2$

31. $x + 2y = -6$

32. $x + 2y = 8$

33. $y = -\frac{2}{3}x + 4$

34. $y = \frac{3}{2}x + 1$

35. $8x - 4y = 12$

36. $6x - 3y = 9$

37. $6y + 2x = 8$

38. $8y + 2x = -4$

Solve by graphing. Label all axes, and show where each solution is located on the graph.

39. *Health insurance.* The number of full-time workers in the United States without health insurance continues to grow. This number, n, can be approximated by

$$n = 0.9t + 19,$$

where n is in millions and t is the number of years since 2001. Graph the equation and use the graph to estimate the number of uninsured full-time workers in 2010.
Source: *The New York Times*, 9/30/03

40. *Value of a color copier.* The value of Dupliographic's color copier is given by

$$v = -0.68t + 3.4,$$

where v is the value, in thousands of dollars, t years from the date of purchase. Graph the equation and use the graph to estimate the value of the copier after $2\frac{1}{2}$ yr.

41. *Increasing life expectancy.* A smoker is 15 times more likely to die from lung cancer than a nonsmoker. An ex-smoker who stopped smoking t years ago is w times more likely to die from lung cancer than a nonsmoker, where

$$w = 15 - t.$$

Graph the equation and use the graph to estimate how much more likely it is for Sandy to die from lung cancer than Polly, if Polly never smoked and Sandy quit $2\frac{1}{2}$ years ago.

Source: Data from *Body Clock* by Dr. Martin Hughes, p. 60. New York: Facts on File, Inc.

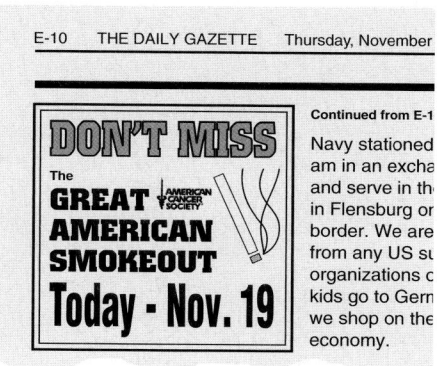

42. *Price of printing.*   The price p, in cents, of a photocopied and bound lab manual is given by $p = \frac{7}{2}n + 20$, where n is the number of pages in the manual. Graph the equation and use the graph to estimate the cost of a 25-page manual.

43. *Value of computer software.*   The value v of a shopkeeper's inventory software program, in hundreds of dollars, is given by $v = -\frac{3}{4}t + 6$, where t is the number of years since the shopkeeper first bought the program. Graph the equation and use the graph to estimate what the program is worth 4 yr after it was first purchased.

44. *Bottled water.*   The number of gallons of bottled water w consumed by an average American in one year is given by $w = 1.1t + 7.3$, where t is the number of years since 1990. Graph the equation and use the graph to predict the number of gallons consumed per person in 2008.

Source: Beverage Marketing Corporation

45. *Cost of college.*   The cost T, in hundreds of dollars, of tuition and fees at many community colleges can be approximated by $T = \frac{3}{4}c + 1$, where c is the number of credits for which a student registers. Graph the equation and use the graph to estimate the cost of tuition and

fees when a student registers for 4 three-credit courses.

46. *Cost of college.*   The cost C, in thousands of dollars, of a year at a private four-year college (all expenses) can be approximated by $C = \frac{6}{5}t + 21$, where t is the number of years since 1995. Graph the equation and use the graph to estimate the cost of a year at a private four-year college in 2009.

Source: Based on information in *Statistical Abstract of the United States*, 2002

47. *Record temperature drop.*   On January 22, 1943, the temperature T, in degrees Fahrenheit, in Spearfish, South Dakota, could be approximated by $T = -2m + 54$, where m is the number of minutes since 9:00 A.M. that morning. Graph the equation and use the graph to estimate the temperature at 9:15 A.M.

Source: Based on information from the National Oceanic and Atmospheric Administration

48. *Coffee consumption.*   The number of gallons of coffee n consumed each year by the average U.S. consumer can be approximated by $n = \frac{6}{5}d + 20$, where d is the number of years since 1995. Graph the equation and use the graph to estimate what the average coffee consumption was in 2004.

Source: Based on information in *Statistical Abstract of the United States*, 2002

49. The equations $3x + 4y = 8$ and $y = -\frac{3}{4}x + 2$ are equivalent. Which equation would be easier to graph and why?

50. Suppose that a linear equation is graphed by plotting three points and that the three points line up with each other. Does this *guarantee* that the equation is being correctly graphed? Why or why not?

SKILL MAINTENANCE

Solve and check. [2.2]

51. $5x + 3 \cdot 0 = 12$

52. $2x - 5 \cdot 0 = 9$

53. $7 \cdot 0 - 4y = 10$

Solve. [2.3]

54. $pq + p = w$, for p

55. $Ax + By = C$, for y

56. $A = \dfrac{T + Q}{2}$, for Q

SYNTHESIS

57. Janice consistently makes the mistake of plotting the *x*-coordinate of an ordered pair using the *y*-axis, and the *y*-coordinate using the *x*-axis. How will Janice's incorrect graph compare with the appropriate graph?

58. Explain how the graph in Example 8 can be used to determine when the cost of a shipment will reach $50.

59. *Bicycling.* Long Beach Island in New Jersey is a long, narrow, flat island. For exercise, Lauren routinely bikes to the northern tip of the island and back. Because of the steady wind, she uses one gear going north and another for her return. Lauren's bike has 14 gears and the sum of the two gears used on her ride is always 18. Write and graph an equation that represents the different pairings of gears that Lauren uses. Note that there are no fractional gears on a bicycle.

In Exercises 60–63, try to find an equation for the graph shown.

60.

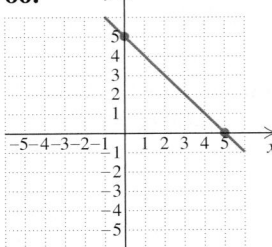

61.

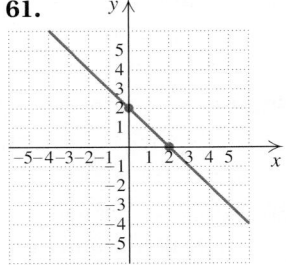

62.

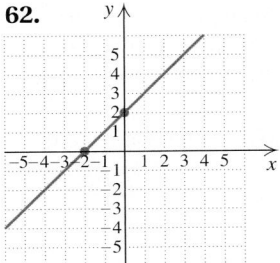

63.

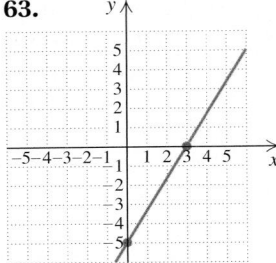

64. Translate to an equation:

    d dimes and *n* nickels total $1.75.

Then graph the equation and use the graph to determine three different combinations of dimes and nickels that total $1.75 (see also Exercise 77).

65. Translate to an equation:

    d $25 dinners and *l* $5 lunches total $225.

Then graph the equation and use the graph to determine three different combinations of lunches and dinners that total $225 (see also Exercise 77).

Use the suggested x-values $-3, -2, -1, 0, 1, 2,$ *and* 3 *to graph each equation.*

66. $y = |x|$

Aha! **67.** $y = -|x|$

Aha! **68.** $y = |x| - 2$

69. $y = -|x| + 2$

70. $y = |x| + 3$

 For Exercises 71–76, use a graphing calculator to graph the equation. Use a $[-10, 10, -10, 10]$ *window.*

71. $y = -2.8x + 3.5$

72. $y = 4.5x + 2.1$

73. $y = 2.8x - 3.5$

74. $y = -4.5x - 2.1$

75. $y = x^2 + 4x + 1$

76. $y = -x^2 + 4x - 7$

77. Study the graph of Exercise 64 or 65. Does *every* point on the graph represent a solution of the associated problem? Why or why not?

3.3   Graphing and Intercepts

Intercepts   •   Using Intercepts to Graph   •   Graphing
Horizontal or Vertical Lines

Unless a line is horizontal or vertical, it will cross both axes. Knowing where the
axes are crossed gives us another way of graphing linear equations.

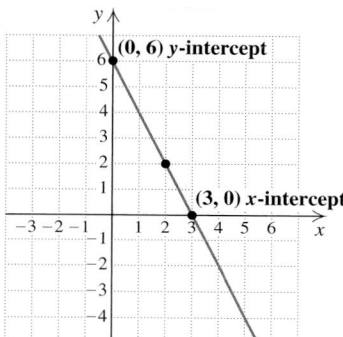

Intercepts

In Example 5 of Section 3.2, we graphed $4x + 2y = 12$ by plotting the points
$(3, 0)$, $(0, 6)$, and $(2, 2)$ and then drawing the line.

The point at which a graph crosses the y-axis is called the **y-intercept**.
In the figure at left, the y-intercept is $(0, 6)$. The x-coordinate of a y-intercept is
always 0.

The point at which a graph crosses the x-axis is called the **x-intercept**. In
the figure at left, the x-intercept is $(3, 0)$. The y-coordinate of an x-intercept is
always 0.

It is possible for the graph of a curve to have more than one y-intercept or
more than one x-intercept.

EXAMPLE 1   For the graph shown below, **(a)** give the coordinates of any x-intercepts and
(b) give the coordinates of any y-intercepts.

Study Skills _____

Create Your Own Glossary

Understanding the meaning of
mathematical terminology is
essential for success in any math
course. To assist with this, try
writing your own glossary of
important words toward the back
of your notebook. Often, just the
act of writing out a word's
definition can help you remember
what the word means.

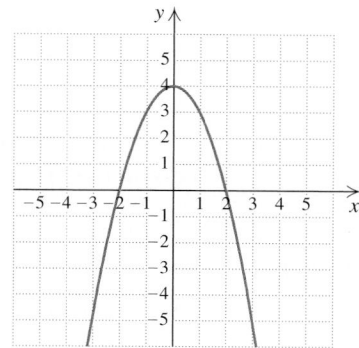

Solution

a) The x-intercepts are the points at which the graph crosses the x-axis. For the
graph shown, the x-intercepts are $(-2, 0)$ and $(2, 0)$.

b) The y-intercept is the point at which the graph crosses the y-axis. For the
graph shown, the y-intercept is $(0, 4)$.

Using Intercepts to Graph

It is important to know how to locate a graph's intercepts from the equation being graphed.

To Find Intercepts

To find the *y*-intercept(s) of an equation's graph, replace *x* with 0 and solve for *y*.

To find the *x*-intercept(s) of an equation's graph, replace *y* with 0 and solve for *x*.

EXAMPLE 2   Find the *y*-intercept and the *x*-intercept of the graph of $2x + 4y = 20$.

Solution   To find the *y*-intercept, we let $x = 0$ and solve for *y*:

$$2 \cdot 0 + 4y = 20 \qquad \text{Replacing } x \text{ with } 0$$
$$4y = 20$$
$$y = 5.$$

Thus the *y*-intercept is $(0, 5)$.
    To find the *x*-intercept, we let $y = 0$ and solve for *x*:

$$2x + 4 \cdot 0 = 20 \qquad \text{Replacing } y \text{ with } 0$$
$$2x = 20$$
$$x = 10.$$

Thus the *x*-intercept is $(10, 0)$.

Since two points are sufficient to graph a line, intercepts can be used to graph linear equations.

EXAMPLE 3   Graph $2x + 4y = 20$ using intercepts.

Solution   In Example 2, we showed that the *y*-intercept is $(0, 5)$ and the *x*-intercept is $(10, 0)$. Before drawing a line, we plot a third point as a check. We substitute any convenient value for *x* and solve for *y*.
    If we let $x = 5$, then

$$2 \cdot 5 + 4y = 20 \qquad \text{Substituting 5 for } x$$
$$10 + 4y = 20$$
$$4y = 10 \qquad \text{Subtracting 10 from both sides}$$
$$y = \tfrac{10}{4}, \text{ or } 2\tfrac{1}{2}. \qquad \text{Solving for } y$$

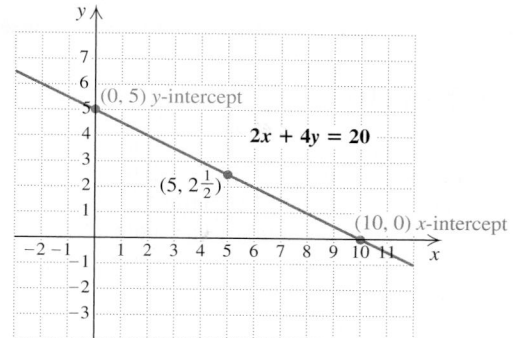

The point $\left(5, 2\frac{1}{2}\right)$ appears to line up with the intercepts, so our work is probably correct. To finish, we draw and label the line.

Note that when we solved for the y-intercept, we replaced x with 0 and simplified $2x + 4y = 20$ to $4y = 20$. Thus, to find the y-intercept, we can momentarily ignore the x-term and solve the remaining equation.

In a similar manner, when we solved for the x-intercept, we simplified $2x + 4y = 20$ to $2x = 20$. Thus, to find the x-intercept, we can momentarily ignore the y-term and then solve this remaining equation.

EXAMPLE 4   Graph $3x - 2y = 60$ using intercepts.

Solution   To find the y-intercept, we let $x = 0$. This amounts to temporarily ignoring the x-term and then solving:

$$-2y = 60 \qquad \text{For } x = 0, \text{ we have } 3 \cdot 0 - 2y, \text{ or simply } -2y.$$
$$y = -30.$$

The y-intercept is $(0, -30)$.

To find the x-intercept, we let $y = 0$. This amounts to temporarily disregarding the y-term and then solving:

$$3x = 60 \qquad \text{For } y = 0, \text{ we have } 3x - 2 \cdot 0, \text{ or simply } 3x.$$
$$x = 20.$$

The x-intercept is $(20, 0)$.

To find a third point, we replace x with 4 and solve for y:

$$3 \cdot 4 - 2y = 60 \qquad \text{Numbers other than 4 can be used for } x.$$
$$12 - 2y = 60$$
$$-2y = 48$$
$$y = -24. \qquad \text{This means that } (4, -24) \text{ is on the graph.}$$

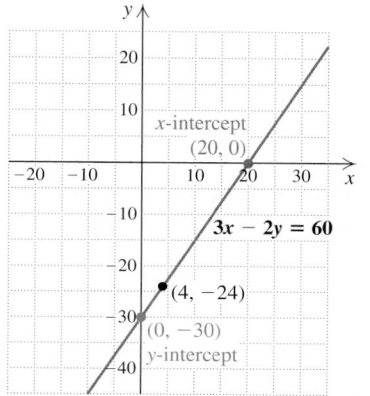

In order for us to graph all three points, the y-axis of our graph must go down to at least -30 and the x-axis must go up to at least 20. Using a scale of 5 units per square allows us to display both intercepts and $(4, -24)$, as well as the origin.

The point $(4, -24)$ appears to line up with the intercepts, so we draw and label the line, as shown at left.

 technology connection

When an equation has been entered into a graphing calculator, we may not be able to see both intercepts. For example, if $y = -0.8x + 17$ is graphed in the $[-10, 10, -10, 10]$ window, neither intercept is visible.

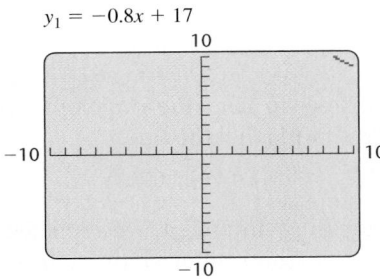

$y_1 = -0.8x + 17$

To better view the intercepts, we can change the window dimensions or we can zoom out. The ZOOM

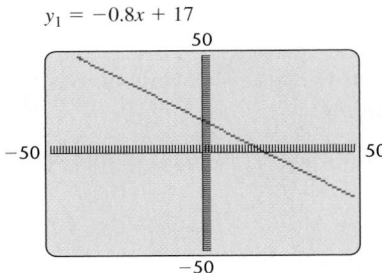

$y_1 = -0.8x + 17$

feature allows us to reduce or magnify a graph or a portion of a graph. Before zooming, the ZOOM *factors* must be set in the memory of the ZOOM key. If we zoom out with factors set at 5, both intercepts are visible but the axes are heavily drawn, as shown in the preceding figure.

This suggests that the *scales* of the axes should be changed. To do this, we use the WINDOW menu and set Xscl to 5 and Yscl to 5. The resulting graph has tick marks 5 units apart and clearly shows both intercepts. Other choices for Xscl and Yscl can also be made.

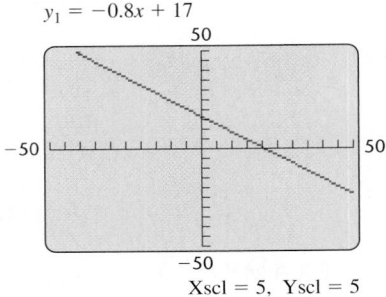

$y_1 = -0.8x + 17$

Xscl = 5, Yscl = 5

Graph each equation so that both intercepts can be easily viewed. Zoom or adjust the window settings so that tick marks can be clearly seen on both axes.

1. $y = -0.72x - 15$ **2.** $y - 2.13x = 27$
3. $5x + 6y = 84$ **4.** $2x - 7y = 150$
5. $19x - 17y = 200$ **6.** $6x + 5y = 159$

Graphing Horizontal or Vertical Lines

The equations graphed in Examples 3 and 4 are both in the form $Ax + By = C$. We have already stated that any equation in the form $Ax + By = C$ is linear, provided A and B are not both zero. What if A or B (but not both) is zero? We will find that when A is zero, there is no x-term and the graph is a horizontal line. We will also find that when B is zero, there is no y-term and the graph is a vertical line.

EXAMPLE 5   Graph: $y = 3$.

Solution   We can regard the equation $y = 3$ as $0 \cdot x + y = 3$. No matter what number we choose for x, we find that y must be 3 if the equation is to be solved. Consider the following table.

Student Notes _____

Many students draw horizontal lines when they should be drawing vertical lines and vice versa. To keep this straight, locate the correct number on the axis whose label is given and then draw a line perpendicular to that axis. Thus, to graph $x = 2$, we locate 2 on the x-axis and then draw a line perpendicular to that axis at that point. Note that the graph of $x = 2$ on a plane is a line, whereas the graph of $x = 2$ on a number line is a point.

$$y = 3$$

Choose any number for x. →

x	y	(x, y)
-2	3	$(-2, 3)$
0	3	$(0, 3)$
4	3	$(4, 3)$

y must be 3.

All pairs will have 3 as the y-coordinate.

When we plot the ordered pairs $(-2, 3)$, $(0, 3)$, and $(4, 3)$ and connect the points, we obtain a horizontal line. Any ordered pair of the form $(x, 3)$ is a solution, so the line is parallel to the x-axis with y-intercept $(0, 3)$.

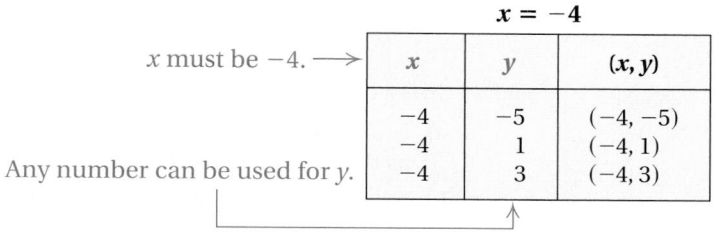

EXAMPLE 6

Graph: $x = -4$.

Solution   We can regard the equation $x = -4$ as $x + 0 \cdot y = -4$. We make up a table with all -4's in the x-column.

$$x = -4$$

x must be -4. →

x	y	(x, y)
-4	-5	$(-4, -5)$
-4	1	$(-4, 1)$
-4	3	$(-4, 3)$

Any number can be used for y.

All pairs will have -4 as the x-coordinate.

When we plot the ordered pairs $(-4, -5)$, $(-4, 1)$, and $(-4, 3)$ and connect them, we obtain a vertical line. Any ordered pair of the form $(-4, y)$ is a solution. The line is parallel to the y-axis with x-intercept $(-4, 0)$.

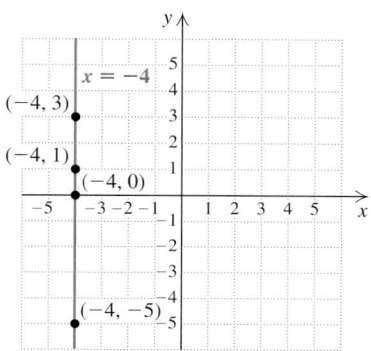

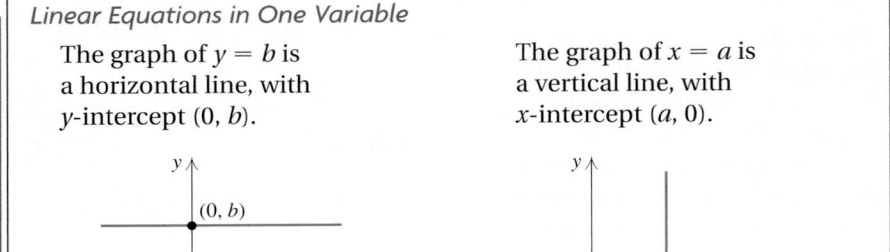

Linear Equations in One Variable

The graph of $y = b$ is a horizontal line, with y-intercept $(0, b)$.

The graph of $x = a$ is a vertical line, with x-intercept $(a, 0)$.

EXAMPLE 7 Write an equation for each graph.

a)

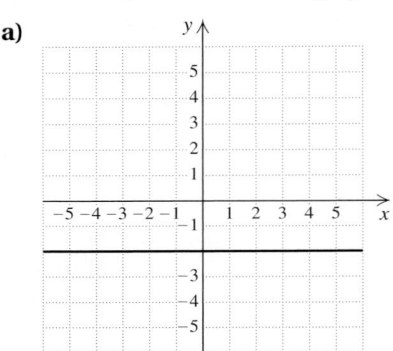

b)

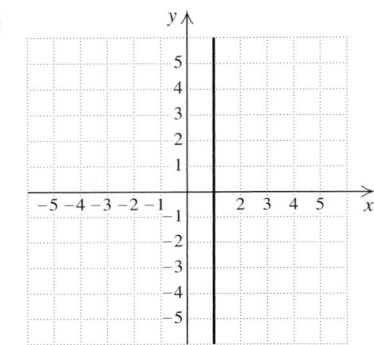

Solution

a) Note that every point on the horizontal line passing through $(0, -2)$ has -2 as the y-coordinate. Thus the equation of the line is $y = -2$.

b) Note that every point on the vertical line passing through $(1, 0)$ has 1 as the x-coordinate. Thus the equation of the line is $x = 1$.

Exercise Set

3.3

↝ Concept Reinforcement *In each of Exercises 1–6, match the phrase with the most appropriate choice from the column on the right.*

1. _____ A vertical line

2. _____ A horizontal line

3. _____ A *y*-intercept

4. _____ An *x*-intercept

5. _____ A third point as a check

6. _____ Use a scale of 10 units per square.

a) $2x + 5y = 100$

b) $(3, -2)$

c) $(1, 0)$

d) $(0, 2)$

e) $y = 3$

f) $x = -4$

For Exercises 7–12, list **(a)** the coordinates of the *y*-intercept and **(b)** the coordinates of all *x*-intercepts.

7.
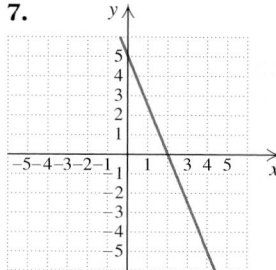

8.
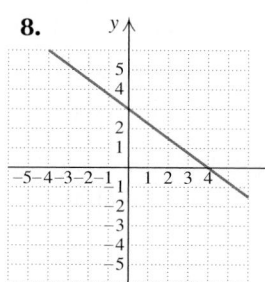

9.
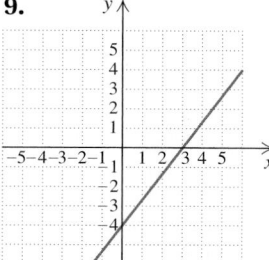

10.
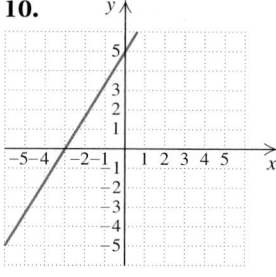

11.
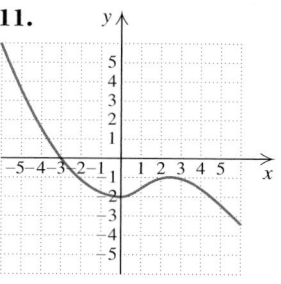

12.
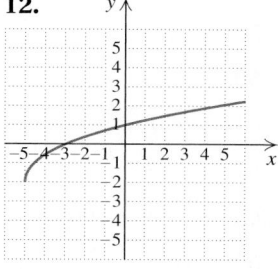

For Exercises 13–22, list **(a)** the coordinates of any *y*-intercept and **(b)** the coordinates of any *x*-intercept. *Do not graph.*

13. $5x + 3y = 15$

14. $5x + 2y = 20$

15. $7x - 2y = 28$

16. $4x - 3y = 24$

17. $-4x + 3y = 150$

18. $-2x + 3y = 80$

Aha! **19.** $y = 9$

20. $x = 8$

21. $x = -7$

22. $y = -1$

Find the intercepts. Then graph.

23. $3x + 2y = 12$

24. $x + 2y = 6$

25. $x + 3y = 6$

26. $6x + 9y = 36$

27. $-x + 2y = 8$

28. $-x + 3y = 9$

29. $3x + y = 9$

30. $2x - y = 8$

31. $y = 2x - 6$

32. $y = -3x + 6$

33. $3x - 9 = 3y$

34. $5x - 10 = 5y$

35. $2x - 3y = 6$

36. $2x - 5y = 10$

37. $4x + 5y = 20$

38. $6x + 2y = 12$

39. $3x + 2y = 8$

40. $x - 1 = y$

41. $2x + 4y = 6$

42. $5x - 6y = 18$

43. $5x + 3y = 180$

44. $10x + 7y = 210$

45. $y = -30 + 3x$

46. $y = -40 + 5x$

47. $-4x = 20y + 80$

48. $60 = 20x - 3y$

49. $y - 3x = 0$

50. $x + 2y = 0$

Graph.

51. $y = 5$

52. $y = 2$

53. $x = 4$

54. $x = 6$

55. $y = -2$

56. $y = -4$

57. $x = -1$

58. $x = -6$

59. $x = 18$

60. $x = -20$

61. $y = 0$

62. $y = \frac{3}{2}$

63. $x = -\frac{5}{2}$

64. $x = 0$

65. $-5y = -300$

66. $12x = -480$

67. $35 + 7y = 0$

68. $-3x - 24 = 0$

Write an equation for each graph.

69.

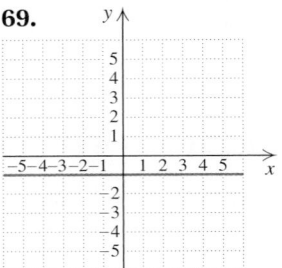

70.

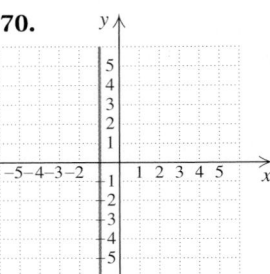

71.

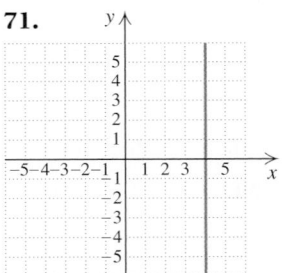

72.

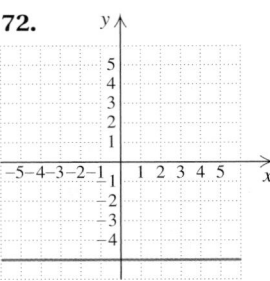

73.

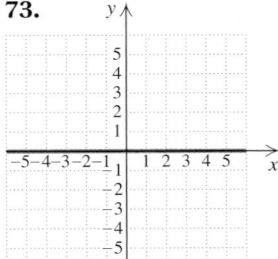

74.

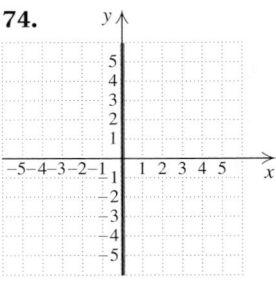

75. Explain in your own words why the graph of $y = 8$ is a horizontal line.

76. Explain in your own words why the graph of $x = -4$ is a vertical line.

SKILL MAINTENANCE

Translate to an algebraic expression. [1.1]

77. 7 less than d

78. 5 more than w

79. The sum of 2 and a number

80. The product of 3 and a number

81. Twice the sum of two numbers

82. Half of the sum of two numbers

SYNTHESIS

83. Describe what the graph of $x + y = C$ will look like for any choice of C.

84. If the graph of a linear equation has one point that is both the x- and the y-intercepts, what is that point? Why?

85. Write an equation for the x-axis.

86. Write an equation of the line parallel to the x-axis and passing through $(3, 5)$.

87. Write an equation of the line parallel to the y-axis and passing through $(-2, 7)$.

88. Find the coordinates of the point of intersection of the graphs of $y = x$ and $y = 6$.

89. Find the coordinates of the point of intersection of the graphs of the equations $x = -3$ and $y = x$.

90. Write an equation of the line shown in Exercise 7.

91. Write an equation of the line shown in Exercise 10.

92. Find the value of C such that the graph of $3x + C = 5y$ has an x-intercept of $(-4, 0)$.

93. Find the value of C such that the graph of $4x = C - 3y$ has a y-intercept of $(0, -8)$.

94. For A and B nonzero, the graphs of $Ax + D = C$ and $By + D = C$ will be parallel to an axis. Explain why.

In Exercises 95–100, find the intercepts of each equation algebraically. Then adjust the window and scale so that the intercepts can be checked graphically with no further window adjustments.

95. $3x + 2y = 50$

96. $2x - 7y = 80$

97. $y = 0.2x - 9$

98. $y = 1.3x - 15$

99. $25x - 20y = 1$

100. $50x + 25y = 1$

3.4 Rates

Rates of Change • Visualizing Rates

Rates of Change

Because graphs make use of two axes, they allow us to visualize how two quantities change with respect to each other. A number accompanied by units is used to represent this type of change and is referred to as a *rate*.

> **Rate**
>
> A *rate* is a ratio that indicates how two quantities change with respect to each other.

Rates occur often in everyday life:

A town that grows by 3400 residents over a period of 2 yr has a *growth rate* of $\frac{3400}{2}$, or 1700, residents per year.

A person running 150 m in 20 sec is moving at a *rate* of $\frac{150}{20}$, or 7.5, m/sec (meters per second).

A class of 25 students pays a total of $93.75 to visit a museum. The *rate* is $\frac{\$93.75}{25}$, or $3.75, per student.

Caution! To calculate a rate, it is important to keep track of the units being used.

EXAMPLE 1   On January 3, Nell rented a Ford Focus with a full tank of gas and 9312 mi on the odometer. On January 7, she returned the car with 9630 mi on the odometer.* If the rental agency charged Nell $108 for the rental and needed 12 gal of gas to fill up the gas tank, find the following rates:

a) The car's rate of gas consumption, in miles per gallon

b) The average cost of the rental, in dollars per day

c) The car's rate of travel, in miles per day

Solution

a) The rate of gas consumption, in miles per gallon, is found by dividing the number of miles traveled by the number of gallons used for that amount of driving:

$$\text{Rate, in miles per gallon} = \frac{9630 \text{ mi} - 9312 \text{ mi}}{12 \text{ gal}} \qquad \text{The word "per" indicates division.}$$

$$= \frac{318 \text{ mi}}{12 \text{ gal}}$$

$$= 26.5 \text{ mi/gal} \qquad \text{Dividing}$$

$$= 26.5 \text{ miles per gallon.}$$

b) The average cost of the rental, in dollars per day, is found by dividing the cost of the rental by the number of days:

$$\text{Rate, in dollars per day} = \frac{108 \text{ dollars}}{4 \text{ days}} \qquad \begin{array}{l} \text{From January 3 to} \\ \text{January 7 is} \\ 7 - 3 = 4 \text{ days.} \end{array}$$

$$= 27 \text{ dollars/day}$$

$$= \$27 \text{ per day.}$$

c) The car's rate of travel, in miles per day, is found by dividing the number of miles traveled by the number of days:

$$\text{Rate, in miles per day} = \frac{318 \text{ mi}}{4 \text{ days}} \qquad \begin{array}{l} 9630 \text{ mi} - 9312 \text{ mi} = 318 \text{ mi;} \\ \text{From January 3 to January 7} \\ \text{is } 7 - 3 = 4 \text{ days.} \end{array}$$

$$= 79.5 \text{ mi/day}$$

$$= 79.5 \text{ mi per day.}$$

Many problems involve a rate of travel, or *speed*. The **speed** of an object is found by dividing the distance traveled by the time required to travel that distance.

---

*For all rental problems, assume that the pickup time was later in the day than the return time so that no late fees were applied.

EXAMPLE 2

Transportation.   An Atlantic City Express bus makes regular trips between Paramus and Atlantic City, New Jersey. At 6:00 P.M., the bus is at mileage marker 40 on the Garden State Parkway, and at 8:00 P.M. it is at marker 170. Find the average speed of the bus.

Solution   Speed is the distance traveled divided by the time spent traveling:

$$\text{Bus speed} = \frac{\text{Distance traveled}}{\text{Time spent traveling}}$$

$$= \frac{\text{Change in mileage}}{\text{Change in time}}$$

$$= \frac{130 \text{ mi}}{2 \text{ hr}} \qquad \begin{array}{l} 170 \text{ mi} - 40 \text{ mi} = 130 \text{ mi};\\ 8{:}00 \text{ P.M.} - 6{:}00 \text{ P.M.} = 2 \text{ hr} \end{array}$$

$$= 65 \frac{\text{mi}}{\text{hr}}$$

$$= 65 \text{ miles per hour.} \qquad \begin{array}{l} \text{This } \textit{average} \text{ speed does not}\\ \text{indicate by how much the bus}\\ \text{speed may vary along the route.} \end{array}$$

Visualizing Rates

Graphs allow us to visualize a rate of change. As a rule, the quantity listed in the numerator appears on the vertical axis and the quantity listed in the denominator appears on the horizontal axis.

EXAMPLE 3

Phone lines.   Between 1992 and 2002, the number of U.S. telephone lines (land and cellular combined) increased at a rate of approximately 17.5 million lines per year.* In 1992, there were about 155 million U.S. telephone lines. Draw a graph to represent this information.

Solution   To label the axes, note that the rate is given in millions of telephone lines per year. Thus we list *Number of telephone lines, in millions*, on the vertical axis and *Year* on the horizontal axis.

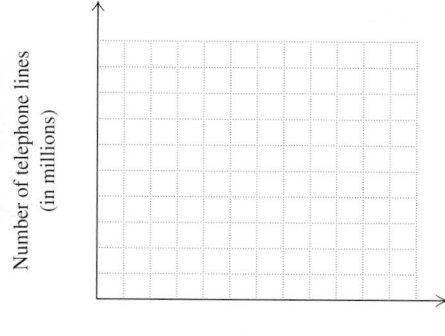

17.5 million per year is 17.5 million/yr; millions of customers is the vertical axis; year is the horizontal axis.

*Based on information from the Federal Communications Commission and the Cellular Telecommunications and Internet Associations

Next, we select a scale for each axis that allows us to plot the given information. If we count by increments of 40 million on the vertical axis, we can easily reach 155 million and beyond—allowing for growth. On the horizontal axis, we list years, making certain that both 1992 and 2002 are included (see the figure on the left below).

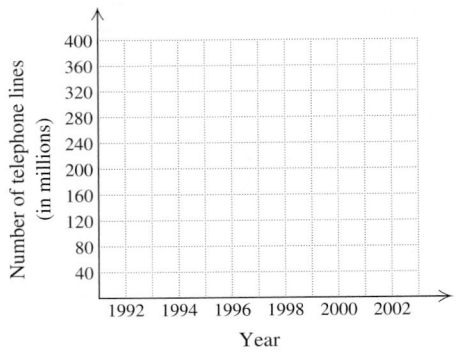
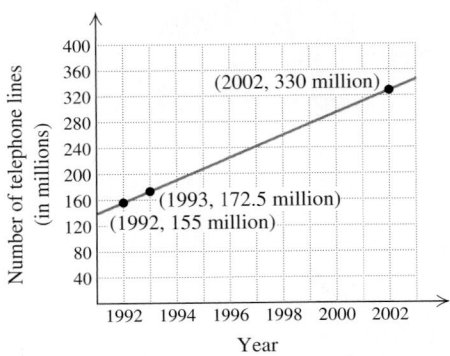

To display the given information, we plot the end point that corresponds to (1992, 155 million). Then, to display the rate of growth, we move from that point to a second point that represents 17.5 million more telephone lines one year later. The coordinates of this point are (1992 + 1, 155 + 17.5 million), or (1993, 172.5 million). To find the coordinates of the year 2002, note that 2002 is 10 years after 1992. Assuming a constant growth rate for those years, we calculate and plot (1992 + 10, 155 + 10 · 17.5 million), or (2002, 330 million). Finally, we draw a line through the three points.

EXAMPLE 4   Haircutting.   Sylvia's World of Hair has a graph displaying data from a recent day of work.

a) What rate can be determined from the graph?

b) What is that rate?

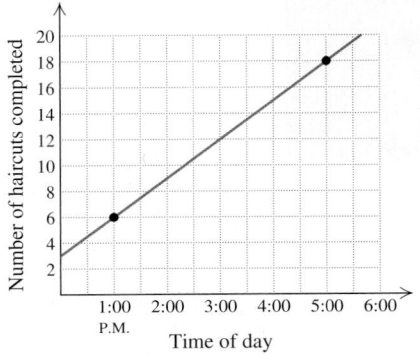

Solution

a) Because the vertical axis shows the number of haircuts completed and the horizontal axis lists the time in hour-long increments, we can find the rate *Number of haircuts per hour*.

b) The points (1:00, 6 haircuts) and (5:00, 18 haircuts) are both on the graph. This tells us that in the 4 hr between 1:00 and 5:00, there were $18 - 6 = 12$ haircuts completed. Thus the rate is

$$\frac{18 \text{ haircuts} - 6 \text{ haircuts}}{5:00 - 1:00} = \frac{12 \text{ haircuts}}{4 \text{ hours}}$$

$$= 3 \text{ haircuts per hour.}$$

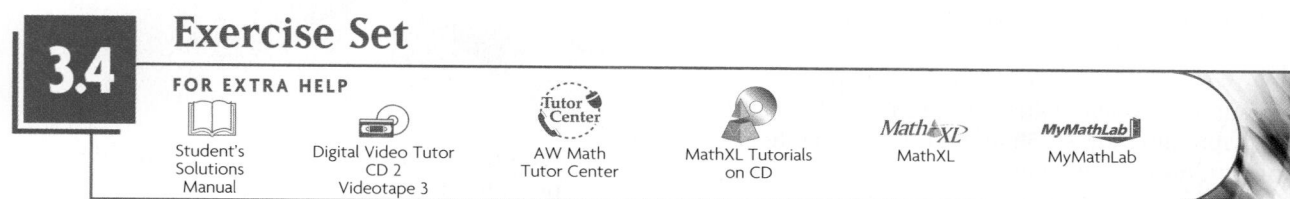

3.4 Exercise Set

FOR EXTRA HELP

Student's Solutions Manual | Digital Video Tutor CD 2 Videotape 3 | Tutor Center AW Math Tutor Center | MathXL Tutorials on CD | MathXL MathXL | MyMathLab MyMathLab

Solve. For Exercises 1–8, round answers to the nearest cent. For Exercises 1 and 2, assume that the pickup time was later in the day than the return time so that no late fees were applied.

1. *Van rentals.* Late on June 5, Deb rented a Dodge Caravan with a full tank of gas and 13,741 mi on the odometer. On June 8, she returned the van with 14,014 mi on the odometer. The rental agency charged Deb $118 for the rental and needed 13 gal of gas to fill up the tank.
  a) Find the van's rate of gas consumption, in miles per gallon.
  b) Find the average cost of the rental, in dollars per day.
  c) Find the rate of travel, in miles per day.
  d) Find the rental rate, in cents per mile.

2. *Car rentals.* On February 10, Oscar rented a Chevy Blazer with a full tank of gas and 13,091 mi on the odometer. On February 12, he returned the vehicle with 13,322 mi on the odometer. The rental agency charged $92 for the rental and needed 14 gal of gas to fill the tank.
  a) Find the Blazer's rate of gas consumption, in miles per gallon.

b) Find the average cost of the rental, in dollars per day.
  c) Find the rate of travel, in miles per day.
  d) Find the rental rate, in cents per mile.

3. *Bicycle rentals.* At 2:00, Perry rented a mountain bike from The Slick Rock Cyclery. He returned the bike at 5:00, after cycling 18 mi. Perry paid $12 for the rental.
  a) Find Perry's average speed, in miles per hour.
  b) Find the rental rate, in dollars per hour.
  c) Find the rental rate, in dollars per mile.

4. *Bicycle rentals.* At 9:00, Jodi rented a mountain bike from The Bike Rack. She returned the bicycle at 11:00, after cycling 14 mi. Jodi paid $15 for the rental.
  a) Find Jodi's average speed, in miles per hour.
  b) Find the rental rate, in dollars per hour.
  c) Find the rental rate, in dollars per mile.

5. *Temporary help.* A typist from Jobsite Services, Inc., reports to AKA International for work at 9:00 A.M. and leaves at 5:00 P.M. after having typed from the end of page 12 to the end of page 48 of a prospectus. AKA International pays $128 for the typist's services.

a) Find the rate of pay, in dollars per hour.
b) Find the average typing rate, in number of pages per hour.
c) Find the rate of pay, in dollars per page.

6. *Temporary help.* A typist for Kelly Services reports to 3E's Properties for work at 10:00 A.M. and leaves at 6:00 P.M. after having typed from the end of page 8 to the end of page 50 of a proposal. 3E's pays $120 for the typist's services.

a) Find the rate of pay, in dollars per hour.
b) Find the average typing rate, in number of pages per hour.
c) Find the rate of pay, in dollars per page.

7. *Two-year-college tuition.* The average tuition at a public two-year college was $1327 in 1999 and approximately $1359 in 2001. Find the rate at which tuition was increasing.
Source: *Statistical Abstract of the United States*, 2002

8. *Four-year-college tuition.* The average tuition at a public four-year college was $3640 in 1999 and approximately $3983 in 2001. Find the rate at which tuition was increasing.
Source: *Statistical Abstract of the United States*, 2002

9. *Elevators.* At 2:38, Serge entered an elevator on the 34th floor of the Regency Hotel. At 2:40, he stepped off at the 5th floor.

a) Find the elevator's average rate of travel, in number of floors per minute.
b) Find the elevator's average rate of travel, in seconds per floor.

10. *Snow removal.* By 1:00 P.M., Erin had already shoveled 2 driveways, and by 6:00 P.M., the number was up to 7.

a) Find Erin's shoveling rate, in number of driveways per hour.
b) Find Erin's shoveling rate, in hours per driveway.

11. *Mountaineering.* As part of an ill-fated expedition to climb Mt. Everest in 1996, author Jon Krakauer departed "The Balcony," elevation

27,600 ft, at 7:00 A.M. and reached the summit, elevation 29,028 ft, at 1:25 P.M.
Source: Krakauer, Jon, *Into Thin Air, the Illustrated Edition*. New York: Random House, 1998

a) Find Krakauer's average rate of ascent, in feet per minute.
b) Find Krakauer's average rate of ascent, in minutes per foot.

12. *Mountaineering.* The fastest ascent of Mt. Everest was accomplished by Lakpa Gelu Sherpa of Nepal in 2003. Lakpa Gelu Sherpa climbed from base camp, elevation 17,552 ft, to the summit, elevation 29,028 ft, in 10 hr 57 min.
Source: *Guinness Book of World Records* 2004 Edition

a) Find Lakpa Gelu Sherpa's rate of ascent, in feet per minute.
b) Find Lakpa Gelu Sherpa's rate of ascent, in minutes per foot.

In Exercises 13–22, draw a linear graph to represent the given information. Be sure to label and number the axes appropriately (see Example 3).

13. *Healthcare costs.* In 2003, the average copayment for a prescription drug on an insurance company's preferred list was $19, and the figure was rising at a rate of $2 per year.
Source: *Daily Reporter*, Greenfield IN, 9/9/03

14. *Healthcare costs.* In 2003, the average cost for health insurance for a family was about $9000, and the figure was rising at a rate of about $875 per year.
Source: *Daily Reporter*, Greenfield IN, 9/9/03

15. *Law enforcement.*   In 2000, there were approximately 26 million crimes reported in the United States, and the figure was dropping at a rate of about 1.2 million per year.
Source: Based on data from the Bureau of Justice Statistics

16. *Fire fighting.*   In 2000, there were approximately 1.7 million fires in the United States, and the figure was dropping at a rate of about 0.1 million per year.
Source: Based on data from the *Statistical Abstract of the United States,* 2002

17. *Train travel.*   At 3:00 P.M., the Boston–Washington Metroliner had traveled 230 mi and was cruising at a rate of 90 miles per hour.

18. *Plane travel.*   At 4:00 P.M., the Seattle–Los Angeles shuttle had traveled 400 mi and was cruising at a rate of 300 miles per hour.

19. *Wages.*   By 2:00 P.M., Diane had earned $50. She continued earning money at a rate of $15 per hour.

20. *Wages.*   By 3:00 P.M., Arnie had earned $70. He continued earning money at a rate of $12 per hour.

21. *Telephone bills.*   Roberta's phone bill was already $7.50 when she made a call for which she was charged at a rate of $0.10 per minute.

22. *Telephone bills.*   At 3:00 P.M., Larry's phone bill was $6.50 and increasing at a rate of 7¢ per minute.

In Exercises 23–32, use the graph provided to calculate a rate of change in which the units of the horizontal axis are used in the denominator.

23. *Hairdresser.*   Eve's Custom Cuts has a graph displaying data from a recent day of work. At what rate does Eve work?

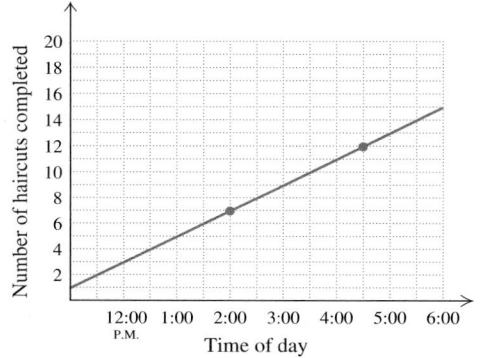

24. *Manicures.*   The following graph shows data from a recent day's work at the O'Hara School of Cosmetology. At what rate do they work?

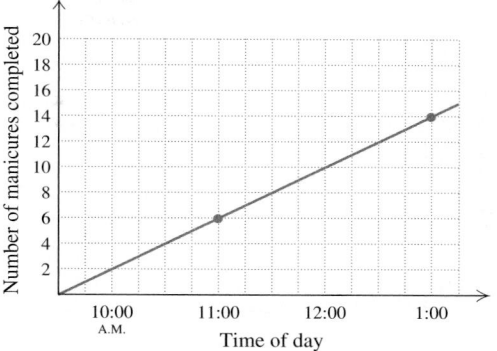

25. *Train travel.*   The following graph shows data from a recent train ride from Chicago to St. Louis. At what rate did the train travel?

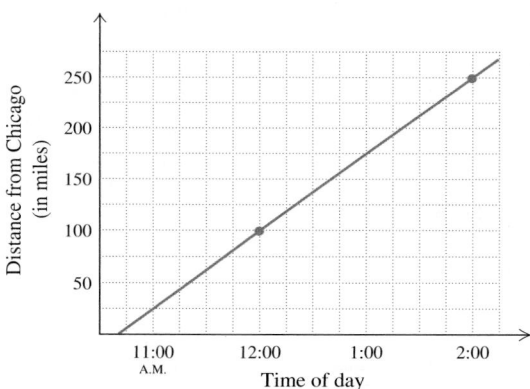

26. *Train travel.*   The following graph shows data from a recent train ride from Denver to Kansas City. At what rate did the train travel?

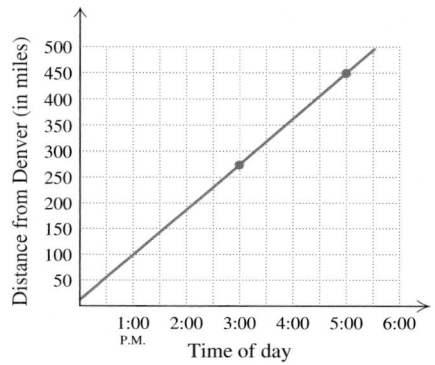

27. *Cost of a telephone call.* The following graph shows data from a recent MCI phone call between San Francisco, CA, and Pittsburgh, PA. At what rate was the customer being billed?

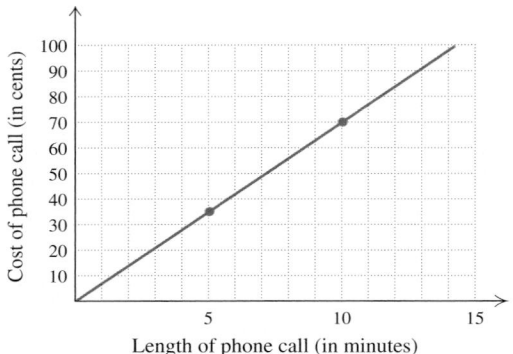

28. *Cost of a telephone call.* The following graph shows data from a recent AT&T phone call between Burlington, VT, and Austin, TX. At what rate was the customer being billed?

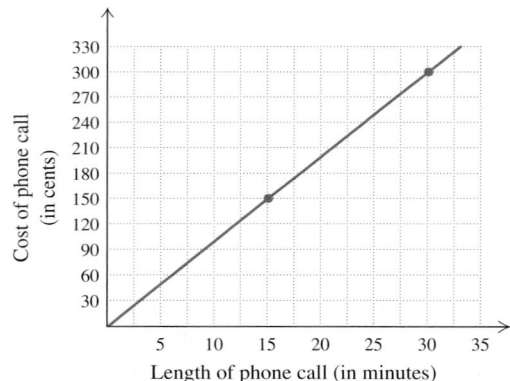

29. *Depreciation of an office machine.* Data regarding the value of a particular color copier is represented in the following graph. At what rate is the value changing?

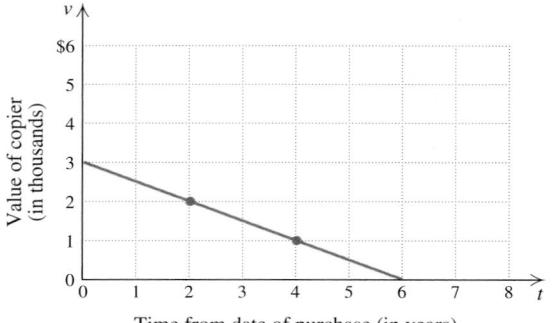

30. *NASA spending.* Spending on the National Aeronautics and Space Administration (NASA) during the late 1990s is represented in the following graph. At what rate was the amount spent on NASA changing?

Source: Based on information in the *Statistical Abstract of the United States*, 1999, 2002

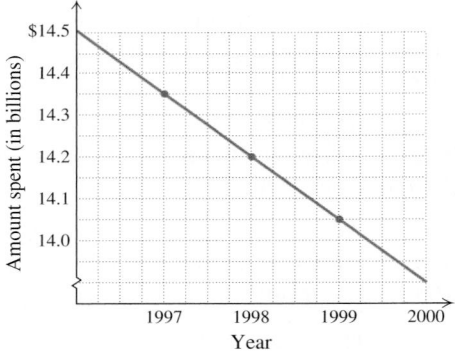

31. *Gas mileage.* The following graph shows data for a Honda Odyssey driven on interstate highways. At what rate was the vehicle consuming gas?

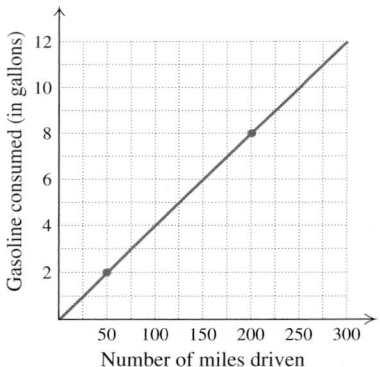

32. *Gas mileage.* The following graph shows data for a Ford Explorer driven on city streets. At what rate was the vehicle consuming gas?

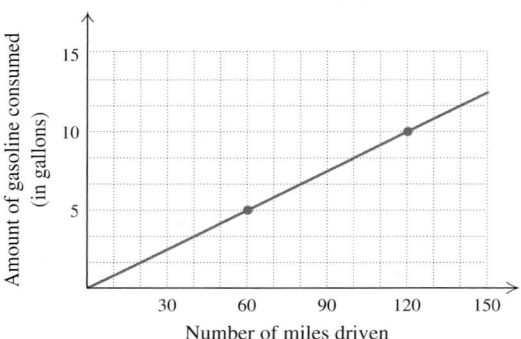

In each of Exercises 33–38, match the description with the most appropriate graph from the choices below. Scales are intentionally omitted. Assume that of the three sports listed, swimming is the slowest and biking is the fastest.

33. _____ Robin trains for triathlons by running, biking, and then swimming every Saturday.

34. _____ Gene trains for triathlons by biking, running, and then swimming every Sunday.

35. _____ Shirley trains for triathlons by swimming, biking, and then running every Sunday.

36. _____ Evan trains for triathlons by swimming, running, and then biking every Saturday.

37. _____ Angie trains for triathlons by biking, swimming, and then running every Sunday.

38. _____ Mick trains for triathlons by running, swimming, and then biking every Saturday.

a)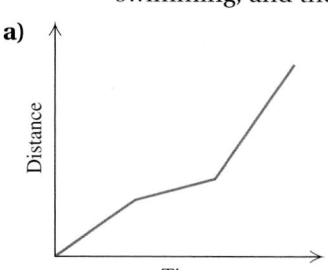

b)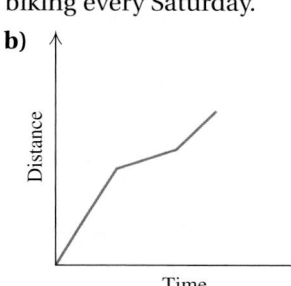

c)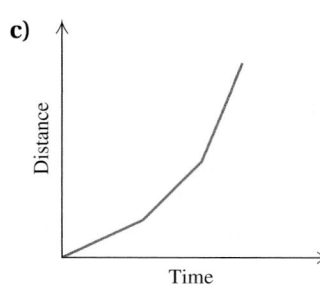

d)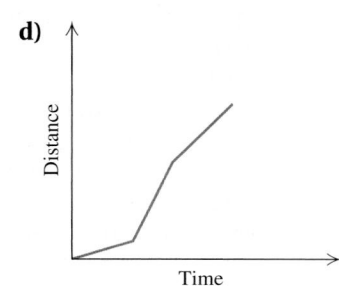

e)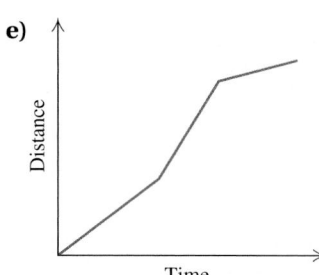

f)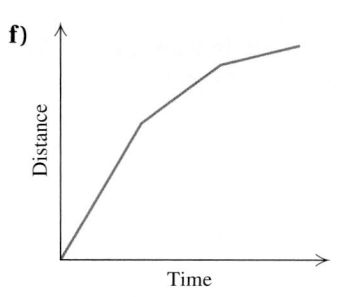

39. What does a negative rate of travel indicate? Explain.

40. Explain how to convert from kilometers per hour to meters per second.

SKILL MAINTENANCE

41. $-2 - (-7)$ [1.6]

42. $-9 - (-3)$ [1.6]

43. $\dfrac{5 - (-4)}{-2 - 7}$ [1.8]

44. $\dfrac{8 - (-4)}{2 - 11}$ [1.8]

45. $\dfrac{-4 - 8}{7 - (-2)}$ [1.8]

46. $\dfrac{-5 - 3}{6 - (-4)}$ [1.8]

SYNTHESIS

47. Write an exercise similar to Exercises 23–32 for a classmate to solve. Design the problem so that the solution is "The plane was traveling at a rate of 300 miles per hour."

48. Write an exercise similar to Exercises 1–12 for a classmate to solve. Design the problem so that the solution is "The motorcycle's rate of gas consumption was 65 miles per gallon."

49. *Aviation.*   A Boeing 737 climbs from sea level to a cruising altitude of 31,500 ft at a rate of 6300 ft/min. After cruising for 3 min, the jet is forced to land, descending at a rate of 3500 ft/min. Represent the flight with a graph in which altitude is measured on the vertical axis and time on the horizontal axis.

50. *Wages with commissions.*   Each salesperson at Mike's Bikes is paid $140 a week plus 13% of all sales up to $2000, and then 20% on any sales in excess of $2000. Draw a graph in which sales are measured on the horizontal axis and wages on the vertical axis. Then use the graph to estimate the wages paid when a salesperson sells $2700 in merchandise in one week.

51. *Taxi fares.*   The driver of a New York City Yellow Cab recently charged $2 plus 50¢ for each fifth of a mile traveled. Draw a graph that could be used to determine the cost of a fare.

52. *Gas mileage.*   Suppose that a Honda motorcycle goes twice as far as a Honda Odyssey on the same amount of gas (see Exercise 31). Draw a graph that reflects this information.

53. *Navigation.*   In 3 sec, Penny walks 24 ft, to the bow (front) of a tugboat. The boat is cruising at a rate of 5 feet per second. What is Penny's rate of travel with respect to land?

54. *Aviation.*   Tim's F-16 jet is moving forward at a deck speed of 95 mph aboard an aircraft carrier that is traveling 39 mph in the same direction. How fast is the jet traveling, in minutes per mile, with respect to the sea?

55. *Running.*   Annette ran from the 4-km mark to the 7-km mark of a 10-km race in 15.5 min. At this rate, how long would it take Annette to run a 5-mi race?

56. *Running.*   Jerod ran from the 2-mi marker to the finish line of a 5-mi race in 25 min. At this rate, how long would it take Jerod to run a 10-km race?

57. At 3:00 P.M., Catanya and Chad had already made 46 candles. By 5:00 P.M., the total reached 100 candles. Assuming a constant production rate, at what time did they make their 82nd candle?

58. Marcy picks apples twice as fast as Ryan. By 4:30, Ryan had already picked 4 bushels of apples. Fifty minutes later, his total reached $5\frac{1}{2}$ bushels. Find Marcy's picking rate. Give your answer in number of bushels per hour.

CORNER

COLLABORATIVE

Determining Depreciation Rates

Focus: Modeling, graphing, and rates

Time: 30 minutes

Group size: 3

Materials: Graph paper and straightedges

From the minute a new car is driven out of the dealership, it *depreciates*, or drops in value with the passing of time. The Kelley Blue Book Auto Market Report is a periodic listing of the trade-in values of used cars. The data below are taken from two such reports from 2003.

ACTIVITY

1. Each group member should select a different one of the cars listed in the table below as his or her own. Assuming that the values are dropping linearly, each student should draw a line representing the trade-in value of his or her car. Draw all three lines on the same graph. Let the horizontal axis represent the time, in months, since April 2003, and let the vertical axis represent the trade-in value of each car. Decide as a group how many months or dollars each square should represent. Make the drawings as neat as possible.

2. At what *rate* is each car depreciating and how are the different rates illustrated in the graph of part (1)?

3. If one of the three cars had to be sold in April 2004, which one would your group sell and why? Compare answers with other groups.

Car	Trade-in Value in April 2003	Trade-in Value in October 2003
2001 Chevrolet Camaro convertible V6	$14,650	$13,600
2001 Ford Mustang convertible V6	$12,550	$11,450
2001 Mitsubishi Eclipse convertible V6	$15,250	$13,800

3.5   Slope

Rate and Slope   •   Horizontal and Vertical Lines   •
Applications

In Section 3.4, we introduced *rate* as a method of measuring how two quantities change with respect to each other. In this section, we will discuss how rate can be related to the slope of a line.

Rate and Slope

Suppose that a car manufacturer operates two plants: one in Michigan and one in Pennsylvania. Knowing that the Michigan plant produces 3 cars every 2 hours and the Pennsylvania plant produces 5 cars every 4 hours, we can set up tables listing the number of cars produced after various amounts of time.

Michigan Plant	
Hours Elapsed	**Cars Produced**
0	0
2	3
4	6
6	9
8	12

Pennsylvania Plant	
Hours Elapsed	**Cars Produced**
0	0
4	5
8	10
12	15
16	20

By comparing the number of cars produced at each plant over a specified period of time, we can compare the two production rates. For example, the Michigan plant produces 3 cars every 2 hours, so its *rate* is $3 \div 2 = 1\frac{1}{2}$, or $\frac{3}{2}$ cars per hour. Since the Pennsylvania plant produces 5 cars every 4 hours, its rate is $5 \div 4 = 1\frac{1}{4}$, or $\frac{5}{4}$ cars per hour.

Let's now graph the pairs of numbers listed in the tables, using the horizontal axis for time and the vertical axis for the number of cars produced. Note that the rate in the Michigan plant is slightly greater so its graph is slightly steeper.

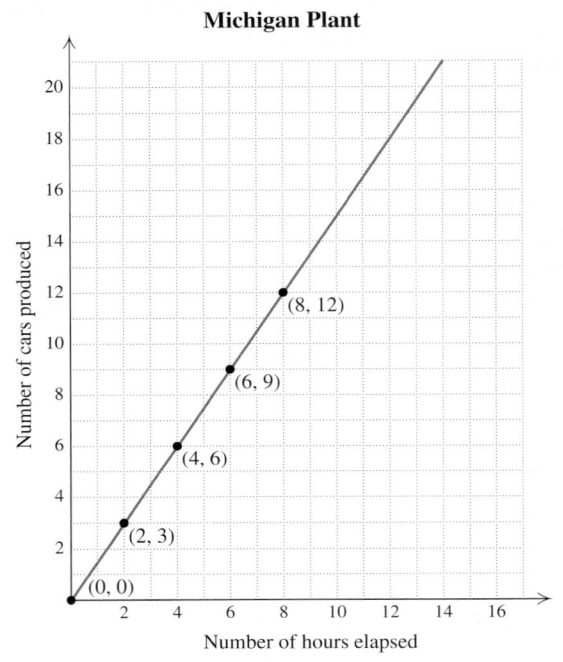

Michigan Plant

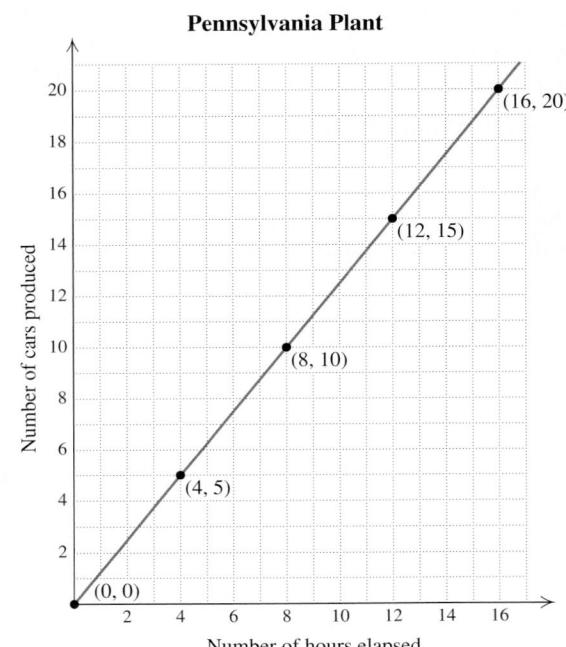

Pennsylvania Plant

The rates $\frac{3}{2}$ and $\frac{5}{4}$ can also be found using the coordinates of any two points that are on the line. For example, we can use the points (6, 9) and (8, 12) to find the production rate for the Michigan plant. To do so, remember that these coordinates tell us that after 6 hr, 9 cars have been produced, and after 8 hr, 12 cars have been produced. In the 2 hr between the 6-hr and 8-hr points, $12 - 9$, or 3 cars were produced. Thus,

$$\text{Michigan production rate} = \frac{\text{change in number of cars produced}}{\text{corresponding change in time}}$$

$$= \frac{12 - 9 \text{ cars}}{8 - 6 \text{ hr}}$$

$$= \frac{3 \text{ cars}}{2 \text{ hr}} = \frac{3}{2} \text{ cars per hour.}$$

Because the line is straight, the same rate is found using *any* pair of points on the line. For example, using (0, 0) and (4, 6), we have

$$\text{Michigan production rate} = \frac{6 - 0 \text{ cars}}{4 - 0 \text{ hr}} = \frac{6 \text{ cars}}{4 \text{ hr}} = \frac{3}{2} \text{ cars per hour.}$$

Note that the rate is always the vertical change divided by the corresponding horizontal change.

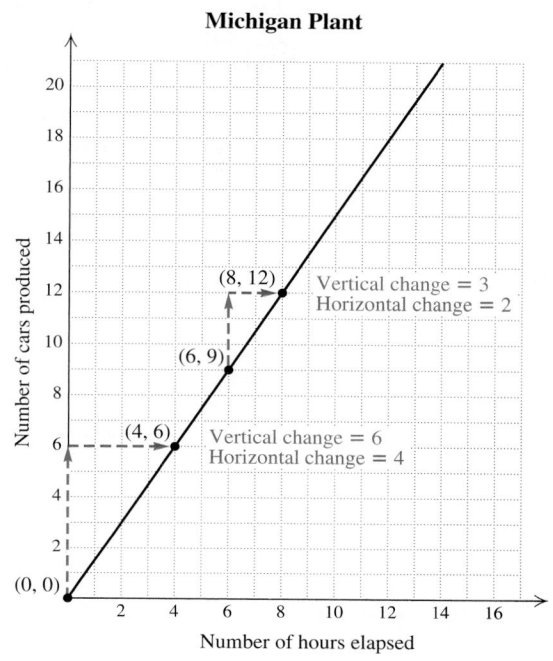

Michigan Plant

EXAMPLE 1   Use the graph of car production at the Pennsylvania plant to find the rate of production.

Solution   We can use any two points on the line, such as (12, 15) and (16, 20):

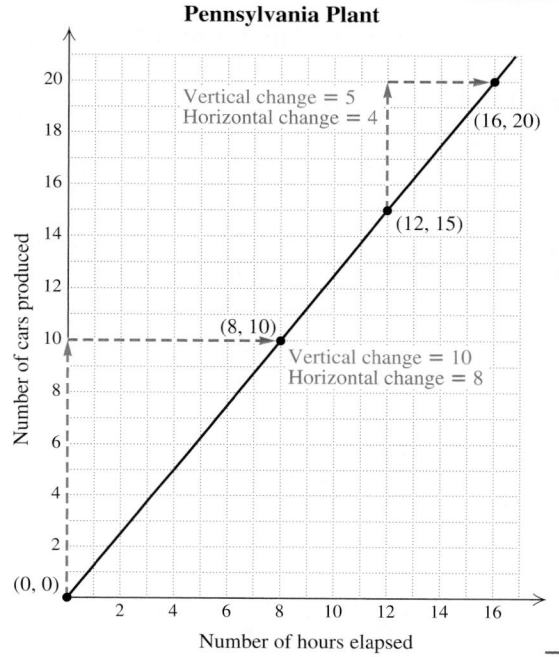

Pennsylvania Plant

$$\begin{array}{l}\text{Pennsylvania} \\ \text{production rate}\end{array} = \frac{\text{change in number of cars produced}}{\text{corresponding change in time}}$$

$$= \frac{20 - 15 \text{ cars}}{16 - 12 \text{ hr}}$$

$$= \frac{5 \text{ cars}}{4 \text{ hr}}$$

$$= \frac{5}{4} \text{ cars per hour.}$$

As a check, we can use another pair of points, like (0, 0) and (8, 10):

$$\begin{array}{l}\text{Pennsylvania} \\ \text{production rate}\end{array} = \frac{10 - 0 \text{ cars}}{8 - 0 \text{ hr}}$$

$$= \frac{10 \text{ cars}}{8 \text{ hr}}$$

$$= \frac{5}{4} \text{ cars per hour.}$$

When the axes of a graph are simply labeled x and y, it is still useful to know the ratio of vertical change to horizontal change. This ratio is a measure of a line's slant, or **slope**, and is the rate at which y is changing with respect to x.

Consider a line passing through (2, 3) and (6, 5), as shown below. We find the ratio of vertical change, or *rise*, to horizontal change, or *run*, as follows:

$$\text{Ratio of vertical change to horizontal change} = \frac{\text{change in } y}{\text{change in } x} = \frac{\text{rise}}{\text{run}}$$

$$= \frac{5-3}{6-2}$$

$$= \frac{2}{4}, \text{ or } \frac{1}{2}.$$

Note that these calculations can be performed without viewing a graph.

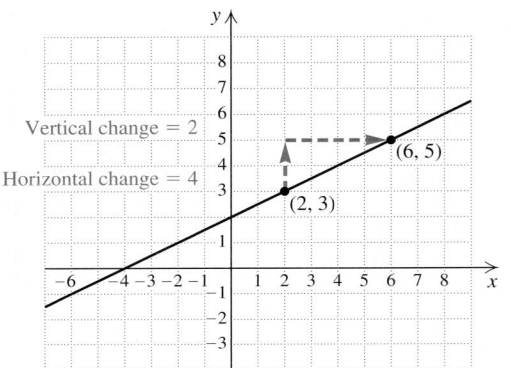

Vertical change = 2
Horizontal change = 4

Thus the y-coordinates of points on this line increase at a rate of 2 units for every 4-unit increase in x, 1 unit for every 2-unit increase in x, or $\frac{1}{2}$ unit for every 1-unit increase in x. The slope of the line is $\frac{1}{2}$.

In the box below, the *subscripts* 1 and 2 are used to distinguish two arbitrary points, point 1 and point 2, from each other. The slightly lowered 1's and 2's are not exponents but are used to denote x-values (or y-values) that differ from each other.

Slope

The *slope* of the line containing points (x_1, y_1) and (x_2, y_2) is given by

$$m = \frac{\text{change in } y}{\text{change in } x} = \frac{\text{rise}}{\text{run}} = \frac{y_2 - y_1}{x_2 - x_1}.$$

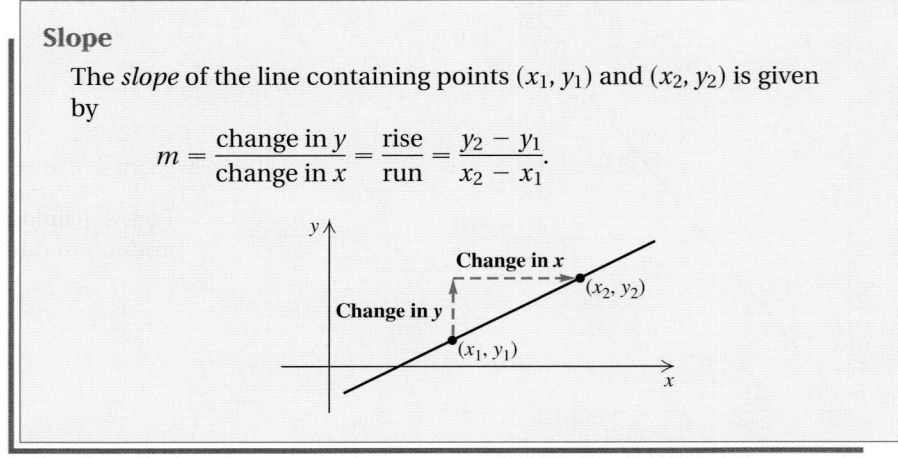

EXAMPLE 2 Graph the line containing the points $(-4, 3)$ and $(2, -6)$ and find the slope.

Solution The graph is shown below. From $(-4, 3)$ to $(2, -6)$, the change in y, or rise, is $-6 - 3$, or -9. The change in x, or run, is $2 - (-4)$, or 6. Thus,

$$\text{Slope} = \frac{\text{change in } y}{\text{change in } x}$$

$$= \frac{\text{rise}}{\text{run}}$$

$$= \frac{-6 - 3}{2 - (-4)}$$

$$= \frac{-9}{6}$$

$$= -\frac{9}{6}, \text{ or } -\frac{3}{2}.$$

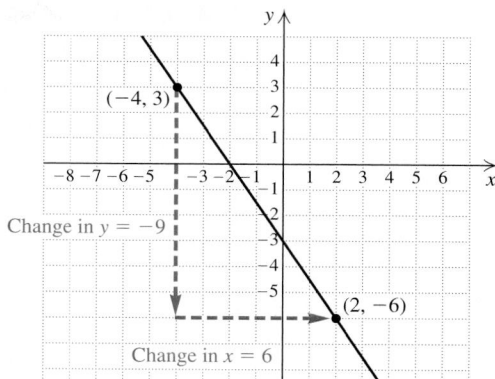

Student Notes

You may wonder which point should be regarded as (x_1, y_1) and which should be (x_2, y_2). To see that the math works out the same either way, perform both calculations on your own.

Caution! When we use the formula

$$m = \frac{y_2 - y_1}{x_2 - x_1},$$

it makes no difference which point is considered (x_1, y_1). What matters is that we subtract the y-coordinates in the same order that we subtract the x-coordinates.

To illustrate, we reverse *both* of the subtractions in Example 2. The slope is still $-\frac{3}{2}$:

$$\text{Slope} = \frac{\text{change in } y}{\text{change in } x} = \frac{3 - (-6)}{-4 - 2} = \frac{9}{-6} = -\frac{3}{2}.$$

As shown in the graphs below, a line with positive slope slants up from left to right, and a line with negative slope slants down from left to right. The larger the absolute value of the slope, the steeper the line.

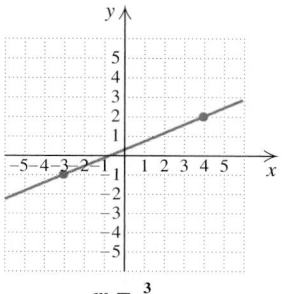

$m = \frac{3}{7}$

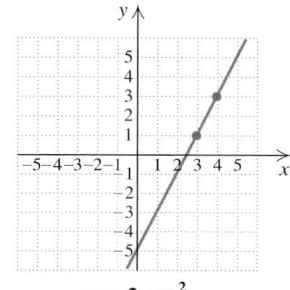

$m = 2, \text{ or } \frac{2}{1}$

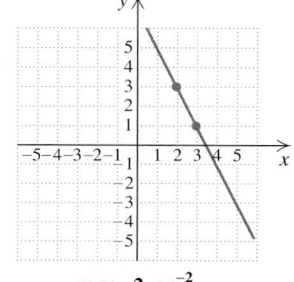

$m = -2, \text{ or } \frac{-2}{1}$

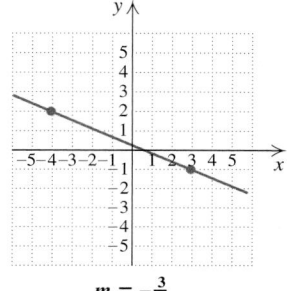

$m = -\frac{3}{7}$

Horizontal and Vertical Lines

What about the slope of a horizontal or a vertical line?

EXAMPLE 3   Find the slope of the line $y = 4$.

Solution   Consider the points (2, 4) and $(-3, 4)$, which are on the line. The change in y, or the rise, is $4 - 4$, or 0. The change in x, or the run, is $-3 - 2$, or -5. Thus,

$$m = \frac{4 - 4}{-3 - 2}$$

$$= \frac{0}{-5}$$

$$= 0.$$

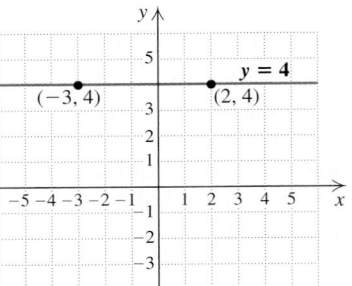

Any two points on a horizontal line have the same y-coordinate. Thus the change in y is 0, so the slope is 0.

A horizontal line has slope 0.

EXAMPLE 4   Find the slope of the line $x = -3$.

Solution   Consider the points $(-3, 4)$ and $(-3, -2)$, which are on the line. The change in y, or the rise, is $-2 - 4$, or -6. The change in x, or the run, is $-3 - (-3)$, or 0. Thus,

$$m = \frac{-2 - 4}{-3 - (-3)}$$

$$= \frac{-6}{0}. \quad \text{(undefined)}$$

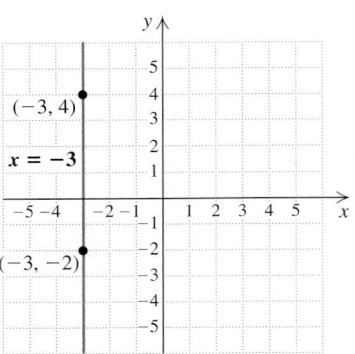

Since division by 0 is not defined, the slope of this line is not defined. The answer to a problem of this type is "The slope of this line is undefined."

The slope of a vertical line is undefined.

Applications

We have seen that slope has many real-world applications, ranging from car speed to production rate. Some applications use slope to measure steepness. For example, numbers like 2%, 3%, and 6% are often used to represent the **grade** of a road, a measure of a road's steepness. That is, a 3% grade means that for every horizontal distance of 100 ft, the road rises or drops 3 ft. The concept of grade also occurs in skiing or snowboarding, where a 7% grade is considered very tame, but a 70% grade is considered steep.

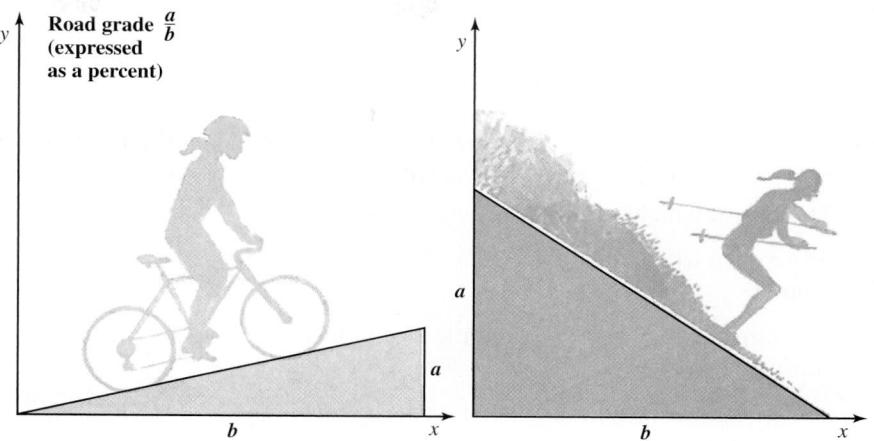

EXAMPLE 5   Skiing.   Among the steepest skiable terrain in North America, the Headwall on Mount Washington, in New Hampshire, drops 720 ft over a horizontal distance of 900 ft. Find the grade of the Headwall.

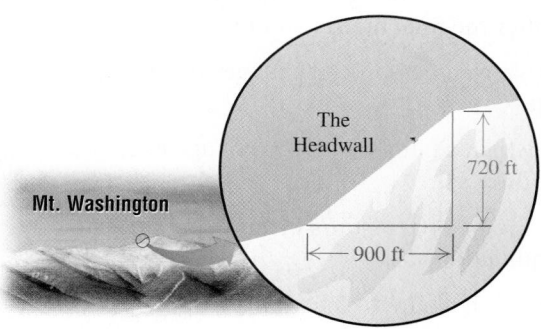

Solution   The grade of the Headwall is its slope, expressed as a percent:

$$m = \frac{720}{900}$$

$$= \frac{8}{10}$$

$$= 80\%.$$

Grade is slope expressed as a percent.

Carpenters use slope when designing stairs, ramps, or roof pitches. Another application occurs in the engineering of a dam—the force or strength of a river depends on how much the river drops over a specified distance.

Exercise Set

3.5

Concept Reinforcement *State whether each of the following rates is positive, negative, or zero.*

1. The rate at which a teenager's height changes

2. The rate at which an elderly person's height changes

3. The rate at which a pond's water level changes during a drought

4. The rate at which a pond's water level changes during the rainy season

5. The rate at which the number of people in attendance at a basketball game changes in the moments before the opening tipoff

6. The rate at which the number of people in attendance at a basketball game changes in the moments after the final buzzer sounds

7. The rate at which a person's IQ changes during his or her sleep

8. The rate at which a gift shop's sales change as the holidays approach

9. The rate at which a bookstore's inventory changes during a liquidation sale

10. The rate at which a buoy's elevation changes while floating for an evening in a lake

11. Find the rate of change of the U.S. population.
    Source: Based on information in the *Statistical Abstract of the United States*, 2002

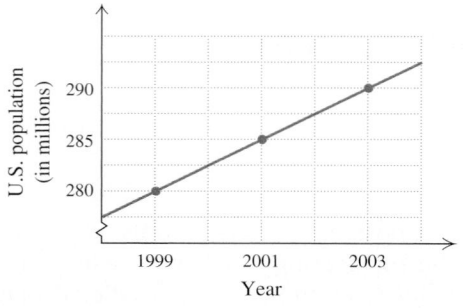

12. Find the rate at which a runner burns calories.

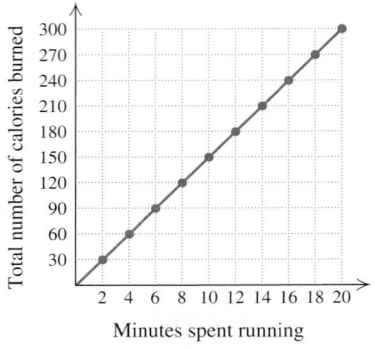

13. Find the rate of change in the percentage of high school students who admitted to cheating on an exam at least once in the past year.
    Source: Based on information from Josephson Institute of Ethics, as reported in *The New York Times*, 10/4/03

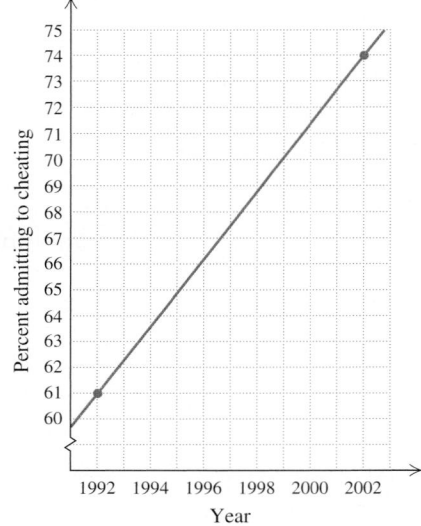

14. Find the rate of change in the percentage of college students who admitted to taking

material from the Internet and claiming it as their own.

Source: Based on information from Donald L. McCabe, as reported in *The New York Times*, 10/4/03

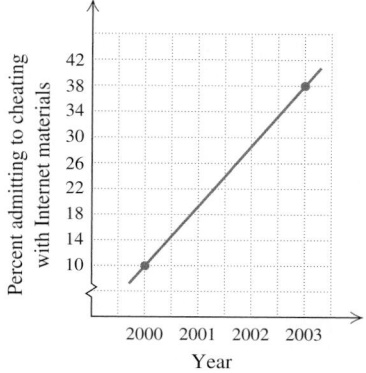

15. Find the rate of change in SAT math scores with respect to family income.

Source: Based on data from 1999 college-bound seniors in Massachusetts

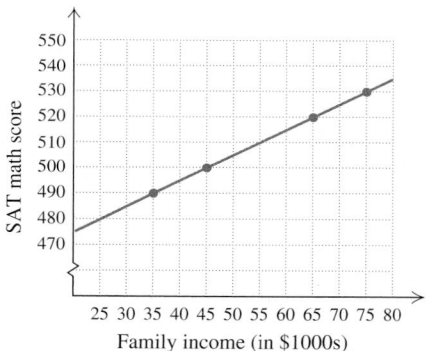

16. Find the rate of change in SAT verbal scores with respect to family income.

Source: Based on data from 1999 college-bound seniors in Massachusetts

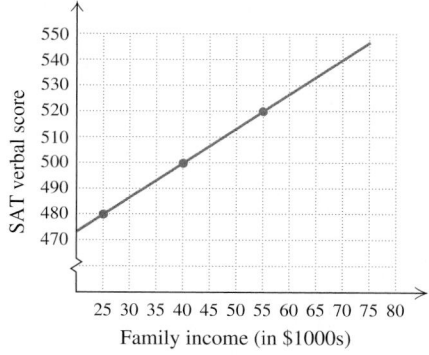

17. Find the rate of change in the temperature in Spearfish, South Dakota, on January 22, 1943, as shown below.

Source: National Oceanic and Atmospheric Administration

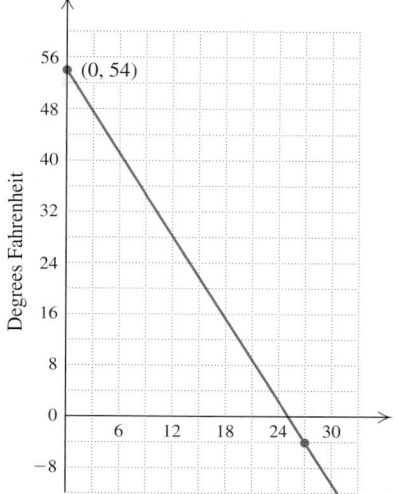

Number of minutes after 9 A.M.

18. Find the rate of change in the number of crimes reported in the United States.

Source: Based on statistics from the U.S. Bureau of Justice

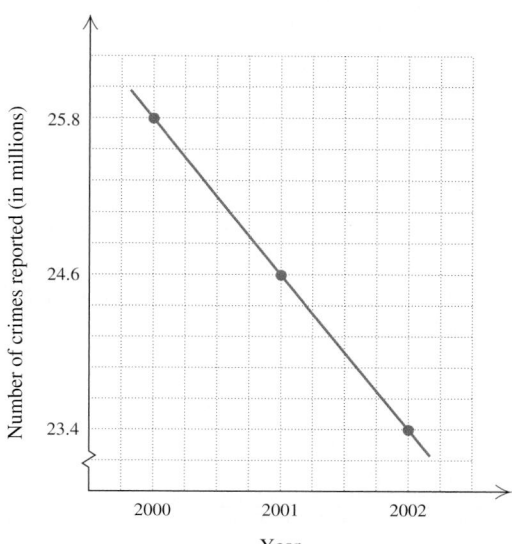

Find the slope, if it is defined, of each line. If the slope is undefined, state this.

19.

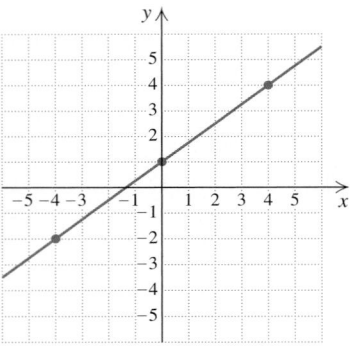

20.

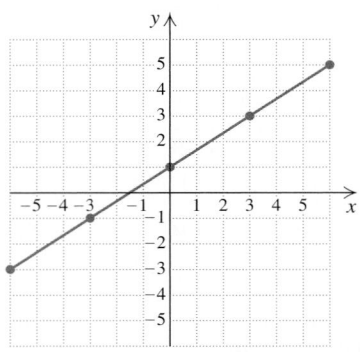

21.

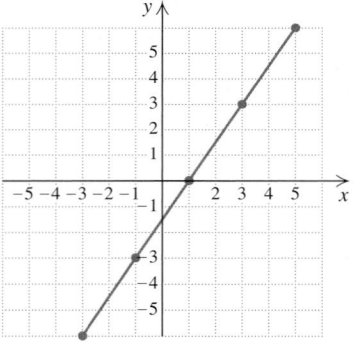

22.

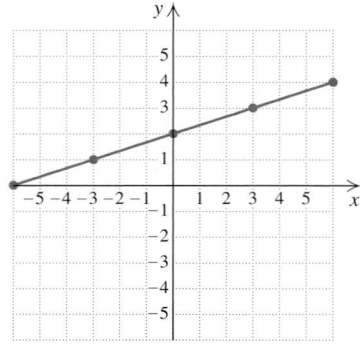

23.

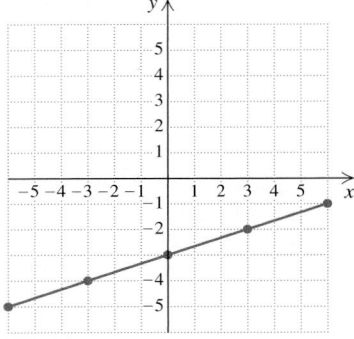

24.

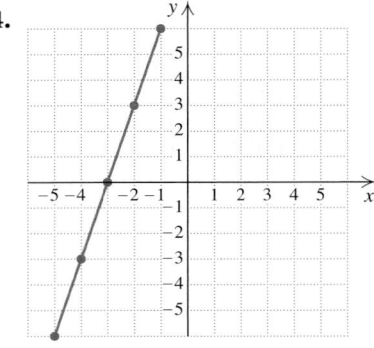

25.

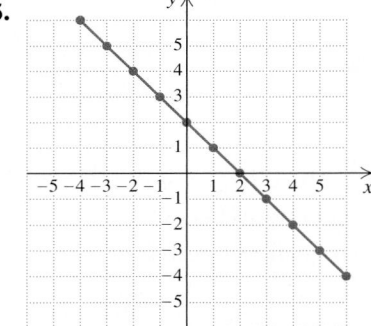

26.

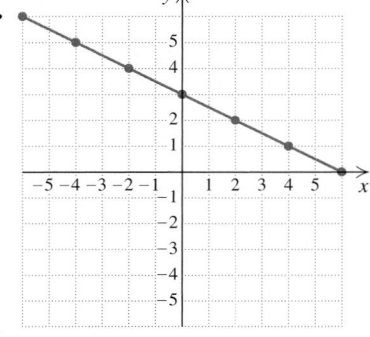

27.

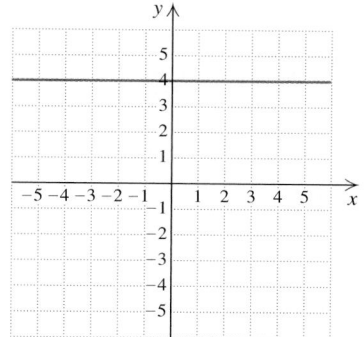

28.

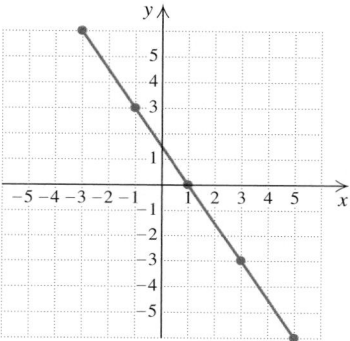

29.

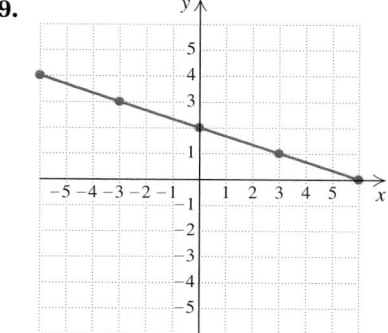

30.

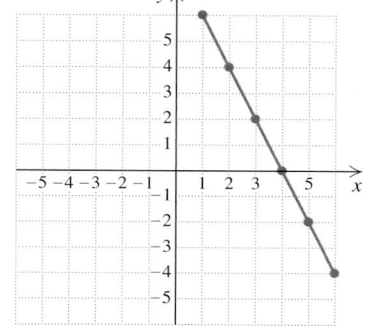

31.

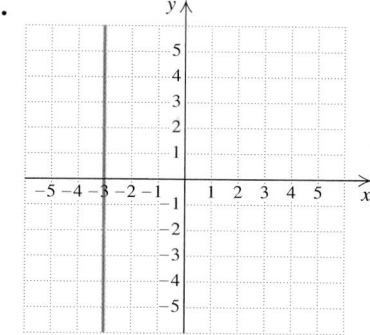

32.

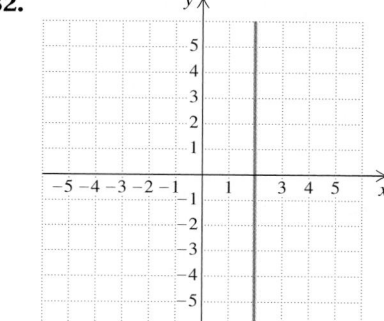

33.

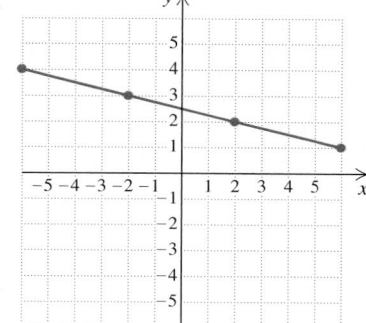

34.

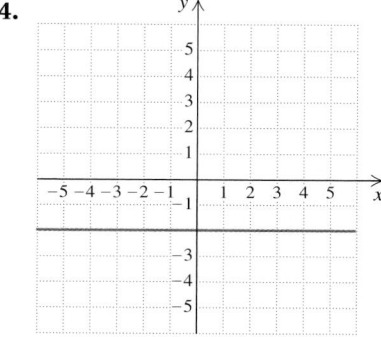

35.

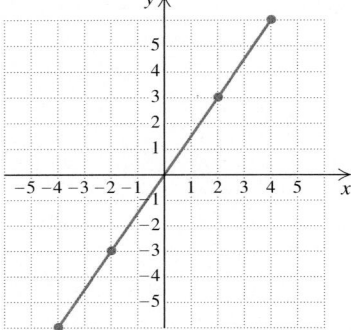

36.

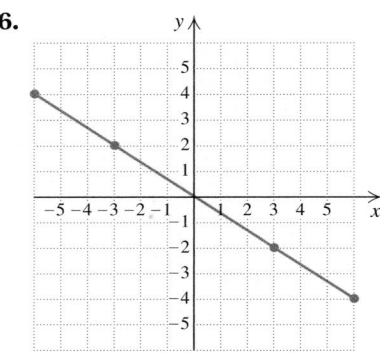

37.

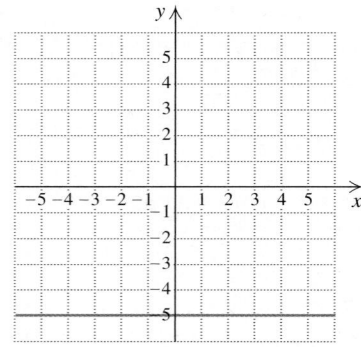

38.

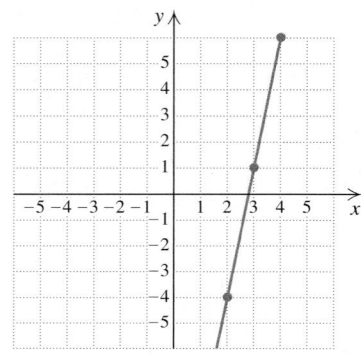

39.

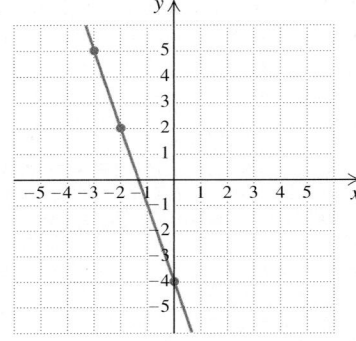

40.

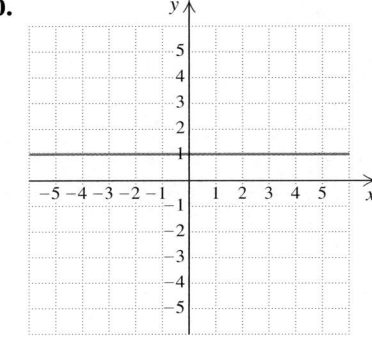

Find the slope of the line containing each given pair of points. If the slope is undefined, state this.

41. (1, 2) and (5, 8)

42. (2, 1) and (6, 9)

43. (−2, 4) and (3, 0)

44. (−4, 2) and (2, −3)

45. (−4, 0) and (5, 7)

46. (3, 0) and (6, 2)

47. (0, 8) and (−3, 10)

48. (0, 9) and (−5, 0)

49. (−2, 3) and (−6, 5)

50. (−2, 4) and (6, −7)

Aha! **51.** $\left(-2, \frac{1}{2}\right)$ and $\left(-5, \frac{1}{2}\right)$

52. (−5, −1) and (2, 3)

53. (3, 4) and (9, −7)

54. (−10, 3) and (−10, 4)

55. (6, −4) and (6, 5)

56. (5, −2) and (−4, −2)

Find the slope of each line whose equation is given. If the slope is undefined, state this.

57. $x = -3$

58. $x = -4$

59. $y = 4$

60. $y = 17$

61. $x = 9$

62. $x = 6$

63. $y = -9$

64. $y = -4$

65. *Surveying.*   Tucked between two ski areas, Vermont Route 108 rises 106 m over a horizontal distance of 1325 m. What is the grade of the road?

66. *Navigation.*   Capital Rapids drops 54 ft vertically over a horizontal distance of 1080 ft. What is the slope of the rapids?

67. *Architecture.*   To meet federal standards, a wheelchair ramp cannot rise more than 1 ft over a horizontal distance of 12 ft. Express this slope as a grade.

68. *Engineering.*   At one point, Yellowstone's Beartooth Highway rises 315 ft over a horizontal distance of 4500 ft. Find the grade of the road.

69. *Carpentry.*   Find the slope (or pitch) of the roof.

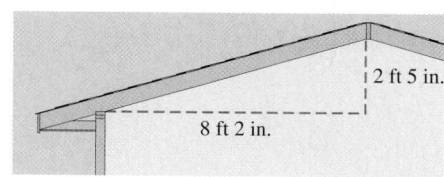

2 ft 5 in.

8 ft 2 in.

70. *Exercise.*   Find the slope (or grade) of the treadmill.

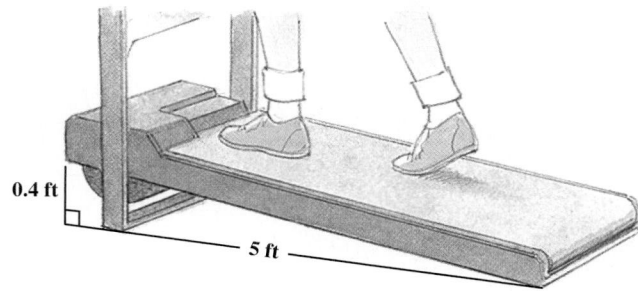

0.4 ft

5 ft

71. *Surveying.*   From a base elevation of 9600 ft, Longs Peak, Colorado, rises to a summit elevation of 14,255 ft over a horizontal distance of 15,840 ft. Find the average grade of Longs Peak.

72. *Construction.*   Public buildings regularly include steps with 7-in. risers and 11-in. treads. Find the grade of such a stairway.

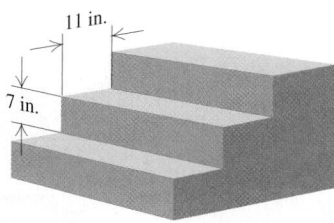

11 in.

7 in.

73. Explain why the order in which coordinates are subtracted to find slope does not matter so long as *y*-coordinates and *x*-coordinates are subtracted in the same order.

74. If one line has a slope of -3 and another has a slope of 2, which line is steeper? Why?

SKILL MAINTENANCE

Solve. [2.3]

75. $ax + by = c$, for y

76. $rx - mn = p$, for r

77. $ax - by = c$, for y

78. $rs + nt = q$, for t

Evaluate. [1.8]

79. $\frac{2}{3}x - 5$, for $x = 12$

80. $\frac{3}{5}x - 7$, for $x = 15$

SYNTHESIS

81. The points $(-4, -3)$, $(1, 4)$, $(4, 2)$, and $(-1, -5)$ are vertices of a quadrilateral. Use slopes to explain why the quadrilateral is a parallelogram.

82. Which is steeper and why: a ski slope that is 50° or one with a grade of 100%?

83. The plans below are for a skateboard "Fun Box". For the ramps labeled A, find the slope or grade.
    Source: www.heckler.com

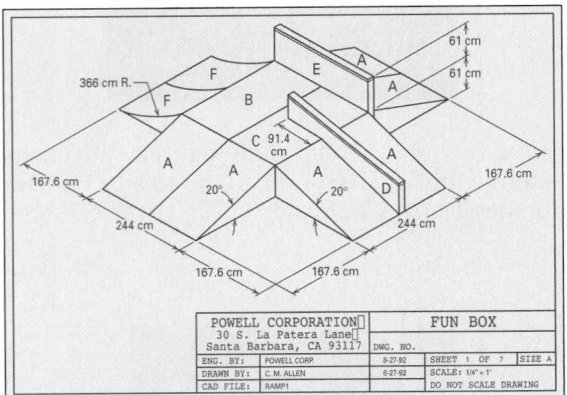

84. A line passes through $(4, -7)$ and never enters the first quadrant. What numbers could the line have for its slope?

85. A line passes through $(2, 5)$ and never enters the second quadrant. What numbers could the line have for its slope?

86. *Architecture.*   Architects often use the equation $x + y = 18$ to determine the height y, in inches, of the riser of a step when the tread is x inches wide. Express the slope of stairs designed with this equation without using the variable y.

In Exercises 87 and 88, the slope of the line is $-\frac{2}{3}$, but the numbering on one axis is missing. How many units should each tick mark on that unnumbered axis represent?

87.

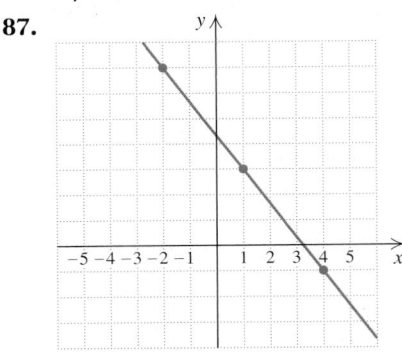

88.

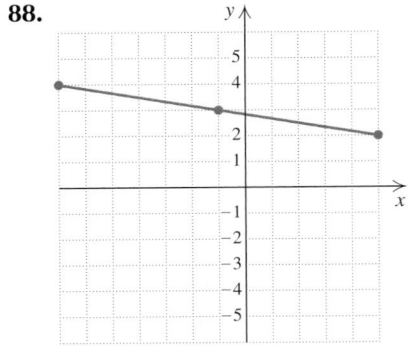

3.6 | Slope–Intercept Form

Using the *y*-intercept and the Slope to Graph a Line   •
Equations in Slope–Intercept Form   •   Graphing and
Slope–Intercept Form   •   Parallel and Perpendicular Lines

CONNECTING THE CONCEPTS

In Chapters 1 and 2, we simplified expressions and solved equations. In Sections 3.1–3.3 of this chapter, we learned how a graph can be used to represent the solutions of an equation in two variables, like $y = 4 - x$ or $2x + 5y = 15$.

Graphs are used in many important applications. Sections 3.4 and 3.5 have shown us that the slope of a line can be used to represent the rate at which the quantity measured on the vertical axis changes with respect to the quantity measured on the horizontal axis.

In Section 3.6, we return to the task of graphing equations. This time, however, our understanding of rates and slope will provide us with the tools necessary to develop shortcuts that will streamline our work.

If we know the slope and the *y*-intercept of a line, it is possible to graph the line. In this section, we will discover that a line's slope and *y*-intercept can be determined directly from the line's equation, provided the equation is written in a certain form.

Using the *y*-intercept and the Slope to Graph a Line

Let's modify the car production situation that first appeared in Section 3.5. Suppose that as a new workshift begins, 4 cars have already been produced. At the Michigan plant, 3 cars were being produced every 2 hours, a rate of $\frac{3}{2}$ cars per hour. If this rate remains the same regardless of how many cars have already been produced, the table and graph shown here can be made.

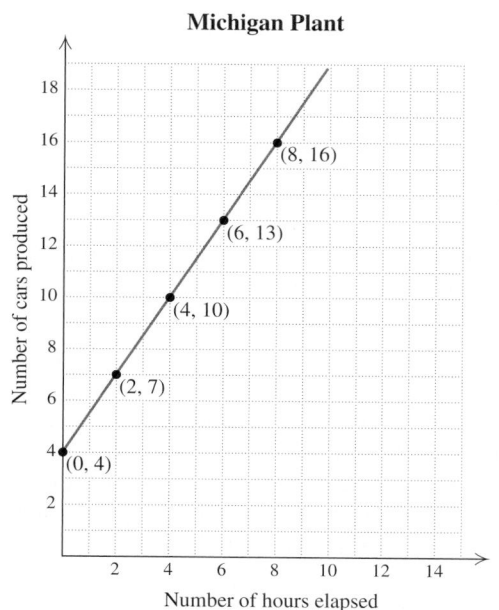

Michigan Plant

Michigan Plant	
Hours Elapsed	**Cars Produced**
0	4
2	7
4	10
6	13
8	16

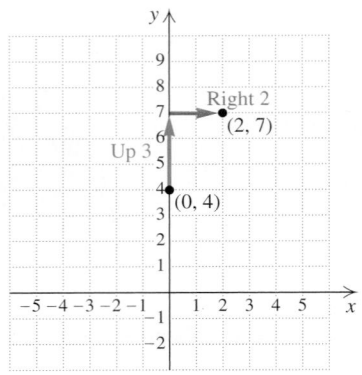

To confirm that the production rate is still $\frac{3}{2}$, we calculate the slope. Recall that

$$\text{Slope} = \frac{\text{change in } y}{\text{change in } x} = \frac{\text{rise}}{\text{run}} = \frac{y_2 - y_1}{x_2 - x_1},$$

where (x_1, y_1) and (x_2, y_2) are any two points on the graphed line. Here we select $(0, 4)$ and $(2, 7)$:

$$\text{Slope} = \frac{\text{change in } y}{\text{change in } x} = \frac{7 - 4}{2 - 0} = \frac{3}{2}.$$

Knowing that the slope is $\frac{3}{2}$, we could have drawn the graph by plotting $(0, 4)$ and from there moving *up* 3 units and *to the right* 2 units. This would have located the point $(2, 7)$. Using $(0, 4)$ and $(2, 7)$, we can then draw the line. This is the method used in the next example.

EXAMPLE 1

Draw a line that has slope $\frac{1}{4}$ and y-intercept $(0, 2)$.

Solution We plot $(0, 2)$ and from there move *up* 1 unit and *to the right* 4 units. This locates the point $(4, 3)$. We plot $(4, 3)$ and draw a line passing through $(0, 2)$ and $(4, 3)$, as shown on the right below.

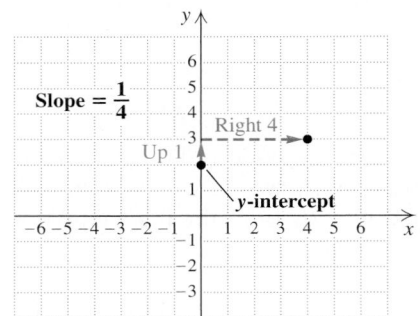

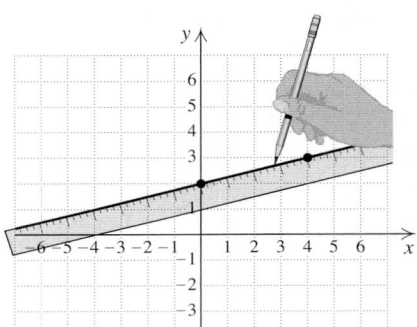

Equations in Slope–Intercept Form

It is not difficult to find the slope and the y-intercept of a line from its equation. Recall from Section 3.3 that to find the y-intercept of an equation's graph, we replace x with 0 and solve the resulting equation for y. For example, to find the y-intercept of the graph of $y = 2x + 3$, we replace x with 0 and solve as follows:

$$y = 2x + 3$$
$$= 2 \cdot 0 + 3 = 0 + 3 = 3. \qquad \text{The } y\text{-intercept is } (0, 3).$$

The y-intercept of the graph of $y = 2x + 3$ is $(0, 3)$. It can be similarly shown that the graph of $y = mx + b$ has the y-intercept $(0, b)$.

To calculate the slope of the graph of $y = 2x + 3$, we need two ordered pairs that are solutions of the equation. The y-intercept $(0, 3)$ is one pair; a second pair, $(1, 5)$, can be found by substituting 1 for x. We then have

$$\text{Slope} = \frac{\text{change in } y}{\text{change in } x} = \frac{5 - 3}{1 - 0} = \frac{2}{1} = 2.$$

Note that the slope, 2, is also the x-coefficient in $y = 2x + 3$. It can be similarly shown that the graph of any equation of the form $y = mx + b$ has slope m (see Exercise 89).

The Slope–Intercept Equation

The equation $y = mx + b$ is called the *slope–intercept equation*. The equation represents a line of slope m with y-intercept $(0, b)$.

The equation of any nonvertical line can be written in this form.

EXAMPLE 2   Find the slope and the y-intercept of each line whose equation is given.

a) $y = \frac{4}{5}x - 8$

b) $2x + y = 5$

c) $3x - 4y = 7$

Solution

a) We rewrite $y = \frac{4}{5}x - 8$ as $y = \frac{4}{5}x + (-8)$. Now we simply read the slope and the y-intercept from the equation:

$$y = \frac{4}{5}x + (-8).$$

The slope is $\frac{4}{5}$. The y-intercept is $(0, -8)$.

b) We first solve for y to find an equivalent equation in the form $y = mx + b$:

$$2x + y = 5$$
$$y = -2x + 5. \qquad \text{Adding } -2x \text{ to both sides}$$

The slope is -2. The y-intercept is $(0, 5)$.

c) We rewrite the equation in the form $y = mx + b$:

$$3x - 4y = 7$$
$$-4y = -3x + 7 \qquad \text{Adding } -3x \text{ to both sides}$$
$$y = -\tfrac{1}{4}(-3x + 7) \qquad \text{Multiplying both sides by } -\tfrac{1}{4}$$
$$y = \tfrac{3}{4}x - \tfrac{7}{4}. \qquad \text{Using the distributive law}$$

The slope is $\frac{3}{4}$. The y-intercept is $\left(0, -\frac{7}{4}\right)$.

EXAMPLE 3   A line has slope $-\frac{12}{5}$ and y-intercept $(0, 11)$. Find an equation of the line.

Solution   We use the slope–intercept equation, substituting $-\frac{12}{5}$ for m and 11 for b:

$$y = mx + b = -\tfrac{12}{5}x + 11.$$

The desired equation is $y = -\frac{12}{5}x + 11$.

EXAMPLE 4

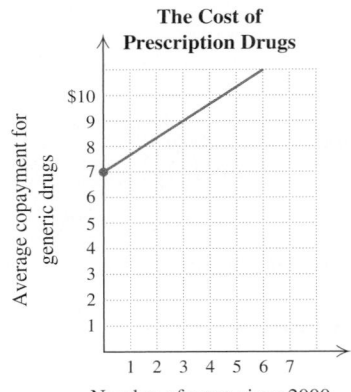

The Cost of Prescription Drugs

Average copayment for generic drugs

Number of years since 2000

Determine an equation for the graph shown at left (*Source*: Based on information from The Kaiser Family Foundation; Health Research and Education Trust, reported in *The Daily Reporter*, Greenfield IN, 9/9/03).

Solution   To write an equation for a line, we can use slope–intercept form, provided the slope and the y-intercept are known. From the graph, we see that $(0, 7)$ is the y-intercept. Looking closely, we see that the line passes through $(3, 9)$. We can either count squares on the graph or use the formula to calculate the slope:

$$m = \frac{\text{change in } y}{\text{change in } x} = \frac{9 - 7}{3 - 0} = \frac{2}{3}.$$

The desired equation is

$$y = \frac{2}{3}x + 7, \qquad \text{Using } \tfrac{2}{3} \text{ for } m \text{ and 7 for } b$$

where y is the average cost, in dollars, of a copayment for generic drugs x years after 2000.

Graphing and Slope–Intercept Form

In Example 1, we drew a graph, knowing only the slope and the y-intercept. In Example 2, we determined the slope and the y-intercept of a line by examining its equation. We now combine the two procedures to develop a quick way to graph a linear equation.

EXAMPLE 5

Graph: **(a)** $y = \frac{3}{4}x + 5$; **(b)** $2x + 3y = 3$.

Solution

a) From the equation $y = \frac{3}{4}x + 5$, we see that the slope of the graph is $\frac{3}{4}$ and the y-intercept is $(0, 5)$. We plot $(0, 5)$ and then consider the slope, $\frac{3}{4}$. Starting at $(0, 5)$, we plot a second point by moving *up* 3 units (since the numerator is *positive* and corresponds to the change in y) and *to the right* 4 units (since the denominator is *positive* and corresponds to the change in x). We reach a new point, $(4, 8)$.

We can also rewrite the slope as $\frac{-3}{-4}$. We again start at the y-intercept, $(0, 5)$, but move *down* 3 units (since the numerator is *negative* and corresponds to the change in y) and *to the left* 4 units (since the denominator is *negative* and corresponds to the change in x). We reach another point, $(-4, 2)$. Once two or three points have been plotted, the line representing all solutions of $y = \frac{3}{4}x + 5$ can be drawn.

Student Notes

Graphing an equation in slope–intercept form can be done rather easily. First, identify and plot the y-intercept. Then identify the slope and use the slope to plot a second point. Finally, use a straightedge to draw a line passing through the two points.

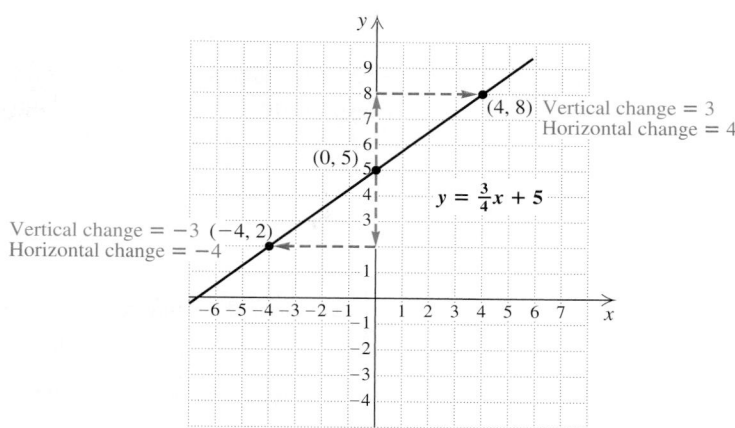

technology connection

Using a standard $[-10, 10, -10, 10]$ window, graph the equations $y_1 = \frac{2}{3}x + 1$, $y_2 = \frac{3}{8}x + 1$, $y_3 = \frac{2}{3}x + 5$, and $y_4 = \frac{3}{8}x + 5$. If you can, use your graphing calculator in the MODE that graphs equations *simultaneously*. Once all lines have been drawn, try to decide which equation corresponds to each line. After matching equations with lines, you can check your matches by using TRACE and the up and down arrow keys to move from one line to the next. The number of the equation will appear in a corner of the screen.

1. Graph $y_1 = -\frac{3}{4}x - 2$, $y_2 = -\frac{1}{5}x - 2$, $y_3 = -\frac{3}{4}x - 5$, and $y_4 = -\frac{1}{5}x - 5$ using the SIMULTANEOUS mode. Then match each line with the corresponding equation. Check using TRACE.

b) To graph $2x + 3y = 3$, we first rewrite it in slope–intercept form:

$$2x + 3y = 3$$
$$3y = -2x + 3 \qquad \text{Adding } -2x \text{ to both sides}$$
$$y = \tfrac{1}{3}(-2x + 3) \qquad \text{Multiplying both sides by } \tfrac{1}{3}$$
$$y = -\tfrac{2}{3}x + 1. \qquad \text{Using the distributive law}$$

To graph $y = -\frac{2}{3}x + 1$, we first plot the y-intercept, $(0, 1)$. We can think of the slope as $\frac{-2}{3}$. Starting at $(0, 1)$ and using the slope, we find a second point by moving *down* 2 units (since the numerator is *negative*) and *to the right* 3 units (since the denominator is *positive*). We plot the new point, $(3, -1)$. In a similar manner, we can move from the point $(3, -1)$ to locate a third point, $(6, -3)$. The line can then be drawn.

Since $-\frac{2}{3} = \frac{2}{-3}$, an alternative approach is to again plot $(0, 1)$, but this time move *up* 2 units (since the numerator is *positive*) and *to the left* 3 units (since the denominator is *negative*). This leads to another point on the graph, $(-3, 3)$.

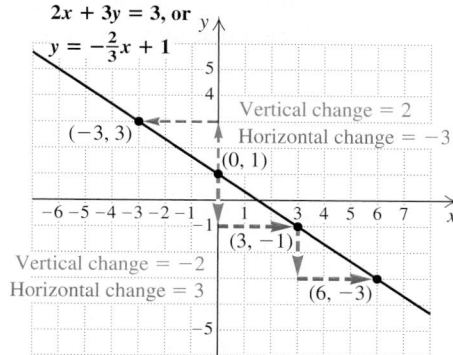

It is important to be able to use both $\frac{2}{-3}$ and $\frac{-2}{3}$ to draw the graph.

Slope–intercept form allows us to quickly determine the slope of a line by simply inspecting its equation. This can be especially helpful when attempting to decide whether two lines are parallel or perpendicular.

Parallel and Perpendicular Lines

Two lines are parallel if they lie in the same plane and do not intersect no matter how far they are extended. If two lines are vertical, they are parallel. How can we tell if nonvertical lines are parallel? The answer is simple: We look at their slopes.

Slope and Parallel Lines

Two lines are parallel if they have the same slope.

EXAMPLE 6   Determine whether the graphs of $y = -3x + 4$ and $6x + 2y = -10$ are parallel.

Solution   When two lines have the same slope but different y-intercepts, they are parallel.

One of the two equations given,

$$y = -3x + 4,$$

represents a line with slope -3 and y-intercept $(0, 4)$. To find the slope of the other line, we need to rewrite

$$6x + 2y = -10$$

in slope–intercept form:

$$6x + 2y = -10$$
$$2y = -6x - 10 \qquad \text{Adding } -6x \text{ to both sides}$$
$$y = -3x - 5. \qquad \text{The slope is } -3 \text{ and the } y\text{-intercept is } (0, -5).$$

Since both lines have slope -3 but different y-intercepts, the graphs are parallel. There is no need for us to actually graph either equation.

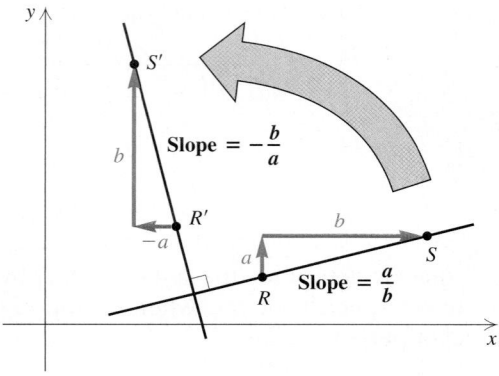

Two lines are perpendicular if they intersect at a right angle. If one line is vertical and another is horizontal, they are perpendicular. There are other instances in which two lines are perpendicular.

Consider a line $\overleftrightarrow{RS}$, as shown at left, with slope a/b. Then think of rotating the figure 90° to get a line $\overleftrightarrow{R'S'}$ perpendicular to $\overleftrightarrow{RS}$. For the new line, the rise and the run are interchanged, but the run is now negative. Thus the slope of the new line is $-b/a$. Let's multiply the slopes:

$$\frac{a}{b}\left(-\frac{b}{a}\right) = -1.$$

This can help us determine which lines are perpendicular.

> **Slope and Perpendicular Lines**
>
> Two lines are perpendicular if the product of their slopes is -1 or if one line is vertical and the other is horizontal.

Thus, if one line has slope m ($m \neq 0$), the slope of a line perpendicular to it is $-1/m$. That is, we take the reciprocal of m ($m \neq 0$) and change the sign.

EXAMPLE 7   Determine whether the graphs of $2x + y = 8$ and $y = \frac{1}{2}x + 7$ are perpendicular.

Solution   First, we find the slope of each line. The second equation,

$$y = \tfrac{1}{2}x + 7,$$

is in slope-intercept form. It represents a line with slope $\frac{1}{2}$.

To find the slope of the other line, we rewrite

$$2x + y = 8$$

in slope–intercept form:

$$2x + y = 8$$
$$y = -2x + 8. \qquad \text{Adding } -2x \text{ to both sides}$$

The slope of the line is -2.

The lines are perpendicular if the product of their slopes is -1. Since

$$\frac{1}{2}(-2) = -1,$$

the graphs are perpendicular.

EXAMPLE 8   Write a slope–intercept equation for the line whose graph is described.

a) Parallel to the graph of $2x - 3y = 7$, with y-intercept $(0, -1)$

b) Perpendicular to the graph of $2x - 3y = 7$, with y-intercept $(0, -1)$

Solution   We begin by determining the slope of the line represented by $2x - 3y = 7$:

$$2x - 3y = 7$$
$$-3y = -2x + 7 \qquad \text{Adding } -2x \text{ to both sides}$$
$$y = \tfrac{2}{3}x - \tfrac{7}{3}. \qquad \text{Dividing both sides by } -3$$

The slope is $\frac{2}{3}$.

a) A line parallel to the graph of $2x - 3y = 7$ has a slope of $\frac{2}{3}$. Since the y-intercept is $(0, -1)$, the slope–intercept equation is

$$y = \tfrac{2}{3}x - 1.$$

b) A line perpendicular to the graph of $2x - 3y = 7$ has a slope that is the negative reciprocal of $\frac{2}{3}$, or $-\frac{3}{2}$. Since the y-intercept is $(0, -1)$, the slope–intercept equation is

$$y = -\tfrac{3}{2}x - 1.$$

Exercise Set

3.6

⤷ *Concept Reinforcement   In each of Exercises 1–6,
match the phrase with the most appropriate choice from
the column on the right.*

1. ____ The y-intercept of the graph of $y = 2x - 3$

2. ____ The y-intercept of the graph of $y = 3x - 2$

3. ____ The slope of the graph of $y = 3x - 2$

4. ____ The slope of the graph of $y = 2x - 3$

5. ____ The slope of the graph of $y = \frac{2}{3}x + 3$

6. ____ The y-intercept of the graph of $y = \frac{2}{3}x + \frac{3}{4}$

a) $\left(0, \frac{3}{4}\right)$

b) 2

c) $(0, -3)$

d) $\frac{2}{3}$

e) $(0, -2)$

f) 3

Draw a line that has the given slope and y-intercept.

7. Slope $\frac{2}{5}$; y-intercept $(0, 1)$

8. Slope $\frac{3}{5}$; y-intercept $(0, -1)$

9. Slope $\frac{5}{3}$; y-intercept $(0, -2)$

10. Slope $\frac{5}{2}$; y-intercept $(0, 1)$

11. Slope $-\frac{3}{4}$; y-intercept $(0, 5)$

12. Slope $-\frac{4}{5}$; y-intercept $(0, 6)$

13. Slope 2; y-intercept $(0, -4)$

14. Slope -2; y-intercept $(0, -3)$

15. Slope -3; y-intercept $(0, 2)$

16. Slope 3; y-intercept $(0, 4)$

*Find the slope and the y-intercept of each line whose
equation is given.*

17. $y = -\frac{2}{7}x + 5$

18. $y = -\frac{3}{8}x + 4$

19. $y = \frac{5}{8}x + 3$

20. $y = \frac{4}{5}x + 1$

21. $y = \frac{9}{5}x - 4$

22. $y = \frac{7}{4}x - 5$

23. $-3x + y = 7$

24. $-4x + y = 7$

25. $5x + 2y = 8$

26. $3x + 4y = 12$

Aha! **27.** $y = 4$

28. $y - 3 = 5$

29. $2x - 5y = -8$

30. $5x - 6y = 9$

*Find the slope–intercept equation for the line with the
indicated slope and y-intercept.*

31. Slope 3; y-intercept $(0, 7)$

32. Slope -4; y-intercept $(0, -2)$

33. Slope $\frac{7}{8}$; y-intercept $(0, -1)$

34. Slope $\frac{5}{7}$; y-intercept $(0, 4)$

35. Slope $-\frac{5}{3}$; y-intercept $(0, -8)$

36. Slope $\frac{3}{4}$; y-intercept $(0, 23)$

Aha! **37.** Slope 0; y-intercept $(0, 3)$

38. Slope 7; y-intercept $(0, 0)$

Determine an equation for each graph shown.

39.

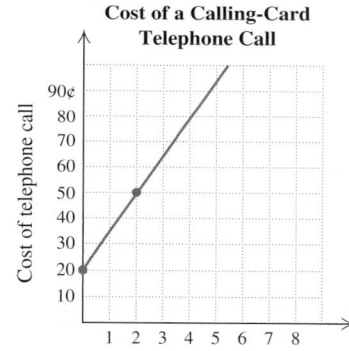

U.S. Bottled Water Consumption

Number of gallons per person

Number of years since 1990

Based on information from the International Bottled Water Association 2002

40.

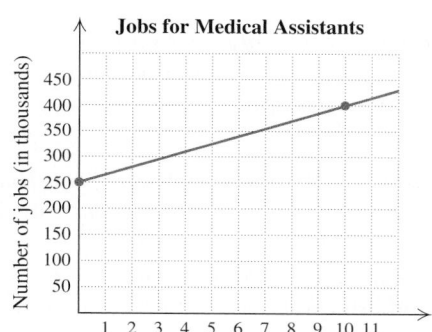

Cost of a Calling-Card Telephone Call

Cost of telephone call

Number of minutes

41.

Jobs for Medical Assistants

Number of jobs (in thousands)

Number of years since 1998

Estimates based on information from *Handbook of U.S. Labor*

42.

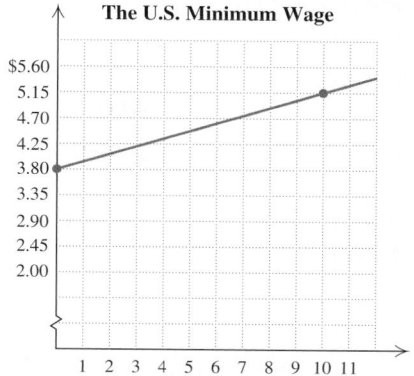

The U.S. Minimum Wage

Number of years since 1990

Based on information from United for a Fair Economy

Graph.

43. $y = \frac{3}{5}x + 2$ **44.** $y = -\frac{3}{5}x - 1$

45. $y = -\frac{3}{5}x + 1$ **46.** $y = \frac{3}{5}x - 2$

47. $y = \frac{5}{3}x + 3$ **48.** $y = \frac{5}{3}x - 2$

49. $y = -\frac{3}{2}x - 2$ **50.** $y = -\frac{4}{3}x + 3$

51. $2x + y = 1$ **52.** $3x + y = 2$

53. $3x + y = 0$ **54.** $2x + y = 0$

55. $2x + 3y = 9$ **56.** $4x + 5y = 15$

57. $x - 4y = 12$ **58.** $x + 5y = 20$

Determine whether each pair of equations represents parallel lines.

59. $y = \frac{2}{3}x + 7,$ **60.** $y = -\frac{5}{4}x + 1,$
$\quad\;\; y = \frac{2}{3}x - 5$ $y = \frac{5}{4}x + 3$

61. $y = 2x - 5,$ **62.** $y = -3x + 1,$
$\quad\;\; 4x + 2y = 9$ $6x + 2y = 8$

63. $3x + 4y = 8,$ **64.** $3x = 5y - 2,$
$\quad\;\; 7 - 12y = 9x$ $10y = 4 - 6x$

Determine whether each pair of equations represents perpendicular lines.

65. $y = 4x - 5,$ **66.** $2x - 5y = -3,$
$\quad\;\; 4y = 8 - x$ $2x + 5y = 4$

67. $x - 2y = 5,$ **68.** $y = -x + 7,$
$\quad\;\; 2x + 4y = 8$ $y - x = 3$

69. $2x + 3y = 1,$ **70.** $y = 5 - 3x,$
$\quad\;\; 3x - 2y = 1$ $3x - y = 8$

Write a slope–intercept equation of the line whose graph is described.

71. Parallel to the graph of $y = 5x - 7$; y-intercept $(0, 11)$

72. Parallel to the graph of $2x - y = 1$; y-intercept $(0, -3)$

73. Perpendicular to the graph of $2x + y = 0$; y-intercept $(0, 0)$

74. Perpendicular to the graph of $y = \frac{1}{3}x + 7$; y-intercept $(0, 5)$

Aha! **75.** Parallel to the graph of $y = x$; y-intercept $(0, 3)$

Aha! **76.** Perpendicular to the graph of $y = x$; y-intercept $(0, 0)$

77. Perpendicular to the graph of $x + y = 3$; y-intercept $(0, -4)$

78. Parallel to the graph of $3x + 2y = 5$; y-intercept $(0, -1)$

79. Can a horizontal line be graphed using the method of Example 5? Why or why not?

80. Can a vertical line be graphed using the method of Example 5? Why or why not?

SKILL MAINTENANCE

Solve. [2.3]

81. $y - k = m(x - h)$, for y

82. $y - 9 = -2(x + 4)$, for y

Simplify. [1.6]

83. $-5 - (-7)$ **84.** $7 - (-9)$

85. $-3 - 6$ **86.** $-2 - 8$

SYNTHESIS

87. Explain how it is possible for an incorrect graph to be drawn, even after plotting three points that line up.

88. Which would you prefer, and why: graphing an equation of the form $y = mx + b$ or graphing an equation of the form $Ax + By = C$?

89. Show that the slope of the line given by $y = mx + b$ is m. (*Hint*: Substitute both 0 and 1 for x to find two pairs of coordinates. Then use the formula, Slope = change in y/change in x.)

90. Write an equation of the line with the same slope as the line given by $5x + 2y = 8$ and the same y-intercept as the line given by $3x - 7y = 10$.

91. Write an equation of the line parallel to the line given by $2x - 6y = 10$ and having the same y-intercept as the line given by $9x + 6y = 18$.

92. Write an equation of the line parallel to the line given by $3x - 2y = 8$ and having the same y-intercept as the line given by $2y + 3x = -4$.

93. Write an equation of the line perpendicular to the line given by $3x - 5y = 8$ and having the same y-intercept as the line given by $2x + 4y = 12$.

94. Write an equation of the line perpendicular to the line given by $2x + 3y = 7$ and having the same y-intercept as the line given by $5x + 2y = 10$.

95. Write an equation of the line perpendicular to the line given by $3x - 2y = 9$ and having the same y-intercept as the line given by $2x + 5y = 0$.

96. Write an equation of the line perpendicular to the line given by $2x + 5y = 6$ that passes through $(2, 6)$. (*Hint*: Draw a graph.)

97. *Aerobic exercise.* The formula $T = -\frac{3}{4}a + 165$ can be used to determine the *target heart rate*, in beats per minute, for a person, a years old, participating in aerobic exercise. Graph the equation and interpret the significance of its slope.

3.7 Point–Slope Form

Writing Equations in Point–Slope Form   •   Graphing and Point–Slope Form   •   Estimations and Predictions Using Two Points

There are many applications in which a slope—or a rate of change—and an ordered pair are known. When the ordered pair is the y-intercept, an equation in slope–intercept form can be easily produced. When the ordered pair represents a point other than the y-intercept, a different form, known as *point–slope form*, is more convenient.

Writing Equations in Point–Slope Form

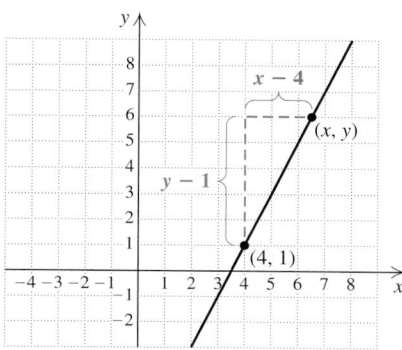

Consider a line with slope 2 passing through the point (4, 1), as shown in the figure. In order for a point (x, y) to be on the line, the coordinates x and y must be solutions of the slope equation

$$\frac{y - 1}{x - 4} = 2.$$

Take a moment to examine this equation. Pairs like (5, 3) and (3, −1) are solutions, since

$$\frac{3 - 1}{5 - 4} = 2 \quad \text{and} \quad \frac{-1 - 1}{3 - 4} = 2.$$

Note, however, that (4, 1) is not itself a solution of the equation:

$$\frac{1 - 1}{4 - 4} \neq 2.$$

To avoid this difficulty, we can use the multiplication principle:

$$(x - 4) \cdot \frac{y - 1}{x - 4} = 2(x - 4) \qquad \text{Multiplying both sides by } x - 4$$

$$y - 1 = 2(x - 4). \qquad \text{Removing a factor equal to 1: } \frac{x - 4}{x - 4} = 1$$

This is considered **point–slope form** for the line shown above. A point–slope equation can be written any time a line's slope and a point on the line are known.

The Point–Slope Equation

The equation $y - y_1 = m(x - x_1)$ is called the *point–slope equation* for the line with slope m that contains the point (x_1, y_1).

Point–slope form is especially useful in more advanced mathematics courses, where problems similar to the following often arise.

EXAMPLE 1   Write a point–slope equation for the line with slope $\frac{1}{5}$ that contains the point $(7, 2)$.

Solution   We substitute $\frac{1}{5}$ for m, 7 for x_1, and 2 for y_1:

$$y - y_1 = m(x - x_1) \qquad \text{Using the point–slope equation}$$
$$y - 2 = \tfrac{1}{5}(x - 7). \qquad \text{Substituting}$$

EXAMPLE 2   Write a point–slope equation for the line with slope $-\frac{4}{3}$ that contains the point $(1, -6)$.

Solution   We substitute $-\frac{4}{3}$ for m, 1 for x_1, and -6 for y_1:

$$y - y_1 = m(x - x_1) \qquad \text{Using the point–slope equation}$$
$$y - (-6) = -\tfrac{4}{3}(x - 1). \qquad \text{Substituting}$$

EXAMPLE 3   Write the slope–intercept equation for the line with slope 2 that contains the point $(3, 1)$.

Student Notes

There are several forms in which a line's equation can be written. For instance, as shown in Example 3, $y - 1 = 2(x - 3), y - 1 = 2x - 6$, and $y = 2x - 5$ all are equations for the same line.

Solution   There are two parts to this solution. First, we write an equation in point–slope form:

$$y - y_1 = m(x - x_1)$$
$$y - 1 = 2(x - 3). \qquad \text{Substituting}$$

Next, we find an equivalent equation of the form $y = mx + b$:

$$y - 1 = 2(x - 3)$$
$$y - 1 = 2x - 6 \qquad \text{Using the distributive law}$$
$$y = 2x - 5. \qquad \text{Adding 1 to both sides to get slope–intercept form}$$

EXAMPLE 4   Consider the line given by the equation $8y = 7x - 24$.

a) Write the slope–intercept equation for a parallel line passing through $(-1, 2)$.

b) Write the slope–intercept equation for a perpendicular line passing through $(-1, 2)$.

Solution   Both parts (a) and (b) require us to find the slope of the line given by $8y = 7x - 24$. To do so, we solve for y to find slope–intercept form:

$$8y = 7x - 24$$
$$y = \tfrac{7}{8}x - 3. \qquad \text{Multiplying both sides by } \tfrac{1}{8}$$
$$\underline{\qquad\qquad} \text{The slope is } \tfrac{7}{8}.$$

a) The slope of any parallel line will be $\frac{7}{8}$. The point–slope equation yields

$$y - 2 = \frac{7}{8}[x - (-1)]$$ Substituting $\frac{7}{8}$ for the slope and $(-1, 2)$ for the point

$$y - 2 = \frac{7}{8}[x + 1]$$

$$y = \frac{7}{8}x + \frac{7}{8} + 2$$ Using the distributive law and adding 2 to both sides

$$y = \frac{7}{8}x + \frac{23}{8}.$$

b) The slope of a perpendicular line is given by the opposite of the reciprocal of $\frac{7}{8}$, or $-\frac{8}{7}$. The point–slope equation yields

$$y - 2 = -\frac{8}{7}[x - (-1)]$$ Substituting $-\frac{8}{7}$ for the slope and $(-1, 2)$ for the point

$$y - 2 = -\frac{8}{7}[x + 1]$$

$$y = -\frac{8}{7}x - \frac{8}{7} + 2$$ Using the distributive law and adding 2 to both sides

$$y = -\frac{8}{7}x + \frac{6}{7}.$$

Graphing and Point–Slope Form

When we know a line's slope and a point that is on the line, we can draw the graph, much as we did in Section 3.6. For example, the information given in the statement of Example 3 is sufficient for drawing a graph.

EXAMPLE 5   Graph the line with slope 2 that passes through (3, 1).

Solution   We plot (3, 1), move *up* 2 and *to the right* 1 $\left(\text{since } 2 = \frac{2}{1}\right)$, and draw the line.

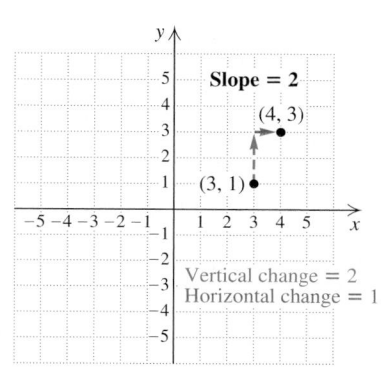

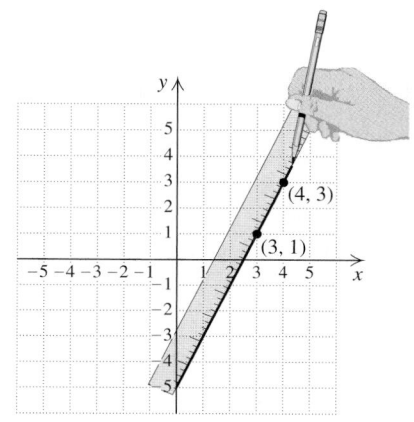

EXAMPLE 6   Graph: $y - 2 = 3(x - 4)$.

Solution   Since $y - 2 = 3(x - 4)$ is in point–slope form, we know that the line has slope 3, or $\frac{3}{1}$, and passes through the point $(4, 2)$. We plot $(4, 2)$ and then find a second point by moving *up* 3 units and *to the right* 1 unit. The line can then be drawn, as shown below.

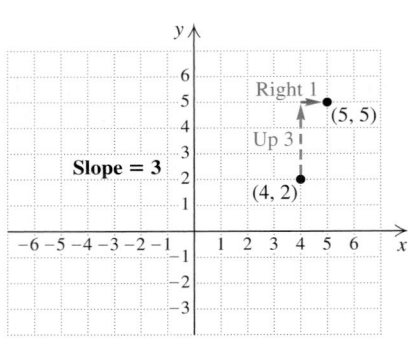 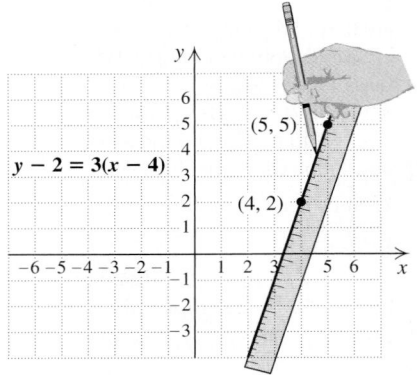

EXAMPLE 7   Graph: $y + 4 = -\frac{5}{2}(x + 3)$.

Solution   Once we have written the equation in point–slope form, $y - y_1 = m(x - x_1)$, we can proceed much as we did in Example 6. To find an equivalent equation in point–slope form, we subtract opposites instead of adding:

$$y + 4 = -\frac{5}{2}(x + 3)$$
$$y - (-4) = -\frac{5}{2}(x - (-3)).$$ Subtracting a negative instead of adding a positive. This is now in point–slope form.

From this last equation, $y - (-4) = -\frac{5}{2}(x - (-3))$, we see that the line passes through $(-3, -4)$ and has slope $-\frac{5}{2}$, or $\frac{5}{-2}$.

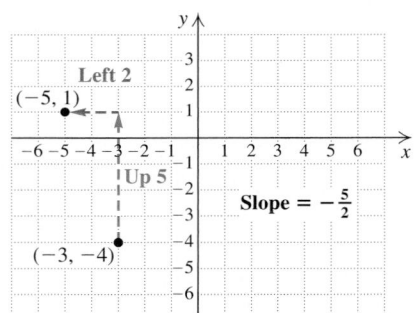 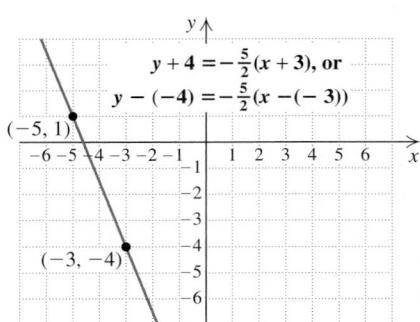

Estimations and Predictions Using Two Points

It is possible to use line graphs to estimate real-life quantities that are not already known. To do so, we calculate the coordinates of an unknown point by using two points with known coordinates. When the unknown point is located *between* the two points, this process is called **interpolation**.* Sometimes a graph passing through the known points is *extended* to predict future values. Making predictions in this manner is called **extrapolation**.* In statistics, methods exist for using a set of several points to interpolate or extrapolate values using curves other than lines.

EXAMPLE 8

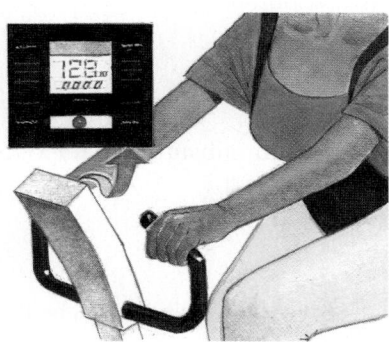

Aerobic exercise.   A person's target heart rate is the number of beats per minute that bring the most aerobic benefit to his or her heart. The target heart rate for a 20-year-old is 150 beats per minute and for a 60-year-old, 120 beats per minute.

a) Graph the given data and calculate the target heart rate for a 36-year-old.

b) Calculate the target heart rate for a 75-year-old.

Solution

a) We first draw a horizontal axis for "Age" and a vertical axis for "Target heart rate." Next, we number the axes, using a scale that will permit us to view both the given and the desired data. The given information allows us to then plot (20, 150) and (60, 120).

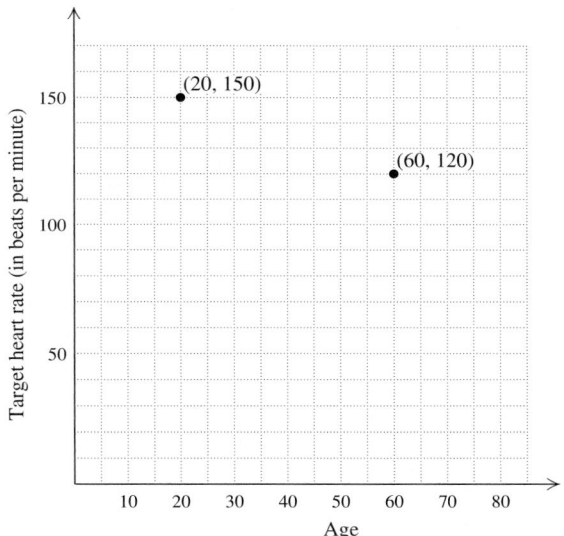

Next, we draw a line passing through both points. The slope of the line is given by

$$m = \frac{\text{change in } y}{\text{change in } x} = \frac{150 - 120 \text{ beats per minute}}{20 - 60 \text{ years}}$$

$$= \frac{30 \text{ beats per minute}}{-40 \text{ years}} = -\frac{3}{4} \text{ beat per minute per year.}$$

---

*Both interpolation and extrapolation can be performed using more than two known points and using curves other than lines.

The target heart rate drops at a rate of $\frac{3}{4}$ beat per minute for each year that we age. We can use either of the given points to write a point–slope equation for the line. Let's use (20, 150) and then use algebra to write an equivalent equation in slope–intercept form:

$$y - 150 = -\frac{3}{4}(x - 20) \qquad \text{This is a point–slope equation.}$$

$$y - 150 = -\frac{3}{4}x + 15 \qquad \text{Using the distributive law}$$

$$y = -\frac{3}{4}x + 165. \qquad \begin{array}{l}\text{Adding 150 to both sides; this is}\\\text{slope–intercept form.}\end{array}$$

To calculate the target heart rate for a 36-year-old, we substitute 36 for x in the slope–intercept equation:

$$y = -\frac{3}{4} \cdot 36 + 165 = -27 + 165 = 138.$$

As the graph confirms, the target heart rate for a 36-year-old is about 138 beats per minute.

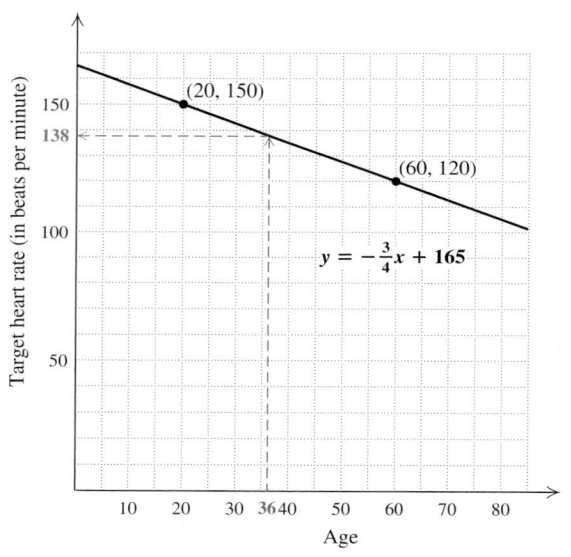

Because 36 is *between* the given values of 20 and 60, we are *interpolating* here.

b) To calculate the target heart rate for a 75-year-old, we again substitute for x in the slope–intercept equation:

$$y = -\frac{3}{4} \cdot 75 + 165 = -56.25 + 165 = 108.75 \approx 109.$$

As the graph confirms, the target heart rate for a 75-year-old is about 109 beats per minute.

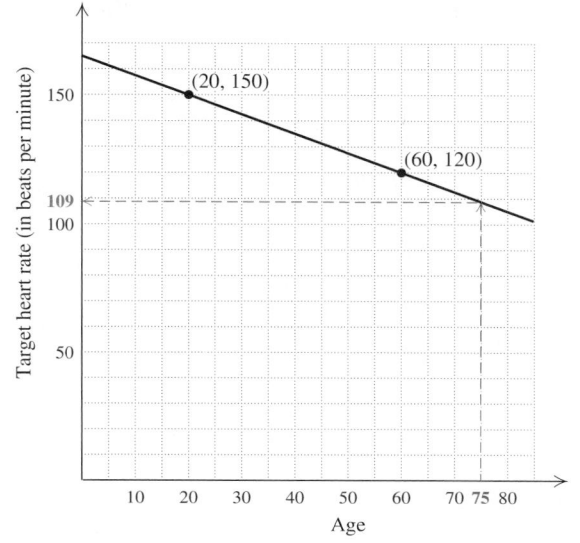

Because 75 is *beyond* the given values, we are *extrapolating* here.

CONNECTING THE CONCEPTS

We have now studied the slope–intercept, point–slope, and standard forms of a linear equation. These are the most common ways in which linear equations are written. Depending on what information we are given and what information we are seeking, one form may be more useful than the others. A referenced summary is given below.

Slope–intercept form, $y = mx + b$	• Useful when an equation is needed and the slope and y-intercept are given. See Example 3 on p. 203. • Useful when a line's slope and y-intercept are needed. See Example 2 on p. 203.
Standard form, $Ax + By = C$	• Allows for easy calculation of intercepts. See Example 2 on p. 170. • Will prove useful in future work. See Sections 8.1–8.2.
Point–slope form, $y - y_1 = m(x - x_1)$	• Useful when an equation is needed and the slope and a point on the line are given. See Example 1 on p. 212. • Will prove useful in future work with curves and tangents in calculus.

Exercise Set

3.7

Concept Reinforcement   *In each of Exercises 1–8, match the given information about a line with the appropriate equation from the column on the right.*

1. _____ Slope 5; includes (2, 3)

2. _____ Slope -5; includes (2, 3)

3. _____ Slope -5; includes $(-2, -3)$

4. _____ Slope 5; includes $(-2, -3)$

5. _____ Slope 5; includes (3, 2)

6. _____ Slope -5; includes (3, 2)

7. _____ Slope -5; includes $(-3, -2)$

8. _____ Slope 5; includes $(-3, -2)$

a) $y + 3 = 5(x + 2)$

b) $y - 2 = 5(x - 3)$

c) $y + 2 = 5(x + 3)$

d) $y - 3 = -5(x - 2)$

e) $y + 3 = -5(x + 2)$

f) $y + 2 = -5(x + 3)$

g) $y - 3 = 5(x - 2)$

h) $y - 2 = -5(x - 3)$

In each of Exercises 9–12, match the graph with the appropriate equation from the column on the right.

9.

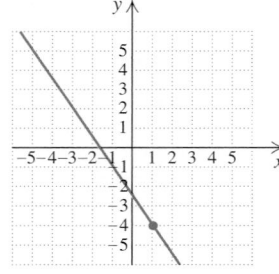

a) $y - 4 = -\frac{3}{2}(x + 1)$

b) $y - 4 = \frac{3}{2}(x + 1)$

c) $y + 4 = -\frac{3}{2}(x - 1)$

d) $y + 4 = \frac{3}{2}(x - 1)$

10.

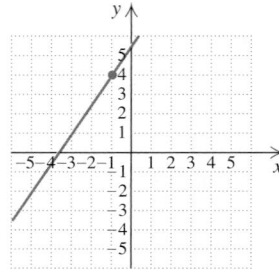

11.

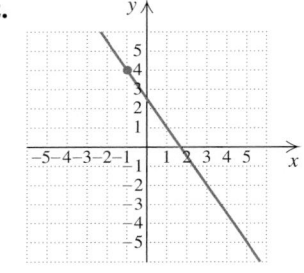

12.

Write a point–slope equation for the line with the given slope that contains the given point.

13. $m = 5$; (6, 2)

14. $m = 4$; (3, 5)

15. $m = -4$; (3, 1)

16. $m = -5$; (6, 2)

17. $m = \frac{3}{2}$; (5, -4)

18. $m = \frac{4}{3}$; (7, -1)

19. $m = \frac{5}{4}$; (-2, 6)

20. $m = \frac{7}{2}$; (-3, 4)

21. $m = -2$; (-4, -1)

22. $m = -3$; (-2, -5)

23. $m = 1$; (-2, 8)

24. $m = -1$; (-3, 6)

Write the slope–intercept equation for the line with the given slope that contains the given point.

25. $m = 2$; (5, 7)

26. $m = 3$; (6, 2)

27. $m = \frac{7}{4}$; (4, -2)

28. $m = \frac{8}{3}$; (3, -4)

29. $m = -3$; (-1, 6)

30. $m = -2$; (-1, 3)

31. $m = -4$; $(-2, -1)$ **32.** $m = -5$; $(-1, -4)$

Aha! **33.** $m = -\frac{5}{6}$; $(0, 4)$ **34.** $m = -\frac{3}{4}$; $(0, 5)$

Write an equation of the line that contains the specified point and is parallel to the indicated line.

35. $(4, 7)$, $x + 2y = 6$

36. $(1, 3)$, $3x - y = 7$

Aha! **37.** $(0, -7)$, $y = 2x + 1$

38. $(-7, 0)$, $5x + 2y = 6$

39. $(2, -6)$, $5x - 3y = 8$

40. $(-4, -5)$, $2x + y = -3$

Write an equation of the line that contains the specified point and is perpendicular to the indicated line.

41. $(3, -2)$, $3x + 6y = 5$

42. $(-3, -5)$, $4x - 2y = 4$

43. $(-4, -7)$, $3x - 5y = 6$

44. $(-4, 5)$, $7x - 2y = 1$

Aha! **45.** $(0, 6)$, $2x - 5 = y$

46. $(4, 0)$, $x - 3y = 0$

47. Graph the line with slope $\frac{4}{3}$ that passes through the point $(1, 2)$.

48. Graph the line with slope $\frac{2}{5}$ that passes through the point $(3, 4)$.

49. Graph the line with slope $-\frac{3}{4}$ that passes through the point $(2, 5)$.

50. Graph the line with slope $-\frac{3}{2}$ that passes through the point $(1, 4)$.

Graph.

51. $y - 2 = \frac{1}{2}(x - 1)$ **52.** $y - 5 = \frac{1}{3}(x - 2)$

53. $y - 1 = -\frac{1}{2}(x - 3)$ **54.** $y - 1 = -\frac{1}{4}(x - 3)$

55. $y + 4 = 3(x + 1)$ **56.** $y + 3 = 2(x + 1)$

57. $y + 3 = -(x + 2)$ **58.** $y + 3 = -1(x - 4)$

59. $y + 1 = -\frac{3}{5}(x + 2)$ **60.** $y + 2 = -\frac{2}{3}(x + 1)$

In Exercises 61–68, graph the given data and determine an equation for the related line (as in Example 8). Then use the equation to answer parts (a) and (b).

61. *Birth rate among teenagers.*   The birth rate among teenagers, measured in births per 1000 females age 15–19, fell steadily from 62.1 in 1991 to 45.9 in 2001.
Source: National Center for Health Statistics; 2001 figure is preliminary.

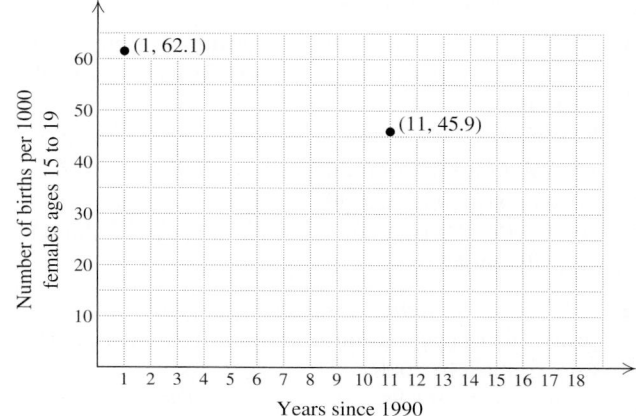

a) Calculate the birth rate among teenagers in 1999.

b) Predict the birth rate among teenagers in 2008.

62. *Food-stamp program participation.*   Participation in the U.S. food-stamp program grew from approximately 17.1 million people in 2000 to approximately 19.1 million in 2002.
Source: U.S. Department of Agriculture

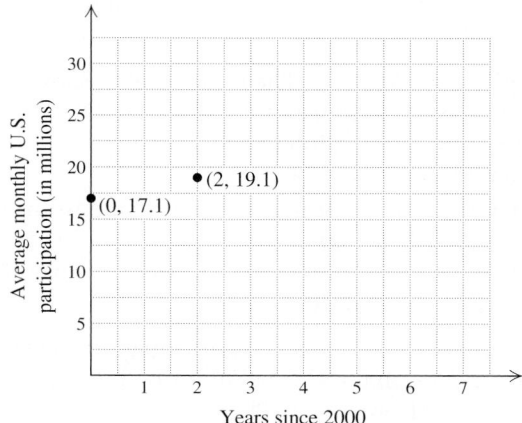

a) Calculate the number of participants in 2001.

b) Predict the number of participants in 2007.

63. *Cigarette smoking.* The percentage of people age 26–34 who smoke has dropped from 45.7% in 1985 to 32.5% in 1998.
Source: *The World Almanac and Book of Facts 2000,* 1999

    a) Calculate the percentage of people age 26–34 who smoked in 1990.

    b) Predict the percentage of people age 26–34 who will smoke in 2008.

64. *Cigarette smoking.* The percentage of people age 18–25 who smoke has changed from 47.4% in 1985 to 41.8% in 1998.
Source: *The World Almanac and Book of Facts 2000,* 1999

    a) Calculate the percentage of people age 18–25 who smoked in 1990.

    b) Predict the percentage of people age 18–25 who will smoke in 2008.

65. *College enrollment.* U.S. college enrollment has grown from approximately 60.3 million in 1990 to 68.3 million in 2000.
Source: *The World Almanac and Book of Facts 2000,* 1999

    a) Calculate the U.S. college enrollment for 1996.

    b) Predict the U.S. college enrollment for 2005.

66. *High school enrollment.* U.S. high school enrollment has changed from approximately 12.5 million in 1990 to 14.9 million in 2000.
Source: *The World Almanac and Book of Facts 2000,* 1999

    a) Calculate the U.S. high school enrollment for 1996.

    b) Predict the U.S. high school enrollment for 2005.

67. *Aging population.* The number of U.S. residents over the age of 65 was approximately 31 million in 1990 and 35.6 million in 2002.
Source: U.S. Bureau of the Census

Aha!   **a)** Calculate the number of U.S. residents over the age of 65 in 1996.

    b) Predict the number of U.S. residents over the age of 65 in 2010.

68. *Urban population.* The percentage of the U.S. population that resides in metropolitan areas increased from about 78% in 1980 to about 80% in 1996.

    a) Calculate the percentage of the U.S. population residing in metropolitan areas in 1992.

    b) Predict the percentage of the U.S. population residing in metropolitan areas in 2008.

Write the slope–intercept equation for the line containing the given pair of points.

69. $(1, 5)$ and $(4, 2)$ **70.** $(3, 7)$ and $(4, 8)$

71. $(-3, 1)$ and $(3, 5)$ **72.** $(-2, 3)$ and $(2, 5)$

73. $(5, 0)$ and $(0, -2)$ **74.** $(-2, 0)$ and $(0, 3)$

75. $(-2, -4)$ and $(2, -1)$ **76.** $(-3, 5)$ and $(-1, -3)$

77. Can equations for horizontal or vertical lines be written in point–slope form? Why or why not?

78. Describe a situation in which it is easier to graph the equation of a line in point–slope form than in slope–intercept form.

SKILL MAINTENANCE

Simplify. [1.8]

79. $(-5)^3$ **80.** $(-2)^6$

81. $3 \cdot 2^4 - 5 \cdot 2^3$ **82.** $5 \cdot 3^2 - 7 \cdot 3$

83. $(-2)^3(-3)^2$ **84.** $(5 - 7)^2(3 - 2 \cdot 2)$

SYNTHESIS

85. Describe a procedure that can be used to write the slope–intercept equation for any nonvertical line passing through two given points.

86. Any nonvertical line has many equations in point–slope form, but only one in slope–intercept form. Why is this?

Graph.

Aha! **87.** $y - 3 = 0(x - 52)$ **88.** $y + 4 = 0(x + 93)$

Write the slope–intercept equation for each line shown.

89.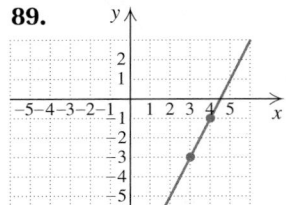

90.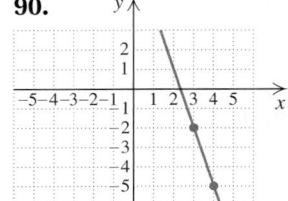

91.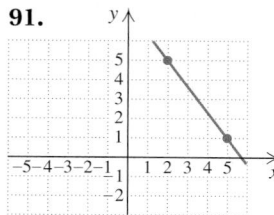

92.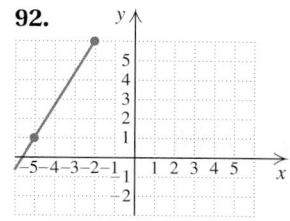

93. Write a point–slope equation of the line passing through $(-4, 7)$ that is parallel to the line given by $2x + 3y = 11$.

94. Write a point–slope equation of the line passing through $(3, -1)$ that is parallel to the line given by $4x - 5y = 9$.

Aha! **95.** Write an equation of the line parallel to the line given by $y = 3 - 4x$ that passes through $(0, 7)$.

96. Write the slope–intercept equation of the line that has the same y-intercept as the line $x - 3y = 6$ and contains the point $(5, -1)$.

97. Write the slope–intercept equation of the line that contains the point $(-1, 5)$ and is parallel to the line passing through $(2, 7)$ and $(-1, -3)$.

98. Write the slope–intercept equation of the line that has x-intercept $(-2, 0)$ and is parallel to $4x - 8y = 12$.

99. Why is slope–intercept form more useful than point–slope form when using a graphing calculator? How can point–slope form be modified so that it is more easily used with graphing calculators?

CORNER

Follow the Bouncing Ball

Focus: Graphing and problem solving

Time: 25 minutes

Group size: 3

Materials: Each group will need a rubber ball, graph paper, and a tape measure.

Does a rubber ball always rebound a fixed percentage of the height from which it is dropped? The following activity attempts to answer this. Please be sure to read all steps before beginning.

ACTIVITY

1. One group member should hold a rubber ball 5 ft above the floor. A second group member should measure this height for accuracy. The ball should then be dropped and caught at the peak of its bounce. The second group member should measure this rebound height and the third group member should record the measurement.

2. Repeat part (1) two more times from the same height. Then find the average of the three rebound heights and use that number to form the ordered pair (original height, rebound height).

3. Repeat parts (1) and (2) four more times at heights of 4 ft, 3 ft, 2 ft, and 1 ft to find five ordered pairs. Graph the five pairs on graph paper and draw a straight line starting at $(0, 0)$ that comes as close as possible to all six points.

4. Use the graph in part (3) to predict the rebound height for a ball dropped from a height of 7 ft. Then perform the drop and check your prediction.

5. Repeat part (4), dropping the ball from a "new" height of the group's choice.

6. Does it appear that a rubber ball will always rebound a fixed percentage of its original height? Why or why not?

7. Compare results from other class groups. What conclusions can you draw?

COLLABORATIVE

3 Study Summary

Information can be communicated quickly and effectively in a visual form through the use of graphs. **Bar graphs, circle graphs**, and **line graphs** are three of the most commonly used types of graphs (pp. 148–150)

When information is given as **ordered pairs**, like $(-3, 2)$ or $(5, 17)$, the pairs can be **plotted** or **graphed** using a **coordinate system** that uses two **axes**, which are most often labeled x and y. The axes intersect at the **origin**, $(0, 0)$ (p. 151).

When an equation contains two variables, any solution is an ordered pair. When all ordered pairs that are solutions of an equation are graphed, we say that we have **graphed the equation** (p. 160). Some equations are said to be **linear**, like $y = 2x - 7$ or $2x + 3y = 12$, because their graphs are lines (p. 160). Other equations have **nonlinear graphs** (p. 165). When a graph crosses the y-axis, the point at which the axis is crossed is called a *y***-intercept.** Similarly, an *x***-intercept** is any point at which a graph crosses the x-axis (p. 169).

A line's slant is measured as its **slope.** Slope measures the **rate** at which the quantity measured on the vertical axis changes with respect to the quantity measured on the horizontal axis (p. 190).

$$\text{Slope} = m = \frac{\text{change in } y}{\text{change in } x} = \frac{\text{rise}}{\text{run}} = \frac{y_2 - y_1}{x_2 - x_1}$$

Any equation that can be written in **standard form**, $Ax + By = C$, is linear (p. 160). An equation in **slope–intercept form**, $y = mx + b$, represents a line with slope m and y-intercept $(0, b)$ (p. 203). An equation in **point–slope form**, $y - y_1 = m(x - x_1)$, represents a line with slope m passing through (x_1, y_1) (p. 211).

Parallel lines have the same slope. **Perpendicular lines** have slopes that are opposite reciprocals; the product of the slopes is -1 (pp. 206, 207).

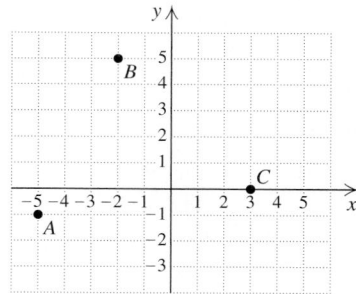

3 Review Exercises

Classify each statement as either true or false.

1. Not every ordered pair lies in one of the four quadrants. [3.1]

2. The equation of a vertical line cannot be written in slope–intercept form. [3.6]

3. Equations for lines written in slope–intercept form appear in the form $Ax + By = C$. [3.6]

4. Every horizontal line has an x-intercept. [3.3]

5. A line's slope is a measure of rate. [3.5]

6. A positive rate of inflation means prices are rising. [3.4]

7. Any two points on a line can be used to determine the line's slope. [3.5]

8. Knowing a line's slope is enough to write the equation of the line. [3.6]

9. Parallel lines that are not vertical have the same slope. [3.6]

10. Knowing two points on a line is enough to write the equation of the line. [3.7]

The following circle graph shows a breakdown of the charities to which Americans donated in a recent year. Use the graph for Exercises 11 and 12.
Source: American Association of Fund-Raising Counsel Trust for Philanthropy Giving USA 2003

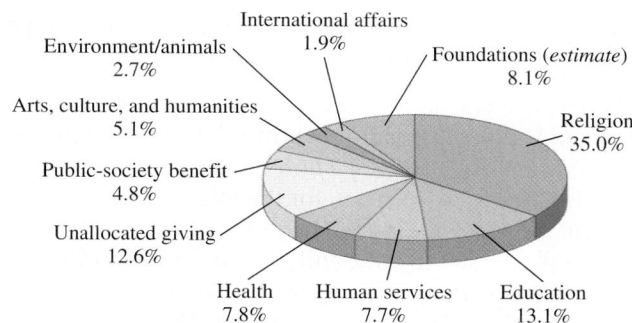

International affairs 1.9%
Environment/animals 2.7%
Foundations (*estimate*) 8.1%
Arts, culture, and humanities 5.1%
Religion 35.0%
Public-society benefit 4.8%
Unallocated giving 12.6%
Health 7.8%
Human services 7.7%
Education 13.1%

All figures are rounded. Total may not be 100%.

11. The citizens of Pittsford are typical of Americans in general with regard to charitable contributions. If the citizens donated a total of $2 million to charities, how much was given to the environment/animals? [3.1]

12. About 5% of the Velez's income of $70,000 goes to charity. If their giving habits are typical, approximate the amount that they contribute to human services. [3.1]

Plot each point. [3.1]

13. $(5, -1)$ 14. $(2, 3)$ 15. $(-4, 0)$

In which quadrant is each point located? [3.1]

16. $(2, -6)$ 17. $(-16.5, -20.3)$

18. $(-3, 12)$

Find the coordinates of each point in the figure. [3.1]

19. *A*   20. *B*   21. *C*

22. Use a grid 10 squares wide and 10 squares high to plot $(-65, -2)$, $(-10, 6)$, and $(25, 7)$. Choose the scale carefully. [3.1]

23. Determine whether the equation $y = 2x - 5$ has *each* ordered pair as a solution: **(a)** $(3, 1)$; **(b)** $(-3, 1)$. [3.2]

24. Show that the ordered pairs $(0, -3)$ and $(2, 1)$ are solutions of the equation $2x - y = 3$. Then use the graph of the two points to determine another solution. Answers may vary. [3.2]

Graph.

25. $y = x - 5$ [3.2]

26. $y = -\frac{1}{4}x$ [3.2]

27. $y = -x + 4$ [3.2]

28. $4x + y = 3$ [3.2]

29. $4x + 5 = 3$ [3.3]

30. $5x - 2y = 10$ [3.3]

31. *Meal service.*   At 8:30 A.M., the Colchester Boy Scouts had served 47 people at their annual pancake breakfast. By 9:15, the total served had reached 67.

   a) Find the Boy Scouts' serving rate, in number of meals per minute. [3.4]

   b) Find the Boy Scouts' serving rate, in minutes per meal. [3.4]

32. *Gas mileage.*   The following graph shows data for a Ford Explorer driven on city streets. At what rate was the vehicle consuming gas? [3.4]

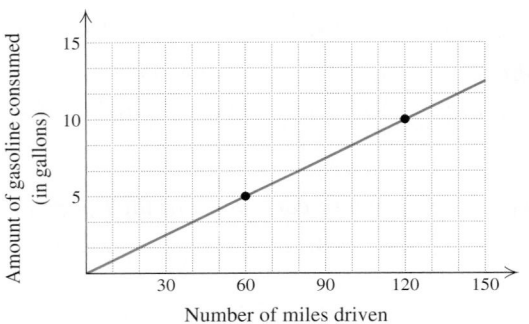

Find the slope of each line. [3.5]

33.

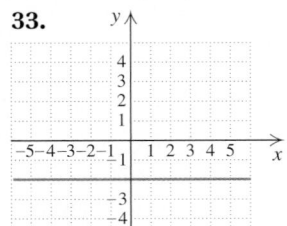

34.

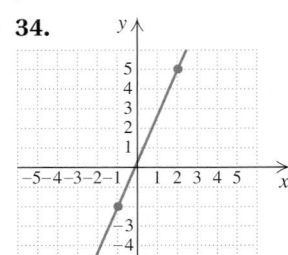

35.

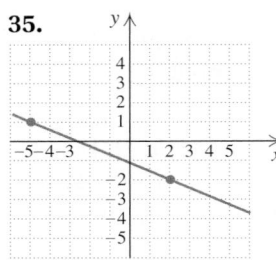

Find the slope of the line containing the given pair of points. [3.5]

36. $(6, 8)$ and $(-2, -4)$

37. $(5, 1)$ and $(-1, 1)$

38. $(-3, 0)$ and $(-3, 5)$

39. $(-8.3, 4.6)$ and $(-9.9, 1.4)$

40. Lombard Street in San Francisco, sometimes called the world's most crooked street, at one point drops 35 ft over a 125-ft horizontal distance. Find the grade. [3.5]

41. Find the *x*-intercept and the *y*-intercept of the line given by $3x + 2y = 18$. [3.3]

42. Find the slope and the *y*-intercept of the line given by $2x + 4y = 20$. [3.6]

Determine whether each pair of lines is parallel, perpendicular, or neither. [3.6]

43. $y + 5 = -x,$
   $x - y = 2$

44. $3x - 5 = 7y,$
   $7y - 3x = 7$

45. Write the slope–intercept equation of the line with slope $-\frac{3}{4}$ and *y*-intercept $(0, 6)$. [3.6]

46. Write a point–slope equation for the line with slope $-\frac{1}{2}$ that contains the point $(3, 6)$. [3.7]

47. The average annual cost of health insurance for a single person rose from about \$2500 in 2000 to about \$3400 in 2003. Graph the data and determine an equation for the related line. Then **(a)** calculate the cost of health insurance for a single person in 2002 and **(b)** predict the cost of health insurance for a single person in 2008. [3.7]
Source: The Kaiser Family Foundation and the Health Research and Education Trust

48. Write the slope–intercept equation for the line with slope 4 that contains the point $(-3, -7)$. [3.7]

49. Write the slope–intercept equation for the line that is perpendicular to the line $3x - 5y = 9$ and that contains the point $(2, -5)$. [3.7]

Graph.

50. $y = \frac{2}{3}x - 5$ [3.6]

51. $2x + y = 4$ [3.3]

52. $y = 6$ [3.3]

53. $x = -2$ [3.3]

54. $y + 2 = -\frac{1}{2}(x - 3)$ [3.7]

SYNTHESIS

55. Can two perpendicular lines share the same y-intercept? Why or why not? [3.3], [3.6]

56. Explain why the first coordinate of the y-intercept is always 0. [3.3]

57. Find the value of m in $y = mx + 3$ such that $(-2, 5)$ is on the graph. [3.2]

58. Find the value of b in $y = -5x + b$ such that $(3, 4)$ is on the graph. [3.2]

59. Find the area and the perimeter of a rectangle for which $(-2, 2)$, $(7, 2)$, and $(7, -3)$ are three of the vertices. [3.1]

60. Find three solutions of $y = 4 - |x|$. [3.2]

3 Chapter Test

Use of tax dollars. *The following pie chart shows how federal income tax dollars were spent in 2003.*
Source: Based on information from www.taxfoundation.com

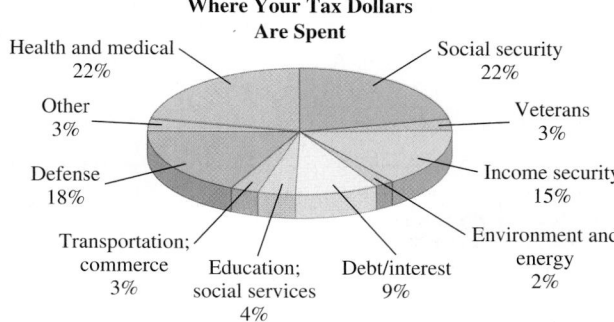

Where Your Tax Dollars Are Spent

Health and medical 22%
Social security 22%
Other 3%
Veterans 3%
Defense 18%
Income security 15%
Transportation; commerce 3%
Education; social services 4%
Debt/interest 9%
Environment and energy 2%

1. Debbie pays 15% of her taxable income of $34,200 in taxes. How much of her income will go to education and social services?

2. Larry pays 10% of his taxable income of $29,000 in taxes. How much of his income will go to social security?

In which quadrant is each point located?

3. $\left(-\frac{1}{2}, 7\right)$

4. $(-5, -6)$

Find the coordinates of each point in the figure.

5. A

6. B

7. C

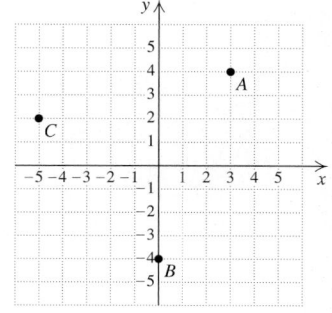

Graph.

8. $y = 2x - 1$

9. $2x - 4y = -8$

10. $y + 1 = 6$

11. $y = \frac{3}{4}x$

12. $y = 7$

13. $2x - y = 3$

14. $x = -1$

Find the x- and y-intercepts. Do not graph.

15. $5x - 3y = 30$

16. $x = 10 - 4y$

Find the slope of the line containing each pair of points.

17. $(4, -1)$ and $(6, 8)$

18. $(-3, -5)$ and $(9, 2)$

19. *Running.* Ted reached the 3-km mark of a race at 2:15 P.M. and the 6-km mark at 2:24 P.M. What is his running rate?

20. At one point Filbert Street, the steepest street in San Francisco, drops 63 ft over a horizontal distance of 200 ft. Find the road grade.

21. Find the slope and the y-intercept of the line given by $y - 3x = 7$.

Graph.

22. $y = \frac{1}{4}x - 2$

23. $y + 4 = -\frac{1}{2}(x - 3)$

Determine without graphing whether each pair of lines is parallel, perpendicular, or neither.

24. $4y + 2 = 3x,$
$-3x + 4y = -12$

25. $y = -2x + 5,$
$2y - x = 6$

26. Write a point–slope equation for the line of slope -3 that contains the point $(6, 8)$.

27. Write the slope–intercept equation of the line that is perpendicular to the line $x - y = 3$ and that contains the point $(3, 7)$.

28. The average urban commuter was stuck in traffic 16 hours in 1982 and 46 hours in 2002.
Source: Urban Mobility Report

a) Calculate the number of hours the average urban commuter was stuck in traffic in 1998.

b) Predict the number of hours the average urban commuter will be stuck in traffic in 2010.

SYNTHESIS

29. Write an equation of the line that is parallel to the graph of $2x - 5y = 6$ and has the same y-intercept as the graph of $3x + y = 9$.

30. A diagonal of a square connects the points $(-3, -1)$ and $(2, 4)$. Find the area and the perimeter of the square.

31. List the coordinates of three other points that are on the same line as $(-2, 14)$ and $(17, -5)$. Answers may vary.

1–3 Cumulative Review

1. Evaluate $\frac{x}{2y}$ for $x = 70$ and $y = 5$. [1.1]

2. Multiply: $3(4x - 5y + 7)$. [1.2]

3. Factor: $15x - 9y + 3$. [1.2]

4. Find the prime factorization of 42. [1.3]

5. Find decimal notation: $\frac{9}{20}$. [1.4]

6. Find the absolute value: $|-4|$. [1.4]

7. Find the opposite of $-\frac{1}{4}$. [1.6]

8. Find the reciprocal of $-\frac{1}{4}$. [1.7]

9. Combine like terms: $2x - 5y + (-3x) + 4y$. [1.6]

10. Find decimal notation: 78.5%. [2.4]

Simplify.

11. $\frac{3}{5} - \frac{5}{12}$ [1.3]

12. $3.4 + (-0.8)$ [1.5]

13. $(-2)(-1.4)(2.6)$ [1.7]

14. $\frac{3}{8} \div \left(-\frac{9}{10}\right)$ [1.7]

15. $1 - [32 \div (4 + 2^2)]$ [1.8]

16. $-5 + 16 \div 2 \cdot 4$ [1.8]

17. $y - (3y + 7)$ [1.8]

18. $3(x - 1) - 2[x - (2x + 7)]$ [1.8]

Solve.

19. $2.6 = 3.8 + x$ [2.1]

20. $\frac{2}{7}x = -6$ [2.1]

21. $5x - 8 = 37$ [2.2]

22. $\dfrac{2}{3} = \dfrac{-m}{10}$ [2.1]

23. $5.4 - 1.9x = 0.8x$ [2.2]

24. $x - \frac{7}{8} = \frac{3}{4}$ [2.1]

25. $2(2 - 3x) = 3(5x + 7)$ [2.2]

26. $\frac{1}{4}x - \frac{2}{3} = \frac{3}{4} + \frac{1}{3}x$ [2.2]

27. $y + 5 - 3y = 5y - 9$ [2.2]

28. $x - 28 < 20 - 2x$ [2.6]

29. $2(x + 2) \geq 5(2x + 3)$ [2.6]

30. Solve $A = 2\pi rh + \pi r^2$ for h. [2.3]

31. In which quadrant is the point $(3, -1)$ located? [3.1]

32. Graph on a number line: $-1 < x \leq 2$. [2.6]

33. Use a grid 10 squares wide and 10 squares high to plot $(-150, -40)$, $(40, -7)$, and $(0, 6)$. Choose the scale carefully. [3.1]

Graph.

34. $x = 3$ [3.3]

35. $2x - 5y = 10$ [3.3]

36. $y = -2x + 1$ [3.2]

37. $y = \frac{2}{3}x$ [3.2]

38. $y = -\frac{3}{4}x + 2$ [3.6]

Find the coordinates of the x- and y-intercepts. Do not graph.

39. $2x - 7y = 21$ [3.3]

40. $y = 4x + 5$ [3.3]

Solve.

41. *Donating blood.*   Each year 8 million Americans donate blood. This is 5% of those healthy enough to do so. How many Americans are eligible to donate blood? [2.4]
Source: *Indianapolis Star,* 10/6/96

42. *Blood types.*   There were recently 117 million Americans with either O-positive or O-negative blood. Those with O-positive blood outnumbered those with O-negative blood by 85.8 million. How many Americans had O-negative blood? [2.5]

43. Tina paid $126 for a cordless drill, including a 5% sales tax. How much did the drill itself cost? [2.4]

44. A 143-m wire is cut into three pieces. The second is 3 m longer than the first. The third is four fifths as long as the first. How long is each piece? [2.5]

45. Tori's contract stipulates that she cannot work more than 40 hr per week. For the first 4 days of one week, she worked 7, 10, 9, and 6 hr. How many hours can she work the fifth day without violating her contract? [2.7]

46. *Fund raising.*   The following graph shows data from a recent road race held to raise money for Amnesty International. At what rate was money raised? [3.4]

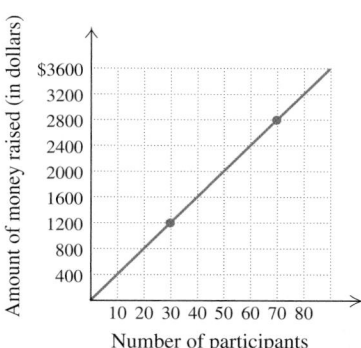

47. Find the slope of the line containing the points $(-4, 1)$ and $(2, -1)$. [3.5]

48. Write an equation of the line with slope $\frac{2}{7}$ and y-intercept $(0, -4)$. [3.6]

49. Find the slope and the y-intercept of the line given by $2x + 6y = 18$. [3.6]

50. Write the slope–intercept equation of the line parallel to the line $x + 2y = 3$ and containing the point $(2, 4)$. [3.7]

51. The percentage of farm operators completing four-year college degrees or higher rose from 15.2% in 1991 to 20.4% in 2001. [3.7]
Source: USDA Economic Research Service

  a) Calculate the percentage of farmers with four-year degrees or higher in 1996.
  b) Estimate the percentage of farmers with four-year degrees or higher in 2009.

SYNTHESIS

52. *Yearly earnings.*   Paula's salary at the end of a year is $26,780. This reflects a 4% salary increase in February and then a 3% cost-of-living adjustment in June. What was her salary at the beginning of the year? [2.4]

Solve. Label any contradictions or identities.

53. $4|x| - 13 = 3$ [1.4], [2.2]

54. $4(x + 2) = 9(x - 2) + 16$ [2.2]

55. $2(x + 3) + 4 = 0$ [2.2]

56. $\dfrac{2 + 5x}{4} = \dfrac{11}{28} + \dfrac{8x + 3}{7}$ [2.2]

57. $5(7 + x) = (x + 6)5$ [2.2]

58. Solve $p = \dfrac{2}{m + Q}$ for Q. [2.3]

59. The points $(-3, 0)$, $(0, 7)$, $(3, 0)$, and $(0, -7)$ are vertices of a parallelogram. Find four equations of lines that intersect to form the parallelogram. [3.6]

4

Polynomials

AN APPLICATION

In computer science, 1 kilobyte (KB) is 1×2^{10}, or approximately 1000, bytes of memory. The TI-84 Plus Silver Edition graphing calculator has 1.5 megabytes (MB) of "FLASH ROM," where 1 MB = 1000 KB. How many bytes of FLASH ROM does this calculator have?

This problem appears as Exercise 121 in Section 4.1.

Damian Isla
AI PROGRAMMER
Seattle, Washington

No video game would ever be developed without math. My job is to simulate the thought processes of computer-controlled characters, and I do this using techniques from trigonometry, vector geometry, calculus, and logic.

O ur work in Chapter 3 concentrated on using graphs to represent solutions of equations in two variables. Here in Chapter 4 we will focus on finding equivalent expressions, not on solving equations.

Algebraic expressions such as $16t^2$, $5a^2 - 3ab$, and $3x^2 - 7x + 5$ are called polynomials. Polynomials occur frequently in applications and appear in most branches of mathematics. Thus learning to add, subtract, multiply, and divide polynomials is an important part of nearly every course in elementary algebra.

4.1 Exponents and Their Properties

Multiplying Powers with Like Bases • Dividing Powers with Like Bases • Zero as an Exponent • Raising a Power to a Power • Raising a Product or a Quotient to a Power

In Section 4.2, we begin our study of polynomials. Before doing so, however, we must develop some rules for manipulating exponents.

Multiplying Powers with Like Bases

Recall from Section 1.8 that an expression like a^3 means $a \cdot a \cdot a$. We can use this fact to find the product of two expressions that have the same base:

$$a^3 \cdot a^2 = (a \cdot a \cdot a)(a \cdot a) \qquad \text{There are three factors in } a^3 \text{ and two factors in } a^2.$$
$$= a \cdot a \cdot a \cdot a \cdot a \qquad \text{Using an associative law}$$
$$= a^5.$$

Note that the exponent in a^5 is the sum of the exponents in $a^3 \cdot a^2$. That is, $3 + 2 = 5$. Similarly,

$$b^4 \cdot b^3 = (b \cdot b \cdot b \cdot b)(b \cdot b \cdot b)$$
$$= b^7, \quad \text{where } 4 + 3 = 7.$$

Adding the exponents gives the correct result.

The Product Rule

For any number a and any positive integers m and n,

$$a^m \cdot a^n = a^{m+n}.$$

(To multiply powers with the same base, keep the base and add the exponents.)

EXAMPLE 1   Multiply and simplify each of the following. (Here "simplify" means express the product as one base to a power whenever possible.)

a) $x^2 \cdot x^9$

b) $5^3 \cdot 5^8 \cdot 5^1$

c) $(r + s)^7 (r + s)^6$

d) $(a^3 b^2)(a^3 b^5)$

Solution

a) $x^2 \cdot x^9 = x^{2+9}$ Adding exponents: $a^m \cdot a^n = a^{m+n}$

$\quad\quad\quad\quad = x^{11}$

b) $5^3 \cdot 5^8 \cdot 5^1 = 5^{3+8+1}$ Adding exponents

$\quad\quad\quad\quad\quad\quad = 5^{12}$ ——— | *Caution!* $5^{12} \neq 5 \cdot 12.$ |

c) $(r + s)^7 (r + s)^6 = (r + s)^{7+6}$ The base here is $r + s$.

$\quad\quad\quad\quad\quad\quad\quad\quad = (r + s)^{13}$ ——— | *Caution!* $(r + s)^{13} \neq r^{13} + s^{13}.$ |

d) $(a^3 b^2)(a^3 b^5) = a^3 b^2 a^3 b^5$ Using an associative law

$\quad\quad\quad\quad\quad\quad\quad = a^3 a^3 b^2 b^5$ Using a commutative law

$\quad\quad\quad\quad\quad\quad\quad = a^6 b^7$ Adding exponents

Dividing Powers with Like Bases

Recall that any expression that is divided or multiplied by 1 is unchanged. This, together with the fact that anything (besides 0) divided by itself is 1, can lead to a rule for division:

$$\frac{a^5}{a^2} = \frac{a \cdot a \cdot a \cdot a \cdot a}{a \cdot a}$$

$$= \frac{a \cdot a \cdot a}{1} \cdot \frac{a \cdot a}{a \cdot a}$$

$$= \frac{a \cdot a \cdot a}{1} \cdot 1$$

$$= a \cdot a \cdot a = a^3.$$

Note that the exponent in a^3 is the difference of the exponents in a^5/a^2. Similarly,

$$\frac{x^4}{x^3} = \frac{x \cdot x \cdot x \cdot x}{x \cdot x \cdot x} = \frac{x}{1} \cdot \frac{x \cdot x \cdot x}{x \cdot x \cdot x} = \frac{x}{1} \cdot 1 = x^1, \quad \text{or } x.$$

Subtracting the exponents gives the correct result.

The Quotient Rule

For any nonzero number a and any positive integers m and n for which $m > n$,

$$\frac{a^m}{a^n} = a^{m-n}.$$

(To divide powers with the same base, subtract the exponent of the denominator from the exponent of the numerator.)

EXAMPLE 2   Divide and simplify. (Here "simplify" means express the quotient as one base to a power whenever possible.)

a) $\dfrac{x^8}{x^2}$ **b)** $\dfrac{7^9}{7^4}$ **c)** $\dfrac{(5a)^{12}}{(5a)^4}$ **d)** $\dfrac{4p^5q^7}{6p^2q}$

Solution

a) $\dfrac{x^8}{x^2} = x^{8-2}$ Subtracting exponents: $\dfrac{a^m}{a^n} = a^{m-n}$

$\quad\quad = x^6$

b) $\dfrac{7^9}{7^4} = 7^{9-4}$

$\quad\quad = 7^5$

> *Caution!*   When we are using the quotient rule, the base is unchanged:
>
> $$\frac{7^9}{7^4} \neq 1^5, \quad \text{rather} \quad \frac{7^9}{7^4} = 7^5.$$

c) $\dfrac{(5a)^{12}}{(5a)^4} = (5a)^{12-4} = (5a)^8$ The base here is $5a$.

d) $\dfrac{4p^5q^7}{6p^2q} = \dfrac{4}{6} \cdot \dfrac{p^5}{p^2} \cdot \dfrac{q^7}{q^1}$ Note that the 4 and 6 are factors, not exponents!

$\quad\quad = \dfrac{2}{3} \cdot p^{5-2} \cdot q^{7-1} = \dfrac{2}{3}p^3q^6$ Using the quotient rule twice; simplifying

---

Zero as an Exponent

The quotient rule can be used to help determine what 0 should mean when it appears as an exponent. Consider a^4/a^4, where a is nonzero. Since the numerator and the denominator are the same,

$$\frac{a^4}{a^4} = 1.$$

On the other hand, using the quotient rule would give us

$$\frac{a^4}{a^4} = a^{4-4} = a^0.$$ Subtracting exponents

Since $a^0 = a^4/a^4 = 1$, this suggests that $a^0 = 1$ for any nonzero value of a.

> **The Exponent Zero**
>
> For any real number a, $a \neq 0$,
>
> $$a^0 = 1.$$
>
> (Any nonzero number raised to the 0 power is 1.)

Note that in the above box, 0^0 is not defined. For this text, we will assume that expressions like a^m do not represent 0^0.

EXAMPLE 3   Simplify: **(a)** 1948^0; **(b)** $(-9)^0$; **(c)** $(3x)^0$; **(d)** $(-1)9^0$; **(e)** -9^0.

Solution

a) $1948^0 = 1$ Any nonzero number raised to the 0 power is 1.

b) $(-9)^0 = 1$ Any nonzero number raised to the 0 power is 1. The base here is -9.

c) $(3x)^0 = 1$, for any $x \neq 0$.   The parentheses indicate that the base is $3x$.

d) Recall that, unless there are calculations within parentheses, exponents are calculated before multiplication:

$$(-1)9^0 = (-1)1 = -1.$$ The base here is 9.

e) -9^0 is read "the opposite of 9^0" and is equivalent to $(-1)9^0$:

$$-9^0 = (-1)9^0 = (-1)1 = -1.$$

Note from parts (b), (d), and (e) that $-9^0 = (-1)9^0$ and $-9^0 \neq (-9)^0$.

> *Caution!* $-9^0 \neq (-9)^0$, and, in general, $-a^n \neq (-a)^n$.

Raising a Power to a Power

Consider an expression like $(7^2)^4$:

$$
\begin{aligned}
(7^2)^4 &= (7^2)(7^2)(7^2)(7^2) & &\text{There are four factors of } 7^2. \\
&= (7 \cdot 7)(7 \cdot 7)(7 \cdot 7)(7 \cdot 7) & &\text{We could also use the} \\
& & &\text{product rule.} \\
&= 7 \cdot 7 \cdot 7 \cdot 7 \cdot 7 \cdot 7 \cdot 7 \cdot 7 & &\text{Using an associative law} \\
&= 7^8.
\end{aligned}
$$

Note that the exponent in 7^8 is the product of the exponents in $(7^2)^4$. Similarly,

$$
\begin{aligned}
(y^5)^3 &= y^5 \cdot y^5 \cdot y^5 & &\text{There are three factors of } y^5. \\
&= (y \cdot y \cdot y \cdot y \cdot y)(y \cdot y \cdot y \cdot y \cdot y)(y \cdot y \cdot y \cdot y \cdot y) \\
&= y^{15}.
\end{aligned}
$$

Once again, we get the same result if we multiply exponents:

$$(y^5)^3 = y^{5 \cdot 3} = y^{15}.$$

> **The Power Rule**
>
> For any number a and any whole numbers m and n,
>
> $$(a^m)^n = a^{mn}.$$
>
> (To raise a power to a power, multiply the exponents and leave the base unchanged.)

Remember that for this text we assume that 0^0 is not considered.

EXAMPLE 4   Simplify: **(a)** $(m^2)^5$; **(b)** $(3^5)^4$.

Student Notes

There are several rules for manipulating exponents in this section. One way to remember them all is to replace variables with small numbers and see what the results suggest. For example, multiplying $2^2 \cdot 2^3$ and examining the result is a fine way of reminding yourself that $a^m \cdot a^n = a^{m+n}$.

Solution

a) $(m^2)^5 = m^{2 \cdot 5}$ Multiplying exponents: $(a^m)^n = a^{mn}$

$\qquad = m^{10}$

b) $(3^5)^4 = 3^{5 \cdot 4}$

$\qquad = 3^{20}$

Raising a Product or a Quotient to a Power

When an expression inside parentheses is raised to a power, the inside expression is the base. Let's compare $2a^3$ and $(2a)^3$:

$2a^3 = 2 \cdot a \cdot a \cdot a;$ The base is a. $(2a)^3 = (2a)(2a)(2a)$ The base is $2a$.

$\qquad\qquad\qquad\qquad\qquad\qquad\qquad\quad = (2 \cdot 2 \cdot 2)(a \cdot a \cdot a)$

$\qquad\qquad\qquad\qquad\qquad\qquad\qquad\quad = 2^3 a^3$

$\qquad\qquad\qquad\qquad\qquad\qquad\qquad\quad = 8a^3.$

We see that $2a^3$ and $(2a)^3$ are *not* equivalent. Note too that $(2a)^3$ can be simplified by cubing each factor. This leads to the following rule for raising a product to a power.

> **Raising a Product to a Power**
>
> For any numbers a and b and any whole number n,
>
> $$(ab)^n = a^n b^n.$$
>
> (To raise a product to a power, raise each factor to that power.)

EXAMPLE 5   Simplify: **(a)** $(4a)^3$; **(b)** $(-5x^4)^2$; **(c)** $(a^7 b)^2(a^3 b^4)$.

Solution

a) $(4a)^3 = 4^3 a^3 = 64a^3$ Raising each factor to the third power and simplifying

b) $(-5x^4)^2 = (-5)^2(x^4)^2$ Raising each factor to the second power. Parentheses are important here.

$\qquad\qquad = 25x^8$ Simplifying $(-5)^2$ and using the product rule

c) $(a^7b)^2(a^3b^4) = (a^7)^2b^2a^3b^4$ Raising a product to a power

$\qquad\qquad\qquad = a^{14}b^2a^3b^4$ Multiplying exponents

$\qquad\qquad\qquad = a^{17}b^6$ Adding exponents

> *Caution!*   The rule $(ab)^n = a^nb^n$ applies only to *products* raised to a power, not to sums or differences. For example, $(3 + 4)^2 \neq 3^2 + 4^2$ since $49 \neq 9 + 16$.

There is a similar rule for raising a quotient to a power.

> **Raising a Quotient to a Power**
>
> For any numbers a and b, $b \neq 0$, and any whole number n,
>
> $$\left(\frac{a}{b}\right)^n = \frac{a^n}{b^n}.$$
>
> (To raise a quotient to a power, raise the numerator to the power and divide by the denominator to the power.)

EXAMPLE 6   Simplify: **(a)** $\left(\dfrac{x}{5}\right)^2$; **(b)** $\left(\dfrac{5}{a^4}\right)^3$; **(c)** $\left(\dfrac{3a^4}{b^3}\right)^2$.

Solution

a) $\left(\dfrac{x}{5}\right)^2 = \dfrac{x^2}{5^2} = \dfrac{x^2}{25}$ Squaring the numerator and the denominator

b) $\left(\dfrac{5}{a^4}\right)^3 = \dfrac{5^3}{(a^4)^3}$ Raising a quotient to a power

$\qquad\qquad = \dfrac{125}{a^{4\cdot3}} = \dfrac{125}{a^{12}}$ Using the power rule and simplifying

c) $\left(\dfrac{3a^4}{b^3}\right)^2 = \dfrac{(3a^4)^2}{(b^3)^2}$ Raising a quotient to a power

$\qquad\qquad = \dfrac{3^2(a^4)^2}{b^{3\cdot2}} = \dfrac{9a^8}{b^6}$ Raising a product to a power and using the power rule

In the following summary of definitions and rules, we assume that no denominators are 0 and that 0^0 is not considered.

Definitions and Properties of Exponents

For any whole numbers m and n,

1 as an exponent:	$a^1 = a$
0 as an exponent:	$a^0 = 1$
The Product Rule:	$a^m \cdot a^n = a^{m+n}$
The Quotient Rule:	$\dfrac{a^m}{a^n} = a^{m-n}$
The Power Rule:	$(a^m)^n = a^{mn}$
Raising a product to a power:	$(ab)^n = a^n b^n$
Raising a quotient to a power:	$\left(\dfrac{a}{b}\right)^n = \dfrac{a^n}{b^n}$

Exercise Set

4.1

FOR EXTRA HELP

 Student's Solutions Manual   Digital Video Tutor CD 2 Videotape 4   Tutor Center AW Math Tutor Center    MathXL Tutorials on CD   Math XL MathXL    MyMathLab MyMathLab

Concept Reinforcement   *In each of Exercises 1–8, complete the sentence using the most appropriate phrase from the column on the right.*

1. To raise a power to a power, ____

2. To raise a quotient to a power, ____

3. To raise a product to a power, ____

4. To divide powers with the same base, ____

5. Any nonzero number raised to the 0 power ____

6. To multiply powers with the same base, ____

7. To square a fraction, ____

8. To square a product, ____

a) keep the base and add the exponents.

b) multiply the exponents and leave the base unchanged.

c) square the numerator and square the denominator.

d) square each factor.

e) raise each factor to that power.

f) raise the numerator to the power and divide by the denominator to the power.

g) is one.

h) subtract the exponent of the denominator from the exponent of the numerator.

Simplify. Assume that no denominator is zero and that 0^0 is not considered.

9. $r^4 \cdot r^6$

10. $8^4 \cdot 8^3$

11. $9^5 \cdot 9^3$

12. $n^3 \cdot n^{20}$

13. $a^6 \cdot a$

14. $y^7 \cdot y^9$

15. $8^4 \cdot 8^7$

16. $t^0 \cdot t^{16}$

17. $(3y)^4(3y)^8$

18. $(2t)^8(2t)^{17}$

19. $(5t)(5t)^6$

20. $(8x)^0(8x)^1$

21. $(a^2b^7)(a^3b^2)$

22. $(m-3)^4(m-3)^5$

23. $(x+1)^5(x+1)^7$

24. $(a^8b^3)(a^4b)$

25. $r^3 \cdot r^7 \cdot r^0$

26. $s^4 \cdot s^5 \cdot s^2$

27. $(xy^4)(xy)^3$

28. $(a^3b)(ab)^4$

29. $\dfrac{7^5}{7^2}$

30. $\dfrac{4^7}{4^3}$

31. $\dfrac{x^{15}}{x^3}$

32. $\dfrac{a^{10}}{a^2}$

33. $\dfrac{t^5}{t}$

34. $\dfrac{x^7}{x}$

35. $\dfrac{(5a)^7}{(5a)^6}$

36. $\dfrac{(3m)^9}{(3m)^8}$

Aha!   **37.** $\dfrac{(x+y)^8}{(x+y)^8}$

38. $\dfrac{(a-b)^4}{(a-b)^3}$

39. $\dfrac{6m^5}{8m^2}$

40. $\dfrac{10n^7}{15n^3}$

41. $\dfrac{8a^9b^7}{2a^2b}$

42. $\dfrac{12r^{10}s^7}{4r^2s}$

43. $\dfrac{m^9n^8}{m^0n^4}$

44. $\dfrac{a^{10}b^{12}}{a^2b^0}$

Simplify.

45. x^0 when $x = 13$

46. y^0 when $y = 38$

47. $5x^0$ when $x = -4$

48. $7m^0$ when $m = 1.7$

49. $7^0 + 4^0$

50. $(8+5)^0$

51. $(-3)^1 - (-3)^0$

52. $(-4)^0 - (-4)^1$

Simplify. Assume that no denominator is zero and that 0^0 is not considered.

53. $(x^4)^7$

54. $(a^3)^8$

55. $(5^8)^2$

56. $(2^5)^3$

57. $(m^7)^5$

58. $(n^9)^2$

59. $(t^{20})^4$

60. $(t^3)^9$

61. $(7x)^2$

62. $(5a)^2$

63. $(-2a)^3$

64. $(-3x)^3$

65. $(4m^3)^2$

66. $(5n^4)^2$

67. $(a^2b)^7$

68. $(xy^4)^9$

69. $(x^3y)^2(x^2y^5)$

70. $(a^4b^6)(a^2b)^5$

71. $(2x^5)^3(3x^4)$

72. $(5x^3)^2(2x^7)$

73. $\left(\dfrac{a}{4}\right)^3$

74. $\left(\dfrac{3}{x}\right)^4$

75. $\left(\dfrac{7}{5a}\right)^2$

76. $\left(\dfrac{5x}{2}\right)^3$

77. $\left(\dfrac{a^4}{b^3}\right)^5$

78. $\left(\dfrac{x^5}{y^2}\right)^7$

79. $\left(\dfrac{y^3}{2}\right)^2$

80. $\left(\dfrac{a^5}{2}\right)^3$

81. $\left(\dfrac{x^2y}{z^3}\right)^4$

82. $\left(\dfrac{x^3}{y^2z}\right)^5$

83. $\left(\dfrac{a^3}{-2b^5}\right)^4$

84. $\left(\dfrac{x^5}{-3y^3}\right)^4$

85. $\left(\dfrac{5x^7y}{2z^4}\right)^3$

86. $\left(\dfrac{4a^2b}{3c^7}\right)^3$

Aha!   **87.** $\left(\dfrac{4x^3y^5}{3z^7}\right)^0$

88. $\left(\dfrac{5a^7}{2b^5c}\right)^0$

89. Explain in your own words why $-5^2 \neq (-5)^2$.

90. Under what circumstances should exponents be added?

SKILL MAINTENANCE

Factor. [1.2]

91. $3s - 3r + 3t$

92. $-7x + 7y - 7z$

Combine like terms. [1.6]

93. $9x + 2y - x - 2y$

94. $5a - 7b - 8a + b$

Use the commutative law of addition to write an equivalent expression. [1.2]

95. $3x + 2y$

96. $2xy + 5z$

SYNTHESIS

97. Under what conditions does a^n represent a negative number? Why?

98. Using the quotient rule, explain why 9^0 is 1.

99. Suppose that the width of a square is three times the width of a second square. How do the areas of the squares compare? Why?

$3x$ x

100. Suppose that the width of a cube is twice the width of a second cube. How do the volumes of the cubes compare? Why?

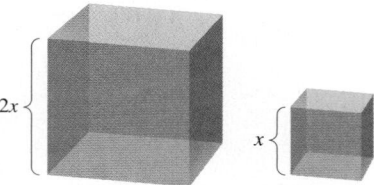

$2x$ x

Find a value of the variable that shows that the two expressions are not *equivalent. Answers may vary.*

101. $(a + 5)^2$; $a^2 + 5^2$ **102.** $3x^2$; $(3x)^2$

103. $\dfrac{a + 7}{7}$; a **104.** $\dfrac{t^6}{t^2}$; t^3

Simplify.

105. $a^{10k} \div a^{2k}$ **106.** $y^{4x} \cdot y^{2x}$

107. $\dfrac{\left(\frac{1}{2}\right)^3\left(\frac{2}{3}\right)^4}{\left(\frac{5}{6}\right)^3}$ **108.** $\dfrac{x^{5t}(x^t)^2}{(x^{3t})^2}$

109. Solve for x:
$$\frac{t^{26}}{t^x} = t^x.$$

Replace ■ *with* $>$, $<$, *or* $=$ *to write a true sentence.*

110. 3^5 ■ 3^4 **111.** 4^2 ■ 4^3

112. 4^3 ■ 5^3 **113.** 4^3 ■ 3^4

114. 9^7 ■ 3^{13} **115.** 25^8 ■ 125^5

Use the fact that $10^3 \approx 2^{10}$ *to estimate each of the following powers of 2. Then compute the power of 2 with a calculator and find the difference between the exact value and the approximation.*

116. 2^{14} **117.** 2^{22}

118. 2^{26} **119.** 2^{31}

In computer science, 1 KB of memory refers to 1 kilobyte, or 1×10^3 *bytes, of memory. This is really an approximation of* 1×2^{10} *bytes (since computer memory uses powers of 2).*

120. The TI-84 Plus graphing calculator has 480 KB of "FLASH ROM." How many bytes is this?

121. The TI-84 Plus Silver Edition graphing calculator has 1.5 MB (megabytes) of FLASH ROM, where 1 MB is 1000 KB (see Exercise 120). How many bytes of FLASH ROM does this calculator have?

4.2   Polynomials

Terms   •   Types of Polynomials   •   Degree and Coefficients   •
Combining Like Terms   •   Evaluating Polynomials and Applications

We now examine an important algebraic expression known as a *polynomial.* Certain polynomials have appeared earlier in this text so you already have some experience working with them.

Terms

At this point, we have seen a variety of algebraic expressions like

$$3a^2b^4, \qquad 2l + 2w, \quad \text{and} \quad 5x^2 + x - 2.$$

Of these, $3a^2b^4$, $2l$, $2w$, $5x^2$, x, and -2 are examples of *terms.* A **term** (see p. 16) can be a number (like -2), a variable (like x), a product of numbers and/or variables (like $3a^2b^4$, $2l$, $2w$, or $5x^2$), or a quotient of numbers and/or variables (like $7/t$ or $(a^2b^3)/(4c)$).*

Types of Polynomials

A term that is a product of constants and/or variables is called a **monomial.**† All of the terms listed above, except for the quotients, are monomials. Other examples of monomials are

$$7, \qquad t, \qquad 23x^2y, \quad \text{and} \quad \tfrac{3}{7}a^5.$$

A **polynomial** is a monomial or a sum of monomials. The following are examples of polynomials:

$$4x + 7, \quad \tfrac{2}{3}t^2, \quad 6a + 7, \quad -5n^2 + m - 1, \quad 42r^5, \quad x, \quad \text{and} \quad 0.$$

The following algebraic expressions are *not* polynomials:

$$\textbf{(1)} \ \frac{x + 3}{x - 4}, \qquad \textbf{(2)} \ 5x^3 - 2x^2 + \frac{1}{x}, \qquad \textbf{(3)} \ \frac{1}{x^3 - 2}.$$

Expressions (1) and (3) are not polynomials because they represent quotients, not sums. Expression (2) is not a polynomial because $1/x$ is not a monomial.

When a polynomial is written as a sum of monomials, each monomial is called a *term of the polynomial.*

---

*Later in this text, expressions like $5x^{3/2}$ and $2a^{-7}b$ will be discussed. Such expressions are also considered terms.

†Note that a term, but not a monomial, can include division by a variable.

EXAMPLE 1

Identify the terms of the polynomial $3t^4 - 5t^6 - 4t + 2$.

Solution   The terms are $3t^4$, $-5t^6$, $-4t$, and 2. We can see this by rewriting all subtractions as additions of opposites:

$$3t^4 - 5t^6 - 4t + 2 = 3t^4 + (-5t^6) + (-4t) + 2.$$

These are the terms of the polynomial.

A polynomial that is composed of two terms is called a **binomial**, whereas those composed of three terms are called **trinomials**. Polynomials with four or more terms have no special name.

Monomials	Binomials	Trinomials	No Special Name
$4x^2$	$2x + 4$	$3t^3 + 4t + 7$	$4x^3 - 5x^2 + xy - 8$
9	$3a^5 + 6bc$	$6x^7 - 8z^2 + 4$	$z^5 + 2z^4 - z^3 + 7z + 3$
$-7a^{19}b^5$	$-9x^7 - 6$	$4x^2 - 6x - \frac{1}{2}$	$4x^6 - 3x^5 + x^4 - x^3 + 2x - 1$

Degree and Coefficients

The **degree of a term** of a polynomial is the number of variable factors in that term. Thus the degree of $7t^2$ is 2 because $7t^2$ has two variable factors: $7t^2 = 7 \cdot t \cdot t$. We will revisit the meaning of degree in Section 4.6 when polynomials in several variables are examined.

EXAMPLE 2

Determine the degree of each term: **(a)** $8x^4$;   **(b)** $3x$;   **(c)** 7.

Solution

a) The degree of $8x^4$ is 4. x^4 represents 4 variable factors: $x \cdot x \cdot x \cdot x$.

b) The degree of $3x$ is 1.   There is 1 variable factor.

c) The degree of 7 is 0.   There is no variable factor.

The part of a term that is a constant factor is the **coefficient** of that term. Thus the coefficient of $3x$ is 3, and the coefficient for the term 7 is simply 7.

EXAMPLE 3

Identify the coefficient of each term in the polynomial

$$4x^3 - 7x^2y + x - 8.$$

Solution

The coefficient of $4x^3$ is 4.

The coefficient of $-7x^2y$ is -7.

The coefficient of the third term is 1, since $x = 1x$.

The coefficient of -8 is simply -8.

The **leading term** of a polynomial is the term of highest degree. Its coefficient is called the **leading coefficient** and its degree is referred to as the **degree of the polynomial**. To see how this terminology is used, consider the polynomial

$$3x^2 - 8x^3 + 5x^4 + 7x - 6.$$

The *terms* are $3x^2, \quad -8x^3, \quad 5x^4, \quad 7x, \quad$ and $\quad -6.$
The *coefficients* are $3, \quad -8, \quad 5, \quad 7, \quad$ and $\quad -6.$
The *degree of each term* is $2, \quad 3, \quad 4, \quad 1, \quad$ and $\quad 0.$
The *leading term* is $5x^4$ and the *leading coefficient* is 5.
The *degree of the polynomial* is 4.

Combining Like Terms

Recall from Section 1.8 that *like*, or *similar*, *terms* are either constant terms or terms containing the same variable(s) raised to the same power(s). To simplify certain polynomials, we can often *combine*, or *collect*, like terms.

EXAMPLE 4   Identify the like terms in $4x^3 + 5x - 7x^2 + 2x^3 + x^2$.

Solution

Like terms: $4x^3$ and $2x^3$ Same variable and exponent
Like terms: $-7x^2$ and x^2 Same variable and exponent

EXAMPLE 5   Combine like terms.

a) $2x^3 - 6x^3$
b) $5x^2 + 7 + 2x^4 + 4x^2 - 11 - 2x^4$
c) $7a^3 - 5a^2 + 9a^3 + a^2$
d) $\frac{2}{3}x^4 - x^3 - \frac{1}{6}x^4 + \frac{2}{5}x^3 - \frac{3}{10}x^3$

Student Notes _____

Remember that when we combine like terms, we are not solving equations, but are forming equivalent expressions.

Solution

a) $2x^3 - 6x^3 = (2 - 6)x^3$ Using the distributive law
$\qquad\qquad = -4x^3$

b) $5x^2 + 7 + 2x^4 + 4x^2 - 11 - 2x^4 = 5x^2 + 4x^2 + 2x^4 - 2x^4 + 7 - 11$
$\qquad = (5 + 4)x^2 + (2 - 2)x^4 + (7 - 11)$ ⎫   These steps are often
$\qquad = 9x^2 + 0x^4 + (-4)$ ⎭   done mentally.
$\qquad = 9x^2 - 4$

c) $7a^3 - 5a^2 + 9a^3 + a^2 = 7a^3 - 5a^2 + 9a^3 + 1a^2$ When a term appears without a coefficient, we can write in 1.

$\qquad\qquad\qquad = 16a^3 - 4a^2$

d) $\frac{2}{3}x^4 - x^3 - \frac{1}{6}x^4 + \frac{2}{5}x^3 - \frac{3}{10}x^3 = \left(\frac{2}{3} - \frac{1}{6}\right)x^4 + \left(-1 + \frac{2}{5} - \frac{3}{10}\right)x^3$
$\qquad\qquad = \left(\frac{4}{6} - \frac{1}{6}\right)x^4 + \left(-\frac{10}{10} + \frac{4}{10} - \frac{3}{10}\right)x^3$
$\qquad\qquad = \frac{3}{6}x^4 - \frac{9}{10}x^3$
$\qquad\qquad = \frac{1}{2}x^4 - \frac{9}{10}x^3$ There are no similar terms, so we are done.

Note in Example 5 that the solutions are written so that the term of highest degree appears first, followed by the term of next highest degree, and so on. This is known as **descending order** and is the form in which answers will normally appear.

Evaluating Polynomials and Applications

When each variable in a polynomial is replaced with a number, the polynomial then represents a number, or *value*, that can be calculated using the rules for order of operations.

EXAMPLE 6   Evaluate $-x^2 + 3x + 9$ for $x = -2$.

Solution   For $x = -2$, we have

$$-x^2 + 3x + 9 = -(-2)^2 + 3(-2) + 9$$ The negative sign in front of x^2 remains.

$$= -(4) + 3(-2) + 9$$
$$= -4 + (-6) + 9$$
$$= -10 + 9 = -1.$$

EXAMPLE 7   Games in a sports league.   In a sports league of n teams in which each team plays every other team twice, the total number of games to be played is given by the polynomial

$$n^2 - n.$$

A girls' soccer league has 10 teams. How many games are played if each team plays every other team twice?

Solution   We evaluate the polynomial for $n = 10$:

$$n^2 - n = 10^2 - 10$$
$$= 100 - 10 = 90.$$

The league plays 90 games.

EXAMPLE 8   Daily accidents.   The average number of accidents per day involving drivers of age r can be approximated by the polynomial

$$0.4r^2 - 40r + 1039.$$

Find the average number of accidents per day involving 20-year-old drivers.

Solution   To find the average number of accidents involving 20-year-olds, we evaluate the polynomial for $r = 20$:

$$0.4r^2 - 40r + 1039 = 0.4(20)^2 - 40 \cdot 20 + 1039$$
$$= 0.4 \cdot 400 - 800 + 1039$$
$$= 160 - 800 + 1039$$
$$= 399.$$

There are, on average, approximately 399 accidents each day involving 20-year-old drivers.

Sometimes, a graph can be used to estimate the value of a polynomial visually.

EXAMPLE 9

Daily accidents. In the following graph, the polynomial from Example 8 has been graphed by evaluating it for several choices of x. Use the graph to estimate the average number of accidents per day involving 30-year-old drivers.

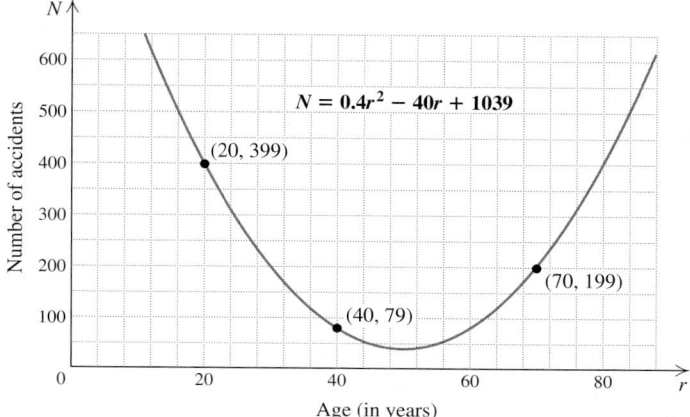

Solution To estimate the number of accidents involving 30-year-olds, we locate 30 on the horizontal axis. From there, we move vertically until we meet the curve at some point. From that point, we move horizontally to the N-axis.

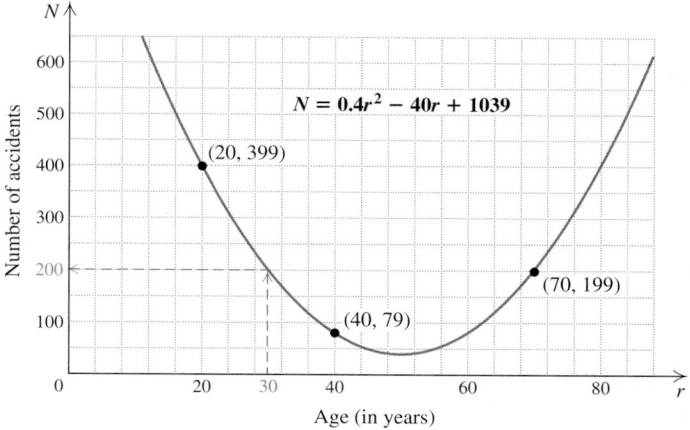

There are, on average, about 200 accidents each day involving 30-year-old drivers. (For $r = 30$, the value of $0.4r^2 - 40r + 1039$ is approximately 200.)

 **technology connection**

One way to evaluate a polynomial is to use the TRACE key. For example, to evaluate $0.4r^2 - 40r + 1039$ in Example 9 for $r = 30$, we can enter the polynomial as
$y = 0.4x^2 - 40x + 1039$. We then use TRACE and enter an x-value of 30. The value of the polynomial appears as y, and the cursor automatically appears at (30, 199). The Value option of the CALC menu works in a similar way.

1. Use TRACE or CALC Value to find the value of $0.4r^2 - 40r + 1039$ for $r = 60$.

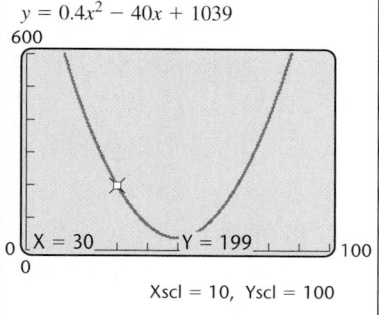

$y = 0.4x^2 - 40x + 1039$

Xscl = 10, Yscl = 100

Exercise Set

4.2

FOR EXTRA HELP

 Student's Solutions Manual    Digital Video Tutor CD 2 Videotape 4   AW Math Tutor Center   MathXL Tutorials on CD   Math XL   MyMathLab

🌿 *Concept Reinforcement   In each of Exercises 1–8, match the description with the most appropriate algebraic expression from the column on the right.*

1. _____ A polynomial with four terms

2. _____ A trinomial written in descending order

3. _____ A polynomial with degree 5

4. _____ A polynomial with 7 as its leading coefficient

5. _____ A binomial with degree 7

6. _____ An expression with two terms that is not a binomial

7. _____ An expression with three terms that is not a trinomial

8. _____ A monomial of degree 0

a) $8x^3 + \dfrac{2}{x^2}$

b) $5x^4 + 3x^3 - 4x + 7$

c) $\dfrac{3}{x} - 6x^2 + 9$

d) $8t - 4t^5$

e) 5

f) $6x^2 + 7x^4 - 2x^3$

g) $4t - 2t^7$

h) $3t^2 + 4t + 7$

Identify the terms of each polynomial.

9. $7x^4 + x^3 - 5x + 8$

10. $5a^3 + 4a^2 - a - 7$

11. $-t^4 + 7t^3 - 3t^2 + 6$

12. $n^5 - 4n^3 + 2n - 8$

Determine the coefficient and the degree of each term in each polynomial.

13. $4x^5 + 7x$

14. $9a^3 - 4a^2$

15. $9t^2 - 3t + 4$

16. $7x^4 + 5x - 3$

17. $7a^4 + 9a + a^3$

18. $6t^5 - 3t^2 - t$

19. $x^4 - x^3 + 4x - 3$

20. $3a^4 - a^3 + a - 9$

*For each of the following polynomials, **(a)** list the degree of each term; **(b)** determine the leading term and the leading coefficient; and **(c)** determine the degree of the polynomial.*

21. $5x - 9x^2 + 3x^6$

22. $2t + 3 + 4t^2$

23. $3a^2 - 7 + 2a^4$

24. $9x^4 + x^2 + x^7 + 4$

25. $8 + 6x^2 - 3x - x^5$

26. $9a - a^4 + 3 + 2a^3$

27. Complete the following table for the polynomial $7x^2 + 8x^5 - 4x^3 + 6 - \frac{1}{2}x^4$.

Term	Coefficient	Degree of the Term	Degree of the Polynomial
		5	
$-\frac{1}{2}x^4$			
	-4		
		2	
	6		

28. Complete the following table for the polynomial $-3x^4 + 6x^3 - 2x^2 + 8x + 7$.

Term	Coefficient	Degree of the Term	Degree of the Polynomial
	-3		
$6x^3$			
		2	
		1	
	7		

Classify each polynomial as a monomial, binomial, trinomial, or none of these.

29. $x^2 - 23x + 17$

30. $-9x^2$

31. $x^3 - 7x^2 + 2x - 4$

32. $t^3 + 4$

33. $8t^2 + 5t$

34. $4x^2 + 12x + 9$

35. 17

36. $2x^4 - 7x^3 + x^2 + x - 6$

Combine like terms. Write all answers in descending order.

37. $7x^2 + 3x + 4x^2$

38. $5a + 7a^2 + 3a$

39. $3a^4 - 2a + 2a + a^4$

40. $9b^5 + 3b^2 - 2b^5 - 3b^2$

41. $2x^2 - 6x + 3x + 4x^2$

42. $3x^4 - 7x + x^4 - 2x$

43. $9x^3 + 2x - 4x^3 + 5 - 3x$

44. $6x^2 + 2x^4 - 2x^2 - x^4 - 4x^2$

45. $10x^2 + 2x^3 - 3x^3 - 4x^2 - 6x^2 - x^4$

46. $8x^5 - x^4 + 2x^5 + 5x^4 - 4x^4 - x^6$

47. $\frac{1}{5}x^4 + 7 - 2x^2 + 3 - \frac{2}{15}x^4 + 2x^2$

48. $\frac{1}{6}x^3 + 3x^2 - \frac{1}{3}x^3 + 7 + x^2 - 10$

49. $5.9x^2 - 2.1x + 6 + 3.4x - 2.5x^2 - 0.5$

50. $7.4x^3 - 4.9x + 2.9 - 3.5x - 4.3 + 1.9x^3$

Evaluate each polynomial for $x = 3$ and for $x = -3$.

51. $-7x + 4$

52. $-5x + 7$

53. $2x^2 - 3x + 7$

54. $4x^2 - 6x + 9$

55. $-2x^3 - 3x^2 + 4x + 2$

56. $-3x^3 + 7x^2 - 4x - 8$

57. $\frac{1}{3}x^4 - 2x^3$

58. $2x^4 - \frac{1}{9}x^3$

59. $-x^4 - x^3 - x^2$

60. $-x^2 - 3x^3 - x^4$

Consumer debt. *The amount of debt, in trillions of dollars, held by U.S. consumers can be estimated by the polynomial*

$$0.15t + 1.42,$$

where t is the number of years since 2000.
Source: Based on information from the Federal Reserve

61. Estimate U.S. consumer debt in 2005.

62. Estimate U.S. consumer debt in 2008.

63. *Skydiving.* During the first 13 sec of a jump, the number of feet that a skydiver falls in t seconds is approximated by the polynomial

$$11.12t^2.$$

Approximately how far has a skydiver fallen 10 sec after jumping from a plane?

64. *Skydiving.* For jumps that exceed 13 sec, the polynomial $173t - 369$ can be used to approximate the distance, in feet, that a skydiver has fallen in t seconds. Approximately how far has a skydiver fallen 20 sec after jumping from a plane?

Circumference. *The circumference of a circle of radius r is given by the polynomial $2\pi r$, where π is an irrational number. For an approximation of π, use 3.14.*

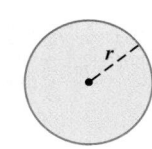

65. Find the circumference of a circle with radius 10 cm.

66. Find the circumference of a circle with radius 5 ft.

Area of a circle. *The area of a circle of radius r is given by the polynomial πr^2. Use 3.14 for π.*

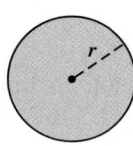

67. Find the area of a circle with radius 7 m.

68. Find the area of a circle with radius 6 ft.

Electricity use.   *The consumption of electricity in the United States, in millions of gigawatt hours, t years after 2000, can be approximated by*

$$0.19t + 3.93.$$

Use the following graph for Exercises 69–72.
Source: Based on information from Cambridge Energy Research Associates

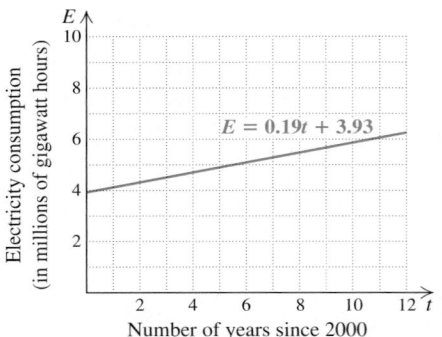

69. Estimate U.S. electricity consumption in 2006.

70. Estimate U.S. electricity consumption in 2008.

71. Estimate U.S. electricity consumption in 2003.

72. Estimate U.S. electricity consumption in 2005.

Memorizing words.   *Participants in a psychology experiment were able to memorize an average of M words in t minutes, where $M = -0.001t^3 + 0.1t^2$. Use the following graph for Exercises 73–76.*

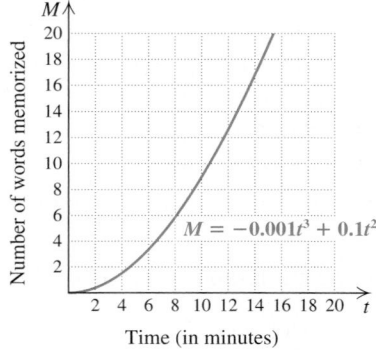

73. Estimate the number of words memorized after 10 min.

74. Estimate the number of words memorized after 14 min.

75. Find the approximate value of M for $t = 8$.

76. Find the approximate value of M for $t = 12$.

Hearing-impaired Americans.   *The number, in millions, of hearing-impaired Americans of age x can be approximated by*

$$-0.00006x^3 + 0.006x^2 - 0.1x + 1.9.$$

Use the following graph for Exercises 77 and 78.
Source: Based on information from American Speech-Language Hearing Association

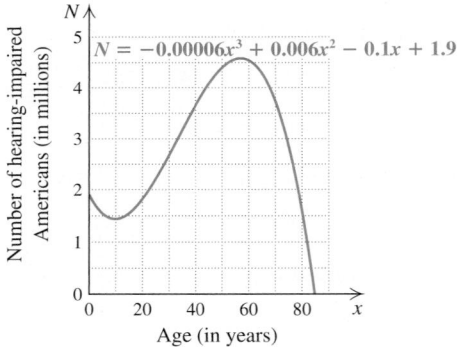

77. Approximate the number of hearing-impaired Americans of ages 20 and 50.

78. Approximate the number of hearing-impaired Americans of ages 30 and 60.

79. Explain how it is possible for a term to not be a monomial.

80. Is it possible to evaluate polynomials without understanding the rules for order of operations? Why or why not?

SKILL MAINTENANCE

Simplify.

81. $-19 + 24$ [1.5]

82. $5 - 14$ [1.6]

Factor. [1.2]

83. $5x + 15$

84. $7a - 21$

85. A family spent $2011 to drive a car one year, during which the car was driven 14,800 mi. The family spent $972 for insurance and $114 for registration and oil. The only other cost was for gasoline. How much did gasoline cost per mile? [3.4]

86. The sum of the page numbers on the facing pages of a book is 549. What are the page numbers? [2.5]

SYNTHESIS

87. Suppose that the coefficients of a polynomial are all integers and the polynomial is evaluated for some integer. Must the value of the polynomial then also be an integer? Why or why not?

88. Is it easier to evaluate a polynomial before or after like terms have been combined? Why?

89. Construct a polynomial in x (meaning that x is the variable) of degree 5 with four terms, with coefficients that are consecutive even integers. Write in descending order.

Revenue, cost, and profit.   Gigabytes Electronics is selling a new type of computer monitor. Total revenue is the total amount of money taken in and total cost is the total amount paid for producing the items. The firm estimates that for the monitor's first year, revenue from the sale of x monitors is

$$250x - 0.5x^2 \text{ dollars,}$$

and the total cost is given by

$$4000 + 0.6x^2 \text{ dollars.}$$

Profit *is the difference between revenue and cost.*

90. Find the profit when 20 monitors are produced and sold.

91. Find the profit when 30 monitors are produced and sold.

Simplify.

92. $\frac{9}{2}x^8 + \frac{1}{9}x^2 + \frac{1}{2}x^9 + \frac{9}{2}x + \frac{9}{2}x^9 + \frac{8}{9}x^2 + \frac{1}{2}x - \frac{1}{2}x^8$

93. $(3x^2)^3 + 4x^2 \cdot 4x^4 - x^4(2x)^2 + ((2x)^2)^3 - 100x^2(x^2)^2$

94. A polynomial in x has degree 3. The coefficient of x^2 is 3 less than the coefficient of x^3. The coefficient of x is three times the coefficient of x^2. The remaining constant is 2 more than the coefficient of x^3. The sum of the coefficients is -4. Find the polynomial.

95. Use the graph for Exercises 77 and 78 to approximate the ages at which 1.5 million Americans of that age are hearing-impaired.

96. *Path of the Olympic arrow.*   The Olympic flame at the 1992 Summer Olympics was lit by a flaming arrow. As the arrow moved d meters horizontally from the archer, its height h, in meters, was approximated by the polynomial

$$-0.0064d^2 + 0.8d + 2.$$

Complete the table for the choices of d given. Then plot the points and draw a graph representing the path of the arrow.

d	$-0.0064d^2 + 0.8d + 2$
0	
30	
60	
90	
120	

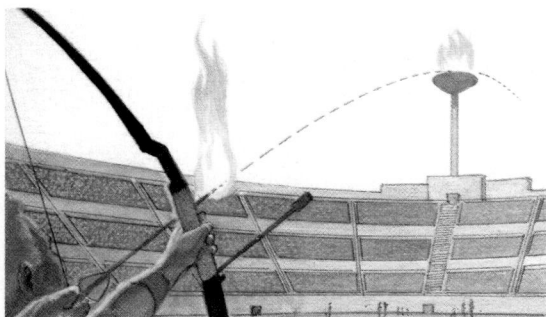

■ *Semester averages.* *Professor Kopecki calculates a student's average for her course using*

$$A = 0.3q + 0.4t + 0.2f + 0.1h,$$

with q, t, f, and h representing a student's quiz average, test average, final exam score, and homework average, respectively. In Exercises 97 and 98, find the given student's course average rounded to the nearest tenth.

97. Mary Lou: quizzes: 60, 85, 72, 91; final exam: 84; tests: 89, 93, 90; homework: 88

98. Nigel: quizzes: 95, 99, 72, 79; final exam: 91; tests: 68, 76, 92; homework: 86

In Exercises 99 and 100, complete the table for the given choices of t. Then plot the points and connect them with a smooth curve representing the graph of the polynomial.

99.

t	$-t^2 + 10t - 18$
3	
4	
5	
6	
7	

100.

t	$-t^2 + 6t - 4$
1	
2	
3	
4	
5	

4.3 Addition and Subtraction of Polynomials

Addition of Polynomials • Opposites of Polynomials •
Subtraction of Polynomials • Problem Solving

Addition of Polynomials

To add two polynomials, we write a plus sign between them and combine like terms.

EXAMPLE 1   Add.

a) $(-5x^3 + 6x - 1) + (4x^3 + 3x^2 + 2)$

b) $\left(\frac{2}{3}x^4 + 3x^2 - 7x + \frac{1}{2}\right) + \left(-\frac{1}{3}x^4 + 5x^3 - 3x^2 + 3x - \frac{1}{2}\right)$

Solution

a) $(-5x^3 + 6x - 1) + (4x^3 + 3x^2 + 2)$

$= (-5 + 4)x^3 + 3x^2 + 6x + (-1 + 2)$ Combining like terms; using the distributive law

$= -x^3 + 3x^2 + 6x + 1$ Note that $-1x^3 = -x^3$.

b) $\left(\frac{2}{3}x^4 + 3x^2 - 7x + \frac{1}{2}\right) + \left(-\frac{1}{3}x^4 + 5x^3 - 3x^2 + 3x - \frac{1}{2}\right)$

$= \left(\frac{2}{3} - \frac{1}{3}\right)x^4 + 5x^3 + (3 - 3)x^2 + (-7 + 3)x + \left(\frac{1}{2} - \frac{1}{2}\right)$ Combining like terms

$= \frac{1}{3}x^4 + 5x^3 - 4x$

After some practice, polynomial addition is often performed mentally.

EXAMPLE 2   Add: $(2 - 3x + x^2) + (-5 + 7x - 3x^2 + x^3)$.

Solution   We have

$$(2 - 3x + x^2) + (-5 + 7x - 3x^2 + x^3)$$
$$= (2 - 5) + (-3 + 7)x + (1 - 3)x^2 + x^3 \qquad \text{You might do}$$
$$\text{this step}$$
$$\text{mentally.}$$
$$= -3 + 4x - 2x^2 + x^3. \qquad \text{Then you would write only this.}$$

Study Skills———

Aim for Mastery

In each exercise set, you may encounter one or more exercises that give you trouble. Do not ignore these problems, but instead work to master them. These are the problems (not the "easy" problems) that you will need to revisit as you progress through the course. Consider marking these exercises with a star so that you can easily locate them when studying or reviewing for a quiz or test.

The polynomials in the last example have the terms arranged according to degree, from least to greatest. Such an arrangement is called *ascending order*. As a rule, answers are written in ascending order when the polynomials in the original problem are given in ascending order. If the polynomials in the original problem are given in descending order, the answer is written in descending order.

We can also add polynomials by writing like terms in columns. Sometimes this makes like terms easier to see.

EXAMPLE 3   Add: $9x^5 - 2x^3 + 6x^2 + 3$ and $5x^4 - 7x^2 + 6$ and $3x^6 - 5x^5 + x^2 + 5$.

Solution   We arrange the polynomials with like terms in columns.

$$
\begin{array}{l}
9x^5 \qquad\quad - 2x^3 + 6x^2 + \;\, 3 \\
\qquad\quad 5x^4 \qquad\quad - 7x^2 + \;\, 6 \qquad \text{We leave spaces for missing terms.} \\
\underline{3x^6 - 5x^5 \qquad\qquad\quad + 1x^2 + \;\, 5} \qquad \text{Writing } x^2 \text{ as } 1x^2 \\
3x^6 + 4x^5 + 5x^4 - 2x^3 \qquad\quad + 14 \qquad \text{Adding}
\end{array}
$$

The answer is $3x^6 + 4x^5 + 5x^4 - 2x^3 + 14$.

CONNECTING THE CONCEPTS

In many ways, polynomials are to algebra as numbers are to arithmetic. Like numbers, polynomials can be added, subtracted, multiplied, and divided. In Sections 4.3–4.7, we examine how these operations are performed. Thus we are again learning how to write equivalent expressions, not how to solve equations.

Note that equations like those in Examples 1–3 are written to show how one expression can be rewritten in an equivalent form. This is very different from solving an equation.

There is much to learn about manipulating polynomials. Not until Section 5.6 will we return to the task of solving equations.

Opposites of Polynomials

In Section 1.8, we used the property of -1 to show that the opposite of a sum is the sum of the opposites. This idea can be extended.

The Opposite of a Polynomial

To find an equivalent polynomial for the *opposite*, or *additive inverse*, of a polynomial, change the sign of every term. This is the same as multiplying the polynomial by -1.

EXAMPLE 4   Write the opposite of $4x^5 - 7x^3 - 8x + \frac{5}{6}$ in two different forms.

Solution

i) $-\left(4x^5 - 7x^3 - 8x + \frac{5}{6}\right)$ This is one representation of the opposite of $4x^5 - 7x^3 - 8x + \frac{5}{6}$.

ii) $-4x^5 + 7x^3 + 8x - \frac{5}{6}$ Changing the sign of every term

Thus, $-\left(4x^5 - 7x^3 - 8x + \frac{5}{6}\right)$ and $-4x^5 + 7x^3 + 8x - \frac{5}{6}$ are equivalent. Both expressions represent the opposite of $4x^5 - 7x^3 - 8x + \frac{5}{6}$.

EXAMPLE 5   Simplify: $-\left(-7x^4 - \frac{5}{9}x^3 + 8x^2 - x + 67\right)$.

Solution   We have

$$-\left(-7x^4 - \tfrac{5}{9}x^3 + 8x^2 - x + 67\right) = 7x^4 + \tfrac{5}{9}x^3 - 8x^2 + x - 67.$$

The same result can be found by multiplying by -1:

$$-\left(-7x^4 - \tfrac{5}{9}x^3 + 8x^2 - x + 67\right)$$
$$= -1(-7x^4) + (-1)\left(-\tfrac{5}{9}x^3\right) + (-1)(8x^2) + (-1)(-x) + (-1)67$$
$$= 7x^4 + \tfrac{5}{9}x^3 - 8x^2 + x - 67.$$

Subtraction of Polynomials

We can now subtract one polynomial from another by adding the opposite of the polynomial being subtracted.

EXAMPLE 6   Subtract: $(9x^5 + x^3 - 2x^2 + 4) - (-2x^5 + x^4 - 4x^3 - 3x^2)$.

Solution

$$(9x^5 + x^3 - 2x^2 + 4) - (-2x^5 + x^4 - 4x^3 - 3x^2)$$

$$= 9x^5 + x^3 - 2x^2 + 4 + 2x^5 - x^4 + 4x^3 + 3x^2 \qquad \text{Adding the opposite}$$

$$= 11x^5 - x^4 + 5x^3 + x^2 + 4 \qquad \text{Combining like terms}$$

EXAMPLE 7   Subtract: $(7x^5 + x^3 - 9x) - (3x^5 - 4x^3 + 5)$.

Solution

$$(7x^5 + x^3 - 9x) - (3x^5 - 4x^3 + 5)$$

$$= 7x^5 + x^3 - 9x + (-3x^5) + 4x^3 - 5 \qquad \text{Adding the opposite}$$

$$= 7x^5 + x^3 - 9x - 3x^5 + 4x^3 - 5 \qquad \text{Try to go directly to this step.}$$

$$= 4x^5 + 5x^3 - 9x - 5 \qquad \text{Combining like terms}$$

To subtract using columns, we first replace the coefficients in the polynomial being subtracted with their opposites. We then add as before.

EXAMPLE 8   Write in columns and subtract: $(5x^2 - 3x + 6) - (9x^2 - 5x - 3)$.

Solution

i) $\begin{aligned} 5x^2 - 3x + 6 \\ -(9x^2 - 5x - 3) \end{aligned}$ Writing similar terms in columns

ii) $\begin{aligned} 5x^2 - 3x + 6 \\ -9x^2 + 5x + 3 \end{aligned}$ Changing signs and removing parentheses

iii) $\begin{aligned} 5x^2 - 3x + 6 \\ \underline{-9x^2 + 5x + 3} \\ -4x^2 + 2x + 9 \end{aligned}$ Adding

---

If you can do so without error, you can arrange the polynomials in columns, mentally find the opposite of each term being subtracted, and write the answer. Lining up like terms is important and may require leaving some blanks.

EXAMPLE 9   Write in columns and subtract: $(x^3 + x^2 + 2x - 12) - (-2x^3 + x^2 - 3x)$.

Solution   We have

$$\begin{aligned} x^3 + x^2 + 2x - 12 \\ \underline{-(-2x^3 + x^2 - 3x \qquad)} \\ 3x^3 \qquad\quad + 5x - 12. \end{aligned}$$ Leaving a blank space for the missing term

Problem Solving

EXAMPLE 10   Find a polynomial for the sum of the areas of rectangles *A*, *B*, *C*, and *D*.

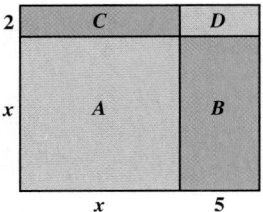

Solution

1. **Familiarize.**   Recall that the area of a rectangle is the product of its length and width.

2. **Translate.**   We translate the problem to mathematical language. The sum of the areas is a sum of products. We find each product and then add:

$$\underbrace{\text{Area of } A}_{x \cdot x} \quad \overset{\text{plus}}{+} \quad \underbrace{\text{area of } B}_{5x} \quad \overset{\text{plus}}{+} \quad \underbrace{\text{area of } C}_{2x} \quad \overset{\text{plus}}{+} \quad \underbrace{\text{area of } D}_{2 \cdot 5.}$$

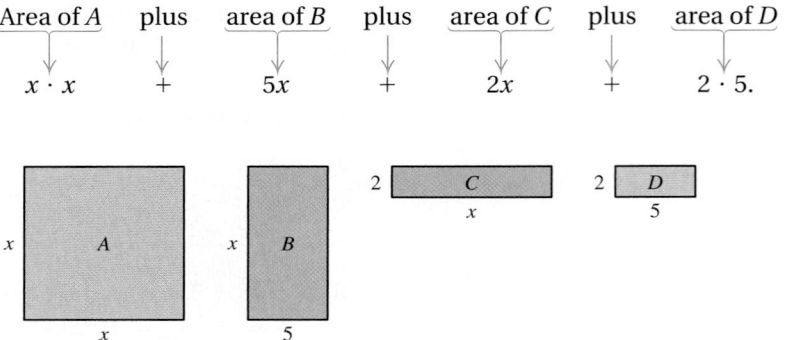

3. **Carry out.**   We simplify $x \cdot x$ and $2 \cdot 5$ and combine like terms:

$$x^2 + 5x + 2x + 10 = x^2 + 7x + 10.$$

4. **Check.**   A partial check is to replace *x* with a number, say 3. Then we evaluate $x^2 + 7x + 10$ and compare that result with an alternative calculation:

$$3^2 + 7 \cdot 3 + 10 = 9 + 21 + 10 = 40.$$

When we substitute 3 for *x* and calculate the total area by regarding the figure as one large rectangle, we should also get 40:

$$\text{Total area} = (x + 5)(x + 2) = (3 + 5)(3 + 2) = 8 \cdot 5 = 40.$$

Our check is only partial, since it is possible for an incorrect answer to equal 40 when evaluated for $x = 3$. This would be unlikely, especially if a second choice of *x*, say $x = 5$, also checks. We leave that check to the student.

5. **State.**   A polynomial for the sum of the areas is $x^2 + 7x + 10$.

EXAMPLE 11 An 8-ft by 8-ft shed is placed on a square lawn x ft on a side. Find a polynomial for the remaining area.

Solution

1. **Familiarize.** We make a drawing of the situation as follows.

2. **Translate.** We reword the problem and translate as follows.

Rewording: Area of lawn − area of shed = area left over

Translating: x ft · x ft − 8 ft · 8 ft = Area left over

3. **Carry out.** We carry out the multiplication:

$$x^2 \text{ ft}^2 - 64 \text{ ft}^2 = \text{Area left over.}$$

4. **Check.** As a partial check, note that the units in the answer are square feet (ft^2), a measure of area, as expected.

5. **State.** The remaining area in the yard is $(x^2 - 64)$ ft^2.

 **technology connection**

To check polynomial addition or subtraction, we can let y_1 = the expression before the addition or subtraction has been performed and y_2 = the simplified sum or difference. If the addition or subtraction is correct, y_1 will equal y_2 and $y_2 - y_1$ will be 0. We enter $y_2 - y_1$ as y_3, using the VARS key. Below is a check of Example 7 in which

$$y_1 = (7x^5 + x^3 - 9x) - (3x^5 - 4x^3 + 5),$$
$$y_2 = 4x^5 + 5x^3 - 9x - 5,$$

and

$$y_3 = y_2 - y_1.$$

We graph only y_3. If indeed y_1 and y_2 are equivalent, then y_3 should equal 0. This means its graph should coincide with the x-axis. The TRACE or TABLE features can confirm

that y_3 is always 0, or we can select y_3 to be drawn bold at the ⟨ Y= ⟩ window.

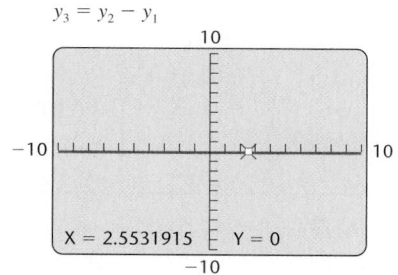

$y_3 = y_2 - y_1$

X = 2.5531915   Y = 0

1. Use a graphing calculator to check Examples 1, 2, and 6.

Exercise Set

4.3

Add.

1. $(3x + 2) + (-5x + 4)$

2. $(5x + 1) + (-9x + 4)$

3. $(-6x + 2) + (x^2 + x - 3)$

4. $(x^2 - 5x + 4) + (8x - 9)$

5. $(7t^2 - 3t + 6) + (2t^2 + 8t - 9)$

6. $(8a^2 + 4a - 7) + (6a^2 - 3a - 1)$

7. $(4m^3 - 4m^2 + m - 5) + (4m^3 + 7m^2 - 4m - 2)$

8. $(5n^3 - n^2 + 4n - 3) + (2n^3 - 4n^2 + 3n - 4)$

9. $(3 + 6a + 7a^2 + 8a^3) + (4 + 7a - a^2 + 6a^3)$

10. $(7 + 4t - 5t^2 + 6t^3) + (2 + t + 6t^2 - 4t^3)$

11. $(9x^8 - 7x^4 + 2x^2 + 5) + (8x^7 + 4x^4 - 2x)$

12. $(4x^5 - 6x^3 - 9x + 1) + (6x^3 + 9x^2 + 9x)$

13. $\left(\frac{1}{4}x^4 + \frac{2}{3}x^3 + \frac{5}{8}x^2 + 9\right) + \left(-\frac{3}{4}x^4 + \frac{3}{8}x^2 - 7\right)$

14. $\left(\frac{1}{3}x^9 + \frac{1}{5}x^5 - \frac{1}{2}x^2 + 7\right) + \left(-\frac{1}{5}x^9 + \frac{1}{4}x^4 - \frac{3}{5}x^5\right)$

15. $(5.3t^2 - 6.4t - 9.1) + (4.2t^3 - 1.8t^2 + 7.3)$

16. $(4.9a^3 + 3.2a^2 - 5.1a) + (2.1a^2 - 3.7a + 4.6)$

17. $\begin{array}{r} -3x^4 + 6x^2 + 2x - 4 \\ -3x^2 + 2x + 4 \\ \hline \end{array}$

18. $\begin{array}{r} -4x^3 + 8x^2 + 3x - 2 \\ -4x^2 + 3x + 2 \\ \hline \end{array}$

19. $\begin{array}{r} 0.15x^4 + 0.10x^3 - 0.9x^2 \\ -0.01x^3 + 0.01x^2 + x \\ 1.25x^4 \qquad + 0.11x^2 \qquad + 0.01 \\ 0.27x^3 \qquad\qquad + 0.99 \\ -0.35x^4 \qquad + 15x^2 - 0.03 \\ \hline \end{array}$

20. $\begin{array}{r} 0.05x^4 + 0.12x^3 - 0.5x^2 \\ -0.02x^3 + 0.02x^2 + 2x \\ 1.5x^4 \qquad + 0.01x^2 \qquad + 0.15 \\ 0.25x^3 \qquad\qquad + 0.85 \\ -0.25x^4 \qquad + 10x^2 - 0.04 \\ \hline \end{array}$

Write the opposite of each polynomial in two different forms, as in Example 4.

21. $-t^3 + 4t^2 - 9$

22. $-4x^3 - 5x^2 + 2x$

23. $12x^4 - 3x^3 + 3$

24. $5a^3 + 2a - 17$

Simplify.

25. $-(8x - 9)$

26. $-(-6x + 5)$

27. $-(3a^4 - 5a^2 + 9)$

28. $-(-6a^3 + 2a^2 - 7)$

29. $-(-4x^4 + 6x^2 + \frac{3}{4}x - 8)$

30. $-(-5x^4 + 4x^3 - x^2 + 0.9)$

Subtract.

31. $(7x + 4) - (-2x + 1)$

32. $(5x + 6) - (-2x + 4)$

33. $(-5t + 6) - (t^2 + 3t - 1)$

34. $(a^2 - 5a + 2) - (3a^2 + 2a - 4)$

35. $(6x^4 + 3x^3 - 1) - (4x^2 - 3x + 3)$

36. $(-4x^2 + 2x) - (3x^3 - 5x^2 + 3)$

37. $(1.2x^3 + 4.5x^2 - 3.8x) - (-3.4x^3 - 4.7x^2 + 23)$

38. $(0.5x^4 - 0.6x^2 + 0.7) - (2.3x^4 + 1.8x - 3.9)$

Aha! **39.** $(7x^3 - 2x^2 + 6) - (7x^3 - 2x^2 + 6)$

40. $(8x^5 + 3x^4 + x - 1) - (8x^5 + 3x^4 - 1)$

41. $(3 + 5a + 3a^2 - a^3) - (2 + 3a - 4a^2 + 2a^3)$

42. $(7 + t - 5t^2 + 2t^3) - (1 + 2t - 4t^2 + 5t^3)$

43. $\left(\frac{5}{8}x^3 - \frac{1}{4}x - \frac{1}{3}\right) - \left(-\frac{1}{8}x^3 + \frac{1}{4}x - \frac{1}{3}\right)$

44. $\left(\frac{1}{5}x^3 + 2x^2 - \frac{3}{10}\right) - \left(-\frac{2}{5}x^3 + 2x^2 + \frac{7}{1000}\right)$

45. $(0.07t^3 - 0.03t^2 + 0.01t) - (0.02t^3 + 0.04t^2 - 1)$

46. $(0.9a^3 + 0.2a - 5) - (0.7a^4 - 0.3a - 0.1)$

47. $x^2 + 5x + 6$ **48.** $x^3 + 3x^2 + 1$
     $\underline{-(x^2 + 2x + 1)}$ $\underline{-(x^3 + x^2 - 5)}$

49. $5x^4 + 6x^3 - 9x^2$
     $\underline{-(-6x^4 - 6x^3 + x^2)}$

50. $4x^4 - 2x^3 + 6x^2$
     $\underline{-(7x^4 + 6x^3 + 7x^2)}$

51. Solve.

   a) Find a polynomial for the sum of the areas of the rectangles shown in the figure.

   b) Find the sum of the areas when $x = 5$ and $x = 7$.

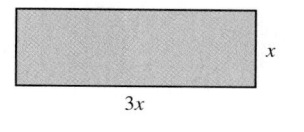 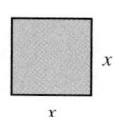
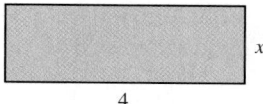 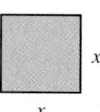

52. Solve.

   a) Find a polynomial for the sum of the areas of the circles shown in the figure.

   b) Find the sum of the areas when $r = 5$ and $r = 11.3$.

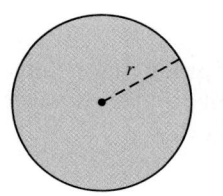 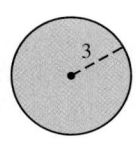 

Find a polynomial for the perimeter of each figure in Exercises 53 and 54.

53.

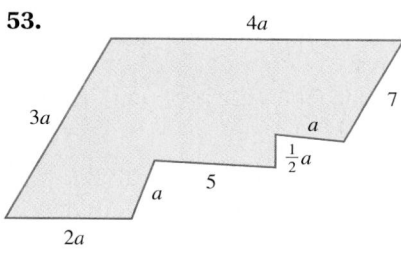

54.

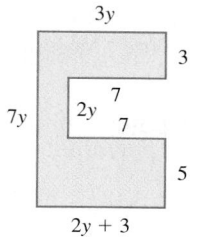

Find two algebraic expressions for the area of each figure. First, regard the figure as one large rectangle, and then regard the figure as a sum of four smaller rectangles.

55.

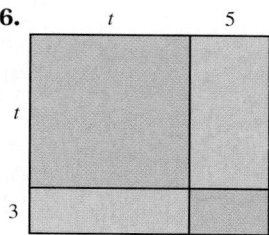

56.

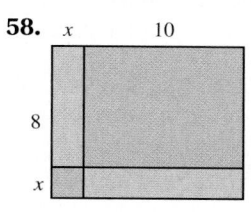

57.

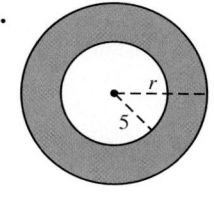

58.

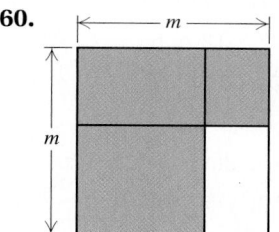

Find a polynomial for the shaded area of each figure.

59.

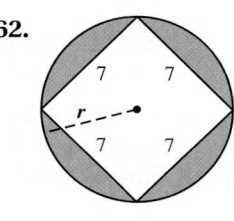

60.

61.

62.

63. A 2-ft by 6-ft bath enclosure is installed in a new bathroom measuring x ft by x ft. Find a polynomial for the remaining floor area.

64. A 5-ft by 7-ft Jacuzzi™ is installed on an outdoor deck measuring y ft by y ft. Find a polynomial for the remaining area of the deck.

65. An 8-ft–wide round hot tub is installed in a patio measuring z ft by z ft. Find a polynomial for the remaining area of the patio.

66. A 10-ft–wide round water trampoline is floating in a pool measuring x ft by x ft. Find a polynomial for the remaining surface area of the pool.

67. A 12-m by 12-m mat includes a circle of diameter d meters for wrestling. Find a polynomial for the area of the mat outside the wrestling circle.

68. A 2-m by 3-m rug is spread inside a tepee that has a diameter of x meters. Find a polynomial for the area of the tepee's floor that is not covered.

69. What advice would you offer to a student who is successful at adding, but not subtracting, polynomials?

70. Is the sum of two trinomials always a trinomial? Why or why not?

SKILL MAINTENANCE

Simplify. [1.8]

71. $5(4 + 3) - 5 \cdot 4 - 5 \cdot 3$

72. $7(2 + 6) - 7 \cdot 2 - 7 \cdot 6$

73. $2(5t + 7) + 3t$ **74.** $3(4t - 5) + 2t$

Solve. [2.6]

75. $2(x + 3) > 5(x - 3) + 7$

76. $7(x - 8) \le 4(x - 5)$

SYNTHESIS

77. What can be concluded about two polynomials whose sum is zero?

78. Which, if any, of the commutative, associative, and distributive laws are needed for adding polynomials? Why?

Simplify.

79. $(6t^2 - 7t) + (3t^2 - 4t + 5) - (9t - 6)$

80. $(3x^2 - 4x + 6) - (-2x^2 + 4) + (-5x - 3)$

81. $(-8y^2 - 4) - (3y + 6) - (2y^2 - y)$

82. $(-4 + x^2 + 2x^3) - (-6 - x + 3x^3) - (-x^2 - 5x^3)$

83. $(345.099x^3 - 6.178x) - (94.508x^3 - 8.99x)$

Find a polynomial for the surface area of the right rectangular solid.

84.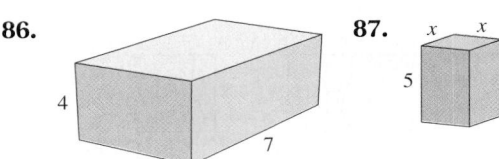

85.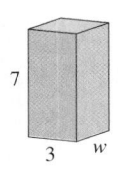

86.

87.

88. Find a polynomial for the total length of all edges in the figure appearing in Exercise 87.

89. Find a polynomial for the total length of all edges in the figure appearing in Exercise 84.

90. *Total profit.*   Hadley Electronics is marketing a new kind of stereo. Total revenue is the total amount of money taken in. The firm determines that when it sells x stereos, its total revenue is given by

$$R = 280x - 0.4x^2.$$

Total cost is the total cost of producing x stereos. Hadley Electronics determines that the total cost of producing x stereos is given by

$$C = 5000 + 0.6x^2.$$

The total profit P is

(Total Revenue) − (Total Cost) = $R - C.$

a) Find a polynomial for total profit.

b) What is the total profit on the production and sale of 75 stereos?

c) What is the total profit on the production and sale of 100 stereos?

91. Does replacing each occurrence of the variable x in $4x^7 - 6x^3 + 2x$ with its opposite result in the opposite of the polynomial? Why or why not?

4.4   Multiplication of Polynomials

Multiplying Monomials • Multiplying a Monomial and
a Polynomial • Multiplying Any Two Polynomials •
Checking by Evaluating

We now multiply polynomials using techniques based largely on the distributive, associative, and commutative laws and the rules for exponents.

Multiplying Monomials

Consider $(3x)(4x)$. We multiply as follows:

$$(3x)(4x) = 3 \cdot x \cdot 4 \cdot x \qquad \text{Using an associative law}$$
$$= 3 \cdot 4 \cdot x \cdot x \qquad \text{Using a commutative law}$$
$$= (3 \cdot 4) \cdot x \cdot x \qquad \text{Using an associative law}$$
$$= 12x^2.$$

To Multiply Monomials

To find an equivalent expression for the product of two monomials, multiply the coefficients and then multiply the variables using the product rule for exponents.

EXAMPLE 1   Multiply: **(a)** $(5x)(6x)$; **(b)** $(3a)(-a)$; **(c)** $(-7x^5)(4x^3)$.

Solution

a) $(5x)(6x) = (5 \cdot 6)(x \cdot x)$ Multiplying the coefficients; multiplying the variables

$\qquad\qquad\quad = 30x^2$ Simplifying

b) $(3a)(-a) = (3a)(-1a)$ Writing $-a$ as $-1a$ can ease calculations.

$\qquad\qquad\quad = (3)(-1)(a \cdot a)$ Using an associative and a commutative law

$\qquad\qquad\quad = -3a^2$

c) $(-7x^5)(4x^3) = (-7 \cdot 4)(x^5 \cdot x^3)$

$\qquad\qquad\qquad = -28x^{5+3}$
$\left.\rule{0pt}{24pt}\right\}$ Using the product rule for exponents
$\qquad\qquad\qquad = -28x^8$

After some practice, you can try writing only the answer.

Student Notes _____

Remember that when we compute $(3 \cdot 5)(2 \cdot 4)$, each factor is used only once, even if we change the order:

$(3 \cdot 5)(2 \cdot 4) = (5 \cdot 2)(3 \cdot 4)$

$\qquad\qquad = 10 \cdot 12 = 120.$

Some students mistakenly "reuse" a factor, as if the distributive law applied.

Multiplying a Monomial and a Polynomial

To find an equivalent expression for the product of a monomial, such as $5x$, and a polynomial, such as $2x^2 - 3x + 4$, we use the distributive law.

EXAMPLE 2   Multiply: **(a)** x and $x + 3$; **(b)** $5x(2x^2 - 3x + 4)$.

Solution

a) $x(x + 3) = x \cdot x + x \cdot 3$ Using the distributive law

$\quad\quad\quad\quad = x^2 + 3x$

b) $5x(2x^2 - 3x + 4) = (5x)(2x^2) - (5x)(3x) + (5x)(4)$ Using the distributive law

$\quad\quad\quad\quad\quad\quad\quad\quad = 10x^3 - 15x^2 + 20x$ Performing the three multiplications

The product in Example 2(a) can be visualized as the area of a rectangle with width x and length $x + 3$.

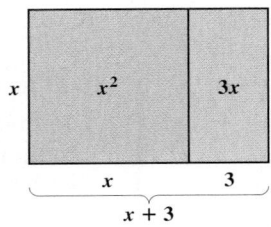

Note that the total area can be expressed as $x(x + 3)$ or, by adding the two smaller areas, $x^2 + 3x$.

> **The Product of a Monomial and a Polynomial**
>
> To multiply a monomial and a polynomial, multiply each term of the polynomial by the monomial.

Try to do this mentally, when possible. Remember that we multiply coefficients and, when the bases match, add exponents.

EXAMPLE 3   Multiply: $2x^2(x^3 - 7x^2 + 10x - 4)$.

Solution

$\qquad\qquad\qquad\qquad$ *Think*: $\underline{2x^2 \cdot x^3} - \underline{2x^2 \cdot 7x^2} + \underline{2x^2 \cdot 10x} - \underline{2x^2 \cdot 4}$

$2x^2(x^3 - 7x^2 + 10x - 4) = 2x^5 \quad - \quad 14x^4 \quad + \quad 20x^3 \quad - \quad 8x^2$

Multiplying Any Two Polynomials

Before considering the product of *any* two polynomials, let's look at products when both polynomials are binomials.

To find an equivalent expression for the product of two binomials, we again begin by using the distributive law. This time, however, it is a *binomial* rather than a monomial that is being distributed.

EXAMPLE 4   Multiply each of the following.

a) $x + 5$ and $x + 4$ 　　　　　　　　**b)** $4x - 3$ and $x - 2$

Solution

a) $(x + 5)\ (x + 4) = (x + 5)\ x + (x + 5)\ 4$ 　　Using the distributive law

$\qquad\qquad\qquad = x(x + 5) + 4(x + 5)$ 　　Using the commutative law for multiplication

$\qquad\qquad\qquad = x \cdot x + x \cdot 5 + 4 \cdot x + 4 \cdot 5$ 　　Using the distributive law (twice)

$\qquad\qquad\qquad = x^2 + 5x + 4x + 20$ 　　Multiplying the monomials

$\qquad\qquad\qquad = x^2 + 9x + 20$ 　　Combining like terms

b) $(4x - 3)\ (x - 2) = (4x - 3)\ x - (4x - 3)\ 2$ 　　Using the distributive law

$\qquad\qquad\qquad = x(4x - 3) - 2(4x - 3)$ 　　Using the commutative law for multiplication. This step is often omitted.

$\qquad\qquad\qquad = x \cdot 4x - x \cdot 3 - 2 \cdot 4x - 2(-3)$ 　　Using the distributive law (twice)

$\qquad\qquad\qquad = 4x^2 - 3x - 8x + 6$ 　　Multiplying the monomials

$\qquad\qquad\qquad = 4x^2 - 11x + 6$ 　　Combining like terms

To visualize the product in Example 4(a), consider a rectangle of length $x + 5$ and width $x + 4$, as shown here.

The total area can be expressed as $(x + 5)(x + 4)$ or, by adding the four smaller areas, $x^2 + 5x + 4x + 20$.

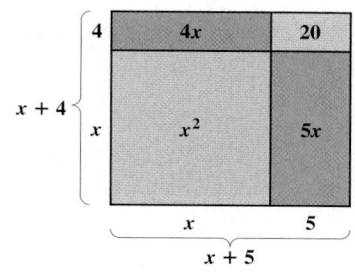

Let's consider the product of a binomial and a trinomial. Again we make repeated use of the distributive law.

EXAMPLE 5   Multiply: $(x^2 + 2x - 3)(x + 4)$.

Solution

$$(x^2 + 2x - 3) \ (x + 4)$$

$$= (x^2 + 2x - 3) \ x + \ (x^2 + 2x - 3) \ 4 \qquad \text{Using the distributive law}$$

$$= x(x^2 + 2x - 3) + 4(x^2 + 2x - 3) \qquad \text{Using the commutative law}$$

$$= x \cdot x^2 + x \cdot 2x - x \cdot 3 + 4 \cdot x^2 + 4 \cdot 2x - 4 \cdot 3 \qquad \text{Using the distributive law (twice)}$$

$$= x^3 + 2x^2 - 3x + 4x^2 + 8x - 12 \qquad \text{Multiplying the monomials}$$

$$= x^3 + 6x^2 + 5x - 12 \qquad \text{Combining like terms}$$

Perhaps you have discovered the following in the preceding examples.

> **The Product of Two Polynomials**
>
> To multiply two polynomials P and Q, select one of the polynomials, say P. Then multiply each term of P by every term of Q and combine like terms.

To use columns for long multiplication, multiply each term in the top row by every term in the bottom row. We write like terms in columns, and then add the results. Such multiplication is like multiplying with whole numbers:

$$
\begin{array}{r}
321 \\
\times \ 12 \\
\hline
642 \\
321 \\
\hline
3852
\end{array}
\qquad
\begin{array}{r}
300 + 20 + 1 \\
\times \qquad\quad 10 + 2 \\
\hline
600 + 40 + 2 \\
3000 + 200 + 10 \\
\hline
3000 + 800 + 50 + 2
\end{array}
\qquad
\begin{array}{l}
\\
\\
\text{Multiplying the top row by 2} \\
\text{Multiplying the top row by 10} \\
\text{Adding}
\end{array}
$$

EXAMPLE 6   Multiply: $(4x^3 - 2x^2 + 3x)(x^2 + 2x)$.

Solution

$$
\left.
\begin{array}{r}
4x^3 - 2x^2 + 3x \\
x^2 + 2x
\end{array}
\right\}
\qquad
\begin{array}{l}
\text{Note that each polynomial is written in} \\
\text{descending order.}
\end{array}
$$

$$
\begin{array}{r}
8x^4 - 4x^3 + 6x^2 \qquad\qquad \text{Multiplying the top row by } 2x \\
4x^5 - 2x^4 + 3x^3 \qquad\qquad\qquad\qquad \text{Multiplying the top row by } x^2 \\
\hline
4x^5 + 6x^4 - \ \ x^3 + 6x^2 \qquad\qquad \text{Combining like terms}
\end{array}
$$

Line up like terms in columns.

If a term is missing, it helps to leave space for it so that like terms can be easily aligned.

EXAMPLE 7   Multiply: $(-2x^2 - 3)(5x^3 - 3x + 4)$.

Solution

$$
\begin{array}{r}
\left.\begin{array}{r}
5x^3 \qquad - 3x + 4 \\
- 2x^2 \qquad - 3
\end{array}\right\} \text{Leaving space for missing terms} \\[4pt]
\hline
\begin{array}{r}
- 15x^3 \qquad + 9x - 12 \\
-10x^5 + 6x^3 - 8x^2
\end{array} \\[2pt]
\hline
-10x^5 - 9x^3 - 8x^2 + 9x - 12
\end{array}
$$

Multiplying by -3
Multiplying by $-2x^2$
Combining like terms

With practice, some steps can be skipped. Sometimes we multiply horizontally, while still aligning like terms.

EXAMPLE 8   Multiply: $(2x^3 + 3x^2 - 4x + 6)(3x + 5)$.

Solution

$$
\begin{aligned}
(2x^3 + 3x^2 - 4x + 6)(3x + 5) &= \overbrace{6x^4 + 9x^3 - 12x^2 + 18x}^{\text{Multiplying by } 3x} \\
&\quad + \underbrace{10x^3 + 15x^2 - 20x + 30}_{\text{Multiplying by } 5} \\
&= 6x^4 + 19x^3 + 3x^2 - 2x + 30
\end{aligned}
$$

Checking by Evaluating

How can we be certain that our multiplication (or addition or subtraction) of polynomials is correct? One check is to simply review our calculations. A different type of check, used in Example 10 of Section 4.3, makes use of the fact that equivalent expressions have the same value when evaluated for the same replacement. Thus a quick, partial, check of Example 8 can be made by selecting a convenient replacement for x (say, 1) and comparing the values of the expressions $(2x^3 + 3x^2 - 4x + 6)(3x + 5)$ and $6x^4 + 19x^3 + 3x^2 - 2x + 30$:

$$
\begin{aligned}
(2x^3 + 3x^2 - 4x + 6)(3x + 5) &= (2 \cdot 1^3 + 3 \cdot 1^2 - 4 \cdot 1 + 6)(3 \cdot 1 + 5) \\
&= (2 + 3 - 4 + 6)(3 + 5) \\
&= 7 \cdot 8 = 56;
\end{aligned}
$$

$$
\begin{aligned}
6x^4 + 19x^3 + 3x^2 - 2x + 30 &= 6 \cdot 1^4 + 19 \cdot 1^3 + 3 \cdot 1^2 - 2 \cdot 1 + 30 \\
&= 6 + 19 + 3 - 2 + 30 \\
&= 28 - 2 + 30 = 56.
\end{aligned}
$$

Since the value of both expressions is 56, the multiplication in Example 8 is very likely correct.

It is possible, by chance, for two expressions that are not equivalent to share the same value when evaluated. For this reason, checking by evaluating is only a partial check. Consult your instructor for the checking approach that he or she prefers.

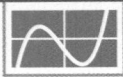 **technology connection**

Tables can also be used to check polynomial multiplication. To illustrate, we can check Example 8 by entering $y_1 = (2x^3 + 3x^2 - 4x + 6)(3x + 5)$ and $y_2 = 6x^4 + 19x^3 + 3x^2 - 2x + 30$.

When ⟨TABLE⟩ is then pressed, we are shown two columns of values—one for y_1 and one for y_2. If our multiplication was correct, the columns of values will match.

X	Y₁	Y₂
−3	36	36
−2	−10	−10
−1	22	22
0	30	30
1	56	56
2	286	286
3	1050	1050

X = −3

1. Form a table and scroll up and down to check Example 7.
2. Check Example 8 using the method discussed in Section 4.3: Let

$$y_1 = (2x^3 + 3x^2 - 4x + 6)(3x + 5),$$
$$y_2 = 6x^4 + 19x^3 + 3x^2 - 2x + 30,$$

and

$$y_3 = y_2 - y_1.$$

Then check that y_3 is always 0.

Exercise Set

4.4

FOR EXTRA HELP

 Student's Solutions Manual

 Digital Video Tutor CD 2 Videotape 4

 AW Math Tutor Center

 MathXL Tutorials on CD

 MathXL

 MyMathLab

↪ *Concept Reinforcement   In each of Exercises 1–4, match the product with the correct result from the column on the right.*

1. _____ $4x^4 \cdot 2x^2$
2. _____ $3x^2 \cdot 2x^4$
3. _____ $4x^3 \cdot 2x^5$
4. _____ $3x^5 \cdot 2x^3$

a) $6x^8$
b) $8x^6$
c) $6x^6$
d) $8x^8$

Multiply.

5. $(4x^3)9$
6. $(5x^4)6$
7. $(-x^3)(x^4)$
8. $(-x^2)(-x)$
9. $(-x^6)(-x^2)$
10. $(-x^5)(x^3)$
11. $(8a^2)(3a^2)$
12. $(7t^5)(4t^3)$
13. $(0.3x^3)(-0.4x^6)$
14. $(-0.1x^6)(0.2x^4)$
15. $\left(-\frac{1}{4}x^4\right)\left(\frac{1}{5}x^8\right)$
16. $\left(-\frac{1}{5}x^3\right)\left(-\frac{1}{3}x\right)$
17. $(-5n^3)(-1)$
18. $19t^2 \cdot 0$

19. $(-4y^5)(6y^2)(-3y^3)$

20. $7x^2(-2x^3)(2x^6)$

21. $2x(4x - 6)$

22. $3x(-x + 5)$

23. $3x(x + 2)$

24. $4x(x + 1)$

25. $(a + 9)3a$

26. $(a - 7)4a$

27. $x^2(x^3 + 1)$

28. $-2x^3(x^2 - 1)$

29. $3x(2x^2 - 6x + 1)$

30. $-4x(2x^3 - 6x^2 - 5x + 1)$

31. $5t^2(3t + 6)$

32. $7t^2(2t + 1)$

33. $-6x^2(x^2 + x)$

34. $-4x^2(x^2 - x)$

35. $\frac{2}{3}a^4\left(6a^5 - 12a^3 - \frac{5}{8}\right)$

36. $\frac{3}{4}t^5\left(8t^6 - 12t^4 + \frac{12}{7}\right)$

37. $(x + 2)(x + 6)$

38. $(x + 5)(x + 2)$

39. $(x + 5)(x - 2)$

40. $(x + 6)(x - 2)$

41. $(a - 6)(a - 7)$

42. $(a - 4)(a - 8)$

43. $(x + 3)(x - 3)$

44. $(x + 6)(x - 6)$

45. $(5 - x)(5 - 2x)$

46. $(3 + x)(6 + 2x)$

47. $\left(t + \frac{3}{2}\right)\left(t + \frac{4}{3}\right)$

48. $\left(a - \frac{2}{5}\right)\left(a + \frac{5}{2}\right)$

49. $\left(\frac{1}{4}a + 2\right)\left(\frac{3}{4}a - 1\right)$

50. $\left(\frac{2}{5}t - 1\right)\left(\frac{3}{5}t + 1\right)$

Draw and label rectangles similar to those following Examples 2 and 4 to illustrate each product.

51. $x(x + 5)$

52. $x(x + 2)$

53. $(x + 1)(x + 2)$

54. $(x + 3)(x + 1)$

55. $(x + 5)(x + 3)$

56. $(x + 4)(x + 6)$

Multiply and check.

57. $(x^2 - x + 5)(x + 1)$

58. $(x^2 + x - 7)(x + 2)$

59. $(2a + 5)(a^2 - 3a + 2)$

60. $(3t + 4)(t^2 - 5t + 1)$

61. $(y^2 - 7)(2y^3 + y + 1)$

62. $(a^2 + 4)(5a^3 - 3a - 1)$

Aha! **63.** $(3x + 2)(5x + 4x + 7)$

64. $(4x - 5x - 3)(1 + 2x^2)$

65. $(x^2 - 3x + 2)(x^2 + x + 1)$

66. $(x^2 + 5x - 1)(x^2 - x + 3)$

67. $(2t^2 - 5t - 4)\left(3t^2 - t + \frac{1}{2}\right)$

68. $\left(5t^2 - t + \frac{1}{2}\right)(2t^2 + t - 4)$

69. $(x + 1)(x^3 + 7x^2 + 5x + 4)$

70. $(x + 2)(x^3 + 5x^2 + 9x + 3)$

71. Is it possible to understand polynomial multiplication without understanding the distributive law? Why or why not?

72. The polynomials

$$(a + b + c + d) \quad \text{and} \quad (r + s + m + p)$$

are multiplied. Without performing the multiplication, determine how many terms the product will contain. Provide a justification for your answer.

SKILL MAINTENANCE

Simplify. [1.8]

73. $5 - 3 \cdot 2 + 7$

74. $4 + 6 \cdot 5 - 3$

75. $(8 - 2)(8 + 2) + 2^2 - 8^2$

76. $(7 - 3)(7 + 3) + 3^2 - 7^2$

SYNTHESIS

77. Under what conditions will the product of two binomials be a trinomial?

78. How can the following figure be used to show that $(x + 3)^2 \neq x^2 + 9$?

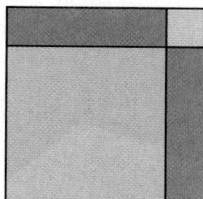

Find a polynomial for the shaded area of each figure.

79.

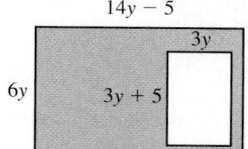

80.

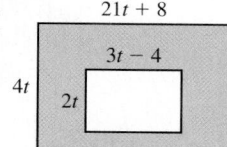

For each figure, determine what the missing number must be in order for the figure to have the given area.

81. Area is $x^2 + 7x + 10$

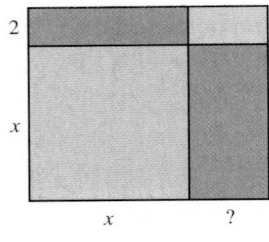

82. Area is $x^2 + 8x + 15$

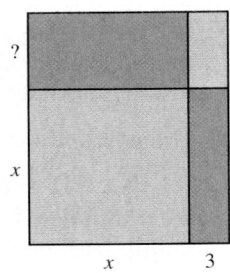

83. A box with a square bottom and no top is to be made from a 12-in.-square piece of cardboard. Squares with side x are cut out of the corners and the sides are folded up. Find the polynomials for the volume and the outside surface area of the box.

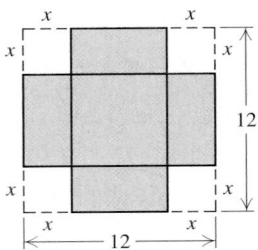

84. An open wooden box is a cube with side x cm. The box, including its bottom, is made of wood that is 1 cm thick. Find a polynomial for the interior volume of the cube.

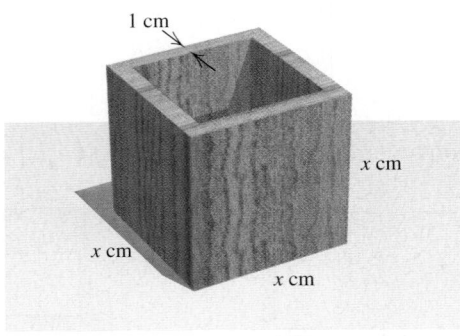

85. Find a polynomial for the volume of the solid shown below.

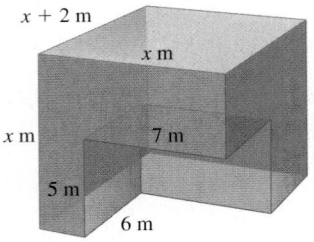

86. A side of a cube is $(x + 2)$ cm long. Find a polynomial for the volume of the cube.

87. A rectangular garden is twice as long as it is wide and is surrounded by a sidewalk that is 4 ft wide (see the figure below). The area of the sidewalk is 256 ft^2. Find the dimensions of the garden.

Compute and simplify.

88. $(x + 3)(x + 6) + (x + 3)(x + 6)$

Aha! **89.** $(x - 2)(x - 7) - (x - 7)(x - 2)$

90. $(x + 5)^2 - (x - 3)^2$

Aha! **91.** Extend the pattern and simplify
$(x - a)(x - b)(x - c)(x - d) \cdots (x - z).$

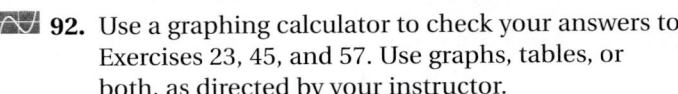 **92.** Use a graphing calculator to check your answers to Exercises 23, 45, and 57. Use graphs, tables, or both, as directed by your instructor.

CORNER

Slick Tricks with Algebra

COLLABORATIVE

Focus: Polynomial multiplication
Time: 15 minutes
Group size: 2

Consider the following dialogue.

Jinny: Cal, let me do a number trick with you. Think of a number between 1 and 7. I'll have you perform some manipulations to this number, you'll tell me the result, and I'll tell you your number.

Cal: OK. I've thought of a number.

Jinny: Good. Write it down so I can't see it. Now double it, and then subtract x from the result.

Cal: Hey, this is algebra!

Jinny: I know. Now square your binomial. After you're through squaring, subtract x^2.

Cal: How did you know I had an x^2? I *thought* this was rigged!

Jinny: It is. Now divide each of the remaining terms by 4 and tell me either your constant term or your x-term. I'll tell you the other term and the number you chose.

Cal: OK. The constant term is 16.

Jinny: Then the other term is $-4x$ and the number you chose was 4.

Cal: You're right! How did you do it?

ACTIVITY

1. Each group member should follow Jinny's instructions. Then determine how Jinny determined Cal's number and the other term.
2. Suppose that, at the end, Cal told Jinny the x-term. How would Jinny have determined Cal's number and the other term?
3. Would Jinny's "trick" work with *any* real number? Why do you think she specified numbers between 1 and 7?

4.5   Special Products

Products of Two Binomials • Multiplying Sums and Differences of Two Terms • Squaring Binomials • Multiplications of Various Types

Certain products of two binomials occur so often that it is helpful to be able to compute them quickly. In this section, we develop methods for doing so.

Products of Two Binomials

In Section 4.4, we found the product $(x + 5)(x + 4)$ by using the distributive law a total of three times (see p. 259). Note that each term in $x + 5$ is multiplied by each term in $x + 4$. To shorten our work, we can go right to this step:

$$(x + 5)(x + 4) = x \cdot x + x \cdot 4 + 5 \cdot x + 5 \cdot 4$$
$$= x^2 + 4x + 5x + 20$$
$$= x^2 + 9x + 20.$$

Note that $x \cdot x$ is found by multiplying the *First* terms of each binomial, $x \cdot 4$ is found by multiplying the *Outer* terms of the two binomials, $5 \cdot x$ is the product of the *Inner* terms of the two binomials, and $5 \cdot 4$ is the product of the *Last* terms of each binomial:

$$\underbrace{\text{First terms}} \quad \underbrace{\text{Outer terms}} \quad \underbrace{\text{Inner terms}} \quad \underbrace{\text{Last terms}}$$

$$(x + 5)(x + 4) = x \cdot x + 4 \cdot x + 5 \cdot x + 5 \cdot 4.$$

To remember this shortcut for multiplying, we use the initials **FOIL**.

The FOIL Method

To multiply two binomials, $A + B$ and $C + D$, multiply the First terms AC, the Outer terms AD, the Inner terms BC, and then the Last terms BD. Then combine like terms, if possible.

$$(A + B)(C + D) = AC + AD + BC + BD$$

1. Multiply First terms: AC.
2. Multiply Outer terms: AD.
3. Multiply Inner terms: BC.
4. Multiply Last terms: BD.

$$\downarrow$$
$$\text{FOIL}$$

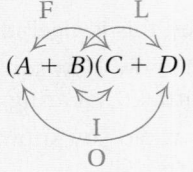

Because addition is commutative, the individual multiplications can be performed in any order. Both FLOI and FIOL yield the same result as FOIL, but FOIL is most easily remembered and most widely used.

EXAMPLE 1   Multiply: $(x + 8)(x^2 + 5)$.

Solution

$$(x + 8)(x^2 + 5) = \overset{\text{F}}{x^3} + \overset{\text{O}}{5x} + \overset{\text{I}}{8x^2} + \overset{\text{L}}{40} \qquad \text{There are no like terms.}$$
$$= x^3 + 8x^2 + 5x + 40 \qquad \text{Writing in descending order}$$

After multiplying, remember to combine any like terms.

EXAMPLE 2   Multiply.

a) $(x + 7)(x + 4)$

b) $(y + 3)(y - 2)$

c) $(4t^3 + 5t)(3t^2 - 2)$

d) $(3 - 4x)(7 - 5x^3)$

Solution

a) $(x + 7)(x + 4) = x^2 + 4x + 7x + 28$ Using FOIL
$= x^2 + 11x + 28$ Combining like terms

b) $(y + 3)(y - 2) = y^2 - 2y + 3y - 6$
$= y^2 + y - 6$

c) $(4t^3 + 5t)(3t^2 - 2) = 12t^5 - 8t^3 + 15t^3 - 10t$ Remember to add exponents when multiplying terms with the same base.

$= 12t^5 + 7t^3 - 10t$

d) $(3 - 4x)(7 - 5x^3) = 21 - 15x^3 - 28x + 20x^4$
$= 21 - 28x - 15x^3 + 20x^4$ In general, if the original binomials are written in *ascending* order, the answer is also written that way.

Multiplying Sums and Differences of Two Terms

Consider the product of the sum and difference of the same two terms, such as
$(x + 5)(x - 5)$.

Since this is the product of two binomials, we can use FOIL. In doing so, we find that the "outer" and "inner" products are opposites:

a) $(x + 5)(x - 5) = x^2 - 5x + 5x - 25$
$= x^2 - 25;$

b) $(3a - 2)(3a + 2) = 9a^2 + 6a - 6a - 4$ The "outer" and "inner" terms "drop out." Their sum is zero.
$= 9a^2 - 4;$

c) $\left(x^3 + \frac{2}{7}\right)\left(x^3 - \frac{2}{7}\right) = x^6 - \frac{2}{7}x^3 + \frac{2}{7}x^3 - \frac{4}{49}$
$= x^6 - \frac{4}{49}.$

Because opposites always add to zero, for products like $(x + 5)(x - 5)$ we can use a shortcut that is faster than FOIL.

The Product of a Sum and Difference

The product of the sum and difference of the same two terms is the square of the first term minus the square of the second term:

$$(A + B)(A - B) = \underline{A^2 - B^2}.$$

This is called a *difference of squares*.

EXAMPLE 3   Multiply.

a) $(x + 4)(x - 4)$

b) $(5 + 2w)(5 - 2w)$

c) $(3a^4 - 5)(3a^4 + 5)$

Solution

$$(A + B)(A - B) = A^2 - B^2$$
$$\downarrow \quad \downarrow \quad \downarrow \qquad \downarrow \quad \downarrow \qquad \downarrow \quad \downarrow$$

a) $(x + 4)(x - 4) = x^2 - 4^2$ Saying the words can help: "The square of the first term, x^2, minus the square of the second, 4^2"

$\qquad\qquad\qquad = x^2 - 16$ Simplifying

b) $(5 + 2w)(5 - 2w) = 5^2 - (2w)^2$

$\qquad\qquad\qquad\qquad = 25 - 4w^2$ Squaring both 5 and $2w$

c) $(3a^4 - 5)(3a^4 + 5) = (3a^4)^2 - 5^2$

$\qquad\qquad\qquad\qquad\quad = 9a^8 - 25$ Using the rules for exponents. Remember to multiply exponents when raising a power to a power.

Squaring Binomials

Consider the square of a binomial, such as $(x + 3)^2$. This can be expressed as $(x + 3)(x + 3)$. Since this is the product of two binomials, we can use FOIL. But again, this product occurs so often that a faster method has been developed. Look for a pattern in the following:

a) $(x + 3)^2 = (x + 3)(x + 3)$

$\qquad\qquad = x^2 + 3x + 3x + 9$

$\qquad\qquad = x^2 + 6x + 9;$

b) $(5 - 3p)^2 = (5 - 3p)(5 - 3p)$

$\qquad\qquad\quad = 25 - 15p - 15p + 9p^2$

$\qquad\qquad\quad = 25 - 30p + 9p^2;$

c) $(a^3 - 7)^2 = (a^3 - 7)(a^3 - 7)$

$\qquad\qquad\quad = a^6 - 7a^3 - 7a^3 + 49$

$\qquad\qquad\quad = a^6 - 14a^3 + 49.$

Perhaps you noticed that in each product the "outer" and "inner" products are identical. The other two terms, the "first" and "last" products, are squares.

> **The Square of a Binomial**
>
> The square of a binomial is the square of the first term, plus twice the product of the two terms, plus the square of the last term:
>
> $$(A + B)^2 = A^2 + 2AB + B^2;$$
> $$(A - B)^2 = A^2 - 2AB + B^2.$$
>
> These are called *perfect-square trinomials.**

EXAMPLE 4   Multiply: **(a)** $(x + 7)^2$;   **(b)** $(t - 5)^2$;   **(c)** $(3a + 0.4)^2$;   **(d)** $(5x - 3x^4)^2$.

Solution

$$(A + B)^2 = A^2 + 2 \cdot A \cdot B + B^2$$

a) $(x + 7)^2 = x^2 + 2 \cdot x \cdot 7 + 7^2$

Saying the words can help: "The square of the first term, x^2, plus twice the product of the terms, $2 \cdot 7x$, plus the square of the second term, 7^2"

$$= x^2 + 14x + 49$$

b) $(t - 5)^2 = t^2 - 2 \cdot t \cdot 5 + 5^2$

$$= t^2 - 10t + 25$$

c) $(3a + 0.4)^2 = (3a)^2 + 2 \cdot 3a \cdot 0.4 + 0.4^2$

$$= 9a^2 + 2.4a + 0.16$$

d) $(5x - 3x^4)^2 = (5x)^2 - 2 \cdot 5x \cdot 3x^4 + (3x^4)^2$

$$= 25x^2 - 30x^5 + 9x^8$$ Using the rules for exponents

> *Caution!*   Although the square of a product is the product of the squares, the square of a sum is *not* the sum of the squares. That is, $(AB)^2 = A^2B^2$, but
>
> The term $2AB$ is missing.
>
> $$(A + B)^2 \neq A^2 + B^2.$$
>
> To confirm this inequality, note that
>
> $$(7 + 5)^2 = 12^2 = 144,$$
>
> whereas
>
> $$7^2 + 5^2 = 49 + 25 = 74, \quad \text{and} \quad 74 \neq 144.$$

---

*In some books, these are called *trinomial squares*.

Geometrically, $(A + B)^2$ can be viewed as the area of a square with sides of length $A + B$:

$$(A + B)(A + B) = (A + B)^2.$$

This is equal to the sum of the areas of the four smaller regions:

$$A^2 + AB + AB + B^2 = A^2 + 2AB + B^2.$$

Thus,

$$(A + B)^2 = A^2 + 2AB + B^2.$$

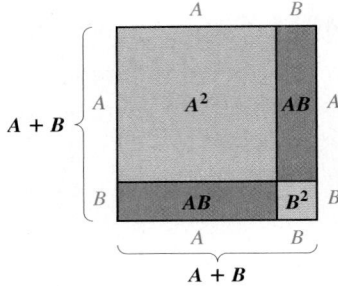

Multiplications of Various Types

Recognizing patterns often helps when new problems are encountered. To simplify a new multiplication problem, always examine what type of product it is so that the best method for finding that product can be used. To do this, ask yourself questions similar to the following.

Multiplying Two Polynomials

1. Is the multiplication the product of a monomial and a polynomial? If so, multiply each term of the polynomial by the monomial.
2. Is the multiplication the product of two binomials? If so:

   a) Is it the product of the sum and difference of the *same* two terms? If so, use the pattern

   $$(A + B)(A - B) = A^2 - B^2.$$

   b) Is the product the square of a binomial? If so, use the pattern

   $$(A + B)(A + B) = (A + B)^2 = A^2 + 2AB + B^2,$$

   or

   $$(A - B)(A - B) = (A - B)^2 = A^2 - 2AB + B^2.$$

   c) If neither (a) nor (b) applies, use FOIL.

3. Is the multiplication the product of two polynomials other than those above? If so, multiply each term of one by every term of the other. Use columns if you wish.

EXAMPLE 5   Multiply.

a) $(x + 3)(x - 3)$

b) $(t + 7)(t - 5)$

c) $(x + 7)(x + 7)$

d) $2x^3(9x^2 + x - 7)$

e) $(p + 3)(p^2 + 2p - 1)$

f) $\left(3x + \frac{1}{4}\right)^2$

Solution

a) $(x + 3)(x - 3) = x^2 - 9$ This is the product of the sum and difference of the same two terms.

b) $(t + 7)(t - 5) = t^2 - 5t + 7t - 35$ Using FOIL
$$= t^2 + 2t - 35$$

c) $(x + 7)(x + 7) = x^2 + 14x + 49$ This is the square of a binomial, $(x + 7)^2$.

d) $2x^3(9x^2 + x - 7) = 18x^5 + 2x^4 - 14x^3$ Multiplying each term of the trinomial by the monomial

e)
$$
\begin{array}{r}
p^2 + 2p - 1 \\
p + 3 \\
\hline
3p^2 + 6p - 3 \\
p^3 + 2p^2 - p \\
\hline
p^3 + 5p^2 + 5p - 3
\end{array}
$$
Using columns to multiply a binomial and a trinomial

Multiplying by 3

Multiplying by p

f) $\left(3x + \frac{1}{4}\right)^2 = 9x^2 + 2(3x)\left(\frac{1}{4}\right) + \frac{1}{16}$ Squaring a binomial
$$= 9x^2 + \frac{3}{2}x + \frac{1}{16}$$

Exercise Set

4.5

FOR EXTRA HELP

 Student's Solutions Manual

 Digital Video Tutor CD 2 Videotape 4

 AW Math Tutor Center

 MathXL Tutorials on CD

Math XL MathXL

MyMathLab MyMathLab

❧ *Concept Reinforcement   Identify each statement as either true or false.*

1. FOIL is simply a memory device for finding the product of two binomials.

2. Once FOIL is used, it is always possible to combine like terms.

3. The square of a binomial cannot be found using FOIL.

4. The square of $A + B$ is not the sum of the squares of A and B.

Multiply.

5. $(x + 3)(x^2 + 5)$

6. $(x^2 - 3)(x - 1)$

7. $(x^3 + 6)(x + 2)$

8. $(x^4 + 2)(x + 12)$

9. $(y + 2)(y - 3)$

10. $(a + 2)(a + 2)$

11. $(3x + 2)(3x + 5)$

12. $(4x + 1)(2x + 7)$

13. $(5x - 4)(x + 2)$

14. $(t - 9)(t + 9)$

15. $(1 + 3t)(2 - 3t)$

16. $(7 - a)(2 + 3a)$

17. $(2x - 5)(x - 4)$

18. $(2x - 1)(3x + 1)$

19. $\left(p - \frac{1}{4}\right)\left(p + \frac{1}{4}\right)$

20. $\left(q + \frac{3}{4}\right)\left(q + \frac{3}{4}\right)$

21. $(x - 0.1)(x + 0.1)$

22. $(x + 0.3)(x - 0.4)$

23. $(-2x + 1)(x + 6)$

24. $(-x + 4)(2x - 5)$

25. $(a + 9)(a + 9)$

26. $(2y + 7)(2y + 7)$

27. $(1 + 3t)(1 - 5t)$

28. $(1 + 2t)(1 - 3t^2)$

29. $(x^2 + 3)(x^3 - 1)$

30. $(x^4 - 3)(2x + 1)$

31. $(3x^2 - 2)(x^4 - 2)$

32. $(x^{10} + 3)(x^{10} - 3)$

33. $(2t^3 + 5)(2t^3 + 3)$

34. $(5t^2 + 1)(2t^2 + 3)$

35. $(8x^3 + 5)(x^2 + 2)$

36. $(4 - 2x)(5 - 2x^2)$

37. $(4x^2 + 3)(x - 3)$

38. $(7x - 2)(2x - 7)$

Multiply. Try to recognize what type of product each multiplication is before multiplying.

39. $(x + 7)(x - 7)$

40. $(x + 1)(x - 1)$

41. $(2x + 1)(2x - 1)$

42. $(x^2 + 1)(x^2 - 1)$

43. $(5m - 2)(5m + 2)$

44. $(3x^4 + 2)(3x^4 - 2)$

45. $(3x^4 - 1)(3x^4 + 1)$

46. $(t^2 - 0.2)(t^2 + 0.2)$

47. $(x^4 + 7)(x^4 - 7)$

48. $(t^3 + 4)(t^3 - 4)$

49. $\left(t - \frac{3}{4}\right)\left(t + \frac{3}{4}\right)$

50. $\left(m - \frac{2}{3}\right)\left(m + \frac{2}{3}\right)$

51. $(x + 2)^2$

52. $(2x - 1)^2$

53. $(7x^3 + 1)^2$

54. $(4x^3 + 1)^2$

55. $\left(a - \frac{2}{5}\right)^2$

56. $\left(t - \frac{1}{5}\right)^2$

57. $(t^3 + 5)^2$

58. $(a^4 + 3)^2$

59. $(2 - 3x^4)^2$

60. $(5 - 2t^3)^2$

61. $(5 + 6t^2)^2$

62. $(3p^2 - p)^2$

63. $(7x - 0.3)^2$

64. $(4a - 0.6)^2$

65. $5a^3(2a^2 - 1)$

66. $9x^3(2x^2 - 5)$

67. $(a - 3)(a^2 + 2a - 4)$

68. $(x^2 - 5)(x^2 + x - 1)$

69. $(3 - 2x^3)^2$

70. $(x - 4x^3)^2$

71. $5x(x^2 + 6x - 2)$

72. $6x(-x^5 + 6x^2 + 9)$

73. $(-t^3 + 1)^2$

74. $(-x^2 + 1)^2$

75. $3t^2(5t^3 - t^2 + t)$

76. $-5x^3(x^2 + 8x - 9)$

77. $(6x^4 - 3)^2$

78. $(8a^3 + 5)^2$

79. $(3x + 2)(4x^2 + 5)$

80. $(2x^2 - 7)(3x^2 + 9)$

81. $(5 - 6x^4)^2$

82. $(3 - 4t^5)^2$

83. $(a + 1)(a^2 - a + 1)$

84. $(x - 5)(x^2 + 5x + 25)$

Find the total area of all shaded rectangles.

85.
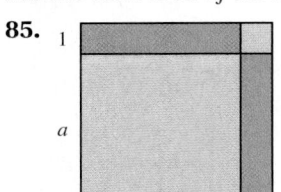

86.
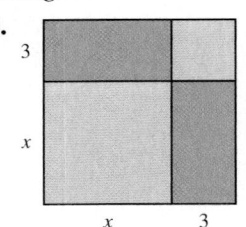

87.
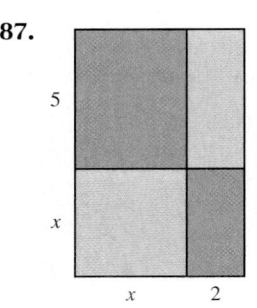

88.
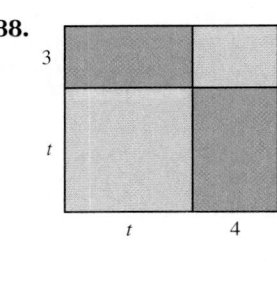

89.
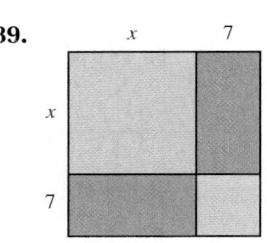

90.
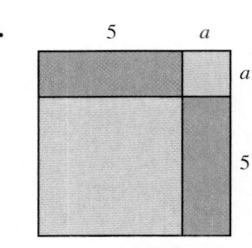

91.
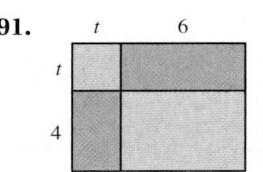

92.
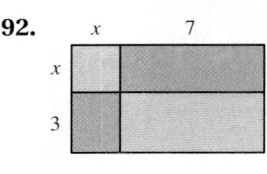

93.
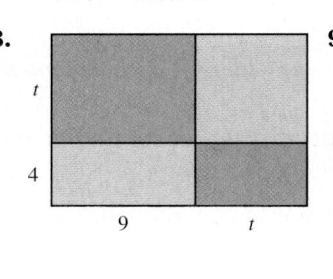

94.
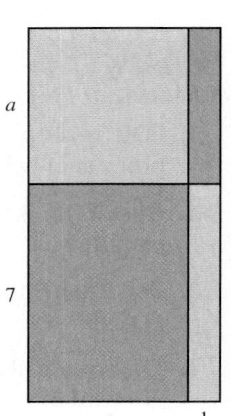

95.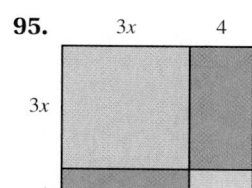

96.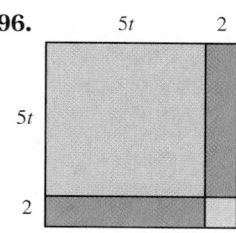

Draw and label rectangles similar to those in Exercises 85–96 to illustrate each of the following.

97. $(x + 5)^2$

98. $(x + 8)^2$

99. $(t + 9)^2$

100. $(a + 12)^2$

101. $(3 + x)^2$

102. $(7 + t)^2$

103. Patti feels that since she can find the product of any two binomials using FOIL, she needn't study the other special products. What advice would you give her?

104. Under what conditions is the product of two binomials a binomial?

SKILL MAINTENANCE

105. *Energy use.*   In an apartment, lamps, an air conditioner, and a television set are all operating at the same time. The lamps take 10 times as many watts as the television set, and the air conditioner takes 40 times as many watts as the television set. The total wattage used in the apartment is 2550 watts. How many watts are used by each appliance? [2.5]

106. In what quadrant is the point $(-3, 4)$ located? [3.1]

Solve. [2.3]

107. $5xy = 8$, for y

108. $3ab = c$, for a

109. $ax - b = c$, for x

110. $st + r = u$, for t

SYNTHESIS

111. Beth claims that by writing $19 \cdot 21$ as $(20 - 1)(20 + 1)$, she can find the product mentally. How is this possible?

112. The product $(A + B)^2$ can be regarded as the sum of the areas of four regions (as shown following Example 4). How might one visually represent $(A + B)^3$? Why?

Multiply.

Aha!   **113.** $(4x^2 + 9)(2x + 3)(2x - 3)$

114. $(9a^2 + 1)(3a - 1)(3a + 1)$

Aha!   **115.** $(3t - 2)^2(3t + 2)^2$

116. $(5a + 1)^2(5a - 1)^2$

117. $(t^3 - 1)^4(t^3 + 1)^4$

118. $(32.41x + 5.37)^2$

Calculate as the difference of squares.

119. 18×22 [*Hint:* $(20 - 2)(20 + 2)$.]

120. 93×107

Solve.

121. $(x + 2)(x - 5) = (x + 1)(x - 3)$

122. $(2x + 5)(x - 4) = (x + 5)(2x - 4)$

Find a polynomial for the total shaded area in each figure.

123.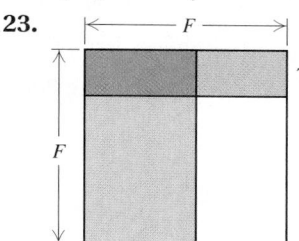

124.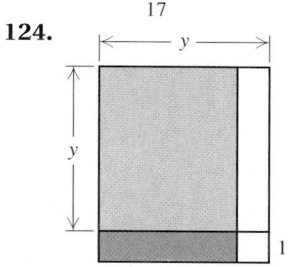

125.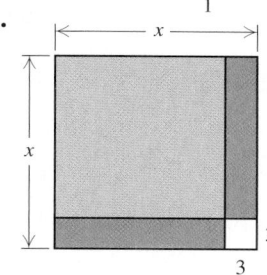

126. Find $(y - 2)^2$ by subtracting the white areas from y^2.

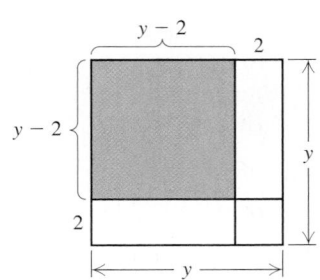

127. Find $(10 - 2x)^2$ by subtracting the white areas from 10^2.

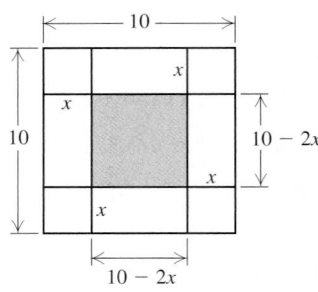

128. Find three consecutive integers for which the sum of the squares is 65 more than three times the square of the smallest integer.

 **129.** Use a graphing calculator and the method developed on p. 262 to check your answers to Exercises 21, 47, and 83.

4.6

Polynomials in Several Variables

Evaluating Polynomials • Like Terms and Degree •
Addition and Subtraction • Multiplication

Thus far, the polynomials that we have studied have had only one variable. Polynomials such as

$$5x + x^2y - 3y + 7, \qquad 9ab^2c - 2a^3b^2 + 8a^2b^3, \quad \text{and} \quad 4m^2 - 9n^2$$

contain two or more variables. In this section, we will add, subtract, multiply, and evaluate such **polynomials in several variables**.

Evaluating Polynomials

To evaluate a polynomial in two or more variables, we substitute numbers for the variables. Then we compute, using the rules for order of operations.

EXAMPLE 1   Evaluate the polynomial $4 + 3x + xy^2 + 8x^3y^3$ for $x = -2$ and $y = 5$.

Solution   We substitute -2 for x and 5 for y:

$$4 + 3x + xy^2 + 8x^3y^3 = 4 + 3(-2) + (-2) \cdot 5^2 + 8(-2)^3 \cdot 5^3$$
$$= 4 - 6 - 50 - 8000 = -8052.$$

EXAMPLE 2   Surface area of a right circular cylinder.   The surface area of a right circular cylinder is given by the polynomial

$$2\pi r h + 2\pi r^2,$$

where h is the height and r is the radius of the base. A 12-oz can has a height of 4.7 in. and a radius of 1.2 in. Approximate its surface area.

Solution   We evaluate the polynomial for $h = 4.7$ in. and $r = 1.2$ in. If 3.14 is used to approximate π, we have

$$\begin{aligned} 2\pi r h + 2\pi r^2 &\approx 2(3.14)(1.2 \text{ in.})(4.7 \text{ in.}) + 2(3.14)(1.2 \text{ in.})^2 \\ &\approx 2(3.14)(1.2 \text{ in.})(4.7 \text{ in.}) + 2(3.14)(1.44 \text{ in}^2) \\ &\approx 35.4192 \text{ in}^2 + 9.0432 \text{ in}^2 \approx 44.4624 \text{ in}^2. \end{aligned}$$

If the π key of a calculator is used, we have

$$\begin{aligned} 2\pi r h + 2\pi r^2 &\approx 2(3.141592654)(1.2 \text{ in.})(4.7 \text{ in.}) + 2(3.141592654)(1.2 \text{ in.})^2 \\ &\approx 44.48495197 \text{ in}^2. \end{aligned}$$

Note that the unit in the answer (square inches) is a unit of area. The surface area is about 44.5 in^2 (square inches).

Like Terms and Degree

Recall that the degree of a monomial is the number of variable factors in the term. For example, the degree of $5x^2$ is 2 because there are two variable factors in $5 \cdot x \cdot x$. Similarly, the degree of $5a^2b^4$ is 6 because there are 6 variable factors in $5 \cdot a \cdot a \cdot b \cdot b \cdot b \cdot b$. Note that 6 can be found by adding the exponents 2 and 4.

As we learned in Section 4.2, the degree of a polynomial is the degree of the term of highest degree.

EXAMPLE 3   Identify the coefficient and the degree of each term and the degree of the polynomial

$$9x^2y^3 - 14xy^2z^3 + xy + 4y + 5x^2 + 7.$$

Solution

Term	Coefficient	Degree	Degree of the Polynomial
$9x^2y^3$	9	5	
$-14xy^2z^3$	-14	6	6
xy	1	2	
$4y$	4	1	
$5x^2$	5	2	
7	7	0	

Note in Example 3 that although both xy and $5x^2$ have degree 2, they are *not* like terms. *Like*, or *similar, terms* either have exactly the same variables with exactly the same exponents or are constants. For example,

$$8a^4b^7 \text{ and } 5b^7a^4 \text{ are like terms}$$

and

$$-17 \text{ and } 3 \text{ are like terms,}$$

but

$$-2x^2y \text{ and } 9xy^2 \text{ are } not \text{ like terms.}$$

As always, combining like terms is based on the distributive law.

EXAMPLE 4   Combine like terms.

a) $9x^2y + 3xy^2 - 5x^2y - xy^2$
b) $7ab - 5ab^2 + 3ab^2 + 6a^3 + 9ab - 11a^3 + b - 1$

Solution

a) $9x^2y + 3xy^2 - 5x^2y - xy^2 = (9 - 5)x^2y + (3 - 1)xy^2$
$$= 4x^2y + 2xy^2 \qquad \text{Try to go directly to this step.}$$

b) $7ab - 5ab^2 + 3ab^2 + 6a^3 + 9ab - 11a^3 + b - 1$
$$= -5a^3 - 2ab^2 + 16ab + b - 1 \qquad \text{We choose to write descending powers of } a. \text{ Other, equivalent, forms can also be used.}$$

Addition and Subtraction

The procedure used for adding polynomials in one variable is used to add polynomials in several variables.

EXAMPLE 5   Add.

a) $(-5x^3 + 3y - 5y^2) + (8x^3 + 4x^2 + 7y^2)$
b) $(5ab^2 - 4a^2b + 5a^3 + 2) + (3ab^2 - 2a^2b + 3a^3b - 5)$

Student Notes _____

Always read the problem carefully. The difference between
$(-5x^3 + 3y - 5y^2) +$
$\qquad\qquad (8x^3 + 4x^2 + 7y^2)$
and
$(-5x^3 + 3y - 5y^2)(8x^3 + 4x^2 + 7y^2)$

is enormous. To avoid wasting time working on an incorrectly copied exercise, be sure to double-check that you have written the correct problem in your notebook.

Solution

a) $(-5x^3 + 3y - 5y^2) + (8x^3 + 4x^2 + 7y^2)$
$$= (-5 + 8)x^3 + 4x^2 + 3y + (-5 + 7)y^2 \qquad \text{Try to do this step mentally.}$$
$$= 3x^3 + 4x^2 + 3y + 2y^2$$

b) $(5ab^2 - 4a^2b + 5a^3 + 2) + (3ab^2 - 2a^2b + 3a^3b - 5)$
$$= 8ab^2 - 6a^2b + 5a^3 + 3a^3b - 3$$

When subtracting a polynomial, remember to find the opposite of each term in that polynomial and then add.

EXAMPLE 6   Subtract: $(4x^2y + x^3y^2 + 3x^2y^3 + 6y) - (4x^2y - 6x^3y^2 + x^2y^2 - 5y)$.

Solution

$$(4x^2y + x^3y^2 + 3x^2y^3 + 6y) - (4x^2y - 6x^3y^2 + x^2y^2 - 5y)$$
$$= 4x^2y + x^3y^2 + 3x^2y^3 + 6y - 4x^2y + 6x^3y^2 - x^2y^2 + 5y$$
$$= 7x^3y^2 + 3x^2y^3 - x^2y^2 + 11y \qquad \text{Combining like terms}$$

Multiplication

To multiply polynomials in several variables, multiply each term of one polynomial by every term of the other, just as we did in Sections 4.4 and 4.5.

EXAMPLE 7   Multiply: $(3x^2y - 2xy + 3y)(xy + 2y)$.

Solution

$$
\begin{array}{r}
3x^2y - 2xy + 3y \\
xy + 2y \\
\hline
6x^2y^2 - 4xy^2 + 6y^2 \\
3x^3y^2 - 2x^2y^2 + 3xy^2 \\
\hline
3x^3y^2 + 4x^2y^2 - xy^2 + 6y^2 \\
\end{array}
$$

Multiplying by $2y$
Multiplying by xy
Adding

The special products discussed in Section 4.5 can speed up our work.

EXAMPLE 8   Multiply.

a) $(p + 5q)(2p - 3q)$ **b)** $(3x + 2y)^2$

c) $(a^3 - 7a^2b)^2$ **d)** $(3x^2y + 2y)(3x^2y - 2y)$

e) $(-2x^3y^2 + 5t)(2x^3y^2 + 5t)$ **f)** $(2x + 3 - 2y)(2x + 3 + 2y)$

Solution

$$\qquad\qquad\qquad \text{F} \qquad \text{O} \qquad \text{I} \qquad \text{L}$$
a) $(p + 5q)(2p - 3q) = 2p^2 - 3pq + 10pq - 15q^2$
$$= 2p^2 + 7pq - 15q^2 \qquad \text{Combining like terms}$$

$$(A + B)^2 = A^2 + 2 \cdot A \cdot B + B^2$$
$$\downarrow \quad\; \downarrow \qquad\quad \downarrow \qquad \downarrow \;\; \downarrow \;\; \downarrow \qquad \downarrow$$
b) $(3x + 2y)^2 = (3x)^2 + 2(3x)(2y) + (2y)^2$ Remembering the pattern for
$$= 9x^2 + 12xy + 4y^2$$
squaring a binomial speeds our work.

$$(A - B)^2 = A^2 - 2 \cdot A \cdot B + B^2$$
$$\downarrow \quad\; \downarrow \qquad\quad \downarrow \qquad \downarrow \;\; \downarrow \;\; \downarrow \qquad \downarrow$$
c) $(a^3 - 7a^2b)^2 = (a^3)^2 - 2(a^3)(7a^2b) + (7a^2b)^2$ Squaring a binomial
$$= a^6 - 14a^5b + 49a^4b^2 \qquad \text{Using the rules for exponents}$$

 **technology connection**

One way to evaluate the polynomial in Example 1 for $x = -2$ and $y = 5$ is to store -2 to X and 5 to Y and enter the polynomial.

```
-2 → X
                    -2
5 → Y
                     5
4+3X+XY²+8X^3Y^3
                  -8052
■
```

Evaluate.

1. $3x^2 - 2y^2 + 4xy + x$, for $x = -6$ and $y = 2.3$
2. $a^2b^2 - 8c^2 + 4abc + 9a$, for $a = 11$, $b = 15$, and $c = -7$

$$(\quad A \quad + \quad B \quad)(\quad A \quad - \quad B \quad) = \quad A^2 \quad - \quad B^2$$

d) $(3x^2y + 2y)(3x^2y - 2y) = (3x^2y)^2 - (2y)^2$ Remembering the pattern for multiplying the sum and difference of two terms speeds our work.

$$= 9x^4y^2 - 4y^2 \quad \text{Using the rules for exponents}$$

e) $(-2x^3y^2 + 5t)(2x^3y^2 + 5t) = (5t - 2x^3y^2)(5t + 2x^3y^2)$ Using the commutative law for addition twice

$$= (5t)^2 - (2x^3y^2)^2 \quad \text{Multiplying the sum and the difference of the same two terms}$$

$$= 25t^2 - 4x^6y^4$$

$$(\quad A \quad - \quad B \quad)(\quad A \quad + \quad B \quad) = \quad A^2 \quad - \quad B^2$$

f) $(\boxed{2x+3} - 2y)(\boxed{2x+3} + 2y) = (\boxed{2x+3})^2 - (2y)^2$ Multiplying a sum and a difference

$$= 4x^2 + 12x + 9 - 4y^2 \quad \text{Squaring a binomial}$$

In Example 8, we recognized patterns that might elude some students, particularly in parts (e) and (f). In part (e), we *can* use FOIL, and in part (f), we *can* use long multiplication, but doing so is much slower. By carefully inspecting a problem before "jumping in," we can save ourselves considerable work. At least one instructor refers to this as "working smart" instead of "working hard."*

*Thanks to Pauline Kirkpatrick of Wharton County Junior College for this language.

Exercise Set

4.6

↪ *Concept Reinforcement*   *Each of the expressions in Exercises 1–8 can be regarded as either* **(a)** *the square of a binomial,* **(b)** *the product of the sum and difference of the same two terms, or* **(c)** *neither* (a) *nor* (b). *Select the appropriate choice for each expression.*

1. $(2x - 7y)^2$
2. $(4x - 9y)(4x + 9y)$
3. $(5a + 6b)(-6b + 5a)$
4. $(7a - 2b)(7a - 2b)$
5. $(r - 3s)(5r + 3s)$
6. $(2x - 7y)(7y - 2x)$
7. $(4x - 9y)(4x - 9y)$
8. $(2r - 3t)^2$

Evaluate each polynomial for $x = 5$ and $y = -2$.

9. $x^2 - 3y^2 + 2xy$
10. $x^2 + 5y^2 - 4xy$

Evaluate each polynomial for $x = 2$, $y = -3$, and $z = -4$.

11. $xyz^2 - z$
12. $xy - xz + yz$

Lung capacity. *The polynomial*

$$0.041h - 0.018A - 2.69$$

can be used to estimate the lung capacity, in liters, of a female with height h, in centimeters, and age A, in years.

13. Find the lung capacity of a 50-year-old woman who is 160 cm tall.

14. Find the lung capacity of a 20-year-old woman who is 165 cm tall.

15. *Male caloric needs.* The number of calories needed each day by a moderately active man who weighs w kilograms, is h centimeters tall, and is a years old can be estimated by the polynomial

$$19.18w + 7h - 9.52a + 92.4.$$

One of the authors of this text is moderately active, weighs 87 kg, is 185 cm tall, and is 59 yr old. What are his daily caloric needs?
Source: Parker, M., *She Does Math*. Mathematical Association of America

16. *Female caloric needs.* The number of calories needed each day by a moderately active woman who weighs w pounds, is h inches tall, and is a years old can be estimated by the polynomial

$$917 + 6w + 6h - 6a.$$

Christine is moderately active, weighs 125 lb, is 64 in. tall, and is 27 yr old. What are her daily caloric needs?
Source: Parker, M., *She Does Math*. Mathematical Association of America

Surface area of a silo. *A silo is a structure that is shaped like a right circular cylinder with a half sphere on top. The surface area of a silo of height h and radius r (including the area of the base) is given by the polynomial $2\pi rh + \pi r^2$. (Note that h is the height of the entire silo.)*

17. A $1\frac{1}{2}$-oz bottle of roll-on deodorant has a height of 4 in. and a radius of $\frac{3}{4}$ in. Find the surface area of the bottle if the bottle is shaped like a silo. Use 3.14 for π.

18. A container of tennis balls is silo-shaped, with a height of $7\frac{1}{2}$ in. and a radius of $1\frac{1}{4}$ in. Find the surface area of the container. Use 3.14 for π.

Altitude of a launched object. *The altitude of an object, in meters, is given by the polynomial*

$$h + vt - 4.9t^2,$$

where h is the height, in meters, at which the launch occurs, v is the initial upward speed (or velocity), in meters per second, and t is the number of seconds for which the object is airborne.

19. A bocce ball is thrown upward with an initial speed of 18 m/sec by a person atop the Leaning Tower of Pisa, which is 50 m above the ground. How high will the ball be 2 sec after it is thrown?

50 m

20. A golf ball is launched upward with an initial speed of 30 m/sec by a golfer atop the Washington Monument, which is 160 m above the ground. How high above the ground will the ball be after 3 sec?

Identify the coefficient and the degree of each term of each polynomial. Then find the degree of each polynomial.

21. $x^3y - 2xy + 3x^2 - 5$

22. $xy^2 - y^2 + 9x^2y + 7$

23. $17x^2y^3 - 3x^3yz - 7$

24. $6 - xy + 8x^2y^2 - y^5$

Combine like terms.

25. $7a + b - 4a - 3b$

26. $8r + s - 5r - 4s$

27. $3x^2y - 2xy^2 + x^2 + 5x$

28. $m^3 + 2m^2n - 3m^2 + 3mn^2$

29. $2u^2v - 3uv^2 + 6u^2v - 2uv^2 + 7u^2$

30. $3x^2 + 6xy + 3y^2 - 5x^2 - 10xy$

31. $5a^2c - 2ab^2 + a^2b - 3ab^2 + a^2c - 2ab^2$

32. $3s^2t + r^2t - 4st^2 - s^2t + 3st^2 - 7r^2t$

Add or subtract, as indicated.

33. $(4x^2 - xy + y^2) + (-x^2 - 3xy + 2y^2)$

34. $(2r^3 + 3rs - 5s^2) - (5r^3 + rs + 4s^2)$

35. $(3a^4 - 5ab + 6ab^2) - (9a^4 + 3ab - ab^2)$

36. $(2r^2t - 5rt + rt^2) - (7r^2t + rt - 5rt^2)$

Aha!   **37.** $(5r^2 - 4rt + t^2) + (-6r^2 - 5rt - t^2) +$ $(-5r^2 + 4rt - t^2)$

38. $(2x^2 - 3xy + y^2) + (-4x^2 - 6xy - y^2) +$ $(4x^2 + 6xy + y^2)$

39. $(x^3 - y^3) - (-2x^3 + x^2y - xy^2 + 2y^3)$

40. $(a^3 + b^3) - (-5a^3 + 2a^2b - ab^2 + 3b^3)$

41. $(2y^4x^2 - 5y^3x) + (5y^4x^2 - y^3x) + (3y^4x^2 - 2y^3x)$

42. $(5a^2b + 7ab) + (9a^2b - 5ab) + (a^2b - 6ab)$

43. Subtract $7x + 3y$ from the sum of $4x + 5y$ and $-5x + 6y$.

44. Subtract $5a + 2b$ from the sum of $2a + b$ and $3a - 4b$.

Multiply.

45. $(3z - u)(2z + 3u)$

46. $(5x + y)(2x - 3y)$

47. $(xy + 7)(xy - 4)$

48. $(ab + 3)(ab - 5)$

49. $(2a - b)(2a + b)$

50. $(a - 3b)(a + 3b)$

51. $(5rt - 2)(3rt + 1)$

52. $(3xy - 1)(4xy + 2)$

53. $(m^3n + 8)(m^3n - 6)$

54. $(3 - c^2d^2)(4 + c^2d^2)$

55. $(6x - 2y)(5x - 3y)$

56. $(7a - 6b)(5a + 4b)$

57. $(pq + 0.1)(-pq + 0.1)$

58. $(rt + 0.2)(-rt + 0.2)$

59. $(x + h)^2$

60. $(r + t)^2$

61. $(4a + 5b)^2$

62. $(3x + 2y)^2$

63. $(c^2 - d)(c^2 + d)$

64. $(p^3 - 5q)(p^3 + 5q)$

65. $(ab + cd^2)(ab - cd^2)$

66. $(xy + pq)(xy - pq)$

Aha!   **67.** $(a + b - c)(a + b + c)$

68. $(x + y + z)(x + y - z)$

69. $[a + b + c][a - (b + c)]$

70. $(a + b + c)(a - b - c)$

Find the total area of each shaded area.

71.
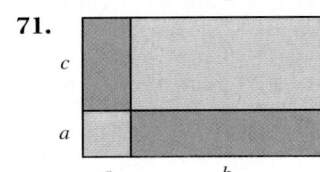

72.
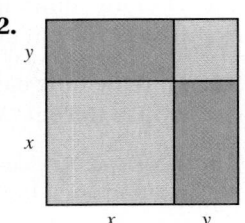

73.
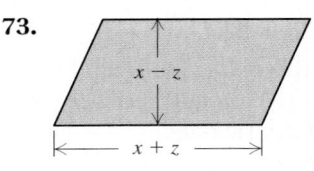

74.
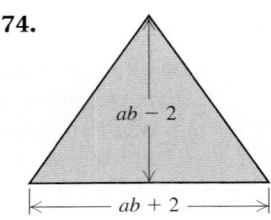

75.
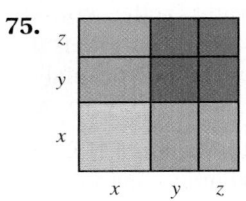

76.
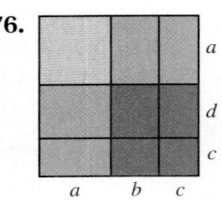

77.
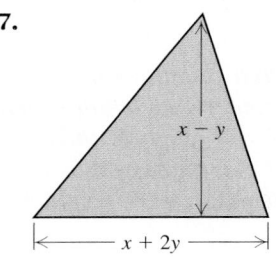

78.
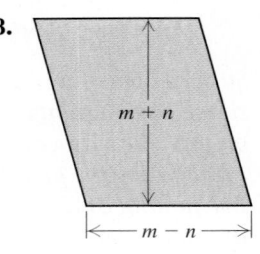

Draw and label rectangles similar to those in Exercises 71, 72, 75, and 76 to illustrate each product.

79. $(r + s)(u + v)$

80. $(m + r)(n + v)$

81. $(a + b + c)(a + d + f)$

82. $(r + s + t)^2$

83. Is it possible for a polynomial in 4 variables to have a degree less than 4? Why or why not?

84. A fourth-degree monomial is multiplied by a third-degree monomial. What is the degree of the product? Explain your reasoning.

SKILL MAINTENANCE

Simplify. [1.8]

85. $5 + \dfrac{7 + 4 + 2 \cdot 5}{3}$

86. $9 - \dfrac{2 + 6 \cdot 3 + 4}{6}$

87. $(4 + 3 \cdot 5 + 8) \div 3 \cdot 3$

88. $(5 + 2 \cdot 7 + 5) \div 2 \cdot 3$

89. $[3 \cdot 5 - 4 \cdot 2 + 7(-3)] \div (-2)$

90. $(7 - 3 \cdot 9 - 2 \cdot 5) \div (-6)$

SYNTHESIS

91. Explain how it is possible for the sum of two trinomials in several variables to be a binomial in one variable.

92. Explain how it is possible for the sum of two trinomials in several variables to be a trinomial in one variable.

Find a polynomial for the shaded area. (Leave results in terms of π where appropriate.)

93.
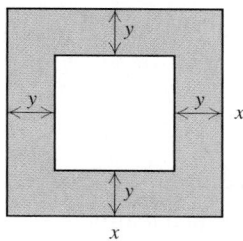

94.
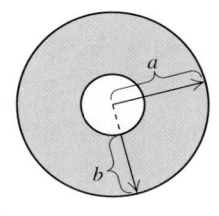

95.
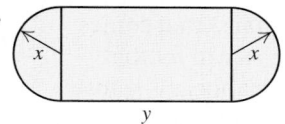

96.
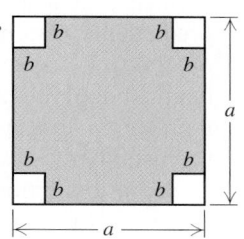

97. Find a polynomial for the total volume of the figure shown.

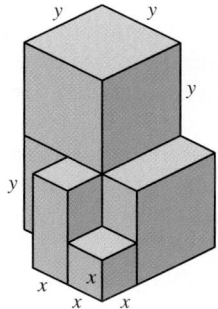

98. Find the shaded area in this figure using each of the approaches given below. Then check that both answers match.

a) Find the shaded area by subtracting the area of the unshaded square from the total area of the figure.

b) Find the shaded area by adding the areas of the three shaded rectangles.

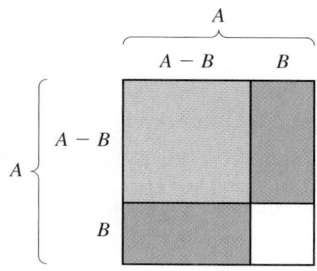

Find a polynomial for the surface area of each solid object shown. (Leave results in terms of π.)

99.
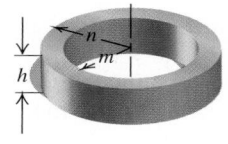

100.
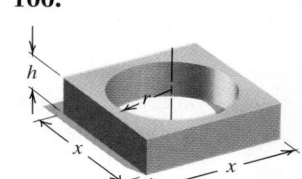

101. The observatory at Danville University is shaped like a silo that is 40 ft high and 30 ft wide (see Exercise 17). The Heavenly Bodies Astronomy Club is to paint the exterior of the observatory using paint that covers 250 ft² per gallon. How many gallons should they purchase? Explain your reasoning.

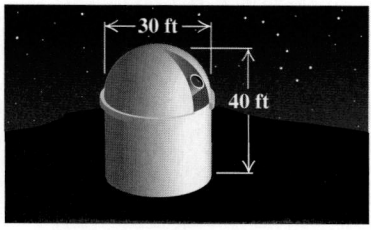

102. Multiply: $(x + a)(x - b)(x - a)(x + b)$.

103. *Interest compounded annually.* An amount of money P that is invested at the yearly interest rate r grows to the amount $P(1 + r)^t$ after t years. Find a polynomial that can be used to determine the amount to which P will grow after 2 yr.

104. *Yearly depreciation.* An investment P that drops in value at the yearly rate r drops in value to

$$P(1 - r)^t$$

after t years. Find a polynomial that can be used to determine the value to which P has dropped after 2 yr.

105. Suppose that $10,400 is invested at 8.5% compounded annually. How much is in the account at the end of 5 yr? (See Exercise 103.)

106. A $90,000 investment in computer hardware is depreciating at a yearly rate of 12.5%. How much is the investment worth after 4 yr? (See Exercise 104.)

CORNER

Finding the Magic Number

Focus: Evaluating polynomials in several variables

Time: 15–25 minutes

Group size: 3

Materials: A coin for each person

When a team nears the end of its schedule in first place, fans begin to discuss the team's "magic number." A team's magic number is the combined number of wins by that team and losses by the second-place team that guarantee the leading team a first-place finish. For example, if the Cubs' magic number is 3 over the Reds, any combination of Cubs wins and Reds losses that totals 3 will guarantee a first-place finish for the Cubs, regardless of how subsequent games are decided. A team's magic number is computed using the polynomial

$$G - P - L + 1,$$

where G is the length of the season, in games, P is the number of games that the leading team has played, and L is the total number of games that the second-place team has lost minus the total number of games that the leading team has lost.

ACTIVITY

1. The standings below are from a fictitious baseball league. Together, the group should calculate the Jaguars' magic number with respect to the Catamounts as well as the Jaguars' magic number with respect to the Wildcats. (Assume that the schedule is 162 games long.)

	W	**L**
Jaguars	92	64
Catamounts	90	66
Wildcats	89	66

2. Each group member should play the role of one of the teams. To simulate each team's remaining games, coin tosses will be performed. If a group member correctly predicts the side (heads or tails) that comes up, the coin toss represents a win for that team. Should the other side appear, the toss represents a loss. Assume that these games are against other (unlisted) teams in the league. Each group member should perform three coin tosses and then update the standings.

3. Recalculate the two magic numbers, using the updated standings from part (2).

4. Slowly—one coin toss at a time—play out the remainder of the season. Record all wins and losses, update the standings, and recalculate the magic numbers each time all three group members have completed a round of coin tosses.

5. Examine the work in part (4) and explain why a magic number of 0 indicates that a team has been eliminated from contention.

COLLABORATIVE

4.7 Division of Polynomials

Dividing by a Monomial • Dividing by a Binomial

In this section, we study division of polynomials. We will find that polynomial division is similar to division in arithmetic.

Dividing by a Monomial

We first consider division by a monomial. When dividing a monomial by a monomial, we use the quotient rule of Section 4.1 to subtract exponents when bases are the same. For example,

$$\frac{15x^{10}}{3x^4} = 5x^{10-4}$$

$$= 5x^6$$

> *Caution!* The coefficients are divided but the exponents are subtracted.

and

$$\frac{42a^2b^5}{-3ab^2} = \frac{42}{-3}a^{2-1}b^{5-2} \qquad \text{Recall that } a^m/a^n = a^{m-n}.$$

$$= -14ab^3.$$

To divide a polynomial by a monomial, we note that since

$$\frac{A}{C} + \frac{B}{C} = \frac{A+B}{C},$$

it follows that

$$\frac{A+B}{C} = \frac{A}{C} + \frac{B}{C}. \qquad \text{Switching the left and right sides of the equation}$$

This is actually how we perform divisions like $86 \div 2$. Although we might simply write

$$\frac{86}{2} = 43,$$

we are really saying

$$\frac{80+6}{2} = \frac{80}{2} + \frac{6}{2} = 40 + 3.$$

Similarly, to divide a polynomial by a monomial, we divide each term by the monomial:

$$\frac{80x^5 + 6x^7}{2x^3} = \frac{80x^5}{2x^3} + \frac{6x^7}{2x^3}$$

$$= \frac{80}{2}x^{5-3} + \frac{6}{2}x^{7-3} \qquad \text{Dividing coefficients and subtracting exponents}$$

$$= 40x^2 + 3x^4.$$

EXAMPLE 1   Divide $x^4 + 15x^3 - 6x^2$ by $3x$.

Solution   We have

$$\frac{x^4 + 15x^3 - 6x^2}{3x} = \frac{x^4}{3x} + \frac{15x^3}{3x} - \frac{6x^2}{3x}$$

$$= \frac{1}{3}x^{4-1} + \frac{15}{3}x^{3-1} - \frac{6}{3}x^{2-1} \qquad \text{Dividing coefficients and subtracting exponents}$$

$$= \frac{1}{3}x^3 + 5x^2 - 2x. \qquad \text{This is the quotient.}$$

To check, we multiply our answer by $3x$, using the distributive law:

$$3x\left(\frac{1}{3}x^3 + 5x^2 - 2x\right) = 3x \cdot \frac{1}{3}x^3 + 3x \cdot 5x^2 - 3x \cdot 2x$$

$$= x^4 + 15x^3 - 6x^2.$$

This is the polynomial that was being divided, so our answer, $\frac{1}{3}x^3 + 5x^2 - 2x$, checks.

EXAMPLE 2   Divide and check: $(10a^5b^4 - 2a^3b^2 + 6a^2b) \div (-2a^2b)$.

Solution   We have

$$\frac{10a^5b^4 - 2a^3b^2 + 6a^2b}{-2a^2b} = \frac{10a^5b^4}{-2a^2b} - \frac{2a^3b^2}{-2a^2b} + \frac{6a^2b}{-2a^2b}$$

$$= -\frac{10}{2}a^{5-2}b^{4-1} - \left(\frac{2}{2}\right)a^{3-2}b^{2-1} + \left(-\frac{6}{2}\right)$$

Dividing coefficients and subtracting exponents

$$= -5a^3b^3 + ab - 3.$$

Check: $-2a^2b(-5a^3b^3 + ab - 3)$

$$= -2a^2b(-5a^3b^3) + (-2a^2b)(ab) + (-2a^2b)(-3)$$

$$= 10a^5b^4 - 2a^3b^2 + 6a^2b$$

Our answer, $-5a^3b^3 + ab - 3$, checks.

Dividing by a Binomial

For divisors with more than one term, we use long division, much as we do in arithmetic. Polynomials are written in descending order and any missing terms in the dividend are written in, using 0 for the coefficients.

EXAMPLE 3   Divide $x^2 + 5x + 6$ by $x + 3$.

Solution   We have

Divide the first term, x^2, by the first term in the divisor: $x^2/x = x$. Ignore the term 3 for the moment.

$$\begin{array}{r} x \\ x + 3\overline{)x^2 + 5x + 6} \\ -(x^2 + 3x) \\ \hline 2x \end{array}$$

— Multiply $x + 3$ by x, using the distributive law

—Subtract by changing signs and adding: $x^2 + 5x - (x^2 + 3x) = 2x$.

Now we "bring down" the next term—in this case, 6. To divide the remainder, $2x + 6$, by $x + 3$, we focus on $2x$ and x.

$$\begin{array}{r} x + 2 \\ x + 3\overline{)x^2 + 5x + 6} \\ -(x^2 + 3x) \\ \hline 2x + 6 \\ -(2x + 6) \\ \hline 0 \end{array}$$

Divide $2x$ by x: $2x/x = 2$.

— Multiply 2 by the divisor, $x + 3$, using the distributive law

— Subtract: $(2x + 6) - (2x + 6) = 0$.

The quotient is $x + 2$. The notation R 0 indicates a remainder of 0, although a remainder of 0 is generally not listed in an answer.

Check: To check, we multiply the quotient by the divisor and add any remainder to see if we get the dividend:

Divisor	Quotient		Remainder		Dividend
$(x + 3)$	$(x + 2)$	$+$	0	$=$	$x^2 + 5x + 6.$

Our answer, $x + 2$, checks.

Student Notes

Long division of polynomials offers many opportunities to make errors. Rather than have one small mistake throw off all your work, we recommend that you double-check each step of your work as you move forward.

EXAMPLE 4   Divide: $(2x^2 + 5x - 1) \div (2x - 1)$.

Solution   We have

Divide the first term by the first term: $2x^2/(2x) = x$.

$$\begin{array}{r} x \\ 2x - 1\overline{)2x^2 + 5x - 1} \\ -(2x^2 - x) \\ \hline 6x \end{array}$$

— Multiply $2x - 1$ by x.

— Subtract by changing signs and adding: $2x^2 + 5x - (2x^2 - x) = 6x$.

Now, we bring down the -1 and divide $6x - 1$ by $2x - 1$.

$$\begin{array}{r} x + 3 \\ 2x - 1\overline{)2x^2 + 5x - 1} \\ -(2x^2 - x) \\ \hline 6x - 1 \\ -(6x - 3) \\ \hline 2 \end{array}$$

Divide $6x$ by $2x$: $6x/(2x) = 3$.

— Multiply 3 by the divisor, $2x - 1$.

— Note that $-1 - (-3) = -1 + 3 = 2$.

The answer is $x + 3$ with R 2.

Another way to write $x + 3$ R 2 is as

$$\text{Quotient} \quad x + 3 + \frac{2}{2x - 1} \quad \begin{matrix} \longleftarrow \text{Remainder} \\ \\ \longleftarrow \text{Divisor} \end{matrix}$$

(This is the way answers will be given at the back of the book.)

Check: To check, we multiply the divisor by the quotient and add the remainder:

$$(2x - 1)(x + 3) + 2 = 2x^2 + 5x - 3 + 2$$
$$= 2x^2 + 5x - 1. \quad \text{Our answer checks.}$$

_____

Our division procedure ends when the degree of the remainder is less than that of the divisor. Check that this was indeed the case in Example 4.

EXAMPLE 5   Divide each of the following.

a) $(x^3 + 1) \div (x + 1)$ **b)** $(x^4 - 3x^2 + 4x - 3) \div (x^2 - 5)$

Solution

a)
$$\begin{array}{r} x^2 - x + 1 \\ x + 1{\overline{\smash{\big)}\,x^3 + 0x^2 + 0x + 1}} \\ \underline{-(x^3 + x^2)} \\ -x^2 + 0x \\ \underline{-(-x^2 - x)} \\ x + 1 \\ \underline{-(x + 1)} \\ 0 \end{array}$$

$\longleftarrow$ Writing in the missing terms

$\longleftarrow$ Subtracting $x^3 + x^2$ from $x^3 + 0x^2$ and bringing down the $0x$

$\longleftarrow$ Subtracting $-x^2 - x$ from $-x^2 + 0x$ and bringing down the 1

The answer is $x^2 - x + 1$.

Check: $(x + 1)(x^2 - x + 1) = x^3 - x^2 + x + x^2 - x + 1$
$$= x^3 + 1.$$

b)
$$\begin{array}{r} x^2 + 2 \\ x^2 - 5{\overline{\smash{\big)}\,x^4 + 0x^3 - 3x^2 + 4x - 3}} \\ \underline{-(x^4 - 5x^2)} \\ 2x^2 + 4x - 3 \\ \underline{-(2x^2 - 10)} \\ 4x + 7 \end{array}$$

Writing in the missing term

Subtracting $x^4 - 5x^2$ from $x^4 - 3x^2$ and bringing down $4x - 3$

$\longleftarrow$ Subtracting $2x^2 - 10$ from $2x^2 + 4x - 3$

Since the remainder, $4x + 7$, is of lower degree than the divisor, the division process stops. The answer is

$$x^2 + 2 + \frac{4x + 7}{x^2 - 5}.$$

Check: $(x^2 - 5)(x^2 + 2) + 4x + 7 = x^4 + 2x^2 - 5x^2 - 10 + 4x + 7$
$$= x^4 - 3x^2 + 4x - 3.$$

_____

Exercise Set

4.7

Divide and check.

1. $\dfrac{32x^5 - 24x}{8}$

2. $\dfrac{12a^4 - 3a^2}{6}$

3. $\dfrac{u - 2u^2 + u^7}{u}$

4. $\dfrac{50x^5 - 7x^4 + x^2}{x}$

5. $(18t^3 - 24t^2 + 6t) \div (3t)$

6. $(20t^3 - 15t^2 + 30t) \div (5t)$

7. $(25x^6 - 20x^4 - 5x^2) \div (-5x^2)$

8. $(16x^6 + 32x^5 - 8x^2) \div (-8x^2)$

9. $(24t^5 - 40t^4 + 6t^3) \div (4t^3)$

10. $(18t^6 - 27t^5 - 3t^3) \div (9t^3)$

11. $\dfrac{8x^2 - 10x + 1}{2}$

12. $\dfrac{9x^2 + 3x - 2}{3}$

13. $\dfrac{4x^7 + 6x^5 + 4x^2}{4x^3}$

14. $\dfrac{10x^6 + 15x^3 + 5x^2}{5x^2}$

15. $\dfrac{9r^2s^2 + 3r^2s - 6rs^2}{-3rs}$

16. $\dfrac{4x^4y - 8x^6y^2 + 12x^8y^6}{4x^4y}$

17. $(x^2 + 4x - 12) \div (x - 2)$

18. $(x^2 - 6x + 8) \div (x - 4)$

19. $(t^2 - 10t - 20) \div (t - 5)$

20. $(t^2 + 8t - 15) \div (t + 4)$

21. $(2x^2 + 11x - 5) \div (x + 6)$

22. $(3x^2 - 2x - 13) \div (x - 2)$

23. $\dfrac{a^3 + 8}{a + 2}$

24. $\dfrac{t^3 + 27}{t + 3}$

25. $\dfrac{t^2 - 13}{t - 4}$

26. $\dfrac{a^2 - 21}{a - 5}$

27. $(3x^2 + 11x - 4) \div (3x - 1)$

28. $(10x^2 + 13x - 3) \div (5x - 1)$

29. $(6a^2 + 17a + 8) \div (2a + 5)$

30. $(10a^2 + 19a + 9) \div (2a + 3)$

31. $\dfrac{2t^3 - 9t^2 + 11t - 3}{2t - 3}$

32. $\dfrac{8t^3 - 22t^2 - 5t + 12}{4t + 3}$

33. $(t^3 - t^2 + t - 1) \div (t + 1)$

34. $(x^3 - x^2 + x - 1) \div (x - 1)$

35. $(t^4 - 2t^2 + 4t - 5) \div (t^2 - 3)$

36. $(t^4 + 4t^2 + 3t - 6) \div (t^2 + 5)$

37. $(4x^4 - 4x^2 - x - 3) \div (2x^2 - 3)$

38. $(6x^4 - 3x^2 + x - 4) \div (2x^2 + 1)$

 **39.** How is the distributive law used when dividing a polynomial by a binomial?

 **40.** On an assignment, Emmy Lou *incorrectly* writes

$$\frac{12x^3 - 6x}{3x} = 4x^2 - 6x.$$

What mistake do you think she is making and how might you convince her that a mistake has been made?

SKILL MAINTENANCE

Simplify.

41. $-4 + (-13)$ [1.5]

42. $-8 + (-15)$ [1.5]

43. $-9 - (-7)$ [1.6]

44. $-2 - (-7)$ [1.6]

45. The perimeter of a rectangular community garden is 640 ft. The length is 15 ft greater than the width. Find the length of the rectangle. [2.5]

46. Solve: $3(2x - 1) = 7x - 5$. [2.2]

47. Graph: $3x - 2y = 12$. [3.2]

48. Plot the points $(4, -1)$, $(0, 5)$, $(-2, 3)$, and $(-3, 0)$. [3.1]

SYNTHESIS

49. Explain how to form trinomials for which division by $x - 5$ results in a remainder of 3.

50. Explain how the quotient of two binomials can have more than two terms.

Divide.

51. $(10x^{9k} - 32x^{6k} + 28x^{3k}) \div (2x^{3k})$

52. $(45a^{8k} + 30a^{6k} - 60a^{4k}) \div (3a^{2k})$

53. $(6t^{3h} + 13t^{2h} - 4t^{h} - 15) \div (2t^{h} + 3)$

54. $(x^4 + a^2) \div (x + a)$

55. $(5a^3 + 8a^2 - 23a - 1) \div (5a^2 - 7a - 2)$

56. $(15y^3 - 30y + 7 - 19y^2) \div (3y^2 - 2 - 5y)$

57. Divide the sum of $4x^5 - 14x^3 - x^2 + 3$ and $2x^5 + 3x^4 + x^3 - 3x^2 + 5x$ by $3x^3 - 2x - 1$.

58. Divide $5x^7 - 3x^4 + 2x^2 - 10x + 2$ by the sum of $(x - 3)^2$ and $5x - 8$.

If the remainder is 0 when one polynomial is divided by another, the divisor is a factor of the dividend. Find the value(s) of c for which $x - 1$ is a factor of each polynomial.

59. $x^2 - 4x + c$

60. $2x^2 - 3cx - 8$

61. $c^2x^2 + 2cx + 1$

4.8   Negative Exponents and Scientific Notation

Negative Integers as Exponents • Scientific Notation • Multiplying and Dividing Using Scientific Notation

We now attach a meaning to negative exponents. Once we understand both positive and negative exponents, we can study a method of writing numbers known as *scientific notation*.

Negative Integers as Exponents

Let's define negative exponents so that the rules that apply to whole-number exponents will hold for all integer exponents. To do so, consider a^{-5} and the rule for adding exponents:

$$a^{-5} = a^{-5} \cdot 1 \qquad \text{Using the identity property of 1}$$

$$= \frac{a^{-5}}{1} \cdot \frac{a^5}{a^5} \qquad \text{Writing 1 as } \frac{a^5}{a^5} \text{ and } a^{-5} \text{ as } \frac{a^{-5}}{1}$$

$$= \frac{a^{-5+5}}{a^5} \qquad \text{Adding exponents}$$

$$= \frac{1}{a^5}. \qquad -5 + 5 = 0 \text{ and } a^0 = 1$$

This leads to our definition of negative exponents.

> **Negative Exponents**
>
> For any real number a that is nonzero and any integer n,
>
> $$a^{-n} = \frac{1}{a^n}.$$
>
> (The numbers a^{-n} and a^n are reciprocals of each other.)

EXAMPLE 1   Express using positive exponents and, if possible, simplify.

a) m^{-3} b) 4^{-2} c) $(-3)^{-2}$ d) ab^{-1}

Solution

a) $m^{-3} = \dfrac{1}{m^3}$ m^{-3} is the reciprocal of m^3.

b) $4^{-2} = \dfrac{1}{4^2} = \dfrac{1}{16}$ 4^{-2} is the reciprocal of 4^2. Note that $4^{-2} \neq 4(-2)$.

c) $(-3)^{-2} = \dfrac{1}{(-3)^2} = \dfrac{1}{(-3)(-3)} = \dfrac{1}{9}$ $\begin{cases} (-3)^{-2} \text{ is the reciprocal of } (-3)^2. \\ \text{Note that } (-3)^{-2} \neq -\dfrac{1}{3^2}. \end{cases}$

d) $ab^{-1} = a\left(\dfrac{1}{b^1}\right) = a\left(\dfrac{1}{b}\right) = \dfrac{a}{b}$ b^{-1} is the reciprocal of b^1.

> *Caution!*   A negative exponent does not, in itself, indicate that an expression is negative. As shown in Example 1,
>
> $$4^{-2} \neq 4(-2) \quad \text{and} \quad (-3)^{-2} \neq -\frac{1}{3^2}.$$

The following is another way to illustrate why negative exponents are defined as they are.

On this side, we divide by 5 at each step.		On this side, the exponents decrease by 1.
	$125 = 5^3$	
	$25 = 5^2$	
	$5 = 5^1$	
	$1 = 5^0$	
	$\dfrac{1}{5} = 5^?$	
	$\dfrac{1}{25} = 5^?$	

To continue the pattern, it follows that

$$\frac{1}{5} = \frac{1}{5^1} = 5^{-1}, \qquad \frac{1}{25} = \frac{1}{5^2} = 5^{-2}, \quad \text{and, in general,} \quad \frac{1}{a^n} = a^{-n}.$$

EXAMPLE 2   Express $\dfrac{1}{x^7}$ using negative exponents.

Solution We know that $\dfrac{1}{a^n} = a^{-n}$. Thus, $\dfrac{1}{x^7} = x^{-7}$.

The rules for powers still hold when exponents are negative.

EXAMPLE 3   Simplify. Do not use negative exponents in the answer.

a) $t^5 \cdot t^{-2}$ **b)** $(5x^{-2}y^3)^{-4}$ **c)** $\dfrac{x^{-4}}{x^{-5}}$

d) $\dfrac{1}{t^{-5}}$ **e)** $\dfrac{s^{-3}}{t^{-5}}$ **f)** $\dfrac{-10x^{-3}y}{5x^2y^5}$

Solution

a) $t^5 \cdot t^{-2} = t^{5+(-2)} = t^3$ Adding exponents

b) $(5x^{-2}y^3)^{-4} = 5^{-4}(x^{-2})^{-4}(y^3)^{-4}$ Raising each factor to the exponent of -4

$\qquad\qquad\qquad\quad = \dfrac{1}{5^4}x^8y^{-12} = \dfrac{x^8}{625y^{12}}$ Multiplying exponents; writing with positive exponents

c) $\dfrac{x^{-4}}{x^{-5}} = x^{-4-(-5)} = x^1 = x$ We subtract exponents even if the exponent in the denominator is negative.

d) Since $\dfrac{1}{a^n} = a^{-n}$, we have $\dfrac{1}{t^{-5}} = t^{-(-5)} = t^5$.

e) $\dfrac{s^{-3}}{t^{-5}} = s^{-3} \cdot \dfrac{1}{t^{-5}} = \dfrac{1}{s^3} \cdot t^5 = \dfrac{t^5}{s^3}$ Using the result from part (d) above

f) $\dfrac{-10x^{-3}y}{5x^2y^5} = \dfrac{-10}{5} \cdot \dfrac{x^{-3}}{x^2} \cdot \dfrac{y^1}{y^5}$ Note that the -10 and 5 are factors.

$\qquad\qquad\quad = -2 \cdot x^{-3-2} \cdot y^{1-5}$ Using the quotient rule twice; simplifying

$\qquad\qquad\quad = -2x^{-5}y^{-4} = \dfrac{-2}{x^5y^4}$

The result from Example 3(e) can be generalized.

> **Factors and Negative Exponents**
>
> For any nonzero real numbers a and b and any integers m and n,
>
> $$\dfrac{a^{-n}}{b^{-m}} = \dfrac{b^m}{a^n}.$$
>
> (A factor can be moved to the other side of the fraction bar if the sign of the exponent is changed.)

EXAMPLE 4   Simplify: $\dfrac{-15x^{-7}}{5y^2z^{-4}}$.

Solution   We can move the factors x^{-7} and z^{-4} to the other side of the fraction bar if we change the sign of each exponent:

$$\frac{-15x^{-7}}{5y^2z^{-4}} = \frac{-3z^4}{y^2x^7}, \quad \text{or} \quad \frac{-3z^4}{x^7y^2}. \qquad \begin{array}{l} \text{When possible, be sure to} \\ \text{simplify the constant factors:} \\ -15/5 = -3. \end{array}$$

Another way to change the sign of the exponent is to take the reciprocal of the base. To understand why this is true, note that

$$\left(\frac{s}{t}\right)^{-5} = \frac{s^{-5}}{t^{-5}} = \frac{t^5}{s^5} = \left(\frac{t}{s}\right)^5.$$

This often provides the easiest way to simplify an expression containing a negative exponent.

Reciprocals and Negative Exponents

For any nonzero real numbers a and b and any integer n,

$$\left(\frac{a}{b}\right)^{-n} = \left(\frac{b}{a}\right)^n.$$

(Any base to a power is equal to the reciprocal of the base raised to the opposite power.)

EXAMPLE 5   Simplify: $\left(\dfrac{x^4}{2y}\right)^{-3}$.

Solution

$$\left(\frac{x^4}{2y}\right)^{-3} = \left(\frac{2y}{x^4}\right)^3 \qquad \begin{array}{l} \text{Taking the reciprocal of the base and changing} \\ \text{the sign of the exponent} \end{array}$$

$$= \frac{(2y)^3}{(x^4)^3} \qquad \begin{array}{l} \text{Raising a quotient to a power by raising both} \\ \text{the numerator and denominator to the power} \end{array}$$

$$= \frac{2^3y^3}{x^{12}} \qquad \begin{array}{l} \text{Raising a product to a power; using the power} \\ \text{rule in the denominator} \end{array}$$

$$= \frac{8y^3}{x^{12}} \qquad \text{Cubing 2}$$

Definitions and Properties of Exponents

The following summary assumes that no denominators are 0 and that 0^0 is not considered. For any integers m and n,

1 as an exponent: $\qquad\qquad a^1 = a$

0 as an exponent: $\qquad\qquad a^0 = 1$

Negative exponents: $\qquad\quad a^{-n} = \dfrac{1}{a^n},$

$$\dfrac{a^{-n}}{b^{-m}} = \dfrac{b^m}{a^n},$$

$$\left(\dfrac{a}{b}\right)^{-n} = \left(\dfrac{b}{a}\right)^n$$

The Product Rule: $\qquad\quad a^m \cdot a^n = a^{m+n}$

The Quotient Rule: $\qquad\quad \dfrac{a^m}{a^n} = a^{m-n}$

The Power Rule: $\qquad\qquad (a^m)^n = a^{mn}$

Raising a product to a power: $\quad (ab)^n = a^n b^n$

Raising a quotient to a power: $\quad \left(\dfrac{a}{b}\right)^n = \dfrac{a^n}{b^n}$

Scientific Notation

When we are working with the very large or very small numbers that frequently occur in science, **scientific notation** provides a useful way of writing numbers. The following are examples of scientific notation.

The mass of the earth:

$$6.0 \times 10^{24} \text{ kilograms (kg)} = 6{,}000{,}000{,}000{,}000{,}000{,}000{,}000{,}000 \text{ kg}$$

The mass of a hydrogen atom:

$$1.7 \times 10^{-24} \text{ g} = 0.0000000000000000000000017 \text{ g}$$

Scientific Notation

Scientific notation for a number is an expression of the type

$$N \times 10^m,$$

where N is at least 1 but less than 10 (that is, $1 \le N < 10$), N is expressed in decimal notation, and m is an integer.

Converting from scientific to decimal notation involves multiplying by a power of 10. Consider the following.

Scientific Notation	*Multiplication*	*Decimal Notation*
4.52×10^2	4.52×100	452.
4.52×10^1	4.52×10	45.2
4.52×10^0	4.52×1	4.52
4.52×10^{-1}	4.52×0.1	0.452
4.52×10^{-2}	4.52×0.01	0.0452

Note that when m, the power of 10, is positive, the decimal point moves right m places in decimal notation. When m is negative, the decimal point moves left $|m|$ places. We generally try to perform this multiplication mentally.

EXAMPLE 6   Convert to decimal notation: **(a)** 7.893×10^5; **(b)** 4.7×10^{-8}.

Solution

a) Since the exponent is positive, the decimal point moves to the right:

$$7.89300. \qquad 7.893 \times 10^5 = 789{,}300 \qquad \text{The decimal point moves}$$

5 places   5 places to the right.

b) Since the exponent is negative, the decimal point moves to the left:

$$0.00000004.7 \qquad 4.7 \times 10^{-8} = 0.000000047 \qquad \text{The decimal point}$$

8 places   moves 8 places to
                                                                         the left.

To convert from decimal to scientific notation, this procedure is reversed.

EXAMPLE 7   Write in scientific notation: **(a)** 83,000; **(b)** 0.0327.

Solution

a) We need to find m such that $83{,}000 = 8.3 \times 10^m$. To change 8.3 to 83,000 requires moving the decimal point 4 places to the right. This can be accomplished by multiplying by 10^4. Thus,

$$83{,}000 = 8.3 \times 10^4. \qquad \text{This is scientific notation.}$$

b) We need to find m such that $0.0327 = 3.27 \times 10^m$. To change 3.27 to 0.0327 requires moving the decimal point 2 places to the left. This can be accomplished by multiplying by 10^{-2}. Thus,

$$0.0327 = 3.27 \times 10^{-2}. \qquad \text{This is scientific notation.}$$

Conversions to and from scientific notation are often made mentally. Remember that positive exponents are used when representing large numbers and negative exponents are used when representing numbers between 0 and 1.

Multiplying and Dividing Using Scientific Notation

Products and quotients of numbers written in scientific notation are found using the rules for exponents.

EXAMPLE 8   Simplify.

a) $(1.8 \times 10^9) \cdot (2.3 \times 10^{-4})$

b) $(3.41 \times 10^5) \div (1.1 \times 10^{-3})$

Solution

a) $(1.8 \times 10^9) \cdot (2.3 \times 10^{-4})$

$\quad = 1.8 \times 2.3 \times 10^9 \times 10^{-4}$ Using the associative and commutative laws

$\quad = 4.14 \times 10^{9+(-4)}$ Adding exponents

$\quad = 4.14 \times 10^5$

b) $(3.41 \times 10^5) \div (1.1 \times 10^{-3})$

$\quad = \dfrac{3.41 \times 10^5}{1.1 \times 10^{-3}}$

$\quad = \dfrac{3.41}{1.1} \times \dfrac{10^5}{10^{-3}}$

$\quad = 3.1 \times 10^{5-(-3)}$ Subtracting exponents

$\quad = 3.1 \times 10^8$

When a problem is stated using scientific notation, we normally use scientific notation for the answer. This often requires an additional conversion.

EXAMPLE 9   Simplify.

a) $(3.1 \times 10^5) \cdot (4.5 \times 10^{-3})$

b) $(7.2 \times 10^{-7}) \div (8.0 \times 10^6)$

Solution

a) We have

$(3.1 \times 10^5) \cdot (4.5 \times 10^{-3}) = 3.1 \times 4.5 \times 10^5 \times 10^{-3}$

$\qquad\qquad\qquad\qquad\qquad\qquad = 13.95 \times 10^2.$

Our answer is not yet in scientific notation because 13.95 is not between 1 and 10. We convert to scientific notation as follows:

$13.95 \times 10^2 = 1.395 \times 10^1 \times 10^2$ Substituting 1.395×10^1 for 13.95

$\qquad\qquad = 1.395 \times 10^3.$ Adding exponents

b) $(7.2 \times 10^{-7}) \div (8.0 \times 10^6) = \dfrac{7.2 \times 10^{-7}}{8.0 \times 10^6} = \dfrac{7.2}{8.0} \times \dfrac{10^{-7}}{10^6}$

$\qquad\qquad\qquad\qquad\qquad\quad = 0.9 \times 10^{-13}$

$\qquad\qquad\qquad\qquad\qquad\quad = 9.0 \times 10^{-1} \times 10^{-13}$ Substituting 9.0×10^{-1} for 0.9

$\qquad\qquad\qquad\qquad\qquad\quad = 9.0 \times 10^{-14}$ Adding exponents

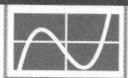 **technology connection**

A key labeled $\boxed{10^x}$, $\boxed{\wedge}$, or $\boxed{EE}$ is used to enter scientific notation into a calculator. Sometimes this is a secondary function, meaning that another key—often labeled SHIFT or **2ND**—must be pressed first.

To check Example 8(a), we press

1.8 $\boxed{EE}$ 9 $\boxed{\times}$ 2.3 $\boxed{EE}$ $\boxed{(-)}$ 4.

When we then press $\boxed{=}$ or **ENTER**, the result 4.14E5 appears. This represents 4.14×10^5. On many calculators, the MODE Sci must be selected in order to display scientific notation.

    ┌─────────────────────────┐
    │ 1.8E9*2.3E−4   │
    │   4.14E5 │
    └─────────────────────────┘

On some calculators, this appears as

or

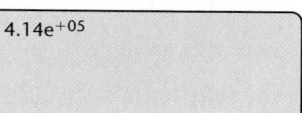

Calculate each of the following.

1. $(3.8 \times 10^9) \cdot (4.5 \times 10^7)$
2. $(2.9 \times 10^{-8}) \div (5.4 \times 10^6)$
3. $(9.2 \times 10^7) \div (2.5 \times 10^{-9})$

Exercise Set

4.8

Express using positive exponents. Then, if possible, simplify.

1. 7^{-2}

2. 2^{-4}

3. 10^{-4}

4. 5^{-3}

5. $(-2)^{-6}$

6. $(-3)^{-4}$

7. x^{-8}

8. t^{-5}

9. xy^{-2}

10. $a^{-3}b$

11. $r^{-5}t$

12. xy^{-9}

13. $\dfrac{1}{t^{-8}}$

14. $\dfrac{1}{z^{-9}}$

15. $\dfrac{1}{h^{-8}}$

16. $\dfrac{1}{a^{-12}}$

17. 7^{-1}

18. 3^{-1}

19. $\left(\dfrac{3}{5}\right)^{-2}$

20. $\left(\dfrac{3}{4}\right)^{-2}$

21. $\left(\dfrac{a}{2}\right)^{-3}$

22. $\left(\dfrac{x}{2}\right)^{-4}$

23. $\left(\dfrac{s}{t}\right)^{-7}$

24. $\left(\dfrac{r}{v}\right)^{-5}$

Express using negative exponents.

25. $\dfrac{1}{6^2}$

26. $\dfrac{1}{5^2}$

27. $\dfrac{1}{t^6}$

28. $\dfrac{1}{y^2}$

29. $\dfrac{1}{a^4}$

30. $\dfrac{1}{t^5}$

31. $\dfrac{1}{p^7}$ **32.** $\dfrac{1}{m^{12}}$ **33.** $\dfrac{1}{5}$

34. $\dfrac{1}{8}$ **35.** $\dfrac{1}{t}$ **36.** $\dfrac{1}{m}$

Simplify. Do not use negative exponents in the answer.

37. $2^{-5} \cdot 2^8$ **38.** $5^{-8} \cdot 5^{10}$ **39.** $x^{-2} \cdot x^{-7}$

40. $x^{-2} \cdot x^{-9}$ **41.** $t^{-3} \cdot t$ **42.** $y^{-5} \cdot y$

43. $(a^{-2})^9$ **44.** $(x^{-5})^6$ **45.** $(t^{-3})^{-6}$

46. $(a^{-4})^{-7}$ **47.** $(t^4)^{-3}$ **48.** $(t^5)^{-2}$

49. $(mn)^{-7}$ **50.** $(ab)^{-9}$ **51.** $(3x^{-4})^2$

52. $(2a^{-5})^3$ **53.** $(5r^{-4}t^3)^2$ **54.** $(4x^5y^{-6})^3$

55. $\dfrac{t^7}{t^{-3}}$ **56.** $\dfrac{x^7}{x^{-2}}$ **57.** $\dfrac{y^{-7}}{y^{-3}}$

58. $\dfrac{z^{-6}}{z^{-2}}$ **59.** $\dfrac{12y^{-4}}{4y^{-9}}$ **60.** $\dfrac{8a^{-6}}{2a^{-10}}$

61. $\dfrac{2x^6}{x}$ **62.** $\dfrac{3x}{x^{-1}}$ **63.** $\dfrac{15a^{-7}}{10b^{-9}}$

64. $\dfrac{12x^{-6}}{8y^{-10}}$ Aha! **65.** $\dfrac{t^{-7}}{t^{-7}}$ **66.** $\dfrac{a^{-5}}{b^{-7}}$

67. $\dfrac{3x^{-5}}{y^{-6}z^{-2}}$ **68.** $\dfrac{4a^{-6}}{b^{-5}c^{-7}}$ **69.** $(3x^4y^5)^{-3}$

70. $(2t^5x^3)^{-4}$ **71.** $(x^{-6}y^{-2})^{-4}$ **72.** $(x^{-2}y^{-7})^{-5}$

73. $(a^{-5}b^7c^{-2})(a^{-3}b^{-2}c^6)$

74. $(x^3y^{-4}z^{-5})(x^{-4}y^{-2}z^9)$

75. $\left(\dfrac{a^4}{3}\right)^{-2}$ **76.** $\left(\dfrac{y^2}{2}\right)^{-2}$ **77.** $\left(\dfrac{m^{-1}}{n^{-4}}\right)^3$

78. $\left(\dfrac{x^2y}{z^{-5}}\right)^3$ **79.** $\left(\dfrac{2a^2}{3b^4}\right)^{-3}$ **80.** $\left(\dfrac{a^2b}{cd^3}\right)^{-5}$

Aha! **81.** $\left(\dfrac{5x^{-2}}{3y^{-2}z}\right)^0$ **82.** $\left(\dfrac{4a^3b^{-2}}{5c^{-3}}\right)^1$ **83.** $\dfrac{-6a^3b^{-5}}{-3a^7b^{-8}}$

84. $\dfrac{12x^{-2}y^4}{-3xy^{-7}}$ **85.** $\dfrac{10x^{-4}yz^7}{8x^7y^{-3}z^{-3}}$ **86.** $\dfrac{9a^6b^{-4}c^7}{27a^{-4}b^5c^9}$

Convert to decimal notation.

87. 7.12×10^4 **88.** 8.92×10^2

89. 8.92×10^{-3} **90.** 7.26×10^{-4}

91. 9.04×10^8 **92.** 1.35×10^7

93. 2.764×10^{-10} **94.** 9.043×10^{-3}

95. 4.209×10^7 **96.** 5.029×10^8

Convert to scientific notation.

97. 490,000   **98.** 71,500

99. 0.00583   **100.** 0.0814

101. 78,000,000,000   **102.** 3,700,000,000,000

103. 0.000000527   **104.** 0.00000000648

105. 0.000000018   **106.** 0.00000000002

107. 1,094,000,000,000,000

108. 1,030,200,000,000,000,000

Multiply or divide, and write scientific notation for the result.

109. $(4 \times 10^7)(2 \times 10^5)$

110. $(1.9 \times 10^8)(3.4 \times 10^{-3})$

111. $(3.8 \times 10^9)(6.5 \times 10^{-2})$

112. $(7.1 \times 10^{-7})(8.6 \times 10^{-5})$

113. $(8.7 \times 10^{-12})(4.5 \times 10^{-5})$

114. $(4.7 \times 10^5)(6.2 \times 10^{-12})$

115. $\dfrac{8.5 \times 10^8}{3.4 \times 10^{-5}}$

116. $\dfrac{5.6 \times 10^{-2}}{2.5 \times 10^5}$

117. $(3.0 \times 10^6) \div (6.0 \times 10^9)$

118. $(1.5 \times 10^{-3}) \div (1.6 \times 10^{-6})$

119. $\dfrac{7.5 \times 10^{-9}}{2.5 \times 10^{12}}$

120. $\dfrac{4.0 \times 10^{-3}}{8.0 \times 10^{20}}$

121. Without performing actual computations, explain why 3^{-29} is smaller than 2^{-29}.

122. What is it about scientific notation that makes it so useful?

SKILL MAINTENANCE

Simplify. [1.8]

123. $(3 - 8)(9 - 12)$

124. $(2 - 9)^2$

125. $7 \cdot 2 + 8^2$

126. $5 \cdot 6 - 3 \cdot 2 \cdot 4$

127. $-3(-2)^2 \div 4 \cdot 5$

128. $-3 - 2 + 7 - 6 \div 2 \cdot 3$

129. Plot the points $(-3, 2)$, $(4, -1)$, $(5, 3)$, and $(-5, -2)$. [3.1]

130. Solve $cx + bt = r$ for t. [2.3]

SYNTHESIS

131. Explain what requirements must be met in order for x^{-n} to represent a negative integer.

132. Explain why scientific notation cannot be used without an understanding of the rules for exponents.

133. Write the reciprocal of 1.25×10^{-6} in scientific notation.

134. Write the reciprocal of 2.5×10^9 in scientific notation.

135. Write $8^{-3} \cdot 32 \div 16^2$ as a power of 2.

136. Write $81^3 \cdot 27 \div 9^2$ as a power of 3.

Simplify each of the following. Use a calculator only where indicated.

Aha! **137.** $\dfrac{125^{-4}(25^2)^4}{125}$

138. $(13^{-12})^2 \cdot 13^{25}$

139. $\dfrac{4.2 \times 10^8[(2.5 \times 10^{-5}) \div (5.0 \times 10^{-9})]}{3.0 \times 10^{-12}}$

140. $\dfrac{27^{-2}(81^2)^3}{9^8}$

141. $\dfrac{7.4 \times 10^{29}}{(5.4 \times 10^{-6})(2.8 \times 10^8)}$

142. $\dfrac{5.8 \times 10^{17}}{(4.0 \times 10^{-13})(2.3 \times 10^4)}$

143. $\dfrac{(7.8 \times 10^7)(8.4 \times 10^{23})}{2.1 \times 10^{-12}}$

144. $\dfrac{(2.5 \times 10^{-8})(6.1 \times 10^{-11})}{1.28 \times 10^{-3}}$

145. Determine whether each of the following is true for all pairs of integers m and n and all positive numbers x and y.
a) $x^m \cdot y^n = (xy)^{mn}$
b) $x^m \cdot y^m = (xy)^{2m}$
c) $(x - y)^m = x^m - y^m$

Write scientific notation for each answer.

146. *Household income.* In 2001, there were about 109.3 million households in the United States. The average income of these households (before taxes) was about \$79,463. Find the total income generated by these households.
Source: *Statistical Abstract of the United States* 2002

147. *Computers.* A gigabyte is a measure of a computer's storage capacity. One gigabyte holds about one billion bytes of information. If a firm's computer network contains 2500 gigabytes of memory, how many bytes are in the network?

148. *River discharge.* The average discharge at the mouth of the Amazon River is 4,200,000 cubic feet per second. How much water is discharged from the Amazon River in 1 yr?

149. *Biology.* A strand of DNA (deoxyribonucleic acid) is about 1.5 m long and 1.3×10^{-10} cm wide. How many times longer is DNA than it is wide?
Source: Human Genome Project Information

150. *Water contamination.* In the United States, 200 million gal of used motor oil is improperly disposed of each year. One gallon of used oil can contaminate one million gallons of drinking water. How many gallons of drinking water can 200 million gallons of oil contaminate?
Source: *The Macmillan Visual Almanac*

Study Summary

Terms were first introduced in Chapter 1 and include expressions like

$$\frac{9}{x^2}, \quad 5a^3, \quad \text{and} \quad -7x^4y^5 \text{ (p. 239).}$$

Of the terms listed, $5a^3$ and $-7x^4y^5$ do not involve division by a variable expression and are examples of **monomials** (p. 239). A **binomial** is the sum or difference of two monomials, such as

$$3x + 5, \quad 2a^4 - 7a^3, \quad \text{or} \quad -x^2y + 6z,$$

and a **trinomial** is formed by adding and/or subtracting three monomials (p. 240). Examples of trinomials include

$$a^2 + 3a - 7, \quad 5x^3 - 4x^2 + 9x, \quad \text{and} \quad 2r^2t + 7rt - r^3.$$

Monomials, binomials, and trinomials are types of **polynomials** (p. 239).

The **degree of a term** in a polynomial is the number of variable factors in the term (p. 240). The **coefficient** of a term is the constant factor in that term (p. 240):

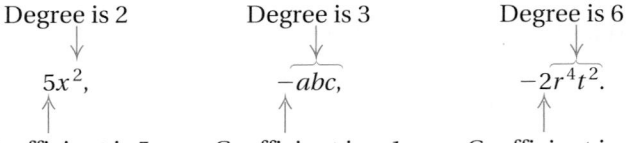

The **leading term of a polynomial** is the term with the highest degree. The **leading coefficient** of a polynomial is the same as the coefficient of the polynomial's leading term, and the **degree of a polynomial** is the degree of the leading term (p. 241). To illustrate, note that

$$5x^2 - abc - 2r^4t^2$$

has degree 6, a leading coefficient of -2, and $-2r^4t^2$ as its leading term. Because this trinomial makes use of more than one variable, it is an example of a **polynomial in several variables** (p. 274).

A polynomial in one variable is written in **descending order** (p. 242) if the degrees of the terms decrease from left to right. Thus,

$$2x^5 + 7x^3 - 5x + 9 \text{ is written in descending order,}$$

whereas

$$2 + 4a + 3a^2 - 2a^5 \text{ is written in \textbf{ascending order} (p. 249).}$$

Polynomials can be added (p. 248) by combining **like terms** (p. 248). To subtract a polynomial (p. 250), we first find its **opposite** (p. 250) and then add. Multiplication (p. 257) is performed by multiplying every term in one polynomial by each term in the other polynomial and, if possible, then combining like terms. Division (p. 283) is performed much like long division of numbers.

Addition: $(3x^2 + 5x - 9) + (4x^2 - 7x) = (3 + 4)x^2 + (5 + (-7))x - 9$
$$= 7x^2 - 2x - 9$$

Subtraction: $(8t^3 + 2t^2 + 3) - (5t^3 + t - 6) = 8t^3 + 2t^2 + 3 + (-5t^3) - t + 6$
$$= (8 - 5)t^3 + 2t^2 - t + (3 + 6)$$
$$= 3t^3 + 2t^2 - t + 9$$

Multiplication: $(x + 4)(x^2 + 3x - 2) = x(x^2 + 3x - 2) + 4(x^2 + 3x - 2)$
$$= x \cdot x^2 + x \cdot 3x + x(-2) + 4 \cdot x^2 + 4 \cdot 3x + 4(-2)$$
$$= x^3 + 3x^2 - 2x + 4x^2 + 12x - 8$$
$$= x^3 + 7x^2 + 10x - 8$$

Division: $(4x^2 + 8x + 7) \div (2x + 3) = 2x + 1 + \dfrac{4}{2x + 3}$

$$
\begin{array}{r}
2x + 1 \\
2x + 3 \overline{)\, 4x^2 + 8x + 7} \\
-(4x^2 + 6x) \\
\hline
2x + 7 \\
-(2x + 3) \\
\hline
4
\end{array}
$$

Important definitions and properties of exponents are listed on p. 292 in Section 4.8.

4 Review Exercises

↪ *Concept Reinforcement*   *Classify each statement as either true or false.*

1. When two polynomials that are written in descending order are added, the result is generally written in ascending order. [4.3]

2. The product of the sum and difference of the same two terms is a difference of squares. [4.5]

3. When a binomial is squared, the result is a perfect-square trinomial. [4.5]

4. FOIL can be used whenever two polynomials are being multiplied. [4.5]

5. The degree of a polynomial can exceed the value of the polynomial's leading coefficient. [4.2]

6. Scientific notation is used only for extremely large numbers. [4.8]

7. FOIL can be used with polynomials in several variables. [4.6]

8. A positive number raised to a negative exponent can never represent a negative number. [4.8]

Simplify. [4.1]

9. $y^7 \cdot y^3 \cdot y$

10. $(3x)^5 \cdot (3x)^9$

11. $t^6 \cdot t^0$

12. $\dfrac{4^5}{4^2}$

13. $\dfrac{(a+b)^4}{(a+b)^4}$

14. $\left(\dfrac{3t^4}{2s^3}\right)^2$

15. $(-2xy^2)^3$

16. $(2x^3)(-3x)^2$

17. $(a^2b)(ab)^5$

Identify the terms of each polynomial. [4.2]

18. $3x^2 + 6x + \frac{1}{2}$

19. $-4y^5 + 7y^2 - 3y - 2$

List the coefficients of the terms in each polynomial. [4.2]

20. $9x^2 - x + 7$

21. $4x^3 + 6x^2 - 5x + \frac{5}{3}$

*For each polynomial, **(a)** list the degree of each term; **(b)** determine the leading term and the leading coefficient; and **(c)** determine the degree of the polynomial.* [4.2]

22. $4t^2 + 6 + 15t^5$

23. $-2x^5 + 7 - 3x^2 + x$

Classify each polynomial as a monomial, a binomial, a trinomial, or none of these. [4.2]

24. $4x^3 - 1$

25. $4 - 9t^3 - 7t^4 + 10t^2$

26. $7y^2$

Combine like terms and write in descending order. [4.2]

27. $3x - x^2 + 4x$

28. $\frac{3}{4}x^3 + 4x^2 - x^3 + 7$

29. $-2x^4 + 16 + 2x^4 + 9 - 3x^5$

30. $3x^2 - 2x + 3 - 5x^2 - 1 - x$

31. $-x + \frac{1}{2} + 14x^4 - 7x^2 - 1 - 4x^4$

Evaluate each polynomial for $x = -1$. [4.2]

32. $9x - 6$

33. $x^2 - 3x + 6$

Add or subtract. [4.3]

34. $(8x^4 - x^3 + x - 4) + (x^5 + 7x^3 - 3x - 5)$

35. $(3x^4 - 5x^3 + 3x^2) + (4x^5 + 4x^3) + (-5x^5 - 5x^2)$

36. $(5x^2 - 4x + 1) - (3x^2 + 7)$

37. $(3x^5 - 4x^4 + 2x^2 + 3) -$
$(2x^5 - 4x^4 + 3x^3 + 4x^2 - 5)$

38. $\begin{array}{l} -\frac{3}{4}x^4 + \frac{1}{2}x^3 \qquad\qquad\qquad + \frac{7}{8} \\ \qquad\quad\; -\frac{1}{4}x^3 - \;\; x^2 - \frac{7}{4}x \\ \underline{+\frac{3}{2}x^4 \qquad\qquad + \frac{2}{3}x^2 \qquad\qquad - \frac{1}{2}} \end{array}$

39. $\begin{array}{l} 2x^5 \qquad\quad -\; x^3 \qquad\quad + x + 3 \\ \underline{-(3x^5 - x^4 + 4x^3 + 2x^2 - x + 3)} \end{array}$

40. The length of a rectangle is 3 m greater than its width.

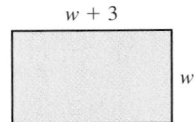

$w + 3$

w

a) Find a polynomial for the perimeter. [4.3]
b) Find a polynomial for the area. [4.4]

Multiply.

41. $3x(-4x^2)$ [4.4]

42. $(7x + 1)^2$ [4.5]

43. $(a - 7)(a + 4)$ [4.5]

44. $(m + 5)(m - 5)$ [4.5]

45. $(4x^2 - 5x + 1)(3x - 2)$ [4.4]

46. $(x - 8)^2$ [4.5]

47. $3t^2(5t^3 - 2t^2 + 4t)$ [4.4]

48. $(a - 9)(a + 9)$ [4.5]

49. $(x - 0.8)(x - 0.5)$ [4.5]

50. $(x^4 - 2x + 3)(x^3 + x - 1)$ [4.4]

51. $(3x - 5)^2$ [4.5]

52. $(2t^2 + 3)(t^2 - 7)$ [4.5]

53. $\left(a - \frac{1}{2}\right)\left(a + \frac{2}{3}\right)$ [4.5]

54. $(-7 + 3x)(7 + 3x)$ [4.5]

55. Evaluate $2 - 5xy + y^2 - 4xy^3 + x^6$ for $x = -1$ and $y = 2$. [4.6]

Identify the coefficient and the degree of each term of each polynomial. Then find the degree of each polynomial. [4.6]

56. $x^5y - 7xy + 9x^2 - 8$

57. $x^2y^5z^9 - y^{40} + x^{13}z^{10}$

Combine like terms. [4.6]

58. $y + w - 2y + 8w - 5$

59. $6m^3 + 3m^2n + 4mn^2 + m^2n - 5mn^2$

Add or subtract. [4.6]

60. $(5x^2 - 7xy + y^2) + (-6x^2 - 3xy - y^2)$

61. $(6x^3y^2 - 4x^2y - 6x) - (-5x^3y^2 + 4x^2y + 6x^2 - 6)$

Multiply. [4.6]

62. $(2x + 5y)(x - 3y)$

63. $\left(3a^4 - \frac{1}{3}b^3\right)^2$

64. Find a polynomial for the shaded area. [4.6]

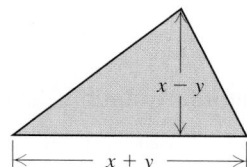

Divide. [4.7]

65. $(10x^3 - x^2 + 8x) \div (2x)$

66. $(6x^3 - 5x^2 - 13x + 13) \div (2x + 3)$

67. $\dfrac{t^4 + t^3 + 2t^2 - t - 3}{t + 1}$

68. Express using a positive exponent: m^{-4}. [4.8]

69. Express using a negative exponent: $\dfrac{1}{a^9}$. [4.8]

Simplify. [4.8]

70. $7^2 \cdot 7^{-4}$

71. $\dfrac{6a^{-5}b}{3a^8b^8}$

72. $(x^3)^{-4}$

73. $(2x^{-3}y)^{-2}$

74. $\left(\dfrac{2x}{y}\right)^{-3}$

75. Convert to decimal notation: 8.3×10^6. [4.8]

76. Convert to scientific notation: 0.0000328. [4.8]

Multiply or divide and write scientific notation for the result. [4.8]

77. $(3.8 \times 10^4)(5.5 \times 10^{-1})$

78. $\dfrac{1.28 \times 10^{-8}}{2.5 \times 10^{-4}}$

79. *Blood donors.* Every 4–6 weeks, one of the authors of this text donates 1.14×10^6 cubic millimeters (two pints) of whole blood, from which platelets are removed and the blood returned to the body. In one cubic millimeter of blood, there are about 2×10^5 platelets. Approximate the number of platelets in a typical donation by this author. [4.8]

SYNTHESIS

80. Explain why $5x^3$ and $(5x)^3$ are not equivalent expressions. [4.1]

81. A binomial is squared and the result, written in descending form, is $x^2 - 6x + 9$. Is it possible to determine what binomial was squared? Why or why not? [4.5]

82. How many terms are there in each of the following? [4.2], [4.5]
   a) $(x - a)(x - b) + (x - a)(x - b)$
   b) $(x + a)(x - b) + (x - a)(x + b)$

83. Simplify:
$$(-3x^5 \cdot 3x^3 - x^6(2x)^2 + (3x^4)^2 + (2x^2)^4 - 20x^2(x^3)^2)^2. \ [4.1], [4.2]$$

84. A polynomial has degree 4. The x^2-term is missing. The coefficient of x^4 is 2 times the coefficient of x^3. The coefficient of x is 3 less than the coefficient of x^4. The remaining coefficient is 7 less than the coefficient of x. The sum of the coefficients is 15. Find the polynomial. [4.2]

Aha! **85.** Multiply: $[(x - 5) - 4x^3][(x - 5) + 4x^3]$. [4.5]

86. Solve: $(x - 7)(x + 10) = (x - 4)(x - 6)$. [2.2], [4.5]

4 Chapter Test

Simplify.

1. $a^4 \cdot a \cdot a^5$

2. $\dfrac{4^5}{4^2}$

3. $\dfrac{(2x)^5}{(2x)^5}$

4. $(x^3)^2$

5. $(-3y^2)^3$

6. $\left(\dfrac{5}{4a^3}\right)^2$

7. $(3x^2)(-2x^5)^3$

8. $(a^3b^2)(ab)^3$

9. Classify $4x^2y - 7y^3$ as a monomial, a binomial, a trinomial, or none of these.

10. Identify the coefficient of each term of the polynomial:
$\frac{1}{3}x^5 - x + 7$.

11. Determine the degree of each term, the leading term and the leading coefficient, and the degree of the polynomial:
$$2t^3 - t + 7t^5 + 4.$$

12. Evaluate $x^2 + 5x - 1$ for $x = -2$.

Combine like terms and write in descending order.

13. $4a^2 - 6 + a^2$

14. $y^2 - 3y - y + \frac{3}{4}y^2$

15. $3 - x^2 + 2x^3 + 5x^2 - 6x - 2x + x^5$

Add or subtract.

16. $(3x^5 + 5x^3 - 5x^2 - 3) + (x^5 + x^4 - 3x^2 + 2x - 4)$

17. $\left(x^4 + \frac{2}{3}x + 5\right) + \left(4x^4 + 5x^2 + \frac{1}{3}x\right)$

18. $(2x^4 + x^3 - 8x^2 - 6x - 3) - (6x^4 - 8x^2 + 2x)$

19. $(x^3 - 0.4x^2 - 12) - (x^5 - 0.3x^3 + 0.4x^2 - 9)$

Multiply.

20. $-2x^2(3x^2 - 3x - 5)$

21. $\left(x - \frac{1}{3}\right)^2$

22. $(5t - 7)(5t + 7)$

23. $(3b + 5)(b - 3)$

24. $(x^6 - 4)(x^8 + 4)$

25. $(8 - y)(6 + 5y)$

26. $(2x + 1)(3x^2 - 5x - 3)$

27. $(8a + 3)^2$

28. Evaluate $2x^2y - 3y^2$ for $x = -3$ and $y = 2$.

29. Combine like terms:
$$2x^3y - y^3 + xy^3 + 8 - 6x^3y - x^2y^2 + 11.$$

30. Subtract:
$$(8a^2b^2 - ab + b^3) - (-6ab^2 - 7ab - ab^3 + 5b^3).$$

31. Multiply: $(3x^5 - y)(3x^5 + y)$.

Divide.

32. $(12x^4 + 9x^3 - 15x^2) \div (3x^2)$

33. $(6x^3 - 8x^2 - 14x + 13) \div (3x + 2)$

34. Express using a positive exponent: 5^{-3}.

35. Express using a negative exponent: $\dfrac{1}{y^8}$.

Simplify.

36. $t^{-4} \cdot t^{-2}$

37. $\dfrac{x^3y^2}{x^8y^{-3}}$

38. $(2a^3b^{-1})^{-4}$

39. $\left(\dfrac{ab}{c}\right)^{-3}$

40. Convert to scientific notation: $3{,}900{,}000{,}000$.

41. Convert to decimal notation: 5×10^{-8}.

Multiply or divide and write scientific notation for the result.

42. $\dfrac{5.6 \times 10^6}{3.2 \times 10^{-11}}$

43. $(2.4 \times 10^5)(5.4 \times 10^{16})$

SYNTHESIS

44. The height of a box is 1 less than its length, and the length is 2 more than its width. Express the volume in terms of the length.

45. Solve: $x^2 + (x - 7)(x + 4) = 2(x - 6)^2$.

46. A CD-ROM can hold about 600 million pieces of information. How many sound files, each needing 40,000 pieces of information, can a 20-pack of CD-ROMs hold? Write scientific notation for the answer.

5

Polynomials and Factoring

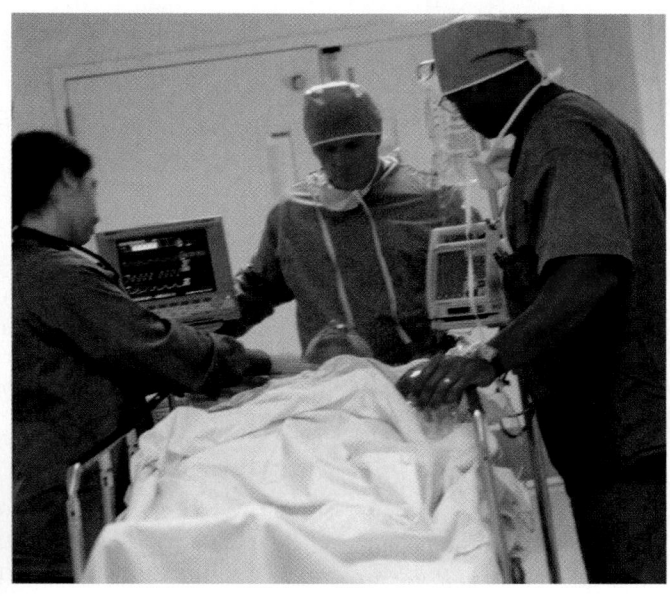

AN APPLICATION

For certain people suffering an extreme allergic reaction, the drug epinephrine (adrenaline) is sometimes prescribed. The number of micrograms N of epinephrine in an adult's bloodstream t minutes after 250 micrograms have been injected can be approximated by $N = -10t^2 + 100t$. How long after an injection will there be about 210 micrograms of epinephrine in the bloodstream?

This problem appears as Example 4 in Section 5.8.

Mary Novinger
EMT
Lima, New York

As an EMT, I use math quite often. In many situations, time is crucial, and quick and accurate calculations of vital signs are essential. I also use math to measure volume and to evaluate burn victims.

*I*n Chapter 1, we learned that factoring *is multiplying reversed. To factor a polynomial is to find an equivalent expression that is a product. In Sections 5.1–5.6, we factor to find equivalent expressions. In Sections 5.7 and 5.8, we use factoring to solve equations, many of which arise from real-world problems. Factoring polynomials requires a solid command of the multiplication methods studied in Chapter 4.*

5.1   Introduction to Factoring

Factoring Monomials   •   Factoring When Terms Have a Common Factor   •   Factoring by Grouping   •   Checking by Evaluating

Just as a number like 15 can be factored as $3 \cdot 5$, a polynomial like $x^2 + 7x$ can be factored as $x(x + 7)$. In both cases, we ask ourselves, "What was multiplied to obtain the given result?" The situation is much like a popular television game show in which an "answer" is given and participants must find the "question" to which the answer corresponds.

> **Factoring**
>
> To *factor* a polynomial is to find an equivalent expression that is a product. An equivalent expression of this type is called a *factorization* of the polynomial.

Factoring Monomials

To factor a monomial, we find two monomials whose product is equivalent to the original monomial. For example, $20x^2$ can be factored as $2 \cdot 10x^2$, $4x \cdot 5x$, or $10x \cdot 2x$, as well as several other ways. To check, we multiply.

EXAMPLE 1   Find three factorizations of $15x^3$.

Solution

a) $15x^3 = (3 \cdot 5)(x \cdot x^2)$ Thinking of how 15 and x^3 can be factored

    $= (3x)(5x^2)$ The factors are $3x$ and $5x^2$; *check*: $3x \cdot 5x^2 = 15x^3$.

b) $15x^3 = (3 \cdot 5)(x^2 \cdot x)$

    $= (3x^2)(5x)$ The factors are $3x^2$ and $5x$; *check*: $3x^2 \cdot 5x = 15x^3$.

c) $15x^3 = ((-5)(-3))x^3$

    $= (-5)(-3x^3)$ The factors are -5 and $-3x^3$; *check*: $(-5)(-3x^3) = 15x^3$.

$(3x)(5x^2)$, $(3x^2)(5x)$, and $(-5)(-3x^3)$ are all factorizations of $15x^3$. Other factorizations exist as well.

———

Recall from Section 1.2 that the word "factor" can be a verb or a noun, depending on the context in which it appears.

Factoring When Terms Have a Common Factor

To multiply a polynomial of two or more terms by a monomial, we use the distributive law: $a(b + c) = ab + ac$. To factor a polynomial with two or more terms of the form $ab + ac$, we use the distributive law with the sides of the equation switched: $ab + ac = a(b + c)$.

Multiply   *Factor*

$3x(x^2 + 2x - 4)$ $3x^3 + 6x^2 - 12x$
$\quad = 3x \cdot x^2 + 3x \cdot 2x - 3x \cdot 4$ $\quad = 3x \cdot x^2 + 3x \cdot 2x - 3x \cdot 4$
$\quad = 3x^3 + 6x^2 - 12x$ $\quad = 3x(x^2 + 2x - 4)$

In the factorization on the right, note that since $3x$ appears as a factor of $3x^3$, $6x^2$, and $-12x$, it is a *common factor* for all the terms of the trinomial $3x^3 + 6x^2 - 12x$.

We have already factored polynomials, although we did not call it factoring at the time. For example, in Section 1.5, we factored when we first learned how to combine like terms (see p. 43).

$$5x + 3x = (5 + 3)x \qquad \text{Here we factored } 5x + 3x.$$
$$= 8x. \qquad\qquad \text{Adding}$$

To *factor* an expression like $10y + 15$, we look to see if both terms have a factor in common. If there *is* a common factor, we can "factor it out" using the distributive law. Factoring each coefficient into its prime factors, we have the following:

The prime factorization of $10y$ is $2 \cdot 5 \cdot y$;
The prime factorization of 15 is $3 \cdot 5$.
$\qquad$ 5 is a common factor.

We generally factor out the *largest* common factor. In this case, that factor is 5 (which is also the *only* common factor here). Thus,

$$10y + 15 = 5 \cdot 2y + 5 \cdot 3 \qquad \text{Try to do this step mentally.}$$
$$= 5(2y + 3). \qquad\qquad \text{Using the distributive law. The student}$$
$$\text{should multiply as a check.}$$

EXAMPLE 2   Factor: $8a - 12$.

Solution

The prime factorization of $8a$ is $2 \cdot 2 \cdot 2 \cdot a$;
The prime factorization of 12 is $2 \cdot 2 \cdot 3$.

Since both factorizations include two factors of 2, the largest common factor is $2 \cdot 2$, or 4:

$$8a - 12 = 4 \cdot 2a - 4 \cdot 3$$

$$8a - 12 = 4(2a - 3). \qquad \text{Try to go directly to this step.}$$

Check: $4(2a - 3) = 4 \cdot 2a - 4 \cdot 3 = 8a - 12$, as expected.

The factorization of $8a - 12$ is $4(2a - 3)$.

> *Caution!* $2 \cdot 2 \cdot 2a - 2 \cdot 2 \cdot 3$ is a factorization of the *terms* of
> $8a - 12$ but not of the polynomial itself. The factorization of $8a - 12$
> is $4(2a - 3)$.

EXAMPLE 3   Factor: $24x^5 + 30x^2$.

Solution

The prime factorization of $24x^5$ is $2 \cdot 2 \cdot 2 \cdot 3 \cdot x \cdot x \cdot x \cdot x \cdot x$.

The prime factorization of $30x^2$ is $2 \cdot 3 \cdot 5 \cdot x \cdot x$.

The largest common factor is $2 \cdot 3 \cdot x \cdot x$, or $6x^2$.

$$24x^5 + 30x^2 = 6x^2 \cdot 4x^3 + 6x^2 \cdot 5 \qquad \text{Factoring each term}$$

$$= 6x^2(4x^3 + 5). \qquad \text{Factoring out } 6x^2$$

Check: $6x^2(4x^3 + 5) = 6x^2 \cdot 4x^3 + 6x^2 \cdot 5 = 24x^5 + 30x^2$, as expected.

The factorization of $24x^5 + 30x^2$ is $6x^2(4x^3 + 5)$.

The largest common factor of a polynomial is the largest common factor of the coefficients times the largest common factor of the variable(s) in all the terms. Suppose in Example 3 that you did not recognize the *largest* common factor, and removed only part of it, as follows:

$$24x^5 + 30x^2 = 2x^2 \cdot 12x^3 + 2x^2 \cdot 15 \qquad 2x^2 \text{ is a common factor.}$$

$$= 2x^2(12x^3 + 15). \qquad 12x^3 + 15 \text{ itself has a common factor.}$$

Note that $12x^3 + 15$ still has a common factor, 3. To find the largest common factor, continue factoring out common factors, as follows, until no more exist:

$$= 2x^2[3(4x^3 + 5)] \qquad \text{Factoring } 12x^3 + 15. \text{ Remember to rewrite the first common factor, } 2x^2.$$

$$= 6x^2(4x^3 + 5). \qquad \text{Using an associative law}$$

Since $4x^3 + 5$ cannot be factored any further, we say that we have factored *completely*. When we are directed simply to factor, it is understood that we should always factor completely.

EXAMPLE 4   Factor: $12x^5 - 15x^4 + 27x^3$.

Solution

The prime factorization of $12x^5$ is $2 \cdot 2 \cdot 3 \cdot x \cdot x \cdot x \cdot x \cdot x$.
The prime factorization of $15x^4$ is $3 \cdot 5 \cdot x \cdot x \cdot x \cdot x$.
The prime factorization of $27x^3$ is $3 \cdot 3 \cdot 3 \cdot x \cdot x \cdot x$.
The largest common factor is $3 \cdot x \cdot x \cdot x$, or $3x^3$.

$$12x^5 - 15x^4 + 27x^3 = 3x^3 \cdot 4x^2 - 3x^3 \cdot 5x + 3x^3 \cdot 9$$
$$= 3x^3(4x^2 - 5x + 9)$$

Since $4x^2 - 5x + 9$ has no common factor, we are done, except for a check:

$$3x^3(4x^2 - 5x + 9) = 3x^3 \cdot 4x^2 - 3x^3 \cdot 5x + 3x^3 \cdot 9$$
$$= 12x^5 - 15x^4 + 27x^3,$$

as expected. The factorization of $12x^5 - 15x^4 + 27x^3$ is $3x^3(4x^2 - 5x + 9)$.

Note in Examples 3 and 4 that the *largest* common variable factor is the *smallest* power of x in the original polynomial.

With practice, we can determine the largest common factor without writing the prime factorization of each term. Then we need only fill in the parentheses. It is customary for the leading coefficient of the polynomial inside the parentheses to be positive.

EXAMPLE 5   Factor: **(a)** $8r^3s^2 + 16rs^3$; **(b)** $-3xy + 6xz - 3x$.

Solution

a) $8r^3s^2 + 16rs^3 = 8rs^2(r^2 + 2s)$ Try to go directly to this step.

The largest common factor is $8rs^2$. $\begin{cases} 8r^3s^2 = 2 \cdot 2 \cdot 2 \cdot r \cdot r^2 \cdot s^2 \\ 16rs^3 = 2 \cdot 2 \cdot 2 \cdot 2 \cdot r \cdot s^2 \cdot s \end{cases}$

Check: $8rs^2(r^2 + 2s) = 8r^3s^2 + 16rs^3$.

b) $-3xy + 6xz - 3x = -3x(y - 2z + 1)$ Note that either $-3x$ or $3x$ can be the largest common factor.

We generally factor out a negative when the first coefficient is negative. The way we factor can depend on the situation in which we are working. We might also factor as follows:

$$-3xy + 6xz - 3x = 3x(-y + 2z - 1).$$

The checks are left to the student.

Student Notes

The 1 in $(y - 2z + 1)$ plays an important role in Example 5(b). Make certain that you understand its significance when using the distributive law.

In some texts, the largest common factor is referred to as the *greatest* common factor. We have avoided this language because, as shown in Example 5, the largest common factor may represent a negative value that is actually *less* than other common factors.

> *Tips for Factoring*
>
> **1.** Factor out the largest common factor, if one exists.
> **2.** The common factor multiplies a polynomial with the same number of terms as the original polynomial.
> **3.** Factoring can always be checked by multiplying. Multiplication should yield the original polynomial.

Factoring by Grouping

Sometimes algebraic expressions contain a common factor with two or more terms.

EXAMPLE 6   Factor: $x^2(x + 1) + 2(x + 1)$.

Solution   The binomial $x + 1$ is a factor of both $x^2(x + 1)$ and $2(x + 1)$. Thus, $x + 1$ is a common factor:

$$x^2(x + 1) + 2(x + 1) = (x + 1)x^2 + (x + 1)2$$

Using a commutative law twice. Try to do this mentally.

$$= (x + 1)(x^2 + 2).$$

Factoring out the common factor, $x + 1$

To check, we could simply reverse the above steps.
The factorization is $(x + 1)(x^2 + 2)$.

In Example 6, the common binomial factor was clearly visible. How do we find such a factor in a polynomial like $5x^3 - x^2 + 15x - 3$? Although there is no factor, other than 1, common to all four terms, $5x^3 - x^2$ and $15x - 3$ can be grouped and factored separately:

$$5x^3 - x^2 = x^2(5x - 1) \quad \text{and} \quad 15x - 3 = 3(5x - 1).$$

Note that $5x^3 - x^2$ and $15x - 3$ share a common factor of $5x - 1$. This means that the original polynomial, $5x^3 - x^2 + 15x - 3$, can be factored:

$$5x^3 - x^2 + 15x - 3 = (5x^3 - x^2) + (15x - 3)$$

Each binomial has a common factor.

$$= x^2(5x - 1) + 3(5x - 1)$$

Factoring each binomial

$$= (5x - 1)(x^2 + 3).$$

Factoring out the common factor, $5x - 1$

Check: $(5x - 1)(x^2 + 3) = 5x \cdot x^2 + 5x \cdot 3 - 1 \cdot x^2 - 1 \cdot 3$
$$= 5x^3 - x^2 + 15x - 3.$$

If a polynomial can be split into groups of terms and the groups share a common factor, then the original polynomial can be factored. This method, known as **factoring by grouping**, can be tried on any polynomial with four or more terms.

EXAMPLE 7 Factor by grouping.

a) $2x^3 + 8x^2 + x + 4$ **b)** $8x^4 + 6x - 28x^3 - 21$

Solution

a) $2x^3 + 8x^2 + x + 4 = (2x^3 + 8x^2) + (x + 4)$

$$= 2x^2(x + 4) + 1(x + 4) \qquad \text{Factoring } 2x^3 + 8x^2 \text{ to find a common binomial factor. Writing the 1 helps with the next step.}$$

Caution! Be sure to include the term 1. The check below shows why it is essential.

$$= (x + 4)(2x^2 + 1) \qquad \text{Factoring out the common factor, } x + 4. \text{ The 1 is essential in the factor } 2x^2 + 1.$$

Check: $(x + 4)(2x^2 + 1) = x \cdot 2x^2 + x \cdot 1 + 4 \cdot 2x^2 + 4 \cdot 1 \qquad \text{Using FOIL}$

$$= 2x^3 + x + 8x^2 + 4$$

$$= 2x^3 + 8x^2 + x + 4. \qquad \text{Using a commutative law}$$

The factorization is $(x + 4)(2x^2 + 1)$.

b) We have a choice of either

$$8x^4 + 6x - 28x^3 - 21 = (8x^4 + 6x) + (-28x^3 - 21)$$

$$= 2x(4x^3 + 3) + 7(-4x^3 - 3)$$

or

$$8x^4 + 6x - 28x^3 - 21 = (8x^4 + 6x) + (-28x^3 - 21)$$

$$= 2x(4x^3 + 3) + (-7)(4x^3 + 3).$$

Because of the common factor $4x^3 + 3$, we choose the latter:

$$8x^4 + 6x - 28x^3 - 21 = 2x(4x^3 + 3) + (-7)(4x^3 + 3)$$

$$= (4x^3 + 3)(2x + (-7)) \qquad \text{Try to do this step mentally.}$$

$$= (4x^3 + 3)(2x - 7). \qquad \text{The common factor } 4x^3 + 3 \text{ was factored out.}$$

Check: $(4x^3 + 3)(2x - 7) = 8x^4 - 28x^3 + 6x - 21$

$$= 8x^4 + 6x - 28x^3 - 21. \qquad \text{This is the original polynomial.}$$

The factorization is $(4x^3 + 3)(2x - 7)$.

Although factoring by grouping can be useful, some polynomials, like $x^3 + x^2 + 2x - 2$, cannot be factored this way. Factoring polynomials of this type is beyond the scope of this text.

Checking by Evaluating

We have seen that one way to check a factorization is to multiply. A second type of check, discussed toward the end of Section 4.4, uses the fact that equivalent expressions have the same value when evaluated for the same replacement. Thus a quick, partial check of Example 7(a) can be made by using a convenient replacement for x (say, 1) and evaluating both $2x^3 + 8x^2 + x + 4$ and $(x + 4)(2x^2 + 1)$:

$$2 \cdot 1^3 + 8 \cdot 1^2 + 1 + 4 = 2 + 8 + 1 + 4$$
$$= 15;$$

$$(1 + 4)(2 \cdot 1^2 + 1) = 5 \cdot 3$$
$$= 15.$$

Since the value of both expressions is the same, the factorization is probably correct.

Keep in mind the possibility that two expressions that are not equivalent may share the same value when evaluated at a certain value. Because of this, unless several values are used (at least one more than the degree of the polynomial, it turns out), evaluating offers only a partial check. Consult with your instructor before making extensive use of this type of check.

 **technology connection**

A partial check of a factorization can be performed using a table or a graph. To check Example 7(a), we let

$$y_1 = 2x^3 + 8x^2 + x + 4 \quad \text{and} \quad y_2 = (x + 4)(2x^2 + 1).$$

Then we set up a table in AUTO mode (see p. 92). If the factorization is correct, the values of y_1 and y_2 will be the same regardless of the table settings used.

ΔTBL = 1

X	Y₁	Y₂
0	4	4
1	15	15
2	54	54
3	133	133
4	264	264
5	459	459
6	730	730

X = 0

We can also graph $y_1 = 2x^3 + 8x^2 + x + 4$ and $y_2 = (x + 4)(2x^2 + 1)$. If the graphs appear to coincide, the factorization is probably correct. The TRACE feature can be used to confirm this.

$$y_1 = 2x^3 + 8x^2 + x + 4,$$
$$y_2 = (x + 4)(2x^2 + 1)$$

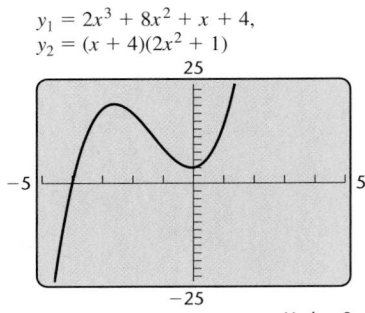

Yscl = 2

Use a table or a graph to determine whether each factorization is correct.

1. $x^2 - 7x - 8 = (x - 8)(x + 1)$
2. $4x^2 - 5x - 6 = (4x + 3)(x - 2)$
3. $5x^2 + 17x - 12 = (5x + 3)(x - 4)$
4. $10x^2 + 37x + 7 = (5x - 1)(2x + 7)$
5. $12x^2 - 17x - 5 = (6x + 1)(2x - 5)$
6. $12x^2 - 17x - 5 = (4x + 1)(3x - 5)$
7. $x^2 - 4 = (x - 2)(x - 2)$
8. $x^2 - 4 = (x + 2)(x - 2)$

Exercise Set

5.1

FOR EXTRA HELP

 Student's Solutions Manual

 Digital Video Tutor CD 3 Videotape 5

AW Math Tutor Center

MathXL Tutorials on CD

Math XL MathXL

MyMathLab MyMathLab

☙ *Concept Reinforcement* *In each of Exercises 1–8, match the phrase with the most appropriate choice from the column on the right.*

1. ____ A factorization of $35a^2b$
2. ____ A factor of $35a^2b$
3. ____ A common factor of $5x + 10$ and $4x + 8$
4. ____ A factorization of $3x^4 - 9x^2$
5. ____ A factorization of $9x^4 - 3x^2$
6. ____ A common factor of $2x + 10$ and $4x + 8$
7. ____ A factor of $3a + 6a^2$
8. ____ A factorization of $3a + 6a^2$

a) $3a(1 + 2a)$
b) $x + 2$
c) $3x^2(3x^2 - 1)$
d) $1 + 2a$
e) $3x^2(x^2 - 3)$
f) $5a^2$
g) 2
h) $7a \cdot 5ab$

Find three factorizations for each monomial. Answers may vary.

9. $10x^3$
10. $6x^3$
11. $-15a^4$
12. $-8t^5$
13. $26x^5$
14. $25x^4$

Factor. Remember to use the largest common factor and to check by multiplying.

15. $7x - 14$
16. $5x - 15$
17. $3t^2 + t$
18. $2t^2 + t$
19. $-4a^2 - 8a$
20. $-6a^2 - 9a$
21. $x^3 + 6x^2$
22. $4x^4 + x^2$
23. $8x^4 - 24x^2$
24. $5x^5 + 10x^3$
25. $2x^2 + 2x - 8$
26. $6x^2 + 3x - 15$
27. $-7a^6 + 10a^4 - 14a^2$
28. $-10t^5 + 15t^4 + 9t^3$
29. $6x^8 + 12x^6 - 24x^4 + 30x^2$
30. $10x^4 - 30x^3 - 50x - 20$
31. $x^5y^5 + x^4y^3 + x^3y^3 - x^2y^2$
32. $x^9y^6 - x^7y^5 + x^4y^4 + x^3y^3$
33. $-5a^3b^4 + 10a^2b^3 - 15a^3b^2$
34. $-21r^5t^4 - 14r^4t^6 + 21r^3t^6$

Factor.

35. $y(y - 2) + 7(y - 2)$
36. $b(b + 5) + 3(b + 5)$
37. $x^2(x + 3) - 7(x + 3)$
38. $3z^2(2z + 9) + (2z + 9)$
39. $y^2(y + 8) + (y + 8)$
40. $x^2(x - 7) - 3(x - 7)$

Factor by grouping, if possible, and check.

41. $x^3 + 3x^2 + 4x + 12$
42. $6z^3 + 3z^2 + 4z + 2$
43. $5a^3 + 15a^2 + 2a + 6$
44. $3a^3 + 2a^2 + 6a + 4$
45. $9x^3 - 12x^2 + 3x - 4$
46. $10x^3 - 25x^2 + 2x - 5$
47. $4t^3 - 20t^2 + 3t - 15$
48. $6a^3 - 8a^2 + 9a - 12$
49. $7x^3 + 2x^2 - 14x - 4$
50. $5x^3 + 4x^2 - 10x - 8$
51. $6a^3 - 7a^2 + 6a - 7$
52. $7t^3 - 5t^2 + 7t - 5$
53. $x^3 + 8x^2 - 3x - 24$
54. $x^3 - x^2 - 2x + 5$
55. $2x^3 + 12x^2 - 5x - 30$
56. $w^3 - 7w^2 + 4w - 28$
57. $p^3 + p^2 - 3p + 10$
58. $x^3 + 7x^2 - 2x - 14$
59. $y^3 + 8y^2 - 2y - 16$
60. $3x^3 + 18x^2 - 5x - 25$
61. $2x^3 - 8x^2 - 9x + 36$
62. $20g^3 - 4g^2 - 25g + 5$

63. In answering a factoring problem, Taylor says the largest common factor is $-5x^2$ and Natasha says the largest common factor is $5x^2$. Can they both be correct? Why or why not?

64. Write a two-sentence paragraph in which the word "factor" is used at least once as a noun and once as a verb.

SKILL MAINTENANCE

Simplify. [4.4]

65. $(x + 3)(x + 5)$
66. $(x + 2)(x + 7)$
67. $(a - 7)(a + 3)$
68. $(a + 5)(a - 8)$
69. $(2x + 5)(3x - 4)$
70. $(3t + 2)(4t - 7)$
71. $(3t - 5)^2$
72. $(2t - 9)^2$

📓 **73.** Marlene recognizes that evaluating provides only a partial check of her factoring. Because of this, she often performs a second check with a different replacement value. Is this a good idea? Why or why not?

📓 **74.** Josh says that for Exercises 15–62 there is no need to print answers at the back of the book. Is he correct in saying this? Why or why not?

Factor, if possible.

75. $4x^5 + 6x^2 + 6x^3 + 9$　　**76.** $x^6 + x^2 + x^4 + 1$

77. $x^{12} + x^7 + x^5 + 1$　　**78.** $x^3 + x^2 - 2x + 2$

Aha! **79.** $5x^5 - 5x^4 + x^3 - x^2 + 3x - 3$

Aha! **80.** $ax^2 + 2ax + 3a + x^2 + 2x + 3$

81. Write a trinomial of degree 7 for which $8x^2y^3$ is the largest common factor. Answers may vary.

5.2 Factoring Trinomials of the Type $x^2 + bx + c$

Constant Term Positive　•　Constant Term Negative　•　Prime Polynomials

Study Skills

Leave a Trail

Students sometimes make the mistake of viewing their supporting work as "scrap" work. Most instructors regard your reasoning as more important than your final answer. Try to organize your supporting work so that your instructor (and you as well) can follow your steps.

We now learn how to factor trinomials like

$$x^2 + 5x + 4 \quad \text{or} \quad x^2 + 3x - 10,$$

for which no common factor exists and the leading coefficient is 1. As preparation for the factoring that follows, compare the following multiplications:

$$
\begin{array}{cccc}
\text{F} & \text{O} & \text{I} & \text{L} \\
\downarrow & \downarrow & \downarrow & \downarrow
\end{array}
$$

$$(x + 2)(x + 5) = x^2 + 5x + 2x + 2 \cdot 5$$
$$= x^2 + \quad 7x \quad + 10;$$

$$(x - 2)(x - 5) = x^2 - 5x - 2x + (-2)(-5)$$
$$= x^2 - \quad 7x \quad + 10;$$

$$(x + 3)(x - 7) = x^2 - 7x + 3x + 3(-7)$$
$$= x^2 - \quad 4x \quad - 21;$$

$$(x - 3)(x + 7) = x^2 + 7x - 3x + (-3)7$$
$$= x^2 + \quad 4x \quad - 21.$$

Note that for all four products:

- The product of the two binomials is a trinomial.
- The coefficient of x in the trinomial is the sum of the constant terms in the binomials.
- The constant term in the trinomial is the product of the constant terms in the binomials.

These observations lead to a method for factoring certain trinomials. The first type we consider has a positive constant term, just as in the first two multiplications above.

Constant Term Positive

To factor a polynomial like $x^2 + 7x + 10$, we think of FOIL in reverse. The x^2 resulted from x times x, which suggests that the first term of each binomial factor is x. Next, we look for numbers p and q such that

$$x^2 + 7x + 10 = (x + p)(x + q).$$

To get the middle term and the last term of the trinomial, we need two numbers, p and q, whose product is 10 and whose sum is 7. Those numbers are 2 and 5. Thus the factorization is

$$(x + 2)(x + 5). \quad \textit{Check:} \ (x + 2)(x + 5) = x^2 + 5x + 2x + 10$$
$$= x^2 + 7x + 10$$

EXAMPLE 1   Factor: $x^2 + 5x + 6$.

A GEOMETRIC APPROACH TO EXAMPLE 1

In Section 4.4, we saw that the product of two binomials can be regarded as the sum of the areas of four rectangles (see p. 259). Thus we can regard the factoring of $x^2 + 5x + 6$ as a search for p and q so that the sum of areas A, B, C, and D is $x^2 + 5x + 6$.

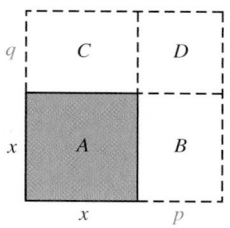

Note that area D is simply the product of p and q. Since p and q are constants, area D is a constant. In order for area D to be 6, p and q must be either 1 and 6 or 2 and 3. We illustrate both below.

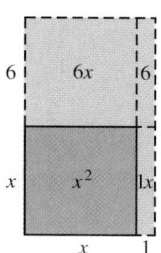 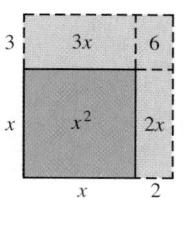

When p and q are 1 and 6, the total area is $x^2 + 7x + 6$, but when p and q are 2 and 3, as shown on the right, the total area is $x^2 + 5x + 6$, as desired. Thus the factorization of $x^2 + 5x + 6$ is $(x + 2)(x + 3)$.

Solution   Think of FOIL in reverse. The first term of each factor is x:

$$(x + \,)(x + \,).$$

To complete the factorization, we need a constant term for each binomial. The constants must have a product of 6 and a sum of 5. We list some pairs of numbers that multiply to 6.

Pairs of Factors of 6	Sums of Factors	
1, 6	7	The numbers
2, 3	5 ←	we seek are 2
−1, −6	−7	and 3.
−2, −3	−5	

Since

$$2 \cdot 3 = 6 \quad \text{and} \quad 2 + 3 = 5,$$

the factorization of $x^2 + 5x + 6$ is $(x + 2)(x + 3)$. To check, we simply multiply the two binomials.

Check: $(x + 2)(x + 3) = x^2 + 3x + 2x + 6$
$$= x^2 + 5x + 6.$$

The product is the original polynomial.

Note that since 5 and 6 are both positive, when factoring $x^2 + 5x + 6$ we need not consider negative factors of 6. Note too that changing the signs of the factors changes only the sign of the sum (see the table above).

At the beginning of this section, we considered the multiplication $(x - 2)(x - 5)$. For this product, the resulting trinomial, $x^2 - 7x + 10$, has a positive constant term but a negative coefficient of x. This is because the *product* of two negative numbers is always positive, whereas the *sum* of two negative numbers is always negative.

To Factor $x^2 + bx + c$ When c Is Positive

When the constant term of a trinomial is positive, look for two numbers with the same sign. The sign is that of the middle term:

$$x^2 - 7x + 10 = (x - 2)(x - 5);$$

$$x^2 + 7x + 10 = (x + 2)(x + 5).$$

EXAMPLE 2

Factor: $y^2 - 8y + 12$.

Solution   Since the constant term is positive and the coefficient of the middle term is negative, we look for a factorization of 12 in which both factors are negative. Their sum must be -8.

Pairs of Factors of 12	Sums of Factors
$-1, -12$	-13
$-2, \ -6$	-8 ←
$-3, \ -4$	-7

We need a sum of -8. The numbers we need are -2 and -6.

The factorization of $y^2 - 8y + 12$ is $(y - 2)(y - 6)$. The check is left to the student.

Constant Term Negative

As we saw in two of the multiplications earlier in this section, the product of two binomials can have a negative constant term:

$$(x + 3)(x - 7) = x^2 - 4x - 21$$

and

$$(x - 3)(x + 7) = x^2 + 4x - 21.$$

It is important to note that when the signs of the constants in the binomials are reversed, only the sign of the middle term of the trinomial changes.

EXAMPLE 3   Factor: $x^2 - 8x - 20$.

Solution   The constant term, -20, must be expressed as the product of a negative number and a positive number. Since the sum of these two numbers must be negative (specifically, -8), the negative number must have the greater absolute value.

Pairs of Factors of -20	Sums of Factors
1, -20	-19
2, -10	-8
4, -5	-1
5, -4	1
10, -2	8
20, -1	19

The numbers we need are 2 and -10.

Because these sums are all positive, for this problem all of the corresponding pairs can be disregarded. Note that in all three pairs, the positive number has the greater absolute value.

The numbers that we are looking for are 2 and -10.

Check: $(x + 2)(x - 10) = x^2 - 10x + 2x - 20$
$$= x^2 - 8x - 20.$$

The factorization of $x^2 - 8x - 20$ is $(x + 2)(x - 10)$.

To Factor $x^2 + bx + c$ When c Is Negative

When the constant term of a trinomial is negative, look for two numbers whose product is negative. One must be positive and the other negative:

$$x^2 - 4x - 21 = (x + 3)(x - 7);$$

$$x^2 + 4x - 21 = (x - 3)(x + 7).$$

Select the two numbers so that the number with the larger absolute value has the same sign as b, the coefficient of the middle term.

EXAMPLE 4   Factor: $t^2 - 24 + 5t$.

Solution   It helps to first write the trinomial in descending order: $t^2 + 5t - 24$. The factorization of the constant term, -24, must have one factor positive and one factor negative. The sum must be 5, so the positive factor must have the larger absolute value. Thus we consider only pairs of factors in which the positive factor has the larger absolute value.

Pairs of Factors of -24	Sums of Factors
$-1, 24$	23
$-2, 12$	10
$-3, \ 8$	5 ←
$-4, \ 6$	2

The numbers we need are -3 and 8.

The factorization is $(t - 3)(t + 8)$. The check is left to the student.

Polynomials in two or more variables, such as $a^2 + 4ab - 21b^2$, are factored in a similar manner.

EXAMPLE 5   Factor: $a^2 + 4ab - 21b^2$.

Solution   It may help to write the trinomial in the equivalent form

$$a^2 + 4ba - 21b^2.$$

We now regard $-21b^2$ as the "constant" term and $4b$ as the "coefficient" of a. Then we try to express $-21b^2$ as a product of two factors whose sum is $4b$. Those factors are $-3b$ and $7b$.

Check: $(a - 3b)(a + 7b) = a^2 + 7ab - 3ba - 21b^2$
$$= a^2 + 4ab - 21b^2.$$

The factorization is $(a - 3b)(a + 7b)$.

Prime Polynomials

EXAMPLE 6   Factor: $x^2 - x + 5$.

Solution   Since 5 has very few factors, we can easily check all possibilities.

Pairs of Factors of 5	Sums of Factors
$5, \ 1$	6
$-5, -1$	-6

Since there are no factors whose sum is -1, the polynomial is *not* factorable into binomials.

In this text, a polynomial like $x^2 - x + 5$ that cannot be factored further is said to be **prime**. In more advanced courses, polynomials like $x^2 - x + 5$ can be factored and are not considered prime.

Often factoring requires two or more steps. Remember, when told to factor, we should *factor completely*. This means that the final factorization should contain only prime polynomials.

EXAMPLE 7

Student Notes _____

Whenever a new set of parentheses is created while factoring, check the expression inside the parentheses to see if it can be factored further.

Factor: $-2x^3 + 20x^2 - 50x$.

Solution *Always* look first for a common factor. Since the leading coefficient is negative, we begin by factoring out $-2x$:

$$-2x^3 + 20x^2 - 50x = -2x(x^2 - 10x + 25).$$

Now consider $x^2 - 10x + 25$. Since the constant term is positive and the coefficient of the middle term is negative, we look for a factorization of 25 in which both factors are negative. Their sum must be -10.

Pairs of Factors of 25	Sums of Factors
$-25, -1$	-26
$-5, -5$	-10 ←

The numbers we need are -5 and -5.

The factorization of $x^2 - 10x + 25$ is $(x - 5)(x - 5)$, or $(x - 5)^2$.

Caution! When factoring involves more than one step, be careful to write out the *entire* factorization.

Check: $-2x(x - 5)(x - 5) = -2x[x^2 - 10x + 25]$ Multiplying binomials
$$= -2x^3 + 20x^2 - 50x.$$ Using the distributive law

The factorization of $-2x^3 + 20x^2 - 50x$ is $-2x(x - 5)(x - 5)$, or $-2x(x - 5)^2$.

Once any common factors have been factored out, the following summary can be used to factor $x^2 + bx + c$.

To Factor $x^2 + bx + c$

1. Find a pair of factors that have c as their product and b as their sum.

   a) If c is positive, its factors will have the same sign as b.
   b) If c is negative, one factor will be positive and the other will be negative. Select the factors such that the factor with the larger absolute value has the same sign as b.

2. Check by multiplying.

Exercise Set

5.2

☙ *Concept Reinforcement   For Exercises 1–6, assume that a polynomial of the form $x^2 + bx + c$ can be factored as $(x + p)(x + q)$. Complete each sentence by replacing each blank with either "positive" or "negative."*

1. If b is positive and c is positive, then p will be _____ and q will be _____.

2. If b is negative and c is positive, then p will be _____ and q will be _____.

3. If p is negative and q is negative, then b must be _____ and c must be _____.

4. If p is positive and q is positive, then b must be _____ and c must be _____.

5. If b, c, and p are all negative, then q must be _____.

6. If b and c are negative and p is positive, then q must be _____.

Factor completely. Remember that you can check by multiplying. If a polynomial is prime, state this.

7. $x^2 + 7x + 6$

8. $x^2 + 6x + 5$

9. $x^2 + 7x + 12$

10. $x^2 + 7x + 10$

11. $x^2 - 6x + 9$

12. $y^2 + 11y + 28$

13. $x^2 + 9x + 14$

14. $a^2 - 11a + 30$

15. $b^2 - 5b + 4$

16. $x^2 + 5x + 4$

17. $a^2 + 4a - 12$

18. $z^2 - 8z + 7$

19. $d^2 - 7d + 10$

20. $x^2 - 8x + 15$

21. $x^2 - 2x - 15$

22. $y^2 - 11y + 10$

23. $x^2 + 2x - 15$

24. $x^2 + x - 42$

25. $3y^2 - 9y - 84$

26. $2x^2 - 14x - 36$

27. $-x^3 + x^2 + 42x$

28. $-x^3 + 6x^2 + 16x$

29. $7x - 60 + x^2$

30. $4y - 45 + y^2$

31. $x^2 - 72 + 6x$

32. $-2x - 99 + x^2$

33. $-5b^2 - 25b + 120$

34. $-c^4 - c^3 + 56c^2$

35. $x^5 - x^4 - 2x^3$

36. $2a^2 - 4a - 70$

37. $x^2 + 2x + 3$

38. $x^2 + x + 1$

39. $50 + 15t + t^2$

40. $27 + 12y + y^2$

41. $x^2 + 20x + 99$

42. $x^2 + 20x + 100$

43. $2x^3 - 40x^2 + 192x$

44. $3x^3 - 63x^2 - 300x$

45. $-4x^2 - 40x - 100$

46. $-2x^2 + 42x + 144$

47. $y^2 - 21y + 108$

48. $x^2 - 25x + 144$

49. $-a^6 - 9a^5 + 90a^4$

50. $-a^4 - a^3 + 132a^2$

51. $t^2 + \frac{2}{3}t + \frac{1}{9}$

52. $x^2 - \frac{2}{5}x + \frac{1}{25}$

53. $11 + w^2 - 4w$

54. $6 - p^2 - 2p$

55. $p^2 + 3pq - 10q^2$

56. $a^2 - 2ab - 3b^2$

57. $m^2 + 5mn + 5n^2$

58. $x^2 - 11xy + 24y^2$

59. $s^2 - 2st - 15t^2$

60. $b^2 + 8bc - 20c^2$

61. $6a^{10} - 30a^9 - 84a^8$

62. $7x^9 - 28x^8 - 35x^7$

63. Without multiplying $(x - 17)(x - 18)$, explain why it cannot possibly be a factorization of $x^2 + 35x + 306$.

64. Marge factors $x^3 - 8x^2 + 15x$ as $(x^2 - 5x)(x - 3)$. Is she wrong? Why or why not? What advice would you offer?

SKILL MAINTENANCE

Solve. [2.2]

65. $3x - 8 = 0$

66. $2x + 7 = 0$

Multiply. [4.4]

67. $(x + 6)(3x + 4)$

68. $(7w + 6)^2$

69. In a recent year, 29,090 people were arrested for counterfeiting. This figure was down 1.2% from the year before. How many people were arrested the year before? [2.4]

70. The first angle of a triangle is four times as large as the second. The measure of the third angle is 30° greater than that of the second. How large are the angles? [2.5]

SYNTHESIS

71. When searching for a factorization, why do we list pairs of numbers with the correct *product* instead of pairs of numbers with the correct *sum*?

72. What is the advantage of writing out the prime factorization of c when factoring $x^2 + bx + c$ with a large value of c?

73. Find all integers b for which $a^2 + ba - 50$ can be factored.

74. Find all integers m for which $y^2 + my + 50$ can be factored.

Factor completely.

75. $y^2 - 0.2y - 0.08$

76. $x^2 + \frac{1}{2}x - \frac{3}{16}$

77. $-\frac{1}{3}a^3 + \frac{1}{3}a^2 + 2a$

78. $-a^7 + \frac{25}{7}a^5 + \frac{30}{7}a^6$

79. $x^{2m} + 11x^m + 28$

80. $-t^{2n} + 7t^n - 10$

Aha! **81.** $(a + 1)x^2 + (a + 1)3x + (a + 1)2$

82. $ax^2 - 5x^2 + 8ax - 40x - (a - 5)9$
(*Hint:* See Exercise 81.)

83. Find the volume of a cube if its surface area is $6x^2 + 36x + 54$ square meters.

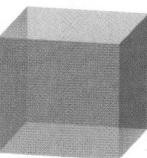

Find a polynomial in factored form for the shaded area in each figure. (Use π in your answers.)

84.

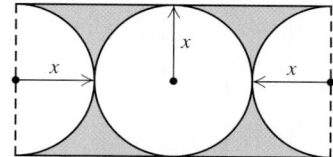

85.

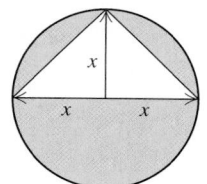

86.

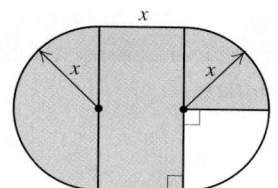

87.

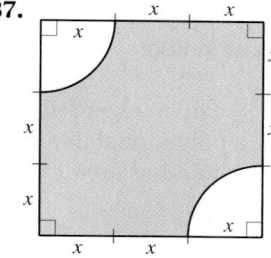

88.

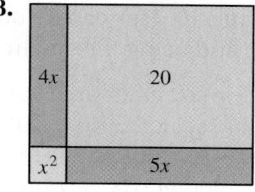

89.

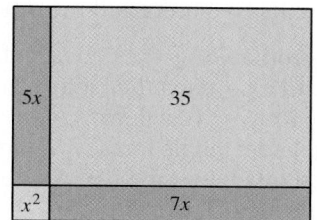

90. A census taker asks a woman, "How many children do you have?"

"Three," she answers.

"What are their ages?"

She responds, "The product of their ages is 36. The sum of their ages is the house number next door."

The math-savvy census taker walks next door, reads the house number, appears puzzled, and returns to the woman, asking, "Is there something you forgot to tell me?"

"Oh yes," says the woman. "I'm sorry. The oldest child is at the park."

The census taker records the three ages, thanks the woman for her time, and leaves.

How old is each child? Explain how you reached this conclusion. (*Hint:* Consider factorizations.)

Source: Adapted from Anita Harnadek, *Classroom Quickies*. Pacific Grove, CA: Critical Thinking Press and Software

CORNER

Visualizing Factoring

COLLABORATIVE

Focus: Visualizing factoring

Time: 20–30 minutes

Group size: 3

Materials: Graph paper and scissors

The product $(x + 2)(x + 3)$ can be regarded as the area of a rectangle with width $x + 2$ and length $x + 3$. Similarly, factoring a polynomial like $x^2 + 5x + 6$ can be thought of as determining the length and the width of a rectangle that has area $x^2 + 5x + 6$. This is the approach used below.

ACTIVITY

1. **a)** To factor $x^2 + 11x + 10$ geometrically, the group needs to cut out shapes like those below to represent x^2, $11x$, and 10. This can be done by either tracing the figures below or by selecting a value for x, say 4, and using the squares on the graph paper to cut out the following:

   x^2:   Using the value selected for x, cut out a square that is x units on each side.

   $11x$:   Using the value selected for x, cut out a rectangle that is 1 unit wide and x units long. Repeat this to form 11 such strips.

   10:   Cut out two rectangles with whole-number dimensions and an area of 10. One should be 2 units by 5 units and the other 1 unit by 10 units.

   b) The group, working together, should then attempt to use one of the two rectangles with area 10, along with all of the other shapes, to piece together one large rectangle. Only one of the rectangles with area 10 will work.

   c) From the large rectangle formed in part (b), use the length and the width to determine the factorization of $x^2 + 11x + 10$. Where do the dimensions of the rectangle representing 10 appear in the factorization?

2. Repeat step (1) above, but this time use the other rectangle with area 10, and use only 7 of the 11 strips, along with the x^2-shape. Piece together the shapes to form one large rectangle. What factorization do the dimensions of this rectangle suggest?

3. Cut out rectangles with area 12 and use the above approach to factor $x^2 + 8x + 12$. What dimensions should be used for the rectangle with area 12?

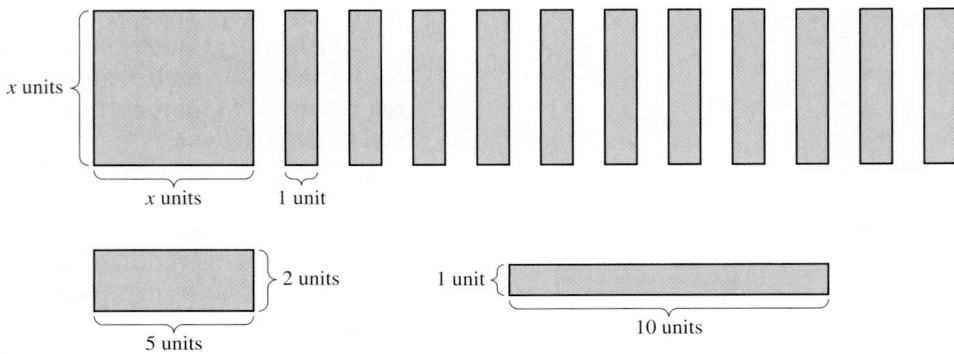

5.3   Factoring Trinomials of the Type $ax^2 + bx + c$

Factoring with FOIL   •   The Grouping Method

In Section 5.2, we learned a FOIL-based method for factoring trinomials of the type $x^2 + bx + c$. Now we learn to factor trinomials in which the leading, or x^2, coefficient is not 1. First we will use another FOIL-based method and then we will use an alternative method that involves factoring by grouping. Use the method that you prefer or the one recommended by your instructor.

Factoring with FOIL

Before factoring trinomials of the type $ax^2 + bx + c$, consider the following:

$$\overset{\text{F}\qquad\text{O}\qquad\text{I}\qquad\text{L}}{(2x + 5)(3x + 4) = 6x^2 + 8x + 15x + 20}$$
$$= 6x^2 + \quad 23x \quad + 20.$$

To factor $6x^2 + 23x + 20$, we reverse the multiplication above and look for two binomials whose product is this trinomial. The product of the First terms must be $6x^2$. The product of the Outer terms plus the product of the Inner terms must be $23x$. The product of the Last terms must be 20. This leads us to

$$(2x + 5)(3x + 4).$$ This factorization is verified above.

How can such a factorization be found without first seeing the corresponding multiplication? Our first approach relies on trial and error and FOIL.

Study Skills

Math Web Sites

If for some reason you feel uneasy about asking an instructor or tutor for help, and have already considered the Addison-Wesley Math Tutor Center (see the Preface and p. 8), you may benefit from visiting the Web sites www.hotmath.com, www.mathforum.org/dr.math, www.sosmath.com, and www.webmath.com. These sites are not specifically for adult learners, but can be very helpful nonetheless.

Source: *The Burlington Free Press*, 4/30/02, p. 8 B

To Factor $ax^2 + bx + c$ Using FOIL

1. Factor out the largest common factor, if one exists.
2. Find two First terms whose product is ax^2:
$$(\blacksquare x + \quad)(\blacksquare x + \quad) = ax^2 + bx + c.$$
   FOIL
3. Find two Last terms whose product is c:
$$(\quad x + \blacksquare)(\quad x + \blacksquare) = ax^2 + bx + c.$$
   FOIL
4. Repeat steps (2) and (3) until a combination is found for which the sum of the Outer and Inner products is bx:
$$(\blacksquare x + \blacksquare)(\blacksquare x + \blacksquare) = ax^2 + bx + c.$$
   I
   O
   FOIL

Always check by multiplying. If no correct combination exists, state that the polynomial is prime.

EXAMPLE 1   Factor: $3x^2 - 10x - 8$.

Solution

1. First, check for a common factor. In this case, there is none (other than 1 or −1).

2. Find two **First** terms whose product is $3x^2$.
   The only possibilities for the **First** terms are $3x$ and x:

   $(3x + \quad)(x + \quad)$.

3. Find two **Last** terms whose product is -8.
   Possible factorizations of -8 are

   $$(-8) \cdot 1, \quad 8 \cdot (-1), \quad (-2) \cdot 4, \quad \text{and} \quad 2 \cdot (-4).$$

   Important!: Since the First terms are not identical, we must also consider the above factors in reverse order:

   $$1 \cdot (-8), \quad (-1) \cdot 8, \quad 4 \cdot (-2), \quad \text{and} \quad (-4) \cdot 2.$$

4. Knowing that all **First** and **Last** products will check, inspect the **O**uter and **I**nner products resulting from steps (2) and (3). Look for the combination in which the sum of the products is the middle term, $-10x$. This may take several tries. Of course, our search normally ends as soon as the correct combination is found. If none exists, we state that the polynomial is prime.

Trial	*Product*	
$(3x - 8)(x + 1)$	$3x^2 + 3x - 8x - 8$	
	$= 3x^2 - 5x - 8$	Wrong middle term
$(3x + 8)(x - 1)$	$3x^2 - 3x + 8x - 8$	
	$= 3x^2 + 5x - 8$	Wrong middle term
$(3x - 2)(x + 4)$	$3x^2 + 12x - 2x - 8$	
	$= 3x^2 + 10x - 8$	Wrong middle term
$(3x + 2)(x - 4)$	$3x^2 - 12x + 2x - 8$	
	$= 3x^2 - 10x - 8$	Correct middle term!
$(3x + 1)(x - 8)$	$3x^2 - 24x + x - 8$	
	$= 3x^2 - 23x - 8$	Wrong middle term
$(3x - 1)(x + 8)$	$3x^2 + 24x - x - 8$	
	$= 3x^2 + 23x - 8$	Wrong middle term
$(3x + 4)(x - 2)$	$3x^2 - 6x + 4x - 8$	
	$= 3x^2 - 2x - 8$	Wrong middle term
$(3x - 4)(x + 2)$	$3x^2 + 6x - 4x - 8$	
	$= 3x^2 + 2x - 8$	Wrong middle term

The correct factorization is $(3x + 2)(x - 4)$. ←

Two observations can be made from Example 1. First, we listed all possible trials even though we usually stop after finding the correct factorization. We did this to show that **each trial differs only in the middle term of the product**. Second, note that as in Section 5.2, **only the sign of the middle term changes when the signs in the binomials are reversed**.

EXAMPLE 2   Factor: $10x^2 + 37x + 7$.

A GEOMETRIC APPROACH TO EXAMPLE 2
The factoring of $10x^2 + 37x + 7$ can be regarded as a search for r and s so that the sum of areas A, B, C, and D is $10x^2 + 37x + 7$.

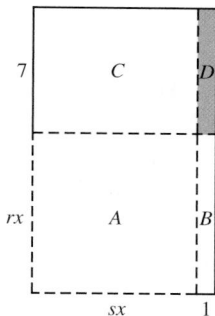

Because A must be $10x^2$, the product rs must be 10. Only when r is 2 and s is 5 will the sum of areas B and C be $37x$ (see below).

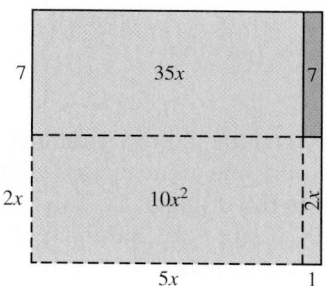

Solution

1. There is no factor (other than 1 or -1) common to all three terms.

2. Because $10x^2$ factors as $10x \cdot x$ or $5x \cdot 2x$, we have two possibilities:

$$(10x + \quad)(x + \quad) \quad \text{or} \quad (5x + \quad)(2x + \quad).$$

3. There are two pairs of factors of 7 and each can be listed two ways:

$$1, 7 \qquad -1, -7$$

and

$$7, 1 \qquad -7, -1.$$

4. From steps (2) and (3), we see that there are 8 possibilities for factorizations. Look for **O**uter and **I**nner products for which the sum is the middle term. Because all coefficients in $10x^2 + 37x + 7$ are positive, we need consider only positive factors of 7.

Trial	*Product*	
$(10x + 1)(x + 7)$	$10x^2 + 70x + 1x + 7$	
	$= 10x^2 + 71x + 7$	Wrong middle term
$(10x + 7)(x + 1)$	$10x^2 + 10x + 7x + 7$	
	$= 10x^2 + 17x + 7$	Wrong middle term
$(5x + 7)(2x + 1)$	$10x^2 + 5x + 14x + 7$	
	$= 10x^2 + 19x + 7$	Wrong middle term
$(5x + 1)(2x + 7)$	$10x^2 + 35x + 2x + 7$	
	$= 10x^2 + 37x + 7$	Correct middle term!

The correct factorization is $(5x + 1)(2x + 7)$.

EXAMPLE 3   Factor: $24x^3 - 76x^2 + 40x$.

Solution

1. First, we factor out the largest common factor, $4x$:

$$4x(6x^2 - 19x + 10).$$

2. Next, we factor $6x^2 - 19x + 10$. Since $6x^2$ can be factored as $3x \cdot 2x$ or $6x \cdot x$, we have two possibilities:

$$(3x + \quad)(2x + \quad) \quad \text{or} \quad (6x + \quad)(x + \quad).$$

3. There are four pairs of factors of 10 and each can be listed two ways:

$$10, 1 \qquad -10, -1 \qquad 5, 2 \qquad -5, -2$$

and

$$1, 10 \qquad -1, -10 \qquad 2, 5 \qquad -2, -5.$$

4. The two possibilities from step (2) and the eight possibilities from step (3) give $2 \cdot 8$, or 16 possibilities for factorizations. We look for **Outer** and **Inner** products resulting from steps (2) and (3) for which the sum is the middle term, $-19x$. Since the sign of the middle term is negative, but the sign of the last term, 10, is positive, the two factors of 10 must both be negative. This means only four pairings from step (3) need be considered. We first try these factors with $(3x + \quad)(2x + \quad)$. If none gives the correct factorization of $6x^2 - 19x + 10$, then we will consider $(6x + \quad)(x + \quad)$.

Trial	*Product*	
$(3x - 10)(2x - 1)$	$6x^2 - 3x - 20x + 10$	
	$= 6x^2 - 23x + 10$	Wrong middle term
$(3x - 1)(2x - 10)$	$6x^2 - 30x - 2x + 10$	
	$= 6x^2 - 32x + 10$	Wrong middle term
$(3x - 5)(2x - 2)$	$6x^2 - 6x - 10x + 10$	
	$= 6x^2 - 16x + 10$	Wrong middle term
$(3x - 2)(2x - 5)$	$6x^2 - 15x - 4x + 10$	
	$= 6x^2 - 19x + 10$	Correct middle term!

Since we have a correct factorization, we need not consider

$$(6x + \quad)(x + \quad).$$

Look again at the possibility $(3x - 5)(2x - 2)$. Without multiplying, we can reject such a possibility. To see why, note that

$$(3x - 5)(2x - 2) = (3x - 5)2(x - 1).$$

The expression $2x - 2$ has a common factor, 2. But we removed the *largest* common factor in step (1). If $2x - 2$ were one of the factors, then 2 would be *another* common factor in addition to the original, $4x$. Thus, $(2x - 2)$ cannot be part of the factorization of $6x^2 - 19x + 10$. Similar reasoning can be used to reject $(3x - 1)(2x - 10)$ as a possible factorization.

Once the largest common factor is factored out, none of the remaining factors can have a common factor.

The factorization of $6x^2 - 19x + 10$ is $(3x - 2)(2x - 5)$, but do not forget the common factor! The factorization of $24x^3 - 76x^2 + 40x$ is

$$4x(3x - 2)(2x - 5).$$

Student Notes

Keep your work organized so that you can see what you have already considered. For example, when factoring $6x^2 - 23x + 10$, we can list all possibilities and cross out those in which a common factor appears:

$(3x - 10)(2x - 1),$

~~$(3x - 1)(2x - 10),$~~

~~$(3x - 5)(2x - 2),$~~

$(3x - 2)(2x - 5),$

~~$(6x - 10)(x - 1),$~~

$(6x - 1)(x - 10),$

$(6x - 5)(x - 2),$

~~$(6x - 2)(x - 5).$~~

By being organized and not erasing, we can see that there are only four possible factorizations.

Tips for Factoring $ax^2 + bx + c$

To factor $ax^2 + bx + c$ $(a > 0)$:

- Always factor out the largest common factor, if one exists.
- Once the largest common factor has been factored out of the original trinomial, no binomial factor can contain a common factor (other than 1 or -1).

(continued)

> - If c is positive, then the signs in both binomial factors must match the sign of b.
> - Reversing the signs in the binomials reverses the sign of the middle term of their product.
> - Organize your work so that you can keep track of which possibilities you have checked.
> - Remember to include the largest common factor—if there is one—in the final factorization.
> - Always check by multiplying.

EXAMPLE 4   Factor: $10x + 8 - 3x^2$.

Solution   An important problem-solving strategy is to find a way to make new problems look like problems we already know how to solve. The factoring tips above apply only to trinomials of the form $ax^2 + bx + c$, with $a > 0$. This leads us to rewrite $10x + 8 - 3x^2$ in descending order:

$$10x + 8 - 3x^2 = -3x^2 + 10x + 8. \qquad \text{Using the commutative law to write descending order}$$

Although $-3x^2 + 10x + 8$ looks similar to the trinomials we have factored, the tips above require a positive leading coefficient. This can be found by factoring out -1:

$$-3x^2 + 10x + 8 = -1(3x^2 - 10x - 8) \qquad \text{Factoring out } -1 \text{ changes the signs of the coefficients.}$$

$$= -1(3x + 2)(x - 4). \qquad \text{Using the result from Example 1}$$

The factorization of $10x + 8 - 3x^2$ is $-1(3x + 2)(x - 4)$.

EXAMPLE 5   Factor: $6p^2 - 13pq - 28q^2$.

Solution   Since no common factor exists, we examine the first term, $6p^2$. There are two possibilities:

$$(2p + \quad)(3p + \quad) \quad \text{or} \quad (6p + \quad)(p + \quad).$$

The last term, $-28q^2$, has the following pairs of factors:

$$28q, -q \qquad 14q, -2q \qquad 7q, -4q,$$

and

$$-28q, \ q \qquad -14q, \ 2q \qquad -7q, \ 4q,$$

as well as each of the pairings reversed.

Some trials, like $(2p + 28q)(3p - q)$ and $(2p + 14q)(3p - 2q)$, cannot be correct because both $(2p + 28q)$ and $(2p + 14q)$ contain a common factor, 2. We try $(2p + 7q)(3p - 4q)$:

$$(2p + 7q)(3p - 4q) = 6p^2 - 8pq + 21pq - 28q^2$$
$$= 6p^2 + 13pq - 28q^2.$$

Our trial is incorrect, but only because of the sign of the middle term. To correctly factor $6p^2 - 13pq - 28q^2$, we simply change the signs in the binomials:

$$(2p - 7q)(3p + 4q) = 6p^2 + 8pq - 21pq - 28q^2$$
$$= 6p^2 - 13pq - 28q^2.$$

The correct factorization is $(2p - 7q)(3p + 4q)$.

The Grouping Method

Another method of factoring trinomials of the type $ax^2 + bx + c$ is known as the *grouping method*. The grouping method relies on rewriting $ax^2 + bx + c$ in the form $ax^2 + px + qx + c$ and then factoring by grouping. To develop this method, consider the following*:

$$(2x + 5)(3x + 4) = 2x \cdot 3x + 2x \cdot 4 + 5 \cdot 3x + 5 \cdot 4 \qquad \text{Using FOIL}$$
$$= 2 \cdot 3 \cdot x^2 + 2 \cdot 4x + 5 \cdot 3x + 5 \cdot 4$$
$$= 2 \cdot 3 \cdot x^2 + (2 \cdot 4 + 5 \cdot 3)x + 5 \cdot 4$$

$$\qquad\qquad \underset{a}{\updownarrow} \qquad\qquad\qquad \underset{b}{\updownarrow} \qquad\qquad \underset{c}{\updownarrow}$$

$$= 6x^2 \qquad + \qquad 23x \qquad + \qquad 20.$$

Note that reversing these steps shows that $6x^2 + 23x + 20$ can be rewritten as $6x^2 + 8x + 15x + 20$ and then factored by grouping. Note that the numbers that add to b (in this case, $2 \cdot 4$ and $5 \cdot 3$), also multiply to ac (in this case, $2 \cdot 3 \cdot 5 \cdot 4$).

To Factor $ax^2 + bx + c$, Using the Grouping Method

1. Factor out the largest common factor, if one exists.
2. Multiply the leading coefficient a and the constant c.
3. Find a pair of factors of ac whose sum is b.
4. Rewrite the middle term, bx, as a sum or difference using the factors found in step (3).
5. Factor by grouping.
6. Include any common factor from step (1) and check by multiplying.

---

*This discussion was inspired by a lecture given by Irene Doo at Austin Community College.

EXAMPLE 6   Factor: $3x^2 - 10x - 8$.

Solution

1. First, we note that there is no common factor (other than 1 or -1).

2. We multiply the leading coefficient, 3, and the constant, -8:

   $$3(-8) = -24.$$

3. We next look for a factorization of -24 in which the sum of the factors is the coefficient of the middle term, -10.

Pairs of Factors of -24	Sums of Factors
1, -24	-23
-1, 24	23
2, -12	-10 ←
-2, 12	10
3, -8	-5
-3, 8	5
4, -6	-2
-4, 6	2

$2 + (-12) = -10$

We normally stop listing pairs of factors once we have found the one we are after.

4. Next, we express the middle term as a sum or difference using the factors found in step (3):

   $$-10x = 2x - 12x.$$

5. We now factor by grouping as follows:

   $3x^2 - 10x - 8 = 3x^2 + 2x - 12x - 8$ Substituting $2x - 12x$ for $-10x$. We could also use $-12x + 2x$.

   $= x(3x + 2) - 4(3x + 2)$ Factoring by grouping; see Section 5.1

   $= (3x + 2)(x - 4).$ Factoring out the common factor, $3x + 2$

6. *Check:* $(3x + 2)(x - 4) = 3x^2 - 12x + 2x - 8 = 3x^2 - 10x - 8$.

   The factorization of $3x^2 - 10x - 8$ is $(3x + 2)(x - 4)$.

EXAMPLE 7   Factor: $8x^3 + 22x^2 - 6x$.

Solution

1. We factor out the largest common factor, $2x$:

   $$8x^3 + 22x^2 - 6x = 2x(4x^2 + 11x - 3).$$

2. To factor $4x^2 + 11x - 3$ by grouping, we multiply the leading coefficient, 4, and the constant term, -3:

   $$4(-3) = -12.$$

3. We next look for factors of -12 that add to 11.

Pairs of Factors of -12	Sums of Factors
1, -12	-11
-1, 12	11
.	.
.	.
.	.

Since $-1 + 12 = 11$, there is no need to list other pairs of factors.

4. We then rewrite the $11x$ in $4x^2 + 11x - 3$ using

$$11x = -1x + 12x, \quad \text{or} \quad 11x = 12x - 1x.$$

5. Next, we factor by grouping:

$$4x^2 + 11x - 3 = 4x^2 - 1x + 12x - 3$$

Rewriting the middle term; $12x - 1x$ could also be used.

$$= x(4x - 1) + 3(4x - 1)$$

Factoring by grouping. Note the common factor, $4x - 1$.

$$= (4x - 1)(x + 3).$$

Factoring out the common factor

6. The factorization of $4x^2 + 11x - 3$ is $(4x - 1)(x + 3)$. But don't forget the common factor, $2x$. The factorization of the original trinomial is

$$2x(4x - 1)(x + 3).$$

Exercise Set

◝ *Concept Reinforcement   In each of Exercises 1–4, match the polynomial with the correct factorization from the column on the right.*

1. ____ $12x^2 + 16x - 3$ **a)** $(7x - 1)(2x + 3)$

2. ____ $14x^2 + 19x - 3$ **b)** $(6x + 1)(2x - 3)$

3. ____ $14x^2 - 19x - 3$ **c)** $(6x - 1)(2x + 3)$

4. ____ $12x^2 - 16x - 3$ **d)** $(7x + 1)(2x - 3)$

Factor completely. If a polynomial is prime, state this.

5. $2x^2 + 7x - 4$

6. $3x^2 + x - 4$

7. $3t^2 + 4t - 15$

8. $5t^2 + t - 18$

9. $6x^2 - 23x + 7$

10. $6x^2 - 13x + 6$

11. $7x^2 + 15x + 2$

12. $6x^2 + 8x + 2$

13. $9a^2 - 6a - 8$

14. $4a^2 - 4a - 15$

15. $6x^2 - 10x - 4$

16. $15x^2 - 19x - 10$

17. $12t^2 - 6t - 6$

18. $18t^2 + 36t - 32$

19. $6 + 19x + 15x^2$

20. $-20 + 31x - 12x^2$

21. $-35x^2 - 34x - 8$

22. $28x^2 + 38x - 6$

23. $4 + 6t^2 - 13t$

24. $9 + 8t^2 - 18t$

25. $25x^2 + 40x + 16$

26. $49t^2 + 42t + 9$

27. $16a^2 + 78a + 27$

28. $24x^2 + 47x - 2$

29. $18t^2 + 24t - 10$

30. $35x^2 - 57x - 44$

31. $-2x^2 + 15 + x$

32. $2t^2 - 19 - 6t$

33. $-6x^2 - 33x - 15$

34. $-12x^2 - 28x + 24$

35. $20x^2 - 25x + 5$

36. $30x^2 - 24x - 54$

37. $12x^2 + 68x - 24$

38. $6x^2 + 21x + 15$

39. $4x + 1 + 3x^2$

40. $-9 + 18x^2 + 21x$

73. $-24a^2 + 34ab - 12b^2$

74. $15a^2 - 5ab - 20b^2$

75. $35x^2 + 34x^3 + 8x^4$

76. $19x^3 - 3x^2 + 14x^4$

77. $18a^7 + 8a^6 + 9a^8$

78. $40a^8 + 16a^7 + 25a^9$

79. Asked to factor $2x^2 - 18x + 36$, Amy *incorrectly* answers
$$2x^2 - 18x + 36 = 2(x^2 + 9x + 18)$$
$$= 2(x + 3)(x + 6).$$
If this were a 10-point quiz question, how many points would you take off? Why?

80. Asked to factor $4x^2 + 28x + 48$, Herb *incorrectly* answers
$$4x^2 + 28x + 48 = (2x + 6)(2x + 8)$$
$$= 2(x + 3)(x + 4).$$
If this were a 10-point quiz question, how many points would you take off? Why?

Factor. Use factoring by grouping even though it would seem reasonable to first combine like terms.

41. $y^2 + 4y - 2y - 8$

42. $x^2 + 5x - 2x - 10$

43. $8t^2 - 6t - 28t + 21$

44. $35t^2 - 40t + 21t - 24$

45. $6x^2 + 4x + 9x + 6$

46. $3x^2 - 2x + 3x - 2$

47. $2t^2 + 6t - t - 3$

48. $5t^2 + 10t - t - 2$

49. $3a^2 - 12a - a + 4$

50. $2a^2 - 10a - a + 5$

Factor completely. If a polynomial is prime, state this.

51. $9t^2 + 14t + 5$

52. $16t^2 + 23t + 7$

53. $-16x^2 - 32x - 7$

54. $-9x^2 - 18x - 5$

55. $10a^2 + 25a - 15$

56. $10x^2 + 30x - 70$

57. $18x^3 + 21x^2 - 9x$

58. $6x^3 - 4x^2 - 10x$

59. $89x + 64 + 25x^2$

60. $47 - 42y + 9y^2$

61. $168x^3 + 45x^2 + 3x$

62. $144x^5 - 168x^4 + 48x^3$

63. $-14t^4 + 19t^3 + 3t^2$

64. $-70a^4 + 68a^3 - 16a^2$

65. $3x + 45x^2 - 18$

66. $2x + 24x^2 - 40$

67. $2a^2 + 5ab + 2b^2$

68. $3p^2 - 16pq - 12q^2$

69. $8s^2 + 18st + 9t^2$

70. $10s^2 + 4st - 6t^2$

71. $18x^2 - 6xy - 24y^2$

72. $-30a^2 - 87ab - 30b^2$

SKILL MAINTENANCE

81. The earth is a sphere (or ball) that is about 40,000 km in circumference. Find the radius of the earth, in kilometers and in miles. Use 3.14 for π. (*Hint*: 1 km ≈ 0.62 mi.) [2.5]

82. The second angle of a triangle is 10° less than twice the first. The third angle is 15° more than four times the first. Find the measure of the second angle. [2.5]

Multiply. [4.5]

83. $(3x + 1)^2$

84. $(5x - 2)^2$

85. $(4t - 5)^2$

86. $(7a + 1)^2$

87. $(5x - 2)(5x + 2)$

88. $(2x - 3)(2x + 3)$

89. $(2t + 7)(2t - 7)$

90. $(4a + 7)(4a - 7)$

91. Which one of the seven tips listed after Example 3 do you find most helpful? Which one do you find least helpful? Explain how you made these determinations.

92. For the trinomial $ax^2 + bx + c$, suppose that a is the product of three different prime factors and c is the product of another two prime factors. How many possible factorizations (like those in Example 1) exist? Explain how you determined your answer.

Factor. If a polynomial is prime, state this.

93. $18x^2y^2 - 3xy - 10$

94. $8x^2y^3 + 10xy^2 + 2y$

95. $9a^2b^3 + 25ab^2 + 16$

96. $-9t^{10} - 12t^5 - 4$

97. $16t^{10} - 8t^5 + 1$

98. $9a^2b^2 - 15ab - 2$

99. $-20x^{2n} - 16x^n - 3$

100. $-15x^{2m} + 26x^m - 8$

101. $a^{2n+1} - 2a^{n+1} + a$

102. $3a^{6n} - 2a^{3n} - 1$

103. $3(a + 1)^{n+1}(a + 3)^2 - 5(a + 1)^n(a + 3)^3$

104. $7(t - 3)^{2n} + 5(t - 3)^n - 2$

5.4 Factoring Perfect-Square Trinomials and Differences of Squares

Recognizing Perfect-Square Trinomials • Factoring Perfect-Square Trinomials • Recognizing Differences of Squares • Factoring Differences of Squares • Factoring Completely

In Section 4.5, we studied some shortcuts for finding the products of certain binomials. Reversing these procedures provides shortcuts for factoring certain polynomials.

Recognizing Perfect-Square Trinomials

Some trinomials are squares of binomials. For example, $x^2 + 10x + 25$ is the square of the binomial $x + 5$. To see this, we can calculate $(x + 5)^2$. It is $x^2 + 2 \cdot x \cdot 5 + 5^2$, or $x^2 + 10x + 25$. A trinomial that is the square of a binomial is called a **perfect-square trinomial**.

In Section 4.5, we considered squaring binomials as a special-product rule:

$$(A + B)^2 = A^2 + 2AB + B^2;$$
$$(A - B)^2 = A^2 - 2AB + B^2.$$

Reading the right sides first, we see that these equations can be used to factor perfect-square trinomials. Note that in order for a trinomial to be the square of a binomial, it must have the following:

1. Two terms, A^2 and B^2, must be squares, such as

$$4, \quad x^2, \quad 81m^2, \quad 16t^2.$$

2. Neither A^2 nor B^2 is being subtracted.

3. The remaining term is either $2 \cdot A \cdot B$ or $-2 \cdot A \cdot B$, where A and B are the square roots of A^2 and B^2.

EXAMPLE 1

Determine whether each of the following is a perfect-square trinomial.

a) $x^2 + 6x + 9$ **b)** $t^2 - 8t - 9$ **c)** $16x^2 + 49 - 56x$

Student Notes

If you're not already quick to recognize the squares that represent $1^2, 2^2, 3^2, ..., 12^2$, this would be a good time to familiarize yourself with these numbers.

Solution

a) To see if $x^2 + 6x + 9$ is a perfect-square trinomial, note that:

1. Two terms, x^2 and 9, are squares.
2. Neither x^2 nor 9 is being subtracted.
3. The remaining term, $6x$, is $2 \cdot x \cdot 3$, where x and 3 are the square roots of x^2 and 9.

Thus, $x^2 + 6x + 9$ *is* a perfect-square trinomial.

b) To see if $t^2 - 8t - 9$ is a perfect-square trinomial, note that:

1. Two terms, t^2 and 9, are squares. But:
2. Since 9 is being subtracted, $t^2 - 8t - 9$ *is not* a perfect-square trinomial.

c) To see if $16x^2 + 49 - 56x$ is a perfect-square trinomial, it helps to first write it in descending order:

$$16x^2 - 56x + 49.$$

Next, note that:

1. Two terms, $16x^2$ and 49, are squares.
2. There is no minus sign before $16x^2$ or 49.
3. Twice the product of the square roots, $2 \cdot 4x \cdot 7$, is $56x$, the opposite of the remaining term, $-56x$.

Thus, $16x^2 + 49 - 56x$ *is* a perfect-square trinomial.

Study Skills

Fill in Your Blanks

Don't hesitate to write out any missing steps that you'd like to see included. For instance, in Example 1(c), we state (in red) that $16x^2$ is a square. To solidify your understanding, you may want to write in $4x \cdot 4x = 16x^2$.

Factoring Perfect-Square Trinomials

Either of the factoring methods from Section 5.3 can be used to factor perfect-square trinomials, but a faster method is to recognize the following patterns.

Factoring a Perfect-Square Trinomial
$$A^2 + 2AB + B^2 = (A + B)^2; \qquad A^2 - 2AB + B^2 = (A - B)^2$$

Each factorization uses the square roots of the squared terms and the sign of the remaining term. To verify these equations, you should compute $(A + B)(A + B)$ and $(A - B)(A - B)$.

EXAMPLE 2

Factor: **(a)** $x^2 + 6x + 9$; **(b)** $t^2 + 49 - 14t$; **(c)** $16x^2 - 40x + 25$.

Solution

a) $x^2 + 6x + 9 = x^2 + 2 \cdot x \cdot 3 + 3^2 = (x + 3)^2$ The sign of the middle term is positive.

$$\underbrace{A^2}_{} + \underbrace{2}_{} \quad \underbrace{A}_{} \quad \underbrace{B}_{} + \underbrace{B^2}_{} = \underbrace{(A + B)^2}_{}$$

b) $t^2 + 49 - 14t = t^2 - 14t + 49$ Using a commutative law to write in descending order

$$= t^2 - 2 \cdot t \cdot 7 + 7^2 = (t - 7)^2$$

$$A^2 - 2 \quad A \quad B + B^2 = (A - B)^2$$

c) $16x^2 - 40x + 25 = (4x)^2 - 2 \cdot 4x \cdot 5 + 5^2 = (4x - 5)^2$ Recall that $(4x)^2 = 16x^2$.

$$A^2 \quad - \; 2 \quad A \quad B + B^2 = (A \; - \; B)^2$$

With practice, it is possible to spot perfect-square trinomials as they occur and factor them quickly.

EXAMPLE 3   Factor: $4p^2 - 12pq + 9q^2$.

Solution   We have

$$4p^2 - 12pq + 9q^2 = (2p)^2 - 2(2p)(3q) + (3q)^2$$ Recognizing the perfect-square trinomial

$$= (2p - 3q)^2.$$ The sign of the middle term is negative.

Check: $(2p - 3q)(2p - 3q) = 4p^2 - 12pq + 9q^2$.

The factorization is $(2p - 3q)^2$.

EXAMPLE 4   Factor: $-75m^3 - 60m^2 - 12m$.

Solution   *Always* look first for a common factor. This time there is one. We factor out $-3m$ so that the leading coefficient of the polynomial inside the parentheses is positive:

$$-75m^3 - 60m^2 - 12m = -3m[25m^2 + 20m + 4]$$ Factoring out the largest common factor

$$= -3m[(5m)^2 + 2(5m)(2) + 2^2]$$ Recognizing the perfect-square trinomial. Try to do this mentally.

$$= -3m(5m + 2)^2.$$

Check: $-3m(5m + 2)^2 = -3m(5m + 2)(5m + 2)$

$$= -3m(25m^2 + 20m + 4)$$

$$= -75m^3 - 60m^2 - 12m.$$

The factorization is $-3m(5m + 2)^2$.

Recognizing Differences of Squares

Some binomials represent the difference of two squares. For example, the binomial $16x^2 - 9$ is a difference of two expressions, $16x^2$ and 9, that are squares. To see this, note that $16x^2 = (4x)^2$ and $9 = 3^2$.

Any expression, like $16x^2 - 9$, that can be written in the form $A^2 - B^2$ is called a **difference of squares**. Note that for a binomial to be a difference of squares, it must have the following.

1. There must be two expressions, both squares, such as

$$25x^2, \quad 9, \quad 4x^2y^2, \quad 1, \quad x^6, \quad 49y^8.$$

2. The terms in the binomial must have different signs.

Note that in order for a term to be a square, its coefficient must be a perfect square and the power(s) of the variable(s) must be even.

EXAMPLE 5   Determine whether each of the following is a difference of squares.

a) $9x^2 - 64$ **b)** $25 - t^3$ **c)** $-4x^{10} + 36$

Solution

a) To see if $9x^2 - 64$ is a difference of squares, note that:
  1. The first expression is a square: $9x^2 = (3x)^2$.
   The second expression is a square: $64 = 8^2$.
  2. The terms have different signs.
  Thus, $9x^2 - 64$ is a difference of squares, $(3x)^2 - 8^2$.

b) To see if $25 - t^3$ is a difference of squares, note that:
  1. The expression t^3 is not a square.
  Thus, $25 - t^3$ is not a difference of squares.

c) To see if $-4x^{10} + 36$ is a difference of squares, note that:
  1. The expressions $4x^{10}$ and 36 are squares: $4x^{10} = (2x^5)^2$ and $36 = 6^2$.
  2. The terms have different signs.
  Thus, $-4x^{10} + 36$ is a difference of squares, $6^2 - (2x^5)^2$. It is often useful to rewrite $-4x^{10} + 36$ in the equivalent form $36 - 4x^{10}$.

Factoring Differences of Squares

To factor a difference of squares, we reverse a pattern from Section 4.5:

> **Factoring a Difference of Squares**
> $$A^2 - B^2 = (A + B)(A - B).$$

Once we have identified the expressions that are playing the roles of A and B, the factorization can be written directly. To verify this equation, simply multiply $(A + B)(A - B)$.

EXAMPLE 6   Factor: **(a)** $x^2 - 4$; **(b)** $m^2 - 9p^2$; **(c)** $9 - 16t^{10}$; **(d)** $50x^2 - 8x^8$.

Solution

a) $x^2 - 4 = x^2 - 2^2 = (x + 2)(x - 2)$

$$A^2 - B^2 = (A + B)(A - B)$$

b) $m^2 - 9p^2 = m^2 - (3p)^2 = (m + 3p)(m - 3p)$

$$A^2 - B^2 = (A + B)(A - B)$$

c) $9 - 16t^{10} = 3^2 - (4t^5)^2$ Using the rules for powers

$$A^2 - B^2$$

$$= (3 + 4t^5)(3 - 4t^5)$$ Try to go directly to this step.

$$(A + B)(A - B)$$

d) *Always* look first for a common factor. This time there is one, $2x^2$:

$50x^2 - 8x^8 = 2x^2(25 - 4x^6)$ Factoring out the common factor

$ = 2x^2[5^2 - (2x^3)^2]$ Recognizing $A^2 - B^2$. Try to do this mentally.

$ = 2x^2(5 + 2x^3)(5 - 2x^3).$ Factoring the difference of squares

Check: $2x^2(5 + 2x^3)(5 - 2x^3) = 2x^2(25 - 4x^6)$

$ = 50x^2 - 8x^8.$

The factorization is $2x^2(5 + 2x^3)(5 - 2x^3)$.

> **Caution!**   Note in Example 6 that a difference of squares is *not* the square of the difference; that is,
>
> $$A^2 - B^2 \neq (A - B)^2.$$ To see this, note that
> $$(A - B)^2 = A^2 - 2AB + B^2.$$

Factoring Completely

Sometimes, as in Examples 4 and 6(d), a *complete* factorization requires two or more steps. Factoring is complete when no factor can be factored further.

EXAMPLE 7   Factor: $p^4 - 16$.

Solution   We have

$$p^4 - 16 = (p^2)^2 - 4^2 \qquad \text{Recognizing } A^2 - B^2$$
$$= (p^2 + 4)(p^2 - 4) \qquad \begin{array}{l} \text{Factoring a difference of} \\ \text{squares. Note that } p^2 - 4 \\ \text{is not prime.} \end{array}$$
$$= (p^2 + 4)(p + 2)(p - 2). \qquad \begin{array}{l} \text{Factoring further: } p^2 - 4 \\ \text{is itself a difference of} \\ \text{squares.} \end{array}$$

Check: $(p^2 + 4)(p + 2)(p - 2) = (p^2 + 4)(p^2 - 4)$
$$= p^4 - 16.$$

The factorization is $(p^2 + 4)(p + 2)(p - 2)$.

Note in Example 7 that the factor $p^2 + 4$ is a *sum* of squares that cannot be factored further.

Caution!   Apart from possibly removing a common factor, you cannot factor a sum of squares. In particular,

$$A^2 + B^2 \neq (A + B)^2.$$

Consider $25x^2 + 100$. Here a sum of squares has a common factor, 25. Factoring, we get $25(x^2 + 4)$, where $x^2 + 4$ is prime.

As you proceed through the exercises, these suggestions may prove helpful.

Tips for Factoring

1. Always look first for a common factor! If there is one, factor it out.
2. Be alert for perfect-square trinomials and for binomials that are differences of squares. Once recognized, they can be factored without trial and error.
3. Always factor completely.
4. Check by multiplying.

Exercise Set

5.4

FOR EXTRA HELP

Student's
Solutions
Manual

Digital Video Tutor
CD 3
Videotape 5

AW Math
Tutor Center

MathXL Tutorials
on CD

Math*XL*
MathXL

MyMathLab
MyMathLab

🐦 *Concept Reinforcement   Identify each of the following as a perfect-square trinomial, a difference of squares, a prime polynomial, or none of these.*

1. $x^2 - 64$

2. $4x^2 + 49$

3. $x^2 - 5x + 4$

4. $t^2 - 100$

5. $a^2 - 8a + 16$

6. $9x^2 + 6x + 1$

7. $-25x^2 - 9$

8. $2t^2 + 10t + 6$

9. $4r^2 + 20r + 25$

10. $16t^2 - 25$

Determine whether each of the following is a perfect-square trinomial.

11. $x^2 + 18x + 81$

12. $x^2 - 16x + 64$

13. $x^2 - 16x - 64$

14. $x^2 - 14x - 49$

15. $x^2 - 3x + 9$

16. $x^2 + 2x + 4$

17. $9x^2 - 36x + 24$

18. $36x^2 - 24x + 16$

Factor completely. Remember to look first for a common factor and to check by multiplying. If a polynomial is prime, state this.

19. $x^2 + 16x + 64$

20. $x^2 + 14x + 49$

21. $x^2 - 14x + 49$

22. $x^2 - 16x + 64$

23. $3x^2 + 6x + 3$

24. $5x^2 - 10x + 5$

25. $4 - 4x + x^2$

26. $4 + x^2 + 4x$

27. $18x^2 + 12x + 2$

28. $25x^2 + 10x + 1$

29. $49 - 56y + 16y^2$

30. $120m + 75 + 48m^2$

31. $-x^5 + 18x^4 - 81x^3$

32. $-2x^2 + 40x - 200$

33. $2x^3 - 4x^2 + 2x$

34. $x^3 + 24x^2 + 144x$

35. $20x^2 + 100x + 125$

36. $12x^2 + 36x + 27$

37. $49 - 42x + 9x^2$

38. $64 - 112x + 49x^2$

39. $16x^2 + 24x + 9$

40. $2a^2 + 28a + 98$

41. $2 + 20x + 50x^2$

42. $9x^2 + 30x + 25$

43. $4p^2 + 12pq + 9q^2$

44. $25m^2 + 20mn + 4n^2$

45. $a^2 - 12ab + 49b^2$

46. $x^2 - 7xy + 9y^2$

47. $-64m^2 - 16mn - n^2$

48. $-81p^2 + 18pq - q^2$

49. $-32s^2 + 80st - 50t^2$

50. $-36a^2 - 96ab - 64b^2$

Determine whether each of the following is a difference of squares.

51. $x^2 + 100$

52. $x^2 - 36$

53. $x^2 - 81$

54. $x^2 + 4$

55. $-26 + 4t^2$

56. $-1 + 49t^2$

Factor completely. Remember to look first for a common factor. If a polynomial is prime, state this.

57. $y^2 - 4$

58. $x^2 - 36$

59. $p^2 - 9$

60. $q^2 + 1$

61. $-49 + t^2$

62. $-64 + m^2$

63. $6a^2 - 54$

64. $x^2 - 8x + 16$

65. $49x^2 - 14x + 1$

66. $3t^2 - 12$

67. $200 - 2t^2$

68. $98 - 8w^2$

69. $-80a^2 + 45$

70. $25x^2 - 4$

71. $5t^2 - 80$

72. $-4t^2 + 64$

73. $8x^2 - 98$

74. $24x^2 - 54$

75. $36x - 49x^3$

76. $16x - 81x^3$

77. $49a^4 - 20$

78. $25a^4 - 9$

79. $t^4 - 1$

80. $x^4 - 16$

81. $-3x^3 + 24x^2 - 48x$

82. $-2a^4 + 36a^3 - 162a^2$

83. $48t^2 - 27$

84. $125t^2 - 45$

85. $a^8 - 2a^7 + a^6$

86. $x^8 - 8x^7 + 16x^6$

87. $7a^2 - 7b^2$

88. $6p^2 - 6q^2$

89. $25x^2 - 4y^2$

90. $16a^2 - 9b^2$

91. $18t^2 - 8s^2$

92. $49x^2 - 16y^2$

93. Explain in your own words how to determine whether a polynomial is a difference of squares.

94. Explain in your own words how to determine whether a polynomial is a perfect-square trinomial.

SKILL MAINTENANCE

95. About 5 L of oxygen can be dissolved in 100 L of water at 0°C. This is 1.6 times the amount that can be dissolved in the same volume of water at 20°C. How much oxygen can be dissolved at 20°C? [2.5]

96. Bonnie is taking a botany course. To get an A, she must average at least 90 after four exams. Bonnie scored 96, 98, and 89 on the first three tests. Determine (in terms of an inequality) what scores on the last test will earn her an A. [2.7]

Simplify. [4.1]

97. $(x^3y^5)(x^9y^7)$

98. $(5a^2b^3)^2$

Graph. [3.2]

99. $y = \frac{3}{2}x - 3$

100. $3x - 5y = 30$

SYNTHESIS

101. Leon concludes that since $x^2 - 9 = (x - 3)(x + 3)$, it must follow that $x^2 + 9 = (x + 3)(x - 3)$. What mistake(s) is he making?

102. Write directions that would enable someone to construct a polynomial that contains a perfect-square trinomial, a difference of squares, and a common factor.

Factor completely. If a polynomial is prime, state this.

103. $x^8 - 2^8$

104. $3x^2 - \frac{1}{3}$

105. $18x^3 - \frac{8}{25}x$

106. $0.49p - p^3$

107. $(y - 5)^4 - z^8$

108. $x^2 - \left(\frac{1}{x}\right)^2$

109. $-x^4 + 8x^2 + 9$

110. $-16x^4 + 96x^2 - 144$

Aha! **111.** $(y + 3)^2 + 2(y + 3) + 1$

112. $49(x + 1)^2 - 42(x + 1) + 9$

113. $27x^3 - 63x^2 - 147x + 343$

114. $a^{2n} - 49b^{2n}$

115. $81 - b^{4k}$

116. $9b^{2n} + 12b^n + 4$

117. Subtract $(x^2 + 1)^2$ from $x^2(x + 1)^2$.

Factor by grouping. Look for a grouping of three terms that is a perfect-square trinomial.

118. $a^2 + 2a + 1 - 9$

119. $y^2 + 6y + 9 - x^2 - 8x - 16$

Find c such that each polynomial is the square of a binomial.

120. $cy^2 + 6y + 1$

121. $cy^2 - 24y + 9$

122. Find the value of a if $x^2 + a^2x + a^2$ factors into $(x + a)^2$.

123. Show that the difference of the squares of two consecutive integers is the sum of the integers. (*Hint*: Use x for the smaller number.)

5.5   Factoring Sums or Differences of Cubes

Factoring a Sum of Two Cubes   •   Factoring a Difference of Two Cubes

It is possible to factor both a sum and a difference of two cubes. To see how, consider the following:

$$(A + B)(A^2 - AB + B^2) = A(A^2 - AB + B^2) + B(A^2 - AB + B^2)$$
$$= A^3 - A^2B + AB^2 + A^2B - AB^2 + B^3$$
$$= A^3 + B^3$$

and

$$(A - B)(A^2 + AB + B^2) = A(A^2 + AB + B^2) - B(A^2 + AB + B^2)$$
$$= A^3 + A^2B + AB^2 - A^2B - AB^2 - B^3$$
$$= A^3 - B^3.$$

These equations show how we can factor a sum or a difference of two cubes.

To Factor a Sum or Difference of Cubes

$$A^3 + B^3 = (A + B)(A^2 - AB + B^2),$$
$$A^3 - B^3 = (A - B)(A^2 + AB + B^2)$$

Remembering this list of cubes may prove helpful when factoring.

N	0.2	0.1	0	1	2	3	4	5	6
N^3	0.008	0.001	0	1	8	27	64	125	216

EXAMPLE 1   Factor: $x^3 - 8$.

Solution   We have

$$x^3 - 8 = x^3 - 2^3 = (x - 2)(x^2 + x \cdot 2 + 2^2).$$
$$A^3 - B^3 = (A - B)(A^2 + A \ B + B^2)$$

This tells us that $x^3 - 8 = (x - 2)(x^2 + 2x + 4)$. Note that we cannot factor $x^2 + 2x + 4$. (It is not a perfect-square trinomial nor can it be factored by trial and error or grouping.) We check by multiplying:

$$(x - 2)(x^2 + 2x + 4) = x(x^2 + 2x + 4) - 2(x^2 + 2x + 4)$$
$$= x^3 + 2x^2 + 4x - 2x^2 - 4x - 8$$
$$= x^3 - 8.$$

The factorization is $(x - 2)(x^2 + 2x + 4)$.

EXAMPLE 2   Factor: $x^3 + 125$.

Solution   We have

$$x^3 + 125 = x^3 + 5^3 = (x + 5)(x^2 - x \cdot 5 + 5^2).$$
$$A^3 + B^3 = (A + B)(A^2 - A \quad B + B^2)$$

Thus, $x^3 + 125 = (x + 5)(x^2 - 5x + 25)$. The check is left to the student.

EXAMPLE 3   Factor: $16a^7b + 54ab^7$.

Solution   We first look for a common factor:

$$16a^7b + 54ab^7 = 2ab[8a^6 + 27b^6]$$
$$= 2ab[(2a^2)^3 + (3b^2)^3] \quad \text{This is of the form } A^3 + B^3,$$
$$\text{where } A = 2a^2 \text{ and } B = 3b^2.$$
$$= 2ab[(2a^2 + 3b^2)(4a^4 - 6a^2b^2 + 9b^4)].$$

We check using the distributive law:

$$2ab(2a^2 + 3b^2)(4a^4 - 6a^2b^2 + 9b^4)$$
$$= 2ab[2a^2(4a^4 - 6a^2b^2 + 9b^4) + 3b^2(4a^4 - 6a^2b^2 + 9b^4)]$$
$$= 2ab[8a^6 - 12a^4b^2 + 18a^2b^4 + 12b^2a^4 - 18a^2b^4 + 27b^6]$$
$$= 2ab[8a^6 + 27b^6]$$
$$= 16a^7b + 54ab^7.$$

We have a check. The factorization is

$$2ab(2a^2 + 3b^2)(4a^4 - 6a^2b^2 + 9b^4).$$

EXAMPLE 4   Factor: $y^3 - 0.001$.

Solution   Since $0.001 = (0.1)^3$, we have a difference of cubes:

$$y^3 - 0.001 = (y - 0.1)(y^2 + 0.1y + 0.01).$$

The check is left to the student.

EXAMPLE 5 Factor: $r^6 - s^6$.

Solution We have

$$r^6 - s^6 = (r^3)^2 - (s^3)^2$$
$$= (r^3 + s^3)(r^3 - s^3).$$ Factoring a difference of two *squares*

Next, we factor the sum and difference of two cubes:

$$r^6 - s^6 = (r + s)(r^2 - rs + s^2)(r - s)(r^2 + rs + s^2).$$

The check is left to the student.

In Example 5, suppose that we first factored $r^6 - s^6$ as a difference of two cubes:

$$(r^2)^3 - (s^2)^3 = (r^2 - s^2)(r^4 + r^2s^2 + s^4)$$
$$= (r + s)(r - s)(r^4 + r^2s^2 + s^4).$$

In this case, we might have missed some factors; $r^4 + r^2s^2 + s^4$ can be factored as $(r^2 - rs + s^2)(r^2 + rs + s^2)$, but we probably would never have suspected that such a factorization exists.

Remember the following about factoring sums or differences of squares and cubes:

Difference of cubes: $A^3 - B^3 = (A - B)(A^2 + AB + B^2),$

Sum of cubes: $A^3 + B^3 = (A + B)(A^2 - AB + B^2),$

Difference of squares: $A^2 - B^2 = (A + B)(A - B).$

Exercise Set

5.5

Factor completely.

1. $t^3 + 8$

2. $p^3 + 27$

3. $a^3 - 64$

4. $w^3 - 1$

5. $z^3 + 125$

6. $x^3 + 1$

7. $8a^3 - 1$

8. $27x^3 - 1$

9. $y^3 - 27$

10. $p^3 - 8$

11. $64 + 125x^3$

12. $8 + 27b^3$

13. $125p^3 - 1$

14. $64w^3 - 1$

15. $27m^3 + 64$

16. $8t^3 + 27$

17. $p^3 - q^3$

18. $a^3 + b^3$

19. $x^3 + \frac{1}{8}$

20. $y^3 + \frac{1}{27}$

21. $2y^3 - 128$

22. $3z^3 - 3$

23. $24a^3 + 3$

24. $54x^3 + 2$

25. $rs^3 - 64r$

26. $ab^3 + 125a$

27. $5x^3 + 40z^3$

28. $2y^3 - 54z^3$

29. $x^3 + 0.001$

30. $y^3 + 0.125$

31. $3z^5 - 3z^2$

32. $2y^4 - 128y$

33. $t^6 + 1$

34. $z^6 - 1$

35. $p^6 - q^6$

36. $x^6 - 64y^6$

37. Dino incorrectly believes that
$$a^3 - b^3 = (a - b)(a^2 + b^2).$$
How could you convince him that he is wrong?

38. Is the following statement true or false and why? If A^3 and B^3 have a common factor, then A and B have a common factor.

SKILL MAINTENANCE

Multiply.

39. $(3x + 5)(3x - 5)$

40. $(3x + 5)^2$

41. $(x - 7)(x + 4)$

42. $(x + 1)(x^2 - x + 1)$

43. In 2005, 55.2 million people will have high-speed Internet access at work. This is 2.3 times the number of people having high-speed access in 2000. (*Source*: Jupiter Media Metrix) How many people had high-speed access in 2000?

44. In January 2001, the Organization of the Petroleum Exporting Countries (OPEC) cut its oil production to 31.3 million barrels per day. This was a decrease of 1.5 million barrels per day. (*Source*: U.S. Department of Energy) What was the daily oil production before the reduction?

SYNTHESIS

45. If $x^3 + c$ is prime, what can you conclude about c? Why?

46. Explain how the geometric model below can be used to verify the formula for factoring $a^3 - b^3$.

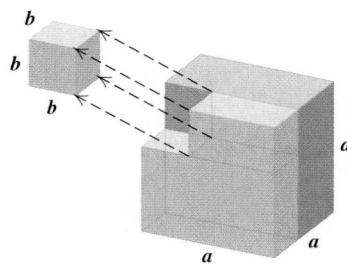

Factor. Assume that variables in exponents represent natural numbers.

47. $125c^6 + 8d^6$

48. $64x^6 + 8t^6$

49. $3x^{3a} - 24y^{3b}$

50. $\frac{8}{27}x^3 - \frac{1}{64}y^3$

51. $\frac{1}{24}x^3y^3 + \frac{1}{3}z^3$

52. $\frac{1}{16}x^{3a} + \frac{1}{2}y^{6a}z^{9b}$

Aha! **53.** $(x + 5)^3 + (x - 5)^3$

54. $x^3 - (x + y)^3$

55. $t^4 - 8t^3 - t + 8$

56. $5x^3y^6 - \frac{5}{8}$

5.6 Factoring: A General Strategy

Choosing the Right Method

CONNECTING THE CONCEPTS

Thus far, each section in this chapter has examined one or two different methods for factoring polynomials. In practice, when the need for factoring a polynomial arises, we must decide on our own which method to use. As preparation for such a situation, we now encounter polynomials of various types, in random order. Regardless of the polynomial we are faced with, the guidelines listed below can always be used.

To Factor a Polynomial

A. Always look for a common factor first. If there is one, factor out the largest common factor. Be sure to include it in your final answer.

B. Then look at the number of terms.

Two terms: Try factoring as a difference of squares first: $A^2 - B^2 = (A - B)(A + B)$. Next, try factoring as a sum or a difference of cubes: $A^3 + B^3 = (A + B)(A^2 - AB + B^2)$ and $A^3 - B^3 = (A - B)(A^2 + AB + B^2)$.

Three terms: If the trinomial is a perfect-square trinomial, factor accordingly: $A^2 + 2AB + B^2 = (A + B)^2$ or $A^2 - 2AB + B^2 = (A - B)^2$. If it is not a perfect-square trinomial, try using FOIL or grouping.

Four terms: Try factoring by grouping.

C. Always *factor completely*. When a factor can itself be factored, be sure to factor it. Remember that some polynomials, like $A^2 + B^2$, are prime.

D. Check by multiplying.

Choosing the Right Method

EXAMPLE 1   Factor: $5t^4 - 80$.

Solution

A. We look for a common factor:

$$5t^4 - 80 = 5(t^4 - 16). \qquad \text{5 is the largest common factor.}$$

B. The factor $t^4 - 16$ is a difference of squares: $(t^2)^2 - 4^2$. We factor it, being careful to rewrite the 5 from step (A):

$$5t^4 - 80 = 5(t^2 + 4)(t^2 - 4). \qquad t^4 - 16 = (t^2 + 4)(t^2 - 4)$$

C. Since $t^2 - 4$ is not prime, we continue factoring:

$$5t^4 - 80 = 5(t^2 + 4)(t^2 - 4) = 5(t^2 + 4)(t - 2)(t + 2)$$

This is a sum of squares with no common factor. It cannot be factored!

D. *Check:* $5(t^2 + 4)(t - 2)(t + 2) = 5(t^2 + 4)(t^2 - 4)$
$$= 5(t^4 - 16) = 5t^4 - 80.$$

The factorization is $5(t^2 + 4)(t - 2)(t + 2)$.

EXAMPLE 2

Study Skills _____

Keep Your Focus

When studying with someone else, it can be very tempting to prolong conversation that has little to do with the mathematics that you need to study. If you see that this may happen, explain to your partner(s) that you enjoy the nonmathematical conversation, but would enjoy it more later—after the math work has been completed.

Factor: $2x^3 + 10x^2 + x + 5$.

Solution

A. We look for a common factor. There is none.
B. Because there are four terms, we try factoring by grouping:

$$2x^3 + 10x^2 + x + 5$$
$$= (2x^3 + 10x^2) + (x + 5) \qquad \text{Separating into two binomials}$$
$$= 2x^2(x + 5) + 1(x + 5) \qquad \text{Factoring out the largest common factor from each binomial. The 1 serves as an aid.}$$

$$= (x + 5)(2x^2 + 1) \qquad \text{Factoring out the common factor, } x + 5$$

C. Nothing can be factored further, so we have factored completely.
D. *Check:* $(x + 5)(2x^2 + 1) = 2x^3 + x + 10x^2 + 5$
$$= 2x^3 + 10x^2 + x + 5.$$

The factorization is $(x + 5)(2x^2 + 1)$.

EXAMPLE 3

Factor: $-x^5 + 2x^4 + 35x^3$.

Solution

A. We note that there is a common factor, $-x^3$:

$$-x^5 + 2x^4 + 35x^3 = -x^3(x^2 - 2x - 35).$$

B. The factor $x^2 - 2x - 35$ is not a perfect-square trinomial. We factor it using trial and error:

$$-x^5 + 2x^4 + 35x^3 = -x^3(x^2 - 2x - 35)$$
$$= -x^3(x - 7)(x + 5).$$

C. Nothing can be factored further, so we have factored completely.
D. *Check:* $-x^3(x - 7)(x + 5) = -x^3(x^2 - 2x - 35)$
$$= -x^5 + 2x^4 + 35x^3.$$

The factorization is $-x^3(x - 7)(x + 5)$.

EXAMPLE 4

Student Notes _____

Quickly checking the leading and constant terms to see if they are squares can save you time. If they aren't both squares, the trinomial can't possibly be a perfect-square trinomial.

Factor: $x^2 - 20x + 100$.

Solution

A. We look first for a common factor. There is none.
B. This polynomial is a perfect-square trinomial. We factor it accordingly:

$$x^2 - 20x + 100 = x^2 - 2 \cdot x \cdot 10 + 10^2 \qquad \text{Try to do this step mentally.}$$

$$= (x - 10)^2.$$

C. Nothing can be factored further, so we have factored completely.
D. *Check:* $(x - 10)(x - 10) = x^2 - 20x + 100$.

The factorization is $(x - 10)(x - 10)$, or $(x - 10)^2$.

EXAMPLE 5

Factor: $6x^2y^4 - 21x^3y^5 + 3x^2y^6$.

Solution

A. We first factor out the largest common factor, $3x^2y^4$:

$$6x^2y^4 - 21x^3y^5 + 3x^2y^6 = 3x^2y^4(2 - 7xy + y^2).$$

B. The constant term in $2 - 7xy + y^2$ is not a square, so we do not have a perfect-square trinomial. Note that x appears only in $-7xy$. If $2 - 7xy + y^2$ could be factored into a form like $(1 - y)(2 - y)$, there would be no x in the middle term. Thus, $2 - 7xy + y^2$ cannot be factored.
C. Nothing can be factored further, so we have factored completely.
D. *Check:* $3x^2y^4(2 - 7xy + y^2) = 6x^2y^4 - 21x^3y^5 + 3x^2y^6$.

The factorization is $3x^2y^4(2 - 7xy + y^2)$.

EXAMPLE 6

Factor: $x^6 - 64$.

Solution

A. We look first for a common factor. There is none.
B. There are two terms, a difference of squares: $(x^3)^2 - (8)^2$. We factor it:

$$x^6 - 64 = (x^3 + 8)(x^3 - 8). \qquad \text{Note that } x^6 = (x^3)^2.$$

C. One factor is a sum of two cubes, and the other factor is a difference of two cubes. We factor both:

$$x^6 - 64 = (x + 2)(x^2 - 2x + 4)(x - 2)(x^2 + 2x + 4).$$

The factorization is complete because no factor can be factored further.
D. *Check:* $(x + 2)(x^2 - 2x + 4)(x - 2)(x^2 + 2x + 4) = (x^3 + 8)(x^3 - 8)$
$$= x^6 - 64.$$

The factorization is $(x + 2)(x^2 - 2x + 4)(x - 2)(x^2 + 2x + 4)$.

EXAMPLE 7

Factor: $25x^2 + 20xy + 4y^2$.

Solution

A. We look first for a common factor. There is none.
B. There are three terms. Note that the first term and the last term are squares: $25x^2 = (5x)^2$ and $4y^2 = (2y)^2$. We see that twice the product of $5x$ and $2y$ is the middle term,

$$2 \cdot 5x \cdot 2y = 20xy,$$

so the trinomial is a perfect square.

To factor, we write a binomial squared:

$$25x^2 + 20xy + 4y^2 = (5x + 2y)^2.$$

C. Nothing can be factored further, so we have factored completely.
D. *Check:* $(5x + 2y)(5x + 2y) = 25x^2 + 20xy + 4y^2.$

The factorization is $(5x + 2y)(5x + 2y)$, or $(5x + 2y)^2$.

EXAMPLE 8

Factor: $p^2q^2 + 7pq + 12.$

Solution

A. We look first for a common factor. There is none.
B. Since only one term is a square, we do not have a perfect-square trinomial. We use trial and error, thinking of the product pq as a single variable:

$$(pq + \quad)(pq + \quad).$$

We factor the last term, 12. All the signs are positive, so we consider only positive factors. Possibilities are 1, 12 and 2, 6 and 3, 4. The pair 3, 4 gives a sum of 7 for the coefficient of the middle term. Thus,

$$p^2q^2 + 7pq + 12 = (pq + 3)(pq + 4).$$

C. Nothing can be factored further, so we have factored completely.
D. *Check:* $(pq + 3)(pq + 4) = p^2q^2 + 7pq + 12.$

The factorization is $(pq + 3)(pq + 4)$.

Compare the variables appearing in Example 7 with those in Example 8. Note that if the leading term contains one variable and a different variable is in the last term, as in Example 7, each binomial contains two variable terms. When two variables appear in the leading term and no variables appear in the last term, as in Example 8, each binomial contains one term that has two variables and one term that is a constant.

EXAMPLE 9

Factor: $a^4 - 16b^4.$

Solution

A. We look first for a common factor. There is none.
B. There are two terms. Since $a^4 = (a^2)^2$ and $16b^4 = (4b^2)^2$, we see that we have a difference of squares. Thus,

$$a^4 - 16b^4 = (a^2 + 4b^2)(a^2 - 4b^2).$$

C. The factor $(a^2 - 4b^2)$ is itself a difference of squares. Thus,

$$a^4 - 16b^4 = (a^2 + 4b^2)(a + 2b)(a - 2b). \qquad \text{Factoring } a^2 - 4b^2$$

D. *Check:* $(a^2 + 4b^2)(a + 2b)(a - 2b) = (a^2 + 4b^2)(a^2 - 4b^2)$
$$= a^4 - 16b^4.$$

The factorization is $(a^2 + 4b^2)(a + 2b)(a - 2b).$

Exercise Set

5.6

🔖 *Concept Reinforcement   In each of Exercises 1–4, complete the sentence.*

1. As a first step when factoring polynomials, always check for a _____.

2. When factoring a trinomial, if two terms are not squares, it cannot be a _____.

3. If a polynomial has four terms and no common factor, it may be possible to factor by _____.

4. It is always possible to check a factorization by _____.

Factor completely. If a polynomial is prime, state this.

5. $10a^2 - 640$

6. $5x^2 - 45$

7. $y^2 + 49 - 14y$

8. $a^2 + 25 + 10a$

9. $2t^2 + 11t + 12$

10. $8t^2 - 18t - 5$

11. $x^3 - 18x^2 + 81x$

12. $x^3 - 24x^2 + 144x$

13. $x^3 - 5x^2 - 25x + 125$

14. $x^3 + 3x^2 - 4x - 12$

15. $27t^3 - 3t$

16. $98t^2 - 18$

17. $9x^3 + 12x^2 - 45x$

18. $20x^3 - 4x^2 - 72x$

19. $t^2 + 25$

20. $x^2 + 4$

21. $6x^2 + 3x - 45$

22. $2t^2 + 6t - 20$

23. $-2a^6 + 8a^5 - 8a^4$

24. $-x^5 - 14x^4 - 49x^3$

25. $5x^5 - 80x$

26. $4x^4 - 64$

27. $t^4 - 9$

28. $9 + t^8$

29. $-x^6 + 2x^5 - 7x^4$

30. $-x^5 + 4x^4 - 3x^3$

31. $x^3 - y^3$

32. $8t^3 + 1$

33. $ax^2 + ay^2$

34. $12n^2 + 24n^3$

35. $36mn - 9m^2n^2$

36. $ab^2 - a^2b$

37. $2\pi rh + 2\pi r^2$

38. $4\pi r^2 + 2\pi r$

Aha! **39.** $(a + b)5a + (a + b)3b$

40. $5c(a^3 + b) - (a^3 + b)$

41. $x^2 + x + xy + y$

42. $n^2 + 2n + np + 2p$

43. $a^2 - 3a + ay - 3y$

44. $2x^2 - 4x + xz - 2z$

45. $3x^2 + 13xy - 10y^2$

46. $-x^2 - y^2 - 2xy$

47. $4b^2 + a^2 - 4ab$

48. $a^2 - 7a - 6$

49. $16x^2 + 24xy + 9y^2$

50. $7p^4 - 7q^4$

51. $t^2 - 8t + 10$

52. $25z^2 + 10zy + y^2$

53. $64t^6 - 1$

54. $m^6 - 1$

55. $4p^2q - pq^2 + 4p^3$

56. $a^2 + ab + 2b^2$

57. $3b^2 + 17ab - 6a^2$

58. $2mn - 360n^2 + m^2$

59. $-12 - x^2y^2 - 8xy$

60. $m^2n^2 - 4mn - 32$

61. $p^2q^2 + 7pq + 6$

62. $a^5b^2 + 3a^4b - 10a^3$

63. $54a^4 + 16ab^3$

64. $54x^3y - 250y^4$

65. $x^6 + x^5y - 2x^4y^2$

66. $2s^6t^2 + 10s^3t^3 + 12t^4$

67. $36a^2 - 15a + \frac{25}{16}$

68. $a^2 + 2a^2bc + a^2b^2c^2$

69. $\frac{1}{81}x^2 - \frac{8}{27}x + \frac{16}{9}$

70. $\frac{1}{4}a^2 + \frac{1}{3}ab + \frac{1}{9}b^2$

71. $1 - 16x^{12}y^{12}$

72. $b^4a - 81a^5$

73. $4a^2b^2 + 12ab + 9$

74. $9c^2 + 6cd + d^2$

75. $a^4 + 8a^2 + 8a^3 + 64a$

76. $t^4 + 7t^2 - 3t^3 - 21t$

🗒 **77.** Kelly factored $16 - 8x + x^2$ as $(x - 4)^2$, while Tony factored it as $(4 - x)^2$. Are they both correct? Why or why not?

🗒 **78.** Describe in your own words a strategy for factoring polynomials.

SKILL MAINTENANCE

79. Show that the pairs $(-1, 11)$, $(0, 7)$, and $(3, -5)$ are solutions of $y = -4x + 7$. [3.2]

Solve. [2.2]

80. $5x - 4 = 0$

81. $3x + 7 = 0$

82. $2x + 9 = 0$

83. $4x - 9 = 0$

84. Graph: $y = -\frac{1}{2}x + 4$. [3.2]

SYNTHESIS

85. There are third-degree polynomials in x that we are not yet able to factor, despite the fact that they are not prime. Explain how such a polynomial could be created.

86. Describe a method that could be used to find a binomial of degree 16 that can be expressed as the product of prime binomial factors.

Factor.

87. $-(x^5 + 7x^3 - 18x)$

88. $18 + a^3 - 9a - 2a^2$

89. $-3a^4 + 15a^2 - 12$

90. $-x^4 + 7x^2 + 18$

Aha! **91.** $y^2(y + 1) - 4y(y + 1) - 21(y + 1)$

92. $y^2(y - 1) - 2y(y - 1) + (y - 1)$

93. $6(x - 1)^2 + 7y(x - 1) - 3y^2$

94. $(y + 4)^2 + 2x(y + 4) + x^2$

95. $2(a + 3)^4 - (a + 3)^3(b - 2) - (a + 3)^2(b - 2)^2$

96. $5(t - 1)^5 - 6(t - 1)^4(s - 1) + (t - 1)^3(s - 1)^2$

CORNER

COLLABORATIVE

*Matching Factorizations**

Focus: Factoring

Time: 20 minutes

Group size: Begin with the entire class. The end result is pairs of students. If there is an odd number of students, the instructor should participate.

Materials: Prepared sheets of paper, pins or tape. On half of the sheets, the instructor writes a polynomial. On the remaining sheets, the instructor writes the factorization of those polynomials. The activity is more interesting if the polynomials and factorizations are similar; for example,

$$x^2 - 2x - 8, \quad (x - 2)(x - 4),$$
$$x^2 - 6x + 8, \quad (x - 1)(x - 8),$$
$$x^2 - 9x + 8, \quad (x + 2)(x - 4).$$

ACTIVITY

1. As class members enter the room, the instructor pins or tapes either a polynomial or a factorization to the back of each student.

Class members are told only whether their sheet of paper contains a polynomial or a factorization. All students should remain quiet and not tell others what is on their sheet of paper.

2. After all students are wearing a sheet of paper, they should mingle with one another, attempting to match up their factorization with the appropriate polynomial or vice versa. They may ask "yes/no" questions of one another that relate to factoring and polynomials. Answers to the questions should be yes or no. For example, a legitimate question might be "Is my last term negative?", "Do my factors have opposite signs?", or "Does my factorization include the factors of 6?"

3. The game is over when all factorization/polynomial pairs have "found" one another.

---

*Thanks to Jann MacInnes of Florida Community College at Jacksonville–Kent Campus for suggesting this activity.

5.7 Solving Polynomial Equations by Factoring

The Principle of Zero Products • Factoring to Solve
Equations • Algebraic–Graphical Connection

CONNECTING THE CONCEPTS

Chapter 4 and Sections 5.1–5.6 have been devoted to finding *equivalent expressions*. Whether we are adding, subtracting, multiplying, dividing, or factoring polynomials, the result of our work is an expression that is equivalent to the original expression.

Here in Section 5.7 we return to the task of *solving equations*. This time, however, the equations will contain a variable raised to a power greater than 1 and will usually have more than one solution. Our ability to factor will play a pivotal role in solving such equations.

Whenever two polynomials are set equal to each other, we have a polynomial equation. The *degree of a polynomial equation* is the same as the highest degree of any term in the equation. Second-degree equations like $4t^2 - 9 = 0$ and $x^2 + 6x + 5 = 0$ are called **quadratic equations**.

Study Skills

Identify the Highlights

If you haven't already tried one, consider using a highlighter as you read. By highlighting sentences or phrases that you find especially important, you will make it easier to review important material in the future. Highlighters are only helpful when used, so be sure to keep your highlighter with you whenever you study.

> **Quadratic Equation**
>
> A *quadratic equation* is an equation equivalent to one of the form
>
> $$ax^2 + bx + c = 0,$$
>
> where a, b, and c are constants, with $a \neq 0$.

In order to solve quadratic equations, we need to develop a new principle.

The Principle of Zero Products

Suppose we are told that the product of two numbers is 6. On the basis of this information, it is impossible to know the value of either number—the product could be $2 \cdot 3$, $6 \cdot 1$, $12 \cdot \frac{1}{2}$, and so on. However, if we are told that the product of two numbers is 0, we know that at least one of the two numbers must itself be 0. For example, if $(x + 3)(x - 2) = 0$, we can conclude that either $x + 3$ is 0 or $x - 2$ is 0.

> **The Principle of Zero Products**
>
> An equation $AB = 0$ is true if and only if $A = 0$ or $B = 0$, or both.
> (A product is 0 if and only if at least one factor is 0.)

EXAMPLE 1   Solve: $(x + 3)(x - 2) = 0$.

Solution   We are told that the product of $x + 3$ and $x - 2$ is 0. In order for a product to be 0, at least one factor must be 0. Therefore, either

$$x + 3 = 0 \quad or \quad x - 2 = 0. \qquad \text{Using the principle of zero products}$$

We solve each equation:

$$\begin{aligned} x + 3 &= 0 \quad or \quad x - 2 = 0 \\ x &= -3 \quad or \qquad x = 2. \end{aligned}$$

Both -3 and 2 should be checked in the original equation.

Check: For -3:

$$\frac{(x + 3)(x - 2) = 0}{\begin{array}{c|c} (-3 + 3)(-3 - 2) & 0 \\ 0(-5) & \\ & 0 \overset{?}{=} 0 \quad \text{TRUE} \end{array}}$$

For 2:

$$\frac{(x + 3)(x - 2) = 0}{\begin{array}{c|c} (2 + 3)(2 - 2) & 0 \\ 5(0) & \\ & 0 \overset{?}{=} 0 \quad \text{TRUE} \end{array}}$$

The solutions are -3 and 2.

When we are using the principle of zero products, the word "or" is meant to emphasize that any one of the factors could be the one that represents 0.

EXAMPLE 2   Solve: $3(5x + 1)(x - 7) = 0$.

Solution   Since the factor 3 is constant, the only way for $3(5x + 1)(x - 7)$ to be 0 is for one of the other factors to be 0. That is,

$$\begin{aligned} 5x + 1 &= 0 \quad or \quad x - 7 = 0 & \text{Using the principle of zero products} \\ 5x &= -1 \quad or \qquad x = 7 & \text{Solving the two equations separately} \\ x &= -\tfrac{1}{5} \quad or \qquad x = 7. & \text{The constant factor 3 is } not \text{ a solution.} \end{aligned}$$

Check: For $-\tfrac{1}{5}$:

$$\frac{3(5x + 1)(x - 7) = 0}{\begin{array}{c|c} 3\big(5\big(-\tfrac{1}{5}\big) + 1\big)\big(-\tfrac{1}{5} - 7\big) & 0 \\ 3(-1 + 1)\big(-7\tfrac{1}{5}\big) & \\ 3(0)\big(-7\tfrac{1}{5}\big) & \\ & 0 \overset{?}{=} 0 \quad \text{TRUE} \end{array}}$$

For 7:

$$\frac{3(5x + 1)(x - 7) = 0}{\begin{array}{c|c} 3(5(7) + 1)(7 - 7) & 0 \\ 3(35 + 1)0 & \\ & 0 \overset{?}{=} 0 \quad \text{TRUE} \end{array}}$$

The solutions are $-\tfrac{1}{5}$ and 7.

The principle of zero products can be used whenever a product equals 0—even if a factor has only one term.

EXAMPLE 3 Solve: $7t(t - 5) = 0$.

Solution We have

$$7 \cdot t(t - 5) = 0 \qquad \text{The factors are 7, } t, \text{ and } t - 5.$$
$$t = 0 \quad or \quad t - 5 = 0 \qquad \text{Using the principle of zero products}$$
$$t = 0 \quad or \qquad t = 5. \qquad \text{Solving. Note that the constant factor, 7, is never 0.}$$

The solutions are 0 and 5. The check is left to the student.

Factoring to Solve Equations

By factoring and using the principle of zero products, we can now solve a variety of quadratic equations.

EXAMPLE 4 Solve: $x^2 + 5x + 6 = 0$.

Solution This equation differs from those solved in Chapter 2. There are no like terms to combine, and there is a squared term. We first factor the polynomial. Then we use the principle of zero products:

$$x^2 + 5x + 6 = 0$$
$$(x + 2)(x + 3) = 0 \qquad \text{Factoring}$$
$$x + 2 = 0 \quad or \quad x + 3 = 0 \qquad \text{Using the principle of zero products}$$
$$x = -2 \quad or \qquad x = -3.$$

Check: For -2:

$$\begin{array}{c|c} x^2 + 5x + 6 = 0 \\ \hline (-2)^2 + 5(-2) + 6 & 0 \\ 4 - 10 + 6 & \\ -6 + 6 & \\ & 0 \overset{?}{=} 0 \quad \text{TRUE} \end{array}$$

For -3:

$$\begin{array}{c|c} x^2 + 5x + 6 = 0 \\ \hline (-3)^2 + 5(-3) + 6 & 0 \\ 9 - 15 + 6 & \\ -6 + 6 & \\ & 0 \overset{?}{=} 0 \quad \text{TRUE} \end{array}$$

The solutions are -2 and -3.

The principle of zero products is used even if the factoring consists of only removing a common factor.

EXAMPLE 5 Solve: $x^2 + 7x = 0$.

Solution Although there is no constant term, because of the x^2-term, the equation is still quadratic. The methods of Chapter 2 are not sufficient, so we try factoring:

$$x^2 + 7x = 0$$
$$x(x + 7) = 0 \qquad \text{Removing the greatest common factor, } x$$
$$x = 0 \quad or \quad x + 7 = 0$$
$$x = 0 \quad or \qquad x = -7.$$

The solutions are 0 and -7. The check is left to the student.

Student Notes _____

Checking for a common factor is an important step that is often overlooked. Also, be sure to remember that only like terms can be combined.

> **Caution!** We *must* have 0 on one side of the equation before the principle of zero products can be used. Get all nonzero terms on one side and 0 on the other.

EXAMPLE 6   Solve: **(a)** $x^2 - 8x = -16$; **(b)** $4t^2 = 25$.

Solution

a) We first add 16 to get 0 on one side:

$$x^2 - 8x = -16$$
$$x^2 - 8x + 16 = 0 \qquad \text{Adding 16 to both sides to get 0 on one side}$$
$$(x - 4)(x - 4) = 0 \qquad \text{Factoring}$$
$$x - 4 = 0 \quad or \quad x - 4 = 0 \qquad \text{Using the principle of zero products}$$
$$x = 4 \quad or \qquad x = 4.$$

There is only one solution, 4. The check is left to the student.

b) We have

$$4t^2 = 25$$
$$4t^2 - 25 = 0 \qquad \text{Subtracting 25 from both sides to get 0 on one side}$$
$$(2t - 5)(2t + 5) = 0 \qquad \text{Factoring a difference of squares}$$
$$2t - 5 = 0 \quad or \quad 2t + 5 = 0$$
$$2t = 5 \quad or \qquad 2t = -5 \qquad \left. \right\} \; \text{Solving the two equations separately}$$
$$t = \tfrac{5}{2} \quad or \qquad t = -\tfrac{5}{2}.$$

The solutions are $\tfrac{5}{2}$ and $-\tfrac{5}{2}$. The check is left to the student.

When solving quadratic equations by factoring, remember that a factorization is not useful unless 0 is on the other side of the equation.

EXAMPLE 7   Solve: $(x + 3)(2x - 1) = 9$.

Solution   Be careful with an equation like this! Since we need 0 on one side, we multiply out the product on the left and then subtract 9 from both sides:

$$(x + 3)(2x - 1) = 9$$
$$2x^2 + 5x - 3 = 9 \qquad \text{Multiplying on the left}$$
$$2x^2 + 5x - 3 - 9 = 9 - 9 \qquad \text{Subtracting 9 from both sides to get 0 on one side}$$
$$2x^2 + 5x - 12 = 0 \qquad \text{Combining like terms}$$
$$(2x - 3)(x + 4) = 0 \qquad \text{Factoring}$$
$$2x - 3 = 0 \quad or \quad x + 4 = 0 \qquad \text{Using the principle of zero products}$$
$$2x = 3 \quad or \qquad x = -4$$
$$x = \tfrac{3}{2} \quad or \qquad x = -4.$$

Check: For $\frac{3}{2}$:

$$\frac{(x + 3)(2x - 1) = 9}{\left(\frac{3}{2} + 3\right)\left(2 \cdot \frac{3}{2} - 1\right) \;\Big|\; 9}$$

$$\left(\frac{9}{2}\right)(2) \;\Big|$$

$$9 \overset{?}{=} 9 \quad \text{TRUE}$$

For -4:

$$\frac{(x + 3)(2x - 1) = 9}{(-4 + 3)(2(-4) - 1) \;\Big|\; 9}$$

$$(-1)(-9) \;\Big|$$

$$9 \overset{?}{=} 9 \quad \text{TRUE}$$

The solutions are $\frac{3}{2}$ and -4.

We can use the principle of zero products to solve polynomials with degree greater than 2, if they can be factored.

EXAMPLE 8   Solve: $3x^3 - 30x = 9x^2$.

Solution   We have

$$3x^3 - 30x = 9x^2$$

$$3x^3 - 9x^2 - 30x = 0 \qquad \text{Getting 0 on one side and writing in descending order}$$

$$3x(x^2 - 3x - 10) = 0 \qquad \text{Factoring out a common factor}$$

$$3x(x + 2)(x - 5) = 0 \qquad \text{Factoring the trinomial}$$

$$3x = 0 \quad or \quad x + 2 = 0 \quad or \quad x - 5 = 0 \qquad \text{Using the principle of zero products}$$

$$x = 0 \quad or \qquad x = -2 \quad or \qquad x = 5.$$

Check:

$$\frac{3x^3 - 30x = 9x^2}{3 \cdot 0^3 - 30 \cdot 0 \;\Big|\; 9 \cdot 0^2}$$

$$0 - 0 \;\Big|\; 9 \cdot 0$$

$$0 \overset{?}{=} 0 \quad \text{TRUE}$$

$$\frac{3x^3 - 30x = 9x^2}{3(-2)^3 - 30(-2) \;\Big|\; 9(-2)^2}$$

$$3(-8) + 60 \;\Big|\; 9 \cdot 4$$

$$-24 + 60 \;\Big|\; 36$$

$$36 \overset{?}{=} 36 \quad \text{TRUE}$$

$$\frac{3x^3 - 30x = 9x^2}{3 \cdot 5^3 - 30 \cdot 5 \;\Big|\; 9 \cdot 5^2}$$

$$3 \cdot 125 - 150 \;\Big|\; 9 \cdot 25$$

$$375 - 150 \;\Big|\; 225$$

$$225 \overset{?}{=} 225 \quad \text{TRUE}$$

The solutions are 0, -2 and 5.

ALGEBRAIC—GRAPHICAL CONNECTION

When graphing a linear equation in Chapter 3, we found the x-intercept by replacing y with 0 and solving for x. This procedure can also be used to find the x-intercepts when graphing an equation of the form $y = ax^2 + bx + c$. Although the details of creating such graphs is left for Chapter 11, we consider them briefly here from the standpoint of finding x-intercepts. The graphs are shaped as shown. Note that each x-intercept represents a solution of $ax^2 + bx + c = 0$.

$$\boxed{\begin{array}{c} y = ax^2 + bx + c \\ a \neq 0 \end{array}}$$

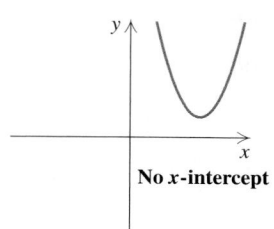

No x-intercept

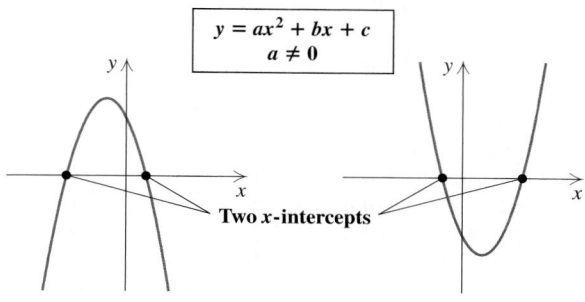

Two x-intercepts

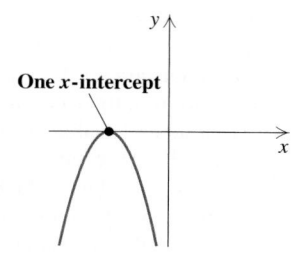

One x-intercept

EXAMPLE 9   Find the x-intercepts for the graph of the equation shown. (The grid is intentionally not included.)

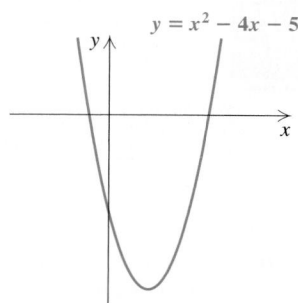

$y = x^2 - 4x - 5$

Solution   To find the x-intercepts, we let $y = 0$ and solve for x:

$$0 = x^2 - 4x - 5 \qquad \text{Substituting 0 for } y$$
$$0 = (x - 5)(x + 1) \qquad \text{Factoring}$$
$$x - 5 = 0 \quad \textit{or} \quad x + 1 = 0 \qquad \text{Using the principle of zero products}$$
$$x = 5 \quad \textit{or} \qquad x = -1. \qquad \text{Solving for } x$$

The x-intercepts are $(5, 0)$ and $(-1, 0)$.

 **technology connection**

A graphing calculator allows us to solve quadratic equations even when an equation cannot be solved by factoring. For example, to solve $x^2 - 3x - 5 = 0$, we can let $y_1 = x^2 - 3x - 5$ and $y_2 = 0$. Selecting a bold line type to the left of y_2 in the ⬚ Y= ⬚ window makes the line easier to see. Using the INTERSECT option of the CALC menu, we select the two graphs in which we are interested, along with a guess. The graphing calculator displays the point of intersection.

An alternate method uses only y_1 and the ZERO or ROOT option of the CALC menu. This option requires you to enter an x-value to the left of the x-intercept as a LEFT BOUND. An x-value to the right of the x-intercept is then entered as a RIGHT BOUND. Finally, a GUESS value between the two bounds is entered and the x-intercept, or ZERO or ROOT, is displayed.

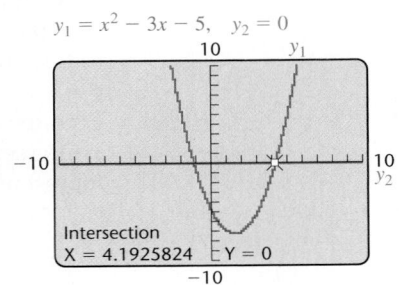

$$y_1 = x^2 - 3x - 5, \quad y_2 = 0$$

Intersection
X = 4.1925824   Y = 0

Use a graphing calculator to find the solutions, if they exist, accurate to two decimal places.

1. $x^2 + 4x - 3 = 0$

2. $x^2 - 5x - 2 = 0$

3. $x^2 + 13.54x + 40.95 = 0$

4. $x^2 - 4.43x + 6.32 = 0$

5. $1.235x^2 - 3.409x = 0$

Exercise Set

5.7

🖐 *Concept Reinforcement   For each of Exercises 1–4, match the phrase with the most appropriate choice from the column on the right.*

1. _____ The name of equations of the type $ax^2 + bx + c = 0$, with $a \neq 0$

a) 2

2. _____ The idea that $A \cdot B = 0$ if and only if $A = 0$ or $B = 0$

b) 0

3. _____ The number of solutions of most quadratic equations

c) Quadratic

4. _____ The number that is on one side of a quadratic equation before the principle of zero products is used

d) The principle of zero products

Solve using the principle of zero products.

5. $(x + 5)(x + 7) = 0$

6. $(x + 1)(x + 2) = 0$

7. $(2x - 9)(x + 4) = 0$

8. $(3x - 5)(x + 1) = 0$

9. $(10x - 9)(4x + 7) = 0$

10. $(2x - 7)(3x + 4) = 0$

11. $x(x + 2) = 0$

12. $t(t + 8) = 0$

13. $\left(\frac{2}{3}x - \frac{12}{11}\right)\left(\frac{7}{4}x - \frac{1}{12}\right) = 0$

14. $\left(\frac{1}{9} - 3x\right)\left(\frac{1}{5} + 2x\right) = 0$

15. $5x(2x + 9) = 0$

16. $12x(4x + 5) = 0$

17. $(20 - 0.4x)(7 - 0.1x) = 0$

18. $(1 - 0.05x)(1 - 0.3x) = 0$

19. $(3x - 2)(x + 5)(x - 1) = 0$

20. $(2x + 1)(x + 3)(x - 5) = 0$

Solve by factoring and using the principle of zero products.

21. $x^2 - 7x + 6 = 0$

22. $x^2 - 6x + 5 = 0$

23. $x^2 + 4x - 21 = 0$

24. $x^2 + 8x + 15 = 0$

25. $x^3 - 3x^2 + 2x = 0$

26. $t^3 - 6t^2 - 7t = 0$

27. $x^2 - 6x = 0$

28. $x^2 + 8x = 0$

29. $6t + t^2 = 0$

30. $3t + t^2 = 0$

31. $9x^2 = 4$

32. $4x^2 = 25$

33. $0 = 25 + x^2 + 10x$

34. $0 = 6x + x^2 + 9$

35. $1 + x^2 = 2x$

36. $x^2 + 16 = 8x$

37. $4t^2 = 8t$

38. $12t = 3t^2$

39. $3x^2 - 7x = 20$

40. $6x^2 - 4x = 10$

41. $2y^2 + 12y = -10$

42. $12y^2 - 5y = 2$

43. $(x - 7)(x + 1) = -16$

44. $(x + 2)(x - 7) = -18$

45. $14z^2 - 3 = 21z - 3$

46. $15z^2 + 7 = 20z + 7$

47. $81x^2 - 5 = 20$

48. $36m^2 - 9 = 40$

49. $(x - 1)(5x + 4) = 2$

50. $(x + 3)(3x + 5) = 7$

51. $x^2 - 2x = 18 + 5x$

52. $3x^2 - 2x = 9 - 8x$

53. $x^2(2x - 1) = 3x$

54. $x^2 = x(10 - 3x^2)$

55. $(2x - 5)(3x^2 + 29x + 56) = 0$

56. $(4x + 9)(15x^2 - 7x - 2) = 0$

57. Use this graph to solve $x^2 + x - 6 = 0$.

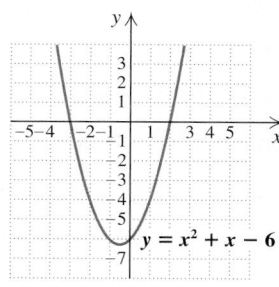

58. Use this graph to solve $x^2 - 3x - 4 = 0$.

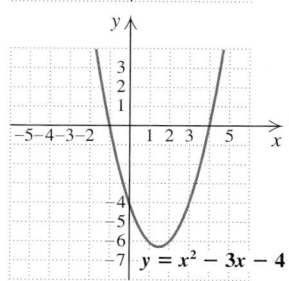

59. Use the following graph to solve $-x^2 + 2x + 3 = 0$.

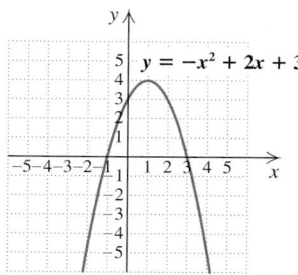

60. Use the following graph to solve $-x^2 - x + 6 = 0$.

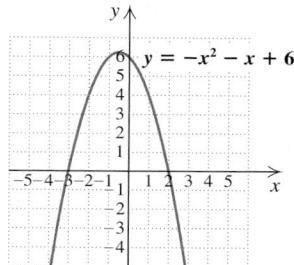
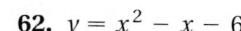

Find the x-intercepts for the graph of each equation. Grids are intentionally not included.

61. $y = x^2 + 3x - 4$

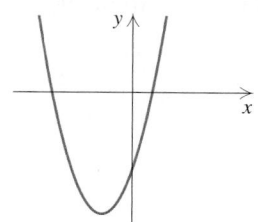

62. $y = x^2 - x - 6$

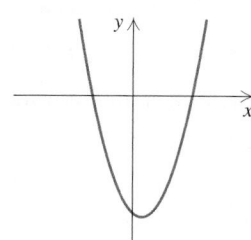

63. $y = x^2 - 2x - 15$

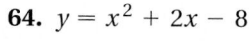
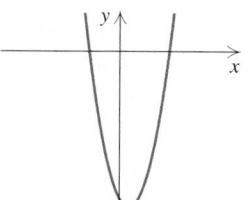

64. $y = x^2 + 2x - 8$

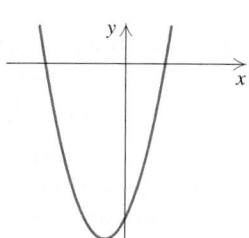

65. $y = 2x^2 + x - 10$ **66.** $y = 2x^2 + 3x - 9$

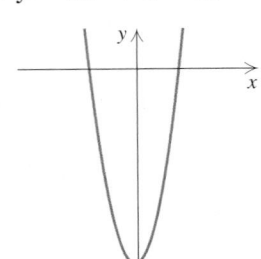

67. The equation $x^2 + 1$ has no real-number solutions. What implications does this have for the graph of $y = x^2 + 1$?

68. What is the difference between a quadratic polynomial and a quadratic equation?

SKILL MAINTENANCE

Translate to an algebraic expression. [1.1]

69. The square of the sum of a and b

70. The sum of the squares of a and b

71. The sum of two consecutive integers

Translate to an inequality. [2.7]

72. 5 more than twice a number is less than 19.

73. 7 less than half of a number exceeds 24.

74. 3 less than a number is at least 34.

SYNTHESIS

75. What is wrong with solving $x^2 = 3x$ by dividing both sides of the equation by x?

76. When the principle of zero products is used to solve a quadratic equation, will there always be two different solutions? Why or why not?

77. Find an equation with integer coefficients that has the given numbers as solutions. For example, 3 and -2 are solutions to $x^2 - x - 6 = 0$.
   a) $-4, 5$ **b)** $-1, 7$ **c)** $\frac{1}{4}, 3$
   d) $\frac{1}{2}, \frac{1}{3}$ **e)** $\frac{2}{3}, \frac{3}{4}$ **f)** $-1, 2, 3$

Solve.

78. $16(x - 1) = x(x + 8)$

79. $a(9 + a) = 4(2a + 5)$

80. $(t - 5)^2 = 2(5 - t)$

81. $-x^2 + \frac{9}{25} = 0$

82. $-x^2 + \frac{25}{36} = 0$

Aha! **83.** $(t + 1)^2 = 9$

84. $\frac{27}{25}x^2 = \frac{1}{3}$

85. For each equation on the left, find an equivalent equation on the right.
   a) $x^2 + 10x - 2 = 0$ $4x^2 + 8x + 36 = 0$
   b) $(x - 6)(x + 3) = 0$ $(2x + 8)(2x - 5) = 0$
   c) $5x^2 - 5 = 0$ $9x^2 - 12x + 24 = 0$
   d) $(2x - 5)(x + 4) = 0$ $(x + 1)(5x - 5) = 0$
   e) $x^2 + 2x + 9 = 0$ $x^2 - 3x - 18 = 0$
   f) $3x^2 - 4x + 8 = 0$ $2x^2 + 20x - 4 = 0$

86. Explain how to construct an equation that has seven solutions.

87. Explain how the graph in Exercise 59 can be used to visualize the solutions of
$$-x^2 + 2x + 3 = -5.$$

Use a graphing calculator to find the solutions of each equation. Round solutions to the nearest hundredth.

88. $x^2 - 9.10x + 15.77 = 0$

89. $-x^2 + 0.63x + 0.22 = 0$

90. $x^2 + 13.74x + 42.00 = 0$

91. $6.4x^2 - 8.45x - 94.06 = 0$

92. $0.84x^2 - 2.30x = 0$

93. $1.23x^2 + 4.63x = 0$

94. $x^2 + 1.80x - 5.69 = 0$

5.8 Solving Applications

Applications • The Pythagorean Theorem

Applications

We can use the five-step problem-solving process and our new methods of solving quadratic equations to solve new types of problems.

EXAMPLE 1

Race numbers.   Roger and Suzanne entered the Twin City Marathon as a two-person team. Their racing bibs were consecutive numbers, the product of which was 156. Find the numbers.

Solution

1. **Familiarize.**   Consecutive numbers are one apart, like 49 and 50. Let $x =$ the first bib number; then $x + 1 =$ the next bib number.

2. **Translate.**   We reword the problem before translating:

   Rewording: The first bib number , times the next bib number , is   156.

   Translating: x $\cdot$ $(x + 1)$ $=$ 156

3. **Carry out.**   We solve the equation as follows:

$$x(x + 1) = 156$$
$$x^2 + x = 156 \qquad \text{Multiplying}$$
$$x^2 + x - 156 = 0 \qquad \text{Subtracting 156 to get 0 on one side}$$
$$(x - 12)(x + 13) = 0 \qquad \text{Factoring}$$
$$x - 12 = 0 \quad or \quad x + 13 = 0 \qquad \text{Using the principle of zero products}$$
$$x = 12 \quad or \qquad x = -13. \qquad \text{Solving each equation}$$

4. **Check.**   The solutions of the equation are 12 and -13. Since bib numbers cannot be negative, -13 must be rejected. On the other hand, if x is 12, then $x + 1$ is 13 and $12 \cdot 13 = 156$. Thus, 12 checks.

5. **State.**   The bib numbers for Suzanne and Roger were 12 and 13.

EXAMPLE 2   Manufacturing.   Wooden Work, Ltd., builds cutting boards that are twice as long as they are wide. The most popular board that Wooden Work makes has an area of 800 cm². What are the dimensions of the board?

Solution

1. **Familiarize.**   We first make a drawing. Recall that the area of any rectangle is Length · Width. We let $x =$ the width of the board, in centimeters. The length is then $2x$, since the board is twice as long as it is wide.

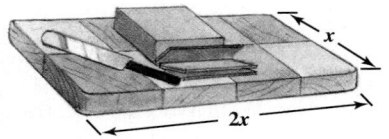

2. **Translate.**   We reword and translate as follows:

Rewording: The area of the rectangle   is   800 cm².

Translating: $2x \cdot x$ =   800

3. **Carry out.**   We solve the equation as follows:

$$2x \cdot x = 800$$
$$2x^2 = 800$$
$$2x^2 - 800 = 0 \qquad \text{Subtracting 800 to get 0 on one side of the equation}$$
$$2(x^2 - 400) = 0 \qquad \text{Removing a common factor of 2}$$
$$2(x - 20)(x + 20) = 0 \qquad \text{Factoring a difference of squares}$$
$$(x - 20)(x + 20) = 0 \qquad \text{Dividing both sides by 2}$$
$$x - 20 = 0 \quad or \quad x + 20 = 0 \qquad \text{Using the principle of zero products}$$
$$x = 20 \quad or \qquad x = -20. \qquad \text{Solving each equation}$$

4. **Check.**   The solutions of the equation are 20 and -20. Since the width must be positive, -20 cannot be a solution. To check 20 cm, we note that if the width is 20 cm, then the length is $2 \cdot 20$ cm $= 40$ cm and the area is 20 cm $\cdot$ 40 cm $= 800$ cm². Thus the solution 20 checks.

5. **State.**   The cutting board is 20 cm wide and 40 cm long.

EXAMPLE 3   Dimensions of a sail.   The mainsail of Stacey's Lightning-styled sailboat has an area of 125 ft². If the sail is 15 ft taller than it is wide, find the height and the width of the sail.

Solution

1. **Familiarize.**   We first make a drawing. The formula for the area of a triangle is Area $= \frac{1}{2} \cdot$ (base) $\cdot$ (height). We let $b =$ the width, in feet, of the triangle's base and $b + 15 =$ the height, in feet.

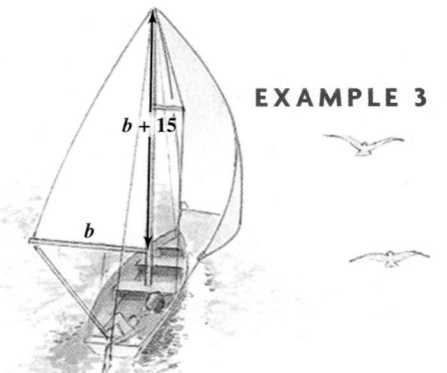

2. Translate.   We reword and translate as follows:

Rewording: $\underbrace{\text{The area of the sail}}$ is $\underbrace{125 \text{ ft}^2}$.

Translating: $\frac{1}{2} \cdot b(b + 15)$ $=$ $125.$

3. Carry out.   We solve the equation as follows:

$$\frac{1}{2} \cdot b \cdot (b + 15) = 125$$
$$\frac{1}{2}(b^2 + 15b) = 125 \qquad \text{Multiplying}$$
$$b^2 + 15b = 250 \qquad \text{Multiplying by 2 to clear fractions}$$
$$b^2 + 15b - 250 = 0 \qquad \text{Subtracting 250 to get 0 on one side}$$
$$(b + 25)(b - 10) = 0 \qquad \text{Factoring}$$
$$b + 25 = 0 \quad or \quad b - 10 = 0 \qquad \text{Using the principle of zero products}$$
$$b = -25 \quad or \qquad b = 10.$$

4. Check.   The width must be positive, so -25 cannot be a solution. Suppose the base is 10 ft. The height would be $10 + 15$, or 25 ft, and the area $\frac{1}{2}(10)(25)$, or 125 ft^2. These numbers check in the original problem.

5. State.   Stacey's mainsail is 25 ft tall and 10 ft wide.

EXAMPLE 4

Medicine.   For certain people suffering an extreme allergic reaction, the drug epinephrine (adrenaline) is sometimes prescribed. The number of micrograms N of epinephrine in an adult's bloodstream t minutes after 250 micrograms have been injected can be approximated by

$$-10t^2 + 100t = N$$

(*Source*: Based on information in Chohan, Naina, Rita M. Doyle, and Patricia Nayle (eds.), *Nursing Handbook*, 21st ed. Springhouse, PA: Springhouse Corporation, 2001). How long after an injection will there be about 210 micrograms of epinephrine in the bloodstream?

Solution

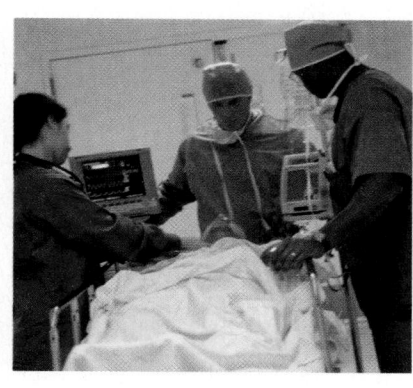

1. Familiarize.   To familiarize ourselves with this problem, we could calculate N for different choices of t. We leave this for the student. We could also speak with a healthcare professional to learn that injections of epinephrine are very quick-acting. Note that there may be two solutions, one on each side of the time at which the drug's effect peaks.

2. Translate.   To find the length of time after injection when 210 micrograms are in the bloodstream, we replace N with 210 in the formula above:

$$-10t^2 + 100t = 210. \qquad \text{Substituting 210 for } N. \text{ This is now an equation in one variable.}$$

technology connection

As a visual check for Example 4, we can either let $y_1 = -10x^2 + 100x$ and $y_2 = 210$, or let $y_1 = -10x^2 + 100x - 210$ and $y_2 = 0$. In either case, the points of intersection occur at $x = 3$ and $x = 7$, as shown.

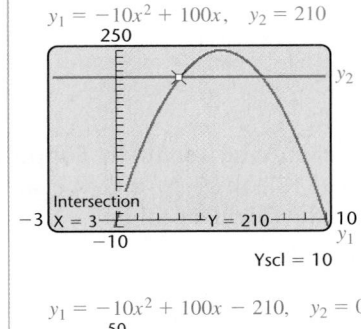

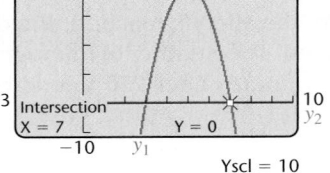

3. Carry out.   We solve the equation as follows:

$$-10t^2 + 100t = 210$$

$$-10t^2 + 100t - 210 = 0 \qquad \text{Subtracting 210 from both sides to get 0 on one side}$$

$$-10(t^2 - 10t + 21) = 0 \qquad \text{Factoring out the largest common factor, } -10$$

$$-10(t - 3)(t - 7) = 0 \qquad \text{Factoring}$$

$$t - 3 = 0 \quad or \quad t - 7 = 0 \qquad \text{Using the principle of zero products}$$

$$t = 3 \quad or \qquad t = 7.$$

4. Check.   Since $-10 \cdot 3^2 + 100 \cdot 3 = -90 + 300 = 210$, the number 3 checks. Since $-10 \cdot 7^2 + 100 \cdot 7 = -490 + 700 = 210$, the number 7 also checks.

5. State.   There will be 210 micrograms of epinephrine in the bloodstream approximately 3 minutes and 7 minutes after injection.

The Pythagorean Theorem

The following problems involve the Pythagorean theorem, which relates the lengths of the sides of a *right* triangle. A triangle is a **right triangle** if it has a 90°, or *right*, angle. The side opposite the 90° angle is called the **hypotenuse**. The other sides are called **legs**.

The Pythagorean Theorem

In any right triangle, if a and b are the lengths of the legs and c is the length of the hypotenuse, then

$$a^2 + b^2 = c^2, \quad \text{or}$$

$$(\text{Leg})^2 + (\text{Other leg})^2 = (\text{Hypotenuse})^2.$$

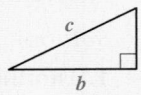

The equation $a^2 + b^2 = c^2$ is called the **Pythagorean equation.***

The Pythagorean theorem is named for the Greek mathematician Pythagoras (569?–500? B.C.). We can think of this relationship as adding areas.

---

*The *converse* of the Pythagorean theorem is also true. That is, if $a^2 + b^2 = c^2$, then the triangle is a right triangle.

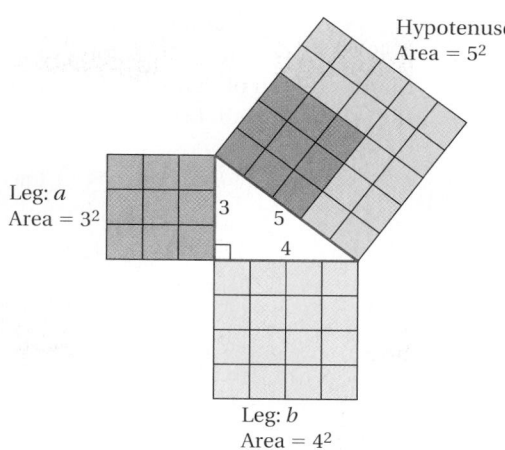

$$a^2 + b^2 = c^2$$
$$3^2 + 4^2 = 5^2$$
$$9 + 16 = 25$$

If we know the lengths of any two sides of a right triangle, we can use the Pythagorean equation to determine the length of the third side.

EXAMPLE 5   *Ladder settings.*   A 13-ft ladder is placed against a building in such a way that the distance from the top of the ladder to the ground is 7 ft more than the distance from the bottom of the ladder to the building. Find both distances.

Solution

1. **Familiarize.**   We first make a drawing. The ladder and the missing distances form the hypotenuse and legs of a right triangle. We let $x =$ the length of the side (leg) across the bottom. Then $x + 7 =$ the length of the other side (leg). The hypotenuse has length 13 ft.

2. **Translate.**   Since a right triangle is formed, we can use the Pythagorean theorem:

$$a^2 + b^2 = c^2$$
$$x^2 + (x + 7)^2 = 13^2. \qquad \text{Substituting}$$

3. **Solve.**   We solve the equation as follows:

$x^2 + (x^2 + 14x + 49) = 169$	Squaring the binomial and 13
$2x^2 + 14x + 49 = 169$	Combining like terms
$2x^2 + 14x - 120 = 0$	Subtracting 169 to get 0 on one side
$2(x^2 + 7x - 60) = 0$	Factoring out a common factor
$2(x + 12)(x - 5) = 0$	Factoring
$x + 12 = 0 \quad or \quad x - 5 = 0$	Using the principle of zero products
$x = -12 \quad or \qquad x = 5.$	

4. **Check.**   The integer -12 cannot be a length of a side because it is negative. When $x = 5$, $x + 7 = 12$, and $5^2 + 12^2 = 13^2$. So 5 checks.

5. **State.**   The distance from the top of the ladder to the ground is 12 ft. The distance from the bottom of the ladder to the building is 5 ft.

EXAMPLE 6   Roadway design.   Elliott Street is 24 ft wide when it ends at Main Street in Brattleboro, Vermont. A 40-ft long diagonal crosswalk allows pedestrians to cross Main Street to or from either corner of Elliott Street (see the figure). Determine the width of Main Street.

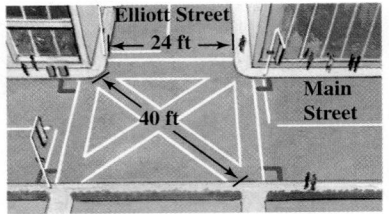

Solution

1. **Familiarize.**   A drawing has already been provided, but we can redraw and label the relevant part. Note that the two streets intersect at a right angle. We let $x =$ the width of Main Street, in feet.

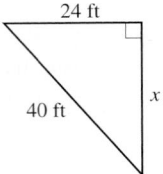

2. **Translate.**   Since a right triangle is formed, we can use the Pythagorean theorem:

$$a^2 + b^2 = c^2$$
$$x^2 + 24^2 = 40^2. \qquad \text{Substituting}$$

3. **Carry out.**   We solve the equation as follows:

$$x^2 + 576 = 1600 \qquad \text{Squaring 24 and 40}$$
$$x^2 - 1024 = 0 \qquad \text{Subtracting 1600 from both sides}$$
$$(x - 32)(x + 32) = 0 \qquad \text{Note that } 1024 = 32^2. \text{ A calculator might be helpful here.}$$
$$x - 32 = 0 \quad or \quad x + 32 = 0 \qquad \text{Using the principle of zero products}$$
$$x = 32 \quad or \qquad x = -32.$$

4. **Check.**   Since the width of a street must be positive, -32 is not a solution. If the width is 32 ft, we have $32^2 + 24^2 = 1024 + 576 = 1600$, which is 40^2. Thus, 32 checks.

5. **State.**   The width of Main Street is 32 ft.

Solve. Use the five-step problem-solving approach.

1. A number is 20 less than its square. Find all such numbers.

2. A number is 2 less than its square. Find all such numbers.

3. One leg of a right triangle is 3 cm longer than the other leg. The length of the hypotenuse is 15 cm. Find the length of each side.

4. One leg of a right triangle is 2 cm shorter than the other leg. The length of the hypotenuse is 10 cm. Find the length of each side.

5. *Locker numbers.* The product of two consecutive locker numbers is 90. Find the locker numbers.

6. *Page numbers.* The product of the page numbers on two facing pages of a book is 210. Find the page numbers.

7. The product of two consecutive odd integers is 255. Find the integers.

8. The product of two consecutive even integers is 224. Find the integers.

9. *Furnishings.* A rectangular table in Arlo's House of Tunes is six times as long as it is wide. If the area of the table is 24 ft^2, find the length and the width of the table.

10. *Framing.* A rectangular picture frame is twice as long as it is wide. If the area of the frame is 288 in^2, find its dimensions.

11. *Design.* The screen of the TI-84 Plus graphing calculator is nearly rectangular. The length of the rectangle is 2 cm more than the width. If the area of the rectangle is 24 cm^2, find the length and the width.

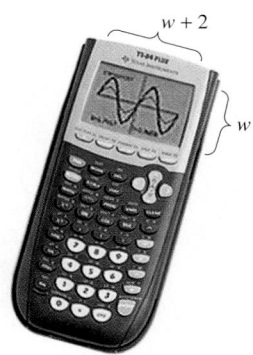

12. *Area of a garden.* The length of a rectangular garden is 4 m greater than the width. The area of the garden is 96 m^2. Find the length and the width.

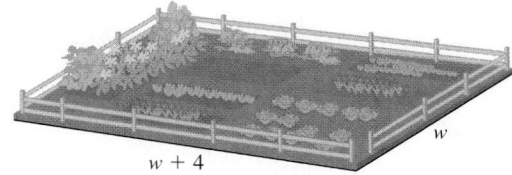

13. *Dimensions of a triangle.* A triangle is 10 cm wider than it is tall. The area is 28 cm^2. Find the height and the base.

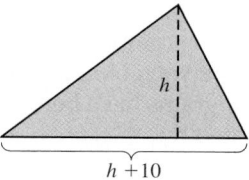

14. *Dimensions of a triangle.*   The height of a triangle is 3 cm less than the length of the base. If the area of the triangle is 35 cm², find the height and the length of the base.

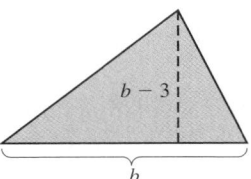

15. *Dimensions of a sail.*   The height of the jib sail on a Lightning sailboat is 5 ft greater than the length of its "foot." If the area of the sail is 42 ft², find the length of the foot and the height of the sail.

16. *Road design.*   A triangular traffic island has a base half as long as its height. Find the base and the height if the island has an area of 64 m².

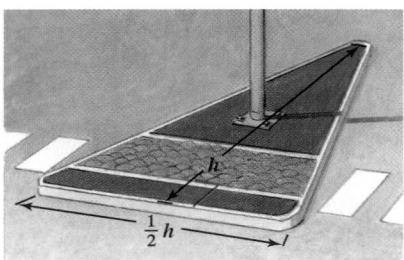

17. *Medicine.*   For many people suffering from constricted bronchial muscles, the drug Albuterol is prescribed. The number of micrograms *A* of Albuterol in a person's bloodstream *t* minutes after 200 micrograms have been inhaled can be approximated by

$$A = -50t^2 + 200t.$$

How long after an inhalation will there be about 150 micrograms of Albuterol in the bloodstream?
Source: Based on information in Chohan, Naina, Rita M. Doyle, and Patricia Nayle (eds.), *Nursing Handbook*, 21st ed. Springhouse, PA: Springhouse Corporation, 2001

18. *Medicine.*   For adults with certain heart conditions, the drug Primacor (milrinone lactate) is prescribed. The number of milligrams *M* of Primacor in the bloodstream of a 132-lb patient *t* hours after a 3-mg dose has been injected can be approximated by

$$M = -\frac{1}{2}t^2 + \frac{5}{2}t.$$

How long after an injection will there be about 2 mg in the bloodstream?
Source: Based on information in Chohan, Naina, Rita M. Doyle, and Patricia Nayle (eds.), *Nursing Handbook*, 21st ed. Springhouse, PA: Springhouse Corporation, 2001

Games in a league's schedule.   In a sports league of x teams in which all teams play each other twice, the total number N of games played is given by

$$x^2 - x = N.$$

Use this formula for Exercises 19 and 20.

19. The Colchester Youth Soccer League plays a total of 240 games, with all teams playing each other twice. How many teams are in the league?

20. A women's softball league plays a total of 132 games. How many teams are in the league?

Number of handshakes.   *The number of possible handshakes H within a group of n people is given by* $H = \frac{1}{2}(n^2 - n)$. *Use this formula for Exercises 21–24.*

21. At a meeting, there are 15 people. How many handshakes are possible?

22. At a party, there are 30 people. How many handshakes are possible?

23. *High-fives.*   After winning the championship, all Los Angeles Lakers teammates exchanged "high-fives." Altogether there were 66 high-fives. How many players were there?

24. *Toasting.*   During a toast at a party, there were 190 "clicks" of glasses. How many people took part in the toast?

25. *Construction.*   The diagonal braces in a lookout tower are 15 ft long and span a distance of 12 ft. How high does each brace reach vertically?

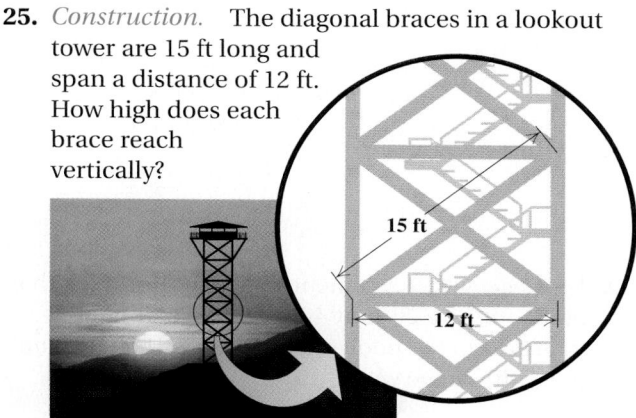

26. *Reach of a ladder.*   Twyla has a 26-ft ladder leaning against her house. If the bottom of the ladder is 10 ft from the base of the house, how high does the ladder reach?

27. *Physical education.*   An outdoor-education ropes course includes a cable that slopes downward from a height of 37 ft to a height of 30 ft. The trees that the cable connects are 24 ft apart. How long is the cable?

28. *Aviation.*   Engine failure forced Geraldine to pilot her Cessna 150 to an emergency landing. To land, Geraldine's plane glided 17,000 ft over a 15,000-ft stretch of deserted highway. From what altitude did the descent begin?

29. *Architecture.*   An architect has allocated a rectangular space of 264 ft² for a square dining room and a 10-ft wide kitchen, as shown in the figure. Find the dimensions of each room.

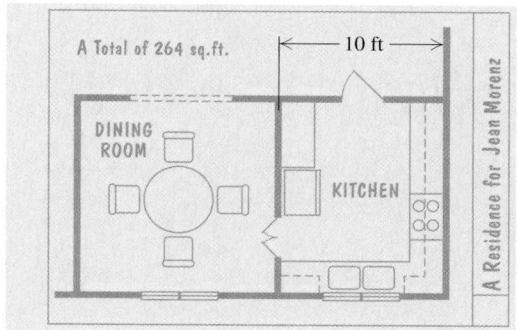

30. *Guy wire.*   The guy wire on a TV antenna is 1 m longer than the height of the antenna. If the guy wire is anchored 3 m from the foot of the antenna, how tall is the antenna?

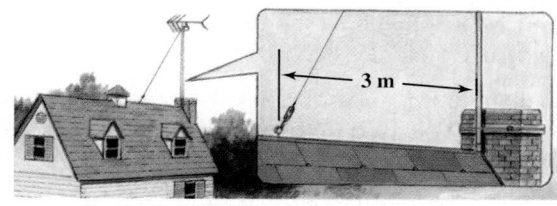

Height of a rocket.   *For Exercises 31–34, assume that a water rocket is launched upward with an initial velocity of 48 ft/sec. Its height h, in feet, after t seconds, is given by* $h = 48t - 16t^2$.

31. Determine the height of the rocket $\frac{1}{2}$ sec after it has been launched.

32. Determine the height of the rocket 1.5 sec after it has been launched.

33. When will the rocket be exactly 32 ft above the ground?

34. When will the rocket crash into the ground?

35. Do we now have the ability to solve *any* problem that translates to a quadratic equation? Why or why not?

36. Write a problem for a classmate to solve such that only one of two solutions of a quadratic equation can be used as an answer.

SKILL MAINTENANCE

Simplify.

37. $-\dfrac{2}{3} \cdot \dfrac{4}{7}$ [1.7]

38. $-\dfrac{4}{5} \cdot \dfrac{2}{9}$ [1.7]

39. $\dfrac{5}{6}\left(\dfrac{-7}{9}\right)$ [1.7]

40. $\dfrac{3}{8}\left(-\dfrac{5}{6}\right)$ [1.7]

41. $-\dfrac{2}{3} + \dfrac{4}{7}$ [1.5]

42. $-\dfrac{4}{5} + \dfrac{2}{9}$ [1.5]

43. $\dfrac{5}{6} + \dfrac{-7}{9}$ [1.5]

44. $\dfrac{3}{8} + \left(-\dfrac{5}{6}\right)$ [1.5]

SYNTHESIS

The converse of the Pythagorean theorem is also true. That is, if $a^2 + b^2 = c^2$, then the triangle is a right triangle (where a and b are the lengths of the legs and c is the length of the hypotenuse). Use this result to answer Exercises 45 and 46.

45. An archaeologist has measuring sticks of 3 ft, 4 ft,
Aha!   and 5 ft. Explain how she could draw a 7-ft by 9-ft rectangle on a piece of land being excavated.

46. Explain how measuring sticks of 5 cm, 12 cm, and 13 cm can be used to draw a right triangle that has two 45° angles.

47. *Sailing.*   The mainsail of a Lightning sailboat is a right triangle in which the hypotenuse is called the leech. If a 24-ft tall mainsail has a leech length of 26 ft and if Dacron® sailcloth costs $10 per square foot, find the cost of a new mainsail.

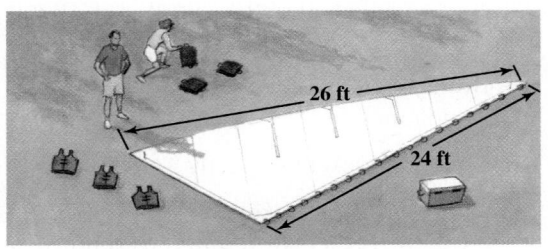

48. *Roofing.*   A *square* of shingles covers 100 ft^2 of surface area. How many squares will be needed to reshingle the house shown?

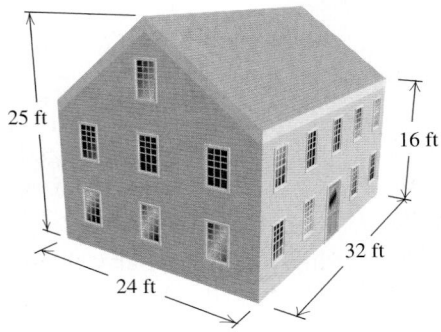

49. Solve for x.

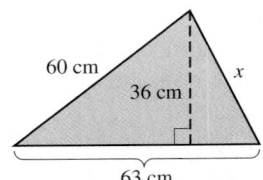

50. *Pool sidewalk.*   A cement walk of uniform width is built around a 20-ft by 40-ft rectangular pool. The total area of the pool and the walk is 1500 ft^2. Find the width of the walk.

51. *Dimensions of an open box.*   A rectangular piece of cardboard is twice as long as it is wide. A 4-cm square is cut out of each corner, and the sides are turned up to make a box with an open top. The volume of the box is 616 cm^3. Find the original dimensions of the cardboard.

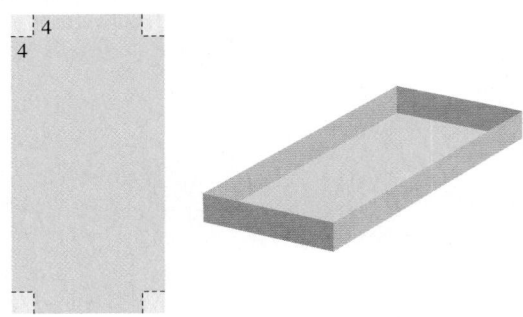

52. Find a polynomial for the shaded area in the figure below.

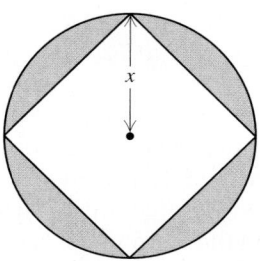

53. *Telephone service.*   Use the information in the figure below to determine the height of the telephone pole.

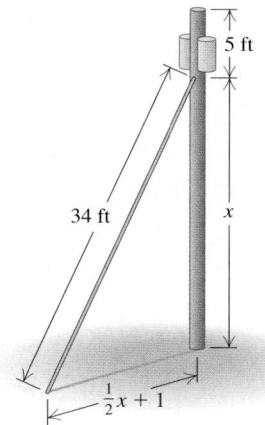

54. *Dimensions of a closed box.*   The total surface area of a closed box is 350 m². The box is 9 m high and has a square base and lid. Find the length of a side of the base.

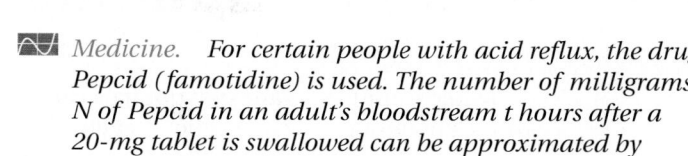
Medicine.   *For certain people with acid reflux, the drug Pepcid (famotidine) is used. The number of milligrams N of Pepcid in an adult's bloodstream t hours after a 20-mg tablet is swallowed can be approximated by*

$$N = -0.009t\,(t - 12)^3.$$

Use a graphing calculator with the window $[-1, 13, -1, 25]$ *and the* TRACE *feature to answer Exercises 55–57.*

Source: Based on information in Chohan, Naina, Rita M. Doyle, and Patricia Nayle (eds.), *Nursing Handbook*, 21st ed. Springhouse, PA: Springhouse Corporation, 2001

55. Approximately how long after a tablet is swallowed will there be 18 mg in the bloodstream?

56. Approximately how long after a tablet is swallowed will there be 10 mg in the bloodstream?

57. Approximately how long after a tablet is swallowed will the peak dosage in the bloodstream occur?

5 Study Summary

Like whole numbers, many polynomials can be expressed as products. The process of writing a polynomial as a product is called **factoring** and the polynomials used in the product are called **factors** (p. 304). In some instances, all terms in the polynomial being factored share a **common factor** that can be factored out by using the distributive law (p. 305). Other times, the polynomial being factored may be a **perfect-square trinomial**, a **difference of squares**, or a **sum** or **difference of cubes** and can be factored using a recognized pattern (pp. 331, 333, and 338). Often trinomials can be factored by either **trial and error**, using FOIL, or **grouping** (pp. 312, 321, and 326). A polynomial that cannot be factored is said to be **prime** (p. 317). In order for a polynomial to be **factored completely**, each factor in the factorization must be prime (p. 334).

$$3x^2 + 12x = 3x(x + 4)$$ ⟵ $3x$ is the largest common factor.
$$ad + ac + bc + bd = (a + b)(c + d)$$ ⟵ Factoring by grouping
$$4t^2 + 20t + 25 = (2t + 5)^2$$ ⟵ Factoring a perfect-square trinomial
$$9x^2 - 1 = (3x - 1)(3x + 1)$$ ⟵ Factoring a difference of squares
$$8m^3 + n^3 = (2m + n)(4m^2 - 2mn + n^2)$$ ⟵ Factoring a sum of cubes
$$y^6 - 125 = (y^2 - 5)(y^4 + 5y^2 + 25)$$ ⟵ Factoring a difference of cubes
$$4t^2 - 21t - 18 = (4t + 3)(t - 6)$$ ⟵ Using trial and error (FOIL) or grouping

(For a general strategy for factoring, see p. 342.)

Certain equations can be solved by factoring and using **the principle of zero products** (p. 348). Often **quadratic equations** can be solved in this manner (p. 348):

$$3x^2 - 4x - 4 = 0$$
$$(3x + 2)(x - 2) = 0$$
$$3x + 2 = 0 \quad or \quad x - 2 = 0$$
$$3x = -2 \quad or \quad x = 2$$
$$x = -\tfrac{2}{3} \quad or \quad x = 2. \quad \text{The solutions are } -\tfrac{2}{3} \text{ and 2.}$$

Quadratic equations sometimes arise from use of the **Pythagorean theorem** (p. 360):

$$a^2 + b^2 = c^2.$$

This indicates 90°.

5 Review Exercises

↪ *Concept Reinforcement* *Classify each statement as either true or false.*

1. Every polynomial has a common factor other than 1 or −1. [5.1]

2. A prime polynomial has no common factor other than 1 or −1. [5.2]

3. Every perfect-square trinomial can be expressed as a binomial squared. [5.4]

4. Every binomial can be regarded as a difference of squares. [5.4]

5. In a right triangle, the hypotenuse is always longer than either leg. [5.8]

6. Every quadratic equation has two different solutions. [5.7]

7. The principle of zero products can be applied whenever a product equals 0. [5.7]

8. The Pythagorean theorem can be applied to any triangle that has an angle measuring at least 90°. [5.8]

Find three factorizations of each monomial. [5.1]

9. $36x^3$

10. $-20x^5$

Factor completely. If a polynomial is prime, state this.

11. $12x^4 - 18x^3$ [5.1]

12. $8a^2 - 12a$ [5.1]

13. $4t^2 - 9$ [5.4]

14. $x^2 + 4x - 12$ [5.2]

15. $x^2 + 14x + 49$ [5.4]

16. $12x^3 + 12x^2 + 3x$ [5.4]

17. $6x^3 + 9x^2 + 2x + 3$ [5.1]

18. $6t^2 - 5t - 1$ [5.3]

19. $25t^2 + 9 - 30t$ [5.4]

20. $48t^2 - 28t + 6$ [5.1]

21. $81a^4 - 1$ [5.4]

22. $9x^3 + 12x^2 - 45x$ [5.3]

23. $2x^3 - 250$ [5.5]

24. $x^4 + 4x^3 - 2x - 8$ [5.1]

25. $a^2b^4 - 36$ [5.4]

26. $-8x^6 + 32x^5 - 4x^4$ [5.1]

27. $75 + 12x^2 - 60x$ [5.4]

28. $a^2 + 4$ [5.4]

29. $-x^3 + x^2 + 30x$ [5.2]

30. $4x^2 - 25$ [5.4]

31. $4z^2 - 6z$ [5.1]

32. $15z + 10z^2$ [5.1]

33. $4t^2 + 13t + 10$ [5.3]

34. $2t^2 - 7t - 4$ [5.3]

35. $7x^3 + 35x^2 + 28x$ [5.2]

36. $5x^3 + 35x^2 + 50x$ [5.2]

37. $18x^2 - 12x + 2$ [5.4]

38. $-3x^2 + 27$ [5.4]

39. $15 - 8x + x^2$ [5.2]

40. $8y^3 + 27x^6$ [5.5]

41. $x^2y^2 + xy - 12$ [5.2]

42. $12a^2 + 84ab + 147b^2$ [5.4]

43. $m^2 + 5m + mt + 5t$ [5.1]

44. $32x^4 - 128y^4z^4$ [5.4]

45. $6m^2 + 2mn + n^2 + 3mn$ [5.1], [5.3]

46. $r^2 - 3rt - 15t^2$ [5.2]

Solve. [5.7]

47. $(x - 7)(x + 8) = 0$

48. $x^3 + 2x^2 - 35x = 0$

49. $9x^2 = 1$

50. $3x^2 + 2 = 5x$

51. $2x^2 - 5x = 12$

52. $(x + 1)(x - 2) = 4$

53. $6t - 8t^2 = 0$

54. $12t^2 - 16t = 0$

55. The square of a number is 12 more than the number. Find all such numbers. [5.8]

56. The formula $x^2 - x = N$ can be used to determine the total number of games played, N, in a league of x teams in which all teams play each other twice. Serena referees for a soccer league in which all teams play each other twice and a total of 90 games is played. How many teams are in the league? [5.8]

57. Find the x-intercepts for the graph of $y = 2x^2 - 3x - 5$. [5.7]

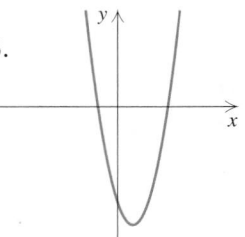

58. A triangular sign is as wide as it is tall. Its area is 800 cm². Find the height and the base. [5.8]

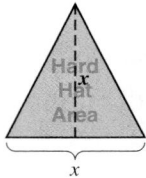

59. Nigel has a 26-ft ladder leaning against his house. If the bottom of the ladder is 10 ft from the base of the house, how high does the ladder reach? [5.8]

SYNTHESIS

60. On a quiz, Edith writes the factorization of $4x^2 - 100$ as $(2x - 10)(2x + 10)$. If this were a 10-point question, how many points would you give Edith? Why? [5.4]

61. How do the equations solved in this chapter differ from those solved in previous chapters? [5.7]

Solve.

62. The pages of a book measure 15 cm by 20 cm. Margins of equal width surround the printing on each page and constitute one half of the area of the page. Find the width of the margins. [5.8]

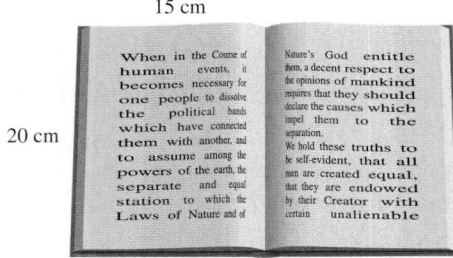

63. The cube of a number is the same as twice the square of the number. Find the number. [5.8]

64. The length of a rectangle is two times its width. When the length is increased by 20 cm and the width is decreased by 1 cm, the area is 160 cm². Find the original length and width. [5.8]

65. The length of each side of a square is increased by 5 cm to form a new square. The area of the new square is $2\frac{1}{4}$ times the area of the original square. Find the area of each square. [5.8]

Solve. [5.7]

66. $(x - 2)2x^2 + x(x - 2) - (x - 2)15 = 0$

Aha! **67.** $x^2 + 25 = 0$

---

5 Chapter Test

1. Find three factorizations of $8x^{4.}$

Factor completely.

2. $x^2 - 7x + 10$

3. $x^2 + 25 - 10x$

4. $6y^2 - 8y^3 + 4y^4$

5. $x^3 + x^2 + 2x + 2$

6. $t^7 - 3t^5$

7. $x^3 + 2x^2 - 3x$

8. $28x - 48 + 10x^2$

9. $4t^2 - 25$

10. $x^2 - x - 12$

11. $-6m^3 - 9m^2 - 3m$

12. $3w^2 - 75$

13. $45r^2 + 60r + 20$

14. $3x^4 - 48$

15. $49t^2 + 36 + 84t$

16. $5x^2 - 26x + 5$

17. $x^4 + 2x^3 - 3x - 6$

18. $2m^6 + 16$

19. $4x^2 - 4x - 15$

20. $6t^3 + 9t^2 - 15t$

21. $3m^2 - 9mn - 30n^2$

Solve.

22. $x^2 + x - 20 = 0$

23. $2x^3 - 7x^2 = 15x$

24. $x(x - 3) = 28$

25. Find the x-intercepts for the graph of $y = 3x^2 - 5x - 8$.

26. The length of a rectangle is 2 m more than the width. The area of the rectangle is 48 m². Find the length and the width.

27. A mason wants to be sure she has a right corner in a building's foundation. She marks a point 3 ft from the corner along one wall and another point 4 ft from the corner along the other wall. If the corner is a right angle, what should the distance be between the two marked points?

SYNTHESIS

28. An open rectangular gutter is made by turning up the sides of a piece of metal 20 in. wide. The area of the cross-section of the gutter is 48 in². Find the possible depths of the gutter.

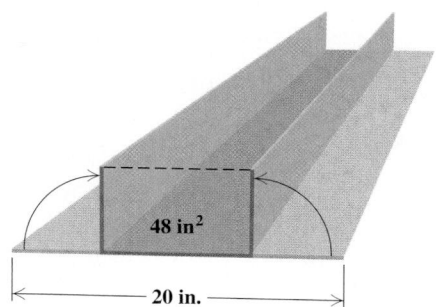

48 in²

20 in.

29. The length of a rectangle is five times its width. When the length is decreased by 3 and the width is increased by 2, the area of the new rectangle is 60. Find the original length and width.

30. Factor: $(a + 3)^2 - 2(a + 3) - 35$.

31. Solve: $20x(x + 2)(x - 1) = 5x^3 - 24x - 14x^2$.

6

Rational Expressions and Equations

AN APPLICATION

In South Africa, the design of every woven handbag, or *gipatsi*, is created by repeating two or more patterns around the bag. If a weaver uses a four-strand, a six-strand, and an eight-strand pattern, what is the smallest number of strands needed in order for all three patterns to repeat a whole number of times?

This problem appears as Exercise 82 in Section 6.3.

Estelle Carlson
HANDWEAVER AND DESIGNER
Los Angeles, California

Weavers use math when designing patterns, from calculating the size of the pattern and number of times it repeats to figuring the number of threads per inch and the weight of the thread. I also use math to convert between English and metric units when dyeing warps and woofs.

*J*ust as fractions are needed to solve certain arithmetic problems, rational expressions *similar to those in the following pages are needed to solve certain algebra problems. We now learn how to simplify, add, subtract, multiply, and* divide rational expressions. These skills will then be used to solve the equations that arise from real-life problems like the one on the preceding page.

6.1 Rational Expressions

Simplifying Rational Expressions • Factors That Are Opposites

Just as a rational number is any number that can be written as a quotient of two integers, a **rational expression** is any expression that can be written as a quotient of two polynomials. The following are examples of rational expressions:

$$\frac{7}{3}, \qquad \frac{5}{x+6}, \qquad \frac{t^2 - 5t + 6}{4t^2 - 7}, \qquad 2x + 7 + \frac{1}{x}.$$

Rational expressions are examples of *algebraic fractions*. They are also examples of *fractional expressions*.

Because rational expressions indicate division, we must be careful to avoid denominators that are 0. When a variable is replaced with a number that produces a denominator of 0, the rational expression is undefined. For example, in the expression

$$\frac{x+3}{x-7},$$

when x is replaced with 7, the denominator is 0, and the expression is undefined:

$$\frac{x+3}{x-7} = \frac{7+3}{7-7} = \frac{10}{0}. \quad \longleftarrow \quad \text{As explained in Chapter 1, division by 0 is undefined.}$$

When x is replaced with a number other than 7—say, 6—the expression *is* defined because the denominator is not zero:

$$\frac{x+3}{x-7} = \frac{6+3}{6-7} = \frac{9}{-1} = -9.$$

EXAMPLE 1 Find all numbers for which the rational expression

$$\frac{x+4}{x^2 - 3x - 10}$$

is undefined.

Solution The value of the numerator has no bearing on whether or not a rational expression is defined. To determine which numbers make the rational expression undefined, we set the *denominator* equal to 0 and solve:

$$x^2 - 3x - 10 = 0 \qquad \text{We must write this equation.}$$

$$(x - 5)(x + 2) = 0 \qquad \text{Factoring}$$

$$x - 5 = 0 \quad or \quad x + 2 = 0 \qquad \text{Using the principle of zero products}$$

$$x = 5 \quad or \qquad x = -2. \qquad \text{Solving each equation}$$

Check:

For $x = 5$:

$$\frac{x + 4}{x^2 - 3x - 10} = \frac{5 + 4}{5^2 - 3 \cdot 5 - 10} \qquad \text{There are no restrictions on the numerator.}$$

$$= \frac{9}{25 - 15 - 10} = \frac{9}{0}. \qquad \text{This expression is undefined, as expected.}$$

For $x = -2$:

$$\frac{x + 4}{x^2 - 3x - 10} = \frac{-2 + 4}{(-2)^2 - 3(-2) - 10}$$

$$= \frac{2}{4 + 6 - 10} = \frac{2}{0}. \qquad \text{This expression is undefined, as expected.}$$

Thus, $\dfrac{x + 4}{x^2 - 3x - 10}$ is undefined for $x = 5$ and $x = -2$.

technology connection

To check Example 1 with a graphing calculator, let $y_1 = x^2 - 3x - 10$ and $y_2 = (x + 4)/(x^2 - 3x - 10)$ or $(x + 4)/y_1$ and use the TABLE feature. Since $x^2 - 3x - 10$ is 0 for $x = 5$, it is impossible to evaluate y_2 for $x = 5$.

TBL START = 0   ΔTBL = 1

X	Y₁	Y₂
0	−10	−.4
1	−12	−.4167
2	−12	−.5
3	−10	−.7
4	−6	−1.333
5	0	ERROR
6	8	1.25

X = 5

Simplifying Rational Expressions

A rational expression is said to be *simplified* when the numerator and the denominator have no factors (other than 1) in common. To simplify a rational expression, we first factor the numerator and the denominator. We then identify factors common to the numerator and the denominator, rewrite the expression as a product of two rational expressions (one of which is equal to 1), and then remove the factor equal to 1. The process is identical to that used in Section 1.3 to simplify $\frac{15}{40}$:

$$\frac{15}{40} = \frac{3 \cdot 5}{8 \cdot 5} \qquad \text{Factoring the numerator and the denominator. Note the common factor, 5.}$$

$$= \frac{3}{8} \cdot \frac{5}{5} \qquad \text{Rewriting as a product of two fractions}$$

$$= \frac{3}{8} \cdot 1 \qquad \frac{5}{5} = 1$$

$$= \frac{3}{8}. \qquad \text{Using the identity property of 1 to remove the factor 1}$$

Similar steps are followed when simplifying rational expressions: We factor and remove a factor equal to 1, using the fact that

$$\frac{ab}{cb} = \frac{a}{c} \cdot \frac{b}{b}.$$

EXAMPLE 2   Simplify: $\dfrac{8x^2}{24x}$.

Solution

$$\dfrac{8x^2}{24x} = \dfrac{8 \cdot x \cdot x}{3 \cdot 8 \cdot x}$$ Factoring the numerator and the denominator. Note the common factor of $8 \cdot x$.

$$= \dfrac{x}{3} \cdot \dfrac{8x}{8x}$$ Rewriting as a product of two rational expressions

$$= \dfrac{x}{3} \cdot 1 \qquad \dfrac{8x}{8x} = 1$$

$$= \dfrac{x}{3}$$ Removing the factor 1

We say that $\dfrac{8x^2}{24x}$ *simplifies* to $\dfrac{x}{3}$.* In the work that follows, we assume that all denominators are nonzero.

EXAMPLE 3   Simplify: $\dfrac{5a + 15}{10}$.

Solution

$$\dfrac{5a + 15}{10} = \dfrac{5(a + 3)}{5 \cdot 2}$$ Factoring the numerator and the denominator. Note the common factor of 5.

$$= \dfrac{5}{5} \cdot \dfrac{a + 3}{2}$$ Rewriting as a product of two rational expressions

$$= 1 \cdot \dfrac{a + 3}{2} \qquad \dfrac{5}{5} = 1$$

$$= \dfrac{a + 3}{2}$$ Removing the factor 1

Sometimes the common factor has two or more terms.

EXAMPLE 4   Simplify.

a) $\dfrac{6x + 12}{7x + 14}$ **b)** $\dfrac{6a^2 + 4a}{2a^2 + 2a}$ **c)** $\dfrac{x^2 + 3x + 2}{x^2 - 1}$

Solution

a) $\dfrac{6x + 12}{7x + 14} = \dfrac{6(x + 2)}{7(x + 2)}$ Factoring the numerator and the denominator. Note the common factor of $x + 2$.

$$= \dfrac{6}{7} \cdot \dfrac{x + 2}{x + 2}$$ Rewriting as a product of two rational expressions

---

*In more advanced courses, we would *not* say that $8x^2/(24x)$ simplifies to $x/3$, but would instead say that $8x^2/(24x)$ simplifies to $x/3$ *with the restriction that $x \neq 0$.*

We can use the TABLE feature as a partial check that rational expressions have been simplified correctly. To check the simplification in Example 4(c),

$$\frac{x^2 + 3x + 2}{x^2 - 1} = \frac{x + 2}{x - 1},$$

we enter $y_1 = (x^2 + 3x + 2)/(x^2 - 1)$ and $y_2 = (x + 2)/(x - 1)$ and select the mode AUTO to look at a table of values of y_1 and y_2. The values should match for all allowable replacements.

X	Y₁	Y₂
−4	.4	.4
−3	.25	.25
−2	0	0
−1	ERROR	−.5
0	−2	−2
1	ERROR	ERROR
2	4	4
X = −4		

The ERROR messages indicate that −1 and 1 are not allowable replacements in y_1 and 1 is not an allowable replacement in y_2. For all other numbers, y_1 and y_2 are the same, so the simplification appears to be correct.

Use the TABLE feature to determine which of the following appear to be correct.

1. $\dfrac{8x^2}{24x} = \dfrac{x}{3}$

2. $\dfrac{5x + 15}{10} = \dfrac{x + 3}{2}$

3. $\dfrac{x + 3}{x} = 3$

4. $\dfrac{x^2 + 3x - 4}{x^2 - 16} = \dfrac{x - 1}{x + 4}$

$$= \frac{6}{7} \cdot 1 \qquad\qquad \frac{x + 2}{x + 2} = 1$$

$$= \frac{6}{7} \qquad\qquad\qquad \text{Removing the factor 1}$$

b) $\dfrac{6a^2 + 4a}{2a^2 + 2a} = \dfrac{2a(3a + 2)}{2a(a + 1)}$ Factoring the numerator and the denominator. Note the common factor of $2a$.

$$= \frac{2a}{2a} \cdot \frac{3a + 2}{a + 1} \qquad \text{Rewriting as a product of two rational expressions}$$

$$= 1 \cdot \frac{3a + 2}{a + 1} \qquad \frac{2a}{2a} = 1$$

$$= \frac{3a + 2}{a + 1} \qquad \begin{array}{l}\text{Removing the factor 1. No common factors of} \\ \text{the numerator and the denominator exist, so} \\ \text{simplification is complete.}\end{array}$$

c) $\dfrac{x^2 + 3x + 2}{x^2 - 1} = \dfrac{(x + 1)(x + 2)}{(x + 1)(x - 1)}$ Factoring; $x + 1$ is the common factor.

$$= \frac{x + 1}{x + 1} \cdot \frac{x + 2}{x - 1} \qquad \text{Rewriting as a product of two rational expressions}$$

$$= 1 \cdot \frac{x + 2}{x - 1} \qquad \frac{x + 1}{x + 1} = 1$$

$$= \frac{x + 2}{x - 1} \qquad \text{Removing the factor 1}$$

Canceling is a shortcut that can be used—and easily *misused*—to simplify rational expressions. As stated in Section 1.3, canceling must be done with care and understanding. Essentially, canceling streamlines the process of removing a factor equal to 1. Example 4(c) could have been streamlined as follows:

$$\frac{x^2 + 3x + 2}{x^2 - 1} = \frac{\cancel{(x + 1)}(x + 2)}{\cancel{(x + 1)}(x - 1)} \qquad \begin{array}{l}\text{When a factor equal to 1 is noted,} \\ \text{it is "canceled":} \dfrac{x + 1}{x + 1} = 1\end{array}$$

$$= \frac{x + 2}{x - 1}. \qquad \text{Simplifying}$$

Caution!   Canceling is often used incorrectly. The following cancellations are *incorrect*:

$$\frac{\cancel{x} + 2}{\cancel{x} + 3}; \qquad \frac{a^2 - \cancel{5}}{\cancel{5}}, \qquad \frac{6\cancel{x}^2 + 5\cancel{x} + 1}{4\cancel{x}^2 - 3\cancel{x}}.$$

Wrong!   Wrong!   Wrong!

None of the above cancellations removes a factor equal to 1. Factors are parts of products. For example, in $x \cdot 2$, x and 2 are factors, but in $x + 2$, x and 2 are terms, *not* factors. If it is not a factor, it cannot be canceled.

EXAMPLE 5   Simplify: $\dfrac{3x^2 - 2x - 1}{x^2 - 3x + 2}$.

Solution   We factor the numerator and the denominator and look for common factors:

$$\frac{3x^2 - 2x - 1}{x^2 - 3x + 2} = \frac{(3x + 1)(x - 1)}{(x - 2)(x - 1)}$$

Try to visualize this as
$$\frac{3x + 1}{x - 2} \cdot \frac{x - 1}{x - 1}.$$

$$= \frac{3x + 1}{x - 2}.$$

Removing a factor equal to 1:
$$\frac{x - 1}{x - 1} = 1$$

When a rational expression is simplified, the result is an equivalent expression. Example 3 says that

$$\frac{5a + 15}{10} \quad \text{is equivalent to} \quad \frac{a + 3}{2}.$$

This result can be partially checked using a replacement for a. For instance, if $a = 2$, then

$$\frac{5a + 15}{10} = \frac{5 \cdot 2 + 15}{10} = \frac{25}{10} = \frac{5}{2}$$

and

> To see why this check is not foolproof, see Exercise 65.

$$\frac{a + 3}{2} = \frac{2 + 3}{2} = \frac{5}{2}.$$

If evaluating both expressions yields differing results, we know that a mistake has been made. For example, if $(5a + 15)/10$ is incorrectly simplified as $(a + 15)/2$ and we evaluate using $a = 2$, we have

$$\frac{5a + 15}{10} = \frac{5 \cdot 2 + 15}{10} = \frac{5}{2}$$

and

Different results

$$\frac{a + 15}{2} = \frac{2 + 15}{2} = \frac{17}{2},$$

which demonstrates that a mistake has been made.

Factors That Are Opposites

Consider

$$\frac{x - 4}{8 - 2x}, \quad \text{or, equivalently,} \quad \frac{x - 4}{2(4 - x)}.$$

At first glance, the numerator and the denominator do not appear to have any common factors. But $x - 4$ and $4 - x$ are opposites, or additive inverses, of each other. Thus we can find a common factor by factoring out -1 in one expression.

EXAMPLE 6   Simplify $\dfrac{x-4}{8-2x}$ and check by evaluating.

Solution   We have

$$\dfrac{x-4}{8-2x} = \dfrac{x-4}{2(4-x)}$$ Factoring

$$= \dfrac{x-4}{2(-1)(x-4)}$$ Note that $4-x = -x+4 = -1(x-4)$.

$$= \dfrac{x-4}{-2(x-4)}$$ Had we originally factored out -2, we could have gone directly to this step.

$$= \dfrac{1}{-2} \cdot \dfrac{x-4}{x-4}$$ Rewriting as a product. It is important to write the 1 in the numerator.

$$= -\dfrac{1}{2}.$$ Removing a factor equal to 1: $(x-4)/(x-4) = 1$

As a partial check, note that for any choice of x other than 4, the value of the rational expression is $-\frac{1}{2}$. For example, if $x = 6$, then

$$\dfrac{x-4}{8-2x} = \dfrac{6-4}{8-2 \cdot 6}$$

$$= \dfrac{2}{-4} = -\dfrac{1}{2}.$$

Exercise Set

6.1

FOR EXTRA HELP

Student's Solutions Manual   Digital Video Tutor CD 3 Videotape 6   AW Math Tutor Center   MathXL Tutorials on CD   MathXL   MyMathLab

↪ *Concept Reinforcement*   *In each of Exercises 1–6, match the rational expression with the list of numbers in the column on the right for which the rational expression is undefined.*

1. ___ $\dfrac{x-5}{(x-2)(x+3)}$

2. ___ $\dfrac{3t}{(t+1)(t-4)}$

a) $-1, 4$

b) $-3, 5$

c) $-\dfrac{4}{3}, \dfrac{1}{2}$

3. ___ $\dfrac{a+7}{a^2-a-12}$

4. ___ $\dfrac{m-3}{m^2-2m-15}$

d) $-3, 4$

e) $-3, 2$

f) $\dfrac{1}{3}, \dfrac{2}{3}$

5. ___ $\dfrac{2t+7}{(2t-1)(3t+4)}$

6. ___ $\dfrac{4x-1}{(3x-1)(3x-2)}$

List all numbers for which each rational expression is undefined.

7. $\dfrac{12}{-7x}$

8. $\dfrac{14}{-5y}$

9. $\dfrac{t-6}{t+8}$

10. $\dfrac{a-8}{a+7}$

11. $\dfrac{a-4}{3a-12}$

12. $\dfrac{x^2-9}{4x-12}$

13. $\dfrac{x^2-25}{x^2-3x-28}$

14. $\dfrac{p^2-9}{p^2-7p+10}$

15. $\dfrac{t^2-4}{2t^3+11t^2-6t}$

16. $\dfrac{x^2-9}{3x^3-x^2-14x}$

Simplify by removing a factor equal to 1. Show all steps.

17. $\dfrac{50a^2b}{40ab^3}$

18. $\dfrac{45x^3y^2}{9x^5y}$

19. $\dfrac{28x^2y}{21x^3y^5}$

20. $\dfrac{12a^5b^6}{18a^3b}$

21. $\dfrac{9x+15}{12x+20}$

22. $\dfrac{14x-7}{10x-5}$

23. $\dfrac{a^2-9}{a^2+4a+3}$

24. $\dfrac{a^2+5a+6}{a^2-9}$

Simplify, if possible. Then check by evaluating, as in Example 6.

25. $\dfrac{36x^6}{24x^9}$

26. $\dfrac{75a^5}{50a^3}$

27. $\dfrac{-2y+6}{-8y}$

28. $\dfrac{4x-12}{6x}$

29. $\dfrac{t^2-16}{t^2+t-20}$

30. $\dfrac{a^2-4}{a^2+5a+6}$

31. $\dfrac{3a^2+9a-12}{6a^2-30a+24}$

32. $\dfrac{2t^2-6t+4}{4t^2+12t-16}$

33. $\dfrac{x^2+8x+16}{x^2-16}$

34. $\dfrac{x^2-25}{x^2-10x+25}$

35. $\dfrac{t^2-1}{t+1}$

36. $\dfrac{a^2-1}{a-1}$

37. $\dfrac{a-2}{a^3-8}$

38. $\dfrac{a^6+27}{a^2+3}$

39. $\dfrac{y^2+4}{y+2}$

40. $\dfrac{x^2+1}{x+1}$

41. $\dfrac{5x^2+20}{10x^2+40}$

42. $\dfrac{6x^2+54}{4x^2+36}$

43. $\dfrac{y^2+6y}{2y^2+13y+6}$

44. $\dfrac{t^2+2t}{2t^2+t-6}$

45. $\dfrac{4x^2-12x+9}{10x^2-11x-6}$

46. $\dfrac{4x^2-4x+1}{6x^2+5x-4}$

47. $\dfrac{x-9}{9-x}$

48. $\dfrac{6-x}{x-6}$

49. $\dfrac{7t-14}{2-t}$

50. $\dfrac{4a-12}{3-a}$

51. $\dfrac{a-b}{3b-3a}$

52. $\dfrac{q-p}{2p-2q}$

53. $\dfrac{3x^2-3y^2}{2y^2-2x^2}$

54. $\dfrac{7a^2-7b^2}{3b^2-3a^2}$

Aha! **55.** $\dfrac{7s^2-28t^2}{28t^2-7s^2}$

56. $\dfrac{9m^2-4n^2}{4n^2-9m^2}$

57. Explain how simplifying is related to the identity property of 1.

58. If a rational expression is undefined for $x=5$ and $x=-3$, what is the degree of the denominator? Why?

SKILL MAINTENANCE

Simplify.

59. $-\dfrac{2}{3}\cdot\dfrac{6}{7}$ [1.7]

60. $\dfrac{5}{9}\left(\dfrac{-6}{11}\right)$ [1.7]

61. $\dfrac{5}{8}\div\left(-\dfrac{1}{6}\right)$ [1.7]

62. $\dfrac{7}{10}\div\left(-\dfrac{8}{15}\right)$ [1.7]

63. $\dfrac{7}{9}-\dfrac{2}{3}\cdot\dfrac{6}{7}$ [1.8]

64. $\dfrac{2}{3}-\left(\dfrac{3}{4}\right)^2$ [1.8]

SYNTHESIS

65. Terry *incorrectly* simplifies

$$\frac{x^2+x-2}{x^2+3x+2} \quad \text{as} \quad \frac{x-1}{x+2}.$$

He then checks his simplification by evaluating both expressions for $x=1$. Use this situation to explain why evaluating is not a foolproof check.

66. How could you convince someone that $a-b$ and $b-a$ are opposites of each other?

Simplify.

67. $\dfrac{16y^4 - x^4}{(x^2 + 4y^2)(x - 2y)}$

68. $\dfrac{(x - 1)(x^4 - 1)(x^2 - 1)}{(x^2 + 1)(x - 1)^2(x^4 - 2x^2 + 1)}$

69. $\dfrac{x^5 - 2x^3 + 4x^2 - 8}{x^7 + 2x^4 - 4x^3 - 8}$

70. $\dfrac{10t^4 - 8t^3 + 15t - 12}{8 - 10t + 12t^2 - 15t^3}$

71. $\dfrac{(t^4 - 1)(t^2 - 9)(t - 9)^2}{(t^4 - 81)(t^2 + 1)(t + 1)^2}$

72. $\dfrac{(t + 2)^3(t^2 + 2t + 1)(t + 1)}{(t + 1)^3(t^2 + 4t + 4)(t + 2)}$

73. $\dfrac{(x^2 - y^2)(x^2 - 2xy + y^2)}{(x + y)^2(x^2 - 4xy - 5y^2)}$

74. $\dfrac{x^4 - y^4}{(y - x)^4}$

75. Select any number x, multiply by 2, add 5, multiply by 5, subtract 25, and divide by 10. What do you get? Explain how this procedure can be used for a number trick.

6.2	**Multiplication and Division**

Multiplication • Division

Multiplication and division of rational expressions are similar to multiplication and division with fractions. In this section, we again assume that all denominators are nonzero.

Multiplication

Recall that to multiply fractions, we multiply numerator times numerator and denominator times denominator. Rational expressions are multiplied in a similar way.

> **The Product of Two Rational Expressions**
>
> To multiply rational expressions, multiply numerators and multiply denominators:
> $$\frac{A}{B} \cdot \frac{C}{D} = \frac{AC}{BD}.$$
> Then factor and simplify the result if possible.

For example,
$$\frac{3}{5} \cdot \frac{8}{11} = \frac{24}{55} \quad \text{and} \quad \frac{x}{3} \cdot \frac{x + 2}{y} = \frac{x(x + 2)}{3y}.$$

Fraction bars are grouping symbols, so parentheses are needed when writing some products. Because we generally simplify, we often leave the product in factored form. There is no need to multiply further.

EXAMPLE 1 Multiply and, if possible, simplify.

a) $\dfrac{5a^3}{4} \cdot \dfrac{2}{5a}$ b) $\dfrac{x^2 + 6x + 9}{x^2 - 4} \cdot \dfrac{x - 2}{x + 3}$ c) $\dfrac{x^2 + x - 2}{15} \cdot \dfrac{5}{2x^2 - 3x + 1}$

Solution

a) $\dfrac{5a^3}{4} \cdot \dfrac{2}{5a} = \dfrac{5a^3(2)}{4(5a)}$ Forming the product of the numerators and the product of the denominators

$= \dfrac{5 \cdot a \cdot a \cdot a \cdot 2}{2 \cdot 2 \cdot 5 \cdot a}$ Factoring the numerator and the denominator

$= \dfrac{5 \cdot a \cdot \cancel{a} \cdot a \cdot 2}{2 \cdot 2 \cdot \cancel{5} \cdot \cancel{a}}$ Removing a factor equal to 1: $\dfrac{2 \cdot 5 \cdot a}{2 \cdot 5 \cdot a} = 1$

$= \dfrac{a^2}{2}$

b) $\dfrac{x^2 + 6x + 9}{x^2 - 4} \cdot \dfrac{x - 2}{x + 3} = \dfrac{(x^2 + 6x + 9)(x - 2)}{(x^2 - 4)(x + 3)}$ Multiplying the numerators and the denominators

$= \dfrac{(x + 3)(x + 3)(x - 2)}{(x + 2)(x - 2)(x + 3)}$ Factoring the numerator and the denominator

$= \dfrac{\cancel{(x + 3)}(x + 3)\cancel{(x - 2)}}{(x + 2)\cancel{(x - 2)}\cancel{(x + 3)}}$ Removing a factor equal to 1: $\dfrac{(x + 3)(x - 2)}{(x + 3)(x - 2)} = 1$

$= \dfrac{x + 3}{x + 2}$

c) $\dfrac{x^2 + x - 2}{15} \cdot \dfrac{5}{2x^2 - 3x + 1} = \dfrac{(x^2 + x - 2)5}{15(2x^2 - 3x + 1)}$ Multiplying the numerators and the denominators

$= \dfrac{(x + 2)(x - 1)5}{5(3)(x - 1)(2x - 1)}$ Factoring the numerator and the denominator. Try to go directly to this step.

$= \dfrac{(x + 2)\cancel{(x - 1)}\cancel{5}}{\cancel{5}(3)\cancel{(x - 1)}(2x - 1)}$ Removing a factor equal to 1: $\dfrac{5(x - 1)}{5(x - 1)} = 1$

$= \dfrac{x + 2}{3(2x - 1)}$

Because our results are often used in addition or subtraction problems that require factored form, there is no need to multiply out the numerator or the denominator.

Division

As with fractions, reciprocals of rational expressions are found by interchanging the numerator and the denominator. For example,

the reciprocal of $\dfrac{2}{7}$ is $\dfrac{7}{2}$,   and   the reciprocal of $\dfrac{3x}{x + 5}$ is $\dfrac{x + 5}{3x}$.

> **The Quotient of Two Rational Expressions**
>
> To divide by a rational expression, multiply by its reciprocal:
>
> $$\frac{A}{B} \div \frac{C}{D} = \frac{A}{B} \cdot \frac{D}{C} = \frac{AD}{BC}.$$
>
> Then factor and, if possible, simplify.

EXAMPLE 2   Divide: **(a)** $\dfrac{x}{5} \div \dfrac{7}{y}$; **(b)** $(x + 2) \div \dfrac{x - 1}{x + 3}$.

Solution

a) $\dfrac{x}{5} \div \dfrac{7}{y} = \dfrac{x}{5} \cdot \dfrac{y}{7}$ Multiplying by the reciprocal of the divisor

$\qquad\quad = \dfrac{xy}{35}$ Multiplying rational expressions

b) $(x + 2) \div \dfrac{x - 1}{x + 3} = \dfrac{x + 2}{1} \cdot \dfrac{x + 3}{x - 1}$ Multiplying by the reciprocal of the divisor. Writing $x + 2$ as $\dfrac{x + 2}{1}$ can be helpful.

$\qquad\qquad\qquad\quad = \dfrac{(x + 2)(x + 3)}{x - 1}$

As usual, we should simplify when possible. Often that requires us to factor one or more polynomials. Our hope is to discover a common factor that appears in both the numerator and the denominator.

EXAMPLE 3   Divide and, if possible, simplify: $\dfrac{x + 1}{x^2 - 1} \div \dfrac{x + 1}{x^2 - 2x + 1}$.

Solution

$$\frac{x + 1}{x^2 - 1} \div \frac{x + 1}{x^2 - 2x + 1} = \frac{x + 1}{x^2 - 1} \cdot \frac{x^2 - 2x + 1}{x + 1}$$ Multiplying by the reciprocal of the divisor

$$= \frac{(x + 1)(x - 1)(x - 1)}{(x + 1)(x - 1)(x + 1)}$$ Multiplying rational expressions and factoring numerators and denominators

$$= \frac{\cancel{(x + 1)}\cancel{(x - 1)}(x - 1)}{\cancel{(x + 1)}\cancel{(x - 1)}(x + 1)}$$ Removing a factor equal to 1: $\dfrac{(x + 1)(x - 1)}{(x + 1)(x - 1)} = 1$

$$= \frac{x - 1}{x + 1}$$

EXAMPLE 4

Divide and, if possible, simplify.

a) $\dfrac{a^2 + 3a + 2}{a^2 + 4} \div (5a^2 + 10a)$

b) $\dfrac{x^2 - 2x - 3}{x^2 - 4} \div \dfrac{x + 1}{x + 5}$

Solution

a) $\dfrac{a^2 + 3a + 2}{a^2 + 4} \div (5a^2 + 10a)$

$= \dfrac{a^2 + 3a + 2}{a^2 + 4} \cdot \dfrac{1}{5a^2 + 10a}$ Multiplying by the reciprocal of the divisor

$= \dfrac{(a + 2)(a + 1)}{(a^2 + 4)5a(a + 2)}$ Multiplying rational expressions and factoring

$= \dfrac{(a + 2)(a + 1)}{(a^2 + 4)5a(a + 2)}$

$= \dfrac{a + 1}{(a^2 + 4)5a}$

Removing a factor equal to 1: $\dfrac{a + 2}{a + 2} = 1$

b) $\dfrac{x^2 - 2x - 3}{x^2 - 4} \div \dfrac{x + 1}{x + 5} = \dfrac{x^2 - 2x - 3}{x^2 - 4} \cdot \dfrac{x + 5}{x + 1}$ Multiplying by the reciprocal of the divisor

$= \dfrac{(x - 3)(x + 1)(x + 5)}{(x - 2)(x + 2)(x + 1)}$ Multiplying rational expressions and factoring

$= \dfrac{(x - 3)(x + 1)(x + 5)}{(x - 2)(x + 2)(x + 1)}$

$= \dfrac{(x - 3)(x + 5)}{(x - 2)(x + 2)}$

Removing a factor equal to 1: $\dfrac{x + 1}{x + 1} = 1$

technology connection

In performing a partial check of Example 4(b), care must be taken in placing parentheses. For example, we enter the original expression in Example 4(b) as $y_1 = ((x^2 - 2x - 3)/(x^2 - 4))/((x + 1)/(x + 5))$ and the simplified expression as $y_2 = ((x - 3)(x + 5))/((x - 2)(x + 2))$. Comparing values of y_1 and y_2, we see that the simplification is probably correct.

X	Y₁	Y₂
−5	ERROR	0
−4	−.5833	−.5833
−3	−2.4	−2.4
−2	ERROR	ERROR
−1	ERROR	5.3333
0	3.75	3.75
1	4	4

X = −5

1. Check Example 4(a).
2. Why are there 3 ERROR messages shown for y_1 on the screen above, and only 1 for y_2?

Exercise Set

6.2

Multiply. Leave each answer in factored form.

1. $\dfrac{7x}{5} \cdot \dfrac{x - 5}{2x + 1}$

2. $\dfrac{3x}{4} \cdot \dfrac{5x + 2}{x - 1}$

3. $\dfrac{a - 4}{a + 6} \cdot \dfrac{a + 2}{a + 6}$

4. $\dfrac{a + 3}{a + 6} \cdot \dfrac{a + 3}{a - 1}$

5. $\dfrac{2x + 3}{4} \cdot \dfrac{x + 1}{x - 5}$

6. $\dfrac{x + 2}{3x - 4} \cdot \dfrac{4}{5x + 6}$

7. $\dfrac{a - 4}{a^2 + 4} \cdot \dfrac{a + 4}{a^2 - 4}$

8. $\dfrac{t + 3}{t^2 - 2} \cdot \dfrac{t + 3}{t^2 - 4}$

9. $\dfrac{x + 6}{3 + x} \cdot \dfrac{x - 1}{x + 1}$

10. $\dfrac{m + 4}{m + 8} \cdot \dfrac{2 + m}{m + 5}$

Multiply and, if possible, simplify.

11. $\dfrac{5a^4}{6a} \cdot \dfrac{2}{a}$

12. $\dfrac{10}{t^7} \cdot \dfrac{3t^2}{25t}$

13. $\dfrac{3c}{d^2} \cdot \dfrac{8d}{6c^3}$

14. $\dfrac{3x^2y}{2} \cdot \dfrac{4}{xy^3}$

15. $\dfrac{x^2 - 3x - 10}{(x-2)^2} \cdot \dfrac{x-2}{x-5}$

16. $\dfrac{t+2}{t-2} \cdot \dfrac{t^2 - 5t + 6}{(t+2)^2}$

17. $\dfrac{a^2 + 25}{a^2 - 4a + 3} \cdot \dfrac{a-5}{a+5}$

18. $\dfrac{x+3}{x^2 + 9} \cdot \dfrac{x^2 + 5x + 4}{x+9}$

19. $\dfrac{a^2 - 9}{a^2} \cdot \dfrac{7a}{a^2 + a - 12}$

20. $\dfrac{x^2 + 10x - 11}{9x} \cdot \dfrac{x^3}{x+11}$

21. $\dfrac{4a^2}{3a^2 - 12a + 12} \cdot \dfrac{3a-6}{2a}$

22. $\dfrac{5v+5}{v-2} \cdot \dfrac{2v^2 - 8v + 8}{v^2 - 1}$

23. $\dfrac{t^2 + 2t - 3}{t^2 + 4t - 5} \cdot \dfrac{t^2 - 3t - 10}{t^2 + 5t + 6}$

24. $\dfrac{x^2 + 5x + 4}{x^2 - 6x + 8} \cdot \dfrac{x^2 + 5x - 14}{x^2 + 8x + 7}$

25. $\dfrac{5a^2 - 180}{10a^2 - 10} \cdot \dfrac{20a + 20}{2a - 12}$

26. $\dfrac{2t^2 - 98}{4t^2 - 4} \cdot \dfrac{8t + 8}{16t - 112}$

Aha! **27.** $\dfrac{x^2 + 4x + 4}{(x-1)^2} \cdot \dfrac{x^2 - 2x + 1}{(x+2)^2}$

28. $\dfrac{x^2 + 7x + 12}{x^2 + 6x + 8} \cdot \dfrac{4 - x^2}{x^2 + x - 6}$

29. $\dfrac{5t^2 + 12t + 4}{t^2 + 4t + 4} \cdot \dfrac{t^2 + 8t + 16}{5t^2 + 22t + 8}$

30. $\dfrac{3y^2 - y - 2}{y^2 - 2y + 1} \cdot \dfrac{y^2 - 4y + 4}{6y^2 - 13y + 2}$

31. $\dfrac{2x^2 - 5x + 3}{6x^2 - 5x - 1} \cdot \dfrac{6x^2 + 13x + 2}{2x^2 + 3x - 9}$

32. $\dfrac{10x^2 - x - 2}{4x^2 - 12x + 5} \cdot \dfrac{4x^2 - 8x + 3}{10x^2 - 11x - 6}$

33. $\dfrac{c^3 + 8}{c^5 - 4c^3} \cdot \dfrac{c^6 - 4c^5 + 4c^4}{c^2 - 2c + 4}$

34. $\dfrac{x^3 - 27}{x^4 - 9x^2} \cdot \dfrac{x^5 - 6x^4 + 9x^3}{x^2 + 3x + 9}$

Find the reciprocal of each expression.

35. $\dfrac{3x}{7}$

36. $\dfrac{3-x}{x^2 + 4}$

37. $a^3 - 8a$

38. $\dfrac{7}{a^2 - b^2}$

Divide and, if possible, simplify.

39. $\dfrac{5}{8} \div \dfrac{3}{7}$

40. $\dfrac{4}{9} \div \dfrac{5}{7}$

41. $\dfrac{x}{4} \div \dfrac{5}{x}$

42. $\dfrac{5}{x} \div \dfrac{x}{12}$

43. $\dfrac{a^5}{b^4} \div \dfrac{a^2}{b}$

44. $\dfrac{x^5}{y^2} \div \dfrac{x^2}{y}$

45. $\dfrac{y+5}{4} \div \dfrac{y}{2}$

46. $\dfrac{a+2}{a-3} \div \dfrac{a-1}{a+3}$

47. $\dfrac{4y - 8}{y + 2} \div \dfrac{y - 2}{y^2 - 4}$

48. $\dfrac{x^2 - 1}{x} \div \dfrac{x+1}{x-1}$

49. $\dfrac{a}{a-b} \div \dfrac{b}{b-a}$

50. $\dfrac{x-y}{6} \div \dfrac{y-x}{3}$

51. $(y^2 - 9) \div \dfrac{y^2 - 2y - 3}{y^2 + 1}$

52. $(x^2 - 5x - 6) \div \dfrac{x^2 - 1}{x + 6}$

53. $\dfrac{-3 + 3x}{16} \div \dfrac{x-1}{5}$

54. $\dfrac{-12 + 4x}{12} \div \dfrac{-6 + 2x}{6}$

55. $\dfrac{a+2}{a-1} \div \dfrac{3a+6}{a-5}$

56. $\dfrac{t-3}{t+2} \div \dfrac{4t-12}{t+1}$

57. $(2x - 1) \div \dfrac{2x^2 - 11x + 5}{4x^2 - 1}$

58. $(a + 7) \div \dfrac{3a^2 + 14a - 49}{a^2 + 8a + 7}$

59. $\dfrac{a^2 - 10a + 25}{2a^2 - a - 21} \div \dfrac{a^2 - a - 20}{3a^2 + 5a - 12}$

60. $\dfrac{3a^2 + 7a - 20}{a^2 - 2a + 1} \div \dfrac{3a^2 - 8a + 5}{a^2 - 5a - 6}$

61. $\dfrac{c^2 + 10c + 21}{c^2 - 2c - 15} \div (5c^2 + 32c - 21)$

62. $\dfrac{z^2 - 2z + 1}{z^2 - 1} \div (4z^2 - z - 3)$

63. $\dfrac{x - y}{x^2 + 2xy + y^2} \div \dfrac{x^2 - y^2}{x^2 - 5xy + 4y^2}$

64. $\dfrac{a^2 - b^2}{a^2 - 4ab + 4b^2} \div \dfrac{a^2 - 3ab + 2b^2}{a - 2b}$

65. $\dfrac{x^3 - 64}{x^3 + 64} \div \dfrac{x^2 - 16}{x^2 - 4x + 16}$

66. $\dfrac{8y^3 - 27}{64y^3 - 1} \div \dfrac{4y^2 - 9}{16y^2 + 4y + 1}$

67. $\dfrac{8a^3 + b^3}{2a^2 + 3ab + b^2} \div \dfrac{8a^2 - 4ab + 2b^2}{4a^2 + 4ab + b^2}$

68. $\dfrac{x^3 + 8y^3}{2x^2 + 5xy + 2y^2} \div \dfrac{x^3 - 2x^2y + 4xy^2}{8x^2 - 2y^2}$

69. Why is it important to insert parentheses when multiplying rational expressions in which the numerators and the denominators contain more than one term?

70. A student claims to be able to divide, but not multiply, rational expressions. Why is this claim difficult to believe?

SKILL MAINTENANCE

Simplify.

71. $\dfrac{3}{4} + \dfrac{5}{6}$ [1.3]

72. $\dfrac{7}{8} + \dfrac{5}{6}$ [1.3]

73. $\dfrac{2}{9} - \dfrac{1}{6}$ [1.3]

74. $\dfrac{3}{10} - \dfrac{7}{15}$ [1.3]

75. $\dfrac{2}{5} - \left(\dfrac{3}{2}\right)^2$ [1.8]

76. $\dfrac{5}{9} + \dfrac{2}{3} \cdot \dfrac{4}{5}$ [1.8]

SYNTHESIS

77. Is the reciprocal of a product the product of the two reciprocals? Why or why not?

78. Explain why the quotient

$$\dfrac{x + 3}{x - 5} \div \dfrac{x - 7}{x + 1}$$

is undefined for $x = 5$, $x = -1$, and $x = 7$, but *is* defined for $x = -3$.

Simplify.

79. $\dfrac{2a^2 - 5ab}{c - 3d} \div (4a^2 - 25b^2)$

80. $(x - 2a) \div \dfrac{a^2x^2 - 4a^4}{a^2x + 2a^3}$

Aha! **81.** $\dfrac{a^2 - 3b}{a^2 + 2b} \cdot \dfrac{a^2 - 2b}{a^2 + 3b} \cdot \dfrac{a^2 + 2b}{a^2 - 3b}$

82. $\dfrac{y^2 - 4xy}{y - x} \div \dfrac{16x^2y^2 - y^4}{4x^2 - 3xy - y^2} \div \dfrac{4}{x^3y^3}$

83. $\dfrac{z^2 - 8z + 16}{z^2 + 8z + 16} \div \dfrac{(z - 4)^5}{(z + 4)^5} \div \dfrac{3z + 12}{z^2 - 16}$

84. $\dfrac{x^2 - x + xy - y}{x^2 + 6x - 7} \div \dfrac{x^2 + 2xy + y^2}{4x + 4y}$

85. $\dfrac{3x + 3y + 3}{9x} \div \dfrac{x^2 + 2xy + y^2 - 1}{x^4 + x^2}$

86. $\dfrac{(t + 2)^3}{(t + 1)^3} \div \dfrac{t^2 + 4t + 4}{t^2 + 2t + 1} \cdot \dfrac{t + 1}{t + 2}$

87. $\dfrac{3y^3 + 6y^2}{y^2 - y - 12} \div \dfrac{y^2 - y}{y^2 - 2y - 8} \cdot \dfrac{y^2 + 5y + 6}{y^2}$

88. $\dfrac{a^4 - 81b^4}{a^2c - 6abc + 9b^2c} \cdot \dfrac{a + 3b}{a^2 + 9b^2} \div \dfrac{a^2 + 6ab + 9b^2}{(a - 3b)^2}$

89. Use a graphing calculator to check that

$$\dfrac{x - 1}{x^2 + 2x + 1} \div \dfrac{x^2 - 1}{x^2 - 5x + 4}$$

is equivalent to

$$\dfrac{x^2 - 5x + 4}{(x + 1)^3}.$$

CORNER

Currency Exchange

COLLABORATIVE

Focus: Least common multiples and proportions
Time: 20 minutes
Group size: 2

Travel between different countries usually necessitates an exchange of currencies. Recently one Canadian dollar was worth 76 U.S. cents. Use this exchange rate for the activity that follows.

ACTIVITY

1. Within each group of two students, one student should play the role of a U.S. citizen planning a visit to Canada. The other student should play the role of a Canadian planning a visit to the United States. Use the exchange rate of one Canadian dollar for 76 U.S. cents.
2. Determine how much Canadian money the U.S. citizen would receive in exchange for $76 U.S. (this should be easy). Then determine how much U.S. money the Canadian would receive in exchange for $76 Canadian. Finally, determine how much the Canadian would receive in exchange for $100 Canadian and how much the U.S. citizen would receive for $100 U.S.

3. The answers to part (2) should indicate that coins smaller than a dollar are needed to exchange $76 Canadian for U.S. funds, or to exchange $100 U.S. for Canadian money. What is the smallest amount of Canadian dollars that can be exchanged for a whole-number amount of U.S. dollars? What is the smallest amount of U.S. dollars that can be exchanged for a whole-number amount of Canadian dollars? (*Hint*: See part 2.)
4. Use the results from part (3) to find two other amounts of U.S. dollars that can be exchanged for a whole-number amount of Canadian dollars. Answers may vary.
5. Find the smallest number a for which both conversions—from a Canadian dollars to U.S. funds and from a U.S. dollars to Canadian funds—use only whole numbers. (*Hint*: Use LCMs and the results of part 2 above.)
6. At one time in 2004, one New Zealand dollar was worth about 60 U.S. cents. Find the smallest number a for which both conversions—from a New Zealand dollars to U.S. funds and from a U.S. dollars to New Zealand funds—use only whole numbers. (*Hint*: See part 5.)

6.3 Addition, Subtraction, and Least Common Denominators

Addition When Denominators Are the Same • Subtraction When Denominators Are the Same • Least Common Multiples and Denominators

Addition When Denominators Are the Same

Recall that to add fractions having the same denominator, like $\frac{2}{7}$ and $\frac{3}{7}$, we add the numerators and keep the common denominator: $\frac{2}{7} + \frac{3}{7} = \frac{5}{7}$. The same procedure is used when rational expressions share a common denominator.

> **The Sum of Two Rational Expressions**
>
> To add when the denominators are the same, add the numerators and keep the common denominator:
>
> $$\frac{A}{B} + \frac{C}{B} = \frac{A + C}{B}.$$

In this section, we again assume that all denominators are nonzero.

EXAMPLE 1   Add. Simplify the result, if possible.

a) $\dfrac{4}{a} + \dfrac{3 + a}{a}$

b) $\dfrac{3x}{x - 5} + \dfrac{2x + 1}{x - 5}$

c) $\dfrac{2x^2 + 3x - 7}{2x + 1} + \dfrac{x^2 + x - 8}{2x + 1}$

d) $\dfrac{x - 5}{x^2 - 9} + \dfrac{2}{x^2 - 9}$

Solution

a) $\dfrac{4}{a} + \dfrac{3 + a}{a} = \dfrac{7 + a}{a}$ When the denominators are alike, add the numerators and keep the common denominator.

b) $\dfrac{3x}{x - 5} + \dfrac{2x + 1}{x - 5} = \dfrac{5x + 1}{x - 5}$ The denominators are alike, so we add the numerators.

c) $\dfrac{2x^2 + 3x - 7}{2x + 1} + \dfrac{x^2 + x - 8}{2x + 1} = \dfrac{(2x^2 + 3x - 7) + (x^2 + x - 8)}{2x + 1}$

$\qquad = \dfrac{3x^2 + 4x - 15}{2x + 1}$ Combining like terms

d) $\dfrac{x - 5}{x^2 - 9} + \dfrac{2}{x^2 - 9} = \dfrac{x - 3}{x^2 - 9}$ Combining like terms in the numerator: $x - 5 + 2 = x - 3$

$\qquad = \dfrac{x - 3}{(x - 3)(x + 3)}$ Factoring

$\qquad = \dfrac{1 \cdot (x - 3)}{(x - 3)(x + 3)}$ Removing a factor equal to 1: $\dfrac{x - 3}{x - 3} = 1$

$\qquad = \dfrac{1}{x + 3}$

Subtraction When Denominators Are the Same

When two fractions have the same denominator, we subtract one numerator from the other and keep the common denominator: $\frac{5}{7} - \frac{2}{7} = \frac{3}{7}$. The same procedure is used with rational expressions.

> ## The Difference of Two Rational Expressions
>
> To subtract when the denominators are the same, subtract the second numerator from the first and keep the common denominator:
>
> $$\frac{A}{B} - \frac{C}{B} = \frac{A - C}{B}.$$

Caution! The fraction bar under a numerator is a grouping symbol, just like parentheses. Thus, when a numerator is subtracted, it is important to subtract *every* term in that numerator.

EXAMPLE 2 Subtract and, if possible, simplify: **(a)** $\dfrac{3x}{x + 2} - \dfrac{x - 5}{x + 2}$; **(b)** $\dfrac{x^2}{x - 4} - \dfrac{x + 12}{x - 4}$.

Solution

a) $\dfrac{3x}{x + 2} - \dfrac{x - 5}{x + 2} = \dfrac{3x - (x - 5)}{x + 2}$ The parentheses are needed to make sure that we subtract both terms.

$= \dfrac{3x - x + 5}{x + 2}$ Removing the parentheses and changing signs (using the distributive law)

$= \dfrac{2x + 5}{x + 2}$ Combining like terms

b) $\dfrac{x^2}{x - 4} - \dfrac{x + 12}{x - 4} = \dfrac{x^2 - (x + 12)}{x - 4}$ Remember the parentheses!

$= \dfrac{x^2 - x - 12}{x - 4}$ Removing parentheses (using the distributive law)

$= \dfrac{(x - 4)(x + 3)}{x - 4}$ Factoring, in hopes of simplifying

$= \dfrac{(x - 4)(x + 3)}{x - 4}$ Removing a factor equal to 1: $\dfrac{x - 4}{x - 4} = 1$

$= x + 3$

Least Common Multiples and Denominators

Thus far, every pair of rational expressions that we have added or subtracted shared a common denominator. To add or subtract rational expressions that have different denominators, we must first find equivalent rational expressions that *do* have a common denominator.

In algebra, we find a common denominator much as we do in arithmetic. Recall that to add $\frac{1}{12}$ and $\frac{7}{30}$, we first identify the smallest number that contains both 12 and 30 as factors. Such a number, the **least common multiple (LCM)** of the denominators, is then used as the **least common denominator (LCD)**.

Let's find the LCM of 12 and 30 using a method that can also be used with polynomials. We begin by writing the prime factorization of 12:

$$12 = 2 \cdot 2 \cdot 3.$$

Next, we write the prime factorization of 30:

$$30 = 2 \cdot 3 \cdot 5.$$

The LCM must include the factors of each number, so it must include each prime factor the greatest number of times that it appears in either of the factorizations. To find the LCM for 12 and 30, we select one factorization, say

$$2 \cdot 2 \cdot 3,$$

and note that because it lacks a factor of 5, it does not contain the entire factorization of 30. If we multiply $2 \cdot 2 \cdot 3$ by 5, every prime factor occurs just often enough to contain both 12 and 30 as factors.

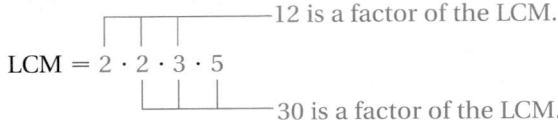

Note that each prime factor—2, 3, and 5—is used the greatest number of times that it appears in either of the individual factorizations. The factor 2 occurs twice and the factors 3 and 5 once each.

To Find the Least Common Denominator (LCD)

1. Write the prime factorization of each denominator.
2. Select one of the factorizations and inspect it to see if it contains the other.

   a) If it does, it represents the LCM of the denominators.
   b) If it does not, multiply that factorization by any factors of the other denominator that it lacks. The final product is the LCM of the denominators.

The LCD is the LCM of the denominators. It should contain each factor the greatest number of times that it occurs in any of the individual factorizations.

Let's finish adding $\dfrac{1}{12}$ and $\dfrac{7}{30}$:

$$\frac{1}{12} + \frac{7}{30} = \frac{1}{2 \cdot 2 \cdot 3} + \frac{7}{2 \cdot 3 \cdot 5}.$$

The least common denominator (LCD) is $2 \cdot 2 \cdot 3 \cdot 5$.

We found above that the LCD is $2 \cdot 2 \cdot 3 \cdot 5$, or 60. To get the LCD, we see that the first denominator needs a factor of 5, and the second denominator needs another factor of 2. To accomplish this, we multiply $\frac{1}{12}$ by 1, using $\frac{5}{5}$, and we multiply $\frac{7}{30}$ by 1, using $\frac{2}{2}$. We can do this because $a \cdot 1 = a$, for any number a:

$$\frac{1}{12} + \frac{7}{30} = \frac{1}{2 \cdot 2 \cdot 3} \cdot \frac{5}{5} + \frac{7}{2 \cdot 3 \cdot 5} \cdot \frac{2}{2} \qquad \frac{5}{5} = 1 \text{ and } \frac{2}{2} = 1$$

$$= \frac{5}{60} + \frac{14}{60} \qquad\qquad \text{Both denominators are now the LCD.}$$

$$= \frac{19}{60}. \qquad\qquad\qquad \text{Adding the numerators and keeping the LCD}$$

Expressions like $\dfrac{5}{36x^2}$ and $\dfrac{7}{24x}$ are added in much the same manner.

EXAMPLE 3   Find the LCD of $\dfrac{5}{36x^2}$ and $\dfrac{7}{24x}$.

Solution

1. We begin by writing the prime factorizations of $36x^2$ and $24x$:

    $36x^2 = 2 \cdot 2 \cdot 3 \cdot 3 \cdot x \cdot x;$

    $24x = 2 \cdot 2 \cdot 2 \cdot 3 \cdot x.$

2. Except for a third factor of 2, the factorization of $36x^2$ contains the entire factorization of $24x$. To find the smallest product that contains both $36x^2$ and $24x$ as factors, we multiply $36x^2$ by a third factor of 2:

    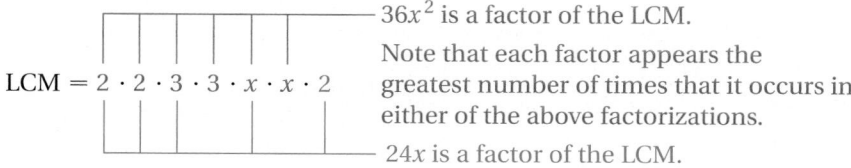

    $36x^2$ is a factor of the LCM.

    Note that each factor appears the greatest number of times that it occurs in either of the above factorizations.

    LCM $= 2 \cdot 2 \cdot 3 \cdot 3 \cdot x \cdot x \cdot 2$

    $24x$ is a factor of the LCM.

    The LCM of the denominators is thus $2^3 \cdot 3^2 \cdot x^2$, or $72x^2$, so the LCD is $72x^2$.

We can now add $\dfrac{5}{36x^2}$ and $\dfrac{7}{24x}$:

$$\frac{5}{36x^2} + \frac{7}{24x} = \frac{5}{2 \cdot 2 \cdot 3 \cdot 3 \cdot x \cdot x} + \frac{7}{2 \cdot 2 \cdot 2 \cdot 3 \cdot x}.$$

In Example 3, we found that the LCD is $2 \cdot 2 \cdot 2 \cdot 3 \cdot 3 \cdot x \cdot x$, or $72x^2$. To obtain equivalent expressions with this LCD, we multiply each expression by 1, using the missing factors of the LCD to write 1:

$$\frac{5}{36x^2} + \frac{7}{24x} = \frac{5}{2 \cdot 2 \cdot 3 \cdot 3 \cdot x \cdot x} \cdot \frac{2}{2} + \frac{7}{2 \cdot 2 \cdot 2 \cdot 3 \cdot x} \cdot \frac{3 \cdot x}{3 \cdot x}$$

The LCD requires another factor of 2.   The LCD requires additional factors of 3 and x.

$$= \frac{10}{72x^2} + \frac{21x}{72x^2} \qquad \text{Both denominators are now the LCD.}$$

$$= \frac{21x + 10}{72x^2}.$$

You now have the "big picture" of why LCMs are needed when adding rational expressions. For the remainder of this section, we will practice finding LCMs and rewriting rational expressions so that they have the LCD as the denominator. In Section 6.4, we will return to the addition and subtraction of rational expressions.

EXAMPLE 4   For each pair of polynomials, find the least common multiple.

a) $15a$ and $35b$

b) $21x^3y^6$ and $7x^5y^2$

c) $x^2 + 5x - 6$ and $x^2 - 1$

Solution

a) We write the prime factorizations and then construct the LCM:

$$15a = 3 \cdot 5 \cdot a$$
$$35b = 5 \cdot 7 \cdot b$$

$$\text{LCM} = 3 \cdot 5 \cdot a \cdot 7 \cdot b$$

$15a$ is a factor of the LCM.

Each factor appears the greatest number of times that it occurs in either of the above factorizations.

$35b$ is a factor of the LCM.

The LCM is $3 \cdot 5 \cdot a \cdot 7 \cdot b$, or $105ab$.

b) $21x^3y^6 = 3 \cdot 7 \cdot x \cdot x \cdot x \cdot y \cdot y \cdot y \cdot y \cdot y \cdot y$ Try to visualize the factors
$7x^5y^2 = 7 \cdot x \cdot x \cdot x \cdot x \cdot x \cdot y \cdot y$ of x and y mentally.

$21x^3y^6$ is a factor of the LCM.

$$\text{LCM} = 3 \cdot 7 \cdot x \cdot x \cdot x \cdot y \cdot y \cdot y \cdot y \cdot y \cdot y \cdot x \cdot x$$

$7x^5y^2$ is a factor of the LCM.

Note that we used the highest power of each factor in $21x^3y^6$ and $7x^5y^2$. The LCM is $21x^5y^6$.

c) $x^2 + 5x - 6 = (x - 1)(x + 6)$

   $x^2 - 1 = (x - 1)(x + 1)$

$$\text{LCM} = (x - 1)(x + 6)(x + 1)$$

   $x^2 + 5x - 6$ is a factor of the LCM.

   $x^2 - 1$ is a factor of the LCM.

The LCM is $(x - 1)(x + 6)(x + 1)$. There is no need to multiply this out.

The above procedure can be used to find the LCM of three or more polynomials as well. We factor each polynomial and then construct the LCM using each factor the greatest number of times that it appears in any one factorization.

EXAMPLE 5   For each group of polynomials, find the LCM.

a) $12x$, $16y$, and $8xyz$

b) $x^2 + 4$, $x + 1$, and 5

Solution

a) $12x = 2 \cdot 2 \cdot 3 \cdot x$

   $16y = 2 \cdot 2 \cdot 2 \cdot 2 \cdot y$

   $8xyz = 2 \cdot 2 \cdot 2 \cdot x \cdot y \cdot z$

$$\text{LCM} = 2 \cdot 2 \cdot 3 \cdot x \cdot 2 \cdot 2 \cdot y \cdot z$$

   $12x$ is a factor of the LCM.

   $16y$ is a factor of the LCM.

   $8xyz$ is a factor of the LCM.

The LCM is $2^4 \cdot 3 \cdot xyz$, or $48xyz$.

b) Since $x^2 + 4$, $x + 1$, and 5 are not factorable, the LCM is their product: $5(x^2 + 4)(x + 1)$.

To add or subtract rational expressions with different denominators, we must be able to write equivalent expressions that have the LCD. Doing so usually requires multiplying each rational expression by a carefully constructed form of 1.

EXAMPLE 6   Find equivalent expressions that have the LCD:

$$\frac{x + 3}{x^2 + 5x - 6}, \qquad \frac{x + 7}{x^2 - 1}.$$

Student Notes

If you prefer, the LCM for a group of three polynomials can be found by finding the LCM of two of them and then finding the LCM of that result and the remaining polynomial.

Solution   From Example 4(c), we know that the LCD is

$$(x + 6)(x - 1)(x + 1).$$

Since

$$x^2 + 5x - 6 = (x + 6)(x - 1),$$

the factor of the LCD that is missing from the first denominator is $x + 1$. We multiply by 1 using $(x + 1)/(x + 1)$:

$$\left.\begin{array}{l} \dfrac{x + 3}{x^2 + 5x - 6} = \dfrac{x + 3}{(x + 6)(x - 1)} \cdot \dfrac{x + 1}{x + 1} \\[3mm] \qquad\qquad = \dfrac{(x + 3)(x + 1)}{(x + 6)(x - 1)(x + 1)}. \end{array}\right\}$$ Finding an equivalent expression that has the least common denominator

For the second expression, we have

$$x^2 - 1 = (x + 1)(x - 1).$$

The factor of the LCD that is missing is $x + 6$. We multiply by 1 using $(x + 6)/(x + 6)$:

$$\left.\begin{array}{l} \dfrac{x + 7}{x^2 - 1} = \dfrac{x + 7}{(x + 1)(x - 1)} \cdot \dfrac{x + 6}{x + 6} \\[3mm] \qquad\quad = \dfrac{(x + 7)(x + 6)}{(x + 1)(x - 1)(x + 6)}. \end{array}\right\}$$ Finding an equivalent expression that has the least common denominator

We leave the results in factored form. In Section 6.4, we will carry out the actual addition and subtraction of such rational expressions.

Exercise Set

6.3

FOR EXTRA HELP

 Student's Solutions Manual   Digital Video Tutor CD 3 Videotape 6   Tutor Center AW Math Tutor Center   MathXL Tutorials on CD   MathXL   MyMathLab

🐦 *Concept Reinforcement*   *Use one or more words to complete each of the following sentences.*

1. To add two rational expressions when the denominators are the same, add _____ and keep the common _____.

2. When a numerator is being subtracted, be sure to treat its fraction bar as a _____ symbol and subtract every _____ in that numerator.

3. The least common multiple of two denominators is usually referred to as the _____ and is abbreviated _____.

4. The least common denominator of two fractions must contain the prime _____ of both _____.

Perform the indicated operation. Simplify, if possible.

5. $\dfrac{6}{x} + \dfrac{4}{x}$

6. $\dfrac{4}{a^2} + \dfrac{9}{a^2}$

7. $\dfrac{x}{12} + \dfrac{2x + 5}{12}$

8. $\dfrac{a}{7} + \dfrac{3a - 4}{7}$

9. $\dfrac{4}{a + 3} + \dfrac{5}{a + 3}$

10. $\dfrac{5}{x + 2} + \dfrac{8}{x + 2}$

11. $\dfrac{8}{a+2} - \dfrac{2}{a+2}$

12. $\dfrac{8}{x+7} - \dfrac{2}{x+7}$

13. $\dfrac{3y+8}{2y} - \dfrac{y+1}{2y}$

14. $\dfrac{5+3t}{4t} - \dfrac{2t+1}{4t}$

15. $\dfrac{7x+8}{x+1} + \dfrac{4x+3}{x+1}$

16. $\dfrac{3a+13}{a+4} + \dfrac{2a+7}{a+4}$

17. $\dfrac{7x+8}{x+1} - \dfrac{4x+3}{x+1}$

18. $\dfrac{3a+13}{a+4} - \dfrac{2a+7}{a+4}$

19. $\dfrac{a^2}{a-4} + \dfrac{a-20}{a-4}$

20. $\dfrac{x^2}{x+5} + \dfrac{7x+10}{x+5}$

21. $\dfrac{x^2}{x-2} - \dfrac{6x-8}{x-2}$

22. $\dfrac{a^2}{a+3} - \dfrac{2a+15}{a+3}$

Aha! **23.** $\dfrac{t^2-5t}{t-1} + \dfrac{5t-t^2}{t-1}$

24. $\dfrac{y^2+6y}{y+2} + \dfrac{2y+12}{y+2}$

25. $\dfrac{x-6}{x^2+5x+6} + \dfrac{9}{x^2+5x+6}$

26. $\dfrac{x-5}{x^2-4x+3} + \dfrac{2}{x^2-4x+3}$

27. $\dfrac{t^2-5t}{t^2+6t+9} + \dfrac{4t-12}{t^2+6t+9}$

28. $\dfrac{y^2-7y}{y^2+8y+16} + \dfrac{6y-20}{y^2+8y+16}$

29. $\dfrac{2x^2+x}{x^2-8x+12} - \dfrac{x^2-2x+10}{x^2-8x+12}$

30. $\dfrac{2x^2+3}{x^2-6x+5} - \dfrac{3+2x^2}{x^2-6x+5}$

31. $\dfrac{3-2x}{x^2-6x+8} + \dfrac{7-3x}{x^2-6x+8}$

32. $\dfrac{1-2t}{t^2-5t+4} + \dfrac{4-3t}{t^2-5t+4}$

33. $\dfrac{x-9}{x^2+3x-4} - \dfrac{2x-5}{x^2+3x-4}$

34. $\dfrac{5-3x}{x^2-2x+1} - \dfrac{x+1}{x^2-2x+1}$

Find the LCM.

35. 15, 27

36. 10, 15

37. 8, 9

38. 12, 15

39. 6, 9, 21

40. 8, 36, 40

Find the LCM.

41. $12x^2, 6x^3$

42. $10t^3, 5t^4$

43. $15a^4b^7, 10a^2b^8$

44. $6a^2b^7, 9a^5b^2$

45. $2(y-3), 6(y-3)$

46. $4(x-1), 8(x-1)$

47. x^2-4, x^2+5x+6

48. x^2+3x+2, x^2-4

49. t^3+4t^2+4t, t^2-4t

50. y^3-y^2, y^4-y^2

51. $10x^2y, 6y^2z, 5xz^3$

52. $8x^3z, 12xy^2, 4y^5z^2$

53. $a+1, (a-1)^2, a^2-1$

54. $x^2-9, x+3, (x-3)^2$

55. m^2-5m+6, m^2-4m+4

56. $2x^2+5x+2, 2x^2-x-1$

57. $6x^3-24x^2+18x, 4x^5-24x^4+20x^3$

58. $9x^3-9x^2-18x, 6x^5-24x^4+24x^3$

59. $2x^3-2, x^2-1$

60. $3a^3+24, a^2-4$

Find equivalent expressions that have the LCD.

61. $\dfrac{3}{10a^3}, \dfrac{b}{5a^6}$

62. $\dfrac{5}{6x^5}, \dfrac{y}{12x^3}$

63. $\dfrac{7}{3x^4y^2}, \dfrac{4}{9xy^3}$

64. $\dfrac{3}{2a^2b}, \dfrac{7}{8ab^2}$

65. $\dfrac{2x}{x^2-4}, \dfrac{4x}{x^2+5x+6}$

66. $\dfrac{5x}{x^2-9}, \dfrac{2x}{x^2+11x+24}$

67. Explain why the product of two numbers is not always their least common multiple.

68. If the LCM of two numbers is their product, what can you conclude about the two numbers?

SKILL MAINTENANCE

Write each number in two equivalent forms. [1.7]

69. $\dfrac{7}{-9}$

70. $-\dfrac{3}{2}$

Simplify. [1.3]

71. $\dfrac{5}{18} - \dfrac{7}{12}$

72. $\dfrac{8}{15} - \dfrac{13}{20}$

Find a polynomial that can represent the shaded area of each figure. [4.6]

73.

74.

SYNTHESIS

75. If the LCM of two third-degree polynomials is a sixth-degree polynomial, what can be concluded about the two polynomials?

76. If the LCM of a binomial and a trinomial is the trinomial, what relationship exists between the two expressions?

Perform the indicated operations. Simplify, if possible.

77. $\dfrac{6x - 1}{x - 1} + \dfrac{3(2x + 5)}{x - 1} + \dfrac{3(2x - 3)}{x - 1}$

78. $\dfrac{2x + 11}{x - 3} \cdot \dfrac{3}{x + 4} + \dfrac{-1}{4 + x} \cdot \dfrac{6x + 3}{x - 3}$

79. $\dfrac{x^2}{3x^2 - 5x - 2} - \dfrac{2x}{3x + 1} \cdot \dfrac{1}{x - 2}$

80. $\dfrac{x + y}{x^2 - y^2} + \dfrac{x - y}{x^2 - y^2} - \dfrac{2x}{x^2 - y^2}$

South African artistry. In South Africa, the design of every woven handbag, or gipatsi *(plural,* sipatsi *) is created by repeating two or more geometric patterns.*

Each pattern encircles the bag, sharing the strands of fabric with any pattern above or below. The length, or period, of each pattern is the number of strands required to construct the pattern. For a gipatsi to be considered beautiful, each individual pattern must fit a whole number of times around the bag.
Source: Gerdes, Paulus, *Women, Art and Geometry in Southern Africa.* Asmara, Eritrea: Africa World Press, Inc., p. 5

81. A weaver is using two patterns to create a gipatsi. Pattern A is 10 strands long, and pattern B is 3 strands long. What is the smallest number of strands that can be used to complete the gipatsi?

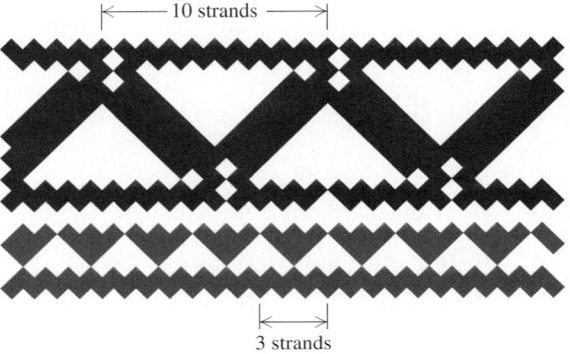

82. A weaver is using a four-strand pattern, a six-strand pattern, and an eight-strand pattern. What is the smallest number of strands that can be used to complete the gipatsi?

83. For technical reasons, the number of strands is generally a multiple of 4. Answer Exercise 81 with this additional requirement in mind.

Find the LCM.

84. 72, 90, 96

85. $6t^2 - 6$, $(3t^2 - 6t + 3)^2$, $8t - 8$

86. $9x^2 - 16$, $6x^2 - x - 12$, $(4x^2 - 12x + 9)^2$

87. *Copiers.* The Brother® DCP-1000 copier can print 10 pages per minute. The Sharp® AL-1540CS copier can print 14 pages per minute. If both machines begin printing at the same instant, how long will it be until they again begin copying a page at exactly the same time?

88. *Running.* Kim and Jed leave the starting point of a fitness loop at the same time. Kim jogs a lap in 6 min and Jed jogs one in 8 min. Assuming they continue to run at the same pace, when will they next meet at the starting place?

89. *Bus schedules.*   Beginning at 5:00 A.M., a hotel shuttle bus leaves Salton Airport every 25 min, and the downtown shuttle bus leaves the airport every 35 min. What time will it be when both shuttles again leave at the same time?

90. *Appliances.*   Dishwashers last an average of 10 yr, clothes washers an average of 14 yr, and refrigerators an average of 20 yr. If an apartment house is equipped with new dishwashers, clothes washers, and refrigerators in 2005, in what year will all three appliances need to be replaced at once?

Source: *Energy Savers: Tips on Saving Energy and Money at Home,* produced for the U.S. Department of Energy by the National Renewable Energy Laboratory, 2003

91. Explain how evaluating can be used to perform a partial check on the result of Example 1(d):

$$\frac{x-5}{x^2-9} + \frac{2}{x^2-9} = \frac{1}{x+3}.$$

92. On p. 390, the second step in finding an LCD is to select one of the factorizations of the denominators. Does it matter which one is selected? Why or why not?

6.4 Addition and Subtraction with Unlike Denominators

> Adding and Subtracting with LCDs   •
> When Factors Are Opposites

Adding and Subtracting with LCDs

We now know how to rewrite two rational expressions in equivalent forms that use the LCD. Once rational expressions share a common denominator, they can be added or subtracted just as in Section 6.3.

To Add or Subtract Rational Expressions Having Different Denominators

1. Find the LCD.
2. Multiply each rational expression by a form of 1 made up of the factors of the LCD missing from that expression's denominator.
3. Add or subtract the numerators, as indicated. Write the sum or difference over the LCD.
4. Simplify, if possible.

EXAMPLE 1 Add: $\dfrac{5x^2}{8} + \dfrac{7x}{12}$.

Study Skills

If a Question Stumps You

Unless you know the material "cold," don't be surprised if a quiz or test includes a question that you feel unsure about. Should this happen, do not get rattled—simply skip the question and continue with the quiz or test, returning to the trouble spot after the other questions have been answered.

Solution

1. First, we find the LCD:

$$\left.\begin{aligned} 8 &= 2 \cdot 2 \cdot 2 \\ 12 &= 2 \cdot 2 \cdot 3 \end{aligned}\right\} \quad \text{LCD} = 2 \cdot 2 \cdot 2 \cdot 3, \text{ or } 24.$$

2. The denominator 8 needs to be multiplied by 3 in order to obtain the LCD. The denominator 12 needs to be multiplied by 2 in order to obtain the LCD. Thus we multiply the first expression by $\frac{3}{3}$ and the second expression by $\frac{2}{2}$ to get the LCD:

$$\begin{aligned} \frac{5x^2}{8} + \frac{7x}{12} &= \frac{5x^2}{2 \cdot 2 \cdot 2} + \frac{7x}{2 \cdot 2 \cdot 3} \\ &= \frac{5x^2}{2 \cdot 2 \cdot 2} \cdot \frac{3}{3} + \frac{7x}{2 \cdot 2 \cdot 3} \cdot \frac{2}{2} \qquad \text{Multiplying each expression by a form of 1 to get the LCD} \\ &= \frac{15x^2}{24} + \frac{14x}{24}. \end{aligned}$$

3. Next, we add the numerators:

$$\frac{15x^2}{24} + \frac{14x}{24} = \frac{15x^2 + 14x}{24}.$$

4. Since $15x^2 + 14x$ and 24 have no common factor,

$$\frac{15x^2 + 14x}{24}$$

cannot be simplified any further.

Subtraction is performed in much the same way.

EXAMPLE 2 Subtract: $\dfrac{7}{8x} - \dfrac{5}{12x^2}$.

Solution We follow the four steps shown above. First, we find the LCD:

$$\left.\begin{aligned} 8x &= 2 \cdot 2 \cdot 2 \cdot x \\ 12x^2 &= 2 \cdot 2 \cdot 3 \cdot x \cdot x \end{aligned}\right\} \quad \text{LCD} = 2 \cdot 2 \cdot 3 \cdot x \cdot x \cdot 2, \text{ or } 24x^2.$$

The denominator $8x$ must be multiplied by $3x$ in order to obtain the LCD. The denominator $12x^2$ must be multiplied by 2 in order to obtain the LCD.

Thus we multiply by $\dfrac{3x}{3x}$ and $\dfrac{2}{2}$ to get the LCD. Then we subtract and, if possible, simplify.

$$\frac{7}{8x} - \frac{5}{12x^2} = \frac{7}{8x} \cdot \frac{3x}{3x} - \frac{5}{12x^2} \cdot \frac{2}{2}$$

$$= \frac{21x}{24x^2} - \frac{10}{24x^2} \longleftarrow$$

> **Caution!** Do not simplify *these* rational expressions or you will lose the LCD.

$$= \frac{21x - 10}{24x^2} \quad \text{This cannot be simplified,}$$
so we are done.

When denominators contain polynomials with two or more terms, the same steps are used.

EXAMPLE 3   Add: $\dfrac{2a}{a^2 - 1} + \dfrac{1}{a^2 + a}$.

Solution   First, we find the LCD:

$$\left.\begin{array}{l} a^2 - 1 = (a - 1)(a + 1) \\ a^2 + a = a(a + 1). \end{array}\right\} \quad \text{LCD} = (a - 1)(a + 1)a$$

We multiply by a form of 1 to get the LCD in each expression:

$$\frac{2a}{a^2 - 1} + \frac{1}{a^2 + a} = \frac{2a}{(a - 1)(a + 1)} \cdot \frac{a}{a} + \frac{1}{a(a + 1)} \cdot \frac{a - 1}{a - 1} \qquad \begin{array}{l} \text{Multiplying by} \\ \dfrac{a}{a} \text{ and } \dfrac{a - 1}{a - 1} \text{ to} \\ \text{get the LCD} \end{array}$$

$$= \frac{2a^2}{(a - 1)(a + 1)a} + \frac{a - 1}{a(a + 1)(a - 1)}$$

$$= \frac{2a^2 + a - 1}{a(a - 1)(a + 1)} \qquad \text{Adding numerators}$$

$$= \frac{(2a - 1)\cancel{(a + 1)}}{a(a - 1)\cancel{(a + 1)}} \left.\begin{array}{l} \\ \\ \end{array}\right\} \quad \begin{array}{l} \text{Simplifying by factoring and} \\ \text{removing a factor equal to 1:} \\ \dfrac{a + 1}{a + 1} = 1 \end{array}$$

$$= \frac{2a - 1}{a(a - 1)}.$$

EXAMPLE 4   Perform the indicated operations.

a) $\dfrac{x + 4}{x - 2} - \dfrac{x - 7}{x + 5}$

b) $\dfrac{t}{t^2 + 11t + 30} + \dfrac{-5}{t^2 + 9t + 20}$

c) $\dfrac{x}{x^2 + 5x + 6} - \dfrac{2}{x^2 + 3x + 2}$

Solution

a) First, we find the LCD. It is just the product of the denominators:

$$\text{LCD} = (x - 2)(x + 5).$$

We multiply by a form of 1 to get the LCD in each expression. Then we subtract and try to simplify.

$$\frac{x + 4}{x - 2} - \frac{x - 7}{x + 5} = \frac{x + 4}{x - 2} \cdot \frac{x + 5}{x + 5} - \frac{x - 7}{x + 5} \cdot \frac{x - 2}{x - 2}$$

$$= \frac{x^2 + 9x + 20}{(x - 2)(x + 5)} - \frac{x^2 - 9x + 14}{(x - 2)(x + 5)} \quad \text{Multiplying out numerators (but not denominators)}$$

$$= \frac{x^2 + 9x + 20 - (x^2 - 9x + 14)}{(x - 2)(x + 5)} \quad \text{When subtracting a numerator with more than one term, parentheses are important.}$$

$$= \frac{x^2 + 9x + 20 - x^2 + 9x - 14}{(x - 2)(x + 5)} \quad \text{Removing parentheses and subtracting every term}$$

$$= \frac{18x + 6}{(x - 2)(x + 5)}$$

Factoring $18x + 6$ as $6(3x + 1)$ would not enable us to simplify our result, so it need not be written in that form.

b) $\dfrac{t}{t^2 + 11t + 30} + \dfrac{-5}{t^2 + 9t + 20}$

$$= \frac{t}{(t + 5)(t + 6)} + \frac{-5}{(t + 5)(t + 4)} \quad \text{Factoring the denominators in order to find the LCD. The LCD is } (t + 5)(t + 6)(t + 4).$$

$$= \frac{t}{(t + 5)(t + 6)} \cdot \frac{t + 4}{t + 4} + \frac{-5}{(t + 5)(t + 4)} \cdot \frac{t + 6}{t + 6} \quad \text{Multiplying to get the LCD}$$

$$= \frac{t^2 + 4t}{(t + 5)(t + 6)(t + 4)} + \frac{-5t - 30}{(t + 5)(t + 6)(t + 4)} \quad \text{Multiplying in each numerator}$$

$$= \frac{t^2 + 4t - 5t - 30}{(t + 5)(t + 6)(t + 4)} \quad \text{Adding numerators}$$

$$= \frac{t^2 - t - 30}{(t + 5)(t + 6)(t + 4)} \quad \text{Combining like terms in the numerator}$$

$$= \frac{\cancel{(t + 5)}(t - 6)}{\cancel{(t + 5)}(t + 6)(t + 4)}$$
$$= \frac{t - 6}{(t + 6)(t + 4)}$$

Always simplify the result, if possible, by removing a factor equal to 1; here $\dfrac{t + 5}{t + 5} = 1$.

Student Notes _____

As you can see, adding or subtracting rational expressions can involve many steps. Therefore, it is important to double-check each step of your work as you work through each problem. Waiting to check your work at the end of the problems is usually a less efficient use of your time.

c) $\dfrac{x}{x^2 + 5x + 6} - \dfrac{2}{x^2 + 3x + 2}$

$$= \dfrac{x}{(x + 2)(x + 3)} - \dfrac{2}{(x + 2)(x + 1)} \qquad \text{Factoring denominators. The LCD is } (x + 2)(x + 3)(x + 1).$$

$$= \dfrac{x}{(x + 2)(x + 3)} \cdot \dfrac{x + 1}{x + 1} - \dfrac{2}{(x + 2)(x + 1)} \cdot \dfrac{x + 3}{x + 3}$$

$$= \dfrac{x^2 + x}{(x + 2)(x + 3)(x + 1)} - \dfrac{2x + 6}{(x + 2)(x + 3)(x + 1)}$$

$$= \dfrac{x^2 + x - (2x + 6)}{(x + 2)(x + 3)(x + 1)} \qquad \text{Don't forget the parentheses!}$$

$$= \dfrac{x^2 + x - 2x - 6}{(x + 2)(x + 3)(x + 1)} \qquad \text{Remember to subtract each term in } 2x + 6.$$

$$= \dfrac{x^2 - x - 6}{(x + 2)(x + 3)(x + 1)} \qquad \text{Combining like terms in the numerator}$$

$$= \dfrac{\cancel{(x + 2)}(x - 3)}{\cancel{(x + 2)}(x + 3)(x + 1)}$$

$$= \dfrac{x - 3}{(x + 3)(x + 1)} \qquad \Biggr\} \text{Factoring and simplifying; } \dfrac{x + 2}{x + 2} = 1$$

When Factors Are Opposites

When one denominator is the opposite of the other, we can first multiply either expression by 1 using $-1/-1$.

EXAMPLE 5   Add: **(a)** $\dfrac{t}{2} + \dfrac{3}{-2}$; **(b)** $\dfrac{x}{x - 5} + \dfrac{7}{5 - x}$.

Solution

a) $\dfrac{t}{2} + \dfrac{3}{-2} = \dfrac{t}{2} + \dfrac{3}{-2} \cdot \dfrac{-1}{-1} \qquad \text{Multiplying by 1 using } \dfrac{-1}{-1}$

$\qquad = \dfrac{t}{2} + \dfrac{-3}{2} \qquad \text{The denominators are now the same.}$

$\qquad = \dfrac{t + (-3)}{2}$

$\qquad = \dfrac{t - 3}{2}$

b) Recall that when an expression of the form $a - b$ is multiplied by -1, the subtraction is reversed: $-1(a - b) = -a + b = b + (-a) = b - a$. Since $x - 5$ and $5 - x$ are opposites, we can find a common denominator by multiplying one of the rational expressions by $-1/-1$. Because polynomials are usually written in descending order, we choose to reverse the subtraction in the second denominator:

$$\frac{x}{x - 5} + \frac{7}{5 - x} = \frac{x}{x - 5} + \frac{7}{5 - x} \cdot \frac{-1}{-1} \qquad \text{Writing 1 as } -1/-1 \text{ and}$$

multiplying to obtain a common denominator

$$= \frac{x}{x - 5} + \frac{-7}{-5 + x}$$

$$= \frac{x}{x - 5} + \frac{-7}{x - 5} \qquad \text{Note that } -5 + x = x + (-5) = x - 5.$$

$$= \frac{x - 7}{x - 5}.$$

Sometimes, after factoring to find the LCD, we find a factor in one denominator that is the opposite of a factor in the other denominator. When this happens, multiplication by $-1/-1$ can again be helpful.

EXAMPLE 6   Perform the indicated operations and simplify.

a) $\dfrac{x}{x^2 - 25} + \dfrac{3}{5 - x}$

b) $\dfrac{x + 9}{x^2 - 4} + \dfrac{6 - x}{4 - x^2} - \dfrac{1 + x}{x^2 - 4}$

Solution

a) $\dfrac{x}{x^2 - 25} + \dfrac{3}{5 - x} = \dfrac{x}{(x - 5)(x + 5)} + \dfrac{3}{5 - x}$ Factoring

$$= \frac{x}{(x - 5)(x + 5)} + \frac{3}{5 - x} \cdot \frac{-1}{-1} \qquad \begin{array}{l} \text{Multiplying by} \\ -1/-1 \text{ changes} \\ 5 - x \text{ to } x - 5. \end{array}$$

$$= \frac{x}{(x - 5)(x + 5)} + \frac{-3}{x - 5} \qquad (5 - x)(-1) = x - 5$$

$$= \frac{x}{(x - 5)(x + 5)} + \frac{-3}{(x - 5)} \cdot \frac{x + 5}{x + 5} \qquad \begin{array}{l} \text{The LCD is} \\ (x - 5)(x + 5). \end{array}$$

$$= \frac{x}{(x - 5)(x + 5)} + \frac{-3x - 15}{(x - 5)(x + 5)}$$

$$= \frac{-2x - 15}{(x - 5)(x + 5)}$$

technology connection

The TABLE feature can be used to check addition or subtraction of rational expressions, much as we did on p. 384 of Section 6.2. Below we check Example 6(a), using $y_1 = x/(x^2 - 25) + 3/(5 - x)$ and $y_2 = (-2x - 15)/((x - 5)(x + 5))$.

ΔTBL = 1

X	Y₁	Y₂
1	.70833	.70833
2	.90476	.90476
3	1.3125	1.3125
4	2.5556	2.5556
5	ERROR	ERROR
6	−2.455	−2.455
7	−1.208	−1.208

X = 1

Because the values for y_1 and y_2 match, we have a check.

b) Since $4 - x^2$ is the opposite of $x^2 - 4$, multiplying the second rational expression by $-1/-1$ will lead to a common denominator:

$$\frac{x + 9}{x^2 - 4} + \frac{6 - x}{4 - x^2} - \frac{1 + x}{x^2 - 4} = \frac{x + 9}{x^2 - 4} + \frac{6 - x}{4 - x^2} \cdot \frac{-1}{-1} - \frac{1 + x}{x^2 - 4}$$

$$= \frac{x + 9}{x^2 - 4} + \frac{x - 6}{x^2 - 4} - \frac{1 + x}{x^2 - 4}$$

$$= \frac{x + 9 + x - 6 - 1 - x}{x^2 - 4} \quad \text{Adding and subtracting numerators}$$

$$= \frac{x + 2}{x^2 - 4}$$

$$= \frac{(x + 2) \cdot 1}{(x + 2)(x - 2)} \quad \left.\begin{array}{c} \\ \\ \end{array}\right\} \text{Simplifying}$$

$$= \frac{1}{x - 2}.$$

Exercise Set

6.4

👈 *Concept Reinforcement* In Exercises 1–4, the four steps for adding rational expressions with different denominators are listed. Fill in the missing word or words for each step.

1. To add or subtract when the denominators are different, first find the _____ .

2. Multiply each rational expression by a form of 1 made up of the factors of the LCD that are _____ from that expression's _____ .

3. Add or subtract the _____ , as indicated. Write the sum or difference over the _____ .

4. _____ , if possible.

Perform the indicated operation. Simplify, if possible.

5. $\dfrac{4}{x} + \dfrac{9}{x^2}$

6. $\dfrac{5}{x} + \dfrac{6}{x^2}$

7. $\dfrac{1}{6r} - \dfrac{3}{8r}$

8. $\dfrac{4}{9t} - \dfrac{7}{6t}$

9. $\dfrac{2}{c^2 d} + \dfrac{7}{cd^3}$

10. $\dfrac{4}{xy^2} + \dfrac{2}{x^2 y}$

11. $\dfrac{-2}{3xy^2} - \dfrac{6}{x^2 y^3}$

12. $\dfrac{8}{9t^3} - \dfrac{5}{6t^2}$

13. $\dfrac{x - 4}{9} + \dfrac{x + 5}{6}$

14. $\dfrac{x + 5}{8} + \dfrac{x - 3}{12}$

15. $\dfrac{x-2}{6} - \dfrac{x+1}{3}$

16. $\dfrac{a+2}{2} - \dfrac{a-4}{4}$

48. $\dfrac{x}{x^2+5x+6} - \dfrac{2}{x^2+3x+2}$

17. $\dfrac{a+4}{16a} + \dfrac{3a+4}{4a^2}$

18. $\dfrac{2a-1}{3a^2} + \dfrac{5a+1}{9a}$

49. $\dfrac{3z}{z^2-4z+4} + \dfrac{10}{z^2+z-6}$

19. $\dfrac{4z-9}{3z} - \dfrac{3z-8}{4z}$

20. $\dfrac{x-1}{4x} - \dfrac{2x+3}{x}$

50. $\dfrac{3}{x^2-9} + \dfrac{2}{x^2-x-6}$

21. $\dfrac{x+y}{xy^2} + \dfrac{3x+y}{x^2y}$

22. $\dfrac{2c-d}{c^2d} + \dfrac{c+d}{cd^2}$

Aha! **51.** $\dfrac{-5}{x^2+17x+16} - \dfrac{0}{x^2+9x+8}$

23. $\dfrac{4x+2t}{3xt^2} - \dfrac{5x-3t}{x^2t}$

24. $\dfrac{5x+3y}{2x^2y} - \dfrac{3x+4y}{xy^2}$

52. $\dfrac{x}{x^2+15x+56} - \dfrac{1}{x^2+13x+42}$

25. $\dfrac{3}{x-2} + \dfrac{3}{x+2}$

26. $\dfrac{5}{x-1} + \dfrac{5}{x+1}$

53. $\dfrac{4x}{5} - \dfrac{x-3}{-5}$

54. $\dfrac{x}{4} - \dfrac{3x-5}{-4}$

27. $\dfrac{5}{x+5} - \dfrac{3}{x-5}$

28. $\dfrac{4}{z-1} - \dfrac{2}{z+1}$

55. $\dfrac{y^2}{y-3} + \dfrac{9}{3-y}$

56. $\dfrac{t^2}{t-2} + \dfrac{4}{2-t}$

29. $\dfrac{3}{x+1} + \dfrac{2}{3x}$

30. $\dfrac{2}{x+5} + \dfrac{3}{4x}$

57. $\dfrac{b-7}{b^2-16} + \dfrac{7-b}{16-b^2}$

58. $\dfrac{a-3}{a^2-25} + \dfrac{a-3}{25-a^2}$

31. $\dfrac{3}{2t^2-2t} - \dfrac{5}{2t-2}$

32. $\dfrac{8}{3t^2-15t} - \dfrac{3}{2t-10}$

59. $\dfrac{4-p}{25-p^2} + \dfrac{p+1}{p-5}$

60. $\dfrac{y+2}{y-7} + \dfrac{3-y}{49-y^2}$

33. $\dfrac{2x}{x^2-16} + \dfrac{x}{x-4}$

34. $\dfrac{4x}{x^2-25} + \dfrac{x}{x+5}$

61. $\dfrac{8x}{16-x^2} - \dfrac{5}{x-4}$

62. $\dfrac{5x}{x^2-9} - \dfrac{4}{3-x}$

35. $\dfrac{6}{z+4} - \dfrac{2}{3z+12}$

36. $\dfrac{t}{t-3} - \dfrac{5}{4t-12}$

63. $\dfrac{a}{a^2-1} + \dfrac{2a}{a-a^2}$

64. $\dfrac{3x+2}{3x+6} + \dfrac{x}{4-x^2}$

37. $\dfrac{3}{x-1} + \dfrac{2}{(x-1)^2}$

38. $\dfrac{2}{x+3} + \dfrac{4}{(x+3)^2}$

65. $\dfrac{4x}{x^2-y^2} - \dfrac{6}{y-x}$

66. $\dfrac{4-a^2}{a^2-9} - \dfrac{a-2}{3-a}$

39. $\dfrac{m-3}{m^3-1} - \dfrac{2}{1-m^3}$

40. $\dfrac{1-6s}{1-s^3} - \dfrac{5}{s^3-1}$

Perform the indicated operations. Simplify, if possible.

67. $\dfrac{x-3}{2-x} - \dfrac{x+3}{x+2} + \dfrac{x+6}{4-x^2}$

41. $\dfrac{3a}{4a-20} + \dfrac{9a}{6a-30}$

42. $\dfrac{4a}{5a-10} + \dfrac{3a}{10a-20}$

68. $\dfrac{t-5}{1-t} - \dfrac{t+4}{t+1} + \dfrac{t+2}{t^2-1}$

Aha! **43.** $\dfrac{x}{x-5} + \dfrac{x}{5-x}$

44. $\dfrac{x+4}{x} + \dfrac{x}{x+4}$

69. $\dfrac{x+5}{x+3} + \dfrac{x+7}{x+2} - \dfrac{7x+19}{(x+3)(x+2)}$

45. $\dfrac{6}{a^2+a-2} + \dfrac{4}{a^2-4a+3}$

70. $\dfrac{2x+5}{x+1} + \dfrac{x+7}{x+5} - \dfrac{5x+17}{(x+1)(x+5)}$

46. $\dfrac{x}{x^2+2x+1} + \dfrac{1}{x^2+5x+4}$

71. $\dfrac{1}{x+y} + \dfrac{1}{x-y} - \dfrac{2x}{x^2-y^2}$

47. $\dfrac{x}{x^2+9x+20} - \dfrac{4}{x^2+7x+12}$

72. $\dfrac{2r}{r^2-s^2} + \dfrac{1}{r+s} - \dfrac{1}{r-s}$

73. What is the advantage of using the *least* common denominator—rather than just *any* common denominator—when adding or subtracting rational expressions?

74. Describe a procedure that can be used to add any two rational expressions.

SKILL MAINTENANCE

Simplify.

75. $-\dfrac{3}{7} \div \dfrac{6}{13}$ [1.7]

76. $\dfrac{5}{12} \div \left(-\dfrac{3}{4}\right)$ [1.7]

77. $\dfrac{\frac{2}{9}}{\frac{5}{3}}$ [1.3]

78. $\dfrac{\frac{7}{10}}{\frac{3}{5}}$ [1.3]

Graph. [3.2]

79. $y = -\dfrac{1}{2}x - 5$

80. $y = \dfrac{1}{2}x - 5$

SYNTHESIS

81. How could you convince someone that
$$\frac{1}{3-x} \quad \text{and} \quad \frac{1}{x-3}$$
are opposites of each other?

82. Are parentheses as important for adding rational expressions as they are for subtracting rational expressions? Why or why not?

Write expressions for the perimeter and the area of each rectangle.

83.

$\dfrac{3}{x+4}$

$\dfrac{2}{x-5}$

84.

$\dfrac{x}{x+4}$

$\dfrac{x}{x+5}$

Perform the indicated operations.

85. $\dfrac{2x+11}{x-3} \cdot \dfrac{3}{x+4} + \dfrac{2x+1}{4+x} \cdot \dfrac{3}{3-x}$

86. $\dfrac{x^2}{3x^2-5x-2} - \dfrac{2x}{3x+1} \cdot \dfrac{1}{x-2}$

Aha! **87.** $\left(\dfrac{x}{x+7} - \dfrac{3}{x+2}\right)\left(\dfrac{x}{x+7} + \dfrac{3}{x+2}\right)$

88. $\dfrac{1}{ay-3a+2xy-6x} - \dfrac{xy+ay}{a^2-4x^2}\left(\dfrac{1}{y-3}\right)^2$

89. $\dfrac{2x^2+5x-3}{2x^2-9x+9} + \dfrac{x+1}{3-2x} + \dfrac{4x^2+8x+3}{x-3} \cdot \dfrac{x+3}{9-4x^2}$

90. $\left(\dfrac{a}{a-b} + \dfrac{b}{a+b}\right)\left(\dfrac{1}{3a+b} + \dfrac{2a+6b}{9a^2-b^2}\right)$

91. Express
$$\frac{a-3b}{a-b}$$
as a sum of two rational expressions with denominators that are opposites of each other. Answers may vary.

92. Use a graphing calculator to check the answer to Exercise 29.

93. Why does the word ERROR appear in the table displayed on p. 403?

6.5   Complex Rational Expressions

Using Division to Simplify   •   Multiplying by the LCD

A **complex rational expression** is a rational expression that has one or more rational expressions within its numerator or denominator. Here are some examples:

$$\dfrac{1+\dfrac{2}{x}}{3}, \quad \dfrac{\dfrac{x+y}{7}}{\dfrac{2x}{x+1}}, \quad \dfrac{\dfrac{4}{3}+\dfrac{1}{5}}{\dfrac{2}{x}-\dfrac{x}{y}}.$$

These are rational expressions within the complex rational expression.

We will consider two methods for simplifying complex rational expressions. Each method offers certain advantages.

Using Division to Simplify (Method 1)

Our first method for simplifying complex rational expressions involves rewriting the expression as a quotient of two rational expressions.

To Simplify a Complex Rational Expression by Dividing

1. Add or subtract, as needed, to get a single rational expression in the numerator.
2. Add or subtract, as needed, to get a single rational expression in the denominator.
3. Divide the numerator by the denominator (invert and multiply).
4. If possible, simplify by removing a factor equal to 1.

The key here is to express a complex rational expression as one rational expression divided by another. We can then proceed as in Section 6.2.

EXAMPLE 1   Simplify: $\dfrac{\dfrac{x}{x-3}}{\dfrac{4}{5x-15}}$.

Solution   Here the numerator and the denominator are already single rational expressions. This allows us to start by dividing (step 3), as in Section 6.2:

$$\frac{\dfrac{x}{x-3}}{\dfrac{4}{5x-15}} = \frac{x}{x-3} \div \frac{4}{5x-15} \qquad \text{Rewriting with a division symbol}$$

$$= \frac{x}{x-3} \cdot \frac{5x-15}{4} \qquad \begin{array}{l}\text{Multiplying by the reciprocal of the}\\\text{divisor (inverting and multiplying)}\end{array}$$

$$= \frac{x}{x-3} \cdot \frac{5(x-3)}{4} \qquad \begin{array}{l}\text{Factoring and removing a factor}\\\text{equal to 1: } \dfrac{x-3}{x-3} = 1\end{array}$$

$$= \frac{5x}{4}.$$

To use this method, we must sometimes add or subtract in the numerator and/or the denominator before we can divide.

EXAMPLE 2   Simplify.

a) $\dfrac{\dfrac{5}{2a} + \dfrac{1}{a}}{\dfrac{1}{4a} - \dfrac{5}{6}}$

b) $\dfrac{\dfrac{x^2}{y} - \dfrac{5}{x}}{xz}$

Solution

a) $\dfrac{\dfrac{5}{2a} + \dfrac{1}{a}}{\dfrac{1}{4a} - \dfrac{5}{6}} = \dfrac{\dfrac{5}{2a} + \dfrac{1}{a} \cdot \dfrac{2}{2}}{\dfrac{1}{4a} \cdot \dfrac{3}{3} - \dfrac{5}{6} \cdot \dfrac{2a}{2a}}$ $\left.\begin{array}{l}\\\\\end{array}\right\} \leftarrow$ Multiplying by 1 to get the LCD, $2a$, for the numerator of the complex rational expression

$\left.\begin{array}{l}\\\\\end{array}\right\} \leftarrow$ Multiplying by 1 to get the LCD, $12a$, for the denominator of the complex rational expression

$$= \dfrac{\dfrac{5}{2a} + \dfrac{2}{2a}}{\dfrac{3}{12a} - \dfrac{10a}{12a}} = \dfrac{\dfrac{7}{2a}}{\dfrac{3-10a}{12a}} \qquad \begin{array}{l}\leftarrow \text{Adding}\\[4pt]\leftarrow \text{Subtracting}\end{array}$$

$$= \frac{7}{2a} \div \frac{3-10a}{12a} \qquad \begin{array}{l}\text{Rewriting with a division symbol.}\\\text{This is often done mentally.}\end{array}$$

$$= \frac{7}{2a} \cdot \frac{12a}{3-10a} \qquad \begin{array}{l}\text{Multiplying by the reciprocal of the}\\\text{divisor (inverting and multiplying)}\end{array}$$

$$= \frac{7}{2a} \cdot \frac{2a \cdot 6}{3-10a} \qquad \text{Removing a factor equal to 1: } \dfrac{2a}{2a} = 1$$

$$= \frac{42}{3-10a}$$

b) $\dfrac{\dfrac{x^2}{y} - \dfrac{5}{x}}{xz} = \dfrac{\dfrac{x^2}{y} \cdot \dfrac{x}{x} - \dfrac{5}{x} \cdot \dfrac{y}{y}}{xz}$ ⟵ Multiplying by 1 to get the LCD, xy, for the numerator of the complex rational expression

$$= \dfrac{\dfrac{x^3}{xy} - \dfrac{5y}{xy}}{xz}$$

$$= \dfrac{\dfrac{x^3 - 5y}{xy}}{xz}$$ ⟵ Subtracting

⟵ If you prefer, write xz as $\dfrac{xz}{1}$.

$$= \dfrac{x^3 - 5y}{xy} \div (xz)$$ Rewriting with a division symbol

$$= \dfrac{x^3 - 5y}{xy} \cdot \dfrac{1}{xz}$$ Multiplying by the reciprocal of the divisor (inverting and multiplying)

$$= \dfrac{x^3 - 5y}{x^2yz}$$

Multiplying by the LCD (Method 2)

A second method for simplifying complex rational expressions relies on multiplying by a carefully chosen expression that is equal to 1. This multiplication by 1 will lead to a simpler, but equivalent, expression.

To Simplify a Complex Rational Expression by Multiplying by the LCD

1. Find the LCD of *all* rational expressions within the complex rational expression.
2. Multiply the complex rational expression by a factor equal to 1. Write 1 as the LCD over itself (LCD/LCD).
3. Simplify. No fractional expressions should remain within the complex rational expression.
4. Factor and, if possible, simplify.

EXAMPLE 3   Simplify: $\dfrac{\dfrac{1}{2} + \dfrac{3}{4}}{\dfrac{5}{6} - \dfrac{3}{8}}$.

Solution

1. Unlike Method 1, in which $\frac{1}{2} + \frac{3}{4}$ would be treated separately from $\frac{5}{6} - \frac{3}{8}$, here we look for the LCD of *all* four fractions. That LCD is 24.

2. We multiply by a form of 1, using the LCD:

$$\frac{\dfrac{1}{2} + \dfrac{3}{4}}{\dfrac{5}{6} - \dfrac{3}{8}} = \frac{\dfrac{1}{2} + \dfrac{3}{4}}{\dfrac{5}{6} - \dfrac{3}{8}} \cdot \frac{24}{24} \qquad \text{Multiplying by a factor equal to 1,}$$

$$\text{using the LCD: } \frac{24}{24} = 1$$

3. Using the distributive law, we perform the multiplication:

$$\frac{\dfrac{1}{2} + \dfrac{3}{4}}{\dfrac{5}{6} - \dfrac{3}{8}} \cdot \frac{24}{24} = \frac{\left(\dfrac{1}{2} + \dfrac{3}{4}\right)24}{\left(\dfrac{5}{6} - \dfrac{3}{8}\right)24} \quad \longleftarrow \text{Multiplying the numerator by 24}$$

Don't forget the parentheses!
$\longleftarrow$ Multiplying the denominator by 24

$$= \frac{\dfrac{1}{2}(24) + \dfrac{3}{4}(24)}{\dfrac{5}{6}(24) - \dfrac{3}{8}(24)} \qquad \text{Using the distributive law}$$

$$= \frac{12 + 18}{20 - 9}, \quad \text{or} \quad \frac{30}{11}. \qquad \text{Simplifying}$$

4. The result, $\frac{30}{11}$, cannot be factored or simplified, so we are done.

Multiplying like this effectively clears fractions in both the top and the bottom of the complex rational expression. In Example 4 we follow, but do not number, the same four steps.

EXAMPLE 4   Simplify.

a) $\dfrac{\dfrac{3}{x} + \dfrac{1}{2x}}{\dfrac{1}{3x} - \dfrac{3}{4x}}$

b) $\dfrac{1 - \dfrac{1}{x}}{1 - \dfrac{1}{x^2}}$

Solution

a) The denominators within the complex expression are x, $2x$, $3x$, and $4x$, so the LCD is $12x$. We multiply by 1 using $(12x)/(12x)$:

$$\frac{\dfrac{3}{x} + \dfrac{1}{2x}}{\dfrac{1}{3x} - \dfrac{3}{4x}} = \frac{\dfrac{3}{x} + \dfrac{1}{2x}}{\dfrac{1}{3x} - \dfrac{3}{4x}} \cdot \frac{12x}{12x} = \frac{\dfrac{3}{x}(12x) + \dfrac{1}{2x}(12x)}{\dfrac{1}{3x}(12x) - \dfrac{3}{4x}(12x)}. \qquad \begin{array}{l}\text{Using the} \\ \text{distributive law}\end{array}$$

When we multiply by $12x$, all fractions in the numerator and the denominator of the complex rational expression are cleared:

$$\frac{\dfrac{3}{x}(12x) + \dfrac{1}{2x}(12x)}{\dfrac{1}{3x}(12x) - \dfrac{3}{4x}(12x)} = \frac{36 + 6}{4 - 9} = -\frac{42}{5}.$$

technology connection

Be careful to place parentheses properly when entering complex rational expressions into a graphing calculator. Remember to enclose the entire numerator of the complex rational expression in one set of parentheses and the entire denominator in another. For example, we enter the expression in Example 4(a) as

$$y_1 = (3/x + 1/(2x))$$
$$/(1/(3x) - 3/(4x)).$$

1. Write Example 4(b) as you would to enter it into a graphing calculator.
2. When must the numerator of a rational expression be enclosed in parentheses? the denominator?

b) $\dfrac{1 - \dfrac{1}{x}}{1 - \dfrac{1}{x^2}} = \dfrac{1 - \dfrac{1}{x}}{1 - \dfrac{1}{x^2}} \cdot \dfrac{x^2}{x^2}$ The LCD is x^2 so we multiply by 1 using x^2/x^2.

$$= \dfrac{1 \cdot x^2 - \dfrac{1}{x} \cdot x^2}{1 \cdot x^2 - \dfrac{1}{x^2} \cdot x^2}$$ Using the distributive law

$$= \dfrac{x^2 - x}{x^2 - 1}$$ All fractions have been cleared within the complex rational expression.

$$= \dfrac{x(x - 1)}{(x + 1)(x - 1)}$$ Factoring and simplifying: $\dfrac{x - 1}{x - 1} = 1$

$$= \dfrac{x}{x + 1}$$

It is important to understand both of the methods studied in this section. Sometimes, as in Example 1, the complex rational expression is either given as—or easily written as—a quotient of two rational expressions. In these cases, Method 1 (using division) is probably the easier method to use. Other times, as in Example 4(a), it is not difficult to find the LCD of all denominators in the complex rational expression. When this occurs, it is usually easier to use Method 2 (multiplying by the LCD). The more practice you get using both methods, the better you will be at selecting the easier method for any given problem.

Exercise Set

6.5

FOR EXTRA HELP

Student's Solutions Manual   Digital Video Tutor CD 3 Videotape 6   Tutor Center AW Math Tutor Center   MathXL Tutorials on CD   MathXL MathXL   MyMathLab MyMathLab

🐾 *Concept Reinforcement*   *In each of Exercises 1–4, use the method listed to match the accompanying expression with the expression in the column on the right that arises when that method is used.*

1. $\dfrac{\dfrac{5}{x^2} + \dfrac{1}{x}}{\dfrac{7}{2} - \dfrac{3}{4x}}$;

Multiplying by the LCD (Method 2)

2. $\dfrac{\dfrac{5}{x^2} + \dfrac{1}{x}}{\dfrac{7}{2} - \dfrac{3}{4x}}$;

Using division to simplify (Method 1)

3.  $\dfrac{\dfrac{4}{5x} - \dfrac{1}{10}}{\dfrac{8}{x^2} + \dfrac{7}{2}}$;

Using division to simplify (Method 1)

4. $\dfrac{\dfrac{4}{5x} - \dfrac{1}{10}}{\dfrac{8}{x^2} + \dfrac{7}{2}}$;

Multiplying by the LCD (Method 2)

a) $\dfrac{\dfrac{5 + x}{x^2}}{\dfrac{14x - 3}{4x}}$

b) $\dfrac{\dfrac{8 - x}{10x}}{\dfrac{16 + 7x^2}{2x^2}}$

c) $\dfrac{\dfrac{4}{5x} \cdot 10x^2 - \dfrac{1}{10} \cdot 10x^2}{\dfrac{8}{x^2} \cdot 10x^2 + \dfrac{7}{2} \cdot 10x^2}$

d) $\dfrac{\dfrac{5}{x^2} \cdot 4x^2 + \dfrac{1}{x} \cdot 4x^2}{\dfrac{7}{2} \cdot 4x^2 - \dfrac{3}{4x} \cdot 4x^2}$

Simplify. Use either method or the method specified by your instructor.

5. $\dfrac{1 + \dfrac{1}{2}}{1 + \dfrac{1}{4}}$

6. $\dfrac{1 + \dfrac{3}{4}}{1 + \dfrac{1}{2}}$

7. $\dfrac{4 + \dfrac{1}{3}}{1 - \dfrac{5}{27}}$

8. $\dfrac{3 + \dfrac{1}{5}}{1 - \dfrac{2}{15}}$

9. $\dfrac{\dfrac{s}{3} + s}{\dfrac{3}{s} + s}$

10. $\dfrac{\dfrac{1}{x} - 5}{\dfrac{1}{x} + 3}$

11. $\dfrac{\dfrac{4}{x}}{\dfrac{3}{x} + \dfrac{2}{x^2}}$

12. $\dfrac{\dfrac{5}{x} - \dfrac{2}{x^2}}{\dfrac{2}{x^2}}$

13. $\dfrac{\dfrac{2a - 5}{3a}}{\dfrac{a - 7}{6a}}$

14. $\dfrac{\dfrac{a + 5}{a^2}}{\dfrac{a - 2}{3a}}$

15. $\dfrac{\dfrac{x}{4} - \dfrac{4}{x}}{\dfrac{1}{4} + \dfrac{1}{x}}$

16. $\dfrac{\dfrac{3}{x} + \dfrac{3}{8}}{\dfrac{x}{8} - \dfrac{3}{x}}$

17. $\dfrac{\dfrac{1}{6} - \dfrac{1}{x}}{\dfrac{6 - x}{6}}$

18. $\dfrac{\dfrac{1}{7} + \dfrac{1}{a}}{\dfrac{7 + a}{7}}$

19. $\dfrac{\dfrac{1}{t^2} + 1}{\dfrac{1}{t} - 1}$

20. $\dfrac{2 + \dfrac{1}{x}}{2 - \dfrac{1}{x^2}}$

21. $\dfrac{\dfrac{x^2}{x^2 - y^2}}{\dfrac{x}{x + y}}$

22. $\dfrac{\dfrac{a^2}{a - 3}}{\dfrac{2a}{a^2 - 9}}$

23. $\dfrac{\dfrac{7}{a^2} + \dfrac{2}{a}}{\dfrac{5}{a^3} - \dfrac{3}{a}}$

24. $\dfrac{\dfrac{5}{x^3} - \dfrac{1}{x^2}}{\dfrac{2}{x} + \dfrac{3}{x^2}}$ *Aha!*

25. $\dfrac{\dfrac{x}{5y^3} + \dfrac{3}{10y}}{\dfrac{3}{10y} + \dfrac{x}{5y^3}}$

26. $\dfrac{\dfrac{a}{6b^3} + \dfrac{4}{9b^2}}{\dfrac{5}{6b} - \dfrac{1}{9b^3}}$

27. $\dfrac{\dfrac{3}{ab^4} + \dfrac{4}{a^3b}}{\dfrac{5}{a^3b} - \dfrac{3}{ab}}$

28. $\dfrac{\dfrac{2}{x^2y} + \dfrac{3}{xy^2}}{\dfrac{3}{xy^2} + \dfrac{2}{x^2y}}$

29. $\dfrac{2 - \dfrac{3}{x^2}}{2 + \dfrac{3}{x^4}}$

30. $\dfrac{3 - \dfrac{2}{a^4}}{2 + \dfrac{3}{a^3}}$

31. $\dfrac{t - \dfrac{2}{t}}{t + \dfrac{5}{t}}$

32. $\dfrac{x + \dfrac{3}{x}}{x - \dfrac{2}{x}}$

33. $\dfrac{\dfrac{1}{a} + \dfrac{1}{b}}{\dfrac{1}{a^3} + \dfrac{1}{b^3}}$

34. $\dfrac{x - y}{\dfrac{1}{x^3} - \dfrac{1}{y^3}}$

35. $\dfrac{3 + \dfrac{4}{ab^3}}{\dfrac{3 + a}{a^2b}}$

36. $\dfrac{5 + \dfrac{3}{x^2y}}{\dfrac{3 + x}{x^3y}}$

37. $\dfrac{t + 5 + \dfrac{3}{t}}{t + 2 + \dfrac{1}{t}}$

38. $\dfrac{a + 3 + \dfrac{2}{a}}{a + 2 + \dfrac{5}{a}}$

39. $\dfrac{x - 2 - \dfrac{1}{x}}{x - 5 - \dfrac{4}{x}}$

40. $\dfrac{x - 3 - \dfrac{2}{x}}{x - 4 - \dfrac{3}{x}}$

41. Is it possible to simplify complex rational expressions without knowing how to divide rational expressions? Why or why not?

42. Why is the distributive law important when simplifying complex rational expressions?

SKILL MAINTENANCE

Solve.

43. $3x - 5 + 2(4x - 1) = 12x - 3$ [2.2]

44. $(x - 1)7 - (x + 1)9 = 4(x + 2)$ [2.2]

45. $\dfrac{3}{4}x - \dfrac{5}{8} = \dfrac{3}{8}x + \dfrac{7}{4}$ [2.2]

46. $\dfrac{5}{9} - \dfrac{2x}{3} = \dfrac{5x}{6} + \dfrac{4}{3}$ [2.2]

47. $x^2 - 7x - 30 = 0$ [5.7]

48. $x^2 + 8x - 20 = 0$ [5.7]

SYNTHESIS

49. Which of the two methods presented would you use to simplify Exercise 28? Why?

50. Which of the two methods presented would you use to simplify Exercise 22? Why?

In Exercises 51–54, find all x-values for which the given expression is undefined.

51. $\dfrac{\dfrac{x - 5}{x - 6}}{\dfrac{x - 7}{x - 8}}$

52. $\dfrac{\dfrac{x + 1}{x + 2}}{\dfrac{x + 3}{x + 4}}$

53. $\dfrac{\dfrac{2x + 3}{5x + 4}}{\dfrac{3}{7} - \dfrac{2x}{21}}$

54. $\dfrac{\dfrac{3x - 5}{2x - 7}}{\dfrac{4x}{5} - \dfrac{8}{15}}$

55. The formula

$$\frac{\dfrac{P\left(1 + \dfrac{i}{12}\right)^2}{\left(1 + \dfrac{i}{12}\right)^2 - 1}}{\dfrac{i}{12}},$$

where P is a loan amount and i is an interest rate, arises in certain business situations. Simplify this expression. (*Hint*: Expand the binomials.)

56. Find the simplified form for the reciprocal of

$$\frac{2}{x - 1} - \frac{1}{3x - 2}.$$

Simplify.

57. $\dfrac{\dfrac{5}{x + 2} - \dfrac{3}{x - 2}}{\dfrac{x}{x - 1} + \dfrac{x}{x + 1}}$

58. $\dfrac{\dfrac{x}{x + 5} + \dfrac{3}{x + 2}}{\dfrac{2}{x + 2} - \dfrac{x}{x + 5}}$

Aha! **59.** $\left[\dfrac{\dfrac{x - 1}{x - 1} - 1}{\dfrac{x + 1}{x - 1} + 1}\right]^5$

60. $1 + \dfrac{1}{1 + \dfrac{1}{1 + \dfrac{1}{x}}}$

61. $\dfrac{\dfrac{z}{1 - \dfrac{z}{2 + 2z}} - 2z}{\dfrac{2z}{5z - 2} - 3}$

62. Under what circumstance(s) will there be no restrictions on the variable appearing in a complex rational expression?

63. Use a graphing calculator to check Example 2(a).

6.6   Solving Rational Equations

Solving a New Type of Equation • A Visual Interpretation

Our study of rational expressions allows us to solve a type of equation that we could not have solved prior to this chapter.

CONNECTING THE CONCEPTS

In Sections 6.1–6.5, we learned how to *simplify expressions* like

$$\frac{4x^2 - 9}{6x + 9}, \quad \frac{t^2 - 1}{8t} \cdot \frac{4t^2}{3}, \quad \text{and} \quad \frac{\dfrac{5}{x}}{\dfrac{2}{x^2} + \dfrac{3}{x}}.$$

In this section, we will return to *solving equations*. These equations will look like

the following:

$$x + \frac{6}{x} = -5 \quad \text{and} \quad \frac{3}{x - 5} + \frac{1}{x + 5} = \frac{2}{x^2 - 25}.$$

As always, be careful not to confuse simplifying an expression with solving an equation. When expressions are simplified, the result is an equivalent expression. When equations are solved, the result is a solution.

Solving a New Type of Equation

A **rational**, or **fractional**, **equation** is an equation containing one or more rational expressions, often with the variable in a denominator. Here are some examples:

$$\frac{2}{3} + \frac{5}{6} = \frac{x}{9}, \qquad t + \frac{7}{t} = -5, \qquad \frac{x^2}{x-1} = \frac{1}{x-1}.$$

To Solve a Rational Equation

1. List any restrictions that exist. Numbers that make a denominator equal 0 cannot possibly be solutions.
2. Clear the equation of fractions by multiplying both sides by the LCM of the denominators.
3. Solve the resulting equation using the addition principle, the multiplication principle, and the principle of zero products, as needed.
4. Check the possible solution(s) in the original equation.

When clearing an equation of fractions, we use the terminology LCM instead of LCD because we are *not* adding or subtracting rational expressions.

EXAMPLE 1   Solve: $\dfrac{x}{6} - \dfrac{x}{8} = \dfrac{1}{12}$.

Solution   Because no variable appears in a denominator, no restrictions exist. The LCM of 6, 8, and 12 is 24, so we multiply both sides by 24:

$$24\left(\frac{x}{6} - \frac{x}{8}\right) = 24 \cdot \frac{1}{12}$$

Using the multiplication principle to multiply both sides by the LCM. Parentheses are important!

$$24 \cdot \frac{x}{6} - 24 \cdot \frac{x}{8} = 24 \cdot \frac{1}{12}$$

Using the distributive law

Be sure to multiply *each* term by the LCM.

$$\left.\begin{array}{c} \dfrac{24x}{6} - \dfrac{24x}{8} = \dfrac{24}{12} \\[2mm] 4x - 3x = 2 \end{array}\right\}$$

$$x = 2.$$

Simplifying. Note that all fractions have been cleared. If fractions remain, we have either made a mistake or have not used the LCM of the denominators.

*Study Skills*_____

Does More Than One Solution Exist?

Keep in mind that many problems—in math and elsewhere—have more than one solution. When asked to solve an equation, we are expected to find any and all solutions of the equation.

Check: $$\dfrac{x}{6} - \dfrac{x}{8} = \dfrac{1}{12}$$

$$
\begin{array}{c|c}
\dfrac{2}{6} - \dfrac{2}{8} & \dfrac{1}{12} \\[2ex]
\dfrac{1}{3} - \dfrac{1}{4} & \\[2ex]
\dfrac{4}{12} - \dfrac{3}{12} & \\[2ex]
\dfrac{1}{12} & \overset{?}{=} \dfrac{1}{12} \quad \text{TRUE}
\end{array}
$$

This checks, so the solution is 2.

Recall that the multiplication principle states that $a = b$ is equivalent to $a \cdot c = b \cdot c$, *provided c is not zero.* Because rational equations often have variables in a denominator, clearing fractions will now require us to multiply both sides by a variable expression. Since a variable expression could represent 0, *multiplying both sides of an equation by a variable expression does not always produce an equivalent equation.* Thus checking each solution in the original equation is essential.

EXAMPLE 2   Solve.

a) $\dfrac{2}{3x} + \dfrac{1}{x} = 10$ **b)** $x + \dfrac{6}{x} = -5$

c) $1 + \dfrac{3x}{x+2} = \dfrac{-6}{x+2}$ **d)** $\dfrac{3}{x-5} + \dfrac{1}{x+5} = \dfrac{2}{x^2-25}$

e) $\dfrac{x^2}{x-1} = \dfrac{1}{x-1}$

Solution

a) Note that x cannot be 0. The LCM is $3x$, so we multiply both sides by $3x$:

$$\dfrac{2}{3x} + \dfrac{1}{x} = 10 \qquad \text{The LCM of } 3x \text{ and } x \text{ is } 3x;\; x \neq 0.$$

$$3x\left(\dfrac{2}{3x} + \dfrac{1}{x}\right) = 3x \cdot 10 \qquad \begin{array}{l}\text{Using the multiplication principle} \\ \text{to multiply both sides by the LCM.} \\ \textit{Don't forget the parentheses!}\end{array}$$

$$\cancel{3x} \cdot \dfrac{2}{\cancel{3x}} + 3\cancel{x} \cdot \dfrac{1}{\cancel{x}} = 3x \cdot 10 \qquad \text{Using the distributive law}$$

$$2 + 3 = 30x \qquad \begin{array}{l}\text{Removing factors equal to 1:} \\ 3x/(3x) = 1 \text{ and } x/x = 1. \text{ This} \\ \text{clears all fractions.}\end{array}$$

$$5 = 30x$$

$$\dfrac{5}{30} = x, \quad \text{so } x = \dfrac{1}{6}. \qquad \begin{array}{l}\text{Dividing both sides by 30, or} \\ \text{multiplying both sides by } 1/30\end{array}$$

Since $\frac{1}{6} \neq 0$, and 0 is the only restricted value, $\frac{1}{6}$ *should* check.

Check:

$$\frac{2}{3x} + \frac{1}{x} = 10$$

$$\begin{array}{c|c} \dfrac{2}{3 \cdot \frac{1}{6}} + \dfrac{1}{\frac{1}{6}} & 10 \\[2ex] \dfrac{2}{\frac{1}{2}} + \dfrac{1}{\frac{1}{6}} & \\[2ex] 2 \cdot \frac{2}{1} + 1 \cdot \frac{6}{1} & \\[1ex] 4 + 6 & \\[1ex] 10 \stackrel{?}{=} 10 & \text{TRUE} \end{array}$$

The solution is $\frac{1}{6}$.

b) Again, note that x cannot be 0. We multiply both sides by x to clear fractions:

$$x + \frac{6}{x} = -5 \qquad \text{We cannot have } x = 0.$$

$$x\left(x + \frac{6}{x}\right) = x(-5) \qquad \begin{array}{l} \text{Multiplying both sides by } x. \\ \textit{Don't forget the parentheses!} \end{array}$$

$$x \cdot x + \not{x} \cdot \frac{6}{\not{x}} = -5x \qquad \text{Using the distributive law}$$

$$x^2 + 6 = -5x \qquad \begin{array}{l} \text{Removing a factor equal to 1: } x/x = 1. \\ \text{We are left with a quadratic equation.} \end{array}$$

$$x^2 + 5x + 6 = 0 \qquad \begin{array}{l} \text{Using the addition principle to add } 5x \text{ to} \\ \text{both sides} \end{array}$$

$$(x + 3)(x + 2) = 0 \qquad \text{Factoring}$$

$$x + 3 = 0 \quad \textit{or} \quad x + 2 = 0 \qquad \begin{array}{l} \text{Using the principle of zero} \\ \text{products} \end{array}$$

$$x = -3 \quad \textit{or} \qquad x = -2. \qquad \begin{array}{l} \text{Since neither solution is 0,} \\ \text{the restriction in the first} \\ \text{step, both answers should} \\ \text{check.} \end{array}$$

Check: For -3:

$$x + \frac{6}{x} = -5$$

$$\begin{array}{c|c} -3 + \dfrac{6}{-3} & -5 \\[2ex] -3 - 2 & \\[1ex] -5 \stackrel{?}{=} -5 & \text{TRUE} \end{array}$$

For -2:

$$x + \frac{6}{x} = -5$$

$$\begin{array}{c|c} -2 + \dfrac{6}{-2} & -5 \\[2ex] -2 - 3 & \\[1ex] -5 \stackrel{?}{=} -5 & \text{TRUE} \end{array}$$

Both of these check, so there are two solutions, -3 and -2.

Student Notes

Not all checking is for finding errors in computation. For these equations, the solution process itself can introduce numbers that do not check.

c) To avoid division by 0, we must have $x + 2 \neq 0$, or $x \neq -2$. With this restriction in mind, we multiply both sides of the equation by the LCM:

$$1 + \frac{3x}{x + 2} = \frac{-6}{x + 2}$$

We cannot have $x = -2$.

$$(x + 2)\left(1 + \frac{3x}{x + 2}\right) = (x + 2)\frac{-6}{x + 2}$$

Multiplying both sides by $x + 2$. Don't forget the parentheses.

$$(x + 2) \cdot 1 + (x + 2)\frac{3x}{x + 2} = (x + 2)\frac{-6}{x + 2}$$

Using the distributive law; removing a factor equal to 1: $(x + 2)/(x + 2) = 1$

$$x + 2 + 3x = -6$$
$$4x + 2 = -6$$
$$4x = -8$$
$$x = -2.$$

Above, we stated that $x \neq -2$.

Because of the above restriction, -2 must be rejected as a solution. The check below simply confirms this.

Check:
$$1 + \frac{3x}{x + 2} = \frac{-6}{x + 2}$$

$$1 + \frac{3(-2)}{-2 + 2} \quad \bigg| \quad \frac{-6}{-2 + 2}$$

$$1 + \frac{-6}{0} \stackrel{?}{=} \frac{-6}{0} \qquad \text{FALSE}$$

The equation has no solution.

d)
$$\frac{3}{x - 5} + \frac{1}{x + 5} = \frac{2}{x^2 - 25}$$

Note that $x \neq 5$ and $x \neq -5$. The LCM of $x - 5$, $x + 5$, and $x^2 - 25$ is $(x - 5)(x + 5)$.

$$(x - 5)(x + 5)\left(\frac{3}{x - 5} + \frac{1}{x + 5}\right) = (x - 5)(x + 5)\frac{2}{(x - 5)(x + 5)}$$

$$\frac{(x - 5)(x + 5)3}{x - 5} + \frac{(x - 5)(x + 5)}{x + 5} = \frac{2(x - 5)(x + 5)}{(x - 5)(x + 5)}$$

Using the distributive law

Removing factors equal to 1: $\dfrac{x - 5}{x - 5} = 1$, $\dfrac{x + 5}{x + 5} = 1$, and $\dfrac{(x - 5)(x + 5)}{(x - 5)(x + 5)} = 1$

$$(x + 5)3 + (x - 5) = 2$$

$$3x + 15 + x - 5 = 2$$

Using the distributive law

$$4x + 10 = 2$$
$$4x = -8$$
$$x = -2$$

$-2 \neq 5$ and $-2 \neq -5$, so -2 *should* check.

We leave it to the student to check that -2 is the solution.

e) To avoid division by 0, we must have $x - 1 \neq 0$, or $x \neq 1$. With this restriction in mind, we multiply both sides by $x - 1$ to clear fractions:

$$\frac{x^2}{x - 1} = \frac{1}{x - 1} \qquad \text{We cannot have } x = 1.$$

$$(x - 1) \cdot \frac{x^2}{x - 1} = (x - 1) \cdot \frac{1}{x - 1} \qquad \text{Multiplying both sides by } x - 1$$

$$x^2 = 1 \qquad \text{Removing a factor equal to 1:} \atop (x - 1)/(x - 1) = 1$$

$$x^2 - 1 = 0 \qquad \text{Subtracting 1 from both sides}$$

$$(x - 1)(x + 1) = 0 \qquad \text{Factoring}$$

$$x - 1 = 0 \quad or \quad x + 1 = 0 \qquad \text{Using the principle of zero products}$$

$$x = 1 \quad or \qquad x = -1. \qquad \text{Above, we stated that } x \neq 1.$$

Because of the above restriction, 1 must be rejected as a solution. The student should check in the original equation that -1 *does* check. The solution is -1.

A Visual Interpretation

 ALGEBRAIC—GRAPHICAL CONNECTION

We can obtain a visual check of the solutions of a rational equation by graphing. For example, consider the equation

$$\frac{x}{4} + \frac{x}{2} = 6.$$

We can examine the solution by graphing the equations

$$y = \frac{x}{4} + \frac{x}{2} \quad \text{and} \quad y = 6$$

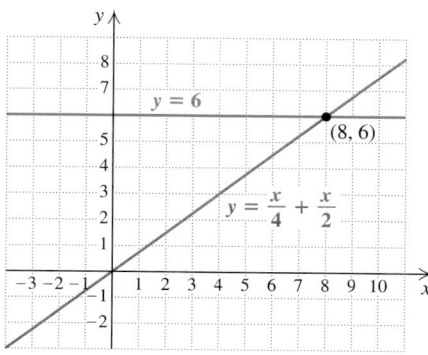

using the same set of axes, as shown at left.

The y-values for each equation will be the same where the graphs intersect. The x-value of that point will yield that value, so it will be the solution of the equation. It appears from the graph that when $x = 8$, the value of $x/4 + x/2$ is 6. We can check by substitution:

$$\frac{x}{4} + \frac{x}{2} = \frac{8}{4} + \frac{8}{2} = 2 + 4 = 6.$$

Thus the solution is 8.

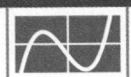

technology connection

We can use a table to check possible solutions of rational equations. Consider the equation in Example 2(e),

$$\frac{x^2}{x-1} = \frac{1}{x-1},$$

and the possible solutions that were found, 1 and -1. To check these solutions, we enter $y_1 = x^2/(x-1)$ and $y_2 = 1/(x-1)$. After setting Indpnt to Ask and Depend to Auto in the TBLSET menu, we display the table and enter $x = 1$. The ERROR messages indicate that 1 is not a solution because it is not an allowable replacement for x in the equation. Next, we enter $x = -1$. Since y_1 and y_2 have the same value, we know that the equation is true when $x = -1$, and thus -1 is a solution.

X	Y₁	Y₂
1	ERROR	ERROR
−1	−.5	−.5

X =

Use a graphing calculator to check the possible solutions of Example 1 and Example 2, parts (a)–(d).

Exercise Set

6.6

FOR EXTRA HELP

 Student's Solutions Manual    Digital Video Tutor CD 3 Videotape 6   Tutor Center AW Math Tutor Center   MathXL Tutorials on CD   MathXL MathXL   MyMathLab MyMathLab

Concept Reinforcement   *Classify each statement as either true or false.*

1. Every rational equation has at least one solution.

2. It is possible for a rational equation to have more than one solution.

3. When both sides of an equation are multiplied by a variable expression, the result is not always an equivalent equation.

4. All the equation-solving principles studied thus far may be needed when solving a rational equation.

Solve. If no solution exists, state this.

5. $\dfrac{3}{5} - \dfrac{5}{8} = \dfrac{x}{20}$

6. $\dfrac{4}{5} - \dfrac{2}{3} = \dfrac{x}{9}$

7. $\dfrac{1}{3} + \dfrac{5}{6} = \dfrac{1}{x}$

8. $\dfrac{3}{5} + \dfrac{1}{8} = \dfrac{1}{x}$

9. $\dfrac{1}{6} + \dfrac{1}{8} = \dfrac{1}{t}$

10. $\dfrac{1}{8} + \dfrac{1}{10} = \dfrac{1}{t}$

11. $x + \dfrac{5}{x} = -6$

12. $x + \dfrac{6}{x} = -7$

13. $\dfrac{x}{6} - \dfrac{6}{x} = 0$

14. $\dfrac{x}{7} - \dfrac{7}{x} = 0$

15. $\dfrac{5}{x} = \dfrac{6}{x} - \dfrac{1}{3}$

16. $\dfrac{4}{t} = \dfrac{5}{t} - \dfrac{1}{2}$

17. $\dfrac{5}{3t} + \dfrac{3}{t} = 1$

18. $\dfrac{3}{4x} + \dfrac{5}{x} = 1$

19. $\dfrac{x-8}{x+3} = \dfrac{1}{4}$

20. $\dfrac{a-4}{a+5} = \dfrac{3}{8}$

21. $x + \dfrac{12}{x} = -7$

22. $x + \dfrac{8}{x} = -9$

23. $\dfrac{2}{x+1} = \dfrac{1}{x-2}$

24. $\dfrac{5}{x-1} = \dfrac{3}{x+2}$

25. $\dfrac{a}{6} - \dfrac{a}{10} = \dfrac{1}{6}$

26. $\dfrac{t}{8} - \dfrac{t}{12} = \dfrac{1}{8}$

27. $\dfrac{x+1}{3} - 1 = \dfrac{x-1}{2}$

28. $\dfrac{x+2}{5} - 1 = \dfrac{x-2}{4}$

29. $\dfrac{4}{t-5} = \dfrac{t-1}{t-5}$

30. $\dfrac{2}{t-9} = \dfrac{t-7}{t-9}$

31. $\dfrac{3}{x+4} = \dfrac{5}{x}$

32. $\dfrac{2}{x+3} = \dfrac{7}{x}$

33. $\dfrac{a-4}{a-1} = \dfrac{a+2}{a-2}$

34. $\dfrac{x-2}{x-3} = \dfrac{x-1}{x+1}$

35. $\dfrac{5}{t-2} + \dfrac{3t}{t-2} = \dfrac{4}{t^2-4t+4}$

36. $\dfrac{4}{t-3} + \dfrac{2t}{t-3} = \dfrac{12}{t^2-6t+9}$

37. $\dfrac{4}{x-3} + \dfrac{2x}{x^2-9} = \dfrac{1}{x+3}$

38. $\dfrac{x}{x+4} - \dfrac{4}{x-4} = \dfrac{x^2+16}{x^2-16}$

39. $\dfrac{5}{y-3} - \dfrac{30}{y^2-9} = 1$

40. $\dfrac{1}{x+3} + \dfrac{1}{x-3} = \dfrac{1}{x^2-9}$

41. $\dfrac{4}{8-a} = \dfrac{4-a}{a-8}$

42. $\dfrac{t+10}{7-t} = \dfrac{3}{t-7}$

Aha! **43.** $\dfrac{-2}{x+2} = \dfrac{x}{x+2}$

44. $\dfrac{3}{2x-6} = \dfrac{x}{2x-6}$

45. When solving rational equations, why do we multiply each side by the LCM of the denominators?

46. Explain the difference between adding rational expressions and solving rational equations.

SKILL MAINTENANCE

47. The sum of two consecutive odd numbers is 276. Find the numbers. [2.5]

48. The length of a rectangular picture window is 3 yd greater than the width. The area of the rectangle is 10 yd². Find the perimeter. [5.8]

49. The height of a triangle is 3 cm longer than its base. If the area of the triangle is 54 cm², find the measurements of the base and the height. [5.8]

50. The product of two consecutive even integers is 48. Find the numbers. [5.8]

51. *Human physiology.* Between June 9 and June 24, Seth's beard grew 0.9 cm. Find the rate at which Seth's beard grows. [3.4]

52. *Gardening.* Between July 7 and July 12, Carla's string beans grew 1.4 in. Find the string beans' growth rate. [3.4]

SYNTHESIS

53. Describe a method that can be used to create rational equations that have no solution.

54. How can a graph be used to determine how many solutions an equation has?

Solve.

55. $1 + \dfrac{x-1}{x-3} = \dfrac{2}{x-3} - x$

56. $\dfrac{4}{y-2} + \dfrac{3}{y^2-4} = \dfrac{5}{y+2} + \dfrac{2y}{y^2-4}$

57. $\dfrac{x}{x^2+3x-4} + \dfrac{x+1}{x^2+6x+8} = \dfrac{2x}{x^2+x-2}$

58. $\dfrac{12-6x}{x^2-4} = \dfrac{3x}{x+2} - \dfrac{3-2x}{2-x}$

59. $\dfrac{x^2}{x^2-4} = \dfrac{x}{x+2} - \dfrac{2x}{2-x}$

60. $7 - \dfrac{a-2}{a+3} = \dfrac{a^2-4}{a+3} + 5$

61. $\dfrac{1}{x-1} + x - 5 = \dfrac{5x-4}{x-1} - 6$

62. $\dfrac{5-3a}{a^2+4a+3} - \dfrac{2a+2}{a+3} = \dfrac{3-a}{a+1}$

63. Use a graphing calculator to check your answers to Exercises 13, 21, 31, and 55.

6.7 Applications Using Rational Equations and Proportions

Problems Involving Work • Problems Involving Motion •
Problems Involving Proportions

In many areas of study, applications involving rates, proportions, or reciprocals translate to rational equations. By using the five steps for problem solving and the lessons of Section 6.6, we can now solve such problems.

Problems Involving Work

EXAMPLE 1   Sorting recyclables.   Cecilia and Aaron work as volunteers at a town's recycling depot. Cecilia can sort a day's accumulation of recyclables in 4 hr, while Aaron requires 6 hr to do the same job. How long would it take them, working together, to sort the recyclables?

Solution

1. **Familiarize.**   We familiarize ourselves with the problem by exploring two common, but *incorrect*, approaches.

   a) One common incorrect approach is to simply add the two times:

   $$4 \text{ hr} + 6 \text{ hr} = 10 \text{ hr}.$$

   Let's think about this. If Cecilia can do the sorting *alone* in 4 hr, then Cecilia and Aaron *together* should take *less* than 4 hr. Thus we reject 10 hr as a solution and reason that the answer must be less than 4 hr.

   b) Another incorrect approach is to assume that Cecilia does half the sorting and Aaron does the other half. Then

   Cecilia sorts $\frac{1}{2}$ of the accumulation in $\frac{1}{2}(4 \text{ hr})$, or 2 hr, and
   Aaron sorts $\frac{1}{2}$ of the accumulation in $\frac{1}{2}(6 \text{ hr})$, or 3 hr.

   This would waste time since Cecilia would finish 1 hr earlier than Aaron. In reality, Cecilia would help Aaron after completing her half, so that Aaron would actually sort less than half of the accumulation. This tells us that the entire job will take them between 2 hr and 3 hr.

   A correct approach is to consider how much of the sorting is finished in 1 hr, 2 hr, 3 hr, and so on. It takes Cecilia 4 hr to sort the recyclables alone, so her *rate* is $\frac{1}{4}$ of the job per hour. It takes Aaron 6 hr to do the sorting alone, so his *rate* is $\frac{1}{6}$ of the job per hour. Working together, they can complete

   $$\frac{1}{4} + \frac{1}{6}, \quad \text{or } \frac{5}{12} \text{ of the sorting in 1 hr.} \qquad \text{Their rate working together is the sum of their individual rates.}$$

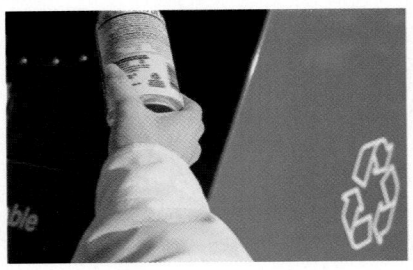

In 2 hr, Cecilia can do $\frac{1}{4} \cdot 2$ of the sorting and Aaron can do $\frac{1}{6} \cdot 2$ of the sorting. Working together, they can complete

$$\frac{1}{4} \cdot 2 + \frac{1}{6} \cdot 2, \quad \text{or } \frac{5}{6} \text{ of the sorting in 2 hr.} \qquad \text{Note that } \frac{5}{12} \cdot 2 = \frac{5}{6}.$$

Continuing this reasoning, we can form a table.

Time	Fraction of the Sorting Completed		
	Cecilia	Aaron	Together
1 hr	$\frac{1}{4}$	$\frac{1}{6}$	$\frac{1}{4} + \frac{1}{6}$, or $\frac{5}{12}$
2 hr	$\frac{1}{4} \cdot 2$	$\frac{1}{6} \cdot 2$	$\left(\frac{1}{4} + \frac{1}{6}\right)2$, or $\frac{5}{12} \cdot 2$, or $\frac{5}{6}$ ← This is too little.
3 hr	$\frac{1}{4} \cdot 3$	$\frac{1}{6} \cdot 3$	$\left(\frac{1}{4} + \frac{1}{6}\right)3$, or $\frac{5}{12} \cdot 3$, or $1\frac{1}{4}$ ← This is too much.
t hr	$\frac{1}{4} \cdot t$	$\frac{1}{6} \cdot t$	$\left(\frac{1}{4} + \frac{1}{6}\right)t$, or $\frac{5}{12} \cdot t$

From the table, we see that if they work 3 hr, the fraction of the sorting that they complete is $1\frac{1}{4}$, which is more of the job than needs to be done. We need to find a number t for which the fraction of the sorting that is completed in t hours is exactly 1, no more and no less.

2. **Translate.** From the table, we see that the time we want is some number t for which

Portion of work done by Cecilia in t hr $\qquad \dfrac{1}{4} \cdot t + \dfrac{1}{6} \cdot t = 1 \qquad$ Portion of work done by Aaron in t hr

or

$$\left(\frac{1}{4} + \frac{1}{6}\right)t = 1 \quad \text{or} \quad \frac{5}{12} \cdot t = 1.$$

Portion of work done together in t hr

3. **Carry out.** We can choose any one of the above equations to solve:

$$\frac{5}{12} \cdot t = 1$$

$$\frac{12}{5} \cdot \frac{5}{12} \cdot t = \frac{12}{5} \cdot 1 \qquad \text{Multiplying both sides by } \tfrac{12}{5}$$

$$t = \frac{12}{5}, \quad \text{or} \quad 2\frac{2}{5} \text{ hr.}$$

4. Check. The check can be done following the pattern used in the table of the *Familiarize* step above:

$$\frac{1}{4} \cdot \frac{12}{5} + \frac{1}{6} \cdot \frac{12}{5} = \frac{3}{5} + \frac{2}{5} = \frac{5}{5} = 1.$$

A second, partial, check is that (as we predicted in step 1) the answer is between 2 hr and 3 hr.

5. State. Together, it takes Cecilia and Aaron $2\frac{2}{5}$ hr to complete the sorting.

EXAMPLE 2

It takes Pepe 9 hr longer than Wendy to rebuild an engine. Working together, they can do the job in 20 hr. How long would it take each, working alone, to rebuild an engine?

Solution

1. **Familiarize.** Unlike Example 1, this problem does not provide us with the times required by the individuals to do the job alone. Let's have $w =$ the number of hours it would take Wendy working alone and $w + 9 =$ the number of hours it would take Pepe working alone.

2. **Translate.** Using the same reasoning as in Example 1, we see that Wendy completes $\dfrac{1}{w}$ of the job in 1 hr and Pepe completes $\dfrac{1}{w + 9}$ of the job in 1 hr.

In 2 hr, Wendy completes $\dfrac{1}{w} \cdot 2$ of the job and Pepe completes $\dfrac{1}{w + 9} \cdot 2$ of the job. We are told that, working together, Wendy and Pepe can complete the entire job in 20 hr. This gives the following:

Portion of job done by Wendy in 20 hr $\underbrace{\dfrac{1}{w} \cdot 20}$ + $\underbrace{\dfrac{1}{w + 9} \cdot 20} = 1,$ Portion of job done by Pepe in 20 hr

or $\dfrac{20}{w} + \dfrac{20}{w + 9} = 1.$

3. **Carry out.** We solve the equation:

$$\frac{20}{w} + \frac{20}{w + 9} = 1$$

$$w(w + 9)\left(\frac{20}{w} + \frac{20}{w + 9}\right) = w(w + 9)1 \qquad \text{Multiplying by the LCD}$$

$$(w + 9)20 + w \cdot 20 = w(w + 9) \qquad \text{Distributing and simplifying}$$

$$40w + 180 = w^2 + 9w$$

$$0 = w^2 - 31w - 180 \qquad \text{Getting 0 on one side}$$

$$0 = (w - 36)(w + 5) \qquad \text{Factoring}$$

$$w - 36 = 0 \quad or \quad w + 5 = 0 \qquad \text{Principle of zero products}$$

$$w = 36 \quad or \qquad w = -5.$$

4. **Check.** Since negative time has no meaning in the problem, -5 is not a solution to the original problem. The number 36 checks since, if Wendy

takes 36 hr alone and Pepe takes $36 + 9 = 45$ hr alone, in 20 hr they would have rebuilt

$$\frac{20}{36} + \frac{20}{45} = \frac{5}{9} + \frac{4}{9} = 1 \text{ complete engine.}$$

5. **State.** It would take Wendy 36 hr to rebuild an engine alone, and Pepe 45 hr.

The Work Principle

Suppose that A requires a units of time to complete a task and B requires b units of time to complete the same task. Then

A works at a rate of $\dfrac{1}{a}$ tasks per unit of time,

B works at a rate of $\dfrac{1}{b}$ tasks per unit of time, and

A and B together work at a rate of $\dfrac{1}{a} + \dfrac{1}{b}$ tasks per unit of time.

If A and B, working together, require t units of time to complete the task, then their rate is $1/t$ and the following equations hold:

$$\frac{1}{a} \cdot t + \frac{1}{b} \cdot t = 1; \quad \left(\frac{1}{a} + \frac{1}{b}\right)t = 1; \quad \frac{t}{a} + \frac{t}{b} = 1; \quad \frac{1}{a} + \frac{1}{b} = \frac{1}{t}.$$

Problems Involving Motion

Problems that deal with distance, speed (or rate), and time are called **motion problems**. Translation of these problems involves the distance formula, $d = r \cdot t$, and/or the equivalent formulas $r = d/t$ and $t = d/r$.

EXAMPLE 3   Flight speed.   Because of a tailwind, a Delta Boeing 777 is able to fly 200 mph faster than a Northwest 777 that is flying *into* the wind. In the same time that it takes the Delta plane to travel 1800 mi, the Northwest plane flies 1200 mi. How fast is each plane traveling?

Solution

1. **Familiarize.** Suppose that the Northwest plane is traveling 300 mph. Then the Delta plane would be flying $300 + 200$, or 500 mph. Thus if r is the speed of the Northwest plane, in miles per hour, then the Delta plane is traveling $r + 200$ mph.

    At 300 mph, the Northwest flight would travel 1200 mi in 1200/300, or 4 hr. At 500 mph, the Delta plane would travel 1800 mi in 1800/500, or $3\frac{3}{5}$ hr. Since both planes spend the same amount of time traveling, and since 4 hr $\neq 3\frac{3}{5}$ hr, we see that our guess of 300 mph is incorrect. Rather than check another guess, we form a table.

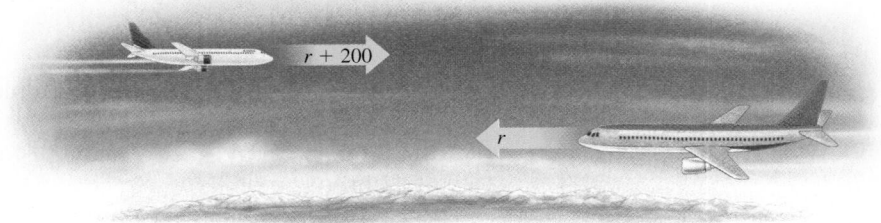

Distance =		Rate ·	Time
	Distance (in miles)	**Speed (in miles per hour)**	**Time (in hours)**
Northwest	1200	r	
Delta	1800	$r + 200$	

2. **Translate.**   Examine how we checked our guess. We found, and then compared, the two flight times. The times were found by dividing the distances, 1200 mi and 1800 mi, by the rates, 300 mph and 500 mph, respectively. Thus the blanks in the table above can be filled, using *time = distance/rate*. This yields a table that uses only one variable.

	Distance (in miles)	**Speed (in miles per hour)**	**Time (in hours)**	
Northwest	1200	r	$1200/r$	The times must
Delta	1800	$r + 200$	$1800/(r + 200)$	be the same.

Since the times must be the same for both planes, we have the equation

$$\frac{1200}{r} = \frac{1800}{r + 200}.$$

Note that $\dfrac{\text{mi}}{\text{mph}} = \dfrac{\text{mi}}{\text{mi/hr}} = \cancel{\text{mi}} \cdot \dfrac{\text{hr}}{\cancel{\text{mi}}} = \text{hr}$, so we are indeed comparing two times.

3. **Carry out.**   To solve the equation, we first multiply both sides by the LCM of the denominators, $r(r + 200)$:

$$r(r + 200) \cdot \frac{1200}{r} = r(r + 200) \cdot \frac{1800}{r + 200}$$

Multiplying both sides by the LCM, $r(r + 200)$. Note that we must have $r \neq 0$ and $r \neq -200$.

$$\begin{aligned}
(r + 200)1200 &= 1800r & \text{Simplifying} \\
1200r + 240{,}000 &= 1800r & \text{Using the distributive law} \\
240{,}000 &= 600r & \text{Subtracting } 1200r \text{ from both sides} \\
400 &= r. & \text{Dividing both sides by 600}
\end{aligned}$$

We now have a possible solution. The speed of the Northwest plane is 400 mph, and the speed of the Delta plane is $400 + 200$, or 600 mph.

4. **Check.** We first reread the problem to confirm that we were to find the speeds. Note that, at 600 mph, the Delta flight would indeed be traveling 200 mph faster than the Northwest flight. At 600 mph, the Delta plane would cover 1800 mi in 1800/600, or 3 hr. If the Northwest plane flies 1200 mi at 400 mph, it is flying for 1200/400, or 3 hr. Since the times are the same, the speeds check.

5. **State.** The Northwest 777 is flying 400 mph and the Delta 777 is flying 600 mph.

In the following example, although the distance is the same in both directions, the key to the translation lies in an additional piece of given information.

EXAMPLE 4 A Hudson River tugboat goes 10 mph in still water. It travels 24 mi upstream and 24 mi back in a total time of 5 hr. What is the speed of the current? (*Sources*: Based on information from the Department of the Interior, U.S. Geological Survey, and *The Tugboat Captain*, Montgomery County Community College)

Solution

1. **Familiarize.** Let's guess that the speed of the current is 4 mph. The tugboat would then be moving $10 - 4 = 6$ mph upstream and $10 + 4 = 14$ mph downstream. The tugboat would require $\frac{24}{6} = 4$ hr to travel 24 mi upstream and $\frac{24}{14} = 1\frac{5}{7}$ hr to travel 24 mi downstream. Since the total time, $4 + 1\frac{5}{7} = 5\frac{5}{7}$ hr, is not the 5 hr mentioned in the problem, we know that our guess is wrong.

   Suppose that the current's speed $= c$ mph. The tugboat's speed would then be $10 - c$ mph going upstream and $10 + c$ mph going downstream. A sketch and table can help display the information.

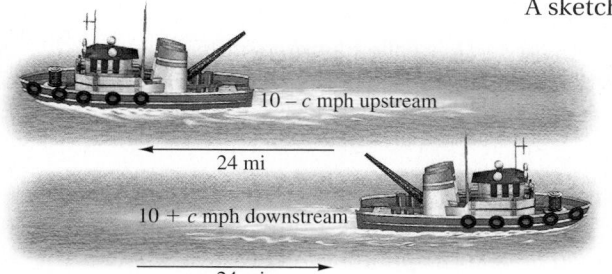

	Distance	Speed	Time
Upstream	24	$10 - c$	t_1
Downstream	24	$10 + c$	t_2

2. **Translate.** From examining our guess, we see that the time traveled can be represented using the formula *Time = Distance/Rate*:

	Distance	Speed	Time
Upstream	24	$10 - c$	$24/(10 - c)$
Downstream	24	$10 + c$	$24/(10 + c)$

   Since the total time upstream and back is 5 hr, we use the last column of the table to form an equation:

   $$\frac{24}{10 - c} + \frac{24}{10 + c} = 5.$$

3. Carry out.   We solve the equation:

$$\frac{24}{10 - c} + \frac{24}{10 + c} = 5$$

$$(10 - c)(10 + c)\left[\frac{24}{10 - c} + \frac{24}{10 + c}\right] = (10 - c)(10 + c)5 \qquad \text{Multiplying by the LCD}$$

$$24(10 + c) + 24(10 - c) = (100 - c^2)5$$

$$480 = 500 - 5c^2 \qquad \text{Simplifying}$$

$$5c^2 - 20 = 0$$

$$5(c^2 - 4) = 0$$

$$5(c - 2)(c + 2) = 0$$

$$c = 2 \quad or \quad c = -2.$$

4. Check.   Since speed cannot be negative in this problem, -2 cannot be a solution. You should confirm that 2 checks in the original problem.

5. State.   The speed of the current is 2 mph.

Problems Involving Proportions

A **ratio** of two quantities is their quotient. For example, 37% is the ratio of 37 to 100, or $\frac{37}{100}$. A **proportion** is an equation stating that two ratios are equal.

> **Proportion**
>
> An equality of ratios, $A/B = C/D$, is called a *proportion*. The numbers within a proportion are said to be *proportional* to each other.

Proportions arise in geometry when we are studying *similar triangles*. If two triangles are **similar**, then their corresponding angles have the same measure and their corresponding sides are proportional. To illustrate, if triangle *ABC* is similar to triangle *RST*, then angles *A* and *R* have the same measure, angles *B* and *S* have the same measure, angles *C* and *T* have the same measure, and

$$\frac{a}{r} = \frac{b}{s} = \frac{c}{t}.$$

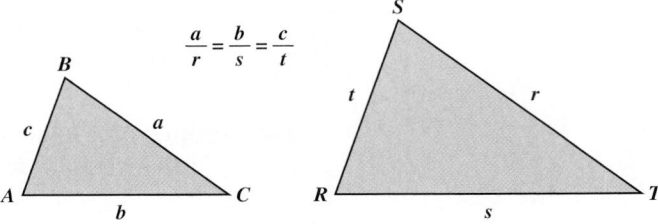

EXAMPLE 5   Similar triangles.   Triangles *ABC* and *XYZ* are similar. Solve for *z* if $x = 10$, $a = 8$, and $c = 5$.

Solution   We make a drawing, write a proportion, and then solve. Note that side *a* is always opposite angle *A*, side *x* is always opposite angle *X*, and so on.

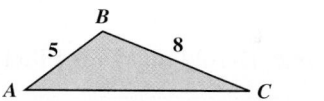 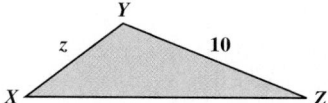

We have

$$\frac{z}{5} = \frac{10}{8}$$ The proportions $\frac{5}{z} = \frac{8}{10}$, $\frac{5}{8} = \frac{z}{10}$, or

$$\frac{8}{5} = \frac{10}{z}$$ could also be used.

$$z = \frac{10}{8} \cdot 5$$ Multiplying both sides by 5

$$z = \frac{50}{8}, \text{ or } 6.25.$$

EXAMPLE 6   Architecture.   A *blueprint* is a scale drawing of a building representing an architect's plans. Emily is adding 12 ft to the length of an apartment and needs to indicate the addition on an existing blueprint. If a 10-ft long bedroom is represented by $2\frac{1}{2}$ in. on the blueprint, how much longer should Emily make the drawing in order to represent the addition?

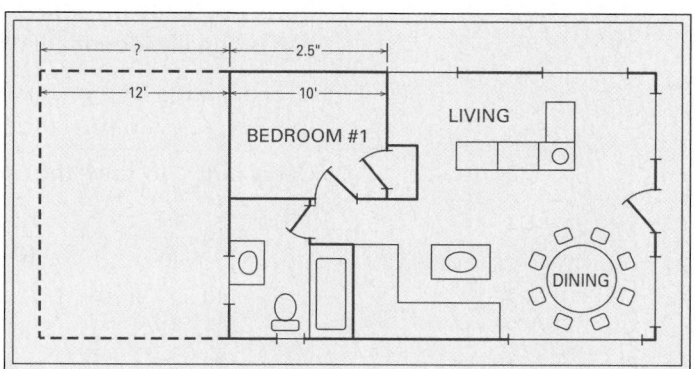

Solution   We let *w* represent the width, in inches, of the addition that Emily is drawing. Because the drawing must be to scale, we have

Inches on drawing $\longrightarrow$ $\dfrac{w}{12} = \dfrac{2.5}{10}$ $\longleftarrow$ Inches on drawing
Feet in real life $\longrightarrow$ $\phantom{\dfrac{w}{12}}$ $\longleftarrow$ Feet in real life

To solve for w, we multiply both sides by the LCM of the denominators, 60:

$$60 \cdot \frac{w}{12} = 60 \cdot \frac{2.5}{10}$$

$$5w = 6 \cdot 2.5 \qquad \text{Simplifying}$$

$$w = \frac{15}{5}, \text{ or } 3.$$

Emily should make the blueprint 3 in. longer.

Proportions can be used to solve a variety of applied problems.

EXAMPLE 7

Environmental science. To determine the number of humpback whales in a pod, a marine biologist, using tail markings, identifies 27 members of the pod. Several weeks later, 40 whales from the pod are randomly sighted. Of the 40 sighted, 12 are from the 27 originally identified. Estimate the number of whales in the pod.

Solution

1. **Familiarize.** If we knew that the 27 whales that were first identified constituted, say, 10% of the pod, we could easily calculate the pod's population from the proportion

$$\frac{27}{W} = \frac{10}{100},$$

where W is the size of the pod's population. Unfortunately, we are *not* told the percentage of the pod identified. We must reread the problem, looking for numbers that could be used to approximate this percentage.

2. **Translate.** Since 12 of the 40 whales that were later sighted were among those originally identified, the ratio 12/40 estimates the percentage of the pod originally identified. We can then translate to a proportion:

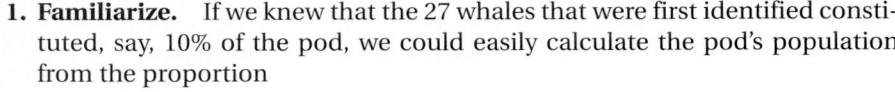

Whales originally identified $\longrightarrow$ $\dfrac{27}{W} = \dfrac{12}{40}.$ $\longleftarrow$ Original whales sighted later
Entire pod $\longrightarrow$ $\phantom{\dfrac{27}{W} = \dfrac{12}{40}}$ $\longleftarrow$ Whales sighted later

3. **Carry out.** To solve the proportion, we multiply by the LCD, 40W:

$$40W \cdot \frac{27}{W} = 40W \cdot \frac{12}{40} \qquad \text{Multiplying both sides by } 40W$$

$$40 \cdot 27 = W \cdot 12 \qquad \begin{array}{l}\text{Removing factors equal to 1:} \\ W/W = 1 \text{ and } 40/40 = 1\end{array}$$

$$\frac{40 \cdot 27}{12} = W \quad \text{or} \quad W = 90. \qquad \text{Dividing both sides by 12}$$

4. **Check.** The check is left to the student.

5. **State.** There are about 90 whales in the pod.

Exercise Set

6.7

Solve.

1. *Volunteerism.* As a "Flynn Spirit," it takes Ned 20 min to fold up all the seats in the Flynn Theater following a performance. Linda requires 30 min to do the job when she volunteers as a Flynn Spirit. How long would it take Ned and Linda to fold up the seats working together?

2. *Construction.* It takes Fontella 4 hr to put up paneling in a room. Omar takes 5 hr to do the same job. How long would it take them, working together, to panel the room?

3. *Carpentry.* By checking work records, a carpenter finds that Juanita can build a small shed in 12 hr. Anton can do the same job in 16 hr. How long would it take if they worked together?

4. *Shoveling.* Vern can shovel the snow from his driveway in 45 min. Nina can do the same job in 60 min. How long would it take Nina and Vern to shovel the driveway if they worked together?

5. *Fax machines.* The Canon Imageclass D680 can fax a year-end report in 7 min while the HP Officejet 6110 can fax the same report in 14 min. How long would it take the two machines, working together, to fax the report? (Assume that the recipient has at least two machines for incoming faxes.)
Source: Manufacturers' marketing brochures

6. *Multifunction copiers.* The Sharp AL-1540CS can copy Charlotte's dissertation in 12 min.

The Brother DCP-1000 can copy the same document in 18 min. If the two machines work together, how long would they take to copy the dissertation?
Source: Manufacturers' marketing brochures

7. *Filling a pool.* The San Paulo community swimming pool can be filled in 12 hr if water enters through a pipe alone or in 30 hr if water enters through a hose alone. If water is entering through both the pipe and the hose, how long will it take to fill the pool?

8. *Filling a tank.* A community water tank can be filled in 18 hr by the town office well alone and in 22 hr by the high school well alone. How long will it take to fill the tank if both wells are working?

9. *Photocopiers.* The HP Officejet 6110 takes twice the time required by the Canon Imageclass D680 to photocopy brochures for the New Bretton Arts Council year-end concert. If working together the two machines can complete the job in 24 min, how long would it take each machine, working alone, to copy the brochures?
Source: Manufacturers' marketing brochures

10. *Computer printers.* The HP Laser Jet 9000 works twice as fast as the Laser Jet 2300. If the machines work together, a university can produce all its staff manuals in 15 hr. Find the time it would take each machine, working alone, to complete the same job.
Source: www.hewlettpackard.com

11. *Forest fires.*   The Erickson Air-Crane helicopter can scoop water and douse a certain forest fire four times as fast as an S-58T helicopter. Working together, the two helicopters can douse the fire in 8 hr. How long would it take each helicopter, working alone, to douse the fire?
Sources: Based on information from www.emergency.com and www.arishelicopters.com

12. *Waxing a car.*   Rosita can wax her car in 2 hr. When she works together with Helga, they can wax the car in 45 min. How long would it take Helga, working by herself, to wax the car?

13. *Newspaper delivery.*   Mariah can deliver papers three times as fast as Stan can. If they work together, it takes them 1 hr. How long would it take each to deliver the papers alone?

14. *Sorting recyclables.*   Together, it takes John and Deb 2 hr 55 min to sort recyclables. Alone, John would require 2 more hr than Deb. How long would it take Deb to do the job alone? (*Hint:* Convert minutes to hours or hours to minutes.)

15. *Paving.*   Together, Larry and Bill require 4 hr 48 min to pave a driveway. Alone, Larry would require 4 hr more than Bill. How long would it take Bill to do the job alone? (*Hint:* Convert minutes to hours.)

16. *Painting.*   Sara takes 3 hr longer to paint a floor than it takes Kate. When they work together, it takes them 2 hr. How long would each take to do the job alone?

17. *Train speeds.*   A B & M freight train is traveling 14 km/h slower than an AMTRAK passenger train. The B & M train travels 330 km in the same time that it takes the AMTRAK train to travel 400 km. Find their speeds. Complete the following table as part of the familiarization.

Distance	*=*	*Rate*	*·*	*Time*

	Distance (in km)	Speed (in km/h)	Time (in hours)
B & M	330		
AMTRAK	400	r	$\dfrac{400}{r}$

18. *Speed of travel.*   A loaded Roadway truck is moving 40 mph faster than a New York Railways freight train. In the time that it takes the train to travel 150 mi, the truck travels 350 mi. Find their speeds. Complete the following table as part of the familiarization.

Distance	*=*	*Rate*	*·*	*Time*

	Distance (in miles)	Speed (in miles per hour)	Time (in hours)
Truck	350	r	$\dfrac{350}{r}$
Train	150		

19. *Driving speed.*   Hillary's Lexus travels 30 mph faster than Bill's Harley. In the same time that Bill travels 75 mi, Hillary travels 120 mi. Find their speeds.

20. *Bicycle speed.*   Hank bicycles 5 km/h slower than Kelly. In the time that it takes Hank to bicycle 42 km, Kelly can bicycle 57 km. How fast does each bicyclist travel?

Aha! **21.** *Tractor speed.*   Manley's tractor is just as fast as Caledonia's. It takes Manley 1 hr more than it takes Caledonia to drive to town. If Manley is 20 mi from town and Caledonia is 15 mi from town, how long does it take Caledonia to drive to town?

22. *Boat speed.*   Tory and Emilio's motorboats both travel at the same speed. Tory pilots her boat 40 km before docking. Emilio continues for another 2 hr, traveling a total of 100 km before docking. How long did it take Tory to navigate the 40 km?

23. *Kayaking.*   The speed of the current in Catamount Creek is 3 mph. Zeno can kayak 4 mi upstream in the same time it takes him to kayak 10 mi downstream. What is the speed of Zeno's kayak in still water?

24. *Boating.*   The current in the Lazy River moves at a rate of 4 mph. Monica's dinghy motors 6 mi upstream in the same time it takes to motor 12 mi downstream. What is the speed of the dinghy in still water?

25. *Moving sidewalks.*   Newark Airport's moving sidewalk moves at a speed of 1.7 ft/sec. Walking on the moving sidewalk, Benny can travel 120 ft forward in the same time it takes to travel 52 ft in the opposite direction. How fast would Benny be walking on a nonmoving sidewalk?

26. *Moving sidewalks.*   The moving sidewalk at O'Hare Airport in Chicago moves 1.8 ft/sec. Walking on the moving sidewalk, Camille travels 105 ft forward in the time it takes to travel 51 ft in the opposite direction. How fast would Camille be walking on a nonmoving sidewalk?

27. *Train speed.*   The speed of the A&M freight train is 14 mph less than the speed of the A&M passenger train. The passenger train travels 400 mi in the same time that the freight train travels 330 mi. Find the speed of each train.

Aha! **28.** *Bus travel.*   A local bus travels 7 mph slower than the express. The express travels 45 mi in the time it takes the local to travel 38 mi. Find the speed of each bus.

29. *Boating.*   Laverne's Mercruiser travels 15 km/h in still water. She motors 140 km downstream in the same time it takes to travel 35 km upstream. What is the speed of the river?

30. *Boating.*   Audrey's paddleboat travels 2 km/h in still water. The boat is paddled 4 km downstream in the same time it takes to go 1 km upstream. What is the speed of the river?

31. *Shipping.*   A barge moves 7 km/h in still water. It travels 45 km upriver and 45 km downriver in a total time of 14 hr. What is the speed of the current?

32. *Aviation.*   A Citation II Jet travels 350 mph in still air and flies 487.5 mi into the wind and 487.5 mi with the wind in a total of 2.8 hr. Find the wind speed.
Source: Eastern Air Charter

Geometry.   *For each pair of similar triangles, find the value of the indicated letter.*

33. *b*
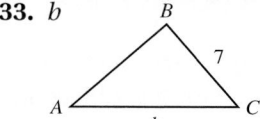
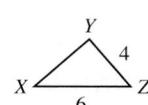

34. *a*

35. *f*

36. *r*

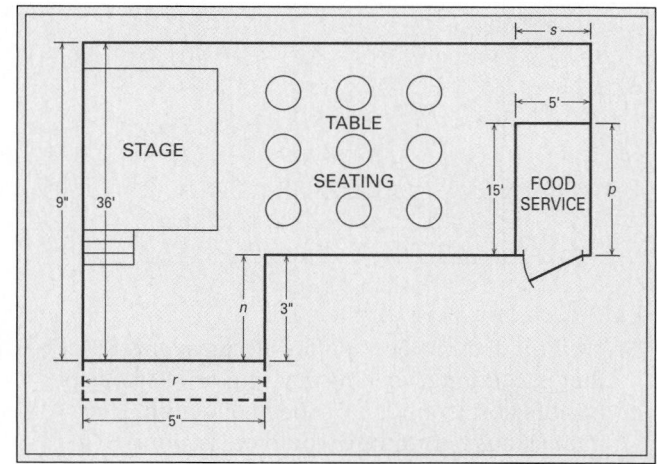

Architecture.   *Use the blueprint below to find the indicated length.*

37. *p*, in inches on blueprint

38. *s*, in inches on blueprint

39. *r*, in feet on actual building

40. *n*, in feet on actual building

Construction.   *Find the indicated length.*
41. *h*

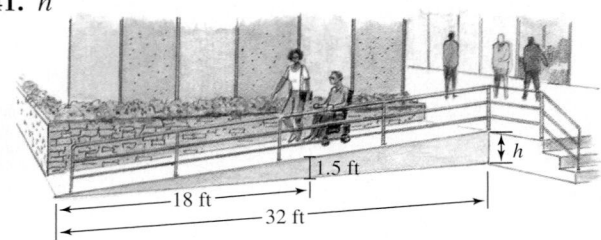

42. *l*

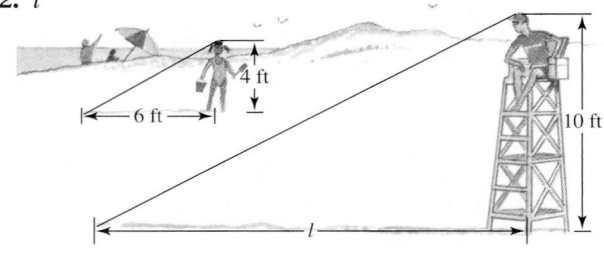

Graphing.   *Find the indicated length.*
43. *r*

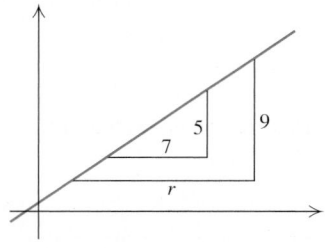

44. *s*

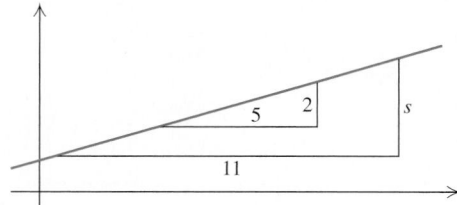

45. *Walking speed.*   For adults with an average stride that is $2\frac{1}{2}$ ft long, walking at a rate of 140 steps per minute corresponds to a speed of 4 mph. How many steps per minute would correspond to a speed of 3 mph?
Source: Runnersworld.com

46. *Burning calories.*   The average 150-lb adult burns about 360 calories walking 4 mi. How far should the average 150-lb adult walk in order to burn 20 calories?
Source: Runnersworld.com

Aha! **47.** *Photography.*   Wanda snapped 234 photos over a period of 14 days. At this rate, how many would she take in 42 days?

48. *Hemoglobin.*   A normal 10-cc specimen of human blood contains 1.2 g of hemoglobin. How much hemoglobin would 16 cc of the same blood contain?

49. *Wages.*   For comparable work, U.S. women earn 77 cents for each dollar earned by a man. If a male sales manager earns $42,000, how much would a female earn for comparable work?
Source: U.S. Bureau of the Census, Public Information Office, 9/26/03

Aha! **50.** *Money.*   The ratio of the weight of copper to the weight of zinc in a U.S. penny is $\frac{1}{39}$. If 50 kg of zinc is being turned into pennies, how much copper is needed?
Source: United States Mint

51. *Firecrackers.*   A sample of 144 firecrackers contained 9 "duds." How many duds would you expect in a sample of 320 firecrackers?

52. *Light bulbs.*   A sample of 184 light bulbs contained 6 defective bulbs. How many defective bulbs would you expect in a sample of 1288 bulbs?

53. *Moose population.*   To determine the size of Pine County's moose population, naturalists catch 69 moose, tag them, and then set them free. Months later, 40 moose are caught, of which 15 have tags. Estimate the size of the moose population.

54. *Deer population.*   To determine the number of deer in the Great Gulf Wilderness, a game warden catches 318 deer, tags them, and lets them loose. Later, 168 deer are caught; 56 of them have tags. Estimate the number of deer in the preserve.

55. *Weight on the moon.*   The ratio of the weight of an object on the moon to the weight of that object on Earth is 0.16 to 1.

a) How much would a 12-ton rocket weigh on the moon?

b) How much would a 180-lb astronaut weigh on the moon?

56. *Weight on Mars.* The ratio of the weight of an object on Mars to the weight of that object on Earth is 0.4 to 1.

    a) How much would a 12-ton rocket weigh on Mars?

    b) How much would a 120-lb astronaut weigh on Mars?

57. Is it correct to assume that two workers will complete a task twice as quickly as one person working alone? Why or why not?

58. If two triangles are exactly the same shape and size, are they similar? Why or why not?

SKILL MAINTENANCE

Graph. [3.2]

59. $y = 2x - 6$

60. $y = -2x + 6$

61. $3x + 2y = 12$

62. $x - 3y = 6$

63. $y = -\dfrac{3}{4}x + 2$

64. $y = \dfrac{2}{5}x - 4$

SYNTHESIS

65. Write a problem similar to Example 1 for a class-mate to solve. Design the problem so that the translation step is

$$\frac{t}{7} + \frac{t}{5} = 1.$$

66. Write a problem similar to Example 2 for a class-mate to solve. Design the problem so that the translation step is

$$\frac{30}{r + 4} = \frac{18}{r}.$$

67. *Car cleaning.* Together, Michelle, Sal, and Kristen can clean and wax a car in 1 hr 20 min. To complete the job alone, Michelle needs twice the time that Sal needs and 2 hr more than Kristen. How long would it take each to clean and wax the car working alone?

68. *Programming.* Rosina, Ng, and Oscar can write a computer program in 3 days. Rosina can write the program in 8 days and Ng can do it in 10 days. How many days will it take Oscar to write the program?

69. *Wiring.* Janet can wire a house in 28 hr. Linus can wire a house in 34 hr. How long will it take Janet and Linus, working together, to wire *two* houses?

70. *Grading.* Alma can grade a batch of placement exams in 3 hr. Kevin can grade a batch in 4 hr. If they work together to grade a batch of exams, what percentage of the exams will have been graded by Alma?

71. *Home maintenance.* Fuel used in lawn mowers and chain saws is made by pouring a 3.2-oz bottle of 2-cycle oil into 160 oz of gasoline. Gus accidentally poured 5.6 oz of 2-cycle oil into 200 oz of gasoline. How much more oil or gasoline should he add in order for the fuel to have the proper ratio of oil to gasoline?

72. *Commuting.* To reach an appointment 50 mi away, Dr. Wright allowed 1 hr. After driving 30 mi, she realized that her speed would have to be increased 15 mph for the remainder of the trip. What was her speed for the first 30 mi?

73. Simplest fraction notation for a rational number is $\frac{9}{17}$. Find an equivalent ratio where the sum of the numerator and the denominator is 104.

74. How soon after 5 o'clock will the hands on a clock first be together?

75. Given that

$$\frac{A}{B} = \frac{C}{D},$$

write three other proportions using *A, B, C,* and *D*.

76. If two triangles are similar, are their areas and perimeters proportional? Why or why not?

77. Are the equations

$$\frac{A + B}{B} = \frac{C + D}{D} \quad \text{and} \quad \frac{A}{B} = \frac{C}{D}$$

equivalent? Why or why not?

COLLABORATIVE

CORNER

Sharing the Workload

Focus: Modeling, estimation, and work problems

Time: 20–25 minutes

Group size: 3

Materials: Paper, pencils, textbooks, and a watch

Many tasks can be done by two people working together. If both people work at the same rate, each does half the task, and the project is completed in half the time. However, when the work rates differ, the faster worker performs more than half of the task.

ACTIVITY

1. The project is to write down (but not answer) Review Exercises 9–42 from Chapter 6 (pp. 436–437) on a sheet of paper. The problems should be spaced apart and written clearly so that they can be used for studying in the future. Two of the members in each group should write down the exercises, one working slowly and one working quickly. The third group member should record the time required for each to write down all 34 exercises.

2. Using the times from step (1), calculate how long it will take the two workers, working together, to complete the task.

3. Next, have the same workers as in step (1)— working at the same speeds as in step (1)— perform the task together. To do this, one person should begin writing with Exercise 9, while the other worker begins with Exercise 42 and lists the problems counting backward. The third worker is again the timekeeper and should observe when the two workers have written all the exercises. To avoid collision, each of the two writers should use a separate sheet of paper.

4. Compare the actual experimental time from part (3) with the time predicted by the model in part (2). List reasons that might account for any discrepancy.

5. Let t_1, t_2, and t_3 represent the times required for the first worker, the second worker, and the two workers together, respectively, to complete a task. Then develop a model that can be used to find t_3 when t_1 and t_2 are known.

Time Required for Worker A, Working Alone	Time Required for Worker B, Working Alone	Estimated Time for the Two Workers, Working Together	Actual Time Required for the Two Workers, Working Together

6 Study Summary

Like the fractions in arithmetic, **rational expressions** can be **simplified, added, subtracted, multiplied**, and **divided** (pp. 374, 375, 381, 383, 388, 389):

$$\frac{A}{B} \cdot \frac{C}{D} = \frac{AC}{BD}; \qquad \frac{A}{B} \div \frac{C}{D} = \frac{A}{B} \cdot \frac{D}{C} = \frac{AD}{BC};$$

$$\frac{A}{B} + \frac{C}{B} = \frac{A+C}{B}; \qquad \frac{A}{B} - \frac{C}{B} = \frac{A-C}{B}.$$

When adding or subtracting rational expressions, we must first make certain that the expressions being added or subtracted have a common denominator. The simplest common denominator to work with is the **least common denominator** (p. 390). To find the least common denominator (LCD), we use the **least common multiple** (LCM) of the denominators (p. 390).

Sometimes rational expressions appear within a rational expression, as in

$$\frac{\dfrac{2}{x+1} - \dfrac{1}{x}}{\dfrac{3}{2x}} \quad \text{or} \quad \frac{5 - \dfrac{t}{t+3}}{t^2}.$$

Such expressions are called **complex rational expressions** and can be simplified either by division or by multiplication involving the least common denominator (pp. 406, 408).

When a rational expression appears in an equation, as in

$$\frac{3}{x+1} + \frac{2}{x} = 7 \quad \text{or} \quad \frac{5}{t} = \frac{7}{9},$$

we have what is called a **rational equation** (p. 413). The multiplication principle is used to solve rational equations and, because it now involves multiplying both sides by a variable expression, checking potential solutions in the original equation is very important. A variety of real-world problems translate to rational equations. Among these are **work problems, motion problems,** and **proportions** (pp. 420, 423, 426).

6 Review Exercises

☞ *Concept Reinforcement* *Classify each statement as either true or false.*

1. Every rational expression can be simplified. [6.1]

2. The expression $(t - 3)/(t^2 - 4)$ is undefined for $t = 2$. [6.1]

3. The expression $(t - 3)/(t^2 - 4)$ is undefined for $t = 3$. [6.1]

4. To multiply rational expressions, a common denominator is never required. [6.2]

5. To divide rational expressions, a common denominator is never required. [6.2]

6. To add rational expressions, a common denominator is never required. [6.3]

7. To subtract rational expressions, a common denominator is never required. [6.3]

8. Every rational equation has at least one solution. [6.6]

List all numbers for which each expression is undefined. [6.1]

9. $\dfrac{17}{-x^2}$

10. $\dfrac{9}{a - 4}$

11. $\dfrac{x - 5}{x^2 - 36}$

12. $\dfrac{x^2 + 3x + 2}{x^2 + x - 30}$

13. $\dfrac{-6}{(t + 2)^2}$

Simplify. [6.1]

14. $\dfrac{4x^2 - 8x}{4x^2 + 4x}$

15. $\dfrac{14x^2 - x - 3}{2x^2 - 7x + 3}$

16. $\dfrac{(y - 5)^2}{y^2 - 25}$

17. $\dfrac{5x^2 - 20y^2}{2y - x}$

Multiply or divide and, if possible, simplify. [6.2]

18. $\dfrac{a^2 - 36}{10a} \cdot \dfrac{2a}{a + 6}$

19. $\dfrac{8t + 8}{2t^2 + t - 1} \cdot \dfrac{t^2 - 1}{t^2 - 2t + 1}$

20. $\dfrac{16 - 8t}{3} \div \dfrac{t - 2}{12t}$

21. $\dfrac{4x^4}{x^2 - 1} \div \dfrac{2x^3}{x^2 - 2x + 1}$

22. $\dfrac{x^2 + 1}{x - 2} \cdot \dfrac{2x + 1}{x + 1}$

23. $(t^2 + 3t - 4) \div \dfrac{t^2 - 1}{t + 4}$

Find the LCM. [6.3]

24. $8a^2b^9, \ 6a^5b^3$

25. $x^2 - x, \ x^5 - x^3, \ x^4$

26. $y^2 - y - 2, \ y^2 - 4$

Add or subtract and, if possible, simplify.

27. $\dfrac{x + 6}{x + 3} + \dfrac{9 - 4x}{x + 3}$ [6.3]

28. $\dfrac{3}{3x - 9} + \dfrac{x - 2}{3 - x}$ [6.4]

29. $\dfrac{6x - 3}{x^2 - x - 12} - \dfrac{2x - 15}{x^2 - x - 12}$ [6.3]

30. $\dfrac{3x - 1}{2x} - \dfrac{x - 3}{x}$ [6.4]

31. $\dfrac{x + 5}{x - 2} - \dfrac{x}{2 - x}$ [6.4]

32. $\dfrac{2a}{a + 1} - \dfrac{4a}{1 - a^2}$ [6.4]

33. $\dfrac{d^2}{d-c} + \dfrac{c^2}{c-d}$ [6.4]

34. $\dfrac{1}{x^2-25} - \dfrac{x-5}{x^2-4x-5}$ [6.4]

35. $\dfrac{3x}{x+2} - \dfrac{x}{x-2} + \dfrac{8}{x^2-4}$ [6.4]

36. $\dfrac{2}{5x} + \dfrac{3}{2x+4}$ [6.4]

Simplify. [6.5]

37. $\dfrac{\dfrac{1}{z}+1}{\dfrac{1}{z^2}-1}$

38. $\dfrac{2+\dfrac{1}{xy^2}}{\dfrac{1+x}{x^4 y}}$

39. $\dfrac{\dfrac{c}{d}-\dfrac{d}{c}}{\dfrac{1}{c}+\dfrac{1}{d}}$

Solve. [6.6]

40. $\dfrac{3}{y} - \dfrac{1}{4} = \dfrac{1}{y}$

41. $\dfrac{5}{x+3} = \dfrac{3}{x+2}$

42. $\dfrac{15}{x} - \dfrac{15}{x+2} = 2$

43. Jackson can sand the oak floors and stairs in a two-story home in 12 hr. Olga can do the same job in 9 hr. How long would it take if they worked together? (Assume that two sanders are available.) [6.7]

44. A research company uses personal computers to process data while the owner is not using the computer. A Pentium 4 3.0-gigahertz processor can process a particular file in 15 sec less time than a Celeron 2.53-gigahertz processor. Working together, the computers can process the file in 18 sec. How long does it take each computer to process the file? [6.7]

45. The distance by highway between Richmond and Waterbury is 70 km, and the distance by rail is 60 km. A car and a train leave Richmond at the same time and arrive in Waterbury at the same time, the car having traveled 15 km/h faster than the train. Find the speed of the car and the speed of the train. [6.7]

46. The Black River's current is 6 mph. A boat travels 50 mi downstream in the same time that it takes to travel 30 mi upstream. What is the speed of the boat in still water? [6.7]

47. To estimate the frog population in a pond, a scientist captures, tags, and releases 24 frogs. Weeks later, 20 frogs are caught and 8 of them are wearing tags. Estimate the frog population of the pond. [6.7]

48. Triangles *ABC* and *XYZ* are similar. Find the value of *x*. [6.7]

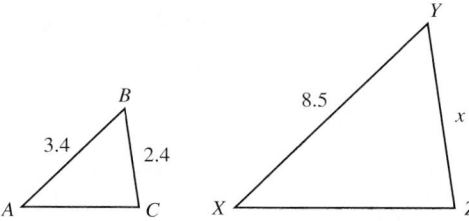

SYNTHESIS

49. For what procedures in this chapter is the LCM of denominators used to clear fractions? [6.1], [6.4], [6.6]

50. A student insists on finding a common denominator by always multiplying the denominators of the expressions being added. How could this approach be improved? [6.3]

Simplify.

51. $\dfrac{2a^2+5a-3}{a^2} \cdot \dfrac{5a^3+30a^2}{2a^2+7a-4} \div \dfrac{a^2+6a}{a^2+7a+12}$ [6.2]

52. $\dfrac{12a}{(a-b)(b-c)} - \dfrac{2a}{(b-a)(c-b)}$ [6.4]

Aha! **53.** $\dfrac{5(x-y)}{(x-y)(x+2y)} - \dfrac{5(x-3y)}{(x+2y)(x-3y)}$ [6.3]

54. It has been over 60 yr since a major-league baseball player went an entire season averaging 4 hits in every 10 at-bats. Suppose that Hideki currently has 153 hits after 395 at-bats. If he is assured 125 more at-bats, what percentage of those must be hits if he is to average 4 hits for every 10 at-bats? [6.7]

6 Chapter Test

List all numbers for which each expression is undefined.

1. $\dfrac{2 - x}{5x}$

2. $\dfrac{5}{x + 8}$

3. $\dfrac{x - 7}{x^2 - 49}$

4. $\dfrac{x^2 + x - 30}{x^2 - 3x + 2}$

5. Simplify: $\dfrac{6x^2 + 17x + 7}{2x^2 + 7x + 3}$.

Multiply or divide and, if possible, simplify.

6. $\dfrac{a^2 - 25}{9a} \cdot \dfrac{6a}{5 - a}$

7. $\dfrac{25y^2 - 1}{9y^2 - 6y} \div \dfrac{5y^2 + 9y - 2}{3y^2 + y - 2}$

8. $\dfrac{4x^2 - 1}{x^2 - 2x + 1} \div \dfrac{x - 2}{x^2 + 1}$

9. $(x^2 + 6x + 9) \cdot \dfrac{(x - 3)^2}{x^2 - 9}$

10. Find the LCM:

$$y^2 - 9, \ y^2 + 10y + 21, \ y^2 + 4y - 21.$$

Add or subtract, and, if possible, simplify.

11. $\dfrac{2 + x}{x^3} + \dfrac{7 - 4x}{x^3}$

12. $\dfrac{5 - t}{t^2 + 1} - \dfrac{t - 3}{t^2 + 1}$

13. $\dfrac{x - 4}{x - 3} + \dfrac{x - 1}{3 - x}$

14. $\dfrac{x - 4}{x - 3} - \dfrac{x - 1}{3 - x}$

15. $\dfrac{7}{t - 2} + \dfrac{4}{t}$

16. $\dfrac{1}{x^2 - 16} - \dfrac{x + 4}{x^2 - 3x - 4}$

17. $\dfrac{1}{x - 1} + \dfrac{4}{x^2 - 1} - \dfrac{2}{x^2 - 2x + 1}$

Simplify.

18. $\dfrac{9 - \dfrac{1}{y^2}}{3 - \dfrac{1}{y}}$

19. $\dfrac{\dfrac{3}{a^2b} - \dfrac{2}{ab^3}}{\dfrac{1}{ab} + \dfrac{2}{a^4b}}$

Solve.

20. $\dfrac{7}{y} - \dfrac{1}{3} = \dfrac{1}{4}$

21. $\dfrac{15}{x} - \dfrac{15}{x - 2} = -2$

22. Kopy Kwik has 2 copiers. One can copy a year-end report in 20 min. The other can copy the same document in 30 min. How long would it take both machines, working together, to copy the report?

23. A recipe for pizza crust calls for $3\frac{1}{2}$ cups of whole wheat flour and $1\frac{1}{4}$ cups of warm water. If 6 cups of whole wheat flour are used, how much water should be used?

24. Craig drives 20 km/h faster than Marilyn. In the same time that Marilyn drives 225 km, Craig drives 325 km. Find the speed of each car.

SYNTHESIS

25. Reggie and Rema work together to mulch the flower beds around an office complex in $2\frac{6}{7}$ hr. Working alone, it would take Reggie 6 hr more than it would take Rema. How long would it take each of them to complete the landscaping working alone?

26. Simplify: $1 - \dfrac{1}{1 - \dfrac{1}{1 - \dfrac{1}{a}}}$.

27. The square of a number is the opposite of the number's reciprocal. Find the number.

1–6 Cumulative Review

1. Use the commutative law of addition to write an expression equivalent to $a + 2b$. [1.2]

2. Write a true sentence using either $<$ or $>$:
$-3.1 \ \blacksquare \ -3.15$. [1.4]

3. Evaluate $(y - 1)^2$ for $y = -4$. [1.8]

4. Simplify: $-4[2(x - 3) - 1]$. [1.8]

Simplify.

5. $-\frac{1}{2} + \frac{3}{8} + (-6) + \frac{3}{4}$ [1.5]

6. $-\frac{72}{108} \div \left(-\frac{2}{3}\right)$ [1.7]

7. $-6.262 \div 1.01$ [1.7]

8. $4 \div (-2) \cdot 2 + 3 \cdot 4$ [1.8]

Solve.

9. $5(x - 2) = 40$ [2.2]

10. $49 = x^2$ [5.6]

11. $-6t = 20$ [2.1]

12. $3x - 4 = 7x + 8$ [2.2]

13. $4(y - 5) = -2(y + 2)$ [2.2]

14. $x^2 + 11x + 10 = 0$ [5.7]

15. $\frac{4}{9}t + \frac{2}{3} = \frac{1}{3}t - \frac{2}{9}$ [2.2]

16. $\frac{4}{x} + x = 5$ [6.6]

17. $6 - y \geq 2y + 8$ [2.6]

18. $\frac{2}{x - 3} = \frac{5}{3x + 1}$ [6.6]

19. $2x^2 + 7x = 4$ [5.7]

20. $4(x + 7) < 5(x - 3)$ [2.6]

21. $\frac{t^2}{t + 5} = \frac{25}{t + 5}$ [6.6]

22. $(2x + 7)(x - 5) = 0$ [5.7]

23. $\frac{2}{x^2 - 9} + \frac{5}{x - 3} = \frac{3}{x + 3}$ [6.6]

Solve each formula. [2.3]

24. $3a - b + 9 = c$, for b

25. $\frac{3}{4}(x + 2y) = z$, for y

Combine like terms. [1.6]

26. $x + 2y - 2z + \frac{1}{2}x - z$

27. $2x^3 - 7 + \frac{3}{7}x^2 - 6x^3 - \frac{4}{7}x^2 + 5$

Graph. [3.2], [3.3]

28. $y = \frac{3}{4}x + 5$

29. $x = -3$

30. $4x + 5y = 20$

31. $y = 6$

32. Find the slope of the line containing the points $(1, 5)$ and $(2, 3)$. [3.5]

33. Find the slope and the y-intercept of the line given by $3x - 6y = 12$. [3.6]

34. Write the slope–intercept equation for the line with slope -3 that contains the point $(2, -5)$. [3.7]

35. A large truck traveling 55 mph gets 7.6 miles per gallon of fuel. The same truck traveling 70 mph gets 6.1 miles per gallon.
Source: Kenworth Truck Co.

a) Graph the data and determine an equation that describes the miles per gallon y in terms of the speed x. [3.7]

b) Use the equation to estimate the miles per gallon for a truck traveling 60 mph. [3.7]

Simplify.

36. $\frac{x^{-5}}{x^{-3}}$ [4.8]

37. $y^2 \cdot y^{-8}$ [4.8]

38. $-(2a^2b^7)^2$ [4.1]

39. Subtract: [4.3]
$(-8y^2 - y + 2) - (y^3 - 6y^2 + y - 5)$.

Multiply.

40. $5(3a - 2b + c)$ [1.2]

41. $(2x^2 - 1)(x^3 + x - 3)$ [4.4]

42. $(6x - 5y)^2$ [4.5]

43. $(x + 8)(2x - 5)$ [4.5]

44. $(2x^3 + 1)(2x^3 - 1)$ [4.5]

Factor.

45. $6x - 2x^2 - 24x^4$ [5.1]

46. $16x^2 - 81$ [5.4]

47. $t^2 - 10t + 24$ [5.2]

48. $8x^2 + 10x + 3$ [5.3]

49. $6x^2 - 28x + 16$ [5.3]

50. $4t^2 - 36$ [5.4]

51. $25t^2 + 40t + 16$ [5.4]

52. $10t^6 - 80$ [5.5]

53. $x^4 + 2x^3 - 3x - 6$ [5.1]

Simplify.

54. $\dfrac{y^2 - 36}{2y + 8} \cdot \dfrac{y + 4}{y - 6}$ [6.2]

55. $\dfrac{x^2 - 1}{x^2 - x - 2} \div \dfrac{x - 1}{x - 2}$ [6.2]

56. $\dfrac{5ab}{a^2 - b^2} + \dfrac{a + b}{a - b}$ [6.4]

57. $\dfrac{x + 2}{4 - x} - \dfrac{x + 3}{x - 4}$ [6.4]

58. $\dfrac{1 + \dfrac{2}{x}}{1 - \dfrac{4}{x^2}}$ [6.5]

59. $\dfrac{\dfrac{1}{t} + 2t}{t - \dfrac{2}{t^2}}$ [6.5]

Divide. [4.7]

60. $\dfrac{18x^4 - 15x^3 + 6x^2 + 12x + 3}{3x^2}$

61. $(15x^4 - 12x^3 + 6x^2 + 2x + 18) \div (x + 3)$

Solve.

62. Linnae has $36 budgeted for stationery. Engraved stationery costs $20 for the first 25 sheets and $0.08 for each additional sheet. How many sheets of stationery can Linnae order and still stay within her budget? [2.7]

63. The price of a box of granola increased 15% to $4.14. What was the price of the granola before the increase? [2.4]

64. If the sides of a square are increased by 2 ft, the area of the original square plus the area of the enlarged square is 452 ft². Find the length of a side of the original square. [5.8]

65. The sum of two consecutive even integers is −554. Find the integers. [2.5]

66. It takes Dina 50 min to shovel 9 in. of snow from her driveway. It takes Nell 75 min to do the same job. How long would it take if they worked together? [6.7]

67. A game warden catches, tags, and then releases 15 zebras. A month later, a sample of 20 zebras is collected and 6 of them have tags. Use this information to estimate the size of the zebra population in that area. [6.7]

68. A 78-in. board is to be cut into two pieces. One piece must be twice as long as the other. How long should the shorter piece be? [2.5]

SYNTHESIS

69. Simplify: $(x + 7)(x - 4) - (x + 8)(x - 5)$. [4.3], [4.5]

70. Solve: $\frac{1}{3}|n| + 8 = 56$. [1.4], [2.2]

Aha! **71.** Multiply: $[4y^3 - (y^2 - 3)][4y^3 + (y^2 - 3)]$. [4.5]

72. Factor: $2a^{32} - 13{,}122b^{40}$. [5.4]

73. Solve: $x(x^2 + 3x - 28) - 12(x^2 + 3x - 28) = 0$. [5.7]

74. Simplify: $-\left|0.875 - \left(-\frac{1}{8}\right) - 8\right|$. [1.4], [1.6]

75. Solve: $\dfrac{2}{x - 3} \cdot \dfrac{3}{x + 3} - \dfrac{4}{x^2 - 7x + 12} = 0$. [6.6]

76. Jesse can peel a bushel of potatoes in 60 min. When Jesse and Priscilla work together, the job takes 20 min. When working together with Jesse, what percentage of the work is performed by Priscilla? [6.7]

77. The truck in Exercise 35 travels 400 mi at 65 mph. On the return trip, the truck travels at 60 mph. If fuel costs $1.95 a gallon, how much less does the return trip cost? [2.5], [3.7]

7

Functions and Graphs

AN APPLICATION

The ultraviolet, or UV, index is a measure issued daily by the National Weather Service that indicates the strength of the sun's rays in a particular locale. For those people whose skin is quite sensitive, a UV rating of 6 will cause sunburn after 10 min (*Source: The Electronic Textbook of Dermatology* found at www.telemedicine.org, January 2004). Given that the number of minutes it takes to burn, t, varies inversely as the UV rating, u, how long will it take a highly sensitive person to burn on a day with a UV rating of 4?

This problem appears as Example 7 in Section 7.5.

Amanda Rainwater
DERMATOLOGIST
Scottsdale, Arizona

Math is an integral part of my practice. I use it to calculate children's prescriptions according to their weight and to formulate solutions for injections. I also need math, of course, to figure costs.

*I*n mathematics, as in many other disciplines, new concepts build on and synthesize topics discussed previously. In this chapter we introduce the concept of a function. As we will see, many functions are described using linear, polynomial, and rational expressions. Functions can also often be visualized graphically, as well as added, subtracted, multiplied, and divided. Near the end of the chapter, we solve formulas using equation-solving techniques studied earlier, and we use function notation when describing direct and inverse variation.

7.1 Introduction to Functions

Correspondences and Functions • Functions and Graphs •
Function Notation and Equations • Applications

We now develop the idea of a *function*—one of the most important concepts in mathematics.

Correspondences and Functions

When forming ordered pairs to graph equations, we often say that the first coordinate of each ordered pair *corresponds* to the second coordinate. In much the same way, a function is a special kind of correspondence between two sets. For example,

To each person in a class	there corresponds	a date of birth.
To each bar code in a store	there corresponds	a price.
To each real number	there corresponds	the cube of that number.

In each example, the first set is called the **domain**. The second set is called the **range**. For any member of the domain, there is *exactly one* member of the range to which it corresponds. This kind of correspondence is called a **function**.

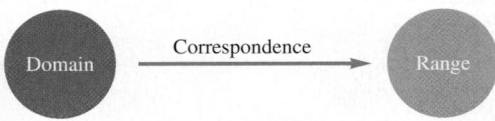

EXAMPLE 1 Determine whether each correspondence is a function.

a)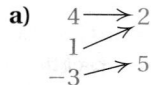

b)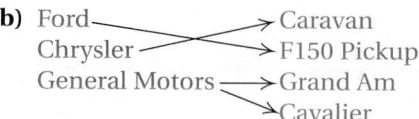

Solution

a) The correspondence *is* a function because each member of the domain corresponds to *exactly one* member of the range.

b) The correspondence *is not* a function because a member of the domain (General Motors) corresponds to more than one member of the range.

> *Function*
>
> A *function* is a correspondence between a first set, called the *domain*, and a second set, called the *range*, such that each member of the domain corresponds to *exactly one* member of the range.

EXAMPLE 2 Determine whether each correspondence is a function.

	Domain	*Correspondence*	*Range*
a)	An elevator full of people	Each person's weight	A set of positive numbers
b)	$\{-2, 0, 1, 2\}$	Each number's square	$\{0, 1, 4\}$
c)	Authors of best-selling books	The titles of books written by each author	A set of book titles

Solution

a) The correspondence *is* a function, because each person has *only one* weight.

b) The correspondence *is* a function, because every number has *only one* square.

c) The correspondence *is not* a function, because some authors have written *more than one* book.

Although the correspondence in Example 2(c) is not a function, it is a *relation*.

> **Relation**
>
> A *relation* is a correspondence between a first set, called the *domain*, and a second set, called the *range*, such that each member of the domain corresponds to *at least one* member of the range.

Functions and Graphs

The functions in Examples 1(a) and 2(b) can be expressed as sets of ordered pairs. Example 1(a) can be written {(−3, 5), (1, 2), (4, 2)} and Example 2(b) can be written {(−2, 4), (0, 0), (1, 1), (2, 4)}. We can graph these functions as follows.

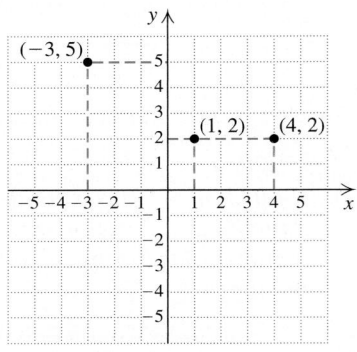

The function {(−3, 5), (1, 2), (4, 2)}
Domain is {−3, 1, 4}
Range is {5, 2}

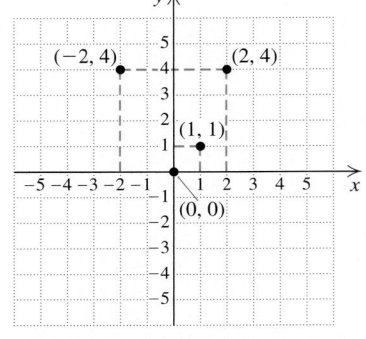

The function {(−2, 4), (0, 0), (1, 1), (2, 4)}
Domain is {−2, 0, 1, 2}
Range is {4, 0, 1}

When a function is given as a set of ordered pairs, the domain is the set of all first coordinates and the range is the set of all second coordinates. Function names are generally represented by lower- or upper-case letters.

EXAMPLE 3   For the function *f* represented below, determine each of the following.

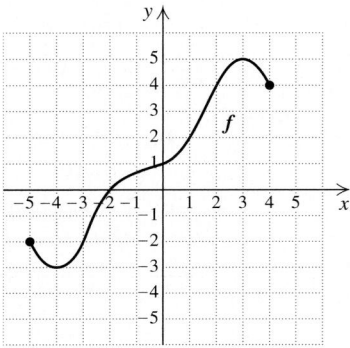

a) What member of the range is paired with 2 (in the domain)

b) What member of the domain is paired with −3 (in the range)

Solution

a) To determine what member of the range is paired with 2, we locate 2 on the horizontal axis. (This is where the domain is located.) Next, we find the point directly above 2 on the graph of *f*. From that point, we can look to the vertical axis to find the corresponding *y*-coordinate, 4. The "input" 2 has the "output" 4.

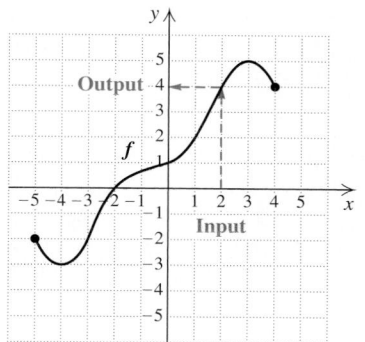

b) To determine what member of the domain is paired with −3, we locate −3 on the vertical axis. (This is where the range is located.) From there we look left and right to the graph of *f* to find any points for which −3 is the second coordinate. One such point exists, (−4, −3). We observe that −4 is the only element of the domain paired with −3.

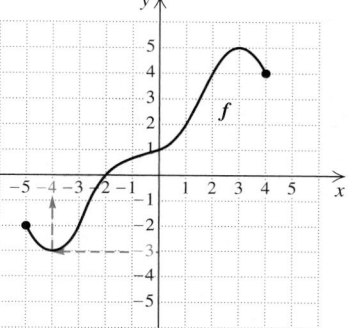

Function Notation and Equations

To understand function notation, it helps to imagine a "function machine." Think of putting a member of the domain (an *input*) into the machine. The machine is programmed to produce the appropriate member of the range (the *output*).

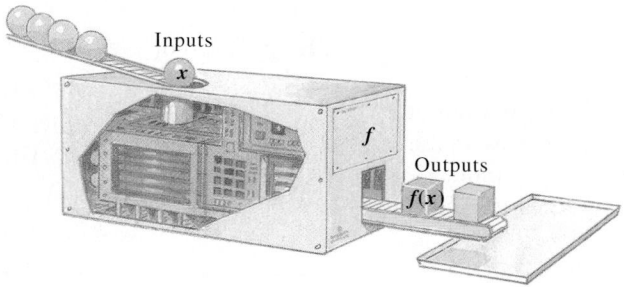

The function pictured has been named *f*. Here *x* represents an arbitrary input, and $f(x)$—read "*f* of *x*," "*f* at *x*," or "the value of *f* at *x*"—represents the corresponding output. In Example 1(a), $f(4)$ is 2, $f(1)$ is 2, and $f(-3)$ is 5. Similarly, in Example 3, we showed that $f(2) = 4$.

Caution! $f(x)$ *does not mean f times x.*

Most functions are described by equations. For example, $f(x) = 2x + 3$ describes the function that takes an input x, multiplies it by 2, and then adds 3.

Input
$$f(x) \quad = \quad 2x \quad + 3$$
Double   Add 3

To calculate the output $f(4)$, we take the input 4, double it, and add 3 to get 11. That is, we substitute 4 into the formula for $f(x)$:

$$f(4) = 2 \cdot 4 + 3$$
$$= 11.$$

Sometimes, in place of $f(x) = 2x + 3$, we write $y = 2x + 3$, where it is understood that the value of y, the *dependent variable*, depends on our choice of x, the *independent variable*. To understand why $f(x)$ notation is so useful, consider two equivalent statements:

a) If $f(x) = 2x + 3$, then $f(4) = 11$.
b) If $\quad y = 2x + 3$, then the value of y is 11 when x is 4.

The notation used in part (a) is far more concise and emphasizes that x is the independent variable.

EXAMPLE 4   Find each indicated function value.

a) $f(5)$, for $f(x) = 3x + 2$
b) $g(-2)$, for $g(r) = 5r^2 + 3r$
c) $h(4)$, for $h(x) = 7$
d) $F(a + 1)$, for $F(x) = 3x + 2$

Solution   Finding function values is much like evaluating an algebraic expression.

a) $f(5) = 3 \cdot 5 + 2 = 17$
b) $g(-2) = 5(-2)^2 + 3(-2)$
$$= 5 \cdot 4 - 6 = 14$$
c) For the function given by $h(x) = 7$, all inputs share the same output, 7. Therefore, $h(4) = 7$. The function h is an example of a *constant function*.
d) $F(a + 1) = 3(a + 1) + 2$
$$= 3a + 3 + 2 = 3a + 5$$

Student Notes _____

In Example 4(d), it is important to note that the parentheses on the left are for function notation, whereas those on the right indicate multiplication.

Note that whether we write $f(x) = 3x + 2$, or $f(t) = 3t + 2$, or $f(\square) = 3\square + 2$, we still have $f(5) = 17$. Thus the independent variable can be thought of as a *dummy variable*. The letter chosen for the dummy variable is not as important as the algebraic manipulations to which it is subjected.

When a function is described by an equation, the domain is often unspecified. In such cases, the domain is the set of all numbers for which function values can be calculated. If an x-value is not in the domain of a function, the graph of that function will not include any point above or below that x-value.

EXAMPLE 5   A typical adult dosage of an antihistamine is 24 mg. Young's rule for determining the dosage size $c(a)$ for a typical child of age a is

$$c(a) = \frac{24a}{a + 12}.^*$$

What should the dosage be for a typical 8-yr-old child?

Solution   We find $c(8)$:

$$c(8) = \frac{24(8)}{8 + 12} = \frac{192}{20} = 9.6.$$

The dosage for a typical 8-yr-old child is 9.6 mg.

Applications

Function notation is often used in formulas. For example, to emphasize that the area A of a circle is a function of its radius r, instead of

$$A = \pi r^2,$$

we can write

$$A(r) = \pi r^2.$$

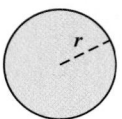

When a function is given as a graph in a problem-solving situation, we are often asked to determine certain quantities on the basis of the graph.

EXAMPLE 6   Elementary school math proficiency.   According to the National Assessment of Educational Progress, the percentage of fourth-graders who are proficient in math has grown from 18% in 1992 to 24% in 2000 and 32% in 2003 (*Source: National Council of Teachers of Mathematics News Bulletin* Jan./Feb. 2004). Estimate the percentage of fourth-graders who showed proficiency in 1996 and predict the percentage who will demonstrate proficiency in 2007.

Source: Olsen, June Looby, Leon J. Ablon, and Anthony Patrick Giangrasso, *Medical Dosage Calculations*, 6th ed.

Solution

1. and **2. Familiarize.** and **Translate.**   The given information enables us to plot and connect three points. We let the horizontal axis represent the year and the vertical axis the percentage of fourth-graders demonstrating mathematical proficiency. We label the function itself *P*.

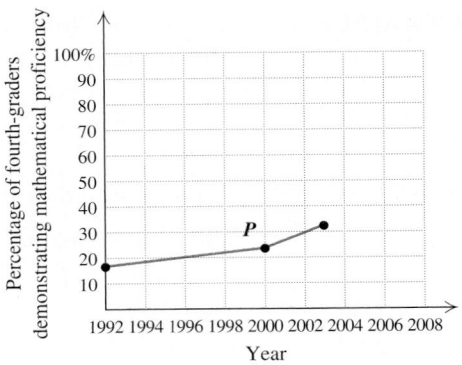

3. Carry out.   To estimate the percentage of fourth-graders showing mathematical proficiency in 1996, we locate the point directly above the year 1996. We then estimate its second coordinate by moving horizontally from that point to the *y*-axis. Although our result is not exact, we see that $P(1996) \approx 20$.

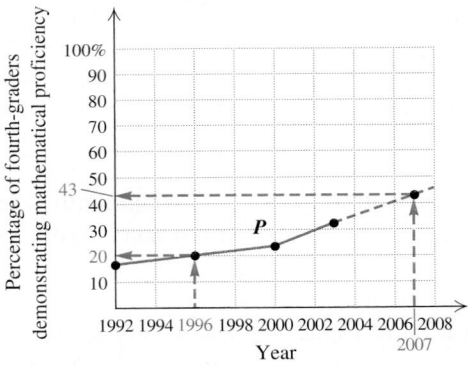

To predict the percentage of fourth-graders showing proficiency in 2007, we extend the graph and extrapolate. It appears that $P(2007) \approx 43$.

4. Check.   A precise check requires consulting an outside information source. Since 20% is between 18% and 24% and 43% is greater than 32%, our estimates seem plausible.

5. State.   In 1996, about 20% of all fourth-graders showed proficiency in math. By 2007, that figure is predicted to grow to 43%.

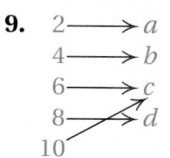

Exercise Set

7.1

FOR EXTRA HELP

 Student's Solutions Manual

 Digital Video Tutor CD 4 Videotape 7

 AW Math Tutor Center

 MathXL Tutorials on CD

Math XL MathXL

MyMathLab MyMathLab

〜 *Concept Reinforcement* *Complete each of the following sentences.*

1. For any function, the set of all inputs, or first values, is called the _____.

2. For any function, the set of all outputs, or second values, is called the _____.

3. In any function, each member of the domain is paired with _____ one member of the range.

4. In a relation, a member of the domain may be paired with _____ than one member of the range.

5. In the notation $f(5) = 8$, the input is _____.

6. In the notation $f(11) = -6$, the output is _____.

7. In the notation $f(x) = c$, the independent variable is _____.

8. The notation $f(3)$ is read _____.

Determine whether each correspondence is a function.

9.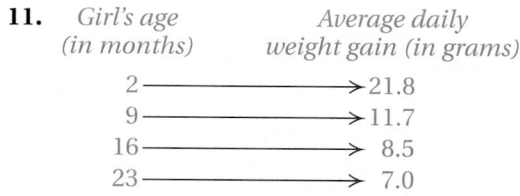

10. 3 ⟶ 9
    6 ⟶ 8
    9 ⟶ 7
    12 ⟶ 6

11. *Girl's age*   *Average daily*
    (in months)   *weight gain (in grams)*

    2 ⟶ 21.8
    9 ⟶ 11.7
    16 ⟶ 8.5
    23 ⟶ 7.0

    Source: *American Family Physician*, December 1993, p. 1435

12. *Boy's age*   *Average daily weight*
    (in months)   *gain (in grams)*

    2 ⟶ 24.3
    9 ⟶ 11.7
    16 ⟶ 8.2
    23 ⟶ 7.0

    Source: *American Family Physician*, December 1993, p. 1435

13. *Celebrity*   *Birthday*

    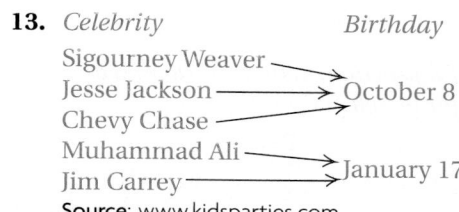

    Source: www.kidsparties.com

14. *Predator*   *Prey*

    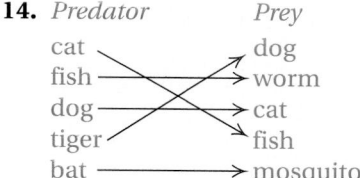

15. *Birthday*   *Celebrity*

    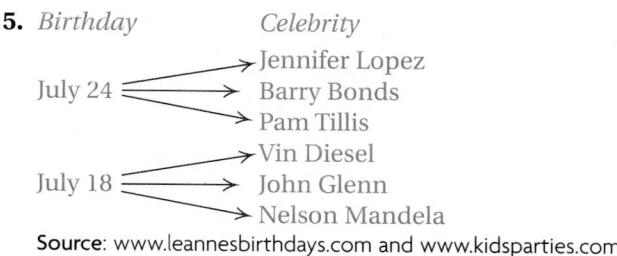

    Source: www.leannesbirthdays.com and www.kidsparties.com

16. *State*   *Neighboring state*

    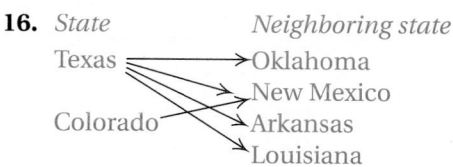

Determine whether each of the following is a function. Identify any relations that are not functions.

Domain	Correspondence	Range
17. A yard full of pumpkins	The price of each pumpkin	A set of prices
18. The members of a rock band	An instrument the person can play	A set of instruments
19. The players on a team	The uniform number of each player	A set of numbers
20. A set of triangles	The area of each triangle	A set of numbers

For each graph of a function, determine **(a)** $f(1)$ *and* **(b)** *any x-values for which* $f(x) = 2$.

21.
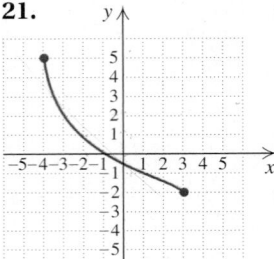

22.
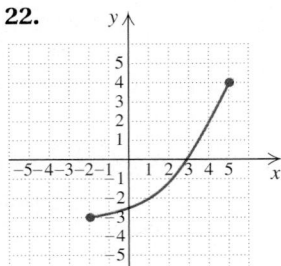

23.
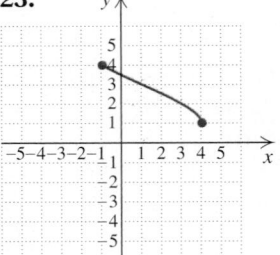

24.
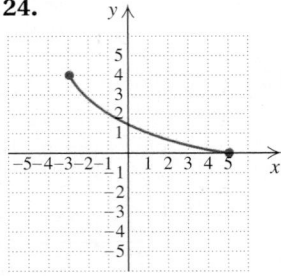

25.
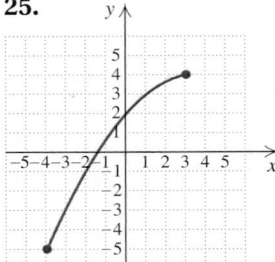

26.
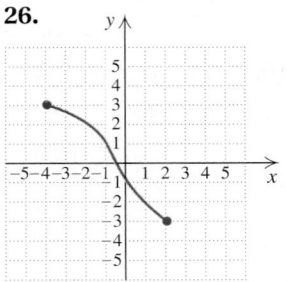

27.
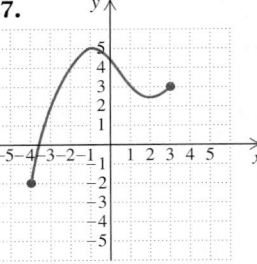

28.
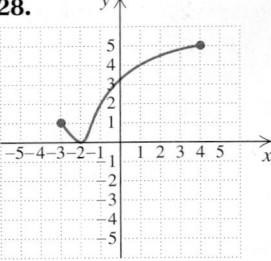

29.
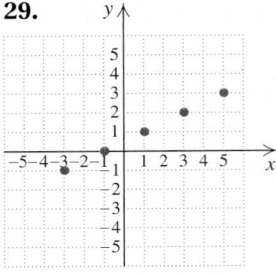

30.
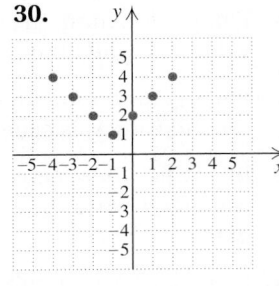

31.
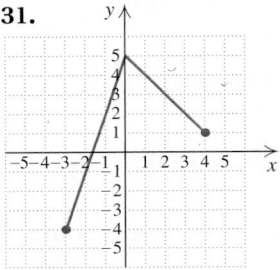

32.
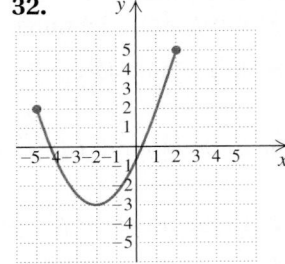

33.
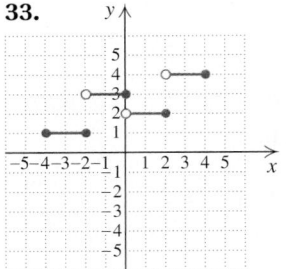

34.
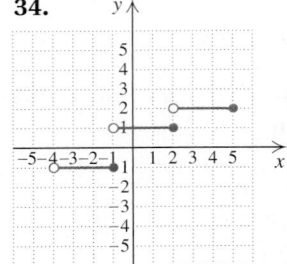

Find the function values.

35. $g(x) = x + 5$

    a) $g(0)$ **b)** $g(-4)$ **c)** $g(-7)$

    d) $g(8)$ **e)** $g(a + 2)$ **f)** $g(a) + 2$

36. $h(x) = x - 2$

    a) $h(4)$ **b)** $h(8)$ **c)** $h(-3)$

    d) $h(-4)$ **e)** $h(a - 1)$ **f)** $h(a) + 3$

37. $f(n) = 5n^2 + 4n$

 a) $f(0)$ **b)** $f(-1)$ **c)** $f(3)$
 d) $f(t)$ **e)** $f(2a)$ **f)** $f(3) - 9$

38. $g(n) = 3n^2 - 2n$

 a) $g(0)$ **b)** $g(-1)$ **c)** $g(3)$
 d) $g(t)$ **e)** $g(2a)$ **f)** $g(3) - 4$

39. $f(x) = \dfrac{x - 3}{2x - 5}$

 a) $f(0)$ **b)** $f(4)$ **c)** $f(-1)$
 d) $f(3)$ **e)** $f(x + 2)$

40. $s(x) = \dfrac{3x - 4}{2x + 5}$

 a) $s(10)$ **b)** $s(2)$ **c)** $s\left(\tfrac{1}{2}\right)$
 d) $s(-1)$ **e)** $s(x + 3)$

The function A described by $A(s) = s^2\dfrac{\sqrt{3}}{4}$ *gives the area of an equilateral triangle with side s.*

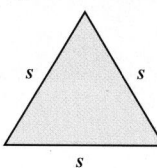

41. Find the area when a side measures 4 cm.

42. Find the area when a side measures 6 in.

The function V described by $V(r) = 4\pi r^2$ *gives the surface area of a sphere with radius r.*

43. Find the surface area when the radius is 3 in.

44. Find the surface area when the radius is 5 cm.

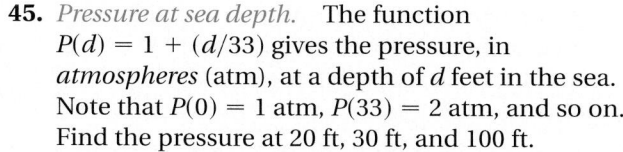

45. *Pressure at sea depth.*   The function $P(d) = 1 + (d/33)$ gives the pressure, in *atmospheres* (atm), at a depth of d feet in the sea. Note that $P(0) = 1$ atm, $P(33) = 2$ atm, and so on. Find the pressure at 20 ft, 30 ft, and 100 ft.

46. *Melting snow.*   The function $W(d) = 0.112d$ approximates the amount, in centimeters, of water that results from d centimeters of snow melting. Find the amount of water that results from snow melting from depths of 16 cm, 25 cm, and 100 cm.

Archaeology.   *The function H described by*
$$H(x) = 2.75x + 71.48$$
can be used to predict the height, in centimeters, of a woman whose humerus *(the bone from the elbow to the shoulder) is x cm long. Predict the height of a woman whose humerus is the length given.*

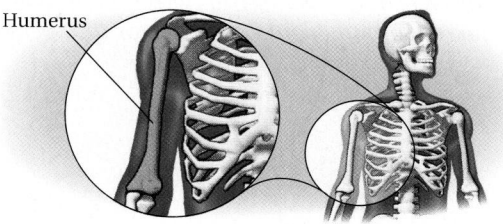

47. 32 cm   **48.** 35 cm

Chemistry.   *The function F described by*
$$F(C) = \tfrac{9}{5}C + 32$$
gives the Fahrenheit temperature corresponding to the Celsius temperature C.

49. Find the Fahrenheit temperature equivalent to $-10°C$.

50. Find the Fahrenheit temperature equivalent to 5°C.

Heart attacks and cholesterol.   *For Exercises 51 and 52, use the following graph, which shows the annual heart attack rate per 10,000 men as a function of blood cholesterol level.**

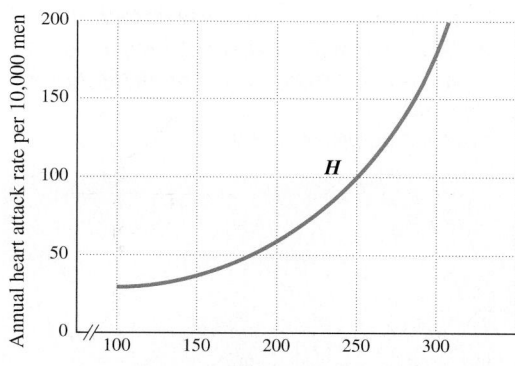

51. Approximate the annual heart attack rate for those men whose blood cholesterol level is 225 mg/dl. That is, find $H(225)$.

---

*Copyright 1989, CSPI. Adapted from *Nutrition Action Healthletter* (1875 Connecticut Avenue, N.W., Suite 300, Washington, DC 20009-5728. $24 for 10 issues).

52. Approximate the annual heart attack rate for those men whose blood cholesterol level is 275 mg/dl. That is, find $H(275)$.

Voting attitudes. *For Exercises 53 and 54, use this graph, which shows the percentage of people responding yes to the question, "If your (political) party nominated a generally well-qualified person for president who happened to be a woman, would you vote for that person?"*
Source: www.gallup.com

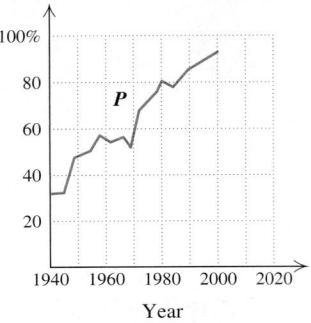

53. Approximate the percentage of Americans willing to vote for a woman for president in 1960. That is, find $P(1960)$.

54. Approximate the percentage of Americans willing to vote for a woman for president in 2000. That is, find $P(2000)$.

Energy-saving lightbulbs. *Compact fluorescent (CFL) lightbulbs can be used in most places that conventional incandescent bulbs are, but they use a fraction of the electricity. The table below lists the CFL wattage and the incandescent wattage required to create the same amount of light.*
Source: Westinghouse Lighting Corporation

Input, CFL Wattage	Output, Wattage of Incandescent Equivalent
7	25
20	75
30	120

55. Use the data in the figure above to draw a graph and to estimate the wattage of an incandescent bulb that creates light equivalent to a 15-watt CFL bulb. Then predict the wattage of an incandescent bulb that creates light equivalent to a 35-watt CFL bulb.

56. Use the graph from Exercise 55 to estimate the wattage of an incandescent bulb that creates light equivalent to a 26-watt CFL bulb. Then predict the wattage of an incandescent bulb that creates light equivalent to a 40-watt CFL bulb.

Blood alcohol level. *The following table can be used to predict the number of drinks required for a person of a specified weight to be considered legally intoxicated (blood alcohol level of 0.08 or above). One 12-oz glass of beer, a 5-oz glass of wine, or a cocktail containing 1 oz of a distilled liquor all count as one drink. Assume that all drinks are consumed within one hour.*

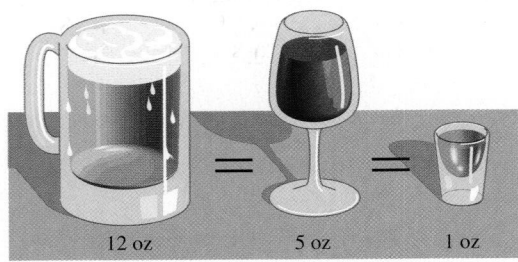

Input, Body Weight (in pounds)	Output, Number of Drinks
100	2.5
160	4
180	4.5
200	5

57. Use the data in the table above to draw a graph and to estimate the number of drinks that a 140-lb person would have to drink to be considered intoxicated. Then predict the number of drinks it would take for a 230-lb person to be considered intoxicated.

58. Use the graph from Exercise 57 to estimate the number of drinks that a 120-lb person would have to drink to be considered intoxicated. Then predict the number of drinks it would take for a 250-lb person to be considered intoxicated.

Incidence of AIDS. *The following table indicates the number of cases of AIDS reported in each of several years.*

Input, Year	Output, Number of Cases Reported
1994	77,103
1996	66,497
1998	48,269
2000	40,282
2002	42,745

Source: U.S. Centers for Disease Control and Prevention

59. Use the data in the table above to draw a graph and to estimate the number of cases of AIDS reported in 1997. Then predict the number of cases that will be reported in 2007.

60. Use the graph from Exercise 59 to estimate the number of cases of AIDS reported in 1999. Then predict the number of cases that will be reported in 2008.

61. *Retailing.* Shoreside Gifts is experiencing constant growth. They recorded a total of $250,000 in sales in 1999 and $285,000 in 2004. Use a graph that displays the store's total sales as a function of time to estimate sales for 2000 and for 2007.

62. Use the graph in Exercise 61 to estimate sales for 2002 and for 2008.

Researchers at Yale University have suggested that the following graphs* may represent three different aspects of love.

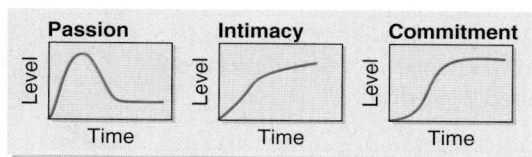

63. In what unit would you measure time if the horizontal length of each graph were ten units? Why?

*From "A Triangular Theory of Love," by R. J. Sternberg, 1986, *Psychological Review,* **93**(2), 119–135. Copyright 1986 by the American Psychological Association, Inc. Reprinted by permission.

64. Do you agree with the researchers that these graphs should be shaped as they are? Why or why not?

SKILL MAINTENANCE

Simplify. [1.8]

65. $\dfrac{10 - 3^2}{9 - 2 \cdot 3}$

66. $\dfrac{2^4 - 10}{6 - 4 \cdot 3}$

67. The surface area of a rectangular solid of length l, width w, and height h is given by $S = 2lh + 2lw + 2wh$. Solve for l. [2.3]

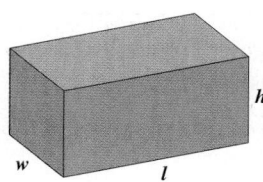

68. Solve the formula in Exercise 67 for w. [2.3]

Solve for y. [2.3]

69. $2x + 3y = 6$

70. $5x - 4y = 8$

SYNTHESIS

71. What would be wrong with using the data for only the years 1994, 1998, and 2002 to answer Exercises 59 and 60?

72. Explain in your own words why every function is a relation, but not every relation is a function.

For Exercises 73 and 74, let $f(x) = 3x^2 - 1$ *and* $g(x) = 2x + 5$.

73. Find $f(g(-4))$ and $g(f(-4))$.

74. Find $f(g(-1))$ and $g(f(-1))$.

75. If f represents the function in Exercise 14, find $f(f(f(f(\text{tiger}))))$.

Pregnancy.   *For Exercises 76–79, use the following graph of a woman's "stress test." This graph shows the size of a pregnant woman's contractions as a function of time.*

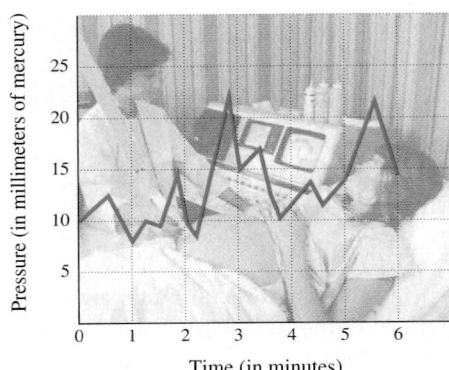

76. How large is the largest contraction that occurred during the test?

77. At what time during the test did the largest contraction occur?

78. On the basis of the information provided, how large a contraction would you expect 60 seconds after the end of the test? Why?

79. What is the frequency of the largest contraction?

80. The *greatest integer function* $f(x) = [\![x]\!]$ is defined as follows: $[\![x]\!]$ is the greatest integer that is less than or equal to x. For example, if $x = 3.74$, then $[\![x]\!] = 3$; and if $x = -0.98$, then $[\![x]\!] = -1$. Graph the greatest integer function for $-5 \leq x \leq 5$. (The notation $f(x) = \text{INT}(x)$ is used in many graphing calculators and computer programs.)

81. Suppose that a function g is such that $g(-1) = -7$ and $g(3) = 8$. Find a formula for g if $g(x)$ is of the form $g(x) = mx + b$, where m and b are constants.

82. *Energy expenditure.*   On the basis of the information given below, what burns more energy: walking $4\frac{1}{2}$ mph for two hours or bicycling 14 mph for one hour?

Approximate Energy Expenditure by a 150-Pound Person in Various Activities

Activity	Calories per Hour
Walking, $2\frac{1}{2}$ mph	210
Bicycling, $5\frac{1}{2}$ mph	210
Walking, $3\frac{3}{4}$ mph	300
Bicycling, 13 mph	660

Source: Based on material prepared by Robert E. Johnson, M.D., Ph.D., and colleagues, University of Illinois.

7.2   Domain and Range

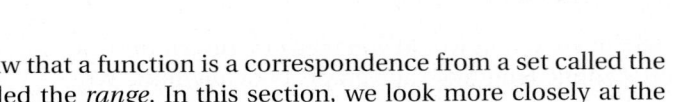

Determining the Domain and the Range • Restrictions on Domain • Functions Defined Piecewise

In Section 7.1, we saw that a function is a correspondence from a set called the *domain* to a set called the *range*. In this section, we look more closely at the concepts of domain and range.

Determining the Domain and the Range

When a function is given as a set of ordered pairs, the domain is the set of all first coordinates and the range is the set of all second coordinates.

EXAMPLE 1   Find the domain and the range for the function f given by

$$f = \{(2, 0), (-1, 5), (8, 0), (-3, 2)\}.$$

Solution   The first coordinates are $2, -1, 8,$ and -3. The second coordinates are $0, 5,$ and 2. Thus we have

Domain of $f = \{2, -1, 8, -3\}$ and

Range of $f = \{0, 5, 2\}$.

We can also determine the domain and the range of a function from its graph.

EXAMPLE 2   Find the domain and the range of the function f below.

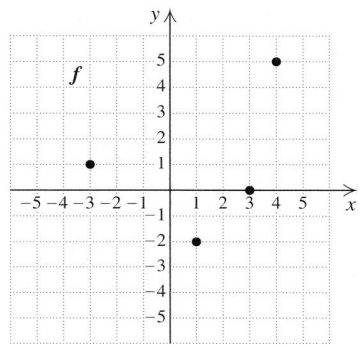

Solution   Here f can be written $\{(-3, 1), (1, -2), (3, 0), (4, 5)\}$. The domain is the set of all first coordinates, $\{-3, 1, 3, 4\}$, and the range is the set of all second coordinates, $\{1, -2, 0, 5\}$.

In Example 2, we could also have found the domain and the range directly, without first writing f, by observing the x- and y-values used in the graph.

EXAMPLE 3   Find the domain and the range of the function f shown here.

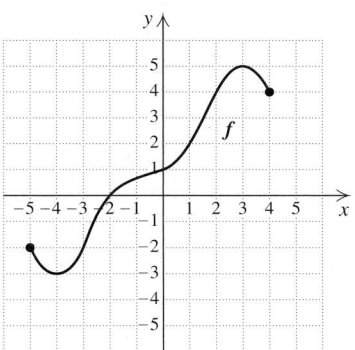

Solution The domain of the function is the set of all x-values that are in the graph. Because there are no breaks in the graph of f, these extend continuously from -5 to 4 and can be viewed as the curve's shadow, or *projection*, on the x-axis. Thus the domain is $\{x\,|\,-5 \le x \le 4\}$ shown below left.

The range of the function is the set of all y-values that are in the graph. These extend continuously from -3 to 5, and can be viewed as the curve's projection on the y-axis. Thus the range is $\{y\,|\,-3 \le y \le 5\}$ shown below right.

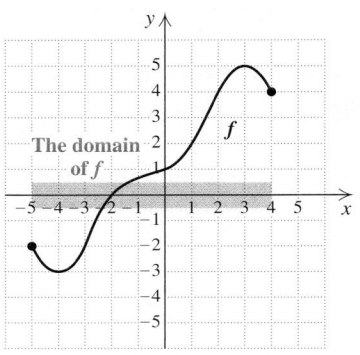
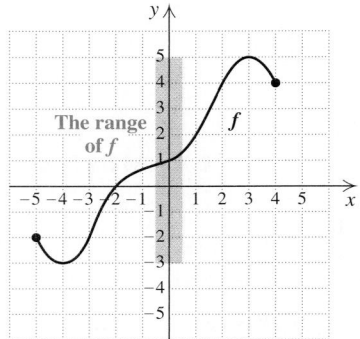

In Example 3, the *endpoints* $(-5, -2)$ and $(4, 4)$ emphasize that the function is not defined for values of x less than -5 or greater than 4.

The graphs of some functions have no endpoints.

EXAMPLE 4 Find the domain and the range of the function f shown here.

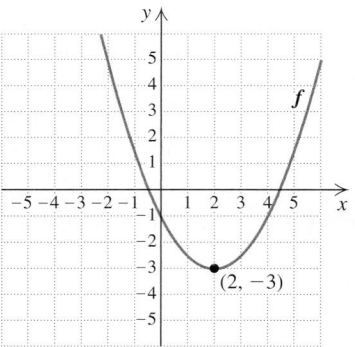

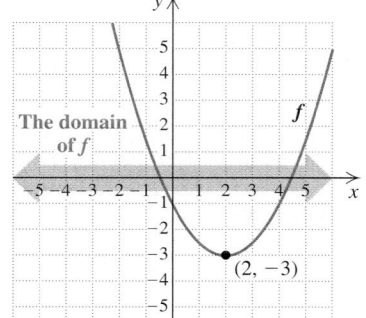

Solution The domain is the set of all x-values that are in the graph. There are no endpoints, which indicates that the graph extends indefinitely. Only a portion of the graph is shown; in fact, it is impossible to show the entire graph. For any x-value, there is a point on the graph. Thus,

Domain of $f = \{x\,|\,x$ is a real number$\}$.

Again, this can be viewed as the curve's projection on the x-axis, as shown at left.

The range of the function is the set of all *y*-values that are in the graph. This can be viewed as the projection of the curve on the *y*-axis. The function has no *y*-values less than −3, and every *y*-value greater than or equal to −3 corresponds to at least one member of the domain. Thus,

Range of $f = \{y \mid y$ is a real number *and* $y \geq -3\}$.

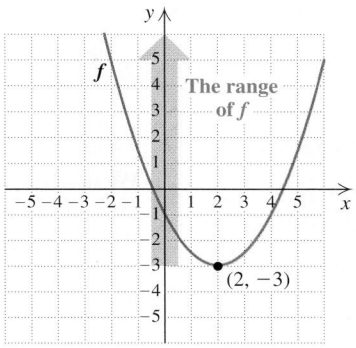

The set of all real numbers is often abbreviated ℝ. Thus, in Example 4, we could write Domain of $f = ℝ$.

A closed dot, as in Example 3, emphasizes that a particular point is on a graph. An open dot, ○, indicates that a particular point is *not* on a graph.

EXAMPLE 5   Find the domain and the range of the function *f* shown here.

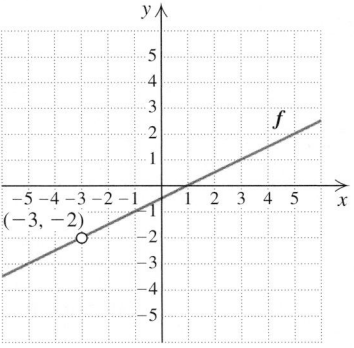

Solution   The domain of *f* is the set of all *x*-values that are in the graph. The open dot in the graph at $(-3, -2)$ indicates that there is no *y*-value that corresponds to $x = -3$; that is, the function is not defined for $x = -3$. Thus, −3 is not in the domain of the function, and

Domain of $f = \{x \mid x$ is a real number *and* $x \neq -3\}$.

There is no function value at $(-3, -2)$, so −2 is not in the range of the function. Thus we have

Range of $f = \{y \mid y$ is a real number *and* $y \neq -2\}$.

When a function is described by an equation, we assume that the domain is the set of all real numbers for which function values can be calculated. If an x-value is not in the domain of a function, the graph of the function will not include any point above or below that x-value.

EXAMPLE 6   For each equation, determine the domain of f.

a) $f(x) = |x|$ **b)** $f(x) = \dfrac{7}{2x - 6}$ **c)** $f(t) = \dfrac{t + 1}{t^2 - 4}$

Solution

a) We ask ourselves, "Is there any number x for which we cannot compute $|x|$?" Since we can find the absolute value of *any* number, the answer is no. Thus the domain of f is $\mathbb{R}$, the set of all real numbers.

b) Is there any number x for which $\dfrac{7}{2x - 6}$ cannot be computed? Since $\dfrac{7}{2x - 6}$ cannot be computed when $2x - 6$ is 0, the answer is yes. To determine what x-value causes the denominator to be 0, we solve an equation:

$$2x - 6 = 0 \qquad \text{Setting the denominator equal to 0}$$
$$2x = 6 \qquad \text{Adding 6 to both sides}$$
$$x = 3. \qquad \text{Dividing both sides by 2}$$

Thus, 3 is *not* in the domain of f, whereas all other real numbers are. The domain of f is $\{x \mid x \text{ is a real number } and \ x \neq 3\}$.

c) The expression $\dfrac{t + 1}{t^2 - 4}$ is undefined when $t^2 - 4 = 0$:

$$t^2 - 4 = 0 \qquad \text{Setting the denominator equal to 0}$$
$$(t + 2)(t - 2) = 0 \qquad \text{Factoring}$$
$$t + 2 = 0 \quad or \quad t - 2 = 0 \qquad \text{Using the principle of zero products}$$
$$t = -2 \quad or \quad t = 2. \qquad \text{Solving; these are the values for which } (t + 1)/(t^2 - 4) \text{ is undefined.}$$

Thus we have

Domain of $f = \{t \mid t \text{ is a real number } and \ t \neq -2 \ and \ t \neq 2\}$.

Note that when the numerator, $t + 1$, is zero, the function value is 0 and *is* defined.

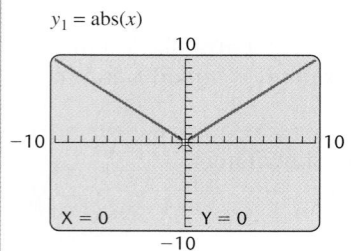

technology connection

To visualize Example 6(a), note that the graph of $y_1 = |x|$ (which is entered $y_1 = \text{abs}(x)$, using the NUM option of the MATH menu) appears without interruption for any piece of the x-axis that we examine.

$y_1 = \text{abs}(x)$

In contrast, the graph of $y_2 = \dfrac{7}{2x - 6}$ in Example 6(b) has a break at $x = 3$.

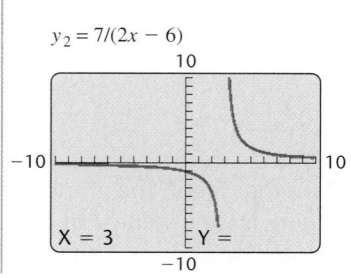

$y_2 = 7/(2x - 6)$

Restrictions on Domain

If a function is used as a model for an application, the problem situation may require restrictions on the domain; for example, length and time are generally nonnegative, and a person's age does not increase indefinitely.

EXAMPLE 7   Prize tee shirts. During intermission at sporting events, it has become common for team mascots to use a powerful slingshot to launch tightly rolled tee shirts into the stands. The height $h(t)$, in feet, of an airborne tee shirt t seconds after being launched can be approximated by

$$h(t) = -15t^2 + 70t + 25.$$

What is the domain of the function?

Solution   The expression $-15t^2 + 70t + 25$ can be evaluated for any number t, so any restrictions on the domain will come from the problem situation.

First, we note that t cannot be negative, since it represents time from launch, so we have $t \geq 0$. If we make a drawing, we also note that the function will not be defined for values of t that make the height negative.

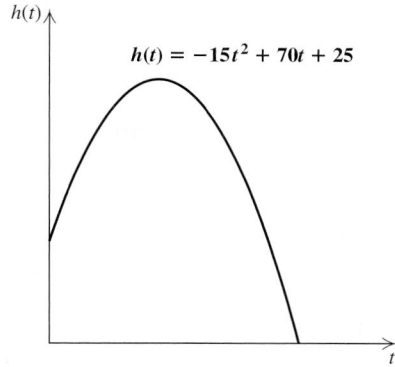

Thus an upper limit for t will be the positive value of t for which $h(t) = 0$. Solving, we obtain

$$h(t) = 0$$

$$-15t^2 + 70t + 25 = 0 \qquad \text{Substituting } -15t^2 + 70t + 25 \text{ for } h(t)$$

$$\left. \begin{array}{l} -5(3t^2 - 14t - 5) = 0 \\ -5(3t + 1)(t - 5) = 0 \end{array} \right\} \qquad \text{Factoring}$$

$$3t + 1 = 0 \quad \text{or} \quad t - 5 = 0 \qquad \text{Using the principle of zero products}$$

$$\left. \begin{array}{rcl} 3t = -1 & \text{or} & t = 5 \\ t = -\dfrac{1}{3} & \text{or} & t = 5. \end{array} \right\} \qquad \text{Solving for } t$$

We already know that $-\frac{1}{3}$ is not in the domain of the function because of the restriction $t \geq 0$ above.

The tee shirt will hit the ground after 5 sec, so we have $t \leq 5$. Putting the two restrictions together, we have $t \geq 0$ *and* $t \leq 5$, so the

Domain of $h = \{t | t$ is a real number *and* $0 \leq t \leq 5\}$.

Functions Defined Piecewise

Some functions are defined by different equations for various parts of their domains. Such functions are said to be **piecewise** defined. For example, the function given by $f(x) = |x|$ is described by

$$f(x) = \begin{cases} x, & \text{if } x \geq 0, \\ -x, & \text{if } x < 0. \end{cases}$$

To evaluate a piecewise-defined function for an input a, we first determine what part of the domain a belongs to. Then we use the appropriate formula for that part of the domain.

EXAMPLE 8   Find each function value for the function f given by

$$f(x) = |x| = \begin{cases} x, & \text{if } x \geq 0, \\ -x, & \text{if } x < 0. \end{cases}$$

a) $f(4)$ **b)** $f(-10)$

Solution

a) Since $4 \geq 0$, we use the equation $f(x) = x$. Thus, $f(4) = 4$.

b) Since $-10 < 0$, we use the equation $f(x) = -x$. Thus,
$f(-10) = -(-10) = 10.$

EXAMPLE 9   Find each function value for the function g given by

$$g(x) = \begin{cases} x + 2, & \text{if } x \leq -2, \\ x^2, & \text{if } -2 < x \leq 5, \\ 3x, & \text{if } x > 5. \end{cases}$$

a) $g(-2)$ **b)** $g(3)$ **c)** $g(10)$

Solution

a) Since $-2 \leq -2$, we use the first equation, $g(x) = x + 2$:

$g(-2) = -2 + 2 = 0.$

b) Since $-2 < 3 \leq 5$, we use the second equation, $g(x) = x^2$:

$g(3) = 3^2 = 9.$

c) Since $10 > 5$, we use the last equation, $g(x) = 3x$:

$g(10) = 3 \cdot 10 = 30.$

Exercise Set

7.2

🔄 *Concept Reinforcement*   *For Exercises 1–6, use the function f given by*

$$f(x) = \begin{cases} x - 5, & \text{if } x < -6, \\ 2x^2, & \text{if } -6 \le x < -1, \\ |x|, & \text{if } -1 \le x < 10, \\ 3x + 1, & \text{if } x \ge 10. \end{cases}$$

Write the letter of the equation that should be used to find each function value. Letters may be used more than once or not at all.

1. ____ $f(0)$

2. ____ $f(15)$

3. ____ $f(10)$

4. ____ $f(-6)$

5. ____ $f(-1)$

6. ____ $f(-3)$

a) $f(x) = x - 5$

b) $f(x) = 2x^2$

c) $f(x) = |x|$

d) $f(x) = 3x + 1$

Find the domain and the range for each function given.

7. $f = \{(2, 8), (9, 3), (-2, 10), (-4, 4)\}$

8. $g = \{(1, 2), (2, 3), (3, 4), (4, 5)\}$

9. $g = \{(0, 0), (4, -2), (-5, 0), (-1, -2)\}$

10. $f = \{(3, 7), (2, 7), (1, 7), (0, 7)\}$

For each graph of a function f, determine the domain and the range of f.

11.

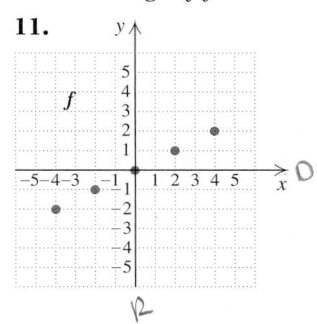

12.

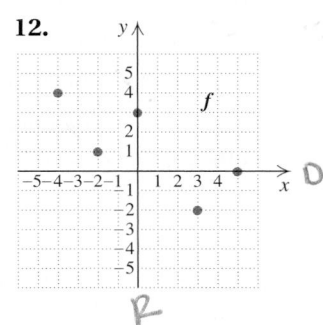

13.

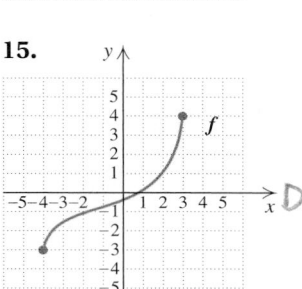

14.

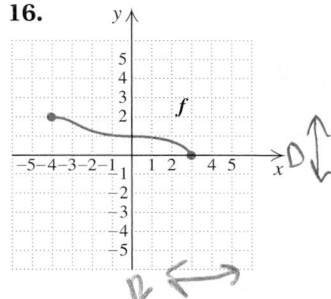

15.

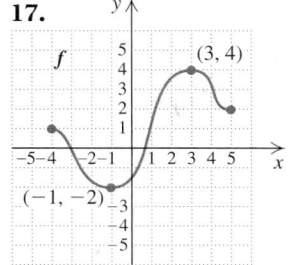

16.

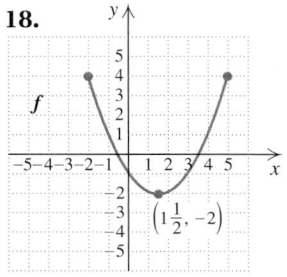

17.

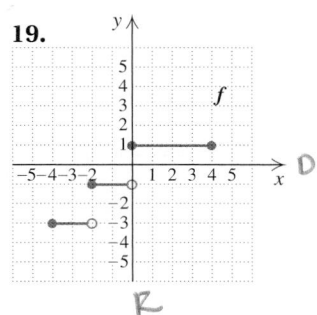

18.

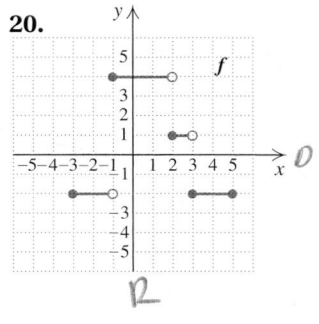

19.

20.

21.

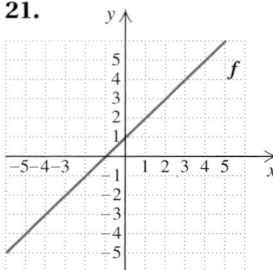

22.

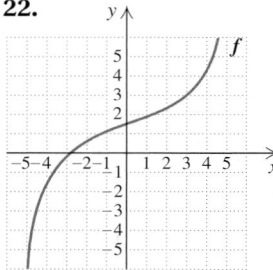

23.

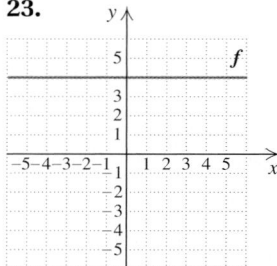

24.

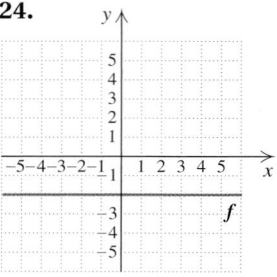

25.

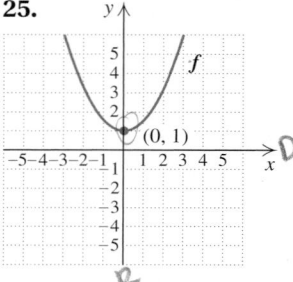

26.

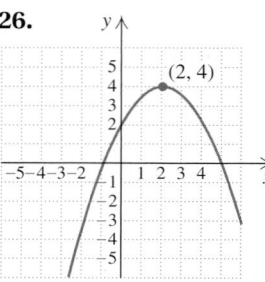

27.

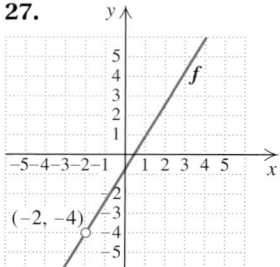

28.

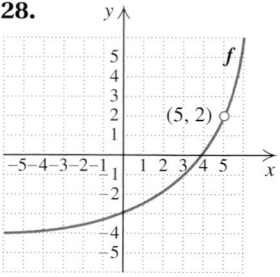

29.

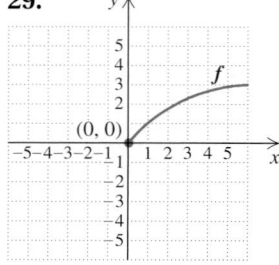

30.

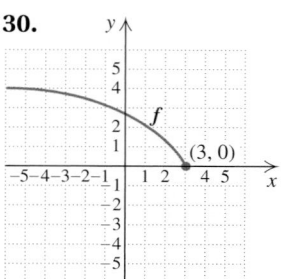

Find the domain of f.

31. $f(x) = \dfrac{5}{x-3}$

32. $f(x) = \dfrac{7}{6-x}$

33. $f(x) = \dfrac{3}{2x-1}$

34. $f(x) = \dfrac{9}{4x+3}$

35. $f(x) = 2x+1$

36. $f(x) = x^2+3$

37. $f(x) = |5-x|$

38. $f(x) = |3x-4|$

39. $f(x) = \dfrac{5}{x^2-9}$

40. $f(x) = \dfrac{x}{x^2-2x+1}$

41. $f(x) = x^2-9$

42. $f(x) = x^2-2x+1$

43. $f(x) = \dfrac{2x-7}{x^2+8x+7}$

44. $f(x) = \dfrac{x+5}{2x^2-x-3}$

45. *Records in the 400-m run.*   The record R for the 400-m run t years after 1930 is given by

$$R(t) = 46.8 - 0.075t.$$

What is the domain of the function?

46. *Records in the 1500-m run.*   The record R for the 1500-m run t years after 1930 is given by

$$R(t) = 3.85 - 0.0075t.$$

What is the domain of the function?

47. *Consumer demand.*   The amount A of coffee that consumers are willing to buy at price p is given by

$$A(p) = -2.5p + 26.5.$$

What is the domain of the function?

48. *Seller's supply.*   The amount A of coffee that suppliers are willing to supply at price p is given by

$$A(p) = 2p - 11.$$

What is the domain of the function?

49. *Pressure at sea depth.*   The pressure P, in atmospheres, at a depth d feet beneath the surface of the ocean is given by

$$P(d) = 0.03d + 1.$$

What is the domain of the function?

50. *Perimeter.*   The perimeter P of an equilateral triangle with sides of length s is given by

$$P(s) = 3s.$$

What is the domain of the function?

51. *Fireworks displays.*   The height h, in feet, of a "weeping willow" fireworks display, t seconds after having been launched from an 80-ft high rooftop, is given by

$$h(t) = -16t^2 + 64t + 80.$$

What is the domain of the function?

52. *Safety flares.*   The height h, in feet, of a safety flare, t seconds after having been launched from a height of 224 ft, is given by

$$h(t) = -16t^2 + 80t + 224.$$

What is the domain of the function?

Find the indicated function values for each function.

53. $f(x) = \begin{cases} x, & \text{if } x < 0, \\ 2x + 1, & \text{if } x \geq 0 \end{cases}$

   a) $f(-5)$ **b)** $f(0)$ **c)** $f(10)$

54. $g(x) = \begin{cases} x - 5, & \text{if } x \leq 5, \\ 3x, & \text{if } x > 5 \end{cases}$

   a) $g(0)$ **b)** $g(5)$ **c)** $g(6)$

55. $G(x) = \begin{cases} x - 5, & \text{if } x < -1, \\ x, & \text{if } -1 \leq x \leq 2, \\ x + 2, & \text{if } x > 2 \end{cases}$

   a) $G(0)$ **b)** $G(2)$ **c)** $G(5)$

56. $F(x) = \begin{cases} 2x, & \text{if } x \leq 0, \\ x, & \text{if } 0 < x \leq 3, \\ -5x, & \text{if } x > 3 \end{cases}$

   a) $F(-1)$ **b)** $F(3)$ **c)** $F(10)$

57. $f(x) = \begin{cases} x^2 - 10, & \text{if } x < -10, \\ x^2, & \text{if } -10 \leq x \leq 10, \\ x^2 + 10, & \text{if } x > 10 \end{cases}$

   a) $f(-10)$ **b)** $f(10)$ **c)** $f(11)$

58. $f(x) = \begin{cases} 2x^2 - 3, & \text{if } x < 2, \\ x^2, & \text{if } 2 \leq x \leq 4, \\ 5x - 7, & \text{if } x > 4 \end{cases}$

   a) $f(0)$ **b)** $f(3)$ **c)** $f(6)$

59. Explain why the domain of the function given by $f(x) = \dfrac{x + 3}{2}$ is $\mathbb{R}$, but the domain of the function given by $g(x) = \dfrac{2}{x + 3}$ is not $\mathbb{R}$.

60. Alayna asserts that for a function described by a set of ordered pairs, the range of the function will always have the same number of elements as there are ordered pairs. Is she correct? Why or why not?

SKILL MAINTENANCE

Graph. [3.6]

61. $y = 2x - 3$ **62.** $y = x + 5$

Find the slope and the y-intercept of each line.

63. $y = \dfrac{2}{3}x - 4$ **64.** $y = -\dfrac{1}{4}x + 6$

65. $y = \dfrac{4}{3}x$ **66.** $y = -5x$

SYNTHESIS

67. Ethan states that $f(x) = \dfrac{x^2}{x}$ and $g(x) = x$ represent the same function. Is he correct? Why or why not?

68. Explain why the domain of a function can be viewed as the projection of its graph on the x-axis.

Sketch the graph of a function for which the domain and range are as given. Graphs may vary.

69. Domain: $\mathbb{R}$; range: $\mathbb{R}$

70. Domain: $\{3, 1, 4\}$; range: $\{0, 5\}$

71. Domain: $\{x \mid 1 \leq x \leq 5\}$; range: $\{y \mid 0 \leq y \leq 2\}$

72. Domain: $\{x \mid x \text{ is a real number } and\ x \neq 1\}$; range: $\{y \mid y \text{ is a real number } and\ y \neq -2\}$

For each graph of a function f, determine the domain and the range of f.

73.

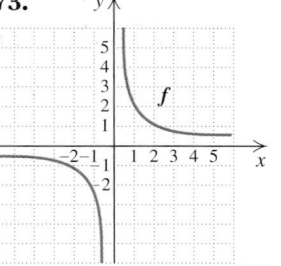

74.

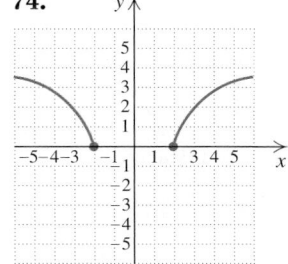

75.

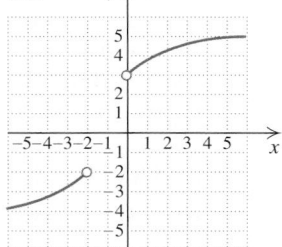

76.

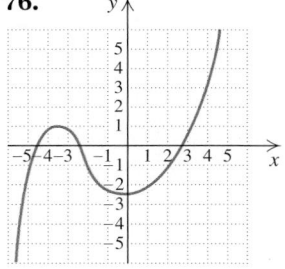

Graph each function on a graphing calculator and estimate its domain and range from the graph.

77. $f(x) = |x - 3|$ **78.** $f(x) = |x| - 3$

79. $f(x) = \dfrac{3}{x - 2}$ **80.** $f(x) = \dfrac{-1}{x + 3}$

81. Use a graphing calculator to estimate the range of the function in Exercise 51.

82. Use a graphing calculator to estimate the range of the function in Exercise 52.

83.–88. *For Exercises 83–88, graph the functions given in each of Exercises 53–58, respectively.*

89. A graphing calculator will interpret an expression like $x \geq 1$ as true or false, depending on the value of x. If the expression is true, the graphing calculator assigns a value of 1 to the expression. If the expression is false, the graphing calculator assigns a value of 0. To graph a piecewise-defined function using a graphing calculator, multiply each part of the definition by its domain, using the TEST menu to enter the inequality symbol. Thus the function in Exercise 53 is entered as $y_1 = x(x < 0) + (2x + 1)(x \geq 0)$. Use a graphing calculator in DOT mode to check your answers to Exercises 83–88.

7.3   Graphs of Functions

The Vertical-Line Test • Linear Functions • Nonlinear Functions

A function can be classified both by the type of equation that is used and by the type of graph it represents. In this section, we will graph a variety of functions that are described by different equations. Before doing so, we examine what types of graphs can represent functions.

The Vertical-Line Test

Recall that a function is a correspondence in which each member of the domain corresponds to *exactly* one member of the range. Thus the correspondence

$$\{(-1, 3), (4, -2), (-4, 3), (-1, 5)\}$$

is not a function because the member -1 of the domain corresponds to the members 3 and 5 of the range. Note on the graph at left that the point $(-1, 5)$ is directly above the point $(-1, 3)$.

Any time two points, such as $(-1, 3)$ and $(-1, 5)$, lie on the same vertical line, the graph containing those points cannot represent a function. This observation is the basis of the *vertical-line test*.

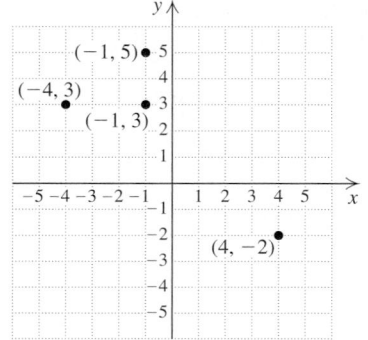

> **The Vertical-Line Test**
>
> If it is possible for a vertical line to cross a graph more than once, then the graph is not the graph of a function.

Study Skills _____

Don't Reach for the Sky

When setting personal study goals, make them reasonable. It is better to set several small, reachable goals than one large one that seems unattainable. For example, you might set a goal of memorizing one formula during each study time, instead of a list of formulas at once. Reaching a goal will encourage you to work toward the next one.

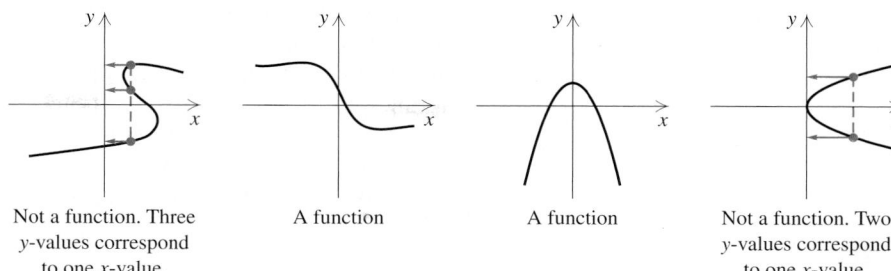

Not a function. Three
y-values correspond
to one x-value.

A function

A function

Not a function. Two
y-values correspond
to one x-value.

Linear Functions

In Chapter 3, we graphed *linear equations*. Here we review such graphs and determine which types of linear graphs represent functions.

Any linear equation can be written in *standard form* $Ax + By = C$. If $B \neq 0$, the equation can also be written in *slope–intercept form*. The *point–slope form* is often used to write equations.

Equations of Lines

Standard form:	$Ax + By = C$
Slope–intercept form:	$y = mx + b$
Point–slope form:	$y - y_1 = m(x - x_1)$

Two points determine a line. If we know that an equation is linear, we can graph the equation by plotting two points that are on the line and drawing the line that goes through those points.

Although any two points can be used to draw a line, some equations can be graphed quickly using their *intercepts*. The point at which a graph crosses the x-axis is its x-intercept; here $y = 0$. Similarly, the point at which a graph crosses the y-axis is its y-intercept; here $x = 0$.

EXAMPLE 1   Graph: $x - 3y = 6$.

Solution   Since $x - 3y = 6$ is in the form $Ax + By = C$, it is a linear equation. We determine the intercepts, plot them, and draw the line.

Let $y = 0$ to find the x-intercept:

$$x - 3y = 6$$
$$x - 3 \cdot 0 = 6 \qquad \text{Substituting 0 for } y$$
$$x = 6. \qquad \text{Solving for } x$$

The x-intercept is $(6, 0)$.

Let $x = 0$ to find the y-intercept:

$$x - 3y = 6$$
$$0 - 3y = 6 \qquad \text{Substituting 0 for } x$$
$$\left. \begin{array}{l} -3y = 6 \\ y = -2. \end{array} \right\} \quad \text{Solving for } y$$

The y-intercept is $(0, -2)$.

We plot $(6, 0)$ and $(0, -2)$ and draw and label the line.

When an equation is written in slope–intercept form $y = mx + b$, the slope of the line is m and the y-intercept is $(0, b)$. Knowing the y-intercept gives us one point on the line, and we can use the slope to determine another point.

Slope is the ratio of the amount of vertical change to the amount of horizontal change, or rise/run. So, for example, a line with slope $\frac{2}{5}$ will rise 2 units for every horizontal run of 5 units. A line with slope -3, or $\frac{-3}{1}$, will rise -3 units, or fall 3 units, for every horizontal run of 1 unit.

EXAMPLE 2

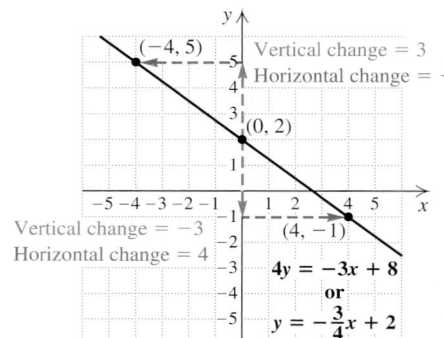

Graph: $4y = -3x + 8$.

Solution   To graph $4y = -3x + 8$, we first rewrite it in slope–intercept form:

$$4y = -3x + 8$$
$$y = \tfrac{1}{4}(-3x + 8) \qquad \text{Multiplying both sides by } \tfrac{1}{4}$$
$$y = -\tfrac{3}{4}x + 2. \qquad \text{Using the distributive law}$$

The slope is $-\frac{3}{4}$ and the y-intercept is $(0, 2)$. We plot $(0, 2)$ and think of the slope as either $\frac{-3}{4}$ or $\frac{3}{-4}$. Using the form $\frac{-3}{4}$, we start at $(0, 2)$ and move *down* 3 units (since the numerator is *negative*) and *to the right* 4 units (since the denominator is *positive*). We plot the new point, $(4, -1)$.

Alternatively, we can think of the slope as $\frac{3}{-4}$. Starting at $(0, 2)$, we move *up* 3 units (since the numerator is *positive*) and *to the left* 4 units (since the denominator is *negative*). This leads to another point on the graph, $(-4, 5)$. Using the points found, we draw and label the graph at left.

The graphs of equations of the form $y = b$ are horizontal lines with a slope of 0, and the graphs of equations of the form $x = a$ are vertical lines. The slope of a vertical line is undefined.

We can use the vertical-line test to determine which types of linear graphs represent functions. Consider the following graphs.

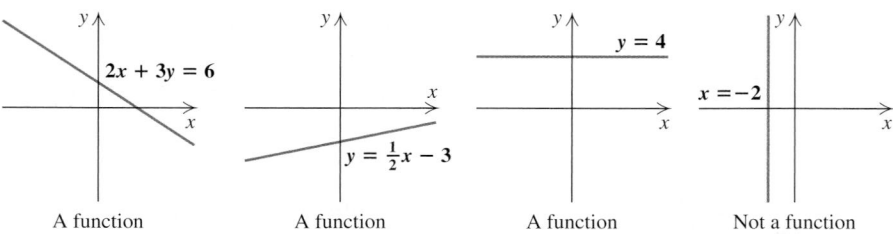

Any vertical line that passes through the graphs of $2x + 3y = 6$, $y = \frac{1}{2}x - 3$, and $y = 4$ will cross the graph only once. However, the vertical line through the point $(-2, 0)$ will cross the graph of $x = -2$ at *every* point. In general, *any* straight line that is not vertical is the graph of a function. A **linear function** is a function described by any linear equation whose graph is not vertical. A horizontal line represents a **constant function**.

> **Linear Function**
>
> A function described by an equation of the form $f(x) = mx + b$ is a *linear function*. Its graph is a straight line with slope m and y-intercept $(0, b)$.
>
> When $m = 0$, the function described by $f(x) = b$ is called a *constant function*. Its graph is a horizontal line through $(0, b)$.

EXAMPLE 3   Graph: $f(x) = 3x + 2$.

Solution   The notations

$$f(x) = 3x + 2 \quad \text{and} \quad y = 3x + 2$$

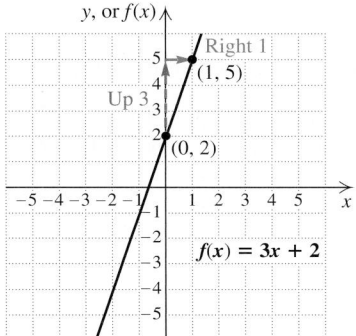

are often used interchangeably. The function notation emphasizes that the second coordinate in each ordered pair is determined by the first coordinate of that pair.

We graph $f(x) = 3x + 2$ in the same way that we would graph $y = 3x + 2$. The vertical axis can be labeled y or $f(x)$. We could use a table of values or, since this is a linear function, use the slope and the y-intercept to graph the function.

Since $f(x) = 3x + 2$ is in the form $f(x) = mx + b$, we can tell from the equation that the slope is 3, or $\frac{3}{1}$, and the y-intercept is $(0, 2)$. We plot $(0, 2)$ and go *up* 3 units and *to the right* 1 unit to determine another point on the line, $(1, 5)$. After we have sketched the line, a third point can be calculated as a check.

EXAMPLE 4   Graph: $f(x) = -3$.

Solution   This is a constant function. For every input x, the output is -3. The graph is a horizontal line.

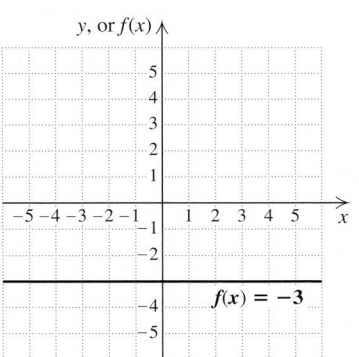

Linear functions are common in today's world.

EXAMPLE 5   Cost projections.   Cleartone Communications charges $50 for a cellular phone and $40 per month for calls made under its Call Anywhere plan. Formulate a mathematical model for the cost. Then use the model to determine the time required for the total cost to reach $250.

Solution

1. **Familiarize.**   The problem describes a situation in which a monthly fee is charged after an initial purchase has been made. After 1 month of service, the total cost will be $50 + $40 = $90. After 2 months, the total cost will be $50 + $40 \cdot 2 = $130. This can be generalized in a model if we let $C(t)$ represent the total cost, in dollars, for t months of service.

2. **Translate.**   We rephrase and translate as follows:

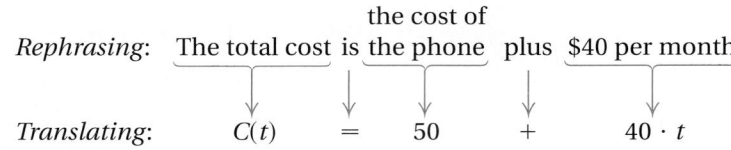

where $t \geq 0$ (since there cannot be a negative number of months).

3. **Carry out.**   The model can be written $C(t) = 40t + 50$. To determine the time required for the total cost to reach $250, we substitute 250 for $C(t)$ and solve for t:

$$C(t) = 40t + 50$$
$$250 = 40t + 50 \qquad \text{Substituting}$$
$$200 = 40t \qquad \text{Subtracting 50 from both sides}$$
$$5 = t. \qquad \text{Dividing both sides by 40}$$

We find that the total cost will reach $250 in 5 months.

4. **Check.**   We evaluate:

$$C(5) = 40 \cdot 5 + 50$$
$$= 200 + 50$$
$$= 250.$$

Our answer checks.

5. **State.**   It takes 5 months for the total cost to reach $250.

EXAMPLE 6   Tattoo removal.   In 1996, an estimated 275,000 Americans visited a doctor for tattoo removal. That figure was expected to grow to 410,000 in 2000 (*Source*: Mike Meyers, staff writer, *Star-Tribune Newspaper of the Twin Cities Minneapolis–St. Paul*, copyright 2000). Assuming constant growth since 1995, how many people will visit a doctor for tattoo removal in 2008?

Solution

1. **Familiarize.**   Constant growth indicates a constant rate of change, so a linear relationship can be assumed. If we let n represent the number of people, in thousands, who visit a doctor for tattoo removal and t the number of years

since 1995, we can form the pairs (1, 275) and (5, 410). After choosing suitable scales on the two axes, we draw the graph.

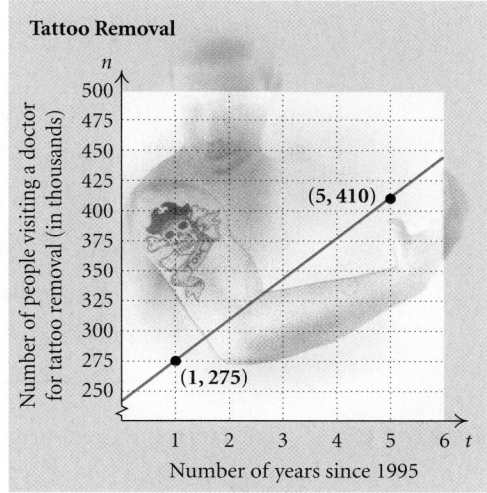

2. **Translate.**   To find an equation relating n and t, we first find the slope of the line. This corresponds to the *growth rate*:

$$m = \frac{410 \text{ thousand people} - 275 \text{ thousand people}}{5 \text{ years} - 1 \text{ year}}$$

$$= \frac{135 \text{ thousand people}}{4 \text{ years}}$$

$$= 33.75 \text{ thousand people per year.}$$

Next, we write point–slope form and solve for n:

$n - 275 = 33.75(t - 1)$	Writing point–slope form
$n - 275 = 33.75t - 33.75$	Using the distributive law
$n = 33.75t + 241.25.$	Adding 275 to both sides

3. **Carry out.**   Using function notation, we have

$$n(t) = 33.75t + 241.25.$$

To predict the number of people who will visit a doctor for tattoo removal in 2008, we find

$n(13) = 33.75 \cdot 13 + 241.25$	2008 is 13 years from 1995.
$= 680.$	This represents 680,000 people.

4. **Check.**   To check, we can repeat our calculations. We could also extend the graph to see if (13, 680) appears to be on the line.

5. **State.**   Assuming constant growth, there will be about 680,000 people visiting a doctor for tattoo removal in 2008.

Since $f(x) = mx + b$ can be evaluated for any choice of x, the domain of all linear functions is $\mathbb{R}$, the set of all real numbers.

The second coordinate of every ordered pair in a constant function $f(x) = b$ is the number b. The range of a constant function thus consists of one number, b. For a nonconstant linear function, the graph extends indefinitely both up and down, so the range is the set of all real numbers, or $\mathbb{R}$.

Domain and Range of a Linear Function

The domain of any linear function $f(x) = mx + b$ is

$\{x \mid x$ is a real number$\}$, or $\mathbb{R}$.

The range of any linear function $f(x) = mx + b$, $m \neq 0$, is

$\{y \mid y$ is a real number$\}$, or $\mathbb{R}$.

The range of any constant function $f(x) = b$ is $\{b\}$.

EXAMPLE 7   Determine the domain and the range of each of the following functions.

a) f, where $f(x) = 2x - 10$ **b)** g, where $g(x) = 4$

Solution

a) Since $f(x) = 2x - 10$ describes a linear function, but not a constant function,

Domain of $f = \mathbb{R}$ and
Range of $f = \mathbb{R}$.

b) The function described by $g(x) = 4$ is a constant function. Thus,

Domain of $g = \mathbb{R}$ and
Range of $g = \{4\}$.

The graphs of nonlinear functions can get quite complex. We will now define several types of nonlinear functions and discuss some of their characteristics. The detailed study of their graphs appears later in this text or in other courses.

Nonlinear Functions

A function for which the graph is not a straight line is a **nonlinear function**. Some important types of nonlinear functions are described below.

Type of function	Description	Example
Absolute-value function	Described by an absolute-value equation	$f(x) = \lvert x \rvert$
Polynomial function	Described by a polynomial equation	$p(x) = x^3 - 4x^2 + 1$
Quadratic function	Described by a polynomial equation of degree 2 (a quadratic equation)	$q(x) = x^2 + 5x + 2$
Rational function	Described by a rational equation	$r(x) = \dfrac{x + 1}{x - 2}$

Note that linear and quadratic functions are special kinds of polynomial functions.

EXAMPLE 8   State whether each equation describes a linear function, an absolute-value function, a general polynomial function, a quadratic function, or a rational function.

a) $f(x) = x^2 - 9$

b) $g(x) = \dfrac{3}{x}$

c) $h(x) = \dfrac{1}{4}x - 16$

d) $v(x) = 4x^4 - 13$

Solution

a) Since f is described by a polynomial equation of degree 2, f is a *quadratic function.*

b) Since g is described by a rational equation, g is a *rational function.*

c) The function h is described by a linear equation, so h is a *linear function.* Note that although $\frac{1}{4}$ is a fraction, there are no variables in a denominator.

d) Since v is described by a polynomial equation, v is a *polynomial function.*

Since the graphs of nonlinear functions are not straight lines, we usually need to calculate more than two or three points to determine the shape of the graph.

EXAMPLE 9   Graph the function given by $f(x) = |x|$, and determine the domain and the range of f.

Solution   We calculate function values for several choices of x and list the results in a table.

$$f(0) = |0| = 0,$$
$$f(1) = |1| = 1,$$
$$f(2) = |2| = 2,$$
$$f(-1) = |-1| = 1,$$
$$f(-2) = |-2| = 2$$

x	$f(x) = |x|$	$(x, f(x))$
0	0	$(0, 0)$
1	1	$(1, 1)$
2	2	$(2, 2)$
-1	1	$(-1, 1)$
-2	2	$(-2, 2)$

When we plot these points, we observe a pattern. The value of the function is 0 when x is 0. Function values increase both as x increases from 0 and as x decreases from 0. The graph of f is V-shaped, with the "point" of the V at the origin.

Because we can find the absolute value of any real number, we have

Domain of $f = \mathbb{R}$.

Because the absolute value of a number is never negative, we have

Range of $f = \{y \mid y \geq 0\}$.

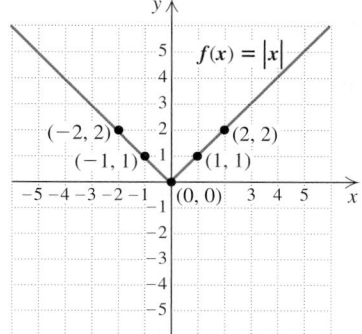

---

technology connection

To check Example 9, we graph $y_1 = |x|$ (which is entered $y_1 = \text{abs}(x)$, using the NUM option of the MATH menu). Note that the graph appears without interruption for any piece of the x-axis that we examine.

$y_1 = \text{abs}(x)$

X = 0, Y = 0

Note, too, that only positive y-values are used.

Graphs of *polynomial functions* generally become more complex as the degree of the polynomial increases. In Chapter 11, we will study in greater detail the graphs of *quadratic functions*, or functions of the form

$$q(x) = ax^2 + bx + c, a \neq 0.$$

Since a polynomial can be evaluated for any real number, the domain of a polynomial is the set of all real numbers.

A *rational function* contains a variable in a denominator; thus its domain may be restricted. Division by zero is undefined, so any values of the variable that make a denominator 0 are not in the domain of the function.

EXAMPLE 10   Determine the domain of *f*.

a) $f(x) = x^3 + 5x^2 - 4x + 1$ **b)** $f(x) = \dfrac{x^2 - 4}{x + 2}$

Solution

a) $f(x) = x^3 + 5x^2 - 4x + 1$ describes a polynomial function. The domain of any polynomial function is $\mathbb{R}$, so the domain of *f* is $\mathbb{R}$.

b) $f(x) = \dfrac{x^2 - 4}{x + 2}$ describes a rational function. Note that $f(x)$ is undefined for $x + 2 = 0$, or, equivalently, for $x = -2$. Thus the domain of $f = \{x | x$ is a real number and $x \neq -2\}$.

CONNECTING THE CONCEPTS

In this section, we have defined several types of functions and described some of their characteristics. They are listed together below for easy reference.

LINEAR FUNCTION

$f(x) = mx + b, m \neq 0$
Graph: straight line;
Domain: $\mathbb{R}$;
Range: $\mathbb{R}$

CONSTANT FUNCTION

$f(x) = b$
Graph: horizontal line;
Domain: $\mathbb{R}$;
Range: $\{b\}$

ABSOLUTE–VALUE FUNCTION

Described by an absolute-value equation
Domain: $\mathbb{R}$

QUADRATIC FUNCTION

$p(x) = ax^2 + bx + c, a \neq 0$
Domain: $\mathbb{R}$

POLYNOMIAL FUNCTION

Described by a polynomial equation
Domain: $\mathbb{R}$

RATIONAL FUNCTION

Described by a rational equation

Domain consists of all real numbers except those that make a denominator 0

↪ *Concept Reinforcement   Answer true or false.*

1. The vertical-line test states that a graph is not that of a function if it contains a vertical line.

2. The graph of a constant function is a horizontal line.

3. The domain of a constant function consists of a single element.

4. The domain of a linear function is the set of all real numbers.

5. Linear functions are typically written in slope–intercept form.

6. Rational functions may have some restrictions on their domains.

Determine whether each of the following is the graph of a function.

7.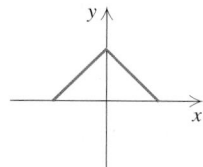

8.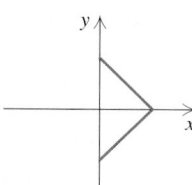

9.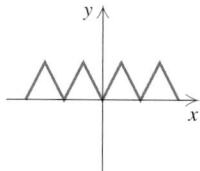

10.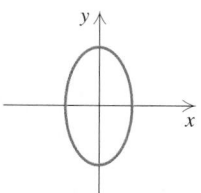

11.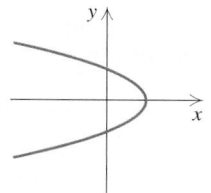

12.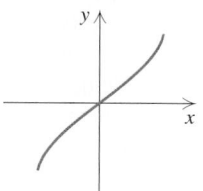

13.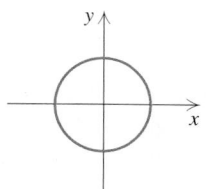

14.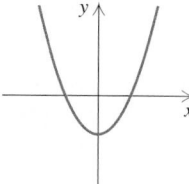

15.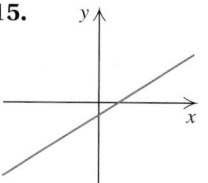

16.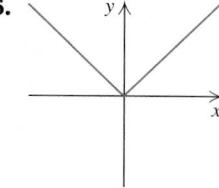

17.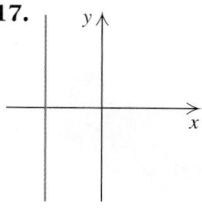

18.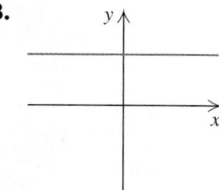

Graph.

19. $x - y = 4$

20. $x + y = 3$

21. $4x - 5y = 20$

22. $2y - 3x = -6$

23. $y = 2x - 1$

24. $y = \frac{1}{3}x + 2$

25. $y = -\frac{2}{3}x + 3$

26. $y = -4x - 2$

27. $3y = 6 - 4x$

28. $5y = 2x - 15$

29. $y = -2$

30. $y = 3$

31. $x = 4$

32. $x = -1$

33. $f(x) = x + 3$

34. $f(x) = 2 - x$

35. $f(x) = \frac{3}{4}x + 1$

36. $f(x) = 3x + 2$

37. $g(x) = 4$

38. $g(x) = -5$

39. *Telephone charges.* Skytone Calling charges $50 for a telephone and $25 per month under its economy plan. Formulate a linear function to model the cost, and determine the amount of time required for the total cost to reach $150.

40. *Cell-phone charges.* The Cellular Connection charges $60 for a cellular phone and $40 per month under its economy plan. Determine the amount of time required for the total cost to reach $260.

41. *Hair growth.* In May, Oscar had his hair cut in a 1-in.-long "buzz cut." His hair grows at a rate of $\frac{1}{2}$ in. per month. Formulate a linear function to model the length of Oscar's hair *t* months after the cut, and determine when his hair will be 3 in. long.

42. *Landscaping.* On Saturday, Shelby Lawncare cut the lawn at Great Harrington Community College to a height of 2 in. Since then, the grass has grown at a rate of $\frac{1}{8}$ in. per day. Formulate a linear function to model the length of the lawn *t* days after having been cut, and determine when the grass will be $3\frac{1}{2}$ in. high.

43. *Cost of a taxi ride.* A taxi ride in Pelham costs $2 plus $0.75 per mile traveled. Formulate a linear function to model the cost of a taxi ride for *d* miles, and determine the length of a taxi ride that cost $4.25.

44. *Natural gas demand.* In 1960, the demand for natural gas was 20 quadrillion joules and was growing at a rate of $\frac{1}{5}$ quadrillion joules per year. Formulate a linear function to model the natural gas demand *t* years after 1960, and determine when the demand was 28 quadrillion joules.

In Exercises 45–54, assume that a constant rate of change exists for each model formed.

45. *Dietary trends.* In 1971, the average American woman consumed 1542 calories per day. By 2000, the figure had risen to 1877 calories per day. Let *C*(*t*) represent the average number of calories consumed per day by an American woman *t* years after 1971.
Source: Centers for Disease Control and Prevention

   a) Find a linear function that fits the data.

   b) Use the function from part (a) to predict the average number of calories consumed per day by an American woman in 2009.

   c) When will the average number of calories consumed per day reach 2000?

46. *Dietary trends.* In 1971, the average American man consumed 2450 calories per day. By 2000, the figure had risen to 2618 calories per day. Let *C*(*t*) represent the average number of calories consumed per day by an American man *t* years after 1971.
Source: Centers for Disease Control and Prevention

   a) Find a linear function that fits the data.

   b) Use the function from part (a) to predict the average number of calories consumed per day by an American man in 2008.

   c) When will the average number of calories consumed per day reach 2750?

47. *Life expectancy of males in the United States.* In 1990, the life expectancy of males was 71.8 yr. In 2000, it was 74.1 yr. Let *E*(*t*) represent life expectancy and *t* the number of years since 1990.
Source: Statistical Abstract of the United States, 2002

   a) Find a linear function that fits the data.

   b) Use the function of part (a) to predict the life expectancy of males in 2009.

48. *Life expectancy of females in the United States.* In 1990, the life expectancy of females was 78.8 yr. In 2000, it was 79.5 yr. Let *E*(*t*) represent life expectancy and *t* the number of years since 1990.
Source: Statistical Abstract of the United States, 2002

   a) Find a linear function that fits the data.

   Aha! **b)** Use the function of part (a) to predict the life expectancy of females in 2010.

49. *PAC contributions.* In 1992, Political Action Committees (PACs) contributed $178.6 million to congressional candidates. In 2002, the figure rose to $282 million. Let $A(t)$ represent the amount of PAC contributions, in millions, and t the number of years since 1992.
Source: Congressional Research Service and Federal Election Commission

a) Find a linear function that fits the data.
b) Use the function of part (a) to predict the amount of PAC contributions in 2008.

50. *Recycling.* In 1996, Americans recycled 57.3 million tons of solid waste. In 2000, the figure grew to 69.9 million tons. Let $N(t)$ represent the number of tons recycled, in millions, and t the number of years since 1996.
Source: Statistical Abstract of the United States, 2002

a) Find a linear function that fits the data.
b) Use the function of part (a) to predict the amount recycled in 2008.

51. *Online travel plans.* In 1999, about 48 million Americans used the Internet to find travel information. By 2003, that number had grown to about 64 million. Let $N(t)$ represent the number of Americans using the Internet for travel information, in millions, t years after 1999.

a) Find a linear function that fits the data.
b) Use the function of part (a) to predict the number of Americans who will find travel information on the Internet in 2009.
c) In what year will 104 million Americans find travel information on the Internet?

52. *Records in the 100-meter run.* In 1991, the record for the 100-m run was 9.86 sec. In 2002, it was 9.78 sec. Let $R(t)$ represent the record in the 100-m run and t the number of years since 1991.
Source: runnersworld.com

a) Find a linear function that fits the data.
b) Use the function of part (a) to predict the record in 2008 and in 2015.
c) When will the record be 9.6 sec?

53. *National Park land.* In 1994, the National Park system consisted of about 74.9 million acres. By 2000, the figure had grown to 78.2 million acres. Let $A(t)$ represent the amount of land in the National Park system, in millions of acres, t years after 1994.
Source: Statistical Abstract of the United States, 2002

a) Find a linear function that fits the data.
b) Use the function of part (a) to predict the amount of land in the National Park system in 2009.

54. *Pressure at sea depth.* The pressure 100 ft beneath the ocean's surface is approximately 4 atm (atmospheres), whereas at a depth of 200 ft, the pressure is about 7 atm.

a) Find a linear function that expresses pressure as a function of depth.
b) Use the function of part (a) to determine the pressure at a depth of 690 ft.

Classify each function as a linear function, an absolute-value function, a quadratic function, another polynomial function, or a rational function, and determine the domain of the function.

55. $f(x) = \dfrac{1}{3}x - 7$ **56.** $g(x) = \dfrac{x}{x + 1}$

57. $p(x) = x^2 + x + 1$ **58.** $t(x) = |x - 7|$

59. $f(t) = \dfrac{12}{3t + 4}$ **60.** $g(n) = 15 - 10n$

61. $f(x) = 0.02x^4 - 0.1x + 1.7$

62. $f(a) = 2|a + 3|$

63. $f(x) = \dfrac{x}{2x - 5}$

64. $g(x) = \dfrac{2x}{3x - 4}$

65. $f(n) = \dfrac{4n - 7}{n^2 + 3n + 2}$

66. $h(x) = \dfrac{x - 5}{2x^2 - 2}$

67. $f(n) = 200 - 0.1n$

68. $g(t) = \dfrac{t^2 - 3t + 7}{8}$

Given the graph of each function, determine the range of f.

69.

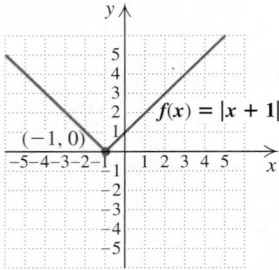

70.

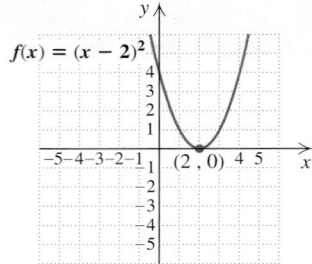

71.

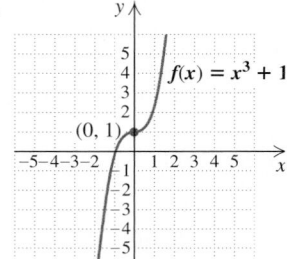

72.

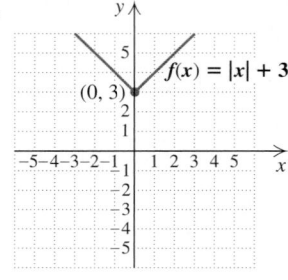

73.

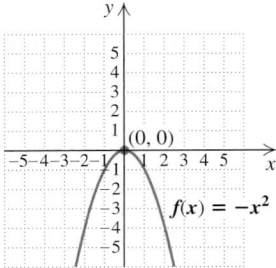

74.

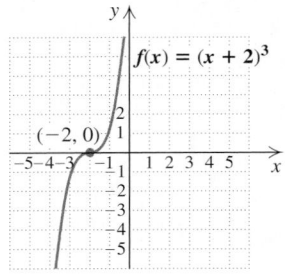

Graph each function and determine its domain and range.

75. $f(x) = x + 3$

76. $f(x) = 2x - 1$

77. $f(x) = -1$

78. $g(x) = 2$

79. $f(x) = |x| + 1$

80. $g(x) = |x - 3|$

81. $g(x) = x^2$

82. $f(x) = x^2 + 2$

83. Explain why the vertical-line test works.

84. Explain why the range of a constant function consists of only one number.

SKILL MAINTENANCE

Perform the indicated operations.

85. $(x^2 + 2x + 7) + (3x^2 - 8)$ [4.3]

86. $(3x^3 - x^2 + x) - (x^3 + 2x - 7)$ [4.3]

87. $(2x + 1)(x - 7)$ [4.5]

88. $(x - 3)(x + 4)$ [4.5]

89. $(x^3 + x^2 - 4x + 7) - (3x^2 - x + 2)$ [4.3]

90. $(2x^2 + x - 3) + (x^3 + 7)$ [4.3]

SYNTHESIS

91. In 1998, Political Action Committees contributed $206.8 million to congressional candidates. Does this information make your answer to Exercise 49(b) seem too low or too high? Why?

92. On the basis of your answers to Exercises 47 and 48, would you predict that at some point in the future the life expectancy of males will exceed that of females? Why or why not?

Given that f(x) = mx + b, classify each of the following as true or false.

93. $f(c + d) = f(c) + f(d)$

94. $f(cd) = f(c)f(d)$

95. $f(kx) = kf(x)$

96. $f(c - d) = f(c) - f(d)$

For Exercises 97–100, assume that a linear equation models each situation.

97. *Temperature conversion.*   Water freezes at 32° Fahrenheit and at 0° Celsius. Water boils at 212°F and at 100°C. What Celsius temperature corresponds to a room temperature of 70°F?

98. *Depreciation of a computer.*   After 6 mos of use, the value of Pearl's computer had dropped to $900. After 8 mos, the value had gone down to $750. How much did the computer cost originally?

99. *Cell-phone charges.*   The total cost of Mel's cell-phone was $230 after 5 mos of service and $390 after 9 mos. What costs had Mel already incurred when his service just began? Assume that Mel's monthly charge is constant.

100. *Operating expenses.*   The total cost for operating Ming's Wings was $7500 after 4 mos and $9250 after 7 mos. Predict the total cost after 10 mos.

101. For a linear function g, $g(3) = -5$ and $g(7) = -1$.
  a) Find an equation for g.
  b) Find $g(-2)$.
  c) Find a such that $g(a) = 75$.

102. When several data points are available and they appear to be nearly collinear, a procedure known as *linear regression* can be used to find an equation for the line that most closely fits the data.
  a) Use a graphing calculator with a LINEAR REGRESSION option and the table that follows to find a linear function that predicts the wattage of an incandescent (conventional)

lightbulb as a function of the wattage of a Compact Fluorescent (CFL) energy-saving bulb of equivalent brightness.

Energy Conservation

CFL Wattage	Incandescent Equivalent
7 W	25 W
20 W	75 W
23 W	90 W
28 W	110 W
30 W	120 W
40 W	135 W

Source: Westinghouse Lighting Corporation

  b) Use the function from part (a) to estimate the incandescent wattage that is equivalent to a 35-watt CFL bulb. Then compare your answer with the corresponding answer to Exercise 55 in Section 7.1. Which answer seems more reliable? Why?

103. Use linear regression (see Exercise 102) to find a linear function that predicts a woman's life expectancy as a function of the year in which she was born. Then use the function to predict the life expectancy in 2010 and compare this with the corresponding answer to Exercise 48 of this exercise set. Which answer seems more reliable? Why?

Life Expectancy of Women

Year, x	Life Expectancy, y (in years)
1920	54.6
1930	61.6
1940	65.2
1950	71.1
1960	73.1
1970	74.7
1980	77.5
1990	78.8
2000	79.5

Sources: Statistical Abstract of the United States 2002 and The World Almanac 1999

7.4 The Algebra of Functions

The Sum, Difference, Product, or Quotient of Two Functions •
Domains and Graphs

We now return to the idea of a function as a machine and examine four ways in which functions can be combined.

The Sum, Difference, Product, or Quotient of Two Functions

Suppose that a is in the domain of two functions, f and g. The input a is paired with $f(a)$ by f and with $g(a)$ by g. The outputs can then be added to get $f(a) + g(a)$.

EXAMPLE 1   Let $f(x) = x + 4$ and $g(x) = x^2 + 1$. Find $f(2) + g(2)$.

Solution   We visualize two function machines. Because 2 is in the domain of each function, we can compute $f(2)$ and $g(2)$.

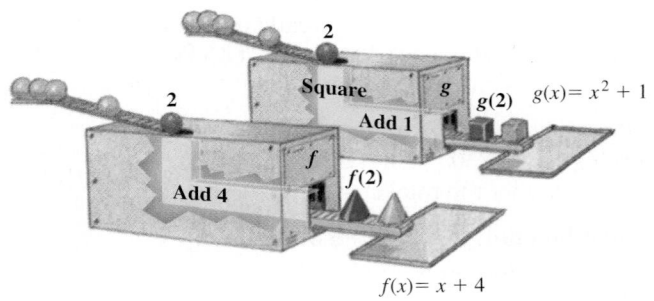

$$g(x) = x^2 + 1$$
$$f(x) = x + 4$$

Since

$$f(2) = 2 + 4 = 6 \quad \text{and} \quad g(2) = 2^2 + 1 = 5,$$

we have

$$f(2) + g(2) = 6 + 5 = 11.$$

In Example 1, suppose that we were to write $f(x) + g(x)$ as $(x + 4) + (x^2 + 1)$, or $f(x) + g(x) = x^2 + x + 5$. This could then be regarded as a "new" function. The notation $(f + g)(x)$ is generally used to denote a function formed in this manner. Similar notations exist for subtraction, multiplication, and division of functions.

> **The Algebra of Functions**
>
> If f and g are functions and x is in the domain of both functions, then:
>
> **1.** $(f + g)(x) = f(x) + g(x)$;
> **2.** $(f - g)(x) = f(x) - g(x)$;
> **3.** $(f \cdot g)(x) = f(x) \cdot g(x)$;
> **4.** $(f/g)(x) = f(x)/g(x)$, provided $g(x) \neq 0$.

EXAMPLE 2   For $f(x) = x^2 - x$ and $g(x) = x + 2$, find the following.

a) $(f + g)(3)$ **b)** $(f - g)(x)$ and $(f - g)(-1)$

c) $(f/g)(x)$ and $(f/g)(-4)$ **d)** $(f \cdot g)(3)$

Solution

a) Since $f(3) = 3^2 - 3 = 6$ and $g(3) = 3 + 2 = 5$, we have

$$(f + g)(3) = f(3) + g(3)$$
$$= 6 + 5 \quad \text{Substituting}$$
$$= 11.$$

Alternatively, we could first find $(f + g)(x)$:

$$(f + g)(x) = f(x) + g(x)$$
$$= x^2 - x + x + 2$$
$$= x^2 + 2. \quad \text{Combining like terms}$$

Thus,

$$(f + g)(3) = 3^2 + 2 = 11. \quad \text{Our results match.}$$

b) We have

$$(f - g)(x) = f(x) - g(x)$$
$$= x^2 - x - (x + 2) \quad \text{Substituting}$$
$$= x^2 - 2x - 2. \quad \begin{array}{l}\text{Removing parentheses and} \\ \text{combining like terms}\end{array}$$

Thus,

$$(f - g)(-1) = (-1)^2 - 2(-1) - 2 \quad \begin{array}{l}\text{Using } (f - g)(x) \text{ is faster than} \\ \text{using } f(x) - g(x).\end{array}$$

$$= 1. \quad \text{Simplifying}$$

c) We have

$$(f/g)(x) = f(x)/g(x)$$
$$= \frac{x^2 - x}{x + 2}. \quad \text{We assume that } x \neq -2.$$

Thus,

$$(f/g)(-4) = \frac{(-4)^2 - (-4)}{-4 + 2} \quad \text{Substituting}$$

$$= \frac{20}{-2} = -10.$$

d) Using our work in part (a), we have

$$(f \cdot g)(3) = f(3) \cdot g(3)$$
$$= 6 \cdot 5$$
$$= 30.$$

Alternatively, we could first find $(f \cdot g)(x)$:

$$(f \cdot g)(x) = f(x) \cdot g(x)$$
$$= (x^2 - x)(x + 2)$$
$$= x^3 + x^2 - 2x. \qquad \text{Multiplying and combining like terms}$$

Then

$$(f \cdot g)(3) = 3^3 + 3^2 - 2 \cdot 3$$
$$= 27 + 9 - 6$$
$$= 30.$$

Domains and Graphs

Although applications involving products and quotients of functions rarely appear in newspapers, situations involving sums or differences of functions often do appear in print. For example, the following graphs are similar to those published by the California Department of Education to promote breakfast programs in which students eat a balanced meal of fruit or juice, toast or cereal, and 2% or whole milk. The combination of carbohydrate, protein, and fat gives a sustained release of energy, delaying the onset of hunger for several hours.

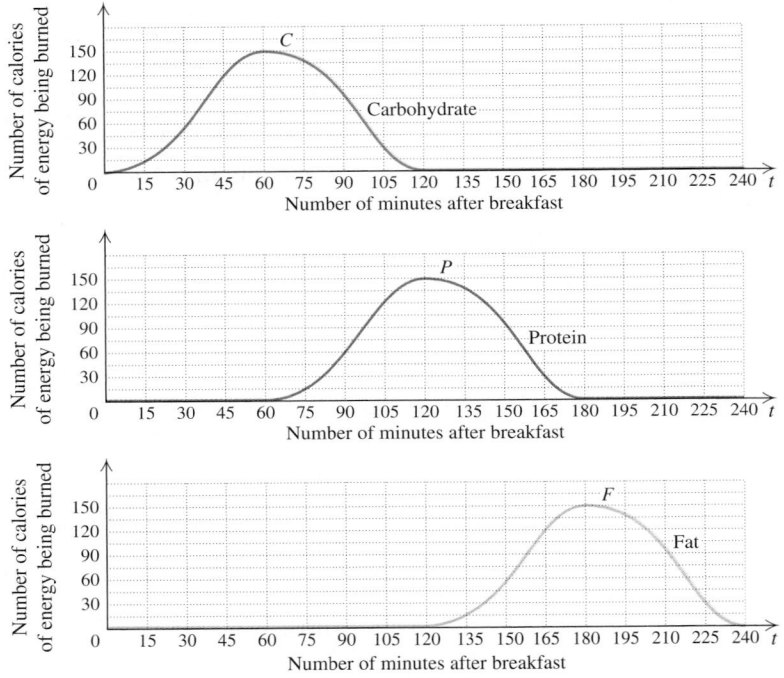

When the three graphs are superimposed, and the calorie expenditures added, it becomes clear that a balanced meal results in a steady, sustained supply of energy.

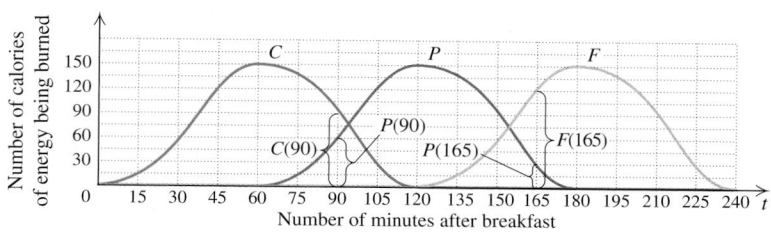

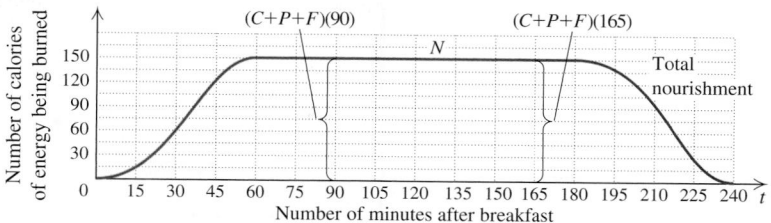

Note that for $t > 120$, we have $C(t) = 0$; for $t < 60$ or $t > 180$, we have $P(t) = 0$; and for $t < 120$, we have $F(t) = 0$. For any point $(t, N(t))$, we have

$$N(t) = (C + P + F)(t) = C(t) + P(t) + F(t).$$

To find $(f + g)(a), (f - g)(a), (f \cdot g)(a)$, or $(f/g)(a)$, we must first be able to find $f(a)$ and $g(a)$. This means a must be in the domain of both f and g.

EXAMPLE 3   Let

$$f(x) = \frac{5}{x} \quad \text{and} \quad g(x) = \frac{2x - 6}{x + 1}.$$

Find the domain of $f + g$, the domain of $f - g$, and the domain of $f \cdot g$.

Solution   Note that because division by 0 is undefined, we have

Domain of $f = \{x \,|\, x$ is a real number *and* $x \neq 0\}$

and

Domain of $g = \{x \,|\, x$ is a real number *and* $x \neq -1\}$.

In order to find $f(a) + g(a), f(a) - g(a)$, or $f(a) \cdot g(a)$, we must know that a is in *both* of the above domains. Thus,

Domain of $f + g =$ Domain of $f - g =$ Domain of $f \cdot g$

$= \{x \,|\, x$ is a real number *and* $x \neq 0$ *and* $x \neq -1\}$.

Suppose in Example 3 that we want to find $(f/g)(3)$. Finding $f(3)$ and $g(3)$ poses no problem:

$$f(3) = \frac{5}{3} \quad \text{and} \quad g(3) = \frac{2 \cdot 3 - 6}{3 + 1} = 0;$$

but then

$$(f/g)(3) = f(3)/g(3)$$
$$= \tfrac{5}{3} \; / \; 0 \; . \qquad \text{Division by 0 is undefined.}$$

Thus, although 3 is in the domain of both f and g, it is not in the domain of f/g.

Determining the Domain

The domain of $f + g, f - g$, or $f \cdot g$ is the set of all values common to the domains of f and g.

The domain of f/g is the set of all values common to the domains of f and g, excluding any values for which $g(x)$ is 0.

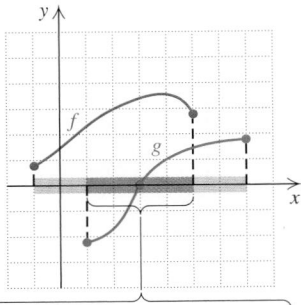

Domain of $f + g, f - g$, and $f \cdot g$

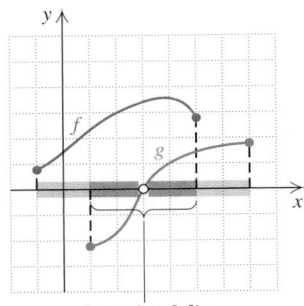

Domain of f/g

EXAMPLE 4

Given $f(x) = 1/x$ and $g(x) = 2x - 7$, find the domains of $f + g, f - g, f \cdot g$, and f/g.

Solution   The domain of f is $\{x \mid x \neq 0\}$ or $\{x \mid x$ is a real number *and* $x \neq 0\}$. The domain of g is $\mathbb{R}$. Thus the domains of $f + g, f - g$, and $f \cdot g$ are the set of all elements common to both the domain of f and the domain of g. We have

the domain of $f + g =$ the domain of $f - g =$ the domain of $f \cdot g$

$$= \{x \mid x \text{ is a real number } and \ x \neq 0\}.$$

To find the domain of f/g, note that

$$(f/g)(x) = \frac{f(x)}{g(x)} = \frac{1/x}{2x - 7} \quad \text{cannot be evaluated if } x = 0 \text{ or if } 2x - 7 = 0.$$

This means the domain of f/g is $\{x \mid x$ is a real number *and* $x \neq 0\}$, *with the additional restriction* that $g(x) \neq 0$. To determine what x-values would make $g(x) = 0$, we solve:

$$2x - 7 = 0 \qquad \text{Replacing } g(x) \text{ with } 2x - 7$$
$$2x = 7$$
$$x = \tfrac{7}{2}.$$

Since $g(x) = 0$ for $x = \tfrac{7}{2}$,

the domain of $f/g = \{x \mid x$ is a real number *and* $x \neq 0$ *and* $x \neq \tfrac{7}{2}\}$.

Student Notes

The concern over a denominator being 0 arises throughout this course. Try to develop the habit of checking for any possible input-values that would create a denominator of 0 whenever you work with functions.

technology connection

A partial check of Example 4 can be performed by setting up a table so the TABLE MINIMUM is 0 and the increment of change (ΔTbl) is 0.7. (Other choices, like 0.1, will also work.) Next, we let $y_1 = 1/x$ and $y_2 = 2x - 7$. Using Y-VARS to write $y_3 = y_1 + y_2$ and $y_4 = y_1/y_2$, we can create the table of values shown here. Note that when x is 3.5, a value for y_3 can be found, but y_4 is undefined. When setting up these functions, you may wish to enter y_1 and y_2 without "selecting" either. Otherwise, the table's columns must be scrolled to display y_3 and y_4.

X	Y₃	Y₄
0	ERROR	ERROR
.7	−4.171	−.2551
1.4	−3.486	−.1701
2.1	−2.324	−.1701
2.8	−1.043	−.2551
3.5	.28571	ERROR
4.2	1.6381	.17007

X = 0

Use a similar approach to partially check Example 3.

Division by 0 is not the only condition that can force restrictions on the domain of a function. In Chapter 10, we will examine functions similar to that given by $f(x) = \sqrt{x}$, for which the concern is taking the square root of a negative number.

Exercise Set

7.4

FOR EXTRA HELP

 Student's Solutions Manual

 Digital Video Tutor CD 4 Videotape 7

Tutor Center AW Math Tutor Center

 MathXL Tutorials on CD

Math XL MathXL

MyMathLab MyMathLab

🔁 *Concept Reinforcement*   *Make each of the following sentences true by selecting the correct word for each blank.*

1. If f and g are functions and x is in the
_____ of both functions, then
  range/domain
$(f + g)(x) = f(x) + g(x)$.

2. One way to compute $(f - g)(2)$ is to
_____ $g(2)$ from $f(2)$.
  erase/subtract

3. One way to compute $(f - g)(2)$ is to simplify
$f(x) - g(x)$ and then _____ the
  evaluate/substitute
result for $x = 2$.

4. The domain of $f + g, f - g$, and $f \cdot g$ is the set of
all values common to the _____
  domains/ranges
of f and g.

5. The domain of f/g is the set of all values common
to the domains of f and g, _____
  including/excluding
any values for which $g(x)$ is 0.

6. The height of $(f + g)(a)$ on a graph is the
_____ of the heights of $f(a)$ and $g(a)$.
  product/sum

Let $f(x) = -3x + 1$ and $g(x) = x^2 + 2$. Find the following.

7. $f(2) + g(2)$

8. $f(-1) + g(-1)$

9. $f(5) - g(5)$

10. $f(4) - g(4)$

11. $f(-1) \cdot g(-1)$

12. $f(-2) \cdot g(-2)$

13. $f(-4)/g(-4)$

14. $f(3)/g(3)$

15. $g(1) - f(1)$

16. $g(2)/f(2)$

17. $(f + g)(x)$

18. $(g - f)(x)$

Let $F(x) = x^2 - 2$ and $G(x) = 5 - x$. Find the following.

19. $(F + G)(x)$

20. $(F + G)(a)$

21. $(F + G)(-4)$

22. $(F + G)(-5)$

23. $(F - G)(3)$

24. $(F - G)(2)$

25. $(F \cdot G)(-3)$

26. $(F \cdot G)(-4)$

27. $(F/G)(x)$

28. $(G - F)(x)$

29. $(F/G)(-2)$

30. $(F/G)(-1)$

In 2004, a study comparing high doses of the cholesterol-lowering drugs Lipitor and Pravachol indicated that patients taking Lipitor were significantly less likely to have heart attacks or require angioplasty or surgery.

In the graph below, $L(t)$ is the percentage of patients on Lipitor (80 mg) and $P(t)$ is the percentage of patients on Pravachol (40 mg) who suffered heart problems or death t years after beginning to take the medication.
Source: *New York Times,* March 9, 2004

31. Use estimates of $P(2)$ and $L(2)$ to estimate $(P - L)(2)$.

32. Use estimates of $P(1)$ and $L(1)$ to estimate $(P - L)(1)$.

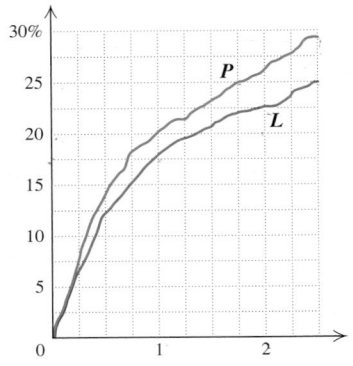

Years of follow-up of patients

Source: New England Journal of Medicine

The following graph shows the number of women, in millions, who had a child the previous year. Here $W(t)$ represents the number of women under 30 who gave birth in year t, $R(t)$ the number of women 30 and older who gave birth in year t, and $N(t)$ the total number of women who gave birth in year t.

33. Use estimates of $R(2000)$ and $W(2000)$ to estimate $N(2000)$.

34. Use estimates of $R(1990)$ and $W(1990)$ to estimate $N(1990)$.

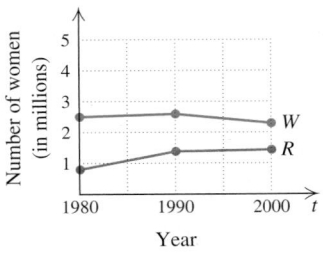

Year

Source: U.S Bureau of the Census

Often function addition is represented by stacking the individual functions directly on top of each other. The graph below indicates how the three major airports servicing New York City have been utilized. The braces indicate the values of the individual functions.

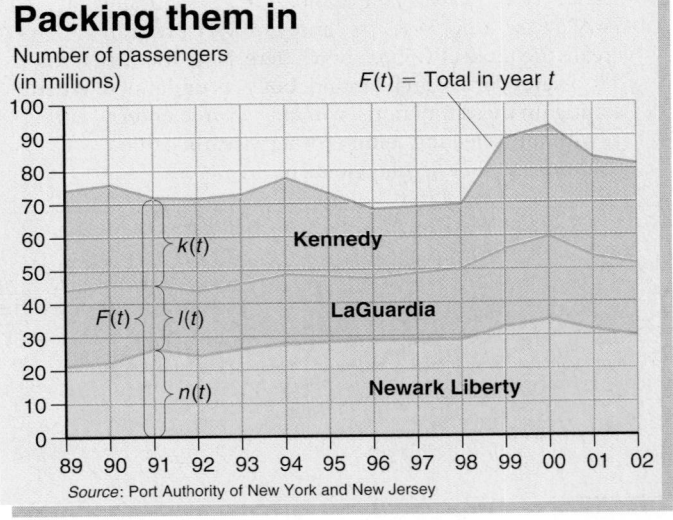

Packing them in

Number of passengers (in millions)

$F(t) = $ Total in year t

Source: Port Authority of New York and New Jersey

35. Estimate $(n + l)('98)$. What does it represent?

36. Estimate $(k + l)('98)$. What does it represent?

37. Estimate $F('02)$. What does it represent?

38. Estimate $F('01)$. What does it represent?

39. Estimate $(F - k)('02)$. What does it represent?

40. Estimate $(F - k)('01)$. What does it represent?

For each pair of functions f and g, determine the domain of the sum, difference, and product of the two functions.

41. $f(x) = x^2,$
$g(x) = 7x - 4$

42. $f(x) = 5x - 1,$
$g(x) = 2x^2$

43. $f(x) = \dfrac{1}{x - 3},$
$g(x) = 4x^3$

44. $f(x) = 3x^2,$
$g(x) = \dfrac{1}{x - 9}$

45. $f(x) = \dfrac{2}{x},$
$g(x) = x^2 - 4$

46. $f(x) = x^3 + 1,$
$g(x) = \dfrac{5}{x}$

47. $f(x) = x + \dfrac{2}{x-1}$,

$g(x) = 3x^3$

48. $f(x) = 9 - x^2$,

$g(x) = \dfrac{3}{x-6} + 2x$

49. $f(x) = \dfrac{3}{x-2}$,

$g(x) = \dfrac{5}{4-x}$

50. $f(x) = \dfrac{5}{x-3}$,

$g(x) = \dfrac{1}{x-2}$

For each pair of functions f and g, determine the domain of f/g.

51. $f(x) = x^4$,

$g(x) = x - 3$

52. $f(x) = 2x^3$,

$g(x) = 5 - x$

53. $f(x) = 3x - 2$,

$g(x) = 2x - 8$

54. $f(x) = 5 + x$,

$g(x) = 6 - 2x$

55. $f(x) = \dfrac{3}{x-4}$,

$g(x) = 5 - x$

56. $f(x) = \dfrac{1}{2-x}$,

$g(x) = 7 - x$

57. $f(x) = \dfrac{2x}{x+1}$,

$g(x) = 2x + 5$

58. $f(x) = \dfrac{7x}{x-2}$,

$g(x) = 3x + 7$

For Exercises 59–66, consider the functions F and G as shown.

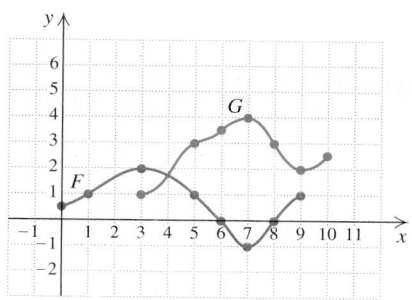

59. Determine $(F + G)(5)$ and $(F + G)(7)$.

60. Determine $(F \cdot G)(6)$ and $(F \cdot G)(9)$.

61. Determine $(G - F)(7)$ and $(G - F)(3)$.

62. Determine $(F/G)(3)$ and $(F/G)(7)$.

63. Find the domains of F, G, $F + G$, and F/G.

64. Find the domains of $F - G$, $F \cdot G$, and G/F.

65. Graph $F + G$.

66. Graph $G - F$.

In the following graph, W(t) represents the number of gallons of whole milk, L(t) the number of gallons of lowfat milk, and S(t) the number of gallons of skim milk consumed by the average American in year t.*

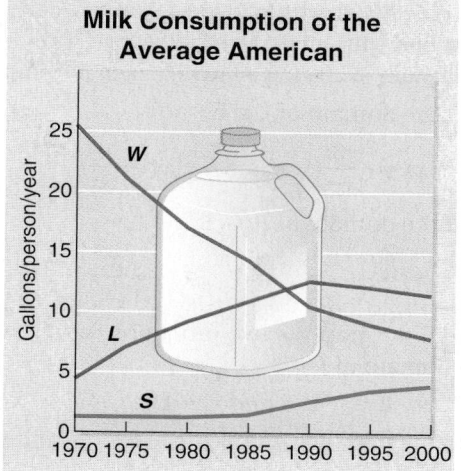

67. From 1970 to 2000, did American milk consumption increase or decrease? Explain how you determined this.

68. Examine the graphs in Exercises 35–40. To what do you attribute the decline in $F(t)$ from 2000 to 2002?

SKILL MAINTENANCE

Solve. [2.3]

69. $4x - 7y = 8$, for x

70. $3x - 8y = 5$, for y

71. $5x + 2y = -3$, for y

72. $6x + 5y = -2$, for x

Translate each of the following. Do not solve. [1.1]

73. Five more than twice a number is 49.

74. Three less than half of some number is 57.

75. The sum of two consecutive integers is 145.

76. The difference between a number and its opposite is 20.

---

*Sources: Copyright 1990, CSPI. Adapted from *Nutrition Action Healthletter* (1875 Connecticut Avenue, N.W., Suite 300, Washington, DC 20009-5728. $24.00 for 10 issues); USDA Agricultural Fact Book 2000, USDA Economic Research Service.

SYNTHESIS

 77. If $f(x) = c$, where c is some positive constant, describe how the graphs of $y = g(x)$ and $y = (f + g)(x)$ will differ.

78. Examine the graphs following Example 2 and explain how they might be modified to represent the absorption of 200 mg of Advil® taken four times a day.

79. Find the domain of f/g, if

$$f(x) = \frac{3x}{2x + 5} \quad \text{and} \quad g(x) = \frac{x^4 - 1}{3x + 9}.$$

80. Find the domain of F/G, if

$$F(x) = \frac{1}{x - 4} \quad \text{and} \quad G(x) = \frac{x^2 - 4}{x - 3}.$$

81. Sketch the graph of two functions f and g such that the domain of f/g is

$$\{x \,|\, -2 \le x \le 3 \text{ and } x \ne 1\}.$$

82. Find the domains of $f + g, f - g, f \cdot g$, and f/g, if
$$f = \{(-2, 1), (-1, 2), (0, 3), (1, 4), (2, 5)\}$$
and
$$g = \{(-4, 4), (-3, 3), (-2, 4), (-1, 0), (0, 5), (1, 6)\}.$$

83. Find the domain of m/n, if
$$m(x) = 3x \text{ for } -1 < x < 5$$
and
$$n(x) = 2x - 3.$$

84. For f and g as defined in Exercise 82, find $(f + g)(-2)$, $(f \cdot g)(0)$, and $(f/g)(1)$.

85. Write equations for two functions f and g such that the domain of $f + g$ is
$$\{x \,|\, x \text{ is a real number } and \; x \ne -2 \; and \; x \ne 5\}.$$

86. Let $y_1 = 2.5x + 1.5$, $y_2 = x - 3$, and $y_3 = y_1/y_2$. Depending on whether the CONNECTED or DOT mode is used, the graph of y_3 appears as follows. Use algebra to determine which graph more accurately represents y_3.

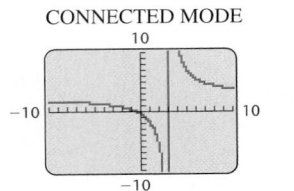 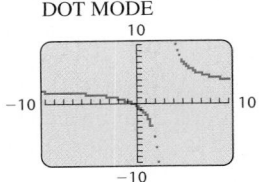

CONNECTED MODE   DOT MODE

87. Using the window $[-5, 5, -1, 9]$, graph $y_1 = 5$, $y_2 = x + 2$, and $y_3 = \sqrt{x}$. Then predict what shape the graphs of $y_1 + y_2$, $y_1 + y_3$, and $y_2 + y_3$ will take. Use a graphing calculator to check each prediction.

88. Use the TABLE feature on a graphing calculator to check your answers to Exercises 45, 47, 55, and 57. (See the Technology Connection on p. 483.)

COLLABORATIVE CORNER

Time On Your Hands

Focus: The algebra of functions

Time: 10–15 minutes

Group size: 2–3

The graph and the data at right chart the average retirement age $R(x)$ and life expectancy $E(x)$ of U.S. citizens in year x.

ACTIVITY

1. Working as a team, perform the appropriate calculations and then graph $E - R$.
2. What does $(E - R)(x)$ represent? In what fields of study or business might the function $E - R$ prove useful?
3. Should E and R really be calculated separately for men and women? Why or why not?

4. What advice would you give to someone considering early retirement?

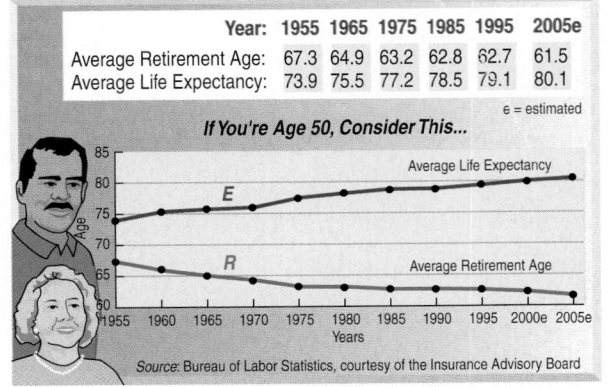

Year:	1955	1965	1975	1985	1995	2005e
Average Retirement Age:	67.3	64.9	63.2	62.8	62.7	61.5
Average Life Expectancy:	73.9	75.5	77.2	78.5	79.1	80.1

e = estimated

If You're Age 50, Consider This...

Source: Bureau of Labor Statistics, courtesy of the Insurance Advisory Board

7.5   Formulas, Applications, and Variation

Formulas • Direct Variation • Inverse Variation •
Joint and Combined Variation

Formulas

Formulas occur frequently as mathematical models. To solve formulas for a
specified letter, we proceed as when solving equations.

EXAMPLE 1   Electronics.   The formula

$$\frac{1}{R} = \frac{1}{r_1} + \frac{1}{r_2}$$

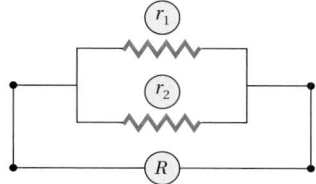

is used by electricians to determine the
resistance R of two resistors r_1 and r_2
connected in parallel.* Solve for r_1.

Solution   We use the same approach as in Section 6.6:

$$Rr_1r_2 \cdot \frac{1}{R} = Rr_1r_2 \cdot \left(\frac{1}{r_1} + \frac{1}{r_2}\right) \qquad \text{Multiplying both sides by the LCD}$$

$$Rr_1r_2 \cdot \frac{1}{R} = Rr_1r_2 \cdot \frac{1}{r_1} + Rr_1r_2 \cdot \frac{1}{r_2} \qquad \text{Multiplying to remove parentheses}$$

$$r_1r_2 = Rr_2 + Rr_1. \qquad \text{Simplifying by removing factors equal to 1: } \frac{R}{R} = 1; \frac{r_1}{r_1} = 1; \frac{r_2}{r_2} = 1$$

At this point it is tempting to multiply by $1/r_2$ to get r_1 alone on the left, *but*
note that there is an r_1 on the right. We must get all the terms involving r_1
on the *same side* of the equation.

$$r_1r_2 - Rr_1 = Rr_2 \qquad \text{Subtracting } Rr_1 \text{ from both sides}$$

$$r_1(r_2 - R) = Rr_2 \qquad \text{Factoring out } r_1 \text{ in order to combine like terms}$$

$$r_1 = \frac{Rr_2}{r_2 - R} \qquad \text{Dividing both sides by } r_2 - R \text{ to get } r_1 \text{ alone}$$

This formula can be used to calculate r_1 whenever R and r_2 are known.

*Recall that the subscripts 1 and 2 merely indicate that r_1 and r_2 are different variables repre-
senting similar quantities.

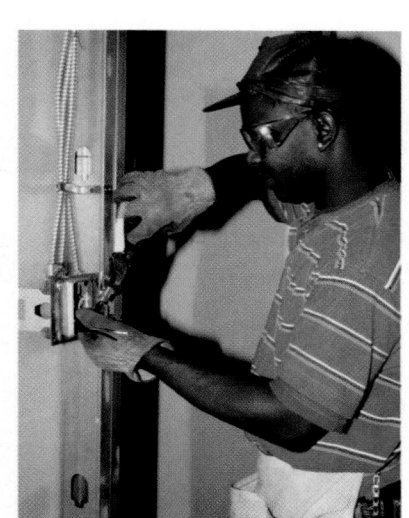

EXAMPLE 2

Astronomy.   The formula

$$\frac{V^2}{R^2} = \frac{2g}{R+h}$$

is used to find a satellite's *escape velocity V*, where R is a planet's radius, h is the satellite's height above the planet, and g is the planet's gravitational constant. Solve for h.

Solution   We first clear fractions by multiplying by the LCD, which is $R^2(R+h)$:

$$\frac{V^2}{R^2} = \frac{2g}{R+h}$$

$$R^2(R+h)\frac{V^2}{R^2} = R^2(R+h)\frac{2g}{R+h}$$

$$\frac{R^2(R+h)V^2}{R^2} = \frac{R^2(R+h)2g}{R+h}$$

$$(R+h)V^2 = R^2 \cdot 2g. \qquad \text{Removing factors equal to 1:}$$
$$\frac{R^2}{R^2} = 1 \text{ and } \frac{R+h}{R+h} = 1$$

Remember: We are solving for h. Although we *could* distribute V^2, since h appears only within the factor $R+h$, it is easier to divide both sides by V^2:

$$\frac{(R+h)V^2}{V^2} = \frac{2R^2g}{V^2} \qquad \text{Dividing both sides by } V^2$$

$$R+h = \frac{2R^2g}{V^2} \qquad \text{Removing a factor equal to 1: } \frac{V^2}{V^2} = 1$$

$$h = \frac{2R^2g}{V^2} - R. \qquad \text{Subtracting } R \text{ from both sides}$$

The last equation can be used to determine the height of a satellite above a planet when the planet's radius and gravitational constant, along with the satellite's escape velocity, are known.

EXAMPLE 3

Acoustics (the Doppler Effect).   The formula

$$f = \frac{sg}{s+v}$$

is used to determine the frequency f of a sound that is moving at velocity v toward a listener who hears the sound as frequency g. Here s is the speed of sound in a particular medium. Solve for s.

Student Notes

The steps used to solve equations are precisely the same steps used to solve formulas. If you feel "rusty" in this regard, study the earlier section in which this type of equation first appeared. Then make sure that you can consistently solve those equations before returning to the work with formulas.

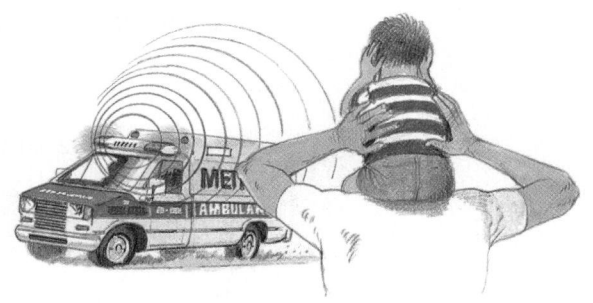

Solution We first clear fractions by multiplying by the LCD, $s + v$:

$$f \cdot (s + v) = \frac{sg}{s + v}(s + v)$$

$$fs + fv = sg. \qquad \text{The variable for which we are solving, } s,$$
$$\text{appears on both sides, forcing us to}$$
$$\text{distribute on the left side.}$$

Next, we must get all terms containing s on one side:

$$fv = sg - fs \qquad \text{Subtracting } fs \text{ from both sides}$$

$$fv = s(g - f) \qquad \text{Factoring out } s. \text{ This is like combining like terms.}$$

$$\frac{fv}{g - f} = s. \qquad \text{Dividing both sides by } g - f$$

Since s is isolated on one side, we have solved for s. This last equation can be used to determine the speed of sound whenever f, v, and g are known.

> **To Solve a Rational Equation for a Specified Variable**
> 1. If necessary, multiply both sides by the LCD to clear fractions.
> 2. Multiply, as needed, to remove parentheses.
> 3. Get all terms with the specified variable alone on one side.
> 4. Factor out the specified variable if it is in more than one term.
> 5. Multiply or divide on both sides to isolate the specified variable.

Variation

To extend our study of formulas and functions, we now examine three real-world situations: direct variation, inverse variation, and combined variation.

Direct Variation

A registered nurse earns $22 per hour. In 1 hr, $22 is earned. In 2 hr, $44 is earned. In 3 hr, $66 is earned, and so on. This gives rise to a set of ordered pairs:

$$(1, 22), (2, 44), (3, 66), (4, 88), \text{ and so on.}$$

Note that the ratio of earnings E to time t is $\frac{22}{1}$ in every case.

If a situation gives rise to pairs of numbers in which the ratio is constant, we say that there is **direct variation**. Here earnings *vary directly* as the time:

We have $\dfrac{E}{t} = 22$, so $E = 22t$ or, using function notation, $E(t) = 22t$.

Direct Variation

When a situation gives rise to a linear function of the form $f(x) = kx$, or $y = kx$, where k is a nonzero constant, we say that there is *direct variation,* that *y varies directly* as *x,* or that *y is proportional to x.* The number k is called the *variation constant,* or *constant of proportionality.*

Note that for $k > 0$, any equation of the form $y = kx$ indicates that as x increases, y increases as well.

EXAMPLE 4   Find the variation constant and an equation of variation if y varies directly as x, and $y = 32$ when $x = 2$.

Solution   We know that $(2, 32)$ is a solution of $y = kx$. Therefore,

$$32 = k \cdot 2 \qquad \text{Substituting}$$

$$\frac{32}{2} = k, \quad \text{or} \quad k = 16. \qquad \text{Solving for } k$$

The variation constant is 16. The equation of variation is $y = 16x$. The notation $y(x) = 16x$ or $f(x) = 16x$ is also used.

EXAMPLE 5   Water from melting snow.   The number of centimeters W of water produced from melting snow varies directly as the number of centimeters S of snow. Meteorologists know that under certain conditions, 150 cm of snow will melt to 16.8 cm of water. The average annual snowfall in Alta, Utah, is 500 in. Assuming the above conditions, how much water will replace the 500 in. of snow?

Alta, Utah

Solution

1. **Familiarize.**   Because of the phrase "W ... varies directly as ... S," we express the amount of water as a function of the amount of snow. Thus, $W(S) = kS$, where k is the variation constant. Knowing that 150 cm of snow becomes 16.8 cm of water, we have $W(150) = 16.8$. Because we are using ratios, it does not matter whether we work in inches or centimeters, provided the same units are used for W and S.

2. **Translate.**   We find the variation constant using the data and then use it to write the equation of variation:

$$W(S) = kS$$
$$W(150) = k \cdot 150 \qquad \text{Replacing } S \text{ with } 150$$
$$16.8 = k \cdot 150 \qquad \text{Replacing } W(150) \text{ with } 16.8$$
$$\frac{16.8}{150} = k \qquad \text{Solving for } k$$
$$0.112 = k. \qquad \text{This is the variation constant.}$$

The equation of variation is $W(S) = 0.112S$. This is the translation.

3. **Carry out.**   To find how much water 500 in. of snow will become, we compute $W(500)$:

$$W(S) = 0.112S$$
$$W(500) = 0.112(500) \qquad \text{Substituting 500 for } S$$
$$W = 56.$$

4. **Check.**   To check, we could reexamine all our calculations. Note that our answer seems reasonable since 500/56 and 150/16.8 are equal.

5. **State.**   Alta's 500 in. of snow will become 56 in. of water.

Inverse Variation

To see what we mean by inverse variation, suppose a bus is traveling 20 mi. At 20 mph, the trip will take 1 hr. At 40 mph, it will take $\frac{1}{2}$ hr. At 60 mph, it will take $\frac{1}{3}$ hr, and so on. This gives rise to pairs of numbers, all having the same product:

$$(20, 1), \left(40, \tfrac{1}{2}\right), \left(60, \tfrac{1}{3}\right), \left(80, \tfrac{1}{4}\right), \quad \text{and so on.}$$

Note that the product of each pair of numbers is 20. Whenever a situation gives rise to pairs of numbers for which the product is constant, we say that there is **inverse variation**. Since $r \cdot t = 20$, the time t, in hours, required for the bus to travel 20 mi at r mph is given by

$$t = \frac{20}{r} \quad \text{or, using function notation,} \quad t(r) = \frac{20}{r}.$$

> **Inverse Variation**
>
> When a situation gives rise to a rational function of the form $f(x) = k/x$, or $y = k/x$, where k is a nonzero constant, we say that there is *inverse variation*, that *y varies inversely as x*, or that *y is inversely proportional to x*. The number k is called the *variation constant*, or *constant of proportionality*.

Note that for $k > 0$, any equation of the form $y = k/x$ indicates that as x increases, y decreases.

EXAMPLE 6   Find the variation constant and an equation of variation if y varies inversely as x, and $y = 32$ when $x = 0.2$.

Solution   We know that (0.2, 32) is a solution of

$$y = \frac{k}{x}.$$

Therefore,

$$32 = \frac{k}{0.2} \qquad \text{Substituting}$$

$$(0.2)32 = k$$

$$6.4 = k. \qquad \text{Solving for } k$$

The variation constant is 6.4. The equation of variation is

$$y = \frac{6.4}{x}.$$

_____

There are many real-life problems that translate to an equation of inverse variation.

EXAMPLE 7   Ultraviolet index.   The ultraviolet, or UV, index is a measure issued daily by the National Weather Service that indicates the strength of the sun's rays in a particular locale. For those people whose skin is quite sensitive, a UV rating of 6 will cause sunburn after 10 min (*Source*: *The Electronic Textbook of Dermatology* found at www.telemedicine.org, January 2004). Given that the number of minutes it takes to burn, t, varies inversely as the UV rating, u, how long will it take a highly sensitive person to burn on a day with a UV rating of 4?

Solution

1. **Familiarize.**   Because of the phrase "… varies inversely as the UV index," we express the amount of time needed to burn as a function of the UV rating: $t(u) = k/u$.

2. **Translate.**   We use the given information to solve for k. Then we use that result to write the equation of variation.

$$t(u) = \frac{k}{u} \qquad \text{Using function notation}$$

$$t(6) = \frac{k}{6} \qquad \text{Replacing } u \text{ with } 6$$

$$10 = \frac{k}{6} \qquad \text{Replacing } t(6) \text{ with } 10$$

$$60 = k \qquad \text{Solving for } k, \text{ the variation constant}$$

The equation of variation is $t(u) = 60/u$. This is the translation.

3. **Carry out.**   To find how long it would take a highly sensitive person to burn on a day with a UV index of 4, we calculate $t(4)$:

$$t(4) = \frac{60}{4} = 15. \qquad t = 15 \text{ when } u = 4$$

4. **Check.**   We could now recheck each step. Note that, as expected, as the UV rating goes *down*, the time it takes to burn goes *up*.

5. **State.**   On a day with a UV rating of 4, a highly sensitive person will begin to burn after 15 min of exposure.

Joint and Combined Variation

When a variable varies directly with more than one other variable, we say that there is *joint variation*. For example, in the formula for the volume of a right circular cylinder, $V = \pi r^2 h$, we say that V varies *jointly* as h and the square of r.

> **Joint Variation**
>
> y varies *jointly* as x and z if, for some nonzero constant k, $y = kxz$.

EXAMPLE 8   Find an equation of variation if y varies jointly as x and z, and $y = 30$ when $x = 2$ and $z = 3$.

Solution   We have

$$y = kxz,$$

so

$$30 = k \cdot 2 \cdot 3 \cdot$$

$$k = 5. \qquad \text{The variation constant is 5.}$$

The equation of variation is $y = 5xz$.

Joint variation is one form of *combined variation*. In general, when a variable varies directly and/or inversely, at the same time, with more than one other variable, there is **combined variation**. Examples 8 and 9 are both examples of combined variation.

EXAMPLE 9   Find an equation of variation if y varies jointly as x and z and inversely as the square of w, and $y = 105$ when $x = 3$, $z = 20$, and $w = 2$.

Solution   The equation of variation is of the form

$$y = k \cdot \frac{xz}{w^2},$$

so, substituting, we have

$$105 = k \cdot \frac{3 \cdot 20}{2^2}$$

$$105 = k \cdot 15$$

$$k = 7.$$

Thus,

$$y = 7 \cdot \frac{xz}{w^2}.$$

Exercise Set

7.5

FOR EXTRA HELP

 Student's Solutions Manual   Digital Video Tutor CD 4 Videotape 7   Tutor Center AW Math Tutor Center    MathXL Tutorials on CD   Math*XL* MathXL   *MyMathLab* MyMathLab

↩ *Concept Reinforcement   Complete each of the following statements.*

1. To solve a formula for a variable appearing in a denominator, we multiply both sides of the equation by the _____.

2. All terms containing the variable being solved for must ultimately be on the _____ side of the equation.

3. If the variable being solved for appears in more than one term on the same side of the equation, it is usually necessary to _____.

4. If y varies directly as x, then $y = kx$ and k is called the _____ of _____.

↩ *Concept Reinforcement   Determine whether each situation represents direct or inverse variation.*

5. Two painters can scrape a house in 9 hr, whereas three painters can scrape the house in 6 hr.

6. Fanny planted 5 bulbs in 20 min and 7 bulbs in 28 min.

7. Glen swam 2 laps in 7 min and 6 laps in 21 min.

8. It took 2 band members 80 min to set up for a show; with 4 members working, it took 40 min.

9. It took 3 hr for 4 volunteers to wrap the campus' collection of Toys for Tots, but only 1.5 hr with 8 volunteers working.

10. Fatuma's air conditioner cooled off 1000 ft^3 in 10 min and 3000 ft^3 in 30 min.

Solve each formula for the specified variable.

11. $f = \dfrac{L}{d}$; d

12. $\dfrac{W_1}{W_2} = \dfrac{d_1}{d_2}$; W_1

13. $s = \dfrac{(v_1 + v_2)t}{2}$; v_1

14. $s = \dfrac{(v_1 + v_2)t}{2}$; t

15. $\dfrac{t}{a} + \dfrac{t}{b} = 1$; b

16. $\dfrac{1}{R} = \dfrac{1}{r_1} + \dfrac{1}{r_2}$; R

17. $I = \dfrac{2V}{R + 2r}$; R

18. $I = \dfrac{2V}{R + 2r}$; r

19. $R = \dfrac{gs}{g + s}$; g

20. $K = \dfrac{rt}{r - t}$; t

21. $I = \dfrac{nE}{R + nr}$; n

22. $I = \dfrac{nE}{R + nr}$; r

23. $\dfrac{1}{p} + \dfrac{1}{q} = \dfrac{1}{f}$; q

24. $\dfrac{1}{p} + \dfrac{1}{q} = \dfrac{1}{f}$; p

25. $S = \dfrac{H}{m(t_1 - t_2)}$; t_1

26. $S = \dfrac{H}{m(t_1 - t_2)}$; H

27. $\dfrac{E}{e} = \dfrac{R + r}{r}$; r

28. $\dfrac{E}{e} = \dfrac{R + r}{R}$; R

29. $S = \dfrac{a}{1 - r}$; r

30. $S = \dfrac{a - ar^n}{1 - r}$; a

Aha! **31.** $c = \dfrac{f}{(a + b)c}$; $a + b$

32. $d = \dfrac{g}{d(c + f)}$; $c + f$

33. *Interest.*   The formula

$$P = \frac{A}{1 + r}$$

is used to determine what principal P should be invested for one year at $(100 \cdot r)\%$ simple interest in order to have A dollars after a year. Solve for r.

34. *Taxable interest.*   The formula

$$I_t = \frac{I_f}{1 - T}$$

gives the *taxable interest rate* I_t equivalent to the *tax-free interest rate* I_f for a person in the $(100 \cdot T)\%$ tax bracket. Solve for T.

35. *Average speed.*   The formula

$$v = \frac{d_2 - d_1}{t_2 - t_1}$$

gives an object's average speed v when that object has traveled d_1 miles in t_1 hours and d_2 miles in t_2 hours. Solve for t_2.

36. *Average acceleration.*   The formula

$$a = \frac{v_2 - v_1}{t_2 - t_1}$$

gives a vehicle's *average acceleration* when its velocity changes from v_1 at time t_1 to v_2 at time t_2. Solve for t_1.

37. *Planetary orbits.*   The formula

$$\frac{x^2}{a^2} + \frac{y^2}{b^2} = 1$$

can be used to plot a planet's elliptical orbit of width $2a$ and length $2b$ (see p. 875 in Section 13.2). Solve for b^2.

38. *Work rate.*   The formula

$$\frac{1}{t} = \frac{1}{a} + \frac{1}{b}$$

gives the total time t required for two workers to complete a job, if the workers' individual times are a and b. Solve for t.

39. *Semester average.*   The formula

$$A = \frac{2Tt + Qq}{2T + Q}$$

gives a student's average A after T tests and Q quizzes, where each test counts as 2 quizzes, t is the test average, and q is the quiz average. Solve for Q.

40. *Astronomy.*   The formula

$$L = \frac{dR}{D - d},$$

where D is the diameter of the sun, d is the diameter of the earth, R is the earth's distance from the sun, and L is some fixed distance, is used in calculating when lunar eclipses occur. Solve for D.

Find the variation constant and an equation of variation if y varies directly as x and the following conditions apply.

41. $y = 28$ when $x = 4$

42. $y = 5$ when $x = 12$

43. $y = 3.4$ when $x = 2$

44. $y = 2$ when $x = 5$

45. $y = 2$ when $x = \frac{1}{3}$

46. $y = 0.9$ when $x = 0.5$

Find the variation constant and an equation of variation in which y varies inversely as x, and the following conditions exist.

47. $y = 3$ when $x = 20$

48. $y = 16$ when $x = 4$

49. $y = 28$ when $x = 4$

50. $y = 9$ when $x = 5$

51. $y = 27$ when $x = \frac{1}{3}$

52. $y = 81$ when $x = \frac{1}{9}$

53. *Use of aluminum cans.* The number N of aluminum cans used each year varies directly as the number of people using the cans. If 250 people use 60,000 cans in one year, how many cans are used each year in Dallas, which has a population of 1,008,000?

54. *Hooke's law.* Hooke's law states that the distance d that a spring is stretched by a hanging object varies directly as the mass m of the object. If the distance is 20 cm when the mass is 3 kg, what is the distance when the mass is 5 kg?

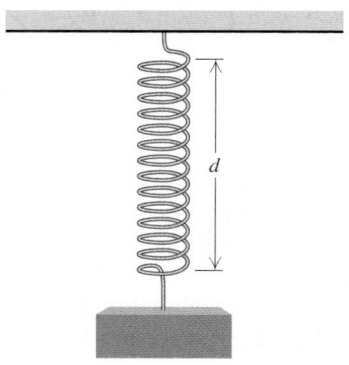

55. *Ohm's law.* The electric current I, in amperes, in a circuit varies directly as the voltage V. When 15 volts are applied, the current is 5 amperes. What is the current when 18 volts are applied?

56. *Pumping rate.* The time t required to empty a tank varies inversely as the rate r of pumping. If a Briggs and Stratton pump can empty a tank in 45 min at the rate of 600 kL/min, how long will it take the pump to empty the tank at 1000 kL/min?

57. *Work rate.* The time T required to do a job varies inversely as the number of people P working. It takes 5 hr for 7 volunteers to pick up rubbish from 1 mi of roadway. How long would it take 10 volunteers to complete the job?

58. *Weekly allowance.* According to Fidelity Investments *Investment Vision Magazine*, the average weekly allowance A of children varies directly as their grade level, G. In a recent year, the average allowance of a 9th-grade student was $13.50 per week. What was the average weekly allowance of a 4th-grade student?
Source: Based on data from www.kidsmoney.org

Aha! **59.** *Mass of water in a human.* The number of kilograms W of water in a human body varies directly as the mass of the body. A 96-kg person contains 64 kg of water. How many kilograms of water are in a 48-kg person?

60. *Weight on Mars.* The weight M of an object on Mars varies directly as its weight E on Earth. A person who weighs 95 lb on Earth weighs 38 lb on Mars. How much would a 100-lb person weigh on Mars?

61. *Bicycling.* The number of calories burned by a bicyclist is directly proportional to the time spent bicycling. At 10 mph, it takes 30 min to burn 150 calories. How long would it take to burn 250 calories when biking 10 mph?
Source: Physical Activity and Health: A Report of the Surgeon General

62. *Wavelength and frequency.* The wavelength W of a radio wave varies inversely as its frequency F. A wave with a frequency of 1200 kilohertz has a length of 300 meters. What is the length of a wave with a frequency of 800 kilohertz?

63. *Ultraviolet index.*   At an ultraviolet, or UV, rating of 4, those people who are less sensitive to the sun will burn in 75 min. Given that the number of minutes it takes to burn, t, varies inversely with the UV rating, u, how long will it take less sensitive people to burn when the UV rating is 14?
Source: *The Electronic Textbook of Dermatology* at www.telemedicine.org

64. *Current and resistance.*   The current I in an electrical conductor varies inversely as the resistance R of the conductor. If the current is $\frac{1}{2}$ ampere when the resistance is 240 ohms, what is the current when the resistance is 540 ohms?

65. *Air pollution.*   The average U.S. household of 2.6 people released 1.1 tons of carbon monoxide into the environment in a recent year. How many tons were released nationally? Use 289,000,000 as the U.S. population.
Sources: Based on data from the U.S. Environmental Protection Agency and the U.S. Bureau of the Census

66. *Relative aperture.*   The relative aperture, or f-stop, of a 23.5-mm lens is directly proportional to the focal length F of the lens. If a lens with a 150-mm focal length has an f-stop of 6.3, find the f-stop of a 23.5-mm lens with a focal length of 80 mm.

Find an equation of variation in which:

67. y varies directly as the square of x, and $y = 6$ when $x = 3$.

68. y varies directly as the square of x, and $y = 0.15$ when $x = 0.1$.

69. y varies inversely as the square of x, and $y = 6$ when $x = 3$.

70. y varies inversely as the square of x, and $y = 0.15$ when $x = 0.1$.

71. y varies jointly as x and the square of z, and $y = 105$ when $x = 14$ and $z = 5$.

72. y varies jointly as x and z and inversely as w, and $y = \frac{3}{2}$ when $x = 2$, $z = 3$, and $w = 4$.

73. y varies jointly as w and the square of x and inversely as z, and $y = 49$ when $w = 3$, $x = 7$, and $z = 12$.

74. y varies directly as x and inversely as w and the square of z, and $y = 4.5$ when $x = 15$, $w = 5$, and $z = 2$.

75. *Electrical safety.*   The amount of time t needed for an electrical shock to stop a 150-lb person's heart from beating varies inversely as the square of the current flowing through the body. It is known that a 0.089-amp current is deadly to a 150-lb person after 3.4 sec. How long would it take a 0.096-amp current to be deadly?
Source: Safety Consulting Services

76. *Stopping distance of a car.*   The stopping distance d of a car after the brakes have been applied varies directly as the square of the speed r. Once the brakes are applied, a car traveling 60 mph can stop in 138 ft. What stopping distance corresponds to a speed of 40 mph?
Source: Based on data from Edmunds.com

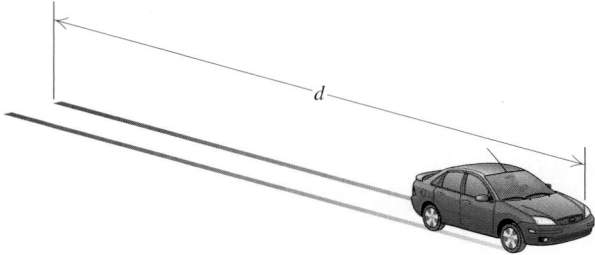

77. *Volume of a gas.*   The volume V of a given mass of a gas varies directly as the temperature T and inversely as the pressure P. If $V = 231$ cm^3 when $T = 300°$K (Kelvin) and $P = 20$ lb/cm^2, what is the volume when $T = 320°$K and $P = 16$ lb/cm^2?

78. *Intensity of a signal.* The intensity I of a television signal varies inversely as the square of the distance d from the transmitter. If the intensity is 25 W/m^2 at a distance of 2 km, what is the intensity 6.25 km from the transmitter?

79. *Atmospheric drag.* Wind resistance, or atmospheric drag, tends to slow down moving objects. Atmospheric drag W varies jointly as an object's surface area A and velocity v. If a car traveling at a speed of 40 mph with a surface area of 37.8 ft^2 experiences a drag of 222 N (Newtons), how fast must a car with 51 ft^2 of surface area travel in order to experience a drag force of 430 N?

80. *Drag force.* The drag force F on a boat varies jointly as the wetted surface area A and the square of the velocity of the boat. If a boat traveling 6.5 mph experiences a drag force of 86 N when the wetted surface area is 41.2 ft^2, find the wetted surface area of a boat traveling 8.2 mph with a drag force of 94 N.

81. Which exercise did you find easier to work: Exercise 13 or Exercise 19? Why?

82. If y varies directly as x, does doubling x cause y to be doubled as well? Why or why not?

SKILL MAINTENANCE

Solve. Label any contradictions or identities.

83. $2x - 5 = 8$ [2.2]

84. $3a + 2 = 7$ [2.2]

85. $\dfrac{1}{x + 1} = \dfrac{3}{x}$ [6.6]

86. $6x^2 = 11x + 35$ [5.7]

87. $3a + 1 = 3(a + 1)$ [2.2]

88. $a + (a - 5) = 2a - 5$ [2.2]

SYNTHESIS

89. Suppose that the number of customer complaints is inversely proportional to the number of employees hired. Will a firm reduce the number of complaints more by expanding from 5 to 10 employees, or from 20 to 25? Explain. Consider using a graph to help justify your answer.

90. Why do you think subscripts are used in Exercises 13 and 25 but not in Exercises 27 and 28?

91. *Escape velocity.* A satellite's escape velocity is 6.5 mi/sec, the radius of the earth is 3960 mi, and the earth's gravitational constant is 32.2 ft/sec^2. How far is the satellite from the surface of the earth? (See Example 2.)

92. The *harmonic mean* of two numbers a and b is a number M such that the reciprocal of M is the average of the reciprocals of a and b. Find a formula for the harmonic mean.

93. *Health-care.* Young's rule for determining the size of a particular child's medicine dosage c is

$$c = \frac{a}{a + 12} \cdot d,$$

where a is the child's age and d is the typical adult dosage. If a child's age is doubled, the dosage increases. Find the ratio of the larger dosage to the smaller dosage. By what percent does the dosage increase?

Source: Olsen, June Looby, Leon J. Ablon, and Anthony Patrick Giangrasso, *Medical Dosage Calculations,* 6th ed.

94. Solve for x:

$$x^2\left(1 - \frac{2pq}{x}\right) = \frac{2p^2q^3 - pq^2x}{-q}.$$

95. *Average acceleration.* The formula

$$a = \frac{\dfrac{d_4 - d_3}{t_4 - t_3} - \dfrac{d_2 - d_1}{t_2 - t_1}}{t_4 - t_2}$$

can be used to approximate average acceleration, where the d's are distances and the t's are the corresponding times. Solve for t_1.

96. If y varies inversely as the cube of x and x is multiplied by 0.5, what is the effect on y?

97. *Intensity of light.* The intensity I of light from a bulb varies directly as the wattage of the bulb and inversely as the square of the distance d from the bulb. If the wattage of a light source and its distance from reading matter are both doubled, how does the intensity change?

98. Describe in words the variation represented by $W = \dfrac{km_1M_1}{d^2}$. Assume k is a constant.

99. *Tension of a musical string.* The tension T on a string in a musical instrument varies jointly as the string's mass per unit length m, the square of its length l, and the square of its fundamental frequency f. A 2-m long string of mass 5 gm/m with a fundamental frequency of 80 has a tension of 100 N. How long should the same string be if its tension is going to be changed to 72 N?

100. *Volume and cost.* A peanut butter jar in the shape of a right circular cylinder is 4 in. high and 3 in. in diameter and sells for $1.20. If we assume that cost is proportional to volume, how much should a jar 6 in. high and 6 in. in diameter cost?

101. *Golf distance finder.* A device used in golf to estimate the distance d to a hole measures the size s that the 7-ft pin *appears* to be in a viewfinder. The viewfinder uses the principle, diagrammed here, that s gets bigger when d gets smaller. If $s = 0.56$ in. when $d = 50$ yd, find an equation of variation that expresses d as a function of s. What is d when $s = 0.40$ in.?

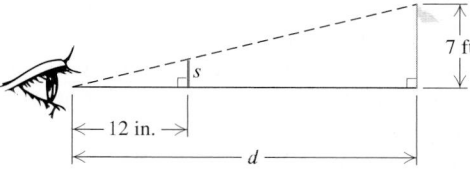

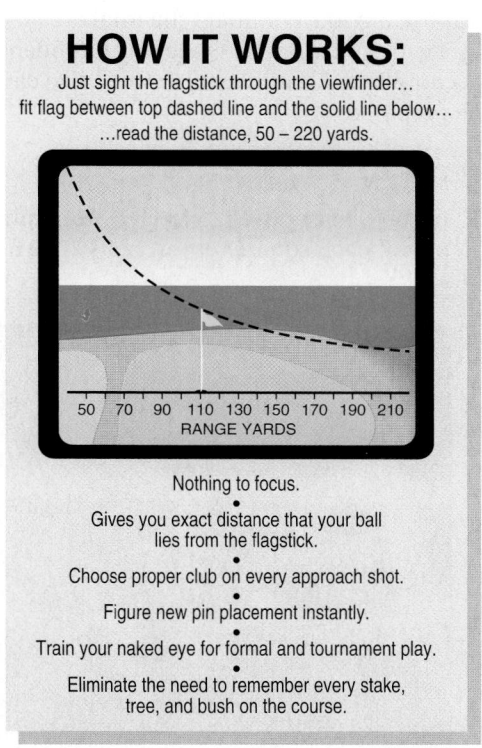

HOW IT WORKS:
Just sight the flagstick through the viewfinder...
fit flag between top dashed line and the solid line below...
...read the distance, 50 – 220 yards.

50 70 90 110 130 150 170 190 210
RANGE YARDS

Nothing to focus.
•
Gives you exact distance that your ball lies from the flagstick.
•
Choose proper club on every approach shot.
•
Figure new pin placement instantly.
•
Train your naked eye for formal and tournament play.
•
Eliminate the need to remember every stake, tree, and bush on the course.

CORNER

COLLABORATIVE

How Many Is a Million?

Focus: Direct variation and estimation

Time: 15 minutes

Group size: 2 or 3 and entire class

The National Park Service's estimates of crowd sizes for static (stationary) mass demonstrations vary directly as the area covered by the crowd. Park Service officials have found that at basic "shoulder-to-shoulder" demonstrations, 1 acre of land (about 45,000 ft^2) holds about 9000 people. Using aerial photographs, officials impose a grid to estimate the total area covered by the demonstrators. Once this has been accomplished, estimates of crowd size can be prepared.

ACTIVITY

1. In the grid imposed on the photograph below, each square represents 10,000 ft^2.

Estimate the size of the crowd photographed. Then compare your group's estimate with those of other groups. What might explain discrepancies between estimates? List ways in which your group's estimate could be made more accurate.

2. Park Service officials use an "acceptable margin of error" of no more than 20%. Using all estimates from part (1) above and allowing for error, find a range of values within which you feel certain that the actual crowd size lies.

3. The Million Man March of 1995 was not a static demonstration because of a periodic turnover of people in attendance (many people stayed for only part of the day's festivities). How might you change your methodology to compensate for this complication?

Study Summary

A **function** (p. 443) is a set of ordered pairs for which no two different pairs have the same first coordinate. The set of all of a function's first coordinates is called the **domain** of that function (p. 442). The set of all of a function's second coordinates is called the **range** of the function (p. 442). When an equation of a function has the variable representing the second coordinate alone on one side of an equation, it appears as the **dependent variable** (p. 446). This means that its value is calculated after a choice has been made for the replacement of the other, **independent**, **variable** (p. 446). Numbers replacing the independent variable are often called **inputs** and the resulting values of the dependent variable are often called **outputs** (p. 445).

One of the reasons functions are important in mathematics is the simplicity of the notation. For example, the notation

$$f(3) = 5 \qquad \text{This is read "} f \text{ of 3 is 5."}$$

means that the value of the function f, or its output, is 5 when the input is 3. The notation

$$f(x) = x^2 + 1$$

means that every input x is associated with an output found by squaring x and adding 1. Here x is the independent variable.

A function can be classified by the type of equation used to define the function (p. 472).

Type of function	Example
Constant function	$f(x) = -6$
Linear function	$f(x) = \frac{2}{3}x - 5$
Absolute-value function	$g(x) = |x - 7|$
Quadratic function	$f(t) = t^2 + 2t + 1$
Polynomial function	$f(x) = x^3 - 7x + 3$
Rational function	$r(x) = \dfrac{x^2 - 7}{2x + 5}$

Graphs provide a useful way to represent functions. The **vertical-line test** (p. 464) provides a quick way to determine whether a graph represents a function: If no vertical line crosses the graph more than once, then the graph is a function. The domain of a function can be viewed as its **projection** (p. 456) on the x-axis, and the range as its projection on the y-axis.

Functions can be added, subtracted, multiplied, and divided as follows:

1. $(f + g)(x) = f(x) + g(x)$
2. $(f - g)(x) = f(x) - g(x)$
3. $(f \cdot g)(x) = f(x) \cdot g(x)$
4. $(f/g)(x) = f(x)/g(x)$, provided $g(x) \neq 0$ (p. 479).

The domain of $f + g, f - g$, and $f \cdot g$ is the set of all values common to the domains of f and g (p. 482).

The domain of f/g is the set of all values common to the domains of f and g, excluding any values for which $g(x)$ is 0 (p. 482).

Functions can be used to describe equations of variation. **Direct variation** produces equations of the form $y = kx$, and **inverse variation** produces equations of the form $y = k/x$ (pp. 490, 492). In both cases, k is the **constant of proportionality**.

Direct variation: $y = kx$;   for $k > 0$, as x increases, y increases.

Inverse variation: $y = k/x$;   for $k > 0$, as x increases, y decreases.

In **combined** and **joint variation**, one variable depends on at least two other variables (p. 493).

7 Review Exercises

⮕ *Concept Reinforcement*   *Classify each of the following as either true or false.*

1. Every function is a relation. [7.1]

2. When we are discussing functions, the notation $f(3)$ does not mean $f \cdot 3$. [7.1]

3. If a graph includes both (9, 5) and (7, 5), it cannot represent a function. [7.1]

4. The domain and the range of a function can be the same set of numbers. [7.2]

5. The horizontal-line test is a quick way to determine whether a graph represents a function. [7.3]

6. In a piecewise-defined function, the function values are determined using more than one rule. [7.2]

7. $(f + g)(x) = f(x) + g(x)$ is not an example of the distributive law when f and g are functions. [7.4]

8. In order for $(f/g)(a)$ to exist, we must have $g(a) \neq 0$. [7.4]

9. If x varies inversely as y, then there exists some constant k for which $x = k/y$. [7.5]

10. If 2 people can decorate for a party in 5 hr, and 10 people can decorate for the same party in 4 hr, the situation represents inverse variation. [7.5]

11. For the following graph of f, determine **(a)** $f(2)$; **(b)** the domain of f; **(c)** any x-values for which $f(x) = 2$; and **(d)** the range of f. [7.1], [7.2]

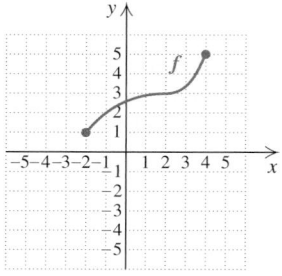

12. Find $g(-3)$ for $g(x) = \dfrac{x}{x+1}$. [7.1]

13. Find $f(2a)$ for $f(x) = x^2 + 2x - 3$. [7.1]

14. The function $A(t) = 0.233t + 5.87$ can be used to estimate the median age of cars in the United States t years after 1990. (In this context, a median age of 3 yr means that half the cars are more than 3 yr old and half are less.) Predict the median age of cars in 2010; that is, find $A(20)$. [7.1]
Source: The Polk Co.

The following table shows the U.S. minimum hourly wage.

Input, Year	Output, U.S. Minimum Hourly Wage
1955	$0.75
1980	3.10
2004	5.15

15. Use the data in the table to draw a graph and to estimate the U.S. minimum hourly wage in 1985. [7.1]

16. Use the graph from Exercise 15 to estimate the U.S. hourly minimum wage in 2009. [7.1]

For each of the graphs in Exercises 17–20, (a) determine whether the graph represents a function and (b) if so, determine the domain and the range of the function.

17.

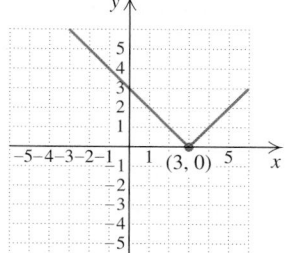

18.

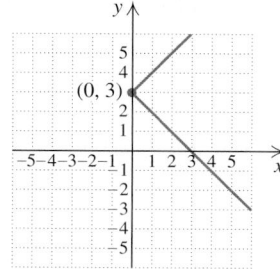

19.

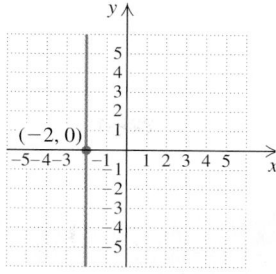

20.

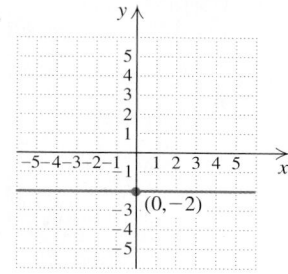

Find the domain of each function.

21. $f(x) = 3x^2 - 7$ [7.2]

22. $g(x) = \dfrac{x^2}{x-1}$ [7.2]

23. $f(t) = \dfrac{1}{t^2 + 5t + 4}$ [7.2]

24. Each salesperson at Knobby's Furniture is paid $P(x)$ dollars, where $P(x) = 0.05x + 200$ and x is the value of the salesperson's sales for the week. What is the domain of the function? [7.2]

25. If a used model 8T lawnmower is returned after purchase, the amount that Turner's Hardware will refund is given by the function

$$r(t) = 850 - 25t,$$

where t is the number of weeks since the date of purchase. What is the domain of the function? [7.2]

26. For the function given by

$$f(x) = \begin{cases} 2 - x, & \text{for } x \le -2, \\ x^2, & \text{for } -2 < x \le 5, \\ x + 10, & \text{for } x > 5, \end{cases}$$

find **(a)** $f(-3)$; **(b)** $f(-2)$; **(c)** $f(4)$; and **(d)** $f(25)$. [7.2]

27. It costs $75 plus $15 a month to join the Family Fitness Center. Formulate a linear function to model the cost for t months of membership, and determine the time required for the cost to reach $180. [7.3]

28. There were 48.5 million international visitors to the United States in 1999 and 41.9 million in 2002. Let v represent the number of international visitors, in millions, t years after 1999. [7.3]
   a) Find a linear function $v(t)$ that fits the data.
   b) Use the function of part (a) to predict the number of visitors in 2001.
   Source: Tourism Industries, International Trade Administration, U.S. Department of Commerce

Classify each function as a linear function, an absolute-value function, a quadratic function, another polynomial function, or a rational function. [7.3]

29. $f(x) = |3x - 7|$

30. $g(x) = 4x^5 - 8x^3 + 7$

31. $p(x) = x^2 + x - 10$

32. $h(n) = 4n - 17$

33. $s(t) = \dfrac{t + 1}{t + 2}$

Graph each function and determine its domain and range. [7.3]

34. $f(x) = 3$

35. $f(x) = 2x + 1$

36. $g(x) = |x + 1|$

Let $g(x) = 3x - 6$ and $h(x) = x^2 + 1$. Find the following.

37. $(g \cdot h)(4)$ [7.4]

38. $(g - h)(-2)$ [7.4]

39. $(g/h)(-1)$ [7.4]

40. The domains of $g + h$ and $g \cdot h$ [7.4]

41. The domain of h/g [7.4]

42. Find an equation of variation in which y varies directly as x, and $y = 30$ when $x = 4$. [7.5]

43. Find an equation of variation in which y varies inversely as x, and $y = 3$ when $x = \frac{1}{4}$. [7.5]

44. Find an equation of variation in which y varies jointly as x and the square of w and inversely as z, and $y = 150$ when $x = 6$, $w = 10$, and $z = 2$. [7.5]

45. The amount of waste generated by a family varies directly as the number of people in the family. The average U.S. family has 3.14 people and generates 13.8 lb of waste daily. How many pounds of waste would be generated daily by a family of 5? **Sources**: Based on data from the U.S. Bureau of the Census and the U.S. Statistical Abstract 2003 [7.5]

46. A warning dye is used by people in lifeboats to aid searching airplanes. The radius r of the circle formed by the dye varies directly as the square root of the volume V. It is found that 4 L of dye will spread to a circle of radius 5 m. How much dye is needed to form a circle with a 20-m radius? [7.5]

SYNTHESIS

47. If two functions have the same domain and range, are the functions identical? Why or why not? [7.2]

48. Jenna believes that 0 is never in the domain of a rational function. Is she correct? Why or why not? [7.2]

49. Homespun Jellies charges $2.49 for each jar of preserves. Shipping charges are $3.75 for handling, plus $0.60 per jar. Find a linear function for determining the cost of purchasing and shipping x jars of preserves. [7.3]

50. Determine the domain and the range of the function graphed below. [7.2]

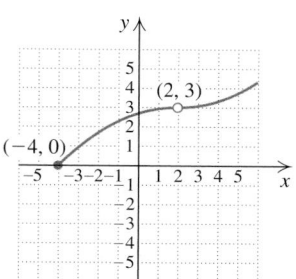

51. Rosewood Graphics recently promised its employees 6% raises each year for the next five years. Amy currently earns $30,000 a year. Can she use a linear function to predict her salary for the next five years? Why or why not? [7.3]

7 Chapter Test

1. For the following graph of f, determine **(a)** $f(-2)$; **(b)** the domain of f; **(c)** any x-value for which $f(x) = \frac{1}{2}$; and **(d)** the range of f.

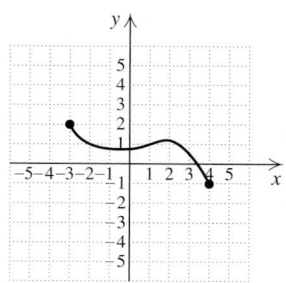

2. Find $f(-2)$ for $f(x) = \dfrac{x}{1 + x}$.

3. There were 43.3 million international visitors to the United States in 1995, and 48.5 million in 1999. Use a graph to estimate the number of international visitors in 1997.
Source: Tourism Industries, International Trade Administration, U.S. Department of Commerce

For each of the following graphs, **(a)** *determine whether the graph represents a function and* **(b)** *if so, determine the domain and the range of the function.*

4.

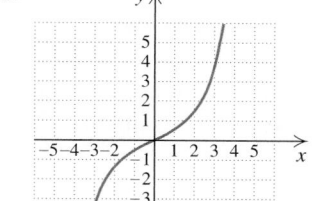

5.

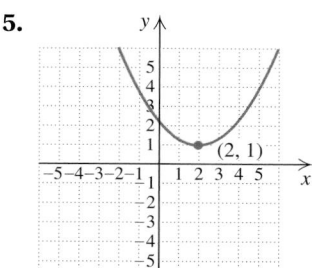

6.

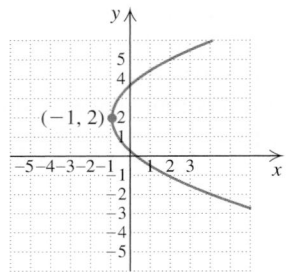

7. The distance d, in miles, that Kerry is from Chicago is given by the function $d(t) = 240 - 60t$, where t is the number of hours since he left Indianapolis. What is the domain of the function?

8. For the function given by
$$f(x) = \begin{cases} x^2, & \text{for } x < 0, \\ 3x - 5, & \text{for } 0 \le x \le 2, \\ x + 7, & \text{for } x > 2, \end{cases}$$
find **(a)** $f(0)$; **(b)** $f(3)$.

9. For a catered party, Jennette's Catering charges $25 per person plus a $75 setup fee. Formulate a linear function to model the cost of a party for n people, and determine the number of people in a party that costs $500.

10. Jon rented a van for one day and drove it 250 mi at a cost of $100. Josh rented the same van for one day and drove it 300 mi at a cost of $115. Let $C(m)$ represent the cost, in dollars, of driving m miles.

a) Find a linear function that fits the data.
b) Use the function to find how much it will cost to rent the van for one day and drive it 500 mi.

Classify each function as a linear function, a quadratic function, another polynomial function, an absolute-value function, or a rational function. Then find the domain of each function.

11. $f(x) = \dfrac{1}{4}x + 7$

12. $g(x) = \dfrac{3}{x - 5}$

13. $p(x) = 4x^2 + 7$

Graph each function and determine its domain and range.

14. $f(x) = \dfrac{1}{3}x - 2$ **15.** $g(x) = x^2 - 1$

16. $h(x) = -\dfrac{1}{2}$

Let $g(x) = \dfrac{1}{x + 4}$ and $h(x) = x - 7$. Find the following.

17. $g(0)$ **18.** $h(-3)$

19. $g(2a)$ **20.** $(g \cdot h)(-1)$

21. The domain of g

22. The domain of h

23. The domain of $g + h$

24. The domain of g/h

25. Find an equation of variation in which y varies directly as x, and $y = 10$ when $x = 20$.

26. The number of workers n needed to clean a stadium after a game varies inversely as the amount of time t allowed for the cleanup. If it takes 25 workers to clean the stadium when there are 6 hr allowed for the job, how many workers are needed if the stadium must be cleaned in 5 hr?

27. The surface area of a balloon varies directly as the square of its radius. The area is 3.4 in^2 when the radius is 5 in. What is the area when the radius is 7 in.?

SYNTHESIS

28. The function $f(t) = 5 + 15t$ can be used to determine a bicycle racer's location, in miles from the starting line, measured t hours after passing the 5-mi mark.

a) How far from the start will the racer be 1 hr and 40 min after passing the 5-mi mark?
b) Assuming a constant rate, how fast is the racer traveling?

29. Given that $f(x) = 5x^2 + 1$ and $g(x) = 4x - 3$, find an expression for $h(x)$ so that the domain of $f/g/h$ is $\left\{x \mid x \text{ is a real number } and \; x \ne \frac{3}{4} \text{ and } x \ne \frac{2}{7}\right\}$.
Answers may vary.

8 Systems of Linear Equations and Problem Solving

AN APPLICATION

King Street Printing recently charged 1.9¢ per sheet of paper, but 2.4¢ per sheet for paper made of recycled fibers. Darren's bill for 150 sheets of paper was $3.41. How many sheets of each type were used?

This problem appears as Exercise 15 in Section 8.3.

Nancy Plunkett
WASTE REDUCTION MANAGER
Williston, Vermont

I regularly use algebra whenever I create spreadsheets to track the costs and performance of recycling and composting programs and to compile survey data. I also use it to convert volumes of different materials to weight and to calculate rates of change in program participation.

*T*he most difficult part of problem solving is almost always translating the problem situation to mathematical language. Once a problem has been translated, the rest is usually straightforward. In this chapter, we study systems of equations *and how to solve them using graphing, substitution, elimination, and matrices. Systems of equations often provide the easiest way to model real-world situations in fields such as psychology, sociology, business, education, engineering, and science.*

8.1	Systems of Equations in Two Variables

Translating   •   Identifying Solutions   •
Solving Systems Graphically

Translating

Problems involving two unknown quantities are often solved most easily if we can first translate the situation to two equations in two unknowns.

EXAMPLE 1

Bottled-water consumption.   Americans are buying more bottled water than ever before. At the time of this writing, the average American buys more bottled water than milk, coffee, or beer. In 2003, the average American purchased 76.4 gal of soft drinks and water. The amount of water was only 4.3 gal less than half of the amount of soft drinks. (*Source*: Based on data from www.beveragemarketing.com) How many gallons of water and how many of soft drinks did the average American buy in 2003?

Solution

1. **Familiarize.** We have already seen problems in which we need to look up certain formulas or the meaning of certain words. Here we simply observe that the words *per person* mean the same thing as the amount consumed by the *average American*.

   Often problems contain information that has no bearing on the situation being discussed. In this case, the fact that more water is bought than milk, coffee, or beer is irrelevant to the question being asked. Instead we focus on the amounts of water and soft drinks consumed. Rather than guess and check, let's proceed to the next step, using *w* to represent the average number of gallons of bottled water purchased and *s* for the average number of gallons of soft drinks purchased in 2003.

2. Translate.   There are two statements to translate. First we look at the total number of gallons purchased by the average American:

Rewording: The amount of water purchased   plus   the amount of soft drinks purchased   was   76.4.

Translating: w $+$ s $=$ 76.4

The second statement compares the two amounts, w and s:

Rewording: The amount of water purchased   was   4.3 gal less than half the amount of soft drinks purchased.

Translating: w $=$ $\dfrac{1}{2}s - 4.3$

We have now translated the problem to a pair, or **system, of equations**:

$$w + s = 76.4,$$
$$w = \frac{1}{2}s - 4.3.$$

> **System of Equations**
>
> A *system of equations* is a set of two or more equations, in two or more variables, for which a common solution is sought.

Problems like Example 1 *can* be solved using one variable; however, as problems become complicated, you will find that using more than one variable (and more than one equation) is often the preferable approach.

EXAMPLE 2   Purchasing.   Recently the Woods County Art Center purchased 120 stamps for $33.90. If the stamps were a combination of 23¢ postcard stamps and 37¢ first-class stamps, how many of each type were bought?

Solution

1. Familiarize.   To familiarize ourselves with this problem, let's guess that the art center bought 60 stamps at 23¢ each and 60 stamps at 37¢ each. The total cost would then be

$$60 \cdot \$0.23 + 60 \cdot \$0.37 = \$13.80 + \$22.20, \text{ or } \$36.00.$$

Since $\$36.00 \neq \$33.90$, our guess is incorrect. Rather than guess again, let's see how algebra can be used to translate the problem.

2. Translate. We let p = the number of postcard stamps and f = the number of first-class stamps. The information can be organized in a table, which will help with the translating.

Type of Stamp	Postcard	First-class	Total	
Number Sold	p	f	120	→ $p + f = 120$
Price	$0.23	$0.37		
Amount	$0.23p	$0.37f	$33.90	→ $0.23p + 0.37f = 33.90$

The first row of the table and the first sentence of the problem indicate that a total of 120 stamps were bought:

$$p + f = 120.$$

Since each postcard stamp cost \$0.23 and p stamps were bought, $0.23p$ represents the amount paid, in dollars, for the postcard stamps. Similarly, $0.37f$ represents the amount paid, in dollars, for the first-class stamps. This leads to a second equation:

$$0.23p + 0.37f = 33.90.$$

Multiplying both sides by 100, we can clear the decimals. This gives the following system of equations as the translation:

$$p + f = 120,$$
$$23p + 37f = 3390.$$

We will complete the solutions of Examples 1 and 2 in Section 8.3.

Identifying Solutions

A *solution* of a system of two equations in two variables is an ordered pair of numbers that makes *both* equations true.

EXAMPLE 3   Determine whether $(-4, 7)$ is a solution of the system

$$x + y = 3,$$
$$5x - y = -27.$$

Solution   As discussed in Chapter 3, unless stated otherwise, we use alphabetical order of the variables. Thus we replace x with -4 and y with 7:

$$\frac{x + y = 3}{-4 + 7 \mid 3}$$
$$3 \overset{?}{=} 3 \quad \text{TRUE}$$

$$\frac{5x - y = -27}{5(-4) - 7 \mid -27}$$
$$-20 - 7 \mid$$
$$-27 \overset{?}{=} -27 \quad \text{TRUE}$$

The pair $(-4, 7)$ makes both equations true, so it is a solution of the system. We can also describe the solution by writing $x = -4$ and $y = 7$. Set notation can also be used to list the solution set $\{(-4, 7)\}$.

Solving Systems Graphically

Recall that the graph of an equation is a drawing that represents its solution set. If we graph the equations in Example 3, we find that $(-4, 7)$ is the only point common to both lines. Thus one way to solve a system of two equations is to graph both equations and identify any points of intersection. The coordinates of each point of intersection represent a solution of that system.

$$x + y = 3,$$
$$5x - y = -27$$

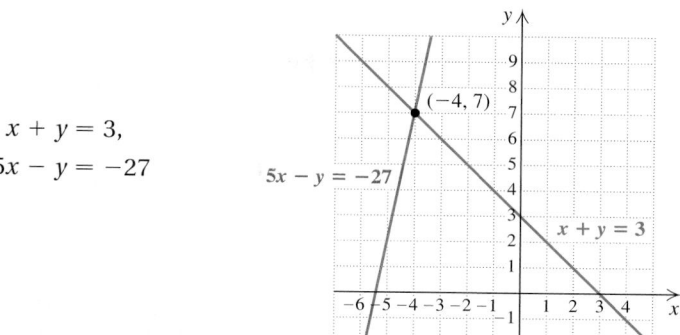

Most pairs of lines have exactly one point in common. We will soon see, however, that this is not always the case.

EXAMPLE 4 Solve each system graphically.

a) $y - x = 1,$
       $y + x = 3$

b) $y = -3x + 5,$
       $y = -3x - 2$

c) $3y - 2x = 6,$
       $-12y + 8x = -24$

Solution

a) We graph each equation using any method studied in Chapter 3. All ordered pairs from line L_1 are solutions of the first equation. All ordered pairs from line L_2 are solutions of the second equation. The point of intersection has coordinates that make *both* equations true. Apparently, $(1, 2)$ is the solution. Graphs are not always accurate, so solving by graphing may yield approximate answers. Our check below shows that $(1, 2)$ is indeed the solution.

$$y - x = 1,$$
$$y + x = 3$$

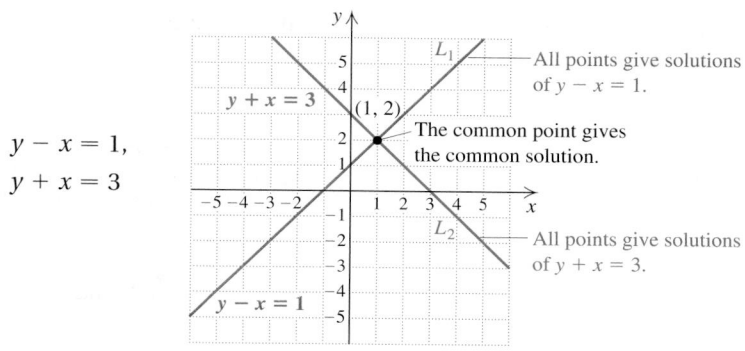

Check:

$$
\begin{array}{c|c}
y - x = 1 \\
\hline
2 - 1 \,\vert\, 1 \\
1 \overset{?}{=} 1 \quad \text{TRUE}
\end{array}
\qquad
\begin{array}{c|c}
y + x = 3 \\
\hline
2 + 1 \,\vert\, 3 \\
3 \overset{?}{=} 3 \quad \text{TRUE}
\end{array}
$$

b) We graph the equations. The lines have the same slope, -3, and different y-intercepts, so they are parallel. There is no point at which they cross, so the system has no solution.

$$y = -3x + 5,$$
$$y = -3x - 2$$

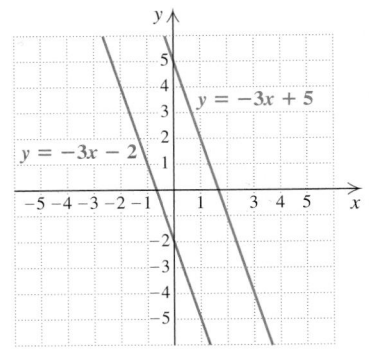

c) We graph the equations and find that the same line is drawn twice. Thus any solution of one equation is a solution of the other. Each equation has an infinite number of solutions, so the system itself has an infinite number of solutions. We check one solution, $(0, 2)$, which is the y-intercept of each equation.

Student Notes _____

Although the system in Example 4(c) is true for an infinite number of ordered pairs, those pairs must be of a certain form. Only pairs that are solutions of $3y - 2x = 6$ or $-12y + 8x = -24$ are solutions of the system. It is incorrect to think that *all* ordered pairs are solutions.

$$3y - 2x = 6,$$
$$-12y + 8x = -24$$

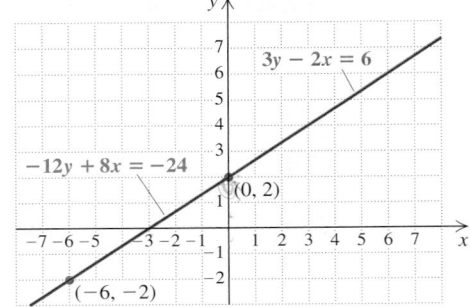

Check:

$3y - 2x = 6$	
$3(2) - 2(0)$	6
$6 - 0$	
$6 \overset{?}{=} 6$	TRUE

$-12y + 8x = -24$	
$-12(2) + 8(0)$	-24
$-24 + 0$	
$-24 \overset{?}{=} -24$	TRUE

You can check that $(-6, -2)$ is another solution of both equations. In fact, any pair that is a solution of one equation is a solution of the other equation as well. Thus the solution set is

$$\{(x, y) \mid 3y - 2x = 6\}$$

or, in words, "the set of all pairs (x, y) for which $3y - 2x = 6$." Since the two equations are equivalent, we could have written instead $\{(x, y) \mid -12y + 8x = -24\}$.

When we graph a system of two linear equations in two variables, one of the following three outcomes will occur.

1. The lines have one point in common, and that point is the only solution of the system (see Example 4a). Any system that has *at least* one solution is said to be **consistent**.

2. The lines are parallel, with no point in common, and the system has no solution (see Example 4b). This type of system is called **inconsistent**.

3. The lines coincide, sharing the same graph. Because every solution of one equation is a solution of the other, the system has an infinite number of solutions (see Example 4c). Since it has at least one solution, this type of system is also consistent.

When one equation in a system can be obtained by multiplying both sides of another equation by a constant, the two equations are said to be **dependent**. Thus the equations in Example 4(c) are dependent, but those in Examples 4(a) and 4(b) are **independent**. For systems of three or more equations, the definitions of dependent and independent will be slightly modified.

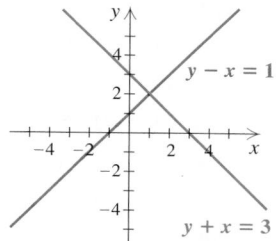

Graphs intersect at one point.
The system is *consistent* and has one solution. Since neither equation is a multiple of the other, they are *independent*.

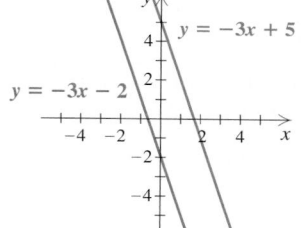

Graphs are parallel.
The system is *inconsistent* because there is no solution. Since the equations are not equivalent, they are *independent*.

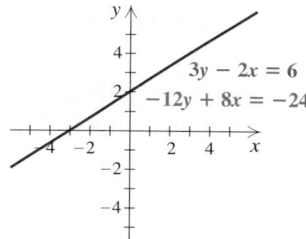

Equations have the same graph.
The system is *consistent* and has an infinite number of solutions. The equations are *dependent* since they are equivalent.

Graphing is helpful when solving systems because it allows us to "see" the solution. It can also be used on systems of nonlinear equations, and in many applications, it provides a satisfactory answer. However, graphing often lacks precision, especially when fraction or decimal solutions are involved. In Section 8.2, we will develop two algebraic methods of solving systems. Both methods produce exact answers.

technology connection

On most graphing calculators, an INTERSECT option allows you to find the coordinates of the intersection directly. This is especially useful when equations contain fractions or decimals or when the coordinates of the intersection are not integers. To illustrate, consider the following system:

$$3.45x + 4.21y = 8.39,$$
$$7.12x - 5.43y = 6.18.$$

After solving for y in each equation, we obtain the graph below. Using INTERSECT, we see that, to the nearest hundredth, the coordinates of the intersection are $(1.47, 0.79)$.

$$y_1 = (8.39 - 3.45x)/4.21,$$
$$y_2 = (6.18 - 7.12x)/(-5.43)$$

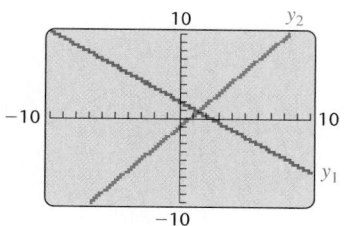

Use a graphing calculator to solve each of the following systems. Make sure that all x- and y-coordinates are correct to the nearest hundredth.

1. $y = -5.43x + 10.89,$
   $y = 6.29x - 7.04$
2. $y = 123.52x + 89.32,$
   $y = -89.22x + 33.76$
3. $2.18x + 7.81y = 13.78,$
   $5.79x - 3.45y = 8.94$
4. $-9.25x - 12.94y = -3.88,$
   $21.83x + 16.33y = 13.69$

Exercise Set

8.1

Concept Reinforcement   Classify each statement as either true or false.

1. Solutions of systems of equations in two variables are ordered pairs.

2. Every system of equations has at least one solution.

3. It is possible for a system of equations to have an infinite number of solutions.

4. The graphs of the equations in a system of two equations may coincide.

5. The graphs of the equations in a system of two equations could be parallel lines.

6. Any system of equations that has at most one solution is said to be consistent.

7. Any system of equations that has more than one solution is said to be inconsistent.

8. If one equation in a system can be obtained by multiplying both sides of another equation in that system by a constant, the two equations are said to be dependent.

Determine whether the ordered pair is a solution of the given system of equations. Remember to use alphabetical order of variables.

9. $(1, 2)$; $4x - y = 2,$
   $10x - 3y = 4$

10. $(-1, -2)$; $2x + y = -4,$
    $x - y = 1$

11. $(2, 5)$; $y = 3x - 1,$
    $2x + y = 4$

12. $(-1, -2)$; $x + 3y = -7,$
    $3x - 2y = 12$

13. $(1, 5)$; $x + y = 6,$
    $y = 2x + 3$

14. $(5, 2)$; $a + b = 7,$
    $2a - 8 = b$

Aha! 15. $(3, 1)$; $3x + 4y = 13,$
    $6x + 8y = 26$

16. $(4, -2)$; $-3x - 2y = -8,$
    $8 = 3x + 2y$

Solve each system graphically. Be sure to check your solution. If a system has an infinite number of solutions, use set-builder notation to write the solution set. If a system has no solution, state this.

17. $x - y = 3,$
    $x + y = 5$

18. $x + y = 4,$
    $x - y = 2$

19. $3x + y = 5,$
    $x - 2y = 4$

20. $2x - y = 4,$
    $5x - y = 13$

21. $4y = x + 8,$
    $3x - 2y = 6$

22. $4x - y = 9,$
    $x - 3y = 16$

23. $x = y - 1,$
    $2x = 3y$

24. $a = 1 + b,$
    $b = 5 - 2a$

25. $x = -3,$
    $y = 2$

26. $x = 4,$
    $y = -5$

27. $t + 2s = -1,$
    $s = t + 10$

28. $b + 2a = 2,$
    $a = -3 - b$

29. $2b + a = 11,$
    $a - b = 5$

30. $y = -\frac{1}{3}x - 1,$
    $4x - 3y = 18$

31. $y = -\frac{1}{4}x + 1,$
    $2y = x - 4$

32. $6x - 2y = 2,$
    $9x - 3y = 1$

33. $y - x = 5,$
    $2x - 2y = 10$

34. $y = -x - 1,$
    $4x - 3y = 24$

35. $y = 3 - x,$
    $2x + 2y = 6$

36. $2x - 3y = 6,$
    $3y - 2x = -6$

37. For the systems in the odd-numbered exercises 17–35, which are consistent?

38. For the systems in the even-numbered exercises 18–36, which are consistent?

39. For the systems in the odd-numbered exercises 17–35, which contain dependent equations?

40. For the systems in the even-numbered exercises 18–36, which contain dependent equations?

Translate each problem situation to a system of equations. Do not attempt to solve, but save for later use.

41. The sum of two numbers is 50. The first number is 25% of the second number. What are the numbers?

42. The sum of two numbers is 40. The first number is 60% of the second number. What are the numbers?

43. *Nontoxic furniture polish.* A nontoxic wood furniture polish can be made by mixing mineral (or olive) oil with vinegar. To make a 16-oz batch for a squirt bottle, Mabel uses an amount of mineral oil that is 4 oz more than twice the amount of vinegar. How much of each ingredient is required?
Sources: Based on information from Chittenden Solid Waste District and *Clean House, Clean Planet* by Karen Logan

44. *Scholastic Aptitude Test.* Many high-school students take the Scholastic Aptitude Test. Each student receives two scores, a *verbal* score and a *math* score. In 2002–2003, the average total score of students was 1026, with the average math score exceeding the verbal score by 12 points. What was the average verbal score and what was the average math score?
Source: College Entrance Examination Board

45. *Geometry.* Two angles are supplementary.* One angle is 3° less than twice the other. Find the measures of the angles.

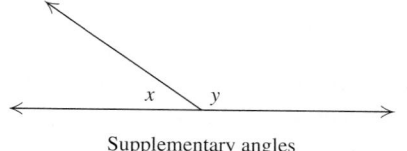

Supplementary angles

46. *Geometry.* Two angles are complementary.† The sum of the measures of the first angle and half the second angle is 64°. Find the measures of the angles.

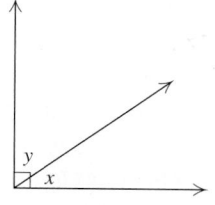

Complementary angles

47. *Basketball scoring.* Wilt Chamberlain once scored 100 points, setting a record for points scored in an NBA game. Chamberlain took only two-point shots and (one-point) foul shots and made a total of 64 shots. How many shots of each type did he make?

48. *Basketball scoring.* The Fenton College Cougars made 40 field goals in a recent basketball game, some 2-pointers and the rest 3-pointers. Altogether the 40 baskets counted for 89 points. How many of each type of field goal was made?

49. *Retail sales.* Paint Town sold 45 paintbrushes, one kind at $8.50 each and another at $9.75 each. In all, $398.75 was taken in for the brushes. How many of each kind were sold?

50. *Retail sales.* Mountainside Fleece sold 40 neckwarmers. Polarfleece neckwarmers sold for $9.90 each and wool ones sold for $12.75 each. In all, $421.65 was taken in for the neckwarmers. How many of each type were sold?

51. *Sales of pharmaceuticals.* In 2004, the Diabetic Express charged $27.06 for a vial of Humulin insulin and $34.39 for a vial of Novolin Velosulin insulin. If a total of $1565.57 was collected for 50 vials of insulin, how many vials of each type were sold?

52. *Fundraising.* The St. Mark's Community Barbecue served 250 dinners. A child's plate cost $3.50 and an adult's plate cost $7.00. A total of $1347.50 was collected. How many of each type of plate was served?

53. *Court dimensions.* The perimeter of a standard basketball court is 288 ft. The length is 44 ft longer than the width. Find the dimensions.

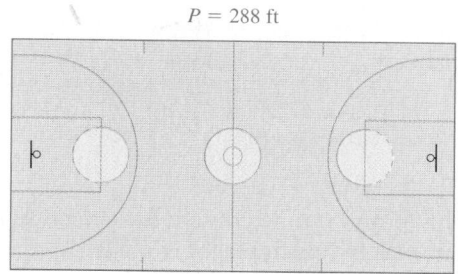

$P = 288$ ft

54. *Court dimensions.* The perimeter of a standard tennis court used for doubles is 228 ft. The width is 42 ft less than the length. Find the dimensions.

55. Write a problem for a classmate to solve that requires writing a system of two equations. Devise the problem so that the solution is "The Lakers made 6 three-point baskets and 31 two-point baskets."

---

*The sum of the measures of two supplementary angles is 180°.
†The sum of the measures of two complementary angles is 90°.

56. Write a problem for a classmate to solve that can be translated into a system of two equations. Devise the problem so that the solution is "Arnie took five 3-credit classes and two 4-credit classes."

SKILL MAINTENANCE

Solve. [2.2]

57. $2(4x - 3) - 7x = 9$ **58.** $6y - 3(5 - 2y) = 4$

59. $4x - 5x = 8x - 9 + 11x$

60. $8x - 2(5 - x) = 7x + 3$

Solve. [2.3]

61. $3x + 4y = 7$, for y

62. $2x - 5y = 9$, for y

SYNTHESIS

Presidential primaries.   For Exercises 63 and 64, consider the following graph showing the results of a poll in which Iowans were asked which Democrat candidate for president they favored.

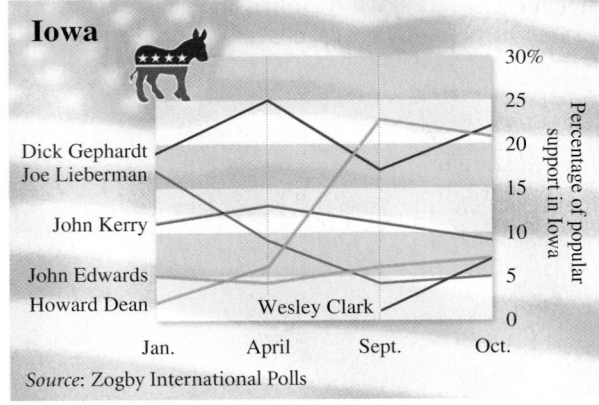

Iowa

Source: Zogby International Polls

63. At what point in time could it have been said that no one was in fourth place? Explain.

64. At what point in time was there no clear leader? Explain how you reach this conclusion.

65. For each of the following conditions, write a system of equations.

a) $(5, 1)$ is a solution.
b) There is no solution.
c) There is an infinite number of solutions.

66. A system of linear equations has $(1, -1)$ and $(-2, 3)$ as solutions. Determine:

a) a third point that is a solution, and
b) how many solutions there are.

67. The solution of the following system is $(4, -5)$. Find A and B.

$$Ax - 6y = 13,$$
$$x - By = -8.$$

Translate to a system of equations. Do not solve.

68. *Ages.*   Burl is twice as old as his son. Ten years ago, Burl was three times as old as his son. How old are they now?

69. *Work experience.*   Lou and Juanita are mathematics professors at a state university. Together, they have 46 years of service. Two years ago, Lou had taught 2.5 times as many years as Juanita. How long has each taught at the university?

70. *Design.*   A piece of posterboard has a perimeter of 156 in. If you cut 6 in. off the width, the length becomes four times the width. What are the dimensions of the original piece of posterboard?

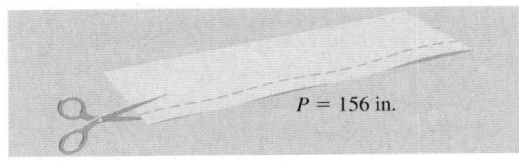

$P = 156$ in.

71. *Nontoxic scouring powder.*   A nontoxic scouring powder is made up of 4 parts baking soda and 1 part vinegar. How much of each ingredient is needed for a 16-oz mixture?

Solve graphically.

72. $y = |x|,$
         $x + 4y = 15$

73. $x - y = 0,$
         $y = x^2$

In Exercises 74–77, use a graphing calculator to solve each system of linear equations for x and y. Round all coordinates to the nearest hundredth.

74. $y = 8.23x + 2.11,$
      $y = -9.11x - 4.66$

75. $y = -3.44x - 7.72,$
      $y = 4.19x - 8.22$

76. $14.12x + 7.32y = 2.98,$
      $21.88x - 6.45y = -7.22$

77. $5.22x - 8.21y = -10.21,$
      $-12.67x + 10.34y = 12.84$

8.2 | Solving by Substitution or Elimination

The Substitution Method • The Elimination Method •
Comparing Methods

The Substitution Method

Algebraic (nongraphical) methods for solving systems are often superior to graphing, especially when fractions are involved. One algebraic method, the *substitution method*, relies on having a variable isolated.

EXAMPLE 1   Solve the system

$$x + y = 4, \quad (1)$$
$$x = y + 1. \quad (2)$$

For easy reference, we have numbered the equations.

Solution   Equation (2) says that x and $y + 1$ name the same number. Thus we can substitute $y + 1$ for x in equation (1):

$$x + y = 4 \qquad \text{Equation (1)}$$
$$(y + 1) + y = 4. \qquad \text{Substituting } y + 1 \text{ for } x$$

We solve this last equation, using methods learned earlier:

$$(y + 1) + y = 4$$
$$2y + 1 = 4 \qquad \text{Removing parentheses and combining like terms}$$
$$2y = 3 \qquad \text{Subtracting 1 from both sides}$$
$$y = \tfrac{3}{2}. \qquad \text{Dividing by 2}$$

We now return to the original pair of equations and substitute $\tfrac{3}{2}$ for y in either equation so that we can solve for x. For this problem, calculations are slightly easier if we use equation (2):

$$x = y + 1 \qquad \text{Equation (2)}$$
$$= \tfrac{3}{2} + 1 \qquad \text{Substituting } \tfrac{3}{2} \text{ for } y$$
$$= \tfrac{3}{2} + \tfrac{2}{2} = \tfrac{5}{2}.$$

We obtain the ordered pair $\left(\tfrac{5}{2}, \tfrac{3}{2}\right)$. A check ensures that it is a solution:

Check:

$$\begin{array}{c|c} x + y = 4 \\ \hline \tfrac{5}{2} + \tfrac{3}{2} & 4 \\ \tfrac{8}{2} & \\ 4 \overset{?}{=} 4 & \text{TRUE} \end{array} \qquad \begin{array}{c|c} x = y + 1 \\ \hline \tfrac{5}{2} & \tfrac{3}{2} + 1 \\ & \tfrac{3}{2} + \tfrac{2}{2} \\ \tfrac{5}{2} \overset{?}{=} \tfrac{5}{2} & \text{TRUE} \end{array}$$

Since $\left(\tfrac{5}{2}, \tfrac{3}{2}\right)$ checks, it is the solution.

The exact solution to Example 1 is difficult to find graphically because it involves fractions. Despite this, the graph shown does serve as a check and provides a visualization of the problem.

Study Skills

Learn from Your Mistakes

Immediately after each quiz or test, write out a step-by-step solution to any questions you missed. Visit your professor during office hours or consult with a tutor for help with problems that are still giving you trouble. Misconceptions tend to resurface if they are not corrected as soon as possible.

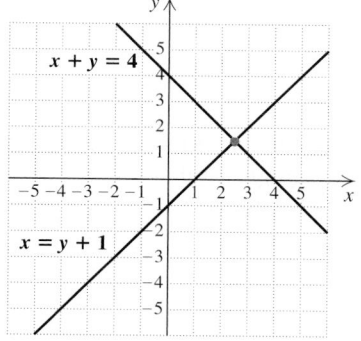

A visualization of Example 1.
Note that the coordinates of the intersection are not obvious.

If neither equation in a system has a variable alone on one side, we first isolate a variable in one equation and then substitute.

EXAMPLE 2   Solve the system

$$2x + y = 6, \qquad (1)$$
$$3x + 4y = 4. \qquad (2)$$

Solution   First, we select an equation and solve for one variable. To isolate y, we can subtract $2x$ from both sides of equation (1):

$$2x + y = 6 \qquad (1)$$
$$y = 6 - 2x. \qquad (3) \qquad \text{Subtracting } 2x \text{ from both sides}$$

Next, we proceed as in Example 1, by substituting:

$$3x + 4(6 - 2x) = 4 \qquad \text{Substituting } 6 - 2x \text{ for } y \text{ in equation (2).}$$
$$\qquad\qquad\qquad\qquad\quad \text{Use parentheses!}$$
$$3x + 24 - 8x = 4 \qquad \text{Distributing to remove parentheses}$$
$$3x - 8x = 4 - 24 \qquad \text{Subtracting 24 from both sides}$$
$$-5x = -20$$
$$x = 4. \qquad \text{Dividing both sides by } -5$$

Next, we substitute 4 for x in either equation (1), (2), or (3). It is easiest to use equation (3) because it has already been solved for y:

$$y = 6 - 2x$$
$$= 6 - 2(4)$$
$$= 6 - 8 = -2.$$

The pair $(4, -2)$ appears to be the solution. We check in equations (1) and (2).

Check:

$2x + y = 6$
$2(4) + (-2) \mid 6$
$8 - 2 \mid$
$6 \overset{?}{=} 6$ TRUE

$3x + 4y = 4$
$3(4) + 4(-2) \mid 4$
$12 - 8 \mid$
$4 \overset{?}{=} 4$ TRUE

Since $(4, -2)$ checks, it is the solution.

A visualization of Example 2

Some systems have no solution, as we saw graphically in Section 8.1. How do we recognize such systems if we are solving by an algebraic method?

EXAMPLE 3   Solve the system

$$y = -3x + 5, \qquad (1)$$
$$y = -3x - 2. \qquad (2)$$

Solution   We solved this system graphically in Example 4(b) of Section 8.1, and found that the lines are parallel and the system has no solution. Let's now

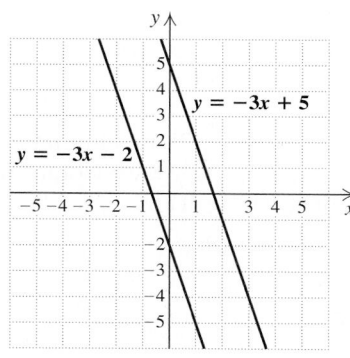

A visualization of Example 3

try to solve the system by substitution. Proceeding as in Example 1, we substitute $-3x - 2$ for y in the first equation:

$$-3x - 2 = -3x + 5 \qquad \text{Substituting } -3x - 2 \text{ for } y \text{ in equation (1)}$$

$$-2 = 5. \qquad \text{Adding } 3x \text{ to both sides; } -2 = 5 \text{ is a contradiction.}$$

When we add $3x$ to get the x-terms on one side, the x-terms drop out and the result is a contradiction—an equation that is always false. When solving algebraically yields a contradiction, we state that the system has no solution.

The Elimination Method

The *elimination method* for solving systems of equations makes use of the *addition principle*: If $a = b$, then $a + c = b + c$. Consider the following system:

$$2x - 3y = 0, \qquad (1)$$
$$-4x + 3y = -1. \qquad (2)$$

To see why the elimination method works well with this system, notice the $-3y$ in one equation and the $3y$ in the other. These terms are opposites. If we add all terms on the left side of the equations, the sum of $-3y$ and $3y$ is 0, so in effect, the variable y is "eliminated."

To use the addition principle with a system, note that according to equation (2), $-4x + 3y$ and -1 are the same number. Thus we can work vertically and add $-4x + 3y$ to the left side of equation (1) and -1 to the right side:

$$2x - 3y = 0 \qquad (1)$$
$$\underline{-4x + 3y = -1} \qquad (2)$$
$$-2x + 0y = -1. \qquad \text{Adding}$$

This eliminates the variable y, and leaves an equation with just one variable, x, for which we solve:

$$-2x = -1$$
$$x = \tfrac{1}{2}.$$

Next, we substitute $\tfrac{1}{2}$ for x in equation (1) and solve for y:

$$2 \cdot \tfrac{1}{2} - 3y = 0 \qquad \text{Substituting. We also could have used equation (2).}$$
$$1 - 3y = 0$$
$$-3y = -1, \text{ so } y = \tfrac{1}{3}.$$

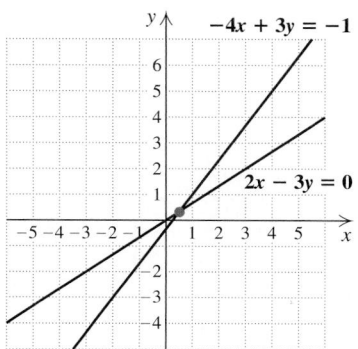

A visualization of the solution of
$$2x - 3y = 0,$$
$$-4x + 3y = -1$$

Check:

$$\begin{array}{c|c} 2x - 3y = 0 \\ \hline 2\left(\tfrac{1}{2}\right) - 3\left(\tfrac{1}{3}\right) & 0 \\ 1 - 1 & \\ 0 \stackrel{?}{=} 0 & \text{TRUE} \end{array} \qquad \begin{array}{c|c} -4x + 3y = -1 \\ \hline -4\left(\tfrac{1}{2}\right) + 3\left(\tfrac{1}{3}\right) & -1 \\ -2 + 1 & \\ -1 \stackrel{?}{=} -1 & \text{TRUE} \end{array}$$

Since $\left(\tfrac{1}{2}, \tfrac{1}{3}\right)$ checks, it is the solution. See also the graph at left.

To eliminate a variable, we must sometimes multiply before adding.

EXAMPLE 4   Solve the system

$$5x + 4y = 22, \quad (1)$$
$$-3x + 8y = 18. \quad (2)$$

Student Notes

It is wise to double-check each step of your work as you go along, rather than checking all steps after reaching the end of a problem. Finding and correcting an error as it occurs will save you time in the long run. One common error is to forget to multiply *both* sides of the equation when you use the multiplication principle.

Solution   If we add the left sides of the two equations, we will not eliminate a variable. However, if the $4y$ in equation (1) were changed to $-8y$, we would. To accomplish this change, we multiply both sides of equation (1) by -2:

$$\begin{array}{ll} -10x - 8y = -44 & \text{Multiplying both sides of equation (1) by } -2 \\ \underline{-3x + 8y = 18} & \\ -13x + 0 = -26 & \text{Adding} \\ x = 2. & \text{Solving for } x \end{array}$$

Then

$$\begin{array}{ll} -3 \cdot 2 + 8y = 18 & \text{Substituting 2 for } x \text{ in equation (2)} \\ -6 + 8y = 18 & \\ \left.\begin{array}{r} 8y = 24 \\ y = 3. \end{array}\right\} & \text{Solving for } y \end{array}$$

We obtain $(2, 3)$, or $x = 2, y = 3$. We leave it to the student to confirm that this checks and is the solution.

Sometimes we must multiply twice in order to make two terms become opposites.

EXAMPLE 5   Solve the system

$$2x + 3y = 17, \quad (1)$$
$$5x + 7y = 29. \quad (2)$$

Solution   We multiply so that the x-terms are eliminated.

$$\begin{array}{lll} 2x + 3y = 17, & \xrightarrow{\text{Multiplying both}}_{\text{sides by 5}} & 10x + 15y = 85 \\ 5x + 7y = 29 & \xrightarrow{\text{Multiplying both}}_{\text{sides by } -2} & \underline{-10x - 14y = -58} \\ & & 0 + y = 27 \quad \text{Adding} \\ & & y = 27 \end{array}$$

Next, we substitute to find x:

$$\begin{array}{ll} 2x + 3 \cdot 27 = 17 & \text{Substituting 27 for } y \text{ in equation (1)} \\ 2x + 81 = 17 & \\ \left.\begin{array}{r} 2x = -64 \\ x = -32. \end{array}\right\} & \text{Solving for } x \end{array}$$

Check:

$$\begin{array}{c|c} \underline{2x + 3y = 17} & \\ 2(-32) + 3(27) & 17 \\ -64 + 81 & \\ \hline 17 \overset{?}{=} 17 & \text{TRUE} \end{array} \qquad \begin{array}{c|c} \underline{5x + 7y = 29} & \\ 5(-32) + 7(27) & 29 \\ -160 + 189 & \\ \hline 29 \overset{?}{=} 29 & \text{TRUE} \end{array}$$

We obtain $(-32, 27)$, or $x = -32, y = 27$, as the solution.

EXAMPLE 6

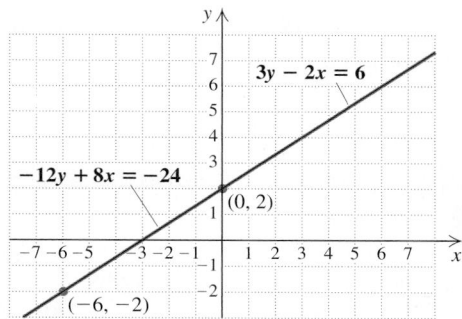

A visualization of Example 6

Solve the system

$$3y - 2x = 6, \qquad (1)$$
$$-12y + 8x = -24. \qquad (2)$$

Solution   We graphed this system in Example 4(c) of Section 8.1, and found that the lines coincide and the system has an infinite number of solutions. Suppose we were to solve this system using the elimination method:

$$
\begin{array}{ll}
12y - 8x = 24 & \text{Multiplying both sides of equation (1) by 4} \\
\underline{-12y + 8x = -24} & \\
0 = 0. & \text{We obtain an identity; } 0 = 0 \text{ is always true.}
\end{array}
$$

Note that both variables have been eliminated and what remains is an identity—that is, an equation that is always true. Any pair that is a solution of equation (1) is also a solution of equation (2). The equations are dependent and the solution set is infinite:

$$\{(x, y) \mid 3y - 2x = 6\}, \quad \text{or equivalently,} \quad \{(x, y) \mid -12y + 8x = -24\}.$$

> **Rules for Special Cases**
>
> When solving a system of two linear equations in two variables:
>
> 1. If an identity is obtained, such as $0 = 0$, then the system has an infinite number of solutions. The equations are dependent and, since a solution exists, the system is consistent.*
> 2. If a contradiction is obtained, such as $0 = 7$, then the system has no solution. The system is inconsistent.

Should decimals or fractions appear, it often helps to *clear* before solving.

EXAMPLE 7

Solve the system

$$0.2x + 0.3y = 1.7,$$
$$\tfrac{1}{7}x + \tfrac{1}{5}y = \tfrac{29}{35}.$$

Solution   We have

$$0.2x + 0.3y = 1.7, \rightarrow \text{Multiplying both sides by 10} \rightarrow 2x + 3y = 17$$
$$\tfrac{1}{7}x + \tfrac{1}{5}y = \tfrac{29}{35} \rightarrow \text{Multiplying both sides by 35} \rightarrow 5x + 7y = 29.$$

We multiplied both sides of the first equation by 10 to clear the decimals. Multiplication by 35, the least common denominator, clears the fractions in the second equation. The problem now happens to be identical to Example 5. The solution is $(-32, 27)$, or $x = -32$, $y = 27$.

---

*Consistent systems and dependent equations are discussed in greater detail in Section 8.4.

Comparing Methods

The following table is a summary that compares the graphical, substitution, and elimination methods for solving systems of equations.

CONNECTING THE CONCEPTS

We now have three different methods for solving systems of equations. Each method has certain strengths and weaknesses, as outlined below.

Method	Strengths	Weaknesses
Graphical	Solutions are displayed graphically. Can be used with any system that can be graphed.	Inexact when solutions involve numbers that are not integers. Solution may not appear on the part of the graph drawn.
Substitution	Yields exact solutions. Easy to use when a variable is alone on one side.	Introduces extensive computations with fractions when solving more complicated systems. Solutions are not displayed graphically.
Elimination	Yields exact solutions. Easy to use when fractions or decimals appear in the system. The preferred method for systems of 3 or more equations in 3 or more variables (see Section 8.4).	Solutions are not displayed graphically.

Before selecting a method to use, try to remember the strengths and weaknesses of each method. If possible, begin solving the system mentally to help discover the method that seems best suited for that particular system. Selecting the "best" method for a problem is a bit like selecting one of three different saws with which to cut a piece of wood. The "best" choice depends on what kind of wood is being cut and what type of cut is being made, as well as your skill level with each saw.

Note that each of the three methods was introduced using a rather simple example. As the examples became more complicated, additional steps were required in order to "turn" the new problem into a more familiar format. This is a common approach in mathematics: We perform one or more steps to make a "new" problem resemble a problem we already know how to solve.

Exercise Set

8.2

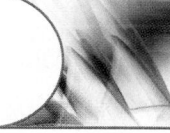

🔁 *Concept Reinforcement   In each of Exercises 1–6, match the system listed with the choice from the column on the right that would be a subsequent step in solving the system.*

1. ___ $-2x + 3y = 7,$
$2x + 5y = -8$

2. ___ $x = 4y - 5,$
$5x + 7y = 2$

3. ___ $3x + 4y = 13,$
$4x - 2y = 5$

4. ___ $8x + 6y = -15,$
$5x - 3y = 8$

5. ___ $y = 4x - 7,$
$6x + 3y = 19$

6. ___ $y = \frac{2}{3}x + 4,$
$y = -\frac{1}{5}x + 4$

a) $3x + 4y = 13,$
$8x - 4y = 10$

b) The lines intersect at (0, 4).

c) $6x + 3(4x - 7) = 19$

d) $8y = -1$

e) $5(4y - 5) + 7y = 2$

f) $8x + 6y = -15,$
$10x - 6y = 16$

For Exercises 7–54, if a system has an infinite number of solutions, use set-builder notation to write the solution set. If a system has no solution, state this.

Solve using the substitution method.

7. $y = 5 - 4x,$
$2x - 3y = 13$

8. $2y + x = 9,$
$x = 3y - 3$

9. $3x + 5y = 3,$
$x = 8 - 4y$

10. $9x - 2y = 3,$
$3x - 6 = y$

11. $3s - 4t = 14,$
$5s + t = 8$

12. $m - 2n = 16,$
$4m + n = 1$

13. $4x - 2y = 6,$
$2x - 3 = y$

14. $t = 4 - 2s,$
$t + 2s = 6$

15. $-5x + t = 11,$
$4x + 12t = 4$

16. $5x + 6y = 14,$
$-3y + x = 7$

17. $2x + 2y = 2,$
$3x - y = 1$

18. $4p - 2q = 16,$
$5p + 7q = 1$

19. $x - 4y = 3,$
$5x + 3y = 4$

20. $2a + 2b = 5,$
$3a - b = 7$

21. $2x - 3 = y,$
$y - 2x = 1$

22. $a - 2b = 3,$
$3a = 6b + 9$

Solve using the elimination method.

23. $x + 3y = 7,$
$-x + 4y = 7$

24. $2x + y = 6,$
$x - y = 3$

25. $2x - y = -3,$
$x + y = 9$

26. $x - 2y = 6,$
$-x + 3y = -4$

27. $9x + 3y = -3,$
$2x - 3y = -8$

28. $6x - 3y = 18,$
$6x + 3y = -12$

29. $5x + 3y = 19,$
$2x - 5y = 11$

30. $3x + 2y = 3,$
$9x - 8y = -2$

31. $5r - 3s = 24,$
$3r + 5s = 28$

32. $5x - 7y = -16,$
$2x + 8y = 26$

33. $6x + 9t = 12,$
$4x + 6t = 5$

34. $10a + 6b = 8,$
$5a + 3b = 2$

35. $\frac{1}{2}x - \frac{1}{6}y = 3,$
$\frac{2}{5}x + \frac{1}{2}y = 2$

36. $\frac{1}{3}x + \frac{1}{5}y = 7,$
$\frac{1}{6}x - \frac{2}{5}y = -4$

37. $\dfrac{x}{2} + \dfrac{y}{3} = \dfrac{7}{6},$
$\dfrac{2x}{3} + \dfrac{3y}{4} = \dfrac{5}{4}$

38. $\dfrac{2x}{3} + \dfrac{3y}{4} = \dfrac{11}{12},$
$\dfrac{x}{3} + \dfrac{7y}{18} = \dfrac{1}{2}$

Aha! **39.** $12x - 6y = -15,$
$-4x + 2y = 5$

40. $8s + 12t = 16,$
$6s + 9t = 12$

41. $0.2a + 0.3b = 1,$
$0.3a - 0.2b = 4$

42. $-0.4x + 0.7y = 1.3,$
$0.7x - 0.3y = 0.5$

Solve using any appropriate method.

43. $a - 2b = 16,$
$b + 3 = 3a$

44. $5x - 9y = 7,$
$7y - 3x = -5$

45. $10x + y = 306,$
$10y + x = 90$

46. $3(a - b) = 15,$
$4a = b + 1$

47. $3y = x - 2,$
$x = 2 + 3y$

48. $x + 2y = 8,$
$x = 4 - 2y$

49. $3s - 7t = 5,$
$7t - 3s = 8$

50. $2s - 13t = 120,$
$-14s + 91t = -840$

51. $0.05x + 0.25y = 22,$
$0.15x + 0.05y = 24$

52. $2.1x - 0.9y = 15,$
$-1.4x + 0.6y = 10$

53. $13a - 7b = 9,$
$2a - 8b = 6$

54. $3a - 12b = 9,$
$14a - 11b = 5$

55. Describe a procedure that can be used to write an inconsistent system of equations.

56. Describe a procedure that can be used to write a system that has an infinite number of solutions.

SKILL MAINTENANCE

Solve. [2.5]

57. The fare for a taxi ride from Johnson Street to Elm Street is $5.20. If the rate of the taxi is $1.00 for the first $\frac{1}{2}$ mi and 30¢ for each additional $\frac{1}{4}$ mi, how far is it from Johnson Street to Elm Street?

58. A student's average after 4 tests is 78.5. What score is needed on the fifth test in order to raise the average to 80?

59. *Home remodeling.*   In a recent year, Americans spent $35 billion to remodel bathrooms and kitchens. Twice as much was spent on kitchens as on bathrooms. How much was spent on each?
Source: *Indianapolis Star*

60. A 480-m wire is cut into three pieces. The second piece is three times as long as the first. The third is four times as long as the second. How long is each piece?

61. *Car rentals.*   Badger Rent-A-Car rents a compact car at a daily rate of $34.95 plus 10¢ per mile. A sales representative is allotted $80 for car rental for one day. How many miles can she travel on the $80 budget?

62. *Car rentals.*   Badger rents midsized cars at a rate of $43.95 plus 10¢ per mile. A tourist has a car-rental budget of $90 for one day. How many miles can he travel on the $90?

SYNTHESIS

63. Some systems are more easily solved by substitution and some are more easily solved by elimination. What guidelines could be used to help someone determine which method to use?

64. Explain how it is possible to solve Exercise 39 mentally.

65. If $(1, 2)$ and $(-3, 4)$ are two solutions of $f(x) = mx + b$, find m and b.

66. If $(0, -3)$ and $\left(-\frac{3}{2}, 6\right)$ are two solutions of $px - qy = -1$, find p and q.

67. Determine a and b for which $(-4, -3)$ is a solution of the system
$$ax + by = -26,$$
$$bx - ay = 7.$$

68. Solve for x and y in terms of a and b:
$$5x + 2y = a,$$
$$x - y = b.$$

Solve.

69. $\dfrac{x + y}{2} - \dfrac{x - y}{5} = 1,$
$\dfrac{x - y}{2} + \dfrac{x + y}{6} = -2,$

70. $3.5x - 2.1y = 106.2,$
$4.1x + 16.7y = -106.28$

Each of the following is a system of nonlinear equations. However, each is reducible to linear, since an appropriate substitution (say, u for $1/x$ and v for $1/y$) yields a linear system. Make such a substitution, solve for the new variables, and then solve for the original variables.

71. $\dfrac{2}{x} + \dfrac{1}{y} = 0,$
$\dfrac{5}{x} + \dfrac{2}{y} = -5$

72. $\dfrac{1}{x} - \dfrac{3}{y} = 2,$
$\dfrac{6}{x} + \dfrac{5}{y} = -34$

73. A student solving the system
$$17x + 19y = 102,$$
$$136x + 152y = 826$$
graphs both equations on a graphing calculator and gets the following screen. The student then (incorrectly) concludes that the equations are dependent and the solution set is infinite. How can algebra be used to convince the student that a mistake has been made?

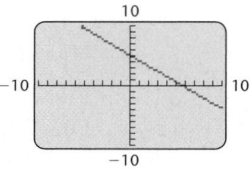

CORNER

How Many Two's? How Many Three's?

Focus: Systems of linear equations

Time: 20 minutes

Group size: 3

The box score at right, from the 2004 NBA All-Star game, contains information on how many field goals and free throws each player attempted and made. For example, the line "Kidd 4-6 3-4 14" means that the East's Jason Kidd made 4 field goals out of 6 attempts and 3 free throws out of 4 attempts, for a total of 14 points. (Each free throw is worth 1 point and each field goal is worth either 2 or 3 points, depending on how far from the basket it was shot.)

ACTIVITY

1. Work as a group to develop a system of two equations in two unknowns that can be used to determine how many 2-pointers and how many 3-pointers were made by the West.
2. Each group member should solve the system from part (1) in a different way: one person

algebraically, one person by making a table and methodically checking all combinations of 2- and 3-pointers, and one person by guesswork. Compare answers when this has been completed.
3. Determine, as a group, how many 2- and 3-pointers the East made.

East (132)
Iverson 1-6 1-4 3, McGrady 5-11 2-4 13, J.O'Neal 7-13 2-4 16, Carter 5-7 0-0 11, Wallace 2-5 0-0 4, Martin 8-10 1-2 17, Kidd 4-6 3-4 14, Magloire 9-16 1-2 19, Artest 3-5 1-2 7, Davis 3-9 0-0 7, Redd 5-12 0-0 13, Pierce 4-8 0-0 8
Totals 56-108 11-22 132

West (136)
Bryant 9-12 0-1 20, Francis 6-9 0-0 13, Garnett 6-14 0-0 12, Duncan 6-11 2-4 14, Yao Ming 8-14 0-0 16, S.O'Neal 12-19 0-1 24, Allen 6-13 3-4 16, Cassell 2-3 0-0 4, Nowitzki 1-3 0-0 4, Stojakovic 2-5 0-0 5, Kirilenko 1-3 0-0 2, Brad Miller 4-5 0-0 8
Totals 63-111 5-10 136

| East | 33 | 31 | 37 | 31 | — | 132 |
| West | 31 | 27 | 45 | 33 | — | 136 |

8.3 Solving Applications: Systems of Two Equations

Total-Value and Mixture Problems • Motion Problems

You are in a much better position to solve problems now that you know how systems of equations can be used. Using systems often makes the translating step easier.

EXAMPLE 1   Bottled-water consumption.   Americans are buying more bottled water than ever before. At the time of this writing, the average American buys more bottled water than milk, coffee, or beer. In 2003, the average American purchased 76.4 gal of soft drinks and water. The amount of water was only 4.3 gal less than

Study Skills _____

Expect to be Challenged

Do not be surprised if your success rate drops some as you work on real-world problems. *This is normal.* Your success rate will increase as you gain experience with these types of problems and use some of the study skills already listed.

half of the amount of soft drinks. (*Source*: Based on data from www.beveragemarketing.com) How many gallons of water and how many of soft drinks did the average American buy in 2003?

Solution The *Familiarize* and *Translate* steps have been done in Example 1 of Section 8.1. The resulting system of equations is

$$w + s = 76.4,$$
$$w = \frac{1}{2}s - 4.3,$$

where w is the number of gallons of water and s is the number of gallons of soft drinks purchased by the average American in 2003.

3. Carry out. We solve the system of equations. Since one equation already has a variable isolated, let's use the substitution method:

$$w + s = 76.4$$

$$\frac{1}{2}s - 4.3 + s = 76.4 \qquad \text{Substituting } \frac{1}{2}s - 4.3 \text{ for } w$$

$$\frac{3}{2}s - 4.3 = 76.4 \qquad \text{Combining like terms}$$

$$\frac{3}{2}s = 80.7 \qquad \text{Adding 4.3 to both sides}$$

$$s = \frac{2}{3} \cdot 80.7 \qquad \text{Multiplying both sides by } \frac{2}{3}: \frac{2}{3} \cdot \frac{3}{2} = 1$$

$$s = 53.8. \qquad \text{Simplifying}$$

Next, using either of the original equations, we substitute and solve for w:

$$w = \frac{1}{2} \cdot 53.8 - 4.3 = 26.9 - 4.3 = 22.6.$$

Student Notes _____

It is very important that you clearly label precisely what each variable represents. Not only will this assist you in writing equations, but it will help you to identify and state solutions.

4. Check. The sum of 53.8 and 22.6 is 76.4, so the total consumed is correct. Since 4.3 less than half of 53.8 is 26.9 − 4.3, or 22.6, the numbers check.

5. State. In 2003, the average American purchased 22.6 gal of bottled water and 53.8 gal of soft drinks.

Total-Value and Mixture Problems

EXAMPLE 2

Purchasing.   Recently the Woods County Art Center purchased 120 stamps for $33.90. If the stamps were a combination of 23¢ postcard stamps and 37¢ first-class stamps, how many of each type were bought?

Solution The *Familiarize* and *Translate* steps were completed in Example 2 of Section 8.1.

3. Carry out. We are to solve the system of equations

$$p + f = 120, \qquad (1)$$
$$23p + 37f = 3390, \qquad (2) \qquad \text{Working in cents rather than dollars}$$

where p is the number of postcard stamps bought and f is the number of first-class stamps bought. Because both equations are in the form

$Ax + By = C$, let's use the elimination method to solve the system. We can eliminate p by multiplying both sides of equation (1) by -23 and adding them to the corresponding sides of equation (2):

$$
\begin{array}{ll}
-23p - 23f = -2760 & \text{Multiplying both sides of equation (1) by } -23 \\
\underline{23p + 37f = 3390} & \\
 14f = 630 & \text{Adding} \\
 f = 45. & \text{Solving for } f
\end{array}
$$

To find p, we substitute 45 for f in equation (1) and then solve for p:

$$
\begin{array}{ll}
p + f = 120 & \text{Equation (1)} \\
p + 45 = 120 & \text{Substituting 45 for } f \\
p = 75. & \text{Solving for } p
\end{array}
$$

We obtain (45, 75), or $f = 45$ and $p = 75$.

4. **Check.** We check in the original problem. Recall that f is the number of first-class stamps and p the number of postcard stamps.

Number of stamps: $f + p = 45 + 75 = 120$
Cost of first-class stamps: $\$0.37f = 0.37 \times 45 = \$16.65$
Cost of postcard stamps: $\$0.23p = 0.23 \times 75 = \underline{\$17.25}$
                                                          Total $= \$33.90$

The numbers check.

5. **State.** The art center bought 45 first-class stamps and 75 postcard stamps.

Example 2 involved two types of items (first-class stamps and postcard stamps), the quantity of each type bought, and the total value of the items. We refer to this type of problem as a *total-value problem*.

EXAMPLE 3   Blending teas.   Sonya's House of Tea sells loose Lapsang Souchong tea for 95¢ an ounce and Assam Gingia for $1.43 an ounce. Sonya wants to make a 20-oz mixture of the two types, called Dragon Blend, that sells for $1.10 an ounce. How much tea of each type should Sonya use?

Solution

1. **Familiarize.**   This problem is similar to Example 2. Rather than postcard stamps and first-class stamps, we have ounces of Assam Gingia and ounces of Lapsang Souchong. Instead of a different price for each type of stamp, we

have a different price per ounce for each type of tea. Finally, rather than knowing the total cost of the stamps, we know the weight and the price per ounce of the mixture. Thus we can find the total value of the blend by multiplying 20 ounces times $1.10, or 110¢ per ounce. Although we could make and check a guess, we proceed to let l = the number of ounces of Lapsang Souchong and a = the number of ounces of Assam Gingia.

2. **Translate.**   Since a 20-oz batch is being made, we must have

$$l + a = 20.$$

To find a second equation, note that the total value of the 20-oz blend must match the combined value of the separate ingredients:

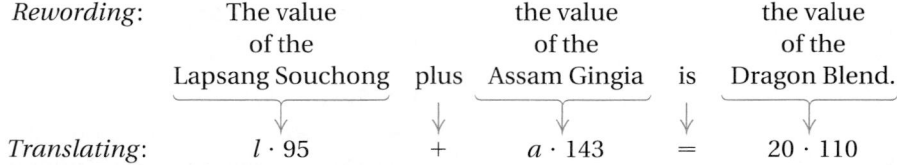

Rewording: The value   the value   the value
                 of the   of the   of the
                 Lapsang Souchong   plus   Assam Gingia   is   Dragon Blend.

Translating: $l \cdot 95$ $+$ $a \cdot 143$ $=$ $20 \cdot 110$

These equations can also be obtained from a table.

	Lapsang Souchong	**Assam Gingia**	**Dragon Blend**	
Number of Ounces	l	a	20	→ $l + a = 20$
Price per Ounce	95¢	143¢	110¢	
Value of Tea	$95l$	$143a$	$20 \cdot 110$, or 2200¢	→ $95l + 143a = 2200$

We have translated to a system of equations:

$$l + \quad a = 20, \qquad (1)$$
$$95l + 143a = 2200. \qquad (2)$$

3. **Carry out.**   We can solve using substitution. When equation (1) is solved for l, we have $l = 20 - a$. Substituting $20 - a$ for l in equation (2), we find a:

$$95(20 - a) + 143a = 2200 \qquad \text{Substituting}$$
$$1900 - 95a + 143a = 2200 \qquad \text{Using the distributive law}$$
$$48a = 300 \qquad \text{Combining like terms; subtracting}$$
$$\text{1900 from both sides}$$
$$a = 6.25. \qquad \text{Dividing both sides by 48}$$

We have $a = 6.25$ and, from equation (1) above, $l + a = 20$. Thus, $l = 13.75$.

4. **Check.**   If 13.75 oz of Lapsang Souchong and 6.25 oz of Assam Gingia are combined, a 20-oz blend will result. The value of 13.75 oz of Lapsang Souchong is 13.75(95¢) or 1306.25¢. The value of 6.25 oz of Assam Gingia is 6.25(143¢), or 893.75¢. Thus the combined value of the blend is 1306.25¢ + 893.75¢, or 2200¢, which is $22. A 20-oz blend priced at $1.10 an ounce would also be worth $22, so our answer checks.

5. **State.**   The Dragon Blend should be made by combining 13.75 oz of Lapsang Souchong with 6.25 oz of Assam Gingia.

EXAMPLE 4

Student loans.   Ranjay's student loans totaled $9600. Part was a Perkins loan made at 5% interest and the rest was a Stafford Loan made at 8% interest. After one year, Ranjay's loans accumulated $633 in interest. What was the original amount of each loan?

Solution

1. **Familiarize.**   We begin with a guess. If $7000 was borrowed at 5% and $2600 was borrowed at 8%, the two loans would total $9600. The interest would then be 0.05($7000), or $350, and 0.08($2600), or $208, for a total of only $558 in interest. Our guess was wrong, but checking the guess familiarized us with the problem. More than $2600 was borrowed at the higher rate.

2. **Translate.**   We let p = the amount of the Perkins loan and f = the amount of the Stafford Loan. Next, we organize a table in which the entries in each column come from the formula for simple interest:

$$Principal \cdot Rate \cdot Time = Interest.$$

	Perkins Loan	Stafford Loan	Total
Principal	p	f	$9600	$\longrightarrow p + f = 9600$
Rate of Interest	5%	8%	
Time	1 yr	1 yr	
Interest	$0.05p$	$0.08f$	$633	$\longrightarrow 0.05p + 0.08f = 633$

The total amount borrowed is found in the first row of the table:

$$p + f = 9600.$$

A second equation, representing the accumulated interest, can be found in the last row:

$$0.05p + 0.08f = 633, \quad or \quad 5p + 8f = 63{,}300. \qquad \text{Clearing decimals}$$

3. **Carry out.**   The system can be solved by elimination:

$$\begin{aligned} p + f &= 9600, \\ 5p + 8f &= 63{,}300. \end{aligned} \quad \begin{array}{c}\longrightarrow \text{Multiplying both} \\ \text{sides by } -5 \end{array} \longrightarrow \begin{aligned} -5p - 5f &= -48{,}000 \\ 5p + 8f &= 63{,}300 \\ \hline 3f &= 15{,}300 \end{aligned}$$

$$\begin{aligned} p + f &= 9600 \longleftarrow \quad f = 5100 \\ p + 5100 &= 9600 \\ p &= 4500. \end{aligned}$$

We find that $p = 4500$ and $f = 5100$.

4. **Check.**   The total amount borrowed is $4500 + $5100, or $9600. The interest on $4500 at 5% for 1 yr is 0.05($4500), or $225. The interest on $5100 at 8% for 1 yr is 0.08($5100), or $408. The total amount of interest is $225 + $408, or $633, so the numbers check.

5. **State.**   The Perkins loan was for $4500 and the Stafford loan was for $5100.

Before proceeding to Example 5, briefly scan Examples 2–4 for similarities. Note that in each case, one of the equations in the system is a simple sum while the other equation represents a sum of products. Example 5 continues this pattern with what is commonly called a *mixture problem*.

Problem-Solving Tip

When solving a problem, see if it is patterned or modeled after a problem that you have already solved.

EXAMPLE 5   Mixing fertilizers.   Sky Meadow Gardening, Inc., carries two brands of fertilizer containing nitrogen and water. "Gently Green" is 5% nitrogen and "Sun Saver" is 15% nitrogen. Sky Meadow Gardening needs to combine the two types of solutions in order to make 90 L of a solution that is 12% nitrogen. How much of each brand should be used?

Solution

1. **Familiarize.**   We make a drawing and then make a guess to gain familiarity with the problem.

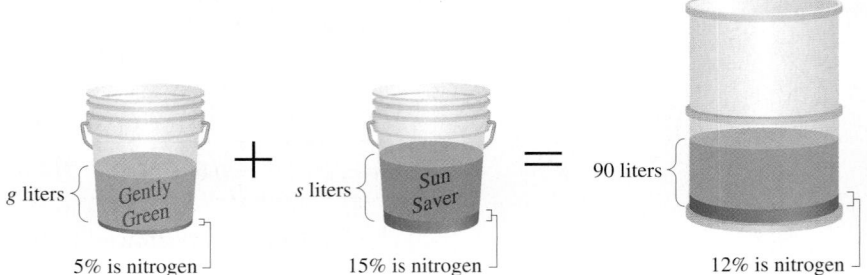

Suppose that 40 L of Gently Green and 50 L of Sun Saver are mixed. The resulting mixture will be the right size, 90 L, but will it be the right strength? To find out, note that 40 L of Gently Green would contribute $0.05(40) = 2$ L of nitrogen to the mixture while 50 L of Sun Saver would contribute $0.15(50) = 7.5$ L of nitrogen to the mixture. The total amount of nitrogen in the mixture would then be $2 + 7.5$, or 9.5 L. But we want 12% of 90, or 10.8 L, to be nitrogen. Thus our guess of 40 L and 50 L is incorrect. Still, checking our guess has familiarized us with the problem.

2. **Translate.**   Let $g =$ the number of liters of Gently Green and $s =$ the number of liters of Sun Saver. The information can be organized in a table.

	Gently Green	Sun Saver	Mixture	
Number of Liters	g	s	90	$\rightarrow g + s = 90$
Percent of Nitrogen	5%	15%	12%	
Amount of Nitrogen	$0.05g$	$0.15s$	0.12×90, or 10.8 liters	$\rightarrow 0.05g + 0.15s = 10.8$

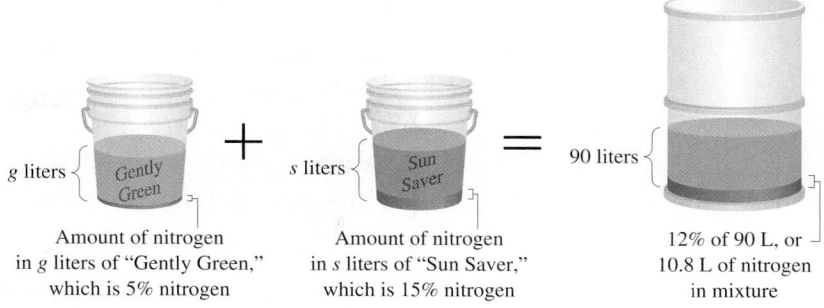

| Amount of nitrogen in g liters of "Gently Green," which is 5% nitrogen | Amount of nitrogen in s liters of "Sun Saver," which is 15% nitrogen | 12% of 90 L, or 10.8 L of nitrogen in mixture |

If we add g and s in the first row, we get one equation. It represents the total amount of mixture: $g + s = 90$.

If we add the amounts of nitrogen listed in the third row, we get a second equation. This equation represents the amount of nitrogen in the mixture: $0.05g + 0.15s = 10.8$.

After clearing decimals, we have translated the problem to the system

$$g + \quad s = 90 \qquad (1)$$
$$5g + 15s = 1080. \qquad (2)$$

3. **Carry out.**   We use the elimination method to solve the system:

$$
\begin{array}{ll}
-5g - 5s = -450 & \text{Multiplying both sides of} \\
\underline{5g + 15s = 1080} & \text{equation (1) by } -5 \\
10s = 630 & \text{Adding} \\
s = 63; & \text{Solving for } s \\
\\
g + 63 = 90 & \text{Substituting into equation (1)} \\
g = 27. & \text{Solving for } g
\end{array}
$$

4. **Check.**   Remember, g is the number of liters of Gently Green and s is the number of liters of Sun Saver.

Total amount of mixture: $g + s = 27 + 63 = 90$

Total amount of nitrogen: 5% of 27 + 15% of 63 = 1.35 + 9.45 = 10.8

Percentage of nitrogen in mixture: $\dfrac{\text{Total amount of nitrogen}}{\text{Total amount of mixture}} = \dfrac{10.8}{90} = 12\%$

The numbers check in the original problem.

5. **State.**   Sky Meadow Gardening should mix 27 L of Gently Green with 63 L of Sun Saver.

Motion Problems

When a problem deals with distance, speed (rate), and time, recall the following.

Distance, Rate, and Time Equations

If r represents rate, t represents time, and d represents distance, then:

$$d = rt, \qquad r = \frac{d}{t}, \quad \text{and} \quad t = \frac{d}{r}.$$

Be sure to remember at least one of these equations. The others can be obtained by multiplying or dividing on both sides as needed.

EXAMPLE 6   *Train travel.*   A Vermont Railways freight train, loaded with logs, leaves Boston, heading to Washington D.C. at a speed of 60 km/h. Two hours later, an Amtrak® Metroliner leaves Boston, bound for Washington D.C., on a parallel track at 90 km/h. At what point will the Metroliner catch up to the freight train?

Solution

1. **Familiarize.**   Let's make a guess—say, 180 km—and check to see if it is correct. The freight train, traveling 60 km/h, would travel 180 km in $\frac{180}{60} = 3$ hr. The Metroliner, traveling 90 km/h, would cover 180 km in $\frac{180}{90} = 2$ hr. Since 3 hr is *not* two hours more than 2 hr, our guess of 180 km is incorrect. Although our guess is wrong, we see that the time that the trains are running and the point at which they meet are both unknown. We let $t =$ the number of hours that the freight train is running before they meet and $d =$ the distance at which the trains meet. Since the freight train has a 2-hr head start, the Metroliner runs for $t - 2$ hours before catching up to the freight train, at which point both trains have traveled the same distance.

60 km/h
d kilometers
t hours

90 km/h
d kilometers
t − 2 hours

Trains meet here

2. **Translate.**   We can organize the information in a chart. Each row is determined by the formula *Distance = Rate · Time*.

	Distance	Rate	Time	
Freight Train	d	60	t	→ $d = 60t$
Metroliner	d	90	$t - 2$	→ $d = 90(t - 2)$

Using *Distance* = *Rate* · *Time* twice, we get two equations:

$$d = 60t, \qquad (1)$$
$$d = 90(t - 2). \qquad (2)$$

3. **Carry out.**   We solve the system using substitution:

$$60t = 90(t - 2) \qquad \text{Substituting } 60t \text{ for } d \text{ in equation (2)}$$
$$60t = 90t - 180$$
$$-30t = -180$$
$$t = 6.$$

The time for the freight train is 6 hr, which means that the time for the Metroliner is 6 − 2, or 4 hr. Remember that it is distance, not time, that the problem asked for. Thus for $t = 6$, we have $d = 60 \cdot 6 = 360$ km.

4. **Check.**   At 60 km/h, the freight train will travel $60 \cdot 6$, or 360 km, in 6 hr. At 90 km/h, the Metroliner will travel $90 \cdot (6 - 2) = 360$ km in 4 hr. The numbers check.

5. **State.**   The freight train will catch up to the Metroliner at a point 360 km from Boston.

EXAMPLE 7

Jet travel.   A Boeing 747-400 jet flies 4 hr west with a 60-mph tailwind. Returning *against* the wind takes 5 hr. Find the speed of the plane with no wind.

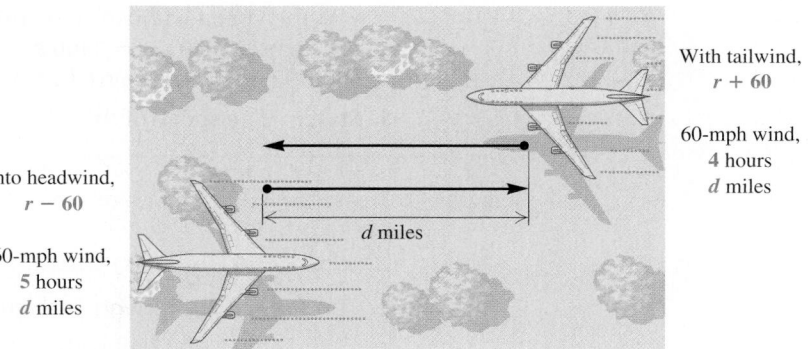

With tailwind,
$r + 60$

60-mph wind,
4 hours
d miles

Into headwind,
$r - 60$

60-mph wind,
5 hours
d miles

d miles

Solution

1. **Familiarize.**   We imagine the situation and make a drawing. Note that the wind *speeds up* the jet on the outbound flight, but *slows down* the jet on the return flight. Since the distances traveled each way must be the same, we can check a guess of the jet's speed with no wind. Suppose the speed of the jet with no wind is 400 mph. The jet would then fly $400 + 60 = 460$ mph with the wind and $400 - 60 = 340$ mph into the wind. In 4 hr, the jet would travel $460 \cdot 4 = 1840$ mi with the wind and $340 \cdot 5 = 1700$ mi against the wind. Since $1840 \neq 1700$, our guess of 400 mph is incorrect. Rather than guess again, let's have $r =$ the speed, in miles per hour, of the jet in still air. Then $r + 60 =$ the jet's speed with the wind and $r - 60 =$ the jet's speed against the wind. We also let $d =$ the distance traveled, in miles.

2. Translate. The information can be organized in a chart. The distances traveled are the same, so we use *Distance* = *Rate* (or *Speed*) · *Time*. Each row of the chart gives an equation.

	Distance	**Rate**	**Time**
With Wind	d	$r + 60$	4
Against Wind	d	$r - 60$	5

$\longrightarrow d = (r + 60)4$

$\longrightarrow d = (r - 60)5$

The two equations constitute a system:

$$d = (r + 60)4, \qquad (1)$$
$$d = (r - 60)5. \qquad (2)$$

3. Carry out. We solve the system using substitution:

$(r - 60)5 = (r + 60)4$ Substituting $(r - 60)5$ for d in equation (1)

$5r - 300 = 4r + 240$ Using the distributive law

$r = 540.$ Solving for r

4. Check. When $r = 540$, the speed with the wind is $540 + 60 = 600$ mph, and the speed against the wind is $540 - 60 = 480$ mph. The distance with the wind, $600 \cdot 4 = 2400$ mi, matches the distance into the wind, $480 \cdot 5 = 2400$ mi, so we have a check.

5. State. The speed of the plane with no wind is 540 mph.

Tips for Solving Motion Problems

1. Draw a diagram using an arrow or arrows to represent distance and the direction of each object in motion.
2. Organize the information in a chart.
3. Look for times, distances, or rates that are the same. These often can lead to an equation.
4. Translating to a system of equations allows for the use of two variables.
5. Always make sure that you have answered the question asked.

	Exercise Set
8.3	

FOR EXTRA HELP

  Student's Solutions Manual    Digital Video Tutor CD 4 Videotape 8   Tutor Center AW Math Tutor Center    MathXL Tutorials on CD    Math XL MathXL    MyMathLab MyMathLab

1.–14. For Exercises 1–14, solve Exercises 41–54 from p. 515.

15. *Printing.*   King Street Printing recently charged 1.9¢ per sheet of paper, but 2.4¢ per sheet for paper made of recycled fibers. Darren's bill for 150 sheets of paper was $3.41. How many sheets of each type were used?

16. *Photocopying.*   Quick Copy recently charged 6¢ a page for copying pages that can be machine-fed and 18¢ a page for copying pages that must be hand-placed on the copier. If Lea's bill for 90 copies was $9.24, how many copies of each type were made?

17. *Lighting.*   Booth Bros. Hardware charges $7.50 for a General Electric Biax Energy Saver light bulb and $5 for an SLi Lighting Cool White Energy Saver bulb. If Paul County Hospital purchased 200 such bulbs for $1150, how many of each type did they purchase?

18. *Office supplies.*   Barlow's Office Supply charges $16.75 for a box of Erase-A-Gel™ pens and $14.25 for a box of Icy™ automatic pencils. If Letsonville Community College purchased 120 such boxes for

$1790, how many boxes of each type did they purchase?

19. *Sales.*   Staples® recently sold a black Apple Stylewriter II ink cartridge for $30.86 and a black HP Designjet 10PS cartridge for $43.58. At the start of a recent fall semester, a total of 50 of these cartridges was sold for a total of $1733.80. How many of each type were purchased?

20. *Sales.*   Staples® recently sold a wirebound graph-paper notebook for $2.50 and a college-ruled note-book made of recycled paper for $2.30. At the start of a recent spring semester, a combination of 50 of these notebooks was sold for a total of $118.60. How many of each type were sold?

21. *Blending coffees.*   The Bean Counter charges $9.00 per pound for Kenyan French Roast coffee and $8.00 per pound for Sumatran coffee. How much of each type should be used to make a 20-lb blend that sells for $8.40 per pound?

22. *Mixed nuts.*   Oh Nuts! sells cashews for $6.75 per pound and Brazil nuts for $5.00 per pound. How much of each type should be used to make a 50-lb mixture that sells for $5.70 per pound?

23. *Catering.*   Casella's Catering is planning a wedding reception. The bride and groom would like to serve a nut mixture containing 25% peanuts. Casella has available mixtures that are either 40% or 10% pea-nuts. How much of each type should be mixed to get a 20-lb mixture that is 25% peanuts?

24. *Ink remover.*   Etch Clean Graphics uses one cleanser that is 25% acid and a second that is 50% acid. How many liters of each should be mixed to get 30 L of a solution that is 40% acid?

25. *Blending granola.*   Deep Thought Granola is 25% nuts and dried fruit. Oat Dream Granola is 10% nuts and dried fruit. How much of Deep

Thought and how much of Oat Dream should be mixed to form a 20-lb batch of granola that is 19% nuts and dried fruit?

26. *Livestock feed.* Soybean meal is 16% protein and corn meal is 9% protein. How many pounds of each should be mixed to get a 350-lb mixture that is 12% protein?

27. *Student loans.* Lomasi's two student loans totaled $12,000. One of her loans was at 6% simple interest and the other at 9%. After one year, Lomasi owed $855 in interest. What was the amount of each loan?

28. *Investments.* An executive nearing retirement made two investments totaling $15,000. In one year, these investments yielded $1432 in simple interest. Part of the money was invested at 9% and the rest at 10%. How much was invested at each rate?

29. *Automotive maintenance.* "Arctic Antifreeze" is 18% alcohol and "Frost No-More" is 10% alcohol. How many liters of each should be mixed to get 20 L of a mixture that is 15% alcohol?

30. *Chemistry.* E-Chem Testing has a solution that is 80% base and another that is 30% base. A technician needs 150 L of a solution that is 62% base. The 150 L will be prepared by mixing the two solutions on hand. How much of each should be used?

31. *Octane ratings.* The octane rating of a gasoline is a measure of the amount of isooctane in the gas. The 2002 Dodge Neon RT requires 91-octane gasoline. How much 87-octane gas and 93-octane gas should Kasey mix in order to make 12 gal of 91-octane gas for her Neon RT?
Sources: Champlain Electric and Petroleum Equipment; Goss Dodge

32. *Octane ratings.* The octane rating of a gasoline is a measure of the amount of isooctane in the gas. The 2005 Chrysler Crossfire requires 93-octane gasoline. How much 87-octane gas and 95-octane gas should Ken mix in order to make 10 gal of 93-octane gas for his Crossfire?
Sources: Champlain Electric and Petroleum Equipment; Freedom Chrysler Plymouth

33. *Food science.* The following bar graph shows the milk fat percentages in three dairy products. How many pounds each of whole milk and cream should be mixed to form 200 lb of milk for cream cheese?

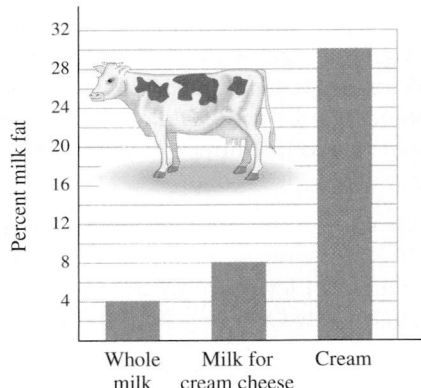

34. *Food science.*   How much lowfat (1% fat) milk and how much whole milk (4% fat) should be mixed to make 5 gal of reduced fat (2% fat) milk?

35. *Train travel.*   A train leaves Danville Junction and travels north at a speed of 75 km/h. Two hours later, an express train leaves on a parallel track and travels north at 125 km/h. How far from the station will they meet?

36. *Car travel.*   Two cars leave Salt Lake City, traveling in opposite directions. One car travels at a speed of 80 km/h and the other at 96 km/h. In how many hours will they be 528 km apart?

37. *Boating.*   Mia's motorboat took 3 hr to make a trip downstream with a 6-mph current. The return trip against the same current took 5 hr. Find the speed of the boat in still water.

38. *Canoeing.*   Alvin paddled for 4 hr with a 6-km/h current to reach a campsite. The return trip against the same current took 10 hr. Find the speed of Alvin's canoe in still water.

39. *Point of no return.*   A plane flying the 3458-mi trip from New York City to London has a 50-mph tailwind. The flight's *point of no return* is the point at which the flight time required to return to New York is the same as the time required to continue to London. If the speed of the plane in still air is 360 mph, how far is New York from the point of no return?

40. *Point of no return.*   A plane is flying the 2553-mi trip from Los Angeles to Honolulu into a 60-mph headwind. If the speed of the plane in still air is 310 mph, how far from Los Angeles is the plane's point of no return? (See Exercise 39.)

41. *Architecture.*   The rectangular ground floor of the John Hancock building has a perimeter of 860 ft. The length is 100 ft more than the width. Find the length and the width.

$x + 100$
x

42. *Real estate.*   The perimeter of a rectangular oceanfront lot is 190 m. The width is one fourth of the length. Find the dimensions.

43. *Real estate.*   In 1996, the Simon Property Group and the DeBartolo Realty Corporation merged to form the largest real estate company in the United States, owning 183 shopping centers in 32 states. Prior to merging, Simon owned twice as many properties as DeBartolo. How many properties did each company own before the merger?

The Indianapolis Star
REAL ESTATE GIANTS MERGE

DeBartolo shareholders would receive 0.68 share of Simon common stock for each share of DeBartolo common stock. Simon also would agree to repay $1.5 billion in DeBartolo debt. At Tuesday's closing price of $ 23.625 a share for common stock, the transaction is valued at roughly $3 billion.

Executives say the proposed company, Simon DeBartolo Group, would be the largest real estate company in the United States, worth $7.5 billion. **Not included in the deal:** DeBartolo's ownership stake in the San Francisco 49ers, or the Indiana Pacers, owned separately by the Simon

44. *Hockey rankings.*   Hockey teams receive 2 points for a win and 1 point for a tie. The Wildcats once won a championship with 60 points. They won 9 more games than they tied. How many wins and how many ties did the Wildcats have?

45. *Radio airplay.*   Roscoe must play 12 commercials during his 1-hr radio show. Each commercial is either 30 sec or 60 sec long. If the total commercial time during that hour is 10 min, how many commercials of each type does Roscoe play?

46. *Video rentals.*   J. P.'s Video rents general-interest films for $3.00 each and children's films for $1.50 each. In one day, a total of $213 was taken in from the rental of 77 videos. How many of each type of video was rented?

47. *Making change.*   Cecilia makes a $9.25 purchase at the bookstore with a $20 bill. The store has no bills and gives her the change in quarters and fifty-cent pieces. There are 30 coins in all. How many of each kind are there?

48. *Teller work.*   Ashford goes to a bank and gets change for a $50 bill consisting of all $5 bills and $1 bills. There are 22 bills in all. How many of each kind are there?

49. In what ways are Examples 3 and 4 similar? In what sense are their systems of equations similar?

50. Write at least three study tips of your own for someone beginning this exercise set.

SKILL MAINTENANCE

Evaluate.

51. $2x - 3y + 12$, for $x = 5$ and $y = 2$ [1.1]

52. $7x - 4y + 9$, for $x = 2$ and $y = 3$ [1.1]

53. $5a - 7b + 3c$, for $a = -2$, $b = 3$, and $c = 1$ [1.1], [1.8]

54. $3a - 8b - 2c$, for $a = -4$, $b = -1$, and $c = 3$ [1.1], [1.8]

55. $4 - 2y + 3z$, for $y = \frac{1}{3}$ and $z = \frac{1}{4}$ [1.3]

56. $3 - 5y + 4z$, for $y = \frac{1}{2}$ and $z = \frac{1}{5}$ [1.3]

SYNTHESIS

57. Suppose that in Example 3 you are asked only for the amount of Assam Gingia needed for the Dragon Blend. Would the method of solving the problem change? Why or why not?

58. Write a problem similar to Example 2 for a classmate to solve. Design the problem so that the solution is "The florist sold 14 hanging plants and 9 flats of petunias."

59. *Recycled paper.*   Unable to purchase 60 reams of paper that contains 20% post-consumer fiber, the Naylor School bought paper that was either 0% post-consumer fiber or 30% post-consumer fiber. How many reams of each should be purchased in order to use the same amount of post-consumer fiber as if the 20% post-consumer fiber paper were available?

60. *Retail.*   Some of the world's best and most expensive coffee is Hawaii's Kona coffee. In order for coffee to be labeled "Kona Blend," it must contain at least 30% Kona beans. Bean Town Roasters has 40 lb of Mexican coffee. How much Kona coffee must they add if they wish to market it as Kona Blend?

61. *Automotive maintenance.*   The radiator in Michelle's car contains 6.3 L of antifreeze and water. This mixture is 30% antifreeze. How much of this mixture should she drain and replace with pure antifreeze so that there will be a mixture of 50% antifreeze?

62. *Exercise.*   Natalie jogs and walks to school each day. She averages 4 km/h walking and 8 km/h

jogging. From home to school is 6 km and Natalie makes the trip in 1 hr. How far does she jog in a trip?

63. *Book sales.*   A limited edition of a book published by a historical society was offered for sale to members. The cost was one book for $12 or two books for $20 (maximum of two per member). The society sold 880 books, for a total of $9840. How many members ordered two books?

64. The tens digit of a two-digit positive integer is 2 more than three times the units digit. If the digits are interchanged, the new number is 13 less than half the given number. Find the given integer. (*Hint*: Let $x =$ the tens-place digit and $y =$ the units-place digit; then $10x + y$ is the number.)

65. *Wood stains.*   Williams' Custom Flooring has 0.5 gal of stain that is 20% brown and 80% neutral. A customer orders 1.5 gal of a stain that is 60% brown and 40% neutral. How much pure brown stain and how much neutral stain should be added to the original 0.5 gal in order to make up the order?*

66. *Train travel.*   A train leaves Union Station for Central Station, 216 km away, at 9 A.M. One hour later, a train leaves Central Station for Union Station. They meet at noon. If the second train had started at 9 A.M. and the first train at 10:30 A.M., they would still have met at noon. Find the speed of each train.

67. *Fuel economy.*   Grady's station wagon gets 18 miles per gallon (mpg) in city driving and 24 mpg in highway driving. The car is driven 465 mi on 23 gal of gasoline. How many miles were driven in the city and how many were driven on the highway?

68. *Biochemistry.*   Industrial biochemists routinely use a machine to mix a buffer of 10% acetone by adding 100% acetone to water. One day, instead of adding 5 L of acetone to create a vat of buffer, a machine added 10 L. How much additional water was needed to bring the concentration down to 10%?

---

*This problem was suggested by Professor Chris Burditt of Yountville, California.

 **69.** See Exercise 65 above. Let $x =$ the amount of pure brown stain added to the original 0.5 gal. Find a function $P(x)$ that can be used to determine the percentage of brown stain in the 1.5-gal mixture. On a graphing calculator, draw the graph of P and use INTERSECT to confirm the answer to Exercise 65.

70. *Gender.* Phil and Phyllis are siblings. Phyllis has twice as many brothers as she has sisters. Phil has the same number of brothers as sisters. How many girls and how many boys are in the family?

8.4 Systems of Equations in Three Variables

Identifying Solutions • Solving Systems in Three Variables • Dependency, Inconsistency, and Geometric Considerations

CONNECTING THE CONCEPTS

As often happens in mathematics, once an idea is thoroughly understood, it can be extended to increasingly more complicated problems. This is precisely the situation for the material in Sections 8.4–8.7: We will extend the elimination method of Section 8.2 to systems of three equations in three unknowns. Although we will not do so in this text, the approach that we use can

be further extended to systems with four equations in four unknowns, five equations in five unknowns, and so on.

Another common occurrence in mathematics is the streamlining of a sequence of steps that are used repeatedly. In Sections 8.6 and 8.7, we develop different notations that streamline the calculations of Sections 8.2 and 8.4.

Some problems translate directly to two equations. Others more naturally call for a translation to three or more equations. In this section, we learn how to solve systems of three linear equations. Later, we will use such systems in problem-solving situations.

Identifying Solutions

A **linear equation in three variables** is an equation equivalent to one in the form $Ax + By + Cz = D$, where A, B, C, and D are real numbers. We refer to the form $Ax + By + Cz = D$ as *standard form* for a linear equation in three variables.

A solution of a system of three equations in three variables is an ordered triple (x, y, z) that makes *all three* equations true.

EXAMPLE 1 Determine whether $\left(\frac{3}{2}, -4, 3\right)$ is a solution of the system

$$4x - 2y - 3z = 5,$$
$$-8x - y + z = -5,$$
$$2x + y + 2z = 5.$$

Solution   We substitute $\left(\frac{3}{2}, -4, 3\right)$ into the three equations, using alphabetical order:

$$
\begin{array}{r|l}
\multicolumn{2}{l}{4x - 2y - 3z = 5} \\
\hline
4 \cdot \frac{3}{2} - 2(-4) - 3 \cdot 3 & 5 \\
6 + 8 - 9 & \\
5 \overset{?}{=} 5 & \text{TRUE}
\end{array}
\qquad
\begin{array}{r|l}
\multicolumn{2}{l}{-8x - y + z = -5} \\
\hline
-8 \cdot \frac{3}{2} - (-4) + 3 & -5 \\
-12 + 4 + 3 & \\
-5 \overset{?}{=} -5 & \text{TRUE}
\end{array}
$$

$$
\begin{array}{r|l}
\multicolumn{2}{l}{2x + y + 2z = 5} \\
\hline
2 \cdot \frac{3}{2} + (-4) + 2 \cdot 3 & 5 \\
3 - 4 + 6 & \\
5 \overset{?}{=} 5 & \text{TRUE}
\end{array}
$$

The triple makes all three equations true, so it is a solution.

Solving Systems in Three Variables

Graphical methods for solving linear equations in three variables are problematic, because a three-dimensional coordinate system is required and the graph of a linear equation in three variables is a plane. The substitution method *can* be used but becomes very cumbersome unless one or more of the equations has only two variables. Fortunately, the elimination method allows us to manipulate a system of three equations in three variables so that a simpler system of two equations in two variables is formed. Once that simpler system has been solved, we can substitute into one of the three original equations and solve for the third variable.

EXAMPLE 2   Solve the following system of equations:

$$
\begin{aligned}
x + y + z &= 4, & (1) \\
x - 2y - z &= 1, & (2) \\
2x - y - 2z &= -1. & (3)
\end{aligned}
$$

Solution   We select *any* two of the three equations and work to get one equation in two variables. Let's add equations (1) and (2):

$$
\begin{array}{rl}
x + y + z = 4 & (1) \\
\underline{x - 2y - z = 1} & (2) \\
2x - y \phantom{{}+ z}= 5. & (4) \qquad \text{Adding to eliminate } z
\end{array}
$$

Next, we select a different pair of equations and eliminate the *same variable* that we did above. Let's use equations (1) and (3) to again eliminate z. Be careful here! A common error is to eliminate a different variable in this step.

$$
\begin{array}{l}
x + y + z = 4, \\
2x - y - 2z = -1
\end{array}
\xrightarrow[\text{of equation (1) by 2}]{\text{Multiplying both sides}}
\begin{array}{rl}
2x + 2y + 2z = 8 \\
\underline{2x - y - 2z = -1} \\
4x + y \phantom{{}+ 2z}= 7 & (5)
\end{array}
$$

Now we solve the resulting system of equations (4) and (5). That solution will give us two of the numbers in the solution of the original system.

$$2x - y = 5 \qquad (4)$$
$$\underline{4x + y = 7} \qquad (5)$$
$$6x = 12 \qquad \text{Adding}$$
$$x = 2$$

Note that we now have two equations in two variables. Had we not eliminated the same variable in both of the above steps, this would not be the case.

We can use either equation (4) or (5) to find y. We choose equation (5):

$$4x + y = 7 \qquad (5)$$
$$4 \cdot 2 + y = 7 \qquad \text{Substituting 2 for } x \text{ in equation (5)}$$
$$8 + y = 7$$
$$y = -1.$$

We now have $x = 2$ and $y = -1$. To find the value for z, we use any of the original three equations and substitute to find the third number, z. Let's use equation (1) and substitute our two numbers in it:

$$x + y + z = 4 \qquad (1)$$
$$2 + (-1) + z = 4 \qquad \text{Substituting 2 for } x \text{ and } -1 \text{ for } y$$
$$1 + z = 4$$
$$z = 3.$$

We have obtained the triple $(2, -1, 3)$. It should check in *all three* equations:

$$\frac{x + y + z = 4}{2 + (-1) + 3 \,\big|\, 4}$$
$$4 \overset{?}{=} 4 \quad \text{TRUE}$$

$$\frac{x - 2y - z = 1}{2 - 2(-1) - 3 \,\big|\, 1}$$
$$1 \overset{?}{=} 1 \quad \text{TRUE}$$

$$\frac{2x - y - 2z = -1}{2 \cdot 2 - (-1) - 2 \cdot 3 \,\big|\, -1}$$
$$-1 \overset{?}{=} -1 \quad \text{TRUE}$$

The solution is $(2, -1, 3)$.

Solving Systems of Three Linear Equations

To use the elimination method to solve systems of three linear equations:

1. Write all equations in the standard form $Ax + By + Cz = D$.
2. Clear any decimals or fractions.
3. Choose a variable to eliminate. Then select two of the three equations and work to get one equation in which the selected variable is eliminated.
4. Next, use a different pair of equations and eliminate the same variable that you did in step (3).
5. Solve the system of equations that resulted from steps (3) and (4).
6. Substitute the solution from step (5) into one of the original three equations and solve for the third variable. Then check.

EXAMPLE 3   Solve the system

$$4x - 2y - 3z = 5, \quad (1)$$
$$-8x - y + z = -5, \quad (2)$$
$$2x + y + 2z = 5. \quad (3)$$

Student Notes _____

Because solving systems of three equations can be lengthy, it is important that you use plenty of paper, work in pencil, and double-check each step as you proceed.

Solution

1., 2. The equations are already in standard form with no fractions or decimals.

3. Next, select a variable to eliminate. We decide on y because the y-terms are opposites of each other in equations (2) and (3). We add:

$$-8x - y + z = -5 \quad (2)$$
$$\underline{2x + y + 2z = 5} \quad (3)$$
$$-6x \qquad + 3z = 0. \quad (4) \qquad \text{Adding}$$

4. We use another pair of equations to create a second equation in x and z. That is, we eliminate the same variable, y, as in step (3). We use equations (1) and (3):

$$4x - 2y - 3z = 5,$$
$$2x + y + 2z = 5$$

$\xrightarrow{\text{Multiplying both sides of equation (3) by 2}}$

$$4x - 2y - 3z = 5$$
$$\underline{4x + 2y + 4z = 10}$$
$$8x \qquad + z = 15. \quad (5)$$

5. Now we solve the resulting system of equations (4) and (5). That allows us to find two parts of the ordered triple.

$$-6x + 3z = 0,$$
$$8x + z = 15$$

$\xrightarrow{\text{Multiplying both sides of equation (5) by } -3}$

$$-6x + 3z = 0$$
$$\underline{-24x - 3z = -45}$$
$$-30x \qquad = -45$$
$$x = \frac{-45}{-30} = \frac{3}{2}$$

We use equation (5) to find z:

$$8x + z = 15$$
$$8 \cdot \frac{3}{2} + z = 15 \qquad \text{Substituting } \frac{3}{2} \text{ for } x$$
$$12 + z = 15$$
$$z = 3.$$

6. Finally, we use any of the original equations and substitute to find the third number, y. We choose equation (3):

$$2x + y + 2z = 5 \quad (3)$$
$$2 \cdot \frac{3}{2} + y + 2 \cdot 3 = 5 \qquad \text{Substituting } \frac{3}{2} \text{ for } x \text{ and } 3 \text{ for } z$$
$$3 + y + 6 = 5$$
$$y + 9 = 5$$
$$y = -4.$$

The solution is $\left(\frac{3}{2}, -4, 3\right)$. The check was performed as Example 1.

Sometimes, certain variables are missing at the outset.

EXAMPLE 4   Solve the system

$$x + y + z = 180, \quad (1)$$
$$x \quad - z = -70, \quad (2)$$
$$2y - z = 0. \quad (3)$$

Solution

1., 2. The equations appear in standard form with no fractions or decimals.

3., 4. Note that there is no y in equation (2). Thus, at the outset, we already have y eliminated from one equation. We need another equation with y eliminated, so we use equations (1) and (3):

$$
\begin{array}{l}
x + y + z = 180, \\
2y - z = 0
\end{array}
\quad
\xrightarrow[\text{of equation (1) by } -2]{\text{Multiplying both sides}}
\quad
\begin{array}{r}
-2x - 2y - 2z = -360 \\
2y - z = 0 \\
\hline
-2x \quad - 3z = -360. \quad (4)
\end{array}
$$

5., 6. Now we solve the resulting system of equations (2) and (4):

$$
\begin{array}{l}
x - z = -70, \\
-2x - 3z = -360
\end{array}
\quad
\xrightarrow[\text{of equation (2) by } 2]{\text{Multiplying both sides}}
\quad
\begin{array}{r}
2x - 2z = -140 \\
-2x - 3z = -360 \\
\hline
-5z = -500 \\
z = 100.
\end{array}
$$

Continuing as in Examples 2 and 3, we get the solution (30, 50, 100). The check is left to the student.

Dependency, Inconsistency, and Geometric Considerations

Each equation in Examples 2, 3, and 4 has a graph that is a plane in three dimensions. The solutions are points common to the planes of each system. Since three planes can have an infinite number of points in common or no points at all in common, we need to generalize the concept of *consistency*.

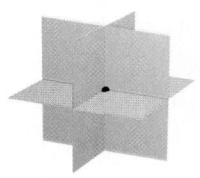 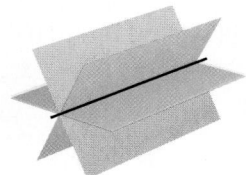  

Planes intersect at one point. System is *consistent* and has one solution.

Planes intersect along a common line. System is *consistent* and has an infinite number of solutions.

Three parallel planes. System is *inconsistent;* it has no solution.

Planes intersect two at a time, with no point common to all three. System is *inconsistent;* it has no solution.

> *Consistency*
>
> A system of equations that has at least one solution is said to be **consistent**.
>
> A system of equations that has no solution is said to be **inconsistent**.

EXAMPLE 5 Solve:

$$y + 3z = 4, \quad (1)$$
$$-x - y + 2z = 0, \quad (2)$$
$$x + 2y + z = 1. \quad (3)$$

Solution The variable x is missing in equation (1). By adding equations (2) and (3), we can find a second equation in which x is missing:

$$-x - y + 2z = 0 \quad (2)$$
$$\underline{x + 2y + z = 1} \quad (3)$$
$$y + 3z = 1. \quad (4) \qquad \text{Adding}$$

Equations (1) and (4) form a system in y and z. We solve as before:

$$y + 3z = 4, \quad \xrightarrow[\text{of equation (1) by } -1]{\text{Multiplying both sides}} \quad -y - 3z = -4$$
$$y + 3z = 1 \qquad\qquad\qquad\qquad \underline{y + 3z = 1}$$
$$\text{This is a contradiction.} \longrightarrow 0 = -3. \qquad \text{Adding}$$

Since we end up with a *false* equation, or contradiction, we know that the system has no solution. It is *inconsistent*.

The notion of *dependency* from Section 8.1 can also be extended.

EXAMPLE 6 Solve:

$$2x + y + z = 3, \quad (1)$$
$$x - 2y - z = 1, \quad (2)$$
$$3x + 4y + 3z = 5. \quad (3)$$

Solution Our plan is to first use equations (1) and (2) to eliminate z. Then we will select another pair of equations and again eliminate z:

$$2x + y + z = 3$$
$$\underline{x - 2y - z = 1}$$
$$3x - y = 4. \quad (4)$$

Next, we use equations (2) and (3) to eliminate z again:

$$x - 2y - z = 1, \quad \xrightarrow[\text{of equation (2) by } 3]{\text{Multiplying both sides}} \quad 3x - 6y - 3z = 3$$
$$3x + 4y + 3z = 5 \qquad\qquad\qquad\qquad \underline{3x + 4y + 3z = 5}$$
$$6x - 2y = 8. \quad (5)$$

We now try to solve the resulting system of equations (4) and (5):

$$3x - y = 4,$$
$$6x - 2y = 8$$

Multiplying both sides of equation (4) by -2

$$-6x + 2y = -8$$
$$\underline{6x - 2y = 8}$$
$$0 = 0. \qquad (6)$$

Equation (6), which is an identity, indicates that equations (1), (2), and (3) are *dependent*. This means that the original system of three equations is equivalent to a system of two equations. One way to see this is to observe that two times equation (1), minus equation (2), is equation (3). Thus removing equation (3) from the system does not affect the solution of the system.* In writing an answer to this problem, we simply state that "the equations are dependent."

Recall that when dependent equations appeared in Section 8.1, the solution sets were always infinite in size and were written in set-builder notation. There, all systems of dependent equations were *consistent*. This is not always the case for systems of three or more equations. The following figures illustrate some possibilities geometrically.

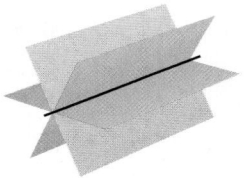

The planes intersect along a common line. The equations are *dependent* and the system is *consistent*. There is an infinite number of solutions.

The planes coincide. The equations are *dependent* and the system is *consistent*. There is an infinite number of solutions.

Two planes coincide. The third plane is parallel. The equations are *dependent* and the system is *inconsistent*. There is no solution.

Exercise Set

8.4

FOR EXTRA HELP

Student's Solutions Manual

Digital Video Tutor CD 4 Videotape 8

AW Math Tutor Center

MathXL Tutorials on CD

Math XL
MathXL

MyMathLab
MyMathLab

↪ *Concept Reinforcement   Classify each statement as either true or false.*

1. $3x + 5y + 4z = 7$ is a linear equation in three variables.

2. It is not difficult to solve a system of three equations in three unknowns by graphing.

3. Every system of three equations in three unknowns has at least one solution.

4. If, when we are solving a system of three equations, a false equation results from adding a multiple of one equation to another, the system is inconsistent.

*A set of equations is dependent if at least one equation can be expressed as a sum of multiples of other equations in that set.

5. If, when we are solving a system of three equations, an identity results from adding a multiple of one equation to another, the equations are dependent.

6. Whenever a system of three equations contains dependent equations, there is an infinite number of solutions.

7. Determine whether $(2, -1, -2)$ is a solution of the system
$$x + y - 2z = 5,$$
$$2x - y - z = 7,$$
$$-x - 2y + 3z = 6.$$

8. Determine whether $(1, -2, 3)$ is a solution of the system
$$x + y + z = 2,$$
$$x - 2y - z = 2,$$
$$3x + 2y + z = 2.$$

Solve each system. If a system's equations are dependent or if there is no solution, state this.

9. $2x - y + z = 10,$
$4x + 2y - 3z = 10,$
$x - 3y + 2z = 8$

10. $x + y + z = 6,$
$2x - y + 3z = 9,$
$-x + 2y + 2z = 9$

11. $x - y + z = 6,$
$2x + 3y + 2z = 2,$
$3x + 5y + 4z = 4$

12. $2x - y - 3z = -1,$
$2x - y + z = -9,$
$x + 2y - 4z = 17$

13. $6x - 4y + 5z = 31,$
$5x + 2y + 2z = 13,$
$x + y + z = 2$

14. $2x - 3y + z = 5,$
$x + 3y + 8z = 22,$
$3x - y + 2z = 12$

15. $x + y + z = 0,$
$2x + 3y + 2z = -3,$
$-x - 2y - z = 1$

16. $3a - 2b + 7c = 13,$
$a + 8b - 6c = -47,$
$7a - 9b - 9c = -3$

17. $2x + y - 3z = -4,$
$4x - 2y + z = 9,$
$3x + 5y - 2z = 5$

18. $4x + y + z = 17,$
$x - 3y + 2z = -8,$
$5x - 2y + 3z = 5$

19. $2x + y + 2z = 11,$
$3x + 2y + 2z = 8,$
$x + 4y + 3z = 0$

20. $2x + y + z = -2,$
$2x - y + 3z = 6,$
$3x - 5y + 4z = 7$

21. $-2x + 8y + 2z = 4,$
$x + 6y + 3z = 4,$
$3x - 2y + z = 0$

22. $x - y + z = 4,$
$5x + 2y - 3z = 2,$
$4x + 3y - 4z = -2$

23. $4x - y - z = 4,$
$2x + y + z = -1,$
$6x - 3y - 2z = 3$

24. $a + 2b + c = 1,$
$7a + 3b - c = -2,$
$a + 5b + 3c = 2$

25. $r + \frac{3}{2}s + 6t = 2,$
$2r - 3s + 3t = 0.5,$
$r + s + t = 1$

26. $5x + 3y + \frac{1}{2}z = \frac{7}{2},$
$0.5x - 0.9y - 0.2z = 0.3,$
$3x - 2.4y + 0.4z = -1$

27. $4a + 9b = 8,$
$8a + 6c = -1,$
$6b + 6c = -1$

28. $3p + 2r = 11,$
$q - 7r = 4,$
$p - 6q = 1$

29. $x + y + z = 57,$
$-2x + y = 3,$
$x - z = 6$

30. $x + y + z = 105,$
$10y - z = 11,$
$2x - 3y = 7$

31. $a - 3c = 6,$
$b + 2c = 2,$
$7a - 3b - 5c = 14$

32. $2a - 3b = 2,$
$7a + 4c = \frac{3}{4},$
$2c - 3b = 1$

Aha! 33. $x + y + z = 83,$
$y = 2x + 3,$
$z = 40 + x$

34. $l + m = 7,$
$3m + 2n = 9,$
$4l + n = 5$

35. $x + z = 0,$
$x + y + 2z = 3,$
$y + z = 2$

36. $x + y = 0,$
$x + z = 1,$
$2x + y + z = 2$

37. $x + y + z = 1,$
$-x + 2y + z = 2,$
$2x - y = -1$

38. $y + z = 1,$
$x + y + z = 1,$
$x + 2y + 2z = 2$

39. Describe a method for writing an inconsistent system of three equations in three variables.

40. Abbie recommends that a frustrated classmate double- and triple-check each step of work when attempting to solve a system of three equations. Is this good advice? Why or why not?

SKILL MAINTENANCE

Translate each sentence to mathematics. [1.1]

41. One number is twice another.

42. The sum of two numbers is three times the first number.

43. The sum of three consecutive numbers is 45.

44. One number plus twice another number is 17.

45. The sum of two numbers is five times a third number.

46. The product of two numbers is twice their sum.

SYNTHESIS

47. Is it possible for a system of three linear equations to have exactly two ordered triples in its solution set? Why or why not?

48. Describe a procedure that could be used to solve a system of four equations in four variables.

Solve.

49. $\dfrac{x+2}{3} - \dfrac{y+4}{2} + \dfrac{z+1}{6} = 0,$

$\dfrac{x-4}{3} + \dfrac{y+1}{4} - \dfrac{z-2}{2} = -1,$

$\dfrac{x+1}{2} + \dfrac{y}{2} + \dfrac{z-1}{4} = \dfrac{3}{4}$

50. $w + x + y + z = 2,$
$w + 2x + 2y + 4z = 1,$
$w - x + y + z = 6,$
$w - 3x - y + z = 2$

51. $w + x - y + z = 0,$
$w - 2x - 2y - z = -5,$
$w - 3x - y + z = 4,$
$2w - x - y + 3z = 7$

For Exercises 52 and 53, let u represent 1/x, v represent 1/y, and w represent 1/z. Solve for u, v, and w, and then solve for x, y, and z.

52. $\dfrac{2}{x} - \dfrac{1}{y} - \dfrac{3}{z} = -1,$

$\dfrac{2}{x} - \dfrac{1}{y} + \dfrac{1}{z} = -9,$

$\dfrac{1}{x} + \dfrac{2}{y} - \dfrac{4}{z} = 17$

53. $\dfrac{2}{x} + \dfrac{2}{y} - \dfrac{3}{z} = 3,$

$\dfrac{1}{x} - \dfrac{2}{y} - \dfrac{3}{z} = 9,$

$\dfrac{7}{x} - \dfrac{2}{y} + \dfrac{9}{z} = -39$

Determine k so that each system is dependent.

54. $x - 3y + 2z = 1,$
$2x + y - z = 3,$
$9x - 6y + 3z = k$

55. $5x - 6y + kz = -5,$
$x + 3y - 2z = 2,$
$2x - y + 4z = -1$

In each case, three solutions of an equation in x, y, and z are given. Find the equation.

56. $Ax + By + Cz = 12;$
$\left(1, \frac{3}{4}, 3\right), \left(\frac{4}{3}, 1, 2\right),$ and $(2, 1, 1)$

57. $z = b - mx - ny;$
$(1, 1, 2), (3, 2, -6),$ and $\left(\frac{3}{2}, 1, 1\right)$

58. Write an inconsistent system of equations that contains dependent equations.

CORNER

Finding the Preferred Approach

COLLABORATIVE

Focus: Systems of three linear equations

Time: 10–15 minutes

Group size: 3

Consider the six steps outlined on p. 183 along with the following system:

$2x + 4y = 3 - 5z,$

$0.3x = 0.2y + 0.7z + 1.4,$

$0.04x + 0.03y = 0.07 + 0.04z.$

ACTIVITY

1. Working independently, each group member should solve the system above. One person should begin by eliminating x, one should first eliminate y, and one should first eliminate z. Write neatly so that others can follow your steps.

2. Once all group members have solved the system, compare your answers. If the answers do not check, exchange notebooks and check each other's work. If a mistake is detected, allow the person who made the mistake to make the repair.

3. Decide as a group which of the three approaches above (if any) ranks as easiest and which (if any) ranks as most difficult. Then compare your rankings with the other groups in the class.

8.5 Solving Applications: Systems of Three Equations

Applications of Three Equations in Three Unknowns

Solving systems of three or more equations is important in many applications. Such systems arise in the natural and social sciences, business, and engineering. In mathematics, purely numerical applications also arise.

EXAMPLE 1   The sum of three numbers is 4. The first number minus twice the second, minus the third is 1. Twice the first number minus the second, minus twice the third is −1. Find the numbers.

Solution

1. **Familiarize.**   There are three statements involving the same three numbers. Let's label these numbers x, y, and z.

2. **Translate.**   We can translate directly as follows.

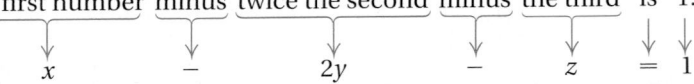

The sum of the three numbers is 4.

$$x + y + z = 4$$

The first number minus twice the second minus the third is 1.

$$x - 2y - z = 1$$

Twice the first number minus the second minus twice the third is −1.

$$2x - y - 2z = -1$$

We now have a system of three equations:

$$
\begin{aligned}
x + y + z &= 4, \\
x - 2y - z &= 1, \\
2x - y - 2z &= -1.
\end{aligned}
$$

3. **Carry out.**   We need to solve the system of equations. Note that we found the solution, $(2, -1, 3)$, in Example 2 of Section 8.4.

4. **Check.**   The first statement of the problem says that the sum of the three numbers is 4. That checks, because $2 + (-1) + 3 = 4$. The second statement says that the first number minus twice the second, minus the third is 1: $2 - 2(-1) - 3 = 1$. That checks. The check of the third statement is left to the student.

5. **State.**   The three numbers are 2, −1, and 3.

EXAMPLE 2   Architecture.   In a triangular cross section of a roof, the largest angle is 70° greater than the smallest angle. The largest angle is twice as large as the remaining angle. Find the measure of each angle.

Solution

1. Familiarize.   The first thing we do is make a drawing, or a sketch.

Since we don't know the size of any angle, we use x, y, and z to represent the three measures, from smallest to largest. Recall that the measures of the angles in any triangle add up to 180°.

2. Translate.   This geometric fact about triangles gives us one equation:

$$x + y + z = 180.$$

Two of the statements can be translated almost directly.

The largest angle	is	70° greater than the smallest angle.
↓	↓	↓
z	=	$x + 70$

The largest angle	is	twice as large as the remaining angle.
↓	↓	↓
z	=	$2y$

We now have a system of three equations:

$$
\begin{aligned}
x + y + z &= 180, \\
x + 70 &= z, \\
2y &= z;
\end{aligned}
\qquad \text{or} \qquad
\begin{aligned}
x + y + z &= 180, \\
x \qquad - z &= -70, \\
2y - z &= 0.
\end{aligned}
\qquad
\begin{array}{l}
\text{Rewriting in} \\
\text{standard form}
\end{array}
$$

3. Carry out.   The system was solved in Example 4 of Section 8.4. The solution is (30, 50, 100).

4. Check.   The sum of the numbers is 180, so that checks. The measure of the largest angle, 100°, is 70° greater than the measure of the smallest angle, 30°, so that checks. The measure of the largest angle is also twice the measure of the remaining angle, 50°. Thus we have a check.

5. State.   The angles in the triangle measure 30°, 50°, and 100°.

EXAMPLE 3   Cholesterol levels.   Recent studies indicate that a child's intake of cholesterol should be no more than 300 mg per day. By eating 1 egg, 1 cupcake, and 1 slice of pizza, a child consumes 302 mg of cholesterol. A child who eats 2 cupcakes and 3 slices of pizza takes in 65 mg of cholesterol. By eating 2 eggs and 1 cupcake, a child consumes 567 mg of cholesterol. How much cholesterol is in each item?

Solution

1. **Familiarize.**   After reading the problem, it becomes clear that an egg contains considerably more cholesterol than the other foods. Let's guess that one egg contains 200 mg of cholesterol and one cupcake contains 50 mg. Because of the second sentence in the problem, it would follow that a slice of pizza contains 52 mg of cholesterol since $200 + 50 + 52 = 302$.

   To see if our guess satisfies the other statements in the problem, we find the amount of cholesterol that 2 cupcakes and 3 slices of pizza would contain: $2 \cdot 50 + 3 \cdot 52 = 256$. Since this does not match the 65 mg listed in the fourth sentence of the problem, our guess was incorrect. Rather than guess again, we examine how we checked our guess and let g, c, and $s =$ the number of milligrams of cholesterol in an egg, a cupcake, and a slice of pizza, respectively.

2. **Translate.**   Rewording some of the sentences, we can translate as follows:

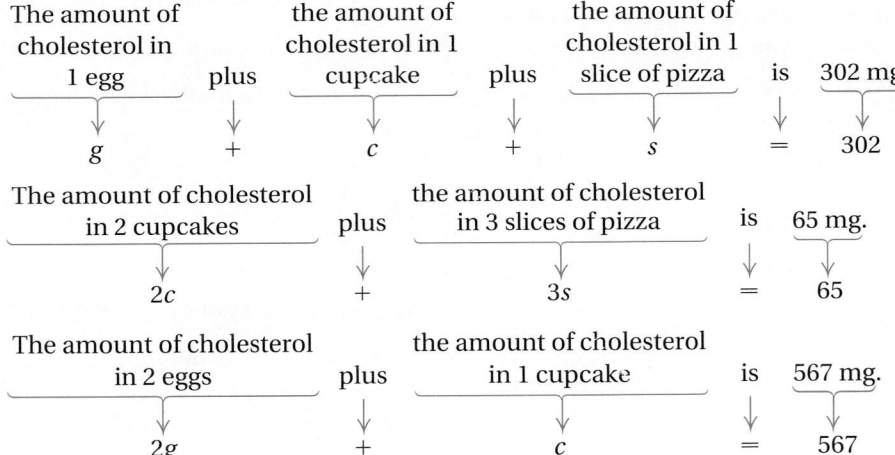

We now have a system of three equations:

$$g + c + s = 302,$$
$$2c + 3s = 65,$$
$$2g + c = 567.$$

3. **Carry out.**   We solve and get $g = 274$, $c = 19$, and $s = 9$.

4. **Check.**   The sum of 274, 19, and 9 is 302 so the total cholesterol in 1 egg, 1 cupcake, and 1 slice of pizza checks. Two cupcakes and three slices of pizza would contain $2 \cdot 19 + 3 \cdot 9 = 65$ mg, while two eggs and one cupcake would contain $2 \cdot 274 + 19 = 567$ mg of cholesterol. The answer checks.

5. **State.**   An egg contains 274 mg of cholesterol, a cupcake contains 19 mg of cholesterol, and a slice of pizza contains 9 mg of cholesterol.

Exercise Set

8.5

Solve.

1. The sum of three numbers is 57. The second is 3 more than the first. The third is 6 more than the first. Find the numbers.

2. The sum of three numbers is 5. The first number minus the second plus the third is 1. The first minus the third is 3 more than the second. Find the numbers.

3. The sum of three numbers is 26. Twice the first minus the second is 2 less than the third. The third is the second minus three times the first. Find the numbers.

4. The sum of three numbers is 105. The third is 11 less than ten times the second. Twice the first is 7 more than three times the second. Find the numbers.

5. *Geometry.* In triangle *ABC*, the measure of angle *B* is three times that of angle *A*. The measure of angle *C* is 20° more than that of angle *A*. Find the angle measures.

6. *Geometry.* In triangle *ABC*, the measure of angle *B* is twice the measure of angle *A*. The measure of angle *C* is 80° more than that of angle *A*. Find the angle measures.

7. *Health insurance.* In 2004, UNICARE® health insurance for adults under the age of 30 cost $121/month for a couple, $107/month for an adult with one child, and $164/month for a couple with one child. On the basis of the information given, find the monthly rate for an individual adult, a spouse, and a child.
Source: UNICARE Life and Health Insurance Company® advertisement

Affordable health insurance is just a phone call away.	
Monthly Rates	Under 30
Applicant and spouse	$121
Applicant and child	107
Family with one child	164
Family with two children	
One child only	

UNICARE

8. *Health insurance.* In 2004, UNICARE® health insurance cost $160/month for a 35–39-year-old adult and spouse, $145/month for a 35–39-year-old adult with one child, and $245/month for a 39-year-old adult with a spouse and child. On the basis of the information given, find the monthly rates for a 39-year-old adult, a 39-year-old's spouse, and a 39-year-old's child.
Source: UNICARE Life and Insurance Company® advertisement

9. *Nutrition.* Most nutritionists now agree that a healthy adult diet should include 25–35 g of fiber each day. A breakfast of 2 bran muffins, 1 banana, and a 1-cup serving of Wheaties® contains 9 g of fiber; a breakfast of 1 bran muffin, 2 bananas, and a 1-cup serving of Wheaties® contains 10.5 g of fiber; and a breakfast of 2 bran muffins and a 1-cup serving of Wheaties® contains 6 g of fiber. How much fiber is in each of these foods?
Sources: usda.gov and InteliHealth.com

10. *Nutrition.* Refer to Exercise 9. A breakfast consisting of 2 pancakes and a 1-cup serving of strawberries contains 4.5 g of fiber, whereas a breakfast of 2 pancakes and a 1-cup serving of Cheerios® contains 4 g of fiber. When a meal consists of 1 pancake, a 1-cup serving of Cheerios®, and a 1-cup serving of strawberries, it contains 7 g of fiber. How much fiber is in each of these foods?
Source: InteliHealth.com

Aha! **11.** *Automobile pricing.* The basic model of a 2004 Jeep Grand Cherokee Laredo (2WD) with a power sunroof cost $25,495. When equipped with 4WD and a sunroof, the vehicle's price rose to $27,465. The cost of the basic model with 4WD was $26,665. Find the basic price, the cost of 4WD, and the cost of a sunroof.

12. *Lens production.* When Sight-Rite's three polishing machines, A, B, and C, are all working, 5700 lenses can be polished in one week. When only A and B are working, 3400 lenses can be polished in one week. When only B and C are working, 4200 lenses can be polished in one week. How many lenses can be polished in a week by each machine?

13. *Welding rates.* Elrod, Dot, and Wendy can weld 74 linear feet per hour when working together. Elrod and Dot together can weld 44 linear feet per hour, while Elrod and Wendy can weld 50 linear feet per hour. How many linear feet per hour can each weld alone?

14. *Telemarketing.* Sven, Tillie, and Isaiah can process 740 telephone orders per day. Sven and Tillie together can process 470 orders, while Tillie and Isaiah together can process 520 orders per day. How many orders can each person process alone?

15. *Coffee prices.* Roz works at a Starbucks® coffee shop where a 12-oz cup of coffee costs $1.40, a 16-oz cup costs $1.60, and a 20-oz cup costs $1.70. During one busy period, Roz served 55 cups of coffee, emptying six 144-oz "brewers" while collecting a total of $85.90. How many cups of each size did Roz fill?

| 12 oz | 16 oz | 20 oz |
| $1.40 | $1.60 | $1.70 |

16. *Advertising.* In a recent year, U.S. companies spent a total of $106.5 billion on newspaper, television, and radio ads. The total amount spent on television and radio ads was $18.7 billion more than the amount spent on newspaper ads alone. The amount spent on newspaper ads was $28.8 billion more than what was spent on radio ads. How much was spent on each form of advertising?
Sources: NAA (newspapers); McCann–Erickson Inc. (television and radio)

17. *Restaurant management.* McDonald's® recently sold small soft drinks for $1, medium soft drinks for $1.15, and large soft drinks for $1.30. During a lunch-time rush, Chris sold 40 soft drinks for a total of $45.25. The number of small and large drinks, combined, was 10 fewer than the number of medium drinks. How many drinks of each size were sold?

18. *Investments.* A business class divided an imaginary investment of $80,000 among three mutual funds. The first fund grew by 10%, the second by 6%, and the third by 15%. Total earnings were $8850. The earnings from the first fund were $750 more than the earnings from the third. How much was invested in each fund?

19. *Nutrition.* A dietician in a hospital prepares meals under the guidance of a physician. Suppose that for a particular patient a physician prescribes a meal to have 800 calories, 55 g of protein, and 220 mg of vitamin C. The dietician prepares a meal of roast beef, baked potatoes, and broccoli according to the data in the following table.

Serving Size	Calories	Protein (in grams)	Vitamin C (in milligrams)
Roast Beef, 3 oz	300	20	0
Baked Potato, 1	100	5	20
Broccoli, 156 g	50	5	100

How many servings of each food are needed in order to satisfy the doctor's orders?

20. *Nutrition.* Repeat Exercise 19 but replace the broccoli with asparagus, for which a 180-g serving contains 50 calories, 5 g of protein, and 44 mg of vitamin C. Which meal would you prefer eating?

21. *World population growth.*   The world population is projected to be 9.1 billion in 2050. At that time, there are expected to be approximately 3 billion more people in Asia than in Africa. The population for the rest of the world will be approximately 0.1 billion more than half the population of Asia. Find the projected populations of Asia, Africa, and the rest of the world in 2050.

Sources: U.S. Bureau of the Census; *Burlington Free Press* 3/23/04

22. *Crying rate.*   The sum of the average number of times a man, a woman, and a one-year-old child cry each month is 56.7. A woman cries 3.9 more times than a man. The average number of times a one-year-old cries per month is 43.3 more than the average number of times combined that a man and a woman cry. What is the average number of times per month that each cries?

23. *Basketball scoring.*   The New York Knicks recently scored a total of 92 points on a combination of 2-point field goals, 3-point field goals, and 1-point foul shots. Altogether, the Knicks made 50 baskets and 19 more 2-pointers than foul shots. How many shots of each kind were made?

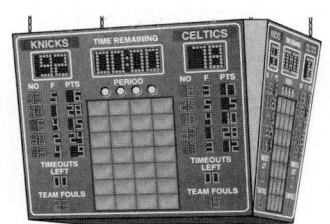

24. *History.*   Find the year in which the first U.S. transcontinental railroad was completed. The following are some facts about the number. The sum of the digits in the year is 24. The ones digit is 1 more than the hundreds digit. Both the tens and the ones digits are multiples of 3.

25. Problems like Exercises 15 and 17 could be classified as total-value problems. How do these problems differ from the total-value problems of Section 8.3?

26. Write a problem for a classmate to solve. Design the problem so that it translates to a system of three equations in three variables.

SKILL MAINTENANCE

Simplify. [1.8]

27. $5(-3) + 7$ **28.** $-4(-6) + 9$

29. $-6(8) + (-7)$ **30.** $7(-9) + (-8)$

31. $-7(2x - 3y + 5z)$ **32.** $-6(4a + 7b - 9c)$

33. $-4(2a + 5b) + 3a + 20b$

34. $3(2x - 7y) + 5x + 21y$

SYNTHESIS

35. Consider Exercise 23. Suppose there were no foul shots made. Would there still be a solution? Why or why not?

36. Consider Exercise 15. Suppose Roz collected $46. Could the problem still be solved? Why or why not?

37. *Health insurance.*   In 2004, UNICARE® health insurance for a 35–39-year-old and his or her spouse cost $160/month. That rate increased to $203/month if a child were included and $243/month if two children were included. The rate dropped to $145/month for just the applicant and one child. Find the separate costs for insuring the applicant, the spouse, the first child, and the second child.

Source: UNICARE Life and Health Insurance Company® advertisement

38. Find a three-digit positive integer such that the sum of all three digits is 14, the tens digit is 2 more than the ones digit, and if the digits are reversed, the number is unchanged.

39. *Ages.*   Tammy's age is the sum of the ages of Carmen and Dennis. Carmen's age is 2 more than the sum of the ages of Dennis and Mark. Dennis's age is four times Mark's age. The sum of all four ages is 42. How old is Tammy?

40. *Ticket revenue.* A magic show's audience of 100 people consists of adults, students, and children. The ticket prices are $10 for adults, $3 for students, and 50¢ for children. The total amount of money taken in is $100. How many adults, students, and children are in attendance? Does there seem to be some information missing? Do some more careful reasoning.

41. *Sharing raffle tickets.* Hal gives Tom as many raffle tickets as Tom first had and Gary as many as Gary first had. In like manner, Tom then gives Hal and Gary as many tickets as each then has. Similarly, Gary gives Hal and Tom as many tickets

as each then has. If each finally has 40 tickets, with how many tickets does Tom begin?

42. Find the sum of the angle measures at the tips of the star in this figure.

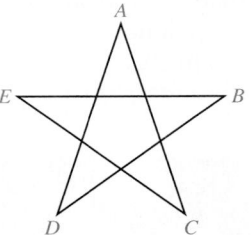

8.6   Elimination Using Matrices

Matrices and Systems • Row-Equivalent Operations

In solving systems of equations, we perform computations with the constants. The variables play no important role until the end. Thus we can simplify writing a system by omitting the variables. For example, the system

$$3x + 4y = 5,$$
$$x - 2y = 1$$

simplifies to

$$\begin{array}{ccc} 3 & 4 & 5 \\ 1 & -2 & 1 \end{array}$$

if we do not write the variables, the operation of addition, and the equals signs.

Matrices and Systems

In the example above, we have written a rectangular array of numbers. Such an array is called a **matrix** (plural, **matrices**). We ordinarily write brackets around matrices. The following are matrices:

$$\begin{bmatrix} -3 & 1 \\ 0 & 5 \end{bmatrix}, \begin{bmatrix} 2 & 0 & -1 & 3 \\ -5 & 2 & 7 & -1 \\ 4 & 5 & 3 & 0 \end{bmatrix}, \begin{bmatrix} 2 & 3 \\ 7 & 15 \\ -2 & 23 \\ 4 & 1 \end{bmatrix}$$

The individual numbers are called *elements* or *entries*.

The **rows** of a matrix are horizontal, and the **columns** are vertical.

$$\begin{bmatrix} 5 & -2 & 2 \\ 1 & 0 & 1 \\ 0 & 1 & 2 \end{bmatrix} \begin{array}{l} \longrightarrow \text{row 1} \\ \longrightarrow \text{row 2} \\ \longrightarrow \text{row 3} \end{array}$$

column 1 column 2 column 3

Let's see how matrices can be used to solve a system.

EXAMPLE 1   Solve the system

$$5x - 4y = -1,$$
$$-2x + 3y = 2.$$

As an aid for understanding, we list the corresponding system in the margin.

$$5x - 4y = -1,$$
$$-2x + 3y = 2$$

Solution   We write a matrix using only coefficients and constants, listing *x*-coefficients in the first column and *y*-coefficients in the second. Note that in each matrix a dashed line separates the coefficients from the constants:

$$\begin{bmatrix} 5 & -4 & \vdots & -1 \\ -2 & 3 & \vdots & 2 \end{bmatrix}.$$

Consult the notes in the margin for further information.

Our goal is to transform

$$\begin{bmatrix} 5 & -4 & \vdots & -1 \\ -2 & 3 & \vdots & 2 \end{bmatrix} \quad \text{into the form} \quad \begin{bmatrix} a & b & \vdots & c \\ 0 & d & \vdots & e \end{bmatrix}.$$

The variables *x* and *y* can then be reinserted to form equations from which we can complete the solution.

We do calculations that are similar to those that we would do if we wrote the entire equations. The first step is to multiply and/or interchange the rows so that each number in the first column below the first number is a multiple of that number. Here that means multiplying Row 2 by 5. This corresponds to multiplying both sides of the second equation by 5.

$$5x - 4y = -1,$$
$$-10x + 15y = 10$$

$$\begin{bmatrix} 5 & -4 & \vdots & -1 \\ -10 & 15 & \vdots & 10 \end{bmatrix}$$

New Row 2 = 5(Row 2 from above)

Next, we multiply the first row by 2, add this to Row 2, and write that result as the "new" Row 2. This corresponds to multiplying the first equation by 2 and adding the result to the second equation in order to eliminate a variable. Write out these computations as necessary—we perform them mentally.

$$5x - 4y = -1,$$
$$7y = 8$$

$$\begin{bmatrix} 5 & -4 & \vdots & -1 \\ 0 & 7 & \vdots & 8 \end{bmatrix}$$

$2(5 \quad -4 \ \vdots \ -1) = (10 \quad -8 \ \vdots \ -2)$ and
$(10 \quad -8 \ \vdots \ -2) + (-10 \quad 15 \ \vdots \ 10) = (0 \quad 7 \ \vdots \ 8)$
New Row 2 = 2(Row 1) + (Row 2)

If we now reinsert the variables, we have

$$5x - 4y = -1, \qquad (1)$$
$$7y = 8. \qquad (2)$$

We can now proceed as before, solving equation (2) for *y*:

$$7y = 8 \qquad (2)$$
$$y = \tfrac{8}{7}.$$

Next, we substitute $\tfrac{8}{7}$ for *y* in equation (1):

$$5x - 4y = -1 \qquad (1)$$
$$5x - 4 \cdot \tfrac{8}{7} = -1 \qquad \text{Substituting } \tfrac{8}{7} \text{ for } y \text{ in equation (1)}$$
$$x = \tfrac{5}{7}. \qquad \text{Solving for } x$$

The solution is $\left(\tfrac{5}{7}, \tfrac{8}{7}\right)$. The check is left to the student.

EXAMPLE 2   Solve the system

$$2x - y + 4z = -3,$$
$$x \qquad - 4z = 5,$$
$$6x - y + 2z = 10.$$

Solution   We first write a matrix, using only the constants. Where there are missing terms, we must write 0's:

$$2x - y + 4z = -3,$$
$$x \qquad - 4z = 5,$$
$$6x - y + 2z = 10$$

$$\begin{bmatrix} 2 & -1 & 4 & | & -3 \\ 1 & 0 & -4 & | & 5 \\ 6 & -1 & 2 & | & 10 \end{bmatrix}$$

Our goal is to transform the matrix to one of the form

$$ax + by + cz = d,$$
$$ey + fz = g,$$
$$hz = i$$

$$\begin{bmatrix} a & b & c & | & d \\ 0 & e & f & | & g \\ 0 & 0 & h & | & i \end{bmatrix}.$$

A matrix of this form can be rewritten as a system of equations that is equivalent to the original system, and from which a solution can be easily found.

The first step is to multiply and/or interchange the rows so that each number in the first column is a multiple of the first number in the first row. In this case, we do so by interchanging Rows 1 and 2:

$$x \qquad - 4z = 5,$$
$$2x - y + 4z = -3,$$
$$6x - y + 2z = 10$$

$$\begin{bmatrix} 1 & 0 & -4 & | & 5 \\ 2 & -1 & 4 & | & -3 \\ 6 & -1 & 2 & | & 10 \end{bmatrix}$$ This corresponds to interchanging the first two equations.

Next, we multiply the first row by -2, add it to the second row, and replace Row 2 with the result:

$$x \qquad - 4z = 5,$$
$$-y + 12z = -13,$$
$$6x - y + 2z = 10$$

$$\begin{bmatrix} 1 & 0 & -4 & | & 5 \\ 0 & -1 & 12 & | & -13 \\ 6 & -1 & 2 & | & 10 \end{bmatrix}.$$ $-2(1 \quad 0 \quad -4 \, | \, 5) = (-2 \quad 0 \quad 8 \, | \, -10)$ and $(-2 \quad 0 \quad 8 \, | \, -10) + (2 \quad -1 \quad 4 \, | \, -3) = (0 \quad -1 \quad 12 \, | \, -13)$

Now we multiply the first row by -6, add it to the third row, and replace Row 3 with the result:

$$x \qquad - 4z = 5,$$
$$-y + 12z = -13,$$
$$-y + 26z = -20$$

$$\begin{bmatrix} 1 & 0 & -4 & | & 5 \\ 0 & -1 & 12 & | & -13 \\ 0 & -1 & 26 & | & -20 \end{bmatrix}.$$ $-6(1 \quad 0 \quad -4 \, | \, 5) = (-6 \quad 0 \quad 24 \, | \, -30)$ and $(-6 \quad 0 \quad 24 \, | \, -30) + (6 \quad -1 \quad 2 \, | \, 10) = (0 \quad -1 \quad 26 \, | \, -20)$

Next, we multiply Row 2 by -1, add it to the third row, and replace Row 3 with the result:

$$x \qquad - 4z = 5,$$
$$-y + 12z = -13,$$
$$14z = -7$$

$$\begin{bmatrix} 1 & 0 & -4 & | & 5 \\ 0 & -1 & 12 & | & -13 \\ 0 & 0 & 14 & | & -7 \end{bmatrix}.$$ $-1(0 \quad -1 \quad 12 \, | \, -13) = (0 \quad 1 \quad -12 \, | \, 13)$ and $(0 \quad 1 \quad -12 \, | \, 13) + (0 \quad -1 \quad 26 \, | \, -20) = (0 \quad 0 \quad 14 \, | \, -7)$

Reinserting the variables gives us

$$x \qquad - 4z = 5,$$
$$-y + 12z = -13,$$
$$14z = -7.$$

We now solve this last equation for z and get $z = -\frac{1}{2}$. Next, we substitute $-\frac{1}{2}$ for z in the preceding equation and solve for y: $-y + 12\left(-\frac{1}{2}\right) = -13$, so $y = 7$. Since there is no y-term in the first equation of this last system, we need only substitute $-\frac{1}{2}$ for z to solve for x: $x - 4\left(-\frac{1}{2}\right) = 5$, so $x = 3$. The solution is $\left(3, 7, -\frac{1}{2}\right)$. The check is left to the student.

The operations used in the preceding example correspond to those used to produce equivalent systems of equations. We call the matrices **row-equivalent** and the operations that produce them **row-equivalent operations**.

Row-Equivalent Operations

Row-Equivalent Operations

Each of the following row-equivalent operations produces a row-equivalent matrix:

a) Interchanging any two rows.
b) Multiplying all elements of a row by a nonzero constant.
c) Replacing a row with the sum of that row and a multiple of another row.

Student Notes

Try to remember that row-equivalent matrices are not *equal*. It is the solutions of the corresponding systems that are the same.

The best overall method for solving systems of equations is by row-equivalent matrices; even computers are programmed to use them. Matrices are part of a branch of mathematics known as linear algebra. They are also studied in many courses in finite mathematics.

technology
connection

Row-equivalent operations can be performed on a graphing calculator. For example, to interchange the first and second rows of the matrix, as in step (1) of Example 2 above, we enter the matrix as matrix **A** and select "rowSwap" from the MATRIX MATH menu. Some graphing calculators will not automatically store the matrix produced using a row-equivalent operation, so when several operations are to be performed in succession, it is helpful to store the result of each operation as it is produced. In the window at right, we see both the matrix produced by the rowSwap operation and the indication that this matrix is stored, using **STO▸**, as matrix **B**.

```
rowSwap([A],1,2)→[B]
[[1   0  -4   5]
 [2  -1   4  -3]
 [6  -1   2  10]]
```

1. Use a graphing calculator to proceed through all the steps in Example 2.

Exercise Set

8.6

↪ *Concept Reinforcement* *Complete each of the following statements.*

1. The rows of a matrix are _____ and the _____ are vertical.

2. Multiplying the numbers in a row of a matrix by a constant corresponds to multiplying both sides of a(n) _____ by a constant.

3. Each number in a matrix is called a(n) _____ or element.

4. The plural of the word matrix is _____.

5. To solve a system using matrices, we can replace any row by the sum of that row and a(n) _____ of another row.

6. In the final step of solving a system of equations, the leftmost column has zeros in all rows except the _____ one.

Solve using matrices.

7. $9x - 2y = 5,$
$3x - 3y = 11$

8. $4x + y = 7,$
$5x - 3y = 13$

9. $x + 4y = 8,$
$3x + 5y = 3$

10. $x + 4y = 5,$
$-3x + 2y = 13$

11. $6x - 2y = 4,$
$7x + y = 13$

12. $3x + 4y = 7,$
$-5x + 2y = 10$

13. $3x + 2y + 2z = 3,$
$x + 2y - z = 5,$
$2x - 4y + z = 0$

14. $4x - y - 3z = 19,$
$8x + y - z = 11,$
$2x + y + 2z = -7$

15. $p - 2q - 3r = 3,$
$2p - q - 2r = 4,$
$4p + 5q + 6r = 4$

16. $x + 2y - 3z = 9,$
$2x - y + 2z = -8,$
$3x - y - 4z = 3$

17. $3p + 2r = 11,$
$q - 7r = 4,$
$p - 6q = 1$

18. $4a + 9b = 8,$
$8a + 6c = -1,$
$6b + 6c = -1$

19. $2x + 2y - 2z - 2w = -10,$
$w + y + z + x = -5,$
$x - y + 4z + 3w = -2,$
$w - 2y + 2z + 3x = -6$

20. $-w - 3y + z + 2x = -8,$
$x + y - z - w = -4,$
$w + y + z + x = 22,$
$x - y - z - w = -14$

Solve using matrices.

21. *Coin value.* A collection of 42 coins consists of dimes and nickels. The total value is \$3.00. How many dimes and how many nickels are there?

22. *Coin value.* A collection of 43 coins consists of dimes and quarters. The total value is \$7.60. How many dimes and how many quarters are there?

23. *Mixed granola.* Grace sells two kinds of granola. One is worth \$4.05 per pound and the other is worth \$2.70 per pound. She wants to blend the two granolas to get a 15-lb mixture worth \$3.15 per pound. How much of each kind of granola should be used?

24. *Trail mix.* Phil mixes nuts worth \$1.60 per pound with oats worth \$1.40 per pound to get 20 lb of trail mix worth \$1.54 per pound. How many pounds of nuts and how many pounds of oats should be used?

25. *Investments.* Elena receives \$212 per year in simple interest from three investments totaling \$2500. Part is invested at 7%, part at 8%, and part at 9%. There is \$1100 more invested at 9% than at 8%. Find the amount invested at each rate.

26. *Investments.* Miguel receives \$306 per year in simple interest from three investments totaling \$3200. Part is invested at 8%, part at 9%, and part at 10%. There is \$1900 more invested at 10% than at 9%. Find the amount invested at each rate.

27. Explain how you can recognize dependent equations when solving with matrices.

28. Explain how you can recognize an inconsistent system when solving with matrices.

SKILL MAINTENANCE

Simplify. [1.8]

29. $5(-3) - (-7)4$

30. $8(-5) - (-2)9$

31. $-2(5 \cdot 3 - 4 \cdot 6) - 3(2 \cdot 7 - 15) + 4(3 \cdot 8 - 5 \cdot 4)$

32. $6(2 \cdot 7 - 3(-4)) - 4(3(-8) - 10) + 5(4 \cdot 3 - (-2)7)$

SYNTHESIS

33. If the matrices

$$\begin{bmatrix} a_1 & b_1 & | & c_1 \\ d_1 & e_1 & | & f_1 \end{bmatrix} \quad \text{and} \quad \begin{bmatrix} a_2 & b_2 & | & c_2 \\ d_2 & e_2 & | & f_2 \end{bmatrix}$$

share the same solution, does it follow that the corresponding entries are all equal to each other ($a_1 = a_2$, $b_1 = b_2$, etc.)? Why or why not?

34. Explain how the row-equivalent operations make use of the addition, multiplication, and distributive properties.

35. The sum of the digits in a four-digit number is 10. Twice the sum of the thousands digit and the tens digit is 1 less than the sum of the other two digits. The tens digit is twice the thousands digit. The ones digit equals the sum of the thousands digit and the hundreds digit. Find the four-digit number.

36. Solve for x and y:

$$ax + by = c,$$
$$dx + ey = f.$$

8.7   Determinants and Cramer's Rule

Determinants of 2 × 2 Matrices   •   Cramer's Rule: 2 × 2
Systems   •   Cramer's Rule: 3 × 3 Systems

Determinants of 2 × 2 Matrices

When a matrix has m rows and n columns, it is called an "m by n" matrix. Thus its *dimensions* are denoted by $m \times n$. If a matrix has the same number of rows and columns, it is called a **square matrix**. Associated with every square matrix is a number called its **determinant**, defined as follows for 2 × 2 matrices.

---

2 × 2 Determinants

The determinant of a two-by-two matrix $\begin{bmatrix} a & c \\ b & d \end{bmatrix}$ is denoted $\begin{vmatrix} a & c \\ b & d \end{vmatrix}$ and is defined as follows:

$$\begin{vmatrix} a & c \\ b & d \end{vmatrix} = ad - bc.$$

EXAMPLE 1 Evaluate: $\begin{vmatrix} 2 & -5 \\ 6 & 7 \end{vmatrix}$.

Solution We multiply and subtract as follows:

$$\begin{vmatrix} 2 & -5 \\ 6 & 7 \end{vmatrix} = 2 \cdot 7 - 6 \cdot (-5) = 14 + 30 = 44.$$

Cramer's Rule: 2 × 2 Systems

One of the many uses for determinants is in solving systems of linear equations in which the number of variables is the same as the number of equations and the constants are not all 0. Let's consider a system of two equations:

$$a_1x + b_1y = c_1,$$
$$a_2x + b_2y = c_2.$$

If we use the elimination method, a series of steps can show that

$$x = \frac{c_1b_2 - c_2b_1}{a_1b_2 - a_2b_1} \quad \text{and} \quad y = \frac{a_1c_2 - a_2c_1}{a_1b_2 - a_2b_1}.$$

These fractions can be rewritten using determinants.

Cramer's Rule: 2 × 2 Systems

The solution of the system

$$a_1x + b_1y = c_1,$$
$$a_2x + b_2y = c_2,$$

if it is unique, is given by

$$x = \frac{\begin{vmatrix} c_1 & b_1 \\ c_2 & b_2 \end{vmatrix}}{\begin{vmatrix} a_1 & b_1 \\ a_2 & b_2 \end{vmatrix}}, \qquad y = \frac{\begin{vmatrix} a_1 & c_1 \\ a_2 & c_2 \end{vmatrix}}{\begin{vmatrix} a_1 & b_1 \\ a_2 & b_2 \end{vmatrix}}.$$

These formulas apply only if the denominator is not 0. If the denominator *is* 0, then one of two things happens:

1. If the denominator is 0 and the numerators are also 0, then the equations in the system are dependent.
2. If the denominator is 0 and at least one numerator is not 0, then the system is inconsistent.

To use Cramer's rule, we find the determinants and compute x and y as shown above. Note that the denominators are identical and the coefficients of x and y appear in the same position as in the original equations. In the numerator of x, the constants c_1 and c_2 replace a_1 and a_2. In the numerator of y, the constants c_1 and c_2 replace b_1 and b_2.

EXAMPLE 2   Solve using Cramer's rule:

$$2x + 5y = 7,$$
$$5x - 2y = -3.$$

Solution   We have

$$x = \frac{\begin{vmatrix} 7 & 5 \\ -3 & -2 \end{vmatrix}}{\begin{vmatrix} 2 & 5 \\ 5 & -2 \end{vmatrix}} \qquad \text{Using Cramer's rule}$$

$$= \frac{7(-2) - (-3)5}{2(-2) - 5 \cdot 5} = -\frac{1}{29}$$

and

$$y = \frac{\begin{vmatrix} 2 & 7 \\ 5 & -3 \end{vmatrix}}{\begin{vmatrix} 2 & 5 \\ 5 & -2 \end{vmatrix}} \qquad \text{Using Cramer's rule}$$

$$= \frac{2(-3) - 5 \cdot 7}{-29} = \frac{41}{29}. \qquad \text{The denominator is the same as in the expression for } x.$$

The solution is $\left(-\frac{1}{29}, \frac{41}{29}\right)$. The check is left to the student.

Cramer's Rule: 3×3 Systems

Cramer's rule can be extended for systems of three linear equations. However, before doing so, we must define what a 3×3 determinant is.

> **3×3 Determinants**
> The determinant of a three-by-three matrix is defined as follows:
>
> $$\begin{vmatrix} a_1 & b_1 & c_1 \\ a_2 & b_2 & c_2 \\ a_3 & b_3 & c_3 \end{vmatrix} = a_1 \overset{\text{Subtract.}}{\begin{vmatrix} b_2 & c_2 \\ b_3 & c_3 \end{vmatrix}} - a_2 \begin{vmatrix} b_1 & c_1 \\ b_3 & c_3 \end{vmatrix} + \overset{\text{Add.}}{a_3} \begin{vmatrix} b_1 & c_1 \\ b_2 & c_2 \end{vmatrix}$$

Note that the a's come from the first column. Note too that the 2×2 determinants above can be obtained by crossing out the row and the column in which the a occurs.

Student Notes

Cramer's rule and the evaluation of determinants rely on patterns. The specific formulas are less important than the patterns that they represent.

For a_1:

$$\begin{vmatrix} \cancel{a_1} & \cancel{b_1} & \cancel{c_1} \\ a_2 & b_2 & c_2 \\ a_3 & b_3 & c_3 \end{vmatrix}$$

For a_2:

$$\begin{vmatrix} a_1 & b_1 & c_1 \\ \cancel{a_2} & \cancel{b_2} & \cancel{c_2} \\ a_3 & b_3 & c_3 \end{vmatrix}$$

For a_3:

$$\begin{vmatrix} a_1 & b_1 & c_1 \\ a_2 & b_2 & c_2 \\ \cancel{a_3} & \cancel{b_3} & \cancel{c_3} \end{vmatrix}$$

EXAMPLE 3   Evaluate:

$$\begin{vmatrix} -1 & 0 & 1 \\ -5 & 1 & -1 \\ 4 & 8 & 1 \end{vmatrix}.$$

Solution   We have

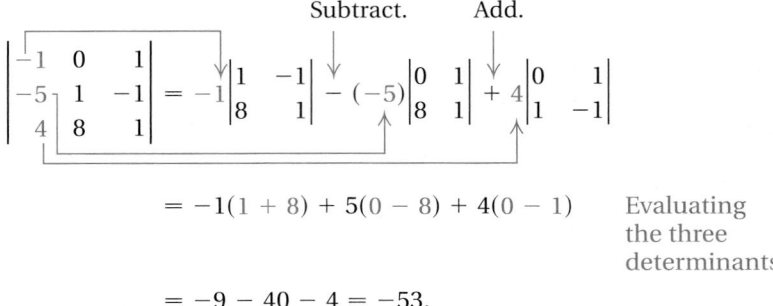

$$= -1(1 + 8) + 5(0 - 8) + 4(0 - 1) \qquad \text{Evaluating the three determinants}$$

$$= -9 - 40 - 4 = -53.$$

Cramer's Rule: 3×3 Systems

The solution of the system

$$a_1 x + b_1 y + c_1 z = d_1,$$
$$a_2 x + b_2 y + c_2 z = d_2,$$
$$a_3 x + b_3 y + c_3 z = d_3$$

can be found using the following determinants:

$$D = \begin{vmatrix} a_1 & b_1 & c_1 \\ a_2 & b_2 & c_2 \\ a_3 & b_3 & c_3 \end{vmatrix}, \qquad D_x = \begin{vmatrix} d_1 & b_1 & c_1 \\ d_2 & b_2 & c_2 \\ d_3 & b_3 & c_3 \end{vmatrix},$$

D contains only coefficients. In D_x, the *d*'s replace the *a*'s.

$$D_y = \begin{vmatrix} a_1 & d_1 & c_1 \\ a_2 & d_2 & c_2 \\ a_3 & d_3 & c_3 \end{vmatrix}, \qquad D_z = \begin{vmatrix} a_1 & b_1 & d_1 \\ a_2 & b_2 & d_2 \\ a_3 & b_3 & d_3 \end{vmatrix}.$$

In D_y, the *d*'s replace the *b*'s. In D_z, the *d*'s replace the *c*'s.

If a unique solution exists, it is given by

$$x = \frac{D_x}{D}, \qquad y = \frac{D_y}{D}, \qquad z = \frac{D_z}{D}.$$

EXAMPLE 4   Solve using Cramer's rule:

$$x - 3y + 7z = 13,$$
$$x + y + z = 1,$$
$$x - 2y + 3z = 4.$$

Determinants can be evaluated on most graphing calculators using **2ND** **MATRIX** . After entering a matrix, we select the determinant operation from the MATRIX MATH menu and enter the name of the matrix. The graphing calculator will return the value of the determinant of the matrix. For example, if

$$\mathbf{A} = \begin{bmatrix} 1 & 6 & -1 \\ -3 & -5 & 3 \\ 0 & 4 & 2 \end{bmatrix},$$

we have

$$\boxed{\begin{array}{l} \det([A]) \\ \qquad\qquad 26 \end{array}}$$

1. Confirm the calculations in Example 4.

Solution   We compute D, D_x, D_y, and D_z:

$$D = \begin{vmatrix} 1 & -3 & 7 \\ 1 & 1 & 1 \\ 1 & -2 & 3 \end{vmatrix} = -10; \qquad D_x = \begin{vmatrix} 13 & -3 & 7 \\ 1 & 1 & 1 \\ 4 & -2 & 3 \end{vmatrix} = 20;$$

$$D_y = \begin{vmatrix} 1 & 13 & 7 \\ 1 & 1 & 1 \\ 1 & 4 & 3 \end{vmatrix} = -6; \qquad D_z = \begin{vmatrix} 1 & -3 & 13 \\ 1 & 1 & 1 \\ 1 & -2 & 4 \end{vmatrix} = -24.$$

Then

$$x = \frac{D_x}{D} = \frac{20}{-10} = -2;$$

$$y = \frac{D_y}{D} = \frac{-6}{-10} = \frac{3}{5};$$

$$z = \frac{D_z}{D} = \frac{-24}{-10} = \frac{12}{5}.$$

The solution is $\left(-2, \frac{3}{5}, \frac{12}{5}\right)$. The check is left to the student.

In Example 4, we need not have evaluated D_z. Once x and y were found, we could have substituted them into one of the equations to find z.

To use Cramer's rule, we divide by D, provided $D \neq 0$. If $D = 0$ and at least one of the other determinants is not 0, then the system is inconsistent. If *all* the determinants are 0, then the equations in the system are dependent.

Exercise Set

8.7

FOR EXTRA HELP

Student's Solutions Manual

Digital Video Tutor CD 4 Videotape 8

AW Math Tutor Center

MathXL Tutorials on CD

MathXL

MyMathLab

↪ *Concept Reinforcement   Classify each of the following as either true or false.*

1. A square matrix has the same number of rows and columns.

2. A 3×4 matrix has 3 rows and 4 columns.

3. Cramer's rule exists only for 2×2 systems.

4. Whenever Cramer's rule yields a denominator that is 0, the system has no solution.

5. Whenever Cramer's rule yields a numerator that is 0, the equations are dependent.

6. Cramer's rule allows us to solve some systems that could not be solved any other way.

Evaluate.

7. $\begin{vmatrix} 5 & 1 \\ 2 & 4 \end{vmatrix}$

8. $\begin{vmatrix} 3 & 2 \\ 2 & -3 \end{vmatrix}$

9. $\begin{vmatrix} 6 & -9 \\ 2 & 3 \end{vmatrix}$

10. $\begin{vmatrix} 3 & 2 \\ -7 & 5 \end{vmatrix}$

11. $\begin{vmatrix} 1 & 4 & 0 \\ 0 & -1 & 2 \\ 3 & -2 & 1 \end{vmatrix}$

12. $\begin{vmatrix} 3 & 0 & -2 \\ 5 & 1 & 2 \\ 2 & 0 & -1 \end{vmatrix}$

13. $\begin{vmatrix} -1 & -2 & -3 \\ 3 & 4 & 2 \\ 0 & 1 & 2 \end{vmatrix}$ **14.** $\begin{vmatrix} 1 & 2 & 2 \\ 2 & 1 & 0 \\ 3 & 3 & 1 \end{vmatrix}$

15. $\begin{vmatrix} -4 & -2 & 3 \\ -3 & 1 & 2 \\ 3 & 4 & -2 \end{vmatrix}$ **16.** $\begin{vmatrix} 2 & -1 & 1 \\ 1 & 2 & -1 \\ 3 & 4 & -3 \end{vmatrix}$

Solve using Cramer's rule.

17. $5x + 8y = 1,$
    $3x + 7y = 5$

18. $3x - 4y = 6,$
    $5x + 9y = 10$

19. $5x - 4y = -3,$
    $7x + 2y = 6$

20. $-2x + 4y = 3,$
    $3x - 7y = 1$

21. $3x - y + 2z = 1,$
    $x - y + 2z = 3,$
    $-2x + 3y + z = 1$

22. $3x + 2y - z = 4,$
    $3x - 2y + z = 5,$
    $4x - 5y - z = -1$

23. $2x - 3y + 5z = 27,$
    $x + 2y - z = -4,$
    $5x - y + 4z = 27$

24. $x - y + 2z = -3,$
    $x + 2y + 3z = 4,$
    $2x + y + z = -3$

25. $r - 2s + 3t = 6,$
    $2r - s - t = -3,$
    $r + s + t = 6$

26. $a - 3c = 6,$
    $b + 2c = 2,$
    $7a - 3b - 5c = 14$

27. What is it about Cramer's rule that makes it useful?

28. Which version of Cramer's rule do you find more useful: the version for 2×2 systems or the version for 3×3 systems? Why?

SKILL MAINTENANCE

Solve. [2.2]

29. $0.5x - 2.34 + 2.4x = 7.8x - 9$

30. $5x + 7x = -144$

31. A piece of wire 32.8 ft long is to be cut into two pieces, and those pieces are each to be bent to make a square. The length of a side of one square is to be 2.2 ft greater than the length of a side of the other. How should the wire be cut? [2.5]

32. *Inventory.*   The Freeport College store paid $1728 for an order of 45 calculators. The store paid $9 for each scientific calculator. The others, all graphing calculators, cost the store $58 each. How many of each type of calculator was ordered? [8.3]

33. *Insulation.*   The Mazzas' attic required three and a half times as much insulation as did the Kranepools'. Together, the two attics required 36 rolls of insulation. How much insulation did each attic require? [8.3]

34. *Sales of food.*   High Flyin' Wings charges $12 for a bucket of chicken wings and $7 for a chicken dinner. After filling 28 orders for buckets and dinners, High Flyin' Wings had collected $281. How many buckets and how many dinners did they sell? [8.3]

SYNTHESIS

35. Cramer's rule states that if $a_1x + b_1y = c_1$ and $a_2x + b_2y = c_2$ are dependent, then
$$\begin{vmatrix} a_1 & b_1 \\ a_2 & b_2 \end{vmatrix} = 0.$$
Explain why this will always happen.

36. Under what conditions can a 3×3 system of linear equations be consistent but unable to be solved using Cramer's rule?

Solve.

37. $\begin{vmatrix} y & -2 \\ 4 & 3 \end{vmatrix} = 44$ **38.** $\begin{vmatrix} 2 & x & -1 \\ -1 & 3 & 2 \\ -2 & 1 & 1 \end{vmatrix} = -12$

39. $\begin{vmatrix} m+1 & -2 \\ m-2 & 1 \end{vmatrix} = 27$

40. Show that an equation of the line through (x_1, y_1) and (x_2, y_2) can be written
$$\begin{vmatrix} x & y & 1 \\ x_1 & y_1 & 1 \\ x_2 & y_2 & 1 \end{vmatrix} = 0.$$

8.8 | Business and Economic Applications

Break-Even Analysis   •   Supply and Demand

*Study Skills*_____

Try to Look Ahead

If you are able to at least skim
through an upcoming section
before your instructor covers that
lesson, you will be better able to
focus on what is being emphasized
in class. Similarly, if you can begin
studying for a quiz or test a day
or two before you really *have* to,
you will reap great rewards for
doing so.

Break-Even Analysis

When a company manufactures x units of a product, it spends money. This is **total cost** and can be thought of as a function C, where $C(x)$ is the total cost of producing x units. When the company sells x units of the product, it takes in money. This is **total revenue** and can be thought of as a function R, where $R(x)$ is the total revenue from the sale of x units. **Total profit** is the money taken in less the money spent, or total revenue minus total cost. Total profit from the production and sale of x units is a function P given by

$$\textbf{Profit = Revenue − Cost,} \quad \text{or} \quad \textbf{\textit{P}(\textit{x}) = \textit{R}(\textit{x}) − \textit{C}(\textit{x}).}$$

If $R(x)$ is greater than $C(x)$, there is a gain and $P(x)$ is positive. If $C(x)$ is greater than $R(x)$, there is a loss and $P(x)$ is negative. When $R(x) = C(x)$, the company breaks even.

There are two kinds of costs. First, there are costs like rent, insurance, machinery, and so on. These costs, which must be paid whether a product is produced or not, are called *fixed costs*. When a product is being produced, there are costs for labor, materials, marketing, and so on. These are called *variable costs*, because they vary according to the amount being produced. The sum of the fixed cost and the variable cost gives the *total cost* of producing a product.

> *Caution!*   Do not confuse "cost" with "price." When we discuss the
> *cost* of an item, we are referring to what it costs to produce the item.
> The *price* of an item is what a consumer pays to purchase the item and
> is used when calculating revenue.

EXAMPLE 1   Manufacturing lamps.   Ergs, Inc., is planning to make a new lamp. Fixed costs will be $90,000, and it will cost $15 to produce each lamp (variable costs). Each lamp sells for $26.

a) Find the total cost $C(x)$ of producing x lamps.

b) Find the total revenue $R(x)$ from the sale of x lamps.

c) Find the total profit $P(x)$ from the production and sale of x lamps.

d) What profit will the company realize from the production and sale of 3000 lamps? of 14,000 lamps?

e) Graph the total-cost, total-revenue, and total-profit functions using the same set of axes. Determine the break-even point.

Solution

a) Total cost is given by

$$C(x) = \text{(Fixed costs) plus (Variable costs)},$$

$$\text{or} \quad C(x) = \quad 90{,}000 \quad + \quad 15x,$$

where x is the number of lamps produced.

b) Total revenue is given by

$$R(x) = 26x. \qquad \text{\$26 times the number of lamps sold.}$$
We assume that every lamp produced is sold.

c) Total profit is given by

$$P(x) = R(x) - C(x) \qquad \text{Profit is revenue minus cost.}$$

$$= 26x - (90{,}000 + 15x)$$

$$= 11x - 90{,}000.$$

d) Profits will be

$$P(3000) = 11 \cdot 3000 - 90{,}000 = -\$57{,}000$$

when 3000 lamps are produced and sold, and

$$P(14{,}000) = 11 \cdot 14{,}000 - 90{,}000 = \$64{,}000$$

when 14,000 lamps are produced and sold. Thus the company loses money if only 3000 lamps are sold, but makes money if 14,000 are sold.

e) The graphs of each of the three functions are shown below:

$$R(x) = 26x, \qquad\qquad \text{This represents the revenue function.}$$

$$C(x) = 90{,}000 + 15x, \qquad \text{This represents the cost function.}$$

$$P(x) = 11x - 90{,}000. \qquad \text{This represents the profit function.}$$

$R(x)$, $C(x)$, and $P(x)$ are all in dollars.

The revenue function has a graph that goes through the origin and has a slope of 26. The cost function has an intercept on the $-axis of 90,000 and has a slope of 15. The profit function has an intercept on the $-axis of $-90,000$ and has a slope of 11. It is shown by the dashed line. The red dashed line shows a "negative" profit, which is a loss. (That is what is known as "being in the red.") The black dashed line shows a "positive" profit, or gain. (That is what is known as "being in the black.")

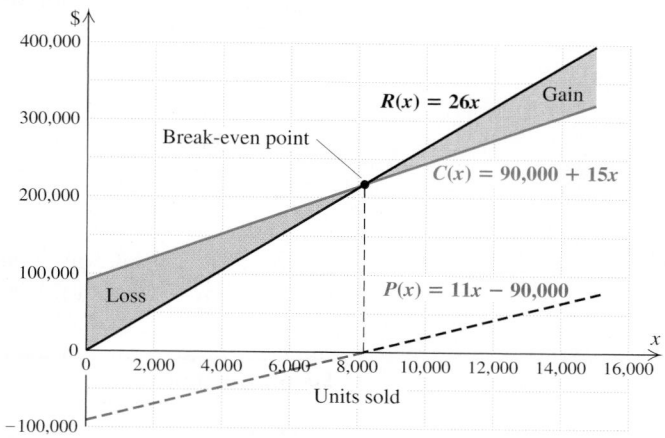

*Student Notes*_____

If you plan to study business or economics, you may want to consult the material in this section when these topics arise in your other courses.

Gains occur where the revenue is greater than the cost. Losses occur where the revenue is less than the cost. The **break-even point** occurs where the graphs of R and C cross. Thus to find the break-even point, we solve a system:

$$R(x) = 26x,$$
$$C(x) = 90{,}000 + 15x.$$

Since both revenue and cost are in *dollars* and they are equal at the break-even point, the system can be rewritten as

$$d = 26x, \qquad (1)$$
$$d = 90{,}000 + 15x \qquad (2)$$

and solved using substitution:

$26x = 90{,}000 + 15x$ Substituting $26x$ for d in equation (2)

$11x = 90{,}000$

$x \approx 8181.8.$

The firm will break even if it produces and sells about 8182 lamps (8181 will yield a tiny loss and 8182 a tiny gain), and takes in a total of $R(8182) = 26 \cdot 8182 = \$212{,}732$ in revenue. Note that the x-coordinate of the break-even point can also be found by solving $P(x) = 0$. The break-even point is (8182 lamps, \$212,732).

Supply and Demand

As the price of coffee varies, the amount sold varies. The table and graph below show that *consumers will demand less as the price goes up.*

Demand Function, D

Price, p, per Kilogram	Quantity, $D(p)$ (in millions of kilograms)
$ 8.00	25
9.00	20
10.00	15
11.00	10
12.00	5

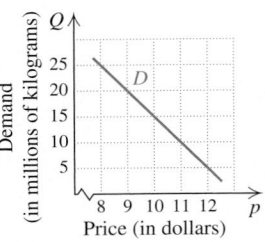

As the price of coffee varies, the amount available varies. The table and graph below show that *sellers will supply more as the price goes up.*

Supply Function, S

Price, p, per Kilogram	Quantity, $S(p)$ (in millions of kilograms)
$ 9.00	5
9.50	10
10.00	15
10.50	20
11.00	25

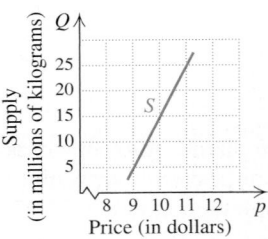

Let's look at the above graphs together. We see that as price increases, demand decreases. As price increases, supply increases. The point of intersection is called the **equilibrium point**. At that price, the amount that the seller will supply is the same amount that the consumer will buy. The situation is analogous to a buyer and a seller negotiating the price of an item. The equilibrium point is the price and quantity that they finally agree on.

Any ordered pair of coordinates from the graph is (price, quantity), because the horizontal axis is the price axis and the vertical axis is the quantity axis. If D is a demand function and S is a supply function, then the equilibrium point is where demand equals supply:

$$D(p) = S(p).$$

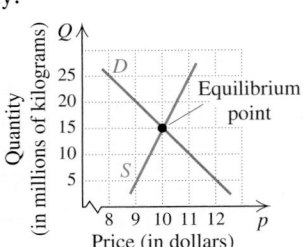

EXAMPLE 2   Find the equilibrium point for the demand and supply functions given:

$$D(p) = 1000 - 60p, \quad (1)$$
$$S(p) = 200 + 4p. \quad (2)$$

Solution   Since both demand and supply are *quantities* and they are equal at the equilibrium point, we rewrite the system as

$$q = 1000 - 60p, \quad (1)$$
$$q = 200 + 4p. \quad (2)$$

We substitute $200 + 4p$ for q in equation (1) and solve:

$$200 + 4p = 1000 - 60p \qquad \text{Substituting } 200 + 4p \text{ for } q \text{ in equation (1)}$$
$$200 + 64p = 1000 \qquad \text{Adding } 60p \text{ to both sides}$$
$$64p = 800 \qquad \text{Adding } -200 \text{ to both sides}$$
$$p = \frac{800}{64} = 12.5.$$

Thus the equilibrium price is $12.50 per unit.

To find the equilibrium quantity, we substitute $12.50 into either $D(p)$ or $S(p)$. We use $S(p)$:

$$S(12.5) = 200 + 4(12.5) = 200 + 50 = 250.$$

Thus the equilibrium quantity is 250 units, and the equilibrium point is ($12.50, 250).

Exercise Set

8.8

FOR EXTRA HELP

Student's Solutions Manual   Digital Video Tutor CD 4 Videotape 8   AW Math Tutor Center   MathXL Tutorials on CD   MathXL   MyMathLab

↪ *Concept Reinforcement*   *In each of Exercises 1–8, match the word or phrase with the most appropriate choice from the column on the right.*

1. ___ Total cost

2. ___ Total revenue

3. ___ Total profit

4. ___ Fixed costs

5. ___ Variable costs

6. ___ Break-even point

7. ___ Equilibrium point

8. ___ Price

a) The amount of money that a company takes in

b) The sum of fixed costs and variable costs

c) The point at which total revenue equals total cost

d) What consumers pay per item

e) The difference between total revenue and total cost

f) What companies spend whether or not a product is produced

g) The point at which supply equals demand

h) The costs that vary according to the number of items produced

For each of the following pairs of total-cost and total-revenue functions, find (**a**) *the total-profit function and* (**b**) *the break-even point.*

9. $C(x) = 45x + 300{,}000$;
$R(x) = 65x$

10. $C(x) = 25x + 270{,}000$;
$R(x) = 70x$

11. $C(x) = 10x + 120{,}000$;
$R(x) = 60x$

12. $C(x) = 30x + 49{,}500$;
$R(x) = 85x$

13. $C(x) = 40x + 22{,}500$;
$R(x) = 85x$

14. $C(x) = 20x + 10{,}000$;
$R(x) = 100x$

15. $C(x) = 22x + 16{,}000$;
$R(x) = 40x$

16. $C(x) = 15x + 75{,}000$;
$R(x) = 55x$

Aha! **17.** $C(x) = 75x + 100{,}000$;
$R(x) = 125x$

18. $C(x) = 20x + 120{,}000$;
$R(x) = 50x$

Find the equilibrium point for each of the following pairs of demand and supply functions.

19. $D(p) = 1000 - 10p$,
$S(p) = 230 + p$

20. $D(p) = 2000 - 60p$,
$S(p) = 460 + 94p$

21. $D(p) = 760 - 13p$,
$S(p) = 430 + 2p$

22. $D(p) = 800 - 43p$,
$S(p) = 210 + 16p$

23. $D(p) = 7500 - 25p$,
$S(p) = 6000 + 5p$

24. $D(p) = 8800 - 30p$,
$S(p) = 7000 + 15p$

25. $D(p) = 1600 - 53p$,
$S(p) = 320 + 75p$

26. $D(p) = 5500 - 40p$,
$S(p) = 1000 + 85p$

Solve.

27. *Computer manufacturing.*   Biz.com Electronics is planning to introduce a new line of computers. The fixed costs for production are $125,300. The variable costs for producing each computer are $450. The revenue from each computer is $800. Find the following.

   a) The total cost $C(x)$ of producing x computers

b) The total revenue $R(x)$ from the sale of x computers

c) The total profit $P(x)$ from the production and sale of x computers

d) The profit or loss from the production and sale of 100 computers; of 400 computers

e) The break-even point

28. *Manufacturing CD players.*   SoundGen, Inc., is planning to manufacture a new type of CD player. The fixed costs for production are $22,500. The variable costs for producing each CD player are estimated to be $40. The revenue from each CD player is to be $85. Find the following.

   a) The total cost $C(x)$ of producing x CD players

   b) The total revenue $R(x)$ from the sale of x CD players

   c) The total profit $P(x)$ from the production and sale of x CD players

   d) The profit or loss from the production and sale of 3000 CD players; of 400 CD players

   e) The break-even point

29. *Manufacturing caps.*   Martina's Custom Printing is planning on adding painter's caps to its product line. For the first year, the fixed costs for setting up production are $16,404. The variable costs for producing a dozen caps are $6.00. The revenue on each dozen caps will be $18.00. Find the following.

   a) The total cost $C(x)$ of producing x dozen caps

   b) The total revenue $R(x)$ from the sale of x dozen caps

   c) The total profit $P(x)$ from the production and sale of x dozen caps

   d) The profit or loss from the production and sale of 3000 dozen caps; of 1000 dozen caps

   e) The break-even point

30. *Sport coat production.*   Sarducci's is planning a new line of sport coats. For the first year, the fixed costs for setting up production are $10,000. The variable costs for producing each coat are $30. The revenue from each coat is to be $80. Find the following.

   a) The total cost $C(x)$ of producing x coats

   b) The total revenue $R(x)$ from the sale of x coats

   c) The total profit $P(x)$ from the production and sale of x coats

   d) The profit or loss from the production and sale of 2000 coats; of 50 coats

   e) The break-even point

31. In Example 1, the slope of the line representing Revenue is the sum of the slopes of the other two lines. This is not a coincidence. Explain why.

32. Variable costs and fixed costs are often compared to the slope and the *y*-intercept, respectively, of an equation for a line. Explain why you feel this analogy is or is not valid.

SKILL MAINTENANCE

Solve. [2.2]

33. $3x - 9 = 27$

34. $4x - 7 = 53$

35. $4x - 5 = 7x - 13$

36. $2x + 9 = 8x - 15$

37. $7 - 2(x - 8) = 14$

38. $6 - 4(3x - 2) = 10$

SYNTHESIS

39. Ian claims that since his fixed costs are $1000, he need sell only 20 birdbaths at $50 each in order to break even. Does this sound plausible? Why or why not?

40. In this section, we examined supply and demand functions for coffee. Does it seem realistic to you for the graph of *D* to have a constant slope? Why or why not?

41. *Yo-yo production.* Bing Boing Hobbies is willing to produce 100 yo-yo's at $2.00 each and 500 yo-yo's at $8.00 each. Research indicates that the public will buy 500 yo-yo's at $1.00 each and 100 yo-yo's at $9.00 each. Find the equilibrium point.

42. *Loudspeaker production.* Fidelity Speakers, Inc., has fixed costs of $15,400 and variable costs of $100 for each pair of speakers produced. If the speakers sell for $250 a pair, how many pairs of speakers must be produced (and sold) in order to have enough profit to cover the fixed costs of two additional facilities? Assume that all fixed costs are identical.

Use a graphing calculator to solve.

43. *Dog food production.* Puppy Love, Inc., will soon begin producing a new line of puppy food. The marketing department predicts that the demand function will be $D(p) = -14.97p + 987.35$ and the supply function will be $S(p) = 98.55p - 5.13$.

a) To the nearest cent, what price per unit should be charged in order to have equilibrium between supply and demand?

b) The production of the puppy food involves $87,985 in fixed costs and $5.15 per unit in variable costs. If the price per unit is the value you found in part (a), how many units must be sold in order to break even?

44. *Computer production.* Number Cruncher Computers, Inc., is planning a new line of computers, each of which will sell for $970. The fixed costs in setting up production are $1,235,580 and the variable costs for each computer are $697.

a) What is the break-even point? (Round to the nearest whole number.)

b) The marketing department at Number Cruncher is not sure that $970 is the best price. Their demand function for the new computers is given by $D(p) = -304.5p + 374,580$ and their supply function is given by $S(p) = 788.7p - 576,504$. To the nearest dollar, what price *p* would result in equilibrium between supply and demand?

8 Study Summary

Because so many real-world problems translate into two or more equations in two or more variables, **systems of equations** are studied in great detail (p. 509). Most of the systems studied in this chapter are **consistent**, meaning that they have at least one solution, although we also studied some **inconsistent** systems, for which there is no solution (p. 513). The equations in the systems we solved were **independent**, except for those cases in which one equation could be written as a multiple and/or sum of the other equation(s) (such equations are called **dependent**) (p. 513).

Three methods—graphing, substitution, and elimination—can be used to solve systems of equations (pp. 511, 517, 519). Of the three methods, graphing is the easiest to visualize.

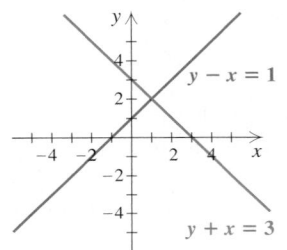

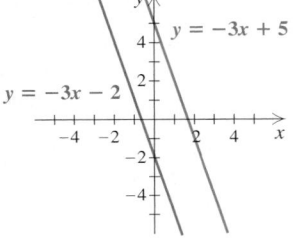

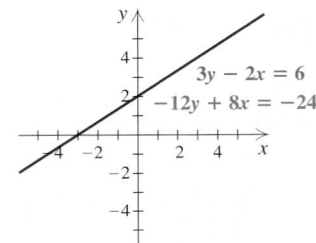

Graphs intersect at one point.
The system is *consistent* and has one solution. Since neither equation is a multiple of the other, they are *independent*.

Graphs are parallel.
The system is *inconsistent* because there is no solution. Since the equations are not equivalent, they are *independent*.

Equations have the same graph.
The system is *consistent* and has an infinite number of solutions. The equations are *dependent* since they are equivalent.

Graphing is especially useful when working with **revenue**, **cost**, and **profit** functions to determine a **break-even point** (p. 567). It is also used when working with **supply** and **demand** functions to determine an **equilibrium point** (p. 568).

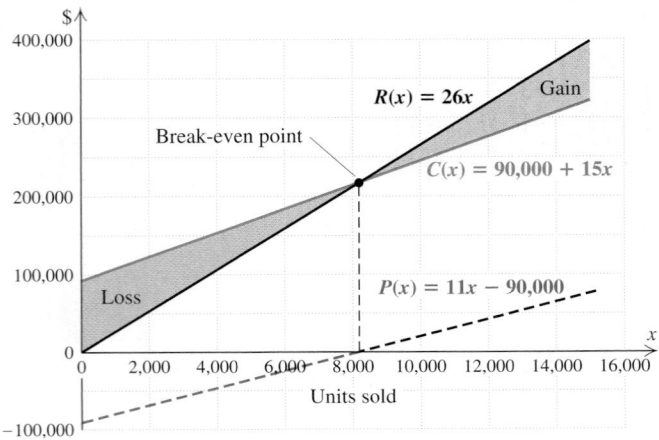

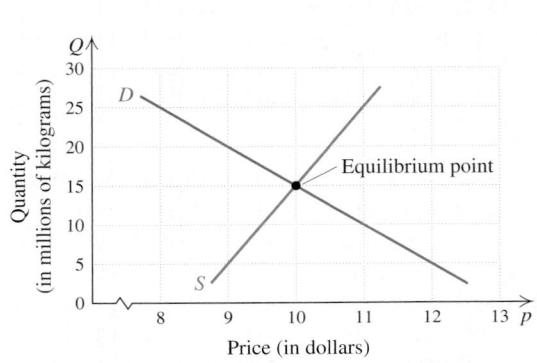

The substitution and elimination methods are the most commonly used methods for solving **total-value**, **mixture**, and **motion problems**.

Total Value

King Street Printing recently charged 1.9¢ per sheet of paper, but 2.4¢ per sheet for paper made of recycled fibers. Darren's bill for 150 sheets of paper was $3.41. How many sheets of each type were used? (Exercise 15, p. 535)

Mixture

Etch Clean Graphics uses one cleanser that is 25% acid and a second that is 50% acid. How many liters of each should be mixed to get 30 L of a solution that is 40% acid? (Exercise 24, p. 535)

Motion

Mia's motorboat took 3 hr to make a trip downstream with a 6-mph current. The return trip against the same current took 5 hr. Find the speed of the boat in still water. (Exercise 37, p. 537)

The elimination method can be extended to systems of three or more equations in three or more variables. One way in which this is accomplished is through the use of **matrices** (singular: **matrix**) (p. 554). A matrix is an array of numbers that are displayed in **rows** and **columns** (p. 554). The individual numbers are called **entries** or **elements** (p. 554).

By using **row-equivalent** operations, we can use matrices to solve systems of equations without needing to continually rewrite all of the variables (p. 557). The solution of systems is streamlined even further when **determinants** are used as part of **Cramer's rule** (pp. 559–562).

Determinant of a 2 × 2 Matrix

$$\begin{vmatrix} a & c \\ b & d \end{vmatrix} = ad - bc$$

Determinant of a 3 × 3 Matrix

$$\begin{vmatrix} a_1 & b_1 & c_1 \\ a_2 & b_2 & c_2 \\ a_3 & b_3 & c_3 \end{vmatrix} = a_1 \begin{vmatrix} b_2 & c_2 \\ b_3 & c_3 \end{vmatrix} - a_2 \begin{vmatrix} b_1 & c_1 \\ b_3 & c_3 \end{vmatrix} + a_3 \begin{vmatrix} b_1 & c_1 \\ b_2 & c_2 \end{vmatrix}$$

Cramer's Rule: 2 × 2 Systems

The solution of the system

$$a_1x + b_1y = c_1,$$
$$a_2x + b_2y = c_2,$$

if it is unique, is given by

$$x = \frac{\begin{vmatrix} c_1 & b_1 \\ c_2 & b_2 \end{vmatrix}}{\begin{vmatrix} a_1 & b_1 \\ a_2 & b_2 \end{vmatrix}}, \qquad y = \frac{\begin{vmatrix} a_1 & c_1 \\ a_2 & c_2 \end{vmatrix}}{\begin{vmatrix} a_1 & b_1 \\ a_2 & b_2 \end{vmatrix}}.$$

Cramer's Rule: 3 × 3 Systems

The solution of the system

$$a_1x + b_1y + c_1z = d_1,$$
$$a_2x + b_2y + c_2z = d_2,$$
$$a_3x + b_3y + c_3z = d_3,$$

if it is unique, is given by

$$x = \frac{\begin{vmatrix} d_1 & b_1 & c_1 \\ d_2 & b_2 & c_2 \\ d_3 & b_3 & c_3 \end{vmatrix}}{\begin{vmatrix} a_1 & b_1 & c_1 \\ a_2 & b_2 & c_2 \\ a_3 & b_3 & c_3 \end{vmatrix}}, \qquad y = \frac{\begin{vmatrix} a_1 & d_1 & c_1 \\ a_2 & d_2 & c_2 \\ a_3 & d_3 & c_3 \end{vmatrix}}{\begin{vmatrix} a_1 & b_1 & c_1 \\ a_2 & b_2 & c_2 \\ a_3 & b_3 & c_3 \end{vmatrix}}, \qquad z = \frac{\begin{vmatrix} a_1 & b_1 & d_1 \\ a_2 & b_2 & d_2 \\ a_3 & b_3 & d_3 \end{vmatrix}}{\begin{vmatrix} a_1 & b_1 & c_1 \\ a_2 & b_2 & c_2 \\ a_3 & b_3 & c_3 \end{vmatrix}}.$$

8 Review Exercises

↪ *Concept Reinforcement*   *Complete each of the following sentences.*

1. The system

   $5x + 3y = 7,$
   $y = 2x + 1$

   is most easily solved using the _____ method. [8.2]

2. The system

   $-2x + 3y = 8,$
   $2x + 2y = 7$

   is most easily solved using the _____ method. [8.2]

3. A weakness in using graphs to solve a system is that when solutions involve fractions or decimals, the graph may yield only a(n)_____ solution. [8.2]

4. When one equation in a system is a multiple of another equation in that system, the equations are said to be _____. [8.1]

5. A system for which there is no solution is said to be _____. [8.1]

6. When using elimination to solve a system of two equations, if an identity is obtained, we know that there is a(n) _____ number of solutions. [8.2]

7. When we are graphing to solve a system of two equations, if there is no solution, the lines will be _____. [8.1]

8. When a matrix has the same number of rows and columns, it is said to be _____. [8.7]

9. Cramer's rule is a formula in which the numerator and the denominator of each fraction is a(n) _____. [8.7]

10. At the break-even point, the value of the profit function is _____. [8.8]

For Exercises 11–19, if a system has an infinite number of solutions, use set-builder notation to write the solution set. If a system has no solution, state this.

Solve graphically. [8.1]

11. $3x + 2y = -4,$
    $y = 3x + 7$

12. $2x + 3y = 12,$
    $4x - y = 10$

Solve using the substitution method. [8.2]

13. $9x - 6y = 2,$
    $x = 4y + 5$

14. $y = x + 2,$
    $y - x = 8$

15. $x - 3y = -2,$
    $7y - 4x = 6$

Solve using the elimination method. [8.2]

16. $8x - 2y = 10,$
    $-4y - 3x = -17$

17. $4x - 7y = 18,$
    $9x + 14y = 40$

18. $3x - 5y = -4,$
    $5x - 3y = 4$

19. $1.5x - 3 = -2y,$
    $3x + 4y = 6$

Solve. [8.3]

20. Luther bought two DVD's and one videocassette for $48. If he had purchased one DVD and two videocassettes, he would have spent $3 less. What is the price of a DVD? What is the price of a videocassette?

21. A freight train leaves Houston at midnight traveling north at a speed of 44 mph. One hour later, a passenger train, going 55 mph, travels north from Houston on a parallel track. How many hours will the passenger train travel before it overtakes the freight train?

22. Yolanda wants 14 L of fruit punch that is 10% juice. At the store, she finds punch that is 15% juice and punch that is 8% juice. How much of each should she purchase?

Solve. If a system's equations are dependent or if there is no solution, state this.

23. $x + 4y + 3z = 2,$
    $2x + y + z = 10,$
    $-x + y + 2z = 8$
    [8.4]

24. $4x + 2y - 6z = 34,$
    $2x + y + 3z = 3,$
    $6x + 3y - 3z = 37$
    [8.4]

25. $2x - 5y - 2z = -4,$
  $7x + 2y - 5z = -6,$
  $-2x + 3y + 2z = 4$
  [8.4]

26. $-5x + 5y = -6,$
  $2x - 2y = 4$
  [8.2]

27. $3x + y = 2,$
  $x + 3y + z = 0,$
  $x + z = 2$ [8.4]

Solve.

28. In triangle ABC, the measure of angle A is four times the measure of angle C, and the measure of angle B is 45° more than the measure of angle C. What are the measures of the angles of the triangle? [8.5]

29. *Nontoxic floor wax.* A nontoxic floor wax can be made from lemon juice and food-grade linseed oil. The amount of oil should be twice the amount of lemon juice. How much of each ingredient is needed to make 32 oz of floor wax? (The mix should be spread with a rag and buffed when dry.) [8.3]

30. *Lumber production.* Denison Lumber can convert logs into either lumber or plywood. In a given day, the mill turns out 42 pallets of plywood and lumber. It makes a profit of $75 on a pallet of lumber and $120 on a pallet of plywood. How many pallets of each type must be produced and sold in order to make a profit of $3735? [8.3]

Solve using matrices. Show your work. [8.6]

31. $3x + 4y = -13,$
  $5x + 6y = 8$

32. $3x - y + z = -1,$
  $2x + 3y + z = 4,$
  $5x + 4y + 2z = 5$

Evaluate. [8.7]

33. $\begin{vmatrix} -2 & 4 \\ -3 & 5 \end{vmatrix}$

34. $\begin{vmatrix} 2 & 3 & 0 \\ 1 & 4 & -2 \\ 2 & -1 & 5 \end{vmatrix}$

Solve using Cramer's rule. Show your work. [8.7]

35. $2x + 3y = 6,$
  $x - 4y = 14$

36. $2x + y + z = -2,$
  $2x - y + 3z = 6,$
  $3x - 5y + 4z = 7$

37. Find the equilibrium point for the demand and supply functions

$$S(p) = 60 + 7p$$

and

$$D(p) = 120 - 13p. \; [8.8]$$

38. Auriel is beginning to produce organic honey. For the first year, the fixed costs for setting up production are $9000. The variable costs for producing each pint of honey are $0.75. The revenue from each pint of honey is $5.25. Find the following. [8.8]

  a) The total cost $C(x)$ of producing x pints of honey

  b) The total revenue $R(x)$ from the sale of x pints of honey

  c) The total profit $P(x)$ from the production and sale of x pints of honey

  d) The profit or loss from the production and sale of 1500 pints of honey; of 5000 pints of honey

  e) The break-even point

SYNTHESIS

39. How would you go about solving a problem that involves four variables? [8.5]

40. Explain how a system of equations can be both dependent and inconsistent. [8.4]

41. Auriel is quitting a job that pays $27,000 a year to make honey (see Exercise 38). How many pints of honey must she produce and sell in order to make the same amount that she made in the job she left? [8.8]

42. Solve graphically:

$$y = x + 2,$$
$$y = x^2 + 2. \; [8.1]$$

43. The graph of $f(x) = ax^2 + bx + c$ contains the points $(-2, 3)$, $(1, 1)$, and $(0, 3)$. Find a, b, and c and give a formula for the function. [8.5]

8 Chapter Test

1. Solve graphically:

$$2x + y = 8,$$
$$y - x = 2.$$

Solve, if possible, using the substitution method.

2. $x + 3y = -8,$
$4x - 3y = 23$

3. $2x + 4y = -6,$
$y = 3x - 9$

Solve, if possible, using the elimination method.

4. $4x - 6y = 3,$
$6x - 4y = -3$

5. $4y + 2x = 18,$
$3x + 6y = 26$

6. The perimeter of a rectangle is 96. The length of the rectangle is 6 less than twice the width. Find the dimensions of the rectangle.

7. Pepperidge Farm® Goldfish is a snack food for which 40% of its calories come from fat. Rold Gold® Pretzels receive 9% of their calories from fat. How many grams of each would be needed to make 620 g of a snack mix for which 15% of the calories are from fat?

Solve. If a system's equations are dependent or if there is no solution, state this.

8. $-3x + y - 2z = 8,$
$-x + 2y - z = 5,$
$2x + y + z = -3$

9. $6x + 2y - 4z = 15,$
$-3x - 4y + 2z = -6,$
$4x - 6y + 3z = 8$

10. $2x + 2y = 0,$
$4x + 4z = 4,$
$2x + y + z = 2$

11. $3x + 3z = 0,$
$2x + 2y = 2,$
$3y + 3z = 3$

Solve using matrices.

12. $7x - 8y = 10,$
$9x + 5y = -2$

13. $x + 3y - 3z = 12,$
$3x - y + 4z = 0,$
$-x + 2y - z = 1$

Evaluate.

14. $\begin{vmatrix} 4 & -2 \\ 3 & 7 \end{vmatrix}$

15. $\begin{vmatrix} 3 & 4 & 2 \\ 2 & -5 & 4 \\ 4 & 5 & -3 \end{vmatrix}$

16. Solve using Cramer's rule:

$$8x - 3y = 5,$$
$$2x + 6y = 3.$$

17. An electrician, a carpenter, and a plumber are hired to work on a house. The electrician earns $21 per hour, the carpenter $19.50 per hour, and the plumber $24 per hour. The first day on the job, they worked a total of 21.5 hr and earned a total of $469.50. If the plumber worked 2 more hours than the carpenter did, how many hours did each work?

18. Find the equilibrium point for the demand and supply functions

$$D(p) = 79 - 8p \quad \text{and} \quad S(p) = 37 + 6p.$$

19. Kick Back, Inc., is producing a new hammock. For the first year, the fixed costs for setting up production are $40,000. The variable costs for producing each hammock are $25. The revenue from each hammock is $70. Find the following.

   a) The total cost $C(x)$ of producing x hammocks
   b) The total revenue $R(x)$ from the sale of x hammocks
   c) The total profit $P(x)$ from the production and sale of x hammocks
   d) The profit or loss from the production and sale of 300 hammocks; of 900 hammocks
   e) The break-even point

SYNTHESIS

20. The graph of the function $f(x) = mx + b$ contains the points $(-1, 3)$ and $(-2, -4)$. Find m and b.

21. At a county fair, an adult's ticket sold for $5.50, a senior citizen's ticket for $4.00, and a child's ticket for $1.50. On opening day, the number of adults' and senior citizens' tickets sold was 30 more than the number of children's tickets sold. The number of adults' tickets sold was 6 more than four times the number of senior citizens' tickets sold. Total receipts from the ticket sales were $11,219.50. How many of each type of ticket were sold?

9

Inequalities and Problem Solving

AN APPLICATION

The yearly U.S. production of crude oil $C(t)$, in millions of barrels, t years after 1990, can be approximated by the equation

$$C(t) = -53.5t + 2683$$

(*Source*: Based on data from the *Statistical Abstract of the United States* 2003). Determine (using an inequality) those years for which domestic production will be less than 1750 million barrels.

This problem appears as Example 4 in Section 9.1.

Melissa Leadley
OILFIELD SERVICES ENGINEER
Bakersfield, California

When trying to find oil and gas, we are always solving for the unknown. We use math to calculate important information like oil saturation and rock permeability, which can influence multimillion dollar decisions on whether to produce an oil well, stimulate it, or abandon it.

*I*nequalities are mathematical sentences containing symbols such as $<$ (is less than). In this chapter, we use the principles for solving inequalities developed in Chapter 2 to solve compound inequalities. We also combine our knowledge of inequalities and systems of equations to solve systems of inequalities.

9.1   Interval Notation and Applications

Solving Inequalities • Interval Notation • Problem Solving

Solving Inequalities

Recall from Chapter 1 that an **inequality** is any sentence containing $<, >, \leq, \geq,$ or $\neq$ (see Section 1.4)—for example,

$$-2 < a, \qquad x > 4, \qquad x + 3 \leq 6, \qquad 6 - 7y \geq 10y - 4, \quad \text{and} \quad 5x \neq 10.$$

Any replacement for the variable that makes an inequality true is called a **solution**. The set of all solutions is called the **solution set**. When all solutions of an inequality are found, we say that we have **solved** the inequality.

    We can use two principles, developed in Chapter 2, to solve inequalities.

> **The Addition Principle for Inequalities**
>
> For any real numbers a, b, and c:
>
> $$a < b \text{ is equivalent to } a + c < b + c;$$
> $$a > b \text{ is equivalent to } a + c > b + c.$$
>
> Similar statements hold for $\leq$ and $\geq$.

> **The Multiplication Principle for Inequalities**
>
> For any real numbers a and b, and for any *positive* number c,
>
> $$a < b \text{ is equivalent to } ac < bc;$$
> $$a > b \text{ is equivalent to } ac > bc.$$
>
> For any real numbers a and b, and for any *negative* number c,
>
> $$a < b \text{ is equivalent to } ac > bc;$$
> $$a > b \text{ is equivalent to } ac < bc.$$
>
> Similar statements hold for $\leq$ and $\geq$.

The *graph* of an inequality is a visual representation of the inequality's solution set. An inequality in one variable can be graphed on a number line. Inequalities in two variables are graphed on a coordinate plane, and appear later in this chapter.

The solutions of the inequality $x < 4$ are graphed on the following number line:

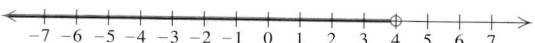

The open dot indicates that 4 is not a solution, and the shading indicates that all real numbers less than 4 are solutions.

We can write the solution set using *set-builder notation* (see Section 2.6):

$\{x \mid x < 4\}$.

This is read

"The set of all x such that x is less than 4."

Interval Notation

Another way to write solutions of an inequality in one variable is to use **interval notation**. Interval notation uses parentheses, ( ), and brackets, [ ].

If a and b are real numbers such that $a < b$, we define the **open interval** **(a, b)** as the set of all numbers x for which $a < x < b$. Thus,

$(a, b) = \{x \mid a < x < b\}$.   Parentheses are used to exclude endpoints.

Its graph excludes the endpoints:

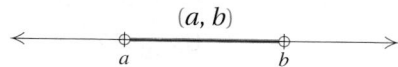

Caution!   Do not confuse the *interval* (a, b) with the *ordered pair* (a, b). The context in which the notation appears usually makes the meaning clear.

The **closed interval [a, b]** is defined as the set of all numbers x for which $a \leq x \leq b$. Thus,

$[a, b] = \{x \mid a \leq x \leq b\}$.   Brackets are used to include endpoints.

Its graph includes the endpoints, as indicated by solid dots*:

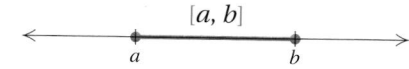

---

*Some books use the graphs ⟨————⟩ and ⊢————⊣ instead of, respectively, ⊕————⊕ and •————• .

There are two kinds of **half-open intervals**, defined as follows:

1. $(a, b] = \{x \mid a < x \leq b\}$. This is open on the left. Its graph is as follows:

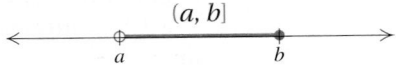

2. $[a, b) = \{x \mid a \leq x < b\}$. This is open on the right. Its graph is as follows:

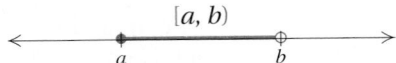

We use the symbols ∞ and $-\infty$ to represent positive and negative infinity, respectively. Thus the notation (a, ∞) represents the set of all real numbers greater than a, and $(-\infty, a)$ represents the set of all real numbers less than a.

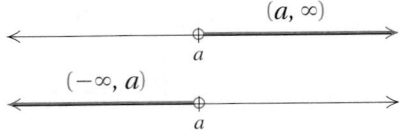

The notations $[a, \infty)$ and $(-\infty, a]$ are used when we want to include the endpoint a.

EXAMPLE 1   Graph $y \geq -2$ on a number line and write the solution set using both set-builder and interval notations.

Solution   Using set-builder notation, we write the solution set as $\{y \mid y \geq -2\}$; using interval notation, we write $[-2, \infty)$. To graph the solution, we shade all numbers to the right of -2 and use a solid dot to indicate that -2 is also a solution.

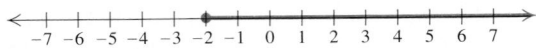

EXAMPLE 2   Solve: $16 - 7y \geq 10y - 4$. Write the solution set in both set-builder and interval notation.

Solution

$$16 - 7y \geq 10y - 4$$
$$-16 + 16 - 7y \geq -16 + 10y - 4 \qquad \text{Adding } -16 \text{ to both sides}$$
$$-7y \geq 10y - 20$$
$$-10y + (-7y) \geq -10y + 10y - 20 \qquad \text{Adding } -10y \text{ to both sides}$$
$$-17y \geq -20$$

The symbol must be reversed.

$$-\tfrac{1}{17} \cdot (-17y) \leq -\tfrac{1}{17} \cdot (-20) \qquad \text{Multiplying both sides by } -\tfrac{1}{17}$$
$$\text{or dividing both sides by } -17$$

$$y \leq \tfrac{20}{17}$$

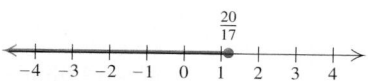

The solution set is $\left\{y \mid y \le \frac{20}{17}\right\}$, or $\left(-\infty, \frac{20}{17}\right]$.

Caution!   Remember that whenever we multiply or divide both sides of an inequality by a negative number, we must reverse the inequality symbol.

We can use interval notation to describe the values for which one function's value is greater than another's.

EXAMPLE 3

Let $f(x) = -3(x + 8) - 5x$ and $g(x) = 4x - 9$. Find all values of x for which $f(x) > g(x)$.

Solution   We are looking for values of x for which $f(x) > g(x)$.

$$f(x) > g(x)$$

$-3(x + 8) - 5x > 4x - 9$	Replacing $f(x)$ with $-3(x + 8) - 5x$ and $g(x)$ with $4x - 9$
$-3x - 24 - 5x > 4x - 9$	Using the distributive law
$-24 - 8x > 4x - 9$	
$-24 - 8x + 8x > 4x - 9 + 8x$	Adding $8x$ to both sides
$-24 > 12x - 9$	
$-24 + 9 > 12x - 9 + 9$	Adding 9 to both sides
$-15 > 12x$	
	The symbol stays the same.
$-\frac{5}{4} > x$	Dividing by 12 and simplifying

The solution set is $\left\{x \mid -\frac{5}{4} > x\right\}$, or $\left\{x \mid x < -\frac{5}{4}\right\}$, or $\left(-\infty, -\frac{5}{4}\right)$.

technology connection

On most calculators, Example 3 can be checked by graphing $y_1 = -3(x + 8) - 5x > 4x - 9$ ($>$ is often found by pressing **2ND** **MATH**). The solution set is then displayed as an interval (shown by a horizontal line 1 unit above the x-axis).

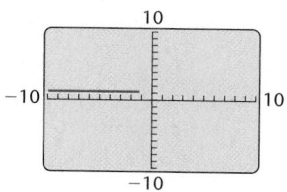

A check can also be made by graphing $y_1 = -3(x + 8) - 5x$ and $y_2 = 4x - 9$ and identifying those x-values for which $y_1 > y_2$.

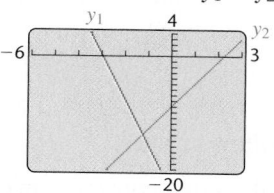

The INTERSECT option helps us find that $y_1 = y_2$ when $x = -1.25$. Note that $y_1 > y_2$ for x-values in the interval $(-\infty, -1.25)$.

Problem Solving

Many problem-solving situations translate to inequalities.

EXAMPLE 4

Domestic oil production. The yearly U.S. production of crude oil $C(t)$, in millions of barrels, t years after 1990, can be approximated by the equation

$$C(t) = -53.5t + 2683$$

(*Source*: Based on data from the *Statistical Abstract of the United States* 2003). Determine (using an inequality) those years for which domestic production will be less than 1750 million barrels.

Solution

1. **Familiarize.** We already have a formula. To become more familiar with it, we might make a substitution for t. Suppose we want to predict production after 20 years, in 2010. We substitute 20 for t:

$$C(20) = -53.5 \cdot 20 + 2683 = 1613.$$

We see that by 2010, production will be less than 1750 million barrels. To predict the exact years in which fewer than 1750 million barrels will be produced, we could check other substitutions. Instead, we proceed to the next step.

2. **Translate.** We are asked to find the years for which U.S. oil production $C(t)$ will be *less than* 1750 million barrels. Thus we have

$$C(t) < 1750.$$

We replace $C(t)$ with $-53.5t + 2683$ to find the times t that solve the inequality:

$$-53.5t + 2683 < 1750. \qquad \text{Substituting}$$

3. **Carry out.** We solve the inequality:

$$-53.5t + 2683 < 1750$$
$$-53.5t < -933 \qquad \text{Adding } -2683 \text{ to both sides}$$
$$t > 17.44. \qquad \text{Dividing both sides by } -53.5, \text{ reversing the symbol, and rounding}$$

4. **Check.** A partial check is to substitute a value for t greater than 17.44. We did that in the *Familiarize* step.

5. **State.** U.S. oil production will fall below 1750 million barrels about 17.4 years after 1990, or in 2007, and will remain below 1750 million for all years after that.

EXAMPLE 5

Job offers. After graduation, Rose had two job offers in sales:

Uptown Fashions: A salary of $600 per month, plus a commission of 4% of sales;

Ergo Designs: A salary of $800 per month, plus a commission of 6% of sales in excess of $10,000.

If sales always exceed $10,000, for what amount of sales would Uptown Fashions provide higher pay?

Solution

1. **Familiarize.**   Listing the given information in a table will be helpful.

Uptown Fashions Monthly Income	Ergo Designs Monthly Income
$600 salary 4% of sales *Total*: $600 + 4% of sales	$800 salary 6% of sales over $10,000 *Total*: $800 + 6% of sales over $10,000

Next, suppose that Rose sold a certain amount—say, $12,000—in one month. Which plan would be better? Working for Uptown, she would earn $600 plus 4% of $12,000, or

$$600 + 0.04(12,000) = \$1080.$$

Since with Ergo Designs commissions are paid only on sales in excess of $10,000, Rose would earn $800 plus 6% of ($12,000 − $10,000), or

$$800 + 0.06(2000) = \$920.$$

This shows that for monthly sales of $12,000, Uptown pays better. Similar calculations will show that for sales of $30,000 a month, Ergo pays better. To determine *all* values for which Uptown pays more money, we must solve an inequality that is based on the calculations above.

2. **Translate.**   We let S = the amount of monthly sales, in dollars, and will assume $S > 10,000$ so that both plans will pay a commission. Examining the calculations in the *Familiarize* step, we see that monthly income from Uptown is $600 + 0.04S$ and from Ergo is $800 + 0.06(S − 10,000)$. We want to find all values of S for which

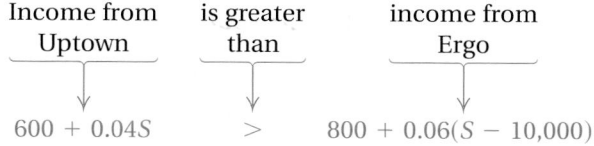

Income from Uptown	is greater than	income from Ergo

$$600 + 0.04S \quad > \quad 800 + 0.06(S − 10,000).$$

3. **Carry out.**   We solve the inequality:

$600 + 0.04S > 800 + 0.06(S − 10,000)$	
$600 + 0.04S > 800 + 0.06S − 600$	Using the distributive law
$600 + 0.04S > 200 + 0.06S$	Combining like terms
$400 > 0.02S$	Subtracting 200 and $0.04S$ from both sides
$20,000 > S$, or $S < 20,000$.	Dividing both sides by 0.02

4. **Check.**   The above steps indicate that income from Uptown Fashions is higher than income from Ergo Designs for sales less than $20,000. In the *Familiarize* step, we saw that for sales of $12,000, Uptown pays more. Since $12,000 < 20,000$, this is a partial check.

5. **State.**   When monthly sales are less than $20,000, Uptown Fashions provides the higher pay.

Exercise Set

9.1

FOR EXTRA HELP

 Student's Solutions Manual

 Digital Video Tutor CD 5 Videotape 9

 AW Math Tutor Center

MathXL Tutorials on CD

Math XL MathXL

MyMathLab MyMathLab

🐦 *Concept Reinforcement*   Classify each of the following as equivalent inequalities, equivalent equations, equivalent expressions, or not equivalent.

1. $x - 7 > -2$, $x > 5$

2. $t + 3 < 1$, $t < 2$

3. $5x + 7 = 6 - 3x$, $8x + 7 = 6$

4. $2(4x + 1)$, $8x + 2$

5. $-4t \le 12$, $t \le -3$

6. $\frac{3}{5}a + \frac{1}{5} = 2$, $3a + 1 = 10$

7. $6a + 9$, $3(2a + 3)$

8. $-4x \ge -8$, $x \ge 2$

9. $-\frac{1}{2}x < 7$, $x > 14$

10. $-\frac{1}{3}t \le -5$, $t \ge 15$

Graph each inequality, and write the solution set using both interval and set-builder notations.

11. $y < 6$

12. $x > 4$

13. $x \ge -4$

14. $t \le 6$

15. $t > -3$

16. $y < -3$

17. $x \le -7$

18. $x \ge -6$

Solve. Then graph.

19. $y - 9 > -18$

20. $y - 8 > -14$

21. $y - 20 \le -6$

22. $x - 11 \le -2$

23. $9t < -81$

24. $8x \ge 24$

25. $-9x \ge -8.1$

26. $-8y \le 3.2$

27. $-\frac{3}{4}x \ge -\frac{5}{8}$

28. $-\frac{5}{6}y \le -\frac{3}{4}$

29. $\dfrac{2x + 7}{5} < -9$

30. $\dfrac{5y + 13}{4} > -2$

31. $\dfrac{3t - 7}{-4} \le 5$

32. $\dfrac{2t - 9}{-3} \ge 7$

33. Let $f(x) = 2x + 1$ and $g(x) = x + 7$. Find all values of x for which $f(x) \ge g(x)$.

34. Let $f(x) = 5 - x$ and $g(x) = 4x - 5$. Find all values of x for which $f(x) \ge g(x)$.

35. Let $f(x) = 7 - 3x$ and $g(x) = 2x - 3$. Find all values of x for which $f(x) \le g(x)$.

36. Let $f(x) = 8x - 9$ and $g(x) = 3x - 11$. Find all values of x for which $f(x) \le g(x)$.

37. Let $f(x) = 2x - 7$ and $g(x) = 5x - 9$. Find all values of x for which $f(x) < g(x)$.

38. Let $f(x) = 0.4x + 5$ and $g(x) = 1.2x - 4$. Find all values of x for which $g(x) \ge f(x)$.

39. Let $f(x) = \frac{3}{8} + 2x$ and $g(x) = 3x - \frac{1}{8}$. Find all values of x for which $g(x) \ge f(x)$.

40. Let $f(x) = 2x + 1$ and $g(x) = -\frac{1}{2}x + 6$. Find all values of x for which $f(x) < g(x)$.

Solve.

41. $4(3y - 2) \ge 9(2y + 5)$

42. $4m + 7 \ge 14(m - 3)$

43. $5(t - 3) + 4t < 2(7 + 2t)$

44. $2(4 + 2x) > 2x + 3(2 - 5x)$

45. $5[3m - (m + 4)] > -2(m - 4)$

46. $8x - 3(3x + 2) - 5 \ge 3(x + 4) - 2x$

47. $19 - (2x + 3) \le 2(x + 3) + x$

48. $13 - (2c + 2) \ge 2(c + 2) + 3c$

49. $\frac{1}{4}(8y + 4) - 17 < -\frac{1}{2}(4y - 8)$

50. $\frac{1}{3}(6x + 24) - 20 > -\frac{1}{4}(12x - 72)$

51. $2[8 - 4(3 - x)] - 2 \ge 8[2(4x - 3) + 7] - 50$

52. $5[3(7 - t) - 4(8 + 2t)] - 20 \le -6[2(6 + 3t) - 4]$

Phone rates. *In Vermont, Verizon recently charged customers* $13.55 *for monthly service plus* 2.2¢ *per minute for local phone calls between* 9 A.M. *and* 9 P.M. *weekdays. The charge for off-peak local calls was* 0.5¢ *per minute. Calls were free after the total monthly charges reached* $39.40.

53. Assume that only peak local calls were made. For how long must a customer speak on the phone if the $39.40 maximum charge is to apply?

54. Assume that only off-peak calls were made. For how long must a customer speak on the phone if the $39.40 maximum charge is to apply?

55. *Checking-account rates.* The Hudson Bank offers two checking-account plans. Their Anywhere plan charges 20¢ per check whereas their Acu-checking plan costs $2 per month plus 12¢ per check. For what numbers of checks per month will the Acu-checking plan cost less?

56. *Moving costs.* Musclebound Movers charges $85 plus $40 an hour to move households across town. Champion Moving charges $60 an hour for cross-town moves. For what lengths of time is Champion more expensive?

57. *Wages.* Toni can be paid in one of two ways:

> *Plan A*: A salary of $400 per month, plus a commission of 8% of gross sales;
>
> *Plan B*: A salary of $610 per month, plus a commission of 5% of gross sales.

For what amount of gross sales should Toni select plan A?

58. *Wages.* Branford can be paid for his masonry work in one of two ways:

> *Plan A*: $300 plus $9.00 per hour;
>
> *Plan B*: Straight $12.50 per hour.

Suppose that the job takes n hours. For what values of n is plan B better for Branford?

59. *Wedding costs.* The Arnold Inn offers two plans for wedding parties. Under plan A, the inn charges $30 for each person in attendance. Under plan B, the inn charges $1300 plus $20 for each person in excess of the first 25 who attend. For what size parties will plan B cost less? (Assume that more than 25 guests will attend.)

60. *Insurance benefits.* Bayside Insurance offers two plans. Under plan A, Giselle would pay the first $50 of her medical bills and 20% of all bills after that. Under plan B, Giselle would pay the first $250 of bills, but only 10% of the rest. For what amount of medical bills will plan B save Giselle money? (Assume that her bills will exceed $250.)

61. *Show business.* Slobberbone receives $750 plus 15% of receipts over $750 for playing a club date. If a club charges a $6 cover charge, how many people must attend in order for the band to receive at least $1200?

62. *Temperature conversion.* The function
$$C(F) = \tfrac{5}{9}(F - 32)$$
can be used to find the Celsius temperature $C(F)$ that corresponds to $F°$ Fahrenheit.
 a) Gold is solid at Celsius temperatures less than 1063°C. Find the Fahrenheit temperatures for which gold is solid.
 b) Silver is solid at Celsius temperatures less than 960.8°C. Find the Fahrenheit temperatures for which silver is solid.

63. *Manufacturing.* Ergs, Inc., is planning to make a new kind of lamp. Fixed costs will be $90,000, and variable costs will be $15 for the production of each lamp. The total-cost function for x lamps is
$$C(x) = 90{,}000 + 15x.$$
The company makes $26 in revenue for each lamp sold. The total-revenue function for x lamps is
$$R(x) = 26x.$$
(See Section 8.8.)
 a) When $R(x) < C(x)$, the company loses money. Find the values of x for which the company loses money.
 b) When $R(x) > C(x)$, the company makes a profit. Find the values of x for which the company makes a profit.

64. *Publishing.* The demand and supply functions for a locally produced poetry book are approximated by
$$D(p) = 2000 - 60p \quad \text{and}$$
$$S(p) = 460 + 94p,$$
where p is the price in dollars (see Section 8.8).

a) Find those values of p for which demand exceeds supply.

b) Find those values of p for which demand is less than supply.

65. Explain in your own words why the inequality symbol must be reversed when both sides of an inequality are multiplied by a negative number.

66. Why isn't roster notation used to write solutions of inequalities?

SKILL MAINTENANCE

Find the domain of f. [7.2]

67. $f(x) = \dfrac{3}{x - 2}$

68. $f(x) = \dfrac{x - 5}{4x + 12}$

69. $f(x) = \dfrac{5x}{7 - 2x}$

70. $f(x) = \dfrac{x + 3}{9 - 4x}$

Simplify. [1.8]

71. $9x - 2(x - 5)$

72. $8x + 7(2x - 1)$

SYNTHESIS

73. A Presto photocopier costs \$510 and an Exact Image photocopier costs \$590. Write a problem that involves the cost of the copiers, the cost per page of photocopies, and the number of copies for which the Presto machine is the more expensive machine to own.

74. Explain how the addition principle can be used to avoid ever needing to multiply or divide both sides of an inequality by a negative number.

Solve for x and y. Assume that a, b, c, d, and m are positive constants.

75. $3ax + 2x \geq 5ax - 4$; assume $a > 1$

76. $6by - 4y \leq 7by + 10$

77. $a(by - 2) \geq b(2y + 5)$; assume $a > 2$

78. $c(6x - 4) < d(3 + 2x)$; assume $3c > d$

79. $c(2 - 5x) + dx > m(4 + 2x)$; assume $5c + 2m < d$

80. $a(3 - 4x) + cx < d(5x + 2)$; assume $c > 4a + 5d$

Determine whether each statement is true or false. If false, give an example that shows this.

81. For any real numbers a, b, c, and d, if $a < b$ and $c < d$, then $a - c < b - d$.

82. For all real numbers x and y, if $x < y$, then $x^2 < y^2$.

83. Are the inequalities

$$x < 3 \quad \text{and} \quad x + \frac{1}{x} < 3 + \frac{1}{x}$$

equivalent? Why or why not?

84. Are the inequalities

$$x < 3 \quad \text{and} \quad 0 \cdot x < 0 \cdot 3$$

equivalent? Why or why not?

Solve. Then graph.

85. $x + 5 \leq 5 + x$

86. $x + 8 < 3 + x$

87. $x^2 > 0$

88. Assume that the graphs of $y_1 = -\frac{1}{2}x + 5$, $y_2 = x - 1$, and $y_3 = 2x - 3$ are as shown below. Solve each inequality, referring only to the figure.

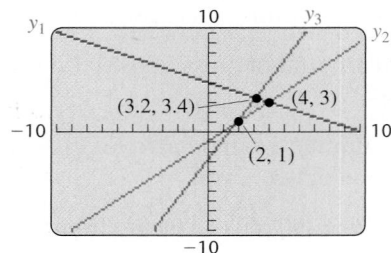

a) $-\frac{1}{2}x + 5 > x - 1$

b) $x - 1 \leq 2x - 3$

c) $2x - 3 \geq -\frac{1}{2}x + 5$

89. Using an approach similar to that in the Technology Connection on p. 581, use a graphing calculator to check your answers to Exercises 19, 35, 51, and 55.

CORNER

Reduce, Reuse, and Recycle

COLLABORATIVE

Focus: Inequalities and problem solving

Time: 15–20 minutes

Group size: 2

In the United States, the amount of solid waste (rubbish) being recycled is slowly catching up to the amount being generated. In 1991, each person generated, on average, 4.3 lb of solid waste every day, of which 0.8 lb was recycled. In 2001, each person generated, on average, 4.4 lb of solid waste, of which 1.3 lb was recycled. (*Sources*: U.S. Bureau of the Census, *Statistical Abstract of the United States* 2003, and EPA Municipal Solid Waste Factbook)

ACTIVITY

Assume that the amount of solid waste being generated and the amount recycled are both increasing linearly. One group member should find a linear function w for which $w(t)$ represents the number of pounds of waste generated per person per day t years after 1991. The other group member should find a linear function r for which $r(t)$ represents the number of pounds recycled per person per day t years after 1991. Finally, working together, the group should determine those years for which the amount recycled will meet or exceed the amount generated.

9.2 Intersections, Unions, and Compound Inequalities

> Intersections of Sets and Conjunctions of Sentences • Unions of Sets and Disjunctions of Sentences • Interval Notation and Domains

Two inequalities joined by the word "and" or the word "or" are called **compound inequalities**. Thus, "$2x - 7 < 3$ *or* $x - 1 > 4$" and "$7x < 9$ *and* $x - 1 > -5$" are two examples of compound inequalities. In order to discuss how to solve compound inequalities, we must first study ways in which sets can be combined.

Intersections of Sets and Conjunctions of Sentences

The **intersection** of two sets A and B is the set of all elements that are common to both A and B. We denote the intersection of sets A and B as

$$A \cap B.$$

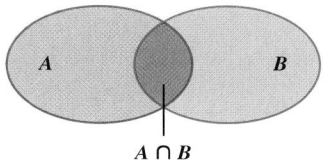

$A \cap B$

The intersection of two sets is represented by the purple region shown in the figure at left. For example, if $A = \{$all students who are more than 5′4″ tall$\}$ and $B = \{$all students who weigh more than 120 lb$\}$, then $A \cap B = \{$all students who are more than 5′4″ tall and weigh more than 120 lb$\}$.

EXAMPLE 1   Find the intersection: $\{1, 2, 3, 4, 5\} \cap \{-2, -1, 0, 1, 2, 3\}$.

Solution   The numbers 1, 2, and 3 are common to both sets, so the intersection is $\{1, 2, 3\}$.

When two or more sentences are joined by the word *and* to make a compound sentence, the new sentence is called a **conjunction** of the sentences. The following is a conjunction of inequalities:

$$-2 < x \quad and \quad x < 1.$$

A number is a solution of a conjunction if it is a solution of *both* of the separate parts. For example, -1 is a solution because it is a solution of $-2 < x$ as well as $x < 1$.

Below we show the graph of $-2 < x$, followed by the graph of $x < 1$, and finally the graph of the conjunction $-2 < x$ and $x < 1$. *Note that the solution set of a conjunction is the intersection of the solution sets of the individual sentences.*

$\{x \mid -2 < x\}$ $(-2, \infty)$

$\{x \mid x < 1\}$ $(-\infty, 1)$

$\{x \mid -2 < x\} \cap \{x \mid x < 1\}$
$= \{x \mid -2 < x \text{ and } x < 1\}$ $(-2, 1)$

Because there are numbers that are both greater than -2 and less than 1, the conjunction $-2 < x$ and $x < 1$ can be abbreviated by $-2 < x < 1$. Thus the interval $(-2, 1)$ can be represented as $\{x \mid -2 < x < 1\}$, the set of all numbers that are *simultaneously* greater than -2 *and* less than 1. Note that for $a < b$,

$$a < x \quad and \quad x < b \quad \text{can be abbreviated} \quad a < x < b;$$

and, equivalently,

$$b > x \quad and \quad x > a \quad \text{can be abbreviated} \quad b > x > a.$$

EXAMPLE 2   Solve and graph: $-1 \le 2x + 5 < 13$.

Solution   This inequality is an abbreviation for the conjunction

$$-1 \le 2x + 5 \quad and \quad 2x + 5 < 13.$$

The word *and* corresponds to set *intersection*. To solve the conjunction, we solve each of the two inequalities separately and then find the intersection of the solution sets:

$$-1 \le 2x + 5 \quad and \quad 2x + 5 < 13$$

$\quad -6 \le 2x \qquad\quad and \qquad\quad 2x < 8 \qquad$ Subtracting 5 from both sides of each inequality

$\quad -3 \le x \qquad\quad and \qquad\quad x < 4. \qquad$ Dividing both sides of each inequality by 2

We now abbreviate the answer:

$$-3 \leq x < 4.$$

The solution set is $\{x \mid -3 \leq x < 4\}$, or, in interval notation, $[-3, 4)$. The graph is the intersection of the two separate solution sets.

$\{x \mid -3 \leq x\}$ $[-3, \infty)$

$\{x \mid x < 4\}$ $(-\infty, 4)$

$\{x \mid -3 \leq x\} \cap \{x \mid x < 4\}$
$= \{x \mid -3 \leq x < 4\}$ $[-3, 4)$

The steps in Example 2 are often combined as follows:

$$-1 \leq 2x + 5 < 13$$
$$-1 - 5 \leq 2x + 5 - 5 < 13 - 5 \qquad \text{Subtracting 5 from all three regions}$$
$$-6 \leq 2x < 8$$
$$-3 \leq x < 4.$$

Such an approach saves some writing and will prove useful in Section 9.3.

> *Caution!*   The abbreviated form of a conjunction, like $-3 \leq x < 4$, can be written only if both inequality symbols point in the same direction. It is *not acceptable* to write a sentence like $-1 > x < 5$ since doing so does not indicate if *both* $-1 > x$ and $x < 5$ must be true or if it is enough for one of the separate inequalities to be true.

EXAMPLE 3   Solve and graph: $2x - 5 \geq -3$ *and* $5x + 2 \geq 17$.

Solution   We first solve each inequality separately, retaining the word *and*:

$$2x - 5 \geq -3 \quad and \quad 5x + 2 \geq 17$$
$$2x \geq 2 \quad and \qquad 5x \geq 15$$
$$x \geq 1 \quad and \qquad x \geq 3.$$

Next, we find the intersection of the two separate solution sets.

$\{x \mid x \geq 1\}$ $[1, \infty)$

$\{x \mid x \geq 3\}$ $[3, \infty)$

$\{x \mid x \geq 1\} \cap \{x \mid x \geq 3\}$
$= \{x \mid x \geq 3\}$ $[3, \infty)$

The numbers common to both sets are those greater than or equal to 3. Thus the solution set is $\{x \mid x \geq 3\}$, or, in interval notation, $[3, \infty)$. You should check that any number in $[3, \infty)$ satisfies the conjunction whereas numbers outside $[3, \infty)$ do not.

> **Mathematical Use of the Word "and"**
>
> The word "and" corresponds to "intersection" and to the symbol " $\cap$ ". Any solution of a conjunction must make each part of the conjunction true.

Sometimes there is no way to solve both parts of a conjunction at once.

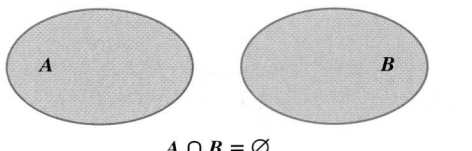

$A \cap B = \varnothing$

When $A \cap B = \varnothing$, A and B are said to be *disjoint*.

EXAMPLE 4   Solve and graph: $2x - 3 > 1 \;$*and*$\; 3x - 1 < 2$.

Solution   We solve each inequality separately:

$$2x - 3 > 1 \quad and \quad 3x - 1 < 2$$
$$2x > 4 \quad and \quad 3x < 3$$
$$x > 2 \quad and \quad x < 1.$$

The solution set is the intersection of the individual inequalities.

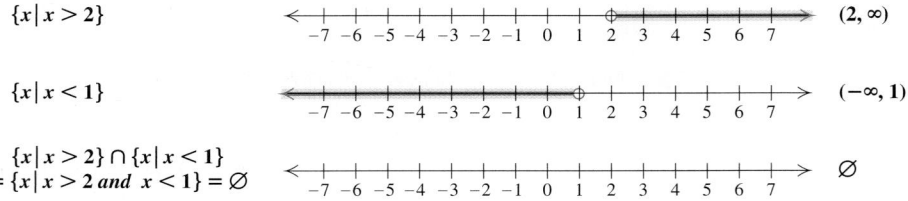

$\{x \mid x > 2\}$ $(2, \infty)$

$\{x \mid x < 1\}$ $(-\infty, 1)$

$\{x \mid x > 2\} \cap \{x \mid x < 1\}$
$= \{x \mid x > 2 \text{ } and \text{ } x < 1\} = \varnothing$ $\varnothing$

Since no number is both greater than 2 and less than 1, the solution set is the empty set, $\varnothing$.

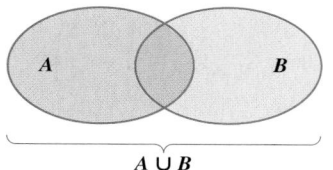

$A \cup B$

Unions of Sets and Disjunctions of Sentences

The **union** of two sets A and B is the collection of elements belonging to A and/or B. We denote the union of A and B by

$$A \cup B.$$

The union of two sets is often pictured as shown at left. For example, if $A =$ {all parents} and $B =$ {all people who are at least 30 yr old}, then $A \cup B =$ {all people who are parents *or* who are at least 30 yr old}. Note that this set includes people who are parents *and* at least 30 yr old.

EXAMPLE 5   Find the union: $\{2, 3, 4\} \cup \{3, 5, 7\}$.

Solution   The numbers in either or both sets are 2, 3, 4, 5, and 7, so the union is $\{2, 3, 4, 5, 7\}$.

Student Notes

Remember that the union or intersection of two sets is itself a set and should be written with set braces.

When two or more sentences are joined by the word *or* to make a compound sentence, the new sentence is called a **disjunction** of the sentences. Here is an example:

$$x < -3 \quad or \quad x > 3.$$

A number is a solution of a disjunction if it is a solution of at least one of the separate parts. For example, -5 is a solution of this disjunction since -5 is a solution of $x < -3$. Below we show the graph of $x < -3$, followed by the graph of $x > 3$, and finally the graph of the disjunction $x < -3$ *or* $x > 3$. *Note that the solution set of a disjunction is the union of the solution sets of the individual sentences.*

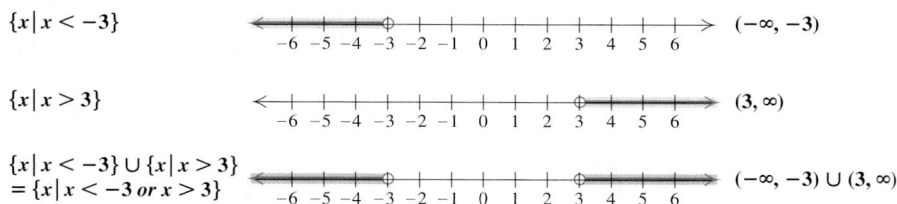

$\{x \mid x < -3\}$ $(-\infty, -3)$

$\{x \mid x > 3\}$ $(3, \infty)$

$\{x \mid x < -3\} \cup \{x \mid x > 3\}$
$= \{x \mid x < -3 \ or \ x > 3\}$ $(-\infty, -3) \cup (3, \infty)$

The solution set of $x < -3$ *or* $x > 3$ is $\{x \mid x < -3 \ or \ x > 3\}$, or, in interval notation, $(-\infty, -3) \cup (3, \infty)$. There is no simpler way to write the solution.

Mathematical Use of the Word "or"

The word "or" corresponds to "union" and to the symbol " $\cup$ ". For a number to be a solution of a disjunction, it must be in *at least one* of the solution sets of the individual sentences.

EXAMPLE 6   Solve and graph: $7 + 2x < -1$ *or* $13 - 5x \leq 3$.

Solution   We solve each inequality separately, retaining the word *or*:

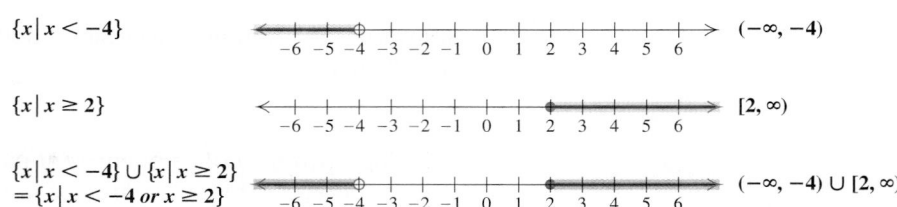

$$7 + 2x < -1 \quad or \quad 13 - 5x \leq 3$$
$$2x < -8 \quad or \quad -5x \leq -10$$

Dividing by a negative and reversing the symbol

$$x < -4 \quad or \quad x \geq 2.$$

To find the solution set of the disjunction, we consider the individual graphs. We graph $x < -4$ and then $x \geq 2$. Then we take the union of the graphs.

$\{x \mid x < -4\}$ $(-\infty, -4)$

$\{x \mid x \geq 2\}$ $[2, \infty)$

$\{x \mid x < -4\} \cup \{x \mid x \geq 2\}$
$= \{x \mid x < -4 \text{ or } x \geq 2\}$ $(-\infty, -4) \cup [2, \infty)$

The solution set is $\{x \mid x < -4 \text{ or } x \geq 2\}$, or $(-\infty, -4) \cup [2, \infty)$.

> *Caution!*   A compound inequality like
>
> $$x < -4 \quad or \quad x \geq 2,$$
>
> as in Example 6, *cannot* be expressed as $2 \leq x < -4$ because to do so would be to say that x is *simultaneously* less than -4 and greater than or equal to 2. No number is both less than -4 *and* greater than 2, but many are less than -4 *or* greater than 2.

EXAMPLE 7   Solve: $-2x - 5 < -2$ *or* $x - 3 < -10$.

Solution   We solve the individual inequalities separately, retaining the word *or*:

$$-2x - 5 < -2 \quad or \quad x - 3 < -10$$
$$-2x < 3 \quad or \quad x < -7$$

Dividing by a negative and reversing the symbol   Keep the word "or."

$$x > -\tfrac{3}{2} \quad or \quad x < -7.$$

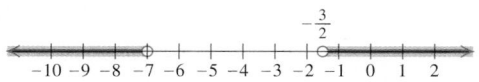

The solution set is $\left\{x \mid x < -7 \text{ or } x > -\tfrac{3}{2}\right\}$, or $(-\infty, -7) \cup \left(-\tfrac{3}{2}, \infty\right)$.

EXAMPLE 8   Solve: $3x - 11 < 4$ *or* $4x + 9 \geq 1$.

Solution   We solve the individual inequalities separately, retaining the word *or*:

$$3x - 11 < 4 \quad or \quad 4x + 9 \geq 1$$
$$3x < 15 \quad or \quad 4x \geq -8$$
$$x < 5 \quad or \quad x \geq -2.$$

— Keep the word "or."

To find the solution set, we first look at the individual graphs.

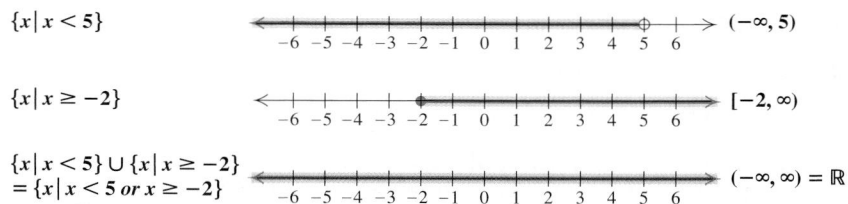

$\{x \mid x < 5\}$ \quad $(-\infty, 5)$

$\{x \mid x \geq -2\}$ \quad $[-2, \infty)$

$\{x \mid x < 5\} \cup \{x \mid x \geq -2\}$
$= \{x \mid x < 5 \ or \ x \geq -2\}$ \quad $(-\infty, \infty) = \mathbb{R}$

Since *all* numbers are less than 5 or greater than or equal to -2, the two sets fill the entire number line. Thus the solution set is $\mathbb{R}$, the set of all real numbers.

Interval Notation and Domains

In Section 7.2, we saw that if $g(x) = \dfrac{5x - 2}{3x - 7}$, then the domain of $g = \{x \mid x$ is a real number *and* $x \neq \frac{7}{3}\}$. We can now represent such a set using interval notation:

$$\{x \mid x \text{ is a real number } and \ x \neq \tfrac{7}{3}\} = \left(-\infty, \tfrac{7}{3}\right) \cup \left(\tfrac{7}{3}, \infty\right).$$

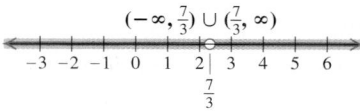

$\left(-\infty, \frac{7}{3}\right) \cup \left(\frac{7}{3}, \infty\right)$

$\frac{7}{3}$

EXAMPLE 9   Use interval notation to write the domain of f if $f(x) = \sqrt{x + 2}$.

Solution   The expression $\sqrt{x + 2}$ is not a real number when $x + 2$ is negative. Thus the domain of f is the set of all x-values for which $x + 2 \geq 0$:

$$x + 2 \geq 0 \qquad x + 2 \text{ cannot be negative.}$$
$$x \geq -2. \qquad \text{Adding } -2 \text{ to both sides}$$

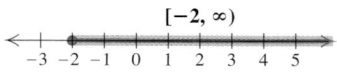

$[-2, \infty)$

We have the domain of $f = \{x \mid x \geq -2\} = [-2, \infty)$.

Exercise Set
9.2

FOR EXTRA HELP

Student's Solutions Manual

Digital Video Tutor CD 5 Videotape 9

Tutor Center

AW Math Tutor Center

MathXL Tutorials on CD

Math$_{XL}$ MathXL

MyMathLab MyMathLab

☙ *Concept Reinforcement* In each of Exercises 1–10, match the set with the most appropriate choice from the column on the right.

1. ____ $\{x \mid x < -2 \ or \ x > 2\}$

2. ____ $\{x \mid x < -2 \ and \ x > 2\}$

3. ____ $\{x \mid x > -2\} \cap \{x \mid x < 2\}$

4. ____ $\{x \mid x \le -2\} \cup \{x \mid x \ge 2\}$

5. ____ $\{x \mid x \le -2\} \cup \{x \mid x \le 2\}$

6. ____ $\{x \mid x \le -2\} \cap \{x \mid x \le 2\}$

7. ____ $\{x \mid x \ge -2\} \cap \{x \mid x \ge 2\}$

8. ____ $\{x \mid x \ge -2\} \cup \{x \mid x \ge 2\}$

9. ____ $\{x \mid x \le 2\} \ and \ \{x \mid x \ge -2\}$

10. ____ $\{x \mid x \le 2\} \ or \ \{x \mid x \ge -2\}$

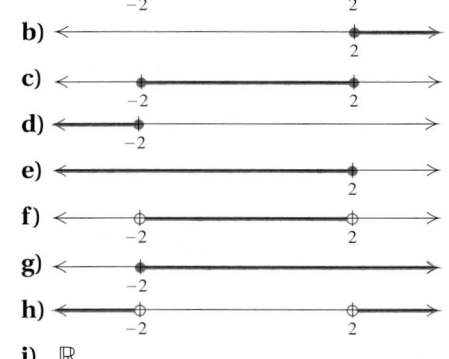

a)

b)

c)

d)

e)

f)

g)

h)

i) $\mathbb{R}$

j) $\varnothing$

Find each indicated intersection or union.

11. $\{5, 9, 11\} \cap \{9, 11, 18\}$

12. $\{2, 4, 8\} \cup \{8, 9, 10\}$

13. $\{0, 5, 10, 15\} \cup \{5, 15, 20\}$

14. $\{2, 5, 9, 13\} \cap \{5, 8, 10\}$

15. $\{a, b, c, d, e, f\} \cap \{b, d, f\}$

16. $\{a, b, c\} \cup \{a, c\}$

17. $\{r, s, t\} \cup \{r, u, t, s, v\}$

18. $\{m, n, o, p\} \cap \{m, o, p\}$

19. $\{3, 6, 9, 12\} \cap \{5, 10, 15\}$

20. $\{1, 5, 9\} \cup \{4, 6, 8\}$

21. $\{3, 5, 7\} \cup \varnothing$

22. $\{3, 5, 7\} \cap \varnothing$

Graph and write interval notation for each compound inequality.

23. $3 < x < 7$

24. $0 \le y \le 4$

25. $-6 \le y \le -2$

26. $-9 \le x < -5$

27. $x < -1 \ or \ x > 4$

28. $x < -5 \ or \ x > 1$

29. $x \le -2 \ or \ x > 1$

30. $x \le -5 \ or \ x > 2$

31. $-4 \le -x < 2$

32. $x > -7 \ and \ x < -2$

33. $x > -2 \ and \ x < 4$

34. $3 > -x \ge -1$

35. $5 > a \ or \ a > 7$

36. $t \ge 2 \ or \ -3 > t$

37. $x \ge 5 \ or \ -x \ge 4$

38. $-x < 3 \ or \ x < -6$

39. $7 > y \ and \ y \ge -3$

40. $6 > -x \ge 0$

41. $x < 7 \ and \ x \ge 3$

42. $x \ge -3 \ and \ x < 3$

Aha! 43. $t < 2 \ or \ t < 5$

44. $t > 4 \ or \ t > -1$

Solve and graph each solution set.

45. $-2 < t + 1 < 8$

46. $-3 < t + 1 \le 5$

47. $2 < x + 3 \ and \ x + 1 \le 5$

48. $-1 < x + 2 \ and \ x - 4 < 3$

49. $-7 \leq 2a - 3 \ and \ 3a + 1 < 7$

50. $-4 \leq 3n + 5 \ and \ 2n - 3 \leq 7$

Aha!   **51.** $x + 7 \leq -2 \ or \ x + 7 \geq -3$

52. $x + 5 < -3 \ or \ x + 5 \geq 4$

53. $5 > \dfrac{x - 3}{4} > 1$

54. $3 \geq \dfrac{x - 1}{2} \geq -4$

55. $-7 \leq 4x + 5 \leq 13$

56. $-4 \leq 2x + 3 \leq 15$

57. $2 \leq f(x) \leq 8$, where $f(x) = 3x - 1$

58. $7 \geq g(x) \geq -2$, where $g(x) = 3x - 5$

59. $-21 \leq f(x) < 0$, where $f(x) = -2x - 7$

60. $4 > g(t) \geq 2$, where $g(t) = -3t - 8$

61. $f(x) \leq 2 \ or \ f(x) \geq 8$, where $f(x) = 3x - 1$

62. $g(x) \leq -2 \ or \ g(x) \geq 10$, where $g(x) = 3x - 5$

63. $f(x) < -3 \ or \ f(x) > 5$, where $f(x) = 2x - 7$

64. $g(x) < -7 \ or \ g(x) > 7$, where $g(x) = 3x + 5$

65. $6 > 2a - 1 \ or \ -4 \leq -3a + 2$

66. $3a - 7 > -10 \ or \ 5a + 2 \leq 22$

67. $a + 3 < -2 \ and \ 3a - 4 < 8$

68. $1 - a < -2 \ and \ 2a + 1 > 9$

69. $3x + 2 < 2 \ or \ 4 - 2x < 14$

70. $2x - 1 > 5 \ or \ 3 - 2x \geq 7$

71. $2t - 7 \leq 5 \ or \ 5 - 2t > 3$

72. $5 - 3a \leq 8 \ or \ 2a + 1 > 7$

For $f(x)$ as given, use interval notation to write the domain of f.

73. $f(x) = \dfrac{9}{x + 8}$

74. $f(x) = \dfrac{2}{x + 3}$

75. $f(x) = \sqrt{x - 6}$

76. $f(x) = \sqrt{x - 2}$

77. $f(x) = \dfrac{x + 3}{2x - 8}$

78. $f(x) = \dfrac{x - 1}{3x + 6}$

79. $f(x) = \sqrt{2x + 7}$

80. $f(x) = \sqrt{8 - 5x}$

81. $f(x) = \sqrt{8 - 2x}$

82. $f(x) = \sqrt{10 - 2x}$

83. Why can the conjunction $2 < x \ and \ x < 5$ be rewritten as $2 < x < 5$, but the disjunction $2 < x$ or $x < 5$ cannot be rewritten as $2 < x < 5$?

84. Can the solution set of a disjunction be empty? Why or why not?

SKILL MAINTENANCE

Graph.

85. $y = 5$ [3.6]

86. $y = -2$ [3.6]

87. $f(x) = |x|$ [7.3]

88. $g(x) = x - 1$ [7.3]

Solve each system graphically. [8.1]

89. $y = x - 3,$
$y = 5$

90. $y = x + 2,$
$y = -3$

SYNTHESIS

91. What can you conclude about a, b, c, and d, if $[a, b] \cup [c, d] = [a, d]$? Why?

92. What can you conclude about a, b, c, and d, if $[a, b] \cap [c, d] = [a, b]$? Why?

93. Use the accompanying graph of $f(x) = 2x - 5$ to solve $-7 < 2x - 5 < 7$.

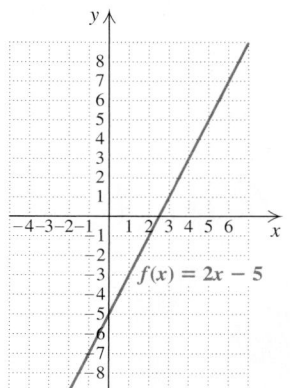

94. Use the accompanying graph of $g(x) = 4 - x$ to solve $4 - x < -2 \ or \ 4 - x > 7$.

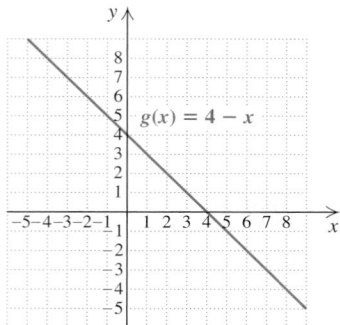

95. *Childless women.* On the basis of trends from the late 1900s, the function given by

$$P(t) = 0.44t + 10.2$$

can be used to estimate the percentage of U.S. women age 40–44, $P(t)$, t years after 1980, who have not given birth. For what years will the percentage of childless 40–44-year-old women be between 19 and 30 percent?

Sources: Based on data from the U.S. Bureau of the Census; *Deseret Morning News*, 11/03

96. *Pressure at sea depth.* The function given by

$$P(d) = 1 + \frac{d}{33}$$

gives the pressure, in atmospheres (atm), at a depth of d feet in the sea. For what depths d is the pressure at least 1 atm and at most 7 atm?

97. *Converting dress sizes.* The function given by

$$f(x) = 2(x + 10)$$

can be used to convert dress sizes x in the United States to dress sizes $f(x)$ in Italy. For what dress sizes in the United States will dress sizes in Italy be between 32 and 46?

98. *Solid-waste generation.* The function given by

$$w(t) = 0.01t + 4.3$$

can be used to estimate the number of pounds of solid waste, $w(t)$, produced daily, on average, by each person in the United States, t years after 1991. For what years will waste production range from 4.5 to 4.75 lb per person per day?

99. *Records in the women's 100-m dash.* Florence Griffith Joyner set a world record of 10.49 sec in the women's 100-m dash in 1988. The function given by

$$R(t) = -0.0433t + 10.49$$

can be used to predict the world record in the women's 100-m dash t years after 1988. Predict (using an inequality) those years for which the world record was between 11.5 and 10.8 sec. (Measure from the middle of 1988.)

Sources: *Guinness Book of World Records* 2004; www.Runnersworld.com

100. *Temperatures of liquids.* The formula

$$C = \tfrac{5}{9}(F - 32)$$

can be used to convert Fahrenheit temperatures F to Celsius temperatures C.

a) Gold is liquid for Celsius temperatures C such that $1063° \leq C < 2660°$. Find a comparable inequality for Fahrenheit temperatures.

b) Silver is liquid for Celsius temperatures C such that $960.8° \leq C < 2180°$. Find a comparable inequality for Fahrenheit temperatures.

101. *Minimizing tolls.* A \$3.00 toll is charged to cross the bridge from Sanibel Island to mainland Florida. A six-month pass, costing \$15.00, reduces the toll to \$0.50. A one-year pass, costing \$150, allows for free crossings. How many crossings per year does it take, on average, for two consecutive six-month passes to be the most economical choice? Assume a constant number of trips per month.

Source: www.leewayinfo.com

Solve and graph.

102. $4a - 2 \leq a + 1 \leq 3a + 4$

103. $4m - 8 > 6m + 5 \ or \ 5m - 8 < -2$

104. $x - 10 < 5x + 6 \leq x + 10$

105. $3x < 4 - 5x < 5 + 3x$

Determine whether each sentence is true or false for all real numbers a, b, and c.

106. If $-b < -a$, then $a < b$.

107. If $a \le c$ and $c \le b$, then $b > a$.

108. If $a < c$ and $b < c$, then $a < b$.

109. If $-a < c$ and $-c > b$, then $a > b$.

For f(x) as given, use interval notation to write the domain of f.

110. $f(x) = \dfrac{\sqrt{5 + 2x}}{x - 1}$

111. $f(x) = \dfrac{\sqrt{3 - 4x}}{x + 7}$

112. Let $y_1 = -1$, $y_2 = 2x + 5$, and $y_3 = 13$. Then use the graphs of y_1, y_2, and y_3 to check the solution to Example 2.

113. Let $y_1 = -2x - 5$, $y_2 = -2$, $y_3 = x - 3$, and $y_4 = -10$. Then use the graphs of y_1, y_2, y_3, and y_4 to check the solution to Example 7.

114. Use a graphing calculator to check your answers to Exercises 43–46 and Exercises 63–66.

115. On many graphing calculators, the TEST key provides access to inequality symbols, while the LOGIC option of that same key accesses the conjunction *and* and the disjunction *or*. Thus, if $y_1 = x > -2$ and $y_2 = x < 4$, Exercise 33 can be checked by forming the expression $y_3 = y_1$ *and* y_2. The interval(s) in the solution set appears as a horizontal line 1 unit above the *x*-axis. (Be careful to "deselect" y_1 and y_2 so that only y_3 is drawn.)

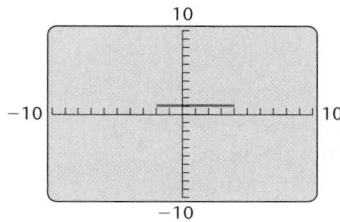

CORNER

Saving on Shipping Costs

COLLABORATIVE

Focus: Compound inequalities and solution sets

Time: 20–30 minutes

Group size: 2–3

For Priority Mail, the United States Postal Service charges (at present, in 2005) $3.85 for the first pound plus $0.51 for each pound or part thereof after the first. For UPS Ground Residential shipping, the first pound costs $6.25 and each additional pound or part thereof costs $0.17.*

_____

*This activity is based on an article by Michael Contino in *Mathematics Teacher*, May 1995.

ACTIVITY

1. One group member should determine the function p, where $p(x)$ represents the cost, in dollars, of mailing x pounds as Priority Mail.
2. One member should determine the function r, where $r(x)$ represents the cost, in dollars, of shipping x pounds using UPS Ground.
3. A third member should graph p and r on the same set of axes.
4. Finally, working together, use the graph to determine those weights for which Priority Mail is less expensive than UPS Ground Residential shipping. Express your answer in both set-builder and interval notations.

9.3 Absolute-Value Equations and Inequalities

Equations with Absolute Value   •   Inequalities with Absolute Value

Equations with Absolute Value

Recall from Section 1.4 that the absolute value of a number a is the number of units that a is from zero. Another definition uses opposites.

> **Absolute Value**
>
> The absolute value of x, denoted $|x|$, is defined as
> $$|x| = \begin{cases} x, & \text{if } x \geq 0, \\ -x, & \text{if } x < 0. \end{cases}$$
>
> (When x is nonnegative, the absolute value of x is x. When x is negative, the absolute value of x is the opposite of x.)

To better understand this definition, suppose x is -5. Then $|x| = |-5| = 5$, and 5 is the opposite of -5. This shows that when x represents a negative number, the absolute value of x is the opposite of x (which is positive).

Since distance is always nonnegative, we can think of a number's absolute value as its distance from zero on a number line.

EXAMPLE 1   Find the solution set: **(a)** $|x| = 4$; **(b)** $|x| = 0$; **(c)** $|x| = -7$.

Solution

a) We interpret $|x| = 4$ to mean that the number x is 4 units from zero on a number line. There are two such numbers, 4 and -4. Thus the solution set is $\{-4, 4\}$.

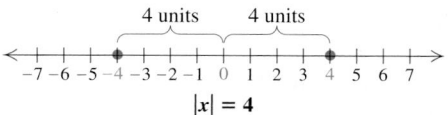

$|x| = 4$

A second way to visualize this problem is to graph $f(x) = |x|$ (see Section 7.3). We also graph $g(x) = 4$. The x-values of the points of intersection are the solutions of $|x| = 4$.

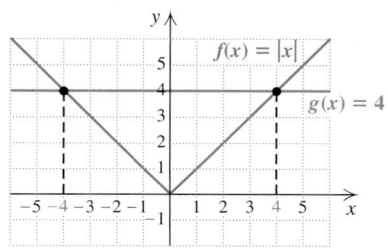

b) We interpret $|x| = 0$ to mean that x is 0 units from zero on a number line. The only number that satisfies this is 0 itself. Thus the solution set is $\{0\}$.

c) Since distance is always nonnegative, it doesn't make sense to talk about a number that is -7 units from zero. Remember: The absolute value of a number is never negative. Thus, $|x| = -7$ has no solution; the solution set is $\varnothing$.

Example 1 leads us to the following principle for solving equations.

> **The Absolute-Value Principle for Equations**
>
> For any positive number p and any algebraic expression X:
>
> **a)** The solutions of $|X| = p$ are those numbers that satisfy $X = -p \text{ or } X = p$.
> **b)** The equation $|X| = 0$ is equivalent to the equation $X = 0$.
> **c)** The equation $|X| = -p$ has no solution.

EXAMPLE 2

Find the solution set: **(a)** $|2x + 5| = 13$; **(b)** $|4 - 7x| = -8$.

Solution

a) We use the absolute-value principle, knowing that $2x + 5$ must be either 13 or -13:

$$|X| = p$$
$$|2x + 5| = 13 \qquad \text{Substituting}$$
$$2x + 5 = -13 \quad or \quad 2x + 5 = 13$$
$$2x = -18 \quad or \quad 2x = 8$$
$$x = -9 \quad or \quad x = 4.$$

Check: For -9:

$$\frac{|2x + 5| = 13}{\begin{array}{c|c} |2(-9) + 5| & 13 \\ |-18 + 5| & \\ |-13| & \\ & 13 \overset{?}{=} 13 \quad \text{TRUE} \end{array}}$$

For 4:

$$\frac{|2x + 5| = 13}{\begin{array}{c|c} |2 \cdot 4 + 5| & 13 \\ |8 + 5| & \\ |13| & \\ & 13 \overset{?}{=} 13 \quad \text{TRUE} \end{array}}$$

The number $2x + 5$ is 13 units from zero if x is replaced with -9 or 4. The solution set is $\{-9, 4\}$.

b) The absolute-value principle reminds us that absolute value is always nonnegative. The equation $|4 - 7x| = -8$ has no solution. The solution set is $\varnothing$.

To use the absolute-value principle, we must be sure that the absolute-value expression is alone on one side of the equation.

---

technology connection

To check Example 2(a), we use the NUM option after pressing **MATH** and let $y_1 = \text{abs}(2x + 5)$ and $y_2 = 13$. Next, using the window $[-12, 8, -2, 18]$, we use the INTERSECT option after pressing **2ND** CALC to find the intersections $(-9, 13)$ and $(4, 13)$. The x-coordinates, -9 and 4, are the solutions.

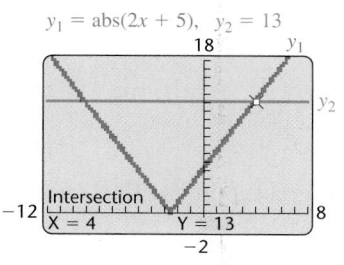

$y_1 = \text{abs}(2x + 5), \quad y_2 = 13$

1. Use a graphing calculator to show that Example 2(b) has no solution.

EXAMPLE 3 Given that $f(x) = 2|x + 3| + 1$, find all x for which $f(x) = 15$.

Solution Since we are looking for $f(x) = 15$, we substitute:

$$f(x) = 15$$

$$2|x + 3| + 1 = 15 \qquad \text{Replacing } f(x) \text{ with } 2|x + 3| + 1$$

$$2|x + 3| = 14 \qquad \text{Subtracting 1 from both sides}$$

$$|x + 3| = 7 \qquad \text{Dividing both sides by 2}$$

$$x + 3 = -7 \quad or \quad x + 3 = 7 \qquad \text{Using the absolute-value principle for equations}$$

$$x = -10 \quad or \qquad x = 4.$$

We leave it to the student to check that $f(-10) = f(4) = 15$. The solution set is $\{-10, 4\}$.

EXAMPLE 4 Solve: $|x - 2| = 3$.

Solution Because this equation is of the form $|a - b| = c$, it can be solved in two different ways.

Method 1. We interpret $|x - 2| = 3$ as stating that the number $x - 2$ is 3 units from zero. Using the absolute-value principle, we replace X with $x - 2$ and p with 3:

$$|X| = p$$

$$|x - 2| = 3$$

$$x - 2 = -3 \quad or \quad x - 2 = 3 \qquad \text{Using the absolute-value principle}$$

$$x = -1 \quad or \qquad x = 5.$$

Method 2. This approach is helpful in calculus. The expressions $|a - b|$ and $|b - a|$ can be used to represent the *distance between a and b* on the number line. For example, the distance between 7 and 8 is given by $|8 - 7|$ or $|7 - 8|$. From this viewpoint, the equation $|x - 2| = 3$ states that the distance between x and 2 is 3 units. We draw a number line and locate all numbers that are 3 units from 2.

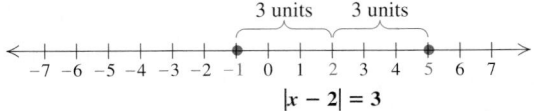

The solutions of $|x - 2| = 3$ are -1 and 5.

Check: The check consists of observing that both methods give the same solutions. The solution set is $\{-1, 5\}$.

Sometimes an equation has two absolute-value expressions. Consider $|a| = |b|$. This means that a and b are the same distance from zero.

If a and b are the same distance from zero, then either they are the same number or they are opposites.

EXAMPLE 5

Solve: $|2x - 3| = |x + 5|$.

Solution   The given equation tells us that $2x - 3$ and $x + 5$ are the same distance from 0. This means that they are either the same number or opposites:

This assumes these numbers are the same.

This assumes these numbers are opposites.

$$2x - 3 = x + 5 \quad or \quad 2x - 3 = -(x + 5)$$
$$x - 3 = 5 \quad or \quad 2x - 3 = -x - 5$$
$$x = 8 \quad or \quad 3x - 3 = -5$$
$$3x = -2$$
$$x = -\tfrac{2}{3}.$$

The check is left to the student. The solutions are 8 and $-\tfrac{2}{3}$ and the solution set is $\left\{-\tfrac{2}{3}, 8\right\}$.

Inequalities with Absolute Value

Our methods for solving equations with absolute value can be adapted for solving inequalities. Inequalities of this sort arise regularly in more advanced courses.

EXAMPLE 6

Solve $|x| < 4$. Then graph.

Solution   The solutions of $|x| < 4$ are all numbers whose *distance from zero is less than* 4. By substituting or by looking at the number line, we can see that numbers like $-3, -2, -1, -\tfrac{1}{2}, -\tfrac{1}{4}, 0, \tfrac{1}{4}, \tfrac{1}{2}, 1, 2,$ and 3 are all solutions. In fact, the solutions are all the numbers between -4 and 4. The solution set is $\{x | -4 < x < 4\}$. In interval notation, the solution set is $(-4, 4)$. The graph is as follows:

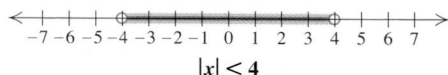

$|x| < 4$

We can also visualize Example 6 by graphing $f(x) = |x|$ and $g(x) = 4$, as in Example 1. The solution set consists of all x-values for which $(x, f(x))$ is below the horizontal line $g(x) = 4$. These x-values comprise the interval $(-4, 4)$.

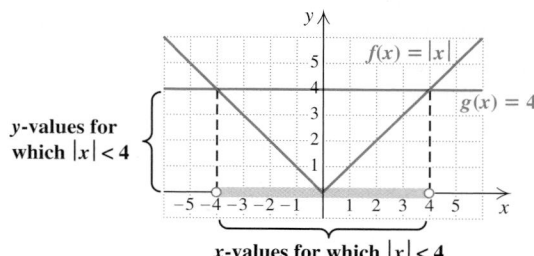

x-values for which $|x| < 4$

EXAMPLE 7

Solve $|x| \geq 4$. Then graph.

Solution   The solutions of $|x| \geq 4$ are all numbers that are at least 4 units from zero—in other words, those numbers x for which $x \leq -4$ *or* $4 \leq x$. The solution set is $\{x \mid x \le -4 \text{ or } x \ge 4\}$. In interval notation, the solution set is $(-\infty, -4] \cup [4, \infty)$. We can check mentally with numbers like $-4.1, -5, 4.1,$ and 5. The graph is as follows:

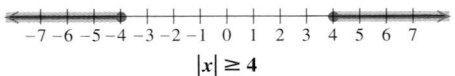

$|x| \ge 4$

As with Examples 1 and 6, Example 7 can be visualized by graphing $f(x) = |x|$ and $g(x) = 4$. The solution set of $|x| \ge 4$ consists of all x-values for which $(x, f(x))$ is on or above the horizontal line $g(x) = 4$. These x-values comprise $(-\infty, -4] \cup [4, \infty)$.

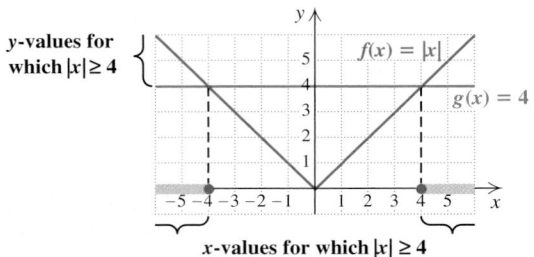

Examples 1, 6, and 7 illustrate three types of problems in which absolute-value symbols appear. The following is a general principle for solving such problems.

Principles for Solving Absolute-Value Problems

For any positive number p and any expression X:

a) The solutions of $|X| = p$ are those numbers that satisfy $X = -p \text{ or } X = p$.

b) The solutions of $|X| < p$ are those numbers that satisfy $-p < X < p$.

c) The solutions of $|X| > p$ are those numbers that satisfy $X < -p \text{ or } p < X$.

Of course, if p is negative, any value of X will satisfy the inequality $|X| > p$ because absolute value is never negative. (The solution set is $\mathbb{R}$.) By the same reasoning, $|X| < p$ has no solution when p is not positive. (The solution set is $\varnothing$). Thus, $|2x - 7| > -3$ is true for any real number x, and $|2x - 7| < -3$ has no solution.

Note that an inequality of the form $|X| < p$ corresponds to a *con*junction, whereas an inequality of the form $|X| > p$ corresponds to a *dis*junction.

EXAMPLE 8

Solve $|3x - 2| < 4$. Then graph.

Solution   The number $3x - 2$ must be less than 4 units from 0. This is of the form $|X| < p$, so part (b) of the principles listed above applies:

$$|X| < p$$

$$|3x - 2| < 4 \qquad \text{Replacing } X \text{ with } 3x - 2 \text{ and } p \text{ with } 4$$

$$-4 < 3x - 2 < 4 \qquad \text{The number } 3x - 2 \text{ must be within 4 units of zero.}$$

$$-2 < \quad 3x \quad < 6 \qquad \text{Adding 2}$$

$$-\tfrac{2}{3} < \quad x \quad < 2. \qquad \text{Multiplying by } \tfrac{1}{3}$$

The solution set is $\left\{x \,|\, -\tfrac{2}{3} < x < 2\right\}$. In interval notation, the solution set is $\left(-\tfrac{2}{3}, 2\right)$. The graph is as follows:

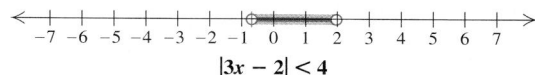

$$|3x - 2| < 4$$

EXAMPLE 9

Given that $f(x) = |4x + 2|$, find all x for which $f(x) \geq 6$.

Solution   We have

$$f(x) \geq 6,$$

$$\text{or} \qquad |4x + 2| \geq 6. \qquad \text{Substituting}$$

To solve, we use part (c) of the principles listed above. In this case, X is $4x + 2$ and p is 6:

$$|X| \geq p$$

$$|4x + 2| \geq 6 \qquad \text{Replacing } X \text{ with } 4x + 2 \text{ and } p \text{ with } 6$$

$$4x + 2 \leq -6 \quad or \quad 6 \leq 4x + 2 \qquad \text{The number } 4x + 2 \text{ must be at least 6 units from zero.}$$

$$4x \leq -8 \quad or \quad 4 \leq 4x \qquad \text{Adding } -2$$

$$x \leq -2 \quad or \quad 1 \leq x. \qquad \text{Multiplying by } \tfrac{1}{4}$$

The solution set is $\{x \,|\, x \leq -2 \ or \ x \geq 1\}$. In interval notation, the solution is $(-\infty, -2] \cup [1, \infty)$. The graph is as follows:

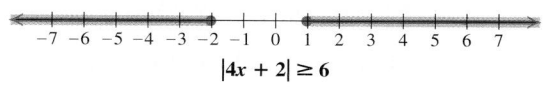

$$|4x + 2| \geq 6$$

Exercise Set
9.3

FOR EXTRA HELP

| Student's Solutions Manual | Digital Video Tutor CD 5 Videotape 9 | AW Math Tutor Center | MathXL Tutorials on CD | MathXL | MyMathLab |

Concept Reinforcement *Classify each of the following as either true or false.*

1. If x is negative, then $|x| = -x$.

2. $|x|$ is never negative.

3. $|x|$ is always positive.

4. The distance between a and b can be expressed as $|a - b|$.

5. The number a is $|a|$ units from 0.

6. There are two solutions of $|3x - 8| = 17$.

7. There is no solution of $|4x + 9| > -5$.

8. All real numbers are solutions of $|2x - 7| < -3$.

Solve.

9. $|x| = 7$

10. $|x| = 9$

Aha! 11. $|x| = -6$

12. $|x| = -3$

13. $|p| = 0$

14. $|y| = 7.3$

15. $|t| = 5.5$

16. $|m| = 0$

17. $|2x - 3| = 4$

18. $|5x + 2| = 7$

19. $|3x - 5| = -8$

20. $|7x - 2| = -9$

21. $|x - 2| = 6$

22. $|x - 3| = 8$

23. $|x - 5| = 3$

24. $|x - 6| = 1$

25. $|x - 7| = 9$

26. $|x - 4| = 5$

27. $|5x| - 3 = 37$

28. $|2y| - 5 = 13$

29. $7|q| - 2 = 9$

30. $7|z| + 2 = 16$

31. $\left|\dfrac{2x - 1}{3}\right| = 5$

32. $\left|\dfrac{4 - 5x}{6}\right| = 3$

33. $|m + 5| + 9 = 16$

34. $|t - 7| + 1 = 4$

35. $5 - 2|3x - 4| = -5$

36. $3|2x - 5| - 7 = -1$

37. Let $f(x) = |2x + 6|$. Find all x for which $f(x) = 8$.

38. Let $f(x) = |2x + 4|$. Find all x for which $f(x) = 10$.

39. Let $f(x) = |x| - 3$. Find all x for which $f(x) = 5.7$.

40. Let $f(x) = |x| + 7$. Find all x for which $f(x) = 18$.

41. Let $f(x) = \left|\dfrac{3x - 2}{5}\right|$. Find all x for which $f(x) = 2$.

42. Let $f(x) = \left|\dfrac{1 - 2x}{3}\right|$. Find all x for which $f(x) = 1$.

Solve.

43. $|x + 4| = |2x - 7|$

44. $|3x + 5| = |x - 6|$

45. $|x + 4| = |x - 3|$

46. $|x - 9| = |x + 6|$

47. $|3a - 1| = |2a + 4|$

48. $|5t + 7| = |4t + 3|$

Aha! 49. $|n - 3| = |3 - n|$

50. $|y - 2| = |2 - y|$

51. $|7 - a| = |a + 5|$

52. $|6 - t| = |t + 7|$

53. $\left|\frac{1}{2}x - 5\right| = \left|\frac{1}{4}x + 3\right|$

54. $\left|2 - \frac{2}{3}x\right| = \left|4 + \frac{7}{8}x\right|$

Solve and graph.

55. $|a| \le 9$

56. $|x| < 2$

57. $|x| > 8$

58. $|a| \ge 3$

59. $|t| > 0$

60. $|t| \ge 1.7$

61. $|x - 1| < 4$

62. $|x - 1| < 3$

63. $|x + 2| \le 6$

64. $|x + 4| \le 1$

65. $|x - 3| + 2 > 7$

66. $|x - 4| + 5 > 2$

Aha! 67. $|2y - 9| > -5$

68. $|3y - 4| > 8$

69. $|3a - 4| + 2 \ge 8$

70. $|2a - 5| + 1 \ge 9$

71. $|y - 3| < 12$

72. $|p - 2| < 3$

73. $9 - |x + 4| \le 5$

74. $12 - |x - 5| \le 9$

75. $|4 - 3y| > 8$

76. $|7 - 2y| < -6$

Aha!
77. $|5 - 4x| < -6$

78. $7 + |4a - 5| \le 26$

79. $\left| \dfrac{2 - 5x}{4} \right| \ge \dfrac{2}{3}$

80. $\left| \dfrac{1 + 3x}{5} \right| > \dfrac{7}{8}$

81. $|m + 3| + 8 \le 14$

82. $|t - 7| + 3 \ge 4$

83. $25 - 2|a + 3| > 19$

84. $30 - 4|a + 2| > 12$

85. Let $f(x) = |2x - 3|$. Find all x for which $f(x) \le 4$.

86. Let $f(x) = |5x + 2|$. Find all x for which $f(x) \le 3$.

87. Let $f(x) = 5 + |3x - 4|$. Find all x for which $f(x) \ge 16$.

88. Let $f(x) = |2 - 9x|$. Find all x for which $f(x) \ge 25$.

89. Let $f(x) = 7 + |2x - 1|$. Find all x for which $f(x) < 16$.

90. Let $f(x) = 5 + |3x + 2|$. Find all x for which $f(x) < 19$.

91. Explain in your own words why -7 is not a solution of $|x| < 5$.

92. Explain in your own words why $[6, \infty)$ is only part of the solution of $|x| \ge 6$.

SKILL MAINTENANCE

Solve using substitution or elimination. [8.2]

93. $2x - 3y = 7,$
$3x + 2y = -10$

94. $3x - 5y = 9,$
$4x - 3y = 1$

95. $x = -2 + 3y,$
$x - 2y = 2$

96. $y = 3 - 4x,$
$2x - y = -9$

Solve graphically. [8.1]

97. $x + 2y = 9,$
$3x - y = -1$

98. $2x + y = 7,$
$-3x - 2y = 10$

SYNTHESIS

99. Is it possible for an equation in x of the form $|ax + b| = c$ to have exactly one solution? Why or why not?

100. Explain why the inequality $|x + 5| \ge 2$ can be interpreted as "the number x is at least 2 units from -5."

101. From the definition of absolute value, $|x| = x$ only when $x \ge 0$. Solve $|3t - 5| = 3t - 5$ using this same reasoning.

Solve.

102. $|3x - 5| = x$

103. $|x + 2| > x$

104. $2 \le |x - 1| \le 5$

105. $|5t - 3| = 2t + 4$

106. $t - 2 \le |t - 3|$

Find an equivalent inequality with absolute value.

107. $-3 < x < 3$

108. $-5 \le y \le 5$

109. $x \le -6 \text{ or } 6 \le x$

110. $x < -4 \text{ or } 4 < x$

111. $x < -8 \text{ or } 2 < x$

112. $-5 < x < 1$

113. x is less than 2 units from 7.

114. x is less than 1 unit from 5.

Write an absolute-value inequality for which the interval shown is the solution.

115.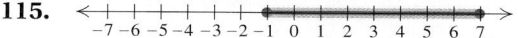

116.

117.

118.

119. *Bungee jumping.* A bungee jumper is bouncing up and down so that her distance d above a river satisfies the inequality $|d - 60 \text{ ft}| \le 10 \text{ ft}$ (see the figure below). If the bridge from which she jumped is 150 ft above the river, how far is the bungee jumper from the bridge at any given time?

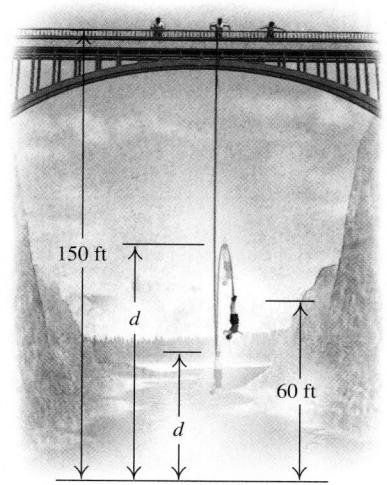

120. *Water level.*   Depending on how dry or wet the weather has been, water in a well will rise and fall. The distance d that a well's water level is below the ground satisfies the inequality $|d - 15| \le 2.5$ (see the figure below).

a) Solve for d.

b) How tall a column of water is in the well at any given time?

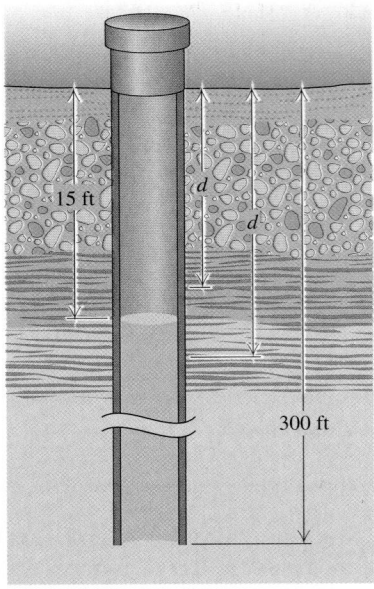

121. Use the accompanying graph of $f(x) = |2x - 6|$ to solve $|2x - 6| \le 4$.

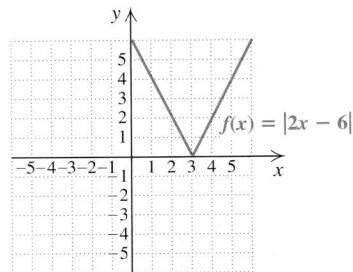

122. Describe a procedure that could be used to solve any equation of the form $g(x) < c$ graphically.

123. Use a graphing calculator to check the solutions to Examples 3 and 5.

124. Use a graphing calculator to check your answers to Exercises 9, 17, 23, 49, 61, 71, 79, and 103.

125. Isabel is using the following graph to solve $|x + 3| < 4$. How can you tell that a mistake has been made in entering $y = \text{abs}(x + 3)$?

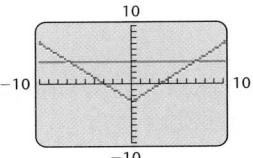

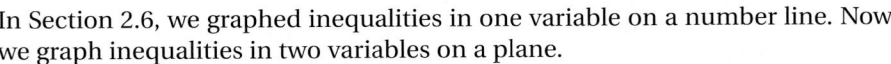

9.4   Inequalities in Two Variables

Graphs of Linear Inequalities   •   Systems of Linear Inequalities

In Section 2.6, we graphed inequalities in one variable on a number line. Now we graph inequalities in two variables on a plane.

Graphs of Linear Inequalities

When the equals sign in a linear equation is replaced with an inequality sign, a **linear inequality** is formed. Solutions of linear inequalities are ordered pairs.

EXAMPLE 1   Determine whether $(-3, 2)$ and $(6, -7)$ are solutions of the inequality $5x - 4y > 13$.

Solution   Below, on the left, we replace x with -3 and y with 2. On the right, we replace x with 6 and y with -7.

$$\begin{array}{c|c} \multicolumn{2}{c}{5x - 4y > 13} \\ \hline 5(-3) - 4 \cdot 2 & 13 \\ -15 - 8 & \\ -23 \overset{?}{>} 13 & \text{FALSE} \end{array}$$

$$\begin{array}{c|c} \multicolumn{2}{c}{5x - 4y > 13} \\ \hline 5(6) - 4(-7) & 13 \\ 30 + 28 & \\ 58 \overset{?}{>} 13 & \text{TRUE} \end{array}$$

Since $-23 > 13$ is false, $(-3, 2)$ is not a solution.

Since $58 > 13$ is true, $(6, -7)$ is a solution.

The graph of a linear equation is a straight line. The graph of a linear inequality is a **half-plane**, with a **boundary** that is a straight line. To find the equation of the boundary line, we simply replace the inequality sign with an equals sign.

EXAMPLE 2   Graph: $y \leq x$.

Solution   We first graph the equation of the boundary, $y = x$. Every solution of $y = x$ is an ordered pair, like $(3, 3)$, in which both coordinates are the same. The graph of $y = x$ is shown on the left below. Since the inequality symbol is $\leq$, the line is drawn solid and is part of the graph of $y \leq x$.

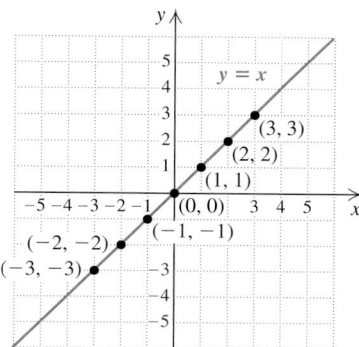 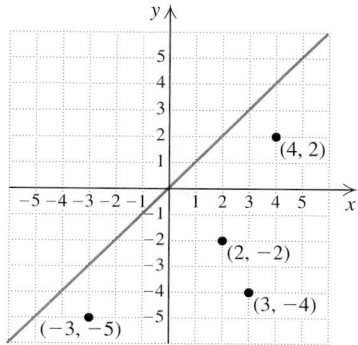

Note that in the graph on the right each ordered pair on the half-plane below $y = x$ contains a y-coordinate that is less than the x-coordinate. All these pairs represent solutions of $y \leq x$. We check one pair, $(4, 2)$, as follows:

$$\begin{array}{c|c} \multicolumn{2}{c}{y \leq x} \\ \hline 2 & 4 \quad \text{TRUE} \end{array}$$

It turns out that *any* point on the same side of $y = x$ as $(4, 2)$ is also a solution. Thus, if one point in a half-plane is a solution, then *all* points in that half-plane are solutions.

We finish drawing the solution set by shading the half-plane below $y = x$. The complete solution set consists of the shaded half-plane as well as the boundary line itself.

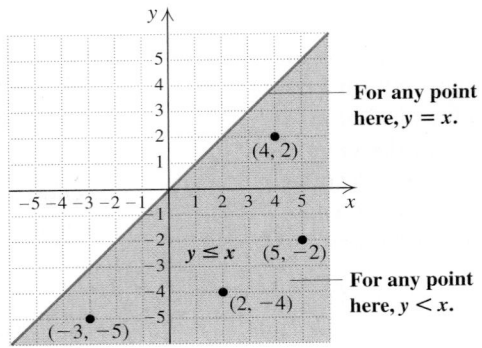

From Example 2, we see that for any inequality of the form $y \leq f(x)$ or $y < f(x)$, we shade *below* the graph of $y = f(x)$.

EXAMPLE 3

Graph: $8x + 3y > 24$.

Solution   First, we sketch the graph of $8x + 3y = 24$. Since the inequality sign is $>$, points on this line do not represent solutions of the inequality, so the line is drawn dashed. Points representing solutions of $8x + 3y > 24$ are in either the half-plane above the line or the half-plane below the line. To determine which, we select a point that is not on the line and determine whether it is a solution of $8x + 3y > 24$. Let's use $(1, 1)$ as this *test point*:

$$
\begin{array}{c|c}
\multicolumn{2}{c}{8x + 3y > 24} \\
\hline
8(1) + 3(1) & 24 \\
8 + 3 & \\
11 \overset{?}{>} 24 & \text{FALSE}
\end{array}
$$

Since $11 > 24$ is *false*, $(1, 1)$ is not a solution. Thus no point in the half-plane containing $(1, 1)$ is a solution. The points in the other half-plane *are* solutions, so we shade that half-plane and obtain the graph shown at right.

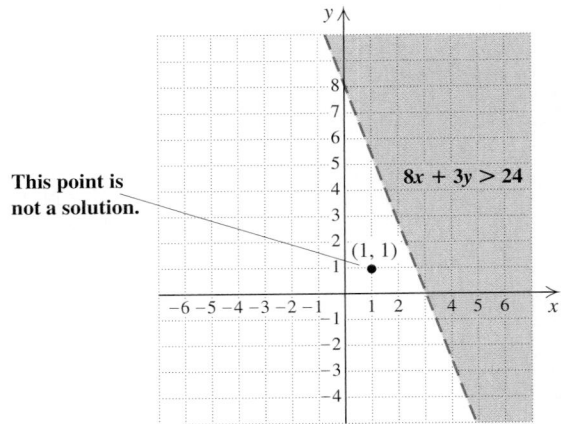

Steps for Graphing Linear Inequalities

1. Replace the inequality sign with an equals sign and graph this line as the boundary. If the inequality symbol is $<$ or $>$, draw the line dashed. If the inequality symbol is $\le$ or $\ge$, draw the line solid.
2. The graph of the inequality consists of a half-plane on one side of the line and, if the line is solid, the line as well.

   a) If the inequality is of the form $y < mx + b$ or $y \le mx + b$, shade *below* the line.
      If the inequality is of the form $y > mx + b$ or $y \ge mx + b$, shade *above* the line.
   b) If y is not isolated, either solve for y and graph as in part (a) or simply graph the boundary and use a test point (as in Example 3). If the test point *is* a solution, shade the half-plane containing the point. If it is not a solution, shade the other half-plane.

EXAMPLE 4   Graph: $6x - 2y < 12$.

Solution   We could graph $6x - 2y = 12$ and use a test point, as in Example 3. Instead, let's solve $6x - 2y < 12$ for y:

$$6x - 2y < 12$$
$$-2y < -6x + 12 \qquad \text{Adding } -6x \text{ to both sides}$$
$$y > 3x - 6. \qquad \text{Dividing both sides by } -2 \text{ and reversing the } < \text{ symbol}$$

The graph consists of the half plane above the dashed boundary line $y = 3x - 6$ (see the graph below).

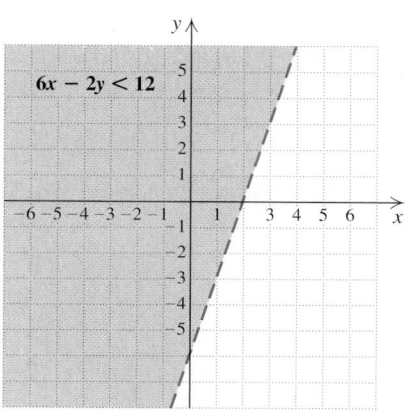

EXAMPLE 5 Graph $x > -3$ on a plane.

Solution There is a missing variable in this inequality. If we graph the inequality on a line, its graph is as follows:

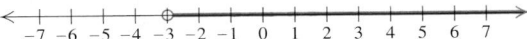

However, we can also write this inequality as $x + 0y > -3$ and graph it on a plane. We can use the same technique as in the examples above. First, we graph the boundary $x = -3$ in the plane, using a dashed line. Then we test some point, say, (2, 5):

$$\frac{x + 0y > -3}{2 + 0 \cdot 5 \mid -3}$$
$$2 \overset{?}{>} -3 \quad \text{TRUE}$$

Since (2, 5) is a solution, all points in the half-plane containing (2, 5) are solutions. We shade that half-plane. Another approach is to simply note that the solutions of $x > -3$ are all pairs with first coordinates greater than -3.

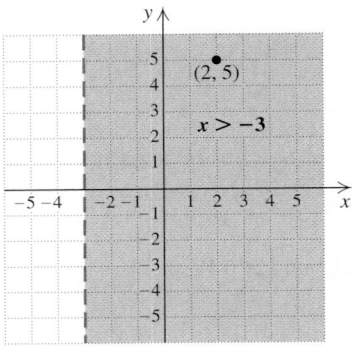

---

technology connection

On most graphing calculators, an inequality like $y < \frac{6}{5}x + 3.49$ can be drawn by entering $(6/5)x + 3.49$ as y_1, moving the cursor to the GraphStyle icon just to the left of y_1, pressing **ENTER** until ◣ appears, and then pressing **GRAPH**.

Many newer calculators have an INEQUALZ program that is accessed using the **APPS** key. Running this program allows us to write inequalities at the **Y=** screen by pressing **ALPHA** and then one of the five keys just below the screen.

Although the graphs should be identical regardless of the method used, on the newer calculators the boundary line appears dashed when < or > is selected.

$$y_1 < (6/5)x + 3.49, \text{ or}$$
$$◣ \, y_1 = (6/5)x + 3.49$$

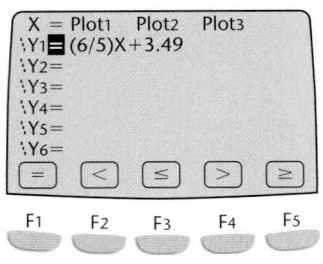

Inequalities containing ≤ or ≥ are handled in a similar manner.

Graph each of the following. Solve for y first if necessary.

1. $y > x + 3.5$
2. $7y \le 2x + 5$
3. $8x - 2y < 11$
4. $11x + 13y + 4 \ge 0$

EXAMPLE 6   Graph $y \leq 4$ on a plane.

Solution   The inequality is of the form $y \leq mx + b$ (with $m = 0$), so we shade below the solid horizontal line representing $y = 4$.

This inequality can also be graphed by drawing $y = 4$ and testing a point above or below the line. The student should check that this results in a graph identical to the one at right.

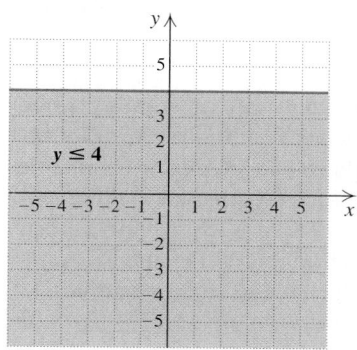

Systems of Linear Inequalities

To graph a system of equations, we graph the individual equations and then find the intersection of the individual graphs. We do the same thing for a system of inequalities, that is, we graph each inequality and find the intersection of the individual graphs.

EXAMPLE 7   Graph the system

$$x + y \leq 4,$$
$$x - y < 4.$$

Solution   To graph $x + y \leq 4$, we graph $x + y = 4$ using a solid line. Since the test point $(0, 0)$ *is* a solution and $(0, 0)$ is below the line, we shade the half-plane below the graph red. The arrows near the ends of the line are another way of indicating the half-plane containing solutions.

Next, we graph $x - y < 4$. We graph $x - y = 4$ using a dashed line and consider $(0, 0)$ as a test point. Again, $(0, 0)$ is a solution, so we shade that side of the line blue. The solution set of the system is the region that is shaded purple (both red and blue) and part of the line $x + y = 4$.

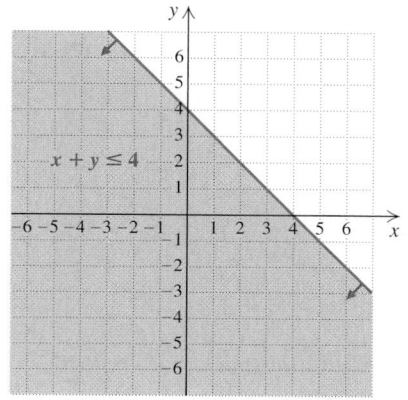
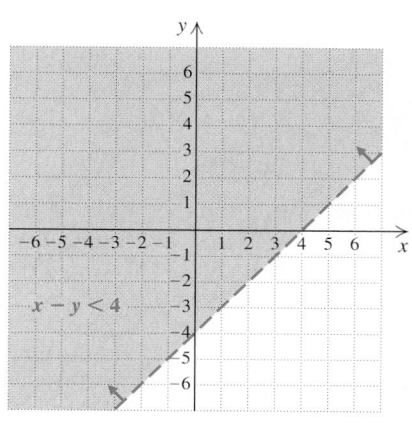
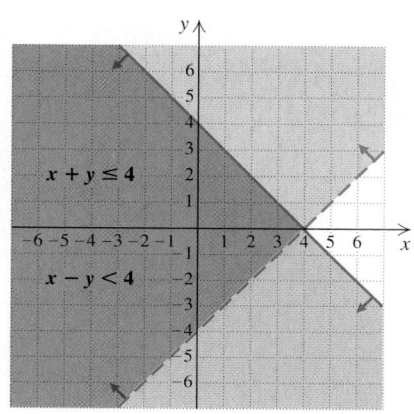

EXAMPLE 8

Student Notes

If you don't use differently colored pencils or pens to shade different regions, consider using a pencil to make slashes that tilt in different directions in each region. You may also find it useful to attach arrows to the lines, as in the examples shown.

Graph: $-2 < x \leq 3$.

Solution   This is a system of inequalities:

$$-2 < x,$$
$$x \leq 3.$$

We graph the equation $-2 = x$, and see that the graph of the first inequality is the half-plane to the right of the boundary $-2 = x$. It is shaded red.

We graph the second inequality, starting with the boundary $x = 3$, and find that its graph is the line and also the half-plane to its left. It is shaded blue.

The solution set of the system is the region that is the intersection of the individual graphs. Since it is shaded both blue and red, it appears to be purple. All points in this region have x-coordinates that are greater than -2 but do not exceed 3.

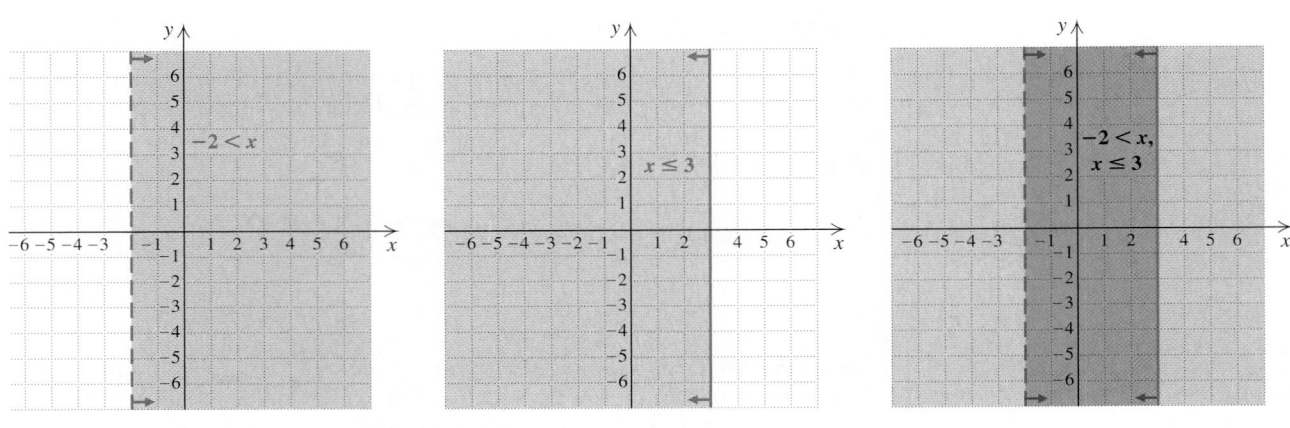

A system of inequalities may have a graph that consists of a polygon and its interior. In Section 9.5, we will have use for the corners, or *vertices* (singular, *vertex*), of such a graph.

EXAMPLE 9

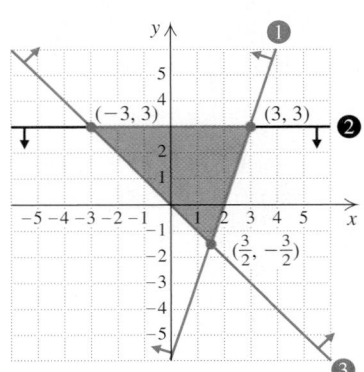

Graph the system of inequalities. Find the coordinates of any vertices formed.

$$6x - 2y \leq 12, \quad (1)$$
$$y - 3 \leq 0, \quad (2)$$
$$x + y \geq 0 \quad (3)$$

Solution   We graph the boundaries

$$6x - 2y = 12,$$
$$y - 3 = 0,$$
and $$x + y = 0$$

using solid lines. The regions for each inequality are indicated by the arrows near the ends of the lines. We note where the regions overlap and shade the region of solutions purple.

To find the vertices, we solve three different systems of two equations. The system of boundary equations from inequalities (1) and (2) is

$$6x - 2y = 12,$$
$$y - 3 = 0.$$

Solving, we obtain the vertex $(3, 3)$.

The system of boundary equations from inequalities (1) and (3) is

$$6x - 2y = 12,$$
$$x + y = 0.$$

Solving, we obtain the vertex $\left(\frac{3}{2}, -\frac{3}{2}\right)$.

The system of boundary equations from inequalities (2) and (3) is

$$y - 3 = 0,$$
$$x + y = 0.$$

Solving, we obtain the vertex $(-3, 3)$.

![technology connection]

Systems of inequalities can be graphed by solving for y and then graphing each inequality as in the Technology Connection on p. 610. To graph systems directly using the INEQUALZ application, enter the correct inequalities, press ⌈GRAPH⌉, and then press **ALPHA** and Shades (⌈F1⌉ or ⌈F2⌉). At the SHADES menu, select Ineq Intersection to see the final graph. To find the vertices, or points of intersection, select PoI-Trace from the graph menu.

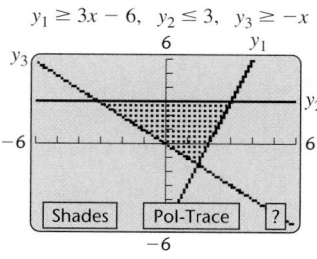

$y_1 \geq 3x - 6, \quad y_2 \leq 3, \quad y_3 \geq -x$

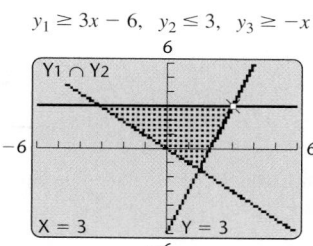

$y_1 \geq 3x - 6, \quad y_2 \leq 3, \quad y_3 \geq -x$

1. Use a graphing calculator to check the solution of Example 7.

CONNECTING THE CONCEPTS

We have now solved a variety of equations, inequalities, systems of equations, and systems of inequalities. In each case, there are different ways to represent the solution. Below is a list of the different types of problems we have solved, along with illustrations of each type.

Type	Example	Solution	Graph
Linear equations in one variable	$2x - 8 = 3(x + 5)$	A number	
Linear inequalities in one variable	$-3x + 5 > 2$	A set of numbers; an interval	
Linear equations in two variables	$2x + y = 7$	A set of ordered pairs; a line	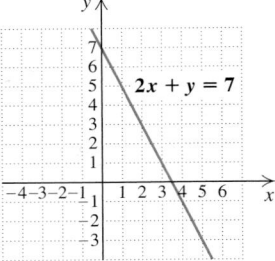
Linear inequalities in two variables	$x + y \geq 4$	A set of ordered pairs; a half-plane	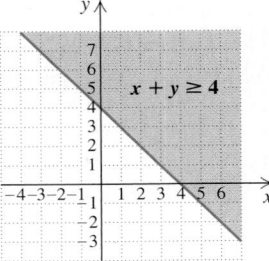
System of equations in two variables	$x + y = 3,$ $5x - y = -27$	An ordered pair or a (possibly empty) set of ordered pairs	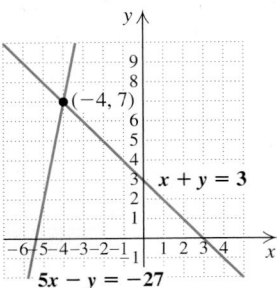
System of inequalities in two variables	$6x - 2y \leq 12,$ $y - 3 \leq 0,$ $x + y \geq 0$	A set of ordered pairs; a region of a plane	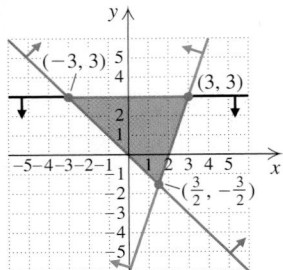

Keeping in mind how these solutions vary and what their graphs look like will help you as you progress further in this book and in mathematics in general.

Exercise Set

9.4

☙ *Concept Reinforcement*   *In each of Exercises 1–6, match the phrase with the most appropriate choice from the column on the right.*

1. ____ A solution of a linear inequality

2. ____ The graph of a linear inequality

3. ____ The graph of a system of linear inequalities

4. ____ Often a convenient test point

5. ____ The name for the corners of a graph of a system of linear inequalities

6. ____ A dashed line

a) $(0, 0)$

b) Vertices

c) A half-plane

d) The intersection of two or more half-planes

e) An ordered pair that satisfies the inequality

f) Indicates the line is not part of the solution

Determine whether each ordered pair is a solution of the given inequality.

7. $(-4, 2)$; $2x + 3y < -1$

8. $(3, -6)$; $4x + 2y \leq -2$

9. $(8, 14)$; $2y - 3x \geq 9$

10. $(7, 20)$; $3x - y > -1$

Graph on a plane.

11. $y > \frac{1}{2}x$

12. $y > 2x$

13. $y \geq x - 3$

14. $y < x + 3$

15. $y \leq x + 5$

16. $y > x - 2$

17. $x - y \leq 4$

18. $x + y < 4$

19. $2x + 3y < 6$

20. $3x + 4y \leq 12$

21. $2y - x \leq 4$

22. $2y - 3x > 6$

23. $2x - 2y \geq 8 + 2y$

24. $3x - 2 \leq 5x + y$

25. $y \geq 3$

26. $x < -5$

27. $x \leq 6$

28. $y > -3$

29. $-2 < y < 7$

30. $-4 < y < -1$

31. $-4 \leq x \leq 2$

32. $-3 \leq y \leq 4$

33. $0 \leq y \leq 3$

34. $0 \leq x \leq 6$

Graph each system.

35. $y > -x,$
    $y < x + 2$

36. $y < x,$
    $y > -x + 1$

37. $y \geq x,$
    $y \geq 2x - 4$

38. $y \geq x,$
    $y \leq -x + 4$

39. $y \leq -3,$
    $x \geq -1$

40. $y \geq -3,$
    $x \geq 1$

41. $x > -4,$
    $y < -2x + 3$

42. $x < 3,$
    $y > -3x + 2$

43. $y \leq 5,$
    $y \geq -x + 4$

44. $y \geq -2,$
    $y \geq x + 3$

45. $x + y \leq 6,$
    $x - y \leq 4$

46. $x + y < 1,$
    $x - y < 2$

47. $y + 3x > 0,$
    $y + 3x < 2$

48. $y - 2x \geq 1,$
    $y - 2x \leq 3$

Graph each system of inequalities. Find the coordinates of any vertices formed.

49. $y \leq 2x - 3,$
    $y \geq -2x + 1,$
    $x \leq 5$

50. $2y - x \leq 2,$
    $y - 3x \geq -4,$
    $y \geq -1$

51. $x + 2y \leq 12,$
    $2x + y \leq 12,$
    $x \geq 0,$
    $y \geq 0$

52. $x - y \leq 2,$
    $x + 2y \geq 8,$
    $y \leq 4$

53. $8x + 5y \leq 40,$
    $x + 2y \leq 8,$
    $x \geq 0,$
    $y \geq 0$

54. $4y - 3x \geq -12,$
    $4y + 3x \geq -36,$
    $y \leq 0,$
    $x \leq 0$

55. $y - x \geq 2,$
    $y - x \leq 4,$
    $2 \leq x \leq 5$

56. $3x + 4y \geq 12,$
    $5x + 6y \leq 30,$
    $1 \leq x \leq 3$

57. In Example 7, is the point $(4, 0)$ part of the solution set? Why or why not?

58. When graphing linear inequalities, Ron makes a habit of always shading above the line when the symbol $\geq$ is used. Is this wise? Why or why not?

SKILL MAINTENANCE

Solve.

59. *Catering.* Sandy's Catering needs to provide 10 lb of mixed nuts for a wedding reception. Peanuts cost $2.50 per pound and fancy nuts cost $7 per pound. If $40 has been allocated for nuts, how many pounds of each type should be mixed? [8.3]

60. *Household waste.* The Hendersons generate two and a half times as much trash as their neighbors, the Savickis. Together, the two households produce 14 bags of trash each month. How much trash does each household produce? [8.3]

61. *Paid admissions.* There were 203 tickets sold for a volleyball game. For activity-card holders the price was $2.50, and for noncard holders the price was $4. The total amount of money collected was $620. How many of each type of ticket were sold? [8.3]

62. *Paid admissions.* There were 200 tickets sold for a women's basketball game. Tickets for students were $4 each and for adults were $6 each. The total amount collected was $1060. How many of each type of ticket were sold? [8.3]

63. *Landscaping.* Grass seed is being spread on a triangular traffic island. If the grass seed can cover an area of 200 ft^2 and the island's base is 16 ft long, how tall a triangle can the seed fill? [2.5]

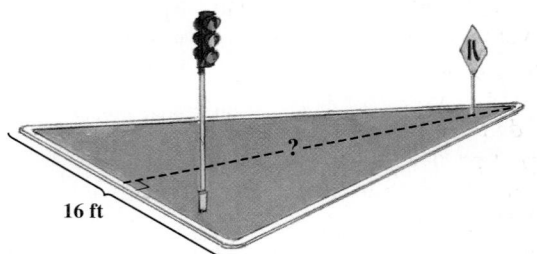

16 ft

64. *Interest rate.* What rate of interest is required in order for a principal of $1280 to earn $17.60 in half a year? [2.4]

SYNTHESIS

65. Explain how a system of linear inequalities could have a solution set containing exactly one pair.

66. Do all systems of linear inequalities have solutions? Why or why not?

Graph.

67. $x + y > 8,$
$x + y \leq -2$

68. $x + y \geq 1,$
$-x + y \geq 2,$
$x \geq -2,$
$y \geq 2,$
$y \leq 4,$
$x \leq 2$

69. $x - 2y \leq 0,$
$-2x + y \leq 2,$
$x \leq 2,$
$y \leq 2,$
$x + y \leq 4$

70. Write four systems of four inequalities that describe a 2-unit by 2-unit square that has (0, 0) as one of the vertices.

71. *Luggage size.* Unless an additional fee is paid, most major airlines will not check any luggage for which the sum of the item's length, width, and height exceeds 62 in. The U.S. Postal Service will ship a package only if the sum of the package's length and girth (distance around its midsection) does not exceed 130 in. Video Promotions is ordering several 30-in. long cases that will be both mailed and checked as luggage. Using w and h for width and height (in inches), respectively, write and graph an inequality that represents all acceptable combinations of width and height.
Sources: U.S. Postal Service; www.case2go.com

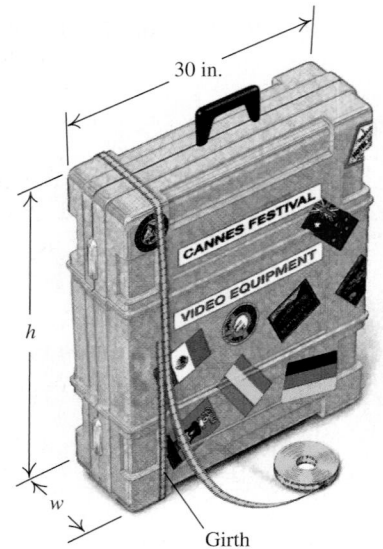

30 in.

h

w

Girth

72. *Hockey wins and losses.*   The Skating Stars figure that they need at least 60 points for the season in order to make the playoffs. A win is worth 2 points and a tie is worth 1 point. Graph a system of inequalities that describes the situation. (*Hint*: Let $w =$ the number of wins and $t =$ the number of ties.)

73. *Elevators.*   Many elevators have a capacity of 1 metric ton (1000 kg). Suppose that c children, each weighing 35 kg, and a adults, each 75 kg, are on an elevator. Graph a system of inequalities that indicates when the elevator is overloaded.

74. *Widths of a basketball floor.*   Sizes of basketball floors vary due to building sizes and other constraints such as cost. The length L is to be at most 94 ft and the width W is to be at most 50 ft. Graph a system of inequalities that describes the possible dimensions of a basketball floor.

75. Use a graphing calculator to graph each inequality.
   a) $3x + 6y > 2$
   b) $x - 5y \leq 10$
   c) $13x - 25y + 10 \leq 0$
   d) $2x + 5y > 0$

76. Use a graphing calculator to check your answers to Exercises 35–48. Then use INTERSECT to determine any point(s) of intersection.

CORNER

COLLABORATIVE

How Old Is Old Enough?

***Focus*:** Linear inequalities

***Time*:** 15–25 minutes

***Group size*:** 2

It is not unusual for the ages of a bride and groom to differ significantly. Yet is it possible for the difference in age to be too great? In answer to this question, the following rule of thumb has emerged: *The younger spouse's age should be at least seven more than half the age of the older spouse.* (*Source*: http://home.earthlink.net/~mybrainhurts/2002_06_01_archive.html)

ACTIVITY

1. Let $b =$ the age of the bride, in years, and $g =$ the age of the groom, in years. One group member should write an equation for

calculating the bride's minimum age if the groom's age is known. The other group member should write an equation for finding the groom's minimum age if the bride's age is known. The equations should look similar.

2. Convert each equation into an inequality by selecting the appropriate symbol from $<$, $>$, $\leq$, and $\geq$. Be sure to reflect the rule of thumb stated above.

3. Graph both inequalities from step 2 as a system of linear inequalities. What does the solution set represent?

4. If your group feels that a minimum or maximum age for marriage should exist, adjust your graph accordingly.

5. Compare your finished graph with those of other groups.

9.5 Applications Using Linear Programming

Objective Functions and Constraints   •   Linear Programming

There are many real-world situations in which we need to find a greatest value (a maximum) or a least value (a minimum). For example, most businesses like to know how to make the *most* profit with the *least* expense possible. Some such problems can be solved using systems of inequalities.

Objective Functions and Constraints

Often a quantity we wish to maximize depends on two or more other quantities. For example, a gardener's profits P might depend on the number of shrubs s and the number of trees t that are planted. If the gardener makes a $10 profit from each shrub and an $18 profit from each tree, the total profit, in dollars, is given by the **objective function**

$$P = 10s + 18t.$$

Thus the gardener might be tempted to simply plant lots of trees since they yield the greater profit. This would be a good idea were it not for the fact that the number of trees and shrubs planted—and thus the total profit—is subject to the demands, or **constraints**, of the situation. For example, to improve drainage, the gardener might be required to plant at least 3 shrubs. Thus the objective function would be subject to the *constraint*

$$s \geq 3.$$

Because of the limited space, the gardener might also be required to plant no more than 10 plants. This would subject the objective function to a *second* constraint:

$$s + t \leq 10.$$

Finally, the gardener might be told to spend no more than $700 on the plants. If the shrubs cost $40 each and the trees cost $100 each, the objective function is subject to a *third* constraint:

The cost of the shrubs plus the cost of the trees cannot exceed $700.

$$40s \qquad + \qquad 100t \qquad \leq \qquad 700$$

In short, the gardener wishes to maximize the objective function

$$P = 10s + 18t,$$

subject to the constraints

$$s \geq 3,$$
$$s + t \leq 10,$$
$$40s + 100t \leq 700,$$
$$\left.\begin{array}{l} s \geq 0, \\ t \geq 0. \end{array}\right\}$$ Because the number of trees and shrubs cannot be negative

These constraints form a system of linear inequalities that can be graphed.

Linear Programming

The gardener's problem is "How many shrubs and trees should be planted, subject to the constraints listed, in order to maximize profit?" To solve such a problem, we use an important result from a branch of mathematics known as **linear programming**.

The Corner Principle

Suppose that an objective function $F = ax + by + c$ depends on x and y (with a, b, and c constant). Suppose also that F is subject to constraints on x and y, which form a system of linear inequalities. If F has a minimum or a maximum value, then it can be found as follows:

1. Graph the system of inequalities and find the vertices.
2. Find the value of the objective function at each vertex. The greatest and the least of those values are the maximum and the minimum of the function, respectively.
3. The ordered pair at which the maximum or minimum occurs indicates the choice of (x, y) for which that maximum or minimum occurs.

   This result was proven during World War II, when linear programming was developed to help with shipping troops and supplies from North America to Europe.

EXAMPLE 1   Solve the gardener's problem discussed above.

Solution   We are asked to maximize $P = 10s + 18t$, subject to the constraints

$$s \geq 3,$$
$$s + t \leq 10,$$
$$40s + 100t \leq 700,$$
$$s \geq 0,$$
$$t \geq 0.$$

We graph the system, using the techniques of Section 9.4. The portion of the graph that is shaded represents all pairs that satisfy the constraints. It is sometimes called the *feasible region*.

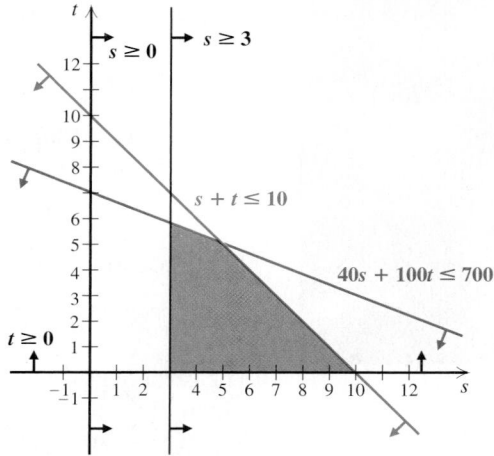

According to the corner principle, P is maximized at one of the vertices of the shaded region. To determine the coordinates of the vertices, we solve the following systems:

$$\left.\begin{array}{r} 40s + 100t = 700, \\ s = 3; \end{array}\right\}$$ The student can verify that the solution of this system is (3, 5.8).

$$\left.\begin{array}{r} s + t = 10, \\ 40s + 100t = 700; \end{array}\right\}$$ The student can verify that the solution of this system is (5, 5).

$$\left.\begin{array}{r} s + t = 10, \\ t = 0; \end{array}\right\}$$ The solution of this system is (10, 0).

$$\left.\begin{array}{r} t = 0, \\ s = 3. \end{array}\right\}$$ The solution of this system is (3, 0).

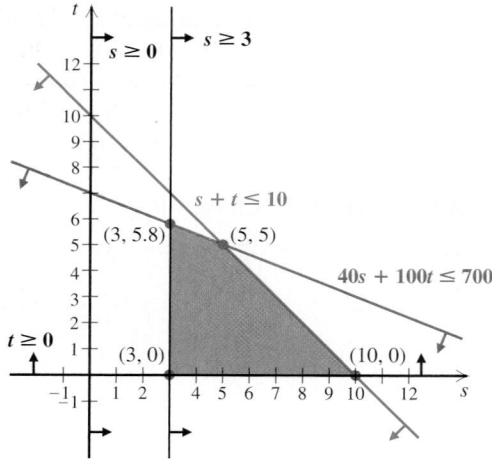

We now find the value of P at each vertex.

Vertex (s, t)	Profit $P = 10s + 18t$	
(3, 5.8)	$10(3) + 18(5.8) = 134.4$	
(5, 5)	$10(5) + 18(5) = 140$	← Maximum
(10, 0)	$10(10) + 18(0) = 100$	
(3, 0)	$10(3) + 18(0) = 30$	← Minimum

The greatest value of P occurs at (5, 5). Thus profit is maximized at \$140 if the gardener plants 5 shrubs and 5 trees. Incidentally, we have also shown that profit is minimized at \$30 if 3 shrubs and 0 trees are planted.

EXAMPLE 2

Test scores.   Corinna is taking a test in which multiple-choice questions are worth 10 points each and short-answer questions are worth 15 points each. It takes her 3 min to answer each multiple-choice question and 6 min to answer each short-answer question. The total time allowed is 60 min, and no more than 16 questions can be answered. Assuming that all her answers are correct, how many items of each type should Corinna answer in order to get the best score?

Solution

1. Familiarize.   Tabulating information will help us to see the picture.

Type	Number of Points for Each	Time Required for Each	Number Answered
Multiple-choice	10	3 min	x
Short-answer	15	6 min	y
Total time: 60 min			
Total number of items: 16 or fewer			

Note that we use x to represent the number of multiple-choice questions and y to represent the number of short-answer questions that are answered.

2. Translate.   In this case, it helps to extend the table.

Type	Number of Points for Each	Time Required for Each	Number Answered	Total Time for Each Type	Total Points for Each Type
Multiple-choice	10	3 min	x	$3x$	$10x$
Short-answer	15	6 min	y	$6y$	$15y$
Total			$x + y \leq 16$	$3x + 6y \leq 60$	$10x + 15y$

Because no more than 16 items may be answered

Because the time cannot exceed 60 min

This is what we want to maximize: the total score on the test.

Let T represent the total score. We express this as a function of the same variables used in the constraints—in this case, x and y:

$$T = 10x + 15y.$$

We wish to maximize T subject to the constraints listed above:

$$x + y \leq 16,$$
$$3x + 6y \leq 60,$$
$$\left.\begin{matrix} x \geq 0, \\ y \geq 0. \end{matrix}\right\}$$ We include this because the number of questions answered cannot be negative.

3. **Carry out.** The mathematical manipulation consists of graphing the system and evaluating T at each vertex. The graph is as follows:

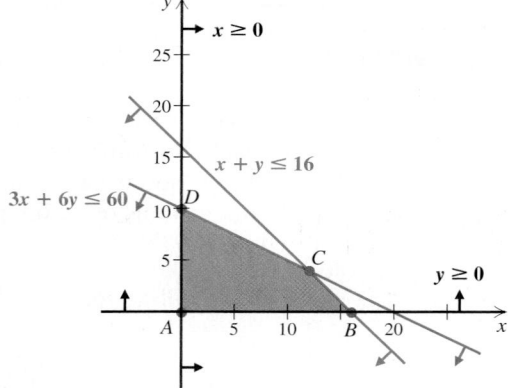

We find the coordinates of each vertex by solving a system of two linear equations. The coordinates of point A are obviously $(0, 0)$. To find the coordinates of point C, we solve the system

$$x + y = 16, \quad (1)$$
$$3x + 6y = 60, \quad (2)$$

as follows:

$$-3x - 3y = -48 \qquad \text{Multiplying both sides of equation (1) by } -3$$
$$\underline{3x + 6y = 60}$$
$$3y = 12 \qquad \text{Adding}$$
$$y = 4.$$

Then we find that $x = 12$. Thus the coordinates of vertex C are $(12, 4)$. Point B is the x-intercept of the line given by $x + y = 16$, so B is $(16, 0)$. Point D is the y-intercept of $3x + 6y = 60$, so D is $(0, 10)$. Computing the test score for each ordered pair, we obtain the table at left.

The greatest score in the table is 180, obtained when 12 multiple-choice and 4 short-answer questions are answered.

4. **Check.** We can check that $T \leq 180$ for any other pair in the shaded region. This is left to the student.

5. **State.** In order to maximize her score, Corinna should answer 12 multiple-choice questions and 4 short-answer questions.

Vertex (x, y)	Score $T = 10x + 15y$
A $(0, 0)$	0
B $(16, 0)$	160
C $(12, 4)$	180
D $(0, 10)$	150

Exercise Set

9.5

↘ *Concept Reinforcement*   *Complete each of the following sentences.*

1. In linear programing, a function is maximized or _____ subject to the demands of a given situation.

2. In linear programming, the function on which the demands are placed is known as the _____ function.

3. In linear programming, the demands arising from the given situation are known as _____.

4. To solve a linear programming problem, we make use of the _____ principle.

5. The shaded portion of a graph that represents all points that satisfy a problem's constraints is known as the _____ region.

6. In linear programming, the corners of the shaded portion of the graph are referred to as _____.

Find the maximum and the minimum values of each objective function and the values of x and y at which they occur.

7. $F = 2x + 14y$,
   subject to
   $5x + 3y \leq 34$,
   $3x + 5y \leq 30$,
   $x \geq 0$,
   $y \geq 0$

8. $G = 7x + 8y$,
   subject to
   $3x + 2y \leq 12$,
   $2y - x \leq 4$,
   $x \geq 0$,
   $y \geq 0$

9. $P = 8x - y + 20$,
   subject to
   $6x + 8y \leq 48$,
   $0 \leq y \leq 4$,
   $0 \leq x \leq 7$

10. $Q = 24x - 3y + 52$,
    subject to
    $5x + 4y \leq 20$,
    $0 \leq y \leq 4$,
    $0 \leq x \leq 3$

11. $F = 2y - 3x$,
    subject to
    $y \leq 2x + 1$,
    $y \geq -2x + 3$,
    $x \leq 3$

12. $G = 5x + 2y + 4$,
    subject to
    $y \leq 2x + 1$,
    $y \geq -x + 3$,
    $x \leq 5$

13. *Gas mileage.* Roschelle owns a car and a moped. She has at most 12 gal of gasoline to be used between the car and the moped. The car's tank holds at most 18 gal and the moped's 3 gal. The mileage for the car is 20 mpg and for the moped is 100 mpg. How many gallons of gasoline should each vehicle use if Roschelle wants to travel as far as possible? What is the maximum number of miles?

14. *Lunch-time profits.* Elrod's lunch cart sells burritos and chili. To stay in business, Elrod must sell at least 10 orders of chili and 30 burritos each day. Because of limited space, no more than 40 orders of chili or 70 burritos can be made. The total number of orders cannot exceed 90. If profit is $1.65 per chili order and $1.05 per burrito, how many of each item should Elrod sell in order to maximize profit?

15. *Milling.* Johnson Lumber can convert logs into either lumber or plywood. In a given week, the mill can turn out 400 units of production, of which 100 units of lumber and 150 units of plywood are required by regular customers. The profit on a unit of lumber is $20 and on a unit of plywood is $30. How many units of each type should the mill produce in order to maximize profit?

16. *Cycle production.* Yawaka manufactures motorcycles and bicycles. In order to stay in business, the company determines that the number of bicycles made cannot exceed 3 times the number of motorcycles made. Yawaka lacks the facilities to produce more than 60 motorcycles or more than 120 bicycles. The total production of motorcycles and bicycles cannot exceed 160. The profit on a motorcycle is $1340 and on a bicycle is $200. Find the number of each that should be manufactured in order to maximize profit.

Aha! **17.** *Investing.*   Rosa is planning to invest up to $40,000 in corporate or municipal bonds, or both. She must invest from $6000 to $22,000 in corporate bonds, and she does not want to invest more than $30,000 in municipal bonds. The interest on corporate bonds is 8% and on municipal bonds is $7\frac{1}{2}\%$. This is simple interest for one year. How much should Rosa invest in each type of bond in order to earn the most interest? What is the maximum interest?

18. *Investing.*   Jamaal is planning to invest up to $22,000 in City Bank or the Southwick Credit Union, or both. He wants to invest at least $2000 but no more than $14,000 in City Bank. Because of insurance limitations, he will invest no more than $15,000 in the Southwick Credit Union. The interest in City Bank is 6% and in the credit union is $6\frac{1}{2}\%$. This is simple interest for one year. How much should Jamaal invest in each bank in order to earn the most interest? What is the maximum interest?

19. *Test scores.*   Phil is about to take a test that contains matching questions worth 10 points each and essay questions worth 25 points each. He must do at least 3 matching questions, but time restricts doing more than 12. Phil must do at least 4 essays, but time restricts doing more than 15. If no more than 20 questions can be answered, how many of each type should Phil do in order to maximize his score? What is this maximum score?

20. *Test scores.*   Edy is about to take a test that contains short-answer questions worth 4 points each and word problems worth 7 points each. Edy must do at least 5 short-answer questions, but time restricts doing more than 10. She must do at least 3 word problems, but time restricts doing more than 10. Edy can do no more than 18 questions in total. How many of each type of question must Edy do in order to maximize her score? What is this maximum score?

21. *Grape growing.*   Auggie's vineyard consists of 240 acres upon which he wishes to plant Merlot and Cabernet grapes. Profit per acre of Merlot is $400 and profit per acre of Cabernet is $300. Furthermore, the total number of hours of labor available during the harvest season is 3200. Each acre of Merlot requires 20 hr of labor and each acre of Cabernet requires 10 hr of labor. Determine how the land should be divided between Merlot and Cabernet in order to maximize profit.

22. *Coffee blending.*   The Coffee Peddler has 1440 lb of Sumatran coffee and 700 lb of Kona coffee. A batch of Hawaiian Blend requires 8 lb of Kona and 12 lb of Sumatran, and yields a profit of $90. A batch of Classic Blend requires 4 lb of Kona and 16 lb of Sumatran, and yields a $55 profit. How many batches of each kind should be made in order to maximize profit? What is the maximum profit? (*Hint*: Organize the information in a table.)

23. *Textile production.*   It takes Cosmic Stitching 2 hr of cutting and 4 hr of sewing to make a knit suit. To make a worsted suit, it takes 4 hr of cutting and 2 hr of sewing. At most 20 hr per day are available for cutting and at most 16 hr per day are available for sewing. The profit on a knit suit is $68 and on a worsted suit is $62. How many of each kind of suit should be made in order to maximize profit?

24. *Biscuit production.*   The Broad St. Biscuit Factory makes two types of biscuits, Biscuit Jumbos and Mitimite Biscuits. The oven can cook at most 200 biscuits per hour. Jumbos each require 2 oz of flour, Mitimites require 1 oz of flour, and there is at most 1440 oz of flour available. The income from Jumbos is $1.00 and from Mitimites is $0.80. How many of each type of biscuit should be made in order to maximize income? What is the maximum income?

25. Before a student begins work in this section, what three sections of the text would you suggest he or she study? Why?

26. What does the use of the word "constraint" in this section have in common with the use of the word in everyday speech?

SKILL MAINTENANCE

Evaluate. [1.8]

27. $5x^3 - 4x^2 - 7x + 2$, for $x = -2$

28. $6t^3 - 3t^2 + 5t$, for $t = 2$

Simplify. [1.2]

29. $3(2x - 5) + 4(x + 5)$

30. $4(5t - 7) + 6(t + 8)$

31. $6x - 3(x + 2)$

32. $8t - 2(3t - 1)$

SYNTHESIS

33. Explain how Exercises 17 and 18 can be answered by logical reasoning without linear programming.

34. Write a linear programming problem for a classmate to solve. Devise the problem so that profit must be maximized subject to at least two (nontrivial) constraints.

35. *Airplane production.*   Alpha Tours has two types of airplanes, the T3 and the S5, and contracts requiring accommodations for a minimum of 2000 first-class, 1500 tourist-class, and 2400 economy-class passengers. The T3 costs $30 per mile to operate and can accommodate 40 first-class, 40 tourist-class, and 120 economy-class passengers, whereas the S5 costs $25 per mile to operate and can accommodate 80 first-class, 30 tourist-class, and 40 economy-class passengers. How many of each type of airplane should be used in order to minimize the operating cost?

36. *Airplane production.*   A new airplane, the T4, is now available, having an operating cost of $37.50 per mile and accommodating 40 first-class, 40 tourist-class, and 80 economy-class passengers. If the T3 of Exercise 35 were replaced with the T4, how many S5's and how many T4's would be needed in order to minimize the operating cost?

37. *Furniture production.*   P. J. Edward Furniture Design produces chairs and sofas. The chairs require 20 ft of wood, 1 lb of foam rubber, and 2 sq yd of fabric. The sofas require 100 ft of wood, 50 lb of foam rubber, and 20 sq yd of fabric. The company has 1900 ft of wood, 500 lb of foam rubber, and 240 sq yd of fabric. The chairs can be sold for $80 each and the sofas for $1200 each. How many of each should be produced in order to maximize income?

9 Study Summary

Our focus in this chapter has been on **inequalities** and their **solutions** (p. 578). Because most inequalities have an infinite number of solutions, **set-builder notation** and **interval notation** are used to write **solution sets** (p. 579). To solve an inequality, principles similar to those used for solving equations are used (p. 578).

The Addition Principle for Inequalities

For any real numbers a, b, and c:

$$a < b \text{ is equivalent to } a + c < b + c; \qquad a > b \text{ is equivalent to } a + c > b + c.$$

The Multiplication Principle for Inequalities

For any real numbers a and b, and for any positive number c,

$$a < b \text{ is equivalent to } ac < bc; \qquad a > b \text{ is equivalent to } ac > bc.$$

For any real numbers a and b, and for any *negative* number c,

$$a < b \text{ is equivalent to } ac > bc; \qquad a > b \text{ is equivalent to } ac < bc.$$

Similar statements hold for $\leq$ and $\geq$.

These principles can be used, for example, to show that the solution set of $-2x + 7 > 1$ is $\{x \mid x < 3\}$ or $(-\infty, 3)$.

The solution of a **conjunction** like $x + 5 \leq 7$ *and* $2x \geq -3$ is an **intersection** of sets (pp. 587–588). The solution of a **disjunction** like $3x < 12$ *or* $x - 9 > 5$ is a **union** of sets (p. 591). Both conjunctions and disjunctions are **compound inequalities** (p. 587). We use compound inequalities to solve **absolute-value** problems like $|5x + 2| < 9$ (pp. 598–602).

The Absolute-Value Principles for Equations and Inequalities

For any positive number p and any algebraic expression X:

a) The solutions of $|X| = p$ are those numbers that satisfy $X = -p$ *or* $X = p$.
b) The solutions of $|X| < p$ are those numbers that satisfy $-p < X < p$.
c) The solutions of $|X| > p$ are those numbers that satisfy $X < -p$ *or* $p < X$.

If $|X| = 0$, then $X = 0$. If p is negative, then $|X| = p$ and $|X| < p$ have no solution, and any value of X will satisfy $|X| > p$.

Linear inequalities like $2x + 3y \leq 10$ are solved by graphing the **boundary** $2x + 3y = 10$ (pp. 606–607). To determine which **half-plane** is part of the solution set, we use a convenient **test point** (pp. 607, 608). When more than one linear inequality is to be solved at a time, we have a **system of linear inequalities** (p. 611). The solution of such a system requires finding the intersection of two or more half-planes.

Maximization or minimization problems in **linear programming** require us to identify a **feasible region** by solving a system of linear inequalities that represent the **constraints** of the problem (pp. 618–620). The corners, or **vertices**, of the region are then tested in the problem's **objective function** (pp. 612, 618). The **corner principle** states that if a maximum or minimum exists, then it will occur at one of the vertices (p. 619).

9 Review Exercises

↪ *Concept Reinforcement*   *Classify each of the following as either true or false.*

1. The addition and multiplication principles for inequalities are used to write equivalent inequalities. [9.1]

2. It is always true that if $a > b$, then $ac > bc$. [9.1]

3. The solution of $|3x - 5| \leq 8$ is a closed interval. [9.3]

4. The inequality $2 < 5x + 1 < 9$ is equivalent to $2 < 5x + 1$ *or* $5x + 1 < 9$. [9.2]

5. The solution set of a disjunction is the union of two solution sets. [9.2]

6. The equation $|x| = -p$ has no solution when p is positive. [9.3]

7. $|f(x)| > 3$ is equivalent to $f(x) < -3$ *or* $f(x) > 3$. [9.3]

8. A test point is used to determine whether the line in a linear inequality is drawn solid or dashed. [9.4]

9. The graph of a system of linear inequalities is always a half-plane. [9.4]

10. The corner principle states that every objective function has a maximum or minimum value. [9.5]

Graph each inequality and write the solution set using both set-builder and interval notations. [9.1]

11. $x \leq -2$

12. $a + 7 \leq -14$

13. $y - 5 \geq -12$

14. $4y > -15$

15. $-0.3y < 9$

16. $-6x - 5 < 4$

17. $-\frac{1}{2}x - \frac{1}{4} > \frac{1}{2} - \frac{1}{4}x$

18. $0.3y - 7 < 2.6y + 15$

19. $-2(x - 5) \geq 6(x + 7) - 12$

20. Let $f(x) = 3x - 5$ and $g(x) = 11 - x$. Find all values of x for which $f(x) \leq g(x)$. [9.1]

Solve. [9.1]

21. Rose can choose between two summer jobs. She can work as a checker in a discount store for $8.40 an hour, or she can mow lawns for $12.00 an hour. In order to mow lawns, she must buy a $450 lawnmower. For how many hours must Rose work in order for the mowing to be more profitable than checking?

22. Clay is going to invest $9000, part at 3% and the rest at 3.5%. What is the most he can invest at 3% and still be guaranteed $300 in interest each year?

23. Find the intersection:

$$\{1, 2, 5, 6, 9\} \cap \{1, 3, 5, 9\}. \text{ [9.2]}$$

24. Find the union:

$$\{1, 2, 5, 6, 9\} \cup \{1, 3, 5, 9\}. \text{ [9.2]}$$

Graph and write interval notation. [9.2]

25. $x \le 3 \text{ and } x > -5$

26. $x \le 3 \text{ or } x > -5$

Solve and graph each solution set. [9.2]

27. $-4 < x + 8 \le 5$

28. $-15 < -4x - 5 < 0$

29. $3x < -9 \text{ or } -5x < -5$

30. $2x + 5 < -17 \text{ or } -4x + 10 \le 34$

31. $2x + 7 \le -5 \text{ or } x + 7 \ge 15$

32. $f(x) < -5 \text{ or } f(x) > 5, \text{ where } f(x) = 3 - 5x$

For f(x) as given, use interval notation to write the domain of f. [9.2]

33. $f(x) = \dfrac{2x}{x - 8}$

34. $f(x) = \sqrt{x + 5}$

35. $f(x) = \sqrt{8 - 3x}$

Solve. [9.3]

36. $|x| = 5$

37. $|t| \ge 3.5$

38. $|x - 3| = 7$

39. $|2x + 5| < 12$

40. $|3x - 4| \ge 15$

41. $|2x + 5| = |x - 9|$

42. $|5n + 6| = -8$

43. $\left| \dfrac{x + 4}{6} \right| \le 2$

44. $2|x - 5| - 7 > 3$

45. Let $f(x) = |3x - 5|$. Find all x for which $f(x) < 0$. [9.3]

46. Graph $x - 2y \ge 6$ on a plane. [9.4]

Graph each system of inequalities. Find the coordinates of any vertices formed. [9.4]

47. $x + 3y > -1,$
    $x + 3y < 4$

48. $x - 3y \le 3,$
    $x + 3y \ge 9,$
    $y \le 6$

49. Find the maximum and the minimum values of

$$F = 3x + y + 4$$

subject to

$$y \le 2x + 1,$$
$$x \le 7,$$
$$y \ge 3. \qquad \text{[9.5]}$$

50. Custom Computers has two manufacturing plants. The Oregon plant cannot produce more than 60 computers a week, while the Ohio plant cannot produce more than 120 computers a week. The Electronics Outpost sells at least 160 Custom computers each week. It costs \$40 to ship a computer to The Electronics Outpost from the Oregon plant and \$25 to ship from the Ohio plant. How many computers should be shipped from each plant in order to minimize cost? [9.5]

SYNTHESIS

51. Explain in your own words why $|X| = p$ has two solutions when p is positive and no solution when p is negative. [9.3]

52. Explain why the graph of the solution of a system of linear inequalities is the intersection, not the union, of the individual graphs. [9.4]

53. Solve: $|2x + 5| \le |x + 3|$. [9.3]

54. Classify as true or false: If $x < 3$, then $x^2 < 9$. If false, give an example showing why. [9.1]

55. Just-For-Fun manufactures marbles with a 1.1-cm diameter and a ± 0.03-cm manufacturing tolerance, or allowable variation in diameter. Write the tolerance as an inequality with absolute value. [9.3]

9 Chapter Test

Graph each inequality and write the solution set using both set-builder and interval notations.

1. $x - 2 < 10$

2. $-0.6y < 30$

3. $-4y - 3 \geq 5$

4. $3a - 5 \leq -2a + 6$

5. $3(7 - x) < 2x + 5$

6. $-8(2x + 3) + 6(4 - 5x) \geq 2(1 - 7x) - 4(4 + 6x)$

7. Let $f(x) = -5x - 1$ and $g(x) = -9x + 3$. Find all values of x for which $f(x) > g(x)$.

8. Lia can rent a van for either $40 with unlimited mileage or $30 with 100 free miles and an extra charge of 15¢ for each mile over 100. For what numbers of miles traveled would the unlimited mileage plan save Lia money?

9. A refrigeration repair company charges $80 for the first half-hour of work and $60 for each additional hour. Blue Mountain Camp has budgeted $200 to repair its walk-in cooler. For what lengths of a service call will the budget not be exceeded?

10. Find the intersection:

$\{1, 3, 5, 7, 9\} \cap \{3, 5, 11, 13\}$.

11. Find the union:

$\{1, 3, 5, 7, 9\} \cup \{3, 5, 11, 13\}$.

12. Write the domain of f using interval notation if $f(x) = \sqrt{8 - 2x}$.

Solve and graph each solution set.

13. $-2 < x - 3 < 5$

14. $-11 \leq -5t - 2 < 0$

15. $3x - 2 < 7$ or $x - 2 > 4$

16. $-3x > 12$ or $4x > -10$

17. $-\frac{1}{3} \leq \frac{1}{6}x - 1 < \frac{1}{4}$

18. $|x| = 13$

19. $|a| > 7$

20. $|3x - 1| < 7$

21. $|-5t - 3| \geq 10$

22. $|2 - 5x| = -12$

23. $g(x) < -3$ or $g(x) > 3$, where $g(x) = 4 - 2x$

24. Let $f(x) = |x + 10|$ and $g(x) = |x - 12|$. Find all values of x for which $f(x) = g(x)$.

Graph the system of inequalities. Find the coordinates of any vertices formed.

25. $x + y \geq 3,$
$x - y \geq 5$

26. $2y - x \geq -7,$
$2y + 3x \leq 15,$
$y \leq 0,$
$x \leq 0$

27. Find the maximum and the minimum values of

$F = 5x + 3y$

subject to

$x + y \leq 15,$
$1 \leq x \leq 6,$
$0 \leq y \leq 12.$

28. Sassy Salon makes $12 on each manicure and $18 on each haircut. A manicure takes 30 minutes and a haircut takes 50 minutes, and there are 5 stylists who each work 6 hours a day. If the salon can schedule 50 appointments a day, how many should be manicures and how many haircuts in order to maximize profit? What is the maximum profit?

SYNTHESIS

Solve. Write the solution set using interval notation.

29. $|2x - 5| \leq 7$ and $|x - 2| \geq 2$

30. $7x < 8 - 3x < 6 + 7x$

31. Write an absolute-value inequality for which the interval shown is the solution.

1–9 Cumulative Review

1. Evaluate
$$\frac{2x - y^2}{x + y}$$
for $x = 3$ and $y = -4$. [1.8]

2. Convert to scientific notation: 391,000,000. [4.8]

3. Determine the slope and the y-intercept for the line given by $7x - 4y = 12$. [3.6]

4. Find an equation for the line that passes through the points $(-1, 7)$ and $(2, -3)$. [3.7]

5. Solve the system
$$5x - 2y = -23,$$
$$3x + 4y = 7. \text{ [8.2]}$$

6. Solve the system
$$-3x + 4y + z = -5,$$
$$x - 3y - z = 6,$$
$$2x + 3y + 5z = -8. \text{ [8.4]}$$

7. Folsom Elementary School sold 45 pizzas for a fundraiser. Small pizzas sold for $7.00 each and large pizzas for $10.00 each. The total amount of funds received from the sale was $402. How many of each size pizza were sold? [8.3]

8. The sum of three numbers is 20. The first number is 3 less than twice the third number. The second number minus the third number is -7. What are the numbers? [8.5]

9. Trex Company makes decking material from waste wood fibers and reclaimed polyethylene. Its sales rose from $49.2 million in 1998 to $191 million in 2003. Calculate the rate at which sales were rising. [3.4]
Source: U.S. Securities and Exchange Commission

10. In 1989, the average length of a visit to a physician in an HMO was 15.4 min; and in 1998, it was 17.9 min. Let V represent the average length of a visit t years after 1989. [7.3]
Sources: Rutgers University Study and National Center for Health Statistics
   a) Find a linear function $V(t)$ that fits the data.
   b) Use the function of part (a) to predict the average length of a visit in 2005.

11. If
$$f(x) = \frac{x - 2}{x - 5},$$
find **(a)** $f(3)$ and **(b)** the domain of f. [7.1], [7.2]

Solve.

12. $8x = 1 + 16x^2$ [5.7]

13. $144 = 49y^2$ [5.7]

14. $20 > 2 - 6x$ [9.1]

15. $\frac{1}{3}x - \frac{1}{5} \geq \frac{1}{5}x - \frac{1}{3}$ [9.1]

16. $-8 < x + 2 < 15$ [9.2]

17. $3x - 2 < -6 \text{ or } x + 3 > 9$ [9.2]

18. $|x| > 6.4$ [9.3]

19. $|3x - 2| \leq 14$ [9.3]

20. $\frac{8}{n} - \frac{2}{n} = 3$ [6.6]

21. $\frac{6}{x - 5} = \frac{2}{2x}$ [6.6]

22. $\frac{3x}{x - 2} - \frac{6}{x + 2} = \frac{24}{x^2 - 4}$ [6.6]

23. $\frac{3x^2}{x + 2} + \frac{5x - 22}{x - 2} = \frac{-48}{x^2 - 4}$ [6.6]

24. Let $f(x) = |3x - 5|$. Find all values of x for which $f(x) = 2$. [9.3]

25. Write the domain of f using interval notation if $f(x) = \sqrt{x - 9}$. [9.2]

Solve.

26. $5m - 3n = 4m + 12$, for n [2.3]

27. $P = \dfrac{4a}{a + b}$, for a [7.5]

Graph on a plane.

28. $4x \geq 5y + 12$ [9.4]

29. $y = \frac{1}{3}x - 2$ [3.6]

Perform the indicated operations and simplify.

30. $(2x^2 - 8x + 7) + (6x - 3x^3 + 9x^2 - 4)$ [4.3]

31. $(8x^3y^2)(-3xy^2)$ [4.6]

32. $(3a + b - 2c) - (-4b + 3c - 2a)$ [4.6]

33. $(5x^2 - 2x + 1)(3x^2 + x - 2)$ [4.4]

34. $(3x^2 + y)^2$ [4.6]

35. $(2x^2 - y)(2x^2 + y)$ [4.6]

36. $(-5m^3n^2 - 3mn^3) + (-4m^2n^2 + 4m^3n^2) - (2mn^3 - 3m^2n^2)$ [4.6]

37. $\dfrac{y^2 - 36}{2y + 8} \cdot \dfrac{y + 4}{y + 6}$ [6.2]

38. $\dfrac{x^4 - 1}{x^2 - x - 2} \div \dfrac{x^2 + 1}{x - 2}$ [6.2]

39. $\dfrac{5ab}{a^2 - b^2} + \dfrac{a + b}{a - b}$ [6.4]

40. $\dfrac{2}{m + 1} + \dfrac{3}{m - 5} - \dfrac{m^2 - 1}{m^2 - 4m - 5}$ [6.4]

41. $y - \dfrac{2}{3y}$ [6.4]

42. Simplify: $\dfrac{\dfrac{1}{x} - \dfrac{1}{y}}{x + y}$. [6.5]

43. Divide: $(9x^3 + 5x^2 + 2) \div (x + 2)$. [4.7]

Factor.

44. $4x^3 - 14x^2$ [5.1]

45. $x^2 + 8x - 84$ [5.2]

46. $16y^2 - 81$ [5.4]

47. $64x^3 + 8$ [5.5]

48. $t^2 - 16t + 64$ [5.4]

49. $x^6 - x^2$ [5.4]

50. $0.027b^3 - 0.008c^3$ [5.5]

51. $20x^2 - 7x - 3$ [5.3]

52. $3t^2 + 17t - 28$ [5.3]

53. $x^5 - x^3y + x^2y - y^2$ [5.1]

54. If $f(x) = x^2 - 4$ and $g(x) = x^2 - 7x + 10$, find the domain of f/g. [7.4]

55. A digital data circuit can transmit a particular set of data in 4 sec. An analog phone circuit can transmit the same data in 20 sec. How long would it take, working together, for both circuits to transmit the data? [6.7]

56. The floor area of a rental trailer is rectangular. The length is 3 ft more than the width. A rug of area 54 ft^2 exactly fills the floor of the trailer. Find the perimeter of the trailer. [5.8]

57. The sum of the squares of three consecutive even integers is equal to 8 more than three times the square of the second number. Find the integers. [5.8]

58. *Logging.* The volume of wood V in a tree trunk varies jointly as the height h and the square of the girth g (girth is distance around). If the volume is 35 ft^3 when the height is 20 ft and the girth is 5 ft, what is the height when the volume is 85.75 ft^3 and the girth is 7 ft? [7.5]

SYNTHESIS

59. Multiply: $(x - 4)^3$. [4.4]

60. Solve: $x^4 + 225 = 34x^2$. [5.7]

Solve.

61. $4 \leq |3 - x| \leq 6$ [9.2], [9.3]

62. $\dfrac{18}{x - 9} + \dfrac{10}{x + 5} = \dfrac{28x}{x^2 - 4x - 45}$ [6.6]

63. $16x^3 = x$ [5.7]

10

Exponents and Radicals

AN APPLICATION

The function given by $f(x) = 262 \cdot 2^{x/12}$ can be used to determine the frequency, in cycles per second, of a musical note that is x half-steps above a piano's middle C. Show that the G that is 7 half-steps (a perfect 5th) above middle C has a frequency that is about 1.5 times that of middle C.

This problem appears as Exercise 113 in Section 10.2.

R. J. Malloy
GUITAR MAKER
Englewood, Florida

Music is mathematics with a soul. From determining what thickness and length of a string will produce the vibrations of a pure C note to determining how a new bracing system will affect the tone of a guitar, I use math constantly. Even the shape of a violin is determined by algebraic formulas that are 500 years old and based on the speed of sound through solid materials.

*I*n this chapter, we learn about square roots, cube roots, fourth roots, and so on. These roots are studied in connection with the manipulation of radical expressions and the solution of real-world applications. Exponents that are fractions are also studied and used to ease some of our work with radicals. The chapter closes with an examination of the complex-number system.

10.1 Radical Expressions and Functions

Square Roots and Square-Root Functions • Expressions of the Form $\sqrt{a^2}$ • Cube Roots • Odd and Even nth Roots

Study Skills

Advance Planning Pays Off

The best way to prepare for a final exam is to do so over a period of at least two weeks. First review each chapter, studying the words, formulas, problems, properties, and procedures in the Study Summaries and Review Exercises. Then retake your quizzes and tests. If you miss any questions, spend extra time reviewing the corresponding topics. Watch the videotapes or DVDs that accompany the text or use the accompanying tutorial software. Also consider participating in a study group or attending a tutoring or review session.

In this section, we consider roots, such as square roots and cube roots. We look at the symbolism that is used and ways in which symbols can be manipulated to get equivalent expressions. All of this will be important in problem solving.

Square Roots and Square-Root Functions

When a number is multiplied by itself, we say that the number is squared. Often we need to know what number was squared in order to produce some value a. If such a number can be found, we call that number a *square root* of a.

> **Square Root**
> The number c is a *square root* of a if $c^2 = a$.

For example,

9 has -3 and 3 as square roots because $(-3)^2 = 9$ and $3^2 = 9$.

25 has -5 and 5 as square roots because $(-5)^2 = 25$ and $5^2 = 25$.

-4 does not have a real-number square root because there is no real number c for which $c^2 = -4$.

Note that every positive number has two square roots, whereas 0 has only itself as a square root. Negative numbers do not have real-number square roots, although later in this chapter we introduce the *complex-number* system in which such square roots do exist.

EXAMPLE 1   Find the two square roots of 64.

Solution   The square roots are 8 and -8, because $8^2 = 64$ and $(-8)^2 = 64$.

Whenever we refer to *the* square root of a number, we mean the nonnegative square root of that number. This is often referred to as the *principal square root* of the number.

Student Notes _____

It is important to remember the difference between *the* square root of 9 and *a* square root of 9. *A* square root of 9 means either 3 or -3, whereas *the* square root of 9, denoted $\sqrt{9}$, means the principal square root of 9, or 3.

> **Principal Square Root**
>
> The *principal square root* of a nonnegative number is its nonnegative square root. The symbol $\sqrt{}$ is called a *radical sign* and is used to indicate the principal square root of the number over which it appears.

EXAMPLE 2   Simplify each of the following.

a) $\sqrt{25}$

b) $\sqrt{\dfrac{25}{64}}$

c) $-\sqrt{64}$

d) $\sqrt{0.0049}$

Solution

a) $\sqrt{25} = 5$ $\sqrt{}$ indicates the principal square root. Note that $\sqrt{25} \neq -5$.

b) $\sqrt{\dfrac{25}{64}} = \dfrac{5}{8}$ Since $\left(\dfrac{5}{8}\right)^2 = \dfrac{25}{64}$

c) $-\sqrt{64} = -8$ Since $\sqrt{64} = 8$, $-\sqrt{64} = -8$.

d) $\sqrt{0.0049} = 0.07$ $(0.07)(0.07) = 0.0049$. Note too that $0.0049 = \dfrac{49}{10,000}$ and $\sqrt{49/10,000} = \dfrac{7}{100}$.

In addition to being read as "the principal square root of *a*," $\sqrt{a}$ is also read as "the square root of *a*," "root *a*," or "radical *a*." Any expression in which a radical sign appears is called a *radical expression*. The following are radical expressions:

$$\sqrt{5}, \qquad \sqrt{a}, \qquad -\sqrt{3x}, \qquad \sqrt{\dfrac{y^2 + 7}{y}}, \qquad \sqrt{x} + 8.$$

The expression under the radical sign is called the **radicand**. In the expressions above, the radicands are 5, a, $3x$, $(y^2 + 7)/y$, and x.

All but the most basic calculators give values for square roots. These values are, for the most part, approximations. For example, on many calculators, if you enter 5 and then press $\boxed{\checkmark}$, a number like

2.23606798

appears, depending on how the calculator rounds. (On some calculators, the $\boxed{\checkmark}$ key is pressed first.) The exact value of $\sqrt{5}$, for example, is not given by any repeating or terminating decimal. In general, for any whole number a that is not a perfect square, $\sqrt{a}$ is a nonterminating, nonrepeating decimal. We discussed such *irrational numbers* in Chapter 1.

The square-root function, given by

$$f(x) = \sqrt{x},$$

has the interval $[0, \infty)$ as its domain. We can draw its graph by selecting convenient values for x and calculating the corresponding outputs. Once these ordered pairs have been graphed, a smooth curve can be drawn.

$f(x) = \sqrt{x}$

x	$\sqrt{x}$	$(x, f(x))$
0	0	$(0, 0)$
1	1	$(1, 1)$
4	2	$(4, 2)$
9	3	$(9, 3)$

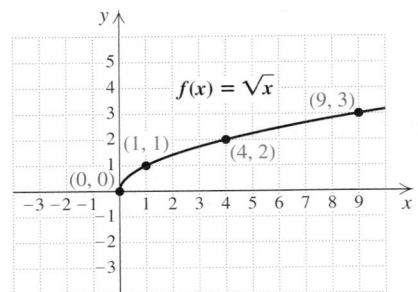

EXAMPLE 3   For each function, find the indicated function value.

a) $f(x) = \sqrt{3x - 2};\ f(1)$ **b)** $g(z) = -\sqrt{6z + 4};\ g(3)$

Solution

a) $f(1) = \sqrt{3 \cdot 1 - 2}$ Substituting

$ = \sqrt{1} = 1$ Simplifying

b) $g(3) = -\sqrt{6 \cdot 3 + 4}$ Substituting

$ = -\sqrt{22}$ Simplifying. This is the most exact way to write the answer.

$ \approx -4.69041576$ Using a calculator to write an approximate answer

Expressions of the Form $\sqrt{a^2}$

It is tempting to write $\sqrt{a^2} = a$, but the next example shows that, as a rule, this is untrue.

EXAMPLE 4   Evaluate $\sqrt{a^2}$ for the following values: **(a)** 5; **(b)** 0; **(c)** -5.

Solution

a) $\sqrt{5^2} = \sqrt{25} = 5$

$\underline{}$ Same

b) $\sqrt{0^2} = \sqrt{0} = 0$

$\underline{}$ Same

c) $\sqrt{(-5)^2} = \sqrt{25} = 5$

$\underline{}$ Opposites   Note that $\sqrt{(-5)^2} \neq -5$.

You may have noticed that evaluating $\sqrt{a^2}$ is just like evaluating $|a|$.

> **Simplifying** $\sqrt{a^2}$
>
> For any real number a,
> $$\sqrt{a^2} = |a|.$$
> (The principal square root of a^2 is the absolute value of a.)

When a radicand is the square of a variable expression, like $(x + 5)^2$ or $36t^2$, absolute-value signs are needed when simplifying. We use absolute-value signs unless we know that the expression being squared is nonnegative. This assures that our result is never negative.

EXAMPLE 5 Simplify each expression. Assume that the variable can represent any real number.

a) $\sqrt{(x + 1)^2}$ **b)** $\sqrt{x^2 - 8x + 16}$

c) $\sqrt{a^8}$ **d)** $\sqrt{t^6}$

Solution

a) $\sqrt{(x + 1)^2} = |x + 1|$ Since $x + 1$ might be negative (for example, if $x = -3$), absolute-value notation is necessary.

b) $\sqrt{x^2 - 8x + 16} = \sqrt{(x - 4)^2} = |x - 4|$ Since $x - 4$ might be negative, absolute-value notation is necessary.

c) Note that $(a^4)^2 = a^8$ and that a^4 is never negative. Thus,
$$\sqrt{a^8} = a^4.$$ Absolute-value notation is unnecessary here.

d) Note that $(t^3)^2 = t^6$. Thus,
$$\sqrt{t^6} = |t^3|.$$ Since t^3 might be negative, absolute-value notation is necessary.

---

technology connection

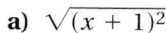

To see the necessity of the absolute-value signs, let y_1 represent the left side and y_2 the right side of each of the following equations. Then use a graph or table to determine whether these equations are true.

1. $\sqrt{x^2} \overset{?}{=} x$

2. $\sqrt{x^2} \overset{?}{=} |x|$

3. $x \overset{?}{=} |x|$

EXAMPLE 6 Simplify each expression. Assume that no radicands were formed by raising negative quantities to even powers.

a) $\sqrt{y^2}$ **b)** $\sqrt{a^{10}}$ **c)** $\sqrt{9x^2 - 6x + 1}$

Solution

a) $\sqrt{y^2} = y$ We are assuming that y is nonnegative, so no absolute-value notation is necessary. When y *is* negative, $\sqrt{y^2} \neq y$.

b) $\sqrt{a^{10}} = a^5$ Assuming that a^5 is nonnegative. Note that $(a^5)^2 = a^{10}$.

c) $\sqrt{9x^2 - 6x + 1} = \sqrt{(3x - 1)^2} = 3x - 1$ Assuming that $3x - 1$ is nonnegative

Cube Roots

We often need to know what number was cubed in order to produce a certain value. When such a number is found, we say that we have found a *cube root*. For example,

2 is the cube root of 8 because $2^3 = 2 \cdot 2 \cdot 2 = 8$;

-4 is the cube root of -64 because $(-4)^3 = (-4)(-4)(-4) = -64$.

> **Cube Root**
>
> The number c is the *cube root* of a if $c^3 = a$. In symbols, we write $\sqrt[3]{a}$ to denote the cube root of a.

Because each real number has only one cube root, a function can be formed. The cube-root function, given by

$$f(x) = \sqrt[3]{x},$$

has $\mathbb{R}$ as its domain and $\mathbb{R}$ as its range. To draw its graph, we select convenient values for x and calculate the corresponding outputs. Once these ordered pairs have been graphed, a smooth curve is drawn. Note that the cube root of a positive number is positive, and the cube root of a negative number is negative.

$$f(x) = \sqrt[3]{x}$$

x	$\sqrt[3]{x}$	$(x, f(x))$
0	0	$(0, 0)$
1	1	$(1, 1)$
8	2	$(8, 2)$
-1	-1	$(-1, -1)$
-8	-2	$(-8, -2)$

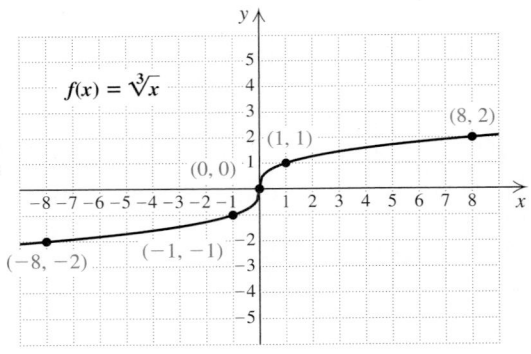

EXAMPLE 7   For each function, find the indicated function value.

a) $f(y) = \sqrt[3]{y}$; $f(125)$

b) $g(x) = \sqrt[3]{x - 1}$; $g(-26)$

Solution

a) $f(125) = \sqrt[3]{125} = 5$ Since $5 \cdot 5 \cdot 5 = 125$

b) $g(-26) = \sqrt[3]{-26 - 1}$

$\qquad\quad = \sqrt[3]{-27}$

$\qquad\quad = -3$ Since $(-3)(-3)(-3) = -27$

EXAMPLE 8   Simplify: $\sqrt[3]{-8y^3}$.

Solution

$$\sqrt[3]{-8y^3} = -2y \qquad \text{Since } (-2y)(-2y)(-2y) = -8y^3$$

Odd and Even *n*th Roots

The fourth root of a number a is the number c for which $c^4 = a$. There are also 5th roots, 6th roots, and so on. We write $\sqrt[n]{a}$ for the nth root. The number n is called the *index* (plural, *indices*). When the index is 2, we do not write it.

EXAMPLE 9   Find each of the following.

a) $\sqrt[5]{32}$

b) $\sqrt[5]{-32}$

c) $-\sqrt[5]{32}$

d) $-\sqrt[5]{-32}$

Solution

a) $\sqrt[5]{32} = 2$ \qquad\qquad Since $2^5 = 32$

b) $\sqrt[5]{-32} = -2$ \qquad Since $(-2)^5 = -32$

c) $-\sqrt[5]{32} = -2$ \qquad Taking the opposite of $\sqrt[5]{32}$

d) $-\sqrt[5]{-32} = -(-2) = 2$ \qquad Taking the opposite of $\sqrt[5]{-32}$

Note that every number has exactly one real root when n is odd. Odd roots of positive numbers are positive and odd roots of negative numbers are negative. Absolute-value signs are not used when finding odd roots.

EXAMPLE 10   Find each of the following.

a) $\sqrt[7]{x^7}$

b) $\sqrt[9]{(t-1)^9}$

Solution

a) $\sqrt[7]{x^7} = x$

b) $\sqrt[9]{(t-1)^9} = t - 1$

When the index n is even, we say that we are taking an *even root*. Every positive real number has two real nth roots when n is even—one positive and one negative. For example, the fourth roots of 16 are -2 and 2. Negative numbers do not have real nth roots when n is even.

When n is even, the notation $\sqrt[n]{a}$ indicates the nonnegative nth root. Thus, to write even nth roots, absolute-value signs are often required.

EXAMPLE 11 Simplify each expression, if possible. Assume that variables can represent any real number.

a) $\sqrt[4]{16}$ **b)** $-\sqrt[4]{16}$

c) $\sqrt[4]{-16}$ **d)** $\sqrt[4]{81x^4}$

e) $\sqrt[6]{(y+7)^6}$

Solution

a) $\sqrt[4]{16} = 2$ Since $2^4 = 16$

b) $-\sqrt[4]{16} = -2$ Taking the opposite of $\sqrt[4]{16}$

c) $\sqrt[4]{-16}$ cannot be simplified. $\sqrt[4]{-16}$ is not a real number.

d) $\sqrt[4]{81x^4} = |3x|$, or $3|x|$ Use absolute-value notation since x could represent a negative number.

e) $\sqrt[6]{(y+7)^6} = |y+7|$ Use absolute-value notation since $y + 7$ is negative for $y < -7$.

We summarize as follows.

Simplifying nth Roots

n	a	$\sqrt[n]{a}$	$\sqrt[n]{a^n}$
Even	Positive	Positive	a
	Negative	Not a real number	$-a$
Odd	Positive	Positive	a
	Negative	Negative	a

EXAMPLE 12 Determine the domain of $g(x) = \sqrt[6]{7 - 3x}$.

Solution Since the index is even, the radicand, $7 - 3x$, must be nonnegative. We solve the inequality:

$$7 - 3x \geq 0 \qquad \text{We cannot find the 6th root of a negative number.}$$

$$-3x \geq -7$$

$$x \leq \tfrac{7}{3}. \qquad \text{Multiplying both sides by } -\tfrac{1}{3} \text{ and reversing the inequality}$$

Thus,

$$\text{Domain of } g = \left\{ x \,\middle|\, x \leq \tfrac{7}{3} \right\}$$
$$= \left(-\infty, \tfrac{7}{3}\right].$$

technology connection

To enter cube or higher roots on a graphing calculator, select options 4 or 5 of the **MATH** menu. The characters $6\sqrt[x]{\ }$ indicate the sixth root.

1. Use a **TABLE** or **GRAPH** and **TRACE** to check the solution of Example 12.

Exercise Set

10.1

↪ *Concept Reinforcement   Select the appropriate word to complete each of the following.*

1. Every positive number has _____ square root(s).
one/two

2. The principal square root is never _____.
negative/positive

3. For any _____ number a, we have
negative/positive
$\sqrt{a^2} = a$.

4. For any _____ number a, we have
negative/positive
$\sqrt{a^2} = -a$.

5. If a is a whole number that is not a perfect square, then $\sqrt{a}$ is a(n) _____ number.
irrational/rational

6. The domain of the function f given by $f(x) = \sqrt[3]{x}$ is the set of all _____ numbers.
whole/real/positive

7. If $\sqrt[4]{x}$ is a real number, then x must be
_____.
negative/positive/nonnegative

8. If $\sqrt[3]{x}$ is negative, then x must be _____.
negative/positive

For each number, find all of its square roots.

9. 49

10. 81

11. 144

12. 9

13. 400

14. 2500

15. 900

16. 225

Simplify.

17. $-\sqrt{\dfrac{36}{49}}$

18. $-\sqrt{\dfrac{361}{9}}$

19. $\sqrt{441}$

20. $\sqrt{196}$

21. $-\sqrt{\dfrac{16}{81}}$

22. $-\sqrt{\dfrac{81}{144}}$

23. $\sqrt{0.04}$

24. $\sqrt{0.36}$

25. $-\sqrt{0.0025}$

26. $\sqrt{0.0144}$

Identify the radicand and the index for each expression.

27. $5\sqrt{p^2 + 4}$

28. $-7\sqrt{y^2 - 8}$

29. $x^2 y^3 \sqrt[3]{\dfrac{x}{y + 4}}$

30. $a^2 b^3 \sqrt[3]{\dfrac{a}{a^2 - b}}$

For each function, find the specified function value, if it exists.

31. $f(t) = \sqrt{5t - 10};\ f(6), f(2), f(1), f(-1)$

32. $g(x) = \sqrt{x^2 - 25};\ g(-6), g(3), g(6), g(13)$

33. $t(x) = -\sqrt{2x + 1};\ t(4), t(0), t(-1), t\left(-\tfrac{1}{2}\right)$

34. $p(z) = \sqrt{2z^2 - 20};\ p(4), p(3), p(-5), p(0)$

35. $f(t) = \sqrt{t^2 + 1};\ f(0), f(-1), f(-10)$

36. $g(x) = -\sqrt{(x + 1)^2};\ g(-3), g(4), g(-5)$

37. $g(x) = \sqrt{x^3 + 9};\ g(-2), g(-3), g(3)$

38. $f(t) = \sqrt{t^3 - 10};\ f(2), f(3), f(4)$

Simplify. Remember to use absolute-value notation when necessary. If a root cannot be simplified, state this.

39. $\sqrt{36x^2}$

40. $\sqrt{25t^2}$

41. $\sqrt{(-6b)^2}$

42. $\sqrt{(-7c)^2}$

43. $\sqrt{(8 - t)^2}$

44. $\sqrt{(a + 3)^2}$

45. $\sqrt{y^2 + 16y + 64}$

46. $\sqrt{x^2 - 4x + 4}$

47. $\sqrt{4x^2 + 28x + 49}$

48. $\sqrt{9x^2 - 30x + 25}$

49. $\sqrt[4]{256}$

50. $-\sqrt[4]{625}$

51. $\sqrt[5]{-1}$

52. $-\sqrt[5]{3^5}$

53. $\sqrt[5]{-\dfrac{32}{243}}$

54. $\sqrt[5]{-\dfrac{1}{32}}$

55. $\sqrt[6]{x^6}$

56. $\sqrt[8]{y^8}$

57. $\sqrt[4]{(6a)^4}$

58. $\sqrt[4]{(7b)^4}$

59. $\sqrt[10]{(-6)^{10}}$

60. $\sqrt[12]{(-10)^{12}}$

61. $\sqrt[414]{(a + b)^{414}}$

62. $\sqrt[1976]{(2a + b)^{1976}}$

63. $\sqrt{a^{22}}$

64. $\sqrt{x^{10}}$

65. $\sqrt{-25}$

66. $\sqrt{-16}$

Simplify. Assume that no radicands were formed by raising negative quantities to even powers.

67. $\sqrt{16x^2}$

68. $\sqrt{25t^2}$

69. $\sqrt{(3t)^2}$

70. $\sqrt{(7c)^2}$

71. $\sqrt{(a+1)^2}$

72. $\sqrt{(5+b)^2}$

73. $\sqrt{4x^2+8x+4}$

74. $\sqrt{9x^2+36x+36}$

75. $\sqrt{9t^2-12t+4}$

76. $\sqrt{25t^2-20t+4}$

77. $\sqrt[3]{27}$

78. $-\sqrt[3]{64}$

79. $\sqrt[4]{16x^4}$

80. $\sqrt[4]{81x^4}$

81. $\sqrt[3]{-216}$

82. $-\sqrt[3]{-100,000}$

83. $-\sqrt[3]{-125y^3}$

84. $-\sqrt[3]{-64x^3}$

85. $\sqrt{t^{18}}$

86. $\sqrt{a^{14}}$

87. $\sqrt{(x-2)^8}$

88. $\sqrt{(x+3)^{10}}$

For each function, find the specified function value, if it exists.

89. $f(x) = \sqrt[3]{x+1}$; $f(7), f(26), f(-9), f(-65)$

90. $g(x) = -\sqrt[3]{2x-1}$; $g(0), g(-62), g(-13), g(63)$

91. $g(t) = \sqrt[4]{t-3}$; $g(19), g(-13), g(1), g(84)$

92. $f(t) = \sqrt[4]{t+1}$; $f(0), f(15), f(-82), f(80)$

Determine the domain of each function described.

93. $f(x) = \sqrt{x-6}$

94. $g(x) = \sqrt{x+8}$

95. $g(t) = \sqrt[4]{t+8}$

96. $f(x) = \sqrt[4]{x-9}$

97. $g(x) = \sqrt[4]{2x-10}$

98. $g(t) = \sqrt[3]{2t-6}$

99. $f(t) = \sqrt[5]{8-3t}$

100. $f(t) = \sqrt[6]{4-3t}$

101. $h(z) = -\sqrt[6]{5z+2}$

102. $d(x) = -\sqrt[4]{7x-5}$

Aha! **103.** $f(t) = 7 + \sqrt[8]{t^8}$

104. $g(t) = 9 + \sqrt[6]{t^6}$

105. Explain how to write the negative square root of a number using radical notation.

106. Does the square root of a number's absolute value always exist? Why or why not?

SKILL MAINTENANCE

Simplify. Do not use negative exponents in your answer. [4.1], [4.8]

107. $(a^3b^2c^5)^3$

108. $(5a^7b^8)(2a^3b)$

109. $(2a^{-2}b^3c^{-4})^{-3}$

110. $(5x^{-3}y^{-1}z^2)^{-2}$

111. $\dfrac{8x^{-2}y^5}{4x^{-6}z^{-2}}$

112. $\dfrac{10a^{-6}b^{-7}}{2a^{-2}c^{-3}}$

SYNTHESIS

113. Under what conditions does the nth root of x^3 exist? Explain your reasoning.

114. Under what conditions does the nth root of x^2 exist? Explain your reasoning.

115. *Dairy farming.* As a calf grows, it needs more milk for nourishment. The number of pounds of milk, M, required by a calf weighing x pounds can be estimated using the formula

$$M = -5 + \sqrt{6.7x - 444}.$$

Estimate the number of pounds of milk required by a calf of the given weight: **(a)** 300 lb; **(b)** 100 lb; **(c)** 200 lb; **(d)** 400 lb.
Source: www.ext.vt.edu

116. *Spaces in a parking lot.* A parking lot has attendants to park the cars. The number N of stalls needed for waiting cars before attendants can get to them is given by the formula $N = 2.5\sqrt{A}$, where A is the number of arrivals in peak hours. Find the number of spaces needed for the given number of arrivals in peak hours: **(a)** 25; **(b)** 36; **(c)** 49; **(d)** 64.

Determine the domain of each function described. Then draw the graph of each function.

117. $f(x) = \sqrt{x+5}$

118. $g(x) = \sqrt{x} + 5$

119. $g(x) = \sqrt{x} - 2$

120. $f(x) = \sqrt{x-2}$

121. Find the domain of f if

$$f(x) = \frac{\sqrt{x+3}}{\sqrt[4]{2-x}}.$$

122. Find the domain of g if

$$g(x) = \frac{\sqrt[4]{5-x}}{\sqrt[6]{x+4}}.$$

123. Find the domain of F if $F(x) = \dfrac{x}{\sqrt{x^2-5x-6}}$.

124. Use a graphing calculator to check your answers to Exercises 39, 45, and 57. On some graphing calculators, a MATH key is needed to enter higher roots.

125. Use a graphing calculator to check your answers to Exercises 117 and 118. (See Exercise 124.)

10.2   Rational Numbers as Exponents

Rational Exponents • Negative Rational Exponents •
Laws of Exponents • Simplifying Radical Expressions

In Chapter 1, we considered the natural numbers as exponents. Our discussion of exponents was expanded in Chapter 4 to include all integers. We now expand the study still further—to include all rational numbers. This will give meaning to expressions like $a^{1/3}$, $7^{-1/2}$, and $(3x)^{4/5}$. Such notation will help us simplify certain radical expressions.

Rational Exponents

Consider $a^{1/2} \cdot a^{1/2}$. If we still want to add exponents when multiplying, it must follow that $a^{1/2} \cdot a^{1/2} = a^{1/2+1/2}$, or a^1. This suggests that $a^{1/2}$ is a square root of a. Similarly, $a^{1/3} \cdot a^{1/3} \cdot a^{1/3} = a^{1/3+1/3+1/3}$, or a^1, so $a^{1/3}$ should mean $\sqrt[3]{a}$.

$$a^{1/n} = \sqrt[n]{a}$$

$a^{1/n}$ means $\sqrt[n]{a}$. When a is nonnegative, n can be any natural number greater than 1. When a is negative, n must be odd.

Note that the denominator of the exponent becomes the index and the base becomes the radicand.

EXAMPLE 1   Write an equivalent expression using radical notation.

a) $x^{1/2}$
b) $(-8)^{1/3}$
c) $(abc)^{1/5}$
d) $(25x^{16})^{1/2}$

Solution

a) $x^{1/2} = \sqrt{x}$
b) $(-8)^{1/3} = \sqrt[3]{-8} = -2$ The denominator of the exponent becomes the index. The base becomes the radicand. Recall that for square roots, the index 2 is understood without being written.
c) $(abc)^{1/5} = \sqrt[5]{abc}$
d) $(25x^{16})^{1/2} = 25^{1/2}x^8 = \sqrt{25} \cdot x^8 = 5x^8$

EXAMPLE 2   Write an equivalent expression using exponential notation.

a) $\sqrt[5]{9xy}$ b) $\sqrt[7]{\dfrac{x^3y}{4}}$ c) $\sqrt{5x}$

Solution   Parentheses are required to indicate the base.

a) $\sqrt[5]{9xy} = (9xy)^{1/5}$

b) $\sqrt[7]{\dfrac{x^3y}{4}} = \left(\dfrac{x^3y}{4}\right)^{1/7}$

> The index becomes the denominator of the exponent. The radicand becomes the base.

c) $\sqrt{5x} = (5x)^{1/2}$ The index 2 is understood without being written. We assume $x \geq 0$.

How shall we define $a^{2/3}$? If the property for multiplying exponents is to hold, we must have $a^{2/3} = (a^{1/3})^2$ and $a^{2/3} = (a^2)^{1/3}$. This would suggest that $a^{2/3} = (\sqrt[3]{a})^2$ and $a^{2/3} = \sqrt[3]{a^2}$. We make our definition accordingly.

Positive Rational Exponents

For any natural numbers m and n ($n \neq 1$) and any real number a for which $\sqrt[n]{a}$ exists,

$$a^{m/n} \quad \text{means} \quad (\sqrt[n]{a})^m, \quad \text{or} \quad \sqrt[n]{a^m}.$$

EXAMPLE 3   Write an equivalent expression using radical notation and simplify.

a) $27^{2/3}$

b) $25^{3/2}$

Solution

a) $27^{2/3}$ means $(\sqrt[3]{27})^2$ or, equivalently, $\sqrt[3]{27^2}$. Let's see which is easier to simplify:

$$(\sqrt[3]{27})^2 = 3^2 \qquad\qquad \sqrt[3]{27^2} = \sqrt[3]{729}$$
$$= 9; \qquad\qquad\qquad\qquad = 9.$$

The simplification on the left is probably easier for most people.

b) $25^{3/2}$ means $(\sqrt[2]{25})^3$ or, equivalently, $\sqrt[2]{25^3}$ (the index 2 is normally omitted). Since $\sqrt{25}$ is more commonly known than $\sqrt{25^3}$, we use that form:

$$25^{3/2} = (\sqrt{25})^3 = 5^3 = 125.$$

Student Notes _____

It is important to remember both meanings of $a^{m/n}$. When the root of the base a is known, $(\sqrt[n]{a})^m$ is generally easier to work with. When it is not known, $\sqrt[n]{a^m}$ is often more convenient.

EXAMPLE 4   Write an equivalent expression using exponential notation.

a) $\sqrt[3]{9^4}$

b) $(\sqrt[4]{7xy})^5$

Solution

a) $\sqrt[3]{9^4} = 9^{4/3}$

b) $(\sqrt[4]{7xy})^5 = (7xy)^{5/4}$

> The index becomes the denominator of the fraction that is the exponent.

Negative Rational Exponents

Recall from Section 4.8 that $x^{-2} = 1/x^2$. Negative rational exponents behave similarly.

To approximate $7^{2/3}$, we enter
7 ⌒ (2/3).

1. Why are the parentheses needed above?
2. Compare the graphs of $y_1 = x^{1/2}$, $y_2 = x$, and $y_3 = x^{3/2}$ and determine those x-values for which $y_1 > y_3$.

Negative Rational Exponents

For any rational number m/n and any nonzero real number a for which $a^{m/n}$ exists,

$$a^{-m/n} \text{ means } \frac{1}{a^{m/n}}.$$

Caution!   A negative exponent does not indicate that the expression in which it appears is negative: $a^{-1} \neq -a$.

EXAMPLE 5   Write an equivalent expression with positive exponents and, if possible, simplify.

a) $9^{-1/2}$

b) $(5xy)^{-4/5}$

c) $64^{-2/3}$

d) $4x^{-2/3}y^{1/5}$

e) $\left(\dfrac{3r}{7s}\right)^{-5/2}$

Solution

a) $9^{-1/2} = \dfrac{1}{9^{1/2}}$ $9^{-1/2}$ is the reciprocal of $9^{1/2}$.

Since $9^{1/2} = \sqrt{9} = 3$, the answer simplifies to $\dfrac{1}{3}$.

b) $(5xy)^{-4/5} = \dfrac{1}{(5xy)^{4/5}}$ $(5xy)^{-4/5}$ is the reciprocal of $(5xy)^{4/5}$.

c) $64^{-2/3} = \dfrac{1}{64^{2/3}}$ $64^{-2/3}$ is the reciprocal of $64^{2/3}$.

Since $64^{2/3} = \left(\sqrt[3]{64}\right)^2 = 4^2 = 16$, the answer simplifies to $\dfrac{1}{16}$.

d) $4x^{-2/3}y^{1/5} = 4 \cdot \dfrac{1}{x^{2/3}} \cdot y^{1/5} = \dfrac{4y^{1/5}}{x^{2/3}}$

e) In Section 4.8, we found that $(a/b)^{-n} = (b/a)^n$. This property holds for *any* negative exponent:

$$\left(\frac{3r}{7s}\right)^{-5/2} = \left(\frac{7s}{3r}\right)^{5/2}.$$ Writing the reciprocal of the base and changing the sign of the exponent

Laws of Exponents

The same laws hold for rational exponents as for integer exponents.

Laws of Exponents

For any real numbers a and b and any rational exponents m and n for which a^m, a^n, and b^m are defined:

1. $a^m \cdot a^n = a^{m+n}$ In multiplying, add exponents if the bases are the same.

2. $\dfrac{a^m}{a^n} = a^{m-n}$ In dividing, subtract exponents if the bases are the same. (Assume $a \neq 0$.)

3. $(a^m)^n = a^{m \cdot n}$ To raise a power to a power, multiply the exponents.

4. $(ab)^m = a^m b^m$ To raise a product to a power, raise each factor to the power and multiply.

EXAMPLE 6   Use the laws of exponents to simplify.

a) $3^{1/5} \cdot 3^{3/5}$

b) $\dfrac{a^{1/4}}{a^{1/2}}$

c) $(7.2^{2/3})^{3/4}$

d) $(a^{-1/3}b^{2/5})^{1/2}$

Solution

a) $3^{1/5} \cdot 3^{3/5} = 3^{1/5+3/5} = 3^{4/5}$ Adding exponents

b) $\dfrac{a^{1/4}}{a^{1/2}} = a^{1/4-1/2} = a^{1/4-2/4}$ Subtracting exponents after finding a common denominator

$$= a^{-1/4}, \text{ or } \frac{1}{a^{1/4}}$$ $a^{-1/4}$ is the reciprocal of $a^{1/4}$.

c) $(7.2^{2/3})^{3/4} = 7.2^{(2/3)(3/4)} = 7.2^{6/12}$ Multiplying exponents

$$= 7.2^{1/2}$$ Using arithmetic to simplify the exponent

d) $(a^{-1/3}b^{2/5})^{1/2} = a^{(-1/3)(1/2)} \cdot b^{(2/5)(1/2)}$ Raising a product to a power and multiplying exponents

$$= a^{-1/6}b^{1/5}, \text{ or } \frac{b^{1/5}}{a^{1/6}}$$

Simplifying Radical Expressions

Many radical expressions contain radicands or factors of radicands that are powers. When these powers and the index share a common factor, rational exponents can be used to simplify the expression.

> **To Simplify Radical Expressions**
> 1. Convert radical expressions to exponential expressions.
> 2. Use arithmetic and the laws of exponents to simplify.
> 3. Convert back to radical notation as needed.

EXAMPLE 7   Use rational exponents to simplify. Do not use exponents that are fractions in the final answer.

a) $\sqrt[6]{(5x)^3}$

b) $\sqrt[5]{t^{20}}$

c) $\left(\sqrt[3]{ab^2c}\right)^{12}$

d) $\sqrt{\sqrt[3]{x}}$

Solution

a) $\sqrt[6]{(5x)^3} = (5x)^{3/6}$ Converting to exponential notation

$= (5x)^{1/2}$ Simplifying the exponent

$= \sqrt{5x}$ Returning to radical notation

b) $\sqrt[5]{t^{20}} = t^{20/5}$ Converting to exponential notation

$= t^4$ Simplifying the exponent

c) $\left(\sqrt[3]{ab^2c}\right)^{12} = (ab^2c)^{12/3}$ Converting to exponential notation

$= (ab^2c)^4$ Simplifying the exponent

$= a^4b^8c^4$ Using the laws of exponents

d) $\sqrt{\sqrt[3]{x}} = \sqrt{x^{1/3}}$ Converting the radicand to exponential notation

$= (x^{1/3})^{1/2}$ Try to go directly to this step.

$= x^{1/6}$ Using the laws of exponents

$= \sqrt[6]{x}$ Returning to radical notation

technology connection

One way to check Example 7(a) is to let $y_1 = (5x)^{3/6}$ and $y_2 = \sqrt{5x}$. Then use GRAPH or TABLE to see if $y_1 = y_2$. An alternative is to let $y_3 = y_2 - y_1$ and see if $y_3 = 0$. Check Example 7(a) using one of these two methods.

1. Why are rational exponents especially useful when working on a graphing calculator?

Exercise Set
10.2

FOR EXTRA HELP

Student's Solutions Manual | Digital Video Tutor CD 5 Videotape 10 | Tutor Center AW Math Tutor Center | MathXL Tutorials on CD | MathXL MathXL |  MyMathLab MyMathLab

↪ *Concept Reinforcement   In each of Exercises 1–8, match the expression with the equivalent expression from the column on the right.*

1. ____ $x^{2/5}$

2. ____ $x^{5/2}$

3. ____ $x^{-5/2}$

4. ____ $x^{-2/5}$

5. ____ $x^{1/5} \cdot x^{2/5}$

6. ____ $(x^{1/5})^{5/2}$

7. ____ $\sqrt[5]{x^4}$

8. ____ $\left(\sqrt[4]{x}\right)^5$

a) $x^{3/5}$

b) $\left(\sqrt[5]{x}\right)^4$

c) $\sqrt{x^5}$

d) $x^{1/2}$

e) $\dfrac{1}{\left(\sqrt{x}\right)^5}$

f) $\sqrt[4]{x^5}$

g) $\sqrt[5]{x^2}$

h) $\dfrac{1}{\left(\sqrt[5]{x}\right)^2}$

Note: Assume for all exercises that even roots are of nonnegative quantities and that all denominators are nonzero.

Write an equivalent expression using radical notation and, if possible, simplify.

9. $x^{1/6}$

10. $y^{1/5}$

11. $16^{1/2}$

12. $8^{1/3}$

13. $81^{1/4}$

14. $64^{1/6}$

15. $9^{1/2}$

16. $25^{1/2}$

17. $(xyz)^{1/3}$

18. $(ab)^{1/4}$

19. $(a^2b^2)^{1/5}$

20. $(x^3y^3)^{1/4}$

21. $t^{2/5}$

22. $b^{3/2}$

23. $16^{3/4}$

24. $4^{7/2}$

25. $27^{4/3}$

26. $9^{5/2}$

27. $(81x)^{3/4}$

28. $(125a)^{2/3}$

29. $(25x^4)^{3/2}$

30. $(9y^6)^{3/2}$

Write an equivalent expression using exponential notation.

31. $\sqrt[3]{20}$

32. $\sqrt[3]{19}$

33. $\sqrt{17}$

34. $\sqrt{6}$

35. $\sqrt{x^3}$

36. $\sqrt{a^5}$

37. $\sqrt[5]{m^2}$

38. $\sqrt[5]{n^4}$

39. $\sqrt[4]{cd}$

40. $\sqrt[5]{xy}$

41. $\sqrt[5]{xy^2z}$

42. $\sqrt[7]{x^3y^2z^2}$

43. $\left(\sqrt{3mn}\right)^3$

44. $\left(\sqrt[3]{7xy}\right)^4$

45. $\left(\sqrt[7]{8x^2y}\right)^5$

46. $\left(\sqrt[6]{2a^5b}\right)^7$

47. $\dfrac{2x}{\sqrt[3]{z^2}}$

48. $\dfrac{3a}{\sqrt[5]{c^2}}$

Write an equivalent expression with positive exponents and, if possible, simplify.

49. $x^{-1/3}$

50. $y^{-1/4}$

51. $(2rs)^{-3/4}$

52. $(5xy)^{-5/6}$

53. $\left(\dfrac{1}{16}\right)^{-3/4}$

54. $\left(\dfrac{1}{8}\right)^{-2/3}$

55. $\dfrac{2c}{a^{-3/5}}$

56. $\dfrac{3b}{a^{-5/7}}$

57. $5x^{-2/3}y^{4/5}z$

58. $2a^{3/4}b^{-1/2}c^{2/3}$

59. $3^{-5/2}a^3b^{-7/3}$

60. $2^{-1/3}x^4y^{-2/7}$

61. $\left(\dfrac{2ab}{3c}\right)^{-5/6}$

62. $\left(\dfrac{7x}{8yz}\right)^{-3/5}$

63. $\dfrac{6a}{\sqrt[4]{b}}$

64. $\dfrac{7x}{\sqrt[3]{z}}$

Use the laws of exponents to simplify. Do not use negative exponents in any answers.

65. $7^{3/4} \cdot 7^{1/8}$

66. $11^{2/3} \cdot 11^{1/2}$

67. $\dfrac{3^{5/8}}{3^{-1/8}}$

68. $\dfrac{8^{7/11}}{8^{-2/11}}$

69. $\dfrac{5.2^{-1/6}}{5.2^{-2/3}}$

70. $\dfrac{2.3^{-3/10}}{2.3^{-1/5}}$

71. $(10^{3/5})^{2/5}$

72. $(5^{5/4})^{3/7}$

73. $a^{2/3} \cdot a^{5/4}$

74. $x^{3/4} \cdot x^{1/3}$

Aha! **75.** $(64^{3/4})^{4/3}$

76. $(27^{-2/3})^{3/2}$

77. $(m^{2/3}n^{-1/4})^{1/2}$

78. $(x^{-1/3}y^{2/5})^{1/4}$

Use rational exponents to simplify. Do not use fraction exponents in the final answer.

79. $\sqrt[6]{x^4}$

80. $\sqrt[6]{a^2}$

81. $\sqrt[4]{a^{12}}$

82. $\sqrt[3]{x^{15}}$

83. $\sqrt[5]{a^{10}}$

84. $\sqrt[6]{x^{18}}$

85. $\left(\sqrt[7]{xy}\right)^{14}$

86. $\left(\sqrt[3]{ab}\right)^{15}$

87. $\sqrt[4]{(7a)^2}$

88. $\sqrt[8]{(3x)^2}$

89. $\left(\sqrt[8]{2x}\right)^6$

90. $\left(\sqrt[10]{3a}\right)^5$

91. $\sqrt[3]{\sqrt[6]{a}}$

92. $\sqrt[4]{\sqrt{x}}$

93. $\sqrt[4]{(xy)^{12}}$

94. $\sqrt{(ab)^6}$

95. $\left(\sqrt[5]{a^2b^4}\right)^{15}$

96. $\left(\sqrt[3]{x^2y^5}\right)^{12}$

97. $\sqrt[3]{\sqrt[4]{xy}}$

98. $\sqrt[5]{\sqrt{2a}}$

99. If $f(x) = (x + 5)^{1/2}(x + 7)^{-1/2}$, find the domain of f. Explain how you found your answer.

100. Explain why $\sqrt[3]{x^6} = x^2$ for any value of x, whereas $\sqrt[2]{x^6} = x^3$ only when $x \geq 0$.

SKILL MAINTENANCE

Simplify.

101. $3x(x^3 - 2x^2) + 4x^2(2x^2 + 5x)$ [4.3], [4.4]

102. $5t^3(2t^2 - 4t) - 3t^4(t^2 - 6t)$ [4.3], [4.4]

103. $(3a - 4b)(5a + 3b)$ [4.5]

104. $(7x - y)^2$ [4.5]

105. *Real estate taxes.* For homes under $100,000, the property transfer tax in Vermont is 0.5% of the selling price. Find the selling price of a home that had a transfer tax of $467.50. [2.4]

106. What numbers are their own squares? [5.8]

SYNTHESIS

107. Let $f(x) = 5x^{-1/3}$. Under what condition will we have $f(x) > 0$? Why?

108. If $g(x) = x^{3/n}$, in what way does the domain of g depend on whether n is odd or even?

Use rational exponents to simplify.

109. $\sqrt{x\sqrt[3]{x^2}}$ **110.** $\sqrt[4]{\sqrt[3]{8x^3y^6}}$

111. $\sqrt[12]{p^2 + 2pq + q^2}$

Music. The function given by $f(x) = k2^{x/12}$ can be used to determine the frequency, in cycles per second, of a musical note that is x half-steps above a note with frequency k.*

112. The frequency of concert A for a trumpet is 440 cycles per second. Find the frequency of the A that is two octaves (24 half-steps) above concert A (few trumpeters can reach this note.)

113. Show that the G that is 7 half-steps (a "perfect fifth") above middle C (262 cycles per second) has a frequency that is about 1.5 times that of middle C.

*This application was inspired by information provided by Dr. Homer B. Tilton of Pima Community College East.

114. Show that the C sharp that is 4 half-steps (a "major third") above concert A (see Exercise 112) has a frequency that is about 25% greater than that of concert A.

115. *Baseball.* The statistician Bill James has found that a baseball team's winning percentage P can be approximated by

$$P = \frac{r^{1.83}}{r^{1.83} + \sigma^{1.83}},$$

where r is the total number of runs scored by that team and σ is the total number of runs scored by their opponents. During a recent season, the San Francisco Giants scored 799 runs and their opponents scored 749 runs. Use James's formula to predict the Giants' winning percentage (the team actually won 55.6% of their games).
Source: M. Bittinger, *One Man's Journey Through Mathematics.* Boston: Addison-Wesley, 2004

116. *Road pavement messages.* In a psychological study, it was determined that the proper length L of the letters of a word printed on pavement is given by

$$L = \frac{0.000169d^{2.27}}{h},$$

where d is the distance of a car from the lettering and h is the height of the eye above the surface of the road. All units are in meters. This formula says that if a person is h meters above the surface of the road and is to be able to recognize a message d meters away, that message will be the most recognizable if the length of the letters is L. Find L to the nearest tenth of a meter, given d and h.

a) $h = 1$ m, $d = 60$ m
b) $h = 0.9906$ m, $d = 75$ m
c) $h = 2.4$ m, $d = 80$ m
d) $h = 1.1$ m, $d = 100$ m

117. *Dating fossils.* The function $r(t) = 10^{-12}2^{-t/5700}$ expresses the ratio of carbon isotopes to carbon atoms in a fossil that is t years old. What ratio of carbon isotopes to carbon atoms would be present in a 1900-year-old bone?

118. *Physics.* The equation $m = m_0(1 - v^2c^{-2})^{-1/2}$, developed by Albert Einstein, is used to determine the mass m of an object that is moving v meters per second and has mass m_0 before the motion begins. The constant c is the speed of light, approximately 3×10^8 m/sec. Suppose that a particle with mass 8 mg is accelerated to a speed of $\frac{9}{5} \times 10^8$ m/sec. Without using a calculator, find the new mass of the particle.

119. Using a graphing calculator, select the **MODE** SIMUL and the FORMAT ExprOff. Then graph
$$y_1 = x^{1/2}, \qquad y_2 = 3x^{2/5},$$
$$y_3 = x^{4/7}, \quad \text{and} \quad y_4 = \tfrac{1}{5}x^{3/4}.$$
Looking only at coordinates, match each graph with its equation.

CORNER

Are Equivalent Fractions Equivalent Exponents?

COLLABORATIVE

Focus: Functions and rational exponents

Time: 10–20 minutes

Group size: 3

Materials: Graph paper

In arithmetic, we have seen that $\frac{1}{3}, \frac{1}{6} \cdot 2$, and $2 \cdot \frac{1}{6}$ all represent the same number. Interestingly,

$$f(x) = x^{1/3},$$
$$g(x) = (x^{1/6})^2, \quad \text{and}$$
$$h(x) = (x^2)^{1/6}$$

represent three *different* functions.

ACTIVITY

1. Selecting a variety of values for x and using the definition of positive rational exponents, one group member should graph f, a second group member should graph g, and a third group member should graph h. Be sure to check whether negative x-values are in the domain of the function.

2. Compare the three graphs and check each other's work. How and why do the graphs differ?

3. Decide as a group which graph, if any, would best represent the graph of $k(x) = x^{2/6}$. Then be prepared to explain your reasoning to the entire class. (*Hint*: Study the definition of $a^{m/n}$ on p. 446 carefully.)

10.3   Multiplying Radical Expressions

Multiplying Radical Expressions • Simplifying by Factoring • Multiplying and Simplifying

Multiplying Radical Expressions

Note that $\sqrt{4}\,\sqrt{25} = 2 \cdot 5 = 10$. Also $\sqrt{4 \cdot 25} = \sqrt{100} = 10$. Likewise,

$$\sqrt[3]{27}\,\sqrt[3]{8} = 3 \cdot 2 = 6 \quad \text{and} \quad \sqrt[3]{27 \cdot 8} = \sqrt[3]{216} = 6.$$

These examples suggest the following.

The Product Rule for Radicals

For any real numbers $\sqrt[n]{a}$ and $\sqrt[n]{b}$,

$$\sqrt[n]{a} \cdot \sqrt[n]{b} = \sqrt[n]{a \cdot b}.$$

(The product of two nth roots is the nth root of the product of the two radicands.)

Rational exponents can be used to derive this rule:

$$\sqrt[n]{a} \cdot \sqrt[n]{b} = a^{1/n} \cdot b^{1/n} = (a \cdot b)^{1/n} = \sqrt[n]{a \cdot b}.$$

EXAMPLE 1   Multiply.

a) $\sqrt{3} \cdot \sqrt{5}$ **b)** $\sqrt{x+3}\,\sqrt{x-3}$

c) $\sqrt[3]{4} \cdot \sqrt[3]{5}$ **d)** $\sqrt[4]{\dfrac{y}{5}} \cdot \sqrt[4]{\dfrac{7}{x}}$

Solution

a) When no index is written, roots are understood to be square roots with an unwritten index of two. We apply the product rule:

$$\sqrt{3} \cdot \sqrt{5} = \sqrt{3 \cdot 5}$$
$$= \sqrt{15}.$$

b) $\sqrt{x+3}\,\sqrt{x-3} = \sqrt{(x+3)(x-3)}$ The product of two square roots is the square root of the product.
$$= \sqrt{x^2 - 9}$$

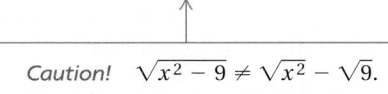

Caution! $\sqrt{x^2 - 9} \neq \sqrt{x^2} - \sqrt{9}.$

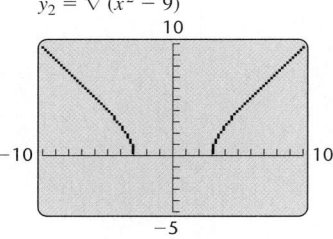
c) Both $\sqrt[3]{4}$ and $\sqrt[3]{5}$ have indices of three, so to multiply we can use the product rule:

$$\sqrt[3]{4} \cdot \sqrt[3]{5} = \sqrt[3]{4 \cdot 5} = \sqrt[3]{20}.$$

d) $\sqrt[4]{\dfrac{y}{5}} \cdot \sqrt[4]{\dfrac{7}{x}} = \sqrt[4]{\dfrac{y}{5} \cdot \dfrac{7}{x}} = \sqrt[4]{\dfrac{7y}{5x}}$ In Section 7.4, we discuss other ways to write answers like this.

> **Caution!**   The product rule for radicals applies only when radicals have the same index:
>
> $$\sqrt[n]{a} \cdot \sqrt[m]{b} \neq \sqrt[nm]{a \cdot b}.$$

Simplifying by Factoring

The number p is a *perfect square* if there exists a rational number q for which $q^2 = p$. We say that p is a *perfect cube* if $q^3 = p$ for some rational number q. In general, p is a *perfect nth power* if $q^n = p$ for some rational number q. Thus, 16 and $\frac{1}{10,000}$ are both perfect 4th powers since $2^4 = 16$ and $\left(\frac{1}{10}\right)^4 = \frac{1}{10,000}$.

The product rule allows us to simplify $\sqrt[n]{ab}$ whenever ab contains a factor that is a perfect nth power.

> **Using the Product Rule to Simplify**
> $$\sqrt[n]{ab} = \sqrt[n]{a} \cdot \sqrt[n]{b}.$$
> ($\sqrt[n]{a}$ and $\sqrt[n]{b}$ must both be real numbers.)

To illustrate, suppose we wish to simplify $\sqrt{20}$. Since this is a *square* root, we check to see if there is a factor of 20 that is a perfect square. There is one, 4, so we express 20 as $4 \cdot 5$ and use the product rule:

$$\sqrt{20} = \sqrt{4 \cdot 5} \qquad \text{Factoring the radicand (4 is a perfect square)}$$
$$= \sqrt{4} \cdot \sqrt{5} \qquad \text{Factoring into two radicals}$$
$$= 2\sqrt{5}. \qquad \text{Finding the square root of 4}$$

> **To Simplify a Radical Expression with Index n by Factoring**
> **1.** Express the radicand as a product in which one factor is the largest perfect nth power possible.
> **2.** Take the nth root of each factor.
> **3.** Simplification is complete when no radicand has a factor that is a perfect nth power.

It is often safe to assume that a radicand does not represent a negative number raised to an even power. We will henceforth make this assumption—unless functions are involved—and discontinue use of absolute-value notation when taking even roots.

EXAMPLE 2   Simplify by factoring: **(a)** $\sqrt{200}$; **(b)** $\sqrt{18x^2y}$; **(c)** $\sqrt[3]{72}$; **(d)** $\sqrt[4]{162x^6}$.

Solution

a) $\sqrt{200} = \sqrt{100 \cdot 2}$ 100 is the largest perfect-square factor of 200.

$\qquad = \sqrt{100} \cdot \sqrt{2} = 10\sqrt{2}$

b) $\sqrt{18x^2y} = \sqrt{9 \cdot 2 \cdot x^2 \cdot y}$ $9x^2$ is the largest perfect-square factor of $18x^2y$.

$\qquad = \sqrt{9x^2} \cdot \sqrt{2y}$ Factoring into two radicals

$\qquad = 3x\sqrt{2y}$ Taking the square root of $9x^2$

c) $\sqrt[3]{72} = \sqrt[3]{8 \cdot 9}$ 8 is the largest perfect-cube (third-power) factor of 72.

$\qquad = \sqrt[3]{8} \cdot \sqrt[3]{9} = 2\sqrt[3]{9}$

Let's look at this example another way. We write a complete factorization and look for triples of factors. Each triple of factors makes a cube:

$\sqrt[3]{72} = \sqrt[3]{\underline{2 \cdot 2 \cdot 2} \cdot 3 \cdot 3}$ Each triple of factors is a cube.

$\qquad = 2\sqrt[3]{3 \cdot 3} = 2\sqrt[3]{9}.$

d) $\sqrt[4]{162x^6} = \sqrt[4]{81 \cdot 2 \cdot x^4 \cdot x^2}$ $81 \cdot x^4$ is the largest perfect fourth-power factor of $162x^6$.

$\qquad = \sqrt[4]{81x^4} \cdot \sqrt[4]{2x^2}$ Factoring into two radicals

$\qquad = 3x\sqrt[4]{2x^2}$ Taking fourth roots

Let's look at this example another way. We write a complete factorization and look for quadruples of factors. Each quadruple makes a perfect fourth power:

$\sqrt[4]{162x^6} = \sqrt[4]{\underline{3 \cdot 3 \cdot 3 \cdot 3} \cdot 2 \cdot \underline{x \cdot x \cdot x \cdot x} \cdot x \cdot x}$ Each quadruple of factors is a power of 4.

$\qquad = 3 \cdot x \cdot \sqrt[4]{2 \cdot x \cdot x}$

$\qquad = 3x\sqrt[4]{2x^2}.$

EXAMPLE 3   If $f(x) = \sqrt{3x^2 - 6x + 3}$, find a simplified form for $f(x)$. Assume that x can be any real number.

Solution

$$f(x) = \sqrt{3x^2 - 6x + 3}$$

$$\left. \begin{aligned} &= \sqrt{3(x^2 - 2x + 1)} \\ &= \sqrt{(x - 1)^2 \cdot 3} \end{aligned} \right\} \quad \text{Factoring the radicand; } x^2 - 2x + 1 \text{ is a perfect square.}$$

$$= \sqrt{(x - 1)^2} \cdot \sqrt{3} \quad \text{Factoring into two radicals}$$

$$= |x - 1|\sqrt{3} \quad \text{Finding the square root of } (x - 1)^2$$

EXAMPLE 4   Simplify: **(a)** $\sqrt{x^7y^{11}z^9}$; **(b)** $\sqrt[3]{16a^7b^{14}}$.

Solution

a) There are many ways to factor $x^7y^{11}z^9$. Because of the square root (index of 2), we identify the largest exponents that are multiples of 2:

$$\sqrt{x^7y^{11}z^9} = \sqrt{x^6 \cdot x \cdot y^{10} \cdot y \cdot z^8 \cdot z} \qquad \text{Using the largest even powers of } x, y, \text{ and } z$$

$$= \sqrt{x^6}\,\sqrt{y^{10}}\,\sqrt{z^8}\,\sqrt{xyz} \qquad \text{Factoring into several radicals}$$

$$= x^{6/2}\,y^{10/2}\,z^{8/2}\sqrt{xyz} \qquad \text{Converting to rational exponents}$$

$$= x^3y^5z^4\sqrt{xyz}.$$

Check: $(x^3y^5z^4\sqrt{xyz})^2 = (x^3)^2(y^5)^2(z^4)^2(\sqrt{xyz})^2$
$$= x^6 \cdot y^{10} \cdot z^8 \cdot xyz = x^7y^{11}z^9$$

Our check shows that $x^3y^5z^4\sqrt{xyz}$ is the square root of $x^7y^{11}z^9$.

b) There are many ways to factor $16a^7b^{14}$. Because of the cube root (index of 3), we identify factors with the largest exponents that are multiples of 3:

$$\sqrt[3]{16a^7b^{14}} = \sqrt[3]{8 \cdot 2 \cdot a^6 \cdot a \cdot b^{12} \cdot b^2} \qquad \text{Using the largest perfect-cube factors; 6 and 12 are multiples of 3}$$

$$= \sqrt[3]{8}\,\sqrt[3]{a^6}\,\sqrt[3]{b^{12}}\,\sqrt[3]{2ab^2} \qquad \text{Factoring into several radicals}$$

$$= 2\;a^{6/3}\,b^{12/3}\sqrt[3]{2ab^2} \qquad \text{Converting to rational exponents}$$

$$= 2a^2b^4\sqrt[3]{2ab^2}$$

As a check, let's redo the problem using a complete factorization of the radicand:

$$\sqrt[3]{16a^7b^{14}} = \sqrt[3]{\underline{2 \cdot 2 \cdot 2} \cdot 2 \cdot \underline{a \cdot a \cdot a} \cdot a \cdot a \cdot a \cdot a \cdot \underline{b \cdot b \cdot b} \cdot \underline{b \cdot b \cdot b} \cdot \underline{b \cdot b \cdot b} \cdot \underline{b \cdot b \cdot b} \cdot b \cdot b}$$

Each triple of factors makes a cube.

$$= 2 \cdot a \cdot a \cdot b \cdot b \cdot b \cdot b \cdot \sqrt[3]{2 \cdot a \cdot b \cdot b}$$

$$= 2a^2b^4\sqrt[3]{2ab^2}. \qquad \text{Our answer checks.}$$

> *Remember*: To simplify an *n*th root, identify factors in the radicand with exponents that are multiples of *n*.

Multiplying and Simplifying

We have used the product rule for radicals to find products and also to simplify radical expressions. For some radical expressions, it is possible to do both: First find a product and then simplify.

EXAMPLE 5   Multiply and simplify.

a) $\sqrt{15}\ \sqrt{6}$ b) $3\sqrt[3]{25}\cdot 2\sqrt[3]{5}$ c) $\sqrt[4]{8x^3y^5}\ \sqrt[4]{4x^2y^3}$

Solution

a) $\sqrt{15}\ \sqrt{6} = \sqrt{15\cdot 6}$ Multiplying radicands

$\phantom{\sqrt{15}\ \sqrt{6}} = \sqrt{90} = \sqrt{9\cdot 10}$ 9 is a perfect square.

$\phantom{\sqrt{15}\ \sqrt{6}} = 3\sqrt{10}$

b) $3\sqrt[3]{25}\cdot 2\sqrt[3]{5} = 3\cdot 2\cdot\sqrt[3]{25\cdot 5}$ Using a commutative law; multiplying radicands

$\phantom{3\sqrt[3]{25}\cdot 2\sqrt[3]{5}} = 6\cdot\sqrt[3]{125}$ 125 is a perfect cube.

$\phantom{3\sqrt[3]{25}\cdot 2\sqrt[3]{5}} = 6\cdot 5,\text{ or }30$

c) $\sqrt[4]{8x^3y^5}\ \sqrt[4]{4x^2y^3} = \sqrt[4]{32x^5y^8}$ Multiplying radicands

$\phantom{\sqrt[4]{8x^3y^5}\ \sqrt[4]{4x^2y^3}} = \sqrt[4]{16x^4y^8\cdot 2x}$ Identifying perfect fourth-power factors

$\phantom{\sqrt[4]{8x^3y^5}\ \sqrt[4]{4x^2y^3}} = \sqrt[4]{16}\ \sqrt[4]{x^4}\ \sqrt[4]{y^8}\ \sqrt[4]{2x}$ Factoring into radicals

$\phantom{\sqrt[4]{8x^3y^5}\ \sqrt[4]{4x^2y^3}} = 2xy^2\sqrt[4]{2x}$ Finding the fourth roots; assume $x \geq 0$.

Student Notes _____

To multiply $\sqrt{x}\cdot\sqrt{x}$, remember what $\sqrt{x}$ represents and go directly to the product, x. Too often students unnecessarily write $\sqrt{x}\cdot\sqrt{x} = \sqrt{x^2} = x$.

The checks are left to the student.

Exercise Set

10.3

↪ *Concept Reinforcement*   *Classify each of the following statements as either true or false.*

1. For any real numbers $\sqrt[n]{a}$ and $\sqrt[n]{b}$,
$\sqrt[n]{a}\cdot\sqrt[n]{b} = \sqrt[n]{ab}$.

2. For any real numbers $\sqrt[n]{a}$ and $\sqrt[n]{b}$,
$\sqrt[n]{a} + \sqrt[n]{b} = \sqrt[n]{a + b}$.

3. For any real numbers $\sqrt[n]{a}$ and $\sqrt[m]{b}$,
$\sqrt[n]{a}\cdot\sqrt[m]{b} = \sqrt[nm]{ab}$.

4. For $x > 0$, $\sqrt{x^2 - 9} = x - 3$.

5. The expression $\sqrt[3]{X}$ is not simplified if X contains a factor that is a perfect cube.

6. It is often possible to simplify $\sqrt{A\cdot B}$ even though $\sqrt{A}$ and $\sqrt{B}$ cannot be simplified.

Multiply.

7. $\sqrt{5}\ \sqrt{7}$ **8.** $\sqrt{10}\ \sqrt{7}$

9. $\sqrt[3]{7}\ \sqrt[3]{2}$ **10.** $\sqrt[3]{2}\ \sqrt[3]{5}$

11. $\sqrt[4]{6}\ \sqrt[4]{3}$ **12.** $\sqrt[4]{8}\ \sqrt[4]{9}$

13. $\sqrt{2x}\ \sqrt{13y}$ **14.** $\sqrt{5a}\ \sqrt{6b}$

15. $\sqrt[5]{8y^3}\ \sqrt[5]{10y}$ **16.** $\sqrt[5]{9t^2}\ \sqrt[5]{2t}$

17. $\sqrt{y - b}\ \sqrt{y + b}$ **18.** $\sqrt{x - a}\ \sqrt{x + a}$

19. $\sqrt[3]{0.7y}\ \sqrt[3]{0.3y}$ **20.** $\sqrt[3]{0.5x}\ \sqrt[3]{0.2x}$

21. $\sqrt[5]{x - 2}\ \sqrt[5]{(x - 2)^2}$

22. $\sqrt[4]{x - 1}\ \sqrt[4]{x^2 + x + 1}$

23. $\sqrt{\dfrac{7}{t}}\sqrt{\dfrac{s}{11}}$

24. $\sqrt{\dfrac{x}{6}}\sqrt{\dfrac{7}{y}}$

25. $\sqrt[7]{\dfrac{x-3}{4}}\sqrt[7]{\dfrac{5}{x+2}}$

26. $\sqrt[6]{\dfrac{a}{b-2}}\sqrt[6]{\dfrac{3}{b+2}}$

Simplify by factoring.

27. $\sqrt{18}$

28. $\sqrt{50}$

29. $\sqrt{27}$

30. $\sqrt{45}$

31. $\sqrt{8}$

32. $\sqrt{75}$

33. $\sqrt{198}$

34. $\sqrt{325}$

35. $\sqrt{36a^4b}$

36. $\sqrt{175y^8}$

37. $\sqrt[3]{8x^3y^2}$

38. $\sqrt[3]{27ab^6}$

39. $\sqrt[3]{-16x^6}$

40. $\sqrt[3]{-32a^6}$

Find a simplified form of $f(x)$. Assume that x can be any real number.

41. $f(x) = \sqrt[3]{125x^5}$

42. $f(x) = \sqrt[3]{16x^6}$

43. $f(x) = \sqrt{49(x-3)^2}$

44. $f(x) = \sqrt{81(x-1)^2}$

45. $f(x) = \sqrt{5x^2 - 10x + 5}$

46. $f(x) = \sqrt{2x^2 + 8x + 8}$

Simplify. Assume that no radicands were formed by raising negative numbers to even powers.

47. $\sqrt{a^6b^7}$

48. $\sqrt{x^6y^9}$

49. $\sqrt[3]{x^5y^6z^{10}}$

50. $\sqrt[3]{a^6b^7c^{13}}$

51. $\sqrt[5]{-32a^7b^{11}}$

52. $\sqrt[4]{16x^5y^{11}}$

53. $\sqrt[5]{x^{13}y^8z^{17}}$

54. $\sqrt[5]{a^6b^8c^9}$

55. $\sqrt[3]{-80a^{14}}$

56. $\sqrt[4]{810x^9}$

Multiply and simplify.

57. $\sqrt{6}\sqrt{3}$

58. $\sqrt{15}\sqrt{5}$

59. $\sqrt{15}\sqrt{21}$

60. $\sqrt{10}\sqrt{14}$

61. $\sqrt[3]{9}\sqrt[3]{3}$

62. $\sqrt[3]{2}\sqrt[3]{4}$

Aha! **63.** $\sqrt{18a^3}\sqrt{18a^3}$

64. $\sqrt{75x^7}\sqrt{75x^7}$

65. $\sqrt[3]{5a^2}\sqrt[3]{2a}$

66. $\sqrt[3]{7x}\sqrt[3]{3x^2}$

67. $\sqrt{2x^5}\sqrt{10x^2}$

68. $\sqrt{5a^7}\sqrt{15a^3}$

69. $\sqrt[3]{s^2t^4}\sqrt[3]{s^4t^6}$

70. $\sqrt[3]{x^2y^4}\sqrt[3]{x^2y^6}$

71. $\sqrt[3]{(x+5)^2}\sqrt[3]{(x+5)^4}$

72. $\sqrt[3]{(a-b)^5}\sqrt[3]{(a-b)^7}$

73. $\sqrt[4]{20a^3b^7}\sqrt[4]{4a^2b^5}$

74. $\sqrt[4]{9x^7y^2}\sqrt[4]{9x^2y^9}$

75. $\sqrt[5]{x^3(y+z)^6}\sqrt[5]{x^3(y+z)^4}$

76. $\sqrt[5]{a^3(b-c)^4}\sqrt[5]{a^7(b-c)^4}$

77. Why do we need to know how to multiply radical expressions before learning how to simplify radical expressions?

78. Why is it incorrect to say that, in general, $\sqrt{x^2} = x$?

SKILL MAINTENANCE

Perform the indicated operation and, if possible, simplify. [6.4]

79. $\dfrac{3x}{16y} + \dfrac{5y}{64x}$

80. $\dfrac{2}{a^3b^4} + \dfrac{6}{a^4b}$

81. $\dfrac{4}{x^2-9} - \dfrac{7}{2x-6}$

82. $\dfrac{8}{x^2-25} - \dfrac{3}{2x-10}$

Simplify. [4.1]

83. $\dfrac{9a^4b^7}{3a^2b^5}$

84. $\dfrac{12a^2b^7}{4ab^2}$

SYNTHESIS

85. Explain why it is true that $\sqrt[n]{ab} = \sqrt[n]{a} \cdot \sqrt[n]{b}$.

86. Is the equation $\sqrt{(2x+3)^8} = (2x+3)^4$ always, sometimes, or never true? Why?

87. *Radar range.* The function given by

$$R(x) = \frac{1}{2}\sqrt[4]{\frac{x \cdot 3.0 \times 10^6}{\pi^2}}$$

can be used to determine the maximum range $R(x)$, in miles, of an ARSR-3 surveillance radar with a peak power of x watts. Determine the maximum radar range when the peak power is 5×10^4 watts.

Source: Introduction to RADAR Techniques, Federal Aviation Administration, 1988

88. *Speed of a skidding car.* Police can estimate the speed at which a car was traveling by measuring its skid marks. The function given by

$$r(L) = 2\sqrt{5L}$$

can be used, where L is the length of a skid mark, in feet, and $r(L)$ is the speed, in miles per hour. Find the exact speed and an estimate (to the nearest tenth mile per hour) for the speed of a car that left skid marks **(a)** 20 ft long; **(b)** 70 ft long; **(c)** 90 ft long. See also Exercise 102.

89. *Wind chill temperature.* When the temperature is T degrees Celsius and the wind speed is v meters per second, the *wind chill temperature*, T_w, is the temperature (with no wind) that it feels like. Here is a formula for finding wind chill temperature:

$$T_w = 33 - \frac{\left(10.45 + 10\sqrt{v} - v\right)\left(33 - T\right)}{22}.$$

Estimate the wind chill temperature (to the nearest tenth of a degree) for the given actual temperatures and wind speeds.
a) $T = 7°C,\ v = 8\ \text{m/sec}$
b) $T = 0°C,\ v = 12\ \text{m/sec}$
c) $T = -5°C,\ v = 14\ \text{m/sec}$
d) $T = -23°C,\ v = 15\ \text{m/sec}$

Simplify. Assume that all variables are nonnegative.
90. $\left(\sqrt{r^3 t}\right)^7$

91. $\left(\sqrt[3]{25x^4}\right)^4$

92. $\left(\sqrt[3]{a^2 b^4}\right)^5$

93. $\left(\sqrt{a^3 b^5}\right)^7$

Draw and compare the graphs of each group of equations.
94. $f(x) = \sqrt{x^2 - 2x + 1}$,
$g(x) = x - 1$,
$h(x) = |x - 1|$

95. $f(x) = \sqrt{x^2 + 2x + 1}$,
$g(x) = x + 1$,
$h(x) = |x + 1|$

96. If $f(t) = \sqrt{t^2 - 3t - 4}$, what is the domain of f?

97. What is the domain of g, if $g(x) = \sqrt{x^2 - 6x + 8}$?

Solve.
98. $\sqrt[3]{5x^{k+1}}\ \sqrt[3]{25x^k} = 5x^7$, for k

99. $\sqrt[5]{4a^{3k+2}}\ \sqrt[5]{8a^{6-k}} = 2a^4$, for k

100. Use a graphing calculator to check your answers to Exercises 21 and 41.

101. Rony is puzzled. When he uses a graphing calculator to graph $y = \sqrt{x} \cdot \sqrt{x}$, he gets the following screen. Explain why Rony did not get the complete line $y = x$.

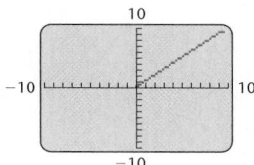

102. Does a car traveling twice as fast as another car leave a skid mark that is twice as long? (See Exercise 88.) Why or why not?

10.4 Dividing Radical Expressions

Dividing and Simplifying   •   Rationalizing Denominators and
Numerators (Part 1)

Dividing and Simplifying

Just as the root of a product can be expressed as the product of two roots, the
root of a quotient can be expressed as the quotient of two roots. For example,

$$\sqrt[3]{\frac{27}{8}} = \frac{3}{2} \quad \text{and} \quad \frac{\sqrt[3]{27}}{\sqrt[3]{8}} = \frac{3}{2}.$$

This example suggests the following.

The Quotient Rule for Radicals

For any real numbers $\sqrt[n]{a}$ and $\sqrt[n]{b}$, $b \neq 0$,

$$\sqrt[n]{\frac{a}{b}} = \frac{\sqrt[n]{a}}{\sqrt[n]{b}}.$$

Remember that an nth root is simplified when its radicand has no factors
that are perfect nth powers. Recall too that we assume that no radicands represent negative quantities raised to an even power.

EXAMPLE 1   Simplify by taking the roots of the numerator and the denominator.

a) $\sqrt[3]{\dfrac{27}{125}}$ b) $\sqrt{\dfrac{25}{y^2}}$

Solution

a) $\sqrt[3]{\dfrac{27}{125}} = \dfrac{\sqrt[3]{27}}{\sqrt[3]{125}} = \dfrac{3}{5}$ Taking the cube roots of the numerator and
the denominator

b) $\sqrt{\dfrac{25}{y^2}} = \dfrac{\sqrt{25}}{\sqrt{y^2}} = \dfrac{5}{y}$ Taking the square roots of the numerator and the
denominator. Assume $y > 0$.

As in Section 10.3, any radical expressions appearing in the answers should
be simplified as much as possible.

EXAMPLE 2 Simplify: **(a)** $\sqrt{\dfrac{16x^3}{y^8}}$; **(b)** $\sqrt[3]{\dfrac{27y^{14}}{8x^3}}$.

Solution

a) $\sqrt{\dfrac{16x^3}{y^8}} = \dfrac{\sqrt{16x^3}}{\sqrt{y^8}}$

$= \dfrac{\sqrt{16x^2 \cdot x}}{\sqrt{y^8}}$

$= \dfrac{4x\sqrt{x}}{y^4}$ Simplifying the numerator and the denominator

b) $\sqrt[3]{\dfrac{27y^{14}}{8x^3}} = \dfrac{\sqrt[3]{27y^{14}}}{\sqrt[3]{8x^3}}$

$= \dfrac{\sqrt[3]{27y^{12}y^2}}{\sqrt[3]{8x^3}}$ y^{12} is the largest perfect-cube factor of y^{14}.

$= \dfrac{\sqrt[3]{27y^{12}} \, \sqrt[3]{y^2}}{\sqrt[3]{8x^3}}$

$= \dfrac{3y^4\sqrt[3]{y^2}}{2x}$ Simplifying the numerator and the denominator

If we read from right to left, the quotient rule tells us that to divide two radical expressions that have the same index, we can divide the radicands.

EXAMPLE 3 Divide and, if possible, simplify.

a) $\dfrac{\sqrt{80}}{\sqrt{5}}$ **b)** $\dfrac{5\sqrt[3]{32}}{\sqrt[3]{2}}$

c) $\dfrac{\sqrt{72xy}}{2\sqrt{2}}$ **d)** $\dfrac{\sqrt[4]{18a^9b^5}}{\sqrt[4]{3b}}$

Student Notes

When writing radical signs, pay careful attention to what is included as the radicand. Each of the following represents a *different* number:

$$\sqrt{\dfrac{5 \cdot 2}{3}}, \quad \dfrac{\sqrt{5 \cdot 2}}{3}, \quad \dfrac{\sqrt{5} \cdot 2}{3}.$$

Solution

a) $\dfrac{\sqrt{80}}{\sqrt{5}} = \sqrt{\dfrac{80}{5}} = \sqrt{16} = 4$

> Because the indices match, we can divide the radicands.

b) $\dfrac{5\sqrt[3]{32}}{\sqrt[3]{2}} = 5\sqrt[3]{\dfrac{32}{2}} = 5\sqrt[3]{16}$

$= 5\sqrt[3]{8 \cdot 2}$ 8 is the largest perfect-cube factor of 16.

$= 5\sqrt[3]{8} \, \sqrt[3]{2} = 5 \cdot 2\sqrt[3]{2}$

$= 10\sqrt[3]{2}$

c) $\dfrac{\sqrt{72xy}}{2\sqrt{2}} = \dfrac{1}{2}\sqrt{\dfrac{72xy}{2}}$

> Because the indices match, we can divide the radicands.

$= \dfrac{1}{2}\sqrt{36xy} = \dfrac{1}{2} \cdot 6\sqrt{xy}$

$= 3\sqrt{xy}$

d) $\dfrac{\sqrt[4]{18a^9b^5}}{\sqrt[4]{3b}} = \sqrt[4]{\dfrac{18a^9b^5}{3b}}$

$= \sqrt[4]{6a^9b^4} = \sqrt[4]{a^8b^4}\,\sqrt[4]{6a}$

Note that 8 is the largest power less than 9 that is a multiple of the index 4.

$= a^2b\sqrt[4]{6a}$

Partial check: $(a^2b)^4 = a^8b^4$

Rationalizing Denominators and Numerators (Part 1)*

When a radical expression appears in a denominator, it can be useful to find an equivalent expression in which the denominator no longer contains a radical.† The procedure for finding such an expression is called **rationalizing the denominator**. We carry this out by multiplying by 1 in either of two ways.

One way is to multiply by 1 *under* the radical to make the denominator of the radicand a perfect power.

EXAMPLE 4   Rationalize each denominator.

a) $\sqrt{\dfrac{7}{3}}$

b) $\sqrt[3]{\dfrac{5}{16}}$

Solution

a) We multiply by 1 under the radical, using $\frac{3}{3}$. We do this so that the denominator of the radicand will be a perfect square:

$\sqrt{\dfrac{7}{3}} = \sqrt{\dfrac{7}{3} \cdot \dfrac{3}{3}}$ Multiplying by 1 under the radical

$= \sqrt{\dfrac{21}{9}}$ The denominator, 9, is now a perfect square.

$= \dfrac{\sqrt{21}}{\sqrt{9}}$ Using the quotient rule for radicals

$= \dfrac{\sqrt{21}}{3}.$

---

*Denominators and numerators with two terms are rationalized in Section 10.5.
†See Exercise 73 on p. 664.

b) Note that $16 = 4^2$. Thus, to make the denominator a perfect cube, we multiply under the radical by $\frac{4}{4}$:

$$\sqrt[3]{\frac{5}{16}} = \sqrt[3]{\frac{5}{4 \cdot 4} \cdot \frac{4}{4}} \qquad \text{Since the index is 3, we need 3 identical factors in the denominator.}$$

$$= \sqrt[3]{\frac{20}{4^3}} \qquad \text{The denominator is now a perfect cube.}$$

$$= \frac{\sqrt[3]{20}}{\sqrt[3]{4^3}}$$

$$= \frac{\sqrt[3]{20}}{4}.$$

Another way to rationalize a denominator is to multiply by 1 *outside* the radical.

EXAMPLE 5   Rationalize each denominator.

a) $\sqrt{\dfrac{4}{5b}}$

b) $\dfrac{\sqrt[3]{a}}{\sqrt[3]{9x}}$

c) $\dfrac{3x}{\sqrt[5]{2x^2y^3}}$

Solution

a) We rewrite the expression as a quotient of two radicals. Then we simplify and multiply by 1:

$$\sqrt{\frac{4}{5b}} = \frac{\sqrt{4}}{\sqrt{5b}} = \frac{2}{\sqrt{5b}} \qquad \text{We assume } b > 0.$$

$$= \frac{2}{\sqrt{5b}} \cdot \frac{\sqrt{5b}}{\sqrt{5b}} \qquad \text{Multiplying by 1}$$

$$= \frac{2\sqrt{5b}}{(\sqrt{5b})^2} \qquad \text{Try to do this step mentally.}$$

$$= \frac{2\sqrt{5b}}{5b}.$$

b) To rationalize the denominator $\sqrt[3]{9x}$, note that $9x$ is $3 \cdot 3 \cdot x$. In order for this radicand to be a cube, we need another factor of 3 and two more factors of x. Thus we multiply by 1, using $\sqrt[3]{3x^2}/\sqrt[3]{3x^2}$:

$$\frac{\sqrt[3]{a}}{\sqrt[3]{9x}} = \frac{\sqrt[3]{a}}{\sqrt[3]{9x}} \cdot \frac{\sqrt[3]{3x^2}}{\sqrt[3]{3x^2}} \qquad \text{Multiplying by 1}$$

$$= \frac{\sqrt[3]{3ax^2}}{\sqrt[3]{27x^3}} \longleftarrow \text{This radicand is now a perfect cube.}$$

$$= \frac{\sqrt[3]{3ax^2}}{3x}.$$

c) To change the radicand $2x^2y^3$ into a perfect fifth power, we need four more factors of 2, three more factors of x, and two more factors of y. Thus we multiply by 1, using $\sqrt[5]{2^4x^3y^2}/\sqrt[5]{2^4x^3y^2}$, or $\sqrt[5]{16x^3y^2}/\sqrt[5]{16x^3y^2}$:

$$\frac{3x}{\sqrt[5]{2x^2y^3}} = \frac{3x}{\sqrt[5]{2x^2y^3}} \cdot \frac{\sqrt[5]{16x^3y^2}}{\sqrt[5]{16x^3y^2}} \qquad \text{Multiplying by 1}$$

$$= \frac{3x\sqrt[5]{16x^3y^2}}{\sqrt[5]{32x^5y^5}} \qquad \text{This radicand is now a perfect fifth power.}$$

$$= \frac{3x\sqrt[5]{16x^3y^2}}{2xy} = \frac{3\sqrt[5]{16x^3y^2}}{2y}. \qquad \text{Always simplify if possible.}$$

Sometimes in calculus it is necessary to rationalize a numerator. To do so, we multiply by 1 to make the radicand in the *numerator* a perfect power.

EXAMPLE 6   Rationalize each numerator: **(a)** $\sqrt{\dfrac{7}{5}}$; **(b)** $\dfrac{\sqrt[3]{4a^2}}{\sqrt[3]{5b}}$.

Solution

a) $\sqrt{\dfrac{7}{5}} = \sqrt{\dfrac{7}{5} \cdot \dfrac{7}{7}}$ Multiplying by 1 under the radical. We also could have multiplied by $\sqrt{7}/\sqrt{7}$ outside the radical.

$= \sqrt{\dfrac{49}{35}}$ The numerator is now a perfect square.

$= \dfrac{\sqrt{49}}{\sqrt{35}}$ Using the quotient rule for radicals

$= \dfrac{7}{\sqrt{35}}$

b) $\dfrac{\sqrt[3]{4a^2}}{\sqrt[3]{5b}} = \dfrac{\sqrt[3]{4a^2}}{\sqrt[3]{5b}} \cdot \dfrac{\sqrt[3]{2a}}{\sqrt[3]{2a}}$ Multiplying by 1

$= \dfrac{\sqrt[3]{8a^3}}{\sqrt[3]{10ba}}$ This radicand is now a perfect cube.

$= \dfrac{2a}{\sqrt[3]{10ab}}$

In Section 10.5, we will discuss rationalizing denominators and numerators in which two terms appear.

Exercise Set

10.4

FOR EXTRA HELP

Student's Solutions Manual

Digital Video Tutor CD 5 Videotape 10

AW Math Tutor Center

MathXL Tutorials on CD

Math XL
MathXL

MyMathLab
MyMathLab

↪ *Concept Reinforcement   In each of Exercises 1–8, match the expression with an equivalent expression from the column on the right. Assume a, b > 0.*

1. ___ $\sqrt[3]{\dfrac{a^2}{b^6}}$

2. ___ $\dfrac{\sqrt[3]{a^6}}{\sqrt[3]{b^9}}$

3. ___ $\sqrt[5]{\dfrac{a^6}{b^4}}$

4. ___ $\sqrt{\dfrac{a}{b^3}}$

5. ___ $\dfrac{\sqrt[5]{a^2}}{\sqrt[5]{b^2}}$

6. ___ $\dfrac{\sqrt{5a^4}}{\sqrt{5a^3}}$

7. ___ $\dfrac{\sqrt[5]{a^2}}{\sqrt[5]{b^3}}$

8. ___ $\sqrt[4]{\dfrac{16a^6}{a^2}}$

a) $\dfrac{\sqrt[5]{a^2}\sqrt[5]{b^2}}{\sqrt[5]{b^5}}$

b) $\dfrac{a^2}{b^3}$

c) $\sqrt{\dfrac{a \cdot b}{b^3 \cdot b}}$

d) $\sqrt{a}$

e) $\dfrac{\sqrt[3]{a^2}}{b^2}$

f) $\sqrt[5]{\dfrac{a^6 b}{b^4 \cdot b}}$

g) $2a$

h) $\dfrac{\sqrt[5]{a^2 b^3}}{\sqrt[5]{b^5}}$

Divide and, if possible, simplify. Assume all variables represent positive numbers.

27. $\dfrac{\sqrt{35x}}{\sqrt{7x}}$

28. $\dfrac{\sqrt{28y}}{\sqrt{4y}}$

29. $\dfrac{\sqrt[3]{270}}{\sqrt[3]{10}}$

30. $\dfrac{\sqrt[3]{40}}{\sqrt[3]{5}}$

31. $\dfrac{\sqrt{40xy^3}}{\sqrt{8x}}$

32. $\dfrac{\sqrt{56ab^3}}{\sqrt{7a}}$

33. $\dfrac{\sqrt[3]{96a^4b^2}}{\sqrt[3]{12a^2b}}$

34. $\dfrac{\sqrt[3]{189x^5y^7}}{\sqrt[3]{7x^2y^2}}$

35. $\dfrac{\sqrt{100ab}}{5\sqrt{2}}$

36. $\dfrac{\sqrt{75ab}}{3\sqrt{3}}$

37. $\dfrac{\sqrt[4]{48x^9y^{13}}}{\sqrt[4]{3xy^{-2}}}$

38. $\dfrac{\sqrt[5]{64a^{11}b^{28}}}{\sqrt[5]{2ab^{-2}}}$

39. $\dfrac{\sqrt[3]{x^3 - y^3}}{\sqrt[3]{x - y}}$ ◄―

40. $\dfrac{\sqrt[3]{r^3 + s^3}}{\sqrt[3]{r + s}}$

Hint: Factor and then simplify.

Simplify by taking the roots of the numerator and the denominator. Assume all variables represent positive numbers.

9. $\sqrt{\dfrac{36}{25}}$

10. $\sqrt{\dfrac{100}{81}}$

11. $\sqrt[3]{\dfrac{64}{27}}$

12. $\sqrt[3]{\dfrac{343}{1000}}$

13. $\sqrt{\dfrac{49}{y^2}}$

14. $\sqrt{\dfrac{121}{x^2}}$

15. $\sqrt{\dfrac{36y^3}{x^4}}$

16. $\sqrt{\dfrac{25a^5}{b^6}}$

17. $\sqrt[3]{\dfrac{27a^4}{8b^3}}$

18. $\sqrt[3]{\dfrac{64x^7}{216y^6}}$

19. $\sqrt[4]{\dfrac{16a^4}{b^4c^8}}$

20. $\sqrt[4]{\dfrac{81x^4}{y^8z^4}}$

21. $\sqrt[4]{\dfrac{a^5b^8}{c^{10}}}$

22. $\sqrt[4]{\dfrac{x^9y^{12}}{z^6}}$

23. $\sqrt[5]{\dfrac{32x^6}{y^{11}}}$

24. $\sqrt[5]{\dfrac{243a^9}{b^{13}}}$

25. $\sqrt[6]{\dfrac{x^6y^8}{z^{15}}}$

26. $\sqrt[6]{\dfrac{a^9b^{12}}{c^{13}}}$

Rationalize each $\boxed{denominator}$ *Assume all variables represent positive numbers.*

41. $\sqrt{\dfrac{3}{2}}$

42. $\sqrt{\dfrac{6}{7}}$

43. $\dfrac{2\sqrt{5}}{7\sqrt{3}}$

44. $\dfrac{3\sqrt{5}}{2\sqrt{2}}$

45. $\sqrt[3]{\dfrac{16}{9}}$

46. $\sqrt[3]{\dfrac{2}{9}}$

47. $\dfrac{\sqrt[3]{3a}}{\sqrt[3]{5c}}$

48. $\dfrac{\sqrt[3]{7x}}{\sqrt[3]{3y}}$

49. $\dfrac{\sqrt[3]{5y^4}}{\sqrt[3]{6x^4}}$

50. $\dfrac{\sqrt[3]{3a^4}}{\sqrt[3]{7b^2}}$

51. $\sqrt[3]{\dfrac{2}{x^2y}}$

52. $\sqrt[3]{\dfrac{5}{ab^2}}$

53. $\sqrt{\dfrac{7a}{18}}$

54. $\sqrt{\dfrac{3x}{10}}$

55. $\sqrt{\dfrac{9}{20x^2y}}$

56. $\sqrt{\dfrac{7}{32a^2b}}$

Aha! 57. $\sqrt{\dfrac{10ab^2}{72a^3b}}$

58. $\sqrt{\dfrac{21x^2y}{75xy^5}}$

Rationalize each numerator. Assume all variables represent positive numbers.

59. $\dfrac{\sqrt{5}}{\sqrt{7x}}$

60. $\dfrac{\sqrt{10}}{\sqrt{3x}}$

61. $\sqrt{\dfrac{14}{21}}$

62. $\sqrt{\dfrac{12}{15}}$

63. $\dfrac{4\sqrt{13}}{3\sqrt{7}}$

64. $\dfrac{5\sqrt{21}}{2\sqrt{5}}$

65. $\dfrac{\sqrt[3]{7}}{\sqrt[3]{2}}$

66. $\dfrac{\sqrt[3]{5}}{\sqrt[3]{4}}$

67. $\sqrt{\dfrac{7x}{3y}}$

68. $\sqrt{\dfrac{7a}{6b}}$

69. $\sqrt[3]{\dfrac{2a^5}{5b}}$

70. $\sqrt[3]{\dfrac{2a^4}{7b}}$

71. $\sqrt{\dfrac{x^3y}{2}}$

72. $\sqrt{\dfrac{ab^5}{3}}$

73. Explain why it is easier to approximate

$$\dfrac{\sqrt{2}}{2} \quad \text{than} \quad \dfrac{1}{\sqrt{2}}$$

if no calculator is available and $\sqrt{2} \approx 1.414213562$.

74. A student *incorrectly* claims that

$$\dfrac{5 + \sqrt{2}}{\sqrt{18}} = \dfrac{5 + \sqrt{1}}{\sqrt{9}} = \dfrac{5 + 1}{3}.$$

How could you convince the student that a mistake has been made? How would you explain the correct way of rationalizing the denominator?

SKILL MAINTENANCE

Multiply. [6.2]

75. $\dfrac{3}{x - 5} \cdot \dfrac{x - 1}{x + 5}$

76. $\dfrac{7}{x + 4} \cdot \dfrac{x - 2}{x - 4}$

Simplify.

77. $\dfrac{a^2 - 8a + 7}{a^2 - 49}$ [6.1]

78. $\dfrac{t^2 + 9t - 22}{t^2 - 4}$ [6.1]

79. $(5a^3b^4)^3$ [4.1]

80. $(3x^4)^2(5xy^3)^2$ [4.1]

SYNTHESIS

81. Is the quotient of two irrational numbers always an irrational number? Why or why not?

82. Is it possible to understand how to rationalize a denominator without knowing how to multiply rational expressions? Why or why not?

83. *Pendulums.* The *period* of a pendulum is the time it takes to complete one cycle, swinging to and fro. For a pendulum that is L centimeters long, the period T is given by the formula

$$T = 2\pi \sqrt{\dfrac{L}{980}},$$

where T is in seconds. Find, to the nearest hundredth of a second, the period of a pendulum of length **(a)** 65 cm; **(b)** 98 cm; **(c)** 120 cm. Use a calculator's $\boxed{\pi}$ key if possible.

Perform the indicated operations.

84. $\dfrac{7\sqrt{a^2b}\,\sqrt{25xy}}{5\sqrt{a^{-4}b^{-1}}\,\sqrt{49x^{-1}y^{-3}}}$

85. $\dfrac{(\sqrt[3]{81mn^2})^2}{(\sqrt[3]{mn})^2}$

86. $\dfrac{\sqrt{44x^2y^9z}\,\sqrt{22y^9z^6}}{(\sqrt{11xy^8z^2})^2}$

87. $\sqrt{a^2 - 3} - \dfrac{a^2}{\sqrt{a^2 - 3}}$

88. $5\sqrt{\dfrac{x}{y}} + 4\sqrt{\dfrac{y}{x}} - \dfrac{3}{\sqrt{xy}}$

89. Provide a reason for each step in the following derivation of the quotient rule:

$$\sqrt[n]{\dfrac{a}{b}} = \left(\dfrac{a}{b}\right)^{1/n} \quad \underline{\hspace{3cm}}$$

$$= \dfrac{a^{1/n}}{b^{1/n}} \quad \underline{\hspace{3cm}}$$

$$= \dfrac{\sqrt[n]{a}}{\sqrt[n]{b}} \quad \underline{\hspace{3cm}}$$

90. Show that $\dfrac{\sqrt[n]{a}}{\sqrt[n]{b}}$ is the nth root of $\dfrac{a}{b}$ by raising it to the nth power and simplifying.

91. Let $f(x) = \sqrt{18x^3}$ and $g(x) = \sqrt{2x}$. Find $(f/g)(x)$ and specify the domain of f/g.

92. Let $f(t) = \sqrt{2t}$ and $g(t) = \sqrt{50t^3}$. Find $(f/g)(t)$ and specify the domain of f/g.

93. Let $f(x) = \sqrt{x^2 - 9}$ and $g(x) = \sqrt{x - 3}$. Find $(f/g)(x)$ and specify the domain of f/g.

10.5 Expressions Containing Several Radical Terms

Adding and Subtracting Radical Expressions   •   Products and Quotients of Two or More Radical Terms   •   Rationalizing Denominators and Numerators (Part 2)   •   Terms with Differing Indices

Radical expressions like $6\sqrt{7} + 4\sqrt{7}$ or $\left(\sqrt{a} + \sqrt{b}\right)\left(\sqrt{a} - \sqrt{b}\right)$ contain more than one *radical term* and can sometimes be simplified.

Adding and Subtracting Radical Expressions

When two radical expressions have the same indices and radicands, they are said to be **like radicals**. Like radicals can be combined (added or subtracted) in much the same way that we combined like terms earlier in this text.

EXAMPLE 1   Simplify by combining like radical terms.

a) $6\sqrt{7} + 4\sqrt{7}$

b) $\sqrt[3]{2} - 7x\sqrt[3]{2} + 5\sqrt[3]{2}$

c) $6\sqrt[5]{4x} + 3\sqrt[5]{4x} - \sqrt[3]{4x}$

Solution

a) $6\sqrt{7} + 4\sqrt{7} = (6 + 4)\sqrt{7}$ Using the distributive law (factoring out $\sqrt{7}$)

$\phantom{6\sqrt{7} + 4\sqrt{7}} = 10\sqrt{7}$ You can think: 6 square roots of 7 plus 4 square roots of 7 results in 10 square roots of 7.

b) $\sqrt[3]{2} - 7x\sqrt[3]{2} + 5\sqrt[3]{2} = (1 - 7x + 5)\sqrt[3]{2}$ Factoring out $\sqrt[3]{2}$

$\phantom{\sqrt[3]{2} - 7x\sqrt[3]{2} + 5\sqrt[3]{2}} = (6 - 7x)\sqrt[3]{2}$ These parentheses are important!

c) $6\sqrt[5]{4x} + 3\sqrt[5]{4x} - \sqrt[3]{4x} = (6 + 3)\sqrt[5]{4x} - \sqrt[3]{4x}$ Try to do this step mentally.

$\phantom{6\sqrt[5]{4x} + 3\sqrt[5]{4x} - \sqrt[3]{4x}} = 9\sqrt[5]{4x} - \sqrt[3]{4x}$ Because the indices differ, we are done.

Our ability to simplify radical expressions can help us to find like radicals even when, at first, it may appear that none exists.

EXAMPLE 2 Simplify by combining like radical terms, if possible.

a) $3\sqrt{8} - 5\sqrt{2}$

b) $9\sqrt{5} - 4\sqrt{3}$

c) $\sqrt[3]{2x^6y^4} + 7\sqrt[3]{2y}$

Solution

a) $3\sqrt{8} - 5\sqrt{2} = 3\sqrt{4 \cdot 2} - 5\sqrt{2}$
$\left.\begin{array}{l} = 3\sqrt{4} \cdot \sqrt{2} - 5\sqrt{2} \\ = 3 \cdot 2 \cdot \sqrt{2} - 5\sqrt{2} \end{array}\right\}$ Simplifying $\sqrt{8}$
$= 6\sqrt{2} - 5\sqrt{2}$
$= \sqrt{2}$ Combining like radicals

b) $9\sqrt{5} - 4\sqrt{3}$ cannot be simplified.

c) $\sqrt[3]{2x^6y^4} + 7\sqrt[3]{2y} = \sqrt[3]{x^6y^3 \cdot 2y} + 7\sqrt[3]{2y}$
$\left.\begin{array}{l} = \sqrt[3]{x^6y^3} \cdot \sqrt[3]{2y} + 7\sqrt[3]{2y} \\ = x^2y \cdot \sqrt[3]{2y} + 7\sqrt[3]{2y} \end{array}\right\}$ Simplifying $\sqrt[3]{2x^6y^4}$
$= (x^2y + 7)\sqrt[3]{2y}$ Factoring to combine like radical terms

Products and Quotients of Two or More Radical Terms

Radical expressions often contain factors that have more than one term. Multiplying such expressions is similar to finding products of polynomials. Some products will yield like radical terms, which we can now combine.

EXAMPLE 3 Multiply.

a) $\sqrt{3}(x - \sqrt{5})$ **b)** $\sqrt[3]{y}(\sqrt[3]{y^2} + \sqrt[3]{2})$

c) $(4\sqrt{3} + \sqrt{2})(\sqrt{3} - 5\sqrt{2})$ **d)** $(\sqrt{a} + \sqrt{b})(\sqrt{a} - \sqrt{b})$

Solution

a) $\sqrt{3}(x - \sqrt{5}) = \sqrt{3} \cdot x - \sqrt{3} \cdot \sqrt{5}$ Using the distributive law
$= x\sqrt{3} - \sqrt{15}$ Multiplying radicals

b) $\sqrt[3]{y}(\sqrt[3]{y^2} + \sqrt[3]{2}) = \sqrt[3]{y} \cdot \sqrt[3]{y^2} + \sqrt[3]{y} \cdot \sqrt[3]{2}$ Using the distributive law
$= \sqrt[3]{y^3} + \sqrt[3]{2y}$ Multiplying radicals
$= y + \sqrt[3]{2y}$ Simplifying $\sqrt[3]{y^3}$

$$\overset{\text{F} \qquad\qquad \text{O} \qquad\quad \text{I} \qquad\quad \text{L}}{}$$

c) $(4\sqrt{3} + \sqrt{2})(\sqrt{3} - 5\sqrt{2}) = 4(\sqrt{3})^2 - 20\sqrt{3} \cdot \sqrt{2} + \sqrt{2} \cdot \sqrt{3} - 5(\sqrt{2})^2$
$= 4 \cdot 3 - 20\sqrt{6} + \sqrt{6} - 5 \cdot 2$ Multiplying radicals
$= 12 - 20\sqrt{6} + \sqrt{6} - 10$
$= 2 - 19\sqrt{6}$ Combining like terms

d) $(\sqrt{a} + \sqrt{b})(\sqrt{a} - \sqrt{b}) = (\sqrt{a})^2 - \sqrt{a}\sqrt{b} + \sqrt{a}\sqrt{b} - (\sqrt{b})^2$ Using FOIL

$$= a - b$$ Combining like terms

In Example 3(d) above, you may have noticed that since the outer and inner products in FOIL are opposites, the result, $a - b$, is not itself a radical expression. Pairs of radical terms, like $\sqrt{a} + \sqrt{b}$ and $\sqrt{a} - \sqrt{b}$, are called **conjugates**. In general, the conjugate of $a\sqrt{b} + c\sqrt{d}$ is $a\sqrt{b} - c\sqrt{d}$. The product of conjugates contains no radical expressions.

Rationalizing Denominators and Numerators (Part 2)

The use of conjugates allows us to rationalize denominators or numerators with two terms.

EXAMPLE 4   Rationalize each denominator: **(a)** $\dfrac{4}{\sqrt{3} + x}$; **(b)** $\dfrac{4 + \sqrt{2}}{\sqrt{5} - \sqrt{2}}$.

Solution

a) $\dfrac{4}{\sqrt{3} + x} = \dfrac{4}{\sqrt{3} + x} \cdot \dfrac{\sqrt{3} - x}{\sqrt{3} - x}$ Multiplying by 1, using the conjugate of $\sqrt{3} + x$, which is $\sqrt{3} - x$

$$= \dfrac{4(\sqrt{3} - x)}{(\sqrt{3} + x)(\sqrt{3} - x)}$$ Multiplying numerators and denominators

$$= \dfrac{4(\sqrt{3} - x)}{(\sqrt{3})^2 - x^2}$$ Using FOIL in the denominator

$$= \dfrac{4\sqrt{3} - 4x}{3 - x^2}$$ Simplifying. No radicals remain in the denominator.

b) $\dfrac{4 + \sqrt{2}}{\sqrt{5} - \sqrt{2}} = \dfrac{4 + \sqrt{2}}{\sqrt{5} - \sqrt{2}} \cdot \dfrac{\sqrt{5} + \sqrt{2}}{\sqrt{5} + \sqrt{2}}$ Multiplying by 1, using the conjugate of $\sqrt{5} - \sqrt{2}$, which is $\sqrt{5} + \sqrt{2}$

$$= \dfrac{(4 + \sqrt{2})(\sqrt{5} + \sqrt{2})}{(\sqrt{5} - \sqrt{2})(\sqrt{5} + \sqrt{2})}$$ Multiplying numerators and denominators

$$= \dfrac{4\sqrt{5} + 4\sqrt{2} + \sqrt{2}\sqrt{5} + (\sqrt{2})^2}{(\sqrt{5})^2 - (\sqrt{2})^2}$$ Using FOIL

$$= \dfrac{4\sqrt{5} + 4\sqrt{2} + \sqrt{10} + 2}{5 - 2}$$ Squaring in the denominator and the numerator

$$= \dfrac{4\sqrt{5} + 4\sqrt{2} + \sqrt{10} + 2}{3}$$ No radicals remain in the denominator.

To rationalize a numerator with more than one term, we use the conjugate of the numerator.

EXAMPLE 5 Rationalize the numerator: $\dfrac{4 + \sqrt{2}}{\sqrt{5} - \sqrt{2}}$.

Solution

$$\dfrac{4 + \sqrt{2}}{\sqrt{5} - \sqrt{2}} = \dfrac{4 + \sqrt{2}}{\sqrt{5} - \sqrt{2}} \cdot \dfrac{4 - \sqrt{2}}{4 - \sqrt{2}}$$ Multiplying by 1, using the conjugate of $4 + \sqrt{2}$, which is $4 - \sqrt{2}$

$$= \dfrac{16 - \left(\sqrt{2}\right)^2}{4\sqrt{5} - \sqrt{5}\sqrt{2} - 4\sqrt{2} + \left(\sqrt{2}\right)^2}$$

$$= \dfrac{14}{4\sqrt{5} - \sqrt{10} - 4\sqrt{2} + 2}$$

Terms with Differing Indices

To multiply or divide radical terms with different indices, we can convert to exponential notation, use the rules for exponents, and then convert back to radical notation.

EXAMPLE 6 Divide and, if possible, simplify: $\dfrac{\sqrt[4]{(x + y)^3}}{\sqrt{x + y}}$.

Student Notes

Expressions similar to the one in Example 6 are most easily simplified by rewriting the expression using exponents in place of radicals. After simplifying, remember to write your final result in radical notation. In general, if a problem is presented in one form, it is expected that the final result be presented in the same form.

Solution

$$\dfrac{\sqrt[4]{(x + y)^3}}{\sqrt{x + y}} = \dfrac{(x + y)^{3/4}}{(x + y)^{1/2}}$$ Converting to exponential notation

$$= (x + y)^{3/4 - 1/2}$$ Since the bases are identical, we can subtract exponents: $\frac{3}{4} - \frac{1}{2} = \frac{3}{4} - \frac{2}{4} = \frac{1}{4}$.

$$\left.\begin{array}{l} = (x + y)^{1/4} \\ = \sqrt[4]{x + y} \end{array}\right\}$$ Converting back to radical notation

The steps used in Example 6 can be used in a variety of situations.

To Simplify Products or Quotients with Differing Indices

 1. Convert all radical expressions to exponential notation.
 2. When the bases are identical, subtract exponents to divide and add exponents to multiply. This may require finding a common denominator.
 3. Convert back to radical notation and, if possible, simplify.

EXAMPLE 7   Multiply and simplify: $\sqrt{x^3}\,\sqrt[3]{x}$.

Solution

$$\sqrt{x^3}\,\sqrt[3]{x} = x^{3/2}\cdot x^{1/3} \qquad \text{Converting to exponential notation}$$

$$= x^{11/6} \qquad \text{Adding exponents: } \tfrac{3}{2}+\tfrac{1}{3}=\tfrac{9}{6}+\tfrac{2}{6}$$

$$= \sqrt[6]{x^{11}} \qquad \text{Converting back to radical notation}$$

$$\left.\begin{array}{l} = \sqrt[6]{x^6}\,\sqrt[6]{x^5} \\[4pt] = x\sqrt[6]{x^5} \end{array}\right\} \quad \text{Simplifying}$$

EXAMPLE 8   If $f(x)=\sqrt[3]{x^2}$ and $g(x)=\sqrt{x}+\sqrt[4]{x}$, find $(f\cdot g)(x)$.

Solution   Recall from Section 7.4 that $(f\cdot g)(x)=f(x)\cdot g(x)$. Thus,

$$(f\cdot g)(x)=\sqrt[3]{x^2}\left(\sqrt{x}+\sqrt[4]{x}\right) \qquad \begin{array}{l}x \text{ is assumed to be}\\ \text{nonnegative.}\end{array}$$

$$= x^{2/3}(x^{1/2}+x^{1/4}) \qquad \begin{array}{l}\text{Converting to exponential}\\ \text{notation}\end{array}$$

$$= x^{2/3}\cdot x^{1/2}+x^{2/3}\cdot x^{1/4} \qquad \text{Using the distributive law}$$

$$= x^{2/3+1/2}+x^{2/3+1/4} \qquad \text{Adding exponents:}$$

$$= x^{7/6}+x^{11/12} \qquad \tfrac{2}{3}+\tfrac{1}{2}=\tfrac{4}{6}+\tfrac{3}{6};\tfrac{2}{3}+\tfrac{1}{4}=\tfrac{8}{12}+\tfrac{3}{12}$$

$$= \sqrt[6]{x^7}+\sqrt[12]{x^{11}} \qquad \begin{array}{l}\text{Converting back to radical}\\ \text{notation}\end{array}$$

$$\left.\begin{array}{l} = \sqrt[6]{x^6}\,\sqrt[6]{x}+\sqrt[12]{x^{11}} \\[4pt] = x\sqrt[6]{x}+\sqrt[12]{x^{11}} \end{array}\right\} \quad \text{Simplifying}$$

If factors are raised to powers that share a common denominator, we can write the final result as a single radical expression.

EXAMPLE 9   Divide and, if possible, simplify: $\dfrac{\sqrt[3]{a^2b^4}}{\sqrt{ab}}$.

Solution

$$\frac{\sqrt[3]{a^2b^4}}{\sqrt{ab}}=\frac{(a^2b^4)^{1/3}}{(ab)^{1/2}} \qquad \text{Converting to exponential notation}$$

$$=\frac{a^{2/3}b^{4/3}}{a^{1/2}b^{1/2}} \qquad \text{Using the product and power rules}$$

$$=a^{2/3-1/2}b^{4/3-1/2} \qquad \text{Subtracting exponents}$$

$$=a^{1/6}b^{5/6}$$

$$=\sqrt[6]{a}\,\sqrt[6]{b^5} \qquad \text{Converting to radical notation}$$

$$=\sqrt[6]{ab^5} \qquad \text{Using the product rule for radicals}$$

Exercise Set

10.5

FOR EXTRA HELP

 Student's Solutions Manual  Digital Video Tutor CD 5 Videotape 10  AW Math Tutor Center MathXL Tutorials on CD MathXL MyMathLab

🌱 **Concept Reinforcement** *For each of Exercises 1–6, fill in the blanks by selecting from the following words (which may be used more than once):*

radicand(s), indices, conjugate(s), base(s), denominator(s), numerator(s).

1. To add radical expressions, the _____ and the _____ must be the same.

2. To multiply radical expressions, the _____ must be the same.

3. To find a product by adding exponents, the _____ must be the same.

4. To add rational expressions, the _____ must be the same.

5. To rationalize the _____ of $\dfrac{\sqrt{a + 0.1} - \sqrt{a}}{0.1}$, we multiply by a form of 1, using the _____ of $\sqrt{a + 0.1} - \sqrt{a}$, or $\sqrt{a + 0.1} + \sqrt{a}$, to write 1.

6. To find a quotient by subtracting exponents, the _____ must be the same.

Add or subtract. Simplify by combining like radical terms, if possible. Assume that all variables and radicands represent positive real numbers.

7. $2\sqrt{5} + 7\sqrt{5}$

8. $4\sqrt{7} + 2\sqrt{7}$

9. $7\sqrt[3]{4} - 5\sqrt[3]{4}$

10. $14\sqrt[5]{2} - 6\sqrt[5]{2}$

11. $\sqrt[3]{y} + 9\sqrt[3]{y}$

12. $9\sqrt[4]{t} - 3\sqrt[4]{t}$

13. $8\sqrt{2} - 6\sqrt{2} + 5\sqrt{2}$

14. $\sqrt{6} + 8\sqrt{6} - 3\sqrt{6}$

15. $9\sqrt[3]{7} - \sqrt{3} + 4\sqrt[3]{7} + 2\sqrt{3}$

16. $5\sqrt{7} - 8\sqrt[4]{11} + \sqrt{7} + 9\sqrt[4]{11}$

17. $4\sqrt{27} - 3\sqrt{3}$

18. $9\sqrt{50} - 4\sqrt{2}$

19. $3\sqrt{45} + 7\sqrt{20}$

20. $5\sqrt{12} + 16\sqrt{27}$

21. $3\sqrt[3]{16} + \sqrt[3]{54}$

22. $\sqrt[3]{27} - 5\sqrt[3]{8}$

23. $\sqrt{5a} + 2\sqrt{45a^3}$

24. $4\sqrt{3x^3} - \sqrt{12x}$

25. $\sqrt[3]{6x^4} + \sqrt[3]{48x}$

26. $\sqrt[3]{54x} - \sqrt[3]{2x^4}$

27. $\sqrt{4a - 4} + \sqrt{a - 1}$

28. $\sqrt{9y + 27} + \sqrt{y + 3}$

29. $\sqrt{x^3 - x^2} + \sqrt{9x - 9}$

30. $\sqrt{4x - 4} - \sqrt{x^3 - x^2}$

Multiply. Assume all variables represent nonnegative real numbers.

31. $\sqrt{3}(4 + \sqrt{3})$

32. $\sqrt{7}(3 - \sqrt{7})$

33. $3\sqrt{5}(\sqrt{5} - \sqrt{2})$

34. $4\sqrt{2}(\sqrt{3} - \sqrt{5})$

35. $\sqrt{2}(3\sqrt{10} - 2\sqrt{2})$

36. $\sqrt{3}(2\sqrt{5} - 3\sqrt{4})$

37. $\sqrt[3]{3}(\sqrt[3]{9} - 4\sqrt[3]{21})$

38. $\sqrt[3]{2}(\sqrt[3]{4} - 2\sqrt[3]{32})$

39. $\sqrt[3]{a}(\sqrt[3]{a^2} + \sqrt[3]{24a^2})$

40. $\sqrt[3]{x}(\sqrt[3]{3x^2} - \sqrt[3]{81x^2})$

41. $(2 + \sqrt{6})(5 - \sqrt{6})$

42. $(4 - \sqrt{5})(2 + \sqrt{5})$

43. $(\sqrt{2} + \sqrt{7})(\sqrt{3} - \sqrt{7})$

44. $(\sqrt{7} - \sqrt{2})(\sqrt{5} + \sqrt{2})$

45. $(3 - \sqrt{5})(3 + \sqrt{5})$

46. $(6 - \sqrt{7})(6 + \sqrt{7})$

47. $(\sqrt{6} + \sqrt{8})(\sqrt{6} - \sqrt{8})$

48. $(\sqrt{5} + \sqrt{3})(\sqrt{5} - \sqrt{3})$

49. $\left(3\sqrt{7} + 2\sqrt{5}\right)\left(2\sqrt{7} - 4\sqrt{5}\right)$

50. $\left(4\sqrt{5} - 3\sqrt{2}\right)\left(2\sqrt{5} + 4\sqrt{2}\right)$

51. $\left(2 + \sqrt{3}\right)^2$

52. $\left(3 + \sqrt{7}\right)^2$

53. $\left(\sqrt{3} - \sqrt{2}\right)^2$

54. $\left(\sqrt{5} - \sqrt{3}\right)^2$

55. $\left(\sqrt{2t} + \sqrt{5}\right)^2$

56. $\left(\sqrt{3x} - \sqrt{2}\right)^2$

57. $\left(3 - \sqrt{x + 5}\right)^2$

58. $\left(4 + \sqrt{x - 3}\right)^2$

59. $\left(2\sqrt[4]{7} - \sqrt[4]{6}\right)\left(3\sqrt[4]{9} + 2\sqrt[4]{5}\right)$

60. $\left(4\sqrt[3]{3} + \sqrt[3]{10}\right)\left(2\sqrt[3]{7} + 5\sqrt[3]{6}\right)$

Rationalize each denominator.

61. $\dfrac{5}{4 - \sqrt{3}}$

62. $\dfrac{3}{4 - \sqrt{7}}$

63. $\dfrac{2 + \sqrt{5}}{6 + \sqrt{3}}$

64. $\dfrac{1 + \sqrt{2}}{3 + \sqrt{5}}$

65. $\dfrac{\sqrt{a}}{\sqrt{a} + \sqrt{b}}$

66. $\dfrac{\sqrt{z}}{\sqrt{x} - \sqrt{z}}$

Aha! **67.** $\dfrac{\sqrt{7} - \sqrt{3}}{\sqrt{3} - \sqrt{7}}$

68. $\dfrac{\sqrt{7} + \sqrt{5}}{\sqrt{5} + \sqrt{2}}$

69. $\dfrac{3\sqrt{2} - \sqrt{7}}{4\sqrt{2} + 2\sqrt{5}}$

70. $\dfrac{5\sqrt{3} - \sqrt{11}}{2\sqrt{3} - 5\sqrt{2}}$

Rationalize each numerator. If possible, simplify your result.

71. $\dfrac{\sqrt{7} + 2}{5}$

72. $\dfrac{\sqrt{3} + 1}{4}$

73. $\dfrac{\sqrt{6} - 2}{\sqrt{3} + 7}$

74. $\dfrac{\sqrt{10} + 4}{\sqrt{2} - 3}$

75. $\dfrac{\sqrt{x} - \sqrt{y}}{\sqrt{x} + \sqrt{y}}$

76. $\dfrac{\sqrt{a} + \sqrt{b}}{\sqrt{a} - \sqrt{b}}$

77. $\dfrac{\sqrt{a + h} - \sqrt{a}}{h}$

78. $\dfrac{\sqrt{x - h} - \sqrt{x}}{h}$

Perform the indicated operation and simplify. Assume all variables represent positive real numbers.

79. $\sqrt{a}\sqrt[4]{a^3}$

80. $\sqrt[3]{x^2}\sqrt[6]{x^5}$

81. $\sqrt[5]{b^2}\sqrt{b^3}$

82. $\sqrt[4]{a^3}\sqrt[3]{a^2}$

83. $\sqrt{xy^3}\sqrt[3]{x^2y}$

84. $\sqrt[5]{a^3b}\sqrt{ab}$

85. $\sqrt[4]{9ab^3}\sqrt{3a^4b}$

86. $\sqrt{2x^3y^3}\sqrt[3]{4xy^2}$

87. $\sqrt{a^4b^3c^4}\sqrt[3]{ab^2c}$

88. $\sqrt[3]{xy^2z}\sqrt{x^3yz^2}$

89. $\dfrac{\sqrt[3]{a^2}}{\sqrt[4]{a}}$

90. $\dfrac{\sqrt[3]{x^2}}{\sqrt[5]{x}}$

91. $\dfrac{\sqrt[4]{x^2y^3}}{\sqrt[3]{xy}}$

92. $\dfrac{\sqrt[5]{a^4b}}{\sqrt[3]{ab}}$

93. $\dfrac{\sqrt{ab^3}}{\sqrt[5]{a^2b^3}}$

94. $\dfrac{\sqrt[5]{x^3y^4}}{\sqrt{xy}}$

95. $\dfrac{\sqrt[4]{(3x - 1)^3}}{\sqrt[5]{(3x - 1)^3}}$

96. $\dfrac{\sqrt[3]{(2 + 5x)^2}}{\sqrt[4]{2 + 5x}}$

97. $\dfrac{\sqrt[3]{(2x + 1)^2}}{\sqrt[5]{(2x + 1)^2}}$

98. $\dfrac{\sqrt[4]{(5 + 3x)^3}}{\sqrt[3]{(5 + 3x)^2}}$

99. $\sqrt[3]{x^2y}\left(\sqrt{xy} - \sqrt[5]{xy^3}\right)$

100. $\sqrt[4]{a^2b}\left(\sqrt[3]{a^2b} - \sqrt[5]{a^2b^2}\right)$

101. $\left(m + \sqrt[3]{n^2}\right)\left(2m + \sqrt[4]{n}\right)$

102. $\left(r - \sqrt[4]{s^3}\right)\left(3r - \sqrt[5]{s}\right)$

In Exercises 103–106, $f(x)$ and $g(x)$ are as given. Find $(f \cdot g)(x)$. Assume all variables represent nonnegative real numbers.

103. $f(x) = \sqrt[4]{x},\ g(x) = \sqrt[4]{2x} - \sqrt[4]{x^{11}}$

104. $f(x) = \sqrt[4]{x^7} + \sqrt[4]{3x^2},\ g(x) = \sqrt[4]{x}$

105. $f(x) = x + \sqrt{7},\ g(x) = x - \sqrt{7}$

106. $f(x) = x - \sqrt{2},\ g(x) = x + \sqrt{6}$

Let $f(x) = x^2$. Find each of the following.

107. $f\left(5 + \sqrt{2}\right)$

108. $f\left(7 + \sqrt{3}\right)$

109. $f\left(\sqrt{3} - \sqrt{5}\right)$

110. $f\left(\sqrt{6} - \sqrt{3}\right)$

111. In what way(s) is combining like radical terms similar to combining like terms that are monomials?

112. Why do we need to know how to multiply radical expressions before learning how to add them?

SKILL MAINTENANCE

Solve.

113. $\dfrac{12x}{x - 4} - \dfrac{3x^2}{x + 4} = \dfrac{384}{x^2 - 16}$ [6.6]

114. $\dfrac{2}{3} + \dfrac{1}{t} = \dfrac{4}{5}$ [6.6]

115. $5x^2 - 6x + 1 = 0$ [5.7]

116. $7t^2 - 8t + 1 = 0$ [5.7]

117. The sum of a number and its square is 20. Find the number. [5.8]

118. The width of a rectangle is one-fourth the length. The area is twice the perimeter. Find the dimensions of the rectangle. [5.8]

SYNTHESIS

119. Ramon *incorrectly* writes
$$\sqrt[5]{x^2} \cdot \sqrt{x^3} = x^{2/5} \cdot x^{3/2} = \sqrt[5]{x^3}.$$
What mistake do you suspect he is making?

120. After examining the expression $\sqrt[4]{25xy^3}\,\sqrt{5x^4y}$, Dyan (correctly) concludes that x and y are both nonnegative. Explain how she could reach this conclusion.

Find a simplified form for $f(x)$. Assume $x \geq 0$.

121. $f(x) = \sqrt{20x^2 + 4x^3} - 3x\sqrt{45 + 9x} + \sqrt{5x^2 + x^3}$

122. $f(x) = \sqrt{x^3 - x^2} + \sqrt{9x^3 - 9x^2} - \sqrt{4x^3 - 4x^2}$

123. $f(x) = \sqrt[4]{x^5 - x^4} + 3\sqrt[4]{x^9 - x^8}$

124. $f(x) = \sqrt[4]{16x^4 + 16x^5} - 2\sqrt[4]{x^8 + x^9}$

Simplify.

125. $\frac{1}{2}\sqrt{36a^5bc^4} - \frac{1}{2}\sqrt[3]{64a^4bc^6} + \frac{1}{6}\sqrt{144a^3bc^6}$

126. $7x\sqrt{(x + y)^3} - 5xy\sqrt{x + y} - 2y\sqrt{(x + y)^3}$

127. $\sqrt{27a^5(b + 1)}\,\sqrt[3]{81a(b + 1)^4}$

128. $\sqrt{8x(y + z)^5}\,\sqrt[3]{4x^2(y + z)^2}$

129. $\dfrac{\dfrac{1}{\sqrt{w}} - \sqrt{w}}{\dfrac{\sqrt{w} + 1}{\sqrt{w}}}$

130. $\dfrac{1}{4 + \sqrt{3}} + \dfrac{1}{\sqrt{3}} + \dfrac{1}{\sqrt{3} - 4}$

Express each of the following as the product of two radical expressions.

131. $x - 5$

132. $y - 7$

133. $x - a$

Multiply.

134. $\sqrt{9 + 3\sqrt{5}}\,\sqrt{9 - 3\sqrt{5}}$

135. $\left(\sqrt{x + 2} - \sqrt{x - 2}\right)^2$

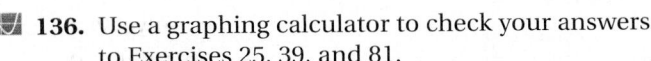 **136.** Use a graphing calculator to check your answers to Exercises 25, 39, and 81.

10.6 Solving Radical Equations

The Principle of Powers   •   Equations with Two Radical Terms

CONNECTING THE CONCEPTS

In Sections 10.1–10.5, we learned how to manipulate radical expressions as well as expressions containing rational exponents. We performed this work to find *equivalent expressions.*

Now that we know how to work with radicals and rational exponents, we can learn how to solve a new type of equation. As in our earlier work with

equations, finding *equivalent equations* will be part of our strategy. What is different, however, is that now we will use an equation-solving step that does not always produce equivalent equations. Checking solutions will therefore be more important than ever.

The Principle of Powers

A **radical equation** is an equation in which the variable appears in a radicand. Examples are

$$\sqrt[3]{2x} + 1 = 5, \quad \sqrt{a} + \sqrt{a - 2} = 7, \quad \text{and} \quad 4 - \sqrt{3x + 1} = \sqrt{6 - x}.$$

To solve such equations, we need a new principle. Suppose $a = b$ is true. If we square both sides, we get another true equation: $a^2 = b^2$. This can be generalized.

> **The Principle of Powers**
>
> If $a = b$, then $a^n = b^n$ for any exponent n.

Note that the principle of powers is an "if–then" statement. The statement obtained by interchanging the two parts of the sentence—"if $a^n = b^n$ for some exponent n, then $a = b$"—is *not always true.* For example, "if $x = 3$, then $x^2 = 9$" is true, but the statement "if $x^2 = 9$, then $x = 3$" is *not* true when x is replaced with -3. For this reason, when both sides of an equation are raised to an even exponent, it is essential to check the answer(s) in the *original* equation.

EXAMPLE 1   Solve: $\sqrt{x} - 3 = 4$.

Solution   Before using the principle of powers, we need to isolate the radical term:

$$\sqrt{x} - 3 = 4$$
$$\sqrt{x} = 7 \qquad \text{Isolating the radical by adding 3 to both sides}$$
$$(\sqrt{x})^2 = 7^2 \qquad \text{Using the principle of powers}$$
$$x = 49.$$

Check: $\dfrac{\sqrt{x} - 3 = 4}{\sqrt{49} - 3 \;\Big|\; 4}$
           $7 - 3 \;\Big|$
                     $4 \overset{?}{=} 4$ TRUE

The solution is 49.

EXAMPLE 2   Solve: $\sqrt{x} - 5 = -7$.

Solution

$$\sqrt{x} - 5 = -7$$
$$\sqrt{x} = -2 \qquad \text{Isolating the radical by adding 5 to both sides}$$

The equation $\sqrt{x} = -2$ has no solution because the principal square root of a number is never negative. We continue as in Example 1 for comparison.

$$\left(\sqrt{x}\right)^2 = (-2)^2 \qquad \text{Using the principle of powers}$$
$$x = 4$$

Check: $\dfrac{\sqrt{x} - 5 = -7}{\sqrt{4} - 5 \;\Big|\; -7}$
           $2 - 5 \;\Big|$
                     $-3 \overset{?}{=} -7$ FALSE

The number 4 does not check. Thus $\sqrt{x} - 5 = -7$ has no solution.

Caution!   Raising both sides of an equation to an even power may not produce an equivalent equation. In this case, a check is essential.

Note in Example 2 that $x = 4$ has solution 4, but that $\sqrt{x} - 5 = -7$ has *no* solution. Thus the equations $x = 4$ and $\sqrt{x} - 5 = -7$ are *not* equivalent.

To Solve an Equation with a Radical Term

1. Isolate the radical term on one side of the equation.
2. Use the principle of powers and solve the resulting equation.
3. Check any possible solution in the original equation.

EXAMPLE 3

Solve: $x = \sqrt{x + 7} + 5$.

Solution

$$x = \sqrt{x + 7} + 5$$

$$x - 5 = \sqrt{x + 7}$$ Isolating the radical by subtracting 5 from both sides

$$\left.\begin{array}{r} (x - 5)^2 = \left(\sqrt{x + 7}\right)^2 \\ x^2 - 10x + 25 = x + 7 \end{array}\right\}$$ Using the principle of powers; squaring both sides

$$x^2 - 11x + 18 = 0$$ Adding $-x - 7$ to both sides to write the quadratic equation in standard form

$$(x - 9)(x - 2) = 0$$ Factoring

$$x = 9 \quad or \quad x = 2$$ Using the principle of zero products

The possible solutions are 9 and 2. Let's check.

Check:

For 9:

$$\begin{array}{c|c} x = \sqrt{x + 7} + 5 \\ \hline 9 & \sqrt{9 + 7} + 5 \\ 9 \overset{?}{=} 9 \end{array}$$ TRUE

For 2:

$$\begin{array}{c|c} x = \sqrt{x + 7} + 5 \\ \hline 2 & \sqrt{2 + 7} + 5 \\ 2 \overset{?}{=} 8 \end{array}$$ FALSE

Since 9 checks but 2 does not, the solution is 9.

technology
connection

To solve Example 3, we can graph $y_1 = x$ and $y_2 = (x + 7)^{1/2} + 5$ and then use the INTERSECT option of the CALC menu to find the point of intersection. The intersection occurs at $x = 9$. Note that there is no intersection when $x = 2$, as predicted in the check of Example 3.

$y_1 = x, \ y_2 = (x + 7)^{1/2} + 5$

1. Use a graphing calculator to solve Examples 1, 2, 4, 5, and 6. Compare your answers with those found using the algebraic methods shown.

It is important to isolate a radical term before using the principle of powers. Suppose in Example 3 that both sides of the equation were squared *before* isolating the radical. We then would have had the expression $\left(\sqrt{x + 7} + 5\right)^2$ or $x + 7 + 10\sqrt{x + 7} + 25$ on the right side, and the radical would have remained in the problem.

EXAMPLE 4

Solve: $(2x + 1)^{1/3} + 5 = 0$.

Solution   We need not use radical notation to solve:

$$(2x + 1)^{1/3} + 5 = 0$$

$$(2x + 1)^{1/3} = -5$$ Subtracting 5 from both sides

$$[(2x + 1)^{1/3}]^3 = (-5)^3$$ Cubing both sides

$$(2x + 1)^1 = (-5)^3$$ Multiplying exponents. Try to do this mentally.

$$2x + 1 = -125$$

$$2x = -126$$ Subtracting 1 from both sides

$$x = -63.$$

Because both sides were raised to an *odd* power, it is not *essential* that we check the answer. Students can confirm that -63 checks and is the solution.

Equations with Two Radical Terms

A strategy for solving equations with two or more radical terms is as follows.

> *To Solve an Equation with Two or More Radical Terms*
>   1. Isolate one of the radical terms.
>   2. Use the principle of powers.
>   3. If a radical remains, perform steps (1) and (2) again.
>   4. Solve the resulting equation.
>   5. Check possible solutions in the original equation.

EXAMPLE 5   Solve: $\sqrt{2x - 5} = 1 + \sqrt{x - 3}$.

Solution

$$\sqrt{2x - 5} = 1 + \sqrt{x - 3}$$

$$\left(\sqrt{2x - 5}\right)^2 = \left(1 + \sqrt{x - 3}\right)^2 \qquad \text{One radical is already isolated.}$$
$$\text{We square both sides.}$$

> This is like squaring a binomial. We square 1, then find twice the product of 1 and $\sqrt{x - 3}$, and then the square of $\sqrt{x - 3}$. Study this carefully.

$$2x - 5 = 1 + 2\sqrt{x - 3} + \left(\sqrt{x - 3}\right)^2$$

$$2x - 5 = 1 + 2\sqrt{x - 3} + (x - 3)$$

$$x - 3 = 2\sqrt{x - 3} \qquad \text{Isolating the remaining radical term}$$

$$(x - 3)^2 = \left(2\sqrt{x - 3}\right)^2 \qquad \text{Squaring both sides}$$

$$x^2 - 6x + 9 = 4(x - 3) \qquad \text{Remember to square both the 2 and the } \sqrt{x - 3} \text{ on the right side.}$$

$$x^2 - 6x + 9 = 4x - 12$$

$$x^2 - 10x + 21 = 0$$

$$(x - 7)(x - 3) = 0 \qquad \text{Factoring}$$

$$x = 7 \quad or \quad x = 3 \qquad \text{Using the principle of zero products}$$

We leave it to the student to show that 7 and 3 both check and are the solutions.

> *Caution!*   A common error in solving equations like
> $$\sqrt{2x - 5} = 1 + \sqrt{x - 3}$$
> is to obtain $1 + (x - 3)$ as the square of the right side. This is wrong because $(A + B)^2 \neq A^2 + B^2$. For example,
> $$\left.\begin{array}{c} (1 + 2)^2 \neq 1^2 + 2^2 \\ 3^2 \neq 1 + 4 \\ 9 \neq 5. \end{array}\right\} \quad \text{See Example 5 for the correct expansion of } \left(1 + \sqrt{x - 3}\right)^2.$$

EXAMPLE 6   Let $f(x) = \sqrt{x + 5} - \sqrt{x - 7}$. Find all x-values for which $f(x) = 2$.

Solution   We must have $f(x) = 2$, or

$$\sqrt{x + 5} - \sqrt{x - 7} = 2. \qquad \text{Substituting for } f(x)$$

To solve, we isolate one radical term and square both sides:

$$\sqrt{x + 5} = 2 + \sqrt{x - 7} \qquad \text{Adding } \sqrt{x - 7} \text{ to both sides.}$$
This isolates one of the radical terms.

$$\left(\sqrt{x + 5}\right)^2 = \left(2 + \sqrt{x - 7}\right)^2 \qquad \text{Using the principle of powers (squaring both sides)}$$

$$x + 5 = 4 + 4\sqrt{x - 7} + (x - 7) \qquad \text{Using } (A + B)^2 = A^2 + 2AB + B^2$$

$$5 = 4\sqrt{x - 7} - 3 \qquad \text{Adding } -x \text{ to both sides and combining like terms}$$

$$8 = 4\sqrt{x - 7} \qquad \text{Isolating the remaining radical term}$$

$$2 = \sqrt{x - 7}$$
$$2^2 = \left(\sqrt{x - 7}\right)^2 \qquad \text{Squaring both sides}$$
$$4 = x - 7$$
$$11 = x.$$

Student Notes

Be careful when checking answers. You don't want to discard a correct answer because of a careless mistake.

Check: $f(11) = \sqrt{11 + 5} - \sqrt{11 - 7}$
$$= \sqrt{16} - \sqrt{4}$$
$$= 4 - 2 = 2.$$

We have $f(x) = 2$ when $x = 11$.

Exercise Set
10.6

☛ *Concept Reinforcement*   *Classify each of the following as either true or false.*

1. If $x^2 = 25$, then $x = 5$.

2. If $t = 7$, then $t^2 = 49$.

3. If $\sqrt{x} = 3$, then $\left(\sqrt{x}\right)^2 = 3^2$.

4. If $x^2 = 36$, then $x = 6$.

5. $\sqrt{x} - 8 = 7$ is equivalent to $\sqrt{x} = 15$.

6. $\sqrt{t} + 5 = 8$ is equivalent to $\sqrt{t} = 3$.

Solve.

7. $\sqrt{5x - 2} = 7$

8. $\sqrt{3x - 2} = 6$

9. $\sqrt{3x} + 1 = 6$

10. $\sqrt{2x} - 1 = 2$

11. $\sqrt{y + 1} - 5 = 8$

12. $\sqrt{x - 2} - 7 = -4$

13. $\sqrt{x - 7} + 3 = 10$

14. $\sqrt{y + 4} + 6 = 7$

15. $\sqrt[3]{x + 5} = 2$

16. $\sqrt[3]{x - 2} = 3$

17. $\sqrt[4]{y - 1} = 3$

18. $\sqrt[4]{x + 3} = 2$

19. $3\sqrt{x} = x$

20. $8\sqrt{y} = y$

21. $2y^{1/2} - 7 = 9$

22. $3x^{1/2} + 12 = 9$

23. $\sqrt[3]{x} = -3$

24. $\sqrt[3]{y} = -4$

25. $t^{1/3} - 2 = 3$

26. $x^{1/4} - 2 = 1$

Aha! **27.** $(y - 3)^{1/2} = -2$

28. $(x + 2)^{1/2} = -4$

29. $\sqrt[4]{3x + 1} - 4 = -1$

30. $\sqrt[4]{2x + 3} - 5 = -2$

31. $(x + 7)^{1/3} = 4$

32. $(y - 7)^{1/4} = 3$

33. $\sqrt[3]{3y + 6} + 7 = 8$

34. $\sqrt[3]{6x + 9} + 5 = 2$

35. $\sqrt{3t + 4} = \sqrt{4t + 3}$

36. $\sqrt{2t - 7} = \sqrt{3t - 12}$

37. $3(4 - t)^{1/4} = 6^{1/4}$

38. $2(1 - x)^{1/3} = 4^{1/3}$

39. $3 + \sqrt{5 - x} = x$

40. $x = \sqrt{x - 1} + 3$

41. $\sqrt{4x - 3} = 2 + \sqrt{2x - 5}$

42. $3 + \sqrt{z - 6} = \sqrt{z + 9}$

43. $\sqrt{20 - x} + 8 = \sqrt{9 - x} + 11$

44. $4 + \sqrt{10 - x} = 6 + \sqrt{4 - x}$

45. $\sqrt{x + 2} + \sqrt{3x + 4} = 2$

46. $\sqrt{6x + 7} - \sqrt{3x + 3} = 1$

47. If $f(x) = \sqrt{x} + \sqrt{x - 9}$, find any x for which $f(x) = 1$.

48. If $g(x) = \sqrt{x} + \sqrt{x - 5}$, find any x for which $g(x) = 5$.

49. If $f(t) = \sqrt{t - 2} - \sqrt{4t + 1}$, find any t for which $f(t) = -3$.

50. If $g(t) = \sqrt{2t + 7} - \sqrt{t + 15}$, find any t for which $g(t) = -1$.

51. If $f(x) = \sqrt{2x - 3}$ and $g(x) = \sqrt{x + 7} - 2$, find any x for which $f(x) = g(x)$.

52. If $f(x) = 2\sqrt{3x + 6}$ and $g(x) = 5 + \sqrt{4x + 9}$, find any x for which $f(x) = g(x)$.

53. If $f(t) = 4 - \sqrt{t - 3}$ and $g(t) = (t + 5)^{1/2}$, find any t for which $f(t) = g(t)$.

54. If $f(t) = 7 + \sqrt{2t - 5}$ and $g(t) = 3(t + 1)^{1/2}$, find any t for which $f(t) = g(t)$.

55. Explain in your own words why it is important to check your answers when using the principle of powers.

56. The principle of powers is an "if–then" statement that becomes false when the sentence parts are interchanged. Give an example of another such if–then statement from everyday life (answers will vary).

SKILL MAINTENANCE

57. The base of a triangle is 2 in. longer than the height. The area is $31\frac{1}{2}$ in². Find the height and the base. [5.8]

58. During a one-hour television show, there were 12 commercials. Some of the commercials were 30 sec long and the others were 60 sec long. If the number of 30-sec commercials was 6 less than the total number of minutes of commercial time during the show, how many 60-sec commercials were used? [8.3]

Graph.

59. $f(x) = \dfrac{2}{3}x - 5$ [7.3]

60. $g(x) = 3x + 4$ [7.3]

61. $F(x) < -2x + 4$ [9.4]

62. $G(x) > -3x + 2$ [9.4]

SYNTHESIS

63. Describe a procedure that could be used to create radical equations that have no solution.

64. Is checking essential when the principle of powers is used with an odd power n? Why or why not?

Steel manufacturing.   In the production of steel and other metals, the temperature of the molten metal is so great that conventional thermometers melt. Instead, sound is transmitted across the surface of the metal to a receiver on the far side and the speed of the sound is measured. The formula

$$S(t) = 1087.7\sqrt{\frac{9t + 2617}{2457}}$$

gives the speed of sound S(t), in feet per second, at a temperature of t degrees Celsius.

65. Find the temperature of a blast furnace where sound travels 1880 ft/sec.

66. Find the temperature of a blast furnace where sound travels 1502.3 ft/sec.

67. Solve the above equation for t.

Automotive repair.   For an engine with a displacement of 2.8 L, the function given by

$$d(n) = 0.75\sqrt{2.8n}$$

can be used to determine the diameter size of the carburetor's opening, in millimeters. Here n is the number of rpm's at which the engine achieves peak performance.
Source: macdizzy.com

68. If a carburetor's opening is 81 mm, for what number of rpm's will the engine produce peak power?

69. If a carburetor's opening is 84 mm, for what number of rpm's will the engine produce peak power?

Escape velocity.   A formula for the escape velocity v of a satellite is

$$v = \sqrt{2gr}\,\sqrt{\frac{h}{r+h}},$$

where g is the force of gravity, r is the planet or star's radius, and h is the height of the satellite above the planet or star's surface.

70. Solve for h.

71. Solve for r.

Sighting to the horizon.   The function $D(h) = 1.2\sqrt{h}$ can be used to approximate the distance D, in miles, that a person can see to the horizon from a height h, in feet.

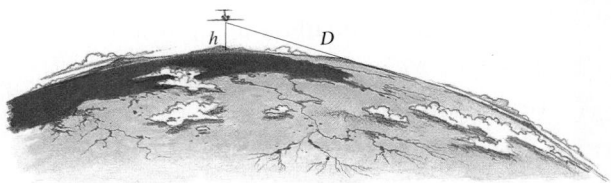

72. How far above sea level must a pilot fly in order to see a horizon that is 180 mi away?

73. How high above sea level must a sailor climb in order to see 10.2 mi out to sea?

Solve.

74. $\left(\dfrac{z}{4} - 5\right)^{2/3} = \dfrac{1}{25}$

75. $\dfrac{x + \sqrt{x+1}}{x - \sqrt{x+1}} = \dfrac{5}{11}$

76. $\sqrt{\sqrt{\sqrt{y} + 49}} = 7$

77. $(z^2 + 17)^{3/4} = 27$

78. $x^2 - 5x - \sqrt{x^2 - 5x - 2} = 4$
    (*Hint*: Let $u = x^2 - 5x - 2$.)

79. $\sqrt{8 - b} = b\sqrt{8 - b}$

Without graphing, determine the x-intercepts of the graphs given by each of the following.

80. $f(x) = \sqrt{x - 2} - \sqrt{x + 2} + 2$

81. $g(x) = 6x^{1/2} + 6x^{-1/2} - 37$

82. $f(x) = (x^2 + 30x)^{1/2} - x - (5x)^{1/2}$

83. Use a graphing calculator to check your answers to Exercises 9, 15, and 31.

84. Saul is trying to solve Exercise 73 using a graphing calculator. Without resorting to trial and error, how can he determine a suitable viewing window for finding the solution?

85. Use a graphing calculator to check your answers to Exercises 27, 35, and 41.

CORNER

Tailgater Alert

Focus: Radical equations and problem solving
Time: 15–25 minutes
Group size: 2–3
Materials: Calculators or square-root tables

The faster a car is traveling, the more distance it needs to stop. Thus it is important for drivers to allow sufficient space between their vehicle and the vehicle in front of them. Police recommend that for each 10 mph of speed, a driver allow 1 car length. Thus a driver going 30 mph should have at least 3 car lengths between his or her vehicle and the one in front.

   In Exercise Set 10.3, the function $r(L) = 2\sqrt{5L}$ was used to find the speed, in miles per hour, that a car was traveling when it left skid marks L feet long.

ACTIVITY

1. Each group member should estimate the length of a car in which he or she frequently travels. (Each should use a different length, if possible.)
2. Using a calculator as needed, each group member should complete the table below.

Column 1 gives a car's speed s, and column 2 lists the minimum amount of space between cars traveling s miles per hour, as recommended by police. Column 3 is the speed that a vehicle *could* travel were it forced to stop in the distance listed in column 2, using the above function.

Column 1 s (in miles per hour)	Column 2 $L(s)$ (in feet)	Column 3 $r(L)$ (in miles per hour)
20		
30		
40		
50		
60		
70		

3. Determine whether there are any speeds at which the "1 car length per 10 mph" guideline might not suffice. On what reasoning do you base your answer? Compare tables to determine how car length affects the results. What recommendations would your group make to a new driver?

10.7   Geometric Applications

Using the Pythagorean Theorem   •   Two Special Triangles

Using the Pythagorean Theorem

There are many kinds of problems that involve powers and roots. Many also involve right triangles and the Pythagorean theorem, which we studied in Section 5.8 and restate here.

Study Skills _____

Making Sketches

One need not be an artist to make highly useful mathematical sketches. That said, it is important to make sure that your sketches are drawn accurately enough to represent the relative sizes within each shape. For example, if one side of a triangle is clearly the longest, make sure your drawing reflects this.

The Pythagorean Theorem*

In any right triangle, if a and b are the lengths of the legs and c is the length of the hypotenuse, then

$$a^2 + b^2 = c^2.$$

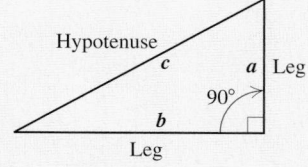

In using the Pythagorean theorem, we often make use of the following principle.

The Principle of Square Roots

For any nonnegative real number n,

If $x^2 = n$,   then $x = \sqrt{n}$ or $x = -\sqrt{n}$.

For most real-world applications involving length or distance, $-\sqrt{n}$ is not needed.

EXAMPLE 1

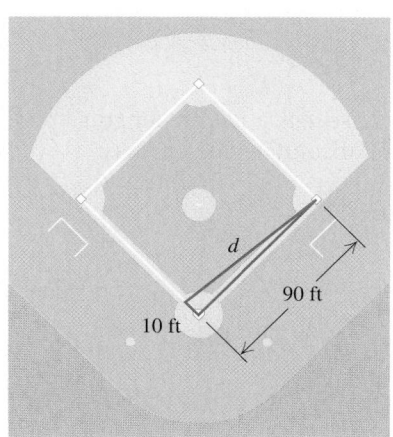

Baseball.   A baseball diamond is actually a square 90 ft on a side. Suppose a catcher fields a ball while standing on the third-base line 10 ft from home plate. How far is the catcher's throw to first base? Give an exact answer and an approximation to three decimal places.

Solution   We make a drawing and let d = the distance, in feet, to first base. Note that a right triangle is formed in which the leg from home plate to first base measures 90 ft and the leg from home plate to where the catcher fields the ball measures 10 ft.

We substitute these values into the Pythagorean theorem to find d:

$$d^2 = 90^2 + 10^2$$
$$d^2 = 8100 + 100$$
$$d^2 = 8200.$$

We now use the principle of square roots: If $d^2 = 8200$, then $d = \sqrt{8200}$ or $d = -\sqrt{8200}$. Since d represents a length, it follows that d is the positive square root of 8200:

$$d = \sqrt{8200} \text{ ft}$$ This is an exact answer.

$$d \approx 90.554 \text{ ft.}$$ Using a calculator for an approximation

*The converse of the Pythagorean theorem also holds. That is, if a, b, and c are the lengths of the sides of a triangle and $a^2 + b^2 = c^2$, then the triangle is a right triangle.

EXAMPLE 2

Guy wires. The base of a 40-ft-long guy wire is located 15 ft from the telephone pole that it is anchoring. How high up the pole does the guy wire reach? Give an exact answer and an approximation to three decimal places.

Solution We make a drawing and let h = the height, in feet, to which the guy wire reaches. A right triangle is formed in which one leg measures 15 ft and the hypotenuse measures 40 ft. Using the Pythagorean theorem, we have

$$h^2 + 15^2 = 40^2$$
$$h^2 + 225 = 1600$$
$$h^2 = 1375$$
$$h = \sqrt{1375}.$$

Exact answer:
$$h = \sqrt{1375} \text{ ft}$$

Approximation:
$$h \approx 37.081 \text{ ft} \qquad \text{Using a calculator}$$

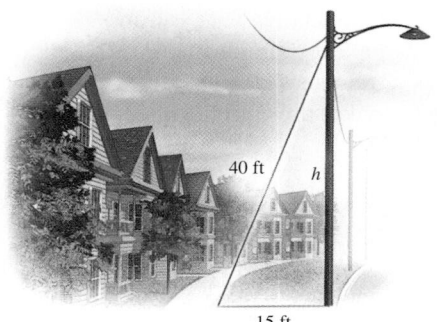

Two Special Triangles

When both legs of a right triangle are the same size, we call the triangle an *isosceles right triangle*, as shown at left. If one leg of an isosceles right triangle has length a, we can find a formula for the length of the hypotenuse as follows:

$$c^2 = a^2 + b^2$$
$$c^2 = a^2 + a^2 \qquad \text{Because the triangle is isosceles, both legs}$$
are the same size: $a = b$.
$$c^2 = 2a^2. \qquad \text{Combining like terms}$$

Next, we use the principle of square roots. Because a, b, and c are lengths, there is no need to consider negative square roots or absolute values. Thus,

$$c = \sqrt{2a^2} \qquad \text{Using the principle of square roots}$$
$$c = \sqrt{a^2 \cdot 2} = a\sqrt{2}.$$

EXAMPLE 3

One leg of an isosceles right triangle measures 7 cm. Find the length of the hypotenuse. Give an exact answer and an approximation to three decimal places.

Solution We substitute:

$$c = a\sqrt{2} \qquad \text{This equation is worth memorizing.}$$
$$c = 7\sqrt{2}.$$

Exact answer: $\qquad c = 7\sqrt{2} \text{ cm}$

Approximation: $\quad c \approx 9.899 \text{ cm} \qquad \text{Using a calculator}$

When the hypotenuse of an isosceles right triangle is known, the lengths of the legs can be found.

EXAMPLE 4

The hypotenuse of an isosceles right triangle is 5 ft long. Find the length of a leg. Give an exact answer and an approximation to three decimal places.

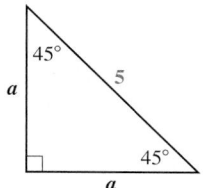

Solution   We replace c with 5 and solve for a:

$$5 = a\sqrt{2} \qquad \text{Substituting 5 for } c \text{ in } c = a\sqrt{2}$$

$$\frac{5}{\sqrt{2}} = a \qquad \text{Dividing both sides by } \sqrt{2}$$

$$\frac{5\sqrt{2}}{2} = a. \qquad \text{Rationalize the denominator if desired.}$$

Exact answer: $a = \dfrac{5}{\sqrt{2}}$ ft, or $\dfrac{5\sqrt{2}}{2}$ ft

Approximation: $a \approx 3.536$ ft   Using a calculator

A second special triangle is known as a 30°–60°–90° right triangle, so named because of the measures of its angles. Note that in an equilateral triangle, all sides have the same length and all angles are 60°. An altitude, drawn dashed in the figure, bisects, or splits in half, one angle and one side. Two 30°–60°–90° right triangles are thus formed.

Because of the way in which the altitude is drawn, if a represents the length of the shorter leg in a 30°–60°–90° right triangle, then $2a$ represents the length of the hypotenuse. We have

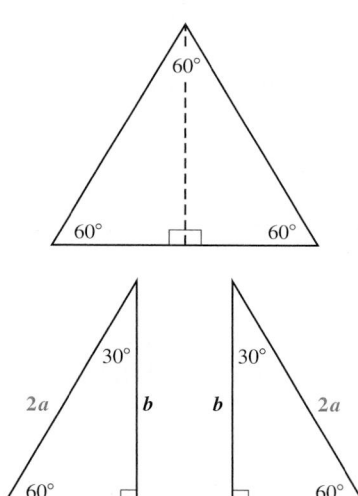

$$a^2 + b^2 = (2a)^2 \qquad \text{Using the Pythagorean theorem}$$

$$a^2 + b^2 = 4a^2$$

$$b^2 = 3a^2 \qquad \text{Subtracting } a^2 \text{ from both sides}$$

$$b = \sqrt{3a^2}$$

$$b = \sqrt{a^2 \cdot 3}$$

$$b = a\sqrt{3}.$$

EXAMPLE 5

The shorter leg of a 30°–60°–90° right triangle measures 8 in. Find the lengths of the other sides. Give exact answers and, where appropriate, an approximation to three decimal places.

Solution   The hypotenuse is twice as long as the shorter leg, so we have

$$c = 2a \qquad\qquad \text{This relationship is worth memorizing.}$$

$$= 2 \cdot 8 = 16 \text{ in.} \qquad \text{This is the length of the hypotenuse.}$$

The length of the longer leg is the length of the shorter leg times $\sqrt{3}$. This gives us

$$b = a\sqrt{3} \qquad\qquad \text{This is also worth memorizing.}$$

$$= 8\sqrt{3} \text{ in.} \qquad \text{This is the length of the longer leg.}$$

Exact answer: $c = 16$ in., $b = 8\sqrt{3}$ in.

Approximation: $b \approx 13.856$ in.

EXAMPLE 6   The length of the longer leg of a 30°–60°–90° right triangle is 14 cm. Find the length of the hypotenuse. Give an exact answer and an approximation to three decimal places.

Solution   The length of the hypotenuse is twice the length of the shorter leg. We first find a, the length of the shorter leg, by using the length of the longer leg:

$$14 = a\sqrt{3} \qquad \text{Substituting 14 for } b \text{ in } b = a\sqrt{3}$$

$$\frac{14}{\sqrt{3}} = a. \qquad \text{Dividing by } \sqrt{3}$$

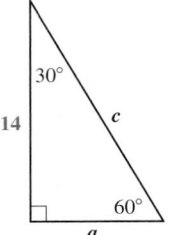

Since the hypotenuse is twice as long as the shorter leg, we have

$$c = 2a$$

$$= 2 \cdot \frac{14}{\sqrt{3}} \qquad \text{Substituting}$$

$$= \frac{28}{\sqrt{3}} \text{ cm.}$$

Exact answer: $c = \dfrac{28}{\sqrt{3}}$ cm, or $\dfrac{28\sqrt{3}}{3}$ cm if the denominator is rationalized.

Approximation: $c \approx 16.166$ cm

Student Notes

Perhaps the easiest way to remember the important results listed in the adjacent box is to write out, on your own, the derivations shown on pp. 484 and 485.

Lengths Within Isosceles and 30°–60°–90° Right Triangles

The length of the hypotenuse in an isosceles right triangle is the length of a leg times $\sqrt{2}$.

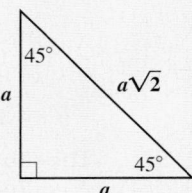

The length of the longer leg in a 30°–60°–90° right triangle is the length of the shorter leg times $\sqrt{3}$. The hypotenuse is twice as long as the shorter leg.

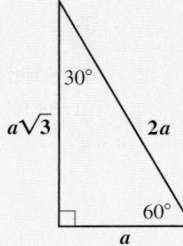

Exercise Set

10.7

↪ *Concept Reinforcement*   *Complete each of the following sentences.*

1. In any _____ triangle, the square of the length of the _____ is the sum of the squares of the lengths of the legs.

2. The shortest side of a right triangle is always one of the two _____.

3. The principle of _____ _____ states that if $x^2 = n$, then $x = \sqrt{n}$ or $x = -\sqrt{n}$.

4. In a(n) _____ right triangle, both legs have the same length.

5. In a(n) ____–____–____ right triangle, the hypotenuse is twice as long as the shorter _____.

6. If both legs have measure a, then the hypotenuse measures _____.

In a right triangle, find the length of the side not given. Give an exact answer and, where appropriate, an approximation to three decimal places.

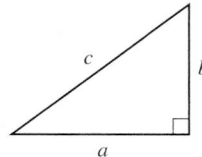

7. $a = 5, b = 3$ **8.** $a = 8, b = 10$

Aha! **9.** $a = 9, b = 9$ **10.** $a = 10, b = 10$

11. $b = 12, c = 13$ **12.** $a = 5, c = 12$

In Exercises 13–18, give an exact answer and, where appropriate, an approximation to three decimal places.

13. A right triangle's hypotenuse is 8 m and one leg is $4\sqrt{3}$ m. Find the length of the other leg.

14. A right triangle's hypotenuse is 6 cm and one leg is $\sqrt{5}$ cm. Find the length of the other leg.

15. The hypotenuse of a right triangle is $\sqrt{20}$ in. and one leg measures 1 in. Find the length of the other leg.

16. The hypotenuse of a right triangle is $\sqrt{15}$ ft and one leg measures 2 ft. Find the length of the other leg.

Aha! **17.** One leg in a right triangle is 1 m and the hypotenuse measures $\sqrt{2}$ m. Find the length of the other leg.

18. One leg of a right triangle is 1 yd and the hypotenuse measures 2 yd. Find the length of the other leg.

In Exercises 19–28, give an exact answer and, where appropriate, an approximation to three decimal places.

19. *Bicycling.*   Clare routinely bicycles across a rectangular parking lot on her way to class. If the lot is 200 ft long and 150 ft wide, how far does Clare travel when she rides across the lot diagonally?

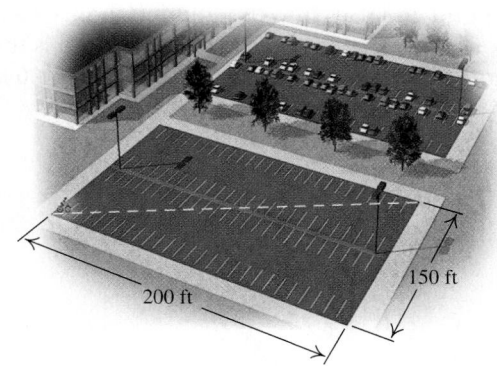

150 ft

200 ft

20. *Guy wire.*   How long is a guy wire if it reaches from the top of a 15-ft pole to a point on the ground 10 ft from the pole?

21. *Softball.*   A slow-pitch softball diamond is actually a square 65 ft on a side. How far is it from home plate to second base?

22. *Baseball.*   Suppose the catcher in Example 1 makes a throw to second base from the same location. How far is that throw?

23. *Television sets.* What does it mean to refer to a 20-in. TV set or a 25-in. TV set? Such units refer to the diagonal of the screen. A 20-in. TV set has a width of 16 in. What is its height?

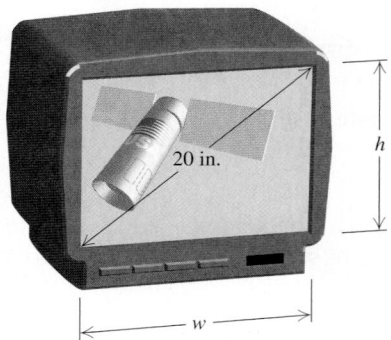

24. *Television sets.* A 25-in. TV set has a screen with a height of 15 in. What is its width? (See Exercise 23.)

25. *Speaker placement.* A stereo receiver is in a corner of a 12-ft by 14-ft room. Speaker wire will run under a rug, diagonally, to a speaker in the far corner. If 4 ft of slack is required on each end, how long a piece of wire should be purchased?

26. *Distance over water.* To determine the width of a pond, a surveyor locates two stakes at either end of the pond and uses instrumentation to place a third stake so that the distance across the pond is the length of a hypotenuse. If the third stake is 90 m from one stake and 70 m from the other, what is the distance across the pond?

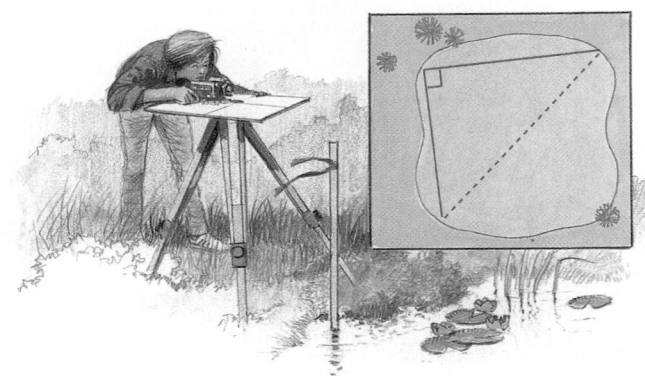

27. *Walking.* Students at Compkin Community College have worn a path that cuts diagonally across the campus "quad." If the quad is actually a rectangle that Marissa measured to be 70 paces long and 40 paces wide, how many paces will Marissa save by using the diagonal path?

28. *Crosswalks.* The diagonal crosswalk at the intersection of State St. and Main St. is the hypotenuse of a triangle in which the crosswalks across State St. and Main St. are the legs. If State St. is 28 ft wide and Main St. is 40 ft wide, how much shorter is the distance traveled by pedestrians using the diagonal crosswalk?

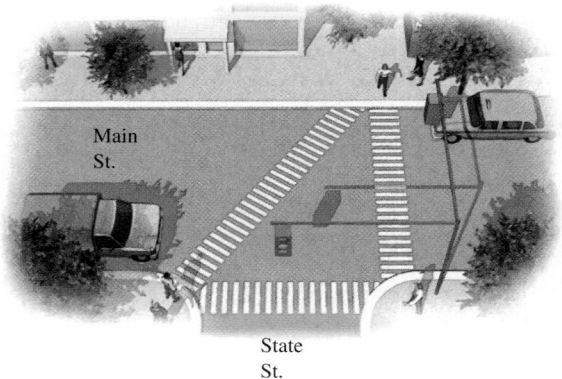

For each triangle, find the missing length(s). Give an exact answer and, where appropriate, an approximation to three decimal places.

29.

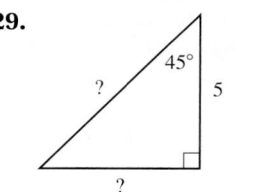

30.

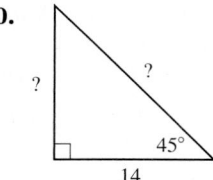

31.

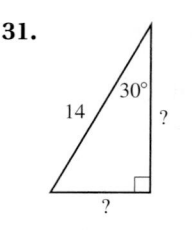

32.

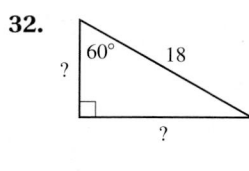

33.

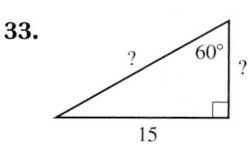

34.

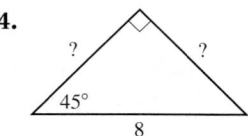

35.

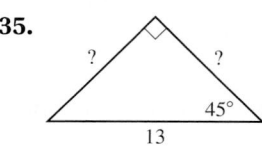

36.

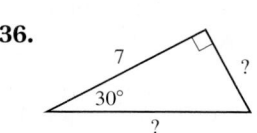

37.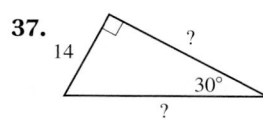

38.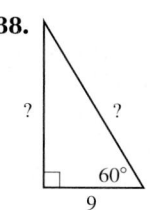

39.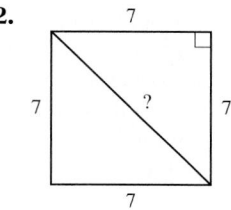

40.

41.

42.

43.

44.

In Exercises 45–50, give an exact answer and, where appropriate, an approximation to three decimal places.

45. *Bridge expansion.* During the summer heat, a 2-mi bridge expands 2 ft in length. If we assume that the bulge occurs straight up the middle, how high is the bulge? (The answer may surprise you. Most bridges have expansion spaces to avoid such buckling.)

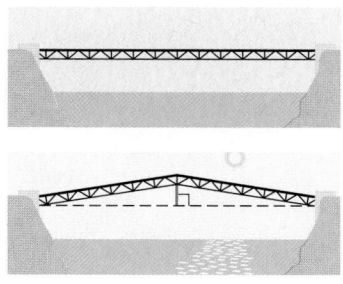

46. Triangle *ABC* has sides of lengths 25 ft, 25 ft, and 30 ft. Triangle *PQR* has sides of lengths 25 ft, 25 ft, and 40 ft. Which triangle, if either, has the greater area and by how much?

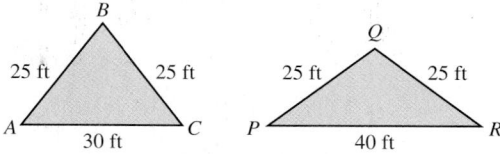

47. *Camping tent.*   The entrance to a pup tent is the shape of an equilateral triangle. If the base of the tent is 4 ft wide, how tall is the tent?

48. The length and the width of a rectangle are given by consecutive integers. The area of the rectangle is 90 cm^2. Find the length of a diagonal of the rectangle.

49. Find all points on the *y*-axis of a Cartesian coordinate system that are 5 units from the point (3, 0).

50. Find all points on the *x*-axis of a Cartesian coordinate system that are 5 units from the point (0, 4).

51. Write a problem for a classmate to solve in which the solution is: "The height of the tepee is $5\sqrt{3}$ yd."

52. Write a problem for a classmate to solve in which the solution is: "The height of the window is $15\sqrt{3}$ ft."

SKILL MAINTENANCE

Simplify. [4.1]

53. $47(-1)^{19}$

54. $(-5)(-1)^{13}$

Factor. [5.6]

55. $x^3 - 9x$

56. $7a^3 - 28a$

Solve. [9.3]

57. $|3x - 5| = 7$

58. $|2x - 3| = |x + 7|$

SYNTHESIS

59. Are there any right triangles, other than those with sides measuring 3, 4, and 5, that have consecutive numbers for the lengths of the sides? Why or why not?

60. If a 30°–60°–90° triangle and an isosceles right triangle have the same perimeter, which will have the greater area? Why?

61. The perimeter of a regular hexagon is 72 cm. Determine the area of the shaded region shown.

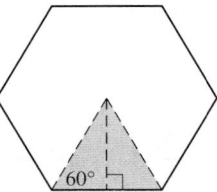

62. If the perimeter of a regular hexagon is 120 ft, what is its area? (*Hint*: See Exercise 61.)

63. Each side of a regular octagon has length s. Find a formula for the distance d between the parallel sides of the octagon.

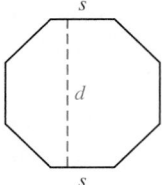

64. *Roofing.*   Kit's home, which is 24 ft wide and 32 ft long, needs a new roof. By counting clapboards that are 4 in. apart, Kit determines that the peak of the roof is 6 ft higher than the sides. If one packet of shingles covers 100 square feet, how many packets will the job require?

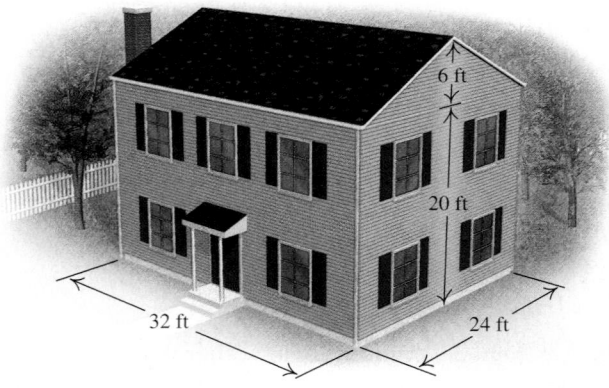

65. *Painting.*   (Refer to Exercise 64.) A gallon of Benjamin Moore® exterior acrylic paint covers 450–500 square feet. If Kit's house has dimensions as shown above, how many gallons of paint should be bought to paint the house? What assumption(s) is made in your answer?

66. *Contracting.*   Oxford Builders has an extension cord on their generator that permits them to work, with electricity, anywhere in a circular area of 3850 ft². Find the dimensions of the largest square room they could work on without having to relocate the generator to reach each corner of the floor plan.

67. *Contracting.*   Cleary Construction has a hose attached to their insulation blower that permits them to work, with electricity, anywhere in a circular area of 6160 ft². Find the dimensions of the largest square room with 12-ft ceilings in which they could reach all corners with the hose while leaving the blower centrally located. Assume that the blower sits on the floor.

68. A cube measures 5 cm on each side. How long is the diagonal that connects two opposite corners of the cube? Give an exact answer.

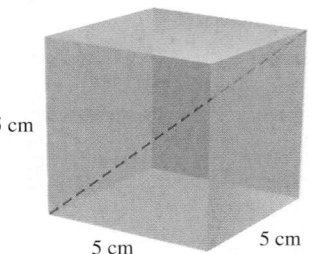

10.8 | The Complex Numbers

Imaginary and Complex Numbers • Addition and Subtraction • Multiplication • Conjugates and Division • Powers of i

Imaginary and Complex Numbers

Negative numbers do not have square roots in the real-number system. However, a larger number system that contains the real-number system is designed so that negative numbers *do* have square roots. That system is called the **complex-number system**, and it will allow us to solve equations like $x^2 + 1 = 0$. The complex-number system makes use of i, a number that is, by definition, a square root of -1.

> **The Number i**
>
> i is the unique number for which $i = \sqrt{-1}$ and $i^2 = -1$.

We can now define the square root of a negative number as follows:

$$\sqrt{-p} = \sqrt{-1}\,\sqrt{p} = i\sqrt{p} \text{ or } \sqrt{p}i, \text{ for any positive number } p.$$

EXAMPLE 1   Express in terms of i: **(a)** $\sqrt{-7}$;   **(b)** $\sqrt{-16}$;   **(c)** $-\sqrt{-13}$;   **(d)** $-\sqrt{-50}$.

Solution

a) $\sqrt{-7} = \sqrt{-1 \cdot 7} = \sqrt{-1} \cdot \sqrt{7} = i\sqrt{7}$, or $\sqrt{7}i$ | i is *not* under the radical. |

b) $\sqrt{-16} = \sqrt{-1 \cdot 16} = \sqrt{-1} \cdot \sqrt{16} = i \cdot 4 = 4i$

c) $-\sqrt{-13} = -\sqrt{-1 \cdot 13} = -\sqrt{-1} \cdot \sqrt{13} = -i\sqrt{13}$, or $-\sqrt{13}i$

d) $-\sqrt{-50} = -\sqrt{-1} \cdot \sqrt{25} \cdot \sqrt{2} = -i \cdot 5 \cdot \sqrt{2} = -5i\sqrt{2}$, or $-5\sqrt{2}i$

> **Imaginary Numbers**
>
> An *imaginary number* is a number that can be written in the form $a + bi$, where a and b are real numbers and $b \neq 0$.

Don't let the name "imaginary" fool you. Imaginary numbers appear in fields such as engineering and the physical sciences. The following are examples of imaginary numbers:

$5 + 4i$,   Here $a = 5, b = 4$.

$\sqrt{3} - \pi i$,   Here $a = \sqrt{3}, b = -\pi$.

$\sqrt{7}i$ Here $a = 0, b = \sqrt{7}$.

The union of the set of all imaginary numbers and the set of all real numbers is the set of all **complex numbers**.

> **Complex Numbers**
>
> A *complex number* is any number that can be written in the form $a + bi$, where a and b are real numbers. (Note that a and b both can be 0.)

The following are examples of complex numbers:

$7 + 3i$ (here $a \neq 0, b \neq 0$); $\qquad$ $4i$ (here $a = 0, b \neq 0$);

8 (here $a \neq 0, b = 0$); $\qquad$ 0 (here $a = 0, b = 0$).

Complex numbers like $17i$ or $4i$, in which $a = 0$ and $b \neq 0$, are imaginary numbers with no real part. Such numbers are called *pure imaginary numbers*.

Note that when $b = 0$, we have $a + 0i = a$, so every real number is a complex number. The relationships among various real and complex numbers are shown below.

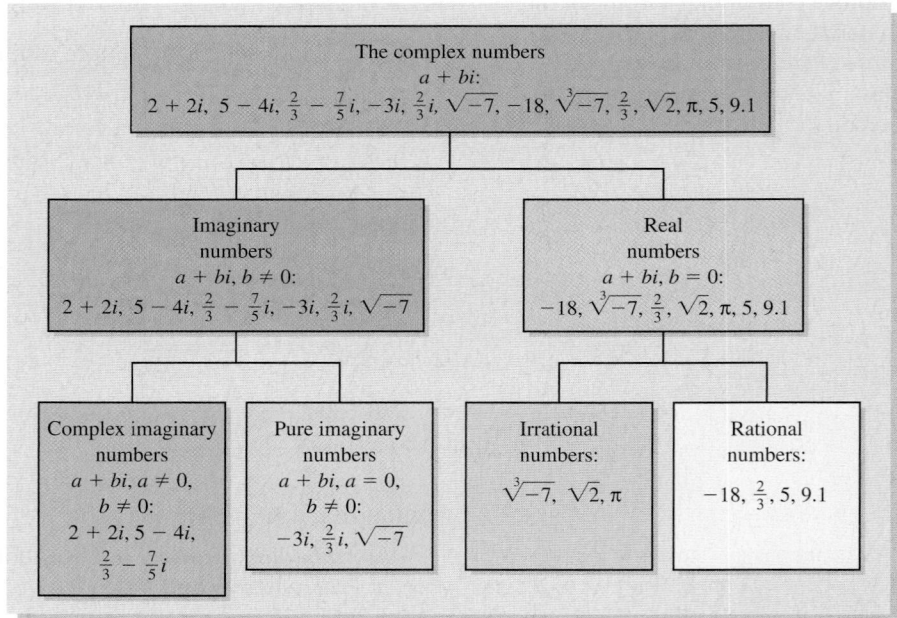

Note that although $\sqrt{-7}$ and $\sqrt[3]{-7}$ are both complex numbers, $\sqrt{-7}$ is imaginary whereas $\sqrt[3]{-7}$ is real.

Addition and Subtraction

The complex numbers obey the commutative, associative, and distributive laws. Thus we can add and subtract them as we do binomials.

EXAMPLE 2   Add or subtract and simplify.

a) $(8 + 6i) + (3 + 2i)$ **b)** $(4 + 5i) - (6 - 3i)$

Solution

a) $(8 + 6i) + (3 + 2i) = (8 + 3) + (6i + 2i)$ Combining the real parts
and the imaginary parts

$$= 11 + (6 + 2)i = 11 + 8i$$

b) $(4 + 5i) - (6 - 3i) = (4 - 6) + [5i - (-3i)]$ Note that the 6 and the
$-3i$ are *both* being
subtracted.

$$= -2 + 8i$$

Student Notes _____

The rule developed in Section 10.3, $\sqrt[n]{a} \cdot \sqrt[n]{b} = \sqrt[n]{a \cdot b}$, does *not* apply when n is 2 and either a or b is negative. Indeed this condition is stated on p. 651 when it is specified that $\sqrt[n]{a}$ and $\sqrt[n]{b}$ are both *real* numbers.

Multiplication

To multiply square roots of negative real numbers, we first express them in terms of i. For example,

$$\sqrt{-2} \cdot \sqrt{-5} = \sqrt{-1} \cdot \sqrt{2} \cdot \sqrt{-1} \cdot \sqrt{5}$$
$$= i \cdot \sqrt{2} \cdot i \cdot \sqrt{5}$$
$$= i^2 \cdot \sqrt{10}$$
$$= -1\sqrt{10} = -\sqrt{10} \text{ is correct!}$$

> *Caution!*   With complex numbers, simply multiplying radicands is *incorrect* when both radicands are negative: $\sqrt{-2} \cdot \sqrt{-5} \neq \sqrt{10}$.

With this in mind, we can now multiply complex numbers.

EXAMPLE 3   Multiply and simplify. When possible, write answers in the form $a + bi$.

a) $\sqrt{-16} \cdot \sqrt{-25}$ **b)** $\sqrt{-5} \cdot \sqrt{-7}$ **c)** $-3i \cdot 8i$

d) $-4i(3 - 5i)$ **e)** $(1 + 2i)(1 + 3i)$

Solution

a) $\sqrt{-16} \cdot \sqrt{-25} = \sqrt{-1} \cdot \sqrt{16} \cdot \sqrt{-1} \cdot \sqrt{25}$
$$= i \cdot 4 \cdot i \cdot 5$$
$$= i^2 \cdot 20$$
$$= -1 \cdot 20 \qquad i^2 = -1$$
$$= -20$$

b) $\sqrt{-5} \cdot \sqrt{-7} = \sqrt{-1} \cdot \sqrt{5} \cdot \sqrt{-1} \cdot \sqrt{7}$ Try to do this step mentally.
$$= i \cdot \sqrt{5} \cdot i \cdot \sqrt{7}$$
$$= i^2 \cdot \sqrt{35}$$
$$= -1 \cdot \sqrt{35} \qquad i^2 = -1$$
$$= -\sqrt{35}$$

c) $-3i \cdot 8i = -24 \cdot i^2$

$\qquad\qquad = -24 \cdot (-1) \qquad i^2 = -1$

$\qquad\qquad = 24$

d) $-4i(3 - 5i) = -4i \cdot 3 + (-4i)(-5i) \qquad$ Using the distributive law

$\qquad\qquad\qquad = -12i + 20i^2$

$\qquad\qquad\qquad = -12i - 20 \qquad\qquad\qquad i^2 = -1$

$\qquad\qquad\qquad = -20 - 12i \qquad\qquad\qquad$ Writing in the form $a + bi$

e) $(1 + 2i)(1 + 3i) = 1 + 3i + 2i + 6i^2 \qquad$ Multiplying each term of $1 + 3i$ by each term of $1 + 2i$ (FOIL)

$\qquad\qquad\qquad\quad = 1 + 3i + 2i - 6 \qquad i^2 = -1$

$\qquad\qquad\qquad\quad = -5 + 5i \qquad\qquad\qquad$ Combining like terms

Conjugates and Division

Recall that the conjugate of $4 + \sqrt{2}$ is $4 - \sqrt{2}$.

Conjugates of complex numbers are defined in a similar manner.

Conjugate of a Complex Number

The *conjugate* of a complex number $a + bi$ is $a - bi$, and the *conjugate* of $a - bi$ is $a + bi$.

EXAMPLE 4   Find the conjugate of each number.

a) $-3 + 7i$ **b)** $14 - 5i$ **c)** $4i$

Solution

a) $-3 + 7i$ The conjugate is $-3 - 7i$.

b) $14 - 5i$ The conjugate is $14 + 5i$.

c) $4i$ The conjugate is $-4i$. Note that $4i = 0 + 4i$.

The product of a complex number and its conjugate is a real number.

EXAMPLE 5   Multiply: $(5 + 7i)(5 - 7i)$.

Solution

$\qquad (5 + 7i)(5 - 7i) = 5^2 - (7i)^2 \qquad$ Using $(A + B)(A - B) = A^2 - B^2$

$\qquad\qquad\qquad\qquad\quad = 25 - 49i^2$

$\qquad\qquad\qquad\qquad\quad = 25 - 49(-1) \qquad i^2 = -1$

$\qquad\qquad\qquad\qquad\quad = 25 + 49 = 74$

Conjugates are used when dividing complex numbers. The procedure is much like that used to rationalize denominators in Section 10.5.

EXAMPLE 6   Divide and simplify to the form $a + bi$.

a) $\dfrac{-5 + 9i}{1 - 2i}$

b) $\dfrac{7 + 3i}{5i}$

Solution

a) To divide and simplify $(-5 + 9i)/(1 - 2i)$, we multiply by 1, using the conjugate of the denominator to form 1:

$$\dfrac{-5 + 9i}{1 - 2i} = \dfrac{-5 + 9i}{1 - 2i} \cdot \dfrac{1 + 2i}{1 + 2i} \qquad \begin{array}{l}\text{Multiplying by 1 using the} \\ \text{conjugate of the denominator} \\ \text{in the symbol for 1}\end{array}$$

$$= \dfrac{(-5 + 9i)(1 + 2i)}{(1 - 2i)(1 + 2i)} \qquad \begin{array}{l}\text{Multiplying numerators;} \\ \text{multiplying denominators}\end{array}$$

$$= \dfrac{-5 - 10i + 9i + 18i^2}{1^2 - 4i^2} \qquad \text{Using FOIL}$$

$$= \dfrac{-5 - i - 18}{1 - 4(-1)} \qquad i^2 = -1$$

$$\left. \begin{array}{l} = \dfrac{-23 - i}{5} \\[2ex] = -\dfrac{23}{5} - \dfrac{1}{5}i \end{array} \right\} \qquad \begin{array}{l}\text{Writing in the form } a + bi; \\ \text{note that } \dfrac{X + Y}{Z} = \dfrac{X}{Z} + \dfrac{Y}{Z}\end{array}$$

b) The conjugate of $5i$ is $-5i$, so we *could* multiply by $-5i/(-5i)$. However, when the denominator is a pure imaginary number, it is easiest if we multiply by i/i:

$$\dfrac{7 + 3i}{5i} = \dfrac{7 + 3i}{5i} \cdot \dfrac{i}{i} \qquad \begin{array}{l}\text{Multiplying by 1 using } i/i. \text{ We can also use} \\ \text{the conjugate of } 5i \text{ to write } -5i/(-5i).\end{array}$$

$$= \dfrac{7i + 3i^2}{5i^2} \qquad \text{Multiplying}$$

$$= \dfrac{7i + 3(-1)}{5(-1)} \qquad i^2 = -1$$

$$= \dfrac{7i - 3}{-5}$$

$$= \dfrac{-3}{-5} + \dfrac{7}{-5}i, \text{ or } \dfrac{3}{5} - \dfrac{7}{5}i. \qquad \text{Writing in the form } a + bi$$

Powers of i

Answers to problems involving complex numbers are generally written in the form $a + bi$. In the following discussion, we show why there is no need to use powers of i (other than 1) when writing answers.

Recall that -1 raised to an *even* power is 1, and -1 raised to an *odd* power is -1. Simplifying powers of i can then be done by using the fact that $i^2 = -1$ and expressing the given power of i in terms of i^2. Consider the following:

$$i^2 = -1, \longleftarrow$$
$$i^3 = i^2 \cdot i = (-1)i = -i,$$
$$i^4 = (i^2)^2 = (-1)^2 = 1,$$
$$i^5 = i^4 \cdot i = (i^2)^2 \cdot i = (-1)^2 \cdot i = i,$$
$$i^6 = (i^2)^3 = (-1)^3 = -1. \longleftarrow \text{The pattern is now repeating.}$$

The powers of i cycle themselves through the values i, -1, $-i$, and 1. Even powers of i are -1 or 1 whereas odd powers of i are i or $-i$.

EXAMPLE 7   Simplify: **(a)** i^{18}; **(b)** i^{24}.

Solution

a) $i^{18} = (i^2)^9$ Using the power rule

$\quad\quad = (-1)^9 = -1$ Raising -1 to a power

b) $i^{24} = (i^2)^{12}$

$\quad\quad = (-1)^{12} = 1$

To simplify i^n when n is odd, we rewrite i^n as $i^{n-1} \cdot i$.

EXAMPLE 8   Simplify: **(a)** i^{29}; **(b)** i^{75}.

Solution

a) $i^{29} = i^{28}i^1$ Using the product rule. This is a key step
$\quad\quad\quad\quad\quad\quad$ when i is raised to an odd power.

$\quad\quad = (i^2)^{14}i$ Using the power rule

$\quad\quad = (-1)^{14}i$

$\quad\quad = 1 \cdot i = i$

b) $i^{75} = i^{74}i^1$ Using the product rule

$\quad\quad = (i^2)^{37}i$ Using the power rule

$\quad\quad = (-1)^{37}i$

$\quad\quad = -1 \cdot i = -i$

Exercise Set

10.8

Concept Reinforcement   *Classify each statement as either true or false.*

1. Imaginary numbers are so named because they have no real-world applications.

2. Every real number is imaginary, but not every imaginary number is real.

3. Every imaginary number is a complex number, but not every complex number is imaginary.

4. Every real number is a complex number, but not every complex number is real.

5. Addition and subtraction of complex numbers has much in common with addition and subtraction of polynomials.

6. The product of a complex number and its conjugate is always a real number.

7. The square of a complex number is always a real number.

8. The quotient of two complex numbers is always a complex number.

Express in terms of i.

9. $\sqrt{-36}$

10. $\sqrt{-25}$

11. $\sqrt{-13}$

12. $\sqrt{-19}$

13. $\sqrt{-18}$

14. $\sqrt{-98}$

15. $\sqrt{-3}$

16. $\sqrt{-4}$

17. $\sqrt{-81}$

18. $\sqrt{-27}$

19. $-\sqrt{-300}$

20. $-\sqrt{-75}$

21. $6 - \sqrt{-84}$

22. $4 - \sqrt{-60}$

23. $-\sqrt{-76} + \sqrt{-125}$

24. $\sqrt{-4} + \sqrt{-12}$

25. $\sqrt{-18} - \sqrt{-100}$

26. $\sqrt{-72} - \sqrt{-25}$

Perform the indicated operation and simplify. Write each answer in the form $a + bi$.

27. $(6 + 7i) + (5 + 3i)$

28. $(4 - 5i) + (3 + 9i)$

29. $(9 + 8i) - (5 + 3i)$

30. $(9 + 7i) - (2 + 4i)$

31. $(7 - 4i) - (5 - 3i)$

32. $(5 - 3i) - (9 + 2i)$

33. $(-5 - i) - (7 + 4i)$

34. $(-2 + 6i) - (-7 + i)$

35. $7i \cdot 6i$

36. $6i \cdot 9i$

37. $(-4i)(-6i)$

38. $7i \cdot (-8i)$

39. $\sqrt{-36}\sqrt{-9}$

40. $\sqrt{-49}\sqrt{-25}$

41. $\sqrt{-5}\sqrt{-2}$

42. $\sqrt{-6}\sqrt{-7}$

43. $\sqrt{-6}\sqrt{-21}$

44. $\sqrt{-15}\sqrt{-10}$

45. $5i(2 + 6i)$

46. $2i(7 + 3i)$

47. $-7i(3 - 4i)$

48. $-4i(6 - 5i)$

49. $(1 + i)(3 + 2i)$

50. $(1 + 5i)(4 + 3i)$

51. $(6 - 5i)(3 + 4i)$

52. $(5 - 6i)(2 + 5i)$

53. $(7 - 2i)(2 - 6i)$

54. $(-4 + 5i)(3 - 4i)$

55. $(-2 + 3i)(-2 + 5i)$

56. $(-3 + 6i)(-3 + 4i)$

57. $(-5 - 4i)(3 + 7i)$

58. $(2 + 9i)(-3 - 5i)$

59. $(4 - 2i)^2$

60. $(1 - 2i)^2$

61. $(2 + 3i)^2$

62. $(3 + 2i)^2$

63. $(-2 + 3i)^2$

64. $(-5 - 2i)^2$

65. $\dfrac{7}{4 + i}$

66. $\dfrac{3}{8 + i}$

67. $\dfrac{2}{3 - 2i}$

68. $\dfrac{4}{2 - 3i}$

69. $\dfrac{2i}{5 + 3i}$

70. $\dfrac{3i}{4 + 2i}$

71. $\dfrac{5}{6i}$

72. $\dfrac{4}{7i}$

73. $\dfrac{5 - 3i}{4i}$

74. $\dfrac{2 + 7i}{5i}$

Aha! **75.** $\dfrac{7i + 14}{7i}$

76. $\dfrac{6i + 3}{3i}$

77. $\dfrac{4 + 5i}{3 - 7i}$

78. $\dfrac{5 + 3i}{7 - 4i}$

79. $\dfrac{2 + 3i}{2 + 5i}$

80. $\dfrac{3 + 2i}{4 + 3i}$

81. $\dfrac{3 - 2i}{4 + 3i}$

82. $\dfrac{5 - 2i}{3 + 6i}$

Simplify.

83. i^7 **84.** i^{11} **85.** i^{24}

86. i^{35} **87.** i^{42} **88.** i^{64}

89. i^9 **90.** $(-i)^{71}$ **91.** $(-i)^6$

92. $(-i)^4$ **93.** $(5i)^3$ **94.** $(-3i)^5$

95. $i^2 + i^4$ **96.** $5i^5 + 4i^3$

97. Is the product of two imaginary numbers always an imaginary number? Why or why not?

98. In what way(s) are conjugates of complex numbers similar to the conjugates used in Section 7.5?

SKILL MAINTENANCE

Graph. [7.3]

99. $f(x) = 3x - 5$ **100.** $g(x) = -2x + 6$

101. $F(x) = x^2$ **102.** $G(x) = |x|$

Solve.

103. $28 = 3x^2 - 17x$ [5.7]

104. $|3x + 7| < 22$ [9.3]

SYNTHESIS

105. Is the set of real numbers a subset of the set of complex numbers? Why or why not?

106. Is the union of the set of imaginary numbers and the set of real numbers the set of complex numbers? Why or why not?

A function g is given by

$$g(z) = \dfrac{z^4 - z^2}{z - 1}.$$

107. Find $g(3i)$.

108. Find $g(1 + i)$.

109. Find $g(5i - 1)$.

110. Find $g(2 - 3i)$.

111. Evaluate

$$\dfrac{1}{w - w^2} \quad \text{for} \quad w = \dfrac{1 - i}{10}.$$

Simplify.

112. $\dfrac{i^5 + i^6 + i^7 + i^8}{(1 - i)^4}$

113. $(1 - i)^3(1 + i)^3$

114. $\dfrac{5 - \sqrt{5}i}{\sqrt{5}i}$

115. $\dfrac{6}{1 + \dfrac{3}{i}}$

116. $\left(\dfrac{1}{2} - \dfrac{1}{3}i\right)^2 - \left(\dfrac{1}{2} + \dfrac{1}{3}i\right)^2$

117. $\dfrac{i - i^{38}}{1 + i}$

10 Study Summary

Radical expressions involve **square roots**, **cube roots**, and, more generally, **nth roots** (pp. 634, 638, and 639):

c is a square root of a if $c^2 = a$;

c is a cube root of a if $c^3 = a$.

The notation $\sqrt{a}$ indicates the **principal** square root of a, the notation $\sqrt[3]{a}$ indicates the cube root of a, and $\sqrt[n]{a}$ indicates the nth root of a (p. 639). For these and other **radical expressions**, we use the following vocabulary (pp. 635, 639):

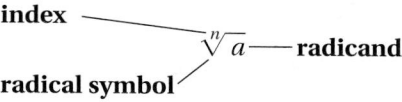

Radical notation can also be written using **rational exponents** (p. 643):

$a^{1/n}$ means $\sqrt[n]{a}$. When a is nonnegative, n can be any natural number greater than 1. When a is negative, n must be odd.

For any natural numbers m and n ($n \neq 1$), and any real number a,

$a^{m/n}$ means $\left(\sqrt[n]{a}\right)^m$ or $\sqrt[n]{a^m}$.

When a is negative, n must be odd.

Radical expressions can be added, subtracted, multiplied, and divided. For multiplication and division, the product and quotient rules, respectively, are used (pp. 651, 658):

The Product Rule for Radicals

For any real numbers $\sqrt[n]{a}$ and $\sqrt[n]{b}$, $\sqrt[n]{a}\,\sqrt[n]{b} = \sqrt[n]{a \cdot b}$.

The Quotient Rule for Radicals

For any real numbers $\sqrt[n]{a}$ and $\sqrt[n]{b}$, $b \neq 0$, $\sqrt[n]{\dfrac{a}{b}} = \dfrac{\sqrt[n]{a}}{\sqrt[n]{b}}$.

When working with radical expressions, it is understood that all results should be written in simplified form. Rational exponents can be especially useful when simplifying certain products or quotients:

Some Ways to Simplify Radical Expressions

1. *Simplifying by factoring.* Factor the radicand and look for factors raised to powers that are divisible by the index.

   Example: $\sqrt[3]{a^6 b} = \sqrt[3]{a^6}\, \sqrt[3]{b} = a^2 \sqrt[3]{b}$

2. *Using rational exponents to simplify.* Convert to exponential notation and then use arithmetic and the laws of exponents to simplify the exponents. Then convert back to radical notation as needed.

   Example: $\sqrt[3]{p} \cdot \sqrt[4]{q^3} = p^{1/3} \cdot q^{3/4}$
   $$= p^{4/12} \cdot q^{9/12}$$
   $$= \sqrt[12]{p^4 q^9}$$

3. *Combining like radical terms.*

   Example: $\sqrt{8} + 3\sqrt{2} = \sqrt{4} \cdot \sqrt{2} + 3\sqrt{2}$
   $$= 2\sqrt{2} + 3\sqrt{2} = 5\sqrt{2}$$

There sometimes arises a need to **rationalize** a numerator or a denominator (pp. 660, 662). This is accomplished by multiplying the original expression by a form of 1 that is written using the **conjugate** of the portion of the fraction being rationalized (p. 667):

$$\frac{3}{2 + \sqrt{5}} = \frac{3}{2 + \sqrt{5}} \cdot \frac{2 - \sqrt{5}}{2 - \sqrt{5}} \longleftarrow \text{This is the conjugate of } 2 + \sqrt{5}.$$
$$= \frac{6 - 3\sqrt{5}}{4 - 5} = \frac{6 - 3\sqrt{5}}{-1} = -6 + 3\sqrt{5}.$$

Radical equations are solved using the **principle of powers** (p. 673): If $a = b$, then $a^n = b^n$ for any exponent n. Because the principle of powers does not always produce equivalent equations, it is important that the solutions be checked in the original equation.

To solve an equation with a radical term:

1. Isolate the radical term on one side of the equation.
2. Use the principle of powers and solve the resulting equation.
3. Check any possible solution in the original equation.

To solve an equation with two or more radical terms:

1. Isolate one of the radical terms.
2. Use the principle of powers.
3. If a radical remains, repeat steps (1) and (2).
4. Solve the resulting equation.
5. Check possible solutions in the original equation.

Radical notation arises frequently when we are using the **Pythagorean theorem** and in work with **isosceles** or **30°–60°–90° right triangles** (pp. 681, 684):

The Pythagorean Theorem

For any right triangle with legs of lengths a and b and hypotenuse of length c,

$$a^2 + b^2 = c^2.$$

Special Triangles

The length of the hypotenuse in an isosceles right triangle is the length of a leg times $\sqrt{2}$.

The length of the longer leg in a 30°–60°–90° right triangle is the length of the shorter leg times $\sqrt{3}$. The hypotenuse is twice as long as the shorter leg.

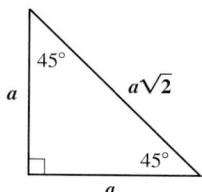

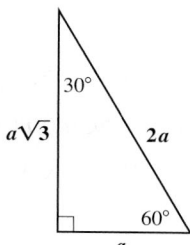

Real numbers and **imaginary** numbers make up the **complex** numbers (pp. 689, 690). Addition, subtraction, multiplication, and division of complex numbers make use of the facts that $i = \sqrt{-1}$ and the product of a complex number and its **conjugate** is a real number (p. 692):

$$(3 + 2i) + (4 - 7i) = 7 - 5i;$$

$$(8 + 6i) - (5 + 2i) = 3 + 4i;$$

$$(2 + 3i)(4 - i) = 8 - 2i + 12i - 3i^2$$
$$= 8 + 10i - 3(-1) = 11 + 10i;$$

$$\frac{1 - 4i}{3 - 2i} = \frac{1 - 4i}{3 - 2i} \cdot \frac{3 + 2i}{3 + 2i} \longleftarrow \text{The conjugate of } 3 - 2i$$

$$= \frac{3 + 2i - 12i - 8i^2}{9 + 6i - 6i - 4i^2}$$

$$= \frac{3 - 10i - 8(-1)}{9 - 4(-1)} = \frac{11 - 10i}{13} = \frac{11}{13} - \frac{10}{13}i.$$

10 Review Exercises

🔖 *Concept Reinforcement*   *Classify each of the following as either true or false.*

1. $\sqrt{ab} = \sqrt{a} \cdot \sqrt{b}$ for any real numbers $\sqrt{a}$ and $\sqrt{b}$. [10.3]

2. $\sqrt{a^2} = a$, for any real number a. [10.1]

3. $\sqrt[3]{a^3} = a$, for any real number a. [10.1]

4. $x^{2/5}$ means $\sqrt[5]{x^2}$ and $(\sqrt[5]{x})^2$. [10.2]

5. A hypotenuse is never shorter than either leg. [10.7]

6. $i^{13} = (i^2)^6 i = i$. [10.8]

7. Some radical equations have no solution. [10.6]

8. If $f(x) = \sqrt{x - 5}$, then the domain of f is the set of all nonnegative real numbers. [10.1]

Simplify. [10.1]

9. $\sqrt{\dfrac{49}{9}}$

10. $-\sqrt{0.25}$

Let $f(x) = \sqrt{2x - 7}$. Find the following. [10.1]

11. $f(16)$

12. The domain of f

Simplify. Assume that each variable can represent any real number.

13. $\sqrt{25t^2}$ [10.1]

14. $\sqrt{(c + 8)^2}$ [10.1]

15. $\sqrt{x^2 - 6x + 9}$ [10.1]

16. $\sqrt{4x^2 + 4x + 1}$ [10.1]

17. $\sqrt[5]{-32}$ [10.1]

18. $\sqrt[3]{-\dfrac{64x^6}{27}}$ [10.4]

19. $\sqrt[4]{x^{12}y^8}$ [10.3]

20. $\sqrt[6]{64x^{12}}$ [10.3]

21. Write an equivalent expression using exponential notation: $(\sqrt[3]{5ab})^4$. [10.2]

22. Write an equivalent expression using radical notation: $(16a^6)^{3/4}$. [10.2]

Use rational exponents to simplify. Assume x, $y \geq 0$. [10.2]

23. $\sqrt{x^6 y^{10}}$

24. $(\sqrt[6]{x^2 y})^2$

Simplify. Do not use negative exponents in the answers. [10.2]

25. $(x^{-2/3})^{3/5}$

26. $\dfrac{7^{-1/3}}{7^{-1/2}}$

27. If $f(x) = \sqrt{25(x - 6)^2}$, find a simplified form for $f(x)$. [10.3]

Perform the indicated operation and, if possible, simplify. Write all answers using radical notation.

28. $\sqrt{2x}\sqrt{3y}$ [10.3]

29. $\sqrt[3]{a^5 b}\sqrt[3]{27b}$ [10.3]

30. $\sqrt[3]{-24x^{10}y^8}\sqrt[3]{18x^7y^4}$ [10.3]

31. $\dfrac{\sqrt[3]{60xy^3}}{\sqrt[3]{10x}}$ [10.4]

32. $\dfrac{\sqrt{75x}}{2\sqrt{3}}$ [10.4]

33. $\sqrt[4]{\dfrac{48a^{11}}{c^8}}$ [10.4]

34. $5\sqrt[3]{x} + 2\sqrt[3]{x}$ [10.5]

35. $2\sqrt{75} - 9\sqrt{3}$ [10.5]

36. $\sqrt[3]{8x^4} + \sqrt[3]{xy^6}$ [10.5]

37. $\sqrt{50} + 2\sqrt{18} + \sqrt{32}$ [10.5]

38. $(\sqrt{3} - 3\sqrt{8})(\sqrt{5} + 2\sqrt{8})$ [10.5]

39. $\sqrt[4]{x}\sqrt{x}$ [10.5]

40. $\dfrac{\sqrt[3]{x^2}}{\sqrt[4]{x}}$ [10.5]

41. If $f(x) = x^2$, find $f(a - \sqrt{2})$. [10.5]

42. Rationalize the denominator:
$$\dfrac{4\sqrt{5}}{\sqrt{2} + \sqrt{3}}. \text{ [10.5]}$$

43. Rationalize the numerator of the expression in Exercise 42. [10.5]

Solve. [10.6]

44. $\sqrt{y+6} - 2 = 3$

45. $(x+1)^{1/3} = -5$

46. $1 + \sqrt{x} = \sqrt{3x-3}$

47. If $f(x) = \sqrt[4]{x+2}$, find a such that $f(a) = 2$. [10.6]

Solve. Give an exact answer and, where appropriate, an approximation to three decimal places. [10.7]

48. The diagonal of a square has length 10 cm. Find the length of a side of the square.

49. A skate-park jump has a ramp that is 6 ft long and is 2 ft high. How long is its base?

6 ft   2 ft

?

50. Find the missing lengths. Give exact answers and, where appropriate, an approximation to three decimal places.

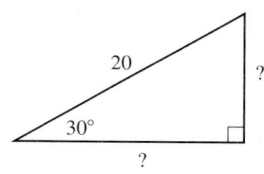

20

?

30°

?

51. Express in terms of i and simplify: $-\sqrt{-8}$. [10.8]

52. Add: $(-4 + 3i) + (2 - 12i)$. [10.8]

53. Subtract: $(9 - 7i) - (3 - 8i)$. [10.8]

Simplify. [10.8]

54. $(2 + 5i)(2 - 5i)$

55. i^{18}

56. Solve:
$$\sqrt{11x + \sqrt{6+x}} = 6. \quad [10.6]$$

57. Simplify:
$$\frac{2}{1-3i} - \frac{3}{4+2i}. \quad [10.8]$$

58. Simplify: $(6 - 3i)(2 - i)$. [10.8]

59. Divide and simplify to the form $a + bi$:
$$\frac{7-2i}{3+4i}. \quad [10.8]$$

60. What makes some complex numbers real and others imaginary? [10.8]

SYNTHESIS

61. Explain why $\sqrt[n]{x^n} = |x|$ when n is even, but $\sqrt[n]{x^n} = x$ when n is odd. [10.1]

62. Write a quotient of two imaginary numbers that is a real number (answers may vary). [10.8]

10 Chapter Test

Simplify. Assume that variables can represent any real number.

1. $\sqrt{50}$

2. $\sqrt[3]{-\dfrac{8}{x^6}}$

3. $\sqrt{81a^2}$

4. $\sqrt{x^2 - 8x + 16}$

5. $\sqrt[5]{x^{12}y^8}$

6. $\sqrt{\dfrac{25x^2}{36y^4}}$

7. $\sqrt[3]{3z}\,\sqrt[3]{5y^2}$

8. $\dfrac{\sqrt[5]{x^3y^4}}{\sqrt[5]{xy^2}}$

9. $\sqrt[4]{x^3y^2}\,\sqrt{xy}$

10. $\dfrac{\sqrt[5]{a^2}}{\sqrt[4]{a}}$

11. $8\sqrt{2} - 2\sqrt{2}$

12. $\sqrt{x^4y} + \sqrt{9y^3}$

13. $\left(7 + \sqrt{x}\right)\left(2 - 3\sqrt{x}\right)$

14. Write an equivalent expression using exponential notation: $\sqrt{7xy}$.

15. Write an equivalent expression using radical notation: $(4a^3b)^{5/6}$.

16. If $f(x) = \sqrt{2x - 10}$, determine the domain of f.

17. If $f(x) = x^2$, find $f(5 + \sqrt{2})$.

18. Rationalize the denominator:
$$\dfrac{\sqrt{3}}{5 + \sqrt{2}}.$$

Solve.

19. $x = \sqrt{2x - 5} + 4$

20. $\sqrt{x} = \sqrt{x + 1} - 5$

Solve. Give exact answers and approximations to three decimal places.

21. One leg of a 30°–60°–90° right triangle is 7 cm long. Find the possible lengths of the other leg.

22. A referee jogs diagonally from one corner of a 50-ft by 90-ft basketball court to the far corner. How far does she jog?

23. Express in terms of i and simplify: $\sqrt{-50}$.

24. Subtract: $(9 + 8i) - (-3 + 6i)$.

25. Multiply: $\sqrt{-16}\,\sqrt{-36}$.

26. Multiply. Write the answer in the form $a + bi$.
$$(4 - i)^2$$

27. Divide and simplify to the form $a + bi$:
$$\dfrac{-2 + i}{3 - 5i}.$$

28. Simplify: i^{37}.

SYNTHESIS

29. Solve:
$$\sqrt{2x - 2} + \sqrt{7x + 4} = \sqrt{13x + 10}.$$

30. Simplify:
$$\dfrac{1 - 4i}{4i(1 + 4i)^{-1}}.$$

31. Drake's Discount Shoe Center has two locations. The sign at the original location is shaped like an isosceles right triangle. The sign at the newer location is shaped like a 30°–60°–90° triangle. The hypotenuse of each sign measures 6 ft. Which sign has the greater area and by how much? (Round to three decimal places.)

11

Quadratic Functions and Equations

AN APPLICATION

An athlete's hang time T, in seconds, is related to vertical leap V, in inches, by the formula $V = 48T^2$. The NBA's Steve Francis reportedly has a vertical leap of 45 in. What is his hang time?

This problem appears as Exercise 37 in Section 11.3.

Elizabeth Bluebird
PHYSICAL THERAPIST
Oklahoma City, Oklahoma

A treadmill test helps me to set up an exercise program for people with heart disease. I use an algebraic formula to figure a single-stage treadmill test using variables such as age, heart rate and speed. I also use math every day when measuring a person's range of motion—the arc or angle through which a body part moves. Knowing this angle, I can help make a diagnosis and show progress in my patient's therapy.

*I*n translating problem situations to mathematics, we often obtain a function or equation containing a second-degree polynomial in one variable. Such functions or equations are said to be quadratic. In this chapter, we examine a variety of equations, inequalities, and applications for which we will need to solve quadratic equations or graph quadratic functions.

11.1   Quadratic Equations

The Principle of Square Roots • Completing the Square • Problem Solving

 ALGEBRAIC—GRAPHICAL CONNECTION

Let's reexamine the graphical connections to the algebraic equation-solving concepts we have studied before.

In Chapter 7, we introduced the graph of a quadratic function given by

$$f(x) = ax^2 + bx + c, \quad a \neq 0.$$

For example, the graphs of $f(x) = x^2 + 6x + 8$ and $g(x) = 0$ are shown below.

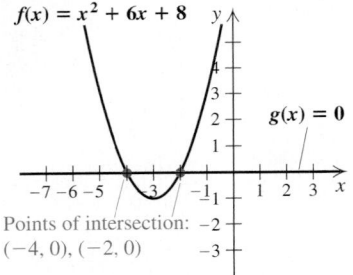

Note that $(-4, 0)$ and $(-2, 0)$ are the points of intersection of the graphs of $f(x) = x^2 + 6x + 8$ and $g(x) = 0$ (the x-axis). In Sections 11.6 and 11.7, we will develop efficient ways to graph quadratic functions. For now, the graphs simply aid in visualizing solutions.

In Chapter 5, we solved equations like $x^2 + 6x + 8 = 0$ by factoring:

$$x^2 + 6x + 8 = 0$$

$$(x + 4)(x + 2) = 0 \qquad \text{Factoring}$$

$$x + 4 = 0 \quad or \quad x + 2 = 0 \qquad \text{Using the principle of zero products}$$

$$x = -4 \quad or \qquad x = -2.$$

Note that -4 and -2 are the first coordinates of the points of intersection (or the x-intercepts) above.

In this section and the next, we develop algebraic methods for solving *any* quadratic equation, whether it is factorable or not.

EXAMPLE 1   Solve: $x^2 = 25$.

Solution   We have

$$x^2 = 25$$
$$x^2 - 25 = 0 \qquad \text{Writing in standard form}$$
$$(x - 5)(x + 5) = 0 \qquad \text{Factoring}$$
$$x - 5 = 0 \quad or \quad x + 5 = 0 \qquad \text{Using the principle of zero products}$$
$$x = 5 \quad or \qquad x = -5.$$

The solutions are 5 and -5. A graph in which $f(x) = x^2$ represents the left side of the original equation and $g(x) = 25$ represents the right side provides a check (see the figure at left). Of course, we can also check by substituting 5 and -5 into the original equation.

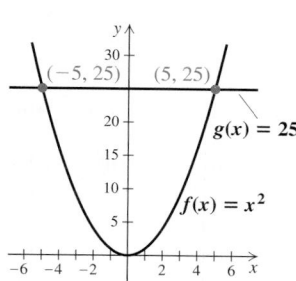

A visualization of Example 1

The Principle of Square Roots

Let's reconsider $x^2 = 25$. We know from Chapter 10 that the number 25 has two real-number square roots, 5 and -5, the solutions of the equation in Example 1. Thus we see that square roots can provide quick solutions for equations of the type $x^2 = k$.

> **The Principle of Square Roots**
> For any real number k, if $x^2 = k$, then $x = \sqrt{k}$ or $x = -\sqrt{k}$.

EXAMPLE 2   Solve: $3x^2 = 6$. Give exact solutions and approximations to three decimal places.

Solution   We have

$$3x^2 = 6$$
$$x^2 = 2 \qquad \text{Isolating } x^2$$
$$x = \sqrt{2} \quad or \quad x = -\sqrt{2}. \qquad \text{Using the principle of square roots}$$

We often use the symbol $\pm\sqrt{2}$ to represent both of the solutions.

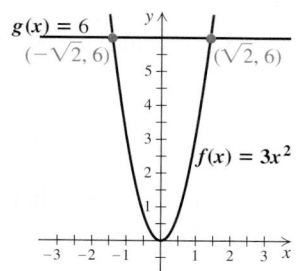

A visualization of Example 2

Check: For $\sqrt{2}$:

$$\frac{3x^2 = 6}{3(\sqrt{2})^2 \;\Big|\; 6}$$
$$3 \cdot 2$$
$$6 \overset{?}{=} 6 \quad \text{TRUE}$$

For $-\sqrt{2}$:

$$\frac{3x^2 = 6}{3(-\sqrt{2})^2 \;\Big|\; 6}$$
$$3 \cdot 2$$
$$6 \overset{?}{=} 6 \quad \text{TRUE}$$

The solutions are $\sqrt{2}$ and $-\sqrt{2}$, or $\pm\sqrt{2}$, which round to 1.414 and -1.414.

EXAMPLE 3   Solve: $-5x^2 + 2 = 0$.

Solution   We have

$$-5x^2 + 2 = 0$$

$$x^2 = \frac{2}{5} \qquad \text{Isolating } x^2$$

$$x = \sqrt{\frac{2}{5}} \quad or \quad x = -\sqrt{\frac{2}{5}}. \qquad \text{Using the principle of square roots}$$

The solutions are $\sqrt{\dfrac{2}{5}}$ and $-\sqrt{\dfrac{2}{5}}$. This can also be written as $\pm\sqrt{\dfrac{2}{5}}$. In the days before calculators, it was standard practice to rationalize denominators when simplifying answers. If we rationalize the denominator, the solutions are $\pm\dfrac{\sqrt{10}}{5}$. The checks are left to the student.

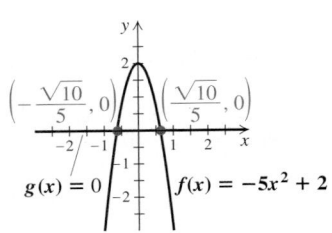

A visualization of Example 3

Sometimes we get solutions that are imaginary numbers.

EXAMPLE 4   Solve: $4x^2 + 9 = 0$.

Solution   We have

$$4x^2 + 9 = 0$$

$$x^2 = -\frac{9}{4} \qquad \text{Isolating } x^2$$

$$x = \sqrt{-\frac{9}{4}} \quad or \quad x = -\sqrt{-\frac{9}{4}} \qquad \text{Using the principle of square roots}$$

$$x = \sqrt{\tfrac{9}{4}}\,\sqrt{-1} \quad or \quad x = -\sqrt{\tfrac{9}{4}}\,\sqrt{-1}$$

$$x = \tfrac{3}{2}i \qquad\qquad or \quad x = -\tfrac{3}{2}i. \qquad \text{Recall that } \sqrt{-1} = i.$$

Check: For $\tfrac{3}{2}i$:

$$\begin{array}{c|c} 4x^2 + 9 = 0 & \\ \hline 4\left(\tfrac{3}{2}i\right)^2 + 9 & 0 \\ 4 \cdot \tfrac{9}{4} \cdot i^2 + 9 & \\ 9(-1) + 9 & \\ & 0 \overset{?}{=} 0 \quad \text{TRUE} \end{array}$$

For $-\tfrac{3}{2}i$:

$$\begin{array}{c|c} 4x^2 + 9 = 0 & \\ \hline 4\left(-\tfrac{3}{2}i\right)^2 + 9 & 0 \\ 4 \cdot \tfrac{9}{4} \cdot i^2 + 9 & \\ 9(-1) + 9 & \\ & 0 \overset{?}{=} 0 \quad \text{TRUE} \end{array}$$

The solutions are $\tfrac{3}{2}i$ and $-\tfrac{3}{2}i$, or $\pm\tfrac{3}{2}i$. The graph at left confirms that there are no real-number solutions.

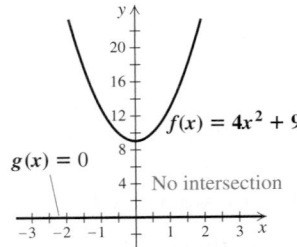

A visualization of Example 4

The principle of square roots can be restated in a more general form that pertains to more complicated algebraic expressions than just x.

> **The Principle of Square Roots (Generalized Form)**
> For any real number k and any algebraic expression X,
>
> $$\text{If } X^2 = k, \quad \text{then} \quad X = \sqrt{k} \quad \text{or} \quad X = -\sqrt{k}.$$

EXAMPLE 5   Let $f(x) = (x - 2)^2$. Find all x-values for which $f(x) = 7$.

Solution   We are asked to find all x-values for which

$$f(x) = 7,$$

or

$$(x - 2)^2 = 7. \qquad \text{Substituting } (x - 2)^2 \text{ for } f(x)$$

The generalized principle of square roots gives us

$$x - 2 = \sqrt{7} \qquad or \quad x - 2 = -\sqrt{7} \qquad \text{Using the principle of square roots}$$

$$x = 2 + \sqrt{7} \quad or \qquad x = 2 - \sqrt{7}.$$

Check: $f\left(2 + \sqrt{7}\right) = \left(2 + \sqrt{7} - 2\right)^2 = \left(\sqrt{7}\right)^2 = 7.$

Similarly,

$$f\left(2 - \sqrt{7}\right) = \left(2 - \sqrt{7} - 2\right)^2 = \left(-\sqrt{7}\right)^2 = 7.$$

The solutions are $2 + \sqrt{7}$ and $2 - \sqrt{7}$, or simply $2 \pm \sqrt{7}$.

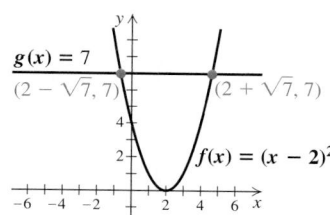

A visualization of Example 5

Example 5 is of the form $(x - a)^2 = c$, where a and c are constants. Sometimes we must factor in order to obtain this form.

EXAMPLE 6   Solve: $x^2 + 6x + 9 = 2$.

Solution   We have

$$x^2 + 6x + 9 = 2 \qquad \text{The left side is the square of a binomial.}$$
$$(x + 3)^2 = 2 \qquad \text{Factoring}$$
$$x + 3 = \sqrt{2} \qquad or \quad x + 3 = -\sqrt{2} \qquad \text{Using the principle of square roots}$$
$$x = -3 + \sqrt{2} \quad or \qquad x = -3 - \sqrt{2}. \qquad \text{Adding } -3 \text{ to both sides}$$

The solutions are $-3 + \sqrt{2}$ and $-3 - \sqrt{2}$, or $-3 \pm \sqrt{2}$. The checks are left to the student.

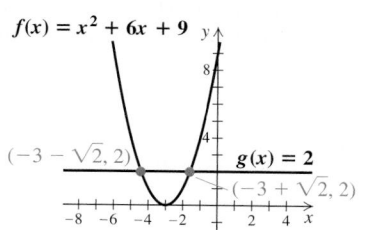

A visualization of Example 6

Completing the Square

Not all quadratic equations can be solved as we did Examples 1–6. By using a method called *completing the square*, we can use the principle of square roots to solve *any* quadratic equation.

EXAMPLE 7   Solve: $x^2 + 6x + 4 = 0$.

Solution   We have

$$x^2 + 6x + 4 = 0$$

$$x^2 + 6x \quad\;\; = -4 \qquad\qquad \text{Subtracting 4 from both sides}$$

$$x^2 + 6x + 9 = -4 + 9 \qquad \text{Adding 9 to both sides. We explain this shortly.}$$

$$(x + 3)^2 = 5 \qquad\qquad\; \text{Factoring the perfect-square trinomial}$$

$$x + 3 = \pm\sqrt{5} \qquad\quad \text{Using the principle of square roots. Remember that } \pm\sqrt{5} \text{ represents two numbers.}$$

$$x = -3 \pm \sqrt{5}. \qquad \text{Adding } -3 \text{ to both sides}$$

Check: For $-3 + \sqrt{5}$:

$$\begin{array}{c|c} x^2 + 6x + 4 = 0 \\ \hline \left(-3 + \sqrt{5}\right)^2 + 6\left(-3 + \sqrt{5}\right) + 4 & 0 \\ 9 - 6\sqrt{5} + 5 - 18 + 6\sqrt{5} + 4 \\ 9 + 5 - 18 + 4 - 6\sqrt{5} + 6\sqrt{5} \\ 0 \overset{?}{=} 0 \quad \text{\small TRUE} \end{array}$$

For $-3 - \sqrt{5}$:

$$\begin{array}{c|c} x^2 + 6x + 4 = 0 \\ \hline \left(-3 - \sqrt{5}\right)^2 + 6\left(-3 - \sqrt{5}\right) + 4 & 0 \\ 9 + 6\sqrt{5} + 5 - 18 - 6\sqrt{5} + 4 \\ 9 + 5 - 18 + 4 + 6\sqrt{5} - 6\sqrt{5} \\ 0 \overset{?}{=} 0 \quad \text{\small TRUE} \end{array}$$

The solutions are $-3 + \sqrt{5}$ and $-3 - \sqrt{5}$, or $-3 \pm \sqrt{5}$.

technology connection

One way to check Example 7 is to store $-3 + \sqrt{5}$ as x using the **STO** key. We can then evaluate $x^2 + 6x + 4$ by entering $x^2 + 6x + 4$ and pressing **ENTER**.

1. Check Example 7 using the method described above.

In Example 7, we chose to add 9 to both sides because it creates a perfect-square trinomial on the left side. The 9 was determined by taking half of the coefficient of x and squaring it—that is,

$$\left(\tfrac{1}{2} \cdot 6\right)^2 = 3^2, \quad \text{or} \quad 9.$$

To help see why this procedure works, examine the following drawings.

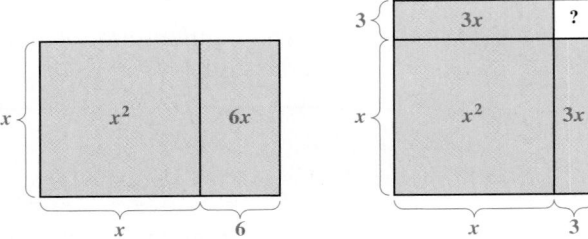

Note that the shaded areas in both figures represent the same area, $x^2 + 6x$. However, only the figure on the right, in which the $6x$ is halved, can be converted into a square with the addition of a constant term. The constant 9 is the "missing" piece that *completes* the square.

To complete the square for $x^2 + bx$, we add $(b/2)^2$.

Example 8, which follows, provides practice in finding numbers that complete the square. We will then use this skill to solve equations.

EXAMPLE 8   Replace the blanks in each equation with constants to form a true equation.

a) $x^2 + 14x + \underline{\quad} = (x + \underline{\quad})^2$
b) $x^2 - 5x + \underline{\quad} = (x - \underline{\quad})^2$
c) $x^2 + \frac{3}{4}x + \underline{\quad} = (x + \underline{\quad})^2$

Solution   We take half of the coefficient of x and square it.

a) Half of 14 is 7, and $7^2 = 49$. Thus, $x^2 + 14x + 49$ is a perfect-square trinomial and is equivalent to $(x + 7)^2$. We have

$$x^2 + 14x + 49 = (x + 7)^2.$$

b) Half of -5 is $-\frac{5}{2}$, and $\left(-\frac{5}{2}\right)^2 = \frac{25}{4}$. Thus, $x^2 - 5x + \frac{25}{4}$ is a perfect-square trinomial and is equivalent to $\left(x - \frac{5}{2}\right)^2$. We have

$$x^2 - 5x + \frac{25}{4} = \left(x - \frac{5}{2}\right)^2.$$

c) Half of $\frac{3}{4}$ is $\frac{3}{8}$, and $\left(\frac{3}{8}\right)^2 = \frac{9}{64}$. Thus, $x^2 + \frac{3}{4}x + \frac{9}{64}$ is a perfect-square trinomial and is equivalent to $\left(x + \frac{3}{8}\right)^2$. We have

$$x^2 + \frac{3}{4}x + \frac{9}{64} = \left(x + \frac{3}{8}\right)^2.$$

Student Notes

In problems like Examples 8(b) and (c), it is best to avoid decimal notation. Most students have an easier time recognizing $\frac{9}{64}$ as $\left(\frac{3}{8}\right)^2$ than regarding 0.140625 as 0.375^2.

We can now use the method of completing the square to solve equations similar to Example 7.

EXAMPLE 9   Solve: **(a)** $x^2 - 8x - 7 = 0$; **(b)** $x^2 + 5x - 3 = 0$.

Solution

a) $x^2 - 8x - 7 = 0$

$x^2 - 8x \qquad = 7$ Adding 7 to both sides. We can now complete the square on the left side.

$x^2 - 8x + 16 = 7 + 16$ Adding 16 to both sides to complete the square: $\frac{1}{2}(-8) = -4$, and $(-4)^2 = 16$

$(x - 4)^2 = 23$ Factoring and simplifying

$x - 4 = \pm\sqrt{23}$ Using the principle of square roots

$x = 4 \pm \sqrt{23}$ Adding 4 to both sides

The solutions are $4 - \sqrt{23}$ and $4 + \sqrt{23}$, or $4 \pm \sqrt{23}$. The checks are left to the student.

b) $x^2 + 5x - 3 = 0$

$x^2 + 5x = 3$ Adding 3 to both sides

$x^2 + 5x + \dfrac{25}{4} = 3 + \dfrac{25}{4}$ Completing the square: $\frac{1}{2} \cdot 5 = \frac{5}{2}$, and $\left(\frac{5}{2}\right)^2 = \frac{25}{4}$

$\left(x + \dfrac{5}{2}\right)^2 = \dfrac{37}{4}$ Factoring and simplifying

$x + \dfrac{5}{2} = \pm\dfrac{\sqrt{37}}{2}$ Using the principle of square roots and the quotient rule for radicals

$x = -\dfrac{5}{2} \pm \dfrac{\sqrt{37}}{2}$, or $\dfrac{-5 \pm \sqrt{37}}{2}$ Adding $-\frac{5}{2}$ to both sides

The checks are left to the student. The solutions are $-\dfrac{5}{2} \pm \dfrac{\sqrt{37}}{2}$, or $\dfrac{-5 \pm \sqrt{37}}{2}$. This can be written as $-\dfrac{5}{2} - \dfrac{\sqrt{37}}{2}$ and $-\dfrac{5}{2} + \dfrac{\sqrt{37}}{2}$, or $\dfrac{-5 - \sqrt{37}}{2}$ and $\dfrac{-5 + \sqrt{37}}{2}$.

Before we complete the square, the x^2-coefficient must be 1. When it is not 1, we divide both sides of the equation by whatever that coefficient may be.

EXAMPLE 10   Find the x-intercepts of the function given by $f(x) = 3x^2 + 7x - 2$.

Solution   The value of $f(x)$ must be 0 at any x-intercepts. Thus,

$f(x) = 0$ We set $f(x)$ equal to 0.

$3x^2 + 7x - 2 = 0$ Substituting

$3x^2 + 7x = 2$ Adding 2 to both sides

$x^2 + \dfrac{7}{3}x = \dfrac{2}{3}$ Dividing both sides by 3

$x^2 + \dfrac{7}{3}x + \dfrac{49}{36} = \dfrac{2}{3} + \dfrac{49}{36}$ Completing the square: $\left(\frac{1}{2} \cdot \frac{7}{3}\right)^2 = \frac{49}{36}$

$\left(x + \dfrac{7}{6}\right)^2 = \dfrac{73}{36}$ Factoring and simplifying

$x + \dfrac{7}{6} = \pm\dfrac{\sqrt{73}}{6}$ Using the principle of square roots and the quotient rule for radicals

$x = -\dfrac{7}{6} \pm \dfrac{\sqrt{73}}{6}$, or $\dfrac{-7 \pm \sqrt{73}}{6}$.   Adding $-\frac{7}{6}$ to both sides

The x-intercepts are

$\left(-\dfrac{7}{6} - \dfrac{\sqrt{73}}{6}, 0\right)$ and $\left(-\dfrac{7}{6} + \dfrac{\sqrt{73}}{6}, 0\right)$, or

$\left(\dfrac{-7 - \sqrt{73}}{6}, 0\right)$ and $\left(\dfrac{-7 + \sqrt{73}}{6}, 0\right)$.

The checks are left to the student.

$f(x) = 3x^2 + 7x - 2$

$\left(\dfrac{-7 - \sqrt{73}}{6}, 0\right)$ $\left(\dfrac{-7 + \sqrt{73}}{6}, 0\right)$

A visualization of Example 10

The procedure used in Example 10 is important because it can be used to solve *any* quadratic equation.

To Solve a Quadratic Equation in x by Completing the Square

1. Isolate the terms with variables on one side of the equation, and arrange them in descending order.
2. Divide both sides by the coefficient of x^2 if that coefficient is not 1.
3. Complete the square by taking half of the coefficient of x and adding its square to both sides.
4. Express the trinomial as the square of a binomial (factor the trinomial) and simplify the other side.
5. Use the principle of square roots (find the square roots of both sides).
6. Solve for x by adding or subtracting on both sides.

Problem Solving

After one year, an amount of money P, invested at 4% per year, is worth 104% of P, or $P(1.04)$. If that amount continues to earn 4% interest per year, after the second year the investment will be worth 104% of $P(1.04)$, or $P(1.04)^2$. This is called **compounding interest** since after the first time period, interest is earned on both the initial investment *and* the interest from the first time period. Continuing the above pattern, we see that after the third year, the investment will be worth 104% of $P(1.04)^2$. Generalizing, we have the following.

The Compound-Interest Formula

If an amount of money P is invested at interest rate r, compounded annually, then in t years, it will grow to the amount A given by

$$A = P(1 + r)^t. \qquad (r \text{ is written in decimal notation.})$$

We can use quadratic equations to solve certain interest problems.

EXAMPLE 11   Investment growth.   Rosa invested $4000 at interest rate r, compounded annually. In 2 yr, it grew to $4410. What was the interest rate?

Solution

1. **Familiarize.**   We are already familiar with the compound-interest formula. If we were not, we would need to consult an outside source.

2. **Translate.**   The translation consists of substituting into the formula:

$$A = P(1 + r)^t$$
$$4410 = 4000(1 + r)^2. \qquad \text{Substituting}$$

3. Carry out. We solve for r:

$$4410 = 4000(1 + r)^2$$

$$\frac{4410}{4000} = (1 + r)^2 \qquad \text{Dividing both sides by 4000}$$

$$\frac{441}{400} = (1 + r)^2 \qquad \text{Simplifying}$$

$$\pm\sqrt{\frac{441}{400}} = 1 + r \qquad \text{Using the principle of square roots}$$

$$\pm\frac{21}{20} = 1 + r \qquad \text{Simplifying}$$

$$-\frac{20}{20} \pm \frac{21}{20} = r \qquad \text{Adding } -1, \text{ or } -\frac{20}{20}, \text{ to both sides}$$

$$\frac{1}{20} = r \quad or \quad -\frac{41}{20} = r.$$

4. Check. Since the interest rate cannot be negative, we need only check $\frac{1}{20}$, or 5%. If \$4000 were invested at 5% interest, compounded annually, then in 2 yr it would grow to $4000(1.05)^2$, or \$4410. The number 5% checks.

5. State. The interest rate was 5%.

EXAMPLE 12   *Free-falling objects.* The formula $s = 16t^2$ is used to approximate the distance s, in feet, that an object falls freely from rest in t seconds. Ireland's Cliffs of Moher are 702 ft tall (*Source*: Based on data from 4windstravel.com). How long will it take a stone to fall from the top? Round to the nearest tenth of a second.

Solution

1. Familiarize. We make a drawing to help visualize the problem and agree to disregard air resistance.

2. Translate. We substitute into the formula:

$$s = 16t^2$$
$$702 = 16t^2.$$

3. Carry out. We solve for t:

$$702 = 16t^2$$
$$\frac{702}{16} = t^2$$
$$43.875 = t^2$$
$$\sqrt{43.875} = t \qquad \text{Using the principle of square roots; rejecting the negative square root since } t \text{ cannot be negative in this problem}$$
$$6.6 \approx t. \qquad \text{Using a calculator and rounding to the nearest tenth}$$

4. Check. Since $16(6.6)^2 \approx 697 \approx 702$, our answer checks.

5. State. It takes about 6.6 sec for a stone to fall freely from the Cliffs of Moher.

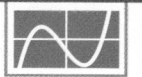 **technology connection**

As we saw in Section 5.7, a graphing calculator can be used to find approximate solutions of any quadratic equation that has real-number solutions.

To check Example 9(a), we graph $y = x^2 - 8x - 7$ and use the ZERO or ROOT option of the CALC menu. When asked for a Left and Right Bound, we enter cursor positions to the left of and to the right of the root. A Guess between the bounds is entered and a value for the root then appears.

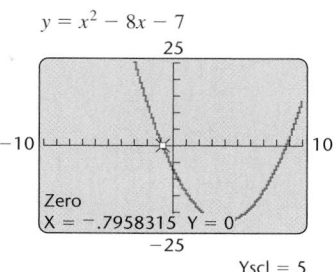

$y = x^2 - 8x - 7$

1. Use a graphing calculator to check the second solution of Example 9(a).
2. Use a graphing calculator to confirm the solutions in Examples 7 and 9(b).
3. Can a graphing calculator be used to find *exact* solutions in Example 10? Why or why not?

4. Use a graphing calculator to confirm that there are no real-number solutions of $x^2 - 6x + 11 = 0$.

Exercise Set

11.1

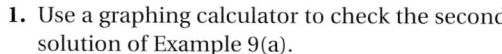 *Concept Reinforcement   Complete each of the following to form true statements.*

1. The principle of square roots states that if $x^2 = k$, then $x = $ _____ or $x = $ _____.

2. If $(x + 5)^2 = 49$, then $x + 5 = $ _____ or $x + 5 = $ _____.

3. If $t^2 + 6t + 9 = 17$, then (_____)$^2 = 17$ and _____ $= \pm\sqrt{17}$.

4. The equations $x^2 + 8x + $ _____ $= 23$ and $x^2 + 8x = 7$ are equivalent.

5. The expressions $t^2 + 10t + $ _____ and $(t + $ _____)^2 are equivalent.

6. The expressions $x^2 - 6x + $ _____ and $(x - $ _____)^2 are equivalent.

Solve.

7. $4x^2 = 20$

8. $7x^2 = 21$

9. $9x^2 + 16 = 0$

10. $25x^2 + 4 = 0$

11. $5t^2 - 7 = 0$

12. $3t^2 - 2 = 0$

13. $(x - 1)^2 = 49$

14. $(x + 2)^2 = 25$

15. $(a - 13)^2 = 18$

16. $(a + 5)^2 = 8$

17. $(x + 1)^2 = -9$

18. $(x - 1)^2 = -49$

19. $\left(y + \frac{3}{4}\right)^2 = \frac{17}{16}$

20. $\left(t + \frac{3}{2}\right)^2 = \frac{7}{2}$

21. $x^2 - 10x + 25 = 64$

22. $x^2 - 6x + 9 = 100$

23. Let $f(x) = (x - 5)^2$. Find x such that $f(x) = 16$.

24. Let $g(x) = (x - 2)^2$. Find x such that $g(x) = 25$.

25. Let $F(t) = (t + 4)^2$. Find t such that $F(t) = 13$.

26. Let $f(t) = (t + 6)^2$. Find t such that $f(t) = 15$.

Aha! 27. Let $g(x) = x^2 + 14x + 49$. Find x such that $g(x) = 49$.

28. Let $F(x) = x^2 + 8x + 16$. Find x such that $F(x) = 9$.

Replace the blanks in each equation with constants to complete the square and form a true equation.

29. $x^2 + 16x + \underline{\quad} = (x + \underline{\quad})^2$

30. $x^2 + 8x + \underline{\quad} = (x + \underline{\quad})^2$

31. $t^2 - 10t + \underline{\quad} = (t - \underline{\quad})^2$

32. $t^2 - 6t + \underline{\quad} = (t - \underline{\quad})^2$

33. $x^2 + 3x + \underline{\quad} = (x + \underline{\quad})^2$

34. $x^2 + 7x + \underline{\quad} = (x + \underline{\quad})^2$

35. $t^2 - 9t + \underline{\quad} = (t - \underline{\quad})^2$

36. $t^2 - 3t + \underline{\quad} = (t - \underline{\quad})^2$

37. $x^2 + \frac{2}{5}x + \underline{\quad} = (x + \underline{\quad})^2$

38. $x^2 + \frac{2}{3}x + \underline{\quad} = (x + \underline{\quad})^2$

39. $t^2 - \frac{5}{6}t + \underline{\quad} = (t - \underline{\quad})^2$

40. $t^2 - \frac{5}{3}t + \underline{\quad} = (t - \underline{\quad})^2$

Solve by completing the square. Show your work.

41. $x^2 + 6x = 7$

42. $x^2 + 8x = 9$

43. $t^2 - 10t = -24$

44. $t^2 - 10t = -21$

45. $x^2 + 10x + 9 = 0$

46. $x^2 + 8x + 7 = 0$

47. $t^2 + 8t - 3 = 0$

48. $t^2 + 6t - 5 = 0$

Complete the square to find the x-intercepts of each function given by the equation listed.

49. $f(x) = x^2 + 6x + 7$

50. $f(x) = x^2 + 5x + 3$

51. $g(x) = x^2 + 12x + 25$

52. $g(x) = x^2 + 4x + 2$

53. $f(x) = x^2 - 10x - 22$

54. $f(x) = x^2 - 8x - 10$

Solve by completing the square. Remember to first divide, as in Example 10, to make sure that the coefficient of x^2 is 1.

55. $9x^2 + 18x = -8$

56. $4x^2 + 8x = -3$

57. $3x^2 - 5x - 2 = 0$

58. $2x^2 - 5x - 3 = 0$

59. $5x^2 + 4x - 3 = 0$

60. $4x^2 + 3x - 5 = 0$

61. Find the x-intercepts of the function given by $f(x) = 4x^2 + 2x - 3$.

62. Find the x-intercepts of the function given by $f(x) = 3x^2 + x - 5$.

63. Find the x-intercepts of the function given by $g(x) = 2x^2 - 3x - 1$.

64. Find the x-intercepts of the function given by $g(x) = 3x^2 - 5x - 1$.

Interest. Use $A = P(1 + r)^t$ to find the interest rate in Exercises 65–70. Refer to Example 11.

65. $2000 grows to $2420 in 2 yr

66. $2560 grows to $2890 in 2 yr

67. $1280 grows to $1805 in 2 yr

68. $1000 grows to $1440 in 2 yr

69. $6250 grows to $6760 in 2 yr

70. $6250 grows to $7290 in 2 yr

Free-falling objects. Use $s = 16t^2$ for Exercises 71–74. Refer to Example 12 and neglect air resistance.

71. Suspended 1053 ft above the water, the bridge over Colorado's Royal Gorge is the world's highest bridge. How long would it take an object to fall freely from the bridge?
Source: *The Guinness Book of Records*

72. The CN Tower in Toronto, at 1815 ft, is the world's tallest self-supporting tower (no guy wires). How long would it take an object to fall freely from the top?
Source: *The Guinness Book of Records*

73. The Sears Tower in Chicago is 1454 ft tall. How long would it take an object to fall freely from the top?

74. The Gateway Arch in St. Louis is 630 ft high. How long would it take an object to fall freely from the top?
Source: www.icivilengineer.com

75. Explain in your own words a sequence of steps that can be used to solve any quadratic equation in the quickest way.

76. Write an interest-rate problem for a classmate to solve. Devise the problem so that the solution is "The loan was made at 7% interest."

SKILL MAINTENANCE

Evaluate. [1.8]

77. $at^2 - bt$, for $a = 3$, $b = 5$, and $t = 4$

78. $mn^2 - mp$, for $m = -2$, $n = 7$, and $p = 3$

Simplify. [10.3]

79. $\sqrt[3]{270}$ **80.** $\sqrt{80}$

Let $f(x) = \sqrt{3x - 5}$. [10.1]

81. Find $f(10)$.   **82.** Find $f(18)$.

SYNTHESIS

83. What would be better: to receive 3% interest every 6 months, or to receive 6% interest every 12 months? Why?

84. Write a problem involving a free-falling object for a classmate to solve (see Example 12). Devise the problem so that the solution is "The object takes about 4.5 sec to fall freely from the top of the structure."

Find b such that each trinomial is a square.

85. $x^2 + bx + 81$ **86.** $x^2 + bx + 49$

87. If $f(x) = 2x^5 - 9x^4 - 66x^3 + 45x^2 + 280x$ and $x^2 - 5$ is a factor of $f(x)$, find all a for which $f(a) = 0$.

88. If $f(x) = \left(x - \frac{1}{3}\right)(x^2 + 6)$ and $g(x) = \left(x - \frac{1}{3}\right)\left(x^2 - \frac{2}{3}\right)$, find all a for which $(f + g)(a) = 0$.

89. *Boating.*   A barge and a fishing boat leave a dock at the same time, traveling at a right angle to each other. The barge travels 7 km/h slower than the fishing boat. After 4 hr, the boats are 68 km apart. Find the speed of each boat.

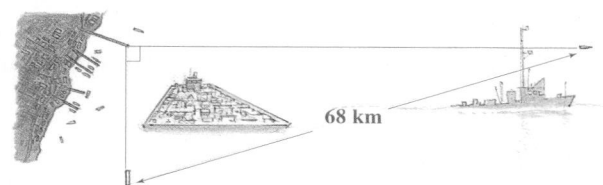

90. Find three consecutive integers such that the square of the first plus the product of the other two is 67.

91. Exercises 15, 23, and 43 can be solved on a graphing calculator without first rewriting in standard form. Simply let y_1 represent the left side of the equation and y_2 the right side. Then use a graphing calculator to determine the x-coordinate of any point of intersection. Use a graphing calculator to solve Exercises 15, 23, and 43 in this manner.

92. Use a graphing calculator to check your answers to Exercises 5, 11, 61, and 63.

93. Example 11 can be solved with a graphing calculator by graphing each side of

$$4410 = 4000(1 + r)^2.$$

How could you determine, from a reading of the problem, a suitable viewing window? What might that window be?

11.2   The Quadratic Formula

Solving Using the Quadratic Formula   •   Approximating Solutions

There are at least two reasons for learning to complete the square. One is to help graph certain equations that appear in this and other chapters. Another is to develop a general formula for solving quadratic equations.

Solving Using the Quadratic Formula

Each time we solve by completing the square, the procedure is the same. In mathematics, when a procedure is repeated many times, a formula can often be developed to speed up our work.

We begin with a quadratic equation in standard form,

$$ax^2 + bx + c = 0,$$

with $a > 0$. For $a < 0$, a slightly different derivation is needed (see Exercise 58), but the result is the same. Let's solve by completing the square. As the steps are performed, compare them with Example 10 on p. 710.

$$ax^2 + bx = -c \qquad \text{Adding } -c \text{ to both sides}$$

$$x^2 + \frac{b}{a}x = -\frac{c}{a} \qquad \text{Dividing both sides by } a$$

Half of $\dfrac{b}{a}$ is $\dfrac{b}{2a}$ and $\left(\dfrac{b}{2a}\right)^2$ is $\dfrac{b^2}{4a^2}$. We add $\dfrac{b^2}{4a^2}$ to both sides:

$$x^2 + \frac{b}{a}x + \frac{b^2}{4a^2} = -\frac{c}{a} + \frac{b^2}{4a^2} \qquad \text{Adding } \frac{b^2}{4a^2} \text{ to complete the square}$$

$$\left(x + \frac{b}{2a}\right)^2 = -\frac{4ac}{4a^2} + \frac{b^2}{4a^2} \qquad \text{Factoring on the left side; finding a common denominator on the right side}$$

$$\left(x + \frac{b}{2a}\right)^2 = \frac{b^2 - 4ac}{4a^2}$$

$$x + \frac{b}{2a} = \pm\frac{\sqrt{b^2 - 4ac}}{2a} \qquad \text{Using the principle of square roots and the quotient rule for radicals; since } a > 0, \sqrt{4a^2} = 2a$$

$$x = \frac{-b \pm \sqrt{b^2 - 4ac}}{2a}. \qquad \text{Adding } -\frac{b}{2a} \text{ to both sides}$$

It is important to remember the quadratic formula and know how to use it.

The Quadratic Formula

The solutions of $ax^2 + bx + c = 0$, $a \neq 0$, are given by

$$x = \frac{-b \pm \sqrt{b^2 - 4ac}}{2a}.$$

EXAMPLE 1   Solve $5x^2 + 8x = -3$ using the quadratic formula.

Solution   We first find standard form and determine a, b, and c:

$$5x^2 + 8x + 3 = 0; \qquad \text{Adding 3 to both sides to get 0 on one side}$$

$$a = 5, \quad b = 8, \quad c = 3.$$

Next, we use the quadratic formula:

$$x = \frac{-b \pm \sqrt{b^2 - 4ac}}{2a}$$

$$x = \frac{-8 \pm \sqrt{8^2 - 4 \cdot 5 \cdot 3}}{2 \cdot 5} \qquad \text{Substituting}$$

$$x = \frac{-8 \pm \sqrt{64 - 60}}{10}$$

Be sure to write the fraction bar all the way across.

$$x = \frac{-8 \pm \sqrt{4}}{10} = \frac{-8 \pm 2}{10}$$

$$x = \frac{-8 + 2}{10} \quad or \quad x = \frac{-8 - 2}{10}$$

$$x = \frac{-6}{10} \qquad or \quad x = \frac{-10}{10}$$

$$x = -\frac{3}{5} \qquad or \quad x = -1.$$

The solutions are $-\frac{3}{5}$ and -1. The checks are left to the student.

Study Skills

Know It "By Heart"

When memorizing something like the quadratic formula, try to first understand and write out the derivation. Doing this two or three times will help you remember the formula.

Because $5x^2 + 8x + 3$ can be factored, the quadratic formula may not have been the fastest way of solving Example 1. However, because the quadratic formula works for *any* quadratic equation, we need not spend too much time struggling to solve a quadratic equation by factoring.

To Solve a Quadratic Equation

1. If the equation can be easily written in the form $ax^2 = p$ or $(x + k)^2 = d$, use the principle of square roots as in Section 11.1.
2. If step (1) does not apply, write the equation in the form $ax^2 + bx + c = 0$.
3. Try factoring and using the principle of zero products.
4. If factoring seems difficult or impossible, use the quadratic formula. Completing the square can also be used, but is slower.

The solutions of a quadratic equation can always be found using the quadratic formula. They cannot always be found by factoring.

Recall that a second-degree polynomial in one variable is said to be quadratic. Similarly, a second-degree polynomial function in one variable is said to be a **quadratic function.**

EXAMPLE 2   For the quadratic function given by $f(x) = 3x^2 - 6x - 4$, find all x for which $f(x) = 0$.

Solution   We substitute and solve for x:

$$f(x) = 0$$

$$3x^2 - 6x - 4 = 0 \qquad \text{Substituting. You can try to solve this by factoring.}$$

$$a = 3; \quad b = -6; \quad c = -4.$$

We then substitute into the quadratic formula:

$$x = \frac{-(-6) \pm \sqrt{(-6)^2 - 4 \cdot 3 \cdot (-4)}}{2 \cdot 3}$$

$$= \frac{6 \pm \sqrt{36 + 48}}{6}$$

$$= \frac{6 \pm \sqrt{84}}{6} \qquad \text{Note that 4 is a perfect-square factor of 84.}$$

$$= \frac{6}{6} \pm \frac{\sqrt{84}}{6}$$

$$= 1 \pm \frac{\sqrt{4}\sqrt{21}}{6} \qquad 84 = 4 \cdot 21$$

$$\left. \begin{array}{l} = 1 \pm \dfrac{2\sqrt{21}}{6} \\[2mm] = 1 \pm \dfrac{\sqrt{21}}{3}. \end{array} \right\} \quad \text{Simplifying}$$

The solutions are $1 - \dfrac{\sqrt{21}}{3}$ and $1 + \dfrac{\sqrt{21}}{3}$. The checks are left to the student.

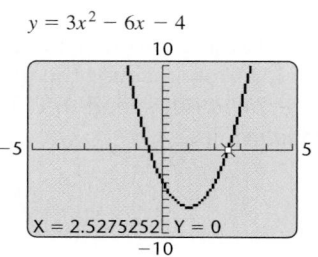

technology connection

To check Example 2 by graphing $y_1 = 3x^2 - 6x - 4$, press `TRACE` and enter $1 + \sqrt{21}/3$. A rational approximation and the y-value 0 should appear.

$y = 3x^2 - 6x - 4$

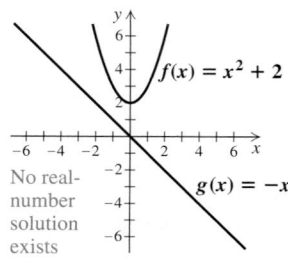

X = 2.5275252   Y = 0

Use this approach to check the other solution of Example 2.

Some quadratic equations have solutions that are imaginary numbers.

EXAMPLE 3   Solve: $x^2 + 2 = -x$.

Solution   We first find standard form:

$$x^2 + x + 2 = 0. \qquad \text{Adding } x \text{ to both sides}$$

Since we cannot factor $x^2 + x + 2$, we use the quadratic formula with $a = 1$, $b = 1$, and $c = 2$:

$$x = \frac{-1 \pm \sqrt{1^2 - 4 \cdot 1 \cdot 2}}{2 \cdot 1} \qquad \text{Substituting}$$

$$= \frac{-1 \pm \sqrt{1 - 8}}{2}$$

$$= \frac{-1 \pm \sqrt{-7}}{2}$$

$$= \frac{-1 \pm i\sqrt{7}}{2}, \text{ or } -\frac{1}{2} \pm \frac{\sqrt{7}}{2}i.$$

$f(x) = x^2 + 2$

$g(x) = -x$

No real-number solution exists

A visualization of Example 3

The solutions are $-\dfrac{1}{2} - \dfrac{\sqrt{7}}{2}i$ and $-\dfrac{1}{2} + \dfrac{\sqrt{7}}{2}i$. The checks are left to the student.

The quadratic formula can be used to solve certain rational equations.

EXAMPLE 4   If $f(x) = 2 + \dfrac{7}{x}$ and $g(x) = \dfrac{4}{x^2}$, find all x for which $f(x) = g(x)$.

Solution   We set $f(x)$ equal to $g(x)$ and solve:

$$f(x) = g(x)$$

$$2 + \frac{7}{x} = \frac{4}{x^2}.$$ Substituting. Note that $x \neq 0$.

This is a rational equation similar to those in Section 6.6. To solve, we multiply both sides by the LCD, x^2:

$$x^2\left(2 + \frac{7}{x}\right) = x^2 \cdot \frac{4}{x^2}$$

$$2x^2 + 7x = 4$$ Simplifying

$$2x^2 + 7x - 4 = 0.$$ Subtracting 4 from both sides

We have

$$a = 2, \quad b = 7, \quad \text{and} \quad c = -4.$$

Substituting then gives us

$$x = \frac{-7 \pm \sqrt{7^2 - 4 \cdot 2 \cdot (-4)}}{2 \cdot 2}$$

$$= \frac{-7 \pm \sqrt{49 + 32}}{4}$$

$$= \frac{-7 \pm \sqrt{81}}{4}$$

$$= \frac{-7 \pm 9}{4}$$

$$x = \frac{2}{4} = \frac{1}{2} \text{ or } x = \frac{-16}{4} = -4.$$ Both answers should check since $x \neq 0$.

You can confirm that $f\left(\tfrac{1}{2}\right) = g\left(\tfrac{1}{2}\right)$ and $f(-4) = g(-4)$. The solutions are $\tfrac{1}{2}$ and -4.

technology connection

We saw in Sections 5.7 and 11.1 how graphing calculators can solve quadratic equations. To determine whether quadratic equations are solved more quickly on a graphing calculator or by using the quadratic formula, solve Examples 2 and 4 both ways. Which method is faster? Which method is more precise? Why?

Approximating Solutions

When the solution of an equation is irrational, a rational-number approximation is often useful. This is often the case in real-world applications similar to those found in Section 11.3.

EXAMPLE 5   Use a calculator to approximate the solutions of Example 2.

Student Notes

It is important that you understand both the rules for order of operations *and* the manner in which your calculator applies those rules.

Solution   On most calculators, one of the following sequences of keystrokes can be used to approximate $1 + \sqrt{21}/3$:

   or

Similar keystrokes can be used to approximate $1 - \sqrt{21}/3$.
The solutions are approximately 2.527525232 and -0.5275252317.

Exercise Set   11.2

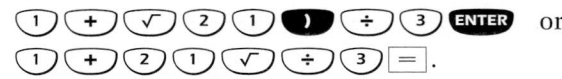

🔖 *Concept Reinforcement*   *Classify each of the following as either true or false.*

1. The quadratic formula can be used to solve *any* quadratic equation.

2. The quadratic formula does not work if solutions are imaginary numbers.

3. Solving by factoring is always slower than using the quadratic formula.

4. The steps used to derive the quadratic formula are the same as those used when solving by completing the square.

5. A quadratic equation can have as many as four solutions.

6. It is possible for a quadratic equation to have no real-number solutions.

Solve.

7. $x^2 + 7x - 3 = 0$

8. $x^2 - 7x + 4 = 0$

9. $3p^2 = 18p - 6$

10. $3u^2 = 8u - 5$

11. $x^2 + x + 1 = 0$

12. $x^2 + x + 2 = 0$

13. $x^2 + 13 = 4x$

14. $x^2 + 13 = 6x$

15. $h^2 + 4 = 6h$

16. $r^2 + 3r = 8$

17. $\dfrac{1}{x^2} - 3 = \dfrac{8}{x}$

18. $\dfrac{9}{x} - 2 = \dfrac{5}{x^2}$

19. $3x + x(x - 2) = 4$

20. $4x + x(x - 3) = 5$

21. $12t^2 + 9t = 1$

22. $15t^2 + 7t = 2$

23. $25x^2 - 20x + 4 = 0$

24. $36x^2 + 84x + 49 = 0$

25. $7x(x + 2) + 5 = 3x(x + 1)$

26. $5x(x - 1) - 7 = 4x(x - 2)$

27. $14(x - 4) - (x + 2) = (x + 2)(x - 4)$

28. $11(x - 2) + (x - 5) = (x + 2)(x - 6)$

29. $5x^2 = 13x + 17$

30. $25x = 3x^2 + 28$

31. $x^2 + 9 = 4x$

32. $x^2 + 7 = 3x$

33. $x^3 - 8 = 0$ (*Hint*: Factor the difference of cubes. Then use the quadratic formula.)

34. $x^3 + 1 = 0$

35. Let $f(x) = 3x^2 - 5x + 2$. Find x such that $f(x) = 0$.

36. Let $g(x) = 4x^2 - 2x - 3$. Find x such that $g(x) = 0$.

37. Let
$$f(x) = \frac{7}{x} + \frac{7}{x+4}.$$
Find all x for which $f(x) = 1$.

38. Let
$$g(x) = \frac{2}{x} + \frac{2}{x+3}.$$
Find all x for which $g(x) = 1$.

39. Let
$$F(x) = \frac{x+3}{x} \quad \text{and} \quad G(x) = \frac{x-4}{3}.$$
Find all x for which $F(x) = G(x)$.

40. Let
$$f(x) = \frac{3-x}{4} \quad \text{and} \quad g(x) = \frac{1}{4x}.$$
Find all x for which $f(x) = g(x)$.

41. Let
$$f(x) = \frac{15-2x}{6} \quad \text{and} \quad g(x) = \frac{3}{x}.$$
Find all x for which $f(x) = g(x)$.

42. Let
$$f(x) = x + 5 \quad \text{and} \quad g(x) = \frac{3}{x-5}.$$
Find all x for which $f(x) = g(x)$.

Solve. Use a calculator to approximate, as precisely as possible, the solutions as rational numbers.

43. $x^2 + 4x - 7 = 0$

44. $x^2 + 6x + 4 = 0$

45. $x^2 - 6x + 4 = 0$

46. $x^2 - 4x + 1 = 0$

47. $2x^2 - 3x - 7 = 0$

48. $3x^2 - 3x - 2 = 0$

49. Are there any equations that can be solved by the quadratic formula but not by completing the square? Why or why not?

50. If you had to choose between remembering the method of completing the square and remembering the quadratic formula, which would you choose? Why?

SKILL MAINTENANCE

51. *Coffee beans.* Twin Cities Roasters has Kenyan coffee for which they pay $6.75 a pound and Kona coffee for which they pay $11.25 a pound. How much of each kind should be mixed in order to obtain a 50-lb mixture that costs them $8.55 a pound? [8.3]

52. *Donuts.* South Street Bakers charges $1.10 for a cream-filled donut and 85¢ for a glazed donut. On a recent Sunday, a total of 90 glazed and cream-filled donuts were sold for $88.00. How many of each type were sold? [8.3]

Simplify.

53. $\sqrt{27a^2b^5} \cdot \sqrt{6a^3b}$ [10.3]

54. $\sqrt{8a^3b} \cdot \sqrt{12ab^5}$ [10.3]

55. $\dfrac{\dfrac{3}{x-1}}{\dfrac{1}{x+1} + \dfrac{2}{x-1}}$ [6.5]

56. $\dfrac{\dfrac{4}{a^2b}}{\dfrac{3}{a} - \dfrac{4}{b^2}}$ [6.5]

SYNTHESIS

57. Suppose you had a large number of quadratic equations to solve and none of the equations had a constant term. Would you use factoring or the quadratic formula to solve these equations? Why?

58. If $a < 0$ and $ax^2 + bx + c = 0$, then $-a$ is positive and the equivalent equation, $-ax^2 - bx - c = 0$, can be solved using the quadratic formula.
   a) Find this solution, replacing a, b, and c in the formula with $-a$, $-b$, and $-c$ from the equation.
   b) How does the result of part (a) indicate that the quadratic formula "works" regardless of the sign of a?

For Exercises 59–61, let

$$f(x) = \frac{x^2}{x-2} + 1 \quad and \quad g(x) = \frac{4x-2}{x-2} + \frac{x+4}{2}.$$

59. Find the *x*-intercepts of the graph of *f*.

60. Find the *x*-intercepts of the graph of *g*.

61. Find all *x* for which $f(x) = g(x)$.

Solve.

62. $x^2 - 0.75x - 0.5 = 0$

63. $z^2 + 0.84z - 0.4 = 0$

64. $\left(1 + \sqrt{3}\right)x^2 - \left(3 + 2\sqrt{3}\right)x + 3 = 0$

65. $\sqrt{2}x^2 + 5x + \sqrt{2} = 0$

66. $ix^2 - 2x + 1 = 0$

67. One solution of $kx^2 + 3x - k = 0$ is -2. Find the other.

68. Use a graphing calculator to solve Exercises 9, 23, and 39.

69. Use a graphing calculator to solve Exercises 15, 31, and 37. Use the method of graphing each side of the equation.

70. Can a graphing calculator be used to solve *any* quadratic equation? Why or why not?

11.3 Applications Involving Quadratic Equations

Solving Problems • Solving Formulas

Solving Problems

As we found in Section 6.7, some problems translate to rational equations. The solution of such rational equations can involve quadratic equations.

EXAMPLE 1

Motorcycle travel. Makita rode her motorcycle 300 mi at a certain average speed. Had she averaged 10 mph more, the trip would have taken 1 hr less. Find the average speed of the motorcycle.

Solution

1. **Familiarize.** We make a drawing, labeling it with the information provided. As in Section 6.7, we can create a table. We let *r* represent the rate, in miles per hour, and *t* the time, in hours, for Makita's trip.

Time *t*   300 miles   Speed *r*

Time *t* − 1   300 miles   Speed *r* + 10

Distance	Speed	Time
300	r	t
300	$r + 10$	$t - 1$

$\longrightarrow \quad r = \dfrac{300}{t}$

$\longrightarrow \quad r + 10 = \dfrac{300}{t-1}$

Recall that the definition of speed, $r = d/t$, relates the three quantities.

2. **Translate.**   From the table, we obtain

$$r = \frac{300}{t} \quad \text{and} \quad r + 10 = \frac{300}{t-1}.$$

3. **Carry out.**   A system of equations has been formed. We substitute for r from the first equation into the second and solve the resulting equation:

$$\frac{300}{t} + 10 = \frac{300}{t-1} \qquad \text{Substituting } 300/t \text{ for } r$$

$$t(t-1) \cdot \left[\frac{300}{t} + 10 \right] = t(t-1) \cdot \frac{300}{t-1} \qquad \text{Multiplying by the LCD}$$

$$\cancel{t}(t-1) \cdot \frac{300}{\cancel{t}} + t(t-1) \cdot 10 = t\cancel{(t-1)} \cdot \frac{300}{\cancel{t-1}} \qquad \text{Using the distributive law and removing factors that equal 1: } \frac{t}{t} = 1; \ \frac{t-1}{t-1} = 1$$

$$\left. \begin{aligned} 300(t-1) + 10(t^2 - t) &= 300t \\ 300t - 300 + 10t^2 - 10t &= 300t \\ 10t^2 - 10t - 300 &= 0 \end{aligned} \right\} \qquad \text{Rewriting in standard form}$$

$$t^2 - t - 30 = 0 \qquad \text{Multiplying by } \tfrac{1}{10} \text{ or dividing by 10}$$

$$(t - 6)(t + 5) = 0 \qquad \text{Factoring}$$

$$t = 6 \quad or \quad t = -5. \qquad \text{Principle of zero products}$$

4. **Check.**   Note that we have solved for t, not r as required. Since negative time has no meaning here, we disregard the -5 and use 6 hr to find r:

$$r = \frac{300 \text{ mi}}{6 \text{ hr}} = 50 \text{ mph.}$$

> *Caution!*   Always make sure that you find the quantity asked for in the problem.

To see if 50 mph checks, we increase the speed 10 mph to 60 mph and see how long the trip would have taken at that speed:

$$t = \frac{d}{r} = \frac{300 \text{ mi}}{60 \text{ mph}} = 5 \text{ hr}.$$ Note that $\text{mi/mph} = \text{mi} \div \frac{\text{mi}}{\text{hr}} =$

$$\bcancel{\text{mi}} \cdot \frac{\text{hr}}{\bcancel{\text{mi}}} = \text{hr}.$$

This is 1 hr less than the trip actually took, so the answer checks.

5. State.   Makita's motorcycle traveled at an average speed of 50 mph.

Solving Formulas

Recall that to solve a formula for a certain letter, we use the principles for solving equations to get that letter alone on one side.

EXAMPLE 2   Period of a pendulum.   The time T required for a pendulum of length l to swing back and forth (complete one period) is given by the formula $T = 2\pi\sqrt{l/g}$, where g is the earth's gravitational constant. Solve for l.

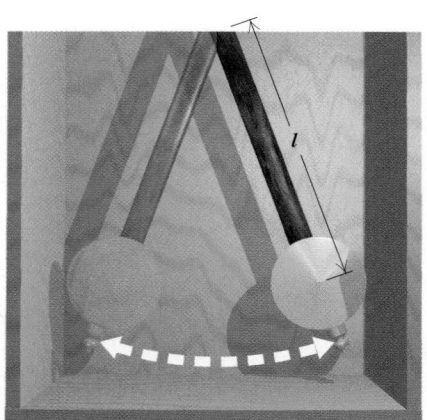

Solution   We have

$$T = 2\pi\sqrt{\frac{l}{g}}$$ This is a radical equation (see Section 10.6).

$$T^2 = \left(2\pi\sqrt{\frac{l}{g}}\right)^2$$ Principle of powers (squaring both sides)

$$T^2 = 2^2\pi^2\frac{l}{g}$$

$$gT^2 = 4\pi^2 l$$ Multiplying both sides by g to clear fractions

$$\frac{gT^2}{4\pi^2} = l.$$ Dividing both sides by $4\pi^2$

We now have l alone on one side and l does not appear on the other side, so the formula is solved for l.

In formulas for which variables represent only nonnegative numbers, there is no need for absolute-value signs when taking square roots.

EXAMPLE 3   Hang time.* An athlete's *hang time* is the amount of time that the athlete can remain airborne when jumping. A formula relating an athlete's vertical leap V, in inches, to hang time T, in seconds, is $V = 48T^2$. Solve for T.

Solution   We have

$$48T^2 = V$$

$$T^2 = \frac{V}{48} \qquad \text{Dividing by 48 to isolate } T^2$$

$$T = \frac{\sqrt{V}}{\sqrt{48}} \qquad \begin{array}{l}\text{Using the principle of square roots} \\ \text{and the quotient rule for radicals.} \\ \text{We assume } V, T \ge 0.\end{array}$$

$$\left.\begin{array}{l} = \dfrac{\sqrt{V}}{\sqrt{16}\,\sqrt{3}} = \dfrac{\sqrt{V}}{4\sqrt{3}} \\[2mm] = \dfrac{\sqrt{V}}{4\sqrt{3}} \cdot \dfrac{\sqrt{3}}{\sqrt{3}} = \dfrac{\sqrt{3V}}{12}. \end{array}\right\} \quad \text{Rationalizing the denominator}$$

EXAMPLE 4   Falling distance. An object tossed downward with an initial speed (velocity) of v_0 will travel a distance of s meters, where $s = 4.9t^2 + v_0 t$ and t is measured in seconds. Solve for t.

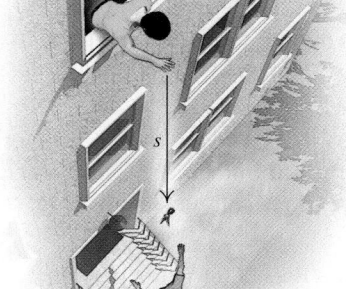

Solution   Since t is squared in one term and raised to the first power in the other term, the equation is quadratic in t.

$$4.9t^2 + v_0 t = s$$

$$4.9t^2 + v_0 t - s = 0 \qquad \text{Writing standard form}$$

$$a = 4.9, \quad b = v_0, \quad c = -s$$

$$t = \frac{-v_0 \pm \sqrt{v_0^2 - 4(4.9)(-s)}}{2(4.9)} \qquad \text{Using the quadratic formula}$$

*This formula is taken from an article by Peter Brancazio, "The Mechanics of a Slam Dunk," *Popular Mechanics*, November 1991. Courtesy of Professor Peter Brancazio, Brooklyn College.

Since the negative square root would yield a negative value for t, we use only the positive root:

$$t = \frac{-v_0 + \sqrt{v_0^2 + 19.6s}}{9.8}.$$

The following list of steps should help you when solving formulas for a given letter. Try to remember that when solving a formula, you use the same approach that you would to solve an equation.

Student Notes

After identifying which numbers to use as a, b, and c, be careful to replace only the *letters* in the quadratic formula.

To Solve a Formula for a Letter—Say, b

1. Clear fractions and use the principle of powers, as needed. Perform these steps until radicals containing b are gone and b is not in any denominator.
2. Combine all like terms.
3. If the only power of b is b^1, the equation can be solved as in Sections 2.3 and 7.5.
4. If b^2 appears but b does not, solve for b^2 and use the principle of square roots to solve for b.
5. If there are terms containing both b and b^2, put the equation in standard form and use the quadratic formula.

Exercise Set

11.3

FOR EXTRA HELP

Student's Solutions Manual | Digital Video Tutor CD 6 Videotape 11 | Tutor Center AW Math Tutor Center | MathXL Tutorials on CD | Math XL MathXL | MyMathLab MyMathLab

Solve.

1. *Car trips.* During the first part of a trip, Trudy's Honda traveled 120 mi at a certain speed. Trudy then drove another 100 mi at a speed that was 10 mph slower. If the total time of Trudy's trip was 4 hr, what was her speed on each part of the trip?

2. *Canoeing.* During the first part of a canoe trip, Tim covered 60 km at a certain speed. He then traveled 24 km at a speed that was 4 km/h slower. If the total time for the trip was 8 hr, what was the speed on each part of the trip?

3. *Car trips.* Petra's Plymouth travels 200 mi averaging a certain speed. If the car had gone 10 mph faster, the trip would have taken 1 hr less. Find Petra's average speed.

4. *Car trips.* Sandi's Subaru travels 280 mi averaging a certain speed. If the car had gone 5 mph faster, the trip would have taken 1 hr less. Find Sandi's average speed.

5. *Air travel.* A Cessna flies 600 mi at a certain speed. A Beechcraft flies 1000 mi at a speed that is 50 mph faster, but takes 1 hr longer. Find the speed of each plane.

6. *Air travel.* A turbo-jet flies 50 mph faster than a super-prop plane. If a turbo-jet goes 2000 mi in 3 hr less time than it takes the super-prop to go 2800 mi, find the speed of each plane.

7. *Bicycling.* Naoki bikes the 40 mi to Hillsboro averaging a certain speed. The return trip is made at a speed that is 6 mph slower. Total time for the round trip is 14 hr. Find Naoki's average speed on each part of the trip.

8. *Car speed.* On a sales trip, Gail drives the 600 mi to Richmond averaging a certain speed. The return trip is made at an average speed that is 10 mph slower. Total time for the round trip is 22 hr. Find Gail's average speed on each part of the trip.

9. *Navigation.* The Hudson River flows at a rate of 3 mph. A patrol boat travels 60 mi upriver and returns in a total time of 9 hr. What is the speed of the boat in still water?

10. *Navigation.* The current in a typical Mississippi River shipping route flows at a rate of 4 mph. In order for a barge to travel 24 mi upriver and then return in a total of 5 hr, approximately how fast must the barge be able to travel in still water?

11. *Filling a pool.* A well and a spring are filling a swimming pool. Together, they can fill the pool in 4 hr. The well, working alone, can fill the pool in 6 hr less time than the spring. How long would the spring take, working alone, to fill the pool?

12. *Filling a tank.* Two pipes are connected to the same tank. Working together, they can fill the tank in 2 hr. The larger pipe, working alone, can fill the tank in 3 hr less time than the smaller one. How long would the smaller one take, working alone, to fill the tank?

13. *Paddleboats.* Ellen paddles 1 mi upstream and 1 mi back in a total time of 1 hr. The speed of the river is 2 mph. Find the speed of Ellen's paddleboat in still water.

14. *Rowing.* Dan rows 10 km upstream and 10 km back in a total time of 3 hr. The speed of the river is 5 km/h. Find Dan's speed in still water.

Solve each formula for the indicated letter. Assume that all variables represent nonnegative numbers.

15. $A = 4\pi r^2$, for r
(Surface area of a sphere of radius r)

16. $A = 6s^2$, for s
(Surface area of a cube with sides of length s)

17. $A = 2\pi r^2 + 2\pi rh$, for r
(Surface area of a right cylindrical solid with radius r and height h)

18. $F = \dfrac{Gm_1m_2}{r^2}$, for r
(Law of gravity)

19. $N = \dfrac{kQ_1Q_2}{s^2}$, for s
(Number of phone calls between two cities)

20. $A = \pi r^2$, for r
(Area of a circle)

21. $T = 2\pi\sqrt{\dfrac{l}{g}}$, for g
(A pendulum formula)

22. $a^2 + b^2 = c^2$, for b
(Pythagorean formula in two dimensions)

23. $a^2 + b^2 + c^2 = d^2$, for c
(Pythagorean formula in three dimensions)

24. $N = \dfrac{k^2 - 3k}{2}$, for k
(Number of diagonals of a polygon with k sides)

25. $s = v_0t + \dfrac{gt^2}{2}$, for t
(A motion formula)

26. $A = \pi r^2 + \pi rs$, for r
(Surface area of a cone)

27. $N = \frac{1}{2}(n^2 - n)$, for n
(Number of games if n teams play each other once)

28. $A = A_0(1 - r)^2$, for r
(A business formula)

29. $V = 3.5\sqrt{h}$, for h
(Distance to horizon from a height)

30. $W = \sqrt{\dfrac{1}{LC}}$, for L
(An electricity formula)

Aha! 31. $at^2 + bt + c = 0$, for t
(An algebraic formula)

32. $A = P_1(1 + r)^2 + P_2(1 + r)$, for r
(Amount in an account when P_1 is invested for 2 yr and P_2 for 1 yr at interest rate r)

Solve.

33. *Falling distance.* (Use $4.9t^2 + v_0t = s$.)

a) A bolt falls off an airplane at an altitude of 500 m. Approximately how long does it take the bolt to reach the ground?

b) A ball is thrown downward at a speed of 30 m/sec from an altitude of 500 m. Approximately how long does it take the ball to reach the ground?

c) Approximately how far will an object fall in 5 sec, when thrown downward at an initial velocity of 30 m/sec from a plane?

34. *Falling distance.* (Use $4.9t^2 + v_0t = s$.)

a) A ring is dropped from a helicopter at an altitude of 75 m. Approximately how long does it take the ring to reach the ground?

b) A coin is tossed downward with an initial velocity of 30 m/sec from an altitude of 75 m. Approximately how long does it take the coin to reach the ground?

c) Approximately how far will an object fall in 2 sec, if thrown downward at an initial velocity of 20 m/sec from a helicopter?

35. *Bungee jumping.* Jesse is tied to one end of a 40-m elasticized (bungee) cord. The other end of the cord is tied to the middle of a bridge. If Jesse jumps off the bridge, for how long will he fall before the cord begins to stretch? (Use $4.9t^2 = s$.)

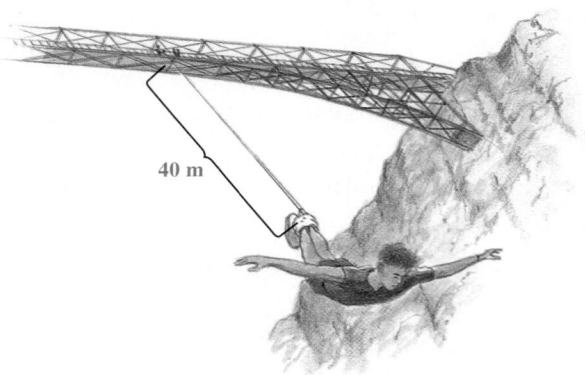

40 m

36. *Bungee jumping.* Sheila is tied to a bungee cord (see Exercise 35) and falls for 2.5 sec before her cord begins to stretch. How long is the bungee cord?

37. *Hang time.* The NBA's Steve Francis reportedly has a vertical leap of 45 in. What is his hang time? (Use $V = 48T^2$.)
Source: www.maximonline.com

38. *League schedules.* In a bowling league, each team plays each of the other teams once. If a total of 66 games is played, how many teams are in the league? (See Exercise 27.)

For Exercises 39 and 40, use $4.9t^2 + v_0t = s$.

39. *Downward speed.* An object thrown downward from a 100-m cliff travels 51.6 m in 3 sec. What was the initial velocity of the object?

40. *Downward speed.* An object thrown downward from a 200-m cliff travels 91.2 m in 4 sec. What was the initial velocity of the object?

For Exercises 41 and 42, use $A = P_1(1 + r)^2 + P_2(1 + r)$. (See Exercise 32.)

41. *Compound interest.* A firm invests $3000 in a savings account for 2 yr. At the beginning of the second year, an additional $1700 is invested. If a total of $5253.70 is in the account at the end of the second year, what is the annual interest rate?

42. *Compound interest.* A business invests $10,000 in a savings account for 2 yr. At the beginning of the second year, an additional $3500 is invested. If a total of $15,569.75 is in the account at the end of the second year, what is the annual interest rate?

43. Marti is tied to a bungee cord that is twice as long as the cord tied to Pedro. Will Marti's fall take twice as long as Pedro's before their cords begin to stretch? Why or why not? (See Exercises 35 and 36.)

44. Under what circumstances would a negative value for t, time, have meaning?

SKILL MAINTENANCE

Evaluate.

45. $b^2 - 4ac$, for $a = 5$, $b = 6$, and $c = 7$ [1.8]

46. $\sqrt{b^2 - 4ac}$, for $a = 3$, $b = 4$, and $c = 5$ [10.8]

Simplify.

47. $\dfrac{x^2 + xy}{2x}$ [6.1]

48. $\dfrac{a^3 - ab^2}{ab}$ [6.1]

49. $\dfrac{3 + \sqrt{45}}{6}$ [10.3]

50. $\dfrac{2 - \sqrt{28}}{10}$ [10.3]

SYNTHESIS

51. Write a problem for a classmate to solve. Devise the problem so that **(a)** the solution is found after solving a rational equation and **(b)** the solution is "The express train travels 90 mph."

52. In what ways do the motion problems of this section (like Example 1) differ from the motion problems in Chapter 6 (see p. 423)?

53. *Biochemistry.* The equation

$$A = 6.5 - \frac{20.4t}{t^2 + 36}$$

is used to calculate the acid level A in a person's blood t minutes after sugar is consumed. Solve for t.

54. *Special relativity.* Einstein found that an object with initial mass m_0 and traveling velocity v has mass

$$m = \frac{m_0}{\sqrt{1 - \dfrac{v^2}{c^2}}},$$

where c is the speed of light. Solve the formula for c.

55. Find a number for which the reciprocal of 1 less than the number is the same as 1 more than the number.

56. *Purchasing.* A discount store bought a quantity of beach towels for $250 and sold all but 15 at a profit of $3.50 per towel. With the total amount received, the manager could buy 4 more than twice as many as were bought before. Find the cost per towel.

57. *Art and aesthetics.* For over 2000 yr, artists, sculptors, and architects have regarded the proportions of a "golden" rectangle as visually appealing. A rectangle of width w and length l is considered "golden" if

$$\frac{w}{l} = \frac{l}{w + l}.$$

Solve for l.

58. *Diagonal of a cube.* Find a formula that expresses the length of the three-dimensional diagonal of a cube as a function of the cube's surface area.

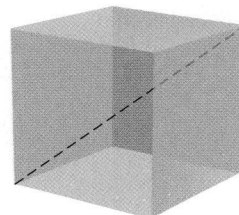

59. Solve for n:
$$mn^4 - r^2pm^3 - r^2n^2 + p = 0.$$

60. *Surface area.* Find a formula that expresses the diameter of a right cylindrical solid as a function of its surface area and its height. (See Exercise 17.)

61. A sphere is inscribed in a cube as shown in the figure below. Express the surface area of the sphere as a function of the surface area S of the cube. (See Exercise 15.)

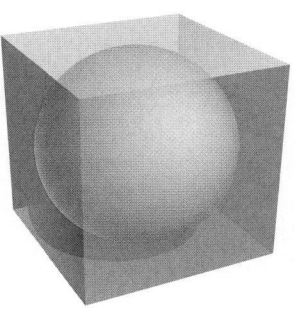

11.4 Studying Solutions of Quadratic Equations

The Discriminant • Writing Equations from Solutions

The Discriminant

It is sometimes enough to know what *type* of number a solution will be, without actually solving the equation. Suppose we want to know if $4x^2 - 5x - 2 = 0$ has rational solutions (and thus can be solved by factoring). Using the quadratic formula, we would have

$$x = \frac{-b \pm \sqrt{b^2 - 4ac}}{2a}$$

$$= \frac{-(-5) \pm \sqrt{(-5)^2 - 4 \cdot 4(-2)}}{2 \cdot 4}.$$

Note that the radicand, $(-5)^2 - 4 \cdot 4 \cdot (-2)$, determines what type of number the solutions will be. Since $(-5)^2 - 4 \cdot 4 \cdot (-2) = 25 - 16(-2) = 25 + 32 = 57$ and since 57 is not a perfect square, we know that the solutions of the equation are not rational numbers. This means that $4x^2 - 5x - 2 = 0$ *cannot* be solved by factoring.

It is $b^2 - 4ac$, known as the **discriminant**, that determines what type of number the solutions of a quadratic equation are. If a, b, and c are rational, then:

- When $b^2 - 4ac$ simplifies to 0, it doesn't matter if we use $+\sqrt{b^2 - 4ac}$ or $-\sqrt{b^2 - 4ac}$; we get the same solution twice. Thus, when the discriminant is 0, there is one *repeated* solution and it is rational.

  Example: $9x^2 + 6x + 1 = 0 \rightarrow b^2 - 4ac = 6^2 - 4 \cdot 9 \cdot 1 = 0.$

- When $b^2 - 4ac$ is positive, there are two different real-number solutions: If $b^2 - 4ac$ is a perfect square, these solutions are rational numbers.

  Example: $6x^2 + 5x + 1 = 0 \rightarrow b^2 - 4ac = 5^2 - 4 \cdot 6 \cdot 1 = 1.$

- When $b^2 - 4ac$ is positive but not a perfect square, there are two irrational solutions and they are conjugates of each other (see p. 469).

  Example: $2x^2 + 4x + 1 = 0 \rightarrow b^2 - 4ac = 4^2 - 4 \cdot 2 \cdot 1 = 8.$

- When the discriminant is negative, there are two imaginary-number solutions and they are complex conjugates of each other.

  Example: $3x^2 + 2x + 1 = 0 \rightarrow b^2 - 4ac = 2^2 - 4 \cdot 3 \cdot 1 = -8.$

Note that any equation for which $b^2 - 4ac$ is a perfect square can be solved by factoring.

Discriminant $b^2 - 4ac$	Nature of Solutions
0	One solution; a rational number

Positive
  Perfect square
  Not a perfect square
	Two different real-number solutions
	  Solutions are rational.
	  Solutions are irrational conjugates.

Negative	Two different imaginary-number solutions (complex conjugates)

EXAMPLE 1   For each equation, determine what type of number the solutions are and how many solutions exist.

a) $9x^2 - 12x + 4 = 0$

b) $x^2 + 5x + 8 = 0$

c) $2x^2 + 7x - 3 = 0$

Solution

a) For $9x^2 - 12x + 4 = 0$, we have

$$a = 9, \quad b = -12, \quad c = 4.$$

We substitute and compute the discriminant:

$$b^2 - 4ac = (-12)^2 - 4 \cdot 9 \cdot 4$$
$$= 144 - 144 = 0.$$

There is exactly one solution, and it is rational. This indicates that $9x^2 - 12x + 4 = 0$ can be solved by factoring.

b) For $x^2 + 5x + 8 = 0$, we have

$$a = 1, \quad b = 5, \quad c = 8.$$

We substitute and compute the discriminant:

$$b^2 - 4ac = 5^2 - 4 \cdot 1 \cdot 8$$
$$= 25 - 32 = -7.$$

Since the discriminant is negative, there are two imaginary-number solutions that are complex conjugates of each other.

c) For $2x^2 + 7x - 3 = 0$, we have

$$a = 2, \quad b = 7, \quad c = -3;$$
$$b^2 - 4ac = 7^2 - 4 \cdot 2(-3)$$
$$= 49 - (-24) = 73.$$

The discriminant is a positive number that is not a perfect square. Thus there are two irrational solutions that are conjugates of each other.

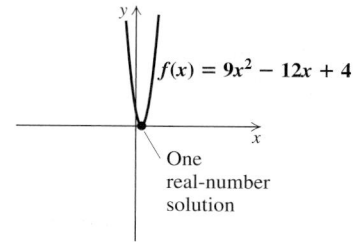

A visualization of part (a)

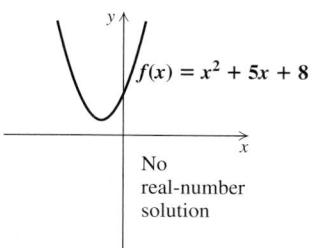

A visualization of part (b)

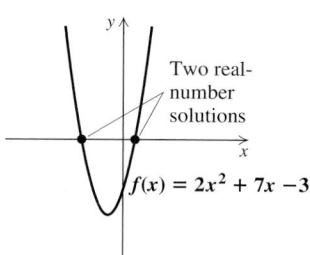

A visualization of part (c)

Discriminants can also be used to determine the number of real-number solutions of $ax^2 + bx + c = 0$. This can be used as an aid in graphing.

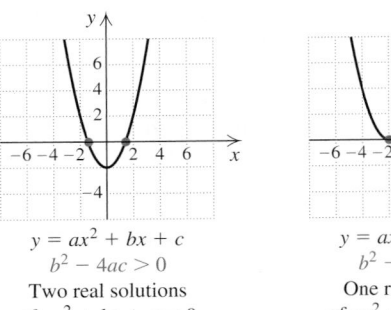

$y = ax^2 + bx + c$
$b^2 - 4ac > 0$
Two real solutions
of $ax^2 + bx + c = 0$
Two x-intercepts

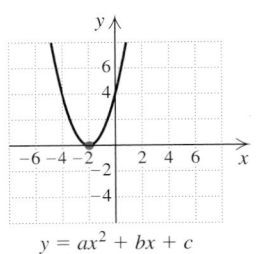

$y = ax^2 + bx + c$
$b^2 - 4ac = 0$
One real solution
of $ax^2 + bx + c = 0$
One x-intercept

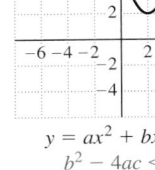

$y = ax^2 + bx + c$
$b^2 - 4ac < 0$
No real solutions
of $ax^2 + bx + c = 0$
No x-intercept

Writing Equations from Solutions

We know by the principle of zero products that $(x - 2)(x + 3) = 0$ has solutions 2 and -3. If we know the solutions of an equation, we can write an equation, using the principle in reverse.

Find an equation for which the given numbers are solutions.

a) 3 and $-\frac{2}{5}$

b) $2i$ and $-2i$

c) $5\sqrt{7}$ and $-5\sqrt{7}$

d) $-4, 0,$ and 1

Solution

a)
$$x = 3 \quad or \quad x = -\frac{2}{5}$$
$$x - 3 = 0 \quad or \quad x + \frac{2}{5} = 0 \qquad \text{Getting 0's on one side}$$
$$(x - 3)\left(x + \frac{2}{5}\right) = 0 \qquad \text{Using the principle of zero products (multiplying)}$$
$$x^2 + \frac{2}{5}x - 3x - 3 \cdot \frac{2}{5} = 0 \qquad \text{Multiplying}$$
$$x^2 - \frac{13}{5}x - \frac{6}{5} = 0 \qquad \text{Combining like terms}$$
$$5x^2 - 13x - 6 = 0 \qquad \text{Multiplying both sides by 5 to clear fractions}$$

Note that multiplying both sides by the LCD, 5, clears the equation of fractions. Had we preferred, we could have multiplied $x + \frac{2}{5} = 0$ by 5, thus clearing fractions *before* using the principle of zero products.

b)
$$x = 2i \quad or \quad x = -2i$$
$$x - 2i = 0 \quad or \quad x + 2i = 0 \qquad \text{Getting 0's on one side}$$
$$(x - 2i)(x + 2i) = 0 \qquad \text{Using the principle of zero products (multiplying)}$$
$$x^2 - (2i)^2 = 0 \qquad \text{Finding the product of a sum and difference}$$
$$x^2 - 4i^2 = 0$$
$$x^2 + 4 = 0 \qquad i^2 = -1$$

EXAMPLE 2

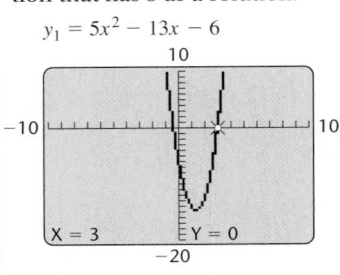

technology connection

To check Example 2(a), we can let $y_1 = 5x^2 - 13x - 6$ and verify that the x-intercepts are 3 and $-\frac{2}{5}$. One way to do this is to press (TRACE) and then enter 3. The cursor then appears on the curve at $x = 3$. Since this is an x-intercept, we know that $5x^2 - 13x - 6 = 0$ is an equation that has 3 as a solution.

$y_1 = 5x^2 - 13x - 6$

X = 3   Y = 0

1. Confirm that $5x^2 - 13x - 6 = 0$ also has $-\frac{2}{5}$ as a solution.
2. Check Example 2(c).
3. Check Example 2(d).

c) $\qquad x = 5\sqrt{7} \quad$ *or* $\qquad x = -5\sqrt{7}$

$\qquad x - 5\sqrt{7} = 0 \qquad$ *or* $\quad x + 5\sqrt{7} = 0 \qquad$ Getting 0's on one side

$\qquad \left(x - 5\sqrt{7}\right)\left(x + 5\sqrt{7}\right) = 0 \qquad$ Using the principle of zero products

$\qquad\qquad x^2 - \left(5\sqrt{7}\right)^2 = 0 \qquad$ Finding the product of a sum and difference

$\qquad\qquad x^2 - 25 \cdot 7 = 0$

$\qquad\qquad\qquad x^2 - 175 = 0$

d) $\qquad x = -4 \quad$ *or* $\quad x = 0 \quad$ *or* $\qquad x = 1$

$\qquad x + 4 = 0 \qquad$ *or* $\quad x = 0 \quad$ *or* $\quad x - 1 = 0 \qquad$ Getting 0's on one side

$\qquad\qquad (x + 4)x(x - 1) = 0 \qquad$ Using the principle of zero products

$\qquad\qquad x(x^2 + 3x - 4) = 0 \qquad$ Multiplying

$\qquad\qquad x^3 + 3x^2 - 4x = 0$

To check any of these equations, we can simply substitute one or more of the given solutions. For example, in Example 2(d) above,

$$(-4)^3 + 3(-4)^2 - 4(-4) = -64 + 3 \cdot 16 + 16$$
$$= -64 + 48 + 16 = 0.$$

The other checks are left to the student.

Exercise Set

11.4

FOR EXTRA HELP

|  Student's Solutions Manual | Digital Video Tutor CD 6 Videotape 11 | Tutor Center AW Math Tutor Center |  MathXL Tutorials on CD | Math*XL* MathXL | MyMathLab MyMathLab |

↪ *Concept Reinforcement   Complete each of the following.*

1. In the quadratic formula, the expression $b^2 - 4ac$ is called the _____.

2. When $b^2 - 4ac$ is 0, there is/are _____ solution(s).

3. When $b^2 - 4ac$ is positive, there is/are _____ solution(s).

4. When $b^2 - 4ac$ is negative, there is/are _____ solution(s).

5. When $b^2 - 4ac$ is a perfect square, the answers are _____ numbers.

6. When $b^2 - 4ac$ is negative, the answer(s) is/are _____ numbers.

For each equation, determine what type of number the solutions are and how many solutions exist.

7. $x^2 - 7x + 5 = 0$

8. $x^2 - 5x + 3 = 0$

9. $x^2 + 3 = 0$

10. $x^2 + 5 = 0$

11. $x^2 - 5 = 0$

12. $x^2 - 3 = 0$

13. $4x^2 + 8x - 5 = 0$

14. $4x^2 - 12x + 9 = 0$

15. $x^2 + 4x + 6 = 0$

16. $x^2 - 2x + 4 = 0$

17. $9t^2 - 48t + 64 = 0$

18. $6t^2 - 19t - 20 = 0$

19. $10x^2 - x - 2 = 0$

20. $6x^2 + 5x - 4 = 0$

Aha! **21.** $9t^2 - 3t = 0$

22. $4m^2 + 7m = 0$

23. $x^2 + 4x = 8$

24. $x^2 + 5x = 9$

25. $2a^2 - 3a = -5$

26. $3a^2 + 5 = 7a$

27. $y^2 + \frac{9}{4} = 4y$

28. $x^2 = \frac{1}{2}x - \frac{3}{5}$

Write a quadratic equation having the given numbers as solutions.

29. $-7, 3$

30. $-6, 4$

31. 3, only solution (*Hint*: It must be a repeated solution.)

32. -5, only solution

33. $-1, -3$

34. $-2, -5$

35. $5, \frac{3}{4}$

36. $4, \frac{2}{3}$

37. $-\frac{1}{4}, -\frac{1}{2}$

38. $\frac{1}{2}, \frac{1}{3}$

39. $2.4, -0.4$

40. $-0.6, 1.4$

41. $-\sqrt{3}, \sqrt{3}$

42. $-\sqrt{7}, \sqrt{7}$

43. $2\sqrt{5}, -2\sqrt{5}$

44. $3\sqrt{2}, -3\sqrt{2}$

45. $4i, -4i$

46. $3i, -3i$

47. $2 - 7i, 2 + 7i$

48. $5 - 2i, 5 + 2i$

49. $3 - \sqrt{14}, 3 + \sqrt{14}$

50. $2 - \sqrt{10}, 2 + \sqrt{10}$

51. $1 - \dfrac{\sqrt{21}}{3}, 1 + \dfrac{\sqrt{21}}{3}$

52. $\dfrac{5}{4} - \dfrac{\sqrt{33}}{4}, \dfrac{5}{4} + \dfrac{\sqrt{33}}{4}$

Write a third-degree equation having the given numbers as solutions.

53. $-2, 1, 5$

54. $-5, 0, 2$

55. $-1, 0, 3$

56. $-2, 2, 3$

57. Under what condition(s) is the discriminant *not* the fastest way to determine how many and what type of solutions exist?

58. Describe a procedure that could be used to write an equation having the first 7 natural numbers as solutions.

SKILL MAINTENANCE

Simplify. [4.1]

59. $(3a^2)^4$

60. $(4x^3)^2$

Find the x-intercepts of the graph of f. [11.1]

61. $f(x) = x^2 - 7x - 8$

62. $f(x) = x^2 - 6x + 8$

63. During a one-hour television show, there were 12 commercials. Some of the commercials were 30 sec long and the others were 60 sec long. The amount of time for 30-sec commercials was 6 min less than the total number of minutes of commercial time during the show. How many 30-sec commercials were used? [8.3]

64. Graph: $y = -\frac{3}{7}x + 4$. [7.3]

SYNTHESIS

65. If we assume that a quadratic equation has integers for coefficients, will the product of the solutions always be a real number? Why or why not?

66. Can a fourth-degree equation have exactly three irrational solutions? Why or why not?

67. The graph of an equation of the form
$$y = ax^2 + bx + c$$
is a curve similar to the one shown below. Determine a, b, and c from the information given.

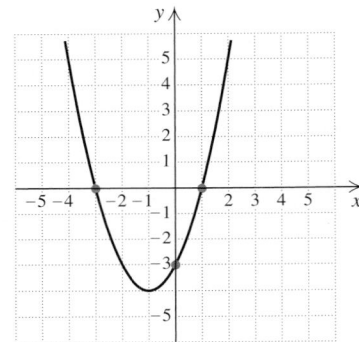

68. Show that the product of the solutions of $ax^2 + bx + c = 0$ is c/a.

For each equation under the given condition, (a) find k and (b) find the other solution.

69. $kx^2 - 2x + k = 0$; one solution is -3

70. $x^2 - kx + 2 = 0$; one solution is $1 + i$

71. $x^2 - (6 + 3i)x + k = 0$; one solution is 3

72. Show that the sum of the solutions of $ax^2 + bx + c = 0$ is $-b/a$.

73. Show that whenever there is just one solution of $ax^2 + bx + c = 0$, that solution is of the form $-b/2a$.

74. Find h and k, where $3x^2 - hx + 4k = 0$, the sum of the solutions is -12, and the product of the solutions is 20. (*Hint*: See Exercises 68 and 72.)

75. Suppose that $f(x) = ax^2 + bx + c$, with $f(-3) = 0$, $f(\frac{1}{2}) = 0$, and $f(0) = -12$. Find a, b, and c.

76. Find an equation for which $2 - \sqrt{3}$, $2 + \sqrt{3}$, $5 - 2i$, and $5 + 2i$ are solutions.

Aha! **77.** Find an equation with integer coefficients for which $1 - \sqrt{5}$ and $3 + 2i$ are two of the solutions.

 78. A discriminant that is a perfect square indicates that factoring can be used to solve the quadratic equation. Why?

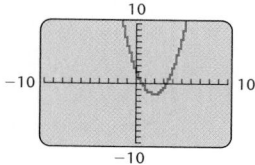 **79.** While solving a quadratic equation of the form $ax^2 + bx + c = 0$ with a graphing calculator, Shawn-Marie gets the following screen. How could the sign of the discriminant help her check the graph?

11.5   Equations Reducible to Quadratic

Recognizing Equations in Quadratic Form   •   Radical and Rational Equations

Recognizing Equations in Quadratic Form

Certain equations that are not really quadratic can be thought of in such a way that they can be solved as quadratic. For example, because the square of x^2 is x^4, the equation $x^4 - 9x^2 + 8 = 0$ is said to be "quadratic in x^2":

$$x^4 \;-\; 9x^2 \;+\; 8 = 0$$

$$(x^2)^2 - 9(x^2) + 8 = 0 \qquad \text{Thinking of } x^4 \text{ as } (x^2)^2$$

$$u^2 \;-\; 9u \;+\; 8 = 0. \qquad \text{To make this clearer, write } u \text{ instead of } x^2.$$

The equation $u^2 - 9u + 8 = 0$ can be solved by factoring or by the quadratic formula. Then, remembering that $u = x^2$, we can solve for x. Equations that can be solved like this are *reducible to quadratic*, or *in quadratic form*.

EXAMPLE 1   Solve: $x^4 - 9x^2 + 8 = 0$.

Solution   Let $u = x^2$. Then we solve by substituting u for x^2 and u^2 for x^4:

$$u^2 - 9u + 8 = 0$$

$$(u - 8)(u - 1) = 0 \qquad\qquad \text{Factoring}$$

$$u - 8 = 0 \quad or \quad u - 1 = 0 \qquad \text{Principle of zero products}$$

$$u = 8 \quad or \qquad\quad u = 1.$$

Student Notes _____

To recognize an equation in quadratic form, note that there will be two terms with variables. The exponent of the first term is twice the exponent of the second term. It is the variable expression in the *second* term that is written as *u*.

We replace u with x^2 and solve these equations:

$$x^2 = 8 \quad or \quad x^2 = 1$$
$$x = \pm\sqrt{8} \quad or \quad x = \pm 1$$
$$x = \pm 2\sqrt{2} \quad or \quad x = \pm 1.$$

To check, note that for both $x = 2\sqrt{2}$ and $-2\sqrt{2}$, we have $x^2 = 8$ and $x^4 = 64$. Similarly, for both $x = 1$ and -1, we have $x^2 = 1$ and $x^4 = 1$. Thus instead of making four checks, we need make only two.

Check:

For $\pm 2\sqrt{2}$:

$$\frac{x^4 - 9x^2 + 8 = 0}{(\pm 2\sqrt{2})^4 - 9(\pm 2\sqrt{2})^2 + 8 \;\Big|\; 0}$$
$$64 - 9 \cdot 8 + 8 \;\Big|$$
$$0 \overset{?}{=} 0 \quad \text{TRUE}$$

For ± 1:

$$\frac{x^4 - 9x^2 + 8 = 0}{(\pm 1)^4 - 9(\pm 1)^2 + 8 \;\Big|\; 0}$$
$$1 - 9 + 8 \;\Big|$$
$$0 \overset{?}{=} 0 \quad \text{TRUE}$$

The solutions are $1, -1, 2\sqrt{2}$, and $-2\sqrt{2}$.

> **Caution!** A common error on problems like Example 1 is to solve for u but forget to solve for x. Remember to solve for the *original* variable!

Example 1 can be solved directly by factoring:

$$x^4 - 9x^2 + 8 = 0$$
$$(x^2 - 1)(x^2 - 8) = 0$$
$$x^2 - 1 = 0 \quad or \quad x^2 - 8 = 0$$
$$x^2 = 1 \quad or \quad x^2 = 8$$
$$x = \pm 1 \quad or \quad x = \pm 2\sqrt{2}.$$

There is nothing wrong with this approach. However, in the examples that follow, you will note that it becomes increasingly difficult to solve the equation without first making a substitution.

Radical and Rational Equations

Sometimes rational equations, radical equations, or equations containing exponents that are fractions are reducible to quadratic. It is especially important that answers to these equations be checked in the original equation.

EXAMPLE 2 Solve: $x - 3\sqrt{x} - 4 = 0$.

Solution This radical equation could be solved using the method discussed in Section 10.6. However, if we note that the square of $\sqrt{x}$ is x, we can regard the equation as "quadratic in $\sqrt{x}$."

We let $u = \sqrt{x}$ and consequently $u^2 = x$:

$$x - 3\sqrt{x} - 4 = 0$$
$$u^2 - 3u - 4 = 0 \qquad \text{Substituting}$$
$$(u - 4)(u + 1) = 0$$
$$u = 4 \quad or \quad u = -1. \qquad \text{Using the principle of zero products}$$

Next, we replace u with $\sqrt{x}$ and solve these equations:

$$\sqrt{x} = 4 \quad or \quad \sqrt{x} = -1.$$

Squaring gives us $x = 16$ or $x = 1$ and also makes checking essential.

Check: For 16:

$$\begin{array}{c|c} x - 3\sqrt{x} - 4 = 0 & \\ \hline 16 - 3\sqrt{16} - 4 & 0 \\ 16 - 3 \cdot 4 - 4 & \\ 0 \stackrel{?}{=} 0 & \text{TRUE} \end{array}$$

For 1:

$$\begin{array}{c|c} x - 3\sqrt{x} - 4 = 0 & \\ \hline 1 - 3\sqrt{1} - 4 & 0 \\ 1 - 3 \cdot 1 - 4 & \\ -6 \stackrel{?}{=} 0 & \text{FALSE} \end{array}$$

The number 16 checks, but 1 does not. Had we noticed that $\sqrt{x} = -1$ has no solution (since principal roots are never negative), we could have solved only the equation $\sqrt{x} = 4$. The solution is 16.

EXAMPLE 3

technology connection

Check Example 3 with a graphing calculator. Use the ZERO, ROOT, or INTERSECT option, if possible.

Find the x-intercepts of the graph of $f(x) = (x^2 - 1)^2 - (x^2 - 1) - 2$.

Solution   The x-intercepts occur where $f(x) = 0$ so we must have

$$(x^2 - 1)^2 - (x^2 - 1) - 2 = 0. \qquad \text{Setting } f(x) \text{ equal to } 0$$

This equation is quadratic in $x^2 - 1$, so we let $u = x^2 - 1$ and $u^2 = (x^2 - 1)^2$:

$$u^2 - u - 2 = 0 \qquad \begin{array}{l}\text{Substituting in}\\ (x^2 - 1)^2 - (x^2 - 1) - 2 = 0\end{array}$$
$$(u - 2)(u + 1) = 0$$
$$u = 2 \quad or \quad u = -1. \qquad \text{Using the principle of zero products}$$

Next, we replace u with $x^2 - 1$ and solve these equations:

$$x^2 - 1 = 2 \qquad or \quad x^2 - 1 = -1$$
$$x^2 = 3 \qquad or \qquad x^2 = 0 \qquad \text{Adding 1 to both sides}$$
$$x = \pm\sqrt{3} \quad or \qquad x = 0. \qquad \begin{array}{l}\text{Using the principle of}\\ \text{square roots}\end{array}$$

The x-intercepts occur at $\left(-\sqrt{3}, 0\right)$, $(0, 0)$, and $\left(\sqrt{3}, 0\right)$.

Sometimes great care must be taken in deciding what substitution to make.

EXAMPLE 4   Solve: $m^{-2} - 6m^{-1} + 4 = 0$.

Solution   Note that the square of m^{-1} is $(m^{-1})^2$, or m^{-2}. This allows us to regard the equation as quadratic in m^{-1}.

We let $u = m^{-1}$ and $u^2 = m^{-2}$:

$$u^2 - 6u + 4 = 0 \qquad \text{Substituting}$$

$$u = \frac{-(-6) \pm \sqrt{(-6)^2 - 4 \cdot 1 \cdot 4}}{2 \cdot 1} \qquad \begin{array}{l}\text{Using the quadratic}\\\text{formula}\end{array}$$

$$\left.\begin{array}{l}u = \dfrac{6 \pm \sqrt{20}}{2} = \dfrac{2 \cdot 3 \pm 2\sqrt{5}}{2}\\[2mm] u = 3 \pm \sqrt{5}.\end{array}\right\} \qquad \text{Simplifying}$$

Next, we replace u with m^{-1} and solve:

$$m^{-1} = 3 \pm \sqrt{5}$$

$$\frac{1}{m} = 3 \pm \sqrt{5} \qquad \text{Recall that } m^{-1} = \frac{1}{m}.$$

$$1 = m(3 \pm \sqrt{5}) \qquad \text{Multiplying both sides by } m$$

$$\frac{1}{3 \pm \sqrt{5}} = m. \qquad \text{Dividing both sides by } 3 \pm \sqrt{5}$$

We can check both solutions as follows.

Check:

For $1/(3 - \sqrt{5})$:

$$m^{-2} - 6m^{-1} + 4 = 0$$

$$\begin{array}{c|c}\left(\dfrac{1}{3 - \sqrt{5}}\right)^{-2} - 6\left(\dfrac{1}{3 - \sqrt{5}}\right)^{-1} + 4 & 0\\[3mm] (3 - \sqrt{5})^2 - 6(3 - \sqrt{5}) + 4 & \\[2mm] 9 - 6\sqrt{5} + 5 - 18 + 6\sqrt{5} + 4 & \\ & 0 \overset{?}{=} 0 \quad \text{TRUE}\end{array}$$

For $1/(3 + \sqrt{5})$:

$$m^{-2} - 6m^{-1} + 4 = 0$$

$$\begin{array}{c|c}\left(\dfrac{1}{3 + \sqrt{5}}\right)^{-2} - 6\left(\dfrac{1}{3 + \sqrt{5}}\right)^{-1} + 4 & 0\\[3mm] (3 + \sqrt{5})^2 - 6(3 + \sqrt{5}) + 4 & \\[2mm] 9 + 6\sqrt{5} + 5 - 18 - 6\sqrt{5} + 4 & \\ & 0 \overset{?}{=} 0 \quad \text{TRUE}\end{array}$$

Both numbers check. The solutions are $1/(3 - \sqrt{5})$ and $1/(3 + \sqrt{5})$, or approximately 1.309016994 and 0.1909830056.

EXAMPLE 5   Solve: $t^{2/5} - t^{1/5} - 2 = 0$.

Solution   Note that the square of $t^{1/5}$ is $(t^{1/5})^2$, or $t^{2/5}$. The equation is therefore quadratic in $t^{1/5}$, so we let $u = t^{1/5}$ and $u^2 = t^{2/5}$:

$$u^2 - u - 2 = 0 \qquad \text{Substituting}$$

$$(u - 2)(u + 1) = 0$$

$$u = 2 \quad or \quad u = -1. \qquad \text{Using the principle of zero products}$$

Now we replace u with $t^{1/5}$ and solve:

$$t^{1/5} = 2 \quad or \quad t^{1/5} = -1$$

$$t = 32 \quad or \quad t = -1. \qquad \begin{array}{l}\text{Principle of powers; raising to the}\\\text{5th power}\end{array}$$

Check:

For 32:

$$\frac{t^{2/5} - t^{1/5} - 2 = 0}{32^{2/5} - 32^{1/5} - 2 \;\Big|\; 0}$$

$$(32^{1/5})^2 - 32^{1/5} - 2$$

$$2^2 - 2 - 2$$

$$0 \overset{?}{=} 0 \quad \text{TRUE}$$

For -1:

$$\frac{t^{2/5} - t^{1/5} - 2 = 0}{(-1)^{2/5} - (-1)^{1/5} - 2 \;\Big|\; 0}$$

$$[(-1)^{1/5}]^2 - (-1)^{1/5} - 2$$

$$(-1)^2 - (-1) - 2$$

$$0 \overset{?}{=} 0 \quad \text{TRUE}$$

Both numbers check. The solutions are 32 and -1.

The following tips may prove useful.

> ## To Solve an Equation That Is Reducible to Quadratic
>
> 1. The equation is quadratic in form if the variable factor in one term is the square of the variable factor in the other variable term.
> 2. Write down any substitutions that you are making.
> 3. Whenever you make a substitution, remember to solve for the variable that is used in the original equation.
> 4. Check possible answers in the original equation.

11.5 Exercise Set

FOR EXTRA HELP

 Student's Solutions Manual

  Digital Video Tutor CD 6 Videotape 11

 AW Math Tutor Center

  MathXL Tutorials on CD

  MathXL MathXL

 MyMathLab MyMathLab

↩ *Concept Reinforcement* In each of Exercises 1–8, match the equation with an appropriate substitution from the column on the right that could be used to reduce the equation to quadratic form.

1. ____ $4x^6 - 2x^3 + 1 = 0$

2. ____ $3x^4 + 4x^2 - 7 = 0$

3. ____ $5x^8 + 2x^4 - 3 = 0$

4. ____ $2x^{2/3} - 5x^{1/3} + 4 = 0$

5. ____ $3x^{4/3} + 4x^{2/3} - 7 = 0$

6. ____ $2x^{-2/3} + x^{-1/3} + 6 = 0$

7. ____ $4x^{-4/3} - 2x^{-2/3} + 3 = 0$

8. ____ $3x^{-4} + 4x^{-2} - 2 = 0$

a) $u = x^{-1/3}$

b) $u = x^{1/3}$

c) $u = x^{-2}$

d) $u = x^2$

e) $u = x^{-2/3}$

f) $u = x^3$

g) $u = x^{2/3}$

h) $u = x^4$

Solve.

9. $x^4 - 5x^2 + 4 = 0$

10. $x^4 - 10x^2 + 9 = 0$

11. $x^4 - 9x^2 + 20 = 0$

12. $x^4 - 12x^2 + 27 = 0$

13. $4t^4 - 19t^2 + 12 = 0$

14. $9t^4 - 14t^2 + 5 = 0$

15. $r - 2\sqrt{r} - 6 = 0$

16. $s - 4\sqrt{s} - 1 = 0$

17. $(x^2 - 7)^2 - 3(x^2 - 7) + 2 = 0$

18. $(x^2 - 1)^2 - 5(x^2 - 1) + 6 = 0$

19. $\left(1 + \sqrt{x}\right)^2 + 5\left(1 + \sqrt{x}\right) + 6 = 0$

20. $\left(3 + \sqrt{x}\right)^2 + 3\left(3 + \sqrt{x}\right) - 10 = 0$

21. $x^{-2} - x^{-1} - 6 = 0$

22. $2x^{-2} - x^{-1} - 1 = 0$

23. $4x^{-2} + x^{-1} - 5 = 0$

24. $m^{-2} + 9m^{-1} - 10 = 0$

25. $t^{2/3} + t^{1/3} - 6 = 0$

26. $w^{2/3} - 2w^{1/3} - 8 = 0$

27. $y^{1/3} - y^{1/6} - 6 = 0$

28. $t^{1/2} + 3t^{1/4} + 2 = 0$

29. $t^{1/3} + 2t^{1/6} = 3$

30. $m^{1/2} + 6 = 5m^{1/4}$

31. $\left(3 - \sqrt{x}\right)^2 - 10\left(3 - \sqrt{x}\right) + 23 = 0$

32. $\left(5 + \sqrt{x}\right)^2 - 12\left(5 + \sqrt{x}\right) + 33 = 0$

33. $16\left(\dfrac{x - 1}{x - 8}\right)^2 + 8\left(\dfrac{x - 1}{x - 8}\right) + 1 = 0$

34. $9\left(\dfrac{x + 2}{x + 3}\right)^2 - 6\left(\dfrac{x + 2}{x + 3}\right) + 1 = 0$

Find all x-intercepts of the given function f. If none exist, state this.

35. $f(x) = 5x + 13\sqrt{x} - 6$

36. $f(x) = 3x + 10\sqrt{x} - 8$

37. $f(x) = (x^2 - 3x)^2 - 10(x^2 - 3x) + 24$

38. $f(x) = (x^2 - 6x)^2 - 2(x^2 - 6x) - 35$

39. $f(x) = x^{2/5} + x^{1/5} - 6$

40. $f(x) = x^{1/2} - x^{1/4} - 6$

Aha! **41.** $f(x) = \left(\dfrac{x^2 + 2}{x}\right)^4 + 7\left(\dfrac{x^2 + 2}{x}\right)^2 + 5$

42. $f(x) = \left(\dfrac{x^2 + 1}{x}\right)^4 + 4\left(\dfrac{x^2 + 1}{x}\right)^2 + 12$

43. To solve $25x^6 - 10x^3 + 1 = 0$, Don lets $u = 5x^3$ and Robin lets $u = x^3$. Can they both be correct? Why or why not?

44. Can the examples and exercises of this section be understood without knowing the rules for exponents? Why or why not?

SKILL MAINTENANCE

Graph. [7.3]

45. $f(x) = \frac{3}{2}x$

46. $f(x) = -\frac{2}{3}x$

47. $f(x) = \dfrac{2}{x}$

48. $f(x) = \dfrac{3}{x}$

49. Hiker's Mix is 18% peanuts and Trail Snax is 45% peanuts. How much of each should be mixed together in order to get 12 lb of a mixture that is 36% peanuts? [8.3]

50. If $g(x) = x^2 - x$, find $g(a + 1)$. [7.1]

SYNTHESIS

51. Describe a procedure that could be used to solve any equation of the form $ax^4 + bx^2 + c = 0$.

52. Describe a procedure that could be used to write an equation that is quadratic in $3x^2 - 1$. Then explain how the procedure could be adjusted to write equations that are quadratic in $3x^2 - 1$ and have no real-number solution.

Solve.

53. $5x^4 - 7x^2 + 1 = 0$

54. $3x^4 + 5x^2 - 1 = 0$

55. $(x^2 - 4x - 2)^2 - 13(x^2 - 4x - 2) + 30 = 0$

56. $(x^2 - 5x - 1)^2 - 18(x^2 - 5x - 1) + 65 = 0$

57. $\dfrac{x}{x - 1} - 6\sqrt{\dfrac{x}{x - 1}} - 40 = 0$

58. $\left(\sqrt{\dfrac{x}{x - 3}}\right)^2 - 24 = 10\sqrt{\dfrac{x}{x - 3}}$

59. $a^5(a^2 - 25) + 13a^3(25 - a^2) + 36a(a^2 - 25) = 0$

60. $a^3 - 26a^{3/2} - 27 = 0$

61. $x^6 - 28x^3 + 27 = 0$

62. $x^6 + 7x^3 - 8 = 0$

63. Use a graphing calculator to check your answers to Exercises 9, 11, 37, and 55.

64. Use a graphing calculator to solve
$$x^4 - x^3 - 13x^2 + x + 12 = 0.$$

65. While trying to solve $0.05x^4 - 0.8 = 0$ with a graphing calculator, Murray gets the following screen. Can Murray solve this equation with a graphing calculator? Why or why not?

$y_1 = .05x^4 - .8$

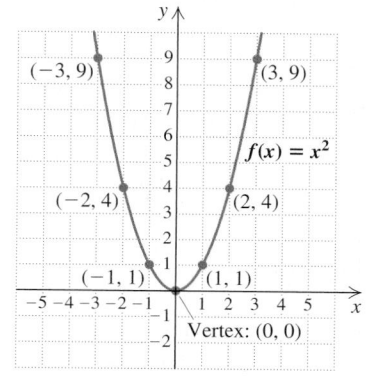

11.6   Quadratic Functions and Their Graphs

The Graph of $f(x) = ax^2$ •   The Graph of $f(x) = a(x - h)^2$ •
The Graph of $f(x) = a(x - h)^2 + k$

We have already used quadratic functions when we solved equations earlier in this chapter. In this section and the next, we learn to graph such functions.

The Graph of $f(x) = ax^2$

The most basic quadratic function is $f(x) = x^2$.

EXAMPLE 1   Graph: $f(x) = x^2$.

Solution   We choose some values for x and compute $f(x)$ for each. Then we plot the ordered pairs and connect them with a smooth curve.

x	$f(x) = x^2$	$(x, f(x))$
-3	9	$(-3, 9)$
-2	4	$(-2, 4)$
-1	1	$(-1, 1)$
0	0	$(0, 0)$
1	1	$(1, 1)$
2	4	$(2, 4)$
3	9	$(3, 9)$

Student Notes

By paying attention to the symmetry of each parabola and the location of the vertex, you save yourself considerable work. Note too that if the x^2-coefficient is a, the x-values 1 unit to the right or left of the vertex are paired with the y-value a units above the vertex. Thus the graph of $y = \frac{3}{2}x^2$ includes the points $\left(-1, \frac{3}{2}\right)$ and $\left(1, \frac{3}{2}\right)$.

All quadratic functions have graphs similar to the one in Example 1. Such curves are called *parabolas*. They are U-shaped and symmetric with respect to a vertical line known as the parabola's *axis of symmetry*. For the graph of $f(x) = x^2$, the y-axis (the vertical line $x = 0$) is the axis of symmetry. Were the paper folded on this line, the two halves of the curve would match. The point $(0, 0)$ is known as the *vertex* of this parabola.

By plotting points, we can compare the graphs of $g(x) = \frac{1}{2}x^2$ and $h(x) = 2x^2$ with the graph of $f(x) = x^2$.

x	$g(x) = \frac{1}{2}x^2$
-3	$\frac{9}{2}$
-2	2
-1	$\frac{1}{2}$
0	0
1	$\frac{1}{2}$
2	2
3	$\frac{9}{2}$

x	$h(x) = 2x^2$
-3	18
-2	8
-1	2
0	0
1	2
2	8
3	18

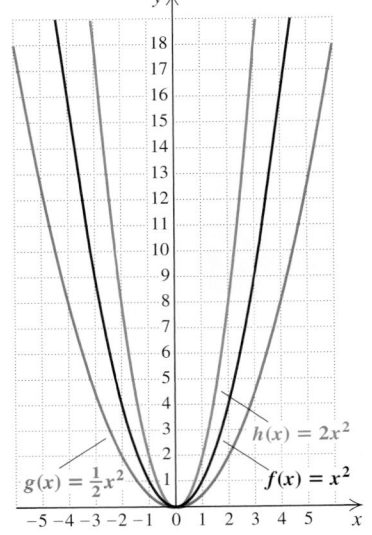

technology connection

To explore the effect of a on the graph of $y = ax^2$, let $y_1 = x^2$, $y_2 = 3x^2$, and $y_3 = \frac{1}{3}x^2$. Graph the equations and use (TRACE) to see how the y-values compare, using △ or ▽ to hop the cursor from one curve to the next.

On many graphing calculators, the (APPS) key will access the Transfrm application. If you run that application and let $y_1 = Ax^2$, the graph becomes interactive and allows for A to be changed while viewing the graph. Selecting the SETTINGS option, after pressing (WINDOW), permits us to adjust the step by which A can be changed.

1. Compare the graphs of $y_1 = \frac{1}{5}x^2$, $y_2 = x^2$, $y_3 = \frac{5}{2}x^2$, $y_4 = -\frac{1}{5}x^2$, $y_5 = -x^2$, and $y_6 = -\frac{5}{2}x^2$.
2. Describe the effect that A has on each graph.

Note that the graph of $g(x) = \frac{1}{2}x^2$ is "wider" than the graph of $\boldsymbol{f(x) = x^2}$, and the graph of $h(x) = 2x^2$ is "narrower." The vertex and the axis of symmetry, however, remain $(0, 0)$ and the line $x = 0$, respectively.

When we consider the graph of $k(x) = -\frac{1}{2}x^2$, we see that the parabola is the same shape as the graph of $g(x) = \frac{1}{2}x^2$, but opens downward. We say that the graphs of k and g are *reflections* of each other across the x-axis.

x	$k(x) = -\frac{1}{2}x^2$
-3	$-\frac{9}{2}$
-2	-2
-1	$-\frac{1}{2}$
0	0
1	$-\frac{1}{2}$
2	-2
3	$-\frac{9}{2}$

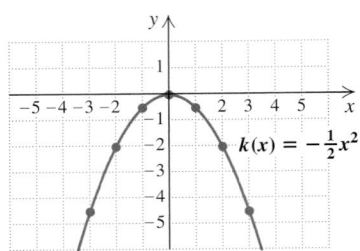

Graphing $f(x) = ax^2$

The graph of $f(x) = ax^2$ is a parabola with $x = 0$ as its axis of symmetry. Its vertex is the origin.

For $a > 0$, the parabola opens upward. For $a < 0$, the parabola opens downward.

If $|a|$ is greater than 1, the parabola is narrower than $y = x^2$.

If $|a|$ is between 0 and 1, the parabola is wider than $y = x^2$.

The Graph of $f(x) = a(x - h)^2$

We *could* next consider graphs of

$$f(x) = ax^2 + bx + c,$$

where b and c are not both 0. In effect, we will do that, but in a disguised form. It turns out to be convenient to first graph $f(x) = a(x - h)^2$, where h is some constant. This allows us to observe similarities to the graphs drawn above.

EXAMPLE 2   Graph: $f(x) = (x - 3)^2$.

Solution   We choose some values for x and compute $f(x)$. It is important to note that when an input here is 3 more than an input for Example 1, the outputs match. We plot the points and draw the curve.

x	$f(x) = (x - 3)^2$
-1	16
0	9
1	4
2	1
3	0
4	1
5	4
6	9

← Vertex

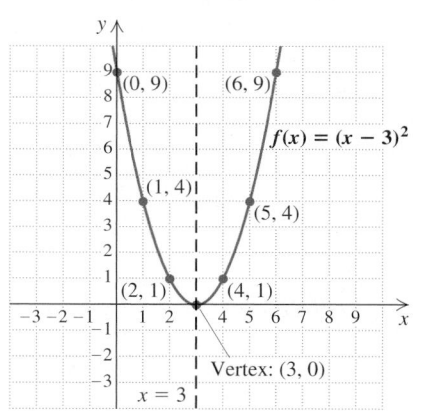

Note that $f(x)$ is smallest when $x - 3$ is 0, that is, for $x = 3$. Thus the line $x = 3$ is now the axis of symmetry and the point $(3, 0)$ is the vertex. Had we recognized earlier that $x = 3$ is the axis of symmetry, we could have computed some values on one side, such as $(4, 1)$, $(5, 4)$, and $(6, 9)$, and then used symmetry to get their mirror images $(2, 1)$, $(1, 4)$, and $(0, 9)$ without further computation.

EXAMPLE 3

To explore the effect of h on the graph of $f(x) = a(x - h)^2$, let $y_1 = 7x^2$ and $y_2 = 7(x - 1)^2$. Graph both y_1 and y_2 and compare y-values, beginning at $x = 1$ and increasing x by one unit at a time. The G-T or HORIZ **MODE** can be used to view a split screen showing both the graph and a table.

Next, let $y_3 = 7(x - 2)^2$ and compare its graph and y-values with those of y_1 and y_2. Then let $y_4 = 7(x + 1)^2$ and $y_5 = 7(x + 2)^2$.

1. Compare graphs and y-values and describe the effect of h on the graph of $f(x) = a(x - h)^2$.
2. If the Transfrm application is available, let $y_1 = A(x - B)^2$ and describe the effect that A and B have on each graph.

Graph: $g(x) = -2(x + 4)^2$.

Solution   We choose some values for x and compute $g(x)$. Note that $g(x)$ is greatest when $x + 4$ is 0, that is, for $x = -4$. Thus the line given by $x = -4$ is the axis of symmetry and the point $(-4, 0)$ is the vertex. We plot some points and draw the curve.

x	$g(x) = -2(x + 4)^2$
-6	-8
-5	-2
-4	0
-3	-2
-2	-8

← Vertex

In Example 2, the graph of $f(x) = (x - 3)^2$ looks just like the graph of $y = x^2$, except that it is moved, or *translated*, 3 units to the right. In Example 3, the graph of $g(x) = -2(x + 4)^2$ looks like the graph of $y = -2x^2$, except that it is shifted 4 units to the left. These results are generalized as follows.

> *Graphing $f(x) = a(x - h)^2$*
>
> The graph of $f(x) = a(x - h)^2$ has the same shape as the graph of $y = ax^2$.
>
> If h is positive, the graph of $y = ax^2$ is shifted h units to the right.
>
> If h is negative, the graph of $y = ax^2$ is shifted $|h|$ units to the left.
>
> The vertex is $(h, 0)$ and the axis of symmetry is $x = h$.

The Graph of $f(x) = a(x - h)^2 + k$

Given a graph of $f(x) = a(x - h)^2$, what happens if we add a constant k? Suppose that we add 2. This increases $f(x)$ by 2, so the curve is moved up. If k is negative, the curve is moved down. The axis of symmetry for the parabola remains $x = h$, but the vertex will be at (h, k), or, equivalently, $(h, f(h))$.

Note that if a parabola opens upward ($a > 0$), the function value, or y-value, at the vertex is a least, or *minimum*, value. That is, it is less than the y-value at any other point on the graph. If the parabola opens downward ($a < 0$), the function value at the vertex is a greatest, or *maximum*, value.

technology connection

To study the effect of k on the graph of $f(x) = a(x - h)^2 + k$, let $y_1 = 7(x - 1)^2$ and $y_2 = 7(x - 1)^2 + 2$. Graph both y_1 and y_2 in the window $[-5, 5, -5, 5]$ and use TRACE or a TABLE to compare the y-values for any given x-value.

1. Let $y_3 = 7(x - 1)^2 - 4$ and compare its graph and y-values with those of y_1 and y_2.
2. Try other values of k, including decimals and fractions. Describe the effect of k on the graph of $f(x) = a(x - h)^2$.
3. If the Transfrm application is available, let $y_1 = A(x - B)^2 + C$ and describe the effect that A, B, and C have on each graph.

Graphs of $f(x) = a(x - h)^2 + k$

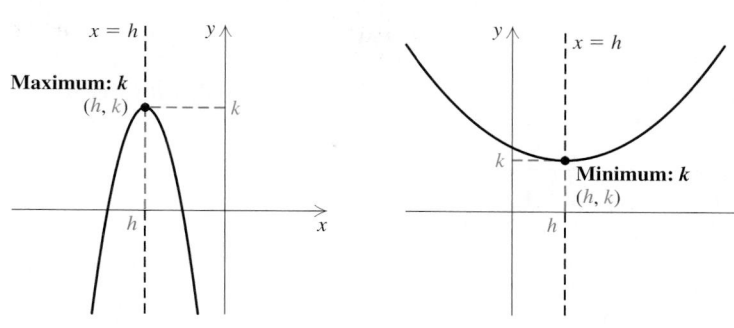

Graphing $f(x) = a(x - h)^2 + k$

The graph of $f(x) = a(x - h)^2 + k$ has the same shape as the graph of $y = a(x - h)^2$.

If k is positive, the graph of $y = a(x - h)^2$ is shifted k units up.

If k is negative, the graph of $y = a(x - h)^2$ is shifted $|k|$ units down.

The vertex is (h, k), and the axis of symmetry is $x = h$.

For $a > 0$, k is the minimum function value. For $a < 0$, k is the maximum function value.

EXAMPLE 4   Graph $g(x) = (x - 3)^2 - 5$, and find the minimum function value.

Solution   The graph will look like that of $f(x) = (x - 3)^2$ (see Example 2) but shifted 5 units down. You can confirm this by plotting some points. For instance, $g(4) = (4 - 3)^2 - 5 = -4$, whereas in Example 2, $f(4) = (4 - 3)^2 = 1$.
The vertex is now $(3, -5)$, and the minimum function value is -5.

x	$g(x) = (x - 3)^2 - 5$
0	4
1	-1
2	-4
3	-5
4	-4
5	-1
6	4

← Vertex

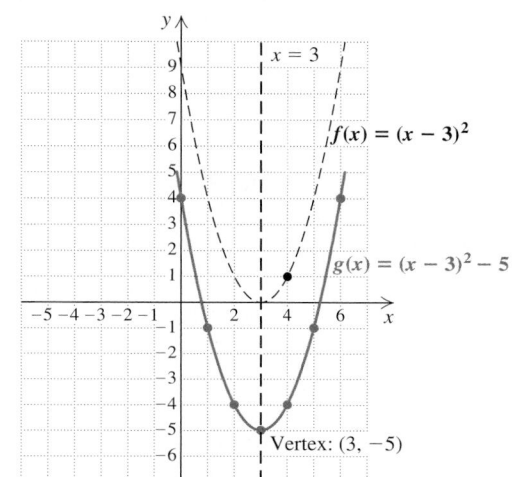

EXAMPLE 5   Graph $h(x) = \frac{1}{2}(x - 3)^2 + 6$, and find the minimum function value.

Solution   The graph looks just like that of $f(x) = \frac{1}{2}x^2$ but moved 3 units to the right and 6 units up. The vertex is (3, 6), and the axis of symmetry is $x = 3$. We draw $f(x) = \frac{1}{2}x^2$ and then shift the curve over and up. The minimum function value is 6. By plotting some points, we have a check.

x	$h(x) = \frac{1}{2}(x - 3)^2 + 6$
0	$10\frac{1}{2}$
1	8
3	6
5	8
6	$10\frac{1}{2}$

←—Vertex

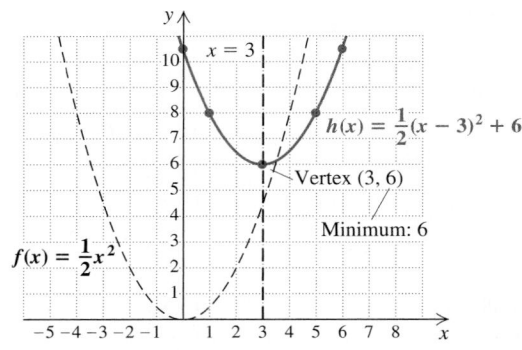

EXAMPLE 6   Graph $y = -2(x + 3)^2 + 5$. Find the vertex, the axis of symmetry, and the maximum or minimum value.

Solution   We first express the equation in the equivalent form

$$y = -2[x - (-3)]^2 + 5.$$

The graph looks like that of $y = -2x^2$ translated 3 units to the left and 5 units up. The vertex is $(-3, 5)$, and the axis of symmetry is $x = -3$. Since -2 is negative, we know that 5, the second coordinate of the vertex, is the maximum y-value.

We compute a few points as needed, selecting convenient x-values on either side of the vertex. The graph is shown here.

x	$y = -2(x + 3)^2 + 5$
-4	3
-3	5
-2	3

←—Vertex

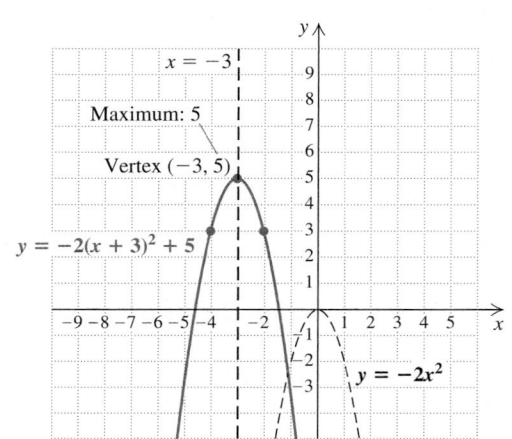

CONNECTING THE CONCEPTS

The ability to graph a function is an important skill. Later in this chapter, as well as in future courses, you will find that graphs of polynomial functions can be used as a tool for solving equations, inequalities, and real-world applications. In the process of learning how to graph quadratic functions, we have developed the ability to reflect or shift (translate) a graph. This skill will prove useful not only in future courses, but in Chapters 12 and 13 as well.

Exercise Set

11.6

FOR EXTRA HELP

 Student's Solutions Manual    Digital Video Tutor CD 6 Videotape 11   AW Math Tutor Center    MathXL Tutorials on CD   *MathXL* MathXL   *MyMathLab* MyMathLab

Concept Reinforcement   *In each of Exercises 1–8, match the equation with the corresponding graph from those shown.*

1. _____ $f(x) = 2(x - 1)^2 + 3$

2. _____ $f(x) = -2(x - 1)^2 + 3$

3. _____ $f(x) = 2(x + 1)^2 + 3$

4. _____ $f(x) = 2(x - 1)^2 - 3$

5. _____ $f(x) = -2(x + 1)^2 + 3$

6. _____ $f(x) = -2(x + 1)^2 - 3$

7. _____ $f(x) = 2(x + 1)^2 - 3$

8. _____ $f(x) = -2(x - 1)^2 - 3$

c)
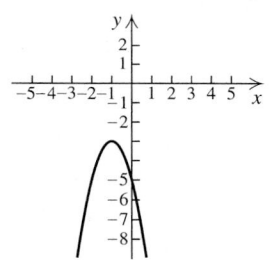

d)
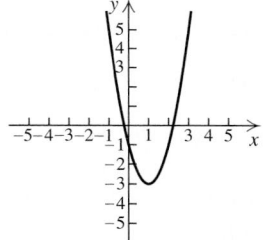

e)
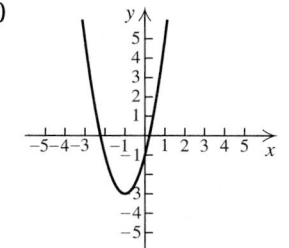

f)
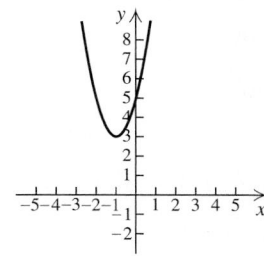

a)
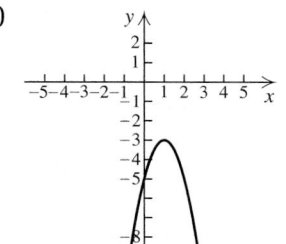

b)
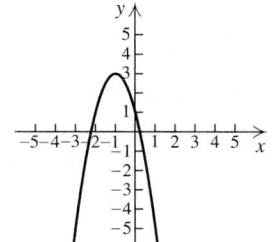

g)
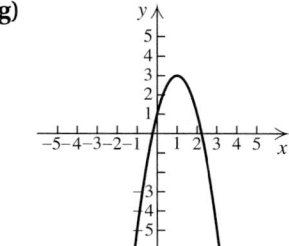

h)
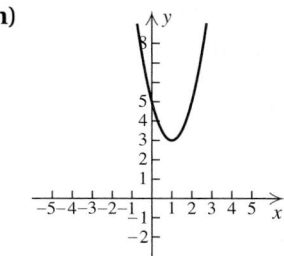

Graph.

9. $f(x) = x^2$

10. $f(x) = -x^2$

11. $f(x) = -2x^2$

12. $f(x) = -3x^2$

13. $g(x) = \frac{1}{3}x^2$

14. $g(x) = \frac{1}{4}x^2$

Aha! **15.** $h(x) = -\frac{1}{3}x^2$

16. $h(x) = -\frac{1}{4}x^2$

17. $f(x) = \frac{5}{2}x^2$

18. $f(x) = \frac{3}{2}x^2$

For each of the following, graph the function, label the vertex, and draw the axis of symmetry.

19. $g(x) = (x + 1)^2$

20. $g(x) = (x + 4)^2$

21. $f(x) = (x - 2)^2$

22. $f(x) = (x - 1)^2$

23. $h(x) = (x - 3)^2$

24. $h(x) = (x - 4)^2$

25. $f(x) = -(x + 1)^2$

26. $f(x) = -(x - 1)^2$

27. $g(x) = -(x - 2)^2$

28. $g(x) = -(x + 4)^2$

29. $f(x) = 2(x + 1)^2$

30. $f(x) = 2(x + 4)^2$

31. $h(x) = -\frac{1}{2}(x - 4)^2$

32. $h(x) = -\frac{3}{2}(x - 2)^2$

33. $f(x) = \frac{1}{2}(x - 1)^2$

34. $f(x) = \frac{1}{3}(x + 2)^2$

35. $f(x) = -2(x + 5)^2$

36. $f(x) = 2(x + 7)^2$

37. $h(x) = -3\left(x - \frac{1}{2}\right)^2$

38. $h(x) = -2\left(x + \frac{1}{2}\right)^2$

For each of the following, graph the function and find the vertex, the axis of symmetry, and the maximum value or the minimum value.

39. $f(x) = (x - 5)^2 + 2$

40. $f(x) = (x + 3)^2 - 2$

41. $f(x) = (x + 1)^2 - 3$

42. $f(x) = (x - 1)^2 + 2$

43. $g(x) = (x + 4)^2 + 1$

44. $g(x) = -(x - 2)^2 - 4$

45. $h(x) = -2(x - 1)^2 - 3$

46. $h(x) = -2(x + 1)^2 + 4$

47. $f(x) = 2(x + 4)^2 + 1$

48. $f(x) = 2(x - 5)^2 - 3$

49. $g(x) = -\frac{3}{2}(x - 1)^2 + 4$

50. $g(x) = \frac{3}{2}(x + 2)^2 - 3$

Without graphing, find the vertex, the axis of symmetry, and the maximum value or the minimum value.

51. $f(x) = 6(x - 8)^2 + 7$

52. $f(x) = 4(x + 5)^2 - 6$

53. $h(x) = -\frac{2}{7}(x + 6)^2 + 11$

54. $h(x) = -\frac{3}{11}(x - 7)^2 - 9$

55. $f(x) = 7\left(x + \frac{1}{4}\right)^2 - 13$

56. $f(x) = 6\left(x - \frac{1}{4}\right)^2 + 15$

57. $f(x) = \sqrt{2}(x + 4.58)^2 + 65\pi$

58. $f(x) = 4\pi(x - 38.2)^2 - \sqrt{34}$

59. Explain, without plotting points, why the graph of $y = x^2 - 4$ looks like the graph of $y = x^2$ translated 4 units down.

60. Explain, without plotting points, why the graph of $y = (x + 2)^2$ looks like the graph of $y = x^2$ translated 2 units to the left.

SKILL MAINTENANCE

Graph using intercepts. [3.3]

61. $2x - 7y = 28$

62. $6x - 3y = 36$

Solve each system. [8.2]

63. $3x + 4y = -19,$
$7x - 6y = -29$

64. $5x + 7y = 9,$
$3x - 4y = -11$

Replace the blanks with constants to form a true equation. [11.1]

65. $x^2 + 5x + \underline{\quad} = \left(x + \underline{\quad}\right)^2$

66. $x^2 - 9x + \underline{\quad} = \left(x - \underline{\quad}\right)^2$

SYNTHESIS

67. Before graphing a quadratic function, Sophie always plots five points. First, she calculates and plots the coordinates of the vertex. Then she plots *four* more points after calculating *two* more ordered pairs. How is this possible?

68. If the graphs of $f(x) = a_1(x - h_1)^2 + k_1$ and $g(x) = a_2(x - h_2)^2 + k_2$ have the same shape, what, if anything, can you conclude about the a's, the h's, and the k's? Why?

Write an equation for a function having a graph with the same shape as the graph of $f(x) = \frac{3}{5}x^2$, but with the given point as the vertex.

69. $(4, 1)$

70. $(2, 6)$

71. $(3, -1)$

72. $(5, -6)$

73. $(-2, -5)$

74. $(-4, -2)$

For each of the following, write the equation of the parabola that has the shape of $f(x) = 2x^2$ or $g(x) = -2x^2$ and has a maximum or minimum value at the specified point.

75. Minimum: $(2, 0)$

76. Minimum: $(-4, 0)$

77. Maximum: $(0, 3)$

78. Maximum: $(3, 8)$

Find an equation for a quadratic function F that satisfies the following conditions.

79. The graph of F is the same shape as the graph of f, where $f(x) = 3(x + 2)^2 + 7$, and $F(x)$ is a minimum at the same point that $g(x) = -2(x - 5)^2 + 1$ is a maximum.

80. The graph of F is the same shape as the graph of f, where $f(x) = -\frac{1}{3}(x - 2)^2 + 7$, and $F(x)$ is a maximum at the same point that $g(x) = 2(x + 4)^2 - 6$ is a minimum.

Functions other than parabolas can be translated. When calculating $f(x)$, if we replace x with $x - h$, where h is a constant, the graph will be moved horizontally. If we replace $f(x)$ with $f(x) + k$, the graph will be moved vertically. Use the graph below for Exercises 81–86.

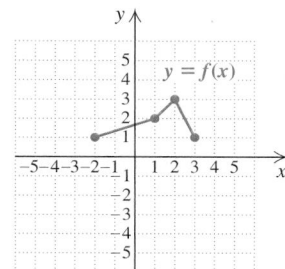

Draw a graph of each of the following.

81. $y = f(x - 1)$

82. $y = f(x + 2)$

83. $y = f(x) + 2$

84. $y = f(x) - 3$

85. $y = f(x + 3) - 2$

86. $y = f(x - 3) + 1$

87. Use the TRACE and/or TABLE features of a graphing calculator to confirm the maximum and minimum values given as answers to Exercises 51, 53, and 55. Be sure to adjust the window appropriately. On many graphing calculators, a maximum or minimum option may be available by using a CALC key.

88. Use a graphing calculator to check your graphs for Exercises 18, 28, and 48.

89. While trying to graph $y = -\frac{1}{2}x^2 + 3x + 1$, Omar gets the following screen. How can Omar tell at a glance that a mistake has been made?

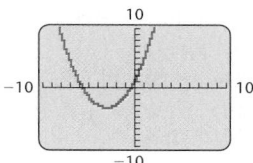

CORNER

Match the Graph

Focus: Graphing quadratic functions

Time: 15–20 minutes

Group size: 6

Materials: Index cards

ACTIVITY

1. On each of six index cards, write one of the following equations:

$y = \frac{1}{2}(x - 3)^2 + 1;$ $y = \frac{1}{2}(x - 1)^2 + 3;$

$y = \frac{1}{2}(x + 1)^2 - 3;$ $y = \frac{1}{2}(x + 3)^2 + 1;$

$y = \frac{1}{2}(x + 3)^2 - 1;$ $y = \frac{1}{2}(x + 1)^2 + 3.$

2. Fold each index card and mix up the six cards in a hat or bag. Then, one by one, each group member should select one of the equations. Do not let anyone see your equation.

3. Each group member should carefully graph the equation selected. Make the graph large enough so that when it is finished, it can be easily viewed by the rest of the group. Be sure to scale the axes and label the vertex, but **do not label the graph with the equation used.**

4. When all group members have drawn a graph, place the graphs in a pile. The group should then match and agree on the correct equation for each graph *with no help from the person who drew the graph.* If a mistake has been made and a graph has no match, determine what its equation *should* be.

5. Compare your group's labeled graphs with those of other groups to reach consensus within the class on the correct label for each graph.

11.7 More About Graphing Quadratic Functions

Completing the Square • Finding Intercepts

Completing the Square

By *completing the square* (see Section 11.1), we can rewrite any polynomial $ax^2 + bx + c$ in the form $a(x - h)^2 + k$. Once that has been done, the procedures discussed in Section 11.6 will enable us to graph any quadratic function.

EXAMPLE 1

Graph: $g(x) = x^2 - 6x + 4$.

Solution We have

$$g(x) = x^2 - 6x + 4$$
$$= (x^2 - 6x) + 4.$$

To complete the square inside the parentheses, we take half the x-coefficient, $\frac{1}{2} \cdot (-6) = -3$, and square it to get $(-3)^2 = 9$. Then we add $9 - 9$ inside the parentheses:

$$g(x) = (x^2 - 6x + 9 - 9) + 4 \qquad \text{The effect is of adding 0.}$$
$$= (x^2 - 6x + 9) + (-9 + 4) \qquad \text{Using the associative law of addition to regroup}$$
$$= (x - 3)^2 - 5. \qquad \text{Factoring and simplifying}$$

This equation appeared as Example 4 of Section 11.6. The graph is that of $f(x) = x^2$ translated right 3 units and down 5 units. The vertex is $(3, -5)$.

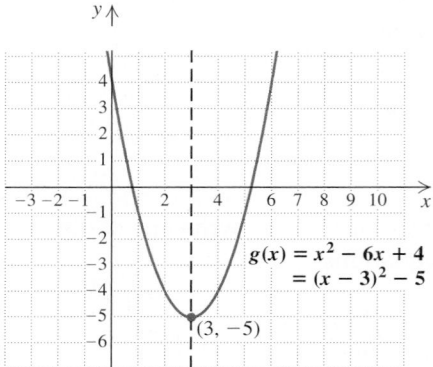

$$g(x) = x^2 - 6x + 4$$
$$= (x - 3)^2 - 5$$

$(3, -5)$

When the leading coefficient is not 1, we factor out that number from the first two terms. Then we complete the square and use the distributive law.

EXAMPLE 2   Graph: $f(x) = 3x^2 + 12x + 13$.

Solution   Since the coefficient of x^2 is not 1, we need to factor out that number—in this case, 3—from the first two terms. Remember that we want the form $f(x) = a(x - h)^2 + k$:

$$f(x) = 3x^2 + 12x + 13$$
$$= 3(x^2 + 4x) + 13.$$

Now we complete the square as before. We take half of the x-coefficient, $\frac{1}{2} \cdot 4 = 2$, and square it: $2^2 = 4$. Then we add $4 - 4$ inside the parentheses:

$$f(x) = 3(x^2 + 4x + 4 - 4) + 13. \qquad \text{Adding } 4 - 4, \text{ or } 0, \text{ inside}$$
$$\text{the parentheses}$$

The distributive law allows us to separate the -4 from the perfect-square trinomial so long as it is multiplied by 3. *This step is critical*:

$$f(x) = 3(x^2 + 4x + 4) + 3(-4) + 13 \qquad \text{This leaves a perfect-square}$$
$$\text{trinomial inside the}$$
$$\text{parentheses.}$$
$$= 3(x + 2)^2 + 1. \qquad \text{Factoring and simplifying}$$

The vertex is $(-2, 1)$, and the axis of symmetry is $x = -2$. The coefficient of x^2 is 3, so the graph is narrow and opens upward. We choose a few x-values on either side of the vertex, compute y-values, and then graph the parabola.

x	$f(x) = 3(x + 2)^2 + 1$	
-2	1	←—Vertex
-3	4	
-1	4	

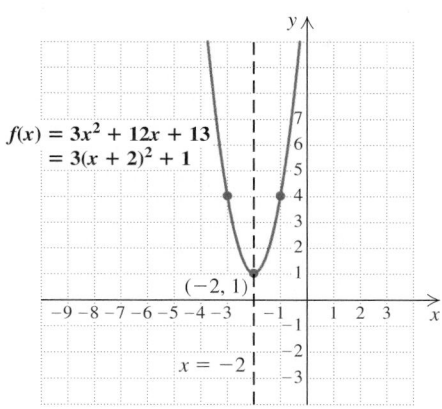

EXAMPLE 3   Graph: $f(x) = -2x^2 + 10x - 7$.

Solution   We first find the vertex by completing the square. To do so, we factor out -2 from the first two terms of the expression. This makes the coefficient of x^2 inside the parentheses 1:

$$f(x) = -2x^2 + 10x - 7$$
$$= -2(x^2 - 5x) - 7.$$

Now we complete the square as before. We take half of the x-coefficient and square it to get $\frac{25}{4}$. Then we add $\frac{25}{4} - \frac{25}{4}$ inside the parentheses:

$$f(x) = -2\left(x^2 - 5x + \frac{25}{4} - \frac{25}{4}\right) - 7$$

$$= -2\left(x^2 - 5x + \frac{25}{4}\right) + (-2)\left(-\frac{25}{4}\right) - 7 \qquad \text{Multiplying by } -2, \text{ using the distributive law, and regrouping}$$

$$= -2\left(x - \frac{5}{2}\right)^2 + \frac{11}{2}. \qquad \text{Factoring and simplifying}$$

The vertex is $\left(\frac{5}{2}, \frac{11}{2}\right)$, and the axis of symmetry is $x = \frac{5}{2}$. The coefficient of x^2, -2, is negative, so the graph opens downward. We plot a few points on either side of the vertex, including the y-intercept, $f(0)$, and graph the parabola.

x	$f(x)$	
$\frac{5}{2}$	$\frac{11}{2}$	← Vertex
0	-7	← y-intercept
1	1	
4	1	

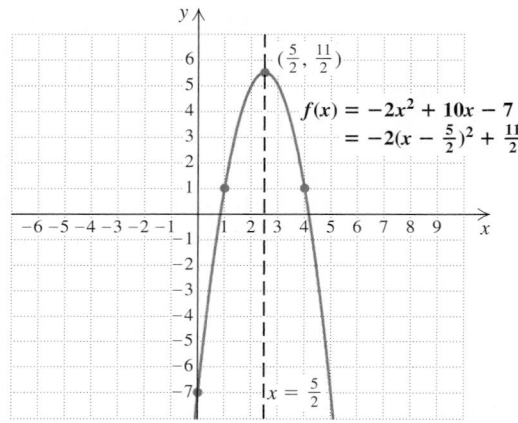

The method used in Examples 1–3 can be generalized to find a formula for locating the vertex. We complete the square as follows:

$$f(x) = ax^2 + bx + c$$

$$= a\left(x^2 + \frac{b}{a}x\right) + c. \qquad \text{Factoring } a \text{ out of the first two terms. Check by multiplying.}$$

Half of the x-coefficient, $\frac{b}{a}$, is $\frac{b}{2a}$. We square it to get $\frac{b^2}{4a^2}$ and add $\frac{b^2}{4a^2} - \frac{b^2}{4a^2}$ inside the parentheses. Then we distribute the a and regroup terms:

$$f(x) = a\left(x^2 + \frac{b}{a}x + \frac{b^2}{4a^2} - \frac{b^2}{4a^2}\right) + c$$

$$= a\left(x^2 + \frac{b}{a}x + \frac{b^2}{4a^2}\right) + a\left(-\frac{b^2}{4a^2}\right) + c \qquad \text{Using the distributive law}$$

$$= a\left(x + \frac{b}{2a}\right)^2 + \frac{-b^2}{4a} + \frac{4ac}{4a} \qquad \text{Factoring and finding a common denominator}$$

$$= a\left[x - \left(-\frac{b}{2a}\right)\right]^2 + \frac{4ac - b^2}{4a}.$$

Student Notes

The easiest way to remember a formula is to understand its derivation. Check with your instructor to determine what, if any, formulas you will be expected to remember.

Thus we have the following.

The Vertex of a Parabola

The vertex of the parabola given by $f(x) = ax^2 + bx + c$ is

$$\left(-\frac{b}{2a}, f\left(-\frac{b}{2a}\right)\right) \quad \text{or} \quad \left(-\frac{b}{2a}, \frac{4ac - b^2}{4a}\right).$$

The x-coordinate of the vertex is $-b/(2a)$. The axis of symmetry is $x = -b/(2a)$. The second coordinate of the vertex is most commonly found by computing $f\left(-\frac{b}{2a}\right)$.

Let's reexamine Example 3 to see how we could have found the vertex directly. From the formula above,

$$\text{the } x\text{-coordinate of the vertex is } -\frac{b}{2a} = -\frac{10}{2(-2)} = \frac{5}{2}.$$

Substituting $\frac{5}{2}$ into $f(x) = -2x^2 + 10x - 7$, we find the second coordinate of the vertex:

$$
\begin{aligned}
f\left(\tfrac{5}{2}\right) &= -2\left(\tfrac{5}{2}\right)^2 + 10\left(\tfrac{5}{2}\right) - 7 \\
&= -2\left(\tfrac{25}{4}\right) + 25 - 7 \\
&= -\tfrac{25}{2} + 18 \\
&= -\tfrac{25}{2} + \tfrac{36}{2} = \tfrac{11}{2}.
\end{aligned}
$$

The vertex is $\left(\frac{5}{2}, \frac{11}{2}\right)$. The axis of symmetry is $x = \frac{5}{2}$.

We have actually developed two methods for finding the vertex. One is by completing the square and the other is by using a formula. You should check to see if your instructor prefers one method over the other or wants you to use both.

Finding Intercepts

For any function f, the y-intercept occurs at $f(0)$. Thus, for $f(x) = ax^2 + bx + c$, the y-intercept is simply $(0, c)$. To find x-intercepts, we look for points where $y = 0$ or $f(x) = 0$. Thus, for $f(x) = ax^2 + bx + c$, the x-intercepts occur at those x-values for which

$$ax^2 + bx + c = 0.$$

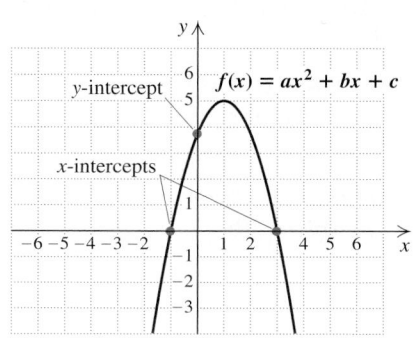

EXAMPLE 4 Find the x- and y-intercepts of the graph of $f(x) = x^2 - 2x - 2$.

Solution The y-intercept is simply $(0, f(0))$, or $(0, -2)$. To find the x-intercepts, we solve the equation

$$0 = x^2 - 2x - 2.$$

We are unable to factor $x^2 - 2x - 2$, so we use the quadratic formula and get $x = 1 \pm \sqrt{3}$. Thus the x-intercepts are $(1 - \sqrt{3}, 0)$ and $(1 + \sqrt{3}, 0)$. If graphing, we would approximate, to get $(-0.7, 0)$ and $(2.7, 0)$.

Exercise Set

11.7

FOR EXTRA HELP

 Student's Solutions Manual

 Digital Video Tutor CD 6 Videotape 11

 Tutor Center AW Math Tutor Center

 MathXL Tutorials on CD

MathXL

MyMathLab

🔁 *Concept Reinforcement* *Complete each of the following.*

1. The expressions $x^2 + 6x + 5$ and $(x^2 + 6x + 9) - \underline{\quad} + 5$ are equivalent.

2. The expressions $x^2 + 8x + 3$ and $(x^2 + 8x + \underline{\quad}) - 16 + 3$ are equivalent.

3. The functions given by $f(x) = 2x^2 + 12x - 7$ and $f(x) = 2(x^2 + 6x + \underline{\quad}) - 18 - 7$ are equivalent.

4. The functions given by $g(x) = 3x^2 - 6x - 5$ and $g(x) = 3(x^2 - 2x + \underline{\quad}) - 3 - 5$ are equivalent.

5. The graph of $f(x) = -2(x - \underline{\quad})^2 + 7$ has its vertex at $(3, 7)$.

6. The graph of $g(x) = -3(x - 4)^2 + \underline{\quad}$ has its vertex at $(4, 1)$.

7. The graph of $f(x) = 2(x - \underline{\quad})^2 + \underline{\quad}$ has its axis of symmetry at $x = \frac{5}{2}$ and has its vertex at $\left(\frac{5}{2}, -4\right)$.

8. The graph of $g(x) = \frac{1}{2}(x - \underline{\quad})^2 + \frac{7}{2}$ has $x = -2$ as its axis of symmetry and $\underline{\quad}$ as its vertex.

For each quadratic function, **(a)** *find the vertex and the axis of symmetry and* **(b)** *graph the function.*

9. $f(x) = x^2 + 4x + 5$

10. $f(x) = x^2 + 2x - 5$

11. $g(x) = x^2 - 6x + 13$

12. $g(x) = x^2 - 4x + 5$

13. $f(x) = x^2 + 8x + 20$

14. $f(x) = x^2 - 10x + 21$

15. $h(x) = 2x^2 - 16x + 25$

16. $h(x) = 2x^2 + 16x + 23$

17. $f(x) = -x^2 + 2x + 5$

18. $f(x) = -x^2 - 2x + 7$

19. $g(x) = x^2 + 3x - 10$

20. $g(x) = x^2 + 5x + 4$

21. $f(x) = 3x^2 - 24x + 50$

22. $f(x) = 4x^2 + 8x - 3$

23. $h(x) = x^2 + 7x$

24. $h(x) = x^2 - 5x$

25. $f(x) = -2x^2 - 4x - 6$

26. $f(x) = -3x^2 + 6x + 2$

27. $g(x) = 2x^2 - 8x + 3$

28. $g(x) = 2x^2 + 5x - 1$

29. $f(x) = -3x^2 + 5x - 2$

30. $f(x) = -3x^2 - 7x + 2$

31. $h(x) = \frac{1}{2}x^2 + 4x + \frac{19}{3}$

32. $h(x) = \frac{1}{2}x^2 - 3x + 2$

Find the x- and y-intercepts. If no x-intercepts exist, state this.

33. $f(x) = x^2 - 6x + 3$

34. $f(x) = x^2 + 5x + 2$

35. $g(x) = -x^2 + 2x + 3$

36. $g(x) = x^2 - 6x + 9$

Aha! **37.** $f(x) = x^2 - 9x$

38. $f(x) = x^2 - 7x$

39. $h(x) = -x^2 + 4x - 4$

40. $h(x) = 4x^2 - 12x + 3$

41. $f(x) = 2x^2 - 4x + 6$

42. $f(x) = x^2 - x + 2$

43. Does the graph of every quadratic function have a *y*-intercept? Why or why not?

44. Is it possible for the graph of a quadratic function to have only one *x*-intercept if the vertex is off the *x*-axis? Why or why not?

SKILL MAINTENANCE

Solve each system.

45. $5x - 3y = 16,$
$4x + 2y = 4$ [8.2]

46. $2x - 5y = 9,$
$5x - 15y = 20$ [8.2]

47. $4a - 5b + c = 3,$
$3a - 4b + 2c = 3,$
$a + b - 7c = -2$
[8.4]

48. $2a - 7b + c = 25,$
$a + 5b - 2c = -18,$
$3a - b + 4c = 14$
[8.4]

Solve. [10.6]

49. $\sqrt{4x - 4} = \sqrt{x + 4} + 1$

50. $\sqrt{5x - 4} + \sqrt{13 - x} = 7$

SYNTHESIS

51. If the graphs of two quadratic functions have the same *x*-intercepts, will they also have the same vertex? Why or why not?

52. Suppose that the graph of $f(x) = ax^2 + bx + c$ has $(x_1, 0)$ and $(x_2, 0)$ as *x*-intercepts. Explain why the graph of $g(x) = -ax^2 - bx - c$ will also have $(x_1, 0)$ and $(x_2, 0)$ as *x*-intercepts.

For each quadratic function, find **(a)** *the maximum or minimum value and* **(b)** *the x- and y-intercepts.*

53. $f(x) = 2.31x^2 - 3.135x - 5.89$

54. $f(x) = -18.8x^2 + 7.92x + 6.18$

55. Graph the function
$$f(x) = x^2 - x - 6.$$
Then use the graph to approximate solutions to each of the following equations.

a) $x^2 - x - 6 = 2$

b) $x^2 - x - 6 = -3$

56. Graph the function
$$f(x) = \frac{x^2}{2} + x - \frac{3}{2}.$$
Then use the graph to approximate solutions to each of the following equations.

a) $\dfrac{x^2}{2} + x - \dfrac{3}{2} = 0$

b) $\dfrac{x^2}{2} + x - \dfrac{3}{2} = 1$

c) $\dfrac{x^2}{2} + x - \dfrac{3}{2} = 2$

Find an equivalent equation of the type
$$f(x) = a(x - h)^2 + k.$$

57. $f(x) = mx^2 - nx + p$

58. $f(x) = 3x^2 + mx + m^2$

59. A quadratic function has $(-1, 0)$ as one of its intercepts and $(3, -5)$ as its vertex. Find an equation for the function.

60. A quadratic function has $(4, 0)$ as one of its intercepts and $(-1, 7)$ as its vertex. Find an equation for the function.

Graph.

61. $f(x) = |x^2 - 1|$

62. $f(x) = |x^2 - 3x - 4|$

63. $f(x) = |2(x - 3)^2 - 5|$

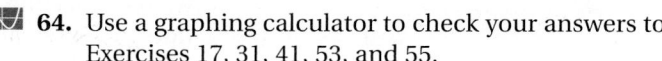 **64.** Use a graphing calculator to check your answers to Exercises 17, 31, 41, 53, and 55.

11.8 Problem Solving and Quadratic Functions

Maximum and Minimum Problems   •   Fitting Quadratic Functions to Data

Let's look now at some of the many situations in which quadratic functions are used for problem solving.

Maximum and Minimum Problems

We have seen that for any quadratic function f, the value of $f(x)$ at the vertex is either a maximum or a minimum. Thus problems in which a quantity must be maximized or minimized can be solved by finding the coordinates of a vertex, assuming the problem can be modeled with a quadratic function.

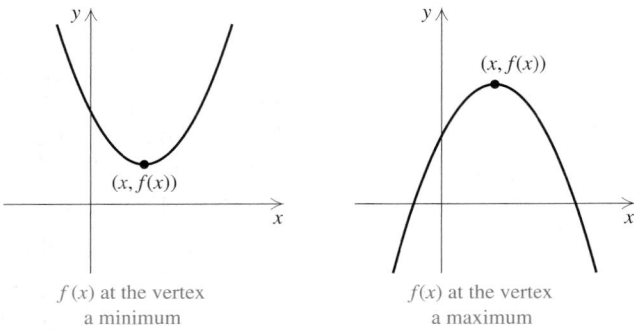

$f(x)$ at the vertex
a minimum

$f(x)$ at the vertex
a maximum

EXAMPLE 1

Newborn calves. The number of pounds of milk per day recommended for a calf that is x weeks old can be approximated by $p(x)$, where $p(x) = -0.2x^2 + 1.3x + 6.2$ (*Source*: C. Chaloux, University of Vermont, 1998). When is a calf's milk consumption greatest and how much milk does it consume at that time?

Solution

1., 2. Familiarize and **Translate.** We are given the function for milk consumption by a calf. Note that it is a quadratic function of x, the calf's age in weeks. Since the coefficient of x^2 is negative, it appears that milk consumption increases and then decreases. The calculator-generated graph at left confirms this.

3. Carry out. We can either complete the square,

$$p(x) = -0.2x^2 + 1.3x + 6.2$$
$$= -0.2(x^2 - 6.5x) + 6.2$$
$$= -0.2(x^2 - 6.5x + 3.25^2 - 3.25^2) + 6.2 \qquad \text{Completing the square;}$$
$$\qquad\qquad\qquad\qquad\qquad\qquad\qquad\qquad\qquad\qquad\qquad\qquad 6.5/2 = 3.25$$
$$= -0.2(x^2 - 6.5x + 3.25^2) + (-0.2)(-3.25^2) + 6.2$$
$$= -0.2(x - 3.25)^2 + 8.3125, \qquad \text{Factoring and simplifying}$$

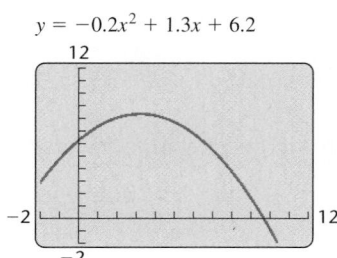

$y = -0.2x^2 + 1.3x + 6.2$

A visualization for Example 1

or we can use $-b/(2a) = -1.3/(-0.4) = 3.25$. Using a calculator, we find that

$$p(3.25) = -0.2(3.25)^2 + 1.3(3.25) + 6.2 = 8.3125.$$

4. **Check.**   Both of the approaches in step (3) indicate that a maximum occurs when $x = 3.25$, or $3\frac{1}{4}$. The graph also serves as a check.

5. **State.**   A calf's milk consumption is greatest when the calf is $3\frac{1}{4}$ weeks old. At that time, it drinks about 8.3 lb of milk per day.

EXAMPLE 2

Swimming area.   A lifeguard has 100 m of roped-together flotation devices with which to cordon off a rectangular swimming area at Lakeside Beach. If the shoreline forms one side of the rectangle, what dimensions will maximize the size of the area for swimming?

Solution

1. **Familiarize.**   We make a drawing and label it, letting $w =$ the width of the rectangle, in meters, and $l =$ the length of the rectangle, in meters.

   Recall that Area $= l \cdot w$ and Perimeter $= 2w + 2l$. Since the beach forms one length of the rectangle, the flotation devices comprise three sides. Thus

   $$2w + l = 100.$$

   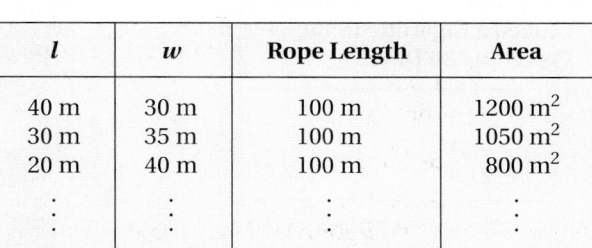

   To get a better feel for the problem, we can look at some possible dimensions for a rectangular area that can be enclosed with 100 m of flotation devices. All possibilities are chosen so that $2w + l = 100$.

technology connection

To generate a table of values for Example 2, let x represent the width of the swimming area, in meters. If l represents the length, in meters, we must have $100 = 2x + l$. Next, solve for l and use that expression for y_1. Then let $y_2 = x \cdot y_1$ (to enter y_1, press **VARS** and select Y-VARS and then FUNCTION and then 1) so that y_2 represents the area. Scroll through the resulting table, adjusting the settings as needed, to determine at what point the area is maximized.

l	w	Rope Length	Area	
40 m	30 m	100 m	1200 m²	What choice of l and w
30 m	35 m	100 m	1050 m²	will maximize A?
20 m	40 m	100 m	800 m²	
⋮	⋮	⋮	⋮	

2. **Translate.**   We have two equations: One guarantees that all 100 m of flotation devices are used; the other expresses area in terms of length and width.

   $$2w + l = 100,$$
   $$A = l \cdot w$$

3. **Carry out.**   We need to express A as a function of l or w but not both. To do so, we solve for l in the first equation to obtain $l = 100 - 2w$. Substituting for l in the second equation, we get a quadratic function:

   $$A = (100 - 2w)w \qquad \text{Substituting for } l$$
   $$= 100w - 2w^2. \qquad \text{This represents a parabola opening downward, so a maximum exists.}$$

Factoring and completing the square, we get

$$A = -2(w^2 - 50w + 625 - 625)$$ We could also use the vertex formula.

$$= -2(w - 25)^2 + 1250.$$ This suggests a maximum of 1250 m^2 when $w = 25$ m.

The maximum area, 1250 m^2, occurs when $w = 25$ m and $l = 100 - 2(25)$, or 50 m.

4. **Check.** Note that 1250 m^2 is greater than any of the values for A found in the *Familiarize* step. To be more certain, we could check values other than those used in that step. For example, if $w = 26$ m, then $l = 100 - 2 \cdot 26 = 48$ m, and $A = 26 \cdot 48 = 1248 \text{ m}^2$. Since 1250 m^2 is greater than 1248 m^2, it appears that we have a maximum.

5. **State.** The largest rectangular area for swimming that can be enclosed is 25 m by 50 m.

Fitting Quadratic Functions to Data

Whenever a certain quadratic function fits a situation, that function can be determined if three inputs and their outputs are known. Each of the given ordered pairs is called a *data point*.

EXAMPLE 3   The decline of teen smoking.   Deadly and less fashionable than ever, teen smoking has declined since 1997. According to the Centers for Disease Control and Prevention, the percentage of high-school students who reported having smoked a cigarette in the preceding 30 days increased from 30.5% in 1993 to 36.4% in 1997, but has declined since then to 21.9% in 2003.

Years After 1993	Percentage of High-School Students Who Smoked a Cigarette in the Preceding 30 Days
0	30.5
4	36.4
10	21.9

Source: Centers for Disease Control and Prevention, *Morbidity and Mortality Weekly Report* 6/18/04

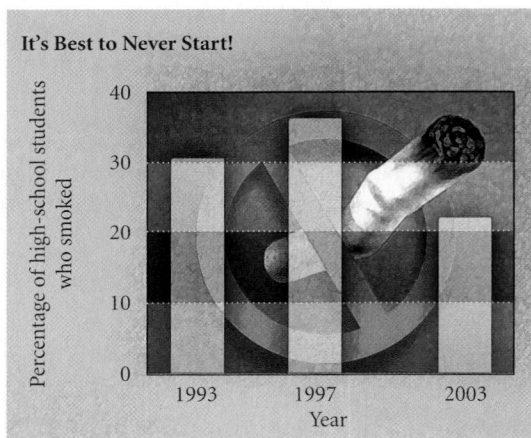

The rise and fall in the percentage of teen smokers suggests that the situation can be modeled by a quadratic function.

a) Use the data points $(0, 30.5)$, $(4, 36.4)$, and $(10, 21.9)$ to find a quadratic function that fits the data.

b) Use the function from part (a) to estimate the percentage of high-school students in 2005 who smoked a cigarette in the preceding 30 days.

Solution

a) We are looking for a function of the form $T(x) = ax^2 + bx + c$, given that $T(0) = 30.5$, $T(4) = 36.4$, and $T(10) = 21.9$. Thus,

$$30.5 = a \cdot 0^2 + b \cdot 0 + c, \qquad \text{Using the data point } (0, 30.5)$$
$$36.4 = a \cdot 4^2 + b \cdot 4 + c, \qquad \text{Using the data point } (4, 36.4)$$
$$21.9 = a \cdot 10^2 + b \cdot 10 + c. \qquad \text{Using the data point } (10, 21.9)$$

After simplifying, we see that we need to solve the system

$$30.5 = c, \qquad\qquad\qquad\quad (\mathbf{1})$$
$$36.4 = 16a + 4b + c, \qquad\;\; (\mathbf{2})$$
$$21.9 = 100a + 10b + c. \quad (\mathbf{3})$$

We know from equation (1) that $c = 30.5$. Substituting that value into equations (2) and (3), we have

$$36.4 = 16a + 4b + 30.5,$$
$$21.9 = 100a + 10b + 30.5.$$

Subtracting 30.5 from both sides of each equation, we have

$$5.9 = 16a + 4b,$$
$$-8.6 = 100a + 10b$$

or

$$\left.\begin{array}{l} 59 = 160a + 40b, \qquad (\mathbf{4}) \\ -86 = 1000a + 100b. \quad (\mathbf{5}) \end{array}\right\} \quad \begin{array}{l}\text{Multiplying both sides} \\ \text{by 10 to clear decimals}\end{array}$$

To solve, we multiply equation (4) by 5 and equation (5) by -2. We then add to eliminate b:

$$\begin{array}{r} 295 = 800a + 200b \\ \underline{172 = -2000a - 200b} \\ 467 = -1200a \end{array}$$

$$-\frac{467}{1200} = a, \quad \text{or} \quad a \approx -0.39. \qquad \text{Converting to decimal notation}$$

Next, we solve for b, using equation (5) above:

$$-86 = 1000\left(-\frac{467}{1200}\right) + 100b$$

$$-86 = -\frac{2335}{6} + 100b$$

$$\frac{1819}{6} = 100b \qquad\qquad\qquad\qquad \text{Adding } \frac{2335}{6} \text{ to both sides and simplifying}$$

$$\frac{1819}{600} = b, \quad \text{or} \quad b \approx 3.03. \qquad \begin{array}{l}\text{Dividing both sides by 100 and} \\ \text{converting to decimal notation}\end{array}$$

We can now write $T(x) = ax^2 + bx + c$ as

$$T(x) = -\frac{467}{1200}x^2 + \frac{1819}{600}x + 30.5 \quad \text{or} \quad T(x) = -0.39x^2 + 3.03x + 30.5.$$

Student Notes

Try to keep the "big picture" in mind on problems like Example 3. Solving a system of three equations is but one part of the solution.

b) To find the percentage of high-school students who will have smoked at least once during 30 days in 2005, we evaluate the function. Note that 2005 is 12 yr after 1993. Thus,

$$T(12) = -\frac{467}{1200} \cdot 12^2 + \frac{1819}{600} \cdot 12 + 30.5$$

$$= 10.84.$$

In 2005, an estimated 10.8% of all high-schoolers smoked at least 1 cigarette during the 30 days preceding the survey in 2005.

 **technology connection**

To use a graphing calculator to fit a quadratic function to the data in Example 3, we first select the EDIT option of the **STAT** key's menu and enter the given data.

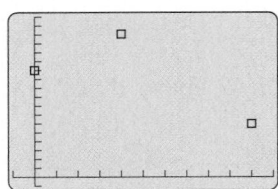

Next, we press **Y=**, **⌃**, and **ENTER** to turn on the plot feature. Using ZoomStat, we obtain the following.

To fit a quadratic function to the data, we press **STAT** **▷** **5** **VARS** **▷** **1** **1** **ENTER**. The first three keystrokes select QuadReg from the STAT CALC menu and display the coefficients a, b, and c of the regression equation $y = ax^2 + bx + c$. The keystrokes **VARS** **▷** **1** **1** copy the regression equation to the equation-editor screen as y_1. We see that the regression equation is $y = -.38916666666668x^2 + 3.0316666666668x + 30.5$. Pressing **ZOOM** **9**, we see the regression equation graphed with the data points.

```
QuadReg
y=ax²+bx+c
a=−.3891666667
b=3.031666667
c=30.5
```

```
Plot1  Plot2  Plot3
\Y1☐−.3891666666
6668x^2+3.031666
6666668x+30.5
\Y2=
\Y3=
\Y4=
\Y5=
```

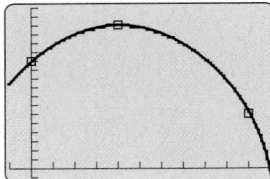

To check Example 3(b), we press **ZOOM** **3** **ENTER** and then **TRACE** **⌃** **1** **2** **ENTER**. The result, 10.84, confirms our earlier answer.

1. Use the above approach to estimate the percentage of high-school students who smoked in 2000.

Exercise Set

11.8

↩ *Concept Reinforcement*   *In each of Exercises 1–6, match the description with the graph that displays that characteristic.*

1. ____ A minimum value of $f(x)$ exists.

2. ____ A maximum value of $f(x)$ exists.

3. ____ No maximum or minimum value of $f(x)$ exists.

4. ____ The data points appear to suggest a linear model.

5. ____ The data points appear to suggest a quadratic model with a maximum.

6. ____ The data points appear to suggest a quadratic model with a minimum.

a)
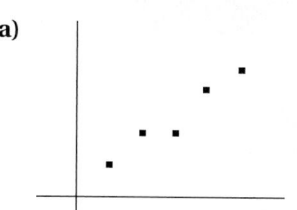

b) $f(x)$
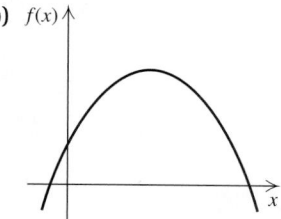

c) $f(x)$
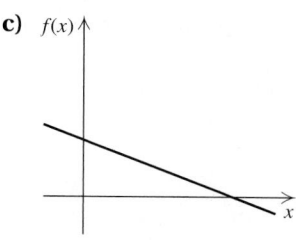

d)
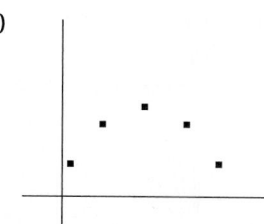

e) $f(x)$
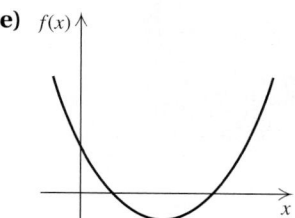

f)
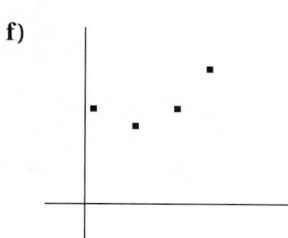

Solve.

7. *Ticket sales.*   The number of tickets sold each day for an upcoming Los Lobos show can be approximated by

$$N(x) = -0.4x^2 + 9x + 11,$$

where x is the number of days since the concert was first announced. When will daily ticket sales peak and how many tickets will be sold that day?

8. *Stock prices.*   The value of a share of I. J. Solar can be represented by $V(x) = x^2 - 6x + 13$, where x is the number of months after January 2004. What is the lowest value $V(x)$ will reach, and when did that occur?

9. *Minimizing cost.*   Sweet Harmony Crafts has determined that when x hundred Dobros are built, the average cost per Dobro can be estimated by

$$C(x) = 0.1x^2 - 0.7x + 2.425,$$

where $C(x)$ is in hundreds of dollars. What is the minimum average cost per Dobro and how many Dobros should be built to achieve that minimum?

10. *Maximizing profit.*   Recall that total profit P is the difference between total revenue R and total cost C. Given $R(x) = 1000x - x^2$ and $C(x) = 3000 + 20x$, find the total profit, the maximum value of the total profit, and the value of x at which it occurs.

11. *Furniture design.*   A furniture builder is designing a rectangular end table with a perimeter of 128 in. What dimensions will yield the maximum area?

12. *Architecture.*   An architect is designing an atrium for a hotel. The atrium is to be rectangular with a perimeter of 720 ft of brass piping. What dimensions will maximize the area of the atrium?

13. *Patio design.*   A stone mason has enough stones to enclose a rectangular patio with 60 ft of perimeter, assuming that the attached house forms one side of the rectangle. What is the maximum area that the mason can enclose? What should the dimensions of the patio be in order to yield this area?

14. *Garden design.*   Ginger is fencing in a rectangular garden, using the side of her house as one side of the rectangle. What is the maximum area that she can enclose with 40 ft of fence? What should the dimensions of the garden be in order to yield this area?

15. *Molding plastics.*   Economite Plastics plans to produce a one-compartment vertical file by bending the long side of an 8-in. by 14-in. sheet of plastic along two lines to form a U shape. How tall should the file be in order to maximize the volume that the file can hold?

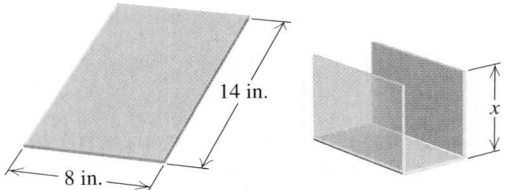

16. *Composting.*   A rectangular compost container is to be formed in a corner of a fenced yard, with 8 ft of chicken wire completing the other two sides of the rectangle. If the chicken wire is 3 ft high, what dimensions of the base will maximize the container's volume?

17. What is the maximum product of two numbers that add to 18? What numbers yield this product?

18. What is the maximum product of two numbers that add to 26? What numbers yield this product?

19. What is the minimum product of two numbers that differ by 8? What are the numbers?

20. What is the minimum product of two numbers that differ by 7? What are the numbers?

Aha! 21. What is the maximum product of two numbers that add to −10? What numbers yield this product?

22. What is the maximum product of two numbers that add to −12? What numbers yield this product?

Choosing models. For the scatterplots and graphs in Exercises 23–34, determine which, if any, of the following functions might be used as a model for the data: Linear, with $f(x) = mx + b$; quadratic, with $f(x) = ax^2 + bx + c, a > 0$; quadratic, with $f(x) = ax^2 + bx + c, a < 0$; neither quadratic nor linear.

23.
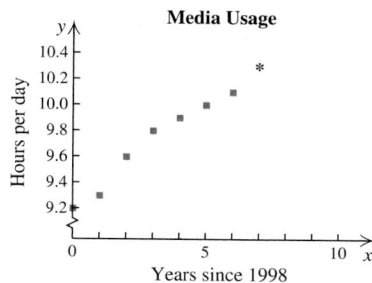

*Estimated
Source: Statistical Abstract of the United States, 2003

24.
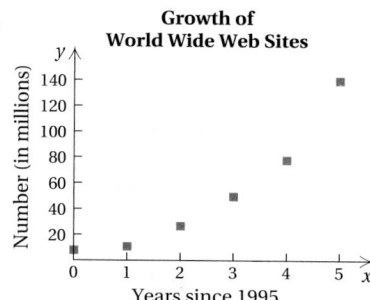

25.
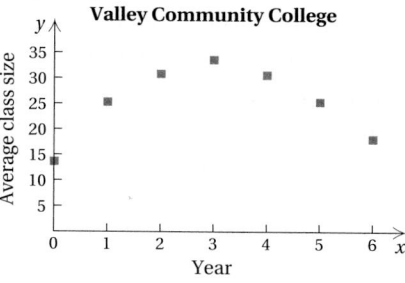

26.
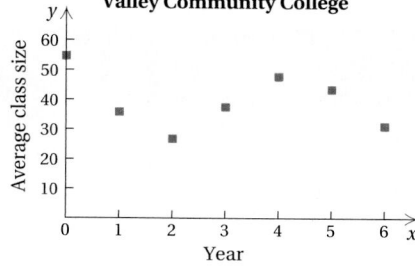

27.

Source: National Highway Traffic Administration

28.
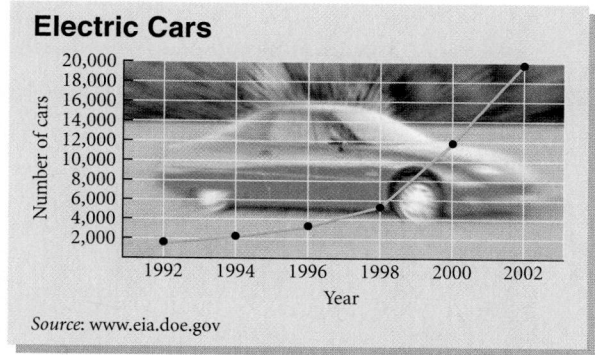

Source: www.eia.doe.gov

29.
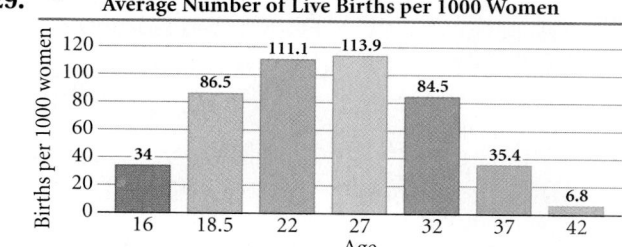

Source: U.S. Centers for Disease Control

30.
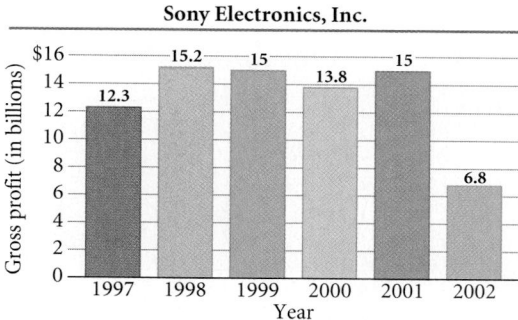

Source: The New York Stock Exchange

31.

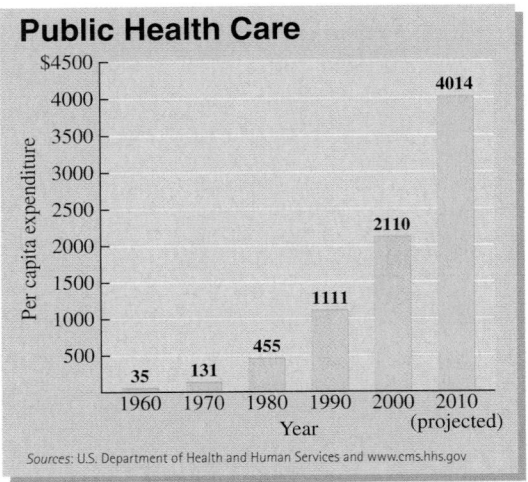

Public Health Care

Sources: U.S. Department of Health and Human Services and www.cms.hhs.gov

32.

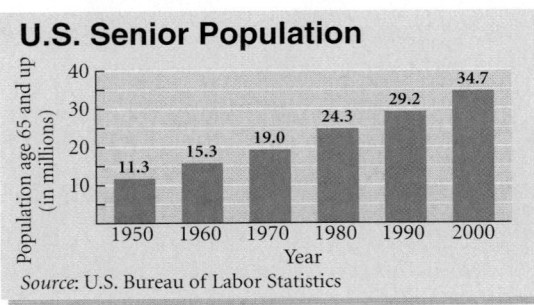

U.S. Senior Population

Source: U.S. Bureau of Labor Statistics

33.

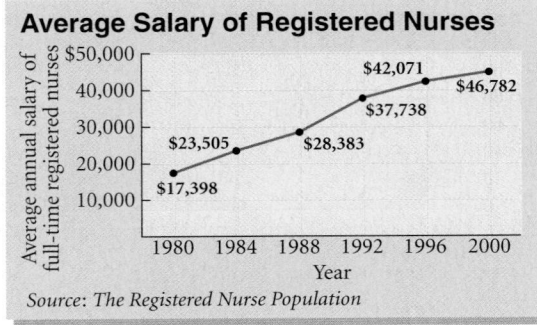

Average Salary of Registered Nurses

Source: *The Registered Nurse Population*

34.

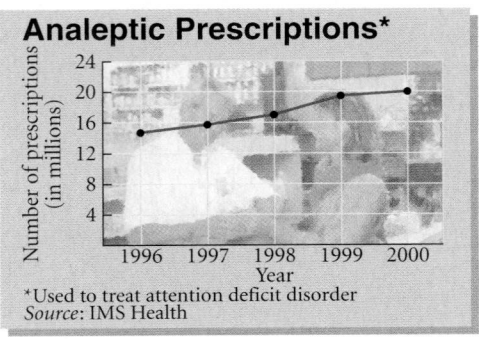

Analeptic Prescriptions*

*Used to treat attention deficit disorder
Source: IMS Health

Find a quadratic function that fits the set of data points.

35. $(1, 4), (-1, -2), (2, 13)$ **36.** $(1, 4), (-1, 6), (-2, 16)$

37. $(2, 0), (4, 3), (12, -5)$ **38.** $(-3, -30), (3, 0), (6, 6)$

39. a) Find a quadratic function that fits the following data.

Travel Speed (in kilometers per hour)	Number of Nighttime Accidents (for every 200 million kilometers driven)
60	400
80	250
100	250

b) Use the function to estimate the number of nighttime accidents that occur at 50 km/h.

40. a) Find a quadratic function that fits the following data.

Travel Speed (in kilometers per hour)	Number of Daytime Accidents (for every 200 million kilometers driven)
60	100
80	130
100	200

b) Use the function to estimate the number of daytime accidents that occur at 50 km/h.

41. *Archery.* The Olympic flame tower at the 1992 Summer Olympics was lit at a height of about 27 m by a flaming arrow that was launched about 63 m from the base of the tower. If the arrow landed about 63 m beyond the tower, find a quadratic function that expresses the height h of the arrow as a function of the distance d that it traveled horizontally.

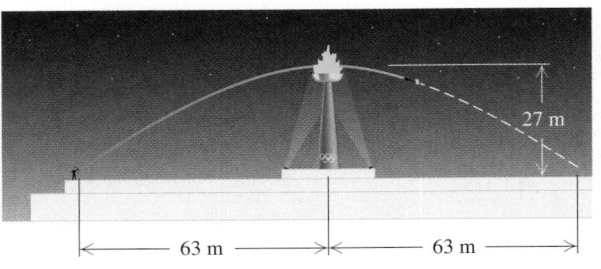

42. *Pizza prices.*   Pizza Unlimited has the following prices for pizzas.

Diameter	Price
8 in.	$ 6.00
12 in.	$ 8.50
16 in.	$11.50

Is price a quadratic function of diameter? It probably should be, because the price should be proportional to the area, and the area is a quadratic function of the diameter. (The area of a circular region is given by $A = \pi r^2$ or $(\pi/4) \cdot d^2$.)

a) Express price as a quadratic function of diameter using the data points $(8, 6)$, $(12, 8.50)$, and $(16, 11.50)$.

b) Use the function to find the price of a 14-in. pizza.

43. Does every nonlinear function have a minimum or maximum value? Why or why not?

44. Explain how the leading coefficient of a quadratic function can be used to determine if a maximum or a minimum function value exists.

SKILL MAINTENANCE

Simplify.

45. $\dfrac{x}{x^2 + 17x + 72} - \dfrac{8}{x^2 + 15x + 56}$ [6.4]

46. $\dfrac{x^2 - 9}{x^2 - 8x + 7} \div \dfrac{x^2 + 6x + 9}{x^2 - 1}$ [6.2]

47. $\dfrac{t^2 - 4}{t^2 - 7t - 8} \cdot \dfrac{t^2 - 64}{t^2 - 5t + 6}$ [6.2]

48. $\dfrac{t}{t^2 - 10t + 21} + \dfrac{t}{t^2 - 49}$ [6.4]

Solve. [9.1]

49. $5x - 9 < 31$

50. $3x - 8 \geq 22$

SYNTHESIS

51. Write a problem for a classmate to solve. Design the problem so that its solution requires finding the minimum or maximum value.

52. Explain what restrictions should be placed on the quadratic functions developed in Exercises 39 and 42 and why such restrictions are needed.

53. *Bridge design.*   The cables supporting a straight-line suspension bridge are nearly parabolic in shape. Suppose that a suspension bridge is being designed with concrete supports 160 ft apart and with vertical cables 30 ft above road level at the midpoint of the bridge and 80 ft above road level at a point 50 ft from the midpoint of the bridge. How long are the longest vertical cables?

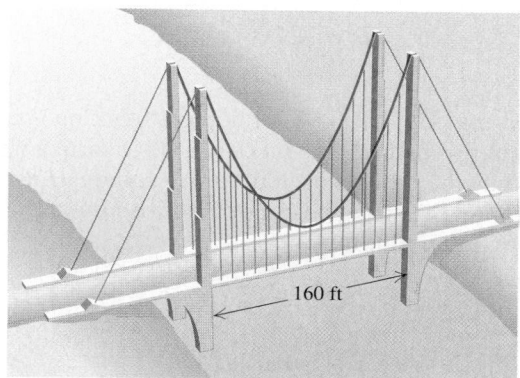

160 ft

54. *Trajectory of a launched object.*   The height above the ground of a launched object is a quadratic function of the time that it is in the air. Suppose that a flare is launched from a cliff 64 ft above sea level. If 3 sec after being launched the flare is again level with the cliff, and if 2 sec after that it lands in the sea, what is the maximum height that the flare will reach?

55. *Cover charges.*   When the owner of Sweet Sounds charges a $10 cover charge, an average of 80 people will attend a show. For each 25¢ increase in admission price, the average number attending decreases by 1. What should the owner charge in order to make the most money?

56. *Crop yield.*   An orange grower finds that she gets an average yield of 40 bushels (bu) per tree when she plants 20 trees on an acre of ground. Each time she adds a tree to an acre, the yield per tree decreases by 1 bu, due to congestion. How many trees per acre should she plant for maximum yield?

57. *Norman window.*   A *Norman window* is a rectangle with a semicircle on top. Big Sky Windows is designing a Norman window that will require 24 ft of trim. What dimensions will allow the maximum amount of light to enter a house?

58. *Minimizing area.*   A 36-in. piece of string is cut into two pieces. One piece is used to form a circle while the other is used to form a square. How should the string be cut so that the sum of the areas is a minimum?

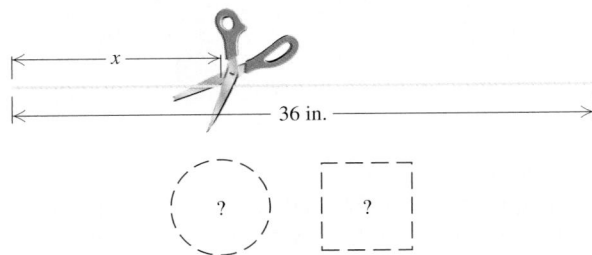

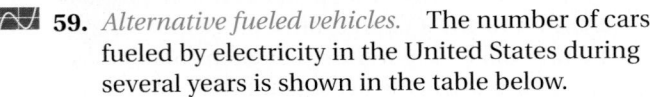

Regression can be used to find the "best"-fitting quadratic function when more than three data points are provided. In Exercises 59 and 60, six data points are given, but the approach used in the Technology Connection on p. 760 still applies.

59. *Alternative fueled vehicles.*   The number of cars fueled by electricity in the United States during several years is shown in the table below.

Year	Number of Cars
1992	1,607
1994	2,224
1996	3,280
1998	5,243
2000	11,834
2002	19,755

Source: www.eia.doe.gov

a) Use regression to find a quadratic function that can be used to estimate the number of cars c that are fueled by electricity x years after 1992.
b) Use the function found in part (a) to predict the number of cars fueled by electricity in 2008.

60. *Hydrology.*   The drawing below shows the cross section of a river. Typically rivers are deepest in the middle, with the depth decreasing to 0 at the edges. A hydrologist measures the depths D, in feet, of a river at distances x, in feet, from one bank. The results are listed in the table below.

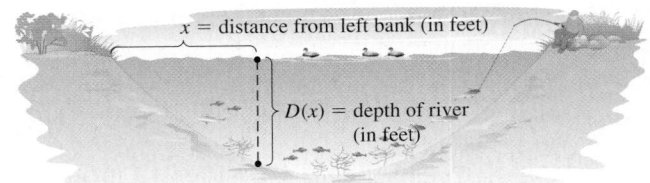

Distance x, from the Left Bank (in feet)	Depth, D, of the River (in feet)
0	0
15	10.2
25	17
50	20
90	7.2
100	0

a) Use regression to find a quadratic function that fits the data.
b) Use the function to estimate the depth of the river 70 ft from the left bank.

CORNER

Quadratic Counter Settings

Focus: Modeling quadratic functions

Time: 20–30 minutes

Group size: 3 or 4

Materials: Graphing calculators are optional.

The Panasonic Portable Stereo System RX-DT680® has a counter for finding locations on an audio cassette. When a fully wound cassette with 45 min of music on a side begins to play, the counter is at 0. After 15 min of music has played, the counter reads 250 and after 35 min, it reads 487. When the 45-min side is finished playing, the counter reads 590.

ACTIVITY

1. The paragraph above describes four ordered pairs of the form (counter number, minutes played). Three pairs are enough to find a function of the form

$$T(n) = an^2 + bn + c,$$

where $T(n)$ represents the time, in minutes, that the tape has run at counter reading n hundred. Each group member should select a different set of three points from the four given and then fit a quadratic function to the data.

2. Of the 3 or 4 functions found in part (1) above, which fits the data "best"? One way to answer this is to see how well each function predicts other pairs. The same counter used above reads 432 after a 45-min tape has played for 30 min. Which function comes closest to predicting this?

3. If a graphing calculator is available with a QUADREG option (see Exercise 59), what function does it fit to the four pairs originally listed?

4. If a class member has access to a Panasonic System RX-DT680, see how well the functions developed above predict the counter readings for a tape that has played for 5 or 10 min.

11.9   Polynomial and Rational Inequalities

Quadratic and Other Polynomial Inequalities   •
Rational Inequalities

Quadratic and Other Polynomial Inequalities

Inequalities like the following are called *polynomial inequalities*:

$$x^3 - 5x > x^2 + 7, \qquad 4x - 3 < 9, \qquad 5x^2 - 3x + 2 \geq 0.$$

Second-degree polynomial inequalities in one variable are called *quadratic inequalities*. To solve polynomial inequalities, we often focus attention on where the outputs of a polynomial function are positive and where they are negative.

EXAMPLE 1   Solve: $x^2 + 3x - 10 > 0$.

Solution   Consider the "related" function $f(x) = x^2 + 3x - 10$ and its graph. Its graph opens upward since the leading coefficient is positive. Thus y-values are positive outside the interval formed by the x-intercepts. To find the intercepts, we set the polynomial equal to 0 and solve:

$$x^2 + 3x - 10 = 0$$
$$(x + 5)(x - 2) = 0$$
$$x + 5 = 0 \quad or \quad x - 2 = 0$$
$$x = -5 \quad or \quad x = 2.$$

Test values can be used to confirm that $f(x)$ is positive outside the interval $[-5, 2]$:

$$f(3) = 3^2 + 3 \cdot 3 - 10 = 9 + 9 - 10 = 8; \quad \text{—Positive}$$
$$f(-6) = (-6)^2 + 3(-6) - 10 = 36 - 18 - 10 = 8.$$

Thus the solution set of the inequality is

$$\{x \mid x < -5 \; or \; x > 2\}, \quad or \quad (-\infty, -5) \cup (2, \infty).$$

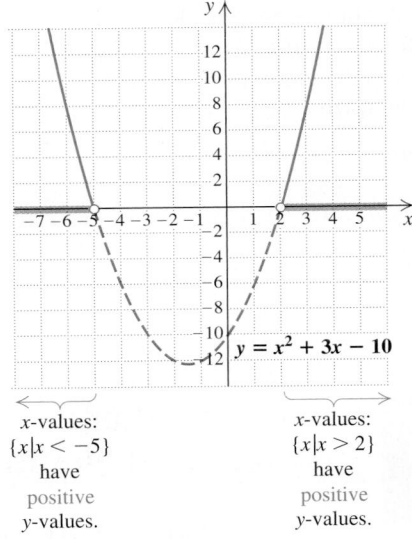

x-values:
$\{x \mid x < -5\}$
have
positive
y-values.

x-values:
$\{x \mid x > 2\}$
have
positive
y-values.

Any inequality with 0 on one side can be solved by considering a graph of the related function and finding intercepts as in Example 1. Sometimes the quadratic formula is needed to find the intercepts.

EXAMPLE 2   Solve: $x^2 - 2x \le 2$.

Solution   We first write the quadratic inequality in standard form, with 0 on one side:

$$x^2 - 2x - 2 \le 0. \qquad \text{This is equivalent to the original inequality.}$$

The graph of $f(x) = x^2 - 2x - 2$ is a parabola opening upward. Values of $f(x)$ are negative for x-values between the x-intercepts. We find the x-intercepts by solving $f(x) = 0$:

$$x = \frac{-b \pm \sqrt{b^2 - 4ac}}{2a}$$

$$= \frac{-(-2) \pm \sqrt{(-2)^2 - 4 \cdot 1(-2)}}{2 \cdot 1}$$

$$= \frac{2 \pm \sqrt{12}}{2} = \frac{2}{2} \pm \frac{2\sqrt{3}}{2} = 1 \pm \sqrt{3}.$$

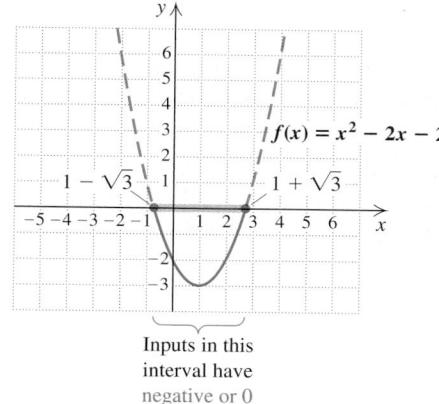

Inputs in this interval have negative or 0 outputs.

At the x-intercepts, $1 - \sqrt{3}$ and $1 + \sqrt{3}$, the value of $f(x)$ is 0. Thus the solution set of the inequality is

$$\left[1 - \sqrt{3}, 1 + \sqrt{3}\right], \quad \text{or} \quad \left\{x \mid 1 - \sqrt{3} \le x \le 1 + \sqrt{3}\right\}.$$

In Example 2, it was not essential to draw the graph. The important information came from finding the x-intercepts and the sign of $f(x)$ on each side of those intercepts. We now solve a third-degree polynomial inequality, without graphing, by locating the x-intercepts, or **zeros**, of f and then using *test points* to determine the sign of $f(x)$ over each interval of the x-axis.

EXAMPLE 3   For $f(x) = 5x^3 + 10x^2 - 15x$, find all x-values for which $f(x) > 0$.

Solution   We first solve the related equation:

$$f(x) = 0$$
$$5x^3 + 10x^2 - 15x = 0 \qquad \text{Substituting}$$
$$5x(x^2 + 2x - 3) = 0$$
$$5x(x + 3)(x - 1) = 0$$
$$5x = 0 \quad \textit{or} \quad x + 3 = 0 \quad \textit{or} \quad x - 1 = 0$$
$$x = 0 \quad \textit{or} \qquad x = -3 \quad \textit{or} \qquad x = 1.$$

The zeros of f are $-3, 0$, and 1. These zeros divide the number line, or x-axis, into four intervals: A, B, C, and D.

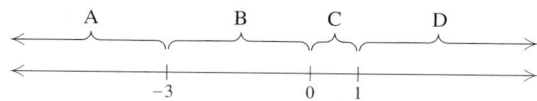

Next, selecting one convenient test value from each interval, we determine the sign of $f(x)$ for that interval. We know that, within each interval, the sign of $f(x)$ cannot change. If it did, there would need to be another zero in that interval.

Student Notes

When we are evaluating test values, there is often no need to do lengthy computations since all we need to determine is the sign of the result.

Using the factored form of $f(x)$ eases the computations:

$$f(x) = 5x(x + 3)(x - 1).$$

For interval A,

$$f(-4) = 5(-4)((-4) + 3)((-4) - 1) \qquad \text{−4 is a convenient value in interval A.}$$

$$= -20(-1)(-5)$$

$$= -100. \qquad f(-4) \text{ is negative.}$$

For interval B,

$$f(-1) = 5(-1)((-1) + 3)((-1) - 1) \qquad \text{−1 is a convenient value in interval B.}$$

$$= -5(2)(-2)$$

$$= 20. \qquad f(-1) \text{ is positive.}$$

For interval C,

$$f(\tfrac{1}{2}) = \underbrace{5 \cdot \tfrac{1}{2}}_{\text{Positive}} \cdot \underbrace{(\tfrac{1}{2} + 3)}_{\text{Positive}} \cdot \underbrace{(\tfrac{1}{2} - 1)}_{\text{Negative}}. \qquad \tfrac{1}{2} \text{ is a convenient value in interval C.}$$

$$\underbrace{}_{\text{Negative}}$$

Only the sign is important. The product is negative, so $f(\tfrac{1}{2})$ is negative.

For interval D,

$$f(2) = \underbrace{5 \cdot 2}_{\text{Positive}} \cdot \underbrace{(2 + 3)}_{\text{Positive}} \cdot \underbrace{(2 - 1)}_{\text{Positive}}.$$

2 is a convenient value in interval D.

$f(2)$ is positive.

Recall that we are looking for all x for which $5x^3 + 10x^2 - 15x > 0$. The calculations above indicate that $f(x)$ is positive for any number in intervals B and D. The solution set of the original inequality is

$$(-3, 0) \cup (1, \infty), \quad \text{or} \quad \{x \mid -3 < x < 0 \text{ or } x > 1\}.$$

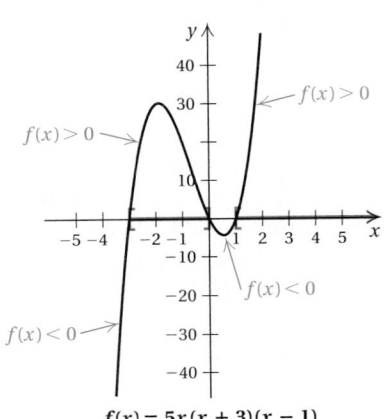

$$f(x) = 5x(x + 3)(x - 1)$$

A visualization of Example 3

The calculations in Example 3 were made simpler by using a factored form of the polynomial. This process was simplified further when, for intervals C and D, we concentrated on only the *sign* of $f(x)$. In the next example, we determine the sign of a polynomial function over each interval by tracking the sign of each factor. By looking at how many positive or negative factors are being multiplied, we will be able to determine the sign of the polynomial function.

EXAMPLE 4   For $f(x) = 4x^3 - 4x$, find all x-values for which $f(x) \leq 0$.

Solution   We first solve the related equation:

$$f(x) = 0$$

$$4x^3 - 4x = 0$$

$$4x(x^2 - 1) = 0$$

$$4x(x + 1)(x - 1) = 0$$

$$4x = 0 \quad \text{or} \quad x + 1 = 0 \quad \text{or} \quad x - 1 = 0$$

$$x = 0 \quad \text{or} \qquad x = -1 \quad \text{or} \qquad x = 1.$$

To solve $2.3x^2 \le 9.11 - 2.94x$, we first rewrite the inequality in the form $2.3x^2 + 2.94x - 9.11 \le 0$ and graph the function $f(x) = 2.3x^2 + 2.94x - 9.11$.

$$y = 2.3x^2 + 2.94x - 9.11$$

To find the values of x for which $f(x) \le 0$, we focus on the region in which the graph lies *on or below* the x-axis. It appears that this region begins somewhere between -3 and -2, and continues to somewhere between 1 and 2. Using the ZERO or ROOT option of CALC, we can find the endpoints of this region. To two decimal places, the endpoints are -2.73 and 1.45. The solution set is approximately $\{x \mid -2.73 \le x \le 1.45\}$.

Had the inequality been $2.3x^2 > 9.11 - 2.94x$, we would look for portions of the graph that lie *above* the x-axis. An approximate solution set of such an inequality would be $\{x \mid x < -2.73 \text{ or } x > 1.45\}$.

Use a graphing calculator to solve each inequality. Round the values of the endpoints to the nearest hundredth.

1. $4.32x^2 - 3.54x - 5.34 \le 0$
2. $7.34x^2 - 16.55x - 3.89 \ge 0$
3. $10.85x^2 + 4.28x + 4.44 > 7.91x^2 + 7.43x + 13.03$
4. $5.79x^3 - 5.68x^2 + 10.68x > 2.11x^3 + 16.90x - 11.69$

The function f has zeros at $-1, 0,$ and 1. Rather than use test values, as in Example 3, let's use the factorization $f(x) = 4x(x + 1)(x - 1)$. The product $4x(x + 1)(x - 1)$ is positive or negative, depending on the signs of $4x$, $x + 1$, and $x - 1$. This is easily determined using a chart.

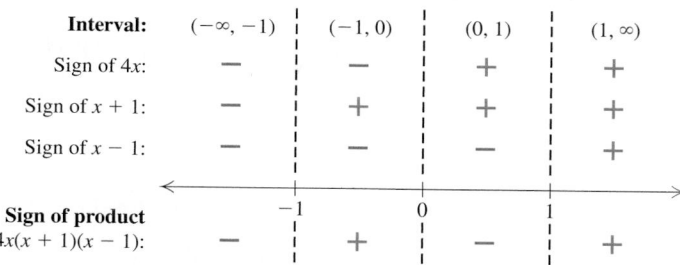

A product is negative when it has an odd number of negative factors. Since the $\le$ sign allows for equality, the endpoints $-1, 0,$ and 1 are solutions. From the chart, we see that the solution set is

$$(-\infty, -1] \cup [0, 1], \text{ or } \{x \mid x \le -1 \text{ or } 0 \le x \le 1\}.$$

To Solve a Polynomial Inequality Using Factors

1. Add or subtract to get 0 on one side and solve the related polynomial equation by factoring.
2. Use the numbers found in step (1) to divide the number line into intervals.
3. Using a test value from each interval, determine the sign of each factor over that interval.
4. Determine the sign of the product of the factors over each interval. Remember that the product of an odd number of negative numbers is negative.
5. Select the interval(s) for which the inequality is satisfied and write set-builder notation or interval notation for the solution set. Include the endpoints of the intervals when $\le$ or $\ge$ is used.

Rational Inequalities

Inequalities involving rational expressions are called **rational inequalities**. Like polynomial inequalities, rational inequalities can be solved using test values. Unlike polynomials, however, rational expressions often have values for which the expression is undefined.

EXAMPLE 5 Solve: $\dfrac{x-3}{x+4} \geq 2$.

Solution We write the related equation by changing the $\geq$ symbol to $=$:

$$\frac{x-3}{x+4} = 2. \qquad \text{Note that } x \neq -4.$$

Next, we solve this related equation:

$$(x+4) \cdot \frac{x-3}{x+4} = (x+4) \cdot 2 \qquad \begin{array}{l}\text{Multiplying both sides} \\ \text{by the LCD, } x+4\end{array}$$

$$x - 3 = 2x + 8$$

$$-11 = x. \qquad \text{Solving for } x$$

In the case of rational inequalities, we must always find any values that make the denominator 0. As noted at the beginning of this example, $x \neq -4$.

Now we use -11 and -4 to divide the number line into intervals:

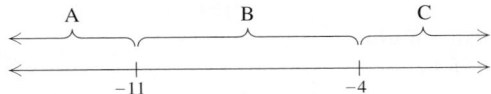

We test a number in each interval to see where the original inequality is satisfied:

$$\frac{x-3}{x+4} \geq 2.$$

A: Test -15, $\quad \dfrac{-15-3}{-15+4} = \dfrac{-18}{-11}$

$$= \frac{18}{11} \not\geq 2 \qquad \begin{array}{l}-15 \text{ } is \text{ } not \text{ a solution, so interval A is} \\ \text{not part of the solution set.}\end{array}$$

B: Test -8, $\quad \dfrac{-8-3}{-8+4} = \dfrac{-11}{-4}$

$$= \frac{11}{4} \geq 2 \qquad \begin{array}{l}-8 \text{ } is \text{ a solution, so interval B is part of} \\ \text{the solution set.}\end{array}$$

C: Test 1, $\quad \dfrac{1-3}{1+4} = \dfrac{-2}{5}$

$$= -\frac{2}{5} \not\geq 2 \qquad \begin{array}{l}1 \text{ } is \text{ } not \text{ a solution, so interval C is not} \\ \text{part of the solution set.}\end{array}$$

The solution set includes interval B. The endpoint -11 is included because the inequality symbol is $\geq$ and -11 is a solution of the related equation. The number -4 is *not* included because $(x-3)/(x+4)$ is undefined for $x = -4$. Thus the solution set of the original inequality is

$$[-11, -4), \quad \text{or} \quad \{x \mid -11 \leq x < -4\}.$$

ALGEBRAIC–GRAPHICAL CONNECTION

Let's compare the algebraic solution of Example 5 to a graphical solution. By graphing $f(x) = (x - 3)/(x + 4)$, we can find the solutions of $(x - 3)/(x + 4) \geq 2$ by sketching the line $y = 2$ and locating all x-values for which $f(x) \geq 2$.

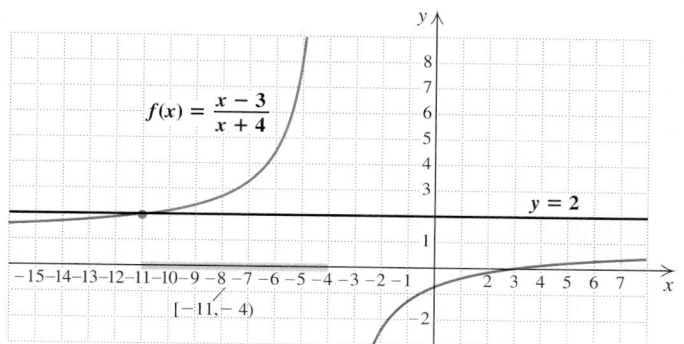

Because graphing rational functions can be very time-consuming, we generally just use test values.

To Solve a Rational Inequality

1. Change the inequality symbol to an equals sign and solve the related equation.
2. Find any replacements for which the rational expression is undefined.
3. Use the numbers found in steps (1) and (2) to divide the number line into intervals.
4. Substitute a test value from each interval into the inequality. If the number is a solution, then the interval to which it belongs is part of the solution set.
5. Select the interval(s) and any endpoints for which the inequality is satisfied and write set-builder or interval notation for the solution set. If the inequality symbol is $\leq$ or $\geq$, then the solutions from step (1) are also included in the solution set. Those numbers found in step (2) should be excluded from the solution set, even if they are solutions from step (1).

Exercise Set
11.9

FOR EXTRA HELP

Student's
Solutions
Manual

Digital Video Tutor
CD 6
Videotape 11

AW Math
Tutor Center

MathXL Tutorials
on CD

MathXL
MathXL

MyMathLab
MyMathLab

Concept Reinforcement   *Classify each of the following as either true or false.*

1. The solution of $(x - 3)(x + 2) \le 0$ is $[-2, 3]$.

2. The solution of $(x + 5)(x - 4) \ge 0$ is $[-5, 4]$.

3. The solution of $(x - 1)(x - 6) > 0$ is $\{x \mid x < 1 \text{ or } x > 6\}$.

4. The solution of $(x + 4)(x + 2) < 0$ is $(-4, -2)$.

5. To solve $\dfrac{x - 5}{x + 4} \ge 0$ using intervals, we divide the number line into the intervals $(-\infty, -4)$, $(-4, 5)$, and $(5, \infty)$.

6. To solve $\dfrac{x + 2}{x - 3} < 0$ using intervals, we divide the number line into the intervals $(-\infty, -2)$, $(-2, 3)$, and $(3, \infty)$.

7. The solution of $\dfrac{3}{x - 5} \le 0$ is $[5, \infty)$.

8. The solution of $\dfrac{-2}{x + 3} \ge 0$ is $(-\infty, -3]$.

Solve.

9. $(x + 4)(x - 3) < 0$

10. $(x - 5)(x + 2) > 0$

11. $(x + 7)(x - 2) \ge 0$

12. $(x - 1)(x + 4) \le 0$

13. $x^2 - x - 2 > 0$

14. $x^2 + x - 2 < 0$

Aha! **15.** $x^2 + 4x + 4 < 0$

16. $x^2 + 6x + 9 < 0$

17. $x^2 - 4x < 12$

18. $x^2 + 6x > -8$

19. $3x(x + 2)(x - 2) < 0$

20. $5x(x + 1)(x - 1) > 0$

21. $(x - 1)(x + 2)(x - 4) \ge 0$

22. $(x + 3)(x + 2)(x - 1) < 0$

23. For $f(x) = x^2 - 1$, find all x-values for which $f(x) \le 3$.

24. For $f(x) = x^2 - 20$, find all x-values for which $f(x) > 5$.

25. For $g(x) = (x - 2)(x - 3)(x + 1)$, find all x-values for which $g(x) > 0$.

26. For $g(x) = (x + 3)(x - 2)(x + 1)$, find all x-values for which $g(x) < 0$.

27. For $F(x) = x^3 - 7x^2 + 10x$, find all x-values for which $F(x) \le 0$.

28. For $G(x) = x^3 - 8x^2 + 12x$, find all x-values for which $G(x) \ge 0$.

Solve.

29. $\dfrac{1}{x + 5} < 0$

30. $\dfrac{1}{x + 4} > 0$

31. $\dfrac{x + 1}{x - 3} \ge 0$

32. $\dfrac{x - 2}{x + 4} \le 0$

33. $\dfrac{x + 1}{x + 6} \ge 1$

34. $\dfrac{x - 1}{x - 2} \le 1$

35. $\dfrac{(x - 2)(x + 1)}{x - 5} \le 0$

36. $\dfrac{(x + 4)(x - 1)}{x + 3} \ge 0$

37. $\dfrac{x}{x + 3} \ge 0$

38. $\dfrac{x - 2}{x} \le 0$

39. $\dfrac{x - 5}{x} < 1$

40. $\dfrac{x}{x - 1} > 2$

41. $\dfrac{x - 1}{(x - 3)(x + 4)} \le 0$

42. $\dfrac{x + 2}{(x - 2)(x + 7)} \ge 0$

43. For $f(x) = \dfrac{5 - 2x}{4x + 3}$, find all x-values for which $f(x) \ge 0$.

44. For $g(x) = \dfrac{2 + 3x}{2x - 4}$, find all x-values for which $g(x) \geq 0$.

45. For $G(x) = \dfrac{1}{x - 2}$, find all x-values for which $G(x) \leq 1$.

46. For $F(x) = \dfrac{1}{x - 3}$, find all x-values for which $F(x) \leq 2$.

47. Explain how any quadratic inequality can be solved by examining a parabola.

48. Describe a method for creating a quadratic inequality for which there is no solution.

SKILL MAINTENANCE

Simplify.

49. $(2a^3b^2c^4)^3$ [4.1]

50. $(5a^4b^7)^2$ [4.1]

51. 2^{-5} [4.8]

52. 3^{-4} [4.8]

53. If $f(x) = 3x^2$, find $f(a + 1)$. [7.1]

54. If $g(x) = 5x - 3$, find $g(a + 2)$. [7.1]

SYNTHESIS

55. Step (5) on p. 773 states that even when the inequality symbol is $\leq$ or $\geq$, the solutions from step (1) are not always part of the solution set. Why?

56. Describe a method that could be used to create quadratic inequalities that have $(-\infty, a] \cup [b, \infty)$ as the solution set. Assume $a < b$.

Find each solution set.

57. $x^2 + 2x < 5$

58. $x^4 + 2x^2 \geq 0$

59. $x^4 + 3x^2 \leq 0$

60. $\left| \dfrac{x + 2}{x - 1} \right| \leq 3$

61. *Total profit.* Derex, Inc., determines that its total-profit function is given by
$$P(x) = -3x^2 + 630x - 6000.$$
a) Find all values of x for which Derex makes a profit.
b) Find all values of x for which Derex loses money.

62. *Height of a thrown object.* The function
$$S(t) = -16t^2 + 32t + 1920$$
gives the height S, in feet, of an object thrown from a cliff that is 1920 ft high. Here t is the time, in seconds, that the object is in the air.
a) For what times does the height exceed 1920 ft?
b) For what times is the height less than 640 ft?

63. *Number of handshakes.* There are n people in a room. The number N of possible handshakes by the people is given by the function
$$N(n) = \dfrac{n(n - 1)}{2}.$$
For what number of people n is $66 \leq N \leq 300$?

64. *Number of diagonals.* A polygon with n sides has D diagonals, where D is given by the function
$$D(n) = \dfrac{n(n - 3)}{2}.$$
Find the number of sides n if
$$27 \leq D \leq 230.$$

Use a graphing calculator to graph each function and find solutions of $f(x) = 0$. Then solve the inequalities $f(x) < 0$ and $f(x) > 0$.

65. $f(x) = x^3 - 2x^2 - 5x + 6$

66. $f(x) = \dfrac{1}{3}x^3 - x + \dfrac{2}{3}$

67. $f(x) = x + \dfrac{1}{x}$

68. $f(x) = x - \sqrt{x}, x \geq 0$

69. $f(x) = \dfrac{x^3 - x^2 - 2x}{x^2 + x - 6}$

70. $f(x) = x^4 - 4x^3 - x^2 + 16x - 12$

71. Use a graphing calculator to solve Exercises 39 and 45 by drawing two curves, one for each side of the inequality.

11 Study Summary

Any equation that can be written in the form $ax^2 + bx + c = 0$, with a, b, and c constant, is said to be **quadratic** (p. 704). There are several methods for solving quadratic equations:

Factoring (p. 704)

Easiest method to use *if* you can factor the polynomial.

$$x^2 - 3x - 10 = 0$$
$$(x + 2)(x - 5) = 0$$
$$x + 2 = 0 \quad or \quad x - 5 = 0$$
$$x = -2 \quad or \quad x = 5$$

Principle of Square Roots (p. 705)

Works only if a perfect-square trinomial is on one side and a constant is on the other side.

$$x^2 - 8x + 16 = 25$$
$$(x - 4)^2 = 25$$
$$x - 4 = -5 \quad or \quad x - 4 = 5$$
$$x = -1 \quad or \quad x = 9$$

Completing the Square (p. 707)

Can be used to solve *any* quadratic equation, but calculations can be lengthy.

$$x^2 + 6x = 1$$
$$x^2 + 6x + \left(\tfrac{6}{2}\right)^2 = 1 + \left(\tfrac{6}{2}\right)^2$$
$$x^2 + 6x + 9 = 1 + 9$$
$$(x + 3)^2 = 10$$
$$x + 3 = \pm\sqrt{10}$$
$$x = -3 \pm \sqrt{10}$$

Quadratic Formula (p. 716)

$$\boxed{\begin{array}{l} \text{If } ax^2 + bx + c = 0, \\ \text{then } x = \dfrac{-b \pm \sqrt{b^2 - 4ac}}{2a}. \end{array}}$$

Is based on completing the square but is quicker to use. Can be used to solve *any* quadratic equation.

$$3x^2 - 2x - 5 = 0$$
$$x = \frac{-(-2) \pm \sqrt{(-2)^2 - 4 \cdot 3(-5)}}{2 \cdot 3}$$
$$x = \frac{2 \pm \sqrt{4 + 60}}{6}$$
$$x = \frac{2 \pm \sqrt{64}}{6}$$
$$x = \frac{2 \pm 8}{6}$$
$$x = \frac{10}{6} = \frac{5}{3} \quad or \quad x = \frac{-6}{6} = -1$$

The **discriminant** of the quadratic formula can be used to find the nature of the solution(s) of a quadratic equation (p. 730):

$$b^2 - 4ac = 0 \rightarrow \text{One solution; a rational number.}$$
$$b^2 - 4ac > 0 \rightarrow \text{Two real solutions, both rational if } b^2 - 4ac \text{ is a perfect square.}$$
$$b^2 - 4ac < 0 \rightarrow \text{Two imaginary-number solutions.}$$

The methods for solving quadratic equations can be applied to equations that are not quadratic but are in **quadratic form** (p. 735):

$$x^4 - 10x^2 + 9 = 0 \qquad \text{Think of } x^2 \text{ as } u \text{ and } x^4 \text{ as } u^2.$$
$$u^2 - 10u + 9 = 0$$
$$(u - 9)(u - 1) = 0$$
$$u - 9 = 0 \quad or \quad u - 1 = 0$$
$$u = 9 \quad or \quad u = 1$$
$$x^2 = 9 \quad or \quad x^2 = 1$$
$$x = \pm 3 \quad or \quad x = \pm 1.$$

The graph of a quadratic function is a **parabola** (p. 742). The graph of $f(x) = ax^2 + bx + c$ opens upward for $a > 0$ and downward for $a < 0$. The graph of an equation of the form $f(x) = a(x - h)^2 + k$ looks like the graph of $y = ax^2$ **translated** $|h|$ units to the right (for $h > 0$) or left (for $h < 0$) and $|k|$ units up (for $k > 0$) or down (for $k < 0$).

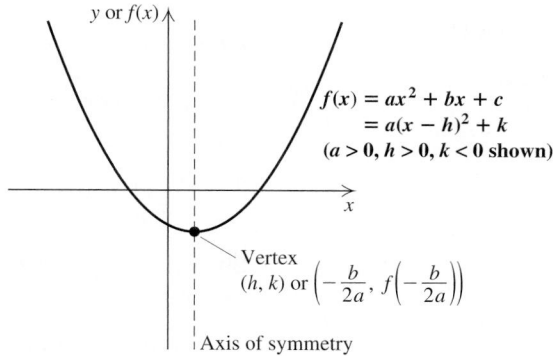

Every parabola has an **axis of symmetry** and **vertex** as shown above (p. 742). For a quadratic function, the second coordinate of the vertex is a **minimum** (for $a > 0$) or **maximum** (for $a < 0$) function value (p. 744).

The x-intercepts, or **zeros**, of a function are used to divide the x-axis into intervals when solving **polynomial** or **rational inequalities** (pp. 767, 771). Rational inequalities also require forming intervals on either side of any x-value(s) for which the denominator is 0. We then use test values from each interval to identify the solution set.

11   Review Exercises

Classify each statement as either true or false.

1. Every quadratic equation has two different solutions. [11.4]

2. Every quadratic equation has at least one solution. [11.4]

3. If an equation cannot be solved by completing the square, it cannot be solved by the quadratic formula. [11.2]

4. A negative discriminant indicates two imaginary-number solutions of a quadratic equation. [11.4]

5. Certain radical or rational equations can be written in quadratic form. [11.5]

6. The graph of $f(x) = 2(x + 3)^2 - 4$ has its vertex at $(3, -4)$. [11.6]

7. The graph of $g(x) = 5x^2$ has $x = 0$ as its axis of symmetry. [11.6]

8. The graph of $f(x) = -2x^2 + 1$ has no minimum value. [11.6]

9. The zeros of $g(x) = x^2 - 9$ are -3 and 3. [11.6]

10. To solve a polynomial inequality, we often must solve a polynomial equation. [11.9]

Solve.

11. $4x^2 - 9 = 0$ [11.1]   12. $8x^2 + 6x = 0$ [11.1]

13. $x^2 - 12x + 36 = 9$ [11.1]   14. $x^2 - 4x + 8 = 0$ [11.2]

15. $x(3x + 4) = 4x(x - 1) + 15$ [11.2]

16. $x^2 + 9x = 1$ [11.2]

17. $x^2 - 5x - 2 = 0$. Use a calculator to approximate the solutions with rational numbers. [11.2]

18. Let $f(x) = 4x^2 - 3x - 1$. Find x such that $f(x) = 0$. [11.2]

Replace the blanks with constants to form a true equation. [11.1]

19. $x^2 - 12x + \underline{\quad} = (x - \underline{\quad})^2$

20. $x^2 + \frac{3}{5}x + \underline{\quad} = (x + \underline{\quad})^2$

21. Solve by completing the square. Show your work.
$x^2 - 6x + 1 = 0$ [11.1]

22. \$2500 grows to \$3025 in 2 yr. Use the formula $A = P(1 + r)^t$ to find the interest rate. [11.1]

23. The Peachtree Center Plaza in Atlanta, Georgia, is 723 ft tall. Use $s = 16t^2$ to approximate how long it would take an object to fall from the top. [11.1]

Solve. [11.3]

24. A corporate pilot must fly from company headquarters to a manufacturing plant and back in 4 hr. The distance between headquarters and the plant is 300 mi. If there is a 20-mph headwind going and a 20-mph tailwind returning, how fast must the plane be able to travel in still air?

25. Working together, Erica and Shawna can answer a day's worth of customer-service e-mails in 4 hr. Working alone, Erica takes 6 hr longer than Shawna. How long would it take Shawna to answer the e-mails alone?

For each equation, determine whether the solutions are real or imaginary. If they are real, specify whether they are rational or irrational. [11.4]

26. $x^2 + 3x - 6 = 0$ 27. $x^2 + 2x + 5 = 0$

28. Write a quadratic equation having the solutions $\sqrt{5}$ and $-\sqrt{5}$. [11.4]

29. Write a quadratic equation having -4 as its only solution. [11.4]

30. Find all x-intercepts of the graph of $f(x) = x^4 - 13x^2 + 36$. [11.5]

Solve. [11.5]

31. $15x^{-2} - 2x^{-1} - 1 = 0$

32. $(x^2 - 4)^2 - (x^2 - 4) - 6 = 0$

33. **a)** Graph: $f(x) = -3(x + 2)^2 + 4$. [11.6]
    b) Label the vertex.
    c) Draw the axis of symmetry.
    d) Find the maximum or the minimum value.

34. For the function given by $f(x) = 2x^2 - 12x + 23$: [11.7]

    a) find the vertex and the axis of symmetry;

    b) graph the function.

35. Find the x- and y-intercepts of

$$f(x) = x^2 - 9x + 14. \text{ [11.7]}$$

36. Solve $N = 3\pi\sqrt{1/p}$ for p. [11.3]

37. Solve $2A + T = 3T^2$ for T. [11.3]

State whether each graph appears to represent a quadratic or linear function. [11.8]

38.

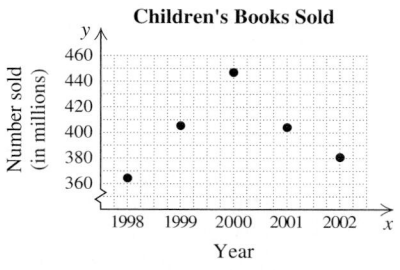

Children's Books Sold

Source: Statistical Abstract of the United States, 2003

39.

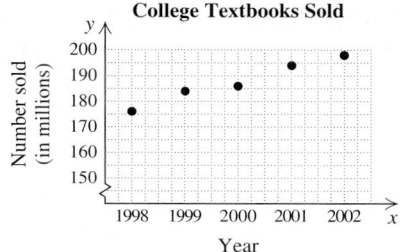

College Textbooks Sold

Source: Statistical Abstract of the United States, 2003

40. Eastgate Consignments wants to build a rectangular area in a corner for children to play in while their parents shop. They have 30 ft of low fencing. What is the maximum area they can enclose? What dimensions will yield this area? [11.8]

41. The following table lists the number of children's books sold (in millions) x years after 1998. (See Exercise 38.) [11.8]

Years Since 1998	Number of Children's Books Sold (in millions)
1	406
2	447
3	404

    a) Find the quadratic function that fits the data.

    b) Use the function to estimate the number of children's books sold in 2002.

Solve. [11.9]

42. $x^3 - 3x > 2x^2$

43. $\dfrac{x - 5}{x + 3} \leq 0$

SYNTHESIS

44. Explain how the x-intercepts of a quadratic function can be used to help find the maximum or minimum value of the function. [11.7], [11.8]

45. Suppose that the quadratic formula is used to solve a quadratic equation. If the discriminant is a perfect square, could factoring have been used to solve the equation? Why or why not? [11.2], [11.4]

46. What is the greatest number of solutions that an equation of the form $ax^4 + bx^2 + c = 0$ can have? Why? [11.5]

47. Discuss two ways in which completing the square was used in this chapter. [11.1], [11.2], [11.7]

48. A quadratic function has x-intercepts at -3 and 5. If the y-intercept is at -7, find an equation for the function. [11.7]

49. Find h and k if, for $3x^2 - hx + 4k = 0$, the sum of the solutions is 20 and the product of the solutions is 80. [11.4]

50. The average of two positive integers is 171. One of the numbers is the square root of the other. Find the integers. [11.5]

11 Chapter Test

Solve.

1. $4x^2 - 11 = 0$

2. $4x(x - 2) - 3x(x + 1) = -18$

3. $x^2 + 2x + 3 = 0$

4. $2x + 5 = x^2$

5. $x^{-2} - x^{-1} = \frac{3}{4}$

6. $x^2 + 3x = 5$. Use a calculator to approximate the solutions with rational numbers.

7. Let $f(x) = 12x^2 - 19x - 21$. Find x such that $f(x) = 0$.

Replace the blanks with constants to form a true equation.

8. $x^2 - 16x + \underline{\quad} = (x - \underline{\quad})^2$

9. $x^2 + \frac{2}{7}x + \underline{\quad} = \left(x + \underline{\quad}\right)^2$

10. Solve by completing the square. Show your work.
$$x^2 + 10x + 15 = 0$$

Solve.

11. The Connecticut River flows at a rate of 4 km/h for the length of a popular scenic route. In order for a cruiser to travel 60 km upriver and then return in a total of 8 hr, how fast must the boat be able to travel in still water?

12. Brock and Ian can assemble a swing set in $1\frac{1}{2}$ hr. Working alone, it takes Ian 4 hr longer than Brock to assemble the swing set. How long would it take Brock, working alone, to assemble the swing set?

13. Determine the type of number that the solutions of $x^2 + 5x + 13 = 0$ will be.

14. Write a quadratic equation having solutions -2 and $\frac{1}{3}$.

15. Find all x-intercepts of the graph of
$$f(x) = x^4 - 15x^2 - 16.$$

16. a) Graph: $f(x) = 4(x - 3)^2 + 5$.
   b) Label the vertex.
   c) Draw the axis of symmetry.
   d) Find the maximum or the minimum function value.

17. For the function $f(x) = 2x^2 + 4x - 6$:
   a) find the vertex and the axis of symmetry;
   b) graph the function.

18. Find the x- and y-intercepts of
$$f(x) = x^2 - x - 6.$$

19. Solve $V = \frac{1}{3}\pi(R^2 + r^2)$ for r. Assume all variables are positive.

20. State whether the graph appears to represent a linear function, a quadratic function, or neither.

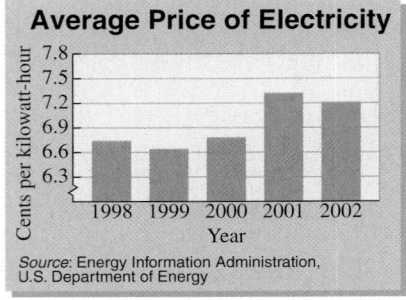

Average Price of Electricity

Source: Energy Information Administration, U.S. Department of Energy

21. Jay's Custom Pickups has determined that when x hundred truck caps are built, the average cost per cap is given by
$$C(x) = 0.2x^2 - 1.3x + 3.4025,$$
where $C(x)$ is in hundreds of dollars. What is the minimum cost per truck cap and how many caps should be built to achieve that minimum?

22. Find the quadratic function that fits the data points $(0, 0)$, $(3, 0)$, and $(5, 2)$.

Solve.

23. $x^2 + 5x < 6$

24. $x - \dfrac{1}{x} \geq 0$

SYNTHESIS

25. One solution of $kx^2 + 3x - k = 0$ is -2. Find the other solution.

26. Find a fourth-degree polynomial equation, with integer coefficients, for which $2 - \sqrt{3}$ and $5 - i$ are solutions.

27. Find a polynomial equation, with integer coefficients, for which 5 is a repeated root and $\sqrt{2}$ and $\sqrt{3}$ are solutions.

12

Exponential and Logarithmic Functions

AN APPLICATION

As more Americans make cell phones their *only* phones, the percentage of phone lines that are land lines is shrinking and can be estimated by

$$P(t) = 63.03(0.95)^t,$$

where t is the number of years since 2000. In what year will the percentage of phones that are land lines drop below 25%?

This problem appears as Exercise 7 in Section 12.7.

Teresa Matos
ENGINEER
Falls Church, Virginia

We use math with everything. Without math, we could not have technology. We use math to quantify cost as well. With knowledge of math, people can do anything.

*T*he functions that we consider in this chapter are interesting not only from a purely intellectual point of view, but also for their rich applications to many fields. We will look at applications such as spread of a disease and population growth, to name just two.

The theory centers on functions with variable exponents (exponential functions). Results follow from those functions, their properties, and the inverses *of those functions.*

12.1 Composite and Inverse Functions

Composite Functions • Inverses and One-to-One Functions •
Finding Formulas for Inverses • Graphing Functions and Their
Inverses • Inverse Functions and Composition

Composite Functions

In the real world, functions frequently occur in which some quantity depends on a variable that, in turn, depends on another variable. For instance, a firm's profits may depend on the number of items the firm produces, which may in turn depend on the number of employees hired. Functions like this are called **composite functions**.

For example, the function g that gives a correspondence between women's shoe sizes in the United States and those in Italy is given by $g(x) = 2x + 24$, where x is the U.S. size and $g(x)$ is the Italian size. Thus a U.S. size 4 corresponds to a shoe size of $g(4) = 2 \cdot 4 + 24$, or 32, in Italy.

A different function gives a correspondence between women's shoe sizes in Italy and those in Britain. This particular function is given by $f(x) = \frac{1}{2}x - 14$, where x is the Italian size and $f(x)$ is the corresponding British size. Thus an Italian size 32 corresponds to a British size $f(32) = \frac{1}{2} \cdot 32 - 14$, or 2.

It seems reasonable to conclude that a U.S. size 4 corresponds to a British size 2 and that some function h describes this correspondence. Can we find a formula for h? If we look at the following tables, we might guess that such a formula is $h(x) = x - 2$, and that is indeed correct. But, for more complicated formulas, we would need to use algebra.

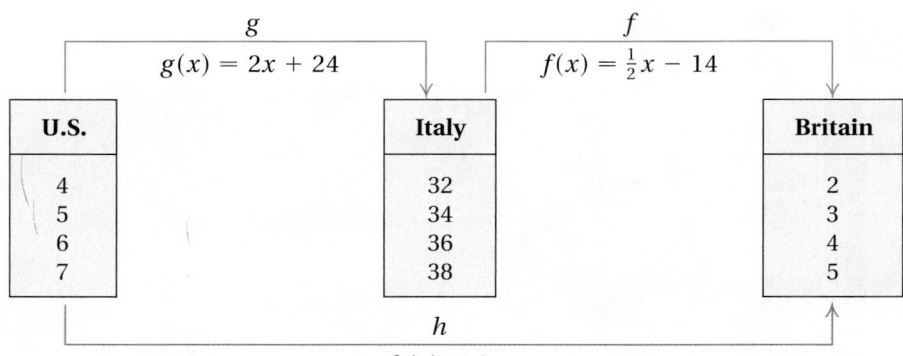

Size x shoes in the United States correspond to size $g(x)$ shoes in Italy, where

$$g(x) = 2x + 24.$$

Size n shoes in Italy correspond to size $f(n)$ shoes in Britain. Similarly, size $g(x)$ shoes in Italy correspond to size $f(g(x))$ shoes in Britain. Since the x in the expression $f(g(x))$ represents a U.S. shoe size, we can find the British shoe size that corresponds to a U.S. size x as follows:

$$f(g(x)) = f(2x + 24) = \tfrac{1}{2} \cdot (2x + 24) - 14 \qquad \text{Using } g(x) \text{ as an input}$$
$$= x + 12 - 14 = x - 2.$$

This gives a formula for h: $h(x) = x - 2$. Thus U.S. size 4 corresponds to British size $h(4) = 4 - 2$, or 2. The function h is the *composition* of f and g and is denoted $f \circ g$ (read "the composition of f and g," "f composed with g," or "f circle g").

Composition of Functions

The *composite function* $f \circ g$, the *composition* of f and g, is defined as

$$(f \circ g)(x) = f(g(x)).$$

We can visualize the composition of functions as follows.

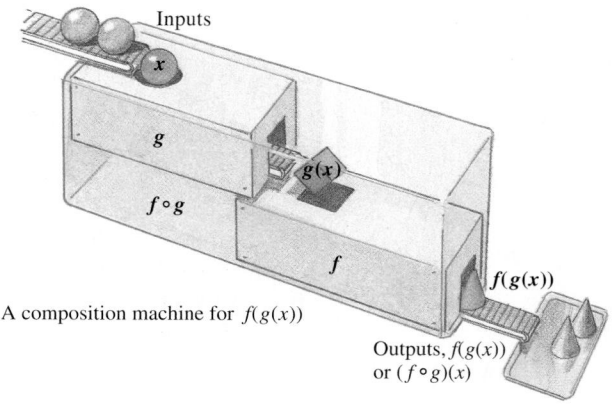

A composition machine for $f(g(x))$

EXAMPLE 1   Given $f(x) = 3x$ and $g(x) = 1 + x^2$:

a) Find $(f \circ g)(5)$ and $(g \circ f)(5)$.   **b)** Find $(f \circ g)(x)$ and $(g \circ f)(x)$.

Solution   Consider each function separately:

$$f(x) = 3x \qquad \qquad \text{This function multiplies each input by 3.}$$

and

$$g(x) = 1 + x^2. \qquad \text{This function adds 1 to the square of each input.}$$

a) To find $(f \circ g)(5)$, we find $g(5)$ and then use that as an input for f:

$$(f \circ g)(5) = f(g(5)) = f(1 + 5^2) \qquad \text{Using } g(x) = 1 + x^2$$
$$= f(26) = 3 \cdot 26 = 78. \qquad \text{Using } f(x) = 3x$$

To find $(g \circ f)(5)$, we find $f(5)$ and then use that as an input for g:

$$(g \circ f)(5) = g(f(5)) = g(3 \cdot 5) \qquad \text{Note that } f(5) = 3 \cdot 5 = 15.$$
$$= g(15) = 1 + 15^2 = 1 + 225 = 226.$$

b) We find $(f \circ g)(x)$ by substituting $g(x)$ for x in the equation for $f(x)$:

$$(f \circ g)(x) = f(g(x)) = f(1 + x^2) \qquad \text{Using } g(x) = 1 + x^2$$
$$= 3 \cdot (1 + x^2) = 3 + 3x^2. \qquad \text{Using } f(x) = 3x$$

To find $(g \circ f)(x)$, we substitute $f(x)$ for x in the equation for $g(x)$:

$$(g \circ f)(x) = g(f(x)) = g(3x) \qquad \text{Substituting } 3x \text{ for } f(x)$$
$$= 1 + (3x)^2 = 1 + 9x^2.$$

As a check, note that $(g \circ f)(5) = 1 + 9 \cdot 5^2 = 1 + 9 \cdot 25 = 226$, as expected from part (a) above.

Example 1 shows that, in general, $(f \circ g)(5) \neq (g \circ f)(5)$ and $(f \circ g)(x) \neq (g \circ f)(x)$.

EXAMPLE 2   Given $f(x) = \sqrt{x}$ and $g(x) = x - 1$, find $(f \circ g)(x)$ and $(g \circ f)(x)$.

Solution

$$(f \circ g)(x) = f(g(x)) = f(x - 1) = \sqrt{x - 1} \qquad \text{Using } g(x) = x - 1$$
$$(g \circ f)(x) = g(f(x)) = g(\sqrt{x}) = \sqrt{x} - 1 \qquad \text{Using } f(x) = \sqrt{x}$$

In fields ranging from chemistry to geology and economics, one needs to recognize how a function can be regarded as the composition of two "simpler" functions. This is sometimes called *de*composition.

EXAMPLE 3   If $h(x) = (7x + 3)^2$, find f and g such that $h(x) = (f \circ g)(x)$.

Solution   We can think of $h(x)$ as the result of first finding $7x + 3$ and then squaring that. This suggests that $g(x) = 7x + 3$ and $f(x) = x^2$. We check by forming the composition:

$$(f \circ g)(x) = f(g(x))$$
$$= f(7x + 3) = (7x + 3)^2 = h(x), \text{ as desired.}$$

This is probably the most "obvious" answer to the question. There are other less obvious answers. For example, if

$$f(x) = (x - 1)^2 \quad \text{and} \quad g(x) = 7x + 4,$$

then

$$(f \circ g)(x) = f(g(x)) = f(7x + 4)$$
$$= (7x + 4 - 1)^2 = (7x + 3)^2 = h(x).$$

Inverses and One-to-One Functions

Let's view the following two functions as relations, or correspondences.

Selected Professions and Their Median Yearly Salary in 2003*

Domain (Set of Inputs)	Range (Set of Outputs)
Registered nurse	→ $49,550
Film and video editor	→ $40,600
Firefighter	→ $37,060
Computer programmer	→ $61,340
Secondary-school teacher	→ $44,580
Architect	→ $57,950

U.S. Senators and Their States

Domain (Set of Inputs)	Range (Set of Outputs)
Feinstein	→ California
Boxer	
Obama	→ Illinois
Durban	
Clinton	→ New York
Schumer	

Suppose we reverse the arrows. We obtain what is called the **inverse relation**. Are these inverse relations functions?

Selected Professions and Their Median Yearly Salary in 2003

Range (Set of Outputs)	Domain (Set of Inputs)
Registered nurse ←	$49,550
Film and video editor ←	$40,600
Firefighter ←	$37,060
Computer programmer ←	$61,340
Secondary-school teacher ←	$44,580
Architect ←	$57,950

U.S. Senators and Their States

Range (Set of Outputs)	Domain (Set of Inputs)
Feinstein ←	California
Boxer ←	
Obama ←	Illinois
Durban ←	
Clinton ←	New York
Schumer ←	

Recall that for each input, a function provides exactly one output. However, a function can have the same output for two or more different inputs. Thus it is possible for different inputs to correspond to the same output. Only when this possibility is *excluded* will the inverse be a function. For the functions listed above, this means the inverse of the "Selected Professions" correspondence is a function, but the inverse of the "U.S. Senator" correspondence is not.

Source: U.S. Bureau of Labor Statistics, May 2003, Occupational Employment and Wage Estimates

In the Selected Professions function, different inputs have different outputs, so it is a **one-to-one function**. In the U.S. Senator function, *Clinton* and *Schumer* are both paired with *New York*. Thus the U.S. Senator function is not one-to-one.

One-To-One Function

A function *f* is *one-to-one* if different inputs have different outputs. That is, if for *a* and *b* in the domain of *f* with $a \neq b$, we have $f(a) \neq f(b)$, then the function *f* is one-to-one. If a function is one-to-one, then its inverse correspondence is also a function.

How can we tell graphically whether a function is one-to-one?

EXAMPLE 4

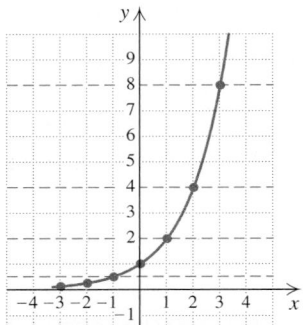

At left is the graph of a function similar to those we will study in Section 9.2. Determine whether the function is one-to-one and thus has an inverse that is a function.

Solution   A function is one-to-one if different inputs have different outputs—that is, if no two *x*-values have the same *y*-value. For this function, we cannot find two *x*-values that have the same *y*-value. Note that this means that no horizontal line can be drawn so that it crosses the graph more than once. The function is one-to-one so its inverse is a function.

The graph of every function must pass the vertical-line test. In order for a function to have an inverse that is a function, it must pass the *horizontal-line test* as well.

The Horizontal-Line Test

If it is impossible to draw a horizontal line that intersects a function's graph more than once, then the function is one-to-one. For every one-to-one function, an inverse function exists.

EXAMPLE 5

Determine whether the function $f(x) = x^2$ is one-to-one and thus has an inverse that is a function.

Solution   The graph of $f(x) = x^2$ is shown here. Many horizontal lines cross the graph more than once. For example, the line $y = 4$ crosses where the first coordinates are -2 and 2. Although these are different inputs, they have the same output. That is, $-2 \neq 2$, but

$$f(-2) = (-2)^2 = 4 = 2^2 = f(2).$$

Thus the function is not one-to-one and no inverse function exists.

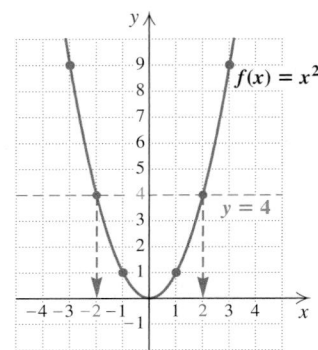

Finding Formulas for Inverses

When the inverse of f is also a function, it is denoted f^{-1} (read "f-inverse").

> *Caution!*   The -1 in f^{-1} is *not* an exponent!

Suppose a function is described by a formula. If its inverse is a function, how do we find a formula for that inverse? For any equation in two variables, if we interchange the variables, we form an equation of the inverse correspondence. If it is a function, we proceed as follows to find a formula for f^{-1}.

> *To Find a Formula for f^{-1}*
>
> First make sure that f is one-to-one. Then:
>
> 1. Replace $f(x)$ with y.
> 2. Interchange x and y. (This gives the inverse function.)
> 3. Solve for y.
> 4. Replace y with $f^{-1}(x)$. (This is inverse function notation.)

EXAMPLE 6

Determine whether each function is one-to-one and if it is, find a formula for $f^{-1}(x)$.

a) $f(x) = x + 2$ **b)** $f(x) = 2x - 3$

Solution

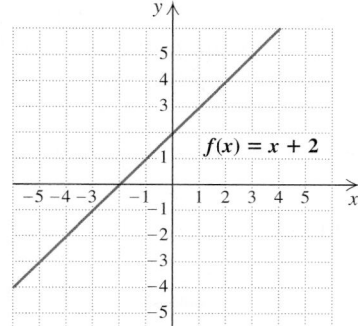

a) The graph of $f(x) = x + 2$ is shown at left. It passes the horizontal-line test, so it is one-to-one. Thus its inverse is a function.

1. Replace $f(x)$ with y: $y = x + 2$.
2. Interchange x and y: $x = y + 2$.   This gives the inverse function.
3. Solve for y: $x - 2 = y$.
4. Replace y with $f^{-1}(x)$: $f^{-1}(x) = x - 2$.   We also "reversed" the equation.

In this case, the function f adds 2 to all inputs. Thus, to "undo" f, the function f^{-1} must subtract 2 from its inputs.

b) The function $f(x) = 2x - 3$ is also linear. Any linear function that is not constant will pass the horizontal-line test. Thus, f is one-to-one.

1. Replace $f(x)$ with y: $y = 2x - 3$.
2. Interchange x and y: $x = 2y - 3$.
3. Solve for y: $x + 3 = 2y$

$$\frac{x + 3}{2} = y.$$

4. Replace y with $f^{-1}(x)$: $f^{-1}(x) = \dfrac{x + 3}{2}$.

In this case, the function f doubles all inputs and then subtracts 3. Thus, to "undo" f, the function f^{-1} adds 3 to each input and then divides by 2.

Graphing Functions and Their Inverses

How do the graphs of a function and its inverse compare?

EXAMPLE 7   Graph $f(x) = 2x - 3$ and $f^{-1}(x) = (x + 3)/2$ on the same set of axes. Then compare.

Solution   The graph of each function follows. Note that the graph of f^{-1} can be drawn by reflecting the graph of f across the line $y = x$. That is, if we graph $f(x) = 2x - 3$ in wet ink and fold the paper along the line $y = x$, the graph of $f^{-1}(x) = (x + 3)/2$ will appear as the impression made by f.

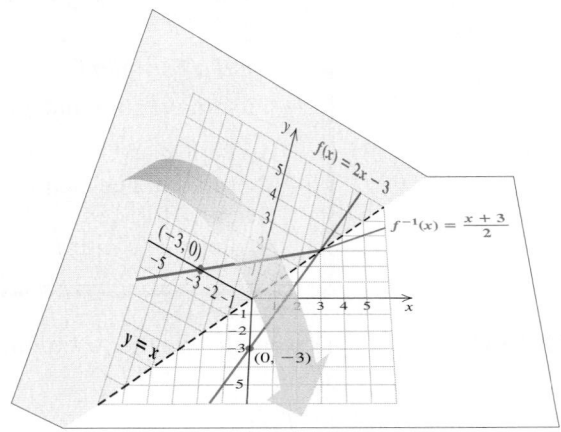

When x and y are interchanged to find a formula for the inverse, we are, in effect, reflecting or flipping the graph of $f(x) = 2x - 3$ across the line $y = x$. For example, when $(0, -3)$, the coordinates of the y-intercept of the graph of f, are reversed, we get $(-3, 0)$, the x-intercept of the graph of f^{-1}.

Visualizing Inverses

  The graph of f^{-1} is a reflection of the graph of f across the line $y = x$.

EXAMPLE 8   Consider $g(x) = x^3 + 2$.

a) Determine whether the function is one-to-one.

b) If it is one-to-one, find a formula for its inverse.

c) Graph the inverse, if it exists.

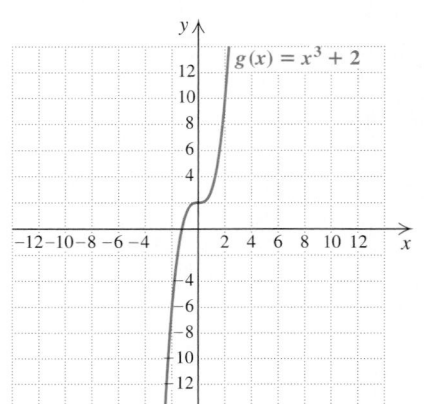

Solution

a) The graph of $g(x) = x^3 + 2$ is shown at right. It passes the horizontal-line test and thus has an inverse.

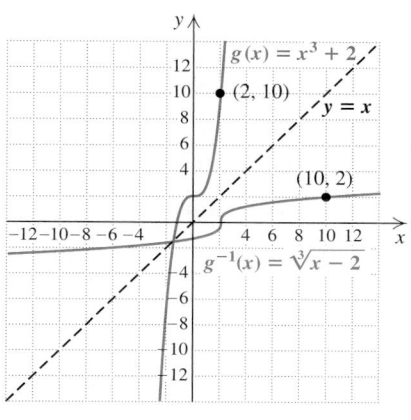

b) 1. Replace $g(x)$ with y: $y = x^3 + 2.$ Using $g(x) = x^3 + 2$
   2. Interchange x and y: $x = y^3 + 2.$
   3. Solve for y: $x - 2 = y^3$
                      $\sqrt[3]{x - 2} = y.$ Each real number has
                                     only one cube root, so
                                     we can solve for y.

   4. Replace y with $g^{-1}(x)$: $g^{-1}(x) = \sqrt[3]{x - 2}.$

c) To find the graph, we reflect the graph of $g(x) = x^3 + 2$ across the line $y = x$, as we did in Example 7. We can also substitute into $g^{-1}(x) = \sqrt[3]{x - 2}$ and plot points. Note that (2, 10) is on the graph of g, whereas (10, 2) is on the graph of g^{-1}. The graphs of g and g^{-1} are shown together at left.

Inverse Functions and Composition

Let's consider inverses of functions in terms of function machines. Suppose that a one-to-one function f is programmed into a machine. If the machine has a reverse switch, when the switch is thrown, the machine performs the inverse function f^{-1}. Inputs then enter at the opposite end, and the entire process is reversed.

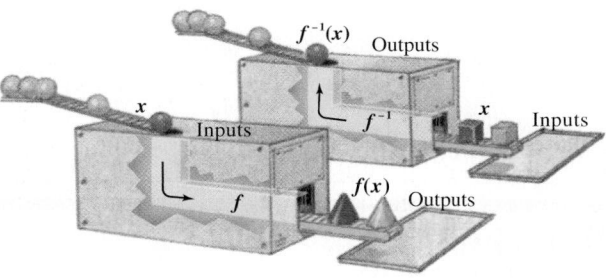

Consider $f(x) = x^3 + 2$ and $f^{-1}(x) = \sqrt[3]{x - 2}$ from Example 8. For the input 3,

$$f(3) = 3^3 + 2 = 27 + 2 = 29.$$

The output is 29. Now we use 29 for the input in the inverse:

$$f^{-1}(29) = \sqrt[3]{29 - 2} = \sqrt[3]{27} = 3.$$

The function f takes 3 to 29. The inverse function f^{-1} takes the number 29 back to 3.

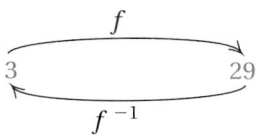

In general, for any output $f(x)$, the function f^{-1} takes that output back to x. Similarly, for any output $f^{-1}(x)$, the function f takes that output back to x.

Composition and Inverses

If a function f is one-to-one, then f^{-1} is the unique function for which

$$(f^{-1} \circ f)(x) = f^{-1}(f(x)) = x \quad \text{and} \quad (f \circ f^{-1})(x) = f(f^{-1}(x)) = x.$$

EXAMPLE 9   Let $f(x) = 2x + 1$. Show that

$$f^{-1}(x) = \frac{x-1}{2}.$$

Solution   We find $(f^{-1} \circ f)(x)$ and $(f \circ f^{-1})(x)$ and check to see that each is x.

$$(f^{-1} \circ f)(x) = f^{-1}(f(x)) = f^{-1}(2x+1)$$

$$= \frac{(2x+1)-1}{2}$$

$$= \frac{2x}{2} = x$$

$$(f \circ f^{-1})(x) = f(f^{-1}(x)) = f\left(\frac{x-1}{2}\right)$$

$$= 2 \cdot \frac{x-1}{2} + 1$$

$$= x - 1 + 1 = x$$

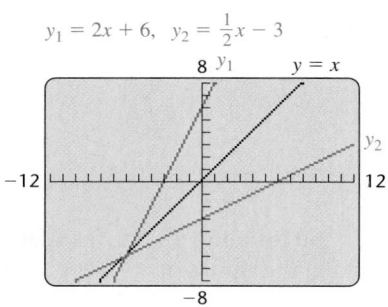

technology connection

To determine whether $y_1 = 2x + 6$ and $y_2 = \frac{1}{2}x - 3$ are inverses of each other, we can graph both functions, along with the line $y = x$, on a "squared" set of axes. It *appears* that y_1 and y_2 are inverses of each other. A more precise check is achieved by selecting the DRAWINV option of the ⟨DRAW⟩ menu. The resulting graph of the inverse of y_1 should coincide with y_2.

$$y_1 = 2x + 6, \quad y_2 = \frac{1}{2}x - 3$$

For a more dependable check, examine a TABLE in which $y_1 = 2x + 6$ and $y_2 = \frac{1}{2} \cdot y_1 - 3$. Note that y_2 "undoes" what y_1 does.

TBL MIN = −3   ΔTBL = 1 $y_2 = \frac{1}{2}y_1 - 3$

X	Y1	Y2
−3	0	−3
−2	2	−2
−1	4	−1
0	6	0
1	8	1
2	10	2
3	12	3

X = 3

1. Use a graphing calculator to check Examples 7, 8, and 9.
2. Will DRAWINV work for *any* choice of y_1? Why or why not?

Exercise Set

12.1

↪ *Concept Reinforcement*   *Classify each statement as either true or false.*

1. The composition of two functions f and g is written $f \circ g$.

2. The notation $(f \circ g)(x)$ means $f(g(x))$.

3. If $f(x) = x^2$ and $g(x) = x + 3$, then $(g \circ f)(x) = (x + 3)^2$.

4. If $f(2) = 15$ and $g(15) = 25$, then $(g \circ f)(2) = 25$.

5. The function f is one-to-one if $f(1) = 1$.

6. For f^{-1} to be a function, f cannot be one-to-one.

7. The function f is the inverse of f^{-1}.

8. If g and h are inverses of each other, then $(g \circ h)(x) = x$.

Find $(f \circ g)(1)$, $(g \circ f)(1)$, $(f \circ g)(x)$, *and* $(g \circ f)(x)$.

9. $f(x) = x^2 + 1$; $g(x) = 2x - 3$

10. $f(x) = 2x + 1$; $g(x) = x^2 - 5$

11. $f(x) = x - 3$; $g(x) = 2x^2 - 7$

12. $f(x) = 3x^2 + 4$; $g(x) = 4x - 1$

13. $f(x) = x + 7$; $g(x) = 1/x^2$

14. $f(x) = 1/x^2$; $g(x) = x + 2$

15. $f(x) = \sqrt{x}$; $g(x) = x + 3$

16. $f(x) = 10 - x$; $g(x) = \sqrt{x}$

17. $f(x) = \sqrt{4x}$; $g(x) = 1/x$

18. $f(x) = \sqrt{x + 3}$; $g(x) = 13/x$

19. $f(x) = x^2 + 4$; $g(x) = \sqrt{x - 1}$

20. $f(x) = x^2 + 8$; $g(x) = \sqrt{x + 17}$

Find $f(x)$ *and* $g(x)$ *such that* $h(x) = (f \circ g)(x)$. *Answers may vary.*

21. $h(x) = (7 + 5x)^2$

22. $h(x) = (3x - 1)^2$

23. $h(x) = \sqrt{2x + 7}$

24. $h(x) = \sqrt{5x + 2}$

25. $h(x) = \dfrac{2}{x - 3}$

26. $h(x) = \dfrac{3}{x} + 4$

Determine whether each function is one-to-one.

27. $f(x) = x - 5$

28. $f(x) = 5 - 2x$

29. $f(x) = x^2 + 1$

30. $f(x) = 1 - x^2$

31.
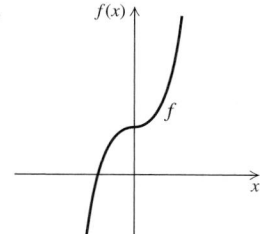

32.
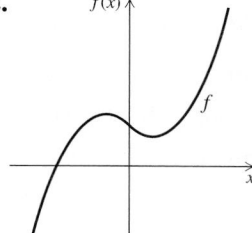

33.
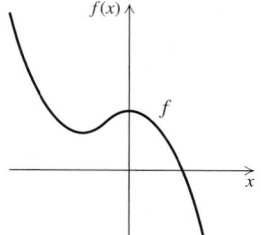

34.
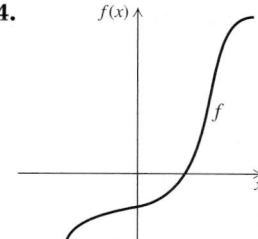

For each function, **(a)** *determine whether it is one-to-one;* **(b)** *if it is one-to-one, find a formula for the inverse.*

35. $f(x) = x + 4$

36. $f(x) = x + 2$

37. $f(x) = 2x$

38. $f(x) = 3x$

39. $g(x) = 3x - 1$

40. $g(x) = 2x - 3$

41. $f(x) = \dfrac{1}{2}x + 1$

42. $f(x) = \dfrac{1}{3}x + 2$

43. $g(x) = x^2 + 5$

44. $g(x) = x^2 - 4$

45. $h(x) = -2x + 4$

46. $h(x) = -3x + 1$

47. $f(x) = \dfrac{1}{x}$

48. $f(x) = \dfrac{3}{x}$

49. $G(x) = 4$

50. $H(x) = 2$

51. $f(x) = \dfrac{2x + 1}{3}$

52. $f(x) = \dfrac{3x + 2}{5}$

53. $f(x) = x^3 - 5$

54. $f(x) = x^3 + 2$

55. $g(x) = (x - 2)^3$

56. $g(x) = (x + 7)^3$

57. $f(x) = \sqrt{x}$

58. $f(x) = \sqrt{x - 1}$

Graph each function and its inverse using the same set of axes.

59. $f(x) = \dfrac{2}{3}x + 4$

60. $g(x) = \dfrac{1}{4}x + 2$

61. $f(x) = x^3 + 1$

62. $f(x) = x^3 - 1$

63. $g(x) = \dfrac{1}{2}x^3$

64. $g(x) = \dfrac{1}{3}x^3$

65. $F(x) = -\sqrt{x}$

66. $f(x) = \sqrt{x}$

67. $f(x) = -x^2, x \geq 0$

68. $f(x) = x^2 - 1, x \leq 0$

69. Let $f(x) = \sqrt[3]{x - 4}$. Show that
$$f^{-1}(x) = x^3 + 4.$$

70. Let $f(x) = 3/(x + 2)$. Show that
$$f^{-1}(x) = \dfrac{3}{x} - 2.$$

71. Let $f(x) = (1 - x)/x$. Show that
$$f^{-1}(x) = \dfrac{1}{x + 1}.$$

72. Let $f(x) = x^3 - 5$. Show that
$$f^{-1}(x) = \sqrt[3]{x + 5}.$$

73. *Dress sizes in the United States and Italy.*   A size-6 dress in the United States is size 36 in Italy. A function that converts dress sizes in the United States to those in Italy is
$$f(x) = 2(x + 12).$$
a) Find the dress sizes in Italy that correspond to sizes 8, 10, 14, and 18 in the United States.
b) Determine whether this function has an inverse that is a function. If so, find a formula for the inverse.
c) Use the inverse function to find dress sizes in the United States that correspond to sizes 40, 44, 52, and 60 in Italy.

74. *Dress sizes in the United States and France.*   A size-6 dress in the United States is size 38 in France. A function that converts dress sizes in the United States to those in France is
$$f(x) = x + 32.$$
a) Find the dress sizes in France that correspond to sizes 8, 10, 14, and 18 in the United States.
b) Determine whether this function has an inverse that is a function. If so, find a formula for the inverse.
c) Use the inverse function to find dress sizes in the United States that correspond to sizes 40, 42, 46, and 50 in France.

75. Is there a one-to-one relationship between the letters and the numbers on the keypad of a telephone? Why or why not?

76. Mathematicians usually try to select "logical" words when forming definitions. Does the term "one-to-one" seem logical? Why or why not?

SKILL MAINTENANCE

Simplify.

77. $(a^5b^4)^2(a^3b^5)$ [4.1]

78. $(x^3y^5)^2(x^4y^2)$ [4.1]

79. $27^{4/3}$ [10.2]

80. $25^{3/2}$ [10.2]

Solve. [2.3]

81. $x = \frac{2}{3}y - 7$, for y

82. $x = 10 - 3y$, for y

SYNTHESIS

83. The function $V(t) = 750(1.2)^t$ is used to predict the value, $V(t)$, of a certain rare stamp t years from 2001. Do not calculate $V^{-1}(t)$, but explain how V^{-1} could be used.

84. An organization determines that the cost per person of chartering a bus is given by the function
$$C(x) = \dfrac{100 + 5x}{x},$$
where x is the number of people in the group and $C(x)$ is in dollars. Determine $C^{-1}(x)$ and explain how this inverse function could be used.

For Exercises 85 and 86, graph the inverse of f.

85.

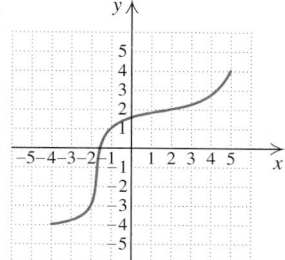

86.

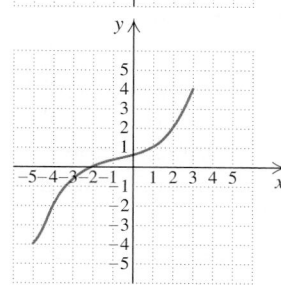

87. *Dress sizes in France and Italy.*   Use the information in Exercises 73 and 74 to find a function for the French dress size that corresponds to a size x dress in Italy.

88. *Dress sizes in Italy and France.*   Use the information in Exercises 73 and 74 to find a function for the Italian dress size that corresponds to a size x dress in France.

89. What relationship exists between the answers to Exercises 87 and 88? Explain how you determined this.

90. Show that function composition is associative by showing that $((f \circ g) \circ h)(x) = (f \circ (g \circ h))(x)$.

91. Show that if $h(x) = (f \circ g)(x)$, then $h^{-1}(x) = (g^{-1} \circ f^{-1})(x)$. (*Hint*: Use Exercise 90.)

Determine whether or not the given pairs of functions are inverses of each other.

92. $f(x) = 0.75x^2 + 2;\ g(x) = \sqrt{\dfrac{4(x-2)}{3}}$

93. $f(x) = 1.4x^3 + 3.2;\ g(x) = \sqrt[3]{\dfrac{x-3.2}{1.4}}$

94. $f(x) = \sqrt{2.5x + 9.25};$
$g(x) = 0.4x^2 - 3.7,\ x \geq 0$

95. $f(x) = 0.8x^{1/2} + 5.23;$
$g(x) = 1.25(x^2 - 5.23),\ x \geq 0$

96. $f(x) = 2.5(x^3 - 7.1);$
$g(x) = \sqrt[3]{0.4x + 7.1}$

97. Match each function in Column A with its inverse from Column B.

Column A

(1) $y = 5x^3 + 10$

(2) $y = (5x + 10)^3$

(3) $y = 5(x + 10)^3$

(4) $y = (5x)^3 + 10$

Column B

A. $y = \dfrac{\sqrt[3]{x} - 10}{5}$

B. $y = \sqrt[3]{\dfrac{x}{5}} - 10$

C. $y = \sqrt[3]{\dfrac{x - 10}{5}}$

D. $y = \dfrac{\sqrt[3]{x - 10}}{5}$

98. Examine the following table. Is it possible that f and g are inverses of each other? Why or why not?

x	$f(x)$	$g(x)$
6	6	6
7	6.5	8
8	7	10
9	7.5	12
10	8	14
11	8.5	16
12	9	18

99. The following window appears on a graphing calculator.

X	Y1	Y2
0	1	-2
1	1.5	0
2	2	2
3	2.5	4
4	3	6
5	3.5	8
6	4	10

X = 0

a) What evidence is there that the functions Y1 and Y2 are inverses of each other?

b) Find equations for Y1 and Y2, assuming that both are linear functions.

c) On the basis of your answer to part (b), are Y1 and Y2 inverses of each other?

12.2 Exponential Functions

Graphing Exponential Functions • Equations with *x* and *y*
Interchanged • Applications of Exponential Functions

CONNECTING THE CONCEPTS

Composite and inverse functions, as shown in Section 12.1, are very useful in and of themselves. The reason they are included in this chapter, however, is that they are needed in order to understand the logarithmic functions that appear in Section 12.3. Here in Section 12.2, we make no reference to composite or inverse functions. Instead, we introduce a new type of function, the *exponential function*, so that we can study both it and its inverse in Sections 12.3–12.7.

Consider the graph below. The rapidly rising curve approximates the graph of an *exponential function*. We now consider such functions and some of their applications.

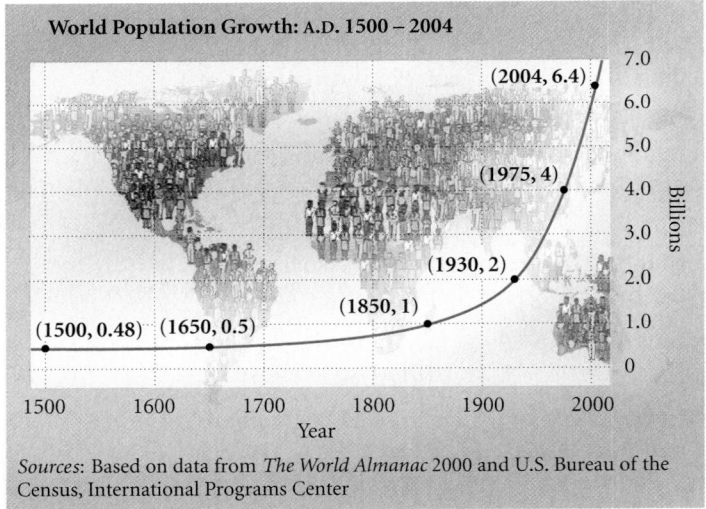

World Population Growth: A.D. 1500 – 2004

(2004, 6.4)
(1975, 4)
(1930, 2)
(1850, 1)
(1500, 0.48) (1650, 0.5)

Sources: Based on data from *The World Almanac* 2000 and U.S. Bureau of the Census, International Programs Center

Graphing Exponential Functions

In Chapter 10, we studied exponential expressions with rational-number exponents, such as

$$5^{1/4}, \qquad 3^{-3/4}, \qquad 7^{2.34}, \qquad 5^{1.73}.$$

For example, $5^{1.73}$, or $5^{173/100}$, represents the 100th root of 5 raised to the 173rd power. What about expressions with irrational exponents, such as $5^{\sqrt{3}}$ or $7^{-\pi}$?

To attach meaning to $5^{\sqrt{3}}$, consider a rational approximation, r, of $\sqrt{3}$. As r gets closer to $\sqrt{3}$, the value of 5^r gets closer to some real number p.

r closes in on $\sqrt{3}$.	5^r closes in on some real number p.
$1.7 < r < 1.8$	$15.426 \approx 5^{1.7} < p < 5^{1.8} \approx 18.119$
$1.73 < r < 1.74$	$16.189 \approx 5^{1.73} < p < 5^{1.74} \approx 16.452$
$1.732 < r < 1.733$	$16.241 \approx 5^{1.732} < p < 5^{1.733} \approx 16.267$

We define $5^{\sqrt{3}}$ to be the number p. To eight decimal places,

$$5^{\sqrt{3}} \approx 16.24245082.$$

Any positive irrational exponent can be interpreted in a similar way. Negative irrational exponents are then defined using reciprocals. Thus, so long as a is positive, a^x has meaning for *any* real number x. All of the laws of exponents still hold, but we will not prove that here. We now define an *exponential function*.

Exponential Function

The function $f(x) = a^x$, where a is a positive constant, $a \neq 1$, is called the *exponential function*, base a.

We require the base a to be positive to avoid imaginary numbers that would result from taking even roots of negative numbers. The restriction $a \neq 1$ is made to exclude the constant function $f(x) = 1^x$, or $f(x) = 1$.

The following are examples of exponential functions:

$$f(x) = 2^x, \qquad f(x) = \left(\tfrac{1}{3}\right)^x, \qquad f(x) = 5^{-3x}. \qquad \text{Note that } 5^{-3x} = (5^{-3})^x.$$

Like polynomial functions, the domain of an exponential function is the set of all real numbers. Unlike polynomial functions, exponential functions have a variable exponent. Because of this, graphs of exponential functions either rise or fall dramatically.

EXAMPLE 1 Graph the exponential function given by $y = f(x) = 2^x$.

Solution We compute some function values, thinking of y as $f(x)$, and list the results in a table. It is a good idea to start by letting $x = 0$.

$$f(0) = 2^0 = 1; \qquad f(-1) = 2^{-1} = \frac{1}{2^1} = \frac{1}{2};$$
$$f(1) = 2^1 = 2;$$
$$f(2) = 2^2 = 4; \qquad f(-2) = 2^{-2} = \frac{1}{2^2} = \frac{1}{4};$$
$$f(3) = 2^3 = 8;$$
$$f(-3) = 2^{-3} = \frac{1}{2^3} = \frac{1}{8}$$

Next, we plot these points and connect them with a smooth curve.

x	y, or $f(x)$
0	1
1	2
2	4
3	8
−1	$\frac{1}{2}$
−2	$\frac{1}{4}$
−3	$\frac{1}{8}$

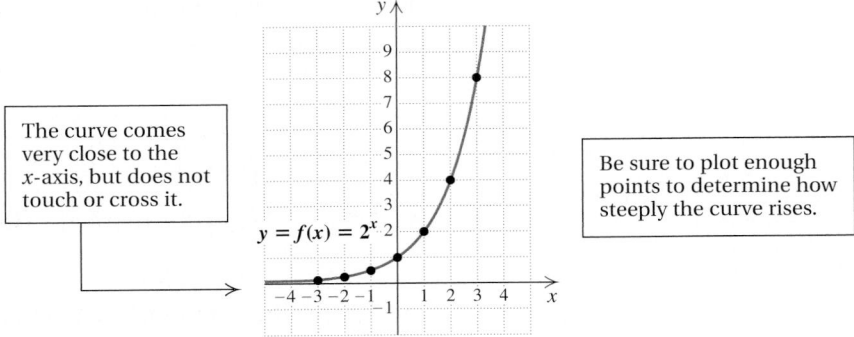

The curve comes very close to the x-axis, but does not touch or cross it.

$y = f(x) = 2^x$

Be sure to plot enough points to determine how steeply the curve rises.

Note that as x increases, the function values increase without bound. As x decreases, the function values decrease, getting very close to 0. The x-axis, or the line $y = 0$, is a horizontal *asymptote*, meaning that the curve gets closer and closer to this line the further we move to the left.

EXAMPLE 2

Graph: $y = f(x) = \left(\frac{1}{2}\right)^x$.

Solution We compute some function values, thinking of y as $f(x)$, and list the results in a table. Before we do this, note that

$$y = f(x) = \left(\tfrac{1}{2}\right)^x = (2^{-1})^x = 2^{-x}.$$

Then we have

$$f(0) = 2^{-0} = 1;$$
$$f(1) = 2^{-1} = \frac{1}{2^1} = \frac{1}{2};$$
$$f(2) = 2^{-2} = \frac{1}{2^2} = \frac{1}{4};$$
$$f(3) = 2^{-3} = \frac{1}{2^3} = \frac{1}{8};$$
$$f(-1) = 2^{-(-1)} = 2^1 = 2;$$
$$f(-2) = 2^{-(-2)} = 2^2 = 4;$$
$$f(-3) = 2^{-(-3)} = 2^3 = 8.$$

x	y, or $f(x)$
0	1
1	$\frac{1}{2}$
2	$\frac{1}{4}$
3	$\frac{1}{8}$
−1	2
−2	4
−3	8

Next, we plot these points and connect them with a smooth curve. This curve is a mirror image, or *reflection*, of the graph of $y = 2^x$ (see Example 1) across the y-axis. The line $y = 0$ is again the asymptote.

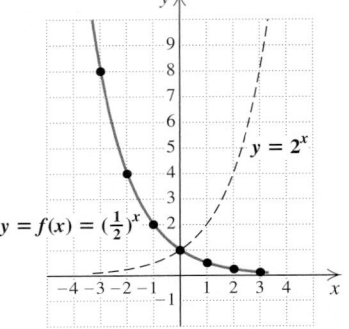

$y = 2^x$

$y = f(x) = \left(\frac{1}{2}\right)^x$

From Examples 1 and 2, we can make the following observations.

A. For $a > 1$, the graph of $f(x) = a^x$ increases from left to right. The greater the value of a, the steeper the curve. (See the figure on the left below.)

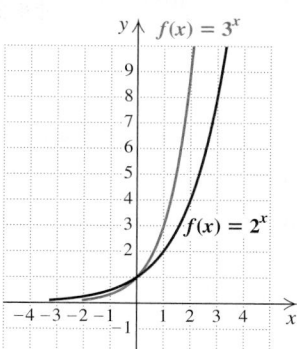 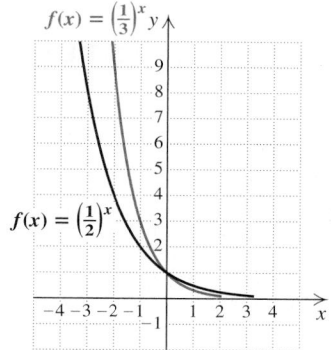

B. For $0 < a < 1$, the graph of $f(x) = a^x$ decreases from left to right. For smaller values of a, the curve becomes steeper. (See the figure on the right above.)
C. All graphs of $f(x) = a^x$ go through the y-intercept $(0, 1)$.
D. All graphs of $f(x) = a^x$ have the x-axis as the asymptote.
E. If $f(x) = a^x$, with $a > 0$, $a \neq 1$, the domain of f is all real numbers, and the range of f is all positive real numbers.
F. For $a > 0$, $a \neq 1$, the function given by $f(x) = a^x$ is one-to-one. Its graph passes the horizontal-line test.

EXAMPLE 3

Graph: $y = f(x) = 2^{x-2}$.

Solution   We construct a table of values. Then we plot the points and connect them with a smooth curve. Here $x - 2$ is the *exponent*.

$$f(0) = 2^{0-2} = 2^{-2} = \frac{1}{4}; \qquad f(-1) = 2^{-1-2} = 2^{-3} = \frac{1}{8};$$

$$f(1) = 2^{1-2} = 2^{-1} = \frac{1}{2}; \qquad f(-2) = 2^{-2-2} = 2^{-4} = \frac{1}{16}$$

$$f(2) = 2^{2-2} = 2^0 = 1;$$
$$f(3) = 2^{3-2} = 2^1 = 2;$$
$$f(4) = 2^{4-2} = 2^2 = 4;$$

x	y, or $f(x)$
0	$\frac{1}{4}$
1	$\frac{1}{2}$
2	1
3	2
4	4
−1	$\frac{1}{8}$
−2	$\frac{1}{16}$

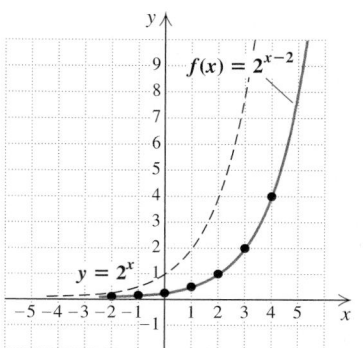

The graph looks just like the graph of $y = 2^x$, but it is translated 2 units to the right. The y-intercept of $y = 2^x$ is $(0, 1)$. The y-intercept of $y = 2^{x-2}$ is $\left(0, \frac{1}{4}\right)$. The line $y = 0$ is again the asymptote.

Student Notes

When using translations, make sure that you are shifting in the correct direction. When in doubt, substitute a value for x and make some calculations.

To practice graphing equations that are translations of each other, use **MODE** SIMUL and (**FORMAT**) ExprOff to graph $y_1 = 2^x$, $y_2 = 2^{x+1}$, $y_3 = 2^{x-1}$, $y_4 = 2^x + 1$, and $y_5 = 2^x - 1$. Use a bold curve for y_1 and then predict which curve represents which equation. Use (**TRACE**) to confirm your predictions. Switching (**FORMAT**) to ExprOn and using (**TRACE**) provides a definitive check (see also Exercise 75).

Equations with *x* and *y* Interchanged

It will be helpful in later work to be able to graph an equation in which the *x* and the *y* in $y = a^x$ are interchanged.

EXAMPLE 4   Graph: $x = 2^y$.

Solution Note that *x* is alone on one side of the equation. To find ordered pairs that are solutions, we choose values for *y* and then compute values for *x*:

For $y = 0$, $x = 2^0 = 1$.

For $y = 1$, $x = 2^1 = 2$.

For $y = 2$, $x = 2^2 = 4$.

For $y = 3$, $x = 2^3 = 8$.

For $y = -1$, $x = 2^{-1} = \dfrac{1}{2}$.

For $y = -2$, $x = 2^{-2} = \dfrac{1}{4}$.

For $y = -3$, $x = 2^{-3} = \dfrac{1}{8}$.

x	*y*
1	0
2	1
4	2
8	3
$\frac{1}{2}$	−1
$\frac{1}{4}$	−2
$\frac{1}{8}$	−3

(1) Choose values for *y*.
(2) Compute values for *x*.

We plot the points and connect them with a smooth curve.

This curve does not touch or cross the *y*-axis, which serves as a vertical asymptote.

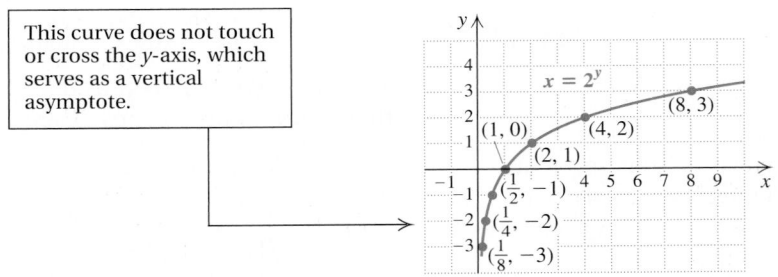

Note too that this curve looks just like the graph of $y = 2^x$, except that it is reflected across the line $y = x$, as shown here.

We have graphed $y = 2^x$ and its inverse, $x = 2^y$.

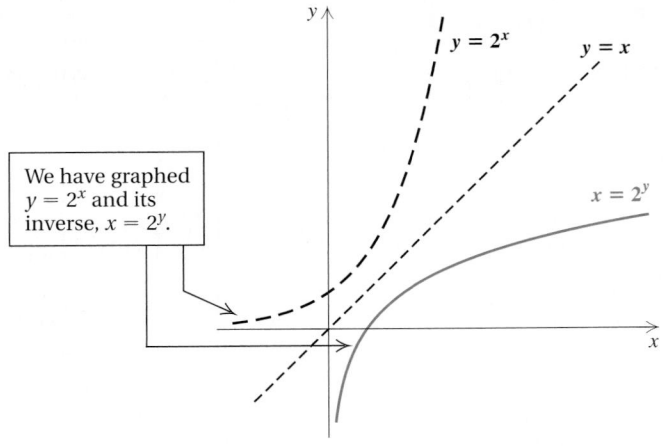

Applications of Exponential Functions

EXAMPLE 5   Interest compounded annually.   The amount of money A that a principal P will be worth after t years at interest rate i, compounded annually, is given by the formula

$$A = P(1 + i)^t. \qquad \text{You might review Example 11 in Section 11.1.}$$

Suppose that $100,000 is invested at 8% interest, compounded annually.

a) Find a function for the amount in the account after t years.
b) Find the amount of money in the account at $t = 0$, $t = 4$, $t = 8$, and $t = 10$.
c) Graph the function.

Solution

a) If $P = \$100{,}000$ and $i = 8\% = 0.08$, we can substitute these values and form the following function:

$$A(t) = \$100{,}000(1 + 0.08)^t \qquad \text{Using } A = P(1 + i)^t$$
$$= \$100{,}000(1.08)^t.$$

b) To find the function values, a calculator with a power key is helpful.

$$A(0) = \$100{,}000(1.08)^0 \qquad\qquad A(8) = \$100{,}000(1.08)^8$$
$$= \$100{,}000(1) \qquad\qquad\qquad \approx \$100{,}000(1.85093021)$$
$$= \$100{,}000 \qquad\qquad\qquad\quad \approx \$185{,}093.02$$

$$A(4) = \$100{,}000(1.08)^4 \qquad\qquad A(10) = \$100{,}000(1.08)^{10}$$
$$= \$100{,}000(1.36048896) \qquad\quad \approx \$100{,}000(2.158924997)$$
$$\approx \$136{,}048.90 \qquad\qquad\qquad\;\; \approx \$215{,}892.50$$

c) We use the function values computed in part (b), and others if we wish, to draw the graph as follows. Note that the axes are scaled differently because of the large numbers.

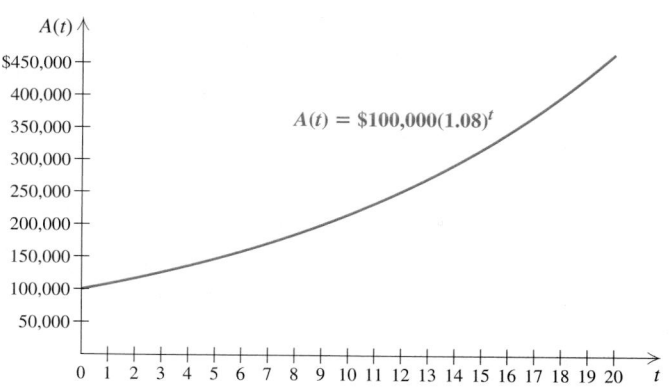

Exercise Set

12.2

↰ *Concept Reinforcement*   *Classify each of the following as either true or false.*

1. The graph of $f(x) = a^x$ always passes through the point (0, 1).

2. The graph of $g(x) = \left(\frac{1}{2}\right)^x$ gets closer and closer to the x-axis as x gets larger and larger.

3. The graph of $f(x) = 2^{x-3}$ looks just like the graph of $y = 2^x$, but it is translated 3 units to the right.

4. The graph of $g(x) = 2^x - 3$ looks just like the graph of $y = 2^x$, but it is translated 3 units up.

5. The graph of $y = 3^x$ gets close to, but never touches, the y-axis.

6. The graph of $x = 3^y$ gets close to, but never touches, the y-axis.

Graph.

7. $y = f(x) = 3^x$

8. $y = f(x) = 2^x$

9. $y = 6^x$

10. $y = 5^x$

11. $y = 2^x + 1$

12. $y = 2^x + 3$

13. $y = 3^x - 2$

14. $y = 3^x - 1$

15. $y = 2^x - 5$

16. $y = 2^x - 4$

17. $y = 2^{x-2}$

18. $y = 2^{x-1}$

19. $y = 2^{x+1}$

20. $y = 2^{x+3}$

21. $y = \left(\frac{1}{4}\right)^x$

22. $y = \left(\frac{1}{5}\right)^x$

23. $y = \left(\frac{1}{3}\right)^x$

24. $y = \left(\frac{1}{2}\right)^x$

25. $y = 2^{x+1} - 3$

26. $y = 2^{x-3} - 1$

27. $x = 6^y$

28. $x = 3^y$

29. $x = 3^{-y}$

30. $x = 2^{-y}$

31. $x = 4^y$

32. $x = 5^y$

33. $x = \left(\frac{4}{3}\right)^y$

34. $x = \left(\frac{3}{2}\right)^y$

Graph each pair of equations on the same set of axes.

35. $y = 3^x$, $x = 3^y$

36. $y = 2^x$, $x = 2^y$

37. $y = \left(\frac{1}{2}\right)^x$, $x = \left(\frac{1}{2}\right)^y$

38. $y = \left(\frac{1}{4}\right)^x$, $x = \left(\frac{1}{4}\right)^y$

Solve.

39. *Population growth.*   The world population $P(t)$, in billions, t years after 1980, can be approximated by
$$P(t) = 4.495(1.015)^t.$$
Source: Based on data from U.S. Bureau of the Census, International Data Base

a) Predict the world population in 2008, 2012, and 2016.

b) Graph the function.

40. *Growth of bacteria.*   The bacteria *Escherichia coli* are commonly found in the human bladder. Suppose that 3000 of the bacteria are present at time $t = 0$. Then t minutes later, the number of bacteria present can be approximated by
$$N(t) = 3000(2)^{t/20}.$$

a) How many bacteria will be present after 10 min? 20 min? 30 min? 40 min? 60 min?

b) Graph the function.

41. *Smoking cessation.*   The percentage of smokers P who receive telephone counseling to quit smoking and are still successful t months later can be approximated by
$$P(t) = 21.4(0.914)^t.$$
Sources: *New England Journal of Medicine*; data from California's Smokers' Hotline

a) Estimate the percentage of smokers receiving telephone counseling who are successful in quitting for 1 month, 3 months, and 1 year.

b) Graph the function.

42. *Smoking cessation.*   The percentage of smokers P who, without telephone counseling, have successfully quit smoking for t months (see Exercise 41) can be approximated by
$$P(t) = 9.02(0.93)^t.$$
Sources: *New England Journal of Medicine*; data from California's Smokers' Hotline

a) Estimate the percentage of smokers not receiving telephone counseling who are successful in quitting for 1 month, 3 months, and 1 year.

b) Graph the function.

43. *Marine biology.* Due to excessive whaling prior to the mid 1970s, the humpback whale is considered an endangered species. The worldwide population of humpbacks, $P(t)$, in thousands, t years after 1900 ($t < 70$) can be approximated by*

$$P(t) = 150(0.960)^t.$$

a) How many humpback whales were alive in 1930? in 1960?

b) Graph the function.

44. *Salvage value.* A photocopier is purchased for $5200. Its value each year is about 80% of the value of the preceding year. Its value, in dollars, after t years is given by the exponential function

$$V(t) = 5200(0.8)^t.$$

a) Find the value of the machine after 0 yr, 1 yr, 2 yr, 5 yr, and 10 yr.

b) Graph the function.

45. *Marine biology.* As a result of preservation efforts in most countries in which whaling was common, the humpback whale population has grown since the 1970s. The worldwide population of humpbacks, $P(t)$, in thousands, t years after 1982 can be approximated by*

$$P(t) = 5.5(1.047)^t.$$

a) How many humpback whales were alive in 1992? in 2004?

b) Graph the function.

46. *Recycling aluminum cans.* It is estimated that $\frac{1}{2}$ of all aluminum cans distributed will be recycled each year. A beverage company distributes 250,000 cans. The number still in use after time t, in years, is given by the exponential function

$$N(t) = 250,000\left(\tfrac{1}{2}\right)^t.$$

Source: The Aluminum Association, Inc., May 2004

a) How many cans are still in use after 0 yr? 1 yr? 4 yr? 10 yr?

b) Graph the function.

47. *Spread of zebra mussels.* Beginning in 1988, infestations of zebra mussels started spreading throughout North American waters.[†] These mussels spread with such speed that water treatment facilities, power plants, and entire

ecosystems can become threatened. The function

$$A(t) = 10 \cdot 34^t$$

can be used to estimate the number of square centimeters of lake bottom that will be covered with mussels t years after an infestation covering 10 cm^2 first occurs.

a) How many square centimeters of lake bottom will be covered with mussels 5 years after an infestation covering 10 cm^2 first appears? 7 years after the infestation first appears?

b) Graph the function.

48. *Cell phones.* The number of cell phones in use in the United States is increasing exponentially. The number N, in millions, in use can be estimated by

$$N(t) = 7.12(1.3)^t,$$

where t is the number of years after 1990.

Source: Cellular Telecommunications and Internet Association

a) Estimate the number of cell phones in use in 1995, 2005, and 2010.

b) Graph the function.

49. Without using a calculator, explain why 2^π must be greater than 8 but less than 16.

50. Suppose that $1000 is invested for 5 yr at 7% interest, compounded annually. In what year will the most interest be earned? Why?

SKILL MAINTENANCE

Simplify.

51. 5^{-2} [4.8]

52. 2^{-5} [4.8]

53. $1000^{2/3}$ [10.2]

54. $25^{-3/2}$ [10.2]

55. $\dfrac{10a^8b^7}{2a^2b^4}$ [4.1]

56. $\dfrac{24x^6y^4}{4x^2y^3}$ [4.1]

SYNTHESIS

57. Examine Exercise 48. Do you believe that the equation for the number of cell phones in use in the United States will be accurate 20 yr from now? Why or why not?

58. Why was it necessary to discuss irrational exponents before graphing exponential functions?

Determine which of the two numbers is larger. Do not use a calculator.

59. $\pi^{1.3}$ or $\pi^{2.4}$

60. $\sqrt{8^3}$ or $8^{\sqrt{3}}$

---

*Based on information from the American Cetacean Society, 2001, and the ASK Archive, 1998.

[†]Many thanks to Dr. Gerald Mackie of the Department of Zoology at the University of Guelph in Ontario for the background information for this exercise.

Graph.

61. $f(x) = 3.8^x$

62. $f(x) = 2.3^x$

63. $y = 2^x + 2^{-x}$

64. $y = \left|\left(\frac{1}{2}\right)^x - 1\right|$

65. $y = |2^x - 2|$

66. $y = 2^{-(x-1)^2}$

67. $y = |2^{x^2} - 1|$

68. $y = 3^x + 3^{-x}$

Graph both equations using the same set of axes.

69. $y = 3^{-(x-1)}$, $x = 3^{-(y-1)}$ **70.** $y = 1^x$, $x = 1^y$

71. *Sales of DVD players.*   As prices of DVD players continue to drop, sales have grown from $171 million in 1997 to $1099 million in 1999 and $2697 million in 2001. After pressing **STAT**, use the ExpReg option in the CALC menu to find an exponential function that models the total sales of DVD players t years after 1997. Then use that function to predict the total sales in 2008.
Source: *Statistical Abstract of the United States,* 2003

72. *Keyboarding speed.*   Ali is studying keyboarding. After he has studied for t hours, Ali's speed, in words per minute, is given by the exponential function

$$S(t) = 200[1 - (0.99)^t].$$

Use a graph and/or table of values to predict Ali's speed after studying for 10 hr, 40 hr, and 80 hr.

73. *Spread of AIDS.*   In 2000, a total of 40,282 cases of AIDS was reported in the United States; in 2001, a total of 41,450 cases; and in 2002, a total of 42,745 cases.

   a) Graph the data points, letting t represent the number of years since 2000.

   b) Which function best fits the data: linear, exponential, or quadratic? Why?

74. Consider any exponential function of the form $f(x) = a^x$ with $a > 1$. Will it always follow that $f(3) - f(2) > f(2) - f(1)$, and, in general, $f(n + 2) - f(n + 1) > f(n + 1) - f(n)$? Why or why not? (*Hint:* Think graphically.)

75. On many graphing calculators, it is possible to enter and graph $y_1 = A \wedge (X - B) + C$ after first pressing **APPS** Transfrm. Use this application to graph $f(x) = 2.5^{x-3} + 2$, $g(x) = 2.5^{x+3} + 2$, $h(x) = 2.5^{x-3} - 2$, and $k(x) = 2.5^{x+3} - 2$.

CORNER

The True Cost of a New Car

Focus: Car loans and exponential functions

Time: 30 minutes

Group size: 2

Materials: Calculators with exponentiation keys

The formula

$$M = \frac{Pr}{1 - (1 + r)^{-n}}$$

is used to determine the payment size, M, when a loan of P dollars is to be repaid in n equally sized monthly payments. Here r represents the monthly interest rate. Loans repaid in this fashion are said to be *amortized* (spread out equally) over a period of n months.

ACTIVITY

1. Suppose one group member is selling the other a car for $2600, financed at 1% interest per month for 24 months. What should be the size of each monthly payment?

2. Suppose both group members are shopping for the same model new car. To save time, each group member visits a different dealer. One dealer offers the car for $13,000 at 10.5% interest (0.00875 monthly interest) for 60 months (no down payment). The other dealer offers the same car for $12,000, but at 12% interest (0.01 monthly interest) for 48 months (no down payment).

   a) Determine the monthly payment size for each offer. Then determine the total amount paid for the car under each offer. How much of each total is interest?

   b) Work together to find the annual interest rate for which the total cost of 60 monthly payments for the $13,000 car would equal the total amount paid for the $12,000 car (as found in part a above).

12.3   Logarithmic Functions

Graphs of Logarithmic Functions   •   Equivalent Equations   •
Solving Certain Logarithmic Equations

We are now ready to study inverses of exponential functions. These functions have many applications and are called *logarithm*, or *logarithmic, functions.*

Graphs of Logarithmic Functions

Consider the exponential function $f(x) = 2^x$. Like all exponential functions, f is one-to-one. Can a formula for f^{-1} be found? To answer this, we use the method of Section 12.1:

1. Replace $f(x)$ with y: $y = 2^x$.

2. Interchange x and y: $x = 2^y$.

3. Solve for y: $y =$ the exponent to which we raise 2 to get x.

4. Replace y with $f^{-1}(x)$: $f^{-1}(x) =$ the exponent to which we raise 2 to get x.

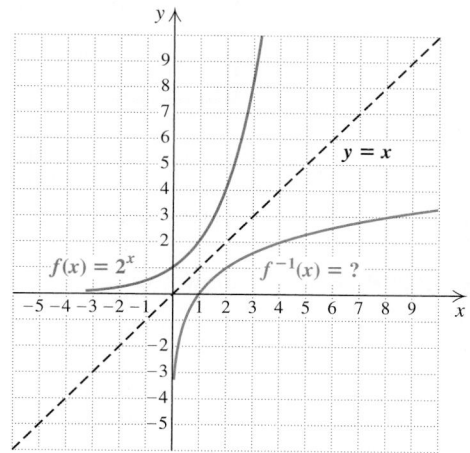

We now define a new symbol to replace the words "the exponent to which we raise 2 to get x":

$\log_2 x$, read "the logarithm, base 2, of x," or "log, base 2, of x," means "the exponent to which we raise 2 to get x."

Thus if $f(x) = 2^x$, then $f^{-1}(x) = \log_2 x$. Note that $f^{-1}(8) = \log_2 8 = 3$, because 3 is *the exponent to which we raise 2 to get* 8.

EXAMPLE 1   Simplify: **(a)** $\log_2 32$; **(b)** $\log_2 1$; **(c)** $\log_2 \frac{1}{8}$.

Solution

a) Think of $\log_2 32$ as the exponent to which we raise 2 to get 32. That exponent is 5. Therefore, $\log_2 32 = 5$.

b) We ask ourselves: "To what exponent do we raise 2 in order to get 1?" That exponent is 0 (recall that $2^0 = 1$). Thus, $\log_2 1 = 0$.

c) To what exponent do we raise 2 in order to get $\frac{1}{8}$? Since $2^{-3} = \frac{1}{8}$, we have $\log_2 \frac{1}{8} = -3$.

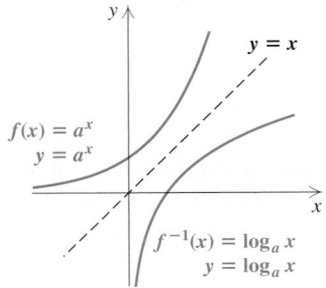

Although numbers like $\log_2 13$ can only be approximated, we must remember that $\log_2 13$ represents *the exponent to which we raise* 2 *to get* 13. That is, $2^{\log_2 13} = 13$. A calculator can be used to show that $\log_2 13 \approx 3.7$ and $2^{3.7} \approx 13$. Later in this chapter, we will use a calculator to find such approximations.

For any exponential function $f(x) = a^x$, the inverse is called a **logarithmic function, base a.** The graph of the inverse can be drawn by reflecting the graph of $f(x) = a^x$ across the line $y = x$. It will be helpful to remember that the inverse of $f(x) = a^x$ is given by $f^{-1}(x) = \log_a x$.

> **The Meaning of $\log_a x$**
>
> For $x > 0$ and a a positive constant other than 1, $\log_a x$ is the exponent to which a must be raised in order to get x. Thus,
>
> $$\log_a x = m \text{ means } a^m = x$$
>
> or equivalently,
>
> $$\log_a x \text{ is that unique exponent for which } a^{\log_a x} = x.$$

It is important to remember that *a logarithm is an exponent.* It might help to repeat several times: "The logarithm, base a, of a number x is the exponent to which a must be raised in order to get x."

EXAMPLE 2   Simplify: $7^{\log_7 85}$.

Solution   Remember that $\log_7 85$ is the exponent to which 7 is raised to get 85. Raising 7 to that exponent, we have

$$7^{\log_7 85} = 85.$$

Because logarithmic and exponential functions are inverses of each other, the result in Example 2 should come as no surprise: If $f(x) = \log_7 x$, then

for $f(x) = \log_7 x$, we have $f^{-1}(x) = 7^x$

and $f^{-1}(f(x)) = f^{-1}(\log_7 x) = 7^{\log_7 x} = x.$

Thus, $f^{-1}(f(85)) = 7^{\log_7 85} = 85.$

The following is a comparison of exponential and logarithmic functions.

Exponential Function	Logarithmic Function
$y = a^x$	$x = a^y$
$f(x) = a^x$	$g(x) = \log_a x$
$a > 0, a \neq 1$	$a > 0, a \neq 1$
The domain is $\mathbb{R}$.	The range is $\mathbb{R}$.
$y > 0$ (Outputs are positive.)	$x > 0$ (Inputs are positive.)
$f^{-1}(x) = \log_a x$	$g^{-1}(x) = a^x$

EXAMPLE 3

Graph: $y = f(x) = \log_5 x$.

Solution   If $y = \log_5 x$, then $5^y = x$. We can find ordered pairs that are solutions by choosing values for y and computing the x-values.

For $y = 0$, $x = 5^0 = 1$.
For $y = 1$, $x = 5^1 = 5$.
For $y = 2$, $x = 5^2 = 25$.
For $y = -1$, $x = 5^{-1} = \frac{1}{5}$.
For $y = -2$, $x = 5^{-2} = \frac{1}{25}$.

(1) Select y.
(2) Compute x.

x, or 5^y	y
1	0
5	1
25	2
$\frac{1}{5}$	-1
$\frac{1}{25}$	-2

This table shows the following:

$\log_5 1 = 0;$
$\log_5 5 = 1;$
$\log_5 25 = 2;$ These can all be checked using the equations above.
$\log_5 \frac{1}{5} = -1;$
$\log_5 \frac{1}{25} = -2.$

We plot the set of ordered pairs and connect the points with a smooth curve. The graphs of $y = 5^x$ and $y = x$ are shown only for reference.

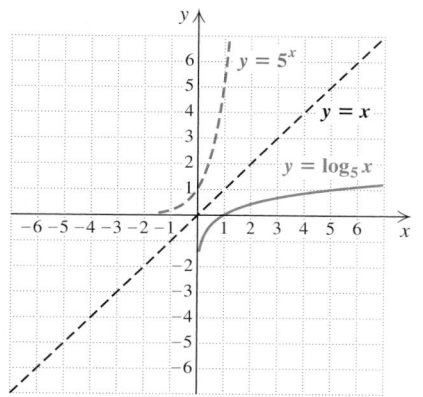

technology connection

To see that $f(x) = 10^x$ and $g(x) = \log_{10} x$ are inverses of each other, let $y_1 = 10^x$ and $y_2 = \log_{10} x = \log x$. Then, using a squared window, compare both graphs. If possible, select DrawInv from the (DRAW) menu and then press (VARS) (▶) (1) (1) (ENTER) to see another representation of f^{-1}. Finally, let $y_3 = y_1(y_2)$ and $y_4 = y_2(y_1)$ to show, using a table or graphs, that, for $x > 0$, $y_3 = y_4 = x$.

Equivalent Equations

We use the definition of logarithm to rewrite a *logarithmic equation* as an equivalent *exponential equation* or the other way around:

$$m = \log_a x \quad \text{is equivalent to} \quad a^m = x.$$

> *Caution!*   **Do not forget this relationship!** It is probably the most important definition in the chapter. Many times this definition will be used to justify a property we are considering.

EXAMPLE 4   Rewrite each as an equivalent exponential equation: **(a)** $y = \log_3 5$; **(b)** $-2 = \log_a 7$; **(c)** $a = \log_b d$.

Solution

a) $y = \log_3 5$ is equivalent to $3^y = 5$ The logarithm is the exponent.

The base remains the base.

b) $-2 = \log_a 7$ is equivalent to $a^{-2} = 7$

c) $a = \log_b d$ is equivalent to $b^a = d$

We also use the definition of logarithm to rewrite an exponential equation as an equivalent logarithmic equation.

EXAMPLE 5   Rewrite each as an equivalent logarithmic equation: **(a)** $8 = 2^x$; **(b)** $y^{-1} = 4$; **(c)** $a^b = c$.

Solution

a) $8 = 2^x$ is equivalent to $x = \log_2 8$ The exponent is the logarithm.

The base remains the base.

b) $y^{-1} = 4$ is equivalent to $-1 = \log_y 4$

c) $a^b = c$ is equivalent to $b = \log_a c$

Solving Certain Logarithmic Equations

Logarithmic equations are often solved by rewriting them as equivalent exponential equations.

EXAMPLE 6   Solve: **(a)** $\log_2 x = -3$; **(b)** $\log_x 16 = 2$.

Solution

a) $\log_2 x = -3$

$\qquad 2^{-3} = x$ Rewriting as an exponential equation

$\qquad \frac{1}{8} = x$ Computing 2^{-3}

Check: $\log_2 \frac{1}{8}$ is the exponent to which 2 is raised to get $\frac{1}{8}$. Since that exponent is -3, we have a check. The solution is $\frac{1}{8}$.

b) $\log_x 16 = 2$

$$x^2 = 16 \qquad \text{Rewriting as an exponential equation}$$
$$x = 4 \quad or \quad x = -4 \qquad \text{Principle of square roots}$$

Check: $\log_4 16 = 2$ because $4^2 = 16$. Thus, 4 is a solution of $\log_x 16 = 2$. Because all logarithmic bases must be positive, -4 cannot be a solution. Logarithmic bases must be positive because logarithms are defined using exponential functions that require positive bases. The solution is 4.

One method for solving certain logarithmic and exponential equations relies on the following property, which results from the fact that exponential functions are one-to-one.

The Principle of Exponential Equality

For any real number b, where $b \neq -1, 0,$ or 1,

$$b^{x_1} = b^{x_2} \quad \text{is equivalent to} \quad x_1 = x_2.$$

(Powers of the same base are equal if and only if the exponents are equal.)

EXAMPLE 7   Solve: **(a)** $\log_{10} 1000 = x$; **(b)** $\log_4 1 = t$.

Solution

a) We rewrite $\log_{10} 1000 = x$ in exponential form and solve:

$$10^x = 1000 \qquad \text{Rewriting as an exponential equation}$$
$$10^x = 10^3 \qquad \text{Writing 1000 as a power of 10}$$
$$x = 3. \qquad \text{Equating exponents}$$

Check: This equation can also be solved directly by determining the exponent to which we raise 10 in order to get 1000. In both cases we find that $\log_{10} 1000 = 3$, so we have a check. The solution is 3.

b) We rewrite $\log_4 1 = t$ in exponential form and solve:

$$4^t = 1 \qquad \text{Rewriting as an exponential equation}$$
$$4^t = 4^0 \qquad \text{Writing 1 as a power of 4. This can be done mentally.}$$
$$t = 0. \qquad \text{Equating exponents}$$

Check: As in part (a), this equation can be solved directly by determining the exponent to which we raise 4 in order to get 1. In both cases we find that $\log_4 1 = 0$, so we have a check. The solution is 0.

Example 7 illustrates an important property of logarithms.

$\log_a 1$

The logarithm, base a, of 1 is always 0: $\log_a 1 = 0$.

This follows from the fact that $a^0 = 1$ is equivalent to the logarithmic equation $\log_a 1 = 0$. Thus, $\log_{10} 1 = 0$, $\log_7 1 = 0$, and so on.

Another property results from the fact that $a^1 = a$. This is equivalent to the equation $\log_a a = 1$.

$\log_a a$

The logarithm, base a, of a is always 1: $\log_a a = 1$.

Thus, $\log_{10} 10 = 1$, $\log_8 8 = 1$, and so on.

Exercise Set

12.3

FOR EXTRA HELP

Student's Solutions Manual | Digital Video Tutor CD 6 Videotape 12 | Tutor Center AW Math Tutor Center | MathXL Tutorials on CD | MathXL MathXL | MyMathLab MyMathLab

🖎 *Concept Reinforcement* *In each of Exercises 1–8,*
match the expression or equation with an equivalent
expression or equation from the column on the right.

1. _____ $\log_5 25$

2. _____ $2^5 = x$

3. _____ $\log_5 5$

4. _____ $\log_2 1$

5. _____ $\log_5 5^x$

6. _____ $\log_x 27 = 5$

7. _____ $8 = 2^x$

8. _____ $x^{-2} = 5$

a) 1

b) x

c) $x^5 = 27$

d) $\log_2 x = 5$

e) $\log_2 8 = x$

f) $\log_x 5 = -2$

g) 2

h) 0

Simplify.

9. $\log_{10} 1000$

10. $\log_{10} 100$

11. $\log_2 16$

12. $\log_2 8$

13. $\log_3 81$

14. $\log_3 27$

15. $\log_4 \frac{1}{16}$

16. $\log_4 \frac{1}{4}$

17. $\log_7 \frac{1}{7}$

18. $\log_7 \frac{1}{49}$

19. $\log_5 625$

20. $\log_5 125$

21. $\log_8 8$

22. $\log_7 1$

23. $\log_8 1$

24. $\log_8 8$

Aha! **25.** $\log_9 9^5$

26. $\log_9 9^{10}$

27. $\log_{10} 0.01$

28. $\log_{10} 0.1$

29. $\log_9 3$

30. $\log_{16} 4$

31. $\log_9 27$

32. $\log_{16} 64$

33. $\log_{1000} 100$

34. $\log_{27} 9$

35. $5^{\log_5 7}$

36. $6^{\log_6 13}$

Graph.

37. $y = \log_{10} x$

38. $y = \log_2 x$

39. $y = \log_3 x$

40. $y = \log_7 x$

41. $f(x) = \log_6 x$ **42.** $f(x) = \log_4 x$

43. $f(x) = \log_{2.5} x$ **44.** $f(x) = \log_{1/2} x$

Graph both functions using the same set of axes.

45. $f(x) = 3^x$, $f^{-1}(x) = \log_3 x$

46. $f(x) = 4^x$, $f^{-1}(x) = \log_4 x$

Rewrite each of the following as an equivalent exponential equation. Do not solve.

47. $t = \log_5 9$ **48.** $h = \log_7 10$

49. $\log_5 25 = 2$ **50.** $\log_6 6 = 1$

51. $\log_{10} 0.1 = -1$ **52.** $\log_{10} 0.01 = -2$

53. $\log_{10} 7 = 0.845$ **54.** $\log_{10} 3 = 0.4771$

55. $\log_c m = 8$ **56.** $\log_b n = 23$

57. $\log_t Q = r$ **58.** $\log_m P = a$

59. $\log_e 0.25 = -1.3863$ **60.** $\log_e 0.989 = -0.0111$

61. $\log_r T = -x$ **62.** $\log_c M = -w$

Rewrite each of the following as an equivalent logarithmic equation. Do not solve.

63. $10^2 = 100$ **64.** $10^4 = 10,000$

65. $4^{-5} = \frac{1}{1024}$ **66.** $5^{-3} = \frac{1}{125}$

67. $16^{3/4} = 8$ **68.** $8^{1/3} = 2$

69. $10^{0.4771} = 3$ **70.** $10^{0.3010} = 2$

71. $z^m = 6$ **72.** $m^n = r$

73. $p^m = V$ **74.** $Q^t = x$

75. $e^3 = 20.0855$ **76.** $e^2 = 7.3891$

77. $e^{-4} = 0.0183$ **78.** $e^{-2} = 0.1353$

Solve.

79. $\log_3 x = 2$ **80.** $\log_4 x = 3$

81. $\log_5 125 = x$ **82.** $\log_4 64 = x$

83. $\log_2 16 = x$ **84.** $\log_3 27 = x$

85. $\log_x 7 = 1$ **86.** $\log_x 8 = 1$

87. $\log_3 x = -2$ **88.** $\log_2 x = -1$

89. $\log_{32} x = \frac{2}{5}$ **90.** $\log_8 x = \frac{2}{3}$

91. Express in words what number is represented by $\log_b c$.

92. Is it true that $2 = b^{\log_b 2}$? Why or why not?

Simplify.

93. $\dfrac{x^{12}}{x^4}$ [4.1]   **94.** $\dfrac{a^{15}}{a^3}$ [4.1]

95. $(a^4 b^6)(a^3 b^2)$ [4.1]   **96.** $(x^3 y^5)(x^2 y^7)$ [4.1]

97. $\dfrac{\dfrac{3}{x} - \dfrac{2}{xy}}{\dfrac{2}{x^2} + \dfrac{1}{xy}}$ [6.5]

98. $\dfrac{\dfrac{4+x}{x^2 + 2x + 1}}{\dfrac{3}{x+1} - \dfrac{2}{x+2}}$ [6.5]

99. Would a manufacturer be pleased or unhappy if sales of a product grew logarithmically? Why?

100. Explain why the number $\log_2 13$ must be between 3 and 4.

101. Graph both equations using the same set of axes:
$$y = \left(\tfrac{3}{2}\right)^x, \qquad y = \log_{3/2} x.$$

Graph.

102. $y = \log_2(x - 1)$

103. $y = \log_3 |x + 1|$

Solve.

104. $|\log_3 x| = 2$

105. $\log_4(3x - 2) = 2$

106. $\log_8(2x + 1) = -1$

107. $\log_{10}(x^2 + 21x) = 2$

Simplify.

108. $\log_{1/4} \frac{1}{64}$

109. $\log_{1/5} 25$

110. $\log_{81} 3 \cdot \log_3 81$

111. $\log_{10}(\log_4(\log_3 81))$

112. $\log_2(\log_2(\log_4 256))$

113. Show that $b^{x_1} = b^{x_2}$ is *not* equivalent to $x_1 = x_2$ for $b = 0$ or $b = 1$.

114. If $\log_b a = x$, does it follow that $\log_a b = 1/x$? Why or why not?

12.4 Properties of Logarithmic Functions

Logarithms of Products • Logarithms of Powers •
Logarithms of Quotients • Using the Properties Together

Logarithmic functions are important in many applications and in more advanced mathematics. We now establish some basic properties that are useful in manipulating expressions involving logarithms. As their proofs reveal, the properties of logarithms are related to the properties of exponents.

Logarithms of Products

The first property we discuss is related to the product rule for exponents: $a^m \cdot a^n = a^{m+n}$. Its proof appears immediately after Example 2.

> **The Product Rule for Logarithms**
>
> For any positive numbers M, N, and a $(a \neq 1)$,
>
> $$\log_a (MN) = \log_a M + \log_a N.$$
>
> (The logarithm of a product is the sum of the logarithms of the factors.)

EXAMPLE 1 Express as an equivalent expression that is a sum of logarithms: $\log_2 (4 \cdot 16)$.

Solution We have

$$\log_2 (4 \cdot 16) = \log_2 4 + \log_2 16. \qquad \text{Using the product rule for logarithms}$$

As a check, note that

$$\log_2 (4 \cdot 16) = \log_2 64 = 6 \qquad 2^6 = 64$$

and that

$$\log_2 4 + \log_2 16 = 2 + 4 = 6. \qquad 2^2 = 4 \text{ and } 2^4 = 16$$

EXAMPLE 2 Express as an equivalent expression that is a single logarithm: $\log_b 7 + \log_b 5$.

Solution We have

$$\log_b 7 + \log_b 5 = \log_b (7 \cdot 5) \qquad \text{Using the product rule for logarithms}$$

$$= \log_b 35.$$

A Proof of the Product Rule. Let $\log_a M = x$ and $\log_a N = y$. Converting to exponential equations, we have $a^x = M$ and $a^y = N$.

Now we multiply the left side of the first exponential equation by the left side of the second equation and similarly multiply the right sides to obtain

$$MN = a^x \cdot a^y, \quad \text{or} \quad MN = a^{x+y}.$$

Converting back to a logarithmic equation, we get

$$\log_a (MN) = x + y.$$

Recalling what x and y represent, we have

$$\log_a (MN) = \log_a M + \log_a N.$$

Logarithms of Powers

The second basic property is related to the power rule for exponents: $(a^m)^n = a^{mn}$. Its proof follows Example 3.

The Power Rule for Logarithms

For any positive numbers M and a ($a \neq 1$), and any real number p,

$$\log_a M^p = p \cdot \log_a M.$$

(The logarithm of a power of M is the exponent times the logarithm of M.)

To better understand the power rule, note that

$$\log_a M^3 = \log_a (M \cdot M \cdot M) = \log_a M + \log_a M + \log_a M = 3 \log_a M.$$

EXAMPLE 3   Use the power rule for logarithms to write an equivalent expression that is a product: **(a)** $\log_a 9^{-5}$; **(b)** $\log_7 \sqrt[3]{x}$.

Solution

a) $\log_a 9^{-5} = -5 \log_a 9$ Using the power rule for logarithms

b) $\log_7 \sqrt[3]{x} = \log_7 x^{1/3}$ Writing exponential notation

$\qquad\qquad = \frac{1}{3} \log_7 x$ Using the power rule for logarithms

---

A Proof of the Power Rule. Let $x = \log_a M$. We then write the equivalent exponential equation, $a^x = M$. Raising both sides to the pth power, we get

$$(a^x)^p = M^p, \quad \text{or} \quad a^{xp} = M^p. \qquad \text{Multiplying exponents}$$

Converting back to a logarithmic equation gives us

$$\log_a M^p = xp.$$

But $x = \log_a M$, so substituting, we have

$$\log_a M^p = (\log_a M)p = p \cdot \log_a M.$$

Student Notes

Without understanding and *remembering* the rules of this section, it will be extremely difficult to solve the equations of Section 12.6.

Logarithms of Quotients

The third property that we study is similar to the quotient rule for exponents: $a^m/a^n = a^{m-n}$.

The Quotient Rule for Logarithms

For any positive numbers M, N, and a $(a \neq 1)$,

$$\log_a \frac{M}{N} = \log_a M - \log_a N.$$

(The logarithm of a quotient is the logarithm of the dividend minus the logarithm of the divisor.)

To better understand the quotient rule, note that

$$\log_a \left(\frac{b^5}{b^3} \right) = \log_a b^2 = 2 \log_a b = 5 \log_a b - 3 \log_a b$$

$$= \log_a b^5 - \log_a b^3.$$

EXAMPLE 4   Express as an equivalent expression that is a difference of logarithms: $\log_t (6/U)$.

Solution

$$\log_t \frac{6}{U} = \log_t 6 - \log_t U \qquad \text{Using the quotient rule for logarithms}$$

EXAMPLE 5   Express as an equivalent expression that is a single logarithm: $\log_b 17 - \log_b 27$.

Solution

$$\log_b 17 - \log_b 27 = \log_b \frac{17}{27} \qquad \begin{array}{l} \text{Using the quotient rule for} \\ \text{logarithms "in reverse"} \end{array}$$

A Proof of the Quotient Rule.   Our proof uses both the product and power rules:

$$\log_a \frac{M}{N} = \log_a MN^{-1} \qquad \text{Rewriting } \frac{M}{N} \text{ as } MN^{-1}$$

$$= \log_a M + \log_a N^{-1} \qquad \begin{array}{l} \text{Using the product rule for} \\ \text{logarithms} \end{array}$$

$$= \log_a M + (-1)\log_a N \qquad \begin{array}{l} \text{Using the power rule for} \\ \text{logarithms} \end{array}$$

$$= \log_a M - \log_a N.$$

Using the Properties Together

EXAMPLE 6   Express as an equivalent expression, using the individual logarithms of x, y, and z.

a) $\log_b \dfrac{x^3}{yz}$

b) $\log_a \sqrt[4]{\dfrac{xy}{z^3}}$

Solution

a) $\log_b \dfrac{x^3}{yz} = \log_b x^3 - \log_b yz$ — Using the quotient rule for logarithms

$= 3 \log_b x - \log_b yz$ — Using the power rule for logarithms

$= 3 \log_b x - (\log_b y + \log_b z)$ — Using the product rule for logarithms. Because of the subtraction, parentheses are essential.

$= 3 \log_b x - \log_b y - \log_b z$ — Using the distributive law

b) $\log_a \sqrt[4]{\dfrac{xy}{z^3}} = \log_a \left(\dfrac{xy}{z^3}\right)^{1/4}$ — Writing exponential notation

$= \dfrac{1}{4} \cdot \log_a \dfrac{xy}{z^3}$ — Using the power rule for logarithms

$= \dfrac{1}{4} \left(\log_a xy - \log_a z^3\right)$ — Using the quotient rule for logarithms. Parentheses are important.

$= \dfrac{1}{4} \left(\log_a x + \log_a y - 3 \log_a z\right)$ — Using the product and power rules for logarithms

---

Caution!   Because the product and quotient rules replace one term with two, it is often best to use the rules within parentheses, as in Example 6.

EXAMPLE 7   Express as an equivalent expression that is a single logarithm.

a) $\dfrac{1}{2} \log_a x - 7 \log_a y + \log_a z$

b) $\log_a \dfrac{b}{\sqrt{x}} + \log_a \sqrt{bx}$

Solution

a) $\dfrac{1}{2} \log_a x - 7 \log_a y + \log_a z$

$= \log_a x^{1/2} - \log_a y^7 + \log_a z$ — Using the power rule for logarithms

$= \left(\log_a \sqrt{x} - \log_a y^7\right) + \log_a z$ — Using parentheses to emphasize the order of operations; $x^{1/2} = \sqrt{x}$

$= \log_a \dfrac{\sqrt{x}}{y^7} + \log_a z$ — Using the quotient rule for logarithms

$= \log_a \dfrac{z\sqrt{x}}{y^7}$ — Using the product rule for logarithms

b) $\log_a \dfrac{b}{\sqrt{x}} + \log_a \sqrt{bx} = \log_a \dfrac{b \cdot \sqrt{bx}}{\sqrt{x}}$ Using the product rule for logarithms

$$= \log_a b\sqrt{b}$$ Removing a factor equal to 1: $\dfrac{\sqrt{x}}{\sqrt{x}} = 1$

$$= \log_a b^{3/2}, \text{ or } \dfrac{3}{2}\log_a b$$ Since $b\sqrt{b} = b^1 \cdot b^{1/2}$

If we know the logarithms of two different numbers (to the same base), the properties allow us to calculate other logarithms.

EXAMPLE 8   Given $\log_a 2 = 0.431$ and $\log_a 3 = 0.683$, calculate a numerical value for each of the following.

a) $\log_a 6$ **b)** $\log_a \frac{2}{3}$ **c)** $\log_a 81$

d) $\log_a \frac{1}{3}$ **e)** $\log_a 2a$ **f)** $\log_a 5$

Solution

a) $\log_a 6 = \log_a(2 \cdot 3) = \log_a 2 + \log_a 3$ Using the product rule for logarithms

$$= 0.431 + 0.683 = 1.114$$

     Check: $a^{1.114} = a^{0.431} \cdot a^{0.683} = 2 \cdot 3 = 6$

b) $\log_a \frac{2}{3} = \log_a 2 - \log_a 3$ Using the quotient rule for logarithms

$$= 0.431 - 0.683 = -0.252$$

c) $\log_a 81 = \log_a 3^4 = 4\log_a 3$ Using the power rule for logarithms

$$= 4(0.683) = 2.732$$

d) $\log_a \frac{1}{3} = \log_a 1 - \log_a 3$ Using the quotient rule for logarithms

$$= 0 - 0.683 = -0.683$$

e) $\log_a 2a = \log_a 2 + \log_a a$ Using the product rule for logarithms

$$= 0.431 + 1 = 1.431$$

f) $\log_a 5$ *cannot be found using these properties.* $(\log_a 5 \neq \log_a 2 + \log_a 3)$

A final property follows from the product rule: Since $\log_a a^k = k\log_a a$, and $\log_a a = 1$, we have $\log_a a^k = k$.

> **The Logarithm of the Base to an Exponent**
>
> For any base a,
>
> $$\log_a a^k = k.$$
>
> (The logarithm, base a, of a to an exponent is the exponent.

This property also follows from the definition of logarithm: k is the exponent to which you raise a in order to get a^k.

EXAMPLE 9   Simplify: **(a)** $\log_3 3^7$; **(b)** $\log_{10} 10^{-5.2}$.

Solution

a) $\log_3 3^7 = 7$ 7 is the exponent to which you raise 3 in order to get 3^7.

b) $\log_{10} 10^{-5.2} = -5.2$

We summarize the properties of logarithms as follows.

For any positive numbers M, N, and a $(a \neq 1)$:

$$\log_a MN = \log_a M + \log_a N; \qquad \log_a M^p = p \cdot \log_a M;$$

$$\log_a \frac{M}{N} = \log_a M - \log_a N; \qquad \log_a a^k = k.$$

Caution!   Keep in mind that, in general,

$$\log_a (M + N) \neq \log_a M + \log_a N, \qquad \log_a MN \neq (\log_a M)(\log_a N),$$

$$\log_a (M - N) \neq \log_a M - \log_a N, \qquad \log_a \frac{M}{N} \neq \frac{\log_a M}{\log_a N}.$$

Exercise Set

12.4

FOR EXTRA HELP

 Student's Solutions Manual    Digital Video Tutor CD 6 Videotape 12   AW Math Tutor Center    MathXL Tutorials on CD   Math XL MathXL   MyMathLab MyMathLab

↪ *Concept Reinforcement   In each of Exercises 1–6, match the expression with an equivalent expression from the column on the right.*

1. ____ $\log_7 20$

2. ____ $\log_7 5^4$

3. ____ $\log_7 \frac{5}{4}$

4. ____ $\log_7 7$

5. ____ $\log_7 1$

6. ____ $\log_7 5 + \log_7 6$

a) $\log_7 5 - \log_7 4$

b) 1

c) 0

d) $\log_7 30$

e) $\log_7 5 + \log_7 4$

f) $4 \log_7 5$

Express as an equivalent expression that is a sum of logarithms.

7. $\log_3 (81 \cdot 27)$

8. $\log_2 (16 \cdot 32)$

9. $\log_4 (64 \cdot 16)$

10. $\log_5 (25 \cdot 125)$

11. $\log_c (rst)$

12. $\log_t (3ab)$

Express as an equivalent expression that is a single logarithm.

13. $\log_a 5 + \log_a 14$

14. $\log_b 65 + \log_b 2$

15. $\log_c t + \log_c y$

16. $\log_t H + \log_t M$

Express as an equivalent expression that is a product.

17. $\log_a r^8$

18. $\log_b t^5$

19. $\log_c y^6$

20. $\log_{10} y^7$

21. $\log_b C^{-3}$

22. $\log_c M^{-5}$

Express as an equivalent expression that is a difference of two logarithms.

23. $\log_2 \frac{25}{13}$

24. $\log_3 \frac{23}{9}$

25. $\log_b \frac{m}{n}$

26. $\log_a \frac{y}{x}$

Express as an equivalent expression that is a single logarithm.

27. $\log_a 17 - \log_a 6$

28. $\log_b 32 - \log_b 7$

29. $\log_b 36 - \log_b 4$

30. $\log_a 26 - \log_a 2$

31. $\log_a 7 - \log_a 18$

32. $\log_b 5 - \log_b 13$

Express as an equivalent expression, using the individual logarithms of w, x, y, and z.

33. $\log_a (xyz)$

34. $\log_a (wxy)$

35. $\log_a (x^3 z^4)$

36. $\log_a (x^2 y^5)$

37. $\log_a (x^2 y^{-2} z)$

38. $\log_a (xy^2 z^{-3})$

39. $\log_a \frac{x^4}{y^3 z}$

40. $\log_a \frac{x^4}{yz^2}$

41. $\log_b \frac{xy^2}{wz^3}$

42. $\log_b \frac{w^2 x}{y^3 z}$

43. $\log_a \sqrt{\frac{x^7}{y^5 z^8}}$

44. $\log_c \sqrt[3]{\frac{x^4}{y^3 z^2}}$

45. $\log_a \sqrt[3]{\frac{x^6 y^3}{a^2 z^7}}$

46. $\log_a \sqrt[4]{\frac{x^8 y^{12}}{a^3 z^5}}$

Express as an equivalent expression that is a single logarithm and, if possible, simplify.

47. $8 \log_a x + 3 \log_a z$

48. $2 \log_b m + \frac{1}{2} \log_b n$

49. $\log_a x^2 - 2 \log_a \sqrt{x}$

50. $\log_a \frac{a}{\sqrt{x}} - \log_a \sqrt{ax}$

51. $\frac{1}{2} \log_a x + 5 \log_a y - 2 \log_a x$

52. $\log_a 2x + 3(\log_a x - \log_a y)$

53. $\log_a (x^2 - 4) - \log_a (x + 2)$

54. $\log_a (2x + 10) - \log_a (x^2 - 25)$

Given $\log_b 3 = 0.792$ and $\log_b 5 = 1.161$. If possible, calculate numerical values for each of the following.

55. $\log_b 15$

56. $\log_b \frac{5}{3}$

57. $\log_b \frac{3}{5}$

58. $\log_b \frac{1}{3}$

59. $\log_b \frac{1}{5}$

60. $\log_b \sqrt{b}$

61. $\log_b \sqrt{b^3}$

62. $\log_b 3b$

63. $\log_b 8$

64. $\log_b 45$

Simplify.

Aha! **65.** $\log_t t^7$

66. $\log_p p^4$

67. $\log_e e^m$

68. $\log_Q Q^{-2}$

69. A student *incorrectly* reasons that

$$\log_b \frac{1}{x} = \log_b \frac{x}{xx}$$

$$= \log_b x - \log_b x + \log_b x = \log_b x.$$

What mistake has the student made?

70. How could you convince someone that

$$\log_a c \neq \log_c a?$$

SKILL MAINTENANCE

Graph. [10.1]

71. $f(x) = \sqrt{x} - 3$

72. $g(x) = \sqrt{x} + 2$

73. $g(x) = \sqrt[3]{x} + 1$

74. $f(x) = \sqrt[3]{x} - 1$

Simplify. [4.1]

75. $(a^3 b^2)^5 (a^2 b^7)$

76. $(x^5 y^3 z^2)(x^2 yz^2)^3$

SYNTHESIS

77. Is it possible to express $\log_b \frac{x}{5}$ as an equivalent expression that is a difference of two logarithms without using the quotient rule? Why or why not?

78. Is it true that $\log_a x + \log_b x = \log_{ab} x$? Why or why not?

Express as an equivalent expression that is a single logarithm and, if possible, simplify.

79. $\log_a (x^8 - y^8) - \log_a (x^2 + y^2)$

80. $\log_a (x + y) + \log_a (x^2 - xy + y^2)$

Express as an equivalent expression that is a sum or difference of logarithms and, if possible, simplify.

81. $\log_a \sqrt{1 - s^2}$

82. $\log_a \dfrac{c - d}{\sqrt{c^2 - d^2}}$

83. If $\log_a x = 2$, $\log_a y = 3$, and $\log_a z = 4$, what is $\log_a \dfrac{\sqrt[3]{x^2 z}}{\sqrt[3]{y^2 z^{-2}}}$?

84. If $\log_a x = 2$, what is $\log_a (1/x)$?

85. If $\log_a x = 2$, what is $\log_{1/a} x$?

Classify each of the following as true or false. Assume a, x, P, and $Q > 0$, $a \neq 1$.

86. $\log_a \left(\dfrac{P}{Q} \right)^x = x \log_a P - \log_a Q$

87. $\log_a (Q + Q^2) = \log_a Q + \log_a (Q + 1)$

 **88.** Use graphs to show that
$$\log x^2 \neq \log x \cdot \log x.$$
(*Note*: log means $\log_{10}$.)

12.5 Common and Natural Logarithms

> Common Logarithms on a Calculator • The Base e and Natural Logarithms on a Calculator • Changing Logarithmic Bases • Graphs of Exponential and Logarithmic Functions, Base e

Any positive number other than 1 can serve as the base of a logarithmic function. However, some numbers are easier to use than others, and there are logarithmic bases that fit into certain applications more naturally than others.

Base-10 logarithms, called **common logarithms**, are useful because they have the same base as our "commonly" used decimal system. Before calculators became widely available, common logarithms helped with tedious calculations. In fact, that is why logarithms were devised.

The logarithmic base most widely used today is an irrational number named e. We will consider e and base e, or *natural*, logarithms later in this section. First we examine common logarithms.

Common Logarithms on a Calculator

Before the advent of scientific calculators, tables were developed to list common logarithms. Today we find common logarithms using calculators.

Here, and in most books, the abbreviation **log**, with no base written, is understood to mean logarithm base 10, or a common logarithm. Thus,

$\log 17$ means $\log_{10} 17$.   It is important to remember this abbreviation.

On most calculators, the key for common logarithms is marked **LOG**. To find the common logarithm of a number, we key in that number and press **LOG**. On most graphing calculators, we press **LOG**, the number, and then **ENTER**.

EXAMPLE 1   Use a scientific calculator to approximate each number to four decimal places.

a) $\log 53{,}128$

b) $\dfrac{\log 6500}{\log 0.007}$

Solution

a) We enter 53,128 and then press **LOG**. We find that

$$\log 53{,}128 \approx 4.7253. \qquad \text{Rounded to four decimal places}$$

b) We enter 6500 and then press **LOG**. Next, we press ÷, enter 0.007, and then press **LOG** =. Be careful not to round until the end:

$$\frac{\log 6500}{\log 0.007} \approx -1.7694. \qquad \text{Rounded to four decimal places}$$

---

technology
connection

To find log 6500/log 0.007 on a graphing calculator, we must use parentheses with care.

1. What keystrokes are needed to create the following?

log(7)/log(3)
1.771243749

The inverse of a logarithmic function is an exponential function. Because of this, on many calculators the **LOG** key doubles as the 10^x key after a **2ND** or SHIFT key is pressed. Calculators lacking a 10^x key may have a key labeled x^y, a^x, or ⌃. Such a key can raise any positive real number to any real-numbered exponent.

EXAMPLE 2   Use a calculator to approximate $10^{3.417}$ to four decimal places.

Solution   We enter 3.417 and then press 10^x. On most graphing calculators, 10^x is pressed first, followed by 3.417 and **ENTER**. Rounding to four decimal places, we have

$$10^{3.417} \approx 2612.1614.$$

The Base *e* and Natural Logarithms on a Calculator

When interest is compounded *n* times a year, the compound interest formula is

$$A = P\left(1 + \frac{r}{n}\right)^{nt},$$

where *A* is the amount that an initial investment *P* will be worth after *t* years at interest rate *r*. Suppose that \$1 is invested at 100% interest for 1 year (no bank would pay this). The preceding formula becomes a function *A* defined in terms of the number of compounding periods *n*:

$$A(n) = \left(1 + \frac{1}{n}\right)^n.$$

Let's find some function values. We round to six decimal places, using a calculator.

technology connection

To visualize the number e, let $y_1 = (1 + 1/x)^x$.

$y_1 = (1 + 1/x)^x$

1. Use $\boxed{\text{TRACE}}$ or $\boxed{\text{TABLE}}$ to confirm that as x gets larger, the number e is more closely approximated.
2. Graph $y_2 = e$ and compare y_1 and y_2 for large values of x.
3. Confirm that 0 is not in the domain of this function. Why?

n	$A(n) = \left(1 + \dfrac{1}{n}\right)^n$
1 (compounded annually)	\$2.00
2 (compounded semiannually)	\$2.25
3	\$2.370370
4 (compounded quarterly)	\$2.441406
5	\$2.488320
100	\$2.704814
365 (compounded daily)	\$2.714567
8760 (compounded hourly)	\$2.718127

The numbers in this table approach a very important number in mathematics, called e. Because e is irrational, its decimal representation does not terminate or repeat.

The Number e

$e \approx 2.7182818284\ldots$

Logarithms base e are called **natural logarithms**, or **Napierian logarithms**, in honor of John Napier (1550–1617), who first "discovered" logarithms.

The abbreviation "ln" is generally used with natural logarithms. Thus,

$\ln 53$ means $\log_e 53$.   It is important to remember this abbreviation.

On most scientific calculators, to find the natural logarithm of a number, we enter that number and press $\boxed{\text{LN}}$. On most graphing calculators, we press $\boxed{\text{LN}}$, the number, and then $\boxed{\text{ENTER}}$.

EXAMPLE 3   Use a scientific calculator to approximate $\ln 4568$ to four decimal places.

Solution   We enter 4568 and then press $\boxed{\text{LN}}$. We find that

$\ln 4568 \approx 8.4268$.   Rounded to four decimal places

On many calculators, the $\boxed{\text{LN}}$ key doubles as the $\boxed{e^x}$ key after a $\boxed{\text{2ND}}$ or $\boxed{\text{SHIFT}}$ key has been pressed.

EXAMPLE 4   Use a calculator to approximate $e^{-1.524}$ to four decimal places.

Solution   We enter -1.524 and then press $\boxed{e^x}$. On most graphing calculators, $\boxed{e^x}$ is pressed first, followed by -1.524 and $\boxed{\text{ENTER}}$. Since $e^{-1.524}$ is irrational, our answer is approximate:

$e^{-1.524} \approx 0.2178$.   Rounded to four decimal places

Changing Logarithmic Bases

Most calculators can find both common logarithms and natural logarithms. To find a logarithm with some other base, a conversion formula is usually needed.

The Change-of-Base Formula

For any logarithmic bases a and b, and any positive number M,

$$\log_b M = \frac{\log_a M}{\log_a b}.$$

(To find the log, base b, of M, we typically compute $\log M/\log b$ or $\ln M/\ln b$.)

Proof. Let $x = \log_b M$. Then,

$b^x = M$	$\log_b M = x$ is equivalent to $b^x = M$.
$\log_a b^x = \log_a M$	Taking the logarithm, base a, on both sides
$x \log_a b = \log_a M$	Using the power rule for logarithms
$x = \dfrac{\log_a M}{\log_a b}.$	Dividing both sides by $\log_a b$

But at the outset we stated that $x = \log_b M$. Thus, by substitution, we have

$$\log_b M = \frac{\log_a M}{\log_a b}. \qquad \text{This is the change-of-base formula.}$$

EXAMPLE 5   Find $\log_5 8$ using the change-of-base formula.

Solution   We use the change-of-base formula with $a = 10$, $b = 5$, and $M = 8$:

$$\log_5 8 = \frac{\log_{10} 8}{\log_{10} 5} \qquad \text{Substituting into } \log_b M = \frac{\log_a M}{\log_a b}$$

$$\approx \frac{0.903089987}{0.6989700043} \qquad \text{Using } \boxed{\text{LOG}} \text{ twice}$$

$$\approx 1.2920. \qquad \text{When using a calculator, it is best not to round before dividing.}$$

To check, note that $\ln 8/\ln 5 \approx 1.2920$. We can also use a calculator to verify that $5^{1.2920} \approx 8$.

EXAMPLE 6   Find $\log_4 31$.

Solution   As shown in the check of Example 5, base e can also be used.

$$\log_4 31 = \frac{\log_e 31}{\log_e 4}$$ Substituting into $\log_b M = \dfrac{\log_a M}{\log_a b}$

$$= \frac{\ln 31}{\ln 4} \approx \frac{3.433987204}{1.386294361}$$ Using **LN** twice

$$\approx 2.4771.$$ *Check:* $4^{2.4771} \approx 31$

Graphs of Exponential and Logarithmic Functions, Base e

EXAMPLE 7   Graph $f(x) = e^x$ and $g(x) = e^{-x}$ and state the domain and the range of f and g.

Solution   We use a calculator with an $\boxed{e^x}$ key to find approximate values of e^x and e^{-x}. Using these values, we can graph the functions.

x	e^x	e^{-x}
0	1	1
1	2.7	0.4
2	7.4	0.1
−1	0.4	2.7
−2	0.1	7.4

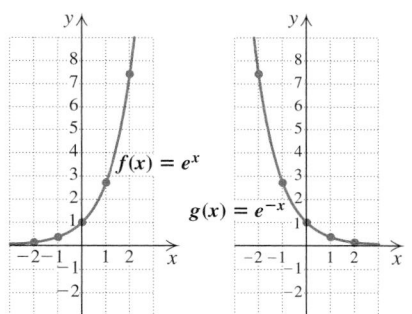

The domain of each function is $\mathbb{R}$ and the range of each function is $(0, \infty)$.

EXAMPLE 8   Graph $f(x) = e^{-x} + 2$ and state the domain and the range of f.

Solution   We find some solutions with a calculator, plot them, and then draw the graph. For example, $f(2) = e^{-2} + 2 \approx 0.1 + 2 \approx 2.1$. The graph is exactly like the graph of $g(x) = e^{-x}$, but is translated up 2 units.

x	$e^{-x} + 2$
0	3
1	2.4
2	2.1
−1	4.7
−2	9.4

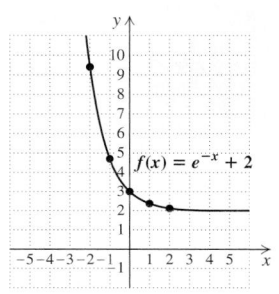

The domain of f is $\mathbb{R}$ and the range is $(2, \infty)$.

EXAMPLE 9   Graph and state the domain and the range of each function.

a) $g(x) = \ln x$ **b)** $f(x) = \ln (x + 3)$

Solution

a) We find some solutions with a calculator and then draw the graph. As expected, the graph is a reflection across the line $y = x$ of the graph of $y = e^x$.

x	$\ln x$
1	0
4	1.4
7	1.9
0.5	-0.7

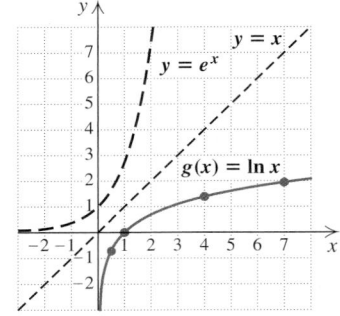

The domain of g is $(0, \infty)$ and the range is $\mathbb{R}$.

b) We find some solutions with a calculator, plot them, and draw the graph.

x	$\ln (x + 3)$
0	1.1
1	1.4
2	1.6
3	1.8
4	1.9
-1	0.7
-2	0
-2.5	-0.7

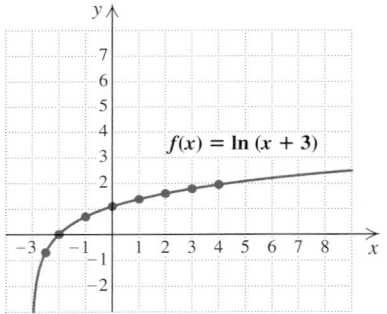

The graph of $y = \ln (x + 3)$ is the graph of $y = \ln x$ translated 3 units to the left. Since $x + 3$ must be positive, the domain is $(-3, \infty)$ and the range is $\mathbb{R}$.

technology connection

Logarithmic functions with bases other than 10 or e can be easily drawn using the change-of-base formula. For example, $y = \log_5 x$ can be written $y = \ln x / \ln 5$.

1. Graph $y = \log_7 x$.
2. Graph $y = \log_5(x + 2)$.
3. Graph $y = \log_7 x + 2$.

Exercise Set

12.5

 Concept Reinforcement   Classify each of the following as either true or false.

1. The expression $\log 23$ means $\log_{10} 23$.

2. The expression $\ln 7$ means $\log_e 7$.

3. The number e is approximately 2.7.

4. The expressions $\log 9$ and $\log 18 / \log 2$ are equivalent.

5. The expressions $\log_2 9$ and $\ln 9 / \ln 2$ are equivalent.

6. The domain of the function given by
$f(x) = \ln (x + 2)$ is $(-2, \infty)$.

7. The range of the function given by $g(x) = e^x$ is
$(0, \infty)$.

8. The range of the function given by $f(x) = \ln x$ is
$(-\infty, \infty)$.

*Use a calculator to find each of the following to the
nearest ten-thousandth.*

9. $\log 6$

10. $\log 5$

11. $\log 72.8$

12. $\log 73.9$

Aha! **13.** $\log 1000$

14. $\log 100$

15. $\log 0.527$

16. $\log 0.493$

17. $\dfrac{\log 8200}{\log 150}$

18. $\dfrac{\log 5700}{\log 90}$

19. $10^{2.3}$

20. $10^{3.4}$

21. $10^{0.173}$

22. $10^{0.247}$

23. $10^{-2.9523}$

24. $10^{-3.2046}$

25. $\ln 5$

26. $\ln 2$

27. $\ln 57$

28. $\ln 30$

29. $\ln 0.0062$

30. $\ln 0.00073$

31. $\dfrac{\ln 2300}{0.08}$

32. $\dfrac{\ln 1900}{0.07}$

33. $e^{2.71}$

34. $e^{3.06}$

35. $e^{-3.49}$

36. $e^{-2.64}$

37. $e^{4.7}$

38. $e^{1.23}$

*Find each of the following logarithms using the
change-of-base formula. Round answers to the nearest
ten-thousandth.*

39. $\log_6 92$

40. $\log_3 78$

41. $\log_2 100$

42. $\log_7 100$

43. $\log_7 65$

44. $\log_5 42$

45. $\log_{0.5} 5$

46. $\log_{0.1} 3$

47. $\log_2 0.2$

48. $\log_2 0.08$

49. $\log_\pi 58$

50. $\log_\pi 200$

*Graph and state the domain and the range of each
function.*

51. $f(x) = e^x$

52. $f(x) = e^{-x}$

53. $f(x) = e^x + 3$

54. $f(x) = e^x + 2$

55. $f(x) = e^x - 2$

56. $f(x) = e^x - 3$

57. $f(x) = 0.5e^x$

58. $f(x) = 2e^x$

59. $f(x) = 0.5e^{2x}$

60. $f(x) = 2e^{-0.5x}$

61. $f(x) = e^{x-3}$

62. $f(x) = e^{x-2}$

63. $f(x) = e^{x+2}$

64. $f(x) = e^{x+3}$

65. $f(x) = -e^x$

66. $f(x) = -e^{-x}$

67. $g(x) = \ln x + 1$

68. $g(x) = \ln x + 3$

69. $g(x) = \ln x - 2$

70. $g(x) = \ln x - 1$

71. $g(x) = 2 \ln x$

72. $g(x) = 3 \ln x$

73. $g(x) = -2 \ln x$

74. $g(x) = -\ln x$

75. $g(x) = \ln (x + 2)$

76. $g(x) = \ln (x + 1)$

77. $g(x) = \ln (x - 1)$

78. $g(x) = \ln (x - 3)$

79. Using a calculator, Zeno *incorrectly* says that
$\log 79$ is between 4 and 5. How could you
convince him, without using a calculator, that
he is mistaken?

80. Examine Exercise 79. What mistake do you believe
Zeno made?

SKILL MAINTENANCE

Solve.

81. $4x^2 - 25 = 0$ [5.7]

82. $5x^2 - 7x = 0$ [5.7]

83. $17x - 15 = 0$ [2.2]

84. $9 - 13x = 0$ [2.2]

85. $x^{1/2} - 6x^{1/4} + 8 = 0$ [11.5]

86. $2y - 7\sqrt{y} + 3 = 0$ [11.5]

SYNTHESIS

87. Explain how the graph of $f(x) = e^x$ could be used
to graph the function given by $g(x) = 1 + \ln x$.

88. Explain how the graph of $f(x) = \ln x$ could be
used to graph the function given by $g(x) = e^{x-1}$.

*Knowing only that $\log 2 \approx 0.301$ and $\log 3 \approx 0.477$,
find each of the following.*

89. $\log_6 81$

90. $\log_9 16$

91. $\log_{12} 36$

92. Find a formula for converting common
logarithms to natural logarithms.

93. Find a formula for converting natural logarithms
to common logarithms.

Solve for x.

94. $\log (275x^2) = 38$

95. $\log (492x) = 5.728$

96. $\dfrac{3.01}{\ln x} = \dfrac{28}{4.31}$

97. $\log 692 + \log x = \log 3450$

For each function given below, (a) determine the domain and the range, (b) set an appropriate window, and (c) draw the graph. Graphs may vary, depending on the scale used.

98. $f(x) = 7.4e^x \ln x$

99. $f(x) = 3.4 \ln x - 0.25e^x$

100. $f(x) = x \ln (x - 2.1)$

101. $f(x) = 2x^3 \ln x$

 102. Use a graphing calculator to check your answers to Exercises 53, 61, and 75.

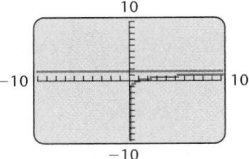 **103.** Use a graphing calculator to check your answers to Exercises 52, 58, and 68.

104. In an attempt to solve $\ln x = 1.5$, Emma gets the following graph. How can Emma tell at a glance that she has made a mistake?

12.6 | Solving Exponential and Logarithmic Equations

Solving Exponential Equations •
Solving Logarithmic Equations

Solving Exponential Equations

Equations with variables in exponents, such as $5^x = 12$ and $2^{7x} = 64$, are called **exponential equations**. In Section 12.3, we solved certain exponential equations by using the principle of exponential equality. We restate that principle below.

> **The Principle of Exponential Equality**
>
> For any real number b, where $b \neq -1, 0,$ or 1,
>
> $$b^x = b^y \quad \text{is equivalent to} \quad x = y.$$
>
> (Powers of the same base are equal if and only if the exponents are equal.)

EXAMPLE 1   Solve: $4^x = 16$.

Solution   Note that $16 = 4^2$. Thus we can write each side as a power of the same base:

$$4^x = 4^2.$$

Since the base is the same, 4, the exponents must be equal. Thus, x must be 2. The solution is 2.

Example 1 is not difficult to solve, but what if the right side of the equation were not a power of 4?

When it seems impossible to write both sides of an equation as powers of the same base, we use the following principle and write an equivalent logarithmic equation.

The Principle of Logarithmic Equality

For any logarithmic base a, and for $x, y > 0$,

$$x = y \quad \text{is equivalent to} \quad \log_a x = \log_a y.$$

(Two expressions are equal if and only if the logarithms of those expressions are equal.)

The principle of logarithmic equality, used together with the power rule for logarithms, allows us to solve equations in which the variable is an exponent.

EXAMPLE 2   Solve: $7^{x-2} = 60$.

Solution   We have

$$7^{x-2} = 60$$

$$\log 7^{x-2} = \log 60 \qquad \text{Using the principle of logarithmic equality to take the common logarithm on both sides. Natural logarithms also would work.}$$

$$(x - 2) \log 7 = \log 60 \qquad \text{Using the power rule for logarithms}$$

$$x - 2 = \frac{\log 60}{\log 7} \longleftarrow \boxed{\textit{Caution!} \quad \text{This is } not \log 60 - \log 7.}$$

$$x = \frac{\log 60}{\log 7} + 2 \qquad \text{Adding 2 to both sides}$$

$$x \approx 4.1041. \qquad \text{Using a calculator and rounding to four decimal places}$$

Since $7^{4.1041-2} \approx 60.0027$, we have a check. We can also note that since $7^{4-2} = 49$, we expect a solution near 4. The solution is $\dfrac{\log 60}{\log 7} + 2$, or approximately 4.1041.

ALGEBRAIC–GRAPHICAL CONNECTION

We can obtain a visual check of the solution of an exponential equation by graphing. For example, the solution of Example 1 can be visualized by graphing $y = 4^x$ and $y = 16$ on the same set of axes. The y-values for both equations are the same where the graphs intersect. The x-value at that point is the solution of the equation. That x-value appears to be 2. Since $4^2 = 16$, we see that this is indeed the case.

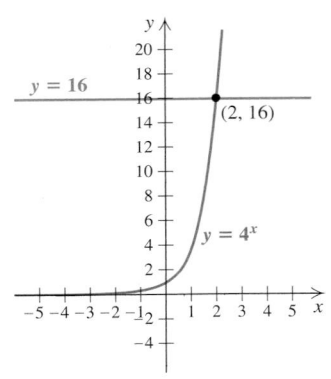

Similarly, the solution of Example 2 can be visualized by graphing $y = 7^{x-2}$ and $y = 60$ and identifying the x-value at the point of intersection. As expected, this value appears to be approximately 4.1.

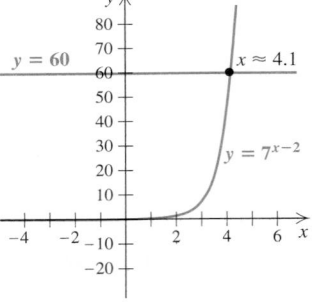

EXAMPLE 3   Solve: $e^{0.06t} = 1500$.

Solution   Since one side is a power of e, it is easiest to take the *natural logarithm* on both sides:

$$\ln e^{0.06t} = \ln 1500 \qquad \text{Taking the natural logarithm on both sides}$$

$$0.06t = \ln 1500 \qquad \text{Finding the logarithm of the base to a power: } \log_a a^k = k. \text{ Logarithmic and exponential functions are inverses of each other.}$$

$$t = \frac{\ln 1500}{0.06} \qquad \text{Dividing both sides by 0.06}$$

$$\approx 121.887. \qquad \text{Using a calculator and rounding to three decimal places}$$

To Solve an Equation of the Form $a^t = b$ for t

1. Take the logarithm (either natural or common) of both sides.
2. Use the power rule for exponents so that the variable is no longer written as an exponent.
3. Divide both sides by the coefficient of the variable to isolate the variable.
4. If appropriate, use a calculator to find an approximate solution in decimal form.

Solving Logarithmic Equations

Equations containing logarithmic expressions are called **logarithmic equations**. We saw in Section 12.3 that certain logarithmic equations can be solved by writing an equivalent exponential equation.

EXAMPLE 4   Solve: **(a)** $\log_4 (8x - 6) = 3$; **(b)** $\ln 5x = 27$.

Solution

a) $\log_4 (8x - 6) = 3$

$$4^3 = 8x - 6 \qquad \text{Writing the equivalent exponential equation}$$

$$64 = 8x - 6$$

$$70 = 8x \qquad \text{Adding 6 to both sides}$$

$$x = \tfrac{70}{8}, \text{ or } \tfrac{35}{4}.$$

Check:
$$\begin{array}{c|c} \log_4 (8x - 6) = 3 \\ \hline \log_4 \left(8 \cdot \tfrac{35}{4} - 6\right) & 3 \\ \log_4 (2 \cdot 35 - 6) & \\ \log_4 64 & \\ 3 \stackrel{?}{=} 3 & \text{TRUE} \end{array}$$

The solution is $\tfrac{35}{4}$.

b) $\ln 5x = 27$ Remember: $\ln 5x$ means $\log_e 5x$.

$$e^{27} = 5x \qquad \text{Writing the equivalent exponential equation}$$

$$\frac{e^{27}}{5} = x \qquad \text{This is a very large number.}$$

The solution is $\dfrac{e^{27}}{5}$. The check is left to the student.

Often the properties for logarithms are needed. The goal is to first write an equivalent equation in which the variable appears in just one logarithmic expression. We then isolate that expression and solve as in Example 4.

EXAMPLE 5   Solve.

a) $\log x + \log (x - 3) = 1$

b) $\log_2 (x + 7) - \log_2 (x - 7) = 3$

c) $\log_7 (x + 1) + \log_7 (x - 1) = \log_7 8$

Solution

a) To increase understanding, we write in the base, 10.

$$\log_{10} x + \log_{10} (x - 3) = 1$$

$$\log_{10} [x(x - 3)] = 1 \qquad \text{Using the product rule for logarithms to obtain a single logarithm}$$

$$x(x - 3) = 10^1 \qquad \text{Writing an equivalent exponential equation}$$

$$x^2 - 3x = 10$$

$$x^2 - 3x - 10 = 0$$

$$(x + 2)(x - 5) = 0 \qquad \text{Factoring}$$

$$x + 2 = 0 \quad or \quad x - 5 = 0 \qquad \text{Using the principle of zero products}$$

$$x = -2 \quad or \qquad x = 5$$

Student Notes

It is essential that you remember the properties of logarithms from Section 12.4. Consider reviewing the properties before attempting to solve equations similar to those in Example 5.

Check:

For −2:

$$\underline{\log x + \log (x - 3) = 1}$$

$$\log (-2) + \log (-2 - 3) \overset{?}{=} 1 \quad \text{FALSE}$$

For 5:

$$\log x + \log (x - 3) = 1$$

$$\log 5 + \log (5 - 3) \; \big| \; 1$$

$$\log 5 + \log 2 \; \big|$$

$$\log 10 \; \big|$$

$$1 \overset{?}{=} 1 \quad \text{TRUE}$$

The number −2 *does not check* because the logarithm of a negative number is undefined. The solution is 5.

b) We have

$$\log_2 (x + 7) - \log_2 (x - 7) = 3$$

$$\log_2 \frac{x + 7}{x - 7} = 3 \qquad \text{Using the quotient rule for logarithms to obtain a single logarithm}$$

$$\frac{x + 7}{x - 7} = 2^3 \qquad \text{Writing an equivalent exponential equation}$$

$$\frac{x + 7}{x - 7} = 8$$

$$x + 7 = 8(x - 7) \qquad \text{Multiplying by the LCD, } x - 7$$

$$x + 7 = 8x - 56 \qquad \text{Using the distributive law}$$

$$63 = 7x$$

$$9 = x. \qquad \text{Dividing by 7}$$

Check:

$$\frac{\log_2 (x + 7) - \log_2 (x - 7) = 3}{\log_2 (9 + 7) - \log_2 (9 - 7) \;\Big|\; 3}$$

$$\log_2 16 - \log_2 2$$

$$4 - 1$$

$$3 \overset{?}{=} 3 \quad \text{TRUE}$$

The solution is 9.

c) We have

$$\log_7 (x + 1) + \log_7 (x - 1) = \log_7 8$$

$$\log_7 [(x + 1)(x - 1)] = \log_7 8 \qquad \text{Using the product rule for logarithms}$$

$$\log_7 (x^2 - 1) = \log_7 8 \qquad \text{Multiplying. Note that both sides are base-7 logarithms.}$$

$$x^2 - 1 = 8 \qquad \text{Using the principle of logarithmic equality. Study this step carefully.}$$

$$x^2 - 9 = 0$$

$$(x - 3)(x + 3) = 0 \qquad \text{Solving the quadratic equation}$$

$$x = 3 \quad or \quad x = -3.$$

We leave it to the student to show that 3 checks but -3 does not. The solution is 3.

technology connection

To solve exponential and logarithmic equations, we can use INTERSECT to determine the x-coordinate at each intersection.

   For example, to solve $e^{0.5x} - 7 = 2x + 6$, we graph $y_1 = e^{0.5x} - 7$ and $y_2 = 2x + 6$ as shown. We then use the INTERSECT option of the (CALC) menu. The x-coordinates at the intersections are approximately -6.48 and 6.52.

   Use a graphing calculator to solve each equation to the nearest hundredth.

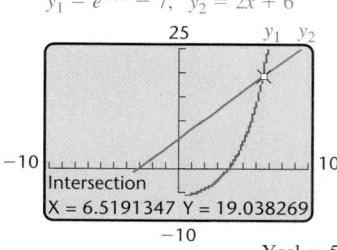

$y_1 = e^{0.5x} - 7, \quad y_2 = 2x + 6$

Intersection
X = 6.5191347   Y = 19.038269

Yscl = 5

1. $e^{7x} = 14$

2. $8e^{0.5x} = 3$

3. $xe^{3x-1} = 5$

4. $4 \ln (x + 3.4) = 2.5$

5. $\ln 3x = 0.5x - 1$

6. $\ln x^2 = -x^2$

Exercise Set

12.6

↪ *Concept Reinforcement In each of Exercises 1–8, match the equation with an equivalent equation from the column on the right that could be the next step in the solution process.*

1. _____ $5^x = 3$

2. _____ $e^{5x} = 3$

3. _____ $\ln x = 3$

4. _____ $\log_x 5 = 3$

5. _____ $\log_5 x + \log_5(x - 2) = 3$

6. _____ $\log_5 x - \log_5(x - 2) = 3$

7. _____ $\ln x - \ln(x - 2) = 3$

8. _____ $\log x + \log(x - 2) = 3$

a) $\ln e^{5x} = \ln 3$

b) $\log_5 (x^2 - 2x) = 3$

c) $\log (x^2 - 2x) = 3$

d) $\log_5 \dfrac{x}{x - 2} = 3$

e) $\log 5^x = \log 3$

f) $e^3 = x$

g) $\ln \dfrac{x}{x - 2} = 3$

h) $x^3 = 5$

Solve. Where appropriate, include approximations to the nearest thousandth.

9. $2^x = 19$

10. $2^x = 15$

11. $8^{x-1} = 17$

12. $4^{x+1} = 13$

13. $e^t = 1000$

14. $e^t = 100$

15. $e^{0.03t} + 2 = 7$

16. $e^{-0.07t} + 3 = 3.08$

17. $5 = 3^{x+1}$

18. $7 = 3^{x-1}$

Aha! **19.** $2^{x+3} = 16$

20. $4^{x+1} = 64$

21. $4.9^x - 87 = 0$

22. $7.2^x - 65 = 0$

23. $19 = 2e^{4x}$

24. $29 = 3e^{2x}$

25. $7 + 3e^{5x} = 13$

26. $4 + 5e^{4x} = 9$

27. $\log_3 x = 4$

Aha! **28.** $\log_2 x = 6$

29. $\log_2 x = -3$

30. $\log_5 x = 3$

31. $\ln x = 5$

32. $\ln x = 4$

33. $\log_8 x = \frac{1}{3}$

34. $\log_4 x = \frac{1}{2}$

35. $\ln 4x = 3$

36. $\ln 3x = 2$

37. $\log x = 2.5$

38. $\log x = 0.5$

39. $\ln (2x + 1) = 4$

40. $\ln (4x - 2) = 3$

Aha! **41.** $\ln x = 1$

42. $\log x = 1$

43. $5 \ln x = -15$

44. $3 \ln x = -3$

45. $\log_2 (8 - 6x) = 5$

46. $\log_5 (2x - 7) = 3$

47. $\log (x - 9) + \log x = 1$

48. $\log (x + 9) + \log x = 1$

49. $\log x - \log (x + 3) = 1$

50. $\log x - \log (x + 7) = -1$

51. $\log_4 (x + 3) = 2 + \log_4 (x - 5)$

52. $\log_2 (x + 3) = 4 + \log_2 (x - 3)$

53. $\log_7 (x + 1) + \log_7 (x + 2) = \log_7 6$

54. $\log_6 (x + 3) + \log_6 (x + 2) = \log_6 20$

55. $\log_5 (x + 4) + \log_5 (x - 4) = \log_5 20$

56. $\log_4 (x + 2) + \log_4 (x - 7) = \log_4 10$

57. $\ln (x + 5) + \ln (x + 1) = \ln 12$

58. $\ln (x - 6) + \ln (x + 3) = \ln 22$

59. $\log_2 (x - 3) + \log_2 (x + 3) = 4$

60. $\log_3 (x - 4) + \log_3 (x + 4) = 2$

61. $\log_{12} (x + 5) - \log_{12} (x - 4) = \log_{12} 3$

62. $\log_6 (x + 7) - \log_6 (x - 2) = \log_6 5$

63. $\log_2 (x - 2) + \log_2 x = 3$

64. $\log_4 (x + 6) - \log_4 x = 2$

65. Could Example 2 have been solved by taking the natural logarithm on both sides? Why or why not?

66. Christina finds that the solution of $\log_3 (x + 4) = 1$ is -1, but rejects -1 as an answer. What mistake is she making?

SKILL MAINTENANCE

67. Find an equation of variation if y varies directly as x, and $y = 7.2$ when $x = 0.8$. [7.5]

68. Find an equation of variation if y varies inversely as x, and $y = 3.5$ when $x = 6.1$. [7.5]

Solve. [11.3]

69. $T = 2\pi\sqrt{L/32}$, for L

70. $E = mc^2$, for c
(Assume E, m, $c > 0$.)

71. Joni can key in a musical score in 2 hr. Miles takes 3 hr to key in the same score. How long would it take them, working together, to key in the score? [6.7]

72. The side exit at the Flynn Theater can empty a capacity crowd in 25 min. The main exit can empty a capacity crowd in 15 min. How long will it take to empty a capacity crowd when both exits are in use? [6.7]

SYNTHESIS

73. Can the principle of logarithmic equality be expanded to include all functions? That is, is the statement "$m = n$ is equivalent to $f(m) = f(n)$" true for any function f? Why or why not?

74. Explain how Exercises 37 and 38 could be solved using the graph of $f(x) = \log x$.

Solve. If no solution exists, state this.

75. $27^x = 81^{2x-3}$

76. $8^x = 16^{3x+9}$

77. $\log_x (\log_3 27) = 3$

78. $\log_6 (\log_2 x) = 0$

79. $x \log \frac{1}{8} = \log 8$

80. $\log_5 \sqrt{x^2 - 9} = 1$

81. $2^{x^2+4x} = \frac{1}{8}$

82. $\log (\log x) = 5$

83. $\log_5 |x| = 4$

84. $\log x^2 = (\log x)^2$

85. $\log \sqrt{2x} = \sqrt{\log 2x}$

86. $1000^{2x+1} = 100^{3x}$

87. $3^{x^2} \cdot 3^{4x} = \frac{1}{27}$

88. $3^{3x} \cdot 3^{x^2} = 81$

89. $\log x^{\log x} = 25$

90. $3^{2x} - 8 \cdot 3^x + 15 = 0$

91. $(81^{x-2})(27^{x+1}) = 9^{2x-3}$

92. $3^{2x} - 3^{2x-1} = 18$

93. Given that $2^y = 16^{x-3}$ and $3^{y+2} = 27^x$, find the value of $x + y$.

94. If $x = (\log_{125} 5)^{\log_5 125}$, what is the value of $\log_3 x$?

95. Find the value of x for which the natural logarithm is the same as the common logarithm.

96. Use a graphing calculator to check your answers to Exercises 3, 29, 41, and 57.

12.7 Applications of Exponential and Logarithmic Functions

Applications of Logarithmic Functions   •
Applications of Exponential Functions

We now consider applications of exponential and logarithmic functions.

Applications of Logarithmic Functions

EXAMPLE 1   Sound levels.   To measure the volume, or "loudness," of a sound, the *decibel* scale is used. The loudness L, in decibels (dB), of a sound is given by

$$L = 10 \cdot \log \frac{I}{I_0},$$

where I is the intensity of the sound, in watts per square meter (W/m^2), and $I_0 = 10^{-12}\ W/m^2$. (I_0 is approximately the intensity of the softest sound that can be heard by the human ear.)

a) It is common for the intensity of sound at live performances of rock music to reach $10^{-1}\ W/m^2$ (even higher close to the stage). How loud, in decibels, is the sound level?

b) The Occupational Safety and Health Administration (OSHA) considers sound levels of 85 dB and above unsafe. What is the intensity of such sounds?

Sarah Fisher puts in her earplugs in the pits at the Indianapolis Motor Speedway on May 19, 2002. She is the youngest woman to qualify for the Indy 500.

Solution

a) To find the loudness, in decibels, we use the above formula:

$$L = 10 \cdot \log \frac{I}{I_0}$$

$$= 10 \cdot \log \frac{10^{-1}}{10^{-12}} \qquad \text{Substituting}$$

$$= 10 \cdot \log 10^{11} \qquad \text{Subtracting exponents}$$

$$= 10 \cdot 11 \qquad\qquad \log 10^a = a$$

$$= 110.$$

The volume of the music is 110 decibels.

b) We substitute and solve for I:

$$L = 10 \cdot \log \frac{I}{I_0}$$

$$85 = 10 \cdot \log \frac{I}{10^{-12}} \qquad \text{Substituting}$$

$$8.5 = \log \frac{I}{10^{-12}} \qquad \text{Dividing both sides by 10}$$

$$8.5 = \log I - \log 10^{-12} \qquad \text{Using the quotient rule for logarithms}$$

$$8.5 = \log I - (-12) \qquad \log 10^a = a$$

$$-3.5 = \log I \qquad \text{Adding } -12 \text{ to both sides}$$

$$10^{-3.5} = I. \qquad \text{Converting to an exponential equation}$$

Earplugs would be recommended for sounds with intensities exceeding $10^{-3.5} \text{ W/m}^2$.

EXAMPLE 2

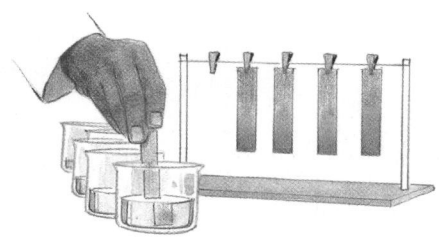

Chemistry: pH of liquids. In chemistry, the pH of a liquid is a measure of its acidity. We calculate pH as follows:

$$\text{pH} = -\log [\text{H}^+],$$

where $[\text{H}^+]$ is the hydrogen ion concentration in moles per liter.

a) The hydrogen ion concentration of human blood is normally about 3.98×10^{-8} moles per liter. Find the pH.

b) The pH of seawater is about 8.3. Find the hydrogen ion concentration.

Solution

a) To find the pH of blood, we use the above formula:

$$\text{pH} = -\log [\text{H}^+]$$

$$= -\log [3.98 \times 10^{-8}]$$

$$\approx -(-7.400117) \qquad \text{Using a calculator}$$

$$\approx 7.4.$$

The pH of human blood is normally about 7.4.

b) We substitute and solve for $[\text{H}^+]$:

$$8.3 = -\log [\text{H}^+] \qquad \text{Using pH} = -\log [\text{H}^+]$$

$$-8.3 = \log [\text{H}^+] \qquad \text{Dividing both sides by } -1$$

$$10^{-8.3} = [\text{H}^+] \qquad \text{Converting to an exponential equation}$$

$$5.01 \times 10^{-9} \approx [\text{H}^+]. \qquad \text{Using a calculator; writing scientific notation}$$

The hydrogen ion concentration of seawater is about 5.01×10^{-9} moles per liter.

Applications of Exponential Functions

EXAMPLE 3   Interest compounded annually.   Suppose that $25,000 is invested at 4% interest, compounded annually. In t years, it will grow to the amount A given by the function

$$A(t) = 25,000(1.04)^t.$$

(See Example 5 in Section 12.2.)

a) How long will it take to accumulate $80,000 in the account?

b) Find the amount of time it takes for the $25,000 to double itself.

Solution

a) We set $A(t) = 80,000$ and solve for t:

$$80,000 = 25,000(1.04)^t$$

$$\frac{80,000}{25,000} = 1.04^t \qquad \text{Dividing both sides by 25,000}$$

$$3.2 = 1.04^t$$

$$\log 3.2 = \log 1.04^t \qquad \text{Taking the common logarithm on both sides}$$

$$\log 3.2 = t \log 1.04 \qquad \text{Using the power rule for logarithms}$$

$$\frac{\log 3.2}{\log 1.04} = t \qquad \text{Dividing both sides by } \log 1.04$$

$$29.7 \approx t. \qquad \text{Using a calculator}$$

Remember that when doing a calculation like this on a calculator, it is best to wait until the end to round off. At an interest rate of 4% per year, it will take about 29.7 yr for $25,000 to grow to $80,000.

b) To find the *doubling time*, we replace $A(t)$ with 50,000 and solve for t:

$$50,000 = 25,000(1.04)^t$$

$$2 = (1.04)^t \qquad \text{Dividing both sides by 25,000}$$

$$\log 2 = \log (1.04)^t \qquad \text{Taking the common logarithm on both sides}$$

$$\log 2 = t \log 1.04 \qquad \text{Using the power rule for logarithms}$$

$$t = \frac{\log 2}{\log 1.04} \approx 17.7. \qquad \text{Dividing both sides by } \log 1.04 \text{ and using a calculator}$$

At an interest rate of 4% per year, the doubling time is about 17.7 yr.

*Student Notes*_____

Study the different steps in the solution of Example 3(b). Note that if 50,000 and 25,000 are replaced with 8000 and 4000, the doubling time is unchanged.

Like investments, populations often grow exponentially.

Exponential Growth

An **exponential growth model** is a function of the form

$$P(t) = P_0 e^{kt}, \quad k > 0,$$

where P_0 is the population at time 0, $P(t)$ is the population at time t, and k is the **exponential growth rate** for the situation. The **doubling time** is the amount of time necessary for the population to double in size.

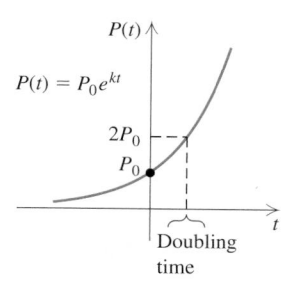

The exponential growth rate is the rate of growth of a population at any *instant* in time. Since the population is continually growing, the percent of total growth after one year will exceed the exponential growth rate.

EXAMPLE 4   Growth of zebra mussel populations.   Zebra mussels, inadvertently imported from Europe, began fouling North American waters in 1988. These mussels are so prolific that lake and river bottoms, as well as water intake pipes, can become blanketed with them, altering an entire ecosystem. In 2000, a portion of the Hudson River contained an average of 10 zebra mussels per square mile. The exponential growth rate was 340% per year.

a) Find the exponential growth function that models the data.

b) Predict the number of mussels per square mile in 2007.

Solution

a) In 2000, at $t = 0$, the population was $10/\text{mi}^2$. We substitute 10 for P_0 and 340%, or 3.4, for k. This gives the exponential growth function

$$P(t) = 10e^{3.4t}.$$

b) In 2007, we have $t = 7$ (since 7 yr have passed since 2000). To find the population in 2007, we compute $P(7)$:

$$P(7) = 10e^{3.4(7)} \qquad \text{Using } P(t) = 10e^{3.4t} \text{ from part (a)}$$
$$= 10e^{23.8}$$
$$\approx 217{,}000{,}000{,}000. \qquad \text{Using a calculator}$$

The population of zebra mussels in the specified portion of the Hudson River will reach approximately 217,000,000,000 per square mile in 2007.

EXAMPLE 5

Spread of a computer virus. The number of computers infected by a virus t hours after it first appears usually increases exponentially. In 2004, the "MyDoom" worm spread from 100 computers to about 100,000 computers in 24 hr. (*Source*: Based on data from IDG News Service)

a) Find the exponential growth rate and the exponential growth function.

b) Assuming exponential growth, estimate how long it took the MyDoom worm to infect 9000 computers.

Solution

a) We use $N(t) = N_0 e^{kt}$, where t is the number of hours since the first 100 computers were infected. Substituting 100 for N_0 gives

$$N(t) = 100e^{kt}.$$

To find the exponential growth rate, k, note that after 24 hr, 100,000 computers had been infected:

$$\left.\begin{array}{l} N(24) = 100e^{k \cdot 24} \\ 100{,}000 = 100e^{24k} \end{array}\right\} \quad \text{Substituting}$$

$$1000 = e^{24k} \qquad \text{Dividing both sides by 100}$$
$$\ln 1000 = \ln e^{24k} \qquad \text{Taking the natural logarithm on both sides}$$
$$\ln 1000 = 24k \qquad \ln e^a = a$$
$$\frac{\ln 1000}{24} = k \qquad \text{Dividing both sides by 24}$$
$$0.288 \approx k. \qquad \text{Using a calculator and rounding}$$

The exponential growth rate is 28.8% and the exponential growth function is given by $N(t) = 100e^{0.288t}$.

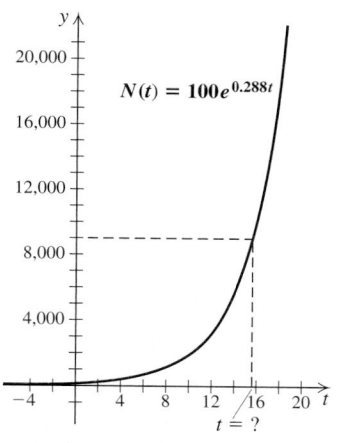

$N(t) = 100e^{0.288t}$

$t = ?$

A visualization of Example 5

b) To estimate how long it took for 9000 computers to be infected, we replace $N(t)$ with 9000 and solve for t:

$$9000 = 100e^{0.288t}$$

$$90 = e^{0.288t} \qquad \text{Dividing both sides by 100}$$

$$\ln 90 = \ln e^{0.288t} \qquad \text{Taking the natural logarithm on both sides}$$

$$\ln 90 = 0.288t \qquad \ln e^a = a$$

$$\frac{\ln 90}{0.288} = t \qquad \text{Dividing both sides by 0.288}$$

$$15.6 \approx t. \qquad \text{Using a calculator}$$

Rounding up to 16, we see that, according to this model, it took about 16 hr for 9000 computers to be infected.

EXAMPLE 6

Interest compounded continuously. When an amount of money P_0 is invested at interest rate k, compounded *continuously*, interest is computed every "instant" and added to the original amount. The balance $P(t)$, after t years, is given by the exponential growth model

$$P(t) = P_0 e^{kt}.$$

a) Suppose that \$30,000 is invested and grows to \$44,754.75 in 5 yr. Find the exponential growth function.

b) What is the doubling time?

Solution

a) We have $P(0) = 30,000$. Thus the exponential growth function is

$$P(t) = 30,000e^{kt}, \quad \text{where } k \text{ must still be determined.}$$

Knowing that for $t = 5$ we have $P(5) = 44,754.75$, it is possible to solve for k:

$$44,754.75 = 30,000e^{k(5)} = 30,000e^{5k}$$

$$\frac{44,754.75}{30,000} = e^{5k} \qquad \text{Dividing both sides by 30,000}$$

$$1.491825 = e^{5k}$$

$$\ln 1.491825 = \ln e^{5k} \qquad \text{Taking the natural logarithm on both sides}$$

$$\ln 1.491825 = 5k \qquad \ln e^a = a$$

$$\frac{\ln 1.491825}{5} = k \qquad \text{Dividing both sides by 5}$$

$$0.08 \approx k. \qquad \text{Using a calculator and rounding}$$

The interest rate is about 0.08, or 8%, compounded continuously. Because interest is being compounded continuously, the yearly interest rate is a bit more than 8%. The exponential growth function is

$$P(t) = 30,000e^{0.08t}.$$

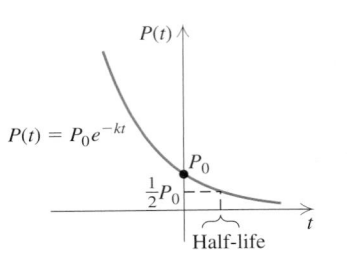

A visualization of Example 6

b) To find the doubling time T, we replace $P(T)$ with 60,000 and solve for T:

$$60{,}000 = 30{,}000e^{0.08T}$$

$$2 = e^{0.08T} \qquad \text{Dividing both sides by 30,000}$$

$$\ln 2 = \ln e^{0.08T} \qquad \text{Taking the natural logarithm on both sides}$$

$$\ln 2 = 0.08T \qquad \ln e^a = a$$

$$\frac{\ln 2}{0.08} = T \qquad \text{Dividing both sides by 0.08}$$

$$8.7 \approx T. \qquad \text{Using a calculator and rounding}$$

Thus the original investment of $30,000 will double in about 8.7 yr.

For any specified interest rate, continuous compounding gives the highest yield and the shortest doubling time.

In some real-life situations, a quantity or population is *decreasing* or *decaying* exponentially.

Exponential Decay

An **exponential decay model** is a function of the form

$$P(t) = P_0 e^{-kt}, \quad k > 0,$$

where P_0 is the quantity present at time 0, $P(t)$ is the amount present at time t, and k is the **decay rate**. The **half-life** is the amount of time necessary for half of the quantity to decay.

EXAMPLE 7

Chaco Canyon, New Mexico

Carbon dating. The radioactive element carbon-14 has a half-life of 5750 yr. The percentage of carbon-14 present in the remains of organic matter can be used to determine the age of that organic matter. Recently, while digging in Chaco Canyon, New Mexico, archaeologists found corn pollen that had lost 38.1% of its carbon-14. The age of this corn pollen was evidence that Indians had been cultivating crops in the Southwest centuries earlier than scientists had thought. What was the age of the pollen? (*Source: American Anthropologist*)

Solution   We first find k. To do so, we use the concept of half-life. When $t = 5750$ (the half-life), $P(t)$ will be half of P_0. Then

$$0.5P_0 = P_0 e^{-k(5750)} \qquad \text{Substituting in } P(t) = P_0 e^{-kt}$$

$$0.5 = e^{-5750k} \qquad \text{Dividing both sides by } P_0$$

$$\ln 0.5 = \ln e^{-5750k} \qquad \text{Taking the natural logarithm on both sides}$$

$$\ln 0.5 = -5750k \qquad \ln e^a = a$$

$$\frac{\ln 0.5}{-5750} = k \qquad \text{Dividing}$$

$$0.00012 \approx k. \qquad \text{Using a calculator and rounding}$$

Now we have a function for the decay of carbon-14:

$$P(t) = P_0 e^{-0.00012t}. \longleftarrow \text{This completes the first part of our solution.}$$

(*Note*: This equation can be used for any subsequent carbon-dating problem.) If the corn pollen has lost 38.1% of its carbon-14 from an initial amount P_0, then $100\% - 38.1\%$, or 61.9%, of P_0 is still present. To find the age t of the pollen, we solve this equation for t:

$$0.619P_0 = P_0 e^{-0.00012t} \qquad \text{We want to find } t \text{ for which } P(t) = 0.619P_0.$$

$$0.619 = e^{-0.00012t} \qquad \text{Dividing both sides by } P_0$$

$$\ln 0.619 = \ln e^{-0.00012t} \qquad \text{Taking the natural logarithm on both sides}$$

$$\ln 0.619 = -0.00012t \qquad \ln e^a = a$$

$$\frac{\ln 0.619}{-0.00012} = t \qquad \text{Dividing}$$

$$4000 \approx t. \qquad \text{Using a calculator}$$

The pollen is about 4000 yr old.

 **Exercise Set**

12.7

FOR EXTRA HELP

 Student's Solutions Manual | Digital Video Tutor CD 6 Videotape 12 | Tutor Center | AW Math Tutor Center | MathXL Tutorials on CD | Math XL MathXL | MyMathLab MyMathLab

Solve.

1. *DVD players.* Yearly sales of DVD players $S(t)$, in millions of dollars, t years after 1997 can be estimated by

$$S(t) = 200 \cdot 2^t.$$

Source: *Statistical Abstract of the United States, 2003*

a) Determine the year in which sales of DVD players first reached $2800 million.
b) What is the doubling time for the yearly sales of DVD players?

2. *Cell phones.* The number of cell phones in use in the United States, in millions, t years after 1990 can be estimated by

$$N(t) = 7.12(1.3)^t.$$

Source: Cellular Telecommunications and Internet Association

a) In what year did the number of cell phones in use first reach 200 million?
b) What is the doubling time for the number of cell phones in use?

3. *Skateboarding.* The number of skateboarders of age x, in thousands, can be approximated by

$$N(x) = 1337(0.9)^x, \qquad 7 \le x \le 60.$$

Sources: Based on figures from the National Sporting Goods Association and *Statistical Abstract of the United States, 2003*

a) Estimate the number of 21-year-old skateboarders.
b) At what age are there only 6300 skateboarders?

4. *Recycling aluminum cans.* Approximately one-half of all aluminum cans distributed will be recycled each year. A beverage company distributes 250,000 cans. The number still in use after t years is given by the function

$$N(t) = 250,000\left(\tfrac{1}{2}\right)^t.$$

Source: The Aluminum Association, Inc., May 2004

a) After how many years will 60,000 cans still be in use?

b) After what amount of time will only 1000 cans still be in use?

5. *Student loan repayment.* A college loan of $29,000 is made at 3% interest, compounded annually. After t years, the amount due, A, is given by the function

$$A(t) = 29,000(1.03)^t.$$

a) After what amount of time will the amount due reach $35,000?

b) Find the doubling time.

6. *Spread of a rumor.* The number of people who have heard a rumor increases exponentially. If all who hear a rumor repeat it to two people a day, and if 20 people start the rumor, the number of people N who have heard the rumor after t days is given by

$$N(t) = 20(3)^t.$$

a) After what amount of time will 1000 people have heard the rumor?

b) What is the doubling time for the number of people who have heard the rumor?

7. *Telephone lines.* As more Americans make cell phones their *only* phones, the percentage of phone lines that are land lines has been shrinking.

The percentage of U.S. phone lines that are land lines $P(t)$, in use t years after 2000, can be estimated by

$$P(t) = 63.03(0.95)^t.$$

Sources: Based on data from Federal Communications Commission; Cellular Telecommunications and Internet Association

a) In what year did/will the percentage of phones that are land lines drop below 50%?

b) In what year will the percentage of phones that are land lines drop below 25%?

8. *Smoking.* The percentage of smokers who received telephone counseling and had successfully quit smoking for t months is given by

$$P(t) = 21.4(0.914)^t.$$

Sources: *New England Journal of Medicine*; data from California's Smoker's Hotline

a) In what month will 15% of those who quit and used telephone counseling still be smoke-free?

b) In what month will 5% of those who quit and used phone counseling still be smoke-free?

9. *Marine biology.* As a result of preservation efforts in countries in which whaling was once common, the humpback whale population has grown since the 1970s. The worldwide population $P(t)$, in thousands, t years after 1982 can be estimated by

$$P(t) = 5.5(1.047)^t.$$

a) In what year will the humpback whale population reach 30,000?

b) Find the doubling time.

10. *World population.* The world population $P(t)$, in billions, t years after 1980 can be approximated by

$$P(t) = 4.495(1.015)^t.$$

Sources: Based on data from U.S. Bureau of the Census; International Data Base

a) In what year will the world population reach 7 billion?

b) Find the doubling time.

Use the pH formula given in Example 2 for Exercises 11–14.

11. *Chemistry.* The hydrogen ion concentration of fresh-brewed coffee is about 1.3×10^{-5} moles per liter. Find the pH.

12. *Chemistry.* The hydrogen ion concentration of milk is about 1.6×10^{-7} moles per liter. Find the pH.

13. *Medicine.*   When the pH of a patient's blood drops below 7.4, a condition called *acidosis* sets in. Acidosis can be deadly when the patient's pH reaches 7.0. What would the hydrogen ion concentration of the patient's blood be at that point?

14. *Medicine.*   When the pH of a patient's blood rises above 7.4, a condition called *alkalosis* sets in. Alkalosis can be deadly when the patient's pH reaches 7.8. What would the hydrogen ion concentration of the patient's blood be at that point?

Use the decibel formula given in Example 1 for Exercises 15–18.

15. *Audiology.*   The intensity of sound in normal conversation is about 3.2×10^{-6} W/m². How loud in decibels is this sound level?

16. *Audiology.*   The intensity of a riveter at work is about 3.2×10^{-3} W/m². How loud in decibels is this sound level?

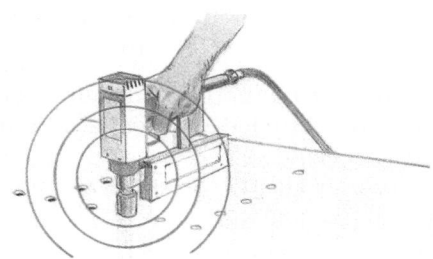

17. *Music.*   The band U2 recently performed and sound measurements of 105 dB were recorded. What is the intensity of such sounds?

18. *Music.*   The band Strange Folk performed in Burlington, VT, and reached sound levels of 111 dB. What is the intensity of such sounds?
Source: Melissa Garrido, *Burlington Free Press*

Use the compound-interest formula in Example 6 for Exercises 19 and 20.

19. *Interest compounded continuously.*   Suppose that P_0 is invested in a savings account where interest is compounded continuously at 2.5% per year.
   a) Express $P(t)$ in terms of P_0 and 0.025.
   b) Suppose that $5000 is invested. What is the balance after 1 yr? after 2 yr?
   c) When will an investment of $5000 double itself?

20. *Interest compounded continuously.*   Suppose that P_0 is invested in a savings account where interest is compounded continuously at 3.1% per year.
   a) Express $P(t)$ in terms of P_0 and 0.031.
   b) Suppose that $1000 is invested. What is the balance after 1 yr? after 2 yr?
   c) When will an investment of $1000 double itself?

21. *Population growth.*   In 2004, the population of the United States was 292.80 million and the exponential growth rate was 0.9% per year.
Source: U.S. Bureau of the Census
   a) Find the exponential growth function.
   b) Predict the U.S. population in 2005.
   c) When will the U.S. population reach 325 million?

22. *World population growth.*   In 2004, the world population was 6.3 billion and the exponential growth rate was 1.1% per year.
Source: U.S. Bureau of the Census
   a) Find the exponential growth function.
   b) Predict the world population in 2009.
   c) When will the world population be 8.0 billion?

23. *iPod sales.*   The number of iPods sold since January 1, 2003, has grown at an exponential growth rate of 10.3% per month. What is the doubling time for iPod sales?
Source: Based on data from iPodlounge.com

24. *Population growth.*   The exponential growth rate of the population of Saudi Arabia is 3.3% per year (one of the highest in the world). What is the doubling time?
Sources: Based on data from U.S. Bureau of the Census; International Data Base 2002

25. *World population.*   The function
$$Y(x) = 67.17 \ln \frac{x}{4.5}$$
can be used to estimate the number of years $Y(x)$ after 1980 required for the world population to reach x billion people.
Sources: Based on data from U.S. Bureau of the Census; International Data Base
   a) In what year will the world population reach 7 billion?
   b) In what year will the world population reach 8 billion?
   c) Graph the function.

26. *Marine biology.* The function

$$Y(x) = 21.77 \ln \frac{x}{5.5}$$

can be used to estimate the number of years $Y(x)$ after 1982 required for the world's humpback whale population to reach x thousand whales.
a) In what year will the whale population reach 15,000?
b) In what year will the whale population reach 25,000?
c) Graph the function.

27. *Forgetting.* Students in an English class took a final exam. They took equivalent forms of the exam at monthly intervals thereafter. The average score $S(t)$, in percent, after t months was found to be given by

$$S(t) = 68 - 20 \log (t + 1), \quad t \geq 0.$$

a) What was the average score when they initially took the test, $t = 0$?
b) What was the average score after 4 months? after 24 months?
c) Graph the function.
d) After what time t was the average score 50%?

28. *Advertising.* A model for advertising response is given by

$$N(a) = 2000 + 500 \log a, \quad a \geq 1,$$

where $N(a)$ is the number of units sold and a is the amount spent on advertising, in thousands of dollars.
a) How many units were sold after spending $1000 ($a = 1$) on advertising?
b) How many units were sold after spending $8000?
c) Graph the function.
d) How much would have to be spent in order to sell 5000 units?

29. *Sexually transmitted disease.* Suppose that in 2000 an outbreak of Herpes infected 17 people at a large university, and that by 2001 the number of those infected had grown to 29.
a) Find an exponential growth function that fits the data.
b) Predict the number of people who will be infected in 2006.

30. *iPod sales.* Sales of iPods have grown exponentially since January 1, 2003, at which time 656,000 units had been sold. Approximately 16 months later, in May 2004, the 3-millionth unit was sold.
Source: Based on data from iPodlounge.com
a) Find an exponential growth function that fits the data.
b) Predict the number of units sold as of April 2005.

31. *Decline in farmland.* The number of acres of farmland in the United States has decreased from 987 million acres in 1990 to 941 million acres in 2002. Assume the number of acres of farmland is decreasing exponentially.
Source: *Statistical Abstract of the United States*, 2003
a) Find the value k, and write an equation for an exponential function that can predict the number of acres of U.S. farmland t years after 1990.
b) Predict the number of acres of farmland in 2008.
c) In what year (theoretically) will there be only 800 million acres of U.S. farmland remaining?

32. *Decline in cases of mumps.* The number of cases of mumps has dropped exponentially from 900 in 1995 to 300 in 2001.
Source: U.S. Centers for Disease Control and Prevention
a) Find the value k, and write an exponential function that can be used to estimate the number of cases t years after 1995.
b) Estimate the number of cases of mumps in 2008.
c) In what year (theoretically) will there be only 1 case of mumps?

33. *Archaeology.* When archaeologists found the Dead Sea Scrolls, they determined that the linen wrapping had lost 22.3% of its carbon-14. How old is the linen wrapping? (See Example 7.)

34. *Archaeology.* In 1996, researchers found an ivory tusk that had lost 18% of its carbon-14. How old was the tusk? (See Example 7.)

35. *Chemistry.* The exponential decay rate of iodine-131 is 9.6% per day. What is its half-life?

36. *Chemistry.* The decay rate of krypton-85 is 6.3% per year. What is its half-life?

37. *Home construction.*   The chemical urea formaldehyde was found in some insulation used in houses built during the mid to late 1960s. Unknown at the time was the fact that urea formaldehyde emitted toxic fumes as it decayed. The half-life of urea formaldehyde is 1 yr. What is its decay rate?

38. *Plumbing.*   Lead pipes and solder are often found in older buildings. Unfortunately, as lead decays, toxic chemicals can get in the water resting in the pipes. The half-life of lead is 22 yr. What is its decay rate?

39. *Value of a sports card.*   Legend has it that because he objected to smoking, and because his first baseball card was issued in cigarette packs, the great shortstop Honus Wagner halted production of his card before many were produced. One of these cards was purchased in 1991 by hockey great Wayne Gretzky (and a partner) for $451,000. The same card was sold in 2000 for $1.1 million. For the following questions, assume that the card's value increases exponentially, as it has for many years.

WAGNER, PITTSBURG

a) Find the exponential growth rate k, and determine an exponential function V that can be used to estimate the dollar value, $V(t)$, of the card t years after 1991.
b) Predict the value of the card in 2006.
c) What is the doubling time for the value of the card?
d) In what year will the value of the card first exceed $3,000,000?

40. *Art masterpieces.*   As of August 2004, the most ever paid for a painting is $104,168,000, paid in 2004 for Pablo Picasso's "Garçon à la Pipe." The same painting sold for $30,000 in 1950.
Source: BBC News, 5/6/04

a) Find the exponential growth rate k, and determine the exponential growth function V, for which $V(t)$ is the painting's value, in millions of dollars, t years after 1950.
b) Estimate the value of the painting in 2009.
c) What is the doubling time for the value of the painting?
d) How long after 1950 will the value of the painting be $1 billion?

41. Write a problem for a classmate to solve in which information is provided and the classmate is asked to find an exponential growth function. Make the problem as realistic as possible.

42. Examine the restriction on t in Exercise 27.
   a) What upper limit might be placed on t?
   b) In practice, would this upper limit ever be enforced? Why or why not?

SKILL MAINTENANCE

Graph. [11.7]

43. $y = x^2 - 8x$ **44.** $y = x^2 - 5x - 6$

45. $f(x) = 3x^2 - 5x - 1$ **46.** $g(x) = 2x^2 - 6x + 3$

Solve by completing the square. [11.1]

47. $x^2 - 8x = 7$ **48.** $x^2 + 10x = 6$

SYNTHESIS

49. Will the model used to predict the number of DVD players in Exercise 1 still be realistic in 2020? Why or why not?

50. *Atmospheric pressure.* Atmospheric pressure P at altitude a is given by

$$P = P_0 e^{-0.00005a},$$

where P_0 is the pressure at sea level ≈ 14.7 lb/in^2 (pounds per square inch). Explain how a barometer, or some other device for measuring atmospheric pressure, can be used to find the height of a skyscraper.

51. *Sports salaries.* As of November 2004, Alex Rodriguez of the New York Yankees has the largest contract in sports history. As part of the 10-year $252-million deal, he will receive $24 million in 2010 (part from the Yankees and part from his former team, the Texas Rangers). How much money would need to be invested in 2004, at 4% interest compounded continuously, in order to have $24 million for Rodriguez in 2010? (This is much like finding what $24 million in 2010 is worth in 2004 dollars.)

Source: *The San Francisco Chronicle*

52. *Supply and demand.* The supply and demand for the sale of stereos by Sound Ideas are given by

$$S(x) = e^x \quad \text{and} \quad D(x) = 162{,}755e^{-x},$$

where $S(x)$ is the price at which the company is willing to supply x stereos and $D(x)$ is the demand price for a quantity of x stereos. Find the equilibrium point. (For reference, see Section 3.8.)

53. Use Exercise 7 to form a model for the percentage of U.S. phone lines that are cellular t years after 2000.

54. Use the model developed in Exercise 53 to predict the percentage of U.S. phone lines that will be cellular in 2020. Does your prediction seem plausible? Why or why not?

55. *Nuclear energy.* Plutonium-239 (Pu-239) is used in nuclear energy plants. The half-life of Pu-239 is 24,360 yr. How long will it take for a fuel rod of Pu-239 to lose 90% of its radioactivity?

Source: *Microsoft Encarta 97 Encyclopedia*

56. *Growth of bacteria.* The bacteria *Escherichia coli* (*E. coli*) are commonly found in the human bladder. Suppose that 3000 of the bacteria are present at time $t = 0$. Then t minutes later, the number of bacteria present is

$$N(t) = 3000(2)^{t/20}.$$

If 100,000,000 bacteria accumulate, a bladder infection can occur. If, at 11:00 A.M., a patient's bladder contains 25,000 *E. coli* bacteria, at what time can infection occur?

57. Show that for exponential growth at rate k, the doubling time T is given by $T = \dfrac{\ln 2}{k}$.

58. Show that for exponential decay at rate k, the half-life T is given by $T = \dfrac{\ln 2}{k}$.

59. *Heart transplants.* In 1967, Dr. Christiaan Barnard of South Africa stunned the world by performing the first heart transplant. Since that time, the operation's popularity has both grown and declined, as shown in the table below.

Year	Number of Heart Transplants Worldwide*
1982	189
1983	318
1984	669
1985	1189
1986	2167
1987	2720
1988	3157
1989	3378
1990	4016
1991	4186
1992	4199
1993	4346
1994	4402
1995	4314
1996	4128
1997	4039
1998	3744
1999	3419
2000	3246
2001	3122
2002	3265
2003	3020

Source: International Society for Heart & Lung Transplantation

a) Using 1982 as $t = 0$, graph the data.

b) Does it appear that an exponential function might have ever served as an appropriate model for these data? If so, for what years would this have been the case?

c) Considering *all* of the data points on your graph, which would be the most appropriate model: a linear, a quadratic, or an exponential function? Why?

CORNER

Investments in Collectibles

COLLABORATIVE

Focus: Exponential-growth models

Time: 30 minutes

Group size: 7

Collecting comic books has long been a popular hobby. It can also be quite profitable since many comic books that originally sold for less than $1 are now worth hundreds or, in some cases, thousands of dollars.

    Collectors often estimate the future value of their collections by examining the value's growth in the past. As with many collectibles, the value of comic books often grows exponentially.

ACTIVITY

1. Suppose that in 2005, each group member invests in the comic books listed in the following table. Looking only at the approximate value in 2005, each student should select $1200 worth of comic books. More than one comic of each type can be selected. The comics chosen will become that student's portfolio.

2. Each group member should select a different one of the comic books. That person should then form an exponential growth model for the value of that comic book using the year of issue and the original cost of the comic book.

3. Using the models developed above, each group member should predict the value of his or her portfolio in 2009. Compare the values. Why, when buying a collector's item, is it important to consider its previous worth?

4. Look at the predicted value of each comic book in 2009. Does an exponential growth model seem appropriate? If possible, find the current value of some of the comic books and compare those values with the predicted values.

Comic Book	Approximate Value in 2005	Year of Issue	Original Cost
Bugs Bunny #28	$ 70	1952	15¢
Josie and the Pussycats #45	140	1969	15¢
Amazing Spider-Man #121	250	1973	25¢
Men In Black #1	45	1990	$2.25
Sonic the Hedgehog #1	35	1993	1.25
G. I. Joe #155	35	1994	1.75
Ultimate Spider-Man #1	90	2000	2.99

Source: Prices based on data kindly provided by Tim Reynolds at Comic Carnival, Indianapolis, IN.

12 Study Summary

Chapter 12 focuses on two special types of functions: **exponential** and **logarithmic** (pp. 794, 803). To understand how these functions relate to each other, it is important to first understand what **composite functions** are and how function **composition** is performed (pp. 782, 783).

The composition of f and g, or $f \circ g$: $(f \circ g)(x) = f(g(x))$

Thus

> if $f(x) = 2x + 1$ and $g(x) = x^2$,
>
> then $(f \circ g)(3) = f(g(3)) = f(9) = 2 \cdot 9 + 1 = 19$.

Composition of functions is developed so we can discuss **inverse functions** (p. 785). To find a function f^{-1}, which is the inverse of f, f must be **one-to-one**, meaning that no two members of f's domain are paired with the same member of the range (p. 786). One quick test for a one-to-one function is the **horizontal-line test**:

A function f is one-to-one if it is not possible to draw a horizontal line that crosses the graph of f at more than one point (p. 786):

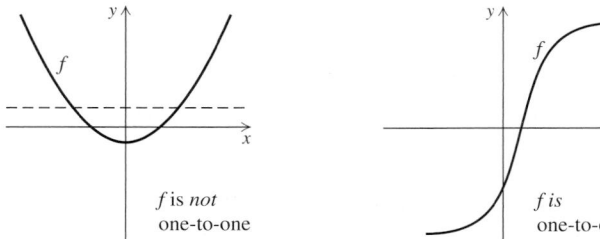

f is *not* one-to-one

f is one-to-one

If f is one-to-one, it is then possible to find its inverse:

1. Replace $f(x)$ with y.
2. Interchange x and y.
3. Solve for y.
4. Replace y with $f^{-1}(x)$.

For any two inverse functions f and f^{-1},

$$(f^{-1} \circ f)(x) = (f \circ f^{-1})(x) = x,$$

or, equivalently,

$$f^{-1}(f(x)) = x \text{ and } f(f^{-1}(x)) = x.$$

Probably the most important pair of inverse functions in mathematics are the exponential and logarithmic functions:

f is an exponential function if $f(x) = a^x$ for $a > 0$, $a \neq 1$.

g is a logarithmic function if $g(x) = \log_a x$ for $a > 0$, $a \neq 1$.

Note that $y = \log_a x$ means $a^y = x$.

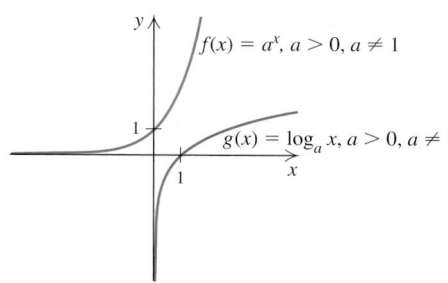

Common (base-10) and **natural** (base-e) logarithms are the most frequently used types of logarithms (pp. 817, 818). Note that $e \approx 2.7$.

Exponential and logarithmic expressions frequently appear in equations. To solve such **exponential** or **logarithmic equations**, we must use certain principles and properties:

Properties of Logarithms

$\log_a (MN) = \log_a M + \log_a N$: $\qquad \log_7 10 = \log_7 5 + \log_7 2$;

$\log_a \dfrac{M}{N} = \log_a M - \log_a N$: $\qquad \log_5 \dfrac{14}{3} = \log_5 14 - \log_5 3$;

$\log_a M^p = p \cdot \log_a M$: $\qquad \log_8 5^{12} = 12 \log_8 5$;

$\log_a 1 = 0$: $\qquad \log_9 1 = 0$;

$\log_a a = 1$: $\qquad \log_4 4 = 1$;

$\log_a a^k = k$: $\qquad \log_3 3^8 = 8$;

$\log M = \log_{10} M$: $\qquad \log 43 = \log_{10} 43$;

$\ln M = \log_e M$: $\qquad \ln 37 = \log_e 37$;

$\log_b M = \dfrac{\log_a M}{\log_a b}$: $\qquad \log_6 31 = \dfrac{\log 31}{\log 6}$

The Principle of Exponential Equality

For any real number b, $b \neq -1$, 0, or 1: $b^x = b^y$ is equivalent to $x = y$.

The Principle of Logarithmic Equality

For any logarithmic base a, and for $x, y > 0$: $x = y$ is equivalent to $\log_a x = \log_a y$.

Thus,

$$25 = 5^x \quad \text{is equivalent to} \quad 5^2 = 5^x, \text{ or } 2 = x,$$

and

$$7^x = 83 \quad \text{is equivalent to} \quad \log 7^x = \log 83, \text{ or } x \log 7 = \log 83, \text{ or } x = \frac{\log 83}{\log 7}.$$

To solve an equation of the form $a^t = b$ for t:

1. Take the logarithm (either natural or common) of both sides.
2. Use the power rule for exponents so that the variable is no longer written as an exponent.
3. Divide both sides by the coefficient of the variable to isolate the variable.
4. If appropriate, use a calculator to find an approximate solution in decimal form.

Exponential and logarithmic equations arise in a wide variety of applications.

Exponential Growth

An **exponential growth model** is a function of the form

$$P(t) = P_0 e^{kt}, \quad k > 0,$$

where P_0 is the population at time 0, $P(t)$ is the population at time t, and k is the **exponential growth rate** for the situation. The **doubling time** is the amount of time necessary for the population to double in size.

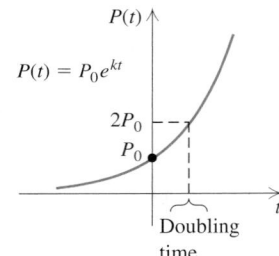

Exponential Decay

An **exponential decay model** is a function of the form

$$P(t) = P_0 e^{-kt}, \quad k > 0,$$

where P_0 is the quantity present at time 0, $P(t)$ is the amount present at time t, and k is the **decay rate**. The **half-life** is the amount of time necessary for half of the quantity to decay.

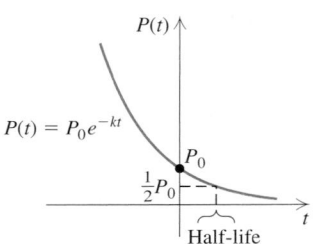

12 Review Exercises

🐦 *Concept Reinforcement   In each of Exercises 1–10, classify the statement as either true or false.*

1. The functions given by $f(x) = e^x$ and $g(x) = \ln x$ are inverses of each other. [12.3]

2. A function's doubling time is the amount of time t for which $f(t) = 2f(0)$. [12.7]

3. A radioactive isotope's half-life is the amount of time t for which $f(t) = \frac{1}{2}f(0)$. [12.7]

4. $\ln(ab) = \ln a - \ln b$ [12.4]

5. $\log x^a = x \ln a$ [12.4]

6. $\log_a \dfrac{m}{n} = \log_a m - \log_a n$ [12.4]

7. For $f(x) = 3^x$, the domain of f is $[0, \infty)$. [12.2]

8. For $g(x) = \log_2 x$, the domain of g is $[0, \infty)$. [12.3]

9. The function F is not one-to-one if $F(-2) = F(5)$. [12.1]

10. The function g is one-to-one if it passes the vertical-line test. [12.1]

11. Find $(f \circ g)(x)$ and $(g \circ f)(x)$ if $f(x) = x^2 + 1$ and $g(x) = 2x - 3$. [12.1]

12. If $h(x) = \sqrt{3 - x}$, find $f(x)$ and $g(x)$ such that $h(x) = (f \circ g)(x)$. Answers may vary. [12.1]

13. Determine whether $f(x) = 4 - x^2$ is one-to-one. [12.1]

Find a formula for the inverse of each function. [12.1]

14. $f(x) = x - 8$

15. $g(x) = \dfrac{3x + 1}{2}$

16. $f(x) = 27x^3$

Graph.

17. $f(x) = 3^x + 1$ [12.2]

18. $x = \left(\frac{1}{4}\right)^y$ [12.2]

19. $y = \log_5 x$ [12.3]

Simplify. [12.3]

20. $\log_3 9$

21. $\log_{10} \frac{1}{100}$

22. $\log_5 5^7$

23. $\log_9 3$

Rewrite as an equivalent logarithmic equation. [12.3]

24. $10^{-2} = \frac{1}{100}$

25. $25^{1/2} = 5$

Rewrite as an equivalent exponential equation. [12.3]

26. $\log_4 16 = x$

27. $\log_8 1 = 0$

Express as an equivalent expression using the individual logarithms of x, y, and z. [12.4]

28. $\log_a x^4 y^2 z^3$

29. $\log_a \dfrac{x^5}{yz^2}$

30. $\log \sqrt[4]{\dfrac{z^2}{x^3 y}}$

Express as an equivalent expression that is a single logarithm and, if possible, simplify. [12.4]

31. $\log_a 7 + \log_a 8$

32. $\log_a 72 - \log_a 12$

33. $\frac{1}{2}\log a - \log b - 2\log c$

34. $\frac{1}{3}[\log_a x - 2\log_a y]$

Simplify. [12.4]

35. $\log_m m$

36. $\log_m 1$

37. $\log_m m^{17}$

Given $\log_a 2 = 1.8301$ and $\log_a 7 = 5.0999$, find each of the following. [12.4]

38. $\log_a 14$

39. $\log_a \frac{2}{7}$

40. $\log_a 28$

41. $\log_a 3.5$

42. $\log_a \sqrt{7}$

43. $\log_a \frac{1}{4}$

📱 *Use a calculator to find each of the following to the nearest ten-thousandth.* [12.5]

44. $\log 75$

45. $10^{1.789}$

46. $\ln 0.05$

47. $e^{-0.98}$

Find each of the following logarithms using the change-of-base formula. Round answers to the nearest ten-thousandth. [12.5]

48. $\log_5 2$

49. $\log_{12} 70$

Graph and state the domain and the range of each function. [12.5]

50. $f(x) = e^x - 1$

51. $g(x) = 0.6 \ln x$

Solve. Where appropriate, include approximations to the nearest ten-thousandth. [12.6]

52. $2^x = 32$

53. $3^x = \frac{1}{9}$

54. $\log_3 x = -4$

55. $\log_x 16 = 4$

56. $\log x = -3$

57. $3 \ln x = -6$

58. $4^{2x-5} = 19$

59. $2^{x^2} \cdot 2^{4x} = 32$

60. $4^x = 8.3$

61. $e^{-0.1t} = 0.03$

62. $2 \ln x = -6$

63. $\log_3 (2x - 5) = 1$

64. $\log_4 x + \log_4 (x - 6) = 2$

65. $\log x + \log (x - 15) = 2$

66. $\log_3 (x - 4) = 3 - \log_3 (x + 4)$

67. In a business class, students were tested at the end of the course with a final exam. They were then tested again 6 months later. The forgetting formula was determined to be

$$S(t) = 82 - 18 \log (t + 1),$$

where t is the time, in months, after taking the final exam. [12.7]

   a) Determine the average score when they first took the exam (when $t = 0$).
   b) What was the average score after 6 months?
   c) After what time was the average score 54?

68. A color photocopier is purchased for $5200. Its value each year is about 80% of its value in the preceding year. Its value in dollars after t years is given by the exponential function

$$V(t) = 5200(0.8)^t. \text{ [12.7]}$$

   a) After what amount of time will the copier's value be $1200?
   b) After what amount of time will the copier's value be half the original value?

69. In 1999, Lucille invested in a building lot that cost $10,000. By 2004, fair-market value for the lot was $19,000. Assume that the value of the lot is increasing exponentially. [12.7]

   a) Find the value k, and write an exponential function that describes the value of Lucille's lot t years after 1999.
   b) Predict the value of the lot in 2009.
   c) In what year will the value of the lot first reach $35,000?

70. The value of Jose's stock market portfolio doubled in 3 yr. What was the exponential growth rate? [12.7]

71. How long will it take $7600 to double itself if it is invested at 4.2%, compounded continuously? [12.7]

72. How old is a skull that has lost 34% of its carbon-14? (Use $P(t) = P_0 e^{-0.00012t}$.) [12.7]

73. What is the pH of a substance if its hydrogen ion concentration is 2.3×10^{-7} moles per liter? (Use pH $= -\log [H^+]$.) [12.7]

74. The intensity of the sound of water at the foot of the Niagara Falls is about 10^{-3} W/m².* How loud in decibels is this sound level? [12.7]

$$\left(\text{Use } L = 10 \cdot \log \frac{I}{10^{-12}}. \right)$$

SYNTHESIS

75. Explain why negative numbers do not have logarithms. [12.3]

76. Explain why taking the natural or common logarithm on each side of an equation produces an equivalent equation. [12.6]

Solve. [12.6]

77. $\ln (\ln x) = 3$

78. $2^{x^2+4x} = \frac{1}{8}$

79. Solve the system:

$$5^{x+y} = 25,$$
$$2^{2x-y} = 64. \text{ [12.6]}$$

Sound and Hearing, Life Science Library. (New York: Time Incorporated, 1965), p. 173.

12 Chapter Test

1. Find $(f \circ g)(x)$ and $(g \circ f)(x)$ if $f(x) = x + x^2$ and $g(x) = 2x + 1$.

2. If
$$h(x) = \frac{1}{2x^2 + 1},$$
find $f(x)$ and $g(x)$ such that $h(x) = (f \circ g)(x)$. Answers may vary.

3. Determine whether $f(x) = |x - 3|$ is one-to-one.

Find a formula for the inverse of each function.

4. $f(x) = 3x + 4$

5. $g(x) = (x + 1)^3$

Graph.

6. $f(x) = 2^x - 3$

7. $g(x) = \log_7 x$

Simplify.

8. $\log_5 125$

9. $\log_{100} 10$

10. $3^{\log_3 18}$

Rewrite as an equivalent logarithmic equation.

11. $4^{-3} = \frac{1}{64}$

12. $256^{1/2} = 16$

Rewrite as an equivalent exponential equation.

13. $m = \log_7 49$

14. $\log_3 81 = 4$

15. Express as an equivalent expression using the individual logarithms of a, b, and c:
$$\log \frac{a^3 b^{1/2}}{c^2}.$$

16. Express as an equivalent expression that is a single logarithm:
$$\tfrac{1}{3} \log_a x + 2 \log_a z.$$

Simplify.

17. $\log_p p$

18. $\log_t t^{23}$

19. $\log_c 1$

Given $\log_a 2 = 0.301$, $\log_a 6 = 0.778$, and $\log_a 7 = 0.845$, find each of the following.

20. $\log_a 14$

21. $\log_a 3$

22. $\log_a 16$

Use a calculator to find each of the following to the nearest ten-thousandth.

23. $\log 12.3$

24. $10^{-0.8}$

25. $\ln 0.035$

26. $e^{4.8}$

27. Find $\log_3 14$ using the change-of-base formula. Round to the nearest ten-thousandth.

Graph and state the domain and the range of each function.

28. $f(x) = e^x + 3$

29. $g(x) = \ln(x - 4)$

Solve. Where appropriate, include approximations to the nearest ten-thousandth.

30. $2^x = \frac{1}{32}$

31. $\log_x 25 = 2$

32. $\log_4 x = \frac{1}{2}$

33. $\log x = 4$

34. $5^{4-3x} = 87$

35. $7^x = 1.2$

36. $\ln x = \frac{1}{4}$

37. $\log(x - 3) + \log(x + 1) = \log 5$

38. The average walking speed R of people living in a city of population P, in thousands, is given by $R = 0.37 \ln P + 0.05$, where R is in feet per second.
   a) The population of Tulsa, Oklahoma, is 878,000. Find the average walking speed.
   b) Baton Rouge, Louisiana, has an average walking speed of about 2.48 ft/sec. Find the population.

39. The population of Nigeria was about 130.5 million in 2002, and the exponential growth rate was 2.4% per year.
   a) Write an exponential function describing the population of Nigeria.
   b) What will the population be in 2007? in 2012?
   c) When will the population be 175 million?
   d) What is the doubling time?

40. The average cost of a year at a private four-year college grew exponentially from \$19,070 in 1997 to \$22,968 in 2003.
Source: National Center for Education Statistics, Digest of Education Statistics, 2002

   a) Find the value k, and write an exponential function that approximates the cost of a year of college t years after 1997.

   b) Predict the cost of a year of college in 2010.

   c) In what year will the average cost of college be \$50,000?

41. An investment with interest compounded continuously doubled itself in 15 yr. What is the interest rate?

42. How old is an animal bone that has lost 43% of its carbon-14? (Use $P(t) = P_0 e^{-0.00012t}$.)

43. The sound of traffic at a busy intersection averages 75 dB. What is the intensity of such a sound?
$$\left(\text{Use } L = 10 \cdot \log \frac{I}{I_0}. \right)$$

44. The hydrogen ion concentration of water is 1.0×10^{-7} moles per liter. What is the pH? (Use pH $= -\log [\text{H}^+]$.)

SYNTHESIS

45. Solve: $\log_5 |2x - 7| = 4$.

46. If $\log_a x = 2$, $\log_a y = 3$, and $\log_a z = 4$, find
$$\log_a \frac{\sqrt[3]{x^2 z}}{\sqrt[3]{y^2 z^{-1}}}.$$

1–12 Cumulative Review

1. Evaluate $\dfrac{x^0 + y}{-z}$ for $x = 6$, $y = 9$, and $z = -5$.
[1.8]

Simplify.

2. $\left| -\frac{5}{2} + \left(-\frac{7}{2} \right) \right|$ [1.8]

3. $(-2x^2 y^{-3})^{-4}$ [4.8]

4. $(-5x^4 y^{-3} z^2)(-4x^2 y^2)$ [4.8]

5. $\dfrac{3x^4 y^6 z^{-2}}{-9x^4 y^2 z^3}$ [4.8]

6. $4x - 3 - 2[5 - 3(2 - x)]$ [1.8]

7. $3^3 + 2^2 - (32 \div 4 - 16 \div 8)$ [1.8]

Solve.

8. $5(2x - 3) = 9 - 5(2 - x)$ [2.2]

9. $4x - 3y = 15,$
$3x + 5y = 4$ [8.2]

10. $x + y - 3z = -1,$
$2x - y + z = 4,$
$-x - y + z = 1$ [8.4]

11. $x(x - 3) = 10$ [5.7]

12. $\dfrac{7}{x^2 - 5x} - \dfrac{2}{x - 5} = \dfrac{4}{x}$ [6.6]

13. $\dfrac{8}{x + 1} + \dfrac{11}{x^2 - x + 1} = \dfrac{24}{x^3 + 1}$ [6.6], [11.2]

14. $\sqrt{4 - 5x} = 2x - 1$ [10.6]

15. $\sqrt[3]{2x} = 1$ [10.6]

16. $3x^2 + 75 = 0$ [11.1]

17. $x - 8\sqrt{x} + 15 = 0$ [11.5]

18. $x^4 - 13x^2 + 36 = 0$ [11.5]

19. $\log_7 x = 1$ [12.3]

20. $\log_x 36 = 2$ [12.3]

21. $9^x = 27$ [12.6]

22. $3^{5x} = 7$ [12.6]

23. $\ln x - \ln (x - 8) = 1$ [12.6]

24. $x^2 + 4x > 5$ [11.9]

25. If $f(x) = x^2 + 6x$, find a such that $f(a) = 11$. [11.2]

26. If $f(x) = |2x - 3|$, find all x for which $f(x) \geq 7$. [9.3]

Solve.

27. $D = \dfrac{ab}{b + a}$, for a [7.5]

28. $\dfrac{1}{p} + \dfrac{1}{q} = \dfrac{1}{f}$, for q [7.5]

29. $M = \dfrac{2}{3}(A + B)$, for B [2.3]

Evaluate. [8.7]

30. $\begin{vmatrix} 6 & -5 \\ 4 & -3 \end{vmatrix}$

31. $\begin{vmatrix} 7 & -6 & 0 \\ -2 & 1 & 2 \\ -1 & 1 & -1 \end{vmatrix}$

32. Find the domain of the function f given by
$$f(x) = \frac{x + 4}{3x^2 - 5x - 2}. \quad [7.2]$$

Solve.

33. *Gasoline consumption.* The number of barrels of gasoline consumed per day in the United States has increased from 7.2 million in 1990 to 8.9 million in 2003.
Source: U.S. Department of Energy, Energy Information Administration

   a) At what rate did gasoline consumption increase from 1990 to 2003? [3.4]
   b) Find a linear function g that fits the data. Let t represent the number of years since 1990. [7.3]
   c) Find an exponential function G that fits the data. [12.7]

34. The perimeter of a rectangular garden is 112 m. The length is 16 m more than the width. Find the length and the width. [2.5]

35. In triangle ABC, the measure of angle B is three times the measure of angle A. The measure of angle C is 105° greater than the measure of angle A. Find the angle measures. [2.5]

36. Good's Candies of Indiana makes all their chocolates by hand. It takes Anne 10 min to coat a tray of candies in chocolate. It takes Clay 12 min to coat a tray of candies. How long would it take Anne and Clay, working together, to coat the candies? [6.7]

37. Joe's Thick and Tasty salad dressing gets 45% of its calories from fat. The Light and Lean dressing gets 20% of its calories from fat. How many ounces of each should be mixed in order to get 15 oz of dressing that gets 30% of its calories from fat? [8.3]

38. A fishing boat with a trolling motor can move at a speed of 5 km/h in still water. The boat travels 42 km downriver in the same time that it takes to travel 12 km upriver. What is the speed of the river? [6.7]

39. What is the minimum product of two numbers whose difference is 14? What are the numbers that yield this product? [11.8]

Students in a biology class just took a final exam. A formula for determining what the average exam grade on a similar test will be t months later is
$$S(t) = 78 - 15 \log (t + 1).$$

40. The average score when the students first took the exam occurs when $t = 0$. Find the students' average score on the final exam. [12.7]

41. What would the average score be on a retest after 4 months? [12.7]

The population of Kenya was 33.7 million in 2002, and the exponential growth rate was 1.0% per year. [12.7]

42. Write an exponential function describing the growth of the population of Kenya.

43. Predict what the population will be in 2008 and in 2014.

44. What is the doubling time of the population?

45. y varies directly as the square of x and inversely as z, and $y = 2$ when $x = 5$ and $z = 100$. What is y when $x = 3$ and $z = 4$? [7.5]

Perform the indicated operations and simplify.

46. $(5p^2q^3 + 6pq - p^2 + p) + (2p^2q^3 + p^2 - 5pq - 9)$ [4.3]

47. $(11x^2 - 6x - 3) - (3x^2 + 5x - 2)$ [4.3]

48. $(3x^2 - 2y)^2$ [4.5]

49. $(5a + 3b)(2a - 3b)$ [4.5]

50. $\dfrac{x^2 + 8x + 16}{2x + 6} \div \dfrac{x^2 + 3x - 4}{x^2 - 9}$ [6.2]

51. $\dfrac{1 + \dfrac{3}{x}}{x - 1 - \dfrac{12}{x}}$ [6.5]

52. $\dfrac{a^2 - a - 6}{a^3 - 27} \cdot \dfrac{a^2 + 3a + 9}{6}$ [6.2]

53. $\dfrac{3}{x + 6} - \dfrac{2}{x^2 - 36} + \dfrac{4}{x - 6}$ [6.4]

Factor.

54. $xy + 2xz - xw$ [5.1]

55. $8 - 125x^3$ [5.5]

56. $6x^2 + 8xy - 8y^2$ [5.3]

57. $x^4 - 4x^3 + 7x - 28$ [5.1]

58. $2m^2 + 12mn + 18n^2$ [5.4]

59. $x^4 - 16y^4$ [5.4]

60. For the function described by
$$h(x) = -3x^2 + 4x + 8,$$
find $h(-2)$. [7.1]

61. Divide: $(x^4 - 5x^3 + 2x^2 - 6) \div (x - 3)$. [4.7]

62. Multiply $(5.2 \times 10^4)(3.5 \times 10^{-6})$. Write scientific notation for the answer. [4.8]

For the radical expressions that follow, assume that all variables represent positive numbers.

63. Divide and simplify:
$$\dfrac{\sqrt[3]{40xy^8}}{\sqrt[3]{5xy}}.$$ [10.4]

64. Multiply and simplify: $\sqrt{7xy^3} \cdot \sqrt{28x^2y}$. [10.3]

65. Write as an equivalent expression without rational exponents: $(27a^6b)^{4/3}$. [10.2]

66. Rationalize the denominator:
$$\dfrac{3 - \sqrt{y}}{2 - \sqrt{y}}.$$ [10.5]

67. Divide and simplify:
$$\dfrac{\sqrt{x + 5}}{\sqrt[5]{x + 5}}.$$ [10.5]

68. Multiply these complex numbers:
$$\left(2 - i\sqrt{3}\right)\left(6 + 2i\sqrt{3}\right).$$ [10.8]

69. Add: $(8 + 2i) + (5 - 3i)$. [10.8]

70. Find the inverse of f if $f(x) = 9 - 2x$. [12.1]

71. Find a linear equation with a graph that contains the points $(0, -8)$ and $(-1, 2)$. [3.7]

72. Find an equation of the line whose graph has a y-intercept of $(0, 7)$ and is perpendicular to the line given by $2x + y = 6$. [3.6]

Graph.

73. $5x = 15 + 3y$ [3.2]

74. $y = 2x^2 - 4x - 1$ [11.7]

75. $y = \log_3 x$ [12.3]

76. $y = 3^x$ [12.2]

77. $-2x - 3y \le 12$ [9.4]

78. Graph: $f(x) = 2(x + 3)^2 + 1$. [11.7]
 a) Label the vertex.
 b) Draw the axis of symmetry.
 c) Find the maximum or minimum value.

79. Graph $f(x) = 2e^x$ and determine the domain and the range. [12.5]

80. Express in terms of logarithms of a, b, and c:
$$\log\left(\dfrac{a^2c^3}{b}\right).$$ [12.4]

81. Express as a single logarithm:
$$3 \log x - \tfrac{1}{2} \log y - 2 \log z.$$ [12.4]

82. Convert to an exponential equation: $\log_a 5 = x$. [12.3]

83. Convert to a logarithmic equation: $x^3 = t$. [12.3]

Find each of the following using a calculator. Round to the nearest ten-thousandth. [12.5]

84. $\log 0.05566$

85. $10^{2.89}$

86. $\ln 12.78$

87. $e^{-1.4}$

SYNTHESIS

Solve.

88. $\dfrac{5}{3x - 3} + \dfrac{10}{3x + 6} = \dfrac{5x}{x^2 + x - 2}$ [6.6]

89. $\log \sqrt{3x} = \sqrt{\log 3x}$ [12.6]

90. A train travels 280 mi at a certain speed. If the speed had been increased by 5 mph, the trip could have been made in 1 hr less time. Find the actual speed. [11.3]

13

Conic Sections

AN APPLICATION

The spotlight on a violin soloist casts an ellipse of light on the floor below her that is 6 ft wide and 10 ft long. Find an equation of that ellipse if the performer is in its center, x is the distance from the performer to the side of the ellipse, and y is the distance from the performer to the top of the ellipse.

This problem appears as Exercise 49 in Section 13.2.

Tony Penna
LIGHTING DESIGNER
Clemson, South Carolina

As a lighting designer, I use math in almost every aspect of my work. Before the lighting instruments are loaded into the theatre, I must determine the angle at which each instrument will be focused at the stage, which requires a great deal of geometry in three dimensions. For each light I use, I must also choose which type of lighting instrument will produce the appropriate sized beam of light.

*T*he ellipse described in the chapter opener is one example of a conic section, meaning that it can be regarded as a cross section of a cone. This chapter presents a variety of applications and equations with graphs that are conic sections. We have already worked with two conic sections, lines *and* parabolas, *in* Chapters 3 and 11.

13.1 Conic Sections: Parabolas and Circles

Parabolas • The Distance and Midpoint Formulas •
Circles

This section and the next two examine curves formed by cross sections of cones. These curves are all graphs of $Ax^2 + By^2 + Cxy + Dx + Ey + F = 0$. The constants A, B, C, D, E, and F determine which of the following shapes will serve as the graph.

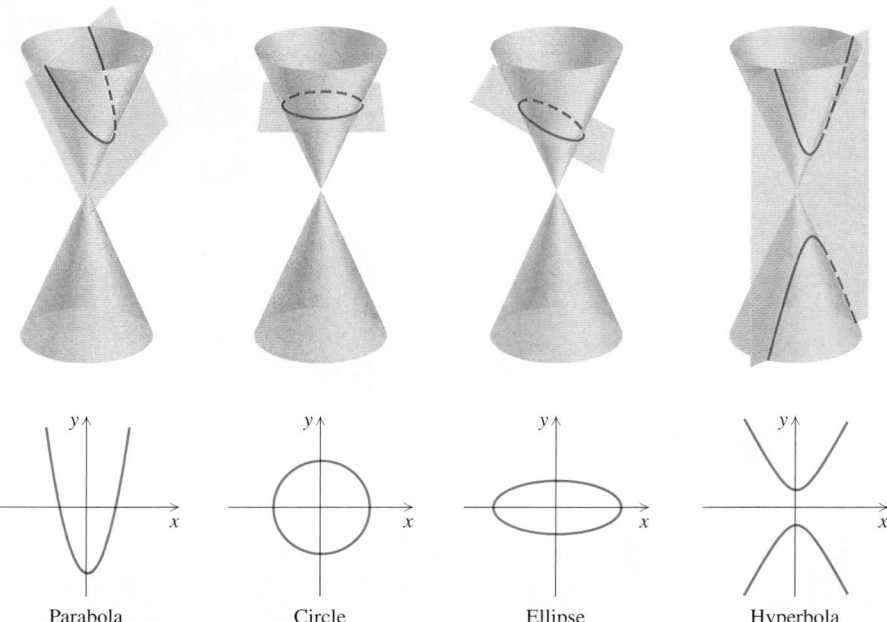

Parabola   Circle   Ellipse   Hyperbola

Parabolas

When a cone is cut as shown in the first figure above, the conic section formed is a **parabola**. Parabolas have many applications in electricity, mechanics, and optics. A cross section of a contact lens or satellite dish is a parabola, and arches that support certain bridges are parabolas.

> **Equation of a Parabola**
>
> A parabola with a vertical axis of symmetry opens upward or downward and has an equation that can be written in the form
>
> $$y = ax^2 + bx + c.$$
>
> A parabola with a horizontal axis of symmetry opens to the right or left and has an equation that can be written in the form
>
> $$x = ay^2 + by + c.$$

Parabolas with equations of the form $f(x) = ax^2 + bx + c$ were graphed in Chapter 11.

EXAMPLE 1   Graph: $y = x^2 - 4x + 9$.

Solution   To locate the vertex, we can use either of two approaches. One way is to complete the square:

$y = (x^2 - 4x) + 9$	Note that half of -4 is -2, and $(-2)^2 = 4$.
$\quad = (x^2 - 4x + 4 - 4) + 9$	Adding and subtracting 4
$\quad = (x^2 - 4x + 4) + (-4 + 9)$	Regrouping
$\quad = (x - 2)^2 + 5.$	Factoring and simplifying

The vertex is $(2, 5)$.

A second way to find the vertex is to recall that the x-coordinate of the vertex of the parabola given by $y = ax^2 + bx + c$ is $-b/(2a)$:

$$x = -\frac{b}{2a} = -\frac{-4}{2(1)} = 2.$$

To find the y-coordinate of the vertex, we substitute 2 for x:

$$y = x^2 - 4x + 9 = 2^2 - 4(2) + 9 = 5.$$

Either way, the vertex is $(2, 5)$. Next, we calculate and plot some points on each side of the vertex. As expected for a positive coefficient of x^2, the graph opens upward.

x	y	
2	5	← Vertex
0	9	← y-intercept
1	6	
3	6	
4	9	

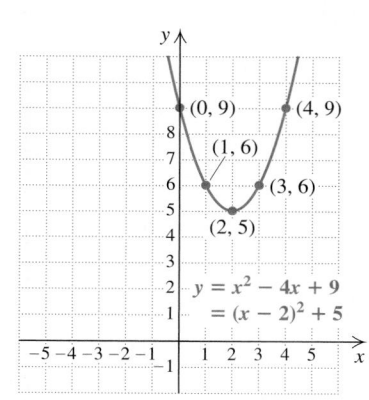

To Graph an Equation of the Form $y = ax^2 + bx + c$

1. Find the vertex (h, k) either by completing the square to find an equivalent equation

$$y = a(x - h)^2 + k,$$

or by using $-b/(2a)$ to find the x-coordinate and substituting to find the y-coordinate.
2. Choose other values for x on each side of the vertex, and compute the corresponding y-values.
3. The graph opens upward for $a > 0$ and downward for $a < 0$.

Equations of the form $x = ay^2 + by + c$ represent horizontal parabolas. These parabolas open to the right for $a > 0$, open to the left for $a < 0$, and have axes of symmetry parallel to the x-axis.

EXAMPLE 2   Graph: $x = y^2 - 4y + 9$.

Solution   This equation is like that in Example 1 but with x and y interchanged. The vertex is $(5, 2)$ instead of $(2, 5)$. To find ordered pairs, we choose values for y on each side of the vertex. Then we compute values for x. Note that the x- and y-values of the table in Example 1 are now switched. You should confirm that, by completing the square, we get $x = (y - 2)^2 + 5$.

x	y
5	2	← Vertex
9	0	← x-intercept
6	1
6	3
9	4

(1) Choose these values for y.

(2) Compute these values for x.

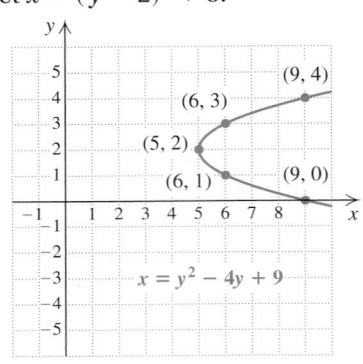

To Graph an Equation of the Form $x = ay^2 + by + c$

1. Find the vertex (h, k) either by completing the square to find an equivalent equation

$$x = a(y - k)^2 + h,$$

or by using $-b/(2a)$ to find the y-coordinate and substituting to find the x-coordinate.
2. Choose other values for y that are above and below the vertex, and compute the corresponding x-values.
3. The graph opens to the right if $a > 0$ and to the left if $a < 0$.

EXAMPLE 3   Graph: $x = -2y^2 + 10y - 7$.

Solution   We find the vertex by completing the square:

$$x = -2y^2 + 10y - 7$$
$$= -2(y^2 - 5y \qquad) - 7$$
$$= -2\left(y^2 - 5y + \tfrac{25}{4}\right) - 7 - (-2)\tfrac{25}{4} \qquad \tfrac{1}{2}(-5) = \tfrac{-5}{2}; \left(\tfrac{-5}{2}\right)^2 = \tfrac{25}{4}; \text{ we}$$
$$\text{add and subtract } (-2)\tfrac{25}{4}.$$
$$= -2\left(y - \tfrac{5}{2}\right)^2 + \tfrac{11}{2}. \qquad \text{Factoring and simplifying}$$

The vertex is $\left(\tfrac{11}{2}, \tfrac{5}{2}\right)$.

For practice, we also find the vertex by first computing its y-coordinate, $-b/(2a)$, and then substituting to find the x-coordinate:

$$y = -\frac{b}{2a} = -\frac{10}{2(-2)} = \frac{5}{2}$$
$$x = -2y^2 + 10y - 7 = -2\left(\tfrac{5}{2}\right)^2 + 10\left(\tfrac{5}{2}\right) - 7$$
$$= \tfrac{11}{2}.$$

To find ordered pairs, we choose values for y on each side of the vertex and then compute values for x. A table is shown below, together with the graph. The graph opens to the left because the y^2-coefficient, -2, is negative.

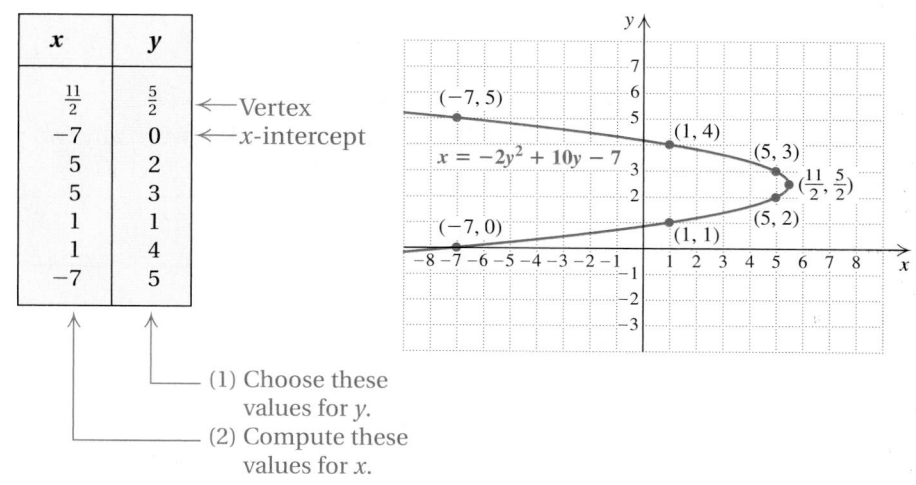

x	y	
$\tfrac{11}{2}$	$\tfrac{5}{2}$	←Vertex
-7	0	←x-intercept
5	2	
5	3	
1	1	
1	4	
-7	5	

(1) Choose these values for y.
(2) Compute these values for x.

The Distance and Midpoint Formulas

If two points are on a horizontal line, they have the same second coordinate. We can find the distance between them by subtracting their first coordinates. This difference may be negative, depending on the order in which we subtract. So, to make sure we get a positive number, we take the absolute value of this difference. The distance between the points (x_1, y_1) and (x_2, y_1) on a horizontal line is thus $|x_2 - x_1|$. Similarly, the distance between the points (x_2, y_1) and (x_2, y_2) on a vertical line is $|y_2 - y_1|$.

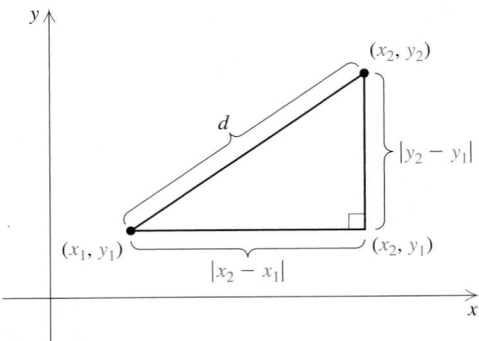

Now consider *any* two points (x_1, y_1) and (x_2, y_2). If $x_1 \neq x_2$ and $y_1 \neq y_2$, these points, along with the point (x_2, y_1), describe a right triangle. The lengths of the legs are $|x_2 - x_1|$ and $|y_2 - y_1|$. We find d, the length of the hypotenuse, by using the Pythagorean theorem:

$$d^2 = |x_2 - x_1|^2 + |y_2 - y_1|^2.$$

Since the square of a number is the same as the square of its opposite, we can replace the absolute-value signs with parentheses:

$$d^2 = (x_2 - x_1)^2 + (y_2 - y_1)^2.$$

Taking the principal square root, we have a formula for distance.

The Distance Formula

The distance d between any two points (x_1, y_1) and (x_2, y_2) is given by

$$d = \sqrt{(x_2 - x_1)^2 + (y_2 - y_1)^2}.$$

EXAMPLE 4   Find the distance between $(5, -1)$ and $(-4, 6)$. Find an exact answer and an approximation to three decimal places.

Solution   We substitute into the distance formula:

$$\begin{aligned} d &= \sqrt{(-4 - 5)^2 + [6 - (-1)]^2} & \text{Substituting} \\ &= \sqrt{(-9)^2 + 7^2} \\ &= \sqrt{130} & \text{This is exact.} \\ &\approx 11.402. & \text{Using a calculator for an approximation} \end{aligned}$$

The distance formula is needed to develop the formula for a circle, which follows, and to verify certain properties of conic sections. It is also needed to verify a formula for the coordinates of the *midpoint* of a segment connecting two points. We state the midpoint formula and leave its proof to the exercises.

Student Notes _____

To help remember the formulas correctly, note that the distance formula (a variation on the Pythagorean theorem) involves both subtraction and addition, whereas the midpoint formula does not include any subtraction.

The Midpoint Formula

If the endpoints of a segment are (x_1, y_1) and (x_2, y_2), then the coordinates of the midpoint are

$$\left(\frac{x_1 + x_2}{2}, \frac{y_1 + y_2}{2} \right).$$

(To locate the midpoint, average the x-coordinates and average the y-coordinates.)

EXAMPLE 5   Find the midpoint of the segment with endpoints $(-2, 3)$ and $(4, -6)$.

Solution   Using the midpoint formula, we obtain

$$\left(\frac{-2 + 4}{2}, \frac{3 + (-6)}{2} \right), \quad \text{or} \quad \left(\frac{2}{2}, \frac{-3}{2} \right), \quad \text{or} \quad \left(1, -\frac{3}{2} \right).$$

Circles

One conic section, the **circle**, is a set of points in a plane that are a fixed distance r, called the **radius** (plural, **radii**), from a fixed point (h, k), called the **center**. Note that the word radius can mean either any segment connecting a point on a circle to the center or the length of such a segment. If (x, y) is on the circle, then by the definition of a circle and the distance formula, it follows that

$$r = \sqrt{(x - h)^2 + (y - k)^2}.$$

Squaring both sides gives the equation of a circle in standard form.

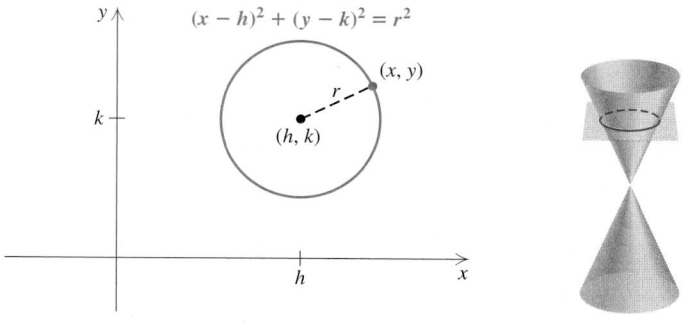

> **Equation of a Circle (Standard Form)**
> The equation of a circle, centered at (h, k), with radius r, is given by
> $$(x - h)^2 + (y - k)^2 = r^2.$$

Note that for $h = 0$ and $k = 0$, the circle is centered at the origin. Otherwise, the circle is translated $|h|$ units horizontally and $|k|$ units vertically.

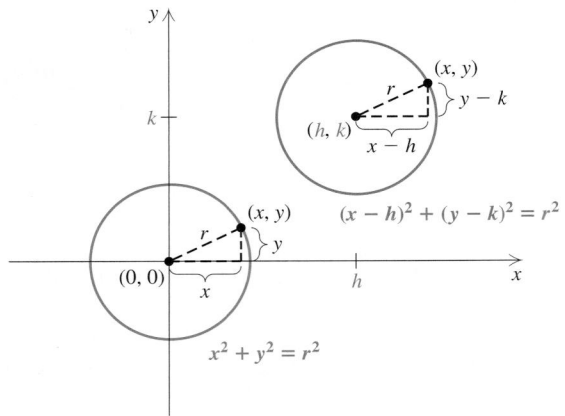

EXAMPLE 6   Find an equation of the circle having center $(4, 5)$ and radius 6.

Solution   Using the standard form, we obtain

$$(x - 4)^2 + (y - 5)^2 = 6^2, \qquad \text{Using } (x - h)^2 + (y - k)^2 = r^2$$

or

$$(x - 4)^2 + (y - 5)^2 = 36.$$

EXAMPLE 7   Find the center and the radius and then graph each circle.

a) $(x - 2)^2 + (y + 3)^2 = 4^2$

b) $x^2 + y^2 + 8x - 2y + 15 = 0$

Solution

a) We write standard form:

$$(x - 2)^2 + [y - (-3)]^2 = 4^2.$$

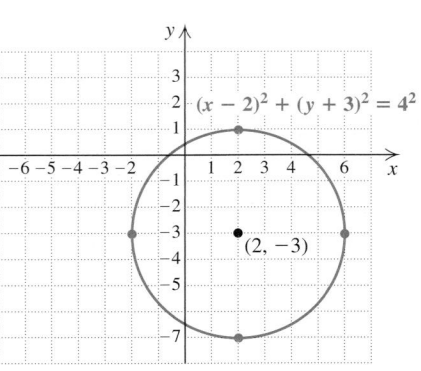

The center is $(2, -3)$ and the radius is 4. To graph, we plot the points $(2, 1)$, $(2, -7)$, $(-2, -3)$, and $(6, -3)$, which are, respectively, 4 units above, below, left, and right of $(2, -3)$. We then either sketch a circle by hand or use a compass.

b) To write the equation $x^2 + y^2 + 8x - 2y + 15 = 0$ in standard form, we complete the square twice, once with $x^2 + 8x$ and once with $y^2 - 2y$:

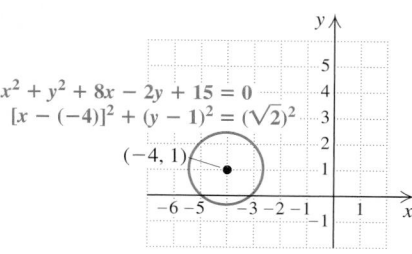

$$x^2 + y^2 + 8x - 2y + 15 = 0$$

$$x^2 + 8x \qquad + y^2 - 2y \qquad = -15$$

Grouping the x-terms and the y-terms; adding -15 to both sides

$$x^2 + 8x + 16 + y^2 - 2y + 1 = -15 + 16 + 1$$

Adding $\left(\frac{8}{2}\right)^2$, or 16, and $\left(-\frac{2}{2}\right)^2$, or 1, to both sides to get standard form

$$(x + 4)^2 + (y - 1)^2 = 2$$

Factoring

$$[x - (-4)]^2 + (y - 1)^2 = \left(\sqrt{2}\right)^2.$$

Writing standard form

The center is $(-4, 1)$ and the radius is $\sqrt{2}$.

technology connection

Most graphing calculators graph only functions, so graphing the equation of a circle usually requires two steps:

1. Solve the equation for y. The result will include a $\pm$ sign in front of a radical.

2. Graph two functions, one for the $+$ sign and the other for the $-$ sign, on the same set of axes.

For example, to graph $(x - 3)^2 + (y + 1)^2 = 16$, solve for $y + 1$ and then y:

$$(y + 1)^2 = 16 - (x - 3)^2$$

$$y + 1 = \pm\sqrt{16 - (x - 3)^2}$$

$$y = -1 \pm \sqrt{16 - (x - 3)^2},$$

or

$$y_1 = -1 + \sqrt{16 - (x - 3)^2}$$

and

$$y_2 = -1 - \sqrt{16 - (x - 3)^2}.$$

When both functions are graphed (in a "squared" window to eliminate distortion), the result is as follows.

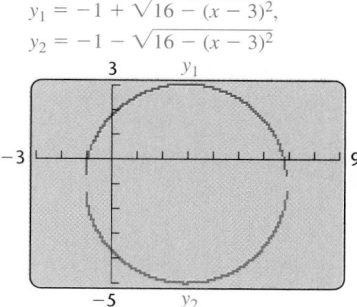

On many calculators, pressing **APPS** and selecting Conics and then Circle accesses a program in which equations in standard form can be graphed directly and then Traced.

Graph each of the following equations.

1. $x^2 + y^2 - 16 = 0$
2. $(x - 1)^2 + (y - 2)^2 = 25$
3. $(x + 3)^2 + (y - 5)^2 = 16$
4. $(x - 5)^2 + (y + 6)^2 = 49$

Exercise Set

13.1

FOR EXTRA HELP

Student's
Solutions
Manual

Digital Video Tutor
CD 7
Videotape 13

AW Math
Tutor Center

MathXL Tutorials
on CD

Math XL
MathXL

MyMathLab
MyMathLab

 Concept Reinforcement   In each of Exercises 1–8, match the equation with the graph of that equation from those shown.

1. _____ $(x - 2)^2 + (y + 5)^2 = 9$

2. _____ $(x + 2)^2 + (y - 5)^2 = 9$

3. _____ $(x - 5)^2 + (y + 2)^2 = 9$

4. _____ $(x + 5)^2 + (y - 2)^2 = 9$

5. _____ $y = (x - 2)^2 - 5$

6. _____ $y = (x - 5)^2 - 2$

7. _____ $x = (y - 2)^2 - 5$

8. _____ $x = (y - 5)^2 - 2$

a)
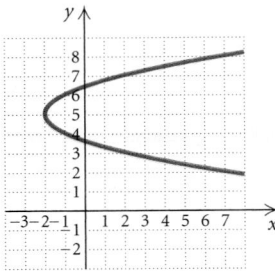

b)
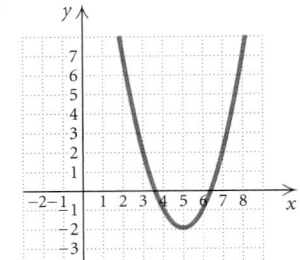

c)
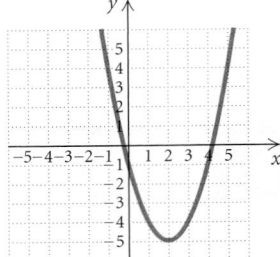

d)
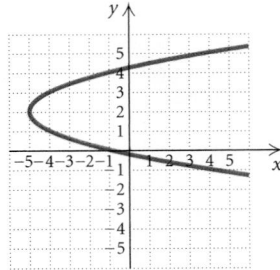

e)
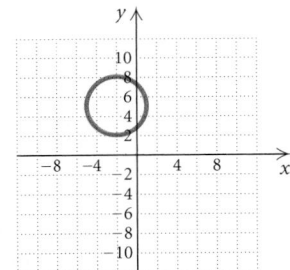

f)
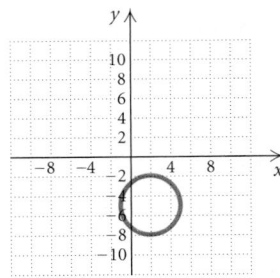

g)
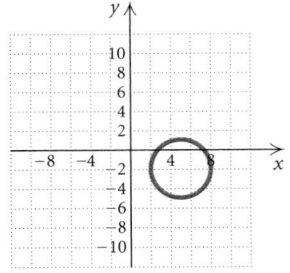

h)
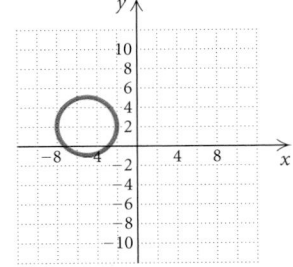

Graph. Be sure to label each vertex.

9. $y = -x^2$

10. $y = 2x^2$

11. $y = -x^2 + 4x - 5$

12. $x = 4 - 3y - y^2$

13. $x = y^2 - 4y + 2$

14. $y = x^2 + 2x + 3$

15. $x = y^2 + 3$

16. $x = 2y^2$

17. $x = -\frac{1}{2}y^2$

18. $x = y^2 - 1$

19. $x = -y^2 - 4y$

20. $x = y^2 + y - 6$

21. $x = 4 - y - y^2$

22. $y = x^2 + 2x + 1$

23. $y = x^2 - 2x + 1$

24. $y = -\frac{1}{2}x^2$

25. $x = -y^2 + 2y - 1$

26. $x = -y^2 - 2y + 3$

27. $x = -2y^2 - 4y + 1$

28. $x = 2y^2 + 4y - 1$

Find the distance between each pair of points. Where appropriate, find an approximation to three decimal places.

29. $(1, 6)$ and $(5, 9)$

30. $(1, 10)$ and $(7, 2)$

31. $(0, -7)$ and $(3, -4)$

32. $(6, 2)$ and $(6, -8)$

33. $(-4, 4)$ and $(6, -6)$

34. $(5, 21)$ and $(-3, 1)$

Aha! **35.** $(8.6, -3.4)$ and $(-9.2, -3.4)$

36. $(5.9, 2)$ and $(3.7, -7.7)$

37. $\left(\frac{5}{7}, \frac{1}{14}\right)$ and $\left(\frac{1}{7}, \frac{11}{14}\right)$

38. $\left(0, \sqrt{7}\right)$ and $\left(\sqrt{6}, 0\right)$

39. $\left(-\sqrt{6}, \sqrt{2}\right)$ and $(0, 0)$

40. $\left(\sqrt{5}, -\sqrt{3}\right)$ and $(0, 0)$

41. $(-4, -2)$ and $(-7, -11)$

42. $(-3, -7)$ and $(-1, -5)$

Find the midpoint of each segment with the given endpoints.

43. $(-7, 6)$ and $(9, 2)$

44. $(6, 7)$ and $(7, -9)$

45. $(2, -1)$ and $(5, 8)$

46. $(-1, 2)$ and $(1, -3)$

47. $(-8, -5)$ and $(6, -1)$

48. $(8, -2)$ and $(-3, 4)$

49. $(-3.4, 8.1)$ and $(2.9, -8.7)$

50. $(4.1, 6.9)$ and $(5.2, -6.9)$

51. $\left(\frac{1}{6}, -\frac{3}{4}\right)$ and $\left(-\frac{1}{3}, \frac{5}{6}\right)$

52. $\left(-\frac{4}{5}, -\frac{2}{3}\right)$ and $\left(\frac{1}{8}, \frac{3}{4}\right)$

53. $\left(\sqrt{2}, -1\right)$ and $\left(\sqrt{3}, 4\right)$

54. $\left(9, 2\sqrt{3}\right)$ and $\left(-4, 5\sqrt{3}\right)$

Find an equation of the circle satisfying the given conditions.

55. Center $(0, 0)$, radius 6

56. Center $(0, 0)$, radius 5

57. Center $(7, 3)$, radius $\sqrt{5}$

58. Center $(5, 6)$, radius $\sqrt{2}$

59. Center $(-4, 3)$, radius $4\sqrt{3}$

60. Center $(-2, 7)$, radius $2\sqrt{5}$

61. Center $(-7, -2)$, radius $5\sqrt{2}$

62. Center $(-5, -8)$, radius $3\sqrt{2}$

63. Center $(0, 0)$, passing through $(-3, 4)$

64. Center $(3, -2)$, passing through $(11, -2)$

65. Center $(-4, 1)$, passing through $(-2, 5)$

66. Center $(-1, -3)$, passing through $(-4, 2)$

Find the center and the radius of each circle. Then graph the circle.

67. $x^2 + y^2 = 64$

68. $x^2 + y^2 = 36$

69. $(x + 1)^2 + (y + 3)^2 = 36$

70. $(x - 2)^2 + (y + 3)^2 = 4$

71. $(x - 4)^2 + (y + 3)^2 = 10$

72. $(x + 5)^2 + (y - 1)^2 = 15$

73. $x^2 + y^2 = 10$

74. $x^2 + y^2 = 7$

75. $(x - 5)^2 + y^2 = \frac{1}{4}$

76. $x^2 + (y - 1)^2 = \frac{1}{25}$

77. $x^2 + y^2 + 8x - 6y - 15 = 0$

78. $x^2 + y^2 + 6x - 4y - 15 = 0$

79. $x^2 + y^2 - 8x + 2y + 13 = 0$

80. $x^2 + y^2 + 6x + 4y + 12 = 0$

81. $x^2 + y^2 + 10y - 75 = 0$

82. $x^2 + y^2 - 8x - 84 = 0$

83. $x^2 + y^2 + 7x - 3y - 10 = 0$

84. $x^2 + y^2 - 21x - 33y + 17 = 0$

85. $36x^2 + 36y^2 = 1$

86. $4x^2 + 4y^2 = 1$

87. Describe a procedure that would use the distance formula to determine whether three points, (x_1, y_1), (x_2, y_2), and (x_3, y_3), are vertices of a right triangle.

88. Does the graph of an equation of a circle include the point that is the center? Why or why not?

SKILL MAINTENANCE

Solve. [6.6]

89. $\dfrac{x}{4} + \dfrac{5}{6} = \dfrac{2}{3}$

90. $\dfrac{t}{6} - \dfrac{1}{9} = \dfrac{7}{12}$

91. A rectangle 10 in. long and 6 in. wide is bordered by a strip of uniform width. If the perimeter of the larger rectangle is twice that of the smaller rectangle, what is the width of the border? [2.5]

92. One airplane flies 60 mph faster than another. To fly a certain distance, the faster plane takes 4 hr and the slower plane takes 4 hr and 24 min. What is the distance? [8.3]

Solve each system. [8.2]

93. $3x - 8y = 5$,
$2x + 6y = 5$

94. $4x - 5y = 9$,
$12x - 10y = 18$

SYNTHESIS

95. Outline a procedure that would use the distance formula to determine whether three points,

(x_1, y_1), (x_2, y_2), and (x_3, y_3), are collinear (lie on the same line).

96. Why does the discussion of the distance formula precede the discussion of circles?

Find an equation of a circle satisfying the given conditions.

97. Center $(3, -5)$ and tangent to (touching at one point) the y-axis

98. Center $(-7, -4)$ and tangent to the x-axis

99. The endpoints of a diameter are $(7, 3)$ and $(-1, -3)$.

100. Center $(-3, 5)$ with a circumference of 8π units

101. Find the point on the y-axis that is equidistant from $(2, 10)$ and $(6, 2)$.

102. Find the point on the x-axis that is equidistant from $(-1, 3)$ and $(-8, -4)$.

103. *Wrestling.* The equation $x^2 + y^2 = \frac{81}{4}$, where x and y represent the number of meters from the center, can be used to draw the outer circle on a wrestling mat used in International, Olympic, and World Championship wrestling. The equation $x^2 + y^2 = 16$ can be used to draw the inner edge of the red zone. Find the area of the red zone.

Source: Based on data from the Government of Western Australia

104. *Snowboarding.* Each side edge of the Salomon Freestyle 500 Pro snowboard is an arc of a circle

with a "running length" of 1180 mm and a "side-cut depth" of 21.5 mm (see the figure below).

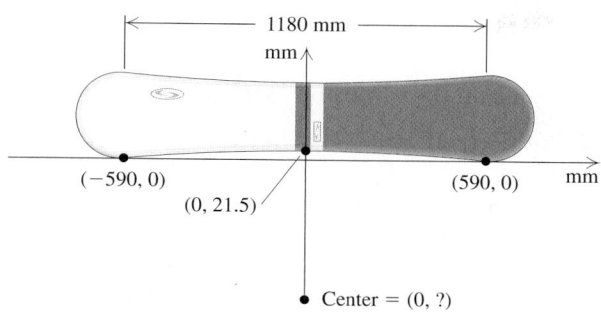

a) Using the coordinates shown, locate the center of the circle. (*Hint*: Equate distances.)
b) What radius is used for the edge of the board?

105. *Snowboarding.*   The Elan Jason Evans 155 snowboard has a running length of 1170 mm and a sidecut depth of 23 mm (see Exercise 104). What radius is used for the edge of this snowboard?

106. *Skiing.*   The Völkl Supersport 5 Star ski, when lying flat and viewed from above, has edges that are arcs of a circle. (Actually, each edge is made of two arcs of slightly different radii. The arc for the rear half of the ski edge has a slightly larger radius.)

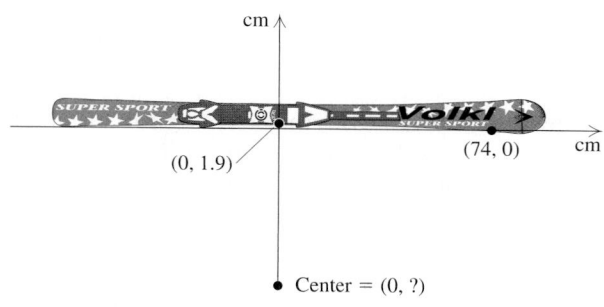

a) Using the coordinates shown, locate the center of the circle. (*Hint*: Equate distances.)

b) What radius is used for the arc passing through (0, 1.9) and (74, 0)?

107. *Doorway construction.*   Ace Carpentry needs to cut an arch for the top of an entranceway. The arch needs to be 8 ft wide and 2 ft high. To draw the arch, the carpenters will use a stretched string with chalk attached at an end as a compass.

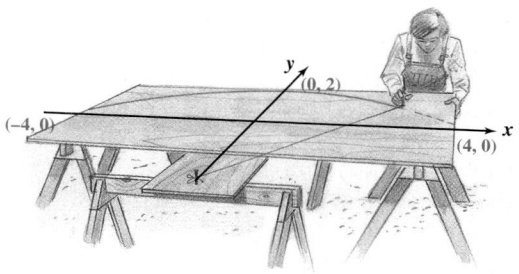

a) Using a coordinate system, locate the center of the circle.
b) What radius should the carpenters use to draw the arch?

108. *Archaeology.*   During an archaeological dig, Martina finds the bowl fragment shown below. What was the original diameter of the bowl?

109. *Ferris wheel design.*   A ferris wheel has a radius of 24.3 ft. Assuming that the center is 30.6 ft off the ground and that the origin is below the center, as in the following figure, find an equation of the circle.

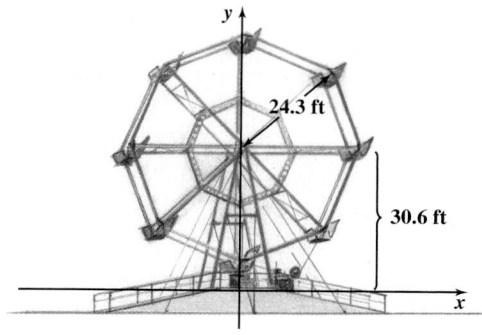

110. Use a graph of the equation $x = y^2 - y - 6$ to approximate to the nearest tenth the solutions of each of the following equations.
a) $y^2 - y - 6 = 2$ (*Hint*: Graph $x = 2$ on the same set of axes as the graph of $x = y^2 - y - 6$.)
b) $y^2 - y - 6 = -3$

111. *Power of a motor.* The horsepower of a certain kind of engine is given by the formula

$$H = \frac{D^2 N}{2.5},$$

where N is the number of cylinders and D is the diameter, in inches, of each piston. Graph this equation, assuming that $N = 6$ (a six-cylinder engine). Let D run from 2.5 to 8.

112. Prove the midpoint formula by showing that

i) the distance from (x_1, y_1) to
$$\left(\frac{x_1 + x_2}{2}, \frac{y_1 + y_2}{2} \right)$$
equals the distance from (x_2, y_2) to
$$\left(\frac{x_1 + x_2}{2}, \frac{y_1 + y_2}{2} \right);$$
and

ii) the points
$$(x_1, y_1), \left(\frac{x_1 + x_2}{2}, \frac{y_1 + y_2}{2} \right),$$
and
$$(x_2, y_2)$$
lie on the same line (see Exercise 95).

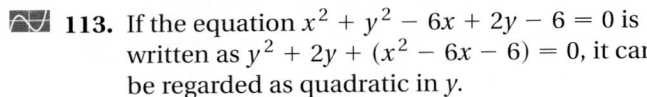 **113.** If the equation $x^2 + y^2 - 6x + 2y - 6 = 0$ is written as $y^2 + 2y + (x^2 - 6x - 6) = 0$, it can be regarded as quadratic in y.

   a) Use the quadratic formula to solve for y.
   b) Show that the graph of your answer to part (a) coincides with the graph in the Technology Connection on p. 863.

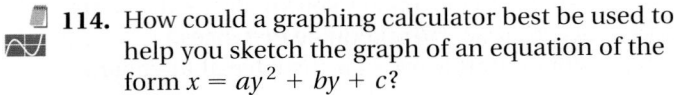 **114.** How could a graphing calculator best be used to help you sketch the graph of an equation of the form $x = ay^2 + by + c$?

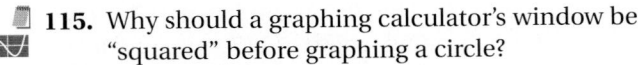 **115.** Why should a graphing calculator's window be "squared" before graphing a circle?

13.2 Conic Sections: Ellipses

Ellipses Centered at (0, 0)   •   Ellipses Centered at (h, k)

When a cone is cut at an angle, as shown below, the conic section formed is an *ellipse*. To draw an ellipse, stick two tacks in a piece of cardboard. Then tie a loose string to the tacks, place a pencil as shown, and draw an oval by moving the pencil while stretching the string tight.

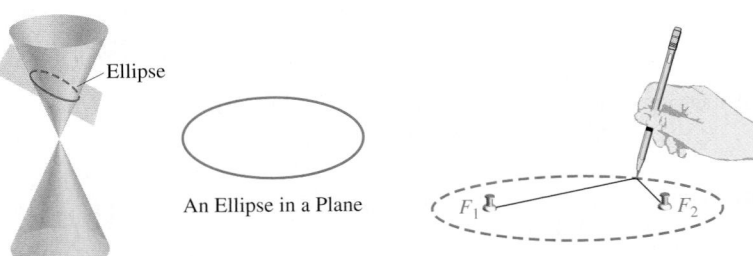

Ellipse

An Ellipse in a Plane

Ellipses Centered at (0, 0)

An **ellipse** is defined as the set of all points in a plane for which the sum of the distances from two fixed points F_1 and F_2 is constant. The points F_1 and F_2 are called **foci** (pronounced fō-sī), the plural of focus. In the figure above, the

Study Skills _____

It is never too early to begin studying for a final exam. If you have at least three days, consider the following:

- Reviewing the highlighted or boxed information in each chapter;
- Studying the Chapter Tests, Review Exercises, Cumulative Reviews, and Study Summaries;
- Re-taking all quizzes and tests that have been returned to you;
- Attending any review sessions being offered;
- Organizing or joining a study group;
- Using the video or software supplements, or asking a tutor or professor about any trouble spots;
- Asking for previous final exams (and answers) for practice.

tacks are at the foci and the length of the string is the constant sum of the distances. The midpoint of the segment F_1F_2 is the **center**. The equation of an ellipse is as follows. Its derivation is left to the exercises.

Equation of an Ellipse Centered at the Origin

The equation of an ellipse centered at the origin and symmetric with respect to both axes is

$$\frac{x^2}{a^2} + \frac{y^2}{b^2} = 1, \quad a, b > 0. \qquad \text{(Standard form)}$$

To graph an ellipse centered at the origin, it helps to first find the intercepts. If we replace x with 0, we can find the y-intercepts:

$$\frac{0^2}{a^2} + \frac{y^2}{b^2} = 1$$

$$\frac{y^2}{b^2} = 1$$

$$y^2 = b^2 \quad \text{or} \quad y = \pm b.$$

Thus the y-intercepts are $(0, b)$ and $(0, -b)$. Similarly, the x-intercepts are $(a, 0)$ and $(-a, 0)$. If $a > b$, the ellipse is said to be horizontal and $(-a, 0)$ and $(a, 0)$ are referred to as the **vertices** (singular, **vertex**). If $b > a$, the ellipse is said to be vertical and $(0, -b)$ and $(0, b)$ are then the vertices.

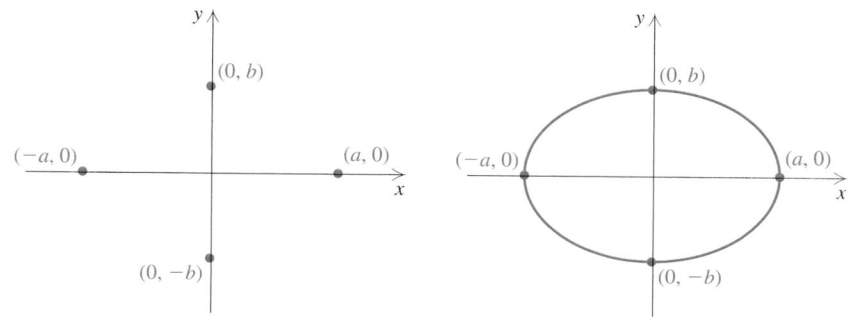

Plotting these four points and drawing an oval-shaped curve, we graph the ellipse. If a more precise graph is desired, we can plot more points.

Using a and b to Graph an Ellipse

For the ellipse

$$\frac{x^2}{a^2} + \frac{y^2}{b^2} = 1,$$

the x-intercepts are $(-a, 0)$ and $(a, 0)$. The y-intercepts are $(0, -b)$ and $(0, b)$. For $a^2 > b^2$, the ellipse is horizontal. For $b^2 > a^2$, the ellipse is vertical.

EXAMPLE 1   Graph the ellipse

$$\frac{x^2}{4} + \frac{y^2}{9} = 1.$$

Solution   Note that

$$\frac{x^2}{4} + \frac{y^2}{9} = \frac{x^2}{2^2} + \frac{y^2}{3^2}.$$ Identifying a and b. Since $b > a$, the ellipse is vertical.

Thus the x-intercepts are $(-2, 0)$ and $(2, 0)$, and the y-intercepts are $(0, -3)$ and $(0, 3)$. We plot these points and connect them with an oval-shaped curve. To plot two other points, we let $x = 1$ and solve for y:

$$\frac{1^2}{4} + \frac{y^2}{9} = 1$$

$$36\left(\frac{1}{4} + \frac{y^2}{9}\right) = 36 \cdot 1$$

$$36 \cdot \frac{1}{4} + 36 \cdot \frac{y^2}{9} = 36$$

$$9 + 4y^2 = 36$$

$$4y^2 = 27$$

$$y^2 = \frac{27}{4}$$

$$y = \pm\sqrt{\frac{27}{4}}$$

$$y \approx \pm 2.6.$$

Thus, $(1, 2.6)$ and $(1, -2.6)$ can also be used to draw the graph. Similarly, the points $(-1, 2.6)$ and $(-1, -2.6)$ should appear on the graph.

EXAMPLE 2   Graph: $4x^2 + 25y^2 = 100$.

Solution   To write the equation in standard form, we divide both sides by 100 to get 1 on the right side:

Student Notes

Note that any equation of the form $Ax^2 + By^2 = C$ can be rewritten as an equivalent equation in standard form. The graph is an ellipse.

$$\frac{4x^2 + 25y^2}{100} = \frac{100}{100}$$ Dividing by 100 to get 1 on the right side

$$\left.\begin{array}{c} \dfrac{4x^2}{100} + \dfrac{25y^2}{100} = 1 \\[2mm] \dfrac{x^2}{25} + \dfrac{y^2}{4} = 1 \end{array}\right\}$$ Simplifying

$$\frac{x^2}{5^2} + \frac{y^2}{2^2} = 1.$$ $a = 5, b = 2$

The x-intercepts are $(-5, 0)$ and $(5, 0)$, and the y-intercepts are $(0, -2)$ and $(0, 2)$. We plot the intercepts and connect them with an oval-shaped curve. Other points can also be computed and plotted.

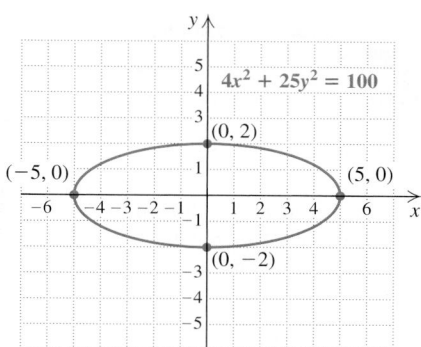

Ellipses Centered at (h, k)

Horizontal and vertical translations, similar to those used in Chapter 8, can be used to graph ellipses that are not centered at the origin.

Equation of an Ellipse Centered at (h, k)

The standard form of a horizontal or vertical ellipse centered at (h, k) is

$$\frac{(x - h)^2}{a^2} + \frac{(y - k)^2}{b^2} = 1.$$

The vertices are $(h + a, k)$ and $(h - a, k)$ if horizontal; $(h, k + b)$ and $(h, k - b)$ if vertical.

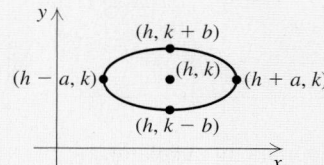

EXAMPLE 3   Graph the ellipse

$$\frac{(x - 1)^2}{4} + \frac{(y + 5)^2}{9} = 1.$$

Solution   Note that

$$\frac{(x - 1)^2}{4} + \frac{(y + 5)^2}{9} = \frac{(x - 1)^2}{2^2} + \frac{(y + 5)^2}{3^2}.$$

Thus, $a = 2$ and $b = 3$. To determine the center of the ellipse, (h, k), note that

$$\frac{(x - 1)^2}{2^2} + \frac{(y + 5)^2}{3^2} = \frac{(x - 1)^2}{2^2} + \frac{(y - (-5))^2}{3^2}.$$

Thus the center is $(1, -5)$. We plot the points 2 units to the left and right of center, as well as the points 3 units above and below center. These are the points $(3, -5)$, $(-1, -5)$, $(1, -2)$, and $(1, -8)$. The graph of the ellipse is shown at left.

Note that this ellipse is the same as the ellipse in Example 1 but translated 1 unit to the right and 5 units down.

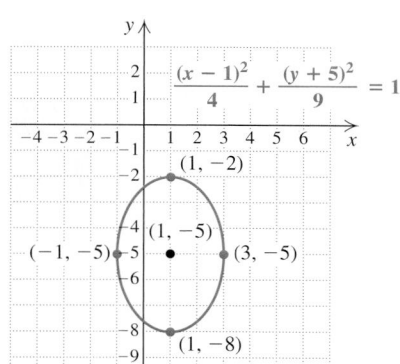

technology connection

Graphing an ellipse on a graphing calculator is much like graphing a circle: We graph it in two pieces after solving for y. To illustrate, let's check Example 2:

$$4x^2 + 25y^2 = 100$$
$$25y^2 = 100 - 4x^2$$
$$y^2 = 4 - \tfrac{4}{25}x^2$$
$$y = \pm\sqrt{4 - \tfrac{4}{25}x^2}.$$

Using a squared window, we have our check:

$$y_1 = -\sqrt{4 - \tfrac{4}{25}x^2}, \quad y_2 = \sqrt{4 - \tfrac{4}{25}x^2}$$

[graph showing ellipse with window from −9 to 9 horizontally and −6 to 6 vertically, labeled y_2 and y_1]

On many calculators, pressing **APPS** and selecting Conics and then Ellipse accesses a program in which equations in Standard Form can be graphed directly.

Ellipses have many applications. Communications satellites move in elliptical orbits with the earth as a focus while the earth itself follows an elliptical path around the sun. A medical instrument, the lithotripter, uses shock waves originating at one focus to crush a kidney stone located at the other focus.

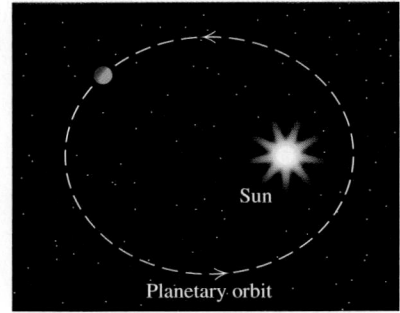

Sun

Planetary orbit

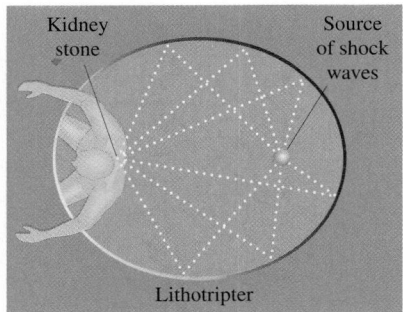

Kidney stone

Source of shock waves

Lithotripter

In some buildings, an ellipsoidal ceiling creates a "whispering gallery" in which a person at one focus can whisper and still be heard clearly at the other focus. This happens because sound waves coming from one focus are all reflected to the other focus. Similarly, light waves bouncing off an ellipsoidal mirror are used in a dentist's or surgeon's reflector light. The light source is located at one focus while the patient's mouth or surgical field is at the other.

Exercise Set

13.2

FOR EXTRA HELP

Student's Solutions Manual

Digital Video Tutor CD 7 Videotape 13

Tutor Center
AW Math Tutor Center

MathXL Tutorials on CD

MathXL
MathXL

MyMathLab
MyMathLab

↘ *Concept Reinforcement Classify each of the following as either true or false.*

1. The graph of $\dfrac{x^2}{9} + \dfrac{y^2}{25} = 1$ includes the points $(-3, 0)$ and $(3, 0)$.

2. The graph of $\dfrac{x^2}{36} + \dfrac{y^2}{25} = 1$ includes the points $(0, -5)$ and $(0, 5)$.

3. The graph of $\dfrac{x^2}{28} + \dfrac{y^2}{48} = 1$ is a vertical ellipse.

4. The graph of $\dfrac{x^2}{30} + \dfrac{y^2}{20} = 1$ is a vertical ellipse.

5. The graph of $\dfrac{x^2}{25} - \dfrac{y^2}{9} = 1$ is a horizontal ellipse.

6. The graph of $\dfrac{-x^2}{20} + \dfrac{y^2}{16} = 1$ is a horizontal ellipse.

7. The graph of $\dfrac{(x+3)^2}{25} + \dfrac{(y-2)^2}{36} = 1$ is an ellipse centered at $(-3, 2)$.

8. The graph of $\dfrac{(x-2)^2}{49} + \dfrac{(y+5)^2}{9} = 1$ is an ellipse centered at $(2, -5)$.

Graph each of the following equations.

9. $\dfrac{x^2}{1} + \dfrac{y^2}{9} = 1$

10. $\dfrac{x^2}{9} + \dfrac{y^2}{1} = 1$

11. $\dfrac{x^2}{25} + \dfrac{y^2}{9} = 1$

12. $\dfrac{x^2}{16} + \dfrac{y^2}{25} = 1$

13. $4x^2 + 9y^2 = 36$

14. $9x^2 + 4y^2 = 36$

15. $16x^2 + 9y^2 = 144$

16. $9x^2 + 16y^2 = 144$

17. $2x^2 + 3y^2 = 6$

18. $5x^2 + 7y^2 = 35$

Aha! **19.** $5x^2 + 5y^2 = 125$

20. $8x^2 + 5y^2 = 80$

21. $3x^2 + 7y^2 - 63 = 0$

22. $3x^2 + 8y^2 - 72 = 0$

23. $8x^2 = 96 - 3y^2$

24. $6y^2 = 24 - 8x^2$

25. $16x^2 + 25y^2 = 1$

26. $9x^2 + 4y^2 = 1$

27. $\dfrac{(x-3)^2}{9} + \dfrac{(y-2)^2}{25} = 1$

28. $\dfrac{(x-2)^2}{25} + \dfrac{(y-4)^2}{9} = 1$

29. $\dfrac{(x+4)^2}{16} + \dfrac{(y-3)^2}{49} = 1$

30. $\dfrac{(x+5)^2}{4} + \dfrac{(y-2)^2}{36} = 1$

31. $12(x-1)^2 + 3(y+4)^2 = 48$
(*Hint*: Divide both sides by 48.)

32. $4(x-6)^2 + 9(y+2)^2 = 36$

Aha! **33.** $4(x+3)^2 + 4(y+1)^2 - 10 = 90$

34. $9(x+6)^2 + (y+2)^2 - 20 = 61$

35. Is the center of an ellipse part of the ellipse itself? Why or why not?

36. Can an ellipse ever be the graph of a function? Why or why not?

SKILL MAINTENANCE

Solve.

37. $\dfrac{3}{x-2} - \dfrac{5}{x-2} = 9$ [6.6]

38. $\dfrac{7}{x+3} - \dfrac{2}{x+3} = 8$ [6.6]

39. $\dfrac{x}{x-4} - \dfrac{3}{x-5} = \dfrac{2}{x-4}$ [11.2]

40. $\dfrac{7}{x-3} - \dfrac{x}{x-2} = \dfrac{4}{x-2}$ [11.2]

41. $9 - \sqrt{2x+1} = 7$ [10.6]

42. $5 - \sqrt{x+3} = 9$ [10.6]

SYNTHESIS

43. An eccentric person builds a pool table in the shape of an ellipse with a hole at one focus and a tiny dot at the other. Guests are amazed at how many bank shots the owner of the pool table makes. Explain why this occurs.

44. Can a circle be considered a special type of ellipse? Why or why not?

Find an equation of an ellipse that contains the following points.

45. $(-9, 0), (9, 0), (0, -11)$, and $(0, 11)$

46. $(-7, 0), (7, 0), (0, -5)$, and $(0, 5)$

47. $(-2, -1), (6, -1), (2, -4)$, and $(2, 2)$

48. $(4, 3), (-6, 3), (-1, -1)$ and $(-1, 7)$

49. *Theatrical lighting.* The spotlight on a violin soloist casts an ellipse of light on the floor below her that is 6 ft wide and 10 ft long. Find an equation of that ellipse if the performer is in its center, x is the distance from the performer to the side of the ellipse, and y is the distance from the performer to the top of the ellipse.

50. *Astronomy.* The maximum distance of the planet Mars from the sun is 2.48×10^8 mi. The minimum distance is 3.46×10^7 mi. The sun is at one focus of the elliptical orbit. Find the distance from the sun to the other focus.

51. Let $(-c, 0)$ and $(c, 0)$ be the foci of an ellipse. Any point $P(x, y)$ is on the ellipse if the sum of the distances from the foci to P is some constant. Use $2a$ to represent this constant.

   a) Show that an equation for the ellipse is given by
   $$\frac{x^2}{a^2} + \frac{y^2}{a^2 - c^2} = 1.$$

   b) Substitute b^2 for $a^2 - c^2$ to get standard form.

52. *President's office.* The Oval Office of the President of the United States is an ellipse 31 ft wide and 38 ft long. Show in a sketch precisely where the President and an adviser could sit to best hear each other using the room's acoustics. (*Hint*: See Exercise 51(b) and the discussion following Example 3.)

53. *Dentistry.* The light source in a dental lamp shines against a reflector that is shaped like a portion of an ellipse in which the light source is one focus of the ellipse. Reflected light enters a patient's mouth at the other focus of the ellipse. If the ellipse from which the reflector was formed is 2 ft wide and 6 ft long, how far should the patient's mouth be from the light source? (*Hint*: See Exercise 51(b).)

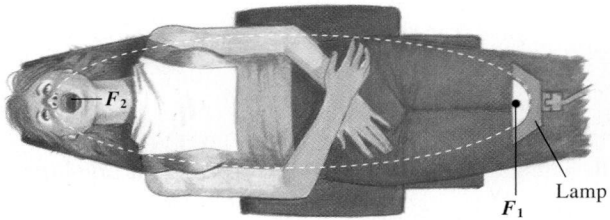

54. *Firefighting.* The size and shape of certain forest fires can be approximated as the union of two "half-ellipses." For the blaze modeled below, the equation of the smaller ellipse—the part of the fire moving *into* the wind—is
$$\frac{x^2}{40,000} + \frac{y^2}{10,000} = 1.$$
The equation of the other ellipse—the part moving *with* the wind—is
$$\frac{x^2}{250,000} + \frac{y^2}{10,000} = 1.$$
Determine the width and the length of the fire.
Source for figure: "Predicting Wind-Driven Wild Land Fire Size and Shape," Hal E. Anderson, Research Paper INT-305, U.S. Department of Agriculture, Forest Service, February 1983

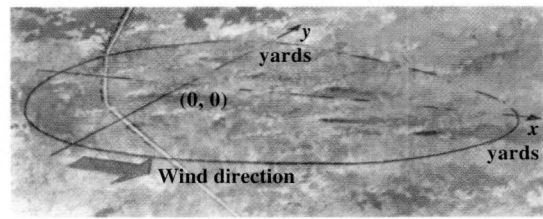

For each of the following equations, complete the square as needed and find an equivalent equation in standard form. Then graph the ellipse.

55. $x^2 - 4x + 4y^2 + 8y - 8 = 0$

56. $4x^2 + 24x + y^2 - 2y - 63 = 0$

57. Use a graphing calculator to check your answers to Exercises 11, 25, 29, and 33.

CORNER

A Cosmic Path

Focus: Ellipses

Time: 20–30 minutes

Group size: 2

Materials: Scientific calculators

In March 1996, the comet Hyakutake came within 21 million mi of the sun, and closer to Earth than any comet in over 500 yr (*Source:* Associated Press newspaper story, 3/20/96). Hyakutake is traveling in an elliptical orbit with the sun at one focus. The comet's average speed is about 100,000 mph (it actually goes much faster near its foci and slower as it gets further from the foci) and one orbit takes about 15,000 yr. (Astronomers estimate the time at 10,000–20,000 yr.)

ACTIVITY

1. The elliptical orbit of Hyakutake is so elongated that the distance traveled in one orbit can be estimated by $4a$ (see the following figure). Use the information above to estimate the distance, in millions of miles, traveled in one orbit. Then determine a.

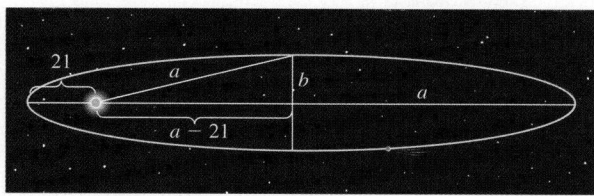

Units are millions of miles.

2. Using the figure above, express b^2 as a function of a. Then solve for b using the value found for a in part (1).
3. Approximately how far will Hyakutake be from the sun at the most distant part of its orbit?
4. Repeat parts (1)–(3), with one group member using the lower estimate of orbit time (10,000 yr) and the other using the upper estimate of orbit time (20,000 yr). By how much do the three answers to part (3) vary?

13.3   Conic Sections: Hyperbolas

Hyperbolas   •   Hyperbolas (Nonstandard Form)   •
Classifying Graphs of Equations

Hyperbolas

A **hyperbola** looks like a pair of parabolas, but the shapes are actually different. A hyperbola has two **vertices** and the line through the vertices is known as the **axis**. The point halfway between the vertices is called the **center**. The two curves that comprise a hyperbola are called **branches**.

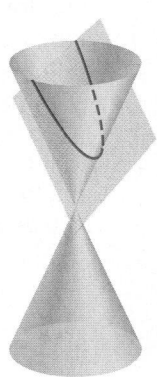

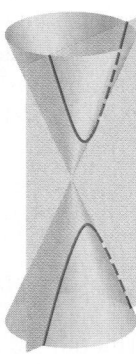
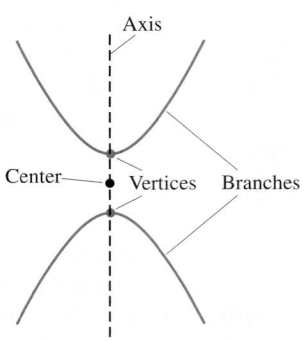

Parabola   Hyperbola in three dimensions   Hyperbola in a plane

---

Equation of a Hyperbola Centered at the Origin

A hyperbola with its center at the origin* has its equation as follows:

$$\frac{x^2}{a^2} - \frac{y^2}{b^2} = 1 \qquad \text{(Horizontal axis);}$$

$$\frac{y^2}{b^2} - \frac{x^2}{a^2} = 1 \qquad \text{(Vertical axis).}$$

Note that both equations have a 1 on the right-hand side and a subtraction symbol between the terms. For the discussion that follows, we assume $a, b > 0$.

---

*Hyperbolas with horizontal or vertical axes and centers *not* at the origin are discussed in Exercises 59–64.

To graph a hyperbola, it helps to begin by graphing two lines called **asymptotes**. Although the asymptotes themselves are not part of the graph, they serve as guidelines for an accurate sketch.

As a hyperbola gets farther away from the origin, it gets closer and closer to its asymptotes. The larger $|x|$ gets, the closer the graph gets to an asymptote. The asymptotes act to "constrain" the graph of a hyperbola. Parabolas are *not* constrained by any asymptotes.

Asymptotes of a Hyperbola

For hyperbolas with equations as shown below, the asymptotes are the lines

$$y = \frac{b}{a}x \quad \text{and} \quad y = -\frac{b}{a}x.$$

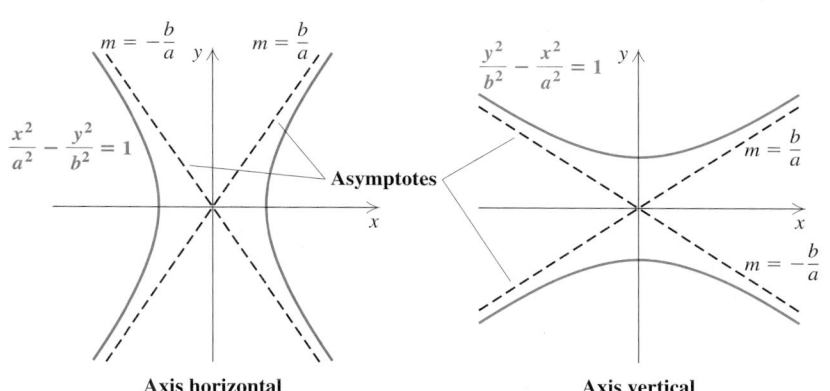

Axis horizontal   Axis vertical

In Section 13.2, we used a and b to determine the width and the length of an ellipse. For hyperbolas, a and b are used to determine the base and the height of a rectangle that can be used as an aid in sketching asymptotes and locating vertices. This is illustrated in the following example.

EXAMPLE 1   Graph: $\dfrac{x^2}{4} - \dfrac{y^2}{9} = 1.$

Solution   Note that

$$\frac{x^2}{4} - \frac{y^2}{9} = \frac{x^2}{2^2} - \frac{y^2}{3^2}, \qquad \text{Identifying } a \text{ and } b$$

so $a = 2$ and $b = 3$. The asymptotes are thus

$$y = \frac{3}{2}x \quad \text{and} \quad y = -\frac{3}{2}x.$$

To help us sketch asymptotes and locate vertices, we use a and b—in this case, 2 and 3—to form the pairs $(-2, 3)$, $(2, 3)$, $(2, -3)$, and $(-2, -3)$. We plot these pairs and lightly sketch a rectangle. The asymptotes pass through the corners and, since this is a horizontal hyperbola, the vertices are where the rectangle intersects the x-axis. Finally, we draw the hyperbola, as shown below.

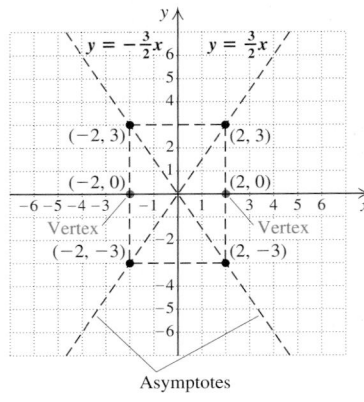 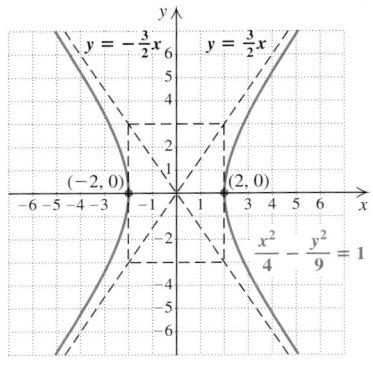

Asymptotes

EXAMPLE 2   Graph: $\dfrac{y^2}{36} - \dfrac{x^2}{4} = 1$.

Solution   Note that

$$\frac{y^2}{36} - \frac{x^2}{4} = \frac{y^2}{6^2} - \frac{x^2}{2^2} = 1.$$

> Whether the hyperbola is horizontal or vertical is determined by the nonnegative term. Here there is a y in this term, so the hyperbola is vertical.

Student Notes

Regarding the orientation of a hyperbola, you may find it helpful to think as follows: "The axis is parallel to the x-axis if $\dfrac{x^2}{a^2}$ is the positive term. The axis is parallel to the y-axis if $\dfrac{y^2}{b^2}$ is the positive term."

Using ± 2 as x-coordinates and ± 6 as y-coordinates, we plot $(2, 6)$, $(2, -6)$, $(-2, 6)$, and $(-2, -6)$, and lightly sketch a rectangle through them. The asymptotes pass through the corners (see the figure on the left below). Since the hyperbola is vertical, its vertices are $(0, 6)$ and $(0, -6)$. Finally, we draw curves through the vertices toward the asymptotes, as shown below.

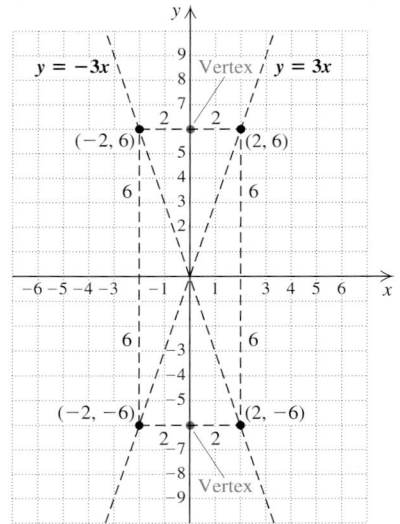 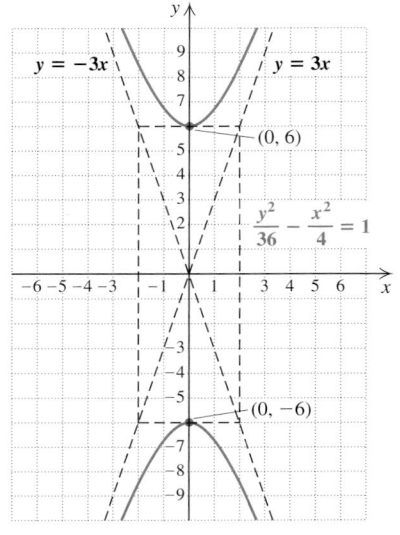

Hyperbolas (Nonstandard Form)

The equations for hyperbolas just examined are the standard ones, but there are other hyperbolas. We consider some of them.

> **Equation of a Hyperbola in Nonstandard Form**
>
> Hyperbolas having the x- and y-axes as asymptotes have equations as follows:
>
> $$xy = c, \quad \text{where } c \text{ is a nonzero constant.}$$

EXAMPLE 3   Graph: $xy = -8$.

Solution   We first solve for y:

$$y = -\frac{8}{x}. \qquad \text{Dividing both sides by } x. \text{ Note that } x \neq 0.$$

Next, we find some solutions, keeping the results in a table. Note that x cannot be 0 and that for large values of $|x|$, y will be close to 0. Thus the x- and y-axes serve as asymptotes. We plot the points and draw two curves.

x	y
2	-4
-2	4
4	-2
-4	2
1	-8
-1	8
8	-1
-8	1

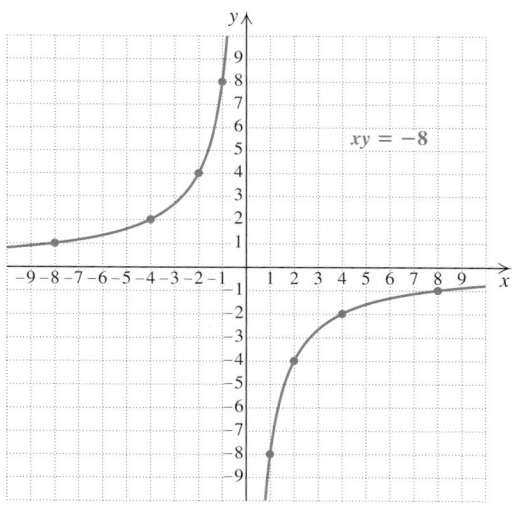

$xy = -8$

Hyperbolas have many applications. A jet breaking the sound barrier creates a sonic boom with a wave front the shape of a cone. The intersection of the cone with the ground is one branch of a hyperbola. Some comets travel in hyperbolic orbits, and a cross section of many lenses is hyperbolic in shape.

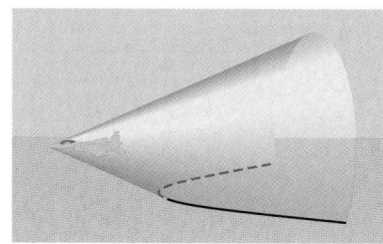

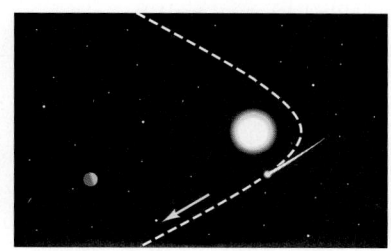

CONNECTING THE CONCEPTS

Recall that the vertical-line test tells us that circles, ellipses, and hyperbolas in standard form do not represent functions. Of the graphs examined in this chapter, only vertical parabolas and hyperbolas similar to the one in Example 3 can represent functions. Because functions are so important, circles and ellipses generally appear in applications of a purely geometric nature, whereas vertical parabolas and nonstandard hyperbolas can be used in applications involving geometry or functions.

In Section 13.4, we return to the challenge of solving real-world problems that translate to a system of equations. There we will find that knowing the general shape of the graph of an equation can help us determine how many solutions, if any, exist.

 technology connection

The procedure used to graph a hyperbola in standard form is similar to that used to draw a circle or an ellipse. Consider the graph of

$$\frac{x^2}{25} - \frac{y^2}{49} = 1.$$

The student should confirm that solving for y yields

$$y_1 = \frac{\sqrt{49x^2 - 1225}}{5}$$

$$= \frac{7}{5}\sqrt{x^2 - 25}$$

and

$$y_2 = \frac{-\sqrt{49x^2 - 1225}}{5}$$

$$= -\frac{7}{5}\sqrt{x^2 - 25},$$

or $y_2 = -y_1.$

When the two pieces are drawn on the same squared window, the result is as shown. The gaps occur where the graph is nearly vertical.

$$y_1 = \tfrac{7}{5}\sqrt{x^2 - 25},$$
$$y_2 = -\tfrac{7}{5}\sqrt{x^2 - 25}$$

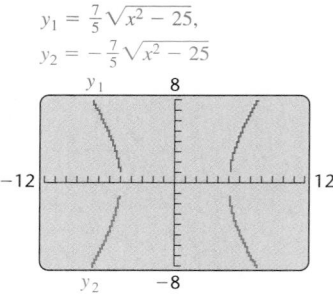

On many calculators, pressing **APPS** and selecting Conics and then Hyperbola accesses a program in which hyperbolas in standard form can be graphed directly.
   Graph.

1. $\dfrac{x^2}{16} - \dfrac{y^2}{60} = 1$ **2.** $16x^2 - 3y^2 = 64$

3. $\dfrac{y^2}{20} - \dfrac{x^2}{64} = 1$ **4.** $45y^2 - 9x^2 = 441$

Classifying Graphs of Equations

We summarize the equations and the graphs of the conic sections studied. The examples resume on p. 685.

PARABOLA

$y = ax^2 + bx + c,\ a > 0$
$\quad = a(x - h)^2 + k$

$y = ax^2 + bx + c,\ a < 0$
$\quad = a(x - h)^2 + k$

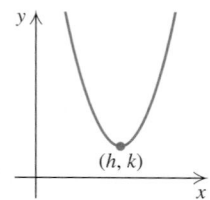

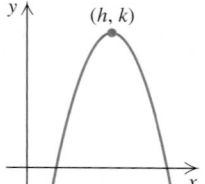

$x = ay^2 + by + c,\ a > 0$
$\quad = a(y - k)^2 + h$

$x = ay^2 + by + c,\ a < 0$
$\quad = a(y - k)^2 + h$

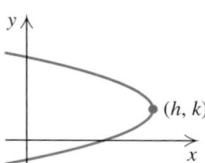

CIRCLE

Center at the origin:
$\quad x^2 + y^2 = r^2$

Center at (h, k):
$\quad (x - h)^2 + (y - k)^2 = r^2$

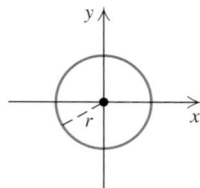

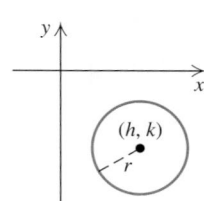

(continued)

CONNECTING THE CONCEPTS

HYPERBOLA

Center at the origin:

$$\frac{x^2}{a^2} - \frac{y^2}{b^2} = 1$$

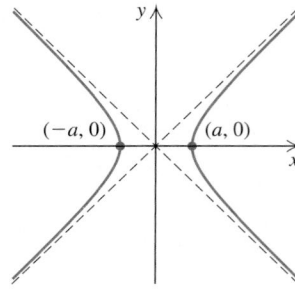

$$\frac{y^2}{b^2} - \frac{x^2}{a^2} = 1$$

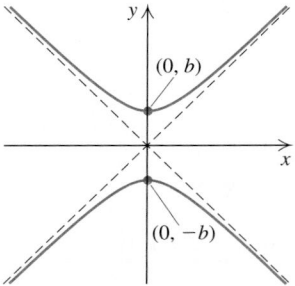

$xy = c, \ c > 0$

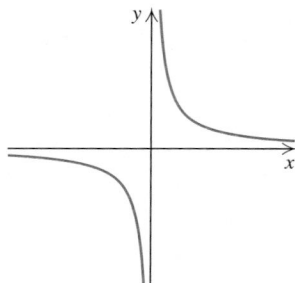

$xy = c, \ c < 0$

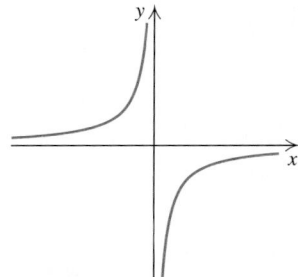

Center at (h, k)*:

$$\frac{(x - h)^2}{a^2} - \frac{(y - k)^2}{b^2} = 1$$

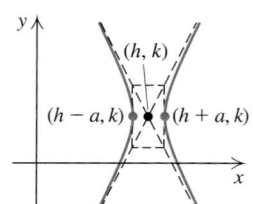

$$\frac{(y - k)^2}{b^2} - \frac{(x - h)^2}{a^2} = 1$$

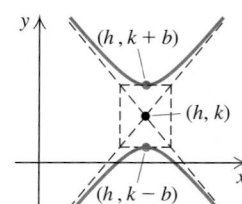

(*continued*)

*See Exercises 59–64.

ELLIPSE

Center at the origin:

$$\frac{x^2}{a^2} + \frac{y^2}{b^2} = 1$$

Center at (h, k):

$$\frac{(x - h)^2}{a^2} + \frac{(y - k)^2}{b^2} = 1$$

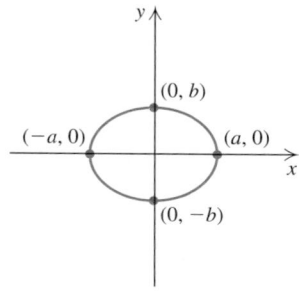

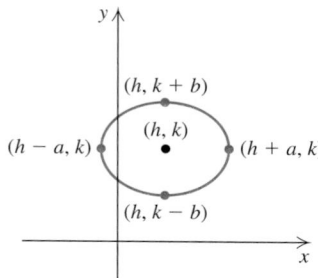

At the beginning of this chapter, we stated that the conic sections represent graphs of $Ax^2 + By^2 + Cxy + Dx + Ey + F = 0$ (we have assumed $C = 0$*):

If A or B (but not both) is 0, the equation can be written in the form $y = ax^2 + bx + c$ or $x = ay^2 + by + c$, and represents a parabola.

If $A = B$, the equation can be written in the form $x^2 + y^2 = c$ or $(x - h)^2 + (y - k)^2 = r^2$, and represents a circle.

If $A \neq B$, but both A and B have the same sign, the equation can be written in the form $b^2x^2 + a^2y^2 = c$ or $b^2(x - h)^2 + a^2(y - k)^2 = c$, and represents an ellipse.

If A and B have opposite signs, the equation can be written in the form $b^2x^2 - a^2y^2 = c$ or $b^2(x - h)^2 - a^2(y - k)^2 = c$, and represents a hyperbola.

Algebraic manipulations may be needed to express an equation in one of the preceding forms.

EXAMPLE 4   Classify the graph of each equation as a circle, an ellipse, a parabola, or a hyperbola.

a) $5x^2 = 20 - 5y^2$

b) $x + 3 + 8y = y^2$

c) $x^2 = y^2 + 4$

d) $x^2 = 16 - 4y^2$

---

*For $C \neq 0$, the graphs are not symmetric with respect to both axes.

Solution

a) We get the terms with variables on one side by adding $5y^2$:

$$5x^2 + 5y^2 = 20.$$

Since x and y are *both* squared, we do not have a parabola. The fact that the squared terms are *added* tells us that we do not have a hyperbola. Do we have a circle? To find out, we need to get $x^2 + y^2$ by itself. We can do that by factoring the 5 out of both terms on the left and then dividing by 5:

$$5(x^2 + y^2) = 20 \qquad \text{Factoring out 5}$$
$$x^2 + y^2 = 4 \qquad \text{Dividing both sides}$$
$$\text{by 5}$$
$$x^2 + y^2 = 2^2. \qquad \text{This is an equation}$$
$$\text{for a circle.}$$

We can see that the graph is a circle with center at the origin and radius 2.

b) The equation $x + 3 + 8y = y^2$ has only one variable squared, so we solve for the other variable:

$$x = y^2 - 8y - 3. \qquad \text{This is an equation}$$
$$\text{for a parabola.}$$

The graph is a horizontal parabola that opens to the right.

c) In $x^2 = y^2 + 4$, both variables are squared, so the graph is not a parabola. We subtract y^2 on both sides and divide by 4 to obtain

$$\frac{x^2}{2^2} - \frac{y^2}{2^2} = 1. \qquad \text{This is an equation for}$$
$$\text{a hyperbola.}$$

The minus sign here indicates that the graph of this equation is a hyperbola. Because it is the x^2-term that is nonnegative, the hyperbola is horizontal.

d) In $x^2 = 16 - 4y^2$, both variables are squared, so the graph cannot be a parabola. We obtain the following equivalent equation:

$$x^2 + 4y^2 = 16.$$

If the coefficients of the terms were the same, we would have the graph of a circle, as in part (a), but they are not. Dividing both sides by 16 yields

$$\frac{x^2}{16} + \frac{y^2}{4} = 1. \qquad \text{This is an equation for}$$
$$\text{an ellipse.}$$

The graph of this equation is a horizontal ellipse.

Exercise Set

13.3

FOR EXTRA HELP

Student's Solutions Manual

Digital Video Tutor CD 7 Videotape 13

AW Math Tutor Center

MathXL Tutorials on CD

Math⧫XL MathXL

MyMathLab MyMathLab

⤻ *Concept Reinforcement* *In each of Exercises 1–8, match the conic section with the equation in the column on the right that represents that type of conic section.*

1. _____ A hyperbola with a horizontal axis

2. _____ A hyperbola with a vertical axis

3. _____ An ellipse with its center not at the origin

4. _____ An ellipse with its center at the origin

5. _____ A circle with its center at the origin

6. _____ A circle with its center not at the origin

7. _____ A parabola opening upward or downward

8. _____ A parabola opening to the right or left

a) $\dfrac{x^2}{10} + \dfrac{y^2}{12} = 1$

b) $(x + 1)^2 + (y - 3)^2 = 30$

c) $y - x^2 = 5$

d) $\dfrac{x^2}{9} - \dfrac{y^2}{10} = 1$

e) $x - 2y^2 = 3$

f) $\dfrac{y^2}{20} - \dfrac{x^2}{35} = 1$

g) $3x^2 + 3y^2 = 75$

h) $\dfrac{(x - 1)^2}{10} + \dfrac{(y - 4)^2}{8} = 1$

Graph each hyperbola. Label all vertices and sketch all asymptotes.

9. $\dfrac{y^2}{16} - \dfrac{x^2}{16} = 1$

10. $\dfrac{x^2}{9} - \dfrac{y^2}{9} = 1$

11. $\dfrac{x^2}{4} - \dfrac{y^2}{25} = 1$

12. $\dfrac{y^2}{16} - \dfrac{x^2}{9} = 1$

13. $\dfrac{y^2}{36} - \dfrac{x^2}{9} = 1$

14. $\dfrac{x^2}{25} - \dfrac{y^2}{36} = 1$

15. $y^2 - x^2 = 25$

16. $x^2 - y^2 = 4$

17. $25x^2 - 16y^2 = 400$

18. $4y^2 - 9x^2 = 36$

Graph.

19. $xy = -5$

20. $xy = 5$

21. $xy = 4$

22. $xy = -9$

23. $xy = -2$

24. $xy = -1$

25. $xy = 1$

26. $xy = 2$

Classify each of the following as the equation of a circle, an ellipse, a parabola, or a hyperbola.

27. $x^2 + y^2 - 6x + 4y - 30 = 0$

28. $y + 9 = 3x^2$

29. $9x^2 + 4y^2 - 36 = 0$

30. $x + 3y = 2y^2 - 1$

31. $4x^2 - 9y^2 - 72 = 0$

32. $y^2 + x^2 = 8$

33. $x^2 + y^2 = 2x + 4y + 4$

34. $2y + 13 + x^2 = 8x - y^2$

35. $4x^2 = 64 - y^2$

36. $y = \dfrac{7}{x}$

37. $x - \dfrac{8}{y} = 0$

38. $x - 4 = y^2 - 3y$

39. $y + 6x = x^2 + 5$

40. $x^2 = 16 + y^2$

41. $9y^2 = 36 + 4x^2$

42. $3x^2 + 5y^2 + x^2 = y^2 + 49$

43. $3x^2 + y^2 - x = 2x^2 - 9x + 10y + 40$

44. $4y^2 + 20x^2 + 1 = 8y - 5x^2$

45. $16x^2 + 5y^2 - 12x^2 + 8y^2 - 3x + 4y = 568$

46. $56x^2 - 17y^2 = 234 - 13x^2 - 38y^2$

47. What does graphing hyperbolas have in common with graphing ellipses?

48. Is it possible for a hyperbola to represent the graph of a function? Why or why not?

SKILL MAINTENANCE

Solve each system. [8.2]

49. $5x + 6y = -12,$
$3x + 9y = 15$

50. $2x + 6y = -6,$
$3x + 5y = 7$

Solve. [5.7]

51. $y^2 - 3 = 6$

52. $x^2 + 3 = 4$

53. The price of a lawn chair, including 5% sales tax, is $36.75. Find the price of the chair before the tax was added. [2.4]

54. A basketball team increases its score by 7 points in each of the two consecutive games after the home opener. If the team scored a total of 228 points in all three games, what was its score in the home opener? [2.5]

SYNTHESIS

55. What is it in the equation of a hyperbola that controls how wide open the branches are? Explain your reasoning.

56. If, in
$$\frac{x^2}{a^2} - \frac{y^2}{b^2} = 1,$$
$a = b$, what are the asymptotes of the graph? Why?

Find an equation of a hyperbola satisfying the given conditions.

57. Having intercepts $(0, 6)$ and $(0, -6)$ and asymptotes $y = 3x$ and $y = -3x$

58. Having intercepts $(8, 0)$ and $(-8, 0)$ and asymptotes $y = 4x$ and $y = -4x$

The standard equations for horizontal or vertical hyperbolas centered at (h, k) are as follows:

$$\frac{(x - h)^2}{a^2} - \frac{(y - k)^2}{b^2} = 1$$

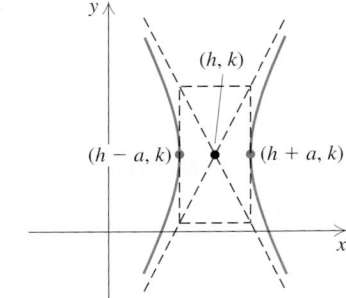

$$\frac{(y - k)^2}{b^2} - \frac{(x - h)^2}{a^2} = 1$$

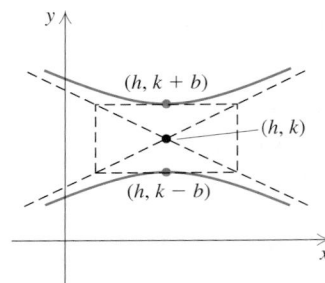

The vertices are as labeled and the asymptotes are
$$y - k = \frac{b}{a}(x - h) \quad and \quad y - k = -\frac{b}{a}(x - h).$$

For each of the following equations of hyperbolas, complete the square, if necessary, and write in standard form. Find the center, the vertices, and the asymptotes. Then graph the hyperbola.

59. $\dfrac{(x - 5)^2}{36} - \dfrac{(y - 2)^2}{25} = 1$

60. $\dfrac{(x - 2)^2}{9} - \dfrac{(y - 1)^2}{4} = 1$

61. $8(y + 3)^2 - 2(x - 4)^2 = 32$

62. $25(x - 4)^2 - 4(y + 5)^2 = 100$

63. $4x^2 - y^2 + 24x + 4y + 28 = 0$

64. $4y^2 - 25x^2 - 8y - 100x - 196 = 0$

65. Use a graphing calculator to check your answers to Exercises 13, 25, 31, and 59.

13.4 Nonlinear Systems of Equations

Systems Involving One Nonlinear Equation • Systems of Two
Nonlinear Equations • Problem Solving

The equations appearing in systems of two equations have thus far in our discussion always been linear. We now consider systems of two equations in
which at least one equation is nonlinear.

Systems Involving One Nonlinear Equation

Suppose that a system consists of an equation of a circle and an equation of a
line. In what ways can the circle and the line intersect? The figures below represent three ways in which the situation can occur. We see that such a system
will have 0, 1, or 2 real solutions.

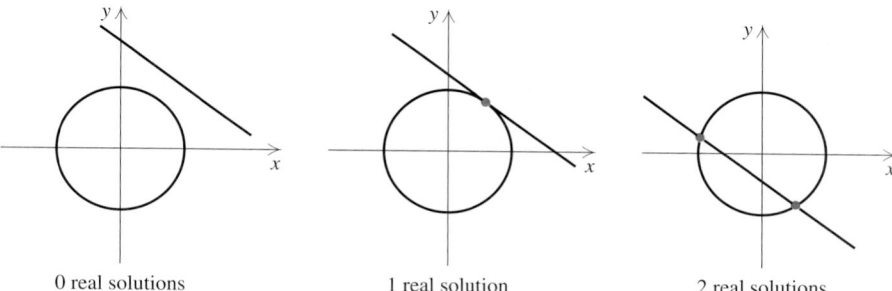

0 real solutions   1 real solution   2 real solutions

Recall that graphing, *elimination*, and *substitution* were all used to solve
systems of linear equations. To solve systems in which one equation is of first
degree and one is of second degree, it is preferable to use the *substitution*
method.

EXAMPLE 1   Solve the system

$$x^2 + y^2 = 25, \quad (1) \qquad \text{(The graph is a circle.)}$$
$$3x - 4y = 0. \quad (2) \qquad \text{(The graph is a line.)}$$

Solution   First, we solve the linear equation, (2), for x:

$$x = \tfrac{4}{3}y. \quad (3) \qquad \text{We could have solved for } y \text{ instead.}$$

Then we substitute $\tfrac{4}{3}y$ for x in equation (1) and solve for y:

$$\left(\tfrac{4}{3}y\right)^2 + y^2 = 25$$
$$\tfrac{16}{9}y^2 + y^2 = 25$$
$$\tfrac{25}{9}y^2 = 25$$
$$y^2 = 9 \qquad \text{Multiplying both sides by } \tfrac{9}{25}$$
$$y = \pm 3. \qquad \text{Using the principle of square roots}$$

Now we substitute these numbers for y in equation (3) and solve for x:

$$\text{for } y = 3, \quad x = \tfrac{4}{3}(3) = 4;$$
$$\text{for } y = -3, \quad x = \tfrac{4}{3}(-3) = -4.$$

Check: For (4, 3):

$$\begin{array}{c|c}
x^2 + y^2 = 25 & \\
\hline
4^2 + 3^2 & 25 \\
16 + 9 & \\
25 \overset{?}{=} 25 & \text{TRUE}
\end{array}
\qquad
\begin{array}{c|c}
3x - 4y = 0 & \\
\hline
3(4) - 4(3) & 0 \\
12 - 12 & \\
0 \overset{?}{=} 0 & \text{TRUE}
\end{array}$$

It is left to the student to confirm that $(-4, -3)$ also checks in both equations.

The pairs (4, 3) and $(-4, -3)$ check, so they are solutions. We can see the solutions in the graph. Intersections occur at (4, 3) and $(-4, -3)$.

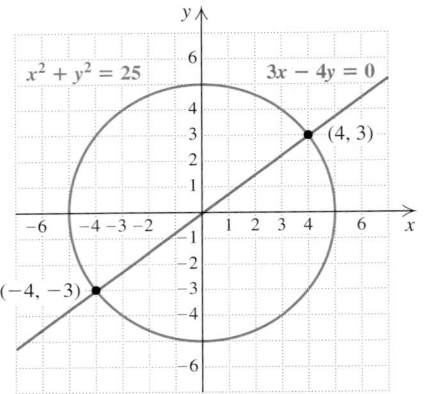

Even if we do not know what the graph of each equation in a system looks like, the algebraic approach of Example 1 can still be used.

EXAMPLE 2   Solve the system

$$y + 3 = 2x, \qquad (1)$$
$$x^2 + 2xy = -1. \qquad (2)$$

Solution   First, we solve the linear equation (1) for y:

$$y = 2x - 3. \qquad (3)$$

Then we substitute $2x - 3$ for y in equation (2) and solve for x:

$$\begin{aligned}
x^2 + 2x(2x - 3) &= -1 \\
x^2 + 4x^2 - 6x &= -1 \\
5x^2 - 6x + 1 &= 0 \\
(5x - 1)(x - 1) &= 0 \qquad && \text{Factoring} \\
5x - 1 = 0 \quad &or \quad x - 1 = 0 \qquad && \text{Using the principle of} \\
& && \text{zero products} \\
x = \tfrac{1}{5} \quad &or \qquad\quad x = 1.
\end{aligned}$$

Now we substitute these numbers for x in equation (3) and solve for y:

for $x = \frac{1}{5}$, $y = 2\left(\frac{1}{5}\right) - 3 = -\frac{13}{5}$;

for $x = 1$, $y = 2(1) - 3 = -1$.

You can confirm that $\left(\frac{1}{5}, -\frac{13}{5}\right)$ and $(1, -1)$ check, so they are both solutions.

EXAMPLE 3   Solve the system

$x + y = 5$,	(1)	(The graph is a line.)
$y = 3 - x^2$.	(2)	(The graph is a parabola.)

Solution   We substitute $3 - x^2$ for y in the first equation:

$$x + 3 - x^2 = 5$$
$$-x^2 + x - 2 = 0 \qquad \text{Adding } -5 \text{ to both sides and rearranging}$$
$$x^2 - x + 2 = 0. \qquad \text{Multiplying both sides by } -1$$

Since $x^2 - x + 2$ does not factor, we need the quadratic formula:

$$x = \frac{-b \pm \sqrt{b^2 - 4ac}}{2a}$$
$$= \frac{-(-1) \pm \sqrt{(-1)^2 - 4 \cdot 1 \cdot 2}}{2(1)} \qquad \text{Substituting}$$
$$= \frac{1 \pm \sqrt{1 - 8}}{2} = \frac{1 \pm \sqrt{-7}}{2} = \frac{1}{2} \pm \frac{\sqrt{7}}{2}i.$$

Solving equation (1) for y gives us $y = 5 - x$. Substituting values for x gives

$$y = 5 - \left(\frac{1}{2} + \frac{\sqrt{7}}{2}i\right) = \frac{9}{2} - \frac{\sqrt{7}}{2}i \quad \text{and}$$
$$y = 5 - \left(\frac{1}{2} - \frac{\sqrt{7}}{2}i\right) = \frac{9}{2} + \frac{\sqrt{7}}{2}i.$$

The solutions are

$$\left(\frac{1}{2} + \frac{\sqrt{7}}{2}i, \frac{9}{2} - \frac{\sqrt{7}}{2}i\right) \quad \text{and} \quad \left(\frac{1}{2} - \frac{\sqrt{7}}{2}i, \frac{9}{2} + \frac{\sqrt{7}}{2}i\right).$$

There are no real-number solutions. Note in the figure at right that the graphs do not intersect. Getting only nonreal solutions tells us that the graphs do not intersect.

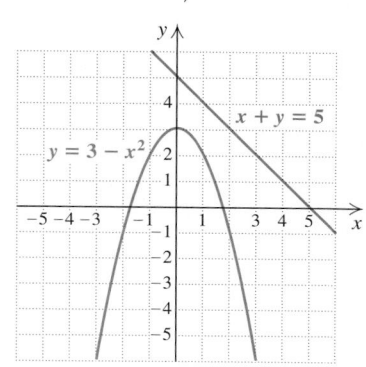

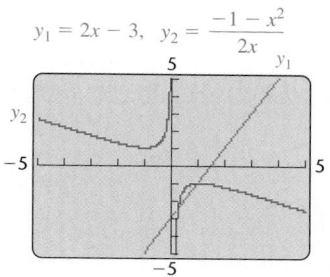

Systems of Two Nonlinear Equations

We now consider systems of two second-degree equations. Graphs of such systems can involve any two conic sections. The following figure shows some ways in which a circle and a hyperbola can intersect.

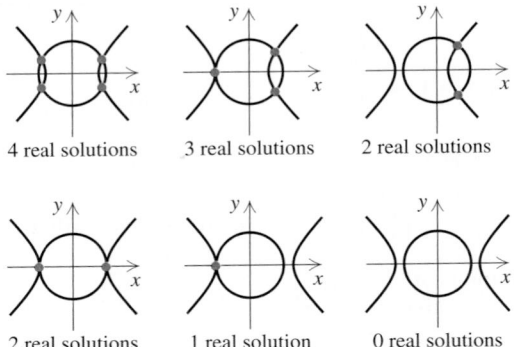

4 real solutions   3 real solutions   2 real solutions

2 real solutions   1 real solution   0 real solutions

To solve systems of two second-degree equations, we either substitute or eliminate. The elimination method is generally better when both equations are of the form $Ax^2 + By^2 = C$. Then we can eliminate an x^2- or y^2-term in a manner similar to the procedure used in Chapter 3.

EXAMPLE 4   Solve the system

$$2x^2 + 5y^2 = 22, \quad (1)$$
$$3x^2 - y^2 = -1. \quad (2)$$

Solution   Here we multiply equation (2) by 5 and then add:

$$
\begin{array}{ll}
2x^2 + 5y^2 = 22 & \\
\underline{15x^2 - 5y^2 = -5} & \text{Multiplying both sides of equation (2) by 5} \\
17x^2 \qquad\quad = 17 & \text{Adding} \\
\qquad x^2 = 1 & \\
\qquad x = \pm 1. &
\end{array}
$$

There is no x-term, and whether x is -1 or 1, we have $x^2 = 1$. Thus we can simultaneously substitute 1 and -1 for x in equation (2):

$$
\left.
\begin{array}{l}
3 \cdot (\pm 1)^2 - y^2 = -1 \\
3 - y^2 = -1 \\
-y^2 = -4
\end{array}
\right\}
\quad
\begin{array}{l}
\text{Since } (-1)^2 = 1^2, \text{ we can evaluate for} \\
x = -1 \text{ and } x = 1 \text{ simultaneously.}
\end{array}
$$

$$y^2 = 4 \quad \text{or} \quad y = \pm 2.$$

Thus, if $x = 1$, then $y = 2$ or $y = -2$; and if $x = -1$, then $y = 2$ or $y = -2$. The four possible solutions are $(1, 2)$, $(1, -2)$, $(-1, 2)$, and $(-1, -2)$.

Check: Since $(2)^2 = (-2)^2$ and $(1)^2 = (-1)^2$, we can check all four pairs at once.

$$
\begin{array}{c|c}
\multicolumn{2}{l}{2x^2 + 5y^2 = 22} \\
\hline
2(\pm 1)^2 + 5(\pm 2)^2 & 22 \\
2 + 20 & \\
22 \overset{?}{=} 22 & \text{TRUE}
\end{array}
\qquad
\begin{array}{c|c}
\multicolumn{2}{l}{3x^2 - y^2 = -1} \\
\hline
3(\pm 1)^2 - (\pm 2)^2 & -1 \\
3 - 4 & \\
-1 \overset{?}{=} -1 & \text{TRUE}
\end{array}
$$

The solutions are $(1, 2)$, $(1, -2)$, $(-1, 2)$, and $(-1, -2)$.

When a product of variables is in one equation and the other equation is of the form $Ax^2 + By^2 = C$, we often solve for a variable in the equation with the product and then use substitution.

EXAMPLE 5   Solve the system

$$x^2 + 4y^2 = 20, \qquad (1)$$
$$xy = 4. \qquad (2)$$

Solution   First, we solve equation (2) for y:

$$y = \frac{4}{x}. \qquad \text{Dividing both sides by } x. \text{ Note that } x \neq 0.$$

Then we substitute $4/x$ for y in equation (1) and solve for x:

$$x^2 + 4\left(\frac{4}{x}\right)^2 = 20$$

$$x^2 + \frac{64}{x^2} = 20$$

$$x^4 + 64 = 20x^2 \qquad \text{Multiplying by } x^2$$

$$x^4 - 20x^2 + 64 = 0 \qquad \begin{array}{l}\text{Obtaining standard} \\ \text{form. This equation is} \\ \text{reducible to quadratic.}\end{array}$$

$$(x^2 - 4)(x^2 - 16) = 0 \qquad \begin{array}{l}\text{Factoring. If you prefer,} \\ \text{let } u = x^2 \text{ and} \\ \text{substitute.}\end{array}$$

$$(x - 2)(x + 2)(x - 4)(x + 4) = 0 \qquad \text{Factoring again}$$

$$x = 2 \quad or \quad x = -2 \quad or \quad x = 4 \quad or \quad x = -4. \qquad \begin{array}{l}\text{Using the prin-} \\ \text{ciple of zero} \\ \text{products}\end{array}$$

Since $y = 4/x$, for $x = 2$, we have $y = 4/2$, or 2. Thus, $(2, 2)$ is a solution. Similarly, $(-2, -2)$, $(4, 1)$, and $(-4, -1)$ are solutions. You can show that all four pairs check.

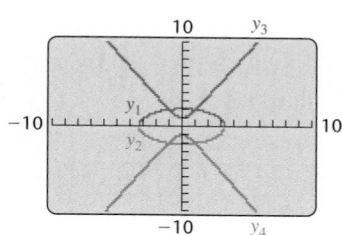

Problem Solving

We now consider applications that can be modeled by a system of equations in which at least one equation is not linear.

EXAMPLE 6   Architecture.   For a college gymnasium, an architect wants to lay out a rectangular piece of land that has a perimeter of 204 m and an area of 2565 m². Find the dimensions of the piece of land.

Solution

1. **Familiarize.**   We draw and label a sketch, letting $l =$ the length and $w =$ the width, both in meters.

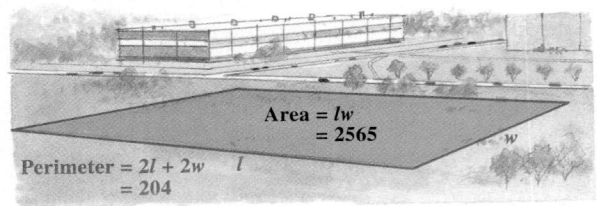

Area = lw
= 2565

Perimeter = $2l + 2w$ l
= 204

2. **Translate.**   We then have the following translation:

   Perimeter: $2w + 2l = 204$;

   Area: $lw = 2565.$

3. **Carry out.**   We solve the system

   $$2w + 2l = 204,$$
   $$lw = 2565.$$

   Solving the second equation for l gives us $l = 2565/w$. Then we substitute $2565/w$ for l in the first equation and solve for w:

   $$2w + 2\left(\frac{2565}{w}\right) = 204$$

   $$2w^2 + 2(2565) = 204w \qquad \text{Multiplying both sides by } w$$

   $$2w^2 - 204w + 2(2565) = 0 \qquad \text{Standard form}$$

   $$w^2 - 102w + 2565 = 0 \qquad \text{Multiplying by } \tfrac{1}{2}$$

   > Factoring could be used instead of the quadratic formula, but the numbers are quite large.

   $$w = \frac{-(-102) \pm \sqrt{(-102)^2 - 4 \cdot 1 \cdot 2565}}{2 \cdot 1}$$

   $$w = \frac{102 \pm \sqrt{144}}{2} = \frac{102 \pm 12}{2}$$

   $$w = 57 \quad or \quad w = 45.$$

   If $w = 57$, then $l = 2565/w = 2565/57 = 45$. If $w = 45$, then $l = 2565/w = 2565/45 = 57$. Since length is usually considered to be longer than width, we have the solution $l = 57$ and $w = 45$, or $(57, 45)$.

4. **Check.**   If $l = 57$ and $w = 45$, the perimeter is $2 \cdot 57 + 2 \cdot 45$, or 204. The area is $57 \cdot 45$, or 2565. The numbers check.

5. **State.**   The length is 57 m and the width is 45 m.

EXAMPLE 7   HDTV dimensions.   High-definition television (HDTV) offers greater clarity than conventional television. The Kaplans' new HDTV screen has an area of 1296 in^2 and has a $\sqrt{3033}$-in. (about 55-in.) diagonal screen. Find the width and the length of the screen.

Solution

1. **Familiarize.**   We make a drawing and label it. Note the right triangle in the figure. We let $l =$ the length and $w =$ the width, both in inches.

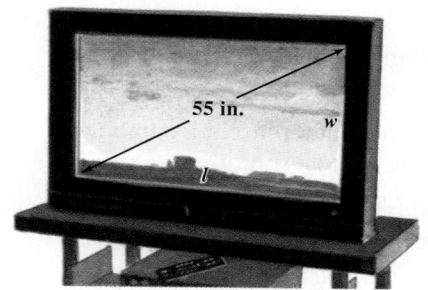

55 in.

2. **Translate.**   We translate to a system of equations:

$$l^2 + w^2 = \sqrt{3033}^2, \qquad \text{Using the Pythagorean theorem}$$
$$lw = 1296. \qquad \text{Using the formula for the area of a rectangle}$$

3. **Carry out.**   We solve the system

$$\left.\begin{array}{l} l^2 + w^2 = 3033, \\ lw = 1296 \end{array}\right\} \quad \text{You should complete the solution of this system.}$$

to get $(48, 27)$, $(27, 48)$, $(-48, -27)$, and $(-27, -48)$.

4. **Check.**   Measurements must be positive and length is usually greater than width, so we check only $(48, 27)$. In the right triangle, $48^2 + 27^2 = 2304 + 729 = 3033$ or $\sqrt{3033}^2$. The area is $48 \cdot 27 = 1296$, so our answer checks.

5. **State.**   The length is 48 in. and the width is 27 in.

Exercise Set

13.4

FOR EXTRA HELP

Student's Solutions Manual Digital Video Tutor CD 7 Videotape 13 AW Math Tutor Center MathXL Tutorials on CD MathXL MyMathLab

↪ *Concept Reinforcement* *Classify each statement as either true or false.*

1. A system of equations that represent a line and an ellipse can have 0, 1, or 2 solutions.

2. A system of equations that represent a parabola and a circle can have up to 4 solutions.

3. A system of equations representing a hyperbola and a circle can have no fewer than 2 solutions.

4. A system of equations representing an ellipse and a line has either 0 or 2 solutions.

5. Systems containing one first-degree equation and one second-degree equation are most easily solved using the substitution method.

6. Systems containing two second-degree equations of the form $Ax^2 + By^2 = C$ are most easily solved using the elimination method.

Solve. Remember that graphs can be used to confirm all real solutions.

7. $x^2 + y^2 = 25,$
   $y - x = 1$

8. $x^2 + y^2 = 100,$
   $y - x = 2$

9. $4x^2 + 9y^2 = 36,$
   $3y + 2x = 6$

10. $9x^2 + 4y^2 = 36,$
    $3x + 2y = 6$

11. $y^2 = x + 3,$
    $2y = x + 4$

12. $y = x^2,$
    $3x = y + 2$

13. $x^2 - xy + 3y^2 = 27,$
    $x - y = 2$

14. $2y^2 + xy + x^2 = 7,$
    $x - 2y = 5$

15. $x^2 + 4y^2 = 25,$
    $x + 2y = 7$

16. $x^2 - y^2 = 16,$
    $x - 2y = 1$

17. $x^2 - xy + 3y^2 = 5,$
    $x - y = 2$

18. $m^2 + 3n^2 = 10,$
    $m - n = 2$

19. $3x + y = 7,$
    $4x^2 + 5y = 24$

20. $2y^2 + xy = 5,$
    $4y + x = 7$

21. $a + b = 7,$
    $ab = 4$

22. $p + q = -6,$
    $pq = -7$

23. $2a + b = 1,$
    $b = 4 - a^2$

24. $4x^2 + 9y^2 = 36,$
    $x + 3y = 3$

25. $a^2 + b^2 = 89,$
    $a - b = 3$

26. $xy = 4,$
    $x + y = 5$

Aha! 27. $y = x^2,$
    $x = y^2$

28. $x^2 + y^2 = 25,$
    $y^2 = x + 5$

29. $x^2 + y^2 = 9,$
    $x^2 - y^2 = 9$

30. $y^2 - 4x^2 = 4,$
    $4x^2 + y^2 = 4$

31. $x^2 + y^2 = 25,$
    $xy = 12$

32. $x^2 - y^2 = 16,$
    $x + y^2 = 4$

33. $x^2 + y^2 = 9,$
    $25x^2 + 16y^2 = 400$

34. $x^2 + y^2 = 4,$
    $9x^2 + 16y^2 = 144$

35. $x^2 + y^2 = 14,$
    $x^2 - y^2 = 4$

36. $x^2 + y^2 = 16,$
    $y^2 - 2x^2 = 10$

37. $x^2 + y^2 = 20,$
    $xy = 8$

38. $x^2 + y^2 = 5,$
    $xy = 2$

39. $x^2 + 4y^2 = 20,$
    $xy = 4$

40. $x^2 + y^2 = 13,$
    $xy = 6$

41. $2xy + 3y^2 = 7,$
    $3xy - 2y^2 = 4$

42. $3xy + x^2 = 34,$
    $2xy - 3x^2 = 8$

43. $4a^2 - 25b^2 = 0,$
    $2a^2 - 10b^2 = 3b + 4$

44. $xy - y^2 = 2,$
    $2xy - 3y^2 = 0$

45. $ab - b^2 = -4,$
    $ab - 2b^2 = -6$

46. $x^2 - y = 5,$
    $x^2 + y^2 = 25$

Solve.

47. *Computer parts.* Dataport Electronics needs a rectangular memory board that has a perimeter of 28 cm and a diagonal of length 10 cm. What should the dimensions of the board be?

48. *Geometry.* A rectangle has an area of 2 yd^2 and a perimeter of 6 yd. Find its dimensions.

49. *Geometry.* A rectangle has an area of 20 in^2 and a perimeter of 18 in. Find its dimensions.

50. *Tile design.* The New World tile company wants to make a new rectangular tile that has a perimeter of 6 in. and a diagonal of length $\sqrt{5}$ in. What should the dimensions of the tile be?

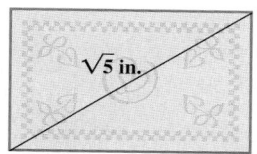

51. *Design of a van.* The cargo area of a delivery van must be 60 ft^2, and the length of a diagonal must accommodate a 13-ft board. Find the dimensions of the cargo area.

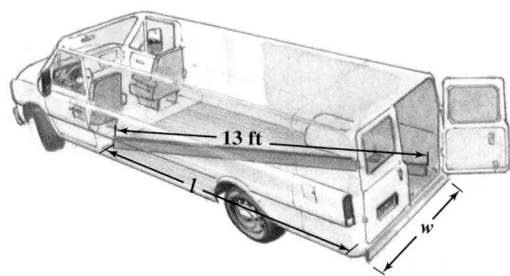

52. *Dimensions of a rug.* The diagonal of a Persian rug is 25 ft. The area of the rug is 300 ft^2. Find the length and the width of the rug.

53. The product of two numbers is 60. The sum of their squares is 136. Find the numbers.

54. *Investments.* A certain amount of money saved for 1 yr at a certain interest rate yielded $225 in interest. If $750 more had been invested and the rate had been 1% less, the interest would have been the same. Find the principal and the rate.

55. *Garden design.* A garden contains two square peanut beds. Find the length of each bed if the sum of their areas is 832 ft^2 and the difference of their areas is 320 ft^2.

56. The area of a rectangle is $\sqrt{3}$ m^2, and the length of a diagonal is 2 m. Find the dimensions.

57. The product of the lengths of the legs of a right triangle is 156. The hypotenuse has length $\sqrt{313}$. Find the lengths of the legs.

58. The area of a rectangle is $\sqrt{2}$ m^2, and the length of a diagonal is $\sqrt{3}$ m. Find the dimensions.

59. How can an understanding of conic sections be helpful when a system of nonlinear equations is being solved algebraically?

60. Suppose a system of equations is comprised of one linear equation and one nonlinear equation. Is it possible for such a system to have three solutions? Why or why not?

SKILL MAINTENANCE

Simplify. [1.8]

61. $(-1)^9(-2)^4$

62. $(-1)^{10}(-2)^5$

Evaluate each of the following. [1.8]

63. $\dfrac{(-1)^k}{k-5}$, for $k = 6$

64. $\dfrac{(-1)^k}{k-5}$, for $k = 9$

65. $\dfrac{n}{2}(3+n)$, for $n = 8$

66. $\dfrac{7(1-r^2)}{1-r}$, for $r = 3$

SYNTHESIS

67. Write a problem that translates to a system of two equations. Design the problem so that at least one equation is nonlinear and so that no real solution exists.

68. Write a problem for a classmate to solve. Devise the problem so that a system of two nonlinear equations with exactly one real solution is solved.

69. Find the equation of a circle that passes through $(-2, 3)$ and $(-4, 1)$ and whose center is on the line $5x + 8y = -2$.

70. Find the equation of an ellipse centered at the origin that passes through the points $(2, -3)$ and $\left(1, \sqrt{13}\right)$.

Solve.

71. $p^2 + q^2 = 13,$
$$\frac{1}{pq} = -\frac{1}{6}$$

72. $a + b = \frac{5}{6},$
$$\frac{a}{b} + \frac{b}{a} = \frac{13}{6}$$

73. *Fence design.* A roll of chain-link fencing contains 100 ft of fence. The fencing is bent at a 90° angle to enclose a rectangular work area of 2475 ft², as shown. Determine the length and the width of the rectangle.

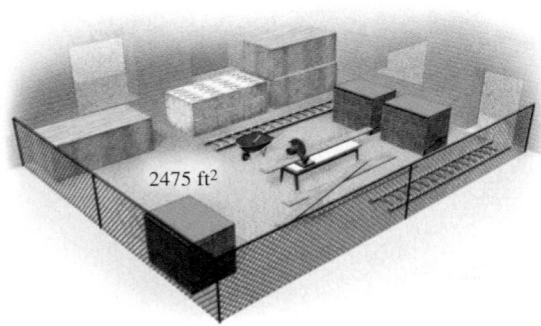

2475 ft²

74. A piece of wire 100 cm long is to be cut into two pieces and those pieces are each to be bent to make a square. The area of one square is to be 144 cm² greater than that of the other. How should the wire be cut?

75. *Box design.* Four squares with sides 5 in. long are cut from the corners of a rectangular metal sheet that has an area of 340 in². The edges are bent up to form an open box with a volume of 350 in³. Find the dimensions of the box.

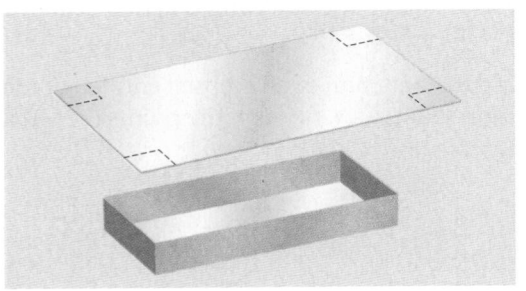

76. *Computer screens.* The ratio of the length to the height of the screen on a computer monitor is 4 to 3. A Dell Inspiron notebook has a 15-in. diagonal screen. Find the dimensions of the screen.

15 in.

77. *HDTV screens.* The ratio of the length to the height of an HDTV screen (see Example 7) is 16 to 9. The Remton Lounge has an HDTV screen with a $\sqrt{4901}$-in. (about 70-in.) diagonal screen. Find the dimensions of the screen.

78. *Railing sales.* Fireside Castings finds that the total revenue R from the sale of x units of railing is given by
$$R = 100x + x^2.$$
Fireside also finds that the total cost C of producing x units of the same product is given by
$$C = 80x + 1500.$$
A break-even point is a value of x for which total revenue is the same as total cost; that is, $R = C$. How many units must be sold to break even?

79. Use a graphing calculator to check your answers to Exercises 13, 25, and 47.

13 Study Summary

The curves formed by cross sections of cones are **conic sections** (p. 856).

Parabola (p. 856):

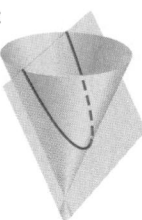

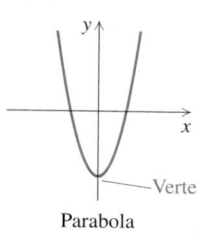

Parabola

$y = ax^2 + bx + c$ (opens upward or downward)
$ = a(x - h)^2 + k$ (vertex at (h, k))

$x = ay^2 + by + c$ (opens right or left)
$ = a(y - k)^2 + h$ (vertex at (h, k))

Circle (p. 861):

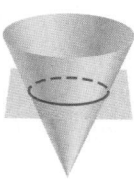

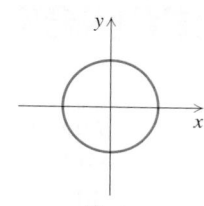

Circle

$x^2 + y^2 = r^2$ (center at $(0, 0)$)

$(x - h)^2 + (y - k)^2 = r^2$ (center at (h, k))

Ellipse (p. 868):

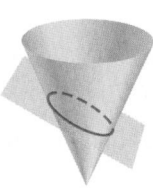

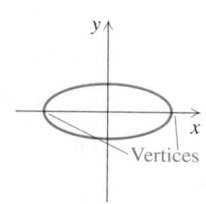

Ellipse

$\dfrac{x^2}{a^2} + \dfrac{y^2}{b^2} = 1$ (center at $(0, 0)$)

$\dfrac{(x - h)^2}{a^2} + \dfrac{(y - k)^2}{b^2} = 1$ (center at (h, k))

Hyperbola (p. 876):

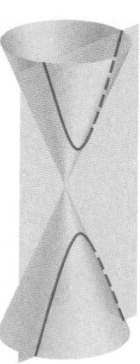

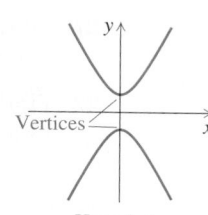

Hyperbola

$\dfrac{x^2}{a^2} - \dfrac{y^2}{b^2} = 1$ (opens right and left)

$\dfrac{y^2}{b^2} - \dfrac{x^2}{a^2} = 1$ (opens upward and downward)

To discuss the conic sections in detail, it is necessary to be able to calculate the **distance** d between any two points (x_1, y_1) and (x_2, y_2):

$$d = \sqrt{(x_2 - x_1)^2 + (y_2 - y_1)^2} \quad \text{(p. 860)}.$$

Using the distance formula, it is possible to develop the formula for the coordinates of the **midpoint** of the segment connecting any two points (x_1, y_1) and (x_2, y_2):

Midpoint: $\left(\dfrac{x_1 + x_2}{2}, \dfrac{y_1 + y_2}{2} \right)$ (p. 861).

Review Exercises

⮌ *Concept Reinforcement*   *Classify each statement as either true or false.*

1. To use the distance formula, one must have an understanding of radical notation. [13.1]

2. The midpoint of the segment connecting (x_1, y_1) and (x_2, y_2) is $(x_1 + x_2, y_1 + y_2)$. [13.1]

3. The center of a circle is part of the circle itself. [13.1]

4. The foci of an ellipse always share the same second coordinate. [13.2]

5. Every parabola that opens upward or downward can represent the graph of a function. [13.1]

6. It is possible for a hyperbola to represent the graph of a function. [13.3]

7. Every system of nonlinear equations has at least one real solution. [13.4]

8. Both substitution and elimination can be used as methods for solving a system of nonlinear equations. [13.4]

Find the distance between each pair of points. Where appropriate, find an approximation to three decimal places. [13.1]

9. $(3, 6)$ and $(7, 6)$

10. $(-1, 1)$ and $(-5, 4)$

11. $(1.4, 3.6)$ and $(4.7, -5.3)$

12. $(2, 3a)$ and $(-1, a)$

Find the midpoint of the segment with the given endpoints. [13.1]

13. $(2, -1)$ and $(7, -1)$

14. $(-1, 10)$ and $(-5, 4)$

15. $\left(1, \sqrt{3}\right)$ and $\left(\frac{1}{2}, -\sqrt{2}\right)$

16. $(2, 3a)$ and $(-1, a)$

Find the center and the radius of each circle. [13.1]

17. $(x + 3)^2 + (y - 2)^2 = 7$

18. $(x - 5)^2 + y^2 = 49$

19. $x^2 + y^2 - 6x - 2y + 1 = 0$

20. $x^2 + y^2 + 8x - 6y = 10$

21. Find an equation of the circle with center $(-4, 3)$ and radius $4\sqrt{3}$. [13.1]

22. Find an equation of the circle with center $(7, -2)$ and radius $2\sqrt{5}$. [13.1]

Classify each equation as a circle, an ellipse, a parabola, or a hyperbola. Then graph.

23. $5x^2 + 5y^2 = 80$ [13.1], [13.3]

24. $9x^2 + 2y^2 = 18$ [13.2], [13.3]

25. $y = -x^2 + 2x - 3$ [13.1], [13.3]

26. $\dfrac{y^2}{9} - \dfrac{x^2}{4} = 1$ [13.3]

27. $xy = 9$ [13.3]

28. $x = y^2 + 2y - 2$ [13.1], [13.3]

29. $\dfrac{(x+1)^2}{3} + (y-3)^2 = 1$ [13.2], [13.3]

30. $x^2 + y^2 + 6x - 8y - 39 = 0$ [13.1], [13.3]

Solve. [13.4]

31. $x^2 - y^2 = 33,$
  $x + y = 11$

32. $x^2 - 2x + 2y^2 = 8,$
  $2x + y = 6$

33. $x^2 - y = 3,$
  $2x - y = 3$

34. $x^2 + y^2 = 25,$
  $x^2 - y^2 = 7$

35. $x^2 - y^2 = 3,$
  $y = x^2 - 3$

36. $x^2 + y^2 = 18,$
  $2x + y = 3$

37. $x^2 + y^2 = 100,$
  $2x^2 - 3y^2 = -120$

38. $x^2 + 2y^2 = 12,$
  $xy = 4$

39. A rectangular bandstand has a perimeter of 38 m and an area of 84 m². What are the dimensions of the bandstand? [13.4]

40. One type of carton used by tableproducts.com exactly fits both a rectangular napkin of area 108 in² and a candle of length 15 in., laid diagonally on top of the napkin. Find the length and the width of the carton. [13.4]

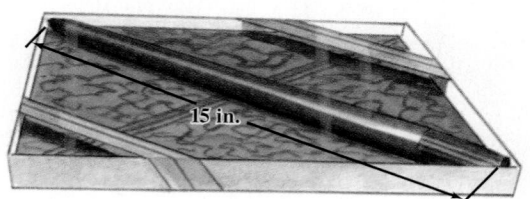

15 in.

41. The perimeter of a square mirror is 12 cm more than the perimeter of another square mirror. Its area exceeds the area of the other by 39 cm². Find the perimeter of each mirror. [13.4]

42. The sum of the areas of two circles is 130π ft². The difference of the circumferences is 16π ft. Find the radius of each circle. [13.4]

SYNTHESIS

43. How does the graph of a hyperbola differ from the graph of a parabola? [13.1], [13.3]

44. Explain why function notation rarely appears in this chapter, and list the graphs discussed for which function notation could be used. [13.1], [13.2], [13.3]

45. Solve: [13.4]
$$4x^2 - x - 3y^2 = 9,$$
$$-x^2 + x + y^2 = 2.$$

46. Find the points whose distance from $(8, 0)$ and from $(-8, 0)$ is 10. [13.1]

47. Find an equation of the circle that passes through $(-2, -4)$, $(5, -5)$, and $(6, 2)$. [13.1], [13.4]

48. Find an equation of the ellipse with the following intercepts: $(-9, 0)$, $(9, 0)$, $(0, -5)$, and $(0, 5)$. [13.2]

49. Find the point on the x-axis that is equidistant from $(-3, 4)$ and $(5, 6)$. [13.1]

13 Chapter Test

Find the distance between each pair of points. Where appropriate, find an approximation to three decimal places.

1. $(5, -1)$ and $(-4, 8)$

2. $(3, -a)$ and $(-3, a)$

Find the midpoint of the segment with the given endpoints.

3. $(4, -1)$ and $(-5, 8)$

4. $(3, -a)$ and $(-3, a)$

Find the center and the radius of each circle.

5. $(x + 5)^2 + (y - 1)^2 = 81$

6. $x^2 + y^2 + 4x - 6y + 4 = 0$

Classify the equation as a circle, an ellipse, a parabola, or a hyperbola. Then graph.

7. $y = x^2 - 4x - 1$

8. $x^2 + y^2 + 2x + 6y + 6 = 0$

9. $\dfrac{x^2}{16} - \dfrac{y^2}{9} = 1$

10. $16x^2 + 4y^2 = 64$

11. $xy = -5$

12. $x = -y^2 + 4y$

Solve.

13. $\dfrac{x^2}{4} + \dfrac{y^2}{9} = 1,$
  $3x + 4y = 12$

14. $x^2 + y^2 = 16,$
  $\dfrac{x^2}{16} - \dfrac{y^2}{9} = 1$

15. $x^2 - 2y^2 = 1,$
  $xy = 6$

16. $x^2 + y^2 = 10,$
  $x^2 = y^2 + 2$

17. A rectangular bookmark with diagonal of length $5\sqrt{5}$ has an area of 22. Find the dimensions of the bookmark.

18. Two squares are such that the sum of their areas is 8 m^2 and the difference of their areas is 2 m^2. Find the length of a side of each square.

19. A rectangular dance floor has a diagonal of length 40 ft and a perimeter of 112 ft. Find the dimensions of the dance floor.

20. Nikki invested a certain amount of money for 1 yr and earned $72 in interest. Erin invested $240 more than Nikki at an interest rate that was $\frac{5}{6}$ of the rate given to Nikki, but she earned the same amount of interest. Find the principal and the interest rate for Nikki's investment.

SYNTHESIS

21. Find an equation of the ellipse passing through $(6, 0)$ and $(6, 6)$ with vertices at $(1, 3)$ and $(11, 3)$.

22. Find the point on the y-axis that is equidistant from $(-3, -5)$ and $(4, -7)$.

23. The sum of two numbers is 36, and the product is 4. Find the sum of the reciprocals of the numbers.

24. *Theatrical production.*   An E.T.C. spotlight for a college's production of *Hamlet* projects an ellipse of light on a stage that is 8 ft wide and 14 ft long. Find an equation of that ellipse if an actor is in its center and x represents the number of feet, horizontally, from the actor to the edge of the ellipse and y represents the number of feet, vertically, from the actor to the edge of the ellipse.

14

Sequences, Series, and the Binomial Theorem

AN APPLICATION

Approximately 534,000 new apartments and houses were built in the United States in 1991. Since then, the number has grown by about 5.35% per year. (*Sources*: Based on data from the U.S. Bureau of the Census and the U.S. Department of Housing and Urban Development) How many new apartments and houses were built in the United States from 1991 through 2004?

This problem appears as Exercise 65 in Section 14.3.

Beverley Dockeray-Ojo
CITY PLANNING DIRECTOR
Atlanta, Georgia

Mathematics is applied in the urban planning field intensively. Specifically used in statistical analysis and demographic projections including population, employment, and income, it is also applied in basic land, floor area, and density calculations. Like many other professions, the numbers tell the story, and provide the basis for problem identification and possible solutions.

*T*he first three sections of this chapter are devoted to sequences *and* series. A sequence is simply an ordered list. For example, when a baseball coach writes a batting order, a sequence is being formed. When the members of a sequence are numbers, they can be added. Such a sum is called a series.

Section 14.4 presents *the* binomial theorem, *which is used to expand expressions of the form* $(a + b)^n$. *Such an expansion is itself a series.*

14.1 Sequences and Series

Sequences • Finding the General Term •
Sums and Series • Sigma Notation

Sequences

Suppose that $10,000 is borrowed at 5%, compounded annually. The value of the loan at the start of years 1, 2, 3, 4, and so on, is

$10,000, $10,500, $11,025, $11,576.25,....

We can regard this as a function that pairs 1 with $10,000, 2 with $10,500, 3 with $11,025, and so on. A **sequence** (or **progression**) is thus a function, where the domain is a set of consecutive positive integers beginning with 1, and the range varies from sequence to sequence.

If we continue computing the amounts in the account forever, we obtain an **infinite sequence**, with function values

$10,000, $10,500, $11,025, $11,576.25, $12,155.06,....

The three dots at the end indicate that the sequence goes on without stopping. If we stop after a certain number of years, we obtain a **finite sequence**:

$10,000, $10,500, $11,025, $11,576.25.

Sequences

An *infinite sequence* is a function having for its domain the set of natural numbers: $\{1, 2, 3, 4, 5, \ldots\}$.

A *finite sequence* is a function having for its domain a set of natural numbers: $\{1, 2, 3, 4, 5, \ldots, n\}$, for some natural number n.

As another example, consider the sequence given by

$$a(n) = 2^n, \quad \text{or} \quad a_n = 2^n.$$

The notation a_n means the same as $a(n)$ but is used more commonly with sequences. Some function values (also called *terms* of the sequence) follow:

$$a_1 = 2^1 = 2,$$
$$a_2 = 2^2 = 4,$$
$$a_3 = 2^3 = 8,$$
$$a_6 = 2^6 = 64.$$

The first term of the sequence is a_1, the fifth term is a_5, and the nth term, or **general term**, is a_n. This sequence can also be denoted in the following ways:

$$2, 4, 8, \ldots;$$

or $2, 4, 8, \ldots, 2^n, \ldots.$ The 2^n emphasizes that the nth term of this sequence is found by raising 2 to the nth power.

EXAMPLE 1   Find the first four terms and the 57th term of the sequence for which the general term is given by $a_n = (-1)^n/(n + 1)$.

Solution   We have

$$a_1 = \frac{(-1)^1}{1 + 1} = -\frac{1}{2},$$
$$a_2 = \frac{(-1)^2}{2 + 1} = \frac{1}{3},$$
$$a_3 = \frac{(-1)^3}{3 + 1} = -\frac{1}{4},$$
$$a_4 = \frac{(-1)^4}{4 + 1} = \frac{1}{5},$$
$$a_{57} = \frac{(-1)^{57}}{57 + 1} = -\frac{1}{58}.$$

Note that the expression $(-1)^n$ causes the signs of the terms to alternate between positive and negative, depending on whether n is even or odd.

technology connection

Sequences are entered and graphed much like functions. The difference is that the SEQUENCE MODE must be selected. You can then enter U_n or V_n using n as the variable. Use this approach to check Example 1 with a table of values for the sequence.

Finding the General Term

When only the first few terms of a sequence are known, it is impossible to be certain what the general term is. Still, a prediction can be made by looking for a pattern.

EXAMPLE 2   For each sequence, predict the general term.

a) $1, 4, 9, 16, 25, \ldots$ **b)** $-1, 2, -4, 8, -16, \ldots$
c) $2, 4, 8, \ldots$

Solution

a) $1, 4, 9, 16, 25, \ldots$

These are squares of consecutive positive integers, so the general term could be n^2.

b) $-1, 2, -4, 8, -16, \ldots$

These are powers of 2 with alternating signs, so the general term may be $(-1)^n[2^{n-1}]$. To check, note that 8 is the fourth term, and

$$(-1)^4[2^{4-1}] = 1 \cdot 2^3$$
$$= 8.$$

c) $2, 4, 8, \ldots$

We regard the pattern as powers of 2, in which case 16 would be the next term and 2^n the general term. The sequence could then be written with more terms as

$$2, 4, 8, 16, 32, 64, 128, \ldots.$$

In part (c) above, suppose that the second term is found by adding 2, the third term by adding 4, the next term by adding 6, and so on. In this case, 14 would be the next term and the sequence would be

$$2, 4, 8, 14, 22, 32, 44, 58, \ldots.$$

This illustrates that the fewer terms we are given, the greater the uncertainty about the *n*th term.

Sums and Series

> **Series**
>
> Given the infinite sequence
>
> $$a_1, \ a_2, \ a_3, \ a_4, \ \ldots, \ a_n, \ldots,$$
>
> the sum of the terms
>
> $$a_1 + a_2 + a_3 + \cdots + a_n + \cdots$$
>
> is called an *infinite series* and is denoted S_∞. A *partial sum* is the sum of the first *n* terms:
>
> $$a_1 + a_2 + a_3 + \cdots + a_n.$$
>
> A partial sum is also called a *finite series* and is denoted S_n.

EXAMPLE 3   For the sequence $-2, 4, -6, 8, -10, 12, -14$, find: **(a)** S_2; **(b)** S_3; **(c)** S_7.

Solution

a) $S_2 = -2 + 4 = 2$ This is the sum of the first 2 terms.

b) $S_3 = -2 + 4 + (-6) = -4$ This is the sum of the first 3 terms.

c) $S_7 = -2 + 4 + (-6) + 8 + (-10) + 12 + (-14) = -8$ This is the sum of the first 7 terms.

Sigma Notation

When the general term of a sequence is known, the Greek letter Σ (capital sigma) can be used to write a series. For example, the sum of the first four terms of the sequence 3, 5, 7, 9, 11, ..., $2k + 1$, ... can be named as follows, using *sigma notation*, or *summation notation*:

$$\sum_{k=1}^{4} (2k + 1). \qquad \begin{array}{l} \text{This represents} \\ (2 \cdot 1 + 1) + (2 \cdot 2 + 1) + (2 \cdot 3 + 1) + (2 \cdot 4 + 1). \end{array}$$

This is read "the sum as k goes from 1 to 4 of $(2k + 1)$." The letter k is called the *index of summation*. The index of summation need not always start at 1.

EXAMPLE 4   Write out and evaluate each sum.

a) $\sum_{k=1}^{5} k^2$ 　　　　　b) $\sum_{k=4}^{6} (-1)^k (2k)$ 　　　　　c) $\sum_{k=0}^{3} (2^k + 5)$

Solution

a) $\sum_{k=1}^{5} k^2 = 1^2 + 2^2 + 3^2 + 4^2 + 5^2 = 1 + 4 + 9 + 16 + 25 = 55$

Evaluate k^2 for all integers from 1 through 5. Then add.

b) $\sum_{k=4}^{6} (-1)^k (2k) = (-1)^4 (2 \cdot 4) + (-1)^5 (2 \cdot 5) + (-1)^6 (2 \cdot 6)$

$= 8 - 10 + 12 = 10$

c) $\sum_{k=0}^{3} (2^k + 5) = (2^0 + 5) + (2^1 + 5) + (2^2 + 5) + (2^3 + 5)$

$= 6 + 7 + 9 + 13 = 35$

Student Notes

A great deal of information is condensed into sigma notation. Be careful to pay attention to what values the index of summation will take on. Evaluate the expression following sigma, the general term, for each value and then add the results.

EXAMPLE 5   Write sigma notation for each sum.

a) $1 + 4 + 9 + 16 + 25$

b) $-1 + 3 - 5 + 7$

c) $3 + 9 + 27 + 81 + \cdots$

Solution

a) $1 + 4 + 9 + 16 + 25$

Note that this is a sum of squares, $1^2 + 2^2 + 3^2 + 4^2 + 5^2$, so the general term is k^2. Sigma notation is

$$\sum_{k=1}^{5} k^2. \qquad \text{The sum starts with } 1^2 \text{ and ends with } 5^2.$$

Answers can vary here. For example, another—perhaps less obvious—way of writing $1 + 4 + 9 + 16 + 25$ is

$$\sum_{k=2}^{6} (k - 1)^2.$$

b) $-1 + 3 - 5 + 7$

Except for the alternating signs, this is the sum of the first four positive odd numbers. It is useful to note that $2k - 1$ is a formula for the kth positive odd number. It is also important to note that since $(-1)^k = 1$ when k is even and $(-1)^k = -1$ when k is odd, the factor $(-1)^k$ can be used to create the alternating signs. The general term is thus $(-1)^k(2k - 1)$, beginning with $k = 1$. Sigma notation is

$$\sum_{k=1}^{4} (-1)^k(2k - 1).$$

To check, we can evaluate $(-1)^k(2k - 1)$ using 1, 2, 3, and 4. Then we can write the sum of the four terms. We leave this to the student.

c) $3 + 9 + 27 + 81 + \cdots$

This is a sum of powers of 3, and it is also an infinite series. We use the symbol ∞ for infinity and write the series using sigma notation:

$$\sum_{k=1}^{\infty} 3^k.$$

Exercise Set

14.1

↪ *Concept Reinforcement*   *In each of Exercises 1–6, match the expression with the most appropriate expression from the column on the right.*

1. ____ $\displaystyle\sum_{k=1}^{4} k^2$

2. ____ $\displaystyle\sum_{k=3}^{6} (-1)^k$

3. ____ $5 + 10 + 15 + 20$

4. ____ $a_n = 5^n$

5. ____ $a_n = 3n + 2$

6. ____ $a_1 + a_2 + a_3$

a) $-1 + 1 + (-1) + 1$

b) $a_2 = 25$

c) $a_2 = 8$

d) $\displaystyle\sum_{k=1}^{4} 5k$

e) S_3

f) $1 + 4 + 9 + 16$

In each of the following, the nth term of a sequence is given. In each case, find the first 4 terms; the 10th term, a_{10}; and the 15th term, a_{15}.

7. $a_n = 2n + 3$

8. $a_n = 5n - 2$

9. $a_n = n^2 + 2$

10. $a_n = \dfrac{n}{n + 1}$

11. $a_n = \dfrac{n^2 - 1}{n^2 + 1}$

12. $a_n = n^2 - 2n$

13. $a_n = \left(-\dfrac{1}{2}\right)^{n-1}$

14. $a_n = n + \dfrac{1}{n}$

15. $a_n = (-1)^n(n + 3)$

16. $a_n = (-1)^n n^2$

17. $a_n = (-1)^n(n^3 - 1)$

18. $a_n = (-1)^{n+1}(3n - 5)$

Find the indicated term of each sequence.

19. $a_n = 2n - 3;\ a_8$

20. $a_n = 3n + 2;\ a_8$

21. $a_n = (3n + 1)(2n - 5);\ a_9$

22. $a_n = (3n + 2)^2; a_6$

23. $a_n = (-1)^{n-1}(3.4n - 17.3); a_{12}$

24. $a_n = (-2)^{n-2}(45.68 - 1.2n); a_{23}$

25. $a_n = 3n^2(9n - 100); a_{11}$

26. $a_n = 4n^2(2n - 39); a_{22}$

27. $a_n = \left(1 + \dfrac{1}{n}\right)^2; a_{20}$

28. $a_n = \left(1 - \dfrac{1}{n}\right)^3; a_{15}$

Look for a pattern and then predict the general term, or nth term, a_n, of each sequence. Answers may vary.

29. $2, 4, 6, 8, 10, \ldots$

30. $1, 3, 5, 7, \ldots$

31. $1, -1, 1, -1, \ldots$

32. $-1, 1, -1, 1, \ldots$

33. $-1, 2, -3, 4, \ldots$

34. $1, -2, 3, -4, \ldots$

35. $3, 5, 7, 9, \ldots$

36. $4, 6, 8, 10, \ldots$

37. $-2, 6, -18, 54, \ldots$

38. $-2, 3, 8, 13, 18, \ldots$

39. $\frac{1}{2}, \frac{2}{3}, \frac{3}{4}, \frac{4}{5}, \frac{5}{6}, \ldots$

40. $1 \cdot 2, 2 \cdot 3, 3 \cdot 4, 4 \cdot 5, \ldots$

41. $5, 25, 125, 625, \ldots$

42. $4, 16, 64, 256, \ldots$

43. $-1, 4, -9, 16, \ldots$

44. $1, -4, 9, -16, \ldots$

Find the indicated partial sum for each sequence.

45. $1, -2, 3, -4, 5, -6, \ldots; S_7$

46. $1, -3, 5, -7, 9, -11, \ldots; S_8$

47. $2, 4, 6, 8, \ldots; S_5$

48. $1, \frac{1}{4}, \frac{1}{9}, \frac{1}{16}, \frac{1}{25}, \ldots; S_5$

Write out and evaluate each sum.

49. $\displaystyle\sum_{k=1}^{5} \dfrac{1}{2k}$

50. $\displaystyle\sum_{k=1}^{6} \dfrac{1}{2k - 1}$

51. $\displaystyle\sum_{k=0}^{4} 3^k$

52. $\displaystyle\sum_{k=4}^{7} \sqrt{2k + 1}$

53. $\displaystyle\sum_{k=2}^{8} \dfrac{k}{k - 1}$

54. $\displaystyle\sum_{k=2}^{5} \dfrac{k - 2}{k + 3}$

55. $\displaystyle\sum_{k=1}^{8} (-1)^{k+1}2^k$

56. $\displaystyle\sum_{k=1}^{7} (-1)^k 4^{k+1}$

57. $\displaystyle\sum_{k=0}^{5} (k^2 - 2k + 3)$

58. $\displaystyle\sum_{k=0}^{5} (k^2 - 3k + 4)$

59. $\displaystyle\sum_{k=3}^{5} \dfrac{(-1)^k}{k(k + 1)}$

60. $\displaystyle\sum_{k=3}^{7} \dfrac{k}{2^k}$

Rewrite each sum using sigma notation. Answers may vary.

61. $\dfrac{2}{3} + \dfrac{3}{4} + \dfrac{4}{5} + \dfrac{5}{6} + \dfrac{6}{7}$

62. $3 + 6 + 9 + 12 + 15$

63. $1 + 4 + 9 + 16 + 25 + 36$

64. $\dfrac{1}{1^2} + \dfrac{1}{2^2} + \dfrac{1}{3^2} + \dfrac{1}{4^2} + \dfrac{1}{5^2}$

65. $4 - 9 + 16 - 25 + \cdots + (-1)^n n^2$

66. $9 - 16 + 25 + \cdots + (-1)^{n+1} n^2$

67. $5 + 10 + 15 + 20 + 25 + \cdots$

68. $7 + 14 + 21 + 28 + 35 + \cdots$

69. $\dfrac{1}{1 \cdot 2} + \dfrac{1}{2 \cdot 3} + \dfrac{1}{3 \cdot 4} + \dfrac{1}{4 \cdot 5} + \cdots$

70. $\dfrac{1}{1 \cdot 2^2} + \dfrac{1}{2 \cdot 3^2} + \dfrac{1}{3 \cdot 4^2} + \dfrac{1}{4 \cdot 5^2} + \cdots$

71. The sequence $1, 4, 9, 16, \ldots$ can be written as $f(x) = x^2$ with the domain the set of all positive integers. Explain how the graph of f would compare with the graph of $y = x^2$.

72. Eric says he expects he will prefer sequences to functions because he dislikes fractions. Will his expectations prove correct? Why or why not?

SKILL MAINTENANCE

Evaluate. [1.1]

73. $\frac{7}{2}(a_1 + a_7)$, for $a_1 = 8$ and $a_7 = 14$

74. $a_1 + (n - 1)d$, for $a_1 = 3$, $n = 6$, and $d = 4$

Multiply. [4.4]

75. $(x + y)^3$

76. $(a - b)^3$

77. $(2a - b)^3$

78. $(2x + y)^3$

SYNTHESIS

79. Explain why the equation

$$\sum_{k=1}^{n} (a_k + b_k) = \sum_{k=1}^{n} a_k + \sum_{k=1}^{n} b_k$$

is true for any positive integer n. What laws are used to justify this result?

80. Consider the sums

$$\sum_{k=1}^{5} 3k^2 \quad \text{and} \quad 3\sum_{k=1}^{5} k^2.$$

  a) Which is easier to evaluate and why?
  b) Is it true that

$$\sum_{k=1}^{n} ca_k = c\sum_{k=1}^{n} a_k?$$

  Why or why not?

Some sequences are given by a recursive *definition. The value of the first term, a_1, is given, and then we are told how to find any subsequent term from the term preceding it. Find the first six terms of each of the following recursively defined sequences.*

81. $a_1 = 1$, $a_{n+1} = 5a_n - 2$

82. $a_1 = 0$, $a_{n+1} = a_n^2 + 3$

83. *Value of a copier.*   The value of a color photocopier is $5200. Its scrap value each year is 75% of its value the year before. Give a sequence that lists the scrap value of the machine at the start of each year for a 10-yr period.

84. *Cell biology.*   A single cell of bacterium divides into two every 15 min. Suppose that the same rate of division is maintained for 4 hr. Give a sequence that lists the number of cells after successive 15-min periods.

85. Find S_{100} and S_{101} for the sequence in which $a_n = (-1)^n$.

Find the first five terms of each sequence; then find S_5.

86. $a_n = \frac{1}{2^n} \log 1000^n$

87. $a_n = i^n$, $i = \sqrt{-1}$

88. Find all values for x that solve the following:

$$\sum_{k=1}^{x} i^k = -1.$$

89. The nth term of a sequence is given by

$$a_n = n^5 - 14n^4 + 6n^3 + 416n^2 - 655n - 1050.$$

Use a graphing calculator with a TABLE feature to determine what term in the sequence is 6144.

90. To define a sequence recursively on a graphing calculator (see Exercises 81 and 82), the SEQ MODE is used. The general term U_n or V_n can often be expressed in terms of U_{n-1} or V_{n-1} by pressing **2ND** **7** or **2ND** **8**. The starting values of U_n, V_n, and n are set as one of the WINDOW variables.

   Use recursion to determine how many handshakes will occur if a group of 50 people shake hands with one another. To develop the recursion formula, begin with a group of 2 and determine how many additional handshakes occur with the arrival of each new group member.

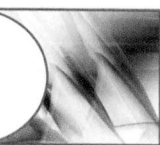

14.2   Arithmetic Sequences and Series

Arithmetic Sequences • Sum of the First n Terms of an
Arithmetic Sequence • Problem Solving

In this section, we concentrate on sequences and series that are said to be arithmetic (pronounced ar-ith-MET-ik).

Arithmetic Sequences

In an **arithmetic sequence** (or **progression**), any term (other than the first) can be found by adding the same number to its preceding term. For example, the sequence $2, 5, 8, 11, 14, 17, \ldots$ is arithmetic because adding 3 to any term produces the next term.

> **Arithmetic Sequence**
>
> A sequence is *arithmetic* if there exists a number d, called the *common difference*, such that $a_{n+1} = a_n + d$ for any integer $n \geq 1$.

EXAMPLE 1   For each arithmetic sequence, identify the first term, a_1, and the common difference, d.

a) $4, 9, 14, 19, 24, \ldots$ b) $27, 20, 13, 6, -1, -8, \ldots$

Solution   To find a_1, we simply use the first term listed. To find d, we choose any term other than a_1 and subtract the preceding term from it.

Sequence	First Term, a_1	Common Difference, d
a) $4, 9, 14, 19, 24, \ldots$	4	$5 \leftarrow 9 - 4 = 5$
b) $27, 20, 13, 6, -1, -8, \ldots$	27	$-7 \leftarrow 20 - 27 = -7$

To find the common difference, we subtracted a_1 from a_2. Had we subtracted a_2 from a_3 or a_3 from a_4, we would have found the same values for d.

Check: As a check, note that when d is added to each term, the result is the next term in the sequence.

a) $4 + 5 = 9, \quad 9 + 5 = 14, \quad 14 + 5 = 19, \quad 19 + 5 = 24$

b) $27 + (-7) = 20, \quad 20 + (-7) = 13, \quad 13 + (-7) = 6, \quad 6 + (-7) = -1,$
    $-1 + (-7) = -8$

To develop a formula for the general, nth, term of any arithmetic sequence, we denote the common difference by d and write out the first few terms:

$a_1,$

$a_2 = a_1 + d,$

$a_3 = a_2 + d = (a_1 + d) + d = a_1 + 2d,$ Substituting $a_1 + d$ for a_2

$a_4 = a_3 + d = (a_1 + 2d) + d = a_1 + 3d.$ Substituting $a_1 + 2d$ for a_3

Note that the coefficient of d in each case is 1 less than the subscript.

Generalizing, we obtain the following formula.

To Find a_n for an Arithmetic Sequence

The nth term of an arithmetic sequence with common difference d is

$$a_n = a_1 + (n - 1)d, \quad \text{for any integer } n \geq 1.$$

EXAMPLE 2   Find the 14th term of the arithmetic sequence 6, 9, 12, 15, ...

Solution   First we note that $a_1 = 6$, $d = 3$, and $n = 14$. Using the formula for the nth term of an arithmetic sequence, we have

$$a_n = a_1 + (n - 1)d$$
$$a_{14} = 6 + (14 - 1) \cdot 3 = 6 + 13 \cdot 3 = 6 + 39 = 45.$$

The 14th term is 45.

EXAMPLE 3   For the sequence in Example 2, which term is 300? That is, find n if $a_n = 300$.

Solution   We substitute into the formula for the nth term of an arithmetic sequence and solve for n:

$$a_n = a_1 + (n - 1)d$$
$$300 = 6 + (n - 1) \cdot 3$$
$$300 = 6 + 3n - 3$$
$$297 = 3n$$
$$99 = n.$$

The term 300 is the 99th term of the sequence.

Given two terms and their places in an arithmetic sequence, we can construct the sequence.

EXAMPLE 4   The 3rd term of an arithmetic sequence is 14, and the 16th term is 79. Find a_1 and d and construct the sequence.

Solution   We know that $a_3 = 14$ and $a_{16} = 79$. Thus we would have to add d 13 times to get from 14 to 79. That is,

$$14 + 13d = 79. \quad \text{a_3 and a_{16} are 13 terms apart; } 16 - 3 = 13$$

Solving $14 + 13d = 79$, we obtain

$$13d = 65 \quad \text{Subtracting 14 from both sides}$$
$$d = 5. \quad \text{Dividing both sides by 13}$$

We subtract d twice from a_3 to get to a_1. Thus,

$$a_1 = 14 - 2 \cdot 5 = 4. \quad \text{a_1 and a_3 are 2 terms apart; } 3 - 1 = 2$$

The sequence is 4, 9, 14, 19, ... Note that we could have subtracted d 15 times from a_{16} in order to find a_1.

In general, d should be subtracted $(n - 1)$ times from a_n in order to find a_1.

Sum of the First n Terms of an Arithmetic Sequence

When the terms of an arithmetic sequence are added, an **arithmetic series** is formed. To develop a formula for computing S_n when the series is arithmetic, we list the first n terms of the sequence as follows:

This is the next-to-last term. If you add d to this term, the result is a_n.

$$a_1, (a_1 + d), (a_1 + 2d), \ldots, (a_n - 2d), (a_n - d), a_n$$

This term is two terms back from the end. If you add d to this term, you get the next-to-last term, $a_n - d$.

Thus, S_n is given by

$$S_n = a_1 + (a_1 + d) + (a_1 + 2d) + \cdots + (a_n - 2d) + (a_n - d) + a_n.$$

Using a commutative law, we have a second equation:

$$S_n = a_n + (a_n - d) + (a_n - 2d) + \cdots + (a_1 + 2d) + (a_1 + d) + a_1.$$

Adding corresponding terms on each side of the above equations, we get

$$2S_n = [a_1 + a_n] + [(a_1 + d) + (a_n - d)] + [(a_1 + 2d) + (a_n - 2d)]$$
$$+ \cdots + [(a_n - 2d) + (a_1 + 2d)] + [(a_n - d) + (a_1 + d)] + [a_n + a_1].$$

This simplifies to

$$2S_n = [a_1 + a_n] + [a_1 + a_n] + [a_1 + a_n]$$
$$+ \cdots + [a_n + a_1] + [a_n + a_1] + [a_n + a_1]. \quad \text{There are n bracketed sums.}$$

Since $[a_1 + a_n]$ is being added n times, it follows that

$$2S_n = n[a_1 + a_n].$$

Dividing both sides by 2 leads to the following formula.

Student Notes

The formula for the sum of an arithmetic sequence is very useful, but remember that it does not work for sequences that are not arithmetic.

> *To Find S_n for an Arithmetic Sequence*
>
> The sum of the first n terms of an arithmetic sequence is given by
>
> $$S_n = \frac{n}{2}(a_1 + a_n).$$

EXAMPLE 5 Find the sum of the first 100 positive even numbers.

Solution The sum is

$$2 + 4 + 6 + \cdots + 198 + 200.$$

This is the sum of the first 100 terms of the arithmetic sequence for which

$$a_1 = 2, \quad n = 100, \quad \text{and} \quad a_n = 200.$$

Substituting in the formula

$$S_n = \frac{n}{2}(a_1 + a_n),$$

we get

$$S_{100} = \frac{100}{2}(2 + 200)$$
$$= 50(202) = 10{,}100.$$

The above formula is useful when we know the first and last terms, a_1 and a_n. To find S_n when a_n is unknown, but a_1, n, and d are known, we can use the formula $a_n = a_1 + (n - 1)d$ to calculate a_n and then proceed as in Example 5.

EXAMPLE 6 Find the sum of the first 15 terms of the arithmetic sequence 4, 7, 10, 13, ...

Solution Note that

$$a_1 = 4, \quad n = 15, \quad \text{and} \quad d = 3.$$

Before using the formula for S_n, we find a_{15}:

$$a_{15} = 4 + (15 - 1)3 \qquad \text{Substituting into the formula for } a_n$$
$$= 4 + 14 \cdot 3 = 46.$$

Thus, knowing that $a_{15} = 46$, we have

$$S_{15} = \tfrac{15}{2}(4 + 46) \qquad \text{Using the formula for } S_n$$
$$= \tfrac{15}{2}(50) = 375.$$

Problem Solving

In problem-solving situations, translation may involve sequences or series. As always, there is often a variety of ways in which a problem can be solved. You should use the approach that is best or easiest for you. In this chapter, however, we will try to emphasize sequences and series and their related formulas.

EXAMPLE 7

Hourly wages.   Chris accepts a job managing a music store, starting with an hourly wage of $14.60, and is promised a raise of 25¢ per hour every 2 months for 5 years. After 5 years of work, what will be Chris's hourly wage?

Solution

1. **Familiarize.**   It helps to write down the hourly wage for several two-month time periods.

   Beginning: 14.60,

   After two months: 14.85,

   After four months: 15.10,

   and so on.

   What appears is a sequence of numbers: 14.60, 14.85, 15.10, ... Since the same amount is added each time, the sequence is arithmetic.

   We list what we know about arithmetic sequences. The pertinent formulas are

   $$a_n = a_1 + (n - 1)d$$

   and

   $$S_n = \frac{n}{2}(a_1 + a_n).$$

   In this case, we are not looking for a sum, so it is probably the first formula that will give us our answer. We want to determine the last term in a sequence. To do so, we need to know a_1, n, and d. From our list above, we see that

   $$a_1 = 14.60 \quad \text{and} \quad d = 0.25.$$

   What is n? That is, how many terms are in the sequence? After 1 year, there have been 6 raises, since Chris gets a raise every 2 months. There are 5 years, so the total number of raises will be $5 \cdot 6$, or 30. Altogether, there will be 31 terms: the original wage and 30 increased rates.

2. **Translate.**   We want to find a_n for the arithmetic sequence in which $a_1 = 14.60$, $n = 31$, and $d = 0.25$.

3. **Carry out.**   Substituting in the formula for a_n gives us

   $$a_{31} = 14.60 + (31 - 1) \cdot 0.25$$
   $$= 22.10.$$

4. **Check.**   We can check by redoing the calculations or we can calculate in a slightly different way for another check. For example, at the end of a year, there will be 6 raises, for a total raise of $1.50. At the end of 5 years, the total raise will be $5 \times \$1.50$, or $7.50. If we add that to the original wage of $14.60, we obtain $22.10. The answer checks.

5. **State.**   After 5 years, Chris's hourly wage will be $22.10.

EXAMPLE 8   Telephone pole storage.   A stack of telephone poles has 30 poles in the bottom row. There are 29 poles in the second row, 28 in the next row, and so on. How many poles are in the stack if there are 5 poles in the top row?

Solution

1. **Familiarize.**   The following figure shows the ends of the poles. There are 30 poles on the bottom and one fewer in each successive row. How many rows will there be?

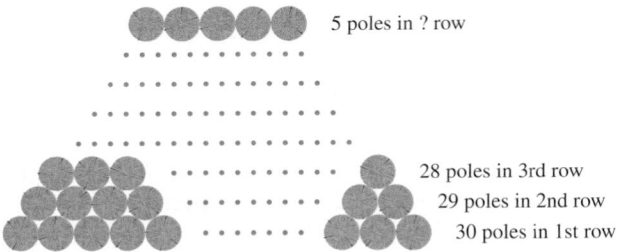

5 poles in ? row

28 poles in 3rd row
29 poles in 2nd row
30 poles in 1st row

Note that there are $30 - 1 = 29$ poles in the 2nd row, $30 - 2 = 28$ poles in the 3rd row, $30 - 3 = 27$ poles in the 4th row, and so on. The pattern leads to $30 - 25 = 5$ poles in the 26th row.

The situation is represented by the equation

$$30 + 29 + 28 + \cdots + 5.$$ There are 26 terms in this series.

Thus we have an arithmetic series. We recall the formula

$$S_n = \frac{n}{2}(a_1 + a_n).$$

2. **Translate.**   We want to find the sum of the first 26 terms of an arithmetic sequence in which $a_1 = 30$ and $a_{26} = 5$.

3. **Carry out.**   Substituting into the above formula gives us

$$S_{26} = \frac{26}{2}(30 + 5)$$
$$= 13 \cdot 35 = 455.$$

4. Check. In this case, we can check the calculations by doing them again. A longer, harder way would be to do the entire addition:

$$30 + 29 + 28 + \cdots + 5.$$

5. State. There are 455 poles in the stack.

Exercise Set

14.2

FOR EXTRA HELP

Student's Solutions Manual | Digital Video Tutor CD 7 Videotape 14 | AW Math Tutor Center | MathXL Tutorials on CD | MathXL | MyMathLab

---

🖐 *Concept Reinforcement   Classify each statement as either true or false.*

1. In an arithmetic sequence, the difference between any two consecutive terms is always the same.

2. In an arithmetic sequence, if $a_9 - a_8 = 4$, then $a_{13} - a_{12} = 4$ as well.

3. In an arithmetic sequence containing 17 terms, the common difference is $a_{17} - a_1$.

4. In an arithmetic sequence, if $a_1 = 8$ and $a_3 = 12$, then a_6 must be 16.

5. The sum of the first 20 terms of an arithmetic sequence can be found by knowing just a_1 and a_{20}.

6. The sum of the first 30 terms of an arithmetic sequence can be found by knowing just a_1 and d, the common difference.

7. The notation S_5 means $a_1 + a_5$.

8. For any arithmetic sequence, $S_9 = S_8 + d$, where d is the common difference.

Find the first term and the common difference.

9. $2, 6, 10, 14, \ldots$

10. $1.06, 1.12, 1.18, 1.24, \ldots$

11. $7, 3, -1, -5, \ldots$

12. $-8, -5, -2, 1, \ldots$

13. $\frac{3}{2}, \frac{9}{4}, 3, \frac{15}{4}, \ldots$

14. $\frac{3}{5}, \frac{1}{10}, -\frac{2}{5}, \ldots$

15. $\$5.12, \$5.24, \$5.36, \$5.48, \ldots$

16. $\$214, \$211, \$208, \$205, \ldots$

17. Find the 15th term of the arithmetic sequence $7, 10, 13, \ldots$.

18. Find the 17th term of the arithmetic sequence $6, 10, 14, \ldots$.

19. Find the 18th term of the arithmetic sequence $8, 2, -4, \ldots$.

20. Find the 14th term of the arithmetic sequence $3, \frac{7}{3}, \frac{5}{3}, \ldots$.

21. Find the 13th term of the arithmetic sequence $\$1200, \$964.32, \$728.64, \ldots$.

22. Find the 10th term of the arithmetic sequence $\$2345.78, \$2967.54, \$3589.30, \ldots$.

23. In the sequence of Exercise 17, what term is 82?

24. In the sequence of Exercise 18, what term is 126?

25. In the sequence of Exercise 19, what term is -328?

26. In the sequence of Exercise 20, what term is -27?

27. Find a_{17} when $a_1 = 2$ and $d = 5$.

28. Find a_{20} when $a_1 = 14$ and $d = -3$.

29. Find a_1 when $d = 4$ and $a_8 = 33$.

30. Find a_1 when $d = 8$ and $a_{11} = 26$.

31. Find n when $a_1 = 5$, $d = -3$, and $a_n = -76$.

32. Find n when $a_1 = 25$, $d = -14$, and $a_n = -507$.

33. For an arithmetic sequence in which $a_{17} = -40$ and $a_{28} = -73$, find a_1 and d. Write the first five terms of the sequence.

34. In an arithmetic sequence, $a_{17} = \frac{25}{3}$ and $a_{32} = \frac{95}{6}$. Find a_1 and d. Write the first five terms of the sequence.

Aha! **35.** Find a_1 and d if $a_{13} = 13$ and $a_{54} = 54$.

36. Find a_1 and d if $a_{12} = 24$ and $a_{25} = 50$.

37. Find the sum of the first 20 terms of the arithmetic series $1 + 5 + 9 + 13 + \cdots$.

38. Find the sum of the first 14 terms of the arithmetic series $11 + 7 + 3 + \cdots$.

39. Find the sum of the first 250 natural numbers.

40. Find the sum of the first 400 natural numbers.

41. Find the sum of the even numbers from 2 to 100, inclusive.

42. Find the sum of the odd numbers from 1 to 99, inclusive.

43. Find the sum of all multiples of 6 from 6 to 102, inclusive.

44. Find the sum of all multiples of 4 that are between 15 and 521.

45. An arithmetic series has $a_1 = 4$ and $d = 5$. Find S_{20}.

46. An arithmetic series has $a_1 = 9$ and $d = -3$. Find S_{32}.

Solve.

47. *Band formations.* The South Brighton Drum and Bugle Corps has 7 marchers in the front row, 9 in the second row, 11 in the third row, and so on, for 15 rows. How many marchers are in the last row? How many marchers are there altogether?

48. *Gardening.* A gardener is planting bulbs near an entrance to a college. She has 39 plants in the front row, 35 in the second row, 31 in the third row, and so on. If the pattern is consistent, how many plants will be in the last row? How many plants will there be altogether?

49. *Archaeology.* Many ancient Mayan pyramids were constructed over a span of several generations. Each layer of the pyramid has a stone perimeter,

enclosing a layer of dirt or debris on which a structure once stood. One drawing of such a pyramid indicates that the perimeter of the bottom layer contains 36 stones, the next level up contains 32 stones, and so on, up to the top row which contains 4 stones. How many stones are in the pyramid?

50. *Telephone pole piles.* How many poles will be in a pile of telephone poles if there are 50 in the first layer, 49 in the second, and so on, until there are 6 in the top layer?

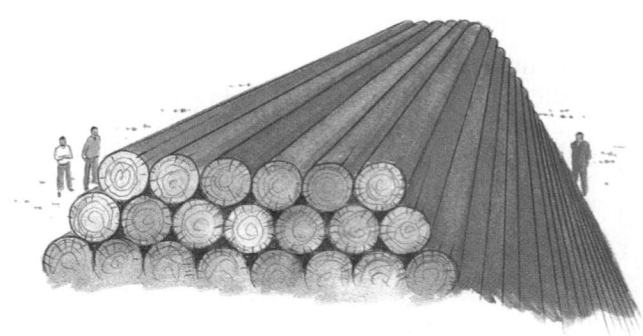

51. *Accumulated savings.* If 10¢ is saved on October 1, another 20¢ on October 2, another 30¢ on October 3, and so on, how much is saved during October? (October has 31 days.)

52. *Accumulated savings.* Renata saves money in an arithmetic sequence: $700 for the first year, another $850 the second, and so on, for 20 yr. How much does she save in all (disregarding interest)?

53. *Auditorium design.* Theaters are often built with more seats per row as the rows move toward the back. The Sanders Amphitheater has 20 seats in the first row, 22 in the second, 24 in the third, and so on, for 19 rows. How many seats are in the amphitheater?

54. *Accumulated savings.* Shirley sets up an investment such that it will return $5000 the first year, $6125 the second year, $7250 the third year, and so on, for 25 yr. How much in all is received from the investment?

55. It is said that as a young child, the mathematician Karl F. Gauss (1777–1855) was able to compute the sum $1 + 2 + 3 + \cdots + 100$ very quickly in his head. Explain how Gauss might have done this and present a formula for the sum of the first n natural numbers. (*Hint*: $1 + 99 = 100$.)

56. Is it true that if every number in a sequence is doubled and then added, the result is the same as if the numbers were first added and the sum then doubled? Why or why not?

SKILL MAINTENANCE

Simplify. [6.4]

57. $\dfrac{3}{10x} + \dfrac{2}{15x}$

58. $\dfrac{2}{9t} + \dfrac{5}{12t}$

Convert to an exponential equation. [12.3]

59. $\log_a P = k$

60. $\ln t = a$

Find an equation of the circle satisfying the given conditions. [13.1]

61. Center $(0, 0)$, radius 9

62. Center $(-2, 5)$, radius $3\sqrt{2}$

SYNTHESIS

63. Write a problem for a classmate to solve. Devise the problem so that its solution requires computing S_{17} for an arithmetic sequence.

64. The sum of the first n terms of an arithmetic sequence is also given by
$$S_n = \frac{n}{2}[2a_1 + (n - 1)d].$$
Use the earlier formulas for a_n and S_n to explain how this equation was developed.

65. A frog is at the bottom of a 100-ft well. With each jump, the frog climbs 4 ft, but then slips back 1 ft. How many jumps does it take for the frog to reach the top of the hole?

66. Find a formula for the sum of the first n consecutive odd numbers starting with 1:
$$1 + 3 + 5 + \cdots + (2n - 1).$$

67. In an arithmetic sequence, $a_1 = \$8760$ and $d = -\$798.23$. Find the first 10 terms of the sequence.

68. Find the sum of the first 10 terms of the sequence given in Exercise 67.

69. Prove that if p, m, and q are consecutive terms in an arithmetic sequence, then
$$m = \frac{p + q}{2}.$$

70. *Straight-line depreciation.* A company buys a color copier for $5200 on January 1 of a given year. The machine is expected to last for 8 yr, at the end of which time its *trade-in*, or *salvage*, *value* will be $1100. If the company figures the decline in value to be the same each year, then the trade-in values, after t years, $0 \le t \le 8$, form an arithmetic sequence given by
$$a_t = C - t\left(\frac{C - S}{N}\right),$$
where C is the original cost of the item, N the years of expected life, and S the salvage value.
   a) Find the formula for a_t for the straight-line depreciation of the copier.
   b) Find the salvage value after 0 yr, 1 yr, 2 yr, 3 yr, 4 yr, 7 yr, and 8 yr.
   c) Find a formula that expresses a_t recursively.

71. Use your answer to Exercise 39 to find the sum of all integers from 501 through 750.

14.3   Geometric Sequences and Series

Geometric Sequences   •   Sum of the First *n* Terms of a
Geometric Sequence   •   Infinite Geometric Series   •
Problem Solving

In an arithmetic sequence, a certain number is added to each term to get the next term. When each term in a sequence is *multiplied* by a certain fixed number to get the next term, the sequence is **geometric**. In this section, we examine geometric sequences (or progressions) and *geometric series*.

Geometric Sequences

Consider the sequence

$$2, 6, 18, 54, 162, \ldots .$$

If we multiply each term by 3, we obtain the next term. The multiplier is called the *common ratio* because it is found by dividing any term by the preceding term.

Geometric Sequence

A sequence is *geometric* if there exists a number *r*, called the *common ratio*, for which

$$\frac{a_{n+1}}{a_n} = r, \quad \text{or} \quad a_{n+1} = a_n \cdot r \quad \text{for any integer } n \geq 1.$$

EXAMPLE 1   For each geometric sequence, find the common ratio.

a) $4, 20, 100, 500, 2500, \ldots$

b) $3, -6, 12, -24, 48, -96, \ldots$

c) $\$5200, \$3900, \$2925, \$2193.75, \ldots$

Solution

Sequence	*Common Ratio*	
a) $4, 20, 100, 500, 2500, \ldots$	5	$\frac{20}{4} = 5, \frac{100}{20} = 5$, and so on
b) $3, -6, 12, -24, 48, -96, \ldots$	-2	$\frac{-6}{3} = -2, \frac{12}{-6} = -2$, and so on
c) $\$5200, \$3900, \$2925, \$2193.75, \ldots$	0.75	$\frac{\$3900}{\$5200} = 0.75, \frac{\$2925}{\$3900} = 0.75$

To develop a formula for the general, or *n*th, term of a geometric sequence, let a_1 be the first term and let r be the common ratio. We write out the first few terms as follows:

$a_1,$

$a_2 = a_1 r,$

$a_3 = a_2 r = (a_1 r)r = a_1 r^2,$ Substituting $a_1 r$ for a_2

$a_4 = a_3 r = (a_1 r^2)r = a_1 r^3.$ Substituting $a_1 r^2$ for a_3

Note that the exponent is 1 less than the subscript.

Generalizing, we obtain the following.

To Find a_n for a Geometric Sequence

The *n*th term of a geometric sequence with common ratio r is given by

$$a_n = a_1 r^{n-1}, \quad \text{for any integer } n \geq 1.$$

EXAMPLE 2 Find the 7th term of the geometric sequence $4, 20, 100, \ldots$.

Solution First, we note that

$$a_1 = 4 \quad \text{and} \quad n = 7.$$

To find the common ratio, we can divide any term (other than the first) by the term preceding it. Since the second term is 20 and the first is 4,

$$r = \frac{20}{4}, \quad \text{or } 5.$$

The formula

$$a_n = a_1 r^{n-1}$$

gives us

$$a_7 = 4 \cdot 5^{7-1} = 4 \cdot 5^6 = 4 \cdot 15{,}625 = 62{,}500.$$

EXAMPLE 3 Find the 10th term of the geometric sequence

$$64, -32, 16, -8, \ldots.$$

Solution First, we note that

$$a_1 = 64, \qquad n = 10, \quad \text{and} \quad r = \frac{-32}{64} = -\frac{1}{2}.$$

Then, using the formula for the *n*th term of a geometric sequence, we have

$$a_{10} = 64 \cdot \left(-\frac{1}{2}\right)^{10-1} = 64 \cdot \left(-\frac{1}{2}\right)^9 = 2^6 \cdot \left(-\frac{1}{2^9}\right) = -\frac{1}{2^3} = -\frac{1}{8}.$$

The 10th term is $-\frac{1}{8}$.

Sum of the First n Terms of a Geometric Sequence

We next develop a formula for S_n when a sequence is geometric:

$$a_1, \ a_1r, \ a_1r^2, \ a_1r^3, \ldots, a_1r^{n-1}, \ldots.$$

The **geometric series** S_n is given by

$$S_n = a_1 + a_1r + a_1r^2 + \cdots + a_1r^{n-2} + a_1r^{n-1}. \tag{1}$$

Multiplying both sides by r gives us

$$rS_n = a_1r + a_1r^2 + a_1r^3 + \cdots + a_1r^{n-1} + a_1r^n. \tag{2}$$

When we subtract corresponding sides of equation (2) from equation (1), the color terms drop out, leaving

$$S_n - rS_n = a_1 - a_1r^n$$
$$S_n(1 - r) = a_1(1 - r^n), \qquad \text{Factoring}$$

or

$$S_n = \frac{a_1(1 - r^n)}{1 - r}. \qquad \text{Dividing both sides by } 1 - r$$

Student Notes

The three determining characteristics of a geometric sequence or series are the first term (a_1), the number of terms (n), and the common ratio (r). Be sure you understand how to use these characteristics to write out a sequence or a series.

To Find S_n for a Geometric Sequence

The sum of the first n terms of a geometric sequence with common ratio r is given by

$$S_n = \frac{a_1(1 - r^n)}{1 - r}, \text{ for any } r \neq 1.$$

EXAMPLE 4   Find the sum of the first 7 terms of the geometric sequence $3, 15, 75, 375, \ldots$.

Solution   First, we note that

$$a_1 = 3, \qquad n = 7, \quad \text{and} \quad r = \frac{15}{3} = 5.$$

Then, substituting in the formula $S_n = \dfrac{a_1(1 - r^n)}{1 - r}$, we have

$$S_7 = \frac{3(1 - 5^7)}{1 - 5} = \frac{3(1 - 78{,}125)}{-4}$$

$$= \frac{3(-78{,}124)}{-4}$$

$$= 58{,}593.$$

Infinite Geometric Series

Suppose we consider the sum of the terms of an infinite geometric sequence, such as $3, 6, 12, 24, 48, \ldots$. We get what is called an **infinite geometric series**:

$$3 + 6 + 12 + 24 + 48 + \cdots.$$

Here, as n increases, the sum of the first n terms, S_n, increases without bound. There are also infinite series that get closer and closer to some specific number. Here is an example:

$$\frac{1}{2} + \frac{1}{4} + \frac{1}{8} + \frac{1}{16} + \cdots + \frac{1}{2^n} + \cdots.$$

Let's consider S_n for the first four values of n:

$$S_1 = \tfrac{1}{2} \qquad\qquad\qquad = \tfrac{1}{2} = 0.5,$$
$$S_2 = \tfrac{1}{2} + \tfrac{1}{4} \qquad\qquad = \tfrac{3}{4} = 0.75,$$
$$S_3 = \tfrac{1}{2} + \tfrac{1}{4} + \tfrac{1}{8} \qquad = \tfrac{7}{8} = 0.875,$$
$$S_4 = \tfrac{1}{2} + \tfrac{1}{4} + \tfrac{1}{8} + \tfrac{1}{16} = \tfrac{15}{16} = 0.9375.$$

> The denominator of each sum is 2^n, where n is the subscript of S. The numerator is $2^n - 1$.

Thus, for this particular series, we have

$$S_n = \frac{2^n - 1}{2^n} = \frac{2^n}{2^n} - \frac{1}{2^n} = 1 - \frac{1}{2^n}.$$

Note that as n gets larger and larger, the value of $1/2^n$ gets closer to 0 and the value of S_n gets closer to 1. We say that 1 is the **limit** of S_n and that 1 is the sum of this infinite geometric series. An infinite geometric series is denoted S_∞. It can be shown (but we will not do it here) that the sum of the terms of an infinite geometric sequence exists if and only if $|r| < 1$ (that is, the common ratio's absolute value is less than 1).

To find a formula for the sum of an infinite geometric series, we first consider the sum of the first n terms:

$$S_n = \frac{a_1(1 - r^n)}{1 - r} = \frac{a_1 - a_1 r^n}{1 - r}. \qquad \text{Using the distributive law}$$

For $|r| < 1$, it follows that the value of r^n gets closer to 0 as n gets larger. (Check this by selecting a number between -1 and 1 and finding larger and larger powers on a calculator.) As r^n gets closer to 0, so too does $a_1 r^n$. Thus, S_n gets closer to $a_1/(1 - r)$.

The Limit of an Infinite Geometric Series

When $|r| < 1$, the limit of an infinite geometric series is given by

$$S_\infty = \frac{a_1}{1 - r}. \qquad \text{(For } |r| \geq 1, \text{ no limit exists.)}$$

EXAMPLE 5   Determine whether each series has a limit. If a limit exists, find it.

a) $1 + 3 + 9 + 27 + \cdots$ **b)** $-35 + 7 - \tfrac{7}{5} + \tfrac{7}{25} + \cdots$

Solution

a) Here $r = 3$, so $|r| = |3| = 3$. Since $|r| \not< 1$, the series does *not* have a limit.

b) Here $r = -\frac{1}{5}$, so $|r| = |-\frac{1}{5}| = \frac{1}{5}$. Since $|r| < 1$, the series *does* have a limit. We find the limit by substituting into the formula for S_∞:

$$S_\infty = \frac{-35}{1 - \left(-\frac{1}{5}\right)} = \frac{-35}{\frac{6}{5}} = -35 \cdot \frac{5}{6} = \frac{-175}{6} = -29\frac{1}{6}.$$

EXAMPLE 6   Find fraction notation for $0.63636363\ldots$.

Solution   We can express this as

$$0.63 + 0.0063 + 0.000063 + \cdots.$$

This is an infinite geometric series, where $a_1 = 0.63$ and $r = 0.01$. Since $|r| < 1$, this series has a limit:

$$S_\infty = \frac{a_1}{1 - r} = \frac{0.63}{1 - 0.01} = \frac{0.63}{0.99} = \frac{63}{99}.$$

Thus fraction notation for $0.63636363\ldots$ is $\frac{63}{99}$, or $\frac{7}{11}$.

Problem Solving

For some problem-solving situations, the translation may involve geometric sequences or series.

EXAMPLE 7   *Daily wages.*   Suppose someone offered you a job for the month of September (30 days) under the following conditions. You will be paid \$0.01 for the first day, \$0.02 for the second, \$0.04 for the third, and so on, doubling your previous day's salary each day. How much would you earn? (Would you take the job? Make a guess before reading further.)

Solution

1. **Familiarize.**   You earn \$0.01 the first day, \$0.01(2) the second day, \$0.01(2)(2) the third day, and so on. Since each day's wages are a constant multiple of the previous day's wages, a geometric sequence is formed.

2. **Translate.**   The amount earned is the geometric series

$$\$0.01 + \$0.01(2) + \$0.01(2^2) + \$0.01(2^3) + \cdots + \$0.01(2^{29}),$$

where $a_1 = \$0.01$, $n = 30$, and $r = 2$.

3. **Carry out.**   Using the formula

$$S_n = \frac{a_1(1 - r^n)}{1 - r},$$

we have

$$S_{30} = \frac{\$0.01(1 - 2^{30})}{1 - 2}$$

$$= \frac{\$0.01(-1,073,741,823)}{-1} \qquad \text{Using a calculator}$$

$$= \$10,737,418.23.$$

4. **Check.** The calculations can be repeated as a check.

5. **State.** The pay exceeds $10.7 million for the month. Most people would probably take the job!

EXAMPLE 8

Loan repayment. Francine's student loan is in the amount of $6000. Interest is to be 9% compounded annually, and the entire amount is to be paid after 10 yr. How much is to be paid back?

Solution

1. **Familiarize.** Suppose we let P represent any principal amount. At the end of one year, the amount owed will be $P + 0.09P$, or $1.09P$. That amount will be the principal for the second year. The amount owed at the end of the second year will be $1.09 \times$ New principal $= 1.09(1.09P)$, or 1.09^2P. Thus the amount owed at the beginning of successive years is as follows:

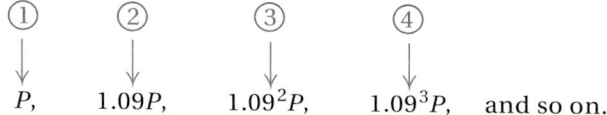

$$P, \quad 1.09P, \quad 1.09^2P, \quad 1.09^3P, \quad \text{and so on.}$$

We have a geometric sequence. The amount owed at the beginning of the 11th year will be the amount owed at the end of the 10th year.

2. **Translate.** We have a geometric sequence with $a_1 = 6000$, $r = 1.09$, and $n = 11$. The appropriate formula is

$$a_n = a_1 r^{n-1}.$$

3. **Carry out.** We substitute and calculate:

$$a_{11} = \$6000(1.09)^{11-1} = \$6000(1.09)^{10}$$

$$\approx \$14{,}204.18. \qquad \text{Using a calculator and rounding to the nearest hundredth}$$

4. **Check.** A check, by repeating the calculations, is left to the student.

5. **State.** Francine will owe $14,204.18 at the end of 10 yr.

EXAMPLE 9

Bungee jumping. A bungee jumper rebounds 60% of the height jumped. Clyde's bungee jump is made using a cord that stretches to 200 ft.

a) After jumping and then rebounding 9 times, how far has Clyde traveled upward (the total rebound distance)?

b) Approximately how far will Clyde have traveled upward (bounced) before coming to rest?

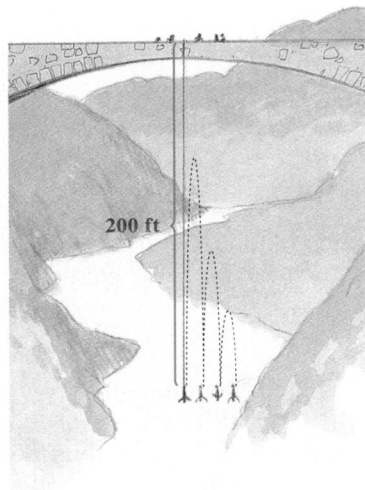

200 ft

Solution

1. **Familiarize.**   Let's do some calculations and look for a pattern.

First fall:	200 ft
First rebound:	0.6×200, or 120 ft
Second fall:	120 ft, or 0.6×200
Second rebound:	0.6×120, or $0.6(0.6 \times 200)$, which is 72 ft
Third fall:	72 ft, or $0.6(0.6 \times 200)$
Third rebound:	0.6×72, or $0.6(0.6(0.6 \times 200))$, which is 43.2 ft

The rebound distances form a geometric sequence:

①   ②   ③   ④

$120, \quad 0.6 \times 120, \quad 0.6^2 \times 120, \quad 0.6^3 \times 120, \ldots.$

2. **Translate.**
   a) The total rebound distance after 9 bounces is the sum of a geometric sequence. The first term is 120 and the common ratio is 0.6. There will be 9 terms, so we can use the formula

   $$S_n = \frac{a_1(1 - r^n)}{1 - r}.$$

   b) Theoretically, Clyde will never stop bouncing. Realistically, the bouncing will eventually stop. To approximate the actual distance bounced, we consider an infinite number of bounces and use the formula

   $$S_\infty = \frac{a_1}{1 - r}. \qquad \text{Since } r = 0.6 \text{ and } |0.6| < 1, \text{ we know that } S_\infty \text{ exists.}$$

3. **Carry out.**
   a) We substitute into the formula and calculate:

   $$S_9 = \frac{120[1 - (0.6)^9]}{1 - 0.6} \approx 297. \qquad \text{Using a calculator}$$

   b) We substitute and calculate:

   $$S_\infty = \frac{120}{1 - 0.6} = 300.$$

4. **Check.**   We can do the calculations again.

5. **State.**
   a) In 9 bounces, Clyde will have traveled upward a total distance of about 297 ft.
   b) Clyde will have traveled upward a total of about 300 ft before coming to rest.

Exercise Set

14.3

↪ *Concept Reinforcement*   *Classify each of the following as an arithmetic sequence, a geometric sequence, an arithmetic series, a geometric series, or none of these.*

1. 2, 6, 18, 54, . . .

2. 3, 5, 7, 9, . . .

3. 1, 6, 11, 16, 21, . . .

4. 5, 15, 45, 135, 405, . . .

5. 4 + 20 + 100 + 500 + 2500 + 12,500

6. 10 + 12 + 14 + 16 + 18 + 20

7. $3 - \frac{3}{2} + \frac{3}{4} - \frac{3}{8} + \frac{3}{16} - \cdots$

8. $1 + \frac{1}{2} + \frac{1}{3} + \frac{1}{4} + \frac{1}{5} + \frac{1}{6} + \cdots$

Find the common ratio for each geometric sequence.

9. 7, 14, 28, 56, . . .

10. 2, 6, 18, 54, . . .

11. 6, −0.6, 0.06, −0.006, . . .

12. −5, −0.5, −0.05, −0.005, . . .

13. $\frac{1}{2}, -\frac{1}{4}, \frac{1}{8}, -\frac{1}{16}, \cdots$

14. $\frac{2}{3}, -\frac{4}{3}, \frac{8}{3}, -\frac{16}{3}, \cdots$

15. $75, 15, 3, \frac{3}{5}, \cdots$

16. $12, -4, \frac{4}{3}, -\frac{4}{9}, \cdots$

17. $\frac{1}{m}, \frac{6}{m^2}, \frac{36}{m^3}, \frac{216}{m^4}, \cdots$

18. $4, \frac{4m}{5}, \frac{4m^2}{25}, \frac{4m^3}{125}, \cdots$

Find the indicated term for each geometric sequence.

19. 3, 6, 12, . . . ; the 7th term

20. 2, 8, 32, . . . ; the 9th term

21. $7, 7\sqrt{2}, 14, \ldots$; the 10th term

22. $4, 4\sqrt{3}, 12, \ldots$; the 8th term

23. $-\frac{8}{243}, \frac{8}{81}, -\frac{8}{27}, \ldots$; the 14th term

24. $\frac{7}{625}, \frac{-7}{125}, \frac{7}{25}, \ldots$; the 13th term

25. \$1000, \$1080, \$1166.40, . . . ; the 12th term

26. \$1000, \$1070, \$1144.90, . . . ; the 11th term

Find the nth, or general, term for each geometric sequence.

27. 1, 5, 25, 125, . . .

28. 2, 4, 8, . . .

29. 1, −1, 1, −1, . . .

30. $\frac{1}{4}, \frac{1}{16}, \frac{1}{64}, \cdots$

31. $\frac{1}{x}, \frac{1}{x^2}, \frac{1}{x^3}, \cdots$

32. $5, \frac{5m}{2}, \frac{5m^2}{4}, \cdots$

For Exercises 33–40, use the formula for S_n to find the indicated sum for each geometric series.

33. S_9 for $6 + 12 + 24 + \cdots$

34. S_6 for $16 - 8 + 4 - \cdots$

35. S_7 for $\frac{1}{18} - \frac{1}{6} + \frac{1}{2} - \cdots$

Aha! **36.** S_5 for $7 + 0.7 + 0.07 + \cdots$

37. S_8 for $1 + x + x^2 + x^3 + \cdots$

38. S_{10} for $1 + x^2 + x^4 + x^6 + \cdots$

39. S_{16} for \$200, \$200(1.06), \$200(1.06)^2, . . .

40. S_{23} for \$1000, \$1000(1.08), \$1000(1.08)^2, . . .

Determine whether each infinite geometric series has a limit. If a limit exists, find it.

41. $16 + 4 + 1 + \cdots$

42. $8 + 4 + 2 + \cdots$

43. $7 + 3 + \frac{9}{7} + \cdots$

44. $12 + 9 + \frac{27}{4} + \cdots$

45. $3 + 15 + 75 + \cdots$

46. $2 + 3 + \frac{9}{2} + \cdots$

47. $4 - 6 + 9 - \frac{27}{2} + \cdots$

48. $-6 + 3 - \frac{3}{2} + \frac{3}{4} - \cdots$

49. $0.43 + 0.0043 + 0.000043 + \cdots$

50. $0.37 + 0.0037 + 0.000037 + \cdots$

51. $\$500(1.02)^{-1} + \$500(1.02)^{-2} + \$500(1.02)^{-3} + \cdots$

52. $\$1000(1.08)^{-1} + \$1000(1.08)^{-2} + \$1000(1.08)^{-3} + \cdots$

Find fraction notation for each infinite sum. (Each can be regarded as an infinite geometric series.)

53. 0.7777...

54. 0.2222...

55. 8.3838...

56. 7.4747...

57. 0.15151515...

58. 0.12121212...

🖩 *Solve. Use a calculator as needed for evaluating formulas.*

59. *Rebound distance.* A ping-pong ball is dropped from a height of 20 ft and always rebounds one-fourth of the distance fallen. How high does it rebound the 6th time?

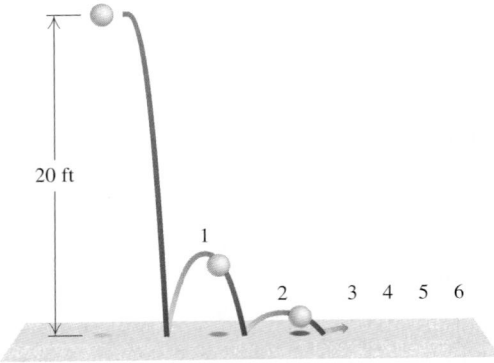

20 ft

60. *Rebound distance.* Approximate the total of the rebound heights of the ball in Exercise 59.

61. *Population growth.* Yorktown has a current population of 100,000, and the population is increasing by 3% each year. What will the population be in 15 yr?

62. *Amount owed.* Gilberto borrows $15,000. The loan is to be repaid in 13 yr at 8.5% interest, compounded annually. How much will be repaid at the end of 13 yr?

63. *Shrinking population.* A population of 5000 fruit flies is dying off at a rate of 4% per minute. How many flies will be alive after 15 min?

64. *Shrinking population.* For the population of fruit flies in Exercise 63, how long will it take for only 1800 fruit flies to remain alive? (*Hint*: Use logarithms.) Round to the nearest minute.

65. *Housing units.* Approximately 534,000 new apartments and houses were built in the United States in 1991. Since then, the number has grown by about 5.35% per year. How many new apartments and houses were built in the United States from 1991 through 2004?
Sources: Based on data from the U.S. Bureau of the Census and the U.S. Department of Housing and Urban Development

66. *Housing units.* Approximately 144,000 new apartments and houses were built in the western United States in 1991. Since then, the number has grown by about 5.0% per year. How many new houses and apartments were built in the western United States from 1991 through 2004?
Sources: Based on data from the U.S. Bureau of the Census and the U.S. Department of Housing and Urban Development

67. *Rebound distance.* A superball dropped from the top of the Washington Monument (556 ft high) rebounds three-fourths of the distance fallen. How far (up and down) will the ball have traveled when it hits the ground for the 6th time?

68. *Rebound distance.* Approximate the total distance that the ball of Exercise 67 will have traveled when it comes to rest.

69. *Stacking paper.* Construction paper is about 0.02 in. thick. Beginning with just one piece, a stack is doubled again and again 10 times. Find the height of the final stack.

70. *Monthly earnings.*   Suppose you accepted a job for the month of February (28 days) under the following conditions. You will be paid $0.01 the first day, $0.02 the second, $0.04 the third, and so on, doubling your previous day's salary each day. How much would you earn?

Aha! **71.** Under what circumstances is it possible for the 5th term of a geometric sequence to be greater than the 4th term but less than the 7th term?

72. When r is negative, a series is said to be *alternating.* Why do you suppose this terminology is used?

SKILL MAINTENANCE

Multiply. [4.4]

73. $(x + y)(x^2 + 2xy + y^2)$

74. $(a - b)(a^2 - 2ab + b^2)$

Solve the system.

75. $5x - 2y = -3,$
$2x + 5y = -24$ [8.2]

76. $x - 2y + 3z = 4,$
$2x - y + z = -1,$
$4x + y + z = 1$ [8.4]

SYNTHESIS

77. Write a problem for a classmate to solve. Devise the problem so that a geometric series is involved and the solution is "The total amount in the bank is $900(1.08)^{40}$, or about $19,550."

78. The infinite series

$$S_\infty = 2 + \frac{1}{2} + \frac{1}{2 \cdot 3} + \frac{1}{2 \cdot 3 \cdot 4} + \frac{1}{2 \cdot 3 \cdot 4 \cdot 5}$$
$$+ \frac{1}{2 \cdot 3 \cdot 4 \cdot 5 \cdot 6} + \cdots$$

is not geometric, but it does have a sum. Using S_1, S_2, S_3, S_4, S_5, and S_6, make a conjecture about the value of S_∞ and explain your reasoning.

Calculate each of the following sums.

79. $\sum_{k=1}^{\infty} 6(0.9)^k$ **80.** $\sum_{k=1}^{\infty} 5(-0.7)^k$

81. Find the sum of the first n terms of
$$x^2 - x^3 + x^4 - x^5 + \cdots.$$

82. Find the sum of the first n terms of
$$1 + x + x^2 + x^3 + \cdots.$$

83. The sides of a square are each 16 cm long. A second square is inscribed by joining the midpoints of the sides, successively. In the second square we repeat the process, inscribing a third square. If this process is continued indefinitely, what is the sum of all of the areas of all the squares? (*Hint*: Use an infinite geometric series.)

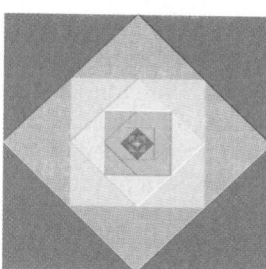

84. Show that $0.999\ldots$ is 1.

85. Using Example 5 and Exercises 41–52, explain how the graph of a geometric sequence can be used to determine whether a geometric series has a limit.

86. To compare the *graphs* of an arithmetic and a geometric sequence, we plot n on the horizontal axis and a_n on the vertical axis. Graph Example 1(a) of Section 14.2 and Example 1(a) of Section 14.3 on the same set of axes. How do the graphs of geometric sequences differ from the graphs of arithmetic sequences?

CORNER

COLLABORATIVE

Bargaining for a Used Car

Focus: Geometric series

Time: 30 minutes

Group size: 2

Materials: Graphing calculators are optional.

ACTIVITY*

1. One group member ("the seller") has a car for sale and is asking $3500. The second ("the buyer") offers $1500. The seller splits the difference ($2000 ÷ 2 = $1000) and lowers the price to $2500. The buyer then splits the difference again ($1000 ÷ 2 = $500) and counters with $2000. Continue in this manner and stop when you are able to agree on the car's selling price to the nearest penny.

2. What should the buyer's initial offer be in order to achieve a purchase price of $2000? (Check several guesses to find the appropriate initial offer.)

———————————————

*This activity is based on the article, "Bargaining Theory, or Zeno's Used Cars," by James C. Kirby, *The College Mathematics Journal*, **27**(4), September 1996.

3. The seller's price in the bargaining above can be modeled recursively (see Exercises 81, 82, and 90 in Section 14.1) by the sequence

$$a_1 = 3500, \qquad a_n = a_{n-1} - \frac{d}{2^{2n-3}},$$

where d is the difference between the initial price and the first offer. Use this recursively defined sequence to solve parts (1) and (2) above either manually or by using the SEQ MODE and the TABLE feature of a graphing calculator.

4. The first four terms in the sequence in part (3) can be written as

$$a_1, \quad a_1 - \frac{d}{2}, \quad a_1 - \frac{d}{2} - \frac{d}{8},$$

$$a_1 - \frac{d}{2} - \frac{d}{8} - \frac{d}{32}.$$

Use the formula for the limit of an infinite geometric series to find a simple algebraic formula for the eventual sale price, P, when the bargaining process from above is followed. Verify the formula by using it to solve parts (1) and (2) above.

14.4   The Binomial Theorem

Binomial Expansion Using Pascal's Triangle •
Binomial Expansion Using Factorial Notation

CONNECTING THE CONCEPTS

Sequences and series occur in many settings, some of which were mentioned in Sections 14.1–14.3. Although you may not have viewed it this way before, the expression $(x + y)^2$ can be regarded as a series: $x^2 + 2xy + y^2$.

In Chapter 4, we found that the expansion of $(x + y)^n$, for powers greater than 2, can be quite time-consuming. The reason for this extends all the way back to Chapter 1 and the rules for the order of operations and the properties of exponents: $(x + y)^n \neq x^n + y^n$. Since the terms in the expansion of $(x + y)^n$ have many uses, we devote this section to two methods that streamline the expansion of this important algebraic expression.

Binomial Expansion Using Pascal's Triangle

Consider the following expanded powers of $(a + b)^n$:

$$(a + b)^0 = 1$$
$$(a + b)^1 = a + b$$
$$(a + b)^2 = a^2 + 2a^1b^1 + b^2$$
$$(a + b)^3 = a^3 + 3a^2b^1 + 3a^1b^2 + b^3$$
$$(a + b)^4 = a^4 + 4a^3b^1 + 6a^2b^2 + 4a^1b^3 + b^4$$
$$(a + b)^5 = a^5 + 5a^4b^1 + 10a^3b^2 + 10a^2b^3 + 5a^1b^4 + b^5.$$

Each expansion is a polynomial. There are some patterns to be noted:

1. There is one more term than the power of the binomial, n. That is, there are $n + 1$ terms in the expansion of $(a + b)^n$.

2. In each term, the sum of the exponents is the power to which the binomial is raised.

3. The exponents of a start with n, the power of the binomial, and decrease to 0 (since $a^0 = 1$, the last term has no factor of a). The first term has no factor of b, so powers of b start with 0 and increase to n.

4. The coefficients start at 1, increase through certain values, and then decrease through these same values back to 1.

Let's study the coefficients further. Suppose we wish to expand $(a + b)^8$. The patterns we noticed above indicate 9 terms in the expansion:

$$a^8 + c_1a^7b + c_2a^6b^2 + c_3a^5b^3 + c_4a^4b^4 + c_5a^3b^5 + c_6a^2b^6 + c_7ab^7 + b^8.$$

Study Skills _____

Make the Most of Each Minute

As your final exam nears, it is essential to make wise use of your time. If it is possible to reschedule haircuts, dentist appointments, and the like until after your exam(s), consider doing so. If you can, consider writing out your hourly schedule of daily activities leading up to the exam(s).

How can we determine the values for the c's? One method seems very simple, but it has some drawbacks. It involves writing down the coefficients in a triangular array as follows. We form what is known as **Pascal's triangle**:

$$
\begin{array}{cccccccccccc}
(a+b)^0\colon & & & & & & 1 & & & & & \\
(a+b)^1\colon & & & & & 1 & & 1 & & & & \\
(a+b)^2\colon & & & & 1 & & 2 & & 1 & & & \\
(a+b)^3\colon & & & 1 & & 3 & & 3 & & 1 & & \\
(a+b)^4\colon & & 1 & & 4 & & 6 & & 4 & & 1 & \\
(a+b)^5\colon & 1 & & 5 & & 10 & & 10 & & 5 & & 1
\end{array}
$$

There are many patterns in the triangle. Find as many as you can.

Perhaps you discovered a way to write the next row of numbers, given the numbers in the row above it. There are always 1's on the outside. Each remaining number is the sum of the two numbers above:

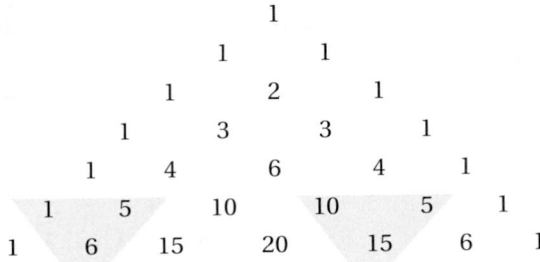

We see that in the bottom (seventh) row

the 1st and last numbers are 1;

the 2nd number is $1 + 5$, or 6;

the 3rd number is $5 + 10$, or 15;

the 4th number is $10 + 10$, or 20;

the 5th number is $10 + 5$, or 15; and

the 6th number is $5 + 1$, or 6.

Thus the expansion of $(a + b)^6$ is

$$(a+b)^6 = 1a^6 + 6a^5b + 15a^4b^2 + 20a^3b^3 + 15a^2b^4 + 6ab^5 + 1b^6.$$

To expand $(a + b)^8$, we complete two more rows of Pascal's triangle:

$$
\begin{array}{ccccccccccccccccc}
 & & & & & & & & 1 & & & & & & & & \\
 & & & & & & & 1 & & 1 & & & & & & & \\
 & & & & & & 1 & & 2 & & 1 & & & & & & \\
 & & & & & 1 & & 3 & & 3 & & 1 & & & & & \\
 & & & & 1 & & 4 & & 6 & & 4 & & 1 & & & & \\
 & & & 1 & & 5 & & 10 & & 10 & & 5 & & 1 & & & \\
 & & 1 & & 6 & & 15 & & 20 & & 15 & & 6 & & 1 & & \\
 & 1 & & 7 & & 21 & & 35 & & 35 & & 21 & & 7 & & 1 & \\
1 & & 8 & & 28 & & 56 & & 70 & & 56 & & 28 & & 8 & & 1
\end{array}
$$

Thus the expansion of $(a + b)^8$ has coefficients found in the 9th row above:

$$(a + b)^8 = 1a^8 + 8a^7b + 28a^6b^2 + 56a^5b^3 + 70a^4b^4 + 56a^3b^5 + 28a^2b^6 + 8ab^7 + 1b^8.$$

We can generalize our results as follows:

The Binomial Theorem (Form 1)

For any binomial $a + b$ and any natural number n,

$$(a + b)^n = c_0a^nb^0 + c_1a^{n-1}b^1 + c_2a^{n-2}b^2 + \cdots + c_{n-1}a^1b^{n-1} + c_na^0b^n,$$

where the numbers $c_0, c_1, c_2, \ldots, c_n$ are from the $(n + 1)$st row of Pascal's triangle.

A proof of the binomial theorem is beyond the scope of this text.

EXAMPLE 1   Expand: $(u - v)^5$.

Solution   Using the binomial theorem, we have $a = u$, $b = -v$, and $n = 5$. We use the 6th row of Pascal's triangle: 1   5   10   10   5   1. Thus,

$$\begin{aligned}
(u - v)^5 &= [u + (-v)]^5 \qquad \text{Rewriting } u - v \text{ as a sum} \\
&= 1(u)^5 + 5(u)^4(-v)^1 + 10(u)^3(-v)^2 + 10(u)^2(-v)^3 \\
&\quad + 5(u)^1(-v)^4 + 1(-v)^5 \\
&= u^5 - 5u^4v + 10u^3v^2 - 10u^2v^3 + 5uv^4 - v^5.
\end{aligned}$$

Note that the signs of the terms alternate between $+$ and $-$. When $-v$ is raised to an odd power, the sign is $-$; when the power is even, the sign is $+$.

EXAMPLE 2   Expand: $\left(2t + \dfrac{3}{t}\right)^6$.

Solution   Note that $a = 2t$, $b = 3/t$, and $n = 6$. We use the 7th row of Pascal's triangle: 1   6   15   20   15   6   1. Thus,

$$\begin{aligned}
\left(2t + \frac{3}{t}\right)^6 &= 1(2t)^6 + 6(2t)^5\left(\frac{3}{t}\right)^1 + 15(2t)^4\left(\frac{3}{t}\right)^2 + 20(2t)^3\left(\frac{3}{t}\right)^3 \\
&\quad + 15(2t)^2\left(\frac{3}{t}\right)^4 + 6(2t)^1\left(\frac{3}{t}\right)^5 + 1\left(\frac{3}{t}\right)^6 \\
&= 64t^6 + 6(32t^5)\left(\frac{3}{t}\right) + 15(16t^4)\left(\frac{9}{t^2}\right) + 20(8t^3)\left(\frac{27}{t^3}\right) \\
&\quad + 15(4t^2)\left(\frac{81}{t^4}\right) + 6(2t)\left(\frac{243}{t^5}\right) + \frac{729}{t^6} \\
&= 64t^6 + 576t^4 + 2160t^2 + 4320 + 4860t^{-2} + 2916t^{-4} \\
&\quad + 729t^{-6}.
\end{aligned}$$

Binomial Expansion Using Factorial Notation

The drawback to using Pascal's triangle is that we must compute all the preceding rows in the table to obtain the row needed for the expansion in which we are interested. The following method avoids this difficulty. It will also enable us to find a specific term—say, the 8th term—without computing all the other terms in the expansion. This method is useful in such courses as finite mathematics, calculus, and statistics.

To develop the method, we need some new notation. Products of successive natural numbers, such as $6 \cdot 5 \cdot 4 \cdot 3 \cdot 2 \cdot 1$ and $8 \cdot 7 \cdot 6 \cdot 5 \cdot 4 \cdot 3 \cdot 2 \cdot 1$, have a special notation. For the product $6 \cdot 5 \cdot 4 \cdot 3 \cdot 2 \cdot 1$, we write 6!, read "6 factorial."

Factorial Notation

For any natural number n,

$$n! = n(n-1)(n-2) \cdots (3)(2)(1).$$

Here are some examples:

$$6! = 6 \cdot 5 \cdot 4 \cdot 3 \cdot 2 \cdot 1 = 720,$$
$$5! = \quad 5 \cdot 4 \cdot 3 \cdot 2 \cdot 1 = 120,$$
$$4! = \quad\quad 4 \cdot 3 \cdot 2 \cdot 1 = \quad 24,$$
$$3! = \quad\quad\quad 3 \cdot 2 \cdot 1 = \quad\quad 6,$$
$$2! = \quad\quad\quad\quad 2 \cdot 1 = \quad\quad 2,$$
$$1! = \quad\quad\quad\quad\quad 1 = \quad\quad 1.$$

We also define 0! to be 1 for reasons explained shortly.

To simplify expressions like

$$\frac{8!}{5! \, 3!},$$

note that

$$8! = 8 \cdot 7 \cdot 6 \cdot 5 \cdot 4 \cdot 3 \cdot 2 \cdot 1 = 8 \cdot 7! = 8 \cdot 7 \cdot 6! = 8 \cdot 7 \cdot 6 \cdot 5!,$$

and so on.

Caution! $\dfrac{6!}{3!} \neq 2!$ To see this, note that

$$\frac{6!}{3!} = \frac{6 \cdot 5 \cdot 4 \cdot \cancel{3} \cdot \cancel{2} \cdot \cancel{1}}{\cancel{3} \cdot \cancel{2} \cdot \cancel{1}} = 6 \cdot 5 \cdot 4.$$

EXAMPLE 3   Simplify: $\dfrac{8!}{5!\,3!}$.

Solution

$$\dfrac{8!}{5!\,3!} = \dfrac{8 \cdot 7 \cdot 6 \cdot 5!}{5! \cdot 3 \cdot 2 \cdot 1} = 8 \cdot 7 \qquad \text{Removing a factor equal to 1:}$$

$$\dfrac{6 \cdot 5!}{5! \cdot 3 \cdot 2} = 1$$

$$= 56$$

Student Notes

It is important to recognize factorial notation as representing a product with descending factors. Thus, $7!$, $7 \cdot 6!$, and $7 \cdot 6 \cdot 5!$ all represent the same product.

The following notation is used in our second formulation of the binomial theorem.

$\dbinom{n}{r}$ **Notation**

For n, r nonnegative integers with $n \geq r$,

$$\dbinom{n}{r}, \quad \text{read "}n\text{ choose }r\text{,"} \quad \text{means} \quad \dfrac{n!}{(n-r)!\,r!}.^*$$

EXAMPLE 4   Simplify: **(a)** $\dbinom{7}{2}$; **(b)** $\dbinom{9}{6}$; **(c)** $\dbinom{6}{6}$.

Solution

a) $\dbinom{7}{2} = \dfrac{7!}{(7-2)!\,2!}$

$$= \dfrac{7!}{5!\,2!} = \dfrac{7 \cdot 6 \cdot 5!}{5! \cdot 2 \cdot 1} = \dfrac{7 \cdot 6}{2} \qquad \text{We can write 7! as } 7 \cdot 6 \cdot 5! \text{ to aid our simplification.}$$

$$= 7 \cdot 3$$

$$= 21$$

b) $\dbinom{9}{6} = \dfrac{9!}{3!\,6!}$

$$= \dfrac{9 \cdot 8 \cdot 7 \cdot 6!}{3 \cdot 2 \cdot 1 \cdot 6!} = \dfrac{9 \cdot 8 \cdot 7}{3 \cdot 2} \qquad \text{Writing 9! as } 9 \cdot 8 \cdot 7 \cdot 6! \text{ to help with simplification}$$

$$= 3 \cdot 4 \cdot 7$$

$$= 84$$

technology connection

The PRB option of the MATH menu provides access to both factorial calculations and nCr. In both cases, a number must be entered first. To find $\dbinom{7}{2}$, we press **7** **MATH**, select PRB and nCr, and press **2** **ENTER**.

7 nCr 2	
	21

1. Find $12!$.

2. Find $\dbinom{8}{3}$ and $\dbinom{12}{5}$.

*In many books and for many calculators, the notation $_nC_r$ is used instead of $\dbinom{n}{r}$.

c) $\dbinom{6}{6} = \dfrac{6!}{0!\,6!} = \dfrac{6!}{1 \cdot 6!}$ Since $0! = 1$

$\phantom{\dbinom{6}{6}} = \dfrac{6!}{6!}$

$\phantom{\dbinom{6}{6}} = 1$

Now we can restate the binomial theorem using our new notation.

The Binomial Theorem (Form 2)

For any binomial $a + b$ and any natural number n,

$$(a + b)^n = \dbinom{n}{0}a^n + \dbinom{n}{1}a^{n-1}b + \dbinom{n}{2}a^{n-2}b^2 + \cdots + \dbinom{n}{n}b^n.$$

EXAMPLE 5   Expand: $(3x + y)^4$.

Solution We use the binomial theorem (Form 2) with $a = 3x$, $b = y$, and $n = 4$:

$$(3x + y)^4 = \dbinom{4}{0}(3x)^4 + \dbinom{4}{1}(3x)^3 y + \dbinom{4}{2}(3x)^2 y^2 + \dbinom{4}{3}(3x)y^3 + \dbinom{4}{4}y^4$$

$$= \dfrac{4!}{4!\,0!}3^4 x^4 + \dfrac{4!}{3!\,1!}3^3 x^3 y + \dfrac{4!}{2!\,2!}3^2 x^2 y^2 + \dfrac{4!}{1!\,3!}3xy^3 + \dfrac{4!}{0!\,4!}y^4$$

$$\left. \begin{aligned} &= 1 \cdot 81x^4 + 4 \cdot 27x^3 y + 6 \cdot 9x^2 y^2 + 4 \cdot 3xy^3 + y^4 \\ &= 81x^4 + 108x^3 y + 54x^2 y^2 + 12xy^3 + y^4. \end{aligned} \right\} \quad \text{Simplifying}$$

EXAMPLE 6   Expand: $(x^2 - 2y)^5$.

Solution   In this case, $a = x^2$, $b = -2y$, and $n = 5$:

$$(x^2 - 2y)^5 = \dbinom{5}{0}(x^2)^5 + \dbinom{5}{1}(x^2)^4(-2y) + \dbinom{5}{2}(x^2)^3(-2y)^2$$

$$+ \dbinom{5}{3}(x^2)^2(-2y)^3 + \dbinom{5}{4}(x^2)(-2y)^4 + \dbinom{5}{5}(-2y)^5$$

$$= \dfrac{5!}{5!\,0!}x^{10} + \dfrac{5!}{4!\,1!}x^8(-2y) + \dfrac{5!}{3!\,2!}x^6(-2y)^2$$

$$+ \dfrac{5!}{2!\,3!}x^4(-2y)^3 + \dfrac{5!}{1!\,4!}x^2(-2y)^4 + \dfrac{5!}{0!\,5!}(-2y)^5$$

$$= x^{10} - 10x^8 y + 40x^6 y^2 - 80x^4 y^3 + 80x^2 y^4 - 32y^5.$$

Note that in the binomial theorem (Form 2), $\binom{n}{0}a^n b^0$ gives us the first term, $\binom{n}{1}a^{n-1}b^1$ gives us the second term, $\binom{n}{2}a^{n-2}b^2$ gives us the third term, and so on. This can be generalized to give a method for finding a specific term without writing the entire expansion.

Finding a Specific Term

When $(a + b)^n$ is expanded and written in descending powers of a, the $(r + 1)$st term is

$$\binom{n}{r}a^{n-r}b^r.$$

EXAMPLE 7   Find the 5th term in the expansion of $(2x - 3y)^7$.

Solution   First, we note that $5 = 4 + 1$. Thus, $r = 4$, $a = 2x$, $b = -3y$, and $n = 7$. Then the 5th term of the expansion is

$$\binom{7}{4}(2x)^{7-4}(-3y)^4, \text{ or } \frac{7!}{3!\,4!}(2x)^3(-3y)^4, \text{ or } 22{,}680x^3 y^4.$$

It is because of the binomial theorem that $\binom{n}{r}$ is called a *binomial coefficient*. We can now explain why 0! is defined to be 1. In the binomial theorem,

$\binom{n}{0}$ must equal 1 when using the definition $\binom{n}{r} = \frac{n!}{(n-r)!\,r!}$.

Thus we must have

$$\binom{n}{0} = \frac{n!}{(n-0)!\,0!} = \frac{n!}{n!\,0!} = 1.$$

This is satisfied only if 0! is defined to be 1.

Exercise Set
14.4

🦢 *Concept Reinforcement*   *Complete each of the following.*

1. The last term in the expansion of $(x + 2)^5$ is _____.

2. The expansion of $(x + y)^7$, when simplified, contains a total of _____ terms.

3. In the expansion of $(a + b)^9$, the exponents in each term add to _____.

4. The expression _____ represents $4 \cdot 3 \cdot 2 \cdot 1$.

5. The expression _____ represents $\dfrac{8!}{3!\,5!}$.

6. In the expansion of $(a + b)^{10}$, the coefficient of $a^8 b^2$ is _____.

7. In the expansion of $(x + y)^9$, the coefficient of $x^2 y^7$ is the same as the coefficient of _____.

8. The notation $\dbinom{10}{4}$ is read _____.

Simplify.

9. $9!$

10. $8!$

11. $11!$

12. $10!$

13. $\dfrac{8!}{6!}$

14. $\dfrac{7!}{4!}$

15. $\dfrac{9!}{5!}$

16. $\dfrac{10!}{7!}$

17. $\dbinom{7}{4}$

18. $\dbinom{8}{2}$

19. $\dbinom{9}{5}$

20. $\dbinom{10}{6}$

21. $\dbinom{30}{3}$

22. $\dbinom{20}{18}$

23. $\dbinom{40}{38}$

24. $\dbinom{35}{2}$

Expand. Use both of the methods shown in this section.

25. $(a - b)^4$

26. $(m + n)^5$

27. $(p + q)^7$

28. $(x - y)^6$

29. $(3c - d)^7$

30. $(x^2 - 3y)^5$

31. $(t^{-2} + 2)^6$

32. $(3c - d)^6$

33. $(x - y)^5$

34. $(x - y)^3$

35. $\left(3s + \dfrac{1}{t}\right)^9$

36. $\left(x + \dfrac{2}{y}\right)^9$

37. $(x^3 - 2y)^5$

38. $(a^2 - b^3)^5$

39. $\left(\sqrt{5} + t\right)^6$

40. $\left(\sqrt{3} - t\right)^4$

41. $\left(\dfrac{1}{\sqrt{x}} - \sqrt{x}\right)^6$

42. $(x^{-2} + x^2)^4$

Find the indicated term for each binomial expression.

43. 3rd, $(a + b)^6$

44. 6th, $(x + y)^7$

45. 12th, $(a - 3)^{14}$

46. 11th, $(x - 2)^{12}$

47. 5th, $\left(2x^3 + \sqrt{y}\right)^8$

48. 4th, $\left(\dfrac{1}{b^2} + c\right)^7$

49. Middle, $(2u + 3v^2)^{10}$

50. Middle two, $\left(\sqrt{x} + \sqrt{3}\right)^5$

Aha! **51.** 9th, $(x - y)^8$

52. 13th, $\left(a - \sqrt{b}\right)^{12}$

📄 **53.** Maya claims that she can calculate mentally the first two and the last two terms of the expansion of $(a + b)^n$ for any whole number n. How do you think she does this?

📄 **54.** Without performing any calculations, explain why the expansions of $(x - y)^8$ and $(y - x)^8$ must be equal.

SKILL MAINTENANCE

Solve. [12.6]

55. $\log_2 x + \log_2 (x - 2) = 3$

56. $\log_3 (x + 2) - \log_3 (x - 2) = 2$

57. $e^t = 280$

58. $\log_5 x^2 = 2$

SYNTHESIS

59. Explain how someone can determine the x^2-term of the expansion of $\left(x - \dfrac{3}{x} \right)^{10}$ without calculating any other terms.

60. Devise two problems requiring the use of the binomial theorem. Design the problems so that one is solved more easily using Form 1 and the other is solved more easily using Form 2. Then explain what makes one form easier to use than the other in each case.

61. Show that there are exactly $\dbinom{5}{3}$ ways of choosing a subset of size 3 from $\{a, b, c, d, e\}$.

62. *Baseball.* At one point in July 2004, Barry Bonds of the San Francisco Giants had a batting average of .360. At that time, if someone were to randomly select 5 of his "at-bats," the probability of Bonds getting exactly 3 hits would be the 3rd term of the binomial expansion of $(0.360 + 0.640)^5$. Find that term and use a calculator to estimate the probability.
Source: www.mlb.com

63. *Widows or divorcees.* The probability that a woman will be either widowed or divorced is 85%. If 8 women are randomly selected, the probability that exactly 5 of them will be either widowed or divorced is the 6th term of the binomial expansion of $(0.15 + 0.85)^8$. Use a calculator to estimate that probability.

64. *Baseball.* In reference to Exercise 62, the probability that Bonds will get *at most* 3 hits is found by adding the last 4 terms of the binomial expansion of $(0.360 + 0.640)^5$. Find these terms and use a calculator to estimate the probability.

65. *Widows or divorcees.* In reference to Exercise 63, the probability that *at least* 6 of the women will be widowed or divorced is found by adding the last three terms of the binomial expansion of $(0.15 + 0.85)^8$. Find these terms and use a calculator to estimate the probability.

66. Find the term of
$$\left(\frac{3x^2}{2} - \frac{1}{3x} \right)^{12}$$
that does not contain x.

67. Prove that
$$\binom{n}{r} = \binom{n}{n - r}$$
for any whole numbers n and r. Assume $r \leq n$.

68. Find the middle term of $(x^2 - 6y^{3/2})^6$.

69. Find the ratio of the 4th term of
$$\left(p^2 - \frac{1}{2} p \sqrt[3]{q} \right)^5$$
to the 3rd term.

70. Find the term containing $\dfrac{1}{x^{1/6}}$ of
$$\left(\sqrt[3]{x} - \frac{1}{\sqrt{x}} \right)^7.$$

Aha! **71.** Multiply: $(x^2 + 2xy + y^2)(x^2 + 2xy + y^2)^2(x + y)$.

72. What is the degree of $(x^3 + 2)^4$?

Study Summary

An ordered list of numbers, such as

$$5, 7, 8, 11, 17 \quad \text{or} \quad 6, 9, 12, 15, 18, 21, \ldots$$

is called a **sequence**, or **progression** (p. 902). The first sequence above ends so it is considered **finite**, whereas the second sequence is **infinite** (p. 902). When the difference between consecutive terms is constant—as in the second sequence above — the sequence is **arithmetic** (p. 909). If d represents the **common difference**, we have

$$a_n = a_1 + (n - 1)d, \qquad \text{To find the } n\text{th term}$$

$$a_{n+1} = a_n + d, \qquad \text{To find the next term when the } n\text{th term is known}$$

$$S_n = \frac{n}{2}(a_1 + a_n). \qquad \text{The sum of the first } n \text{ terms. This is an \textbf{arithmetic series}.}$$

When each term in a sequence is multiplied by a certain fixed number to get the next term, the sequence is **geometric** (p. 918). If r represents this fixed number, or **common ratio**, we have

$$a_n = a_1 r^{n-1}, \qquad \text{To find the } n\text{th term}$$

$$a_{n+1} = a_n \cdot r, \qquad \text{To find the next term when the } n\text{th term is known}$$

$$S_n = \frac{a_1(1 - r^n)}{1 - r}, \qquad \text{The sum of the first } n \text{ terms. This is a \textbf{geometric series}.}$$

$$S_\infty = \frac{a_1}{1 - r}, \, |r| < 1. \qquad \text{The limit of an infinite geometric series with } |r| < 1$$

The expression $\sum\limits_{k=1}^{n} a_k$ indicates a **partial sum**, or **finite series**, using **sigma** or **summation notation** (p. 905). The letter k serves as the **index of summation**.

It is possible to regard expressions of the form $(a + b)^n$ as series for which **Pascal's triangle** can be used to determine the **binomial coefficients** (pp. 930, 935). These coefficients can also be found using the **binomial theorem** (pp. 931, 934). Use of the binomial theorem generally requires understanding **factorial notation** (p. 932).

$$n! = n(n - 1)(n - 2) \cdots 3 \cdot 2 \cdot 1 \qquad \text{Factorial notation}$$

$$\binom{n}{r} = \frac{n!}{(n - r)! \, r!} \qquad \text{Binomial coefficient}$$

$$(a + b)^n = \binom{n}{0}a^n + \binom{n}{1}a^{n-1}b + \binom{n}{2}a^{n-2}b^2 + \cdots + \binom{n}{n}b^n \qquad \text{Binomial theorem}$$

$$\binom{n}{r}a^{n-r}b^r \qquad (r + 1)\text{st term of } (a + b)^n$$

14 Review Exercises

🦆 *Concept Reinforcement* *Classify each of the following as either true or false.*

1. The next term in the arithmetic sequence $10, 15, 20, \ldots$ is 35. [14.2]

2. The next term in the geometric sequence $2, 6, 18, 54, \ldots$ is 162. [14.3]

3. $\sum_{k=1}^{3} k^2$ means $1^2 + 2^2 + 3^2$. [14.1]

4. If $a_n = 3n - 1$, then $a_{17} = 19$. [14.1]

5. The nth term of an arithmetic sequence with common difference d is $a_n = a_1 + (n-1)d$, for any integer $n \geq 1$. [14.2]

6. The nth term of a geometric sequence with common ratio r is given by $a_n = a_1 r^{n-1}$, for any integer $n \geq 1$. [14.3]

7. For any natural number n, $n! = n(n-1)$. [14.4]

8. When simplified, the expansion of $(x + y)^{17}$ has 19 terms. [14.4]

Find the first four terms; the 8th term, a_8; and the 12th term, a_{12}. [14.1]

9. $a_n = 4n - 3$

10. $a_n = \dfrac{n-1}{n^2 + 1}$

Predict the general term. Answers may vary. [14.1]

11. $7, 14, 21, 28, 35, \ldots$

12. $-1, 3, -5, 7, -9, \ldots$

Write out and evaluate each sum. [14.1]

13. $\sum_{k=1}^{5} (-2)^k$

14. $\sum_{k=2}^{7} (1 - 2k)$

Rewrite using sigma notation. [14.1]

15. $4 + 8 + 12 + 16 + 20$

16. $\dfrac{-1}{2} + \dfrac{1}{4} + \dfrac{-1}{8} + \dfrac{1}{16} + \dfrac{-1}{32}$

17. Find the 14th term of the arithmetic sequence $-6, 1, 8, \ldots$. [14.2]

18. An arithmetic sequence has $a_1 = 11$ and $a_{13} = 43$. Find the common difference, d. [14.2]

19. An arithmetic sequence has $a_8 = 20$ and $a_{24} = 40$. Find the first term, a_1, and the common difference, d. [14.2]

20. Find the sum of the first 17 terms of the arithmetic series $-8 + (-11) + (-14) + \cdots$. [14.2]

21. Find the sum of all the multiples of 6 from 12 to 318, inclusive. [14.2]

22. Find the 20th term of the geometric sequence $2, 2\sqrt{2}, 4, \ldots$. [14.3]

23. Find the common ratio of the geometric sequence $2, \frac{4}{3}, \frac{8}{9}, \ldots$. [14.3]

24. Find the nth term of the geometric sequence $-2, 2, -2, \ldots$. [14.3]

25. Find the nth term of the geometric sequence $3, \frac{3}{4}x, \frac{3}{16}x^2, \ldots$. [14.3]

26. Find S_6 for the geometric series
$$3 + 12 + 48 + \cdots. \quad [14.3]$$

27. Find S_{12} for the geometric series
$$3x - 6x + 12x - \cdots. \quad [14.3]$$

Determine whether each infinite geometric series has a limit. If a limit exists, find it. [14.3]

28. $6 + 3 + 1.5 + 0.75 + \cdots$

29. $7 - 4 + \frac{16}{7} - \cdots$

30. $2 + (-2) + 2 + (-2) + \cdots$

31. $0.04 + 0.08 + 0.16 + 0.32 + \cdots$

32. $\$2000 + \$1900 + \$1805 + \$1714.75 + \cdots$

33. Find fraction notation for $0.555555\ldots$. [14.3]

34. Find fraction notation for $1.39393939\ldots$. [14.3]

35. Adam took a job working in a convenience store starting with an hourly wage of $11.50. He was promised a raise of 40¢ per hour every 3 mos for 8 yr. At the end of 8 yr, what will be his hourly wage? [14.2]

36. A stack of poles has 42 poles in the bottom row. There are 41 poles in the second row, 40 poles in the third row, and so on, ending with 1 pole in the top row. How many poles are in the stack? [14.2]

37. Stacey's student loan is in the amount of $12,000. Interest is 4%, compounded annually, and the amount is to be paid off in 7 yr. How much is to be paid back? [14.3]

38. Find the total rebound distance of a ball, given that it is dropped from a height of 12 m and each rebound is one-third of the preceding one. [14.3]

Simplify. [14.4]

39. $7!$

40. $\binom{8}{3}$

41. Find the 3rd term of $(a + b)^{20}$. [14.4]

42. Expand: $(x - 2y)^4$. [14.4]

SYNTHESIS

43. What happens to a_n in a geometric sequence with $|r| < 1$, as n gets larger? Why? [14.3]

44. Compare the two forms of the binomial theorem given in the text. Under what circumstances would one be more useful than the other? [14.4]

45. Find the sum of the first n terms of the geometric series $1 - x + x^2 - x^3 + \cdots$. [14.3]

46. Expand: $(x^{-3} + x^3)^5$. [14.4]

14 Chapter Test

1. Find the first five terms and the 12th term of a sequence with general term $a_n = 6n - 5$.

2. Predict the general term of the sequence
$$\frac{4}{3}, \frac{4}{9}, \frac{4}{27}, \ldots.$$

3. Write out and evaluate:
$$\sum_{k=2}^{6} (3 - 2^k).$$

4. Rewrite using sigma notation:
$$1 + (-8) + 27 + (-64) + 125.$$

5. Find the 13th term, a_{13}, of the arithmetic sequence $9, 4, -1, \ldots$.

Assume arithmetic sequences for Questions 6 and 7.

6. Find the common difference d when $a_1 = 7$ and $a_7 = 9\frac{1}{4}$.

7. Find a_1 and d when $a_5 = 16$ and $a_{10} = -3$.

8. Find the sum of all the multiples of 12 from 24 to 240, inclusive.

9. Find the 6th term of the geometric sequence $72, 18, 4\frac{1}{2}, \ldots$.

10. Find the common ratio of the geometric sequence $22\frac{1}{2}, 15, 10, \ldots$.

11. Find the nth term of the geometric sequence $3, 9, 27, \ldots$.

12. Find the sum of the first nine terms of the geometric series
$$(1 + x) + (2 + 2x) + (4 + 4x) + \cdots.$$

Determine whether each infinite geometric series has a limit. If a limit exists, find it.

13. $0.5 + 0.25 + 0.125 + \cdots$

14. $0.5 + 1 + 2 + 4 + \cdots$

15. $\$1000 + \$80 + \$6.40 + \cdots$

16. Find fraction notation for 0.85858585 ....

17. An auditorium has 31 seats in the first row, 33 seats in the second row, 35 seats in the third row, and so on, for 18 rows. How many seats are in the 17th row?

18. Lindsay's Uncle Ken gave her $100 for her first birthday, $200 for her second birthday, $300 for her third birthday, and so on, until her eighteenth birthday. How much did he give her in all?

19. Each week the price of a $15,000 boat will be reduced 5% of the previous week's price. If we assume that it is not sold, what will be the price after 10 weeks?

20. Find the total rebound distance of a ball that is dropped from a height of 18 m, with each rebound two thirds of the preceding one.

21. Simplify: $\dbinom{12}{9}$.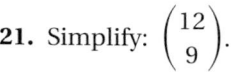

22. Expand: $(x^2 - 3y)^5$.

23. Find the 4th term in the expansion of $(a + x)^{12}$.

SYNTHESIS

24. Find a formula for the sum of the first n even natural numbers:
$$2 + 4 + 6 + \cdots + 2n.$$

25. Find the sum of the first n terms of
$$1 + \frac{1}{x} + \frac{1}{x^2} + \frac{1}{x^3} + \cdots.$$

1-14 Cumulative Review

Simplify.

1. $(-7x^2y^3)(5x^4y^{-7})$ [4.8]

2. $|-3.5 + 9.8|$ [1.8]

3. $y - [3 - 4(5 - 2y) - 3y]$ [1.8]

4. $(10 \cdot 8 - 9 \cdot 7)^2 - 54 \div 9 - 3$ [1.8]

5. Evaluate
$$\frac{ab - ac}{bc}$$
for $a = -2$, $b = 3$, and $c = -4$. [1.8]

Perform the indicated operations to create an equivalent expression. Be sure to simplify your result if possible.

6. $(5a^2 - 3ab - 7b^2) - (2a^2 + 5ab + 8b^2)$ [4.3]

7. $(-3x^2 + 4x^3 - 5x - 1) + (9x^3 - 4x^2 + 7 - x)$ [4.3]

8. $(2a - 1)(3a + 5)$ [4.5]

9. $(3a^2 - 5y)^2$ [4.5]

10. $\dfrac{1}{x - 2} - \dfrac{4}{x^2 - 4} + \dfrac{3}{x + 2}$ [6.4]

11. $\dfrac{x^2 - 6x + 8}{4x + 12} \cdot \dfrac{x + 3}{x^2 - 4}$ [6.2]

12. $\dfrac{3x + 3y}{5x - 5y} \div \dfrac{3x^2 + 3y^2}{5x^3 - 5y^3}$ [6.2]

13. $\dfrac{x - \dfrac{a^2}{x}}{1 + \dfrac{a}{x}}$ [6.5]

Factor, if possible, to form an equivalent expression.

14. $4x^2 - 12x + 9$ [5.4]

15. $27a^3 - 8$ [5.5]

16. $a^3 + 3a^2 - ab - 3b$ [5.1]

17. $15y^4 + 33y^2 - 36$ [5.3]

18. For the function described by
$$f(x) = 3x^2 - 4x,$$
find $f(-2)$. [7.2]

19. Divide:

$$(7x^4 - 5x^3 + x^2 - 4) \div (x - 2). \ [4.7]$$

Solve.

20. $8(x - 1) - 3(x - 2) = 1$ [2.2]

21. $\dfrac{6}{x} + \dfrac{6}{x + 2} = \dfrac{5}{2}$ [6.6]

22. $2x + 1 > 5$ *or* $x - 7 \le 3$ [9.2]

23. $5x + 3y = -2,$
$3x + 5y = 2$ [8.2]

24. $x + y - z = 0,$
$3x + y + z = 6,$
$x - y + 2z = 5$ [8.4]

25. $3\sqrt{x - 1} = 5 - x$ [10.6]

26. $x^4 - 29x^2 + 100 = 0$ [11.5]

27. $x^2 + y^2 = 8,$
$x^2 - y^2 = 2$ [13.4]

28. $4^x = 7$ [12.6]

29. $\log(x^2 - 25) - \log(x + 5) = 3$ [12.6]

30. $\log_5 x = -2$ [12.6]

31. $7^{2x+3} = 49$ [12.6]

32. $|2x - 1| \le 5$ [9.3]

33. $7x^2 + 14 = 0$ [11.1]

34. $x^2 + 4x = 3$ [11.2]

35. $y^2 + 3y > 10$ [11.9]

36. Let $f(x) = x^2 - 2x$. Find a such that $f(a) \le 48$. [11.9]

37. If $f(x) = \sqrt{-x + 4} + 3$ and $g(x) = \sqrt{x - 2} + 3$, find a such that $f(a) = g(a)$. [10.6]

Solve.

38. The perimeter of a rectangular sign is 34 ft. The length of a diagonal is 13 ft. Find the dimensions of the sign. [13.4]

39. A music club offers two types of membership. Limited members pay a fee of $10 a year and can buy CDs for $10 each. Preferred members pay $20 a year and can buy CDs for $7.50 each. For what numbers of annual CD purchases would it be less expensive to be a preferred member? [9.1]

40. Find three consecutive integers whose sum is 198. [2.5]

41. A pentagon with all five sides the same size has a perimeter equal to that of an octagon in which all eight sides are the same size. One side of the pentagon is 2 less than three times one side of the octagon. What is the perimeter of each figure? [8.3]

42. Roger's Organics mixes herbs that cost $2.68 an ounce with herbs that cost $4.60 an ounce to create a seasoning that costs $3.80 an ounce. How many ounces of each herb should be mixed together to make 24 oz of the seasoning? [8.3]

43. An airplane can fly 190 mi with the wind in the same time it takes to fly 160 mi against the wind. The speed of the wind is 30 mph. How fast can the plane fly in still air? [6.7]

44. Bianca can tap the sugar maple trees in Southway Park in 21 hr. Delia can tap the trees in 14 hr. How long would it take them, working together, to tap the trees? [6.7]

45. The centripetal force F of an object moving in a circle varies directly as the square of the velocity v and inversely as the radius r of the circle. If $F = 8$ when $v = 1$ and $r = 10$, what is F when $v = 2$ and $r = 16$? [11.6]

46. The Brighton recreation department plans to fence in a rectangular park next to a river. (Note that no fence will be needed along the river.) What is the area of the largest region that can be fenced in with 200 ft of fencing? [11.8]

Graph.

47. $3x - y = 7$ [3.6]

48. $y - 4 = -\frac{2}{3}(x - 1)$ [3.7]

49. $x^2 + y^2 = 100$ [13.1]

50. $\dfrac{x^2}{36} - \dfrac{y^2}{9} = 1$ [13.3]

51. $y = \log_2 x$ [12.3]

52. $f(x) = 2^x - 3$ [12.2]

53. $2x - 3y < -6$ [9.4]

54. Graph: $f(x) = -2(x - 3)^2 + 1$. [11.7]
   a) Label the vertex.
   b) Draw the axis of symmetry.
   c) Find the maximum or minimum value.

55. Solve $V = P - Prt$ for r. [2.3]

56. Solve $I = \dfrac{R}{R + r}$ for R. [7.5]

57. Find a linear equation whose graph has a y-intercept of $(0, -8)$ and is parallel to the line whose equation is $3x - y = 6$. [3.6]

Find the domain of each function.

58. $f(x) = \sqrt{6 - 8x}$ [9.2]

59. $g(x) = \dfrac{x - 4}{x^2 - 2x + 1}$ [7.2]

60. Multiply $(8.9 \times 10^{-17})(7.6 \times 10^4)$. Write scientific notation for the answer. [4.8]

61. Multiply and simplify: $\sqrt{8x}\,\sqrt{8x^3y}$. [10.3]

62. Simplify: $(25x^{4/3}y^{1/2})^{3/2}$. [10.2]

63. Divide and simplify:
   $$\dfrac{\sqrt[3]{25x}}{\sqrt[3]{5y^2}}.\ [10.5]$$

64. Write an equivalent expression by rationalizing the denominator:
   $$\dfrac{1 - \sqrt{x}}{1 + \sqrt{x}}.\ [10.5]$$

65. Multiply these complex numbers:
   $(3 + 2i)(4 - 7i)$. [10.8]

66. Write a quadratic equation whose solutions are $5\sqrt{2}$ and $-5\sqrt{2}$. [11.4]

67. Find the center and the radius of the circle given by
   $x^2 + y^2 - 4x + 6y - 23 = 0$. [13.1]

68. Write an equivalent expression that is a single logarithm:
   $\frac{2}{3}\log_a x - \frac{1}{2}\log_a y + 5\log_a z$. [12.4]

69. Write an equivalent exponential equation:
   $\log_a c = 5$. [12.3]

Use a calculator to find each of the following. [12.5]

70. log 5677.2

71. $10^{-3.587}$

72. ln 5677.2

73. $e^{-3.587}$

74. The number of personal computers in Mexico has grown exponentially from 0.12 million in 1985 to 5.0 million in 2002. [12.7]
   Source: Center for International Development
   a) Find the exponential growth rate, k, to three decimal places and write an exponential function describing the number of personal computers in Mexico t years after 1985.
   b) Predict the number of personal computers in Mexico in 2010.

75. Find the distance between the points $(-1, -5)$ and $(2, -1)$. [13.1]

76. Find the 21st term of the arithmetic sequence $19, 12, 5, \ldots$. [14.2]

77. Find the sum of the first 25 terms of the arithmetic series $-1 + 2 + 5 + \cdots$. [14.2]

78. Find the general term of the geometric sequence $16, 4, 1, \ldots$. [14.3]

79. Find the 7th term of $(a - 2b)^{10}$. [14.4]

80. Find the sum of the first nine terms of the geometric series $x + 1.5x + 2.25x + \cdots$. [14.3]

81. On Elyse's 9th birthday, her grandmother opened a savings account for her with $500. The account pays 3% interest, compounded annually. If Elyse neither adds to nor withdraws any money from the bank, how much will be in the account on her 18th birthday? [14.3]

SYNTHESIS

Solve.

82. $\dfrac{9}{x} - \dfrac{9}{x + 12} = \dfrac{108}{x^2 + 12x}$ [6.6]

83. $\log_2(\log_3 x) = 2$ [12.6]

84. y varies directly as the cube of x and x is multiplied by 0.5. What is the effect on y? [11.6]

85. Divide these complex numbers:
   $$\dfrac{2\sqrt{6} + 4\sqrt{5}i}{2\sqrt{6} - 4\sqrt{5}i}.\ [10.8]$$

86. Diaphantos, a famous mathematician, spent $\frac{1}{6}$ of his life as a child, $\frac{1}{12}$ as an adolescent, and $\frac{1}{7}$ as a bachelor. Five years after he was married, he had a son who died 4 years before his father at half his father's final age. How long did Diaphantos live? [8.5]

R

Elementary Algebra Review

*T*his chapter is a review of the first six chapters of this text. Each section corresponds to a chapter of the text. For further explanation of the topics in this chapter, refer to the sections or pages referenced in the margin.

R.1   Introduction to Algebraic Expressions

The Real Numbers • Operations on Real Numbers • Algebraic Expressions

The study of algebra requires a thorough understanding of how numbers are manipulated.

The Real Numbers

Numbers can be represented by points on a number line.

$$-3.3 \qquad -\tfrac{3}{4} \qquad \sqrt{8}$$

$$\begin{array}{ccccccc} & -3 & -2 & -1 & 0 & 1 & 2 & 3 \end{array}$$

Some sets of numbers are given specific names.

> **Sets of Numbers**
>
>   **Natural Numbers:** $\{1, 2, 3, \ldots\}$
>
>   **Whole Numbers:** $\{0, 1, 2, 3, \ldots\}$
>
>   **Integers:** $\{\ldots, -3, -2, -1, 0, 1, 2, 3, \ldots\}$
>
>   **Rational numbers:** $\left\{ \dfrac{a}{b} \mid a \text{ and } b \text{ are integers and } b \neq 0 \right\}$

Rational numbers can always be written as **terminating** or **repeating** decimals. **Irrational numbers**, like $\sqrt{2}$ or π, can be thought of as nonterminating and nonrepeating decimals. The set of **real numbers** consists of all rational and irrational numbers, taken together.

We can compare, or **order**, real numbers by their graphs on the number line. For any two numbers, the one to the left is less than the one to the right.

Sentences like $\frac{1}{4} = 0.25$, containing an equals sign, are called **equations**. An **inequality** is a sentence containing > (is greater than), < (is less than), ≥ (is greater than or equal to), or ≤ (is less than or equal to). Equations and inequalities can be true or false.

EXAMPLE 1 Write true or false for each equation or inequality.

a) $-2\frac{1}{3} = -\frac{7}{3}$ b) $1 = -1$ c) $-5 < -2$

d) $-3 \geq 2$ e) $1.1 \leq 1.1$

Solution

a) $-2\frac{1}{3} = -\frac{7}{3}$ is *true* because $-2\frac{1}{3}$ and $-\frac{7}{3}$ represent the same number.

b) $1 = -1$ is a *false* equation.

c) $-5 < -2$ is *true* because -5 is to the left of -2 on the number line.

d) $-3 \geq 2$ is *false* because neither $-3 > 2$ nor $-3 = 2$ is true.

e) $1.1 \leq 1.1$ is *true* because $1.1 = 1.1$ is true.

Absolute value (p. 36)

The distance of a number from 0 is called the **absolute value** of the number. The notation $|-4|$ represents the absolute value of -4. The absolute value of a number is never negative.

EXAMPLE 2 Find the absolute value: (a) $|-4|$; (b) $\left|\frac{11}{3}\right|$; (c) $|0|$.

Solution

a) $|-4| = 4$ since -4 is 4 units from 0.

b) $\left|\frac{11}{3}\right| = \frac{11}{3}$ since $\frac{11}{3}$ is $\frac{11}{3}$ units from 0.

c) $|0| = 0$ since 0 is 0 units from itself.

Operations on Real Numbers

Addition, subtraction, multiplication, and division of real numbers are defined using absolute values.

Addition (Section 1.5)

> **Rules for Addition of Real Numbers**
>
> **1.** *Positive numbers*: Add as usual. The answer is positive.
> **2.** *Negative numbers*: Add absolute values and make the answer negative.
> **3.** *A positive and a negative number*: Subtract absolute values. Then:
>   **a)** If the positive number has the greater absolute value, the answer is positive.
>   **b)** If the negative number has the greater absolute value, the answer is negative.
>   **c)** If the numbers have the same absolute value, the answer is 0.
> **4.** *One number is zero*: The sum is the other number.

Opposite (p. 46)

Every real number has an **opposite**. The opposite of -6 is 6, the opposite of 3.7 is -3.7, and the opposite of 0 is itself. When opposites are added, the result is 0. Finding the opposite of a number is often called "changing its sign."

Subtraction of real numbers is defined in terms of addition and opposites.

Subtraction (Section 1.6)

> **Subtraction of Real Numbers**
>
> To subtract, add the opposite of the number being subtracted.

The rules for multiplication of real numbers are similar to the rules for division.

Multiplication and division
(Section 1.7)

> **Rules for Multiplication and Division**
>
> To multiply or divide two real numbers:
>
> **1.** Using the absolute values, multiply or divide, as indicated.
> **2.** If the signs are the same, the answer is positive.
> **3.** If the signs are different, the answer is negative.

EXAMPLE 3   Perform the indicated operations: **(a)** $-13 + (-9)$; **(b)** $-\frac{4}{5} + \frac{1}{10}$; **(c)** $-6 - (-7.3)$; **(d)** $3(-1.5)$; **(e)** $\left(-\frac{4}{9}\right) \div \left(-\frac{2}{5}\right)$.

Solution

a) $-13 + (-9) = -22$

Two negatives. *Think:* Add the absolute values, 13 and 9, to get 22. Make the answer *negative*, -22.

b) $-\frac{4}{5} + \frac{1}{10} = -\frac{8}{10} + \frac{1}{10} = -\frac{7}{10}$

A negative and a positive. *Think:* The difference of absolute values is $\frac{8}{10} - \frac{1}{10}$, or $\frac{7}{10}$. The negative number has the larger absolute value, so the answer is *negative*, $-\frac{7}{10}$.

c) $-6 - (-7.3) = -6 + 7.3 = 1.3$

Change the subtraction to addition and add the opposite.

d) $3(-1.5) = -4.5$

Think: $3(1.5) = 4.5$. The signs are different, so the answer is negative.

e) $\left(-\frac{4}{9}\right) \div \left(-\frac{2}{5}\right) = \left(-\frac{4}{9}\right) \cdot \left(-\frac{5}{2}\right)$
$= \frac{20}{18} = \frac{10}{9} \cdot \frac{2}{2} = \frac{10}{9}$

Multiplying by the reciprocal. The answer is positive.

Division by 0 (p. 60)

Addition, subtraction, and multiplication are defined for all real numbers, but we cannot **divide by 0**. For example, $\frac{0}{3} = 0 \div 3 = 0$, but $\frac{3}{0} = 3 \div 0$ is **undefined**.

Exponential notation (p. 63)

A product like $2 \cdot 2 \cdot 2 \cdot 2$, in which the factors are the same, is called a **power**. Powers are often written using **exponential notation**:

$$2 \cdot 2 \cdot 2 \cdot 2 = 2^4.$$

There are 4 factors; 4 is the *exponent*.

2 is the *base*.

A number raised to the power of 1 is the number itself; for example, $3^1 = 3$.

An expression containing a series of operations is not necessarily evaluated from left to right. Instead, we perform the operations according to the following rules.

Rules for Order of Operations

1. Calculate within the innermost grouping symbols, ( ), [ ], { }, | |, and above or below fraction bars.
2. Simplify all exponential expressions.
3. Perform all multiplication and division, working from left to right.
4. Perform all addition and subtraction, working from left to right.

EXAMPLE 4

Simplify: $3 - [(4 \times 5) + 12 \div 2^3 \times 6] + 5$.

Solution

$3 - [(4 \times 5) + 12 \div 2^3 \times 6] + 5$

$= 3 - [20 + 12 \div 2^3 \times 6] + 5$ Doing the calculations in the innermost parentheses first

$= 3 - [20 + 12 \div 8 \times 6] + 5$ Working inside the brackets; evaluating 2^3

$= 3 - [20 + 1.5 \times 6] + 5$ $12 \div 8$ is the first multiplication or division working from left to right.

$= 3 - [20 + 9] + 5$ Multiplying

$= 3 - 29 + 5$ Completing the calculations within the brackets

$= -26 + 5$
$= -21$ Adding and subtracting from left to right

Algebraic expression (p. 3)

Constant (p. 2)

Variable (p. 2)

Substitute (p. 3)

Evaluate (p. 3)

Algebraic Expressions

In an **algebraic expression** like $2xt^3$, the number 2 is a **constant** and x and t are **variables**. When numbers are **substituted** for x and t, we say that we are **evaluating** the expression.

Algebraic expressions containing variables can be evaluated by substituting a number for each variable in the expression and following the rules for order of operations.

EXAMPLE 5   The perimeter P of a rectangle of length l and width w is given by the formula $P = 2l + 2w$. Find the perimeter when l is 16 in. and w is 7.5 in.

Solution   We evaluate, substituting 16 in. for l and 7.5 in. for w and carrying out the operations:

$$P = 2l + 2w$$
$$= 2 \cdot 16 + 2 \cdot 7.5$$
$$= 32 + 15$$
$$= 47 \text{ in.}$$

Expressions that represent the same number are said to be **equivalent**. The laws that follow provide methods for writing equivalent expressions.

Laws and Properties of Real Numbers

Commutative laws:	$a + b = b + a$; $ab = ba$
Associative laws:	$a + (b + c) = (a + b) + c$; $a(bc) = (ab)c$
Distributive law:	$a(b + c) = ab + ac$
Identity property of 1:	$1 \cdot a = a \cdot 1 = a$
Identity property of 0:	$a + 0 = 0 + a = a$
Law of opposites:	$a + (-a) = 0$
Multiplicative property of 0:	$0 \cdot a = a \cdot 0 = 0$
Property of -1:	$-1 \cdot a = -a$
Opposite of a sum:	$-(a + b) = -a + (-b)$
$\dfrac{-a}{b} = \dfrac{a}{-b} = -\dfrac{a}{b},$	$\dfrac{-a}{-b} = \dfrac{a}{b}$

Factor (p. 17)

The distributive law can be used to multiply and to **factor** expressions. We factor an expression when we write an equivalent expression that is a product.

EXAMPLE 6   Write an equivalent expression as indicated.

a) Multiply: $-2(5x - 3)$.   **b)** Factor: $5x + 10y + 5$.

Solution

a) $-2(5x - 3) = -2(5x + (-3))$ Adding the opposite

$\qquad\qquad\quad = -2 \cdot 5x + (-2) \cdot (-3)$ Using the distributive law

$\qquad\qquad\quad = (-2 \cdot 5)x + 6$ Using the associative law for multiplication

$\qquad\qquad\quad = -10x + 6$

b) $5x + 10y + 5 = 5 \cdot x + 5 \cdot 2y + 5 \cdot 1$ The common factor is 5.

$\qquad\qquad\qquad\quad = 5(x + 2y + 1)$ Using the distributive law

Factoring can be checked by multiplying:

$$5(x + 2y + 1) = 5 \cdot x + 5 \cdot 2y + 5 \cdot 1 = 5x + 10y + 5.$$

Terms (p. 16)

Combine like terms (p. 43)

The **terms** of an algebraic expression are separated by plus signs. When two terms have variable factors that are exactly the same, the terms are called **like**, or **similar**, **terms**. The distributive law enables us to **combine**, or **collect**, **like terms**.

EXAMPLE 7

Combine like terms: $-5m + 3n - 4n + 10m$.

Solution

$-5m + 3n - 4n + 10m$

$= -5m + 3n + (-4n) + 10m$ Rewriting as addition

$= -5m + 10m + 3n + (-4n)$ Using the commutative law of addition

$= (-5 + 10)m + (3 + (-4))n$ Using the distributive law

$= 5m + (-n)$

$= 5m - n$ Rewriting as subtraction

We can also use the distributive law to help simplify algebraic expressions containing parentheses.

EXAMPLE 8

Simplify: **(a)** $4x - (y - 2x)$; **(b)** $3(t + 2) - 6(t - 1)$.

Solution

a) $4x - (y - 2x) = 4x - y + 2x$ Removing parentheses and changing the sign of every term

$\qquad\qquad\qquad = 6x - y$ Combining like terms

b) $3(t + 2) - 6(t - 1) = 3t + 6 - 6t + 6$ Multiplying each term of $t + 2$ by 3 and each term of $t - 1$ by -6

$\qquad\qquad\qquad\qquad = -3t + 12$ Combining like terms

Value (p. 3)

Solution (p. 6)

We have seen that algebraic expressions can be evaluated for specified numbers. If the expressions on each side of an equation have the same **value** for a given number, then that number is a **solution** of the equation.

EXAMPLE 9   Determine whether each number is a solution of $x - 2 = -5$.

a) 3 **b)** -3

Solution

a) We have:

$$x - 2 = -5 \qquad \text{Writing the equation}$$
$$\overline{3 - 2 \,\big|\, -5} \qquad \text{Substituting 3 for } x$$
$$1 \overset{?}{=} -5 \qquad 1 = -5 \text{ is FALSE}$$

Since $3 - 2 = -5$ is false, 3 is not a solution of $x - 2 = -5$.

b) We have

$$x - 2 = -5$$
$$\overline{-3 - 2 \,\big|\, -5}$$
$$-5 \overset{?}{=} -5 \qquad \text{TRUE}$$

Since $-3 - 2 = -5$ is true, -3 is a solution of $x - 2 = -5$.

Translating to algebraic expressions (p. 4)

Translating to equations (p. 6)

Certain word phrases can be translated to algebraic expressions. These in turn can often be used to translate problems to equations.

EXAMPLE 10   *Time usage.* Translate the following problem to an equation.

The average adult spends 145.6 hr a year shopping for clothes. This is 16 times as many hours as is spent planning for retirement (*Source: Perspective*, September 1996). How many hours a year does the average adult spend planning for retirement?

Solution   We let r represent the number of hours spent planning for retire-ment. We then reword the problem to make the translation more direct.

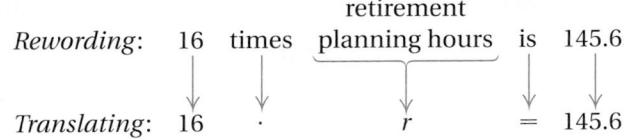

Exercise Set
R.1

Classify each equation or inequality as true or false.

1. $2.3 = 2.31$

2. $-3 \geq -3$

3. $-10 < -1$

4. $0 \leq -1$

5. $5 > 0$

6. $\frac{1}{10} = 0.1$

Find each absolute value.

7. $|4|$

8. $\left|\frac{11}{4}\right|$

9. $|-1.3|$

10. $|-105|$

Simplify.

11. $(-13) + (-12)$

12. $3 - (-2)$

13. $-\frac{1}{3} - \frac{2}{5}$

14. $\frac{3}{8} \div \frac{3}{5}$

15. $4.2 - 10.7$

16. $(-1.3)(2.8)$

17. $-15 + 0$

18. $\left(-\frac{1}{2}\right) + \frac{1}{8}$

19. $0 \div (-10)$

20. $0 - 32$

21. $\left(-\frac{3}{10}\right) + \left(-\frac{1}{5}\right)$

22. $\left(-\frac{4}{7}\right)\left(\frac{7}{4}\right)$

23. $-3.8 + 9.6$

24. $-0.01 + 1$

25. $(-12) \div 4$

26. $(-3)(30)$

27. $32 - (-7)$

28. $-100 + 35$

29. $(-10)(-17.5)$

30. $-10 - 2.68$

31. $(-68) + 36$

32. $175 \div (-25)$

33. $2 + (-3) + 7 + 10$

34. $-5 + (-15) + 13 + (-1)$

35. $3 \cdot (-2) \cdot (-1) \cdot (-1)$

36. $(-6) \cdot (-5) \cdot (-4) \cdot (-3) \cdot (-2) \cdot (-1)$

37. $(-1)^4 + 2^3$

38. $(-1)^5 + 2^4$

39. $2 \times 6 - 3 \times 5$

40. $12 \div 4 + 15 \div 3$

41. $3 - (2 \cdot 4 + 11)$

42. $3 - 2 \cdot 4 + 11$

43. $4 \cdot 5^2$

44. $7 \cdot 2^3$

45. $25 - 8 \times 3 + 1$

46. $12 - 16 \times 5 + 4$

47. $2 - (3^3 + 16 \div (-2)^3)$

48. $-7 - (8 + 10 \times 2^2)$

49. $|6(-3)| + |(-2)(-9)|$

50. $3 - |2 - 7 + 4|$

51. $\dfrac{7000 + (-10)^3}{10^2 \times (2 + 4)}$

52. $\dfrac{3 - 2 \times 6 - 5}{2(3 + 7)^2}$

53. $2 + 8 \div 2 \times 2$

54. $2 + 8 \div (2 \times 2)$

Evaluate.

55. $x - y$, for $x = 10$ and $y = 3$

56. $2m - n$, for $m = 6$ and $n = 11$

57. $-3 - x^2 + 12x$, for $x = 5$

58. $14 + (y - 5)^2 - 12 \div y$, for $y = -2$

59. The area of a parallelogram with base b and height h is bh. Find the area of the parallelogram when the height is 3.5 cm and the base is 8 cm.

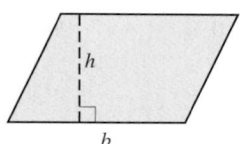

60. The area of a triangle with base b and height h is $\frac{1}{2}bh$. Find the area of the triangle when the height is 2 in. and the base is 6.2 in.

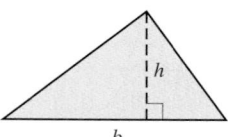

Multiply.

61. $4(2x + 7)$

62. $3(5y + 1)$

63. $5(x - 10)$

64. $4(3x - 2)$

65. $-2(15 - 3x)$

66. $-7(3x - 5)$

67. $2(4a + 6b - 3c)$

68. $5(8p + q - 5r)$

Factor.

69. $8x + 6y$

70. $7p + 14q$

71. $3 + 3w$

72. $4x + 4y$

73. $10x + 50y + 100$

74. $81p + 27q + 36$

Combine like terms.

75. $3p - 2p$

76. $4x + 3x$

77. $4m + 10 - 5m + 12$

78. $3a - 4b - 5b - 6a$

79. $-6x + 7 + 9x$

80. $16r + (-7r) + 3s$

Remove parentheses and simplify.

81. $2p - (7 - 4p)$

82. $4r - (3r + 5)$

83. $2x + 5y - 7(x - y)$

84. $14m - 6(2n - 3m) + n$

85. $6[2a + 4(a - 2b)]$

86. $2[2a + 1 - (3a - 6)]$

87. $3 - 2[5(x - 10y) - (3 + 2y)]$

88. $7 - 4[2(3 - 2x) - 5(4x - 3)]$

Determine whether the given number is a solution of the given equation.

89. $4; 3x - 2 = 10$

90. $12; 100 = 4x + 50$

91. $-3; 4 - x = 1$

92. $-1; 2 = 5 + 3x$

93. $4.6; \dfrac{x}{2} = 2.3$

94. $144; \dfrac{x}{9} = 16$

Translate each problem to an equation. Do not solve.

95. Three times what number is 348?

96. What number added to 256 is 113?

97. *Fast-food calories.* A McDonald's Big Mac® contains 500 calories. This is 69 more calories than a Taco Bell Beef Burrito® provides. How many calories are in a Taco Bell Beef Burrito?

98. *Coca-Cola® consumption.* The average U.S. citizen consumes 296 servings of Coca-Cola each year. This is 7.4 times the international average. What is the international average per capita consumption of Coke?

99. *Vegetable production.* It takes 42 gal of water to produce 1 lb of broccoli. This is twice the amount of water used to produce 1 lb of lettuce. How many gallons of water does it take to produce 1 lb of lettuce?

100. *Sports costs.* The average annual cost for scuba diving is $470. This is $458 more than the average annual cost to play badminton. What is the average annual cost to play badminton?

R.2   Equations, Inequalities, and Problem Solving

Solving Equations and Formulas • Solving Inequalities •
Problem Solving

In this section, we develop a problem-solving approach that can be used to solve many types of problems. Before doing so, we will study some principles used to solve equations and inequalities.

Solving Equations and Formulas

Solution (p. 80)

Any replacement for the variable in an equation that makes the equation true is called a *solution* of the equation. To **solve** an equation means to find all of its solutions.

Equivalent equations (p. 81)

We use the following principles to write **equivalent equations,** or equations with the same solutions.

The Addition and Multiplication Principles for Equations

The addition principle (p. 81)

The Addition Principle

For any real numbers a, b, and c,

$$a = b \quad \text{is equivalent to} \quad a + c = b + c.$$

The multiplication principle (p. 83)

The Multiplication Principle

For any real numbers a, b, and c, with $c \neq 0$,

$$a = b \quad \text{is equivalent to} \quad a \cdot c = b \cdot c.$$

To solve $x + a = b$ for x, we add $-a$ to (or subtract a from) both sides. To solve $ax = b$ for x, we multiply both sides by $\dfrac{1}{a}$ (or divide both sides by a).

To solve an equation like $-3x - 10 = 14$, we first isolate the variable term, $-3x$, using the addition principle. Then we use the multiplication principle to get the variable by itself.

EXAMPLE 1

Solve: $-3x - 10 = 14$.

Solution

$$-3x - 10 = 14$$

$$-3x - 10 + 10 = 14 + 10 \qquad \text{Using the addition principle:}$$
$$\text{Adding 10 to both sides}$$

First isolate the x-term

$$-3x = 24 \qquad \text{Simplifying}$$

$$\frac{-3x}{-3} = \frac{24}{-3} \qquad \text{Dividing both sides by } -3$$

Then isolate x.

$$x = -8 \qquad \text{Simplifying}$$

Check:

$$\frac{-3x - 10 = 14}{}$$
$$\begin{array}{c|c} -3(-8) - 10 & 14 \\ 24 - 10 & \\ 14 \overset{?}{=} 14 & \text{TRUE} \end{array}$$

The solution is -8.

Clearing fractions (p. 93)

Equations are generally easier to solve when they do not contain fractions. The easiest way to clear an equation of fractions is to multiply *every term on both sides* of the equation by the least common denominator.

EXAMPLE 2   Solve: $\frac{5}{2} - \frac{1}{6}t = \frac{2}{3}$.

Solution   The number 6 is the least common denominator, so we multiply both sides by 6.

$$6\left(\frac{5}{2} - \frac{1}{6}t\right) = 6 \cdot \frac{2}{3}$$ Multiplying both sides by 6

$$6 \cdot \frac{5}{2} - 6 \cdot \frac{1}{6}t = 6 \cdot \frac{2}{3}$$ Using the distributive law. Be sure to multiply every term by 6.

$$15 - t = 4$$ The fractions are cleared.

$$15 - t - 15 = 4 - 15$$ Subtracting 15 from both sides

$$-t = -11$$ $15 - t - 15 = 15 + (-t) + (-15)$
$$= -t + 15 + (-15) = -t$$

$$(-1)(-t) = (-1)(-11)$$ Multiplying both sides by -1 to change the sign

$$t = 11$$

Check:

$$\frac{5}{2} - \frac{1}{6}t = \frac{2}{3}$$

$$\begin{array}{c|c} \frac{5}{2} - \frac{1}{6}(11) & \frac{2}{3} \\ \frac{5}{2} - \frac{11}{6} & \\ \frac{15}{6} - \frac{11}{6} & \\ & \frac{2}{3} \overset{?}{=} \frac{2}{3} \quad \text{TRUE} \end{array}$$

The solution is 11.

To solve equations that contain parentheses, we can use the distributive law to first remove the parentheses. If like terms appear in an equation, we combine them and then solve.

EXAMPLE 3   Solve: $1 - 3(4 - x) = 2(x + 5) - 3x$.

Solution

$$1 - 3(4 - x) = 2(x + 5) - 3x$$

$$1 - 12 + 3x = 2x + 10 - 3x$$ Using the distributive law

$$-11 + 3x = -x + 10$$ Combining like terms; $1 - 12 = -11$ and $2x - 3x = -x$

$$-11 + 3x + x = 10$$ Adding x to both sides to get all x-terms on one side

$$-11 + 4x = 10$$ Combining like terms

$$4x = 10 + 11$$ Adding 11 to both sides to isolate the x-term

$$4x = 21$$ Simplifying

$$x = \frac{21}{4}$$ Dividing both sides by 4

Check:

$$
\begin{array}{c|c}
\multicolumn{2}{c}{1 - 3(4 - x) = 2(x + 5) - 3x} \\
\hline
1 - 3\left(4 - \frac{21}{4}\right) & 2\left(\frac{21}{4} + 5\right) - 3\left(\frac{21}{4}\right) \\
1 - 3\left(-\frac{5}{4}\right) & 2\left(\frac{41}{4}\right) - \frac{63}{4} \\
1 + \frac{15}{4} & \frac{82}{4} - \frac{63}{4} \\
\frac{19}{4} & \overset{?}{=} \frac{19}{4} \qquad \text{TRUE}
\end{array}
$$

The solution is $\frac{21}{4}$.

Formulas (Section 2.3)

A **formula** is an equation using two or more letters that represents a relationship between two or more quantities. A formula can be solved for a specified letter using the principles for solving equations.

EXAMPLE 4   The formula

$$A = \frac{a + b + c + d}{4}$$

gives the average A of four test scores a, b, c, and d. Solve for d.

Solution   We have

$$A = \frac{a + b + c + d}{4}$$ We want the letter d alone.

$$4A = a + b + c + d$$ Multiplying by 4 to clear the fraction

$$4A - a - b - c = d.$$ Subtracting $a + b + c$ from (or adding $-a - b - c$ to) both sides. The letter d is now isolated.

We can also write this as $d = 4A - a - b - c$. This formula can be used to determine the test score needed to obtain a specified average if three tests have already been taken.

Solving Inequalities

Solutions of inequalities (p. 127)

A **solution of an inequality** is a replacement of the variable that makes the inequality true.

Graphs of inequalities (p. 128)

The solutions of an inequality in one variable can be **graphed**, or represented by a drawing, on a number line. All points that are solutions are shaded, and dots are used at the endpoints. An open dot indicates an endpoint that is not a solution and a closed dot indicates an endpoint that is a solution.

EXAMPLE 5   Graph each inequality: **(a)** $m \leq 2$; **(b)** $-1 \leq x < 4$.

Solution

a) The solutions of $m \leq 2$ are shown on the number line by shading points to the left of 2 as well as the point at 2. The closed dot at 2 indicates that 2 is a part of the graph (that is, it is a solution of $m \leq 2$).

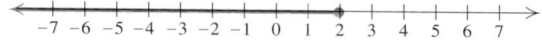

b) In order to be a solution of the inequality $-1 \leq x < 4$, a number must be a solution of both $-1 \leq x$ and $x < 4$. The solutions are shaded on the number line, with an open dot indicating that 4 is not a solution and a closed dot indicating that -1 is a solution.

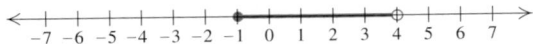

Equivalent inequalities (p. 129)

As with equations, our goal when solving inequalities is to isolate the variable on one side. We use principles that enable us to write **equivalent inequalities**—inequalities having the same solution set. The addition principle is similar to the addition principle for equations; the multiplication principle contains an important difference.

The addition principle for inequalities (p. 129)

The multiplication principle for inequalities (p. 131)

> ### The Addition and Multiplication Principles for Inequalities
>
> #### *The Addition Principle*
>
> For any real numbers a, b, and c,
>
> $\quad a < b$ is equivalent to $a + c < b + c$,   and
> $\quad a > b$ is equivalent to $a + c > b + c$.
>
> #### *The Multiplication Principle*
>
> For any real numbers a and b, and for any *positive* number c,
>
> $\quad a < b$ is equivalent to $ac < bc$,   and
> $\quad a > b$ is equivalent to $ac > bc$.
>
> For any real numbers a and b, and for any *negative* number c,
>
> $\quad a < b$ is equivalent to $ac > bc$,   and
> $\quad a > b$ is equivalent to $ac < bc$.
>
> Similar statements hold for $\leq$ and $\geq$.

Note that when we multiply both sides of an inequality by a negative number, we must reverse the direction of the inequality symbol in order to have an equivalent inequality.

EXAMPLE 6   Solve $-2x \geq 5$ and then graph the solution.

Solution   We have

$$-2x \geq 5$$

$$\frac{-2x}{-2} \leq \frac{5}{-2} \qquad \text{Multiplying by } -\frac{1}{2} \text{ or dividing by } -2$$

$$\underset{}{\qquad\qquad} \text{The symbol must be reversed!}$$

$$x \leq -\frac{5}{2}.$$

Any number less than or equal to $-\frac{5}{2}$ is a solution. The graph is as follows:

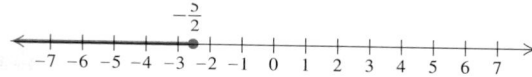

Set-builder notation (p. 130)

In Example 6, note $x \le -\frac{5}{2}$ is an inequality that describes the set of all solutions. Since it is impossible to list all the solutions, we use **set-builder notation**. The solution set of Example 6 is written

$$\left\{x \,\middle|\, x \le -\tfrac{5}{2}\right\},$$

read "the set of all x such that x is less than or equal to $-\frac{5}{2}$." We can use the addition and multiplication principles together to solve inequalities. We can also combine like terms, remove parentheses, and clear fractions and decimals.

EXAMPLE 7

Solve: $2 - 3(x + 5) > 4 - 6(x - 1)$.

Solution   We have

$$2 - 3(x + 5) > 4 - 6(x - 1)$$

$$2 - 3x - 15 > 4 - 6x + 6 \qquad \text{Using the distributive law to remove parentheses}$$

$$-3x - 13 > -6x + 10 \qquad \text{Simplifying}$$

$$-3x + 6x > 10 + 13 \qquad \text{Adding } 6x \text{ and also 13, to get all } x\text{-terms on one side and all other terms on the other side}$$

$$3x > 23 \qquad \text{Combining like terms}$$

$$x > \tfrac{23}{3}. \qquad \text{Multiplying by } \tfrac{1}{3}. \text{ The inequality symbol stays the same because } \tfrac{1}{3} \text{ is positive.}$$

The solution set is $\left\{x \,\middle|\, x > \frac{23}{3}\right\}$.

Problem solving (Section 2.5)

Problem Solving

One of the most important uses of algebra is as a tool for problem solving. The following five steps can be used to help solve problems of many types.

> **Five Steps for Problem Solving in Algebra**
>
> 1. *Familiarize* yourself with the problem.
> 2. *Translate* to mathematical language. (This often means write an equation.)
> 3. *Carry out* some mathematical manipulation. (This often means *solve* an equation.)
> 4. *Check* your possible answer in the original problem.
> 5. *State* the answer clearly.

EXAMPLE 8

Kitchen cabinets.   Cherry kitchen cabinets cost 10% more than oak cabinets. Shelby Custom Cabinets designs a kitchen using $7480 worth of cherry cabinets. How much would the same kitchen cost using oak cabinets?

Solution

Familiarization step (p. 115)

Percent (Section 2.4)

1. **Familiarize.**   The *Familiarize* step is often the most important of the five steps, and may require a significant amount of time. Sometimes it helps to make a drawing or a table, make a guess and check it, or look up further information. For this problem, we could review percent notation. We could also make a guess. Let's suppose that the oak cabinets cost $6500. Then the cherry cabinets would cost 10% more, or an additional $(0.10)($6500) =$ $650. Altogether the cherry cabinets would cost $6500 + $650 = 7150. Since $7150 \neq 7480, our guess is incorrect, but we see that 10% of the price of the oak cabinets must be added to the price of the oak cabinets to get the price of the cherry cabinets. We let $c =$ the cost of the oak cabinets.

2. **Translate.**   What we learned in the *Familiarize* step leads to the translation of the problem to an equation.

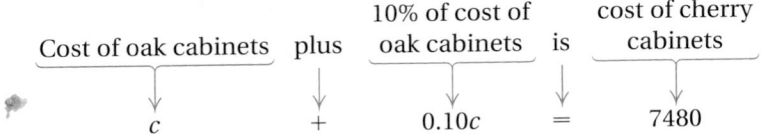

$$
\begin{array}{ccccc}
\underbrace{\text{Cost of oak cabinets}} & \text{plus} & \underbrace{\begin{array}{c}\text{10\% of cost of}\\ \text{oak cabinets}\end{array}} & \text{is} & \underbrace{\begin{array}{c}\text{cost of cherry}\\ \text{cabinets}\end{array}} \\
\downarrow & \downarrow & \downarrow & \downarrow & \downarrow \\
c & + & 0.10c & = & 7480
\end{array}
$$

3. **Carry out.**   We solve the equation:

$$
\begin{aligned}
c + 0.10c &= 7480 \\
1c + 0.10c &= 7480 \qquad &\text{Writing } c \text{ as } 1c \text{ before combining terms} \\
1.10c &= 7480 \qquad &\text{Combining like terms} \\
c &= \frac{7480}{1.10} \qquad &\text{Dividing by 1.10} \\
c &= 6800.
\end{aligned}
$$

4. **Check.**   We check in the wording of the stated problem: Cherry cabinets cost 10% more, so the additional cost is

$$10\% \text{ of } \$6800 = 0.10(\$6800) = \$680.$$

The total cost of the cherry cabinets is then

$$\$6800 + \$680 = \$7480,$$

which is the amount stated in the problem.

5. **State.**   The oak cabinets would cost $6800.

Sometimes the translation of a problem is an inequality.

EXAMPLE 9

Long-distance telephone usage.   Elyse pays a flat rate of 6¢ per minute for long-distance telephone calls. The monthly charge for her local calls is $21.50. How many minutes can she spend calling long distance in a month and not exceed her telephone budget of $50?

Solution

1. Familiarize.   Suppose that Elyse spends 10 hr, or 600 min, making long-distance calls one month. Then her bill would be the local service charge plus the long-distance charges, or

$$\$21.50 + \$0.06(600) = \$57.50.$$

This exceeds $50, so we know that the number of long-distance minutes must be less than 600. We let $m =$ the number of minutes of long-distance calls in a month.

2. Translate.   The *Familiarize* step helps us reword and translate.

	The local service charge	plus	the long-distance charges	cannot exceed	$50.
Rewording:					
Translating:	21.50	+	0.06m	$\leq$	50

Solving applications with inequalities (Section 2.7)

3. Carry out.   We solve the inequality:

$$21.50 + 0.06m \leq 50$$
$$0.06m \leq 28.50 \qquad \text{Subtracting 21.50 from both sides}$$
$$m \leq 475. \qquad \text{Dividing by 0.06. The inequality symbol stays the same.}$$

4. Check.   As a partial check, note that the telephone bill for 475 min of long-distance charges is

$$\$21.50 + \$0.06(475) = \$50.$$

Since this does not exceed the $50 budget, and fewer minutes will cost even less, our answer checks. We also note that 475 is less than 600 min, as noted in the *Familiarize* step.

5. State.   Elyse will not exceed her budget if she talks long distance for no more than 475 min.

Exercise Set

R.2

FOR EXTRA HELP

 Student's Solutions Manual    Digital Video Tutor CD 8 Videotape 15   Tutor Center AW Math Tutor Center    MathXL Tutorials on CD   *Math**XL*** MathXL   **MyMathLab** MyMathLab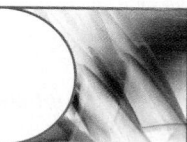

Solve.

1. $y + 5 = 13$

2. $y + 7 = -3$

3. $-3 + m = 9$

4. $-11 = 4 + x$

5. $t + \frac{1}{3} = \frac{1}{4}$

6. $-\frac{2}{3} + p = \frac{1}{6}$

7. $-1.9 = x - 1.1$

8. $x + 4.6 = 1.7$

9. $3y = 13$

10. $2x = 18$

11. $-x = \frac{5}{3}$

12. $-y = -\frac{2}{5}$

13. $-\frac{2}{7}x = -12$

14. $-\frac{1}{4}x = 3$

15. $\dfrac{-t}{5} = 1$

16. $\dfrac{2}{3} = -\dfrac{z}{8}$

17. $3x + 7 = 13$

18. $4x + 3 = -1$

19. $3y - 10 = 15$

20. $12 = 5y - 18$

21. $4x + 7 = 3 - 5x$

22. $2x = 5 + 7x$

23. $2x - 7 = 5x + 1 - x$

24. $a + 7 - 2a = 14 + 7a - 10$

25. $\frac{2}{5} + \frac{1}{3}t = 5$

26. $-\frac{5}{6} + t = \frac{1}{2}$

27. $x + 0.45 = 2.6x$

28. $1.8x + 0.16 = 4.2 - 0.05x$

29. $8(3 - m) + 7 = 47$

30. $2(5 - m) = 5(6 + m)$

31. $4 - (6 + x) = 13$

32. $18 = 9 - (3 - x)$

33. $2 + 3(4 + c) = 1 - 5(6 - c)$

34. $b + (b + 5) - 2(b - 5) = 18 + b$

35. $0.1(a - 0.2) = 1.2 + 2.4a$

36. $\frac{2}{3}\left(\frac{1}{2} - x\right) + \frac{5}{6} = \frac{3}{2}\left(\frac{2}{3}x + 1\right)$

37. $A = lw$, for l

38. $A = lw$, for w

39. $p = 30q$, for q

40. $d = 20t$, for t

41. $I = \dfrac{P}{V}$, for P

42. $b = \dfrac{A}{h}$, for A

43. $q = \dfrac{p + r}{2}$, for p

44. $q = \dfrac{p - r}{2}$, for r

45. $A = \pi r^2 + \pi r^2 h$, for π

46. $ax + by = c$, for a

Determine whether each number is a solution of the given inequality.

47. $x \le -5$

   a) 5   **b)** -5

   c) 0   **d)** -10

48. $y > 0$

   a) -1 **b)** 1

   c) 0   **d)** 100

Solve and graph. Write each answer in set-builder notation.

49. $x + 3 \le 15$

50. $y + 7 < -10$

51. $m - 17 > -5$

52. $x + 9 \ge -8$

53. $2x \ge -3$

54. $-\frac{1}{2}n \le 4$

55. $-5t > 15$

56. $3x > 10$

Solve. Write each answer in set-builder notation.

57. $2y - 7 > 13$

58. $2 - 6y \le 18$

59. $6 - 5a \le a$

60. $4b + 7 > 2 - b$

61. $2(3 + 5x) \ge 7(10 - x)$

62. $2(x + 5) < 8 - 3x$

63. $\frac{2}{3}(6 - x) < \frac{1}{4}(x + 3)$

64. $\frac{2}{3}t + \frac{8}{9} \ge \frac{4}{6} - \frac{1}{4}t$

65. $0.7(2 + x) \ge 1.1x + 5.75$

66. $0.4x + 5.7 \le 2.6 - 3(1.2x - 7)$

Solve. Use the five-step problem-solving process.

67. Three less than the sum of 2 and some number is 6. What is the number?

68. Five times some number is 10 less than the number. What is the number?

69. The sum of two consecutive even integers is 34. Find the numbers.

70. The sum of three consecutive integers is 195. Find the numbers.

71. *Reading.* Leisa is reading a 500-page book. She has twice as many pages to read as she has already finished. How many pages has she already read?

72. *Mowing.* It takes Caleb 50 min to mow his lawn. It will take him three times as many minutes to finish as he has already spent mowing. How long has he already spent mowing?

73. *Perimeter of a rectangle.* The perimeter of a rectangle is 28 cm. The width is 5 cm less than the length. Find the width and the length.

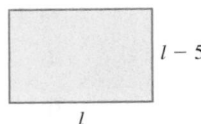

$l - 5$

l

74. *Triangles.* The second angle of a triangle is one third as large as the first. The third angle is 5° more than the first. Find the measure of the second angle.

75. *Water usage.* Rural Water Company charges a monthly service fee of $9.70 plus a volume charge of $2.60 for every hundred cubic feet of water used. How much water was used if the monthly bill is $33.10?

76. *Telephone bills.* Brandon pays $4.95 a month for a long-distance telephone service that offers a flat rate of 7¢ per minute. One month his total long-distance telephone bill was $10.69. How many minutes of long-distance telephone calls were made that month?

77. *Sales prices.* A can of tomatoes is on sale at 20% off for 64¢. What is the normal selling price of the tomatoes?

78. *Plywood.* The price of a piece of plywood rose 5% to $42. What was the original price of the plywood?

R.3   Introduction to Graphing

Points and Ordered Pairs   •   Graphs and Slope   •
Linear Equations

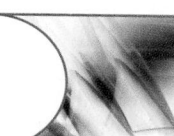

The graph of an equation is a drawing representing the solutions of that equation. Every point on the graph is a solution, and every solution is represented by a point.

Points and Ordered Pairs

Graphing ordered pairs (p. 151)

We can represent, or graph, pairs of numbers such as $(2, -5)$ on a plane. To do so, we use two perpendicular number lines called **axes**. The axes cross at a point called the **origin**. Arrows on the axes show the positive directions.

Coordinates (p. 151)

   The order of the **coordinates**, or numbers in a pair, is important. The **first coordinate** indicates horizontal position and the **second coordinate** indicates vertical position. Such pairs of numbers are called **ordered pairs**. Thus, the ordered pairs $(1, -2)$ and $(-2, 1)$ correspond to different points, as shown in the accompanying figure.

Quadrants (p. 154)

   The axes divide the plane into four regions, or **quadrants**, as indicated by Roman numerals in the figure at right. Points on the axes are not considered to be in any quadrant. The horizontal axis is often labeled the *x*-axis, and the vertical axis the *y*-axis.

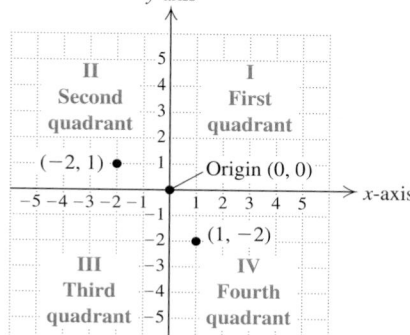

Graphs and Slope

Solutions of equations (p. 159)

When an equation contains two variables, solutions must be ordered pairs. Unless stated otherwise, the first number in each pair replaces the variable that occurs first alphabetically.

EXAMPLE 1

Determine whether $(1, 4)$ is a solution of $y - x = 3$.

Solution We substitute 1 for x and 4 for y since x occurs first alphabetically:

$$\begin{array}{c|c} y - x = 3 \\ \hline 4 - 1 & 3 \\ 3 \overset{?}{=} 3 & \text{TRUE} \end{array}$$

Since $3 = 3$ is true, the pair $(1, 4)$ *is* a solution.

A curve or line that represents all the solutions of an equation is called its **graph**.

EXAMPLE 2

Graph: $y = -2x + 1$.

Solution We select a value for x, calculate the corresponding value of y, and form an ordered pair.

  If $x = 0$, then $y = -2 \cdot 0 + 1 = 1$, and $(0, 1)$ is a solution. Repeating this step, we find other ordered pairs and list the results in a table. We then plot the points corresponding to the pairs. They appear to form a straight line, so we draw a line through the points.

$y = -2x + 1$

x	y	(x, y)
0	1	$(0, 1)$
-1	3	$(-1, 3)$
3	-5	$(3, -5)$
1	-1	$(1, -1)$

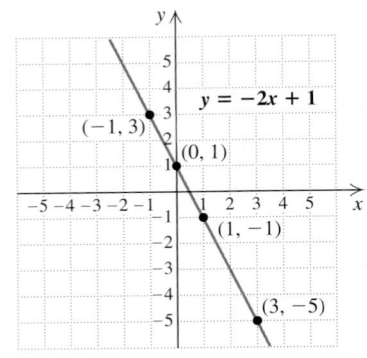

The graph in Example 2 is a straight line. An equation whose graph is a straight line is a **linear equation**. The *rate of change* of y with respect to x is called the **slope** of a graph. A linear graph has constant slope. It can be found using any two points on a line.

Slope (p. 190)

> **Slope**
>
> The *slope* of the line containing points (x_1, y_1) and (x_2, y_2) is given by
>
> $$m = \frac{\text{change in } y}{\text{change in } x} = \frac{\text{rise}}{\text{run}} = \frac{y_2 - y_1}{x_2 - x_1}.$$

EXAMPLE 3   Find the slope of the line containing the points $(-2, 1)$ and $(3, -4)$.

Solution   From $(-2, 1)$ to $(3, -4)$, the change in y, or the rise, is $-4 - 1$, or -5. The change in x, or the run, is $3 - (-2)$, or 5. Thus

$$\text{Slope} = \frac{\text{change in } y}{\text{change in } x} = \frac{\text{rise}}{\text{run}} = \frac{-4 - 1}{3 - (-2)} = \frac{-5}{5} = -1.$$

The slope of a line indicates the direction and steepness of its slant. The larger the absolute value of the slope, the steeper the line. The direction of the slant is indicated by the sign of the slope, as shown in the figures below.

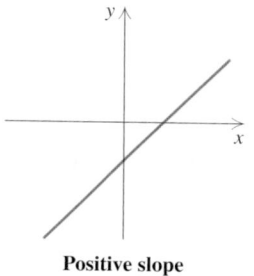

Positive slope

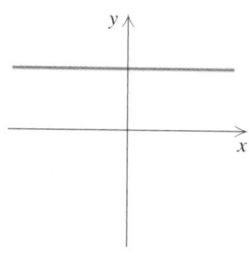

Negative slope

Zero slope

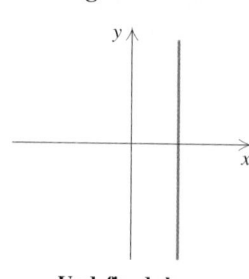

Undefined slope

x-intercept (p. 169)

y-intercept (p. 169)

The **x-intercept** of a line, if it exists, is the point at which the graph crosses the x-axis. To find an x-intercept, we replace y with 0 and calculate x.

The **y-intercept** of a line, if it exists, is the point at which the graph crosses the y-axis. To find a y-intercept, we replace x with 0 and calculate y.

Linear Equations

Any equation that can be written in the **standard form** $Ax + By = C$ is linear. Linear equations can also be written in other forms.

> **Forms of Linear Equations**
>
> | Standard form: | $Ax + By = C$ |
> |---|---|
> | Slope-intercept form: | $y = mx + b$ |
> | Point-slope form: | $y - y_1 = m(x - x_1)$ |

The slope and y-intercept of a line can be read from the slope-intercept form of the line's equation.

> **Slope and y-intercept**
>
>   For the graph of any equation $y = mx + b$,
>
>   - the slope is m, and
>   - the y-intercept is $(0, b)$.

EXAMPLE 4   Find the slope and the y-intercept of the line given by the equation $4x - 3y = 9$.

Solution   We write the equation in slope-intercept form $y = mx + b$:

$$4x - 3y = 9 \qquad \text{We must solve for } y.$$
$$-3y = -4x + 9 \qquad \text{Adding } -4x \text{ to both sides}$$
$$y = \tfrac{4}{3}x - 3. \qquad \text{Dividing both sides by } -3$$

The slope is $\tfrac{4}{3}$ and the y-intercept is $(0, -3)$.

If we know an equation is a straight line, we can plot two points on the line and draw the line through those points. The intercepts are often convenient points to use.

EXAMPLE 5   Graph $2x - 5y = 10$ using intercepts.

Solution   To find the x-intercept, we let $y = 0$ and solve for x:

$$2x - 5 \cdot 0 = 10 \qquad \text{Replacing } y \text{ with } 0$$
$$2x = 10$$
$$x = 5.$$

To find the y-intercept, we let $x = 0$ and solve for y:

$$2 \cdot 0 - 5y = 10 \qquad \text{Replacing } x \text{ with } 0$$
$$-5y = 10$$
$$y = -2.$$

Thus the x-intercept is $(5, 0)$ and the y-intercept is $(0, -2)$. The graph is a line, since $2x - 5y = 10$ is in the form $Ax + By = C$. It passes through these two points.

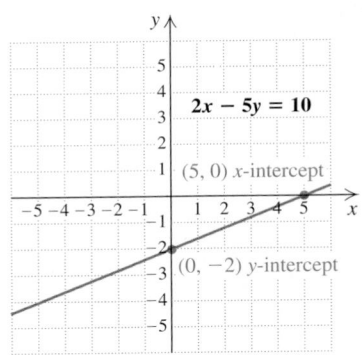

Alternatively, if we know a point on the line and its slope, we can plot the point and "count off" its slope to locate another point on the line.

EXAMPLE 6   Graph: $y = -\dfrac{1}{2}x + 3$.

Solution   The equation is in slope-intercept form, so we can read the slope and *y*-intercept directly from the equation.

Slope: $-\dfrac{1}{2}$

y-intercept: $(0, 3)$

We plot the *y*-intercept and use the slope to find another point.

Another way to write the slope is $\dfrac{-1}{2}$.

This means for a run of 2 units, there is a negative rise, or a fall, of 1 unit. Starting at $(0, 3)$, we move 2 units in the positive horizontal direction and then 1 unit down, to locate the point $(2, 2)$. Then we draw the graph. A third point can be calculated and plotted as a check.

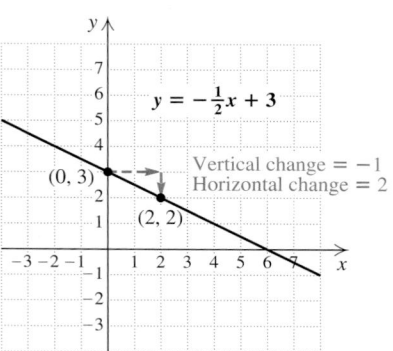

Horizontal and vertical lines intersect only one axis.

Horizontal line (p. 174)

Vertical line (p. 174)

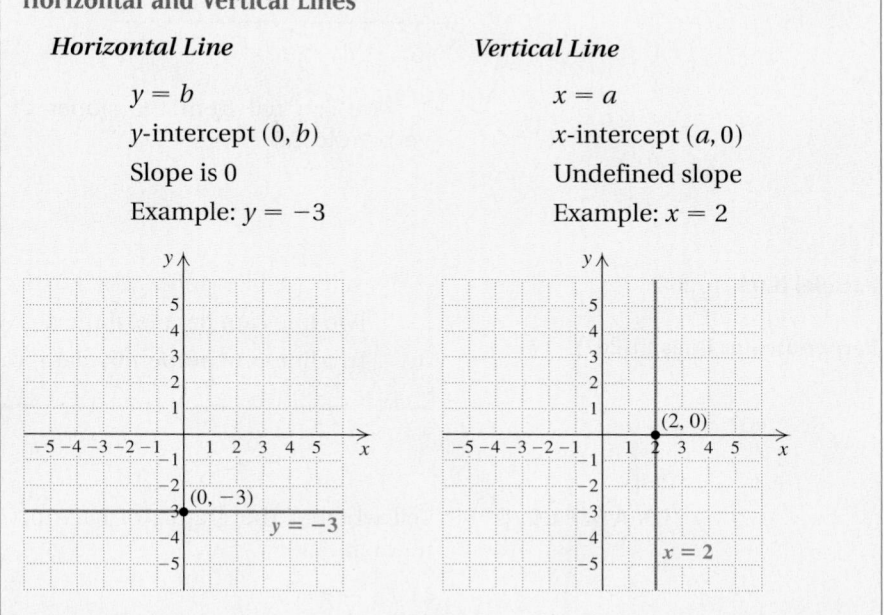

Horizontal and Vertical Lines

Horizontal Line

$y = b$

y-intercept $(0, b)$

Slope is 0

Example: $y = -3$

Vertical Line

$x = a$

x-intercept $(a, 0)$

Undefined slope

Example: $x = 2$

If we know the slope of a line and the coordinates of a point on the line, we can find an equation of the line, using either the slope-intercept equation $y = mx + b$, or the point-slope equation $y - y_1 = m(x - x_1)$.

EXAMPLE 7   Find the slope-intercept equation of a line given the following:

a) The slope is 2, and the y-intercept is $(0, -5)$.

b) The graph contains the points $(-2, 1)$ and $(3, -4)$.

Solution

a) Since the slope and the y-intercept are given, we use the slope-intercept equation:

$$y = mx + b$$
$$y = 2x - 5 \qquad \text{Substituting 2 for } m \text{ and } -5 \text{ for } b$$

b) To use the point-slope equation, we need a point on the line and its slope. The slope can be found from the points given:

$$m = \frac{1 - (-4)}{-2 - 3} = \frac{5}{-5} = -1.$$

Either point can be used for (x_1, y_1). Using $(-2, 1)$, we have

$$y - y_1 = m(x - x_1)$$
$$y - 1 = -1(x - (-2)) \qquad \begin{array}{l}\text{Substituting } -2 \text{ for } x_1, 1 \\ \text{for } y_1, \text{ and } -1 \text{ for } m\end{array}$$
$$y - 1 = -(x + 2)$$
$$y - 1 = -x - 2$$
$$y = -x - 1. \qquad \text{This is in slope-intercept form.}$$

We can tell from the slopes of two lines whether they are parallel or perpendicular.

Parallel lines (p. 206)

Perpendicular lines (p. 207)

> **Parallel and Perpendicular Lines**
> Two lines are parallel if they have the same slope.
> Two lines are perpendicular if the product of the slopes is -1.

EXAMPLE 8   Tell whether the graphs of each pair of lines are parallel, perpendicular, or neither.

a) $2x - y = 7$,
    $y = 2x + 3$

b) $4x - y = 8$,
    $x + 4y = 8$

Solution

a) The slope of $y = 2x + 3$ is 2.
To find the slope of $2x - y = 7$, we solve for y:

$$2x - y = 7$$
$$-y = -2x + 7$$
$$y = 2x - 7.$$

The slope of $2x - y = 7$ is also 2. Since the slopes are equal, the lines are parallel.

b) We solve both equations for y in order to determine the slopes of the lines:

$$4x - y = 8$$
$$-y = -4x + 8$$
$$y = 4x - 8.$$

The slope of $4x - y = 8$ is 4.
For the second line, we have

$$x + 4y = 8$$
$$4y = -x + 8$$
$$y = -\frac{1}{4}x + 2.$$

The slope of $x + 4y = 8$ is $-\frac{1}{4}$. Since $4 \cdot \left(-\frac{1}{4}\right) = -1$, the lines are perpendicular.

Exercise Set

R.3

FOR EXTRA HELP

 Student's Solutions Manual    Digital Video Tutor CD 8 Videotape 15   AW Math Tutor Center    MathXL Tutorials on CD    MathXL    MyMathLab

1. Plot these points.

$(2, -3), (5, 1), (0, 2), (-1, 0),$
$(0, 0), (-2, -5), (-1, 1), (1, -1)$

2. Plot these points.

$(0, -4), (-4, 0), (5, -2), (2, 5),$
$(3, 3), (-3, -1), (-1, 4), (0, 1)$

In which quadrant is each point located?

3. $(2, 1)$

4. $(-2, 5)$

5. $(3, -2.6)$

6. $(-1.7, -5.9)$

7. First coordinates are positive in quadrants ____ and ____.

8. Second coordinates are negative in quadrants ____ and ____.

Determine whether each equation has the given ordered pair as a solution.

9. $y = 2x - 5$; $(1, 3)$

10. $4x + 3y = 8$; $(-1, 4)$

11. $a - 5b = -3$; $(2, 1)$

12. $c = d + 1$; $(1, 2)$

Graph.

13. $y = \frac{1}{3}x + 3$

14. $y = -x - 2$

15. $y = -4x$

16. $y = \frac{3}{4}x + 1$

Find the slope of the line containing each given pair of points.

17. (3, 6) and (2, 5)

18. (−1, 7) and (−3, 4)

19. $\left(-2, -\frac{1}{2}\right)$ and $\left(5, -\frac{1}{2}\right)$

20. (6.8, 7.5) and (6.8, −3.2)

Find the slope and the y-intercept of each equation.

21. $y = 2x - 5$

22. $y = 4x - 8$

23. $2x + y = 1$

24. $x - 2y = 3$

Find the intercepts. Then graph.

25. $3 - y = 2x$

26. $2x + 5y = 10$

27. $y = 3x + 5$

28. $y = -x + 7$

29. $3x - 2y = 6$

30. $2y + 1 = x$

Determine the coordinates of the y-intercept of each equation. Then graph the equation.

31. $y = 2x - 5$

32. $y = -\frac{5}{4}x - 3$

33. $2y + 4x = 6$

34. $3y + x = 4$

Find the slope of each line, and graph.

35. $y = 4$

36. $x = -5$

37. $x = 3$

38. $y = -1$

Find the slope–intercept equation of a line given the conditions.

39. The slope is $\frac{1}{3}$ and the y-intercept is (0, 1).

40. The slope is −1 and the y-intercept is (0, −5).

41. The graph contains the points (0, 3) and (−1, 4).

42. The graph contains the points (5, 1) and (8, 0).

Determine whether each pair of lines is parallel, perpendicular, or neither.

43. $x + y = 5,$
$x - y = 1$

44. $2x + y = 3,$
$y = 4 - 2x$

45. $2x + 3y = 1,$
$2x - 3y = 5$

46. $y = \frac{1}{3}x - 7,$
$y + 3x = 1$

R.4 Polynomials

Exponents • Polynomials • Addition and Subtraction of Polynomials • Multiplication of Polynomials • Division of Polynomials

In this section, we define polynomials and learn to manipulate them. Before doing so, however, we must extend our knowledge of exponents.

Exponents

We know that x^4 means $x \cdot x \cdot x \cdot x$ and that x^1 means x. Exponential notation is also defined for zero and negative exponents.

> **Zero and Negative Exponents**
>
> For any real number a, $a \neq 0$,
>
> $$a^0 = 1 \quad \text{and} \quad a^{-n} = \frac{1}{a^n}.$$

EXAMPLE 1   Simplify: **(a)** $(-36)^0$; **(b)** $(-2x)^0$.

Solution

The exponent zero (p. 233)

a) $(-36)^0 = 1$ since any number (other than 0 itself) raised to the 0 power is 1.

b) $(-2x)^0 = 1$ for any $x \neq 0$.

EXAMPLE 2   Write an equivalent expression using positive exponents:

a) x^{-2} **b)** $\dfrac{1}{x^{-2}}$ **c)** xy^{-1}

Solution

Negative exponents (p. 289)

a) $x^{-2} = \dfrac{1}{x^2}$ x^{-2} is the reciprocal of x^2.

b) $\dfrac{1}{x^{-2}} = x^{-(-2)} = x^2$ The reciprocal of x^{-2} is $x^{-(-2)}$, or x^2.

c) $xy^{-1} = x\left(\dfrac{1}{y^1}\right) = \dfrac{x}{y}$ y^{-1} is the reciprocal of y^1.

The following properties hold for any integers m and n and any real numbers a and b, provided no denominators are 0 and 0^0 is not considered.

Properties of Exponents

The Product Rule: $a^m \cdot a^n = a^{m+n}$

The Quotient Rule: $\dfrac{a^m}{a^n} = a^{m-n}$

The Power Rule: $(a^m)^n = a^{mn}$

Raising a product to a power: $(ab)^n = a^n b^n$

Raising a quotient to a power: $\left(\dfrac{a}{b}\right)^n = \dfrac{a^n}{b^n}$

These properties are often used to simplify exponential expressions.

EXAMPLE 3   Simplify.

a) $(x^2 y^{-1})(xy^{-3})$ **b)** $\dfrac{(3p)^3}{(3p)^{-2}}$

c) $\left(\dfrac{ab^2}{3c^3}\right)^{-4}$

Solution

a) $(x^2y^{-1})(xy^{-3}) = x^2y^{-1}xy^{-3}$ Using an associative law

$= x^2x^1y^{-1}y^{-3}$ Using a commutative law; $x = x^1$

The product rule (p. 230)

$= x^{2+1}y^{-1+(-3)}$ Using the product rule: Adding exponents

$= x^3y^{-4}$, or $\dfrac{x^3}{y^4}$

The quotient rule (p. 232)

b) $\dfrac{(3p)^3}{(3p)^{-2}} = (3p)^{3-(-2)}$ Using the quotient rule: Subtracting exponents

$= (3p)^5$

The power rule (p. 234)

$= 3^5p^5$ Raising each factor to the fifth power

$= 243p^5$

Raising a product to a power (p. 234)

c) $\left(\dfrac{ab^2}{3c^3}\right)^{-4} = \dfrac{(ab^2)^{-4}}{(3c^3)^{-4}}$ Raising the numerator and the denominator to the -4 power

Raising a quotient to a power (p. 235)

$= \dfrac{a^{-4}(b^2)^{-4}}{3^{-4}(c^3)^{-4}}$ Raising each factor to the -4 power

$= \dfrac{a^{-4}b^{-8}}{3^{-4}c^{-12}}$ Multiplying exponents

$= \dfrac{3^4c^{12}}{a^4b^8}$, or $\dfrac{81c^{12}}{a^4b^8}$ Rewriting without negative exponents

Polynomials (Section 4.2)

Polynomials

Algebraic expressions like

$$2x^3 + 3x - 5, \qquad 4x, \qquad -7, \quad \text{and} \quad 2a^3b^2 + ab^3$$

are all examples of **polynomials**. All variables in a polynomial are raised to whole-number powers, and there are no variables in a denominator. The **terms** of a polynomial are separated by addition signs.

Term (p. 239)

A polynomial with one term is called a **monomial**. A polynomial with two terms is called a **binomial**, and one with three terms is called a **trinomial**. The **degree of a term** is the number of variable factors in that term. The **leading term** of a polynomial is the term of highest degree. The **degree of a polynomial** is the degree of the leading term. A polynomial is written in *descending order* when the leading term appears first, followed by the term of next highest degree, and so on.

Types of polynomials (p. 240)

Degree of a term (p. 240)

Leading term (p. 241)

Degree of a polynomial (p. 241)

The number -2 in the term $-2y^3$ is called the **coefficient** of that term. The coefficient of the leading term is the **leading coefficient** of the polynomial. To illustrate this terminology, consider the polynomial

Coefficient (p. 240)

Leading coefficient (p. 241)

$$4y^2 - 8y^5 + y^3 - 6y + 7.$$

The *terms* are $4y^2,$ $-8y^5,$ $y^3,$ $-6y,$ and $7.$

The *coefficients* are $4,$ $-8,$ $1,$ $-6,$ and $7.$

The *degree of each term* is $2,$ $5,$ $3,$ $1,$ and $0.$

The *leading term* is $-8y^5$ and the *leading coefficient* is $-8.$

The *degree of the polynomial* is 5.

Like, or *similar*, *terms* are either constant terms or terms containing the same variable(s) raised to the same power(s). Polynomials containing like terms can be simplified by *combining* those terms.

EXAMPLE 4   Combine like terms: $4x^2y + 2xy - x^2y + xy^2$.

Solution   The like terms are $4x^2y$ and $-x^2y$. Thus we have

$$4x^2y + 2xy - x^2y + xy^2 = 4x^2y - x^2y + 2xy + xy^2$$
$$= 3x^2y + 2xy + xy^2.$$

A polynomial can be evaluated by replacing the variable or variables with a number or numbers.

EXAMPLE 5   Evaluate $-a^2 + 2ab + 5b^2$ for $a = -1$ and $b = 3$.

Solution   We replace a with -1 and b with 3 and calculate the value using the rules for the order of operations:

Evaluating a polynomial (p. 242)

$$-a^2 + 2ab + 5b^2 = -(-1)^2 + 2 \cdot (-1) \cdot 3 + 5 \cdot 3^2$$
$$= -1 - 6 + 45 = 38.$$

Polynomials can be added, subtracted, multiplied, and divided.

Addition and Subtraction of Polynomials

Addition of polynomials
(Section 4.3)

To add two polynomials, we write a plus sign between them and combine like terms.

EXAMPLE 6   Add: $(4x^3 + 3x^2 + 2x - 7) + (-5x^2 + x - 10)$.

Solution

$$(4x^3 + 3x^2 + 2x - 7) + (-5x^2 + x - 10)$$
$$= 4x^3 + (3 - 5)x^2 + (2 + 1)x + (-7 - 10)$$
$$= 4x^3 - 2x^2 + 3x - 17$$

Opposite of a polynomial (p. 250)

In order to subtract polynomials, we must be able to find the *opposite* of a polynomial. To find the opposite of a polynomial, we replace each term with its opposite. This process is also called *changing the sign* of each term. For example, the opposite of

$$3y^4 - 7y^2 - \tfrac{1}{3}y + 17$$

is

$$-\left(3y^4 - 7y^2 - \tfrac{1}{3}y + 17\right) = -3y^4 + 7y^2 + \tfrac{1}{3}y - 17.$$

Subtraction of polynomials
(Section 4.3)

To subtract polynomials, we add the opposite of the polynomial being subtracted.

EXAMPLE 7 Subtract: $(3a^4 - 2a + 7) - (-a^3 + 5a - 1)$.

Solution

$$(3a^4 - 2a + 7) - (-a^3 + 5a - 1)$$
$$= 3a^4 - 2a + 7 + a^3 - 5a + 1 \qquad \text{Adding the opposite}$$
$$= 3a^4 + a^3 - 7a + 8 \qquad \text{Combining like terms}$$

Multiplication of polynomials
(Section 4.4)

Multiplication of Polynomials

To multiply two monomials, we multiply coefficients and then multiply variables using the product rule for exponents. To multiply a monomial and a polynomial, we multiply each term of the polynomial by the monomial, using the distributive property.

EXAMPLE 8 Multiply: $4x^3(3x^4 - 2x^3 + 7x - 5)$.

Solution

$$\textit{Think:} \quad \overbrace{4x^3 \cdot 3x^4}^{} - \overbrace{4x^3 \cdot 2x^3}^{} + \overbrace{4x^3 \cdot 7x}^{} - \overbrace{4x^3 \cdot 5}^{}$$
$$4x^3(3x^4 - 2x^3 + 7x - 5) = 12x^7 \quad - \quad 8x^6 \quad + \quad 28x^4 \quad - \quad 20x^3$$

To multiply any two polynomials P and Q, we select one of the polynomials—say, P. We then multiply each term of P by every term of Q and combine like terms.

EXAMPLE 9 Multiply: $(2a^3 + 3a - 1)(a^2 - 4a)$.

Solution   It is often helpful to use columns for a long multiplication. We multiply each term at the top by every term at the bottom, write like terms in columns, and add the results.

$$
\begin{array}{r}
2a^3 \quad + 3a \; - 1 \\
a^2 - 4a \\
\hline
-8a^4 \qquad - 12a^2 + 4a \\
2a^5 \qquad + 3a^3 \; - a^2 \\
\hline
2a^5 - 8a^4 + 3a^3 - 13a^2 + 4a
\end{array}
$$

Multiplying the top row by $-4a$

Multiplying the top row by a^2

Combining like terms. Be sure that like terms are lined up in columns.

We could multiply two binomials in the same manner in which we multiplied the polynomials in Example 9. However, by observing the pattern of the products formed, we can develop a method of multiplying two binomials more efficiently.

The FOIL Method

To multiply two binomials, $A + B$ and $C + D$, multiply the First terms AC, the Outer terms AD, the Inner terms BC, and then the Last terms BD. Then combine like terms, if possible.

$$(A + B)(C + D) = AC + AD + BC + BD$$

1. Multiply First terms: AC.
2. Multiply Outer terms: AD.
3. Multiply Inner terms: BC.
4. Multiply Last terms: BD.

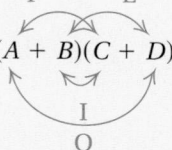

$$\downarrow$$
FOIL

EXAMPLE 10   Multiply: $(3x + 4)(x - 2)$.

Solution

FOIL (p. 266)

$$(3x + 4)(x - 2) = 3x^2 - 6x + 4x - 8$$
$$= 3x^2 - 2x - 8 \qquad \text{Combining like terms}$$

Two types of products of binomials occur so often that specific formulas or methods for computing them have been developed. Products of this type are called *special products*.

Special Products

The product of a sum and difference of the same two terms:

Multiplying sums and differences of two terms (p. 267)

$$(A + B)(A - B) = \underline{A^2 - B^2}$$

This is called a *difference of squares*.

Squaring binomials (p. 269)

The square of a binomial:

$$(A + B)^2 = A^2 + 2AB + B^2$$
$$(A - B)^2 = A^2 - 2AB + B^2$$

EXAMPLE 11   Multiply: **(a)** $(x + 3y)(x - 3y)$; **(b)** $(x^3 + 2)^2$.

Solution

$$(A + B)(A - B) = A^2 - B^2$$

a) $(x + 3y)(x - 3y) = x^2 - (3y)^2$ $A = x$ and $B = 3y$

$$= x^2 - 9y^2$$

$$(A + B)^2 = A^2 + 2 \cdot A \cdot B + B^2$$

b) $(x^3 + 2)^2 = (x^3)^2 + 2 \cdot x^3 \cdot 2 + 2^2$ $A = x^3$ and $B = 2$

$$= x^6 + 4x^3 + 4$$

Division of polynomials
(Section 4.7)

Division of Polynomials

Polynomial division is similar to division in arithmetic. First, let's consider division by a monomial. To divide a polynomial by a monomial, we divide each term by the monomial.

EXAMPLE 12   Divide: $(3x^5 + 8x^3 - 12x) \div 4x$.

Solution   This division can be written

$$\frac{3x^5 + 8x^3 - 12x}{4x} = \frac{3x^5}{4x} + \frac{8x^3}{4x} - \frac{12x}{4x}$$ Dividing each term by $4x$

$$= \frac{3}{4}x^{5-1} + \frac{8}{4}x^{3-1} - \frac{12}{4}x^{1-1}$$ Dividing coefficients and subtracting exponents

$$= \frac{3}{4}x^4 + 2x^2 - 3.$$

To check, we multiply the quotient by $4x$:

$$\left(\tfrac{3}{4}x^4 + 2x^2 - 3\right)4x = 3x^5 + 8x^3 - 12x.$$ The answer checks.

To use long division, we write polynomials in descending order, including terms with 0 coefficients for missing terms. As shown below in Example 13, the procedure ends when the degree of the remainder is less than the degree of the divisor.

EXAMPLE 13   Divide: $(4x^3 - 7x + 1) \div (2x + 1)$.

Solution   The polynomials are already written in descending order, but there is no x^2-term in the dividend. We fill in $0x^2$ for that term.

$$
\begin{array}{r}
2x^2 \\
2x + 1 \overline{)4x^3 + 0x^2 - 7x + 1} \\
\underline{4x^3 + 2x^2} \\
-2x^2
\end{array}
$$

— Divide the first term of the dividend, $4x^3$, by the first term in the divisor, $2x$: $4x^3/(2x) = 2x^2$.
— Multiply $2x^2$ by the divisor, $2x + 1$.
— Subtract: $(4x^3 + 0x^2) - (4x^3 + 2x^2) = -2x^2$.

Then we bring down the next term of the dividend, $-7x$.

$$
\begin{array}{r}
2x^2 - x \\
2x + 1 \overline{)4x^3 + 0x^2 - 7x + 1} \\
\underline{4x^3 + 2x^2} \\
-2x^2 - 7x \\
\underline{-2x^2 - x} \\
-6x
\end{array}
$$

— Divide the first term of $-2x^2 - 7x$ by the first term in the divisor: $-2x^2/(2x) = -x$.
— The $-7x$ has been "brought down."
— Multiply $-x$ by the divisor, $2x + 1$.
— Subtract: $(-2x^2 - 7x) - (-2x^2 - x) = -6x$.

Since the degree of the remainder, $-6x$, is *not* less than the degree of the divisor, we must continue dividing.

$$
\begin{array}{r}
2x^2 - x - 3 \\
2x + 1 \overline{)4x^3 + 0x^2 - 7x + 1} \\
\underline{4x^3 + 2x^2} \\
-2x^2 - 7x \\
\underline{-2x^2 - x} \\
-6x + 1 \\
\underline{-6x - 3} \\
4
\end{array}
$$

— Divide the first term of $-6x + 1$ by the first term in the divisor: $-6x/(2x) = -3$.
— The 1 has been "brought down."
— Multiply -3 by $2x + 1$.
— Subtract.

The answer is $2x^2 - x - 3$ with R4, or

$$
\text{Quotient} \longrightarrow 2x^2 - x - 3 + \frac{4}{2x + 1}.
$$

— Remainder
— Divisor

Check: To check, we can multiply by the divisor and add the remainder:

$$
(2x + 1)(2x^2 - x - 3) + 4 = 4x^3 - 7x - 3 + 4
$$
$$
= 4x^3 - 7x + 1.
$$

Exercise Set

R.4

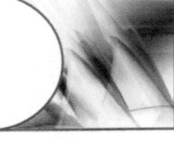
Solve.

1. a^0, for $a = -25$

2. y^0, for $y = 6.97$

3. $4^0 - 4^1$

4. $8^1 - 8^0$

Write an equivalent expression using positive exponents. Then, if possible, simplify.

5. 8^{-2}

6. 2^{-5}

7. $(-2)^{-3}$

8. $(-3)^{-2}$

9. $(ab)^{-2}$

10. ab^{-2}

11. $\dfrac{1}{y^{-10}}$

12. $\dfrac{1}{x^{-t}}$

Write an equivalent expression using negative exponents.

13. $\dfrac{1}{y^4}$

14. $\dfrac{1}{a^2b^3}$

15. $\dfrac{1}{x^t}$

16. $\dfrac{1}{n}$

Simplify.

17. $x^5 \cdot x^8$

18. $a^4 \cdot a^{-2}$

19. $\dfrac{a}{a^{-5}}$

20. $\dfrac{p^{-3}}{p^{-8}}$

21. $\dfrac{(4x)^{10}}{(4x)^2}$

22. $\dfrac{a^2b^9}{a^9b^2}$

23. $(7^8)^5$

24. $(x^3)^{-7}$

25. $(x^{-2}y^{-3})^{-4}$

26. $(-2a^2)^3$

27. $\left(\dfrac{y^2}{4}\right)^3$

28. $\left(\dfrac{ab^2}{c^3}\right)^4$

29. $\left(\dfrac{2p^3}{3q^4}\right)^{-2}$

30. $\left(\dfrac{2}{x}\right)^{-5}$

Identify the terms of each polynomial.

31. $8x^3 - 6x^2 + x - 7$

32. $-a^2b + 4a^2 - 8b + 17$

Determine the coefficient and the degree of each term in each polynomial. Then find the degree of each polynomial.

33. $18x^3 + 36x^9 - 7x + 3$

34. $-8y^7 + y + 19$

35. $-x^2y + 4y^3 - 2xy$

36. $8 - x^2y^4 + y^7$

Determine the leading term and the leading coefficient of each polynomial.

37. $-p^2 + 4 + 8p^4 - 7p$

38. $13 + 20t - 30t^2 - t^3$

Combine like terms. Write each answer in descending order.

39. $3x^3 - x^2 + x^4 + x^2$

40. $5t - 8t^2 + 4t^2$

41. $3 - 2t^2 + 8t - 3t - 5t^2 + 7$

42. $8x^5 - \frac{1}{3} + \frac{4}{5}x + 1 - \frac{1}{2}x$

Evaluate each polynomial for the given replacements of the variables.

43. $3x^2 - 7x + 10$, for $x = -2$

44. $-y + 3y^2 + 2y^3$, for $y = 3$

45. $a^2b^3 + 2b^2 - 6a$, for $a = 2$ and $b = -1$

46. $2pq^3 - 5q^2 + 8p$, for $p = -4$ and $q = -2$

The distance s, in feet, traveled by a body falling freely from rest in t seconds is approximated by

$$s = 16t^2.$$

47. A pebble is dropped into a well and takes 3 sec to hit the water. How far down is the surface of the water?

48. An acorn falls from the top of an oak tree and takes 2 sec to hit the ground. How high is the tree?

Add or subtract, as indicated.

49. $(3x^3 + 2x^2 + 8x) + (x^3 - 5x^2 + 7)$

50. $(-6x^4 + 3x^2 - 16) + (4x^2 + 4x - 7)$

51. $(8y^2 - 2y - 3) - (9y^2 - 7y - 1)$

52. $(4t^2 + 6t - 7) - (t + 5)$

53. $(-x^2y + 2y^2 + y) - (3y^2 + 2x^2y - 7y)$

54. $(ab + x^2y^2) + (2ab - x^2y^2)$

Multiply.

55. $4x^2(3x^3 - 7x + 7)$

56. $a^2b(a^3 + b^2 - ab - 2b)$

57. $(2a + y)(4a + b)$

58. $(x + 7y)(y - 3x)$

59. $(x + 7)(x^2 - 3x + 1)$

60. $(2x - 3)(x^2 - x - 1)$

61. $(x + 7)(x - 7)$

62. $(2x + 1)^2$

63. $(x + y)^2$

64. $(xy + 1)(xy - 1)$

65. $(2x^2 + 7)(3x^2 - 2)$

66. $(x^2 + 2)^2$

67. $(a - 3b)^2$

68. $(1.1x^2 + 5)(0.1x^2 - 2)$

69. $(6a - 5y)(7a + 3y)$

70. $(3p^2 - q^3)^2$

Divide and check.

71. $(3t^5 + 9t^3 - 6t^2 + 15t) \div (-3t)$

72. $(4x^5 + 10x^4 - 16x^2) \div (4x^2)$

73. $(15x^2 - 16x - 15) \div (3x - 5)$

74. $(x^3 - 2x^2 - 14x + 1) \div (x - 5)$

75. $(2x^3 - x^2 + 1) \div (x + 1)$

76. $(2x^3 + 3x^2 - 50) \div (2x - 5)$

77. $(5x^3 + 3x^2 - 5x) \div (x^2 - 1)$

78. $(2x^3 + 3x^2 + 6x + 10) \div (x^2 + 3)$

R.5   Polynomials and Factoring

Common Factors and Factoring by Grouping   •   Factoring Trinomials   •
Factoring Special Forms   •   Solving Polynomial Equations by Factoring

Factor (p. 304)

The reverse of multiplication is factoring. To **factor** a polynomial is to find an equivalent expression that is a product. To factor a monomial, we find two monomials whose product is equivalent to the original monomial. Many monomials have multiple factorizations. For example, three factorizations of $50x^6$ are $5 \cdot 10x^6$, $5x^3 \cdot 10x^3$, and $2x \cdot 25x^5$.

Common factor (p. 305)

Common Factors and Factoring by Grouping

If all the terms in a polynomial share a common factor, that factor can be "factored out" of the polynomial. Whenever you are factoring a polynomial with two or more terms, try to first find the largest common factor of the terms, if one exists.

EXAMPLE 1

Factor: $3x^6 + 15x^4 - 9x^3$.

Solution   The largest factor common to 3, 15, and -9 is 3. The largest power of x common to x^6, x^4, and x^3 is x^3. Thus the largest common factor of the terms of the polynomial is $3x^3$. We factor as follows:

$$3x^6 + 15x^4 - 9x^3 = 3x^3 \cdot x^3 + 3x^3 \cdot 5x - 3x^3 \cdot 3 \qquad \text{Factoring each term}$$

$$= 3x^3(x^3 + 5x - 3). \qquad \text{Factoring out } 3x^3$$

Factorizations can always be checked by multiplying:

$$3x^3(x^3 + 5x - 3) = 3x^6 + 15x^4 - 9x^3.$$

A polynomial with two or more terms can be a common factor.

EXAMPLE 2

Factor: $3x^2(x - 2) + 5(x - 2)$.

Solution   The binomial $x - 2$ is a factor of both $3x^2(x - 2)$ and $5(x - 2)$. Thus we have

$$3x^2(x - 2) + 5(x - 2) = (x - 2)(3x^2 + 5). \qquad \text{Factoring out the common factor, } x - 2$$

Factoring by grouping (p. 308)

If a polynomial with four terms can be split into two groups of terms, and both groups share a common binomial factor, the polynomial can be factored. This method is known as **factoring by grouping**.

EXAMPLE 3

Factor by grouping: $2x^3 + 6x^2 - x - 3$.

Solution   First, we consider the polynomial as two groups of terms, $2x^3 + 6x^2$ and $-x - 3$. Then we factor each group separately:

$$2x^3 + 6x^2 - x - 3 = 2x^2(x + 3) - 1(x + 3) \qquad \text{Factoring out } 2x^2 \text{ and } -1 \text{ to give the common binomial factor, } x + 3$$

$$= (x + 3)(2x^2 - 1).$$

The check is left to the student.

Prime polynomial (p. 317)

Not every polynomial with four terms is factorable by grouping. A polynomial that is not factorable is said to be **prime**.

Factoring trinomials of the type
$x^2 + bx + c$ (Section 5.2)

Factoring Trinomials

Many trinomials that have no common factor can be written as the product of two binomials. We look first at trinomials of the form

$$x^2 + bx + c,$$

for which the leading coefficient is 1.

Factoring trinomials involves a trial-and-error process. In order for the product of two binomials to be $x^2 + bx + c$, the binomials must look like

$$(x + p)(x + q),$$

where p and q are constants that must be determined. For example, to factor $x^2 + 10x + 16$, we must have

$$x^2 + 10x + 16 = (x + p)(x + q)$$
$$= x^2 + (p + q)x + pq. \quad \text{Using FOIL}$$

Therefore,

$$p + q = 10,$$
$$pq = 16.$$

Thus we look for two numbers whose product is 16 and whose sum is 10.

EXAMPLE 4 Factor.

a) $x^2 + 10x + 16$ b) $x^2 - 8x + 15$
c) $x^2 - 2x - 24$ d) $3t^2 - 33st + 84s^2$

Solution

a) The factorization is of the form

$$(x + \quad)(x + \quad).$$

Constant term positive (p. 313)

To find the constant terms, we need a pair of factors whose product is 16 and whose sum is 10. Since 16 is positive, its factors will have the same sign as 10—that is, we need consider only positive factors of 16.

We list the possible factorizations in a table and calculate the sum of each pair of factors.

Pairs of Factors of 16	Sums of Factors
1, 16	17
2, 8	10
4, 4	8

The numbers we seek are 2 and 8.

The factorization of $x^2 + 10x + 16$ is $(x + 2)(x + 8)$. To check, we multiply.

Check: $(x + 2)(x + 8) = x^2 + 8x + 2x + 16 = x^2 + 10x + 16.$

b) For $x^2 - 8x + 15$, c is positive and b is negative. Therefore, the factors of 15 will be negative. Again, we list the possible factorizations in a table.

Pairs of Factors of 15	Sums of Factors
$-1, -15$	-16
$-3, \;\; -5$	-8 ←

The numbers we need are -3 and -5.

The factorization is $(x - 3)(x - 5)$.

Check: $(x - 3)(x - 5) = x^2 - 5x - 3x + 15 = x^2 - 8x + 15$.

Constant term negative (p. 314)

c) For $x^2 - 2x - 24$, c is negative, so one factor of -24 will be negative and one will be positive. Since b is also negative, the negative factor must have the larger absolute value.

Pairs of Factors of -24	Sums of Factors
$1, -24$	-23
$2, -12$	-10
$3, \;\; -8$	-5
$4, \;\; -6$	-2 ←

The numbers we need are 4 and -6.

The factorization is $(x + 4)(x - 6)$.

Check: $(x + 4)(x - 6) = x^2 - 6x + 4x - 24 = x^2 - 2x - 24$.

d) Always look first for a common factor. There is a common factor, 3, which we factor out first:

$$3t^2 - 33st + 84s^2 = 3(t^2 - 11st + 28s^2).$$

Now we consider $t^2 - 11st + 28s^2$. Think of $28s^2$ as the "constant" term c and $-11s$ as the "coefficient" b of the middle term. We try to express $28s^2$ as the product of two factors whose sum is $-11s$. These factors are $-4s$ and $-7s$. Thus the factorization of $t^2 - 11st + 28s^2$ is

$(t - 4s)(t - 7s).$ This is not the entire factorization of $3t^2 - 33st + 84s^2$.

We now include the common factor, 3, and write

$$3t^2 - 33st + 84s^2 = 3(t - 4s)(t - 7s).$$ This is the factorization.

Check: $3(t - 4s)(t - 7s) = 3(t^2 - 11st + 28s^2) = 3t^2 - 33st + 84s^2$.

Factoring trinomials of the type $ax^2 + bx + c$ (Section 5.3)

When the leading coefficient of a trinomial is not 1, the number of trials needed to find a factorization can increase dramatically. We will consider two methods for factoring trinomials of the type $ax^2 + bx + c$: factoring with FOIL and the grouping method.

> *To Factor $ax^2 + bx + c$ Using FOIL*
>
> **1.** Factor out the largest common factor, if one exists.
> **2.** Find two First terms whose product is ax^2:
>
> $$(\square x + \quad)(\square + \quad) = ax^2 + bx + c.$$
> FOIL
>
> **3.** Find two Last terms whose product is c:
>
> $$(\ x + \blacksquare)(\ x + \blacksquare) = ax^2 + bx + c.$$
> FOIL
>
> **4.** Repeat steps (2) and (3) until a combination is found for which the sum of the Outer and Inner products is bx:
>
> $$(\square x + \blacksquare)(\square x + \blacksquare) = ax^2 + bx + c.$$
> I
> O
> FOIL
>
> Always check by multiplying. If no correct combination exists, state that the polynomial is prime.

Factoring with FOIL (p. 321)

EXAMPLE 5   Factor: $20x^3 - 22x^2 - 12x$.

Solution

1. First, we factor out the largest common factor, $2x$:

$$20x^3 - 22x^2 - 12x = 2x(10x^2 - 11x - 6).$$

2. Next, in order to factor the trinomial $10x^2 - 11x - 6$, we search for two terms whose product is $10x^2$. The possibilities are

$$(x + \quad)(10x + \quad) \quad \text{or} \quad (2x + \quad)(5x + \quad).$$

3. There are four pairs of factors of -6. Since the first terms of the binomials are different, the order of the factors is important. So there are eight possibilities for the last terms:

$$1, -6 \quad -1, 6 \quad 2, -3 \quad -2, 3$$

and

$$-6, 1 \quad 6, -1 \quad -3, 2 \quad 3, -2.$$

4. Since each of the eight possibilities from step (3) could be used in either of the two possibilities from step (2), there are $2 \cdot 8$, or 16, possible factorizations. We check the possibilities systematically until we find one that gives the correct factorization. Let's first try factors with $(2x + \quad)(5x + \quad)$.

Trial	Product	
$(2x + 1)(5x - 6)$	$10x^2 - 7x - 6$	⟵ Wrong middle term
$(2x - 1)(5x + 6)$	$10x^2 + 7x - 6$	⟵ Wrong middle term. Note that changing the signs in the binomials changed the sign of middle term in the product.
$(2x + 2)(5x - 3)$	$10x^2 + 4x - 6$	⟵ Wrong middle term. We need not consider $(2x - 2)(5x + 3)$.
$(2x - 6)(5x + 1)$	$10x^2 - 28x - 6$	⟵ Wrong middle term. We need not consider $(2x + 6)(5x - 1)$.
$(2x - 3)(5x + 2)$	$10x^2 - 11x - 6$	⟵ Correct middle term

We can stop when we find a correct factorization. Including the common factor $2x$, we now have

$$20x^3 - 22x^2 - 12x = 2x(2x - 3)(5x + 2).$$

This can be checked by multiplying.

The grouping method (p. 326)

With practice, some of the trials can be skipped or performed mentally.

The second method of factoring trinomials of the type $ax^2 + bx + c$ involves factoring by grouping.

> *To Factor $ax^2 + bx + c$, Using the Grouping Method*
>
> 1. Factor out the largest common factor, if one exists.
> 2. Multiply the leading coefficient a and the constant c.
> 3. Find a pair of factors of ac whose sum is b.
> 4. Rewrite the middle term, bx, as a sum or difference using the factors found in step (3).
> 5. Factor by grouping.
> 6. Always check by multiplying.

EXAMPLE 6   Factor: $7x^2 + 31x + 12$.

Solution

1. There is no common factor (other than 1 or -1).

2. We multiply the leading coefficient, 7, and the constant, 12:

$$7 \cdot 12 = 84.$$

3. We look for a pair of factors of 84 whose sum is 31. Since both 84 and 31 are positive, we need consider only positive factors.

Pairs of Factors of 84	Sums of Factors
1, 84	85
2, 42	44
3, 28	31 ←

$3 + 28 = 31$

4. Next, we rewrite $31x$ using the factors 3 and 28:

$$31x = 3x + 28x.$$

5. We now factor by grouping:

$$7x^2 + 31x + 12 = 7x^2 + 3x + 28x + 12 \qquad \text{Substituting } 3x + 28x \text{ for } 31x$$

$$= x(7x + 3) + 4(7x + 3)$$

$$= (7x + 3)(x + 4). \qquad \text{Factoring out the common factor, } 7x + 3$$

6. *Check:* $(7x + 3)(x + 4) = 7x^2 + 31x + 12.$

Factoring Special Forms

We can factor certain types of polynomials directly, without using trial and error.

Factoring Formulas

Perfect-square trinomial: $A^2 + 2AB + B^2 = (A + B)^2,$
$A^2 - 2AB + B^2 = (A - B)^2$

Difference of squares: $A^2 - B^2 = (A + B)(A - B)$

Sum of cubes: $A^3 + B^3 = (A + B)(A^2 - AB + B^2)$

Difference of cubes: $A^3 - B^3 = (A - B)(A^2 + AB + B^2)$

Before using the factoring formulas, it is important to check carefully that the expression being factored is indeed in one of the forms listed. Note that there is no factoring formula for the sum of two squares.

EXAMPLE 7   Factor: **(a)** $2x^2 - 2$; **(b)** $x^2y^2 + 20xy + 100$; **(c)** $p^3 - 64$; **(d)** $3y^2 + 27$.

Solution

a) We first factor out a common factor, 2:

$$2x^2 - 2 = 2(x^2 - 1).$$

Recognizing and factoring differences of squares (p. 333)

Looking at $x^2 - 1$, we see that it is a difference of squares, with $A = x$ and $B = 1$. The factorization is thus

$$2x^2 - 2 = 2(x^2 - 1) = 2(x + 1)(x - 1).$$

$$\underbrace{\qquad}_{A^2 - B^2} \qquad \underbrace{\qquad}_{(A + B)(A - B)}$$

Recognizing and factoring perfect-square trinomials (pp. 330–331)

b) First, we check for a common factor; there is none. The polynomial is a perfect-square trinomial, since x^2y^2 and 100 are squares; there is no minus sign before either square; and $20xy$ is $2 \cdot xy \cdot 10$, where xy and 10 are square roots of x^2y^2 and 100, respectively. The factorization is thus

$$x^2y^2 + 20xy + 100 = (xy)^2 + 2 \cdot xy \cdot 10 + 10^2 = (xy + 10)^2.$$
$$A^2 + 2 \cdot A \cdot B + B^2 = (A + B)^2$$

Factoring sums or differences of cubes (Section 5.5)

c) This is a difference of cubes, with $A = p$ and $B = 4$:

$$p^3 - 64 = (p)^3 - (4)^3$$
$$= (p - 4)(p^2 + 4p + 16).$$

d) We factor out the common factor, 3:

$$3y^2 + 27 = 3(y^2 + 9).$$

Since $y^2 + 9$ is a sum of squares, no further factorization is possible.

Factoring completely (p. 334)

A polynomial is said to be *factored completely* when no factor can be factored further.

EXAMPLE 8

Factor completely: $x^4 - 1$.

Solution

$$x^4 - 1 = (x^2 + 1)(x^2 - 1)$$ Factoring a difference of squares
$$= (x^2 + 1)(x + 1)(x - 1)$$ The factor $x^2 - 1$ is itself a difference of squares.

Solving Polynomial Equations by Factoring

Polynomial equation (p. 348)

Quadratic equation (p. 348)

A **polynomial equation** is formed by setting two polynomials equal to each other. A **quadratic equation** is a polynomial equation equivalent to one of the form $ax^2 + bx + c = 0$, where $a \neq 0$. Polynomial equations that can be factored can be solved using the principle of zero products.

The principle of zero products (p. 348)

The Principle of Zero Products

An equation $ab = 0$ is true if and only if $a = 0$ or $b = 0$, or both. (A product is 0 if and only if at least one factor is 0.)

If we can write an equation as a product that equals 0, we can try to use the principle of zero products to solve the equation.

EXAMPLE 9   Solve:

a) $x^2 - 11x = 12$

b) $5x^2 + 10x + 5 = 0$

c) $9x^2 = 1$

Solution

a) We must have 0 on one side of the equation before using the principle of zero products:

$$x^2 - 11x = 12$$

$$x^2 - 11x - 12 = 0 \qquad \text{Subtracting 12 from both sides}$$

$$(x - 12)(x + 1) = 0 \qquad \text{Factoring}$$

$$x - 12 = 0 \quad or \quad x + 1 = 0 \qquad \text{Using the principle of zero products}$$

$$x = 12 \quad or \qquad x = -1.$$

The solutions are 12 and -1. The check is left to the student.

b) We have

$$5x^2 + 10x + 5 = 0$$

$$5(x^2 + 2x + 1) = 0 \qquad \text{Factoring out a common factor}$$

$$5(x + 1)(x + 1) = 0 \qquad \text{Factoring completely}$$

$$x + 1 = 0 \quad or \quad x + 1 = 0 \qquad \text{Using the principle of zero products}$$

$$x = -1 \quad or \qquad x = -1.$$

There is only one solution, -1. The check is left to the student.

c) We have

$$9x^2 = 1$$

$$9x^2 - 1 = 0 \qquad \text{Subtracting 1 from both sides to get 0 on one side}$$

$$(3x + 1)(3x - 1) = 0 \qquad \text{Factoring a difference of squares}$$

$$3x + 1 = 0 \quad or \quad 3x - 1 = 0 \qquad \text{Using the principle of zero products}$$

$$3x = -1 \quad or \qquad 3x = 1$$

$$x = -\tfrac{1}{3} \quad or \qquad x = \tfrac{1}{3}.$$

The solutions are $\tfrac{1}{3}$ and $-\tfrac{1}{3}$. The check is left to the student.

___

Quadratic equations can be used to solve problems. One important result that uses squared quantities is the Pythagorean theorem. It relates the lengths of the sides of a **right triangle**, that is, a triangle with a 90° angle. The side opposite the 90° angle is called the **hypotenuse**, and the other sides are called the **legs**.

> **The Pythagorean Theorem**
>
> The sum of the squares of the legs of a right triangle is equal to the square of the hypotenuse:
>
> $$a^2 + b^2 = c^2.$$
>
>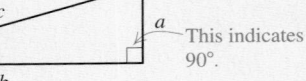
>
> This indicates 90°.

EXAMPLE 10   Swing sets.   The length of a slide on a swing set is 5 ft. The distance from the base of the ladder to the base of the slide is 1 ft more than the height of the ladder. Find the height of the ladder.

Solution

1. **Familiarize.**   We first make a drawing and let $x =$ the height of the ladder, in feet. We know then that the other leg of the triangle is $x + 1$, since it is 1 ft longer than the ladder. The hypotenuse has length 5 ft.

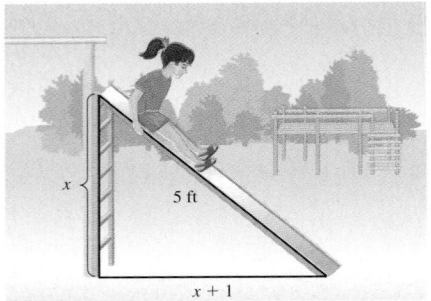

2. **Translate.**   Applying the Pythagorean theorem gives us

$$a^2 + b^2 = c^2$$
$$x^2 + (x + 1)^2 = 5^2. \quad \text{Substituting}$$

3. **Carry out.**   We solve the equation:

$$x^2 + (x + 1)^2 = 5^2$$
$$x^2 + x^2 + 2x + 1 = 25 \qquad \text{Squaring } x + 1; \text{squaring 5}$$
$$2x^2 + 2x + 1 = 25 \qquad \text{Combining like terms}$$
$$2x^2 + 2x - 24 = 0 \qquad \text{Getting 0 on one side}$$
$$2(x^2 + x - 12) = 0 \qquad \text{Factoring out a common factor}$$
$$2(x + 4)(x - 3) = 0 \qquad \text{Factoring a trinomial}$$
$$x + 4 = 0 \quad or \quad x - 3 = 0 \qquad \text{Using the principle of zero products}$$
$$x = -4 \quad or \qquad x = 3.$$

4. **Check.**   We know that the integer -4 is not a solution because the height of the ladder cannot be negative. When $x = 3$, the distance from the base of the ladder to the base of the slide is $x + 1 = 4$, and $3^2 + 4^2 = 5^2$. So the solution 3 checks.

5. **State.**   The ladder is 3 ft high.

Exercise Set

R.5

Factor completely. If a polynomial is prime, state this.

1. $3x^3 + 6x^2 - 9x$

2. $x^2y^4 - 2xy^5 + 3x^3y^6$

3. $y^2 - 6y + 9$

4. $4z^2 - 25$

5. $2p^3(p + 2) + (p + 2)$

6. $6y^2 + y - 1$

7. $16x^2 + 25$

8. $y^3 - 1$

9. $8t^3 + 27$

10. $a^2b^2 + 24ab + 144$

11. $m^2 + 13m + 42$

12. $2x^3 - 6x^2 + x - 3$

13. $x^4 - 81$

14. $x^2 + x + 1$

15. $8x^2 + 22x + 15$

16. $4x^2 - 40x + 100$

17. $x^3 + 2x^2 - x - 2$

18. $(x + 2y)(x - 1) + (x + 2y)(x - 2)$

19. $0.001t^6 - 0.008$

20. $x^2 - 20 - x$

21. $-\frac{1}{16} + x^4$

22. $5x^8 - 5z^{16}$

23. $a^2 + 6a + 9 - y^2$

24. $t^6 - p^6$

25. $5mn + m^2 - 150n^2$

26. $\frac{1}{27} + x^3$

27. $24x^2y - 6y - 10xy$

28. $-3y^2 - 12y - 12$

29. $y^2 + 121 - 22y$

30. $p^2 - m^2 - 2mn - n^2$

Solve.

31. $(x - 2)(x + 7) = 0$

32. $(3x - 5)(7 - 4x) = 0$

33. $8x(4.7 - x) = 0$

34. $(x - 3)(x + 1)(2x - 9) = 0$

35. $x^2 = 100$

36. $8x^2 = 5x$

37. $4x^2 - 18x = 70$

38. $x^2 + 2x + 1 = 0$

39. $2x^2 - 10x = 0$

40. $100x^2 = 81$

41. $(a + 1)(a - 5) = 7$

42. $d(d - 3) = 40$

43. $x^2 + 6x - 55 = 0$

44. $x^2 + 7x - 60 = 0$

45. $\frac{1}{2}x^2 + 5x + \frac{25}{2} = 0$

46. $3 + 10x^2 = 11x$

47. *Landscaping.* A triangular flower garden is 3 ft longer than it is wide. The area of the garden is 20 ft². What are the dimensions of the garden?

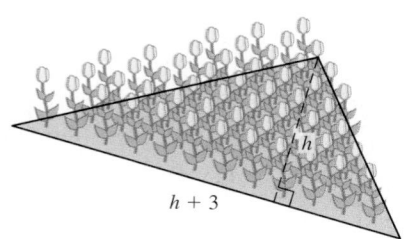

48. *Page numbers.* The product of the page numbers on two facing pages of a book is 156. Find the page numbers.

49. *Right triangles.* The hypotenuse of a right triangle is 17 ft. One leg is 1 ft shorter than twice the length of the other leg. Find the length of the legs.

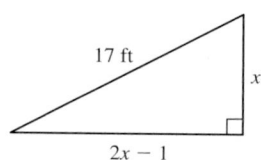

50. *Hiking.*   Jenna hiked 500 ft up a steep incline. Her
   global positioning unit indicated that her horizontal position had changed by 100 ft more than
   her vertical position had changed. What was the
   change in altitude?

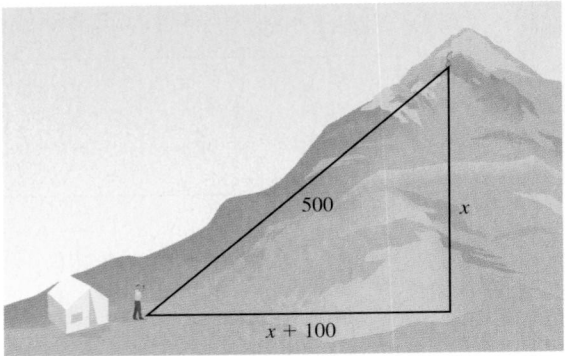

Rational Expressions and Equations

Multiplication and Division of Rational Expressions   •
Addition and Subtraction of Rational Expressions   •
Complex Rational Expressions   •   Solving Rational Equations

Rational expressions (p. 374)

A **rational expression** is a quotient of two polynomials. Because division by 0 is undefined, a rational expression is undefined for any number that will make the denominator 0.

EXAMPLE 1   Find all numbers for which the rational expression

$$\frac{2x + 5}{x^2 - 9x - 10}$$

is undefined.

Solution   We set the denominator equal to 0 and solve:

$$x^2 - 9x - 10 = 0$$

$$(x - 10)(x + 1) = 0 \qquad \text{Factoring}$$

$$x - 10 = 0 \quad or \quad x + 1 = 0 \qquad \text{Using the principle of zero products}$$

$$x = 10 \quad or \qquad x = -1.$$

If x is replaced with 10 or with -1, the denominator is 0. Thus,

$$\frac{2x + 5}{x^2 - 9x - 10} \quad \text{is undefined for } x = 10 \text{ and } x = -1.$$

Multiplication and Division of Rational Expressions

Multiplication and division of rational expressions is similar to multiplication and division with fractions.

The Product and the Quotient of Two Rational Expressions

To multiply two rational expressions, multiply numerators and multiply denominators:

$$\frac{A}{B} \cdot \frac{C}{D} = \frac{AC}{BD}.$$

To divide by a rational expression, multiply by its reciprocal:

$$\frac{A}{B} \div \frac{C}{D} = \frac{A}{B} \cdot \frac{D}{C} = \frac{AD}{BC}.$$

EXAMPLE 2   Simplify: $\dfrac{9x^2 + 12x}{6x^2 - 3x}$.

Solution   We first factor the numerator and the denominator:

$$\frac{9x^2 + 12x}{6x^2 - 3x} = \frac{3x(3x + 4)}{3x(2x - 1)}.$$

We can now write this as a product of two rational expressions using the rule for multiplying rational expressions in reverse. Then we can simplify.

$$\frac{3x(3x + 4)}{3x(2x - 1)} = \frac{3x}{3x} \cdot \frac{3x + 4}{2x - 1} \qquad \text{Rewriting as a product of two rational expressions}$$

$$= 1 \cdot \frac{3x + 4}{2x - 1} \qquad \frac{3x}{3x} = 1$$

$$= \frac{3x + 4}{2x - 1} \qquad \text{Removing the factor 1}$$

Only factors can be removed. Be sure that the numerator and the denominator are factored before you attempt to remove factors equal to 1.

After multiplying or dividing rational expressions, we simplify, if possible.

EXAMPLE 3   Perform each indicated operation and simplify.

a) $\dfrac{x^2 - x - 6}{3x} \cdot \dfrac{12x^3}{x + 2}$

b) $\dfrac{x^2 - 1}{x + 5} \div \dfrac{x^2 + 2x + 1}{2x + 10}$

Solution

Multiplication of rational expressions (p. 381)

a) $\dfrac{x^2 - x - 6}{3x} \cdot \dfrac{12x^3}{x + 2} = \dfrac{(x^2 - x - 6)(12x^3)}{3x(x + 2)}$ Multiplying the numerators and the denominators

$= \dfrac{(x - 3)(x + 2)(3x)(4x^2)}{3x(x + 2)}$ Factoring the numerator. Try to go directly to this step.

$= \dfrac{(x - 3)\cancel{(x + 2)}\cancel{(3x)}(4x^2)}{\cancel{(3x)}\cancel{(x + 2)}}$ Removing a factor equal to 1: $\dfrac{(x + 2)(3x)}{(x + 2)(3x)} = 1$

$= 4x^2(x - 3)$

Division of rational expressions (p. 382)

b) $\dfrac{x^2 - 1}{x + 5} \div \dfrac{x^2 + 2x + 1}{2x + 10} = \dfrac{x^2 - 1}{x + 5} \cdot \dfrac{2x + 10}{x^2 + 2x + 1}$ Multiplying by the reciprocal of the divisor

$= \dfrac{(x + 1)(x - 1)(2)(x + 5)}{(x + 5)(x + 1)(x + 1)}$ Multiplying rational expressions and factoring numerators and denominators

$= \dfrac{\cancel{(x + 1)}(x - 1)(2)\cancel{(x + 5)}}{\cancel{(x + 5)}\cancel{(x + 1)}(x + 1)}$ Removing a factor equal to 1: $\dfrac{(x + 1)(x + 5)}{(x + 1)(x + 5)} = 1$

$= \dfrac{2(x - 1)}{x + 1}$ We leave the numerator in factored form.

Addition and Subtraction of Rational Expressions

Like multiplication and division, addition and subtraction of rational expressions is similar to addition and subtraction of fractions.

> **The Sum and the Difference of Two Rational Expressions**
>
> To add when the denominators are the same, add the numerators and keep the same denominator:
>
> $$\frac{A}{B} + \frac{C}{B} = \frac{A + C}{B}.$$
>
> To subtract when the denominators are the same, subtract the second numerator from the first and keep the same denominator:
>
> $$\frac{A}{B} - \frac{C}{B} = \frac{A - C}{B}.$$

EXAMPLE 4   Add and simplify, if possible:

$$\frac{x - 6}{x^2 - 6x + 5} + \frac{5}{x^2 - 6x + 5}.$$

Solution

Addition of rational expressions
(p. 388)

$$\frac{x-6}{x^2-6x+5} + \frac{5}{x^2-6x+5} = \frac{x-6+5}{x^2-6x+5}$$ Adding numerators

$$= \frac{x-1}{(x-5)(x-1)}$$ Factoring the denominator

$$= \frac{1(x-1)}{(x-5)(x-1)}$$ Removing a factor equal to 1: $\frac{x-1}{x-1} = 1$

$$= \frac{1}{x-5}$$

Least common denominator
(p. 390)

Least common multiple (p. 390)

When two rational expressions do not have a common denominator, we must rewrite them with a common denominator before we can add or subtract them. We generally rewrite them using their **least common denominator** (**LCD**), which is the **least common multiple** (**LCM**) of their denominators.

To Find the Least Common Denominator (LCD)

1. Write the prime factorization of each denominator.
2. Select one of the factorizations and inspect it to see if it contains the other.

  a) If it does, it represents the LCM of the denominators.
  b) If it does not, multiply that factorization by any factors of the other denominator that it lacks. The final product is the LCM of the denominators.

The LCD is the LCM of the denominators. It should contain each factor the greatest number of times that it occurs in any of the individual factorizations.

EXAMPLE 5 Add: $\dfrac{x-3}{x^2-1} + \dfrac{4x^2}{x^2+4x+3}$.

Solution We first find the LCD. We write the prime factorization of each denominator and construct the LCM:

$$x^2 - 1 = (x+1)(x-1);$$
$$x^2 + 4x + 3 = (x+1)(x+3).$$

The LCM must contain both factorizations. We select the factorization of $x^2 - 1$. It does not contain the factor $(x+3)$ from the factorization of $x^2 + 4x + 3$. We multiply $(x+1)(x-1)$ by $(x+3)$:

$$\text{LCM} = (x+1)(x-1)(x+3).$$

The denominator $x^2 - 1 = (x + 1)(x - 1)$ must be multiplied by $x + 3$ in order to obtain the LCD. The denominator $x^2 + 4x + 3 = (x + 1)(x + 3)$ must be multiplied by $x - 1$ in order to obtain the LCD. We multiply each expression by a form of 1 that is made up of these "missing" factors:

$$\frac{x - 3}{x^2 - 1} + \frac{4x^2}{x^2 + 4x + 3} = \frac{x - 3}{(x + 1)(x - 1)} \cdot \frac{x + 3}{x + 3} + \frac{4x^2}{(x + 1)(x + 3)} \cdot \frac{x - 1}{x - 1}$$

$$= \frac{x^2 - 9}{(x + 1)(x - 1)(x + 3)} + \frac{4x^3 - 4x^2}{(x + 1)(x - 1)(x + 3)}$$

$$= \frac{4x^3 - 3x^2 - 9}{(x + 1)(x - 1)(x + 3)}.$$

EXAMPLE 6   Subtract: $\dfrac{x}{x + 2} - \dfrac{2x - 3}{3x - 4}$.

Solution   We have

Subtraction of rational expressions (p. 389)

$$\frac{x}{x + 2} - \frac{2x - 3}{3x - 4}$$

$$= \frac{x}{x + 2} \cdot \frac{3x - 4}{3x - 4} - \frac{2x - 3}{3x - 4} \cdot \frac{x + 2}{x + 2} \qquad \text{The LCD is } (x + 2)(3x - 4)$$

$$= \frac{3x^2 - 4x}{(x + 2)(3x - 4)} - \frac{2x^2 + x - 6}{(x + 2)(3x - 4)} \qquad \begin{array}{l}\text{Multiplying out}\\\text{the numerators}\\\text{(but not the}\\\text{denominators)}\end{array}$$

$$= \frac{3x^2 - 4x - (2x^2 + x - 6)}{(x + 2)(3x - 4)} \qquad \begin{array}{l}\text{Parentheses are}\\\text{important.}\end{array}$$

$$= \frac{3x^2 - 4x - 2x^2 - x + 6}{(x + 2)(3x - 4)} \qquad \begin{array}{l}\text{Removing parentheses}\\\text{in the numerator;}\\\text{subtracting every}\\\text{term}\end{array}$$

$$= \frac{x^2 - 5x + 6}{(x + 2)(3x - 4)}$$

$$= \frac{(x - 2)(x - 3)}{(x + 2)(3x - 4)}. \qquad \begin{array}{l}\text{Factoring the numerator in hopes}\\\text{of simplifying. There are no common}\\\text{factors.}\end{array}$$

The result could be written as either of the last two expressions.

Factors that are opposites (p. 401)

When denominators are opposites, we can find a common denominator by multiplying either rational expression by $-1/-1$.

EXAMPLE 7   Add: $\dfrac{a}{a-b} + \dfrac{5}{b-a}$.

Solution

$$\frac{a}{a-b} + \frac{5}{b-a} = \frac{a}{a-b} + \frac{5}{b-a} \cdot \frac{-1}{-1}$$

Writing 1 as $-1/-1$ and multiplying to obtain a common denominator

$$= \frac{a}{a-b} + \frac{-5}{a-b}$$

$(b-a)(-1) = -b+a = a-b$

$$= \frac{a-5}{a-b}$$

Complex Rational Expressions

Complex rational expressions
(Section 6.5)

A **complex rational expression** is a rational expression that has one or more rational expressions within its numerator or denominator. We will consider two methods for simplifying complex rational expressions. The first involves writing the expression as a quotient of two rational expressions.

> *To Simplify a Complex Rational Expression by Dividing*
> **1.** Add or subtract, as needed, to get a single rational expression in the numerator.
> **2.** Add or subtract, as needed, to get a single rational expression in the denominator.
> **3.** Divide the numerator by the denominator (invert and multiply).
> **4.** If possible, simplify by removing a factor equal to 1.

EXAMPLE 8   Simplify by dividing: $\dfrac{\dfrac{2}{x+1}}{\dfrac{1}{x+2} + \dfrac{1}{x}}$.

Solution

1. There is already a single rational expression in the numerator.

2. We add to get a single rational expression in the denominator:

$$\frac{\dfrac{2}{x+1}}{\dfrac{1}{x+2} + \dfrac{1}{x}} = \frac{\dfrac{2}{x+1}}{\dfrac{1}{x+2} \cdot \dfrac{x}{x} + \dfrac{1}{x} \cdot \dfrac{x+2}{x+2}}$$

Multiplying by 1 to get the LCD, $x(x+2)$, for the denominator

$$= \frac{\dfrac{2}{x+1}}{\dfrac{x}{x(x+2)} + \dfrac{x+2}{x(x+2)}} = \frac{\dfrac{2}{x+1}}{\dfrac{2x+2}{x(x+2)}}.$$

Adding in the denominator

3. Next, we invert and multiply:

$$\frac{\dfrac{2}{x+1}}{\dfrac{2x+2}{x(x+2)}} = \frac{2}{x+1} \div \frac{2x+2}{x(x+2)} = \frac{2}{x+1} \cdot \frac{x(x+2)}{2x+2}.$$

4. Simplifying, we have:

$$\frac{2}{(x+1)} \cdot \frac{x(x+2)}{2x+2} = \frac{2 \cdot x(x+2)}{2(x+1)(x+1)} \qquad \text{Removing a factor equal to 1: } \tfrac{2}{2} = 1$$

$$= \frac{x(x+2)}{(x+1)^2}.$$

A second method for simplifying complex rational expressions involves multiplying by the LCD.

To Simplify a Complex Rational Expression by Multiplying by the LCD

1. Find the LCD of *all* rational expressions within the complex rational expression.
2. Multiply the complex rational expression by a factor equal to 1. Write 1 as the LCD over itself (LCD/LCD).
3. Simplify. No fractional expressions should remain within the complex rational expression.
4. Factor and, if possible, simplify.

EXAMPLE 9   Simplify by multiplying by the LCD: $\dfrac{1 + \dfrac{2}{t}}{\dfrac{4}{t^2} - 1}$.

Solution

1. The denominators *within* the complex rational expression are t and t^2, so the LCD is t^2.

2. We multiply by a form of 1 using t^2/t^2:

$$\frac{1 + \dfrac{2}{t}}{\dfrac{4}{t^2} - 1} = \frac{1 + \dfrac{2}{t}}{\dfrac{4}{t^2} - 1} \cdot \frac{t^2}{t^2}.$$

3. We distribute and simplify:

$$\frac{1 + \dfrac{2}{t}}{\dfrac{4}{t^2} - 1} \cdot \frac{t^2}{t^2} = \frac{1 \cdot t^2 + \dfrac{2}{t} \cdot t^2}{\dfrac{4}{t^2} \cdot t^2 - 1 \cdot t^2}$$

$$= \frac{t^2 + 2t}{4 - t^2}. \qquad \text{No rational expression remains within the numerator or denominator.}$$

4. Finally, we simplify:

$$\frac{t^2 + 2t}{4 - t^2} = \frac{t(t+2)}{(2+t)(2-t)}$$ Factoring and simplifying;

$$\frac{t+2}{t+2} = 1$$

$$= \frac{t}{2-t}.$$

Solving Rational Equations

Solving rational equations
(Section 6.6)

A **rational equation** is an equation containing one or more rational expressions, often with the variable in a denominator.

To Solve a Rational Equation

1. List any restrictions that exist. No possible solution can make a denominator equal 0.
2. Clear the equation of fractions by multiplying both sides by the LCD of all rational expressions in the equation.
3. Solve the resulting equation using the addition principle, the multiplication principle, and the principle of zero products, as needed.
4. Check the possible solution(s) in the original equation.

Because a possible solution in step 3 may make a denominator 0, checking is essential when solving rational equations.

EXAMPLE 10   Solve: $x + \dfrac{10}{x} = 7$.

Solution   First we note that x cannot be 0. The LCD is x, so we multiply both sides by x:

$$x + \frac{10}{x} = 7$$

$$x\left(x + \frac{10}{x}\right) = 7x$$ Don't forget the parentheses!

$$x \cdot x + x \cdot \frac{10}{x} = 7x$$ Using the distributive law

$$x^2 + 10 = 7x$$ We have a quadratic equation.

$$x^2 - 7x + 10 = 0$$ Getting 0 on one side

$$(x - 2)(x - 5) = 0$$ Factoring

$$x - 2 = 0 \quad or \quad x - 5 = 0$$ Using the principle of zero products

$$x = 2 \quad or \qquad x = 5.$$

Check: For 2:

$$x + \frac{10}{x} = 7$$

$$\begin{array}{c|c} 2 + \dfrac{10}{2} & 7 \\ \hline 2 + 5 & \end{array}$$

$$7 \overset{?}{=} 7 \quad \text{TRUE}$$

For 5:

$$x + \frac{10}{x} = 7$$

$$\begin{array}{c|c} 5 + \dfrac{10}{5} & 7 \\ \hline 5 + 2 & \end{array}$$

$$7 \overset{?}{=} 7 \quad \text{TRUE}$$

Both numbers check, so there are two solutions, 2 and 5.

Work problems (p. 420)

Many problems translate to rational equations. **Work problems**, which involve the time that it takes to complete a task, can often be solved using the work principle.

> **The Work Principle**
>
> Suppose that A requires a units of time to complete a task and B requires b units of time to complete the same task. Then
>
> A works at a rate of $\dfrac{1}{a}$ tasks per unit of time,
>
> B works at a rate of $\dfrac{1}{b}$ tasks per unit of time, and
>
> A and B together work at a rate of $\dfrac{1}{a} + \dfrac{1}{b}$ tasks per unit of time.
>
> If A and B, working together, require t units of time to complete the task, then all four of the following equations hold:
>
> $$\frac{1}{a} \cdot t + \frac{1}{b} \cdot t = 1; \quad \left(\frac{1}{a} + \frac{1}{b}\right)t = 1; \quad \frac{t}{a} + \frac{t}{b} = 1; \quad \frac{1}{a} + \frac{1}{b} = \frac{1}{t}.$$

EXAMPLE 11

Drafting. It takes Kerry 30 hr to draw a set of plans for a house. It takes Jesse 45 hr to draw the same set of plans. How long would it take Kerry and Jesse, working together, to draw the set of plans?

Solution

1. **Familiarize.** We could make some guesses to help us understand the problem and then list our results in a table. We could also reason that if Kerry and Jesse each drew half the plans, it would take Kerry 15 hr and Jesse $22\frac{1}{2}$ hr. So the time it takes them working together should be between 15 and $22\frac{1}{2}$ hr. We let $t = $ the time that it takes them to draw the plans, working together.

2. **Translate.** We will use the work principle to translate the problem:

$$\frac{1}{a} \cdot t + \frac{1}{b} \cdot t = 1 \qquad \begin{array}{l} a \text{ is the time that it takes Kerry to draw the plans;} \\ b \text{ is the time that it takes Jesse to draw the plans.} \end{array}$$

$$\frac{t}{30} + \frac{t}{45} = 1.$$

3. **Carry out.**   We solve the equation:

$$\frac{t}{30} + \frac{t}{45} = 1$$

$$90\left(\frac{t}{30} + \frac{t}{45}\right) = 90 \cdot 1 \qquad \text{The LCD is } 2 \cdot 3 \cdot 3 \cdot 5, \text{ or } 90.$$

$$90 \cdot \frac{t}{30} + 90 \cdot \frac{t}{45} = 90$$

$$3t + 2t = 90$$

$$5t = 90$$

$$t = 18.$$

4. **Check.**   We note that, as predicted in the *Familiarize* step, the answer is between 15 and $22\frac{1}{2}$ hr. Also, if each works 18 hr, Kerry will do $\frac{18}{30}$ of the job and Jesse will do $\frac{18}{45}$ of the job, and

$$\frac{18}{30} + \frac{18}{45} = \frac{3}{5} + \frac{2}{5} = 1. \qquad \text{The entire job will be completed.}$$

5. **State.**   Together it will take them 18 hr to draw the plans.

Motion problems (p. 423)

Problems that deal with distance, speed (or rate), and time, or **motion problems**, can often be translated using the distance formula $d = rt$.

EXAMPLE 12

Driving time.   Karen and Eva are each driving to a sales meeting. Because of road conditions, Karen is able to drive 15 mph faster than Eva. In the same time that it takes Karen to travel 120 mi, Eva travels only 90 mi. Find their speeds.

Solution

1. **Familiarize.**   We let $t =$ the time, in hours, that is spent traveling and $r =$ Karen's speed, in mph. Then Eva's speed $= r - 15$. We set up a table.

| | d | = | r | · | t |
	Distance		Speed		Time
Karen	120		r		t
Eva	90		$r - 15$		t

2. **Translate.**   From the distance formula, we have $t = d/r$, so we can replace the times in the table with expressions involving r.

	Distance	Speed	Time
Karen	120	r	$120/r$
Eva	90	$r - 15$	$90/(r - 15)$

Since the times are the same, we have the equation

$$\frac{120}{r} = \frac{90}{r - 15}.$$

3. **Carry out.**   We solve the equation:

$$\frac{120}{r} = \frac{90}{r-15}$$

$$r(r-15)\frac{120}{r} = r(r-15)\frac{90}{r-15} \qquad \text{The LCD is } r(r-15).$$

$$120(r-15) = 90r \qquad\qquad \text{Simplifying}$$

$$120r - 1800 = 90r \qquad\qquad \text{Removing parentheses}$$

$$-1800 = -30r \qquad\qquad \text{Subtracting } 120r$$

$$60 = r. \qquad\qquad\qquad \text{Dividing both sides by } -30$$

4. **Check.**   If $r = 60$, then $r - 15 = 45$. If Karen travels 120 mi at 60 mph, she will have traveled 2 hr. If Eva travels 90 mi at 45 mph, she will also have traveled 2 hr. Since the times are the same, the speeds check.

5. **State.**   Karen is traveling at 60 mph, while Eva is traveling at 45 mph.

Ratio (p. 426)

Proportion (p. 426)

Another type of problem that translates to a rational equation involves proportions. A **ratio** of two quantities is their quotient. A **proportion** is an equation stating that two ratios are equal.

EXAMPLE 13

Baking.   Rob discovers there is $2\frac{1}{2}$ cups of pancake mix left in the box. The directions on the mix indicate that $1\frac{1}{3}$ cups of milk should be added to 2 cups of mix. How much milk should Rob add to the $2\frac{1}{2}$ cups of mix?

Solution   Since the problem translates directly to a proportion, we will not follow all five steps of the problem-solving process. We write the ratio of mix to milk in two ways:

$$\text{Mix} \longrightarrow \frac{2}{1\frac{1}{3}} = \frac{2\frac{1}{2}}{x} \longleftarrow \text{Mix}$$
$$\text{Milk} \longrightarrow \qquad\qquad \longleftarrow \text{Milk}$$

The LCD is $x\left(1\frac{1}{3}\right)$. We solve for x:

$$x\left(1\tfrac{1}{3}\right)\frac{2}{1\frac{1}{3}} = x\left(1\tfrac{1}{3}\right)\frac{2\frac{1}{2}}{x} \qquad \text{Multiplying by the LCD}$$

$$2x = \left(1\tfrac{1}{3}\right)\left(2\tfrac{1}{2}\right) \qquad \text{Simplifying}$$

$$2x = \tfrac{10}{3} \qquad\qquad \text{Converting to fraction notation and multiplying}$$

$$x = \tfrac{5}{3}. \qquad\qquad \text{Multiplying both sides by } \tfrac{1}{2} \text{ and simplifying}$$

Rob needs to add $\frac{5}{3}$ or $1\frac{2}{3}$ cups of milk.

R.6 Exercise Set

FOR EXTRA HELP

Student's
Solutions
Manual

Digital Video Tutor
CD 8
Videotape 15

Tutor
Center
AW Math
Tutor Center

MathXL Tutorials
on CD

Math XL
MathXL

MyMathLab
MyMathLab

List all numbers for which each rational expression is undefined.

1. $\dfrac{x-7}{3x+2}$

2. $\dfrac{10-y}{-6y}$

3. $\dfrac{p^2-1}{p^2-100}$

4. $\dfrac{10x}{x^2+9x+8}$

Simplify by removing a factor equal to 1.

5. $\dfrac{16x^2y}{18xy^2}$

6. $\dfrac{2x+10}{6x+30}$

7. $\dfrac{t^2-2t-8}{t^2-16}$

8. $\dfrac{a^3+2a^2+a}{a^2+4a+3}$

9. $\dfrac{2-x}{x^2-4}$

10. $\dfrac{y-8}{8-y}$

Perform each indicated operation. Then, if possible, simplify.

11. $\dfrac{3x}{x+y} \cdot \dfrac{2x+2y}{x^2}$

12. $\dfrac{5}{x+7} \cdot \dfrac{x+7}{10}$

13. $\dfrac{a^2+2a+1}{a} \div \dfrac{a^2}{a^2-1}$

14. $\dfrac{x}{x+3} + \dfrac{3-x}{x+3}$

15. $\dfrac{2x}{x-7} - \dfrac{x+7}{x-7}$

16. $\dfrac{x}{x+y} \div \dfrac{y}{x+y}$

17. $\dfrac{5}{x} + \dfrac{6}{x^2}$

18. $\dfrac{x^2+4x+3}{x^2+x-2} \cdot \dfrac{x^2+3x+2}{x^2+2x-3}$

19. $\dfrac{2a+b}{a-b} - \dfrac{4}{3a-3b}$

20. $(x^2-16) \div \dfrac{4x+16}{3x^2}$

21. $\dfrac{2-x}{5x^2} \div \dfrac{x^2-4}{3x}$

22. $\dfrac{2x}{x-5} + \dfrac{3}{x+4}$

23. $\dfrac{x^3+2x^2+x}{x^2-4} \cdot \dfrac{x^2-x-2}{x^4+x^3}$

24. $\dfrac{-1}{x^2+7x+10} - \dfrac{3}{x^2+8x+15}$

25. $\dfrac{2}{(x+1)^2} + \dfrac{1}{x+1}$

26. $\dfrac{2x}{x^2-3x} \div (x-3)$

27. $\dfrac{x-y}{2x} \cdot \dfrac{3x^2}{y-x}$

28. $\dfrac{1}{x+y} + \dfrac{2}{x^2+y^2}$

29. $\dfrac{x-2}{x+5} - \dfrac{x+3}{x-4}$

30. $\dfrac{z^2+2z+1}{8z} \div \dfrac{z^2-z-2}{4z^2-4}$

Simplify.

31. $\dfrac{\dfrac{2}{x}-\dfrac{1}{x^2}}{\dfrac{x}{4}}$

32. $\dfrac{\dfrac{x}{3}-\dfrac{3}{x}}{\dfrac{1}{x}+\dfrac{1}{3}}$

33. $\dfrac{\dfrac{3}{x-7}}{\dfrac{4x+3}{x+1}}$

34. $\dfrac{\dfrac{a}{a-b}}{\dfrac{a^2}{a^2-b^2}}$

35. $\dfrac{x-\dfrac{3}{x-2}}{x-\dfrac{12}{x+1}}$

36. $\dfrac{t+\dfrac{1}{t}}{t-\dfrac{2}{t}}$

37. $\dfrac{\dfrac{1}{2}-\dfrac{1}{x}}{\dfrac{2-x}{2}}$

38. $\dfrac{\dfrac{x}{2y^2}+\dfrac{y}{3x^2}}{\dfrac{1}{6xy}+\dfrac{2}{x^2y}}$

Solve.

39. $\dfrac{1}{2} + \dfrac{1}{3} = \dfrac{1}{t}$

40. $\dfrac{1}{4} + \dfrac{1}{t} = \dfrac{1}{3}$

41. $x + \dfrac{1}{x} = 2$

42. $\dfrac{x - 7}{x + 1} = \dfrac{2}{3}$

43. $\dfrac{3}{y + 7} = \dfrac{1}{y - 8}$

44. $\dfrac{x + 1}{x - 2} = \dfrac{3}{x - 2}$

45. $\dfrac{1}{x - 3} - \dfrac{x - 4}{x^2 - 9} = 1$

46. $\dfrac{3}{a + 4} = \dfrac{a - 1}{4 - a}$

47. *Painting.* Quentin can paint the turret on a Queen Anne house in 40 hr. It takes Austin 50 hr to paint the same turret. How long would it take them, working together, to paint the turret?

48. *Building fences.* Lindsay can build a fence in 6 hr. Laura can do the same job in 5 hr. How long will it take them, working together, to build the fence?

49. *Snowmobiling.* Jessica can ride her snowmobile through the fields 20 km/h faster than Josh can ride his through the woods. In the time it takes Jessica to ride 18 km, Josh travels 10 km. Find the speed of each snowmobile.

50. *Bicycling.* Ani bicycles 8 mi and Lia bicycles 12 mi to meet at a park for lunch. Because Ani's trip is mostly uphill, she rides 5 mph slower than Lia. Ani and Lia leave their homes at the same time and arrive at the park at the same time. Find the speed of each bicyclist.

51. *Elk population.* To determine the size of a park's elk population, rangers tag 15 elk and set them free. Months later, 40 elk are caught, of which 12 have tags. Estimate the size of the elk population.

52. *Manufacturing pegs.* A sample of 136 wooden pegs contained 17 defective pegs. How many defective pegs would you expect in a sample of 840 pegs?

Appendixes

A   Mean, Median, and Mode

Mean • Median • Mode

One way to analyze data is to look for a single representative number, called a **center point** or **measure of central tendency**. Those most often used are the **mean** (or **average**), the **median**, and the **mode**.

Mean

Let's first consider the *mean*, or *average*.

> **Mean, or Average**
>
> The *mean*, or *average*, of a set of numbers is the sum of the numbers divided by the number of addends.

EXAMPLE 1   Consider the following data on revenue, in billions of dollars, at McDonald's restaurants in five recent years:

$$\$12.5, \quad \$13.2, \quad \$14.2, \quad \$14.9, \quad \$15.9.$$

What is the mean of the numbers? (*Source*: McDonald's Corporation)

Solution First, we add the numbers:

$$12.5 + 13.2 + 14.2 + 14.9 + 15.9 = 70.7.$$

Then we divide by the number of addends, 5:

$$\frac{(12.5 + 13.2 + 14.2 + 14.9 + 15.9)}{5} = \frac{70.7}{5} = 14.14.$$

The mean, or average, revenue of McDonald's for those five years is $14.14 billion.

Note that $14.14 + 14.14 + 14.14 + 14.14 + 14.14 = 70.7$. If we use this center point, 14.14, repeatedly as the addend, we get the same sum that we do when adding individual data numbers.

Median

The *median* is useful when we wish to de-emphasize extreme scores. For example, suppose five workers in a technology company manufactured the following number of computers during one day's work:

Sarah:	88
Matt:	92
Pat:	66
Jen:	94
Mark:	91

Let's first list the scores in order from smallest to largest:

66 88 **91** 92 94.

↑

Middle number

The middle number—in this case, 91—is the **median**.

Median

Once a set of data has been arranged from smallest to largest, the *median* of the set of data is the middle number if there is an odd number of data numbers. If there is an even number of data numbers, then there are two middle numbers and the median is the *average* of the two middle numbers.

EXAMPLE 2 Find the median of the following set of household incomes:

$76,000, $58,000, $87,000, $32,500, $64,800, $62,500.

Solution   We first rearrange the numbers in order from smallest to largest.

$32,500, \ $58,000, \ $62,500, \ $64,800, \ $76,000, \ $87,000

Median

There is an even number of numbers. We look for the middle two, which are $62,500 and $64,800. In this case, the median is the average of $62,500 and $64,800:

$$\frac{\$62{,}500 + \$64{,}800}{2} = \$63{,}650.$$

Mode

The last center point we consider is called the *mode*. A number that occurs most often in a set of data is sometimes considered a representative number or center point.

> **Mode**
>
> The *mode* of a set of data is the number or numbers that occur most often. If each number occurs the same number of times, there is *no* mode.

EXAMPLE 3   Find the mode of the following data:

23, 24, 27, 18, 19, 27.

Solution   The number that occurs most often is 27. Thus the mode is 27.

EXAMPLE 4   Find the mode of the following data:

83, 84, 84 84, 85, 86, 87, 87, 87, 88, 89, 90.

Solution   There are two numbers that occur most often, 84 and 87. Thus the modes are 84 and 87.

EXAMPLE 5   Find the mode of the following data:

115, 117, 211, 213, 219.

Solution   Each number occurs the same number of times. The set of data has *no* mode.

Exercise Set

A

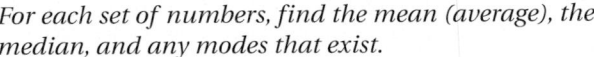

For each set of numbers, find the mean (average), the median, and any modes that exist.

1. 17, 19, 29, 18, 14, 29

2. 72, 83, 85, 88, 92

3. 5, 37, 20, 20, 35, 5, 25

4. 13, 32, 25, 27, 13

5. 4.3, 7.4, 1.2, 5.7, 7.4

6. 13.4, 13.4, 12.6, 42.9

7. 234, 228, 234, 229, 234, 278

8. $29.95, $28.79, $30.95, $29.95

9. *Atlantic storms and hurricanes.* The following bar graph shows the number of Atlantic storms or hurricanes that formed in various months from 1980 to 2000. What is the average number for the 9 months given? the median? the mode?

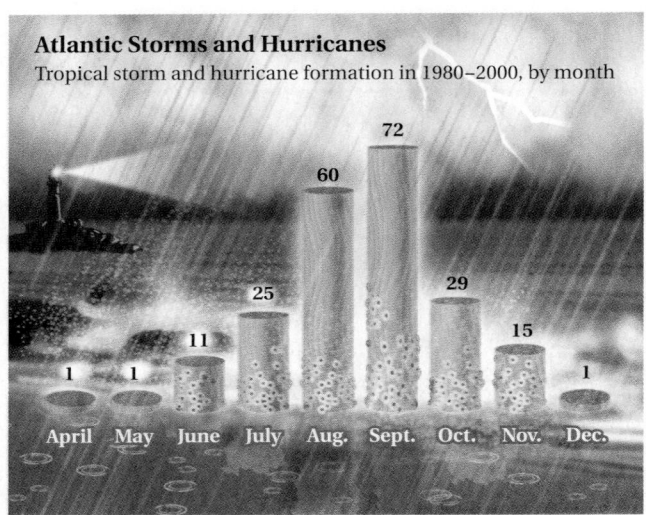

Atlantic Storms and Hurricanes
Tropical storm and hurricane formation in 1980–2000, by month

72
60
25
29
11
15
1 1   1
April May June July Aug. Sept. Oct. Nov. Dec.

Source: Colorado State University

10. *Cheddar cheese prices.* The following prices per pound of sharp cheddar cheese were found at five supermarkets:

   $5.99, $6.79, $5.99, $6.99, $6.79.

What was the average price per pound? the median price? the mode?

11. *Coffee consumption.* The following lists the annual coffee consumption, in number of cups per person, for various countries. Find the mean, the median, and the mode.

Germany	1113
United States	610
Switzerland	1215
France	798
Italy	750

Source: Beverage Marketing Corporation

12. *NBA tall men.* The following is a list of the heights, in inches, of the tallest men in the NBA in a recent year. Find the mean, the median, and the mode.

Shaquille O'Neal	85
Gheorghe Muresan	91
Shawn Bradley	90
Tim Duncan	84
Yao Ming	89
Bruno Sundov	86

Source: National Basketball Association

13. *Salmon prices.* The following prices per pound of Atlantic salmon were found at five fish markets:

   $6.99, $8.49, $8.99, $6.99, $9.49.

What was the average price per pound? the median price? the mode?

14. *PBA scores.*   Chris Barnes rolled scores of 224, 224, 254, and 187 in a recent tournament of the Professional Bowlers Association. What was his average? his median? his mode?
Source: Professional Bowlers Association

15. *Hank Aaron.*   Hank Aaron averaged $34\frac{7}{22}$ home runs per year over a 22-yr career. After 21 yr, Aaron had averaged $35\frac{10}{21}$ home runs per year. How many home runs did Aaron hit in his final year?

16. *Length of pregnancy.*   Marta was pregnant 270 days, 259 days, and 272 days for her first three pregnancies. In order for Marta's average length of pregnancy to equal the worldwide average of 266 days, how long must her fourth pregnancy last?
Source: David Crystal (ed.), *The Cambridge Factfinder.* Cambridge CB2 1RP: Cambridge University Press, 1993, p. 84.

17. The ordered set of data 18, 21, 24, *a*, 36, 37, *b* has a median of 30 and an average of 32. Find *a* and *b*.

18. *Male height.*   Jason's brothers are 174 cm, 180 cm, 179 cm, and 172 cm tall. The average male is 176.5 cm tall. How tall is Jason if he and his brothers have an average height of 176.5 cm?

B   Sets

Naming Sets   •   Membership   •   Subsets   •   Intersections   •   Unions

A **set** is a collection of objects. In mathematics the objects, or **elements**, of a set are generally numbers. This section provides a basic introduction to sets.

Naming Sets

To name the set of whole numbers less than 6, we can use *roster notation*, as follows:

$$\{0, 1, 2, 3, 4, 5\}.$$

The set of real numbers *x* for which *x* is less than 6 cannot be named by listing all its members because there is an infinite number of them. We name such a set using *set-builder notation*, as follows:

$$\{x \mid x < 6\}.$$

This is read

"The set of all *x* such that *x* is less than 6."

See Section 2.6 for more on this notation.

Membership

The symbol $\in$ means *is a member of* or *belongs to*, or *is an element of.* Thus,

$$x \in A$$

means

x is a member of A,   or x belongs to A,   or x is an element of A.

EXAMPLE 1   Classify each of the following as true or false.

a) $1 \in \{1, 2, 3\}$
b) $1 \in \{2, 3\}$
c) $4 \in \{x \mid x \text{ is an even whole number}\}$
d) $5 \in \{x \mid x \text{ is an even whole number}\}$

Solution

a) Since 1 is listed as a member of the set, $1 \in \{1, 2, 3\}$ is true.
b) Since 1 is *not* a member of $\{2, 3\}$, the statement $1 \in \{2, 3\}$ is false.
c) Since 4 is an even whole number, $4 \in \{x \mid x \text{ is an even whole number}\}$ is true.
d) Since 5 is *not* even, $5 \in \{x \mid x \text{ is an even whole number}\}$ is false.

Set membership can be illustrated with a diagram, as shown below.

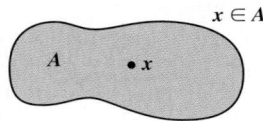

Subsets

If every element of A is also an element of B, then A is a *subset* of B. This is denoted $A \subseteq B$.

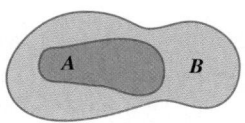

The set of whole numbers is a subset of the set of integers. The set of rational numbers is a subset of the set of real numbers.

EXAMPLE 2   Classify each of the following as true or false.

a) $\{1, 2\} \subseteq \{1, 2, 3, 4\}$
b) $\{p, q, r, w\} \subseteq \{a, p, r, z\}$
c) $\{x \mid x < 6\} \subseteq \{x \mid x \leq 11\}$

Solution

a) Since every element of $\{1, 2\}$ is in the set $\{1, 2, 3, 4\}$, it follows that $\{1, 2\} \subseteq \{1, 2, 3, 4\}$ is true.

b) Since $q \in \{p, q, r, w\}$, but $q \notin \{a, p, r, z\}$, it follows that $\{p, q, r, w\} \subseteq \{a, p, r, z\}$ is false.

c) Since every number that is less than 6 is also less than 11, the statement $\{x \mid x < 6\} \subseteq \{x \mid x \leq 11\}$ is true.

Intersections

The *intersection* of sets A and B, denoted $A \cap B$, is the set of members common to both sets.

EXAMPLE 3   Find each intersection.

a) $\{0, 1, 3, 5, 25\} \cap \{2, 3, 4, 5, 6, 7, 9\}$ b) $\{a, p, q, w\} \cap \{p, q, t\}$

Solution

a) $\{0, 1, 3, 5, 25\} \cap \{2, 3, 4, 5, 6, 7, 9\} = \{3, 5\}$

b) $\{a, p, q, w\} \cap \{p, q, t\} = \{p, q\}$

Set intersection can be illustrated with a diagram, as shown below.

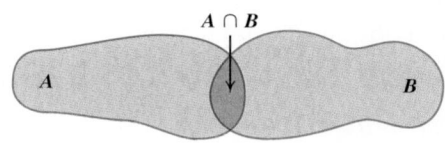

The set without members is known as the *empty set*, and is written $\varnothing$, and sometimes $\{\ \}$. Each of the following is a description of the empty set:

The set of all 12-ft–tall people;

$\{2, 3\} \cap \{5, 6, 7\}$;

$\{x \mid x$ is an even natural number$\} \cap \{x \mid x$ is an odd natural number$\}$.

Unions

Two sets A and B can be combined to form a set that contains the members of both A and B. The new set is called the *union* of A and B, denoted $A \cup B$.

EXAMPLE 4   Find each union.

a) $\{0, 5, 7, 13, 27\} \cup \{0, 2, 3, 4, 5\}$ b) $\{a, c, e, g\} \cup \{b, d, f\}$

Solution

a) $\{0, 5, 7, 13, 27\} \cup \{0, 2, 3, 4, 5\} = \{0, 2, 3, 4, 5, 7, 13, 27\}$
   Note that the 0 and the 5 are *not* listed twice in the solution.

b) $\{a, c, e, g\} \cup \{b, d, f\} = \{a, b, c, d, e, f, g\}$

Set union can be illustrated with a diagram, as shown below.

$A \cup B$ is shaded.

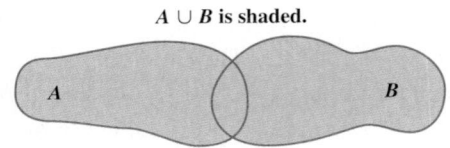

Exercise Set

B

Name each set using the roster method.

1. The set of whole numbers 3 through 7

2. The set of whole numbers 83 through 89

3. The set of odd numbers between 40 and 50

4. The set of multiples of 5 between 10 and 40

5. $\{x \mid$ the square of x is 9$\}$

6. $\{x \mid x$ is the cube of 0.2$\}$

Classify each statement as true or false.

7. $2 \in \{x \mid x$ is an odd number$\}$

8. $7 \in \{x \mid x$ is an odd number$\}$

9. Bruce Springsteen $\in$ The set of all rock stars

10. Apple $\in$ The set of all fruit

11. $-3 \in \{-4, -3, 0, 1\}$

12. $0 \in \{-4, -3, 0, 1\}$

13. $\frac{2}{3} \in \{x \mid x$ is a rational number$\}$

14. Heads $\in$ The set of outcomes of flipping a penny

15. $\{4, 5, 8\} \subseteq \{1, 3, 4, 5, 6, 7, 8, 9\}$

16. The set of vowels $\subseteq$ The set of consonants

17. $\{-1, -2, -3, -4, -5\} \subseteq \{-1, 2, 3, 4, 5\}$

18. The set of integers $\subseteq$ The set of rational numbers

Find each intersection.

19. $\{a, b, c, d, e\} \cap \{c, d, e, f, g\}$

20. $\{a, e, i, o, u\} \cap \{q, u, i, c, k\}$

21. $\{1, 2, 5, 10\} \cap \{0, 1, 7, 10\}$

22. $\{0, 1, 7, 10\} \cap \{0, 1, 2, 5\}$

23. $\{1, 2, 5, 10\} \cap \{3, 4, 7, 8\}$

24. $\{a, e, i, o, u\} \cap \{m, n, f, g, h\}$

Find each union.

25. $\{a, e, i, o, u\} \cup \{q, u, i, c, k\}$

26. $\{a, b, c, d, e\} \cup \{c, d, e, f, g\}$

27. $\{0, 1, 7, 10\} \cup \{0, 1, 2, 5\}$

28. $\{1, 2, 5, 10\} \cup \{0, 1, 7, 10\}$

29. $\{a, e, i, o, u\} \cup \{m, n, f, g, h\}$

30. $\{1, 2, 5, 10\} \cup \{a, b\}$

31. What advantage(s) does set-builder notation have over roster notation?

32. What advantage(s) does roster notation have over set-builder notation?

SYNTHESIS

33. Find the union of the set of integers and the set of whole numbers.

34. Find the intersection of the set of odd integers and the set of even integers.

35. Find the union of the set of rational numbers and the set of irrational numbers.

36. Find the intersection of the set of even integers and the set of positive rational numbers.

37. Find the intersection of the set of rational numbers and the set of irrational numbers.

38. Find the union of the set of negative integers, the set of positive integers, and the set containing 0.

39. For a set A, find each of the following.

    a) $A \cup \varnothing$

    b) $A \cup A$

    c) $A \cap A$

    d) $A \cap \varnothing$

40. A set is *closed* under an operation if, when the operation is performed on its members, the result is in the set. For example, the set of real numbers is closed under the operation of addition since the sum of any two real numbers is a real number.

    a) Is the set of even numbers closed under addition?

    b) Is the set of odd numbers closed under addition?

    c) Is the set $\{0, 1\}$ closed under addition?

    d) Is the set $\{0, 1\}$ closed under multiplication?

    e) Is the set of real numbers closed under multiplication?

    f) Is the set of integers closed under division?

41. Experiment with sets of various types and determine whether the following distributive law for sets is true:

$$A \cap (B \cup C) = (A \cap B) \cup (A \cap C).$$

C   Synthetic Division

Streamlining Long Division • The Remainder Theorem

Streamlining Long Division

To divide a polynomial by a binomial of the type $x - a$, we can streamline the usual procedure to develop a process called *synthetic division*.

Compare the following. In each stage, we attempt to write a bit less than in the previous stage, while retaining enough essentials to solve the problem. At the end, we will return to the usual polynomial notation.

Stage 1

When a polynomial is written in descending order, the coefficients provide the essential information:

$$
\begin{array}{r}
4x^2 + 5x + 11 \\
x - 2\overline{)4x^3 - 3x^2 + x + 7} \\
\underline{4x^3 - 8x^2} \\
5x^2 + x \\
\underline{5x^2 - 10x} \\
11x + 7 \\
\underline{11x - 22} \\
29
\end{array}
$$

$$
\begin{array}{r}
4 + 5 + 11 \\
1 - 2\overline{)4 - 3 + 1 + 7} \\
\underline{4 - 8} \\
5 + 1 \\
\underline{5 - 10} \\
11 + 7 \\
\underline{11 - 22} \\
29
\end{array}
$$

Because the leading coefficient in the divisor is 1, each time we multiply the divisor by a term in the answer, the leading coefficient of that product duplicates a coefficient in the answer. In the next stage, we don't bother to duplicate these numbers. We also show where -2 is used and drop the 1 from the divisor.

Stage 2

$$\begin{array}{r} 4x^2 + 5x + 11 \\ x - 2{\overline{\smash{)}}\,4x^3 - 3x^2 + x + 7} \\ \underline{4x^3 - 8x^2} \\ 5x^2 + x \\ \underline{5x^2 - 10x} \\ 11x + 7 \\ \underline{11x - 22} \\ 29 \end{array}$$

$$\begin{array}{r} 4 + 5 + 11 \\ -2{\overline{\smash{)}}\,4 - 3 + 1 + 7} \\ -8 \longleftarrow \text{ Multiply: } -2 \cdot 4 = -8. \\ 5 + 1 \text{ Subtract: } -3 - (-8) = 5. \\ -10 \longleftarrow \text{ Multiply: } -2 \cdot 5 = -10. \\ 11 + 7 \text{ Subtract: } 1 - (-10) = 11. \\ -22 \longleftarrow \text{ Multiply: } -2 \cdot 11 = -22. \\ 29 \longleftarrow \text{ Subtract: } 7 - (-22) = 29. \end{array}$$

To simplify further, we now reverse the sign of the -2 in the divisor and, in exchange, *add* at each step in the long division.

Stage 3

$$\begin{array}{r} 4x^2 + 5x + 11 \\ x - 2{\overline{\smash{)}}\,4x^3 - 3x^2 + x + 7} \\ \underline{4x^3 - 8x^2} \\ 5x^2 + x \\ \underline{5x^2 - 10x} \\ 11x + 7 \\ \underline{11x - 22} \\ 29 \end{array}$$

$$\begin{array}{r} 4 + 5 + 11 \\ 2{\overline{\smash{)}}\,4 - 3 + 1 + 7} \\ 8 \longleftarrow \text{ Replace the } -2 \text{ with 2.} \\ 5 + 1 \text{ Multiply: } 2 \cdot 4 = 8. \\ 10 \longleftarrow \text{ Add: } -3 + 8 = 5. \\ 11 + 7 \text{ Multiply: } 2 \cdot 5 = 10. \\ 22 \longleftarrow \text{ Add: } 1 + 10 = 11. \\ 29 \longleftarrow \text{ Multiply: } 2 \cdot 11 = 22. \\ \text{ Add: } 7 + 22 = 29. \end{array}$$

The blue numbers can be eliminated if we look at the red numbers instead.

Stage 4

$$\begin{array}{r} 4x^2 + 5x + 11 \\ x - 2{\overline{\smash{)}}\,4x^3 - 3x^2 + x + 7} \\ \underline{4x^3 - 8x^2} \\ 5x^2 + x \\ \underline{5x^2 - 10x} \\ 11x + 7 \\ \underline{11x - 22} \\ 29 \end{array}$$

$$\begin{array}{r|rrr} & 4 & 5 & 11 \\ 2{\overline{\smash{)}}\,4} & -3 & 1 & 7 \\ & 8 & 10 & 22 \\ \hline & 5 & 11 & 29 \end{array}$$

Don't lose sight of how the products 8, 10, and 22 are found. Also, note that the 5 and 11 preceding the remainder 29 coincide with the 5 and 11 following the 4 on the top line. By writing a 4 to the left of 5 on the bottom line, we can eliminate the top line in stage 4 and read our answer from the bottom line. This final stage is commonly called **synthetic division**.

Stage 5

$$
\begin{array}{r}
4 \quad 5 \quad 11 \\
\hline
2)4 \;\; -3 \quad 1 \quad 7 \\
8 \;\; 10 \;\; 22 \\
\hline
5 \;\; 11 \;\; 29
\end{array}
$$

$$
\begin{array}{r|rrr}
2 & 4 & -3 & 1 & 7 \\
& & 8 & 10 & 22 \\
\hline
& 4 & 5 & 11 & 29
\end{array}
$$
⟵ This is the remainder.

This is the zero-degree coefficient.

This is the first-degree coefficient.

This is the second-degree coefficient.

The quotient is $4x^2 + 5x + 11$. The remainder is 29.

Remember that in order for this method to work, the divisor must be of the form $x - a$, that is, a variable minus a constant. The coefficient of the variable must be 1.

EXAMPLE 1   Use synthetic division to divide: $(x^3 + 6x^2 - x - 30) \div (x - 2)$.

Solution

$$
\begin{array}{r|rrrr}
2 & 1 & 6 & -1 & -30 \\
\hline
& 1
\end{array}
$$
Write the 2 of $x - 2$ and the coefficients of the dividend.

Bring down the first coefficient.

$$
\begin{array}{r|rrrr}
2 & 1 & 6 & -1 & -30 \\
& & 2 \\
\hline
& 1 & 8
\end{array}
$$
Multiply 1 by 2 to get 2.

Add 6 and 2.

$$
\begin{array}{r|rrrr}
2 & 1 & 6 & -1 & -30 \\
& & 2 & 16 \\
\hline
& 1 & 8 & 15
\end{array}
$$
Multiply 8 by 2.

Add -1 and 16.

$$
\begin{array}{r|rrrr}
2 & 1 & 6 & -1 & -30 \\
& & 2 & 16 & 30 \\
\hline
& 1 & 8 & 15 & 0
\end{array}
$$
Multiply 15 by 2 and add.

The answer is $x^2 + 8x + 15$ with R 0, or just $x^2 + 8x + 15$.

EXAMPLE 2   Use synthetic division to divide.

a) $(2x^3 + 7x^2 - 5) \div (x + 3)$

b) $(10x^2 - 13x + 3x^3 - 20) \div (4 + x)$

Solution

a) $(2x^3 + 7x^2 - 5) \div (x + 3)$

The dividend has no x-term, so we need to write 0 for its coefficient of x. Note that $x + 3 = x - (-3)$, so we write -3 inside the $\underline{}$.

$$
\begin{array}{r|rrrr}
-3 & 2 & 7 & 0 & -5 \\
   &   & -6 & -3 & 9 \\
\hline
   & 2 & 1 & -3 & \;|\; 4
\end{array}
$$

The answer is $2x^2 + x - 3$, with R 4, or $2x^2 + x - 3 + \dfrac{4}{x + 3}$.

b) We first rewrite $(10x^2 - 13x + 3x^3 - 20) \div (4 + x)$ in descending order:

$$(3x^3 + 10x^2 - 13x - 20) \div (x + 4).$$

Next, we use synthetic division. Note that $x + 4 = x - (-4)$.

$$
\begin{array}{r|rrrr}
-4 & 3 & 10 & -13 & -20 \\
   &   & -12 & 8 & 20 \\
\hline
   & 3 & -2 & -5 & \;|\; 0
\end{array}
$$

The answer is $3x^2 - 2x - 5$.

The Remainder Theorem

Because the remainder is 0, Example 1 shows that $x - 2$ is a factor of $x^3 + 6x^2 - x - 30$ and that we can write $x^3 + 6x^2 - x - 30$ as $(x - 2)(x^2 + 8x + 15)$. Using this result and the principle of zero products, we know that if $f(x) = x^3 + 6x^2 - x - 30$, then $f(2) = 0$ (since $x - 2$ is a factor of $f(x)$). Similarly, from Example 2(b), we know that $x + 4$ is a factor of $g(x) = 10x^2 - 13x + 3x^3 - 20$. This tells us that $g(-4) = 0$. In both examples, the remainder from the division, 0, can serve as a function value. Remarkably, this pattern extends to nonzero remainders. To see this, note that the remainder in Example 2(a) is 4, and if $f(x) = 2x^3 + 7x^2 - 5$, then $f(-3)$ is also 4 (you should check this). The fact that the remainder and the function value coincide is predicted by the remainder theorem, which follows.

> **The Remainder Theorem**
>
> The remainder obtained by dividing $P(x)$ by $x - r$ is $P(r)$.

A proof of this result is outlined in Exercise 23.

EXAMPLE 3   Let $f(x) = 8x^5 - 6x^3 + x - 8$. Use synthetic division to find $f(2)$.

Solution   The remainder theorem tells us that $f(2)$ is the remainder when $f(x)$ is divided by $x - 2$. We use synthetic division to find that remainder:

$$
\begin{array}{r|rrrrrr}
2 & 8 & 0 & -6 & 0 & 1 & -8 \\
  &   & 16 & 32 & 52 & 104 & 210 \\
\hline
  & 8 & 16 & 26 & 52 & 105 & 202 \\
\end{array}
$$

Although the bottom line can be used to find the quotient for the division $(8x^5 - 6x^3 + x - 8) \div (x - 2)$, what we are really interested in is the remainder. It tells us that $f(2) = 202$.

Exercise Set C

Use synthetic division to divide.

1. $(x^3 - 2x^2 + 2x - 7) \div (x + 1)$

2. $(x^3 - 2x^2 + 2x - 7) \div (x - 1)$

3. $(a^2 + 8a + 11) \div (a + 3)$

4. $(a^2 + 8a + 11) \div (a + 5)$

5. $(x^3 - 7x^2 - 13x + 3) \div (x - 2)$

6. $(x^3 - 7x^2 - 13x + 3) \div (x + 2)$

7. $(3x^3 + 7x^2 - 4x + 3) \div (x + 3)$

8. $(3x^3 + 7x^2 - 4x + 3) \div (x - 3)$

9. $(y^3 - 3y + 10) \div (y - 2)$

10. $(x^3 - 2x^2 + 8) \div (x + 2)$

11. $(x^5 - 32) \div (x - 2)$

12. $(y^5 - 1) \div (y - 1)$

13. $\left(3x^3 + 1 - x + 7x^2\right) \div \left(x + \frac{1}{3}\right)$

14. $\left(8x^3 - 1 + 7x - 6x^2\right) \div \left(x - \frac{1}{2}\right)$

Use synthetic division to find the indicated function value.

15. $f(x) = 5x^4 + 12x^3 + 28x + 9; \ f(-3)$

16. $g(x) = 3x^4 - 25x^2 - 18; \ g(3)$

17. $P(x) = 6x^4 - x^3 - 7x^2 + x + 2; \ P(-1)$

18. $F(x) = 3x^4 + 8x^3 + 2x^2 - 7x - 4; \ F(-2)$

19. $f(x) = x^4 - x^3 - 19x^2 + 49x - 30; \ f(4)$

20. $p(x) = x^4 + 7x^3 + 11x^2 - 7x - 12; \ p(2)$

SYNTHESIS

21. Why is it that we *add* when performing synthetic division, but *subtract* when performing long division?

22. Explain how synthetic division could be useful when factoring a polynomial.

23. To prove the remainder theorem, note that any polynomial $P(x)$ can be rewritten as $(x - r) \cdot Q(x) + R$, where $Q(x)$ is the quotient polynomial that arises when $P(x)$ is divided by $x - r$, and R is some constant (the remainder).

  a) How do we know that R must be a constant?

  b) Show that $P(r) = R$ (this says that $P(r)$ is the remainder when $P(x)$ is divided by $x - r$).

24. Let $f(x) = 4x^3 + 16x^2 - 3x - 45$. Find $f(-3)$ and then solve the equation $f(x) = 0$.

25. Let $f(x) = 6x^3 - 13x^2 - 79x + 140$. Find $f(4)$ and then solve the equation $f(x) = 0$.

Nested evaluation. *One way to evaluate a polynomial function like $P(x) = 3x^4 - 5x^3 + 4x^2 - 1$ is to successively factor out x as shown:*

$$P(x) = x(x(x(3x - 5) + 4) + 0) - 1.$$

Computations are then performed using this "nested" form of $P(x)$.

26. Use nested evaluation to find $f(-3)$ in Exercise 24. Note the similarities to the calculations performed with synthetic division.

27. Use nested evaluation to find $f(4)$ in Exercise 25. Note the similarities to the calculations performed with synthetic division.

Tables

TABLE 1   Fraction and Decimal Equivalents

Fraction Notation	$\frac{1}{10}$	$\frac{1}{8}$	$\frac{1}{6}$	$\frac{1}{5}$	$\frac{1}{4}$	$\frac{3}{10}$	$\frac{1}{3}$	$\frac{3}{8}$	$\frac{2}{5}$	$\frac{1}{2}$
Decimal Notation	0.1	0.125	$0.16\overline{6}$	0.2	0.25	0.3	$0.333\overline{3}$	0.375	0.4	0.5
Percent Notation	10%	12.5%, or $12\frac{1}{2}\%$	$16.6\overline{6}\%$, or $16\frac{2}{3}\%$	20%	25%	30%	$33.3\overline{3}\%$, or $33\frac{1}{3}\%$	37.5%, or $37\frac{1}{2}\%$	40%	50%
Fraction Notation	$\frac{3}{5}$	$\frac{5}{8}$	$\frac{2}{3}$	$\frac{7}{10}$	$\frac{3}{4}$	$\frac{4}{5}$	$\frac{5}{6}$	$\frac{7}{8}$	$\frac{9}{10}$	$\frac{1}{1}$
Decimal Notation	0.6	0.625	$0.666\overline{6}$	0.7	0.75	0.8	$0.83\overline{3}$	0.875	0.9	1
Percent Notation	60%	62.5%, or $62\frac{1}{2}\%$	$66.6\overline{6}\%$, or $66\frac{2}{3}\%$	70%	75%	80%	$83.3\overline{3}\%$, or $83\frac{1}{3}\%$	87.5%, or $87\frac{1}{2}\%$	90%	100%

TABLE 2 Squares and Square Roots with Approximations to Three Decimal Places

N	$\sqrt{N}$	N^2	N	$\sqrt{N}$	N^2	N	$\sqrt{N}$	N^2	N	$\sqrt{N}$	N^2
1	1	1	26	5.099	676	51	7.141	2601	76	8.718	5776
2	1.414	4	27	5.196	729	52	7.211	2704	77	8.775	5929
3	1.732	9	28	5.292	784	53	7.280	2809	78	8.832	6084
4	2	16	29	5.385	841	54	7.348	2916	79	8.888	6241
5	2.236	25	30	5.477	900	55	7.416	3025	80	8.944	6400
6	2.449	36	31	5.568	961	56	7.483	3136	81	9	6561
7	2.646	49	32	5.657	1024	57	7.550	3249	82	9.055	6724
8	2.828	64	33	5.745	1089	58	7.616	3364	83	9.110	6889
9	3	81	34	5.831	1156	59	7.681	3481	84	9.165	7056
10	3.162	100	35	5.916	1225	60	7.746	3600	85	9.220	7225
11	3.317	121	36	6	1296	61	7.810	3721	86	9.274	7396
12	3.464	144	37	6.083	1369	62	7.874	3844	87	9.327	7569
13	3.606	169	38	6.164	1444	63	7.937	3969	88	9.381	7744
14	3.742	196	39	6.245	1521	64	8	4096	89	9.434	7921
15	3.873	225	40	6.325	1600	65	8.062	4225	90	9.487	8100
16	4	256	41	6.403	1681	66	8.124	4356	91	9.539	8281
17	4.123	289	42	6.481	1764	67	8.185	4489	92	9.592	8464
18	4.243	324	43	6.557	1849	68	8.246	4624	93	9.644	8649
19	4.359	361	44	6.633	1936	69	8.307	4761	94	9.695	8836
20	4.472	400	45	6.708	2025	70	8.367	4900	95	9.747	9025
21	4.583	441	46	6.782	2116	71	8.426	5041	96	9.798	9216
22	4.690	484	47	6.856	2209	72	8.485	5184	97	9.849	9409
23	4.796	529	48	6.928	2304	73	8.544	5329	98	9.899	9604
24	4.899	576	49	7	2401	74	8.602	5476	99	9.950	9801
25	5	625	50	7.071	2500	75	8.660	5625	100	10	10,000

Glossary

A

Absolute value [1.4]   The distance that a number is from 0 on the number line.

Additive inverse [1.6]   A number's opposite. Two numbers are additive inverses of each other if their sum is zero.

Algebraic expression [1.1]   A number or variable or a collection of numbers and variables on which the operations $+$, $-$, $\cdot$, $\div$, $(\ )^n$, or $\sqrt[n]{(\ )}$ are performed.

Arithmetic sequence [14.2]   A sequence in which the difference between any two successive terms is constant.

Arithmetic series [14.2]   A series for which the associated sequence is arithmetic.

Ascending order [4.3]   A polynomial in one variable written with the terms arranged according to degree, from least to greatest.

Associative law of addition [1.2]   The statement that when three numbers are added, regrouping the addends gives the same sum.

Associative law of multiplication [1.2]   The statement that when three numbers are multiplied, regrouping the factors gives the same product.

Asymptote [13.3]   A line that a graph approaches more and more closely as x increases or as x decreases.

Average [2.7]   Most commonly, the mean of a set of numbers.

Axes [3.1]   Two perpendicular number lines used to identify points in a plane.

Axis of symmetry [11.6]   A line that can be drawn through a graph such that the part of the graph on one side of the line is an exact reflection of the part on the opposite side.

B

Bar graph [3.1]   A graphic display of data using bars proportional in length to the numbers represented.

Base [1.8]   In exponential notation, the number being raised to a power.

Binomial [4.2]   A polynomial composed of two terms.

Branches [13.3]   The two curves that comprise a hyperbola.

Break-even point [8.8]   In business, the point of intersection of the revenue function and the cost function.

C

Circle [13.1]   A set of points in a plane that are a fixed distance r, called the radius, from a fixed point (h, k), called the center.

Circle graph [3.1]   A graphic display of data using sectors of a circle to represent percents.

Circumference [2.3], [4.2]   The distance around a circle.

Closed interval [a, b] [9.1]   The set of all numbers x for which $a \le x \le b$. Thus, $[a, b] = \{x \mid a \le x \le b\}$.

Coefficient [2.1]   The numerical multiplier of a variable.

Combined variation [7.5]   A mathematical relationship in which a variable varies directly and/or inversely, at the same time, with more than one other variable.

Common logarithm [12.5]   A logarithm with base 10.

Commutative law of addition [1.2]   The statement that when two numbers are added, changing the order in which the numbers are added does not affect the sum.

Commutative law of multiplication [1.2]   The statement that when two numbers are multiplied, changing the order in which the numbers are multiplied does not affect the product.

Completing the square [11.1]   Adding a particular constant to an expression so that the resulting sum is a perfect square.

Complex number [10.8]   Any number that can be written as $a + bi$, where a and b are real numbers.

Complex rational expression [6.5]   A rational expression that has one or more rational expressions within its numerator and/or denominator.

Complex-number system [10.8]   A number system that contains the real-number system and is designed so that negative numbers have square roots.

Composite function [12.1]   A function in which a quantity depends on a variable that, in turn, depends on another variable.

Composite number [1.3]   A natural number, other than 1, that is not prime.

Compound inequality [9.2]   A statement in which two or more inequalities are combined using the word *and* or the word *or*.

Compound interest [11.1]   Interest computed on the sum of an original principal and the interest previously accrued by that principal.

Conditional equation [2.2]   An equation that is true for some replacements and false for others.

Conic section [13.1]   A curve formed by the intersection of a plane and a cone.

Conjugates [10.5]   Pairs of radical terms, like $\sqrt{a} + \sqrt{b}$ and $\sqrt{a} - \sqrt{b}$, for which the product does not have a radical term.

Conjunction [9.2]   A sentence in which two statements are joined by the word *and*.

Consecutive numbers [2.5]   Integers that are one unit apart.

Consistent system of equations [8.1]   A system of equations that has at least one solution.

Constant [1.1]   A known number.

Constant function [7.3]   A function given by an equation of the form $f(x) = b$, where b is a real number.

Constant of proportionality [7.5]   The constant in an equation of direct or inverse variation.

Constraint [9.5]   A requirement imposed on a problem.

Contradiction [2.2]   An equation that is never true.

Coordinates [3.1]   The numbers in an ordered pair.

Cube root [10.1]   The number c is called the cube root of a if $c^3 = a$.

D

Data point [3.7]   A given ordered pair of a function, usually found experimentally.

Degree of a polynomial [4.2]   The degree of the term of highest degree in a polynomial.

Degree of a term [4.2]   The number of variable factors in a term.

Demand function [8.8]   A function modeling the relationship between the price of a good and the quantity of that good demanded.

Denominator [1.3]   The number below the fraction bar in a fraction.

Dependent equations [8.1]   The equations in a system are dependent if one equation can be removed without changing the solution set.

Descending order [4.2]   A polynomial in one variable written with the terms arranged according to degree, from greatest to least.

Determinant [8.7]   The determinant of a two-by-two matrix $\begin{bmatrix} a & c \\ b & d \end{bmatrix}$ is denoted $\begin{vmatrix} a & c \\ b & d \end{vmatrix}$ and represents $ad - bc$.

Difference of squares [5.4]   An expression that can be written in the form $a^2 - b^2$.

Direct variation [7.5]   A situation that translates to an equation of the form $y = kx$, with k a constant.

Discriminant [11.4]   The expression $b^2 - 4ac$ from the quadratic formula.

Disjunction [9.2]   A sentence in which two statements are joined by the word *or*.

Distributive law [1.2]   The statement that multiplying a factor by the sum of two numbers gives the same result as multiplying the factor by each of the two numbers and then adding.

Domain [7.2]   The set of all first coordinates of the ordered pairs in a function.

Doubling time [12.7]   The time necessary for a population to double in size.

E

Elimination method [8.2]   An algebraic method that uses the addition principle to solve a system of equations.

Ellipse [13.2]   The set of all points in a plane for which the sum of the distances from two fixed points F_1 and F_2 is constant.

Equation [1.1]   A number sentence with the verb =.

Equation of variation [7.5]   An equation used to represent direct, inverse, or combined variation.

Equilibrium point [8.8]   The point of intersection between the demand function and the supply function.

Equivalent equations [2.1]   Equations with the same solutions.

Equivalent expressions [1.2]   Expressions that have the same value for all allowable replacements.

Equivalent inequalities [2.6]   Inequalities that have the same solution set.

Evaluate [1.1]   To substitute a value for each occurrence of a variable in an expression.

Exponent [1.8]   In expressions of the form a^n, the number n is an exponent. For n a natural number, a^n represents n factors of a.

Exponential decay [12.7]   A decrease in quantity over time that can be modeled by an exponential equation of the form $P(t) = P_0 e^{-kt}$, $k > 0$.

Exponential equation [12.6]   An equation in which a variable appears as an exponent.

Exponential function [12.2]   A function that can be described by an exponential equation.

Exponential growth [12.7]   An increase in quantity over time that can be modeled by an exponential function of the form $P(t) = P_0 e^{kt}$, $k > 0$.

Exponential notation [1.8]   A representation of a number using a base raised to a power.

Extrapolation [3.7]   The process of predicting a future value on the basis of given data.

F

Factor [1.2]   *Verb*: to write an equivalent expression that is a product. *Noun*: a multiplier.

Finite sequence [14.1]   A function having for its domain a set of natural numbers: $\{1, 2, 3, 4, 5, \ldots, n\}$, for some natural number n.

Fixed costs [8.8]   In business, costs that are incurred whether or not a product is produced.

Focus [13.2]   One of two fixed points that determine the points of an ellipse.

FOIL [4.5]   To multiply two binomials by multiplying the First terms, the Outside terms, the Inside terms, and then the Last terms.

Formula [2.3]   An equation that uses numbers or letters to represent a relationship between two or more quantities.

Fraction notation [1.3]   A number written using a numerator and a denominator.

Function [7.5]   A correspondence that assigns to each member of a set called the domain exactly one member of a set called the range.

G

General term of a sequence [14.1]   The nth term, denoted a_n.

Geometric sequence [14.3]   A sequence in which the ratio of every pair of successive terms is constant.

Geometric series [14.3]   A series for which the associated sequence is geometric.

Grade [3.5]   The ratio of the vertical distance a road rises over the horizontal distance it runs, expressed as a percent.

Graph [3.1]   A picture or diagram of the data in a table. A line, curve, or collection of points that represents all the solutions of an equation.

Greatest common factor [5.1]   The common factor of a polynomial with the largest possible coefficient and the largest possible exponent(s).

H

Half-life [12.7]   The amount of time necessary for half of a quantity to decay.

Half-open interval [9.1]   An interval that includes exactly one of two endpoints.

Horizontal-line test [12.1]   If it is impossible to draw a horizontal line that intersects the graph of a function more than once, then that function is one-to-one.

Hyperbola [13.3]   The set of all points P in the plane such that the difference of the distance from P to two fixed points is constant.

Hypotenuse [5.8]   In a right triangle, the side opposite the right angle.

I

Identity [2.2]   An equation that is always true.

Identity property of 0 [1.5]   The statement that the sum of a number and 0 is always the original number.

Identity property of 1 [1.3]   The statement that the product of a number and 1 is always the original number.

Imaginary number [10.8]   A number that can be written in the form $a + bi$, where a and b are real numbers and $b \neq 0$.

Imaginary number i [10.8]   The square root of -1. That is, $i = \sqrt{-1}$ and $i^2 = -1$.

Inconsistent system of equations [8.1]   A system of equations for which there is no solution.

Independent equations [8.1]   Equations that are not dependent.

Index [10.1]   In the radical $\sqrt[n]{a}$, the number n is called the index.

Inequality [1.4]   A mathematical sentence using $<, >, \leq, \geq,$ or $\neq$.

Infinite geometric series [14.3]   The sum of the terms of an infinite geometric sequence.

Infinite sequence [14.1]   A function having for its domain the set of natural numbers: $\{1, 2, 3, 4, 5, \ldots\}$.

Input [7.1]   A member of the domain of a function.

Integers [1.4]   The whole numbers and their opposites.

Interpolation [3.7]   The process of estimating a value between given values.

Intersection of two sets [9.2]   The set of all elements that are common to both sets.

Interval notation [9.1]   The use of a pair of numbers inside parentheses and brackets to represent the set of all numbers between those two numbers. *See also* Closed and Open intervals.

Inverse relation [12.1]   The relation formed by interchanging the members of the domain and the range of a relation.

Inverse variation [7.5]   A situation that translates to an equation of the form $y = k/x$, with k a constant.

Irrational number [1.4]   A real number that cannot be named as a ratio of two integers.

Isosceles right triangle [10.7]   A right triangle in which both legs have the same length.

J

Joint variation [7.5]   A situation that translates to an equation of the form $y = kxz$, with k a constant.

L

Leading coefficient [4.2]   The coefficient of the term of highest degree in a polynomial.

Leading term [4.2]   The term of highest degree in a polynomial.

Least common denominator [6.3]   The least common multiple of the denominators.

Legs [5.8]   In a right triangle, the two sides that form the right angle.

Like radicals [10.5]   Radical expressions that have a common radical factor.

Like terms [1.5]   Terms that have exactly the same variable factors.

Line graph [3.1]   A graph in which quantities are represented as points connected by straight-line segments.

Linear equation [3.2]   Any equation that can be written in the form $y = mx + b$, or $Ax + By = C$, where x and y are variables.

Linear function [7.3]   A function that can be described by an equation of the form $y = mx + b$, where x and y are variables.

Linear inequality [9.4]   An inequality whose related equation is a linear equation.

Linear programming [9.5]   A branch of mathematics involving graphs of inequalities and their constraints.

Logarithmic equation [12.6]   An equation containing a logarithmic expression.

Logarithmic function, base a [12.3]   The inverse of an exponential function with base a.

M

Matrix [8.6]   A rectangular array of numbers.

Maximum value [12.6]   The largest function value (output) achieved by a function.

Mean [2.7]   The sum of a set of numbers divided by the number of addends.

Minimum value [12.6]   The smallest function value (output) achieved by a function.

Monomial [4.2]   A constant, a variable, or a product of a constant and one or more variables.

Motion problem [6.7]   A problem that deals with distance, speed, and time.

Multiplicative inverses [1.3]   Reciprocals; two numbers whose product is 1.

Multiplicative property of zero [1.7]   The statement that the product of 0 and any real number is 0.

N

Natural logarithm [12.5]   A logarithm with base e.

Natural numbers [1.3]   The counting numbers: $1, 2, 3, 4, 5, \ldots$.

Nonlinear function [7.3]   A function whose graph is not a straight line.

Numerator [1.3]   The number above the fraction bar in a fraction.

O

Objective function [9.5]   In linear programming, the function in which the expression being maximized or minimized appears.

One-to-one function [12.1]   A function for which different inputs have different outputs.

Open interval (a, b) [9.1]   The set of all numbers x for which $a < x < b$. Thus, $(a, b) = \{x \mid a < x < b\}$.

Opposite [1.6]   The opposite, or additive inverse, of a number a is written $-a$. Opposites are the same distance from 0 on the number line but on different sides of 0.

Ordered pair [3.1]   A pair of numbers of the form (h, k) for which the order in which the numbers are listed is important.

Origin [3.1]   The point on a graph where the two axes intersect.

Output [7.1]   A member of the range of a function.

P

Parabola [11.6]   A graph of a quadratic function.

Parallel lines [3.6]   Lines that extend indefinitely without intersecting.

Pascal's triangle [14.4]   A triangular array of coefficients of the expansion $(a + b)^n$ for $n = 0, 1, 2, \ldots$.

Perfect square [10.1]   A rational number for which there exists a number a for which $a^2 = p$.

Perfect-square trinomial [5.4]   A trinomial that is the square of a binomial.

Perpendicular lines [3.6]   Lines that form a right angle.

Point–slope equation [3.7]   An equation of the type $y - y_1 = m(x - x_1)$, where x and y are variables.

Polynomial [4.2]   A monomial or a sum of monomials.

Polynomial equation [5.7]   An equation in which two polynomials are set equal to each other.

Polynomial inequality [11.9]   An inequality that is equivalent to an inequality with a polynomial as one side and 0 as the other.

Prime factorization [1.3]   The factorization of a whole number into a product of its prime factors.

Prime number [1.3]   A natural number that has exactly two different factors: the number itself and 1.

Principal square root [10.1]   The nonnegative square root of a number.

Proportion [6.7]   An equation stating that two ratios are equal.

Pure imaginary number [10.8]   A complex number of the form $a + bi$, with $a = 0$ and $b \neq 0$.

Pythagorean theorem [5.8]   In any right triangle, if a and b are the lengths of the legs and c is the length of the hypotenuse, then $a^2 + b^2 = c^2$.

Q

Quadrants [3.1]   The four regions into which the axes divide a plane.

Quadratic equation [5.7]   An equation equivalent to one of the form $ax^2 + bx + c = 0$, where $a \neq 0$.

Quadratic formula [11.2]   The solutions of $ax^2 + bx + c, a \neq 0$, are given by the equation
$$x = \frac{-b \pm \sqrt{b^2 - 4ac}}{2a}.$$

Quadratic function [11.1]   A second-degree polynomial function in one variable.

Quadratic inequality [11.9]   A second-degree polynomial inequality in one variable.

R

Radical equation [10.6]   An equation in which a variable appears in a radicand.

Radical expression [10.1]   An algebraic expression in which a radical sign appears.

Radical sign [10.1]   The symbol $\sqrt{}$.

Radical term [10.5]   A term in which a radical sign appears.

Radicand [10.1]   The expression under the radical sign.

Radius [13.1]   The distance from the center of a circle to a point on the circle. Also, a segment connecting the center to a point on the circle.

Range [7.2]   The set of all second coordinates of the ordered pairs in a function.

Rate [3.4]   A ratio that indicates how two quantities change with respect to each other.

Ratio [6.7]   The ratio of a to b is a/b, also written $a:b$.

Rational equation [6.6]   An equation containing one or more rational expressions.

Rational expression [6.1]   A quotient of two polynomials.

Rational inequality [11.9]   An inequality containing a rational expression.

Rational number [1.4]   A number that can be written in the form $\dfrac{a}{b}$, where a and b are integers and $b \neq 0$.

Rationalizing the denominator [10.4]   A procedure for finding an equivalent expression without a radical in the denominator.

Rationalizing the numerator [10.4]   A procedure for finding an equivalent expression without a radical in the numerator.

Real number [1.4]   Any number that is either rational or irrational.

Reciprocal [1.3]   A multiplicative inverse. Two numbers are reciprocals if their product is 1.

Reflection [11.6]   The mirror image of a graph.

Relation [7.1]   A correspondence between the domain and the range of a function such that each member of the domain corresponds to at least one member of the range.

Repeating decimal [1.4]   A decimal in which a number pattern repeats indefinitely.

Right triangle [5.8]   A triangle that includes a right angle.

Row-equivalent operations [11.6]   Operations used to produce equivalent systems of equations.

S

Scientific notation [4.8]   A number written in the form $N \times 10^{m}$, where m is an integer, $1 \leq N < 10$, and N is expressed in decimal notation.

Sequence [14.1]   A function for which the domain is a set of consecutive positive integers beginning with 1.

Series [14.1]   The sum of specified terms in a sequence.

Set [1.4]   A collection of objects.

Set-builder notation [2.6]   The naming of a set by describing basic characteristics of the elements in the set.

Sigma notation [14.1]   The naming of a sum using the Greek letter Σ (sigma) as part of an abbreviated form.

Similar triangles [10.7]   Triangles in which corresponding sides are proportional.

Simplify   To rewrite an expression in an equivalent, abbreviated, form.

Slope [3.5]   The ratio of the rise to the run for any two points on a line.

Slope–intercept equation [3.6]   An equation of the form $y = mx + b$, where x and y are variables.

Solution [1.1]   A replacement or substitution that makes an equation or inequality true.

Solution set [2.6]   The set of all solutions of an equation, an inequality, or a system of equations or inequalities.

Solve [2.1]   To find all solutions of an equation, an inequality, or a system of equations or inequalities; to find the solution(s) of a problem.

Speed [6.7]   The ratio of distance traveled to the time required to travel that distance.

Square matrix [8.7]   A matrix with the same number of rows and columns.

Square root [10.1]   The number c is a square root of a if $c^2 = a$.

Substitute [1.1]   To replace a variable with a number.

Substitution method [8.2]   An algebraic method for solving systems of equations.

Supply function [11.8]   A function modeling the relationship between the price of a good and the quantity of that good supplied.

System of equations [11.1]   A set of two or more equations that are to be solved simultaneously.

T

Term [1.2]   A number, a variable, or a product or a quotient of numbers and/or variables.

Terminating decimal [1.4]   A decimal that can be written using a finite number of decimal places.

Total cost [8.8]   The amount spent to produce a product.

Total profit [8.8]   The amount taken in less the amount spent, or total revenue minus total cost.

Total revenue [8.8]   The amount taken in from the sale of a product.

Trinomial [4.2]   A polynomial that is composed of three terms.

U

Union of *A* and *B* [9.2]   The set of all elements belonging to either *A* or *B*.

Undefined [1.7]   An expression that has no meaning attached to it.

V

Value [1.1]   The numerical result after a number has been substituted into an expression.

Variable [1.1]   A letter that represents an unknown number.

Variable costs [8.8]   In business, costs that vary according to the amount of products produced.

Variable expression [1.1]   An expression containing a variable.

Vertex [11.6]   The point at which the graph of a quadratic equation crosses its axis of symmetry.

Vertical-line test [7.3]   The statement that a graph represents a function if it is impossible to draw a vertical line that intersects the graph more than once.

W

Whole numbers [1.3]   The natural numbers and 0: $0, 1, 2, 3, \ldots$

X

x-intercept [3.3]   The point at which a graph crosses the x-axis.

Y

y-intercept [3.3]   The point at which a graph crosses the y-axis.

Z

Zeros [11.9]   The x-values for which $f(x)$ is 0, for any function f.

Photo Credits

1, © Peter Turnley/Corbis 7, Corbis 37 (**left**), PhotoDisc 37 (**right**), © Tony Roberts/Corbis 42, PhotoDisc 54, © Peter Turnley/Corbis 79, © Ernest H. Robl 98, © Andrew Holbrooke/Corbis 109, © Antonio N. Rosario/The Image Bank 110, Corbis 116, Appalachian Trail Conference 120, Ryan McVay, PhotoDisc 123, © Kevin Horan, Tony Stone 137, Christine D. Tuff 139, © Ernest H. Robl 141, Brand X Pictures (Getty) 147, 179, 180, Corbis 182, © David Keaton, The Stock Market 187, 199, PhotoDisc 200, Corbis 224, PhotoDisc 229, 238, 246, Corbis 303, 359, Custom Medical Stock Photo, Inc. 364, Corbis 373, 396, © Bachmann, Stock, Boston 420, PhotoDisc 422 Corbis 428, 441, PhotoDisc 474, 425, Corbis 487, PhotoDisc 491, David J. Ellenbogen 493, 497, PhotoDisc 500, National Park Service 504, Corbis 507, DigitalVision, Getty Images 508, PhotoDisc 535, DigitalVision, Getty Images 553, Dex Image, Getty Images 577, 582, PhotoDisc 619, Comstock Images 633, 649, Corbis 679, Brand X 703, NBAE/Getty Images 712, Corbis 714, Bob Thomason, Stone/Getty Images 728, NBAE/Getty Images 781, Corbis 832, ©Reuters/Corbis 835, AP/Wide World Photos 838, 840, Corbis 843 (**left**), From Classic Baseball Cards, by Bert Randolph Sugar, copyright © 1977 by Dover Publishing Co. 843 (**right**), AFP/Getty Images 855, 874, Patrick Wright, Photographic Services, Clemson University 901, 926, Corbis 1004, Ric Ergenbright, Corbis 1006 (**left**), © Corbis Images 1007 (**right**), PBA

Answers

CHAPTER 1

Technology Connection, p. 7

1. 3438   **2.** 47,531

Exercise Set 1.1, pp. 9–12

1. Expression   **3.** Equation   **5.** Equation   **7.** Expression
9. Equation   **11.** Expression   **13.** 36   **15.** 56   **17.** 5
19. 4   **21.** 5   **23.** 6   **25.** 24 ft^2 **27.** 15 cm^2 **29.** 0.368
31. Let r represent Ron's age; $r + 5$, or $5 + r$ **33.** $b + 6$,
or $6 + b$ **35.** $c - 9$ **37.** $6 + q$, or $q + 6$ **39.** Let p
represent Phil's speed; $9p$, or $p \cdot 9$ **41.** $y - x$

43. $x \div w$, or $\dfrac{x}{w}$ **45.** $n - m$

47. Let l represent the length of the box and h represent
the height; $l + h$, or $h + l$ **49.** $9 \cdot 2m$, or $2m \cdot 9$ **51.** Let y
represent "some number"; $\frac{1}{4}y$, or $\frac{y}{4}$ **53.** Let w represent
the number of women attending; 64% of w, or $0.64w$
55. Yes   **57.** No   **59.** Yes   **61.** Yes   **63.** Let x represent
the unknown number; $73 + x = 201$ **65.** Let x represent
the unknown number; $42x = 2352$ **67.** Let s represent the
number of unoccupied squares; $s + 19 = 64$ **69.** Let w
represent the amount of solid waste generated, in millions
of tons; 27% of $w = 56$, or $0.27w = 56$ **71.** (f)   **73.** (d)
75. (g)   **77.** (e)   **79.** ▤   **81.** ▤   **83.** $337.50   **85.** 2
87. 6   **89.** $w + 4$ **91.** $l + w + l + w$, or $2l + 2w$
93. $t + 8$ **95.** ▤

Exercise Set 1.2, pp. 18–19

1. Commutative   **3.** Associative   **5.** Distributive
7. Associative   **9.** Commutative   **11.** $x + 7$ **13.** $c + ab$
15. $3y + 9x$ **17.** $5(1 + a)$ **19.** $a \cdot 2$ **21.** ts **23.** $5 + ba$
25. $(a + 1)5$ **27.** $a + (5 + b)$ **29.** $(r + t) + 7$

31. $ab + (c + d)$ **33.** $8(xy)$ **35.** $(2a)b$ **37.** $(3 \cdot 2)(a + b)$
39. $(r + t) + 6$; $(t + 6) + r$ **41.** $17(ab)$; $b(17a)$
43. $(5 + x) + 2 = (x + 5) + 2$ Commutative law
$\qquad\qquad = x + (5 + 2)$ Associative law
$\qquad\qquad = x + 7$ Simplifying
45. $(m \cdot 3)7 = m(3 \cdot 7)$ Associative law
$\qquad\quad = m \cdot 21$ Simplifying
$\qquad\quad = 21m$ Commutative law
47. $4a + 12$ **49.** $6 + 6x$ **51.** $3x + 3$ **53.** $24 + 8y$
55. $18x + 54$ **57.** $5r + 10 + 15t$ **59.** $2a + 2b$

61. $5x + 5y + 10$ **63.** $x, xyz, 19$ **65.** $2a, \dfrac{a}{b}, 5b$

67. $2(a + b)$ **69.** $7(1 + y)$ **71.** $3(6x + 1)$
73. $5(x + 2 + 3y)$ **75.** $3(4x + 3)$ **77.** $3(a + 3b)$
79. $11(4x + y + 2z)$ **81.** s, t **83.** $3, (x + y)$ **85.** $7, a$
87. $(a - b), (x - y)$ **89.** ▤   **91.** Let k represent Kara's

salary; $2k$ **92.** $\dfrac{1}{2} \cdot m$, or $\dfrac{m}{2}$ **93.** ▤

95. Yes; distributive law   **97.** Yes; distributive law and
commutative law of multiplication   **99.** No; for example,
let $x = 1$ and $y = 2$. Then $30 \cdot 2 + 1 \cdot 15 = 60 + 15 = 75$
and $5[2(1 + 3 \cdot 2)] = 5[2(7)] = 5 \cdot 14 = 70$.   **101.** ▤

Exercise Set 1.3, pp. 28–30

1. Composite   **3.** Prime   **5.** Composite   **7.** Prime
9. Neither   **11.** (b)   **13.** (d)   **15.** $2 \cdot 25$; $5 \cdot 10$; 1, 2, 5, 10,
25, 50   **17.** $3 \cdot 14$; $6 \cdot 7$; 1, 2, 3, 6, 7, 14, 21, 42   **19.** $2 \cdot 13$
21. $2 \cdot 3 \cdot 5$ **23.** $3 \cdot 3 \cdot 3$ **25.** $2 \cdot 3 \cdot 3$ **27.** $2 \cdot 2 \cdot 2 \cdot 5$
29. Prime   **31.** $2 \cdot 3 \cdot 5 \cdot 7$ **33.** $5 \cdot 23$ **35.** $\frac{2}{3}$ **37.** $\frac{2}{7}$
39. $\frac{1}{8}$ **41.** 7   **43.** $\frac{1}{4}$ **45.** 6   **47.** $\frac{21}{25}$ **49.** $\frac{60}{41}$ **51.** $\frac{15}{7}$

53. $\frac{3}{14}$ **55.** $\frac{27}{8}$ **57.** $\frac{1}{2}$ **59.** $\frac{7}{6}$ **61.** $\dfrac{3b}{7a}$ **63.** $\dfrac{7}{a}$ **65.** $\frac{5}{6}$

67. 1   **69.** $\frac{5}{18}$ **71.** 0   **73.** $\frac{35}{18}$ **75.** $\frac{10}{3}$ **77.** 28   **79.** 1
81. $\frac{6}{35}$ **83.** 18   **85.** ▤   **87.** $5(3 + x)$; answers may vary

88. $7 + (b + a)$, or $(a + b) + 7$ **89.** 🖊
91. Row 1: 7, 2, 36, 14, 8, 8; row 2: 9, 18, 2, 10, 12, 21
93. $\frac{2}{5}$ **95.** $\frac{3q}{t}$ **97.** $\frac{6}{25}$ **99.** $\frac{5ap}{2cm}$ **101.** $\frac{23r}{18t}$ **103.** $\frac{28}{45}\text{m}^2$
105. $14\frac{2}{9}\text{m}$ **107.** $27\frac{3}{5}\text{cm}$

Technology Connection, p. 34

1. 2.236067977 **2.** 2.645751311 **3.** 3.605551275
4. 5.196152423 **5.** 6.164414003 **6.** 7.071067812

Exercise Set 1.4, pp. 37–39

1. Repeating **3.** Integer **5.** Rational number
7. Natural number **9.** $-1349, 29,035$ **11.** 950,000,000,
-460 **13.** $2, -6$ **15.** $750, -125$ **17.** Jets: -34; Strikers: 34
19.

21.

23.

25. 0.875 **27.** -0.75

29. $1.\overline{16}$ **31.** $0.\overline{6}$ **33.** -0.5 **35.** $0.1\overline{3}$ **37.** $<$ **39.** $>$
41. $<$ **43.** $<$ **45.** $>$ **47.** $<$ **49.** $<$ **51.** $x < -7$
53. $y \geq -10$ **55.** True **57.** False **59.** True **61.** 58
63. 17 **65.** 5.6 **67.** 329 **69.** $\frac{9}{7}$ **71.** 0 **73.** 8
75. $-83, -4.7, 0, \frac{5}{9}, 8.31, 62$ **77.** $-83, 0, 62$
79. $-83, -4.7, 0, \frac{5}{9}, \pi, \sqrt{17}, 8.31, 62$ **81.** 🖊 **83.** 42
84. $ba + 5$, or $5 + ab$ **85.** 🖊 **87.** 🖊 **89.** $-23, -17, 0, 4$
91. $-\frac{4}{3}, \frac{4}{9}, \frac{4}{8}, \frac{4}{6}, \frac{4}{5}, \frac{4}{3}, \frac{4}{2}$ **93.** $<$ **95.** $=$ **97.** $-7, 7$
99. $-4, -3, 3, 4$ **101.** $\frac{3}{3}$ **103.** $\frac{70}{9}$ **105.** 🖊

Exercise Set 1.5, pp. 44–46

1. (f) **3.** (e) **5.** (b) **7.** -3 **9.** 4 **11.** 0 **13.** -8
15. -27 **17.** -8 **19.** 0 **21.** -41 **23.** 0 **25.** 7
27. -2 **29.** 11 **31.** -33 **33.** 0 **35.** 18 **37.** -45
39. 0 **41.** 20 **43.** -1.7 **45.** -9.1 **47.** $\frac{1}{5}$ **49.** $\frac{-6}{7}$
51. $-\frac{1}{15}$ **53.** $\frac{2}{9}$ **55.** -3 **57.** 0 **59.** The price rose 9¢.
61. Her new balance was $95. **63.** The total gain was
22 yd. **65.** Lyle owes $85. **67.** The elevation of the peak
is 13,796 ft. **69.** $14a$ **71.** $9x$ **73.** $13t$ **75.** $-2m$
77. $-7a$ **79.** $1 - 2x$ **81.** $12x + 17$ **83.** $7r + 8t + 16$
85. $18n + 16$ **87.** 🖊 **89.** $21z + 7y + 14$ **90.** $\frac{28}{3}$ **91.** 🖊
93. $65.25 **95.** $-5y$ **97.** $-7m$ **99.** $-7t, -23$
101. 1 under par

Exercise Set 1.6, pp. 51–54

1. (d) **3.** (f) **5.** (a) **7.** (b) **9.** Four minus ten
11. Two minus negative nine **13.** Nine minus the

opposite of t **15.** The opposite of x minus y
17. Negative three minus the opposite of n **19.** -39
21. 9 **23.** 3.14 **25.** -23 **27.** $\frac{14}{3}$ **29.** -0.101 **31.** 72
33. $-\frac{2}{5}$ **35.** 1 **37.** -7 **39.** -2 **41.** -5 **43.** -6
45. -10 **47.** -6 **49.** 0 **51.** -5 **53.** -10 **55.** 2
57. 0 **59.** 0 **61.** 8 **63.** -11 **65.** 16 **67.** -19
69. -1 **71.** 17 **73.** -5 **75.** -3 **77.** -21 **79.** 5
81. -8 **83.** 10 **85.** -23 **87.** -68 **89.** -58
91. -5.5 **93.** -0.928 **95.** $-\frac{7}{11}$ **97.** $-\frac{4}{5}$ **99.** $\frac{5}{17}$
101. $3.8 - (-5.2); 9$ **103.** $114 - (-79); 193$ **105.** -58
107. 34 **109.** 41 **111.** -62 **113.** -139 **115.** 0
117. $-7x, -4y$ **119.** $9, -5t, -3st$ **121.** $-3x$
123. $-5a + 4$ **125.** $-7n - 9$ **127.** $-6x + 5$
129. $-8t - 7$ **131.** $-12x + 3y + 9$ **133.** $8x + 66$
135. 150°C **137.** 30,384 ft **139.** 116 m **141.** 🖊
143. 432 ft² **144.** $2 \cdot 2 \cdot 2 \cdot 2 \cdot 2 \cdot 3 \cdot 3 \cdot 3$ **145.** 🖊
147. 11:00 P.M., August 14 **149.** False. For example,
let $m = -3$ and $n = -5$. Then $-3 > -5$, but
$-3 + (-5) = -8 \not> 0$. **151.** True. For example, for $m = 4$
and $n = -4$, $4 = -(-4)$ and $4 + (-4) = 0$; for $m = -3$ and
$n = 3$, $-3 = -3$ and $-3 + 3 = 0$.
153. ⊝ ⑨ ⊖ ⊝ ⑦ ENTER

Exercise Set 1.7, pp. 60–62

1. 1 **3.** 0 **5.** 0 **7.** 1 **9.** 1 **11.** -24 **13.** -56
15. -24 **17.** -72 **19.** 42 **21.** 45 **23.** 190 **25.** -144
27. 1200 **29.** 98 **31.** -78 **33.** 21.7 **35.** $-\frac{2}{5}$ **37.** $\frac{1}{12}$
39. -11.13 **41.** $-\frac{5}{12}$ **43.** 252 **45.** 0 **47.** $\frac{1}{28}$ **49.** 150
51. 0 **53.** -720 **55.** $-30,240$ **57.** -7 **59.** -4 **61.** -7
63. 4 **65.** -8 **67.** 2 **69.** -12 **71.** -8 **73.** Undefined
75. -4 **77.** 0 **79.** 0 **81.** $-\frac{8}{3}; \frac{8}{-3}$ **83.** $-\frac{29}{35}; \frac{-29}{35}$
85. $\frac{-7}{3}; \frac{7}{-3}$ **87.** $-\frac{x}{2}; \frac{x}{-2}$ **89.** $-\frac{5}{4}$ **91.** $-\frac{13}{47}$ **93.** $-\frac{1}{10}$
95. $\frac{1}{4.3}$, or $\frac{10}{43}$ **97.** $-\frac{4}{9}$ **99.** Does not exist **101.** $\frac{21}{20}$
103. $\frac{12}{55}$ **105.** -1 **107.** 1 **109.** $-\frac{9}{11}$ **111.** $-\frac{7}{4}$
113. -12 **115.** -3 **117.** 1 **119.** 7 **121.** $-\frac{2}{9}$ **123.** $\frac{1}{10}$
125. $-\frac{7}{6}$ **127.** $\frac{6}{7}$ **129.** $-\frac{14}{15}$ **131.** 🖊 **133.** $\frac{22}{39}$
134. $12x - 2y - 9$ **135.** 🖊 **137.** For 2 and 3, the
reciprocal of the sum is $1/(2 + 3)$ or $1/5$. But
$1/5 \neq 1/2 + 1/3$. **139.** Negative **141.** Negative
143. Negative **145.** (a) m and n have different signs;
(b) either m or n is zero; (c) m and n have the same sign.
147. 🖊

Exercise Set 1.8, pp. 70–72

1. (a) Division; (b) subtraction; (c) addition;
(d) multiplication; (e) subtraction; (f) multiplication
3. 2^3 **5.** x^7 **7.** $(3t)^5$ **9.** 9 **11.** 16 **13.** -16 **15.** 64
17. 625 **19.** 7 **21.** $81t^4$ **23.** $-343x^3$ **25.** 26 **27.** 86
29. 7 **31.** 5 **33.** 1 **35.** 298 **37.** 11 **39.** -36
41. 1291 **43.** 14 **45.** 152 **47.** 36 **49.** 1 **51.** -26

53. -2 **55.** $-\frac{9}{2}$ **57.** -11 **59.** -3 **61.** -15 **63.** 9
65. 30 **67.** 6 **69.** -17 **71.** $-9x - 1$ **73.** $-5 + 6x$
75. $-4a + 3b - 7c$ **77.** $-3x^2 - 5x + 1$ **79.** $2x - 7$
81. $-3a + 9$ **83.** $5x - 6$ **85.** $-3t - 11r$ **87.** $9y - 25z$
89. $x^2 + 2$ **91.** $-t^3 - 2t$ **93.** $37a^2 - 23ab + 35b^2$
95. $-22t^3 - t^2 + 9t$ **97.** $2x - 25$ **99.** 🗒 **101.** Let n
represent the number; $2n + 9$, or $9 + 2n$ **102.** Let m and
n represent the two numbers; $\frac{1}{2}(m + n)$ **103.** 🗒
105. $-6r - 5t + 21$ **107.** $-2x - f$ **109.** 🗒 **111.** True
113. False **115.** 0 **117.** $39{,}000$ **119.** $44x^3$

Review Exercises: Chapter 1, pp. 75–77

1. True **2.** True **3.** False **4.** True **5.** False **6.** False
7. True **8.** False **9.** False **10.** True **11.** 15 **12.** 4
13. -7 **14.** -5 **15.** $z - 7$ **16.** xz **17.** Let m and n
represent the numbers; $mn + 1$, or $1 + mn$ **18.** No
19. Let d represent the number of digital photos taken in
2003, in billions; $29.8 = d + 18.49$ **20.** $t \cdot 3 + 5$
21. $2x + (y + z)$ **22.** $(4x)y$, $4(yx)$, $(4y)x$; answers may vary
23. $18x + 30y$ **24.** $40x + 24y + 16$ **25.** $3(7x + 5y)$
26. $7(5x + 2 + y)$ **27.** $2 \cdot 2 \cdot 13$ **28.** $\frac{5}{12}$ **29.** $\frac{9}{4}$ **30.** $\frac{31}{36}$
31. $\frac{3}{16}$ **32.** $\frac{3}{5}$ **33.** $\frac{72}{25}$ **34.** $-45, 72$
35.

$$\overset{\frac{-1}{3}}{\underset{-5\ -4\ -3\ -2\ -1\ \ 0\ \ 1\ \ 2\ \ 3\ \ 4\ \ 5}{\longleftrightarrow}}$$

36. $x > -3$

37. True **38.** False **39.** -0.875 **40.** 1 **41.** -9
42. -3 **43.** $-\frac{7}{12}$ **44.** 0 **45.** -5 **46.** 5 **47.** $-\frac{7}{5}$
48. -7.9 **49.** 54 **50.** -9.18 **51.** $-\frac{2}{7}$ **52.** -140
53. -7 **54.** -3 **55.** $\frac{3}{4}$ **56.** 92 **57.** 62 **58.** 48
59. 168 **60.** $\frac{21}{8}$ **61.** $\frac{103}{17}$ **62.** $7a - 3b$ **63.** $-2x + 5y$
64. 7 **65.** $-\frac{1}{7}$ **66.** $(2x)^4$ **67.** $-125x^3$ **68.** $-3a + 9$
69. $-2b + 21$ **70.** $-3x + 9$ **71.** $12y - 34$ **72.** $5x + 24$
73. 🗒 The value of a constant never varies. A variable can
represent a variety of numbers. **74.** 🗒 A term is one of
the parts of an expression that is separated from the other
parts by plus signs. A factor is part of a product.
75. 🗒 The distributive law is used in factoring algebraic
expressions, multiplying algebraic expressions, combining
like terms, finding the opposite of a sum, and subtracting
algebraic expressions. **76.** 🗒 A negative number raised
to an even power is positive; a negative number raised to
an odd power is negative. **77.** $25{,}281$
78. (a) $\frac{3}{11}$; (b) $\frac{10}{11}$ **79.** $-\frac{5}{8}$ **80.** -2.1 **81.** (i) **82.** (j)
83. (a) **84.** (h) **85.** (k) **86.** (b) **87.** (c) **88.** (e)
89. (d) **90.** (f) **91.** (g)

Test: Chapter 1, p. 78

1. [1.1] 4 **2.** [1.1] Let x represent the number; $x - 9$
3. [1.1] $240\,\text{ft}^2$ **4.** [1.2] $q + 3p$ **5.** [1.2] $(x \cdot 4) \cdot y$
6. [1.1] No **7.** [1.1] Let p represent the maximum
production capability; $p - 282 = 2518$ **8.** [1.2] $35 - 7x$
9. [1.7] $-5y + 10$ **10.** [1.2] $11(1 - 4x)$

11. [1.2] $7(x + 3 + 2y)$ **12.** [1.3] $2 \cdot 2 \cdot 3 \cdot 5 \cdot 5$
13. [1.3] $\frac{2}{7}$ **14.** [1.4] $<$ **15.** [1.4] $>$ **16.** [1.4] $\frac{9}{4}$
17. [1.4] 2.7 **18.** [1.6] $-\frac{2}{3}$ **19.** [1.7] $-\frac{7}{4}$ **20.** [1.6] 8
21. [1.4] $-2 \geq x$ **22.** [1.6] 7.8 **23.** [1.5] -8
24. [1.6] -2.5 **25.** [1.6] $\frac{7}{8}$ **26.** [1.7] -48 **27.** [1.7] $\frac{3}{16}$
28. [1.7] -6 **29.** [1.7] $\frac{3}{4}$ **30.** [1.7] -9.728 **31.** [1.8] -173
32. [1.6] 15 **33.** [1.8] -4 **34.** [1.8] 448
35. [1.6] $21a + 22y$ **36.** [1.8] $16x^4$ **37.** [1.8] $x + 7$
38. [1.8] $9a - 12b - 7$ **39.** [1.8] $68y - 8$ **40.** [1.1] 5
41. [1.8] $9 - (3 - 4) + 5 = 15$ **42.** [1.8] 15 **43.** [1.8] $4a$
44. [1.8] False

CHAPTER 2

Exercise Set 2.1, pp. 87–88

1. (f) **3.** (a) **5.** (d) **7.** 17 **9.** -11 **11.** -21
13. -31 **15.** 13 **17.** 19 **19.** -4 **21.** $\frac{7}{3}$ **23.** $-\frac{13}{10}$
25. $\frac{41}{24}$ **27.** $-\frac{1}{20}$ **29.** 1.5 **31.** -5 **33.** 14 **35.** 4
37. 12 **39.** -23 **41.** 8 **43.** -7 **45.** 8 **47.** -88
49. 20 **51.** -54 **53.** $\frac{5}{9}$ **55.** 1 **57.** $\frac{9}{2}$ **59.** -7.6
61. -2.5 **63.** -15 **65.** 18 **67.** -6 **69.** -128
71. $-\frac{1}{2}$ **73.** -15 **75.** 12 **77.** 310.756 **79.** 🗒
81. -34 **82.** 41 **83.** 1 **84.** -16 **85.** 🗒 **87.** 9.4
89. 2 **91.** $-13, 13$ **93.** 9000 **95.** 250

Technology Connection, p. 92

1.

X	Y1	
0	5	
1	4	
2	3	
3	2	
4	1	
5	0	
6	−1	
X = 0		

2.

X	Y1	Y2
0	5	17
1	4	13
2	3	9
3	2	5
4	1	1
5	0	−3
6	−1	−7
X = 0		

3. 4; not reliable because, depending on the choice of
ΔTbl, it is easy to scroll past a solution without realizing it.

Exercise Set 2.2, pp. 95–97

1. (c) **3.** (a) **5.** (b) **7.** 8 **9.** 7 **11.** 5 **13.** 14
15. -7 **17.** -11 **19.** -24 **21.** 19 **23.** $\frac{10}{9}$ **25.** 3
27. 15 **29.** -4 **31.** $-\frac{28}{3}$ **33.** All real numbers; identity
35. -3 **37.** 5 **39.** 2 **41.** 0 **43.** 8 **45.** 0 **47.** 10
49. 4 **51.** 0 **53.** 2 **55.** -8 **57.** 2 **59.** No solution;
contradiction **61.** $-\frac{2}{5}$ **63.** $\frac{64}{3}$ **65.** $\frac{2}{5}$ **67.** 3 **69.** -4
71. $1.\overline{6}$ **73.** $-\frac{40}{37}$ **75.** 11 **77.** 6 **79.** $\frac{16}{15}$ **81.** $-\frac{51}{31}$ **83.** 2
85. 🗒 **87.** -7 **88.** 15 **89.** -15 **90.** -28 **91.** 🗒
93. $\dfrac{1136}{909}$, or $1.\overline{2497}$ **95.** No solution; contradiction
97. No solution; contradiction **99.** $\frac{2}{3}$ **101.** 0 **103.** 0
105. -2

Technology Connection, p. 98

1. 72,930

Exercise Set 2.3, pp. 101–104

1. 2 mi **3.** 1423 students **5.** 54,000 Btu's **7.** 255 mg
9. $b = \dfrac{A}{h}$ **11.** $r = \dfrac{d}{t}$ **13.** $P = \dfrac{I}{rt}$ **15.** $m = 65 - H$
17. $l = \dfrac{P - 2w}{2}$, or $l = \dfrac{P}{2} - w$ **19.** $\pi = \dfrac{A}{r^2}$ **21.** $h = \dfrac{2A}{b}$
23. $m = \dfrac{E}{c^2}$ **25.** $d = 2Q - c$ **27.** $b = 3A - a - c$
29. $A = Ms$ **31.** $C = \frac{5}{9}(F - 32)$ **33.** $t = \dfrac{A}{a + b}$
35. $h = \dfrac{2A}{a + b}$ **37.** $L = W - \dfrac{N(R - r)}{400}$, or
$L = \dfrac{400W - NR + Nr}{400}$ **39.** 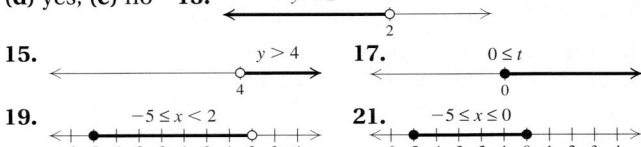 **41.** 0 **42.** 9.18
43. -13 **44.** 65 **45.** **47.** 35 yr **49.** 27 in³
51. $a = \dfrac{w}{c} \cdot d$ **53.** $c = \dfrac{d}{a - b}$ **55.** $a = \dfrac{c}{3 + b + d}$
57. $K = 917 + 13.2276w + 2.3622h - 6a$

Exercise Set 2.4, pp. 109–113

1. (d) **3.** (e) **5.** (c) **7.** (f) **9.** (b) **11.** 0.3 **13.** 0.02
15. 0.77 **17.** 0.09 **19.** 0.6258 **21.** 0.007 **23.** 1.25
25. 64% **27.** 10.6% **29.** 42% **31.** 90% **33.** 0.49%
35. 108% **37.** 230% **39.** 80% **41.** 32% **43.** 25%
45. 24% **47.** $46\frac{2}{3}$, or $\frac{140}{3}$ **49.** 2.5 **51.** 84 **53.** 125%
55. 0.8 **57.** 50% **59.** $198 **61.** $1584 **63.** $528
65. 75 credits **67.** About 626 at-bats **69. (a)** 16%;
(b) $29 **71.** About 72%; about 28% **73.** $280
75. 285 women **77.** $18/hr **79.** 150% **81.** $36
83. $148.50 **85.** About 31.5 lb **87.** 7410 brochures
89. About 165 calories **91.** **93.** Let n represent the
number; $n + 5$ **94.** Let t represent Tino's weight; $t - 4$
95. $8 \cdot 2a$ **96.** Let x and y represent the two numbers;
$xy + 1$ **97.** **99.** 18,500 people **101.** About 5 ft 6 in.
103. About 27% **105.**

Exercise Set 2.5, pp. 122–127

1. 8 **3.** 11 **5.** $85 **7.** $85 **9.** Approximately $62\frac{2}{3}$ mi
11. 190 mi **13.** 1204 and 1205 **15.** 396 and 398
17. 19, 20, 21 **19.** Bride: 102 yr; groom: 83 yr
21. Bathrooms: $11\frac{2}{3}$ billion; kitchens: $23\frac{1}{3}$ billion
23. 140 and 141 **25.** Width: 100 ft; length: 160 ft; area:
16,000 ft² **27.** Width: 50 ft; length: 84 ft **29.** $1\frac{3}{4}$ in. by
$3\frac{1}{2}$ in. **31.** 30°, 90°, 60° **33.** 95° **35.** Bottom: 144 ft;
middle: 72 ft; top: 24 ft **37.** $10\frac{1}{16}$ mi **39.** $128\frac{1}{3}$ mi

41. 65°, 25° **43.** Length: 27.9 cm; width: 21.6 cm
45. $6600 **47.** 1049 points **49.** 160 chirps per minute
51. **53.** < **54.** < **55.** < **56.** > **57.**
59. $37 **61.** 20 **63.** Half-dollars: 5; quarters: 10; dimes:
20; nickels: 60 **65.** 120 apples **67.** 30 games **69.** 76
71. **73.** Width: 23.31 cm; length: 27.56 cm

Exercise Set 2.6, pp. 133–135

1. ≥ **3.** < **5.** Equivalent **7.** Equivalent **9. (a)** Yes;
(b) yes; **(c)** yes; **(d)** no; **(e)** yes **11. (a)** No; **(b)** no; **(c)** yes;
(d) yes; **(e)** no **13.**
15.
17.
19.
21.
23. $\{x \mid x > -4\}$ **25.** $\{x \mid x \le 2\}$ **27.** $\{x \mid x < -1\}$
29. $\{x \mid x \ge 0\}$ **31.** $\{y \mid y > 3\}$,
33. $\{x \mid x \le -21\}$,
35. $\{x \mid x < 17\}$,
37. $\{y \mid y > -6\}$,
39. $\{x \mid x \le 9\}$,
41. $\{y \mid y \le \frac{1}{2}\}$,
43. $\{t \mid t > \frac{5}{8}\}$,
45. $\{x \mid x < 0\}$,
47. $\{t \mid t < 23\}$,
49. $\{x \mid x < 7\}$,
51. $\{x \mid x > -\frac{13}{7}\}$,
53. $\{t \mid t < -3\}$,
55. $\{y \mid y \ge -\frac{2}{7}\}$,
57. $\{y \mid y \ge -\frac{1}{10}\}$,
59. $\{x \mid x > \frac{4}{5}\}$,
61. $\{x \mid x < 9\}$ **63.** $\{y \mid y \ge 4\}$ **65.** $\{t \mid t \le 7\}$
67. $\{y \mid y < -4\}$ **69.** $\{x \mid x > -4\}$ **71.** $\{y \mid y < -\frac{10}{3}\}$
73. $\{x \mid x > -10\}$ **75.** $\{y \mid y < 2\}$ **77.** $\{y \mid y \ge 3\}$
79. $\{x \mid x > -4\}$ **81.** $\{x \mid x > -4\}$ **83.** $\{n \mid n \ge 70\}$
85. $\{x \mid x \le 15\}$ **87.** $\{t \mid t < 14\}$ **89.** $\{y \mid y < 6\}$
91. $\{t \mid t \le -4\}$ **93.** $\{r \mid r > -3\}$ **95.** $\{x \mid x \ge 8\}$
97. $\{x \mid x < \frac{11}{18}\}$ **99.** **101.** Let n represent "some
number"; $3 + n$ **102.** Let x and y represent the two

numbers; $2(x + y)$ **103.** Let x represent the number; $2x - 3$ **104.** Let y represent the number; $2y + 5$ **105.** 📝 **107.** $\{x \mid x \text{ is a real number}\}$ **109.** $\{x \mid x \le \frac{5}{6}\}$ **111.** $\{x \mid x > 7\}$ **113.** $\{x \mid x < \frac{y - b}{a}\}$ **115.** $\{x \mid x \text{ is a real number}\}$

Exercise Set 2.7, pp. 138–142

1. $b \le a$ **3.** $a \le b$ **5.** $b \le a$ **7.** $b < a$ **9.** Let n represent the number; $n \ge 8$ **11.** Let t represent the temperature; $t \le -3$ **13.** Let p represent the price of Pat's PT Cruiser; $p > 21{,}900$ **15.** Let d represent the distance to Normandale Community College; $d \le 15$ **17.** Let n represent the number; $n > -2$ **19.** Let p represent the number of people attending the Million Man March; $400{,}000 < p < 1{,}200{,}000$ **21.** More than 2.5 hr **23.** More than 18 one-way trips per month **25.** Scores greater than or equal to 97 **27.** 8 credits or more **29.** 21 calls or more **31.** Lengths greater than 6 cm **33.** Depths less than 437.5 ft **35.** Blue-book value is greater than or equal to \$10,625 **37.** Lengths greater than or equal to 5 in. **39.** Temperatures greater than 37°C **41.** Heights at least 4 ft **43.** A serving contains at least 16 g of fat. **45.** Dates after September 16 **47.** 14 or fewer copies **49.** Years after 1995 **51.** Mileages less than or equal to 193 **53.** 📝 **55.** 2 **56.** $\frac{1}{2}$ **57.** $-\frac{10}{3}$ **58.** $-\frac{1}{5}$ **59.** 📝 **61.** Temperatures between -15°C and $-9\frac{4}{9}$°C **63.** Lengths less than or equal to 8 cm **65.** They contain at least 7.5 g of fat per serving. **67.** At least \$42 **69.** 📝

Review Exercises: Chapter 2, pp. 144–145

1. True **2.** False **3.** True **4.** True **5.** True **6.** False **7.** True **8.** True **9.** -25 **10.** 7 **11.** -65 **12.** 1 **13.** -5 **14.** 1.11 **15.** $\frac{1}{2}$ **16.** $-\frac{15}{64}$ **17.** $\frac{38}{5}$ **18.** -8 **19.** -5 **20.** $-\frac{1}{3}$ **21.** 4 **22.** 3 **23.** 4 **24.** 16 **25.** 7 **26.** $-\frac{7}{5}$ **27.** 12 **28.** No solution; contradiction **29.** $d = \frac{C}{\pi}$ **30.** $B = \frac{3V}{h}$ **31.** $b = 2A - a$ **32.** 0.009 **33.** 44% **34.** 70% **35.** 140 **36.** Yes **37.** No **38.** Yes **39.**

$$5x - 6 < 2x + 3$$

40.

$$-2 < x \le 5$$

41.

$$t > 0$$

42. $\{t \mid t \ge -\frac{1}{2}\}$ **43.** $\{x \mid x \ge 7\}$ **44.** $\{y \mid y > 3\}$ **45.** $\{y \mid y \le -4\}$ **46.** $\{x \mid x < -11\}$ **47.** $\{y \mid y > -7\}$ **48.** $\{x \mid x > -6\}$ **49.** $\{x \mid x > -\frac{9}{11}\}$ **50.** $\{t \mid t \le -12\}$ **51.** $\{x \mid x \le -8\}$ **52.** 20 bottles **53.** 15 ft, 17 ft **54.** \$126 billion **55.** 57, 59 **56.** Width: 11 cm; length: 17 cm **57.** \$160 **58.** \$46,987.95 **59.** 35°, 85°, 60° **60.** \$105 or less **61.** Widths greater than 17 cm

62. 📝 Multiplying both sides of an equation by *any* nonzero number results in an equivalent equation. When multiplying on both sides of an inequality, the sign of the number being multiplied by must be considered. If the number is positive, the direction of the inequality symbol remains unchanged; if the number is negative, the direction of the inequality symbol must be reversed to produce an equivalent inequality. **63.** 📝 The solutions of an equation can usually each be checked. The solutions of an inequality are normally too numerous to check. Checking a few numbers from the solution set found cannot guarantee that the answer is correct, although if any number does not check, the answer found is incorrect. **64.** 25 hr 39 min **65.** Nile: 6673 km; Amazon: 6440 km **66.** \$18,600 **67.** $-23, 23$ **68.** $-20, 20$ **69.** $a = \frac{y - 3}{2 - b}$

Test: Chapter 2, p. 146

1. [2.1] 9 **2.** [2.1] 15 **3.** [2.1] -3 **4.** [2.1] 49 **5.** [2.1] -12 **6.** [2.2] 2 **7.** [2.1] -8 **8.** [2.1] $-\frac{7}{20}$ **9.** [2.2] 7 **10.** [2.2] $-\frac{5}{3}$ **11.** [2.2] $\frac{23}{3}$ **12.** [2.2] All real numbers; identity **13.** [2.6] $\{x \mid x > -5\}$ **14.** [2.6] $\{x \mid x > -13\}$ **15.** [2.6] $\{x \mid x < \frac{21}{8}\}$ **16.** [2.6] $\{y \mid y \le -13\}$ **17.** [2.6] $\{y \mid y \le -8\}$ **18.** [2.6] $\{x \mid x \le -\frac{1}{20}\}$ **19.** [2.6] $\{x \mid x < -6\}$ **20.** [2.6] $\{x \mid x \le -1\}$ **21.** [2.3] $r = \frac{A}{2\pi h}$ **22.** [2.3] $l = 2w - P$ **23.** [2.4] 2.3 **24.** [2.4] 5.4% **25.** [2.4] 16 **26.** [2.4] 44% **27.** [2.6]

$$y < 4$$

28. [2.6]

$$-2 \le x \le 2$$

29. [2.5] Width: 7 cm; length: 11 cm **30.** [2.5] 60 mi **31.** [2.5] 81 mm, 83 mm, 85 mm **32.** [2.4] \$65 **33.** [2.7] Mileages less than or equal to 525.8 mi **34.** [2.3] $d = \frac{a}{3}$ **35.** [1.4], [2.2] $-15, 15$ **36.** [2.5] 60 tickets

CHAPTER 3

Exercise Set 3.1, pp. 154–158

1. (a) **3.** (b) **5.** 2 drinks **7.** The person weighs more than 200 lb. **9.** About 2,920,000 **11.** About 1,460,000 **13.** About 24.8 million tons **15.** About 2.9 million tons **17.** 120,000,000 phones **19.** 2004

21.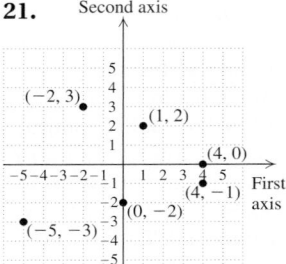

23.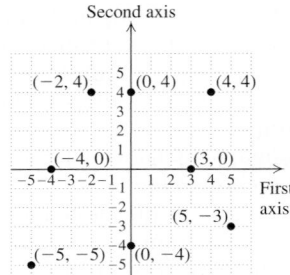

25.

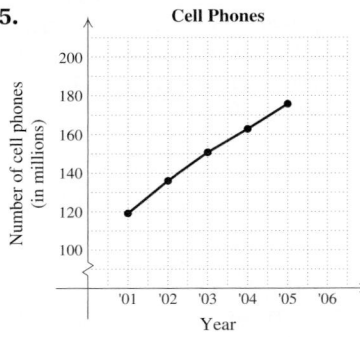

27. $A(-4, 5)$; $B(-3, -3)$; $C(0, 4)$; $D(3, 4)$; $E(3, -4)$
29. $A(4, 1)$; $B(0, -5)$; $C(-4, 0)$; $D(-3, -2)$; $E(3, 0)$
31. 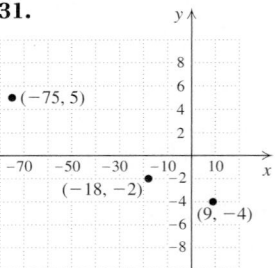   **33.**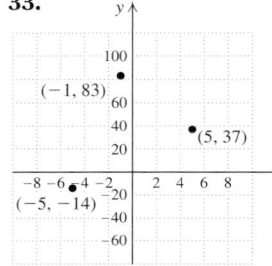

35. 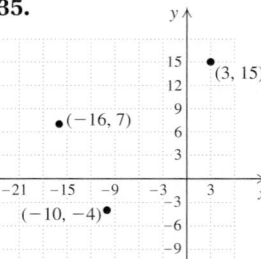   **37.**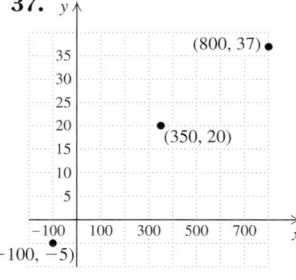

39. 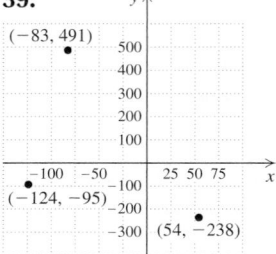   **41.** IV   **43.** III   **45.** I

47. II   **49.** I and IV   **51.** I and III   **53.**   **55.** -18
56. -31 **57.** 6   **58.** 1   **59.** $y = \frac{3}{2}x - 3$
60. $y = \frac{7}{4}x - \frac{7}{2}$ **61.**   **63.** II or IV   **65.** $(-1, -5)$
67. 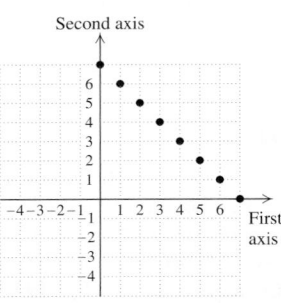   **69.** $\dfrac{65}{2}$ sq units

71. Latitude 27°North; longitude 81° West   **73.**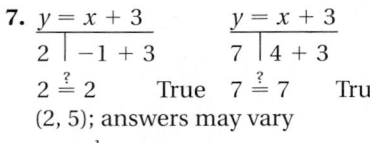

Technology Connection, p. 165

1. $y = -5x + 6.5$ **2.** $y = 3x + 4.5$

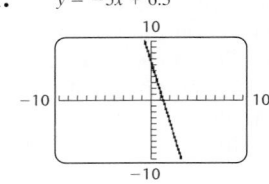

3. $7y - 4x = 22$, or $y = \frac{4}{7}x + \frac{22}{7}$ **4.** $5y + 11x = -20$, or $y = -\frac{11}{5}x - 4$

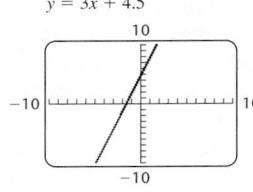

5. $2y - x^2 = 0$, or $y = 0.5x^2$ **6.** $y + x^2 = 8$, or $y = -x^2 + 8$

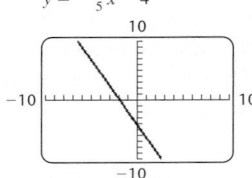

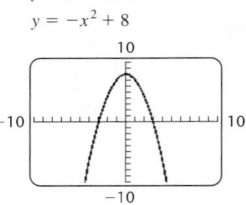

Exercise Set 3.2, pp. 166–168

1. No   **3.** No   **5.** Yes   **7.**

$\begin{array}{c|c} y = x + 3 \\ \hline 2 & -1 + 3 \\ & 2 \overset{?}{=} 2 \quad \text{True} \end{array}$ $\begin{array}{c|c} y = x + 3 \\ \hline 7 & 4 + 3 \\ & 7 \overset{?}{=} 7 \quad \text{True} \end{array}$

$(2, 5)$; answers may vary

9.

$\begin{array}{c|c} y = \frac{1}{2}x + 3 \\ \hline 5 & \frac{1}{2} \cdot 4 + 3 \\ & 2 + 3 \\ & 5 \overset{?}{=} 5 \quad \text{True} \end{array}$ $\begin{array}{c|c} y = \frac{1}{2}x + 3 \\ \hline 2 & \frac{1}{2}(-2) + 3 \\ & -1 + 3 \\ & 2 \overset{?}{=} 2 \quad \text{True} \end{array}$

$(0, 3)$; answers may vary

11.

$y + 3x = 7$	
$1 + 3 \cdot 2$	7
$1 + 6$	

$7 \overset{?}{=} 7$ True

$y + 3x = 7$	
$-5 + 3 \cdot 4$	7
$-5 + 12$	

$7 \overset{?}{=} 7$ True

$(1, 4)$; answers may vary

13.

$4x - 2y = 10$	
$4 \cdot 0 - 2(-5)$	10
$0 + 10$	

$10 \overset{?}{=} 10$ True

$4x - 2y = 10$	
$4 \cdot 4 - 2 \cdot 3$	10
$16 - 6$	

$10 \overset{?}{=} 10$ True

$(2, -1)$; answers may vary

15.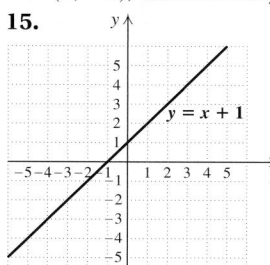

17.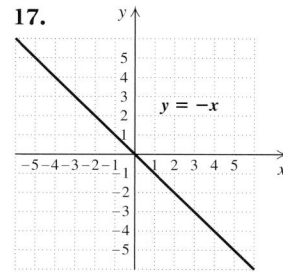

19.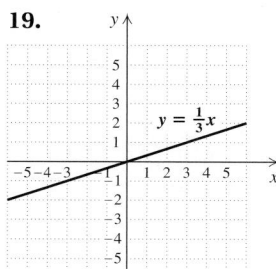

21.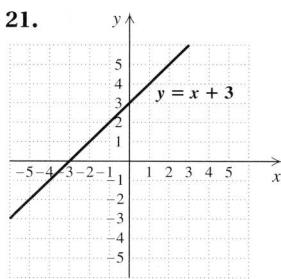

23.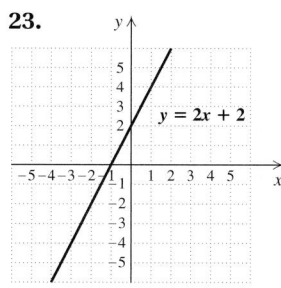

25.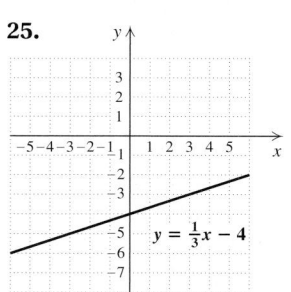

27.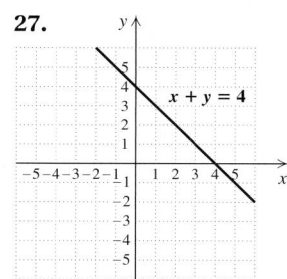

29.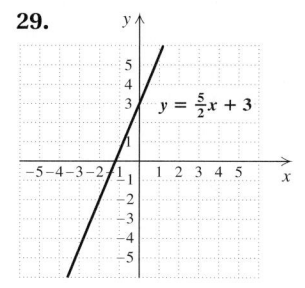

31.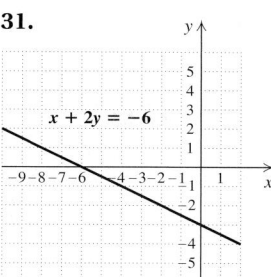

33.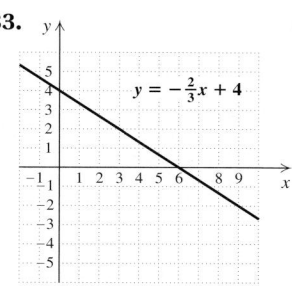

35.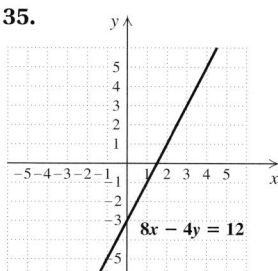

37.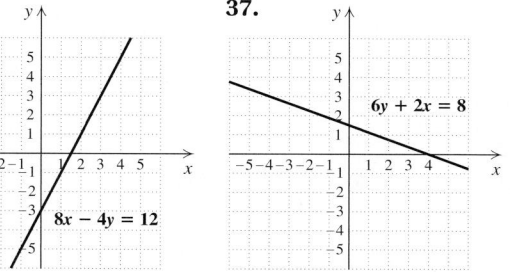

39. 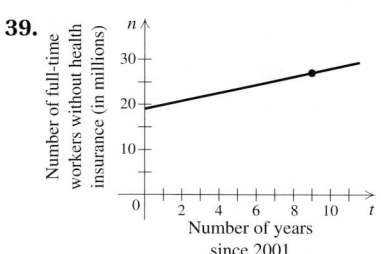   27 million workers

41. 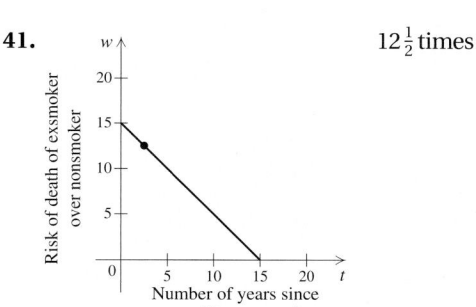 $12\frac{1}{2}$ times

43. 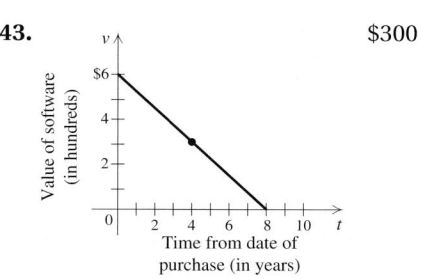 $300

45. 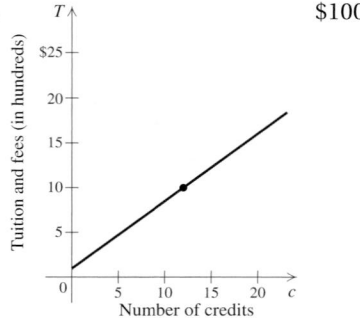 $1000

47. 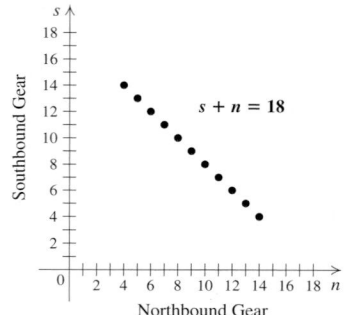   24°F   **49.** 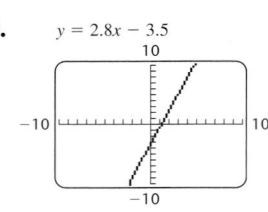   **51.** $\dfrac{12}{5}$

52. $\dfrac{9}{2}$ **53.** $-\dfrac{5}{2}$ **54.** $p = \dfrac{w}{q+1}$ **55.** $y = \dfrac{C - Ax}{B}$

56. $Q = 2A - T$ **57.**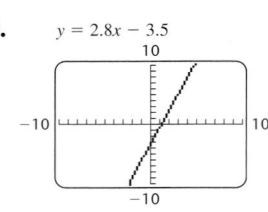

59. $s + n = 18$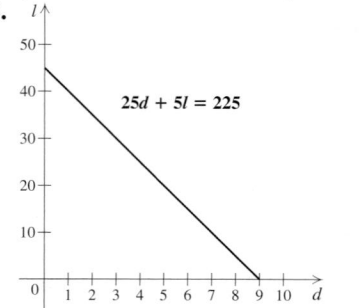

61. $x + y = 2$, or $y = -x + 2$ **63.** $5x - 3y = 15$, or $y = \dfrac{5}{3}x - 5$ **65.**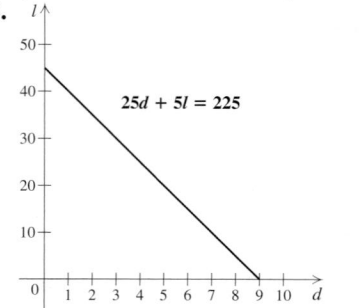

Answers may vary. 1 dinner, 40 lunches;
5 dinners, 20 lunches; 8 dinners, 5 lunches

67. 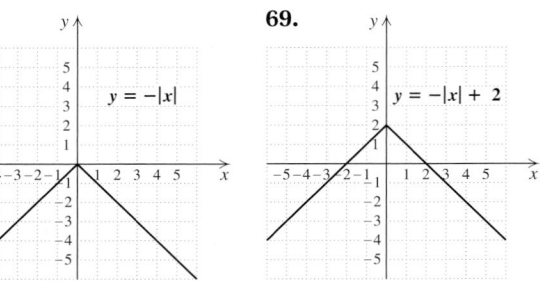 $y = -|x|$

69. 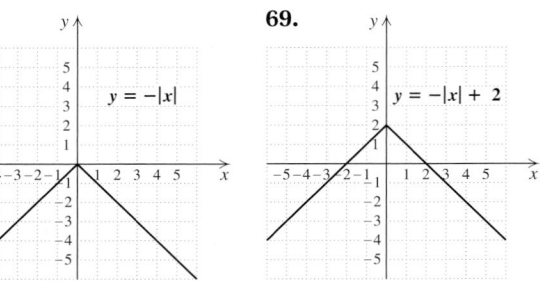 $y = -|x| + 2$

71. $y = -2.8x + 3.5$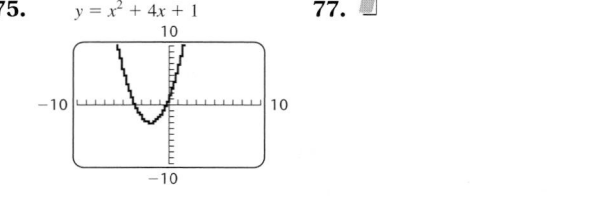

73. $y = 2.8x - 3.5$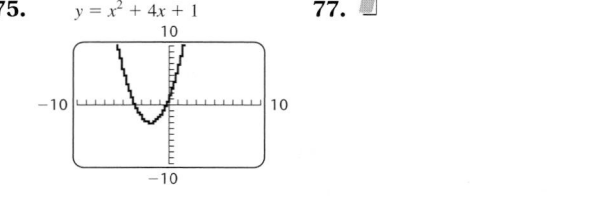

75. $y = x^2 + 4x + 1$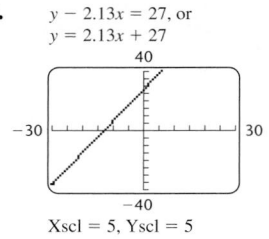

77.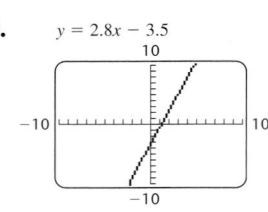

Technology Connection, p. 172

1. $y = -0.72x - 15$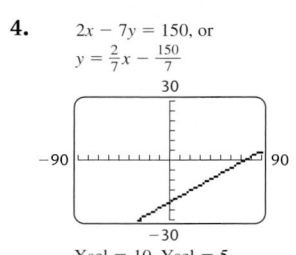
Xscl = 5, Yscl = 5

2. $y - 2.13x = 27$, or $y = 2.13x + 27$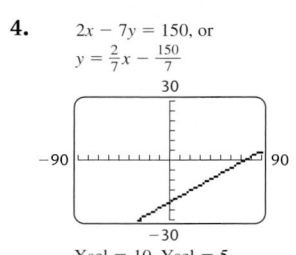
Xscl = 5, Yscl = 5

3. $5x + 6y = 84$, or $y = -\dfrac{5}{6}x + 14$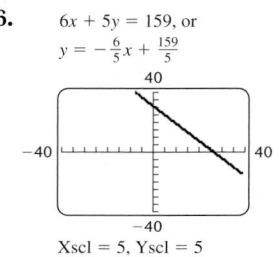
Xscl = 5, Yscl = 5

4. $2x - 7y = 150$, or $y = \dfrac{2}{7}x - \dfrac{150}{7}$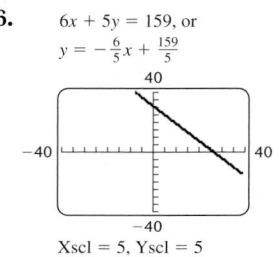
Xscl = 10, Yscl = 5

5. $19x - 17y = 200$, or $y = \dfrac{19}{17}x - \dfrac{200}{17}$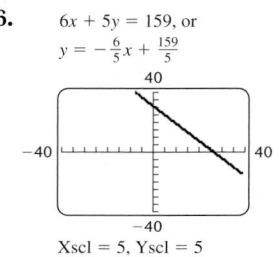

6. $6x + 5y = 159$, or $y = -\dfrac{6}{5}x + \dfrac{159}{5}$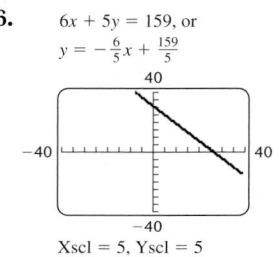
Xscl = 5, Yscl = 5

Exercise Set 3.3, pp. 175–177

1. (f)   **3.** (d)   **5.** (b)   **7. (a)** $(0, 5)$; **(b)** $(2, 0)$
9. (a) $(0, -4)$; **(b)** $(3, 0)$ **11. (a)** $(0, -2)$; **(b)** $(-3, 0)$
13. (a) $(0, 5)$; **(b)** $(3, 0)$ **15. (a)** $(0, -14)$; **(b)** $(4, 0)$
17. (a) $(0, 50)$; **(b)** $\left(-\frac{75}{2}, 0\right)$ **19. (a)** $(0, 9)$; **(b)** none
21. (a) None; **(b)** $(-7, 0)$

23.
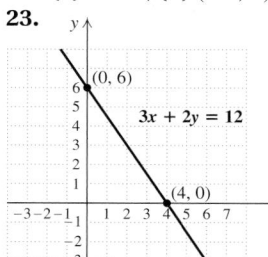

25.
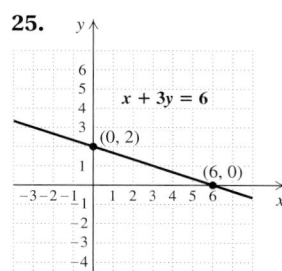

27.
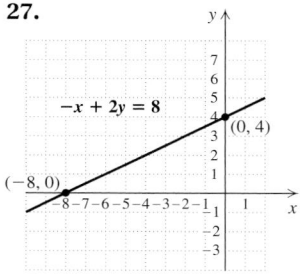

29.
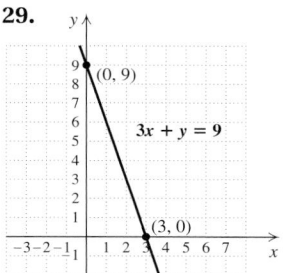

31.
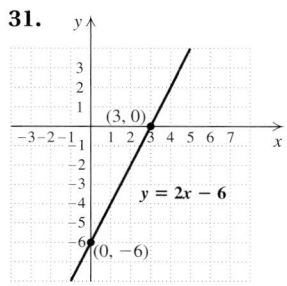

33.
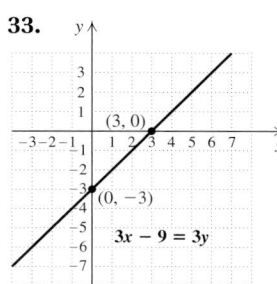

35.
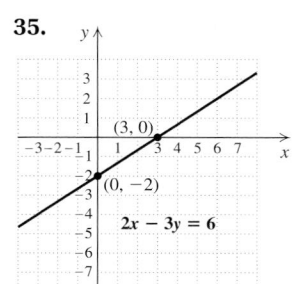

37.
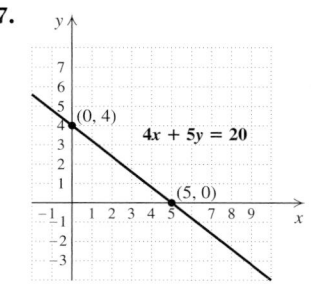

39.
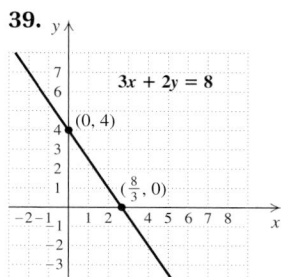

41.
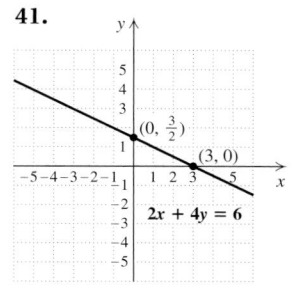

43.
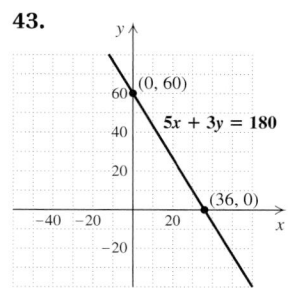

45.
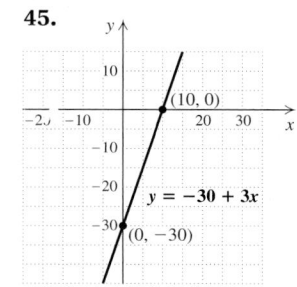

47.
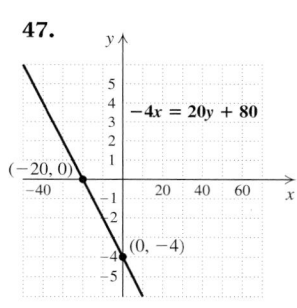

49.
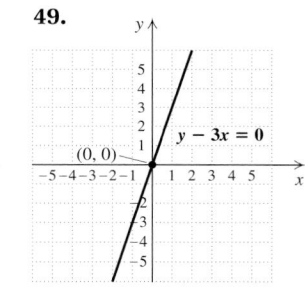

51.
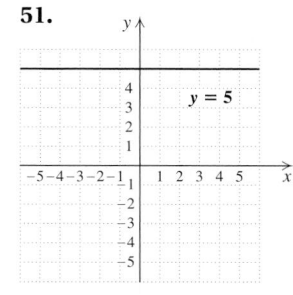

53.
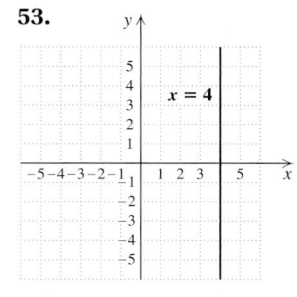

55.
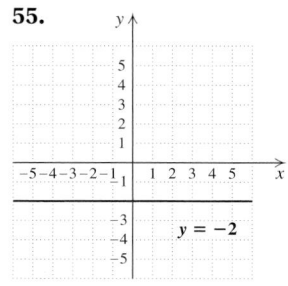

57.
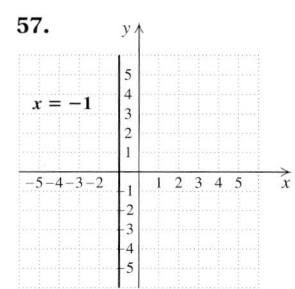

59.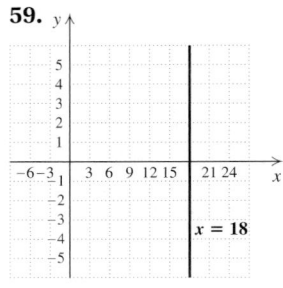

61.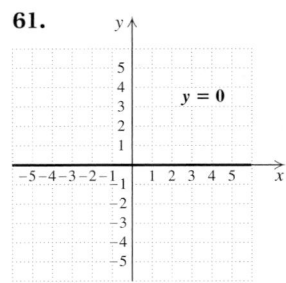

63.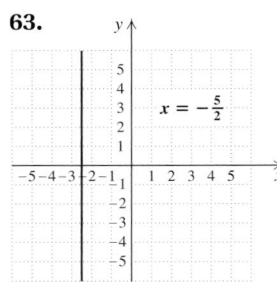

65.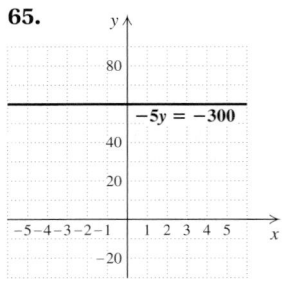

67.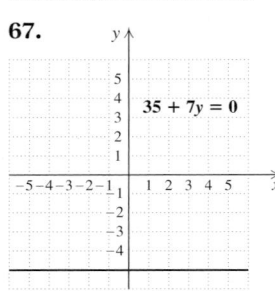

69. $y = -1$ **71.** $x = 4$

73. $y = 0$ **75.** 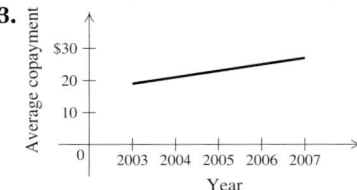 **77.** $d - 7$ **78.** $w + 5$, or $5 + w$
79. Let n represent the number; $2 + n$ **80.** Let n
represent the number; $3n$ **81.** Let x and y represent the
numbers; $2(x + y)$ **82.** Let a and b represent the
numbers; $\frac{1}{2}(a + b)$ **83.** ▨ **85.** $y = 0$ **87.** $x = -2$
89. $(-3, -3)$ **91.** $-5x + 3y = 15$, or $y = \frac{5}{3}x + 5$
93. -24 **95.** $(0, 25)$; $\left(\frac{50}{3}, 0\right)$, or $(16.\overline{6}, 0)$ **97.** $(0, -9)$;
$(45, 0)$ **99.** $\left(0, -\frac{1}{20}\right)$, or $(0, -0.05)$; $\left(\frac{1}{25}, 0\right)$, or $(0.04, 0)$

Exercise Set 3.4, pp. 181–186

1. (a) 21 mpg; (b) \$39.33/day; (c) 91 mi/day; (d) 43¢/mile
3. (a) 6 mph; (b) \$4/hr; (c) \$0.67/mi **5.** (a) \$16/hr;
(b) 4.5 pages/hr; (c) \$3.56/page **7.** \$16/yr
9. (a) 14.5 floors/min; (b) 4.14 sec/floor
11. (a) 3.71 ft/min; (b) 0.27 min/ft
13.

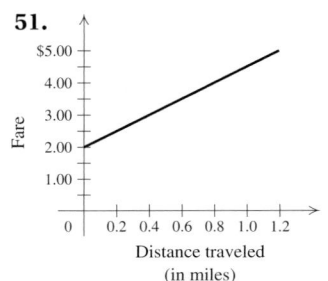

15.

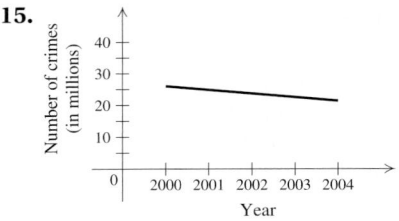

17.

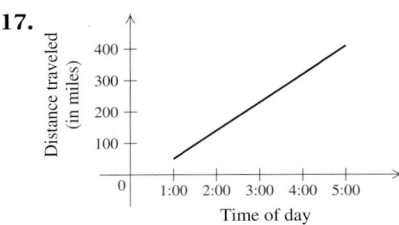

19.

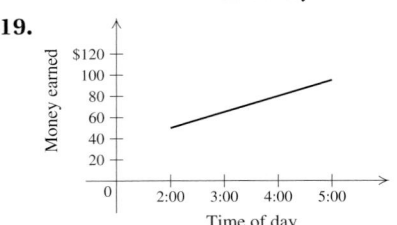

21.

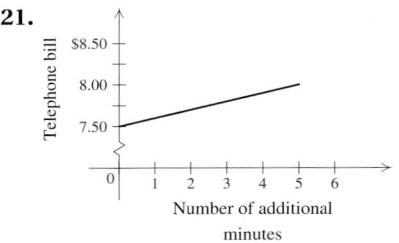

23. 2 haircuts/hr **25.** 75 mi/hr **27.** 7¢/min
29. −\$500/yr **31.** 0.04 gal/mi **33.** (e) **35.** (d)
37. (b) **39.** ▨ **41.** 5 **42.** −6 **43.** −1 **44.** $-\frac{4}{3}$
45. $-\frac{4}{3}$ **46.** $-\frac{4}{5}$ **47.** ▨ **49.**

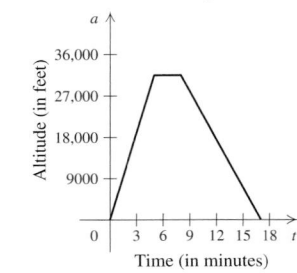

51.   **53.** 13 ft/sec

55. About 41.7 min   **57.** 4:20 P.M.

Exercise Set 3.5, pp. 194–200

1. Positive   **3.** Negative   **5.** Positive   **7.** Zero
9. Negative   **11.** 2.5 million people/yr   **13.** 1.3%/yr
15. 1 point/\$1000 income   **17.** $-2.1°$/min   **19.** $\frac{3}{4}$
21. $\frac{3}{2}$ **23.** $\frac{1}{3}$ **25.** -1 **27.** 0   **29.** $-\frac{1}{3}$ **31.** Undefined
33. $-\frac{1}{4}$ **35.** $\frac{3}{2}$ **37.** 0   **39.** -3 **41.** $\frac{3}{2}$ **43.** $-\frac{4}{5}$ **45.** $\frac{7}{9}$
47. $-\frac{2}{3}$ **49.** $-\frac{1}{2}$ **51.** 0   **53.** $-\frac{11}{6}$ **55.** Undefined
57. Undefined   **59.** 0   **61.** Undefined   **63.** 0   **65.** 8%
67. 8.$\overline{3}$%   **69.** $\frac{29}{98}$, or about 30%   **71.** About 29%   **73.** 🗒
75. $y = \dfrac{c - ax}{b}$ **76.** $r = \dfrac{p + mn}{x}$ **77.** $y = \dfrac{ax - c}{b}$
78. $t = \dfrac{q - rs}{n}$ **79.** 3   **80.** 2   **81.** 🗒
83. 0.364, or 36.4%   **85.** $\left\{ m \mid m \geq \frac{5}{2} \right\}$ **87.** $\frac{1}{2}$

Technology Connection, p. 205

1. $y_1 = -\frac{3}{4}x - 2,\ y_2 = -\frac{1}{5}x - 2,$
$y_3 = -\frac{3}{4}x - 5,\ y_4 = -\frac{1}{5}x - 5$

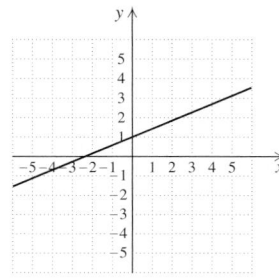

Exercise Set 3.6, pp. 208–210

1. (c)   **3.** (f)   **5.** (d)   **7.**

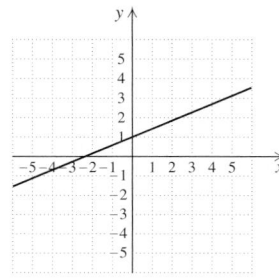

9.

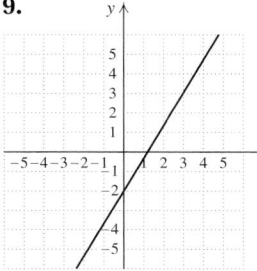

11.

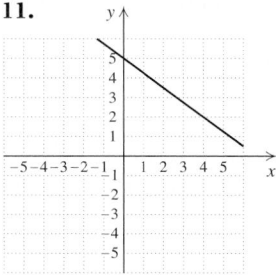

13.

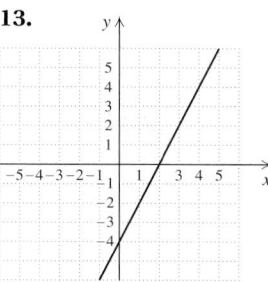

15.

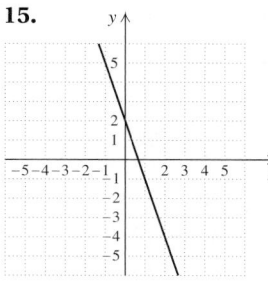

17. $-\frac{2}{7}$; $(0, 5)$ **19.** $\frac{5}{8}$; $(0, 3)$ **21.** $\frac{9}{5}$; $(0, -4)$ **23.** 3; $(0, 7)$
25. $-\frac{5}{2}$; $(0, 4)$ **27.** 0; $(0, 4)$ **29.** $\frac{2}{5}$; $\left(0, \frac{8}{5}\right)$ **31.** $y = 3x + 7$
33. $y = \frac{7}{8}x - 1$ **35.** $y = -\frac{5}{3}x - 8$ **37.** $y = 3$
39. $y = \frac{8}{9}x + 9$, where y is the number of gallons per person and x is the number of years since 1990
41. $y = 15x + 250$, where y is the number of jobs, in thousands, and x is the number of years since 1998

43.

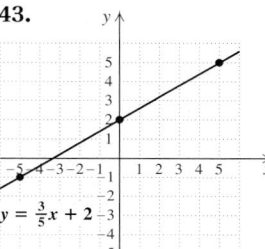

45.

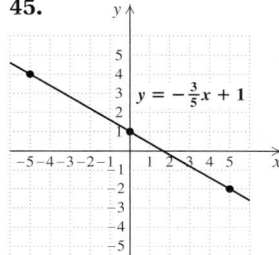

47.

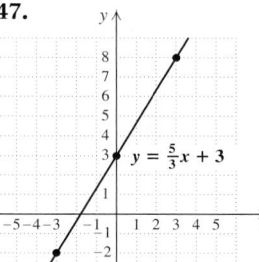

49.

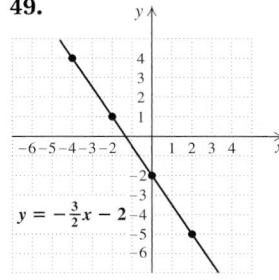

51.

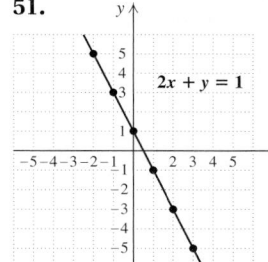

53.

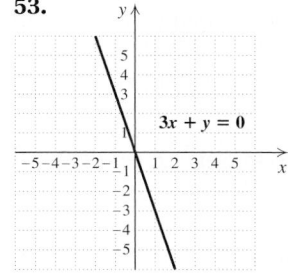

55.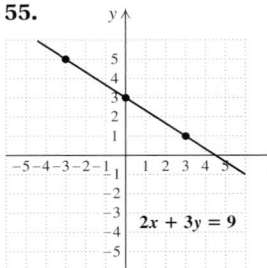

57.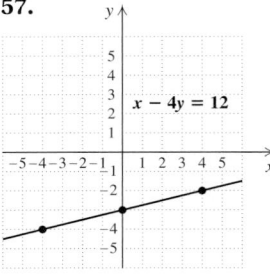

59. Yes **61.** No **63.** Yes **65.** Yes **67.** No **69.** Yes
71. $y = 5x + 11$ **73.** $y = \frac{1}{2}x$ **75.** $y = x + 3$
77. $y = x - 4$ **79.** 🖩 **81.** $y = m(x - h) + k$
82. $y = -2(x + 4) + 9$ **83.** 2 **84.** 16 **85.** -9
86. -10 **87.** 🖩
89. When $x = 0$, $y = b$, so $(0, b)$ is on the line. When
$x = 1$, $y = m + b$, so $(1, m + b)$ is on the line. Then

$$\text{slope} = \frac{(m + b) - b}{1 - 0} = m.$$

91. $y = \frac{1}{3}x + 3$ **93.** $y = -\frac{5}{3}x + 3$ **95.** $y = -\frac{2}{3}x$ **97.** 🖩

Technology Connection, p. 213

1. $y_1 = \frac{3}{4}x + 2$; $y_2 = -\frac{4}{3}x - 1$ **2.** $y_1 = -\frac{2}{5}x - 4$; $y_2 = \frac{5}{2}x + 3$

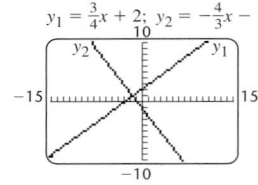 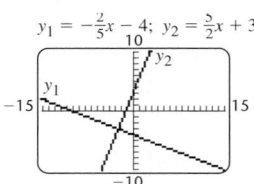

3. $y_1 = \frac{31}{40}x + 2$; $y_2 = -\frac{40}{30}x - 1$ No: $-\frac{40}{30} \neq -\frac{1}{\frac{31}{40}}$

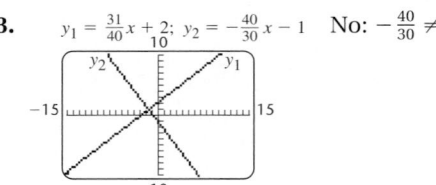

Although the lines appear to be perpendicular, they are
not, because the product of their slopes is not -1:
$$\frac{31}{40}\left(-\frac{40}{30}\right) = -\frac{1240}{1200} \neq -1.$$

Exercise Set 3.7, pp. 218–221

1. (g) **3.** (e) **5.** (b) **7.** (f) **9.** (c) **11.** (d)
13. $y - 2 = 5(x - 6)$ **15.** $y - 1 = -4(x - 3)$
17. $y - (-4) = \frac{3}{2}(x - 5)$ **19.** $y - 6 = \frac{5}{4}(x - (-2))$
21. $y - (-1) = -2(x - (-4))$ **23.** $y - 8 = 1(x - (-2))$
25. $y = 2x - 3$ **27.** $y = \frac{7}{4}x - 9$ **29.** $y = -3x + 3$
31. $y = -4x - 9$ **33.** $y = -\frac{5}{6}x + 4$ **35.** $y = -\frac{1}{2}x + 9$

37. $y = 2x - 7$ **39.** $y = \frac{5}{3}x - \frac{28}{3}$ **41.** $y = 2x - 8$
43. $y = -\frac{5}{3}x - \frac{41}{3}$ **45.** $y = -\frac{1}{2}x + 6$

47.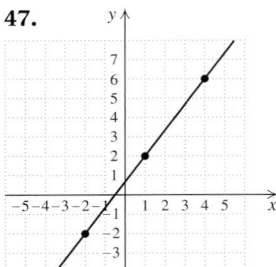

49.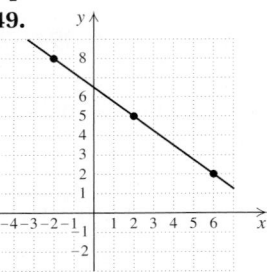

51.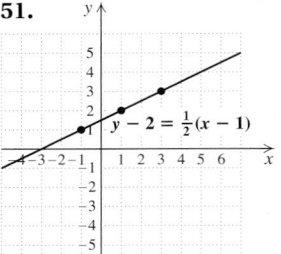

53.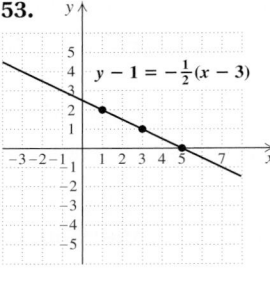

55.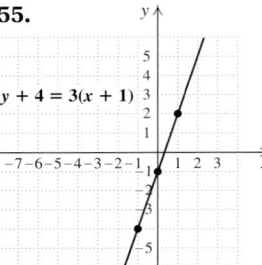

57.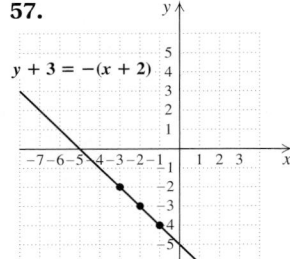

59.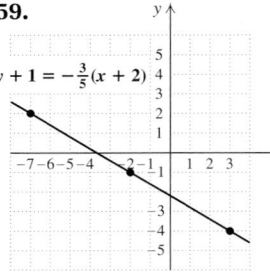

61. (a) 49.14 births per 1000 females; (b) 34.56 births per
1000 females **63.** (a) About 40.6%; (b) about 22.3%
65. (a) 65.1 million students; (b) 72.3 million students
67. (a) 33.3 million residents; (b) 38.7 million residents
69. $y = -x + 6$ **71.** $y = \frac{2}{3}x + 3$ **73.** $y = \frac{2}{5}x - 2$
75. $y = \frac{3}{4}x - \frac{5}{2}$ **77.** 🖩 **79.** -125 **80.** 64 **81.** 8
82. 24 **83.** -72 **84.** -4 **85.** 🖩

87.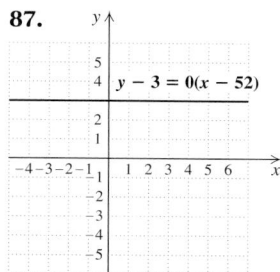

89. $y = 2x - 9$

91. $y = -\frac{4}{3}x + \frac{23}{3}$ **93.** $y - 7 = -\frac{2}{3}(x - (-4))$
95. $y = -4x + 7$ **97.** $y = \frac{10}{3}x + \frac{25}{3}$ **99.** 🗒

Review Exercises: Chapter 3, pp. 223–225

1. True **2.** True **3.** False **4.** False **5.** True **6.** True
7. True **8.** False **9.** True **10.** True **11.** $54,000
12. $269.50 **13.–15.**

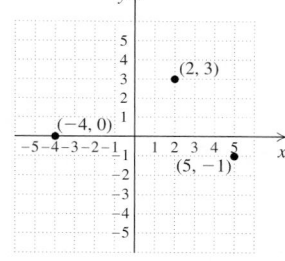

16. IV **17.** III **18.** II **19.** $(-5, -1)$ **20.** $(-2, 5)$
21. $(3, 0)$ **22.** **23.** (a) Yes; (b) no

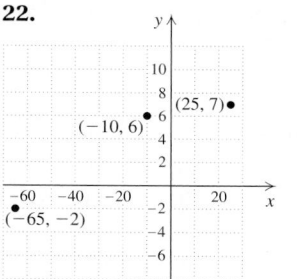

24.
$$\begin{array}{c|c} 2x - y = 3 & 2x - y = 3 \\ \hline 2 \cdot 0 - (-3) \mid 3 & 2 \cdot 2 - 1 \mid 3 \\ 0 + 3 \mid & 4 - 1 \mid \\ 3 \stackrel{?}{=} 3 \text{ True} & 3 \stackrel{?}{=} 3 \text{ True} \end{array}$$
$(-1, -5)$; answers may vary

25. 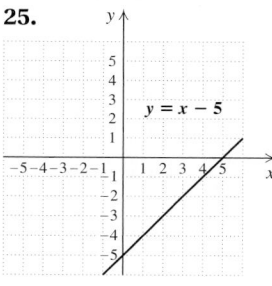 **26.**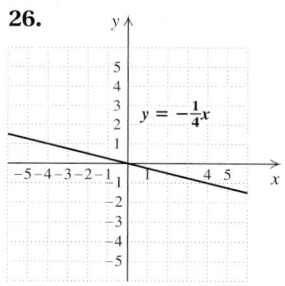

27. 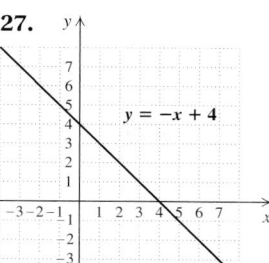 **28.**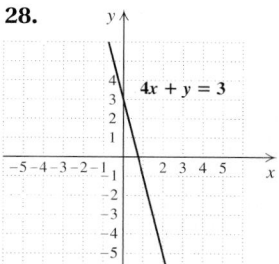

29. 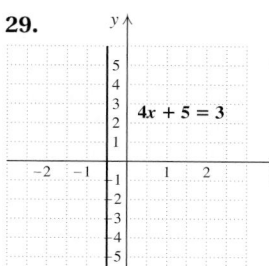 **30.**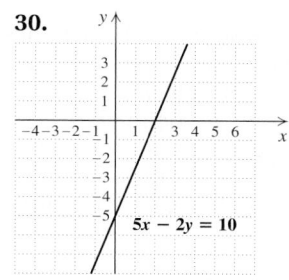

31. (a) $\frac{4}{9}$ meal/min; (b) $2\frac{1}{4}$ min/meal **32.** 12 mpg **33.** 0
34. $\frac{7}{3}$ **35.** $-\frac{3}{7}$ **36.** $\frac{3}{2}$ **37.** 0 **38.** Undefined **39.** 2
40. 28% **41.** x-intercept: $(6, 0)$; y-intercept: $(0, 9)$
42. $-\frac{1}{2}$; $(0, 5)$ **43.** Perpendicular **44.** Parallel
45. $y = -\frac{3}{4}x + 6$ **46.** $y - 6 = -\frac{1}{2}(x - 3)$ **47.** (a) $3100;
(b) $4900 **48.** $y = 4x + 5$ **49.** $y = -\frac{5}{3}x - \frac{5}{3}$
50. 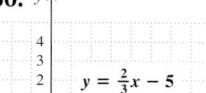 **51.**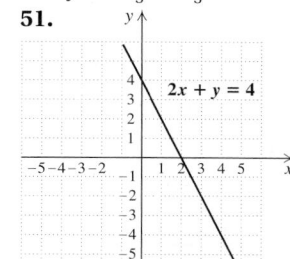

52. 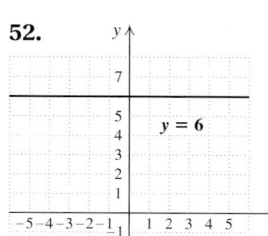 **53.**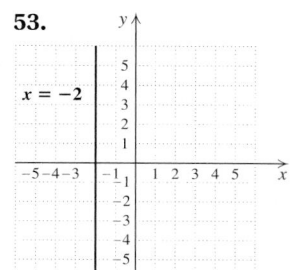

54.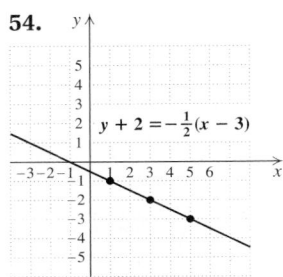

55. 🖉 Two perpendicular lines share the same *y*-intercept if their point of intersection is on the *y*-axis. **56.** 🖉 The *y*-intercept is the point at which the graph crosses the *y*-axis. Since a point on the *y*-axis is neither left nor right of the origin, the first coordinate of the point is 0. **57.** −1 **58.** 19 **59.** Area: 45 sq units; perimeter: 28 units **60.** (0, 4), (1, 3), (−1, 3); answers may vary

Test: Chapter 3, pp. 225–226

1. [3.1] $205.20 **2.** [3.1] $638 **3.** [3.1] II **4.** [3.1] III **5.** [3.1] (3, 4) **6.** [3.1] (0, −4) **7.** [3.1] (−5, 2) **8.** [3.2]

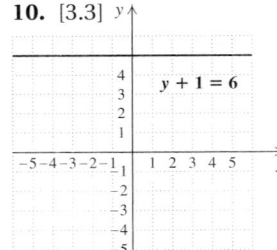

9. [3.3]

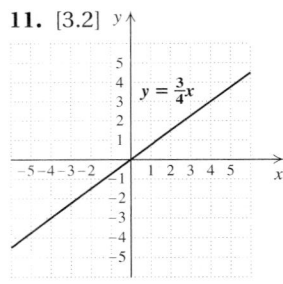

10. [3.3]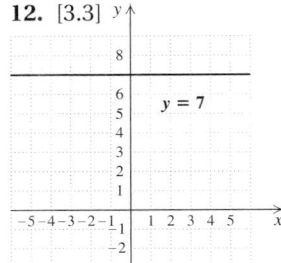

11. [3.2]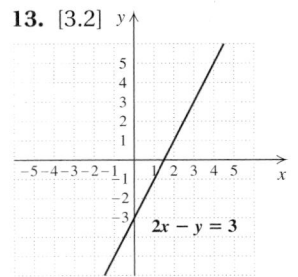

12. [3.3]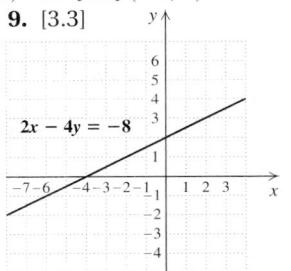

13. [3.2]

14. [3.3]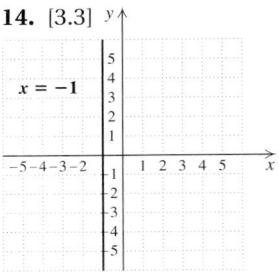

15. [3.3] *x*-intercept: (6, 0); *y*-intercept: (0, −10) **16.** [3.3] *x*-intercept: (10, 0); *y*-intercept: $\left(0, \frac{5}{2}\right)$ **17.** [3.5] $\frac{9}{2}$ **18.** [3.5] $\frac{7}{12}$ **19.** [3.4] $\frac{1}{3}$ km/min **20.** [3.5] 31.5% **21.** [3.6] 3; (0, 7) **22.** [3.6]

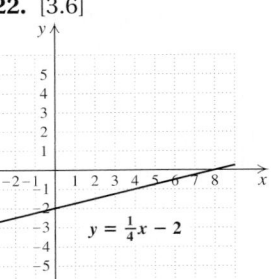

23. [3.7]

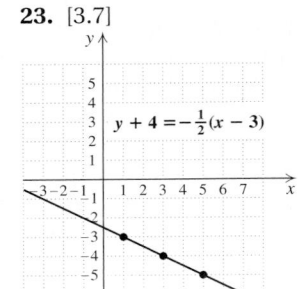

24. [3.6] Parallel **25.** [3.6] Perpendicular **26.** [3.7] $y − 8 = −3(x − 6)$ **27.** [3.7] $y = −x + 10$ **28.** [3.7] **(a)** 40 hr; **(b)** 58 hr **29.** [3.6] $y = \frac{2}{5}x + 9$ **30.** [3.1] Area: 25 sq units; perimeter: 20 units **31.** [3.2], [3.7] (0, 12), (−3, 15), (5, 7)

Cumulative Review: Chapters 1–3, pp. 226–228

1. 7 **2.** $12x − 15y + 21$ **3.** $3(5x − 3y + 1)$ **4.** $2 · 3 · 7$ **5.** 0.45 **6.** 4 **7.** $\frac{1}{4}$ **8.** −4 **9.** $−x − y$ **10.** 0.785 **11.** $\frac{11}{60}$ **12.** 2.6 **13.** 7.28 **14.** $−\frac{5}{12}$ **15.** −3 **16.** 27 **17.** $−2y − 7$ **18.** $5x + 11$ **19.** −1.2 **20.** −21 **21.** 9 **22.** $−\frac{20}{3}$ **23.** 2 **24.** $\frac{13}{8}$ **25.** $−\frac{17}{21}$ **26.** −17 **27.** 2 **28.** $\{x \mid x < 16\}$ **29.** $\{x \mid x \le −\frac{11}{8}\}$ **30.** $h = \dfrac{A − \pi r^2}{2\pi r}$ **31.** IV **32.**

$$-1 < x \le 2$$

33.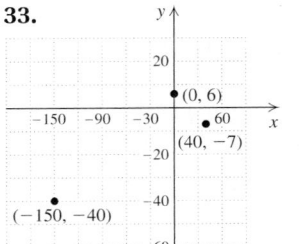

34.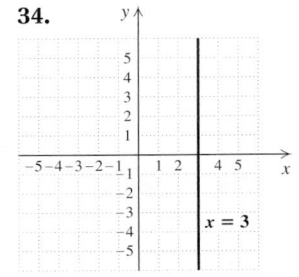

35.

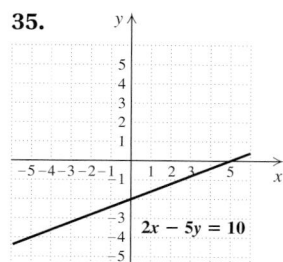

$2x - 5y = 10$

36.

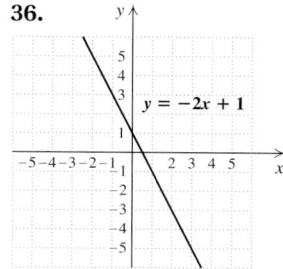

$y = -2x + 1$

37.

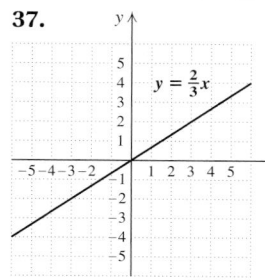

$y = \frac{2}{3}x$

38.

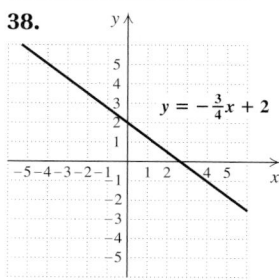

$y = -\frac{3}{4}x + 2$

39. $(10.5, 0); (0, -3)$ **40.** $(-1.25, 0); (0, 5)$
41. 160 million   **42.** 15.6 million   **43.** \$120
44. 50 m, 53 m, 40 m   **45.** No more than 8 hr
46. \$40/person   **47.** $-\frac{1}{3}$ **48.** $y = \frac{2}{7}x - 4$
49. $-\frac{1}{3}; (0, 3)$ **50.** $y = -\frac{1}{2}x + 5$ **51.** **(a)** 17.8%;
(b) 24.56%   **52.** \$25,000   **53.** $-4, 4$ **54.** 2   **55.** -5
56. 3   **57.** No solution; contradiction   **58.** $Q = \dfrac{2 - pm}{p}$
59. $y = -\frac{7}{3}x + 7; y = -\frac{7}{3}x - 7; y = \frac{7}{3}x - 7; y = \frac{7}{3}x + 7$

CHAPTER 4

Exercise Set 4.1, pp. 236–238

1. (b)   **3.** (e)   **5.** (g)   **7.** (c)   **9.** r^{10} **11.** 9^8 **13.** a^7
15. 8^{11} **17.** $(3y)^{12}$ **19.** $(5t)^7$ **21.** $a^5 b^9$ **23.** $(x + 1)^{12}$
25. r^{10} **27.** $x^4 y^7$ **29.** 7^3 **31.** x^{12} **33.** t^4 **35.** $5a$
37. 1   **39.** $\frac{3}{4}m^3$ **41.** $4a^7 b^6$ **43.** $m^9 n^4$ **45.** 1   **47.** 5
49. 2   **51.** -4 **53.** x^{28} **55.** 5^{16} **57.** m^{35} **59.** t^{80}
61. $49x^2$ **63.** $-8a^3$ **65.** $16m^6$ **67.** $a^{14} b^7$ **69.** $x^8 y^7$
71. $24x^{19}$ **73.** $\dfrac{a^3}{64}$ **75.** $\dfrac{49}{25a^2}$ **77.** $\dfrac{a^{20}}{b^{15}}$ **79.** $\dfrac{y^6}{4}$ **81.** $\dfrac{x^8 y^4}{z^{12}}$
83. $\dfrac{a^{12}}{16b^{20}}$ **85.** $\dfrac{125x^{21}y^3}{8z^{12}}$ **87.** 1   **89.** 🖩   **91.** $3(s - r + t)$
92. $-7(x - y + z)$ **93.** $8x$ **94.** $-3a - 6b$ **95.** $2y + 3x$
96. $5z + 2xy$ **97.** 🖩   **99.** 🖩   **101.** Let $a = 1$; then
$(a + 5)^2 = 36$, but $a^2 + 5^2 = 26$.   **103.** Let $a = 0$; then
$\dfrac{a + 7}{7} = 1$, but $a = 0$.   **105.** a^{8k} **107.** $\frac{16}{375}$ **109.** 13
111. $<$ **113.** $<$ **115.** $>$ **117.** 4,000,000; 4,194,304;
194,304   **119.** 2,000,000,000; 2,147,483,648; 147,483,648
121. 1,536,000 bytes, or approximately 1,500,000 bytes

Technology Connection, p. 243

1. 79

Exercise Set 4.2, pp. 244–248

1. (b)   **3.** (d)   **5.** (g)   **7.** (c)   **9.** $7x^4, x^3, -5x, 8$
11. $-t^4, 7t^3, -3t^2, 6$ **13.** Coefficients: 4, 7; degrees: 5, 1
15. Coefficients: 9, -3, 4; degrees: 2, 1, 0
17. Coefficients: 7, 9, 1; degrees: 4, 1, 3   **19.** Coefficients:
1, -1, 4, -3; degrees: 4, 3, 1, 0   **21.** **(a)** 1, 2, 6; **(b)** $3x^6$, 3;
(c) 6   **23.** **(a)** 2, 0, 4; **(b)** $2a^4$, 2; **(c)** 4   **25.** **(a)** 0, 2, 1, 5;
(b) $-x^5$, -1; **(c)** 5
27.

Term	Coefficient	Degree of the Term	Degree of the Polynomial
$8x^5$	8	5	
$-\frac{1}{2}x^4$	$-\frac{1}{2}$	4	
$-4x^3$	-4	3	5
$7x^2$	7	2	
6	6	0	

29. Trinomial   **31.** None of these   **33.** Binomial
35. Monomial   **37.** $11x^2 + 3x$ **39.** $4a^4$ **41.** $6x^2 - 3x$
43. $5x^3 - x + 5$ **45.** $-x^4 - x^3$ **47.** $\frac{1}{15}x^4 + 10$
49. $3.4x^2 + 1.3x + 5.5$ **51.** $-17; 25$ **53.** $16; 34$
55. $-67; 17$ **57.** $-27; 81$ **59.** $-117; -63$
61. \$2.17 trillion   **63.** 1112 ft   **65.** 62.8 cm
67. 153.86 m²   **69.** 5 million gigawatt hours
71. 4.5 million gigawatt hours   **73.** About 9 words
75. About 6   **77.** Approximately 1.8 million;
approximately 4.4 million   **79.** 🖩   **81.** 5   **82.** -9
83. $5(x + 3)$ **84.** $7(a - 3)$ **85.** 6.25¢/mi   **86.** 274 and
275   **87.** 🖩   **89.** $2x^5 + 4x^4 + 6x^3 + 8$; answers may vary
91. \$2510   **93.** $3x^6$ **95.** 5, 13, 80   **97.** 85.0
99.

t	$-t^2 + 10t - 18$
3	3
4	6
5	7
6	6
7	3

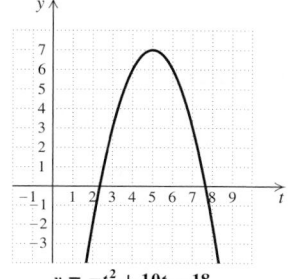

$y = -t^2 + 10t - 18$

Technology Connection, p. 253

1. In each case, let y_1 = the expression before the addition or subtraction has been performed, y_2 = the simplified sum or difference, and $y_3 = y_2 - y_1$; and note that the graph of y_3 coincides with the x-axis. That is, $y_3 = 0$.

Exercise Set 4.3, pp. 254–256

1. $-2x + 6$ **3.** $x^2 - 5x - 1$ **5.** $9t^2 + 5t - 3$
7. $8m^3 + 3m^2 - 3m - 7$ **9.** $7 + 13a + 6a^2 + 14a^3$
11. $9x^8 + 8x^7 - 3x^4 + 2x^2 - 2x + 5$
13. $-\frac{1}{2}x^4 + \frac{2}{3}x^3 + x^2 + 2$
15. $4.2t^3 + 3.5t^2 - 6.4t - 1.8$ **17.** $-3x^4 + 3x^2 + 4x$
19. $1.05x^4 + 0.36x^3 + 14.22x^2 + x + 0.97$
21. $-(-t^3 + 4t^2 - 9); t^3 - 4t^2 + 9$
23. $-(12x^4 - 3x^3 + 3); -12x^4 + 3x^3 - 3$ **25.** $-8x + 9$
27. $-3a^4 + 5a^2 - 9$ **29.** $4x^4 - 6x^2 - \frac{3}{4}x + 8$
31. $9x + 3$ **33.** $-t^2 - 8t + 7$
35. $6x^4 + 3x^3 - 4x^2 + 3x - 4$
37. $4.6x^3 + 9.2x^2 - 3.8x - 23$ **39.** 0
41. $1 + 2a + 7a^2 - 3a^3$ **43.** $\frac{3}{4}x^3 - \frac{1}{2}x$
45. $0.05t^3 - 0.07t^2 + 0.01t + 1$ **47.** $3x + 5$
49. $11x^4 + 12x^3 - 10x^2$ **51.** **(a)** $5x^2 + 4x$; **(b)** 145; 273
53. $\frac{23}{2}a + 12$ **55.** $(r + 11)(r + 9); 9r + 99 + 11r + r^2$
57. $(x + 3)^2; x^2 + 3x + 9 + 3x$ **59.** $\pi r^2 - 25\pi$
61. $xy - 21$ **63.** $(x^2 - 12)$ ft^2 **65.** $(z^2 - 16\pi)$ ft^2
67. $\left(144 - \frac{d^2}{4}\pi\right)$ m^2 **69.** 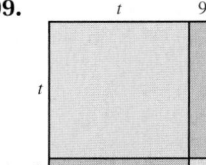   **71.** 0   **72.** 0
73. $13t + 14$ **74.** $14t - 15$ **75.** $\left\{x \mid x < \frac{14}{3}\right\}$
76. $\{x \mid x \le 12\}$ **77.** 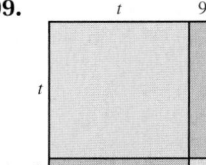   **79.** $9t^2 - 20t + 11$
81. $-10y^2 - 2y - 10$ **83.** $250.591x^3 + 2.812x$
85. $20w + 42$ **87.** $2x^2 + 20x$ **89.** $8x + 24$ **91.**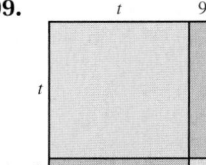

Technology Connection, p. 262

1. Let $y_1 = (-2x^2 - 3)(5x^3 - 3x + 4)$ and $y_2 = -10x^5 - 9x^3 - 8x^2 + 9x - 12$. With the table set in AUTO mode, note that the values in the Y$_1$- and Y$_2$-columns match, regardless of how far we scroll up or down.   **2.** Use TRACE, a table, or a boldly drawn graph to confirm that y_3 is always 0.

Exercise Set 4.4, pp. 262–264

1. (b)   **3.** (d)   **5.** $36x^3$ **7.** $-x^7$ **9.** x^8 **11.** $24a^4$
13. $-0.12x^9$ **15.** $-\frac{1}{20}x^{12}$ **17.** $5n^3$ **19.** $72y^{10}$
21. $8x^2 - 12x$ **23.** $3x^2 + 6x$ **25.** $3a^2 + 27a$
27. $x^5 + x^2$ **29.** $6x^3 - 18x^2 + 3x$ **31.** $15t^3 + 30t^2$
33. $-6x^4 - 6x^3$ **35.** $4a^9 - 8a^7 - \frac{5}{12}a^4$
37. $x^2 + 8x + 12$ **39.** $x^2 + 3x - 10$ **41.** $a^2 - 13a + 42$
43. $x^2 - 9$ **45.** $25 - 15x + 2x^2$ **47.** $t^2 + \frac{17}{6}t + 2$

49. $\frac{3}{16}a^2 + \frac{5}{4}a - 2$ **51.**

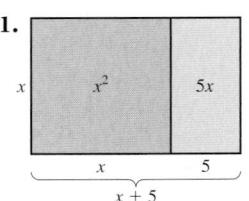

53. **55.**

57. $x^3 + 4x + 5$ **59.** $2a^3 - a^2 - 11a + 10$
61. $2y^5 - 13y^3 + y^2 - 7y - 7$ **63.** $27x^2 + 39x + 14$
65. $x^4 - 2x^3 - x + 2$ **67.** $6t^4 - 17t^3 - 6t^2 + \frac{3}{2}t - 2$
69. $x^4 + 8x^3 + 12x^2 + 9x + 4$ **71.** 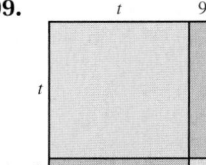   **73.** 6   **74.** 31
75. 0   **76.** 0   **77.** 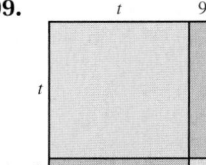   **79.** $75y^2 - 45y$ **81.** 5
83. $V = 4x^3 - 48x^2 + 144x$ in^3; $S = -4x^2 + 144$ in^2
85. $x^3 + 2x^2 - 210$ m^3 **87.** 16 ft by 8 ft   **89.** 0   **91.** 0

Exercise Set 4.5, pp. 271–274

1. True   **3.** False   **5.** $x^3 + 3x^2 + 5x + 15$
7. $x^4 + 2x^3 + 6x + 12$ **9.** $y^2 - y - 6$
11. $9x^2 + 21x + 10$ **13.** $5x^2 + 6x - 8$ **15.** $2 + 3t - 9t^2$
17. $2x^2 - 13x + 20$ **19.** $p^2 - \frac{1}{16}$ **21.** $x^2 - 0.01$
23. $-2x^2 - 11x + 6$ **25.** $a^2 + 18a + 81$
27. $1 - 2t - 15t^2$ **29.** $x^5 + 3x^3 - x^2 - 3$
31. $3x^6 - 2x^4 - 6x^2 + 4$ **33.** $4t^6 + 16t^3 + 15$
35. $8x^5 + 16x^3 + 5x^2 + 10$ **37.** $4x^3 - 12x^2 + 3x - 9$
39. $x^2 - 49$ **41.** $4x^2 - 1$ **43.** $25m^2 - 4$ **45.** $9x^8 - 1$
47. $x^8 - 49$ **49.** $t^2 - \frac{9}{16}$ **51.** $x^2 + 4x + 4$
53. $49x^6 + 14x^3 + 1$ **55.** $a^2 - \frac{4}{5}a + \frac{4}{25}$
57. $t^6 + 10t^3 + 25$ **59.** $4 - 12x^4 + 9x^8$
61. $25 + 60t^2 + 36t^4$ **63.** $49x^2 - 4.2x + 0.09$
65. $10a^5 - 5a^3$ **67.** $a^3 - a^2 - 10a + 12$
69. $9 - 12x^3 + 4x^6$ **71.** $5x^3 + 30x^2 - 10x$
73. $t^6 - 2t^3 + 1$ **75.** $15t^5 - 3t^4 + 3t^3$
77. $36x^8 - 36x^4 + 9$ **79.** $12x^3 + 8x^2 + 15x + 10$
81. $25 - 60x^4 + 36x^8$ **83.** $a^3 + 1$ **85.** $a^2 + 2a + 1$
87. $x^2 + 7x + 10$ **89.** $x^2 + 14x + 49$ **91.** $t^2 + 10t + 24$
93. $t^2 + 13t + 36$ **95.** $9x^2 + 24x + 16$
97. **99.**

101. 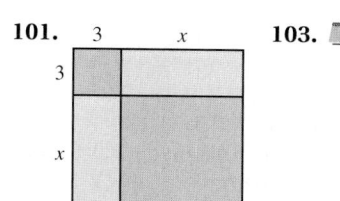 **103.**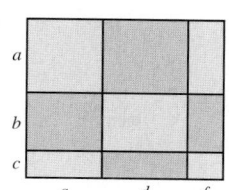

105. Lamps: 500 watts; air conditioner: 2000 watts; television: 50 watts **106.** II **107.** $y = \dfrac{8}{5x}$ **108.** $a = \dfrac{c}{3b}$

109. $x = \dfrac{b + c}{a}$ **110.** $t = \dfrac{u - r}{s}$ **111.**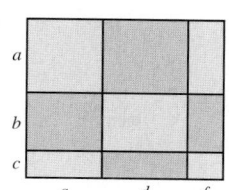

113. $16x^4 - 81$ **115.** $81t^4 - 72t^2 + 16$

117. $t^{24} - 4t^{18} + 6t^{12} - 4t^6 + 1$ **119.** 396 **121.** -7

123. $17F + 7(F - 17)$, $F^2 - (F - 17)(F - 7)$; other equivalent expressions are possible.

125. $x^2 - 9$, $(x - 3)^2 + 3(x - 3) + 3(x - 3)$; other equivalent expressions are possible.

127. $100 - 40x + 4x^2$ **129.**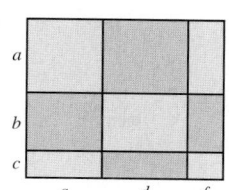

Technology Connection, p. 278

1. 36.22 **2.** 22,312

Exercise Set 4.6, pp. 278–282

1. (a) **3.** (b) **5.** (c) **7.** (a) **9.** -7 **11.** -92

13. 2.97 L **15.** About 2494 calories **17.** 20.60625 in^2

19. 66.4 m

21. Coefficients: 1, -2, 3, -5; degrees: 4, 2, 2, 0; 4

23. Coefficients: 17, -3, -7; degrees: 5, 5, 0; 5

25. $3a - 2b$ **27.** $3x^2y - 2xy^2 + x^2 + 5x$

29. $8u^2v - 5uv^2 + 7u^2$ **31.** $6a^2c - 7ab^2 + a^2b$

33. $3x^2 - 4xy + 3y^2$ **35.** $-6a^4 - 8ab + 7ab^2$

37. $-6r^2 - 5rt - t^2$ **39.** $3x^3 - x^2y + xy^2 - 3y^3$

41. $10y^4x^2 - 8y^3x$ **43.** $-8x + 8y$ **45.** $6z^2 + 7uz - 3u^2$

47. $x^2y^2 + 3xy - 28$ **49.** $4a^2 - b^2$ **51.** $15r^2t^2 - rt - 2$

53. $m^6n^2 + 2m^3n - 48$ **55.** $30x^2 - 28xy + 6y^2$

57. $0.01 - p^2q^2$ **59.** $x^2 + 2xh + h^2$

61. $16a^2 + 40ab + 25b^2$ **63.** $c^4 - d^2$ **65.** $a^2b^2 - c^2d^4$

67. $a^2 + 2ab + b^2 - c^2$ **69.** $a^2 - b^2 - 2bc - c^2$

71. $a^2 + ab + ac + bc$ **73.** $x^2 - z^2$

75. $x^2 + y^2 + z^2 + 2xy + 2xz + 2yz$

77. $\frac{1}{2}x^2 + \frac{1}{2}xy - y^2$

79. We draw a rectangle with dimensions $r + s$ by $u + v$.

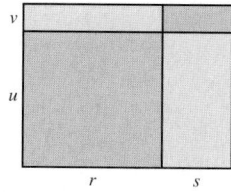

81. 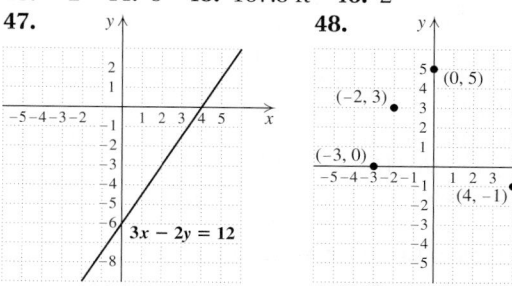 **83.** 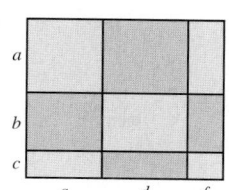 **85.** 12 **86.** 5

87. 27 **88.** 36 **89.** 7 **90.** 5 **91.** **93.** $4xy - 4y^2$

95. $2xy + \pi x^2$ **97.** $x^3 + 2y^3 + x^2y + xy^2$

99. $2\pi nh + 2\pi mh + 2\pi n^2 - 2\pi m^2$ **101.**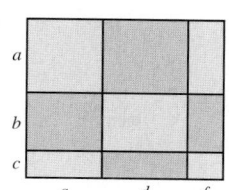

103. $P + 2Pr + Pr^2$ **105.** \$15,638.03

Exercise Set 4.7, pp. 287–288

1. $4x^5 - 3x$ **3.** $1 - 2u + u^6$ **5.** $6t^2 - 8t + 2$

7. $-5x^4 + 4x^2 + 1$ **9.** $6t^2 - 10t + \frac{3}{2}$

11. $4x^2 - 5x + \frac{1}{2}$ **13.** $x^4 + \frac{3}{2}x^2 + \frac{1}{x}$

15. $-3rs - r + 2s$ **17.** $x + 6$ **19.** $t - 5 + \dfrac{-45}{t - 5}$

21. $2x - 1 + \dfrac{1}{x + 6}$ **23.** $a^2 - 2a + 4$ **25.** $t + 4 + \dfrac{3}{t - 4}$

27. $x + 4$ **29.** $3a + 1 + \dfrac{3}{2a + 5}$ **31.** $t^2 - 3t + 1$

33. $t^2 - 2t + 3 + \dfrac{-4}{t + 1}$ **35.** $t^2 + 1 + \dfrac{4t - 2}{t^2 - 3}$

37. $2x^2 + 1 + \dfrac{-x}{2x^2 - 3}$ **39.** 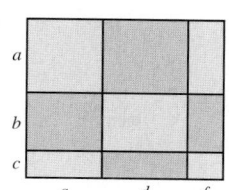 **41.** -17 **42.** -23

43. -2 **44.** 5 **45.** 167.5 ft **46.** 2

47.

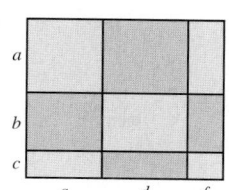

$3x - 2y = 12$

48.

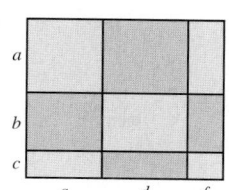

$(0, 5)$; $(-2, 3)$; $(-3, 0)$; $(4, -1)$

49. **51.** $5x^{6k} - 16x^{3k} + 14$ **53.** $3t^{2h} + 2t^h - 5$

55. $a + 3 + \dfrac{5}{5a^2 - 7a - 2}$ **57.** $2x^2 + x - 3$ **59.** 3

61. -1

Technology Connection, p. 295

1. 1.71×10^{17} **2.** $5.\overline{370} \times 10^{-15}$ **3.** 3.68×10^{16}

Exercise Set 4.8, pp. 295–297

1. $\dfrac{1}{7^2} = \dfrac{1}{49}$ **3.** $\dfrac{1}{10^4} = \dfrac{1}{10,000}$ **5.** $\dfrac{1}{(-2)^6} = \dfrac{1}{64}$ **7.** $\dfrac{1}{x^8}$

9. $\dfrac{x}{y^2}$ **11.** $\dfrac{t}{r^5}$ **13.** t^8 **15.** h^8 **17.** $\dfrac{1}{7}$ **19.** $\left(\dfrac{5}{3}\right)^2 = \dfrac{25}{9}$

21. $\left(\dfrac{2}{a}\right)^3 = \dfrac{8}{a^3}$ **23.** $\left(\dfrac{t}{s}\right)^7 = \dfrac{t^7}{s^7}$ **25.** 6^{-2} **27.** t^{-6}

29. a^{-4} **31.** p^{-7} **33.** 5^{-1} **35.** t^{-1} **37.** 2^3, or 8

39. $\dfrac{1}{x^9}$ **41.** $\dfrac{1}{t^2}$ **43.** $\dfrac{1}{a^{18}}$ **45.** t^{18} **47.** $\dfrac{1}{t^{12}}$ **49.** $\dfrac{1}{m^7 n^7}$

51. $\dfrac{9}{x^8}$ **53.** $\dfrac{25t^6}{r^8}$ **55.** t^{10} **57.** $\dfrac{1}{y^4}$ **59.** $3y^5$ **61.** $2x^5$

63. $\dfrac{3b^9}{2a^7}$ **65.** 1 **67.** $\dfrac{3y^6 z^2}{x^5}$ **69.** $\dfrac{1}{27x^{12}y^{15}}$ **71.** $x^{24}y^8$

73. $\dfrac{b^5 c^4}{a^8}$ **75.** $\dfrac{9}{a^8}$ **77.** $\dfrac{n^{12}}{m^3}$ **79.** $\dfrac{27b^{12}}{8a^6}$ **81.** 1 **83.** $\dfrac{2b^3}{a^4}$

85. $\dfrac{5y^4 z^{10}}{4x^{11}}$ **87.** 71,200 **89.** 0.00892 **91.** 904,000,000

93. 0.0000000002764 **95.** 42,090,000 **97.** 4.9×10^5
99. 5.83×10^{-3} **101.** 7.8×10^{10} **103.** 5.27×10^{-7}
105. 1.8×10^{-8} **107.** 1.094×10^{15} **109.** 8×10^{12}
111. 2.47×10^8 **113.** 3.915×10^{-16} **115.** 2.5×10^{13}
117. 5×10^{-4} **119.** 3×10^{-21} **121.** 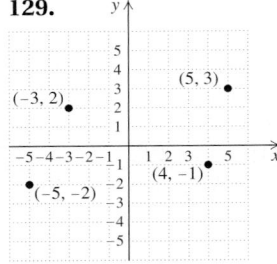 **123.** 15
124. 49 **125.** 78 **126.** 6 **127.** -15 **128.** -7

129.

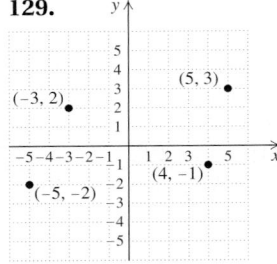

130. $t = \dfrac{r - cx}{b}$ **131.**

133. 8×10^5 **135.** 2^{-12} **137.** 5 **139.** 7×10^{23}
141. $4.894179894 \times 10^{26}$ **143.** 3.12×10^{43}
145. **(a)** False; **(b)** false; **(c)** false
147. About 2.5×10^{12} bytes
149. Approximately 1.15385×10^{12} times

Review Exercises: Chapter 4, pp. 299–301

1. False **2.** True **3.** True **4.** False **5.** True **6.** False
7. True **8.** True **9.** y^{11} **10.** $(3x)^{14}$ **11.** t^6 **12.** 4^3, or
64 **13.** 1 **14.** $\dfrac{9t^8}{4s^6}$ **15.** $-8x^3 y^6$ **16.** $18x^5$ **17.** $a^7 b^6$

18. $3x^2, 6x, \frac{1}{2}$ **19.** $-4y^5, 7y^2, -3y, -2$ **20.** $9, -1, 7$
21. $4, 6, -5, \frac{5}{3}$ **22.** **(a)** $2, 0, 5$; **(b)** $15t^2, 15$; **(c)** 5
23. **(a)** $5, 0, 2, 1$; **(b)** $-2x^5, -2$; **(c)** 5 **24.** Binomial
25. None of these **26.** Monomial **27.** $-x^2 + 7x$
28. $-\frac{1}{4}x^3 + 4x^2 + 7$ **29.** $-3x^5 + 25$
30. $-2x^2 - 3x + 2$ **31.** $10x^4 - 7x^2 - x - \frac{1}{2}$ **32.** -15
33. 10 **34.** $x^5 + 8x^4 + 6x^3 - 2x - 9$
35. $-x^5 + 3x^4 - x^3 - 2x^2$ **36.** $2x^2 - 4x - 6$
37. $x^5 - 3x^3 - 2x^2 + 8$ **38.** $\frac{3}{4}x^4 + \frac{1}{4}x^3 - \frac{1}{3}x^2 - \frac{7}{4}x + \frac{3}{8}$

39. $-x^5 + x^4 - 5x^3 - 2x^2 + 2x$ **40.** **(a)** $4w + 6$;
(b) $w^2 + 3w$ **41.** $-12x^3$ **42.** $49x^2 + 14x + 1$
43. $a^2 - 3a - 28$ **44.** $m^2 - 25$
45. $12x^3 - 23x^2 + 13x - 2$ **46.** $x^2 - 16x + 64$
47. $15t^5 - 6t^4 + 12t^3$ **48.** $a^2 - 81$ **49.** $x^2 - 1.3x + 0.4$
50. $x^7 + x^5 - 3x^4 + 3x^3 - 2x^2 + 5x - 3$
51. $9x^2 - 30x + 25$ **52.** $2t^4 - 11t^2 - 21$
53. $a^2 + \frac{1}{6}a - \frac{1}{3}$ **54.** $9x^2 - 49$ **55.** 49
56. Coefficients: $1, -7, 9, -8$; degrees: $6, 2, 2, 0$; 6
57. Coefficients: $1, -1, 1$; degrees: $16, 40, 23$; 40
58. $-y + 9w - 5$ **59.** $6m^3 + 4m^2 n - mn^2$
60. $-x^2 - 10xy$ **61.** $11x^3 y^2 - 8x^2 y - 6x^2 - 6x + 6$
62. $2x^2 - xy - 15y^2$ **63.** $9a^8 - 2a^4 b^3 + \frac{1}{9}b^6$
64. $\frac{1}{2}x^2 - \frac{1}{2}y^2$ **65.** $5x^2 - \frac{1}{2}x + 4$

66. $3x^2 - 7x + 4 + \dfrac{1}{2x + 3}$ **67.** $t^3 + 2t - 3$ **68.** $\dfrac{1}{m^4}$

69. a^{-9} **70.** $\dfrac{1}{7^2}$, or $\dfrac{1}{49}$ **71.** $\dfrac{2}{a^{13}b^7}$ **72.** $\dfrac{1}{x^{12}}$ **73.** $\dfrac{x^6}{4y^2}$

74. $\dfrac{y^3}{8x^3}$ **75.** 8,300,000 **76.** 3.28×10^{-5} **77.** 2.09×10^4

78. 5.12×10^{-5} **79.** 2.28×10^{11} platelets **80.** In the
expression $5x^3$, the exponent refers only to the x. In the
expression $(5x)^3$, the entire expression $5x$ is the base.
81. It is possible to determine two possibilities for the
binomial that was squared by using the equation
$(A - B)^2 = A^2 - 2AB + B^2$ in reverse. Since, in
$x^2 - 6x + 9$, $A^2 = x^2$ and $B^2 = 9$, or 3^2, the binomial that
was squared was $A - B$, or $x - 3$. If the polynomial is
written $9 - 6x + x^2$, then $A^2 = 9$ and $B^2 = x^2$, so the
binomial that was squared was $3 - x$. We cannot
determine without further information whether the
binomial squared was $x - 3$ or $3 - x$. **82.** **(a)** 4; **(b)** 2
83. $64x^{16}$ **84.** $8x^4 + 4x^3 + 5x - 2$
85. $-16x^6 + x^2 - 10x + 25$ **86.** $\frac{94}{13}$

Test: Chapter 4, p. 302

1. [4.1] a^{10} **2.** [4.1] 4^3, or 64 **3.** [4.1] 1 **4.** [4.1] x^6

5. [4.1] $-27y^6$ **6.** [4.1] $\dfrac{25}{16a^6}$ **7.** [4.1] $-24x^{17}$

8. [4.1] $a^6 b^5$ **9.** [4.2] Binomial **10.** [4.2] $\frac{1}{3}, -1, 7$
11. [4.2] Degrees of terms: $3, 1, 5, 0$; leading term: $7t^5$;
leading coefficient: 7; degree of polynomial: 5
12. [4.2] -7 **13.** [4.2] $5a^2 - 6$ **14.** [4.2] $\frac{7}{4}y^2 - 4y$
15. [4.2] $x^5 + 2x^3 + 4x^2 - 8x + 3$
16. [4.3] $4x^5 + x^4 + 5x^3 - 8x^2 + 2x - 7$
17. [4.3] $5x^4 + 5x^2 + x + 5$
18. [4.3] $-4x^4 + x^3 - 8x - 3$
19. [4.3] $-x^5 + 1.3x^3 - 0.8x^2 - 3$
20. [4.4] $-6x^4 + 6x^3 + 10x^2$ **21.** [4.5] $x^2 - \frac{2}{3}x + \frac{1}{9}$
22. [4.5] $25t^2 - 49$ **23.** [4.5] $3b^2 - 4b - 15$
24. [4.5] $x^{14} - 4x^8 + 4x^6 - 16$ **25.** [4.5] $48 + 34y - 5y^2$
26. [4.4] $6x^3 - 7x^2 - 11x - 3$ **27.** [4.5] $64a^2 + 48a + 9$

28. [4.6] 24   **29.** [4.6] $-4x^3y - x^2y^2 + xy^3 - y^3 + 19$
30. [4.6] $8a^2b^2 + 6ab + 6ab^2 + ab^3 - 4b^3$
31. [4.6] $9x^{10} - y^2$ **32.** [4.7] $4x^2 + 3x - 5$
33. [4.7] $2x^2 - 4x - 2 + \dfrac{17}{3x+2}$ **34.** [4.8] $\dfrac{1}{5^3}$
35. [4.8] y^{-8} **36.** [4.8] $\dfrac{1}{t^6}$ **37.** [4.8] $\dfrac{y^5}{x^5}$ **38.** [4.8] $\dfrac{b^4}{16a^{12}}$
39. [4.8] $\dfrac{c^3}{a^3b^3}$ **40.** [4.8] 3.9×10^9 **41.** [4.8] 0.00000005
42. [4.8] 1.75×10^{17} **43.** [4.8] 1.296×10^{22}
44. [4.4], [4.5] $V = l(l-2)(l-1) = l^3 - 3l^2 + 2l$
45. [2.2], [4.5] $\frac{100}{21}$ **46.** [4.8] 3.0×10^5 sound files

CHAPTER 5

Technology Connection, p. 310

1. Correct   **2.** Correct   **3.** Not correct   **4.** Not correct
5. Not correct   **6.** Correct   **7.** Not correct   **8.** Correct

Exercise Set 5.1, pp. 311–312

1. (h)   **3.** (b)   **5.** (c)   **7.** (d)   **9.** Answers may vary.
$(10x)(x^2), (5x^2)(2x), (-2)(-5x^3)$ **11.** Answers may vary.
$(-15)(a^4), (-5a)(3a^3), (-3a^2)(5a^2)$ **13.** Answers may vary.
$(2x)(13x^4), (13x^5)(2), (-x^3)(-26x^2)$ **15.** $7(x-2)$
17. $t(3t+1)$ **19.** $-4a(a+2)$, or $4a(-a-2)$
21. $x^2(x+6)$ **23.** $8x^2(x^2-3)$ **25.** $2(x^2+x-4)$
27. $-a^2(7a^4 - 10a^2 + 14)$ **29.** $6x^2(x^6 + 2x^4 - 4x^2 + 5)$
31. $x^2y^2(x^3y^3 + x^2y + xy - 1)$
33. $-5a^2b^2(ab^2 - 2b + 3a)$ **35.** $(y-2)(y+7)$
37. $(x+3)(x^2-7)$ **39.** $(y+8)(y^2+1)$
41. $(x+3)(x^2+4)$ **43.** $(a+3)(5a^2+2)$
45. $(3x-4)(3x^2+1)$ **47.** $(t-5)(4t^2+3)$
49. $(7x+2)(x^2-2)$ **51.** $(6a-7)(a^2+1)$
53. $(x+8)(x^2-3)$ **55.** $(x+6)(2x^2-5)$
57. Not factorable by grouping   **59.** $(y+8)(y^2-2)$
61. $(x-4)(2x^2-9)$ **63.** 🖳   **65.** $x^2 + 8x + 15$
66. $x^2 + 9x + 14$ **67.** $a^2 - 4a - 21$ **68.** $a^2 - 3a - 40$
69. $6x^2 + 7x - 20$ **70.** $12t^2 - 13t - 14$
71. $9t^2 - 30t + 25$ **72.** $4t^2 - 36t + 81$ **73.** 🖳
75. $(2x^3 + 3)(2x^2 + 3)$ **77.** $(x^5 + 1)(x^7 + 1)$
79. $(x-1)(5x^4 + x^2 + 3)$ **81.** Answers may vary.
$8x^4y^3 - 24x^2y^4 + 16x^3y^4$

Exercise Set 5.2, pp. 318–319

1. Positive; positive   **3.** Negative; positive   **5.** Positive
7. $(x+6)(x+1)$ **9.** $(x+3)(x+4)$ **11.** $(x-3)^2$
13. $(x+2)(x+7)$ **15.** $(b-4)(b-1)$ **17.** $(a+6)(a-2)$
19. $(d-2)(d-5)$ **21.** $(x-5)(x+3)$ **23.** $(x+5)(x-3)$
25. $3(y+4)(y-7)$ **27.** $-x(x+6)(x-7)$
29. $(x-5)(x+12)$ **31.** $(x-6)(x+12)$

33. $-5(b-3)(b+8)$ **35.** $x^3(x-2)(x+1)$ **37.** Prime
39. $(t+5)(t+10)$ **41.** $(x+9)(x+11)$
43. $2x(x-8)(x-12)$ **45.** $-4(x+5)^2$
47. $(y-9)(y-12)$ **49.** $-a^4(a-6)(a+15)$ **51.** $\left(t + \frac{1}{3}\right)^2$
53. Prime   **55.** $(p+5q)(p-2q)$ **57.** Prime
59. $(s-5t)(s+3t)$ **61.** $6a^8(a+2)(a-7)$ **63.** 🖳
65. $\frac{8}{3}$ **66.** $-\frac{7}{2}$ **67.** $3x^2 + 22x + 24$
68. $49w^2 + 84w + 36$ **69.** 29,443 people
70. 100°, 25°, 55°   **71.** 🖳   **73.** $-5, 5, -23, 23, -49, 49$
75. $(y + 0.2)(y - 0.4)$ **77.** $-\frac{1}{3}a(a-3)(a+2)$
79. $(x^m + 4)(x^m + 7)$ **81.** $(a+1)(x+2)(x+1)$
83. $(x+3)^3$, or $x^3 + 9x^2 + 27x + 27$ cubic meters
85. $x^2(\pi - 1)$ **87.** $x^2\left(9 - \frac{1}{2}\pi\right)$ **89.** $(x+5)(x+7)$

Exercise Set 5.3, pp. 328–330

1. (c)   **3.** (d)   **5.** $(2x-1)(x+4)$ **7.** $(3t-5)(t+3)$
9. $(3x-1)(2x-7)$ **11.** $(7x+1)(x+2)$
13. $(3a+2)(3a-4)$ **15.** $2(3x+1)(x-2)$
17. $6(2t+1)(t-1)$ **19.** $(5x+3)(3x+2)$
21. $-1(7x+4)(5x+2)$, or $-(7x+4)(5x+2)$ **23.** Prime
25. $(5x+4)^2$ **27.** $(8a+3)(2a+9)$ **29.** $2(3t-1)(3t+5)$
31. $-1(x-3)(2x+5)$, or $-(x-3)(2x+5)$
33. $-3(2x+1)(x+5)$ **35.** $5(4x-1)(x-1)$
37. $4(3x-1)(x+6)$ **39.** $(3x+1)(x+1)$
41. $(y+4)(y-2)$ **43.** $(4t-3)(2t-7)$
45. $(3x+2)(2x+3)$ **47.** $(t+3)(2t-1)$
49. $(a-4)(3a-1)$ **51.** $(9t+5)(t+1)$
53. $-1(4x+1)(4x+7)$, or $-(4x+1)(4x+7)$
55. $5(2a-1)(a+3)$ **57.** $3x(3x-1)(2x+3)$
59. $(x+1)(25x+64)$ **61.** $3x(7x+1)(8x+1)$
63. $-t^2(2t-3)(7t+1)$ **65.** $3(5x-3)(3x+2)$
67. $(2a+b)(a+2b)$ **69.** $(2s+3t)(4s+3t)$
71. $6(3x-4y)(x+y)$ **73.** $-2(3a-2b)(4a-3b)$
75. $x^2(4x+7)(2x+5)$ **77.** $a^6(3a+4)(3a+2)$
79. 🖳   **81.** 6369 km, 3949 mi   **82.** 40°   **83.** $9x^2 + 6x + 1$
84. $25x^2 - 20x + 4$ **85.** $16t^2 - 40t + 25$
86. $49a^2 + 14a + 1$ **87.** $25x^2 - 4$ **88.** $4x^2 - 9$
89. $4t^2 - 49$ **90.** $16a^2 - 49$ **91.** 🖳
93. $(3xy + 2)(6xy - 5)$ **95.** Prime   **97.** $(4t^5 - 1)^2$
99. $-1(10x^n + 3)(2x^n + 1)$, or $-(10x^n + 3)(2x^n + 1)$
101. $a(a^n - 1)^2$ **103.** $-2(a+1)^n(a+3)^2(a+6)$

Exercise Set 5.4, pp. 336–337

1. Difference of squares   **3.** None of these   **5.** Perfect-square trinomial   **7.** Prime polynomial   **9.** Perfect-square trinomial   **11.** Yes   **13.** No   **15.** No   **17.** No
19. $(x+8)^2$ **21.** $(x-7)^2$ **23.** $3(x+1)^2$ **25.** $(2-x)^2$, or $(x-2)^2$ **27.** $2(3x+1)^2$ **29.** $(7-4y)^2$, or $(4y-7)^2$
31. $-x^3(x-9)^2$ **33.** $2x(x-1)^2$ **35.** $5(2x+5)^2$
37. $(7-3x)^2$, or $(3x-7)^2$ **39.** $(4x+3)^2$ **41.** $2(1+5x)^2$, or $2(5x+1)^2$ **43.** $(2p+3q)^2$ **45.** Prime
47. $-1(8m+n)^2$, or $-(8m+n)^2$ **49.** $-2(4s-5t)^2$
51. No   **53.** Yes   **55.** No   **57.** $(y+2)(y-2)$

59. $(p + 3)(p - 3)$ **61.** $(7 + t)(-7 + t)$, or $(t + 7)(t - 7)$
63. $6(a + 3)(a - 3)$ **65.** $(7x - 1)^2$ **67.** $2(10 - t)(10 + t)$
69. $-5(4a - 3)(4a + 3)$ **71.** $5(t + 4)(t - 4)$
73. $2(2x + 7)(2x - 7)$ **75.** $x(6 + 7x)(6 - 7x)$ **77.** Prime
79. $(t - 1)(t + 1)(t^2 + 1)$ **81.** $-3x(x - 4)^2$
83. $3(4t + 3)(4t - 3)$ **85.** $a^6(a - 1)^2$ **87.** $7(a + b)(a - b)$
89. $(5x + 2y)(5x - 2y)$ **91.** $2(3t + 2s)(3t - 2s)$ **93.**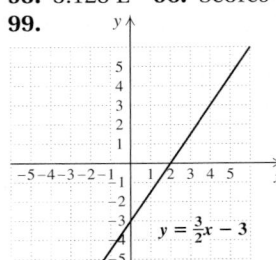
95. 3.125 L   **96.** Scores ≥ 77 **97.** $x^{12}y^{12}$ **98.** $25a^4b^6$
99.

$y = \frac{3}{2}x - 3$

1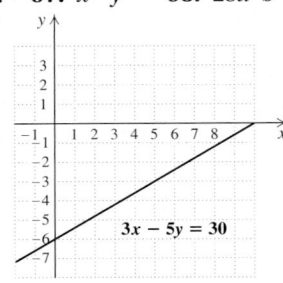
$3x - 5y = 30$

101.   **103.** $(x^4 + 2^4)(x^2 + 2^2)(x + 2)(x - 2)$, or
$(x^4 + 16)(x^2 + 4)(x + 2)(x - 2)$ **105.** $2x\left(3x - \frac{2}{5}\right)\left(3x + \frac{2}{5}\right)$
107. $(y^2 - 10y + 25 + z^4)(y - 5 + z^2)(y - 5 - z^2)$
109. $-1(x^2 + 1)(x + 3)(x - 3)$, or $-(x^2 + 1)(x + 3)(x - 3)$
111. $(y + 4)^2$ **113.** $(3x - 7)^2(3x + 7)$
115. $(9 + b^{2k})(3 + b^k)(3 - b^k)$ **117.** $2x^3 - x^2 - 1$
119. $(y + x + 7)(y - x - 1)$ **121.** 16
123. $(x + 1)^2 - x^2 = [(x + 1) + x][(x + 1) - x] = 2x + 1 = $
$(x + 1) + x$

Exercise Set 5.5, pp. 340–341

1. $(t + 2)(t^2 - 2t + 4)$ **3.** $(a - 4)(a^2 + 4a + 16)$
5. $(z + 5)(z^2 - 5z + 25)$ **7.** $(2a - 1)(4a^2 + 2a + 1)$
9. $(y - 3)(y^2 + 3y + 9)$ **11.** $(4 + 5x)(16 - 20x + 25x^2)$
13. $(5p - 1)(25p^2 + 5p + 1)$
15. $(3m + 4)(9m^2 - 12m + 16)$
17. $(p - q)(p^2 + pq + q^2)$ **19.** $\left(x + \frac{1}{2}\right)\left(x^2 - \frac{1}{2}x + \frac{1}{4}\right)$
21. $2(y - 4)(y^2 + 4y + 16)$ **23.** $3(2a + 1)(4a^2 - 2a + 1)$
25. $r(s - 4)(s^2 + 4s + 16)$
27. $5(x + 2z)(x^2 - 2xz + 4z^2)$
29. $(x + 0.1)(x^2 - 0.1x + 0.01)$
31. $3z^2(z - 1)(z^2 + z + 1)$ **33.** $(t^2 + 1)(t^4 - t^2 + 1)$
35. $(p + q)(p^2 - pq + q^2)(p - q)(p^2 + pq + q^2)$ **37.**
39. $9x^2 - 25$ **40.** $9x^2 + 30x + 25$ **41.** $x^2 - 3x - 28$
42. $x^3 + 1$ **43.** 24 million   **44.** 32.8 million barrels
45.   **47.** $(5c^2 + 2d^2)(25c^4 - 10c^2d^2 + 4d^4)$
49. $3(x^a - 2y^b)(x^{2a} + 2x^ay^b + 4y^{2b})$
51. $\frac{1}{3}\left(\frac{1}{2}xy + z\right)\left(\frac{1}{4}x^2y^2 - \frac{1}{2}xyz + z^2\right)$ **53.** $2x(x^2 + 75)$
55. $(t - 8)(t - 1)(t^2 + t + 1)$

Exercise Set 5.6, pp. 346–347

1. Common factor   **3.** Grouping   **5.** $10(a + 8)(a - 8)$
7. $(y - 7)^2$ **9.** $(2t + 3)(t + 4)$ **11.** $x(x - 9)^2$
13. $(x - 5)^2(x + 5)$ **15.** $3t(3t + 1)(3t - 1)$

17. $3x(3x - 5)(x + 3)$ **19.** Prime   **21.** $3(x + 3)(2x - 5)$
23. $-2a^4(a - 2)^2$ **25.** $5x(x^2 + 4)(x + 2)(x - 2)$
27. $(t^2 + 3)(t^2 - 3)$ **29.** $-x^4(x^2 - 2x + 7)$
31. $(x - y)(x^2 + xy + y^2)$ **33.** $a(x^2 + y^2)$
35. $9mn(4 - mn)$ **37.** $2\pi r(h + r)$ **39.** $(a + b)(5a + 3b)$
41. $(x + 1)(x + y)$ **43.** $(a - 3)(a + y)$
45. $(3x - 2y)(x + 5y)$ **47.** $(a - 2b)^2$ **49.** $(4x + 3y)^2$
51. Prime   **53.** $(2t + 1)(4t^2 - 2t + 1)(2t - 1)(4t^2 + 2t + 1)$
55. $p(4p^2 + 4pq - q^2)$ **57.** $(3b - a)(b + 6a)$
59. $-1(xy + 2)(xy + 6)$, or $-(xy + 2)(xy + 6)$
61. $(pq + 6)(pq + 1)$ **63.** $2a(3a + 2b)(9a^2 - 6ab + 4b^2)$
65. $x^4(x + 2y)(x - y)$ **67.** $\left(6a - \frac{5}{4}\right)^2$ **69.** $\left(\frac{1}{9}x - \frac{4}{3}\right)^2$
71. $(1 + 4x^6y^6)(1 + 2x^3y^3)(1 - 2x^3y^3)$ **73.** $(2ab + 3)^2$
75. $a(a^2 + 8)(a + 8)$ **77.**
79. For $(-1, 11)$:

$$\begin{array}{c|c} y = -4x + 7 \\ \hline 11 & -4(-1) + 7 \\ & 4 + 7 \\ 11 \overset{?}{=} 11 \end{array}$$
True

For $(0, 7)$:

$$\begin{array}{c|c} y = -4x + 7 \\ \hline 7 & -4 \cdot 0 + 7 \\ & 0 + 7 \\ 7 \overset{?}{=} 7 \end{array}$$
True

For $(3, -5)$:

$$\begin{array}{c|c} y = -4x + 7 \\ \hline -5 & -4 \cdot 3 + 7 \\ & -12 + 7 \\ -5 \overset{?}{=} -5 \end{array}$$
True

80. $\frac{4}{5}$ **81.** $-\frac{7}{3}$ **82.** $-\frac{9}{2}$ **83.** $\frac{9}{4}$
84.

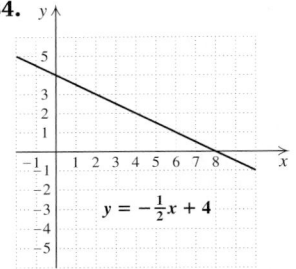
$y = -\frac{1}{2}x + 4$

85.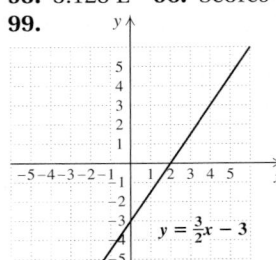

87. $-x(x^2 + 9)(x^2 - 2)$
89. $-3(a + 1)(a - 1)(a + 2)(a - 2)$
91. $(y + 1)(y - 7)(y + 3)$ **93.** $(2x - 2 + 3y)(3x - 3 - y)$
95. $(a + 3)^2(2a + b + 4)(a - b + 5)$

Technology Connection, p. 354

1. $-4.65, 0.65$ **2.** $-0.37, 5.37$ **3.** $-8.98, -4.56$
4. No solution   **5.** $0, 2.76$

Exercise Set 5.7, pp. 354–356

1. (c)   **3.** (a)   **5.** $-7, -5$ **7.** $-4, \frac{9}{2}$ **9.** $-\frac{7}{4}, \frac{9}{10}$
11. $-2, 0$ **13.** $\frac{1}{21}, \frac{18}{11}$ **15.** $-\frac{9}{2}, 0$ **17.** $50, 70$ **19.** $-5, \frac{2}{3}, 1$
21. $1, 6$ **23.** $-7, 3$ **25.** $0, 1, 2$ **27.** $0, 6$ **29.** $-6, 0$
31. $-\frac{2}{3}, \frac{2}{3}$ **33.** -5 **35.** 1 **37.** $0, 2$ **39.** $-\frac{5}{3}, 4$
41. $-5, -1$ **43.** 3 **45.** $0, \frac{3}{2}$ **47.** $-\frac{5}{9}, \frac{5}{9}$ **49.** $-1, \frac{6}{5}$

51. $-2, 9$ **53.** $-1, 0, \frac{3}{2}$ **55.** $-7, -\frac{8}{3}, \frac{5}{2}$ **57.** $-3, 2$
59. $-1, 3$ **61.** $(-4, 0), (1, 0)$ **63.** $(-3, 0), (5, 0)$
65. $\left(-\frac{5}{2}, 0\right), (2, 0)$ **67.** ▨   **69.** $(a + b)^2$ **70.** $a^2 + b^2$
71. Let x represent the first integer; $x + (x + 1)$
72. Let x represent the number; $2x + 5 < 19$
73. Let x represent the number; $\frac{1}{2}x - 7 > 24$
74. Let x represent the number; $x - 3 \geq 34$ **75.** ▨
77. **(a)** $x^2 - x - 20 = 0$; **(b)** $x^2 - 6x - 7 = 0$;
(c) $4x^2 - 13x + 3 = 0$; **(d)** $6x^2 - 5x + 1 = 0$;
(e) $12x^2 - 17x + 6 = 0$; **(f)** $x^3 - 4x^2 + x + 6 = 0$
79. $-5, 4$ **81.** $-\frac{3}{5}, \frac{3}{5}$ **83.** $-4, 2$
85. **(a)** $2x^2 + 20x - 4 = 0$; **(b)** $x^2 - 3x - 18 = 0$;
(c) $(x + 1)(5x - 5) = 0$; **(d)** $(2x + 8)(2x - 5) = 0$;
(e) $4x^2 + 8x + 36 = 0$; **(f)** $9x^2 - 12x + 24 = 0$ **87.** ▨
89. $-0.25, 0.88$ **91.** $-3.23, 4.55$ **93.** $-3.76, 0$

Exercise Set 5.8, pp. 363–367

1. $-4, 5$ **3.** 9 cm, 12 cm, 15 cm   **5.** 9, 10   **7.** -17 and -15; 15 and 17   **9.** Length: 12 ft; width: 2 ft
11. Length: 6 cm; width: 4 cm   **13.** Base: 14 cm; height: 4 cm   **15.** Foot: 7 ft; height: 12 ft   **17.** 1 min, 3 min
19. 16 teams   **21.** 105 handshakes   **23.** 12 players
25. 9 ft   **27.** 25 ft   **29.** Dining room: 12 ft by 12 ft; kitchen: 12 ft by 10 ft   **31.** 20 ft   **33.** 1 sec, 2 sec
35. ▨   **37.** $-\frac{8}{21}$ **38.** $-\frac{8}{45}$ **39.** $-\frac{35}{54}$ **40.** $-\frac{5}{16}$ **41.** $-\frac{2}{21}$
42. $-\frac{26}{45}$ **43.** $\frac{1}{18}$ **44.** $-\frac{11}{24}$ **45.** ▨   **47.** $1200
49. 39 cm   **51.** 15 cm by 30 cm   **53.** 35 ft
55. 2 hr, 4.2 hr   **57.** 3 hr

Review Exercises: Chapter 5, pp. 369–370

1. False   **2.** True   **3.** True   **4.** False   **5.** True   **6.** False
7. True   **8.** False   **9.** Answers may vary.
$(12x)(3x^2), (-9x^2)(-4x), (6x)(6x^2)$ **10.** Answers may vary.
$(-4x^3)(5x^2), (2x^4)(-10x), (-5x)(4x^4)$ **11.** $6x^3(2x - 3)$
12. $4a(2a - 3)$ **13.** $(2t - 3)(2t + 3)$ **14.** $(x - 2)(x + 6)$
15. $(x + 7)^2$ **16.** $3x(2x + 1)^2$ **17.** $(2x + 3)(3x^2 + 1)$
18. $(6t + 1)(t - 1)$ **19.** $(5t - 3)^2$ **20.** $2(24t^2 - 14t + 3)$
21. $(9a^2 + 1)(3a + 1)(3a - 1)$ **22.** $3x(3x - 5)(x + 3)$
23. $2(x - 5)(x^2 + 5x + 25)$ **24.** $(x + 4)(x^3 - 2)$
25. $(ab^2 - 6)(ab^2 + 6)$ **26.** $-4x^4(2x^2 - 8x + 1)$
27. $3(2x - 5)^2$ **28.** Prime   **29.** $-x(x - 6)(x + 5)$
30. $(2x + 5)(2x - 5)$ **31.** $2z(2z - 3)$ **32.** $5z(3 + 2z)$
33. $(4t + 5)(t + 2)$ **34.** $(2t + 1)(t - 4)$
35. $7x(x + 1)(x + 4)$ **36.** $5x(x + 2)(x + 5)$
37. $2(3x - 1)^2$ **38.** $-3(x + 3)(x - 3)$ **39.** $(5 - x)(3 - x)$
40. $(2y + 3x^2)(4y^2 - 6x^2y + 9x^4)$ **41.** $(xy - 3)(xy + 4)$
42. $3(2a + 7b)^2$ **43.** $(m + 5)(m + t)$
44. $32(x^2 + 2y^2z^2)(x^2 - 2y^2z^2)$ **45.** $(2m + n)(3m + n)$
46. Prime   **47.** $-8, 7$ **48.** $-7, 0, 5$ **49.** $-\frac{1}{3}, \frac{1}{3}$ **50.** $\frac{2}{3}, 1$
51. $-\frac{3}{2}, 4$ **52.** $-2, 3$ **53.** $0, \frac{3}{4}$ **54.** $0, \frac{4}{3}$ **55.** $-3, 4$
56. 10 teams   **57.** $(-1, 0), \left(\frac{5}{2}, 0\right)$
58. Height: 40 cm; base: 40 cm   **59.** 24 ft

60. ▨ Answers may vary. Because Edith did not first factor out the largest common factor, 4, her factorization will not be "complete" until she removes a common factor of 2 from each binomial. The answer should be $4(x - 5)(x + 5)$. Awarding 3 to 7 points would seem reasonable.
61. ▨ The equations solved in this chapter have an x^2-term (are quadratic), whereas those solved previously have no x^2-term (are linear). The principle of zero products is used to solve quadratic equations and is not used to solve linear equations.   **62.** 2.5 cm   **63.** 0, 2
64. Length: 12 cm; width: 6 cm   **65.** 100 cm^2, 225 cm^2
66. $-3, 2, \frac{5}{2}$ **67.** No real solution

Test: Chapter 5, pp. 370–371

1. [5.1] $(2x)(4x^3), (-4x^2)(-2x^2), (8x)(x^3)$
2. [5.2] $(x - 2)(x - 5)$ **3.** [5.4] $(x - 5)^2$
4. [5.1] $2y^2(2y^2 - 4y + 3)$ **5.** [5.1] $(x + 1)(x^2 + 2)$
6. [5.1] $t^5(t^2 - 3)$ **7.** [5.2] $x(x + 3)(x - 1)$
8. [5.3] $2(5x - 6)(x + 4)$ **9.** [5.4] $(2t + 5)(2t - 5)$
10. [5.2] $(x - 4)(x + 3)$ **11.** [5.3] $-3m(2m + 1)(m + 1)$
12. [5.4] $3(w + 5)(w - 5)$ **13.** [5.4] $5(3r + 2)^2$
14. [5.4] $3(x^2 + 4)(x + 2)(x - 2)$ **15.** [5.4] $(7t + 6)^2$
16. [5.3] $(5x - 1)(x - 5)$ **17.** [5.1] $(x + 2)(x^3 - 3)$
18. [5.5] $2(m^2 + 2)(m^4 - 2m^2 + 4)$
19. [5.3] $(2x + 3)(2x - 5)$ **20.** [5.3] $3t(2t + 5)(t - 1)$
21. [5.3] $3(m - 5n)(m + 2n)$ **22.** [5.7] $-5, 4$
23. [5.7] $-\frac{3}{2}, 0, 5$ **24.** [5.7] $-4, 7$ **25.** [5.7] $(-1, 0), \left(\frac{8}{3}, 0\right)$
26. [5.8] Length: 8 m; width: 6 m   **27.** [5.8] 5 ft
28. [5.8] 4 in., 6 in.   **29.** [5.8] Width: 3; length: 15
30. [5.2] $(a - 4)(a + 8)$ **31.** [5.7] $-\frac{8}{3}, 0, \frac{2}{5}$

CHAPTER 6

Technology Connection, p. 377

1. Correct   **2.** Correct   **3.** Not correct   **4.** Not correct

Exercise Set 6.1, pp. 379–381

1. (e)   **3.** (d)   **5.** (c)   **7.** 0   **9.** -8 **11.** 4   **13.** $-4, 7$
15. $-6, 0, \frac{1}{2}$ **17.** $\frac{5a}{4b^2}$ **19.** $\frac{4}{3xy^4}$ **21.** $\frac{3}{4}$ **23.** $\frac{a - 3}{a + 1}$
25. $\frac{3}{2x^3}$ **27.** $\frac{y - 3}{4y}$ **29.** $\frac{t + 4}{t + 5}$ **31.** $\frac{a + 4}{2(a - 4)}$ **33.** $\frac{x + 4}{x - 4}$
35. $t - 1$ **37.** $\frac{1}{a^2 + 2a + 4}$ **39.** $\frac{y^2 + 4}{y + 2}$ **41.** $\frac{1}{2}$
43. $\frac{y}{2y + 1}$ **45.** $\frac{2x - 3}{5x + 2}$ **47.** -1 **49.** -7 **51.** $-\frac{1}{3}$
53. $-\frac{3}{2}$ **55.** -1 **57.** ▨   **59.** $-\frac{4}{7}$ **60.** $-\frac{10}{33}$ **61.** $-\frac{15}{4}$
62. $-\frac{21}{16}$ **63.** $\frac{13}{63}$ **64.** $\frac{5}{48}$ **65.** ▨   **67.** $-(2y + x)$

69. $\dfrac{x^3 + 4}{(x^3 + 2)(x^2 + 2)}$ **71.** $\dfrac{(t - 1)(t - 9)^2}{(t + 1)(t^2 + 9)}$

73. $\dfrac{(x - y)^3}{(x + y)^2(x - 5y)}$ **75.** 🖩

Technology Connection, p. 384

1. Let $y_1 = ((x^2 + 3x + 2)/(x^2 + 4))/(5x^2 + 10x)$ and $y_2 = (x + 1)/((x^2 + 4)(5x))$. With the table set in AUTO mode, note that the values in the Y1- and Y2-columns match except for $x = -2$.
2. ERROR messages occur when division by 0 is attempted. Since the simplified expression has no factor of $x + 5$ or $x + 1$ in a denominator, no ERROR message occurs in Y2 for $x = -5$ or -1.

Exercise Set 6.2, pp. 384–386

1. $\dfrac{7x(x - 5)}{5(2x + 1)}$ **3.** $\dfrac{(a - 4)(a + 2)}{(a + 6)^2}$ **5.** $\dfrac{(2x + 3)(x + 1)}{4(x - 5)}$

7. $\dfrac{(a - 4)(a + 4)}{(a^2 + 4)(a^2 - 4)}$ **9.** $\dfrac{(x + 6)(x - 1)}{(3 + x)(x + 1)}$ **11.** $\dfrac{5a^2}{3}$

13. $\dfrac{4}{c^2 d}$ **15.** $\dfrac{x + 2}{x - 2}$ **17.** $\dfrac{(a^2 + 25)(a - 5)}{(a - 3)(a - 1)(a + 5)}$

19. $\dfrac{7(a + 3)}{a(a + 4)}$ **21.** $\dfrac{2a}{a - 2}$ **23.** $\dfrac{t - 5}{t + 5}$ **25.** $\dfrac{5(a + 6)}{a - 1}$

27. 1   **29.** $\dfrac{t + 4}{t + 2}$ **31.** $\dfrac{x + 2}{x + 3}$ **33.** $c(c - 2)$ **35.** $\dfrac{7}{3x}$

37. $\dfrac{1}{a^3 - 8a}$ **39.** $\frac{35}{24}$ **41.** $\dfrac{x^2}{20}$ **43.** $\dfrac{a^3}{b^3}$ **45.** $\dfrac{y + 5}{2y}$

47. $4(y - 2)$ **49.** $-\dfrac{a}{b}$ **51.** $\dfrac{(y + 3)(y^2 + 1)}{y + 1}$ **53.** $\frac{15}{16}$

55. $\dfrac{a - 5}{3(a - 1)}$ **57.** $\dfrac{(2x - 1)(2x + 1)}{x - 5}$ **59.** $\dfrac{(a - 5)(3a - 4)}{(a + 4)(2a - 7)}$

61. $\dfrac{1}{(c - 5)(5c - 3)}$ **63.** $\dfrac{(x - 4y)(x - y)}{(x + y)^3}$

65. $\dfrac{x^2 + 4x + 16}{(x + 4)^2}$ **67.** $\dfrac{(2a + b)^2}{2(a + b)}$ **69.** 🖩   **71.** $\frac{19}{12}$

72. $\frac{41}{24}$ **73.** $\frac{1}{18}$ **74.** $-\frac{1}{6}$ **75.** $-\frac{37}{20}$ **76.** $\frac{49}{45}$ **77.** 🖩

79. $\dfrac{a}{(c - 3d)(2a + 5b)}$ **81.** $\dfrac{a^2 - 2b}{a^2 + 3b}$ **83.** $\dfrac{(z + 4)^3}{3(z - 4)^2}$

85. $\dfrac{x(x^2 + 1)}{3(x + y - 1)}$ **87.** $\dfrac{3(y + 2)^3}{y(y - 1)}$ **89.** 〰️

Exercise Set 6.3, pp. 394–397

1. Numerators; denominator   **3.** Least common denominator; LCD   **5.** $\dfrac{10}{x}$ **7.** $\dfrac{3x + 5}{12}$ **9.** $\dfrac{9}{a + 3}$

11. $\dfrac{6}{a + 2}$ **13.** $\dfrac{2y + 7}{2y}$ **15.** 11   **17.** $\dfrac{3x + 5}{x + 1}$ **19.** $a + 5$

21. $x - 4$ **23.** 0   **25.** $\dfrac{1}{x + 2}$ **27.** $\dfrac{t - 4}{t + 3}$ **29.** $\dfrac{x + 5}{x - 6}$

31. $-\dfrac{5}{x - 4}$, or $\dfrac{5}{4 - x}$ **33.** $-\dfrac{1}{x - 1}$, or $\dfrac{1}{1 - x}$ **35.** 135

37. 72   **39.** 126   **41.** $12x^3$ **43.** $30a^4b^8$ **45.** $6(y - 3)$

47. $(x - 2)(x + 2)(x + 3)$ **49.** $t(t - 4)(t + 2)^2$

51. $30x^2y^2z^3$ **53.** $(a + 1)(a - 1)^2$ **55.** $(m - 3)(m - 2)^2$

57. $12x^3(x - 5)(x - 3)(x - 1)$

59. $2(x + 1)(x - 1)(x^2 + x + 1)$

61. $\dfrac{3a^3}{10a^6}, \dfrac{2b}{10a^6}$ **63.** $\dfrac{21y}{9x^4y^3}, \dfrac{4x^3}{9x^4y^3}$

65. $\dfrac{2x(x + 3)}{(x - 2)(x + 2)(x + 3)}, \dfrac{4x(x - 2)}{(x - 2)(x + 2)(x + 3)}$

67. 🖩   **69.** $-\frac{7}{9}, \frac{-7}{9}$ **70.** $\frac{-3}{2}, \frac{3}{-2}$ **71.** $-\frac{11}{36}$ **72.** $-\frac{7}{60}$

73. $x^2 - 9x + 18$ **74.** $s^2 - \pi r^2$ **75.** 🖩   **77.** $\dfrac{18x + 5}{x - 1}$

79. $\dfrac{x}{3x + 1}$ **81.** 30 strands   **83.** 60 strands

85. $72(t + 1)(t - 1)^4$ **87.** 70 minutes   **89.** 7:55 A.M.

91. 🖩

Exercise Set 6.4, pp. 403–405

1. LCD   **3.** Numerators; LCD   **5.** $\dfrac{4x + 9}{x^2}$ **7.** $-\dfrac{5}{24r}$

9. $\dfrac{2d^2 + 7c}{c^2 d^3}$ **11.** $\dfrac{-2xy - 18}{3x^2 y^3}$ **13.** $\dfrac{5x + 7}{18}$ **15.** $\dfrac{-x - 4}{6}$

17. $\dfrac{a^2 + 16a + 16}{16a^2}$ **19.** $\dfrac{7z - 12}{12z}$ **21.** $\dfrac{x^2 + 4xy + y^2}{x^2 y^2}$

23. $\dfrac{4x^2 - 13xt + 9t^2}{3x^2 t^2}$ **25.** $\dfrac{6x}{(x + 2)(x - 2)}$

27. $\dfrac{2x - 40}{(x - 5)(x + 5)}$ **29.** $\dfrac{11x + 2}{3x(x + 1)}$ **31.** $\dfrac{3 - 5t}{2t(t - 1)}$

33. $\dfrac{x^2 + 6x}{(x - 4)(x + 4)}$ **35.** $\dfrac{16}{3(z + 4)}$ **37.** $\dfrac{3x - 1}{(x - 1)^2}$

39. $\dfrac{1}{m^2 + m + 1}$ **41.** $\dfrac{9a}{4(a - 5)}$ **43.** 0

45. $\dfrac{10}{(a - 3)(a + 2)}$ **47.** $\dfrac{x - 5}{(x + 5)(x + 3)}$

49. $\dfrac{3z^2 + 19z - 20}{(z - 2)^2(z + 3)}$ **51.** $\dfrac{-5}{x^2 + 17x + 16}$ **53.** $\dfrac{5x - 3}{5}$

55. $y + 3$ **57.** $\dfrac{2b - 14}{b^2 - 16}$ **59.** $\dfrac{p^2 + 7p + 1}{(p - 5)(p + 5)}$

61. $\dfrac{13x + 20}{(4 + x)(4 - x)}$ **63.** $\dfrac{-a - 2}{(a + 1)(a - 1)}$, or $\dfrac{a + 2}{(1 + a)(1 - a)}$

65. $\dfrac{10x + 6y}{(x - y)(x + y)}$ **67.** $\dfrac{2x - 3}{2 - x}$ **69.** 2   **71.** 0   **73.** 🖩

75. $-\frac{13}{14}$ **76.** $-\frac{5}{9}$ **77.** $\frac{2}{15}$ **78.** $\frac{7}{6}$

79.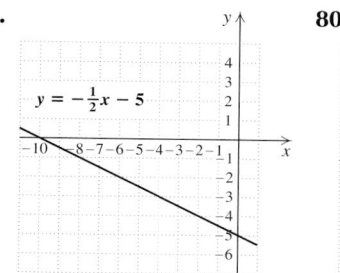

$y = -\frac{1}{2}x - 5$

80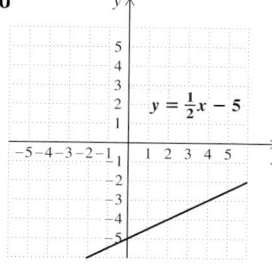

$y = \frac{1}{2}x - 5$

81.

83. Perimeter: $\dfrac{10x - 14}{(x - 5)(x + 4)}$; area: $\dfrac{6}{(x - 5)(x + 4)}$

85. $\dfrac{30}{(x - 3)(x + 4)}$ **87.** $\dfrac{x^4 + 4x^3 - 5x^2 - 126x - 441}{(x + 2)^2(x + 7)^2}$

89. $\dfrac{-x^2 - 3}{(2x - 3)(x - 3)}$ **91.** $\dfrac{a}{a - b} + \dfrac{3b}{b - a}$; answers may vary.

93.

Technology Connection, p. 410

1. $(1 - 1/x)/(1 - 1/x^2)$ **2.** Parentheses are needed to group separate terms into factors. When a fraction bar is replaced with a division sign, we need parentheses to preserve the groupings that had been created by the fraction bar. This holds for denominators and numerators alike.

Exercise Set 6.5, pp. 410–412

1. (d)   **3.** (b)   **5.** $\frac{6}{5}$ **7.** $\frac{117}{22}$ **9.** $\dfrac{4s^2}{9 + 3s^2}$ **11.** $\dfrac{4x}{3x + 2}$

13. $\dfrac{4a - 10}{a - 7}$ **15.** $x - 4$ **17.** $-\dfrac{1}{x}$ **19.** $\dfrac{1 + t^2}{t - t^2}$ **21.** $\dfrac{x}{x - y}$

23. $\dfrac{2a^2 + 7a}{5 - 3a^2}$ **25.** 1   **27.** $\dfrac{3a^2 + 4b^3}{5b^3 - 3a^2b^3}$ **29.** $\dfrac{2x^4 - 3x^2}{2x^4 + 3}$

31. $\dfrac{t^2 - 2}{t^2 + 5}$ **33.** $\dfrac{a^2b^2}{b^2 - ab + a^2}$ **35.** $\dfrac{3a^2b^3 + 4a}{3b^2 + ab^2}$

37. $\dfrac{t^2 + 5t + 3}{(t + 1)^2}$ **39.** $\dfrac{x^2 - 2x - 1}{x^2 - 5x - 4}$ **41.**   **43.** -4

44. -4 **45.** $\frac{19}{3}$ **46.** $-\frac{14}{27}$ **47.** $-3, 10$ **48.** $-10, 2$

49.   **51.** 6, 7, 8   **53.** $-\frac{4}{5}, \frac{9}{2}$ **55.** $\dfrac{P(i + 12)^2}{12(i + 24)}$

57. $\dfrac{(x - 8)(x - 1)(x + 1)}{x^2(x - 2)(x + 2)}$ **59.** 0   **61.** $\dfrac{2z(5z - 2)}{(z + 2)(13z - 6)}$

63.

Exercise Set 6.6, pp. 418–419

1. False   **3.** True   **5.** $-\frac{1}{2}$ **7.** $\frac{6}{7}$ **9.** $\frac{24}{7}$ **11.** $-5, -1$

13. $-6, 6$ **15.** 3   **17.** $\frac{14}{3}$ **19.** $\frac{35}{3}$ **21.** $-4, -3$ **23.** 5

25. $\frac{5}{2}$ **27.** -1 **29.** No solution   **31.** -10 **33.** $\frac{10}{7}$

35. $-2, \frac{7}{3}$ **37.** No solution   **39.** 2   **41.** No solution

43. No solution   **45.**   **47.** 137, 139   **48.** 14 yd

49. Base: 9 cm; height: 12 cm   **50.** $-8, -6; 6, 8$

51. 0.06 cm per day   **52.** 0.28 in. per day   **53.**

55. -2 **57.** $-\frac{1}{6}$ **59.** $-1, 0$ **61.** 4   **63.**

Exercise Set 6.7, pp. 429–433

1. 12 min   **3.** $\frac{48}{7}$ hr, or 6 hr, $51\frac{3}{7}$ min   **5.** $\frac{14}{3}$ min, or $4\frac{2}{3}$ min

7. $8\frac{4}{7}$ hr   **9.** Canon: 36 min; HP: 72 min   **11.** Erickson Air-Crane: 10 hr; S-58T: 40 hr   **13.** Mariah: $\frac{4}{3}$ hr; Stan: 4 hr

15. 8 hr

17.

	Distance (in km)	Speed (in km/h)	Time (in hours)
B & M	330	$r - 14$	$\dfrac{330}{r - 14}$
AMTRAK	400	r	$\dfrac{400}{r}$

AMTRAK: 80 km/h; B & M: 66 km/h

19. Bill: 50 mph; Hillary: 80 mph   **21.** 3 hr   **23.** 7 mph

25. 4.3 ft/sec   **27.** Freight: 66 mph; passenger: 80 mph

29. 9 km/h   **31.** 2 km/h   **33.** 10.5   **35.** $\frac{8}{3}$ **37.** $3\frac{3}{4}$ in.

39. 20 ft   **41.** $2\frac{2}{3}$ ft   **43.** 12.6   **45.** 105 steps per minute

47. 702 photos   **49.** \$32,340   **51.** 20 duds   **53.** 184

55. **(a)** 1.92 T; **(b)** 28.8 lb   **57.**

59. 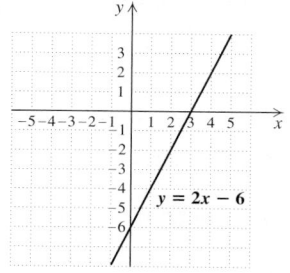   **60.**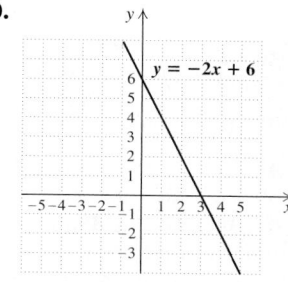

$y = 2x - 6$

$y = -2x + 6$

61. 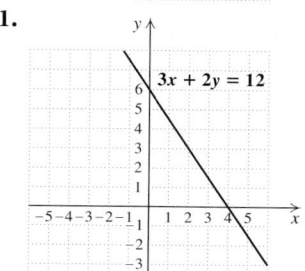   **62.**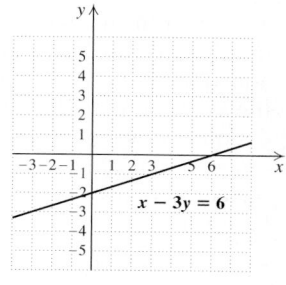

$3x + 2y = 12$

$x - 3y = 6$

63.

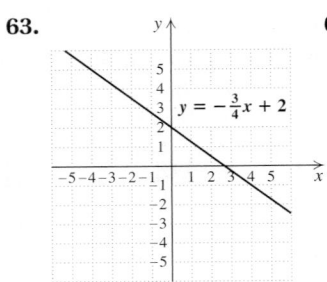

$y = -\frac{3}{4}x + 2$

64.

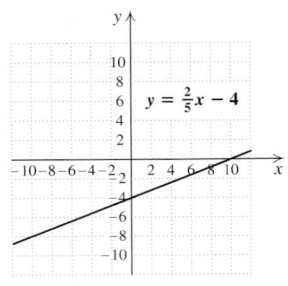

$y = \frac{2}{5}x - 4$

65. 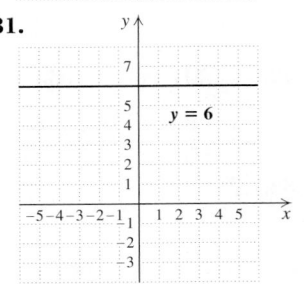 **67.** Michelle: 6 hr; Sal: 3 hr; Kristen: 4 hr
69. $30\frac{22}{31}$ hr **71.** Add 80 oz of gasoline. **73.** $\frac{36}{68}$
75. $\frac{B}{A} = \frac{D}{C}; \frac{A}{C} = \frac{B}{D}; \frac{C}{A} = \frac{D}{B}$ **77.**

Review Exercises: Chapter 6, pp. 436–437

1. False **2.** True **3.** False **4.** True **5.** True **6.** False
7. False **8.** False **9.** 0 **10.** 4 **11.** $-6, 6$ **12.** $-6, 5$
13. -2 **14.** $\frac{x-2}{x+1}$ **15.** $\frac{7x+3}{x-3}$ **16.** $\frac{y-5}{y+5}$
17. $-5(x+2y)$ **18.** $\frac{a-6}{5}$ **19.** $\frac{8(t+1)}{(2t-1)(t-1)}$ **20.** $-32t$
21. $\frac{2x(x-1)}{x+1}$ **22.** $\frac{(x^2+1)(2x+1)}{(x-2)(x+1)}$ **23.** $\frac{(t+4)^2}{t+1}$
24. $24a^5b^9$ **25.** $x^4(x-1)(x+1)$
26. $(y-2)(y+2)(y+1)$ **27.** $\frac{15-3x}{x+3}$ **28.** -1
29. $\frac{4}{x-4}$ **30.** $\frac{x+5}{2x}$ **31.** $\frac{2x+5}{x-2}$ **32.** $\frac{2a}{a-1}$ **33.** $d+c$
34. $\frac{-x^2+x+26}{(x+1)(x-5)(x+5)}$ **35.** $\frac{2(x-2)}{x+2}$ **36.** $\frac{19x+8}{10x(x+2)}$
37. $\frac{z}{1-z}$ **38.** $\frac{x^3(2xy^2+1)}{y(1+x)}$ **39.** $c-d$ **40.** 8 **41.** $-\frac{1}{2}$
42. $-5, 3$ **43.** $5\frac{1}{7}$ hr **44.** Celeron: 45 sec; Pentium 4:
30 sec **45.** Car: 105 km/h; train: 90 km/h **46.** 24 mph
47. 60 **48.** 6 **49.** The LCM of denominators is used
to clear fractions when simplifying a complex rational
expression using the method of multiplying by the LCD,
and when solving rational equations. **50.** Although
multiplying the denominators of the expressions being
added results in a common denominator, it is often not
the *least* common denominator. Using a common
denominator other than the LCD makes the expressions
more complicated, requires additional simplifying after
the addition has been performed, and leaves more room
for error.
51. $\frac{5(a+3)^2}{a}$ **52.** $\frac{10a}{(a-b)(b-c)}$ **53.** 0 **54.** 44%

Test: Chapter 6, p. 438

1. [6.1] 0 **2.** [6.1] -8 **3.** [6.1] $-7, 7$ **4.** [6.1] 1, 2
5. [6.1] $\frac{3x+7}{x+3}$ **6.** [6.2] $\frac{-2(a+5)}{3}$ **7.** [6.2] $\frac{(5y+1)(y+1)}{3y(y+2)}$
8. [6.2] $\frac{(2x+1)(2x-1)(x^2+1)}{(x-1)^2(x-2)}$ **9.** [6.2] $(x+3)(x-3)$
10. [6.3] $(y-3)(y+3)(y+7)$ **11.** [6.3] $\frac{-3x+9}{x^3}$
12. [6.3] $\frac{-2t+8}{t^2+1}$ **13.** [6.4] $\frac{3}{3-x}$ **14.** [6.4] $\frac{2x-5}{x-3}$
15. [6.4] $\frac{11t-8}{t(t-2)}$ **16.** [6.4] $\frac{-x^2-7x-15}{(x+4)(x-4)(x+1)}$
17. [6.4] $\frac{x^2+2x-7}{(x+1)(x-1)^2}$ **18.** [6.5] $\frac{3y+1}{y}$
19. [6.5] $\frac{a^2(3b^2-2a)}{b^2(a^3+2)}$ **20.** [6.6] 12 **21.** [6.6] $-3, 5$
22. [6.7] 12 min **23.** [6.7] $2\frac{1}{7}$ cups **24.** [6.7] Craig:
65 km/h; Marilyn: 45 km/h **25.** [6.7] Rema: 4 hr;
Reggie: 10 hr **26.** [6.5] a **27.** [6.7] -1

Cumulative Review: Chapters 1–6, pp. 439–440

1. $2b+a$ **2.** $>$ **3.** 25 **4.** $-8x+28$ **5.** $-\frac{43}{8}$ **6.** 1
7. -6.2 **8.** 8 **9.** 10 **10.** $-7, 7$ **11.** $-\frac{10}{3}$ **12.** -3
13. $\frac{8}{3}$ **14.** $-10, -1$ **15.** -8 **16.** 1, 4 **17.** $\{y | y \le -\frac{2}{3}\}$
18. -17 **19.** $-4, \frac{1}{2}$ **20.** $\{x | x > 43\}$ **21.** 5 **22.** $-\frac{7}{2}, 5$
23. -13 **24.** $b = 3a - c + 9$ **25.** $y = \frac{4z-3x}{6}$
26. $\frac{3}{2}x + 2y - 3z$ **27.** $-4x^3 - \frac{1}{7}x^2 - 2$

28.

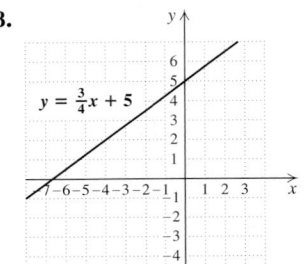

$y = \frac{3}{4}x + 5$

29.

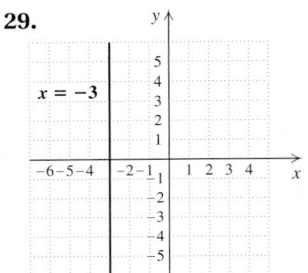

$x = -3$

30.

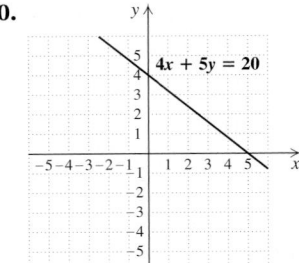

$4x + 5y = 20$

31.

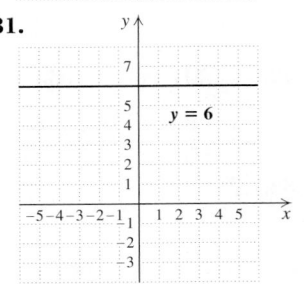

$y = 6$

32. -2 **33.** $\frac{1}{2}$; $(0, -2)$ **34.** $y = -3x + 1$

35. (a) $y = -0.1x + 13.1$; (b) 7.1 miles per gallon   **36.** $\dfrac{1}{x^2}$

37. y^{-6}, or $\dfrac{1}{y^6}$ **38.** $-4a^4b^{14}$ **39.** $-y^3 - 2y^2 - 2y + 7$

40. $15a - 10b + 5c$ **41.** $2x^5 + x^3 - 6x^2 - x + 3$

42. $36x^2 - 60xy + 25y^2$ **43.** $2x^2 + 11x - 40$ **44.** $4x^6 - 1$

45. $2x(3 - x - 12x^3)$ **46.** $(4x + 9)(4x - 9)$

47. $(t - 4)(t - 6)$ **48.** $(4x + 3)(2x + 1)$

49. $2(3x - 2)(x - 4)$ **50.** $4(t + 3)(t - 3)$ **51.** $(5t + 4)^2$

52. $10(t^2 - 2)(t^4 + 2t^2 + 4)$ **53.** $(x + 2)(x^3 - 3)$

54. $\dfrac{y + 6}{2}$ **55.** 1 **56.** $\dfrac{a^2 + 7ab + b^2}{(a + b)(a - b)}$ **57.** $\dfrac{2x + 5}{4 - x}$

58. $\dfrac{x}{x - 2}$ **59.** $\dfrac{t(2t^2 + 1)}{t^3 - 2}$ **60.** $6x^2 - 5x + 2 + \dfrac{4}{x} + \dfrac{1}{x^2}$

61. $15x^3 - 57x^2 + 177x - 529 + \dfrac{1605}{x + 3}$

62. At most 225 sheets   **63.** \$3.60   **64.** 14 ft

65. $-278, -276$ **66.** 30 min   **67.** 50   **68.** 26 in.   **69.** 12

70. $-144, 144$ **71.** $16y^6 - y^4 + 6y^2 - 9$

72. $2(a^{16} + 81b^{20})(a^8 + 9b^{10})(a^4 + 3b^5)(a^4 - 3b^5)$

73. $-7, 4, 12$ **74.** -7 **75.** 18   **76.** $66\frac{2}{3}\%$ **77.** \$8.32

CHAPTER 7

Exercise Set 7.1, pp. 449–454

1. Domain   **3.** Exactly   **5.** 5   **7.** x **9.** Yes   **11.** Yes

13. Yes   **15.** No   **17.** Function   **19.** Function

21. (a) -1; (b) -3 **23.** (a) 3; (b) 3   **25.** (a) 3; (b) 0

27. (a) 3; (b) -3 **29.** (a) 1; (b) 3   **31.** (a) 4; (b) $-1, 3$

33. (a) 2; (b) $\{x \,|\, 0 < x \le 2\}$

35. (a) 5; (b) 1; (c) -2; (d) 13; (e) $a + 7$; (f) $a + 7$

37. (a) 0; (b) 1; (c) 57; (d) $5t^2 + 4t$; (e) $20a^2 + 8a$; (f) 48

39. (a) $\frac{3}{5}$; (b) $\frac{1}{3}$; (c) $\frac{4}{7}$; (d) 0; (e) $\dfrac{x - 1}{2x - 1}$

41. $4\sqrt{3} \text{ cm}^2 \approx 6.93 \text{ cm}^2$ **43.** $36\pi \text{ in}^2 \approx 113.10 \text{ in}^2$

45. $1\frac{20}{33}$ atm; $1\frac{10}{11}$ atm; $4\frac{1}{33}$ atm   **47.** 159.48 cm   **49.** 14°F

51. 75 heart attacks per 10,000 men   **53.** 56%

55. 60 watts; 140 watts

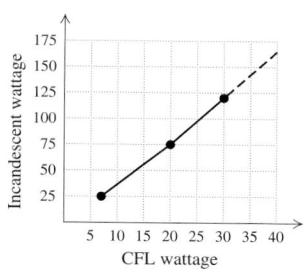

57. 3.5 drinks; 6 drinks

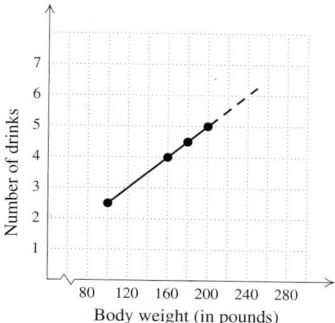

59. 57,000; 50,000

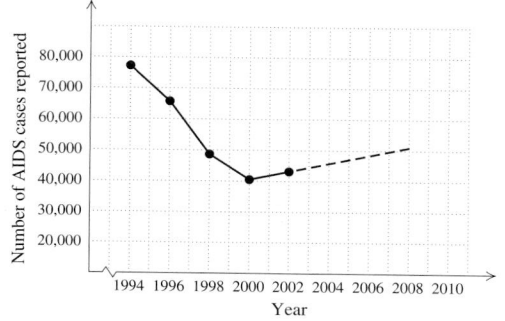

61. \$257,000; \$306,000

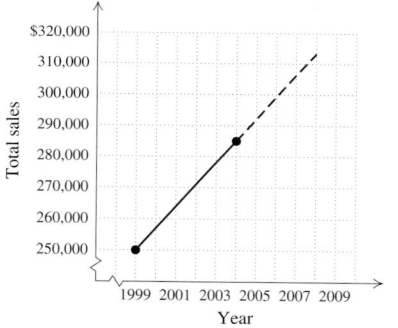

63. 📓   **65.** $\frac{1}{3}$ **66.** -1 **67.** $l = \dfrac{S - 2wh}{2h + 2w}$

68. $w = \dfrac{S - 2lh}{2l + 2h}$ **69.** $y = -\frac{2}{3}x + 2$ **70.** $y = \frac{5}{4}x - 2$

71. 📓   **73.** 26; 99   **75.** Worm   **77.** About 2 min, 50 sec

79. 1 every 3 min   **81.** $g(x) = \frac{15}{4}x - \frac{13}{4}$

Exercise Set 7.2, pp. 461–464

1. (c)   **3.** (d)   **5.** (c)

7. Domain: $\{2, 9, -2, -4\}$; range: $\{8, 3, 10, 4\}$

9. Domain: $\{0, 4, -5, -1\}$; range: $\{0, -2\}$

11. Domain: $\{-4, -2, 0, 2, 4\}$; range: $\{-2, -1, 0, 1, 2\}$

13. Domain: $\{-5, -3, -1, 0, 2, 4\}$; range: $\{-1, 1\}$

15. Domain: $\{x \,|\, -4 \le x \le 3\}$; range: $\{y \,|\, -3 \le y \le 4\}$

17. Domain: $\{x \,|\, -4 \le x \le 5\}$; range: $\{y \,|\, -2 \le y \le 4\}$

19. Domain: $\{x \mid -4 \le x \le 4\}$; range: $\{-3, -1, 1\}$
21. Domain: $\mathbb{R}$; range: $\mathbb{R}$ **23.** Domain: $\mathbb{R}$; range: $\{4\}$
25. Domain: $\mathbb{R}$; range: $\{y \mid y \ge 1\}$
27. Domain: $\{x \mid x$ is a real number *and* $x \ne -2\}$; range: $\{y \mid y$ is a real number *and* $y \ne -4\}$
29. Domain: $\{x \mid x \ge 0\}$; range: $\{y \mid y \ge 0\}$
31. $\{x \mid x$ is a real number *and* $x \ne 3\}$
33. $\{x \mid x$ is a real number *and* $x \ne \frac{1}{2}\}$ **35.** $\mathbb{R}$ **37.** $\mathbb{R}$
39. $\{x \mid x$ is a real number *and* $x \ne 3$ *and* $x \ne -3\}$ **41.** $\mathbb{R}$
43. $\{x \mid x$ is a real number *and* $x \ne -1$ *and* $x \ne -7\}$
45. $\{t \mid 0 \le t < 624\}$ **47.** $\{p \mid \$0 \le p \le \$10.60\}$
49. $\{d \mid d \ge 0\}$ **51.** $\{t \mid 0 \le t \le 5\}$ **53. (a)** -5;
(b) 1; **(c)** 21   **55. (a)** 0; **(b)** 2; **(c)** 7   **57. (a)** 100;
(b) 100; **(c)** 131   **59.**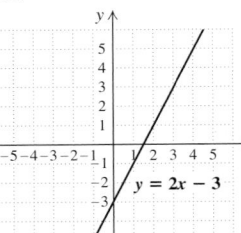

61.   **62.**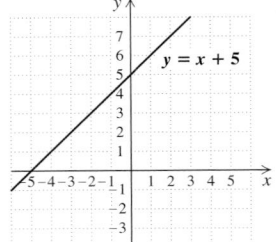

63. Slope: $\frac{2}{3}$; y-intercept: $(0, -4)$
64. Slope: $-\frac{1}{4}$; y-intercept: $(0, 6)$
65. Slope: $\frac{4}{3}$; y-intercept: $(0, 0)$
66. Slope: -5; y-intercept: $(0, 0)$ **67.**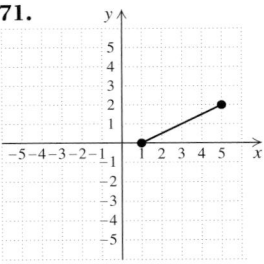

69.   **71.**

73. Domain: $\{x \mid x$ is a real number *and* $x \ne 0\}$; range: $\{y \mid y$ is a real number *and* $y \ne 0\}$
75. Domain: $\{x \mid x < -2$ *or* $x > 0\}$; range: $\{y \mid y < -2$ *or* $y > 3\}$
77. Domain: $\mathbb{R}$; range: $\{y \mid y \ge 0\}$
79. Domain: $\{x \mid x$ is a real number *and* $x \ne 2\}$; range: $\{y \mid y$ is a real number *and* $y \ne 0\}$ **81.** $\{h \mid 0 \le h \le 144\}$
83. 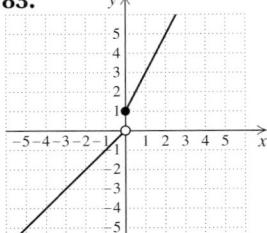   **85.**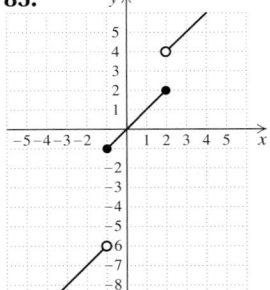

87. 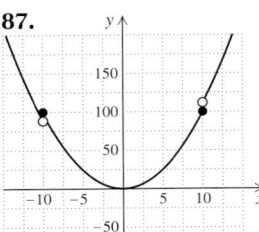 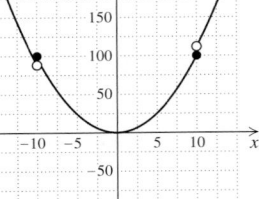   **89.**

Exercise Set 7.3, pp. 473–477

1. False   **3.** False   **5.** True   **7.** Yes   **9.** Yes   **11.** No
13. No   **15.** Yes   **17.** No
19. 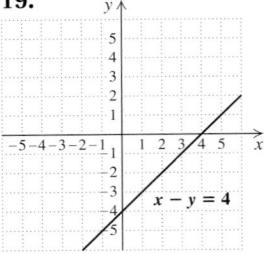   **21.**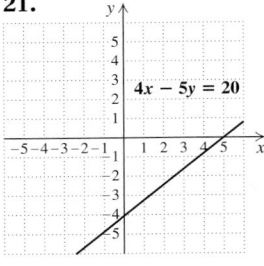

23. 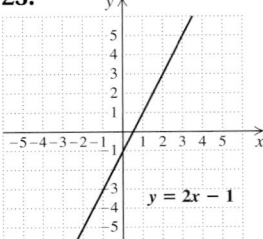   **25.**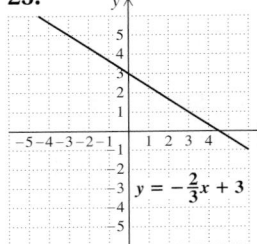

27. 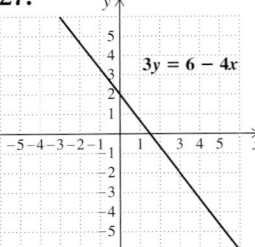   **29.**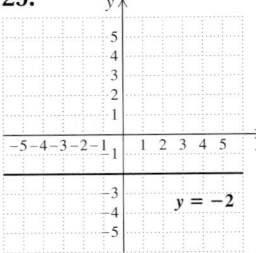

31. 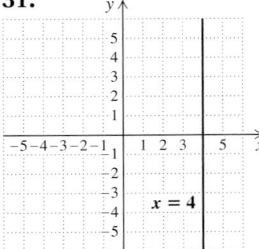   **33.**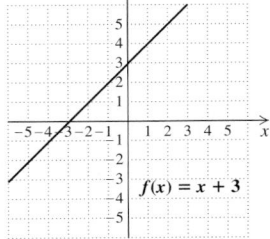

35.

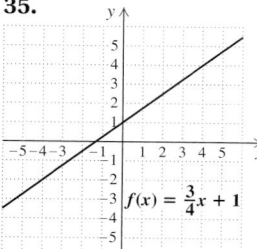

37.

$g(x) = 4$

39. $C(t) = 25t + 50$; 4 months   **41.** $L(t) = \frac{1}{2}t + 1$;
4 months after the cut   **43.** $C(d) = \$0.75d + 2$; 3 mi
45. **(a)** $C(t) = \frac{335}{29}t + 1542$; **(b)** 1981 calories;
(c) approximately 2011
47. **(a)** $E(t) = 0.23t + 71.8$; **(b)** 76.2 yr
49. **(a)** $A(t) = 10.34t + 178.6$; **(b)** \$344 million
51. **(a)** $N(t) = 4t + 48$; **(b)** 88 million Americans; **(c)** 2013
53. **(a)** $A(t) = 0.55t + 74.9$; **(b)** 83.15 million acres
55. Linear function; $\mathbb{R}$ **57.** Quadratic function; $\mathbb{R}$
59. Rational function; $\{t \mid t$ is a real number $and\ t \neq -\frac{4}{3}\}$
61. Polynomial function; $\mathbb{R}$ **63.** Rational function; $\{x \mid x$
is a real number $and\ x \neq \frac{5}{2}\}$ **65.** Rational function; $\{n \mid n$
is a real number $and\ n \neq -1\ and\ n \neq -2\}$ **67.** Linear
function; $\mathbb{R}$ **69.** $\{y \mid y \geq 0\}$ **71.** $\mathbb{R}$ **73.** $\{y \mid y \leq 0\}$
75.

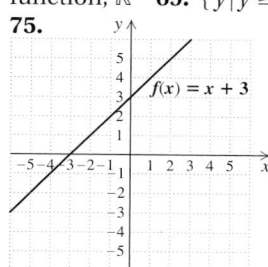

$f(x) = x + 3$

77.

$f(x) = -1$

Domain: $\mathbb{R}$; range: $\mathbb{R}$
79.

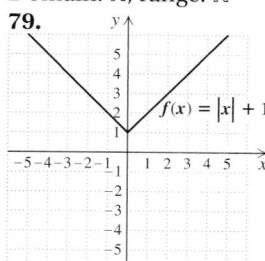

$f(x) = |x| + 1$

81.

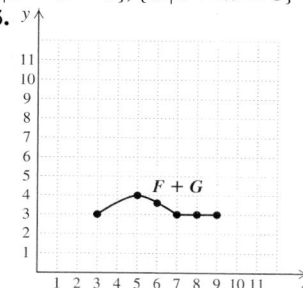

$g(x) = x^2$

Domain: $\mathbb{R}$; range: $\{-1\}$

Domain: $\mathbb{R}$;
range: $\{y \mid y \geq 1\}$

Domain: $\mathbb{R}$;
range: $\{y \mid y \geq 0\}$

83. ▨   **85.** $4x^2 + 2x - 1$
86. $2x^3 - x^2 - x + 7$ **87.** $2x^2 - 13x - 7$
88. $x^2 + x - 12$ **89.** $x^3 - 2x^2 - 3x + 5$
90. $x^3 + 2x^2 + x + 4$ **91.** ▨   **93.** False
95. False   **97.** 21.1°C   **99.** \$30
101. **(a)** $g(x) = x - 8$; **(b)** -10; **(c)** 83
103. $f(x) = 0.299x - 515.3622222$

Exercise Set 7.4, pp. 483–486
1. Domain   **3.** Evaluate   **5.** Excluding   **7.** 1   **9.** -41
11. 12   **13.** $\frac{13}{18}$ **15.** 5   **17.** $x^2 - 3x + 3$ **19.** $x^2 - x + 3$
21. 23   **23.** 5   **25.** 56   **27.** $\frac{x^2 - 2}{5 - x}, x \neq 5$ **29.** $\frac{2}{7}$
31. 4%   **33.** $1.3 + 2.2 = 3.5$ million   **35.** About
50 million; the number of passengers using Newark
Liberty and LaGuardia in 1998   **37.** About 81 million; the
number of passengers using the three airports in 2002
39. About 51 million; the number of passengers using
LaGuardia and Newark Liberty in 2002   **41.** $\mathbb{R}$
43. $\{x \mid x$ is a real number $and\ x \neq 3\}$
45. $\{x \mid x$ is a real number $and\ x \neq 0\}$
47. $\{x \mid x$ is a real number $and\ x \neq 1\}$
49. $\{x \mid x$ is a real number $and\ x \neq 2\ and\ x \neq 4\}$
51. $\{x \mid x$ is a real number $and\ x \neq 3\}$
53. $\{x \mid x$ is a real number $and\ x \neq 4\}$
55. $\{x \mid x$ is a real number $and\ x \neq 4\ and\ x \neq 5\}$
57. $\{x \mid x$ is a real number $and\ x \neq -1\ and\ x \neq -\frac{5}{2}\}$
59. 4; 3   **61.** 5; -1 **63.** $\{x \mid 0 \leq x \leq 9\}$; $\{x \mid 3 \leq x \leq 10\}$;
$\{x \mid 3 \leq x \leq 9\}$; $\{x \mid 3 \leq x \leq 9\}$
65.

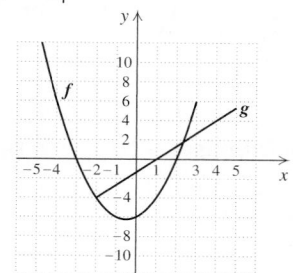

$F + G$

67. ▨   **69.** $x = \frac{7}{4}y + 2$ **70.** $y = \frac{3}{8}x - \frac{5}{8}$
71. $y = -\frac{5}{2}x - \frac{3}{2}$ **72.** $x = -\frac{5}{6}y - \frac{1}{3}$
73. Let n represent the number; $2n + 5 = 49$
74. Let x represent the number; $\frac{1}{2}x - 3 = 57$
75. Let x represent the first integer; $x + (x + 1) = 145$
76. Let n represent the number; $n - (-n) = 20$
77. ▨   **79.** $\{x \mid x$ is a real number $and\ x \neq -\frac{5}{2}\ and$
$x \neq -3\ and\ x \neq 1\ and\ x \neq -1\}$
81. Answers may vary.

83. $\{x\,|\,x$ is a real number $and -1 < x < 5$ $and\ x \neq \frac{3}{2}\}$

85. Answers may vary. $f(x) = \dfrac{1}{x+2}$, $g(x) = \dfrac{1}{x-5}$ **87.**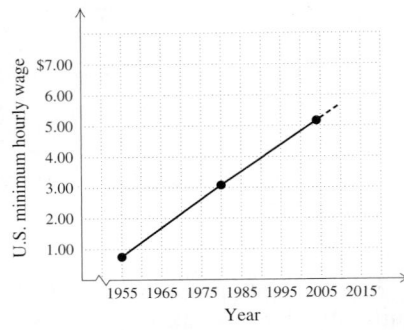

Exercise Set 7.5, pp. 494–499

1. LCD   **3.** Factor   **5.** Inverse   **7.** Direct   **9.** Inverse

11. $d = \dfrac{L}{f}$ **13.** $v_1 = \dfrac{2s}{t} - v_2$, or $\dfrac{2s - tv_2}{t}$ **15.** $b = \dfrac{at}{a-t}$

17. $R = \dfrac{2V}{I} - 2r$, or $\dfrac{2V - 2Ir}{I}$ **19.** $g = \dfrac{Rs}{s-R}$

21. $n = \dfrac{IR}{E - Ir}$ **23.** $q = \dfrac{pf}{p-f}$ **25.** $t_1 = \dfrac{H}{Sm} + t_2$, or

$\dfrac{H + Smt_2}{Sm}$ **27.** $r = \dfrac{Re}{E-e}$ **29.** $r = 1 - \dfrac{a}{S}$, or $\dfrac{S-a}{S}$

31. $a + b = \dfrac{f}{c^2}$ **33.** $r = \dfrac{A}{P} - 1$, or $\dfrac{A-P}{P}$

35. $t_2 = \dfrac{d_2 - d_1}{v} + t_1$, or $\dfrac{d_2 - d_1 + t_1 v}{v}$ **37.** $b^2 = \dfrac{a^2 y^2}{a^2 - x^2}$

39. $Q = \dfrac{2Tt - 2AT}{A - q}$ **41.** $k = 7; y = 7x$

43. $k = 1.7; y = 1.7x$ **45.** $k = 6; y = 6x$

47. $k = 60; y = \dfrac{60}{x}$ **49.** $k = 112; y = \dfrac{112}{x}$

51. $k = 9; y = \dfrac{9}{x}$ **53.** 241,920,000 cans   **55.** 6 amperes

57. 3.5 hr   **59.** 32 kg   **61.** 50 min   **63.** About 21 min

65. 122,269,230 tons   **67.** $y = \frac{2}{3}x^2$ **69.** $y = \dfrac{54}{x^2}$

71. $y = 0.3xz^2$ **73.** $y = \dfrac{4wx^2}{z}$ **75.** About 2.9 sec

77. 308 cm³   **79.** About 57.42 mph   **81.** 🖳   **83.** $\frac{13}{2}$

84. $\frac{5}{3}$ **85.** $-\frac{3}{2}$ **86.** $-\frac{5}{3}, \frac{7}{2}$ **87.** No solution; contradiction

88. All real numbers; identity   **89.** 🖳   **91.** 567 mi

93. Ratio is $\dfrac{a + 12}{a + 6}$; percent increase is $\dfrac{6}{a+6} \cdot 100\%$,

or $\dfrac{600}{a+6}\%$ **95.** $t_1 = t_2 + \dfrac{(d_2 - d_1)(t_4 - t_3)}{a(t_4 - t_2)(t_4 - t_3) + d_3 - d_4}$

97. The intensity is halved.   **99.** About 1.697 m

101. $d(s) = \dfrac{28}{s}$; 70 yd

Review Exercises: Chapter 7, pp. 502–505

1. True   **2.** True   **3.** False   **4.** True   **5.** False   **6.** True
7. True   **8.** True   **9.** True   **10.** False   **11. (a)** 3;
(b) $\{x\,|\,{-2} \le x \le 4\}$; **(c)** -1 **(d)** $\{y\,|\,1 \le y \le 5\}$ **12.** $\frac{3}{2}$
13. $4a^2 + 4a - 3$ **14.** 10.53 yr

15. About \$3.50;

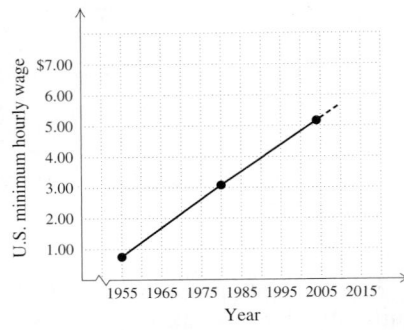

16. About \$5.60   **17. (a)** Yes; **(b)** domain: $\mathbb{R}$;
range: $\{y\,|\,y \ge 0\}$ **18. (a)** No   **19. (a)** No   **20. (a)** Yes;
(b) domain: $\mathbb{R}$; range: $\{-2\}$ **21.** $\mathbb{R}$
22. $\{x\,|\,x$ is a real number $and\ x \neq 1\}$
23. $\{t\,|\,t$ is a real number $and\ t \neq -1\ and\ t \neq -4\}$
24. $\{x\,|\,x \ge 0\}$ **25.** $\{t\,|\,0 \le t \le 34\}$ **26. (a)** 5; **(b)** 4;
(c) 16; **(d)** 35   **27.** $C(t) = 15t + 75$; 7 months
28. (a) $v(t) = -2.2t + 48.5$; **(b)** 44.1 million visitors
29. Absolute-value function   **30.** Polynomial function
31. Quadratic function   **32.** Linear function
33. Rational function

34.

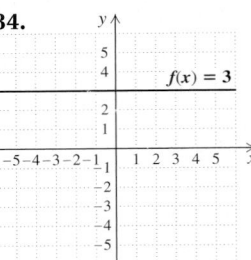

Domain: $\mathbb{R}$; range: $\{3\}$

35.

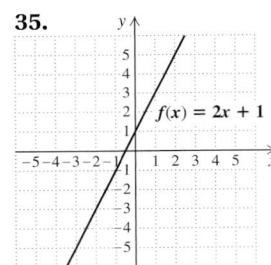

Domain: $\mathbb{R}$; range: $\mathbb{R}$

36.

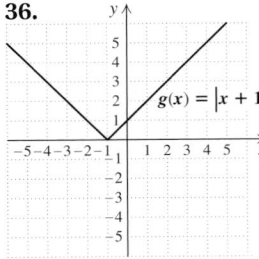

Domain: $\mathbb{R}$; range: $\{y\,|\,y \ge 0\}$
37. 102   **38.** -17 **39.** $-\frac{9}{2}$ **40.** $\mathbb{R}$
41. $\{x\,|\,x$ is a real number $and\ x \neq 2\}$ **42.** $y = \frac{15}{2}x$

43. $y = \dfrac{\frac{3}{4}}{x}$ **44.** $y = \dfrac{1}{2} \cdot \dfrac{xw^2}{z}$ **45.** About 22 lb   **46.** 64 L

47. 🖳 Two functions that have the same domain and
range are not necessarily identical. For example, the
functions $f: \{(-2, 1), (-3, 2)\}$ and $g: \{(-2, 2), (-3, 1)\}$ have
the same domain and range but are different functions.

48. Jenna is not correct. Any value of the variable that makes a denominator 0 is not in the domain; 0 itself may or may not make a denominator 0.
49. $f(x) = 3.09x + 3.75$
50. Domain: $\{x \mid x \geq -4 \text{ and } x \neq 2\}$; range: $\{y \mid y \geq 0 \text{ and } y \neq 3\}$
51. No; the rate of change is not constant. Each year the amount of the raise will be higher since the current year's salary is higher than the previous year's salary.

Test: Chapter 7, pp. 505–506

1. [7.1], [7.2] **(a)** 1; **(b)** $\{x \mid -3 \leq x \leq 4\}$; **(c)** 3; **(d)** $\{y \mid -1 \leq y \leq 2\}$ **2.** [7.1] 2 **3.** [7.1] 46 million
4. (a) [7.3] Yes; **(b)** [7.2] domain: $\mathbb{R}$; range: $\mathbb{R}$
5. (a) [7.3] Yes; **(b)** [7.2] domain: $\mathbb{R}$; range: $\{y \mid y \geq 1\}$
6. (a) [7.3] No **7.** [7.2] $\{t \mid 0 \leq t \leq 4\}$
8. [7.2] **(a)** -5; **(b)** 10
9. [7.3] $c(n) = 25t + 75$; 17 people
10. [7.3] **(a)** $C(m) = 0.3m + 25$; **(b)** $175
11. [7.3] Linear function; $\mathbb{R}$
12. [7.3] Rational function; $\{x \mid x \text{ is a real number } and \, x \neq 5\}$ **13.** [7.3] Quadratic function; $\mathbb{R}$
14. [7.3]

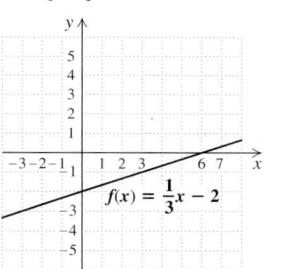

Domain: $\mathbb{R}$; range: $\mathbb{R}$

15. [7.3]

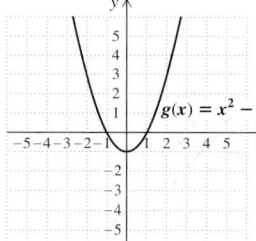

Domain: $\mathbb{R}$; range: $\{y \mid y \geq -1\}$

16. [7.3]

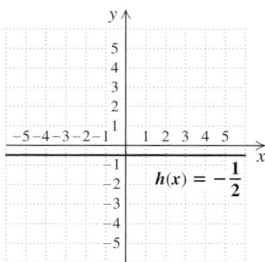

Domain: $\mathbb{R}$; range: $\left\{-\frac{1}{2}\right\}$

17. [7.1] $\frac{1}{4}$ **18.** [7.1] -10 **19.** [7.1] $\dfrac{1}{2a + 4}$
20. [7.4] $-\frac{8}{3}$ **21.** [7.2] $\{x \mid x \text{ is a real number } and \, x \neq -4\}$
22. [7.2] $\mathbb{R}$ **23.** [7.4] $\{x \mid x \text{ is a real number } and \, x \neq -4\}$
24. [7.4] $\{x \mid x \text{ is a real number } and \, x \neq -4 \text{ and } x \neq 7\}$
25. [7.5] $y = \frac{1}{2}x$ **26.** [7.5] 30 workers **27.** [7.5] $\frac{833}{125}$, or 6.664 in^2 **28.** [7.3] **(a)** 30 mi; **(b)** 15 mph
29. [7.4] $h(x) = 7x - 2$

CHAPTER 8

Technology Connection, p. 513

1. $(1.53, 2.58)$ **2.** $(-0.26, 57.06)$ **3.** $(2.23, 1.14)$
4. $(0.87, -0.32)$

Exercise Set 8.1, pp. 514–516

1. True **3.** True **5.** True **7.** False **9.** Yes **11.** No
13. Yes **15.** Yes **17.** $(4, 1)$ **19.** $(2, -1)$ **21.** $(4, 3)$
23. $(-3, -2)$ **25.** $(-3, 2)$ **27.** $(3, -7)$ **29.** $(7, 2)$
31. $(4, 0)$ **33.** No solution **35.** $\{(x, y) \mid y = 3 - x\}$
37. All except 33 **39.** 35 **41.** Let x represent the first number and y the second number; $x + y = 50, x = 0.25y$
43. Let m represent the number of ounces of mineral oil and v the number of ounces of vinegar; $m + v = 16, m = 2v + 4$ **45.** Let x and y represent the angles; $x + y = 180, x = 2y - 3$ **47.** Let x represent the number of two-point shots and y the number of foul shots; $x + y = 64, 2x + y = 100$ **49.** Let x represent the number of $8.50 brushes sold and y the number of $9.75 brushes sold; $x + y = 45, 8.50x + 9.75y = 398.75$
51. Let h represent the number of vials of Humulin sold and n the number of vials of Novolin Velosulin; $h + n = 50, 27.06h + 34.39n = 1565.57$ **53.** Let l represent the length, in feet, and w the width, in feet; $2l + 2w = 288, l = w + 44$ **55.** **57.** 15 **58.** $\frac{19}{12}$
59. $\frac{9}{20}$ **60.** $\frac{13}{3}$ **61.** $y = -\frac{3}{4}x + \frac{7}{4}$ **62.** $y = \frac{2}{5}x - \frac{9}{5}$ **63.**
65. Answers may vary. **(a)** $x + y = 6, x - y = 4$;
(b) $x + y = 1, 2x + 2y = 3$; **(c)** $x + y = 1, 2x + 2y = 2$
67. $A = -\frac{17}{4}, B = -\frac{12}{5}$ **69.** Let x and y represent the number of years that Lou and Juanita have taught at the university, respectively; $x + y = 46, x - 2 = 2.5(y - 2)$
71. Let s and v represent the number of ounces of baking soda and vinegar needed, respectively; $s = 4v, s + v = 16$
73. $(0, 0), (1, 1)$ **75.** $(0.07, -7.95)$ **77.** $(0.00, 1.25)$

Exercise Set 8.2, pp. 523–524

1. (d) **3.** (a) **5.** (c) **7.** $(2, -3)$ **9.** $(-4, 3)$ **11.** $(2, -2)$ **13.** $\{(x, y) \mid 2x - 3 = y\}$ **15.** $(-2, 1)$ **17.** $\left(\frac{1}{2}, \frac{1}{2}\right)$
19. $\left(\frac{25}{23}, -\frac{11}{23}\right)$ **21.** No solution **23.** $(1, 2)$ **25.** $(2, 7)$
27. $(-1, 2)$ **29.** $\left(\frac{128}{31}, -\frac{17}{31}\right)$ **31.** $(6, 2)$ **33.** No solution
35. $\left(\frac{110}{19}, -\frac{12}{19}\right)$ **37.** $(3, -1)$ **39.** $\{(x, y) \mid -4x + 2y = 5\}$
41. $\left(\frac{140}{13}, -\frac{50}{13}\right)$ **43.** $(-2, -9)$ **45.** $(30, 6)$
47. $\{(x, y) \mid x = 2 + 3y\}$ **49.** No solution **51.** $(140, 60)$
53. $\left(\frac{1}{3}, -\frac{2}{3}\right)$ **55.** **57.** 4 mi **58.** 86
59. $11\frac{2}{3}$ billion on bathrooms; $23\frac{1}{3}$ billion on kitchens
60. 30 m, 90 m, 360 m **61.** 450.5 mi **62.** 460.5 mi
63. **65.** $m = -\frac{1}{2}, b = \frac{5}{2}$ **67.** $a = 5, b = 2$
69. $\left(-\frac{32}{17}, \frac{38}{17}\right)$ **71.** $\left(-\frac{1}{5}, \frac{1}{10}\right)$ **73.**

Exercise Set 8.3, pp. 535–539

1. 10, 40   **3.** Mineral oil: 12 oz; vinegar: 4 oz   **5.** 119°, 61°
7. Two-point shots: 36; foul shots: 28   **9.** $8.50-brushes:
32; $9.75-brushes: 13   **11.** Humulin vials: 21; Novolin
Velosulin vials: 29   **13.** Width: 50 ft; length: 94 ft
15. Nonrecycled sheets: 38; recycled sheets: 112
17. General Electric bulbs: 60; SLi bulbs: 140   **19.** HP
cartridges: 15; Apple cartridges: 35   **21.** Kenyan: 8 lb;
Sumatran: 12 lb   **23.** 10 lb of each   **25.** Deep Thought:
12 lb; Oat Dream: 8 lb   **27.** $7500 at 6%; $4500 at 9%
29. Arctic Antifreeze: 12.5 L; Frost No-More: 7.5 L
31. 87-octane: 4 gal; 93-octane: 8 gal   **33.** Whole milk:
$169\frac{3}{13}$ lb; cream: $30\frac{10}{13}$ lb   **35.** 375 km   **37.** 24 mph
39. About 1489 mi   **41.** Length: 265 ft; width: 165 ft
43. Simon: 122 properties; DeBartolo: 61 properties
45. 30-sec commercials: 4; 60-sec commercials: 8
47. Quarters: 17; fifty-cent pieces: 13   **49.** ▨   **51.** 16
52. 11   **53.** −28   **54.** −10   **55.** $\frac{49}{12}$ **56.** $\frac{13}{10}$ **57.** ▨
59. 0%: 20 reams; 30%: 40 reams   **61.** 1.8 L
63. 180 members   **65.** Brown: 0.8 gal; neutral: 0.2 gal
67. City: 261 mi; highway: 204 mi   **69.** $P(x) = \dfrac{0.1 + x}{1.5}$

(This expresses the percent as a decimal quantity.)

Exercise Set 8.4, pp. 545–547

1. True   **3.** False   **5.** True   **7.** No   **9.** (4, 0, 2)
11. (2, −2, 2)   **13.** (3, −2, 1)   **15.** No solution
17. (2, 1, 3)   **19.** (2, −5, 6)   **21.** The equations are
dependent.   **23.** $\left(\frac{1}{2}, 4, -6\right)$ **25.** $\left(\frac{1}{2}, \frac{1}{3}, \frac{1}{6}\right)$ **27.** $\left(\frac{1}{2}, \frac{2}{3}, -\frac{5}{6}\right)$
29. (15, 33, 9)   **31.** (3, 4, −1)   **33.** (10, 23, 50)   **35.** No
solution   **37.** The equations are dependent.   **39.** ▨
41. Let x and y represent the numbers; $x = 2y$ **42.** Let x
and y represent the numbers; $x + y = 3x$ **43.** Let x
represent the first number; $x + (x + 1) + (x + 2) = 45$
44. Let x and y represent the numbers; $x + 2y = 17$
45. Let x, y, and z represent the numbers; $x + y = 5z$
46. Let x and y represent the numbers; $xy = 2(x + y)$
47. ▨   **49.** (1, −1, 2)   **51.** (−3, −1, 0, 4)
53. $\left(-\frac{1}{2}, -1, -\frac{1}{3}\right)$ **55.** 14   **57.** $z = 8 − 2x − 4y$

Exercise Set 8.5, pp. 551–554

1. 16, 19, 22   **3.** 8, 21, −3   **5.** 32°, 96°, 52°   **7.** Individual
adult: $64; spouse: $57; child: $43   **9.** Bran muffin: 1.5 g;
banana: 3 g; 1 cup Wheaties: 3 g   **11.** Basic price: $24,695;
4WD: $1970; sunroof: $800   **13.** Elrod: 20 ft/hr; Dot:
24 ft/hr; Wendy: 30 ft/hr   **15.** 12-oz cups: 17; 16-oz cups:
25; 20-oz cups: 13   **17.** Small: 10; medium: 25; large: 5
19. Roast beef: 2 servings; baked potato: 1 serving;
broccoli: 2 servings   **21.** Asia: 4.8 billion; Africa: 1.8
billion; rest of world: 2.5 billion   **23.** Two-point field
goals: 32; three-point field goals: 5; foul shots: 13
25. ▨   **27.** −8   **28.** 33   **29.** −55   **30.** −71
31. $−14x + 21y − 35z$ **32.** $−24a − 42b + 54c$ **33.** $−5a$

34. $11x$ **35.** ▨   **37.** Applicant: $102; spouse: $58; first
child: $43; second child: $40   **39.** 20 yr   **41.** 35 tickets

Exercise Set 8.6, pp. 558–559

1. Horizontal; columns   **3.** Entry   **5.** Multiple
7. $\left(-\frac{1}{3}, -4\right)$ **9.** (−4, 3)   **11.** $\left(\frac{3}{2}, \frac{5}{2}\right)$ **13.** $\left(2, \frac{1}{2}, -2\right)$
15. (2, −2, 1)   **17.** $\left(4, \frac{1}{2}, -\frac{1}{2}\right)$ **19.** (1, −3, −2, −1)
21. Dimes: 18; nickels: 24   **23.** $4.05-granola: 5 lb;
$2.70-granola: 10 lb   **25.** $400 at 7%; $500 at 8%;
$1600 at 9%   **27.** ▨   **29.** 13   **30.** −22   **31.** 37   **32.** 422
33. ▨   **35.** 1324

Exercise Set 8.7, pp. 563–564

1. True   **3.** False   **5.** False   **7.** 18   **9.** 36   **11.** 27
13. −3   **15.** −5   **17.** (−3, 2)   **19.** $\left(\frac{9}{19}, \frac{51}{38}\right)$
21. $\left(-1, -\frac{6}{7}, \frac{11}{7}\right)$ **23.** (2, −1, 4)   **25.** (1, 2, 3)   **27.** ▨
29. $\frac{333}{245}$ **30.** −12   **31.** One piece: 20.8 ft; other piece: 12 ft
32. Scientific calculators: 18; graphing calculators: 27
33. Mazzas: 28 rolls; Kranepools: 8 rolls   **34.** Buckets:
17; dinners: 11   **35.** ▨   **37.** 12   **39.** 10

Exercise Set 8.8, pp. 569–571

1. (b)   **3.** (e)   **5.** (h)   **7.** (g)
9. (a) $P(x) = 20x − 300,000$; **(b)** (15,000 units, $975,000)
11. (a) $P(x) = 50x − 120,000$; **(b)** (2400 units, $144,000)
13. (a) $P(x) = 45x − 22,500$; **(b)** (500 units, $42,500)
15. (a) $P(x) = 18x − 16,000$; **(b)** (889 units, $35,560)
17. (a) $P(x) = 50x − 100,000$; **(b)** (2000 units, $250,000)
19. ($70, 300)   **21.** ($22, 474)   **23.** ($50, 6250)
25. ($10, 1070)   **27. (a)** $C(x) = 125,300 + 450x$;
(b) $R(x) = 800x$; **(c)** $P(x) = 350x − 125,300$; **(d)** $90,300
loss, $14,700 profit; **(e)** (358 computers, $286,400)
29. (a) $C(x) = 16,404 + 6x$; **(b)** $R(x) = 18x$;
(c) $P(x) = 12x − 16,404$; **(d)** $19,596 profit, $4404 loss;
(e) (1367 dozen caps, $24,606)   **31.** ▨   **33.** 12   **34.** 15
35. $\frac{8}{3}$ **36.** 4   **37.** $\frac{9}{2}$ **38.** $\frac{1}{3}$ **39.** ▨   **41.** ($5, 300 yo-yo's)
43. (a) $8.74; **(b)** 24,509 units

Review Exercises: Chapter 8, pp. 574–575

1. Substitution   **2.** Elimination   **3.** Approximate
4. Dependent   **5.** Inconsistent   **6.** Infinite   **7.** Parallel
8. Square   **9.** Determinant   **10.** Zero   **11.** (−2, 1)
12. (3, 2)   **13.** $\left(-\frac{11}{15}, -\frac{43}{30}\right)$ **14.** No solution   **15.** $\left(-\frac{4}{5}, \frac{2}{5}\right)$
16. $\left(\frac{37}{19}, \frac{53}{19}\right)$ **17.** $\left(\frac{76}{17}, -\frac{2}{119}\right)$ **18.** (2, 2)
19. $\{(x, y) \mid 3x + 4y = 6\}$ **20.** DVD: $17; videocassette: $14
21. 4 hr   **22.** 8% juice: 10 L; 15% juice: 4 L
23. (4, −8, 10)   **24.** The equations are dependent.
25. (2, 0, 4)   **26.** No solution   **27.** $\left(\frac{8}{9}, -\frac{2}{3}, \frac{10}{9}\right)$
28. A: 90°; B: 67.5°; C: 22.5°
29. Oil: $21\frac{1}{3}$ oz; lemon juice: $10\frac{2}{3}$ oz

30. Lumber: 29 pallets; plywood: 13 pallets **31.** $\left(55, -\frac{89}{2}\right)$
32. $(-1, 1, 3)$ **33.** 2 **34.** 9 **35.** $(6, -2)$ **36.** $(-3, 0, 4)$
37. $(\$3, 81)$ **38. (a)** $C(x) = 0.75x + 9000$;
(b) $R(x) = 5.25x$; **(c)** $P(x) = 4.5x - 9000$; **(d)** $2250 loss;
$13,500 profit; **(e)** (2000 pints of honey, $10,500)
39. ▨ To solve a problem involving four variables, go
through the *Familiarize* and *Translate* steps as usual. The
resulting system of equations can be solved using the
elimination method just as for three variables but likely
with more steps. **40.** ▨ A system of equations can be
both dependent and inconsistent if it is equivalent to a
system with fewer equations that has no solution. An
example is a system of three equations in three unknowns
in which two of the equations represent the same plane,
and the third represents a parallel plane. **41.** 8000 pints
42. $(0, 2), (1, 3)$
43. $a = -\frac{2}{3}, b = -\frac{4}{3}, c = 3; f(x) = -\frac{2}{3}x^2 - \frac{4}{3}x + 3$

Test: Chapter 8, p. 576

1. [8.1] $(2, 4)$ **2.** [8.2] $\left(3, -\frac{11}{3}\right)$ **3.** [8.2] $\left(\frac{15}{7}, -\frac{18}{7}\right)$
4. [8.2] $\left(-\frac{3}{2}, -\frac{3}{2}\right)$ **5.** [8.2] No solution **6.** [8.3] Length:
30 units; width: 18 units **7.** [8.3] Pepperidge Farm
Goldfish: 120 g; Rold Gold Pretzels: 500 g **8.** [8.4] The
equations are dependent. **9.** [8.4] $\left(2, -\frac{1}{2}, -1\right)$
10. [8.4] No solution **11.** [8.4] $(0, 1, 0)$
12. [8.6] $\left(\frac{34}{107}, -\frac{104}{107}\right)$ **13.** [8.6] $(3, 1, -2)$ **14.** [8.7] 34
15. [8.7] 133 **16.** [8.7] $\left(\frac{13}{18}, \frac{7}{27}\right)$ **17.** [8.5] Electrician:
3.5 hr; carpenter: 8 hr; plumber: 10 hr **18.** [8.8] $(\$3, 55)$
19. [8.8] **(a)** $C(x) = 25x + 40,000$; **(b)** $R(x) = 70x$;
(c) $P(x) = 45x - 40,000$; **(d)** $26,500 loss, $500 profit;
(e) (889 hammocks, $62,230)
20. [3.6], [8.3] $m = 7, b = 10$ **21.** [8.5] Adults' tickets:
1346; senior citizens' tickets: 335; children's tickets: 1651

CHAPTER 9

Exercise Set 9.1, pp. 584–586

1. Equivalent inequalities **3.** Equivalent equations
5. Not equivalent **7.** Equivalent expressions **9.** Not
equivalent
11. $(-\infty, 6), \{y \mid y < 6\}$
13. $[-4, \infty), \{x \mid x \geq -4\}$
15. $(-3, \infty), \{t \mid t > -3\}$
17. $(-\infty, -7], \{x \mid x \leq -7\}$
19. $\{y \mid y > -9\}$, or $(-9, \infty)$
21. $\{y \mid y \leq 14\}$, or $(-\infty, 14]$
23. $\{t \mid t < -9\}$, or $(-\infty, -9)$
25. $\{x \mid x \leq 0.9\}$, or $(-\infty, 0.9]$

27. $\left\{x \mid x \leq \frac{5}{6}\right\}$, or $\left(-\infty, \frac{5}{6}\right]$
29. $\{x \mid x < -26\}$, or $(-\infty, -26)$
31. $\left\{t \mid t \geq -\frac{13}{3}\right\}$, or $\left[-\frac{13}{3}, \infty\right)$
33. $\{x \mid x \geq 6\}$, or $[6, \infty)$
35. $\{x \mid x \geq 2\}$, or $[2, \infty)$
37. $\left\{x \mid x > \frac{2}{3}\right\}$, or $\left(\frac{2}{3}, \infty\right)$
39. $\left\{x \mid x \geq \frac{1}{2}\right\}$, or $\left[\frac{1}{2}, \infty\right)$

41. $\left\{y \mid y \leq -\frac{53}{6}\right\}$, or $\left(-\infty, -\frac{53}{6}\right]$ **43.** $\left\{t \mid t < \frac{29}{5}\right\}$, or $\left(-\infty, \frac{29}{5}\right)$
45. $\left\{m \mid m > \frac{7}{3}\right\}$, or $\left(\frac{7}{3}, \infty\right)$ **47.** $\{x \mid x \geq 2\}$, or $[2, \infty)$
49. $\{y \mid y < 5\}$, or $(-\infty, 5)$ **51.** $\left\{x \mid x \leq \frac{4}{7}\right\}$, or $\left(-\infty, \frac{4}{7}\right]$
53. For 1175 min or more **55.** More than 25 checks
57. Gross sales greater than $7000
59. Parties of more than 80 **61.** At least 625 people
63. (a) $\left\{x \mid x < 8181\frac{9}{11}\right\}$, or $\{x \mid x \leq 8181\}$;
(b) $\left\{x \mid x > 8181\frac{9}{11}\right\}$, or $\{x \mid x \geq 8182\}$ **65.** ▨
67. $\{x \mid x \text{ is a real number } and \ x \neq 2\}$
68. $\{x \mid x \text{ is a real number } and \ x \neq -3\}$
69. $\left\{x \mid x \text{ is a real number } and \ x \neq \frac{7}{2}\right\}$
70. $\left\{x \mid x \text{ is a real number } and \ x \neq \frac{9}{4}\right\}$ **71.** $7x + 10$
72. $22x - 7$ **73.** ▨ **75.** $\left\{x \mid x \leq \dfrac{2}{a - 1}\right\}$
77. $\left\{y \mid y \geq \dfrac{2a + 5b}{b(a - 2)}\right\}$ **79.** $\left\{x \mid x > \dfrac{4m - 2c}{d - (5c + 2m)}\right\}$
81. False; $2 < 3$ and $4 < 5$, but $2 - 4 = 3 - 5$. **83.** ▨
85. $\mathbb{R}$
87. $\{x \mid x \text{ is a real number } and \ x \neq 0\}$
89.

Exercise Set 9.2, pp. 594–597

1. (h) **3.** (f) **5.** (e) **7.** (b) **9.** (c) **11.** $\{9, 11\}$
13. $\{0, 5, 10, 15, 20\}$ **15.** $\{b, d, f\}$ **17.** $\{r, s, t, u, v\}$
19. $\varnothing$ **21.** $\{3, 5, 7\}$
23. $(3, 7)$
25. $[-6, -2]$
27. $(-\infty, -1) \cup (4, \infty)$
29. $(-\infty, -2] \cup (1, \infty)$
31. $(-2, 4]$
33. $(-2, 4)$

35. 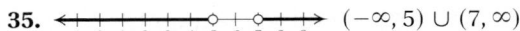 $(-\infty, 5) \cup (7, \infty)$

37. $(-\infty, -4] \cup [5, \infty)$

39. $[-3, 7)$

41. $[3, 7)$

43. $(-\infty, 5)$

45. $\{t \mid -3 < t < 7\}$, or $(-3, 7)$

47. $\{x \mid -1 < x \le 4\}$, or $(-1, 4]$

49. $\{a \mid -2 \le a < 2\}$, or $[-2, 2)$

51. $\mathbb{R}$, or $(-\infty, \infty)$

53. $\{x \mid 7 < x < 23\}$, or $(7, 23)$

55. $\{x \mid -3 \le x \le 2\}$, or $[-3, 2]$

57. $\{x \mid 1 \le x \le 3\}$, or $[1, 3]$

59. $\left\{x \mid -\frac{7}{2} < x \le 7\right\}$, or $\left(-\frac{7}{2}, 7\right]$

61. $\{x \mid x \le 1 \ or \ x \ge 3\}$, or $(-\infty, 1] \cup [3, \infty)$

63. $\{x \mid x < 2 \ or \ x > 6\}$, or $(-\infty, 2) \cup (6, \infty)$

65. $\left\{a \mid a < \frac{7}{2}\right\}$, or $\left(-\infty, \frac{7}{2}\right)$

67. $\{a \mid a < -5\}$, or $(-\infty, -5)$

69. $\mathbb{R}$, or $(-\infty, \infty)$

71. $\{t \mid t \le 6\}$, or $(-\infty, 6]$

73. $(-\infty, -8) \cup (-8, \infty)$ **75.** $[6, \infty)$

77. $(-\infty, 4) \cup (4, \infty)$ **79.** $\left[-\frac{7}{2}, \infty\right)$ **81.** $(-\infty, 4]$

83.

85. 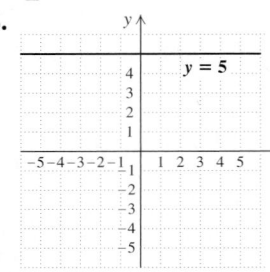 **86.**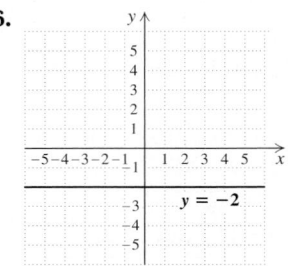

87. 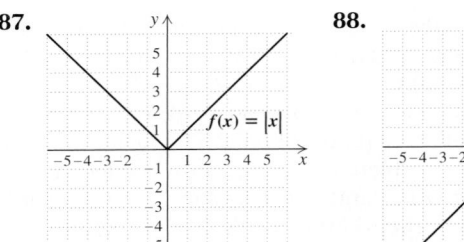 **88.**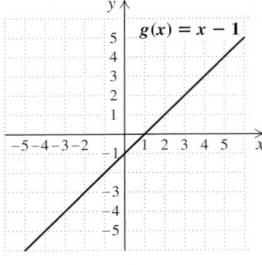

89. $(8, 5)$ **90.** $(-5, -3)$ **91.** ◾ **93.** $(-1, 6)$
95. From 2000 to 2025 **97.** Sizes between 6 and 13
99. $1965 \le y \le 1981$ **101.** Between 12 and 240 trips
103. $\left\{m \mid m < \frac{6}{5}\right\}$, or $\left(-\infty, \frac{6}{5}\right)$

105. $\left\{x \mid -\frac{1}{8} < x < \frac{1}{2}\right\}$, or $\left(-\frac{1}{8}, \frac{1}{2}\right)$

107. False **109.** True **111.** $\left(-\infty, -7\right) \cup \left(-7, \frac{3}{4}\right]$
113. ◾ **115.** ◾

Technology Connection, p. 599

1. The graphs of $y_1 = \text{abs}(4 - 7x)$ and $y_2 = -8$ do not intersect.

Technology Connection, p. 603

1. The x-values on the graph of $y_1 = |4x + 2|$ that are *below* the line $y = 6$ solve the inequality $|4x + 2| < 6$. The x-values on the graph of $y_1 = |4x + 2|$ that are *on* the line $y = 6$ solve the equation $|4x + 2| = 6$. **2.** The x-values on the graph of $y_1 = |3x - 2|$ that are below the line $y = 4$ are in the interval $\left(-\frac{2}{3}, 2\right)$.

Exercise Set 9.3, pp. 604–606

1. True **3.** False **5.** True **7.** False **9.** $\{-7, 7\}$ **11.** $\varnothing$
13. $\{0\}$ **15.** $\{-5.5, 5.5\}$ **17.** $\left\{-\frac{1}{2}, \frac{7}{2}\right\}$ **19.** $\varnothing$ **21.** $\{-4, 8\}$
23. $\{2, 8\}$ **25.** $\{-2, 16\}$ **27.** $\{-8, 8\}$ **29.** $\left\{-\frac{11}{7}, \frac{11}{7}\right\}$
31. $\{-7, 8\}$ **33.** $\{-12, 2\}$ **35.** $\left\{-\frac{1}{3}, 3\right\}$ **37.** $\{-7, 1\}$
39. $\{-8.7, 8.7\}$ **41.** $\left\{-\frac{8}{3}, 4\right\}$ **43.** $\{1, 11\}$ **45.** $\left\{-\frac{1}{2}\right\}$
47. $\left\{-\frac{3}{5}, 5\right\}$ **49.** $\mathbb{R}$ **51.** $\{1\}$ **53.** $\left\{32, \frac{8}{3}\right\}$
55. $\{a \mid -9 \le a \le 9\}$, or $[-9, 9]$

57. $\{x \mid x < -8 \ or \ x > 8\}$, or $(-\infty, -8) \cup (8, \infty)$

59. $\{t \mid t < 0 \ or \ t > 0\}$, or $(-\infty, 0) \cup (0, \infty)$

61. $\{x \mid -3 < x < 5\}$, or $(-3, 5)$

63. $\{x \mid -8 \le x \le 4\}$, or $[-8, 4]$

65. $\{x \mid x < -2 \ or \ x > 8\}$, or $(-\infty, -2) \cup (8, \infty)$

67. $\mathbb{R}$, or $(-\infty, \infty)$

69. $\left\{a \mid a \le -\frac{2}{3} \ or \ a \ge \frac{10}{3}\right\}$, or $\left(-\infty, -\frac{2}{3}\right] \cup \left[\frac{10}{3}, \infty\right)$

71. $\{y \mid -9 < y < 15\}$, or $(-9, 15)$

73. $\{x \mid x \le -8 \ or \ x \ge 0\}$, or $(-\infty, -8] \cup [0, \infty)$

75. $\left\{y \mid y < -\frac{4}{3} \ or \ y > 4\right\}$, or $\left(-\infty, -\frac{4}{3}\right) \cup (4, \infty)$

77. $\varnothing$

79. $\left\{x \mid x \le -\frac{2}{15} \ or \ x \ge \frac{14}{15}\right\}$, or $\left(-\infty, -\frac{2}{15}\right] \cup \left[\frac{14}{15}, \infty\right)$

81. $\{m \mid -9 \le m \le 3\}$, or $[-9, 3]$

83. $\{a \mid -6 < a < 0\}$, or $(-6, 0)$

85. $\left\{x \mid -\frac{1}{2} \le x \le \frac{7}{2}\right\}$, or $\left[-\frac{1}{2}, \frac{7}{2}\right]$

87. $\left\{x \mid x \le -\frac{7}{3} \ or \ x \ge 5\right\}$, or $\left(-\infty, -\frac{7}{3}\right] \cup [5, \infty)$

89. $\{x \mid -4 < x < 5\}$, or $(-4, 5)$

91. 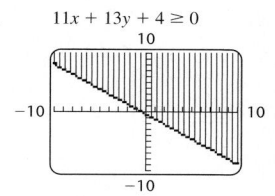   **93.** $\left(-\frac{16}{13}, -\frac{41}{13}\right)$ **94.** $(-2, -3)$ **95.** $(10, 4)$
96. $(-1, 7)$ **97.** $(1, 4)$ **98.** $(24, -41)$ **99.**
101. $\left\{t \mid t \ge \frac{5}{3}\right\}$, or $\left[\frac{5}{3}, \infty\right)$ **103.** $\mathbb{R}$, or $(-\infty, \infty)$ **105.** $\left\{-\frac{1}{7}, \frac{7}{3}\right\}$
107. $|x| < 3$ **109.** $|x| \ge 6$ **111.** $|x + 3| > 5$
113. $|x - 7| < 2$, or $|7 - x| < 2$ **115.** $|x - 3| \le 4$
117. $|x + 4| < 3$ **119.** Between 80 ft and 100 ft
121. $\{x \mid 1 \le x \le 5\}$, or $[1, 5]$ **123.**   **125.**

Technology Connection, p. 610

1. $y > x + 3.5$

2. $7y \le 2x + 5$

3. $8x - 2y < 11$

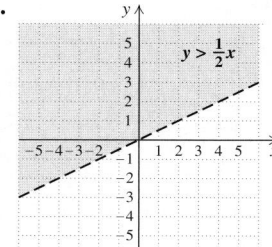

4. $11x + 13y + 4 \ge 0$

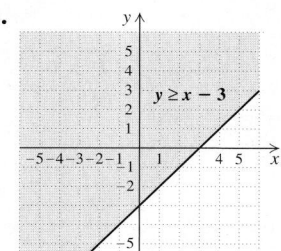

Technology Connection, p. 613

1. $y_1 \le 4 - x, \ y_2 > x - 4$

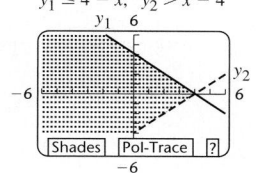

Exercise Set 9.4, pp. 615–617

1. (e)   **3.** (d)   **5.** (b)   **7.** Yes   **9.** No

11. $y > \frac{1}{2}x$

13. $y \ge x - 3$

15. $y \le x + 5$

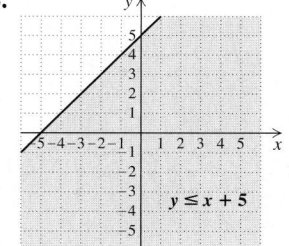

17. $x - y \le 4$

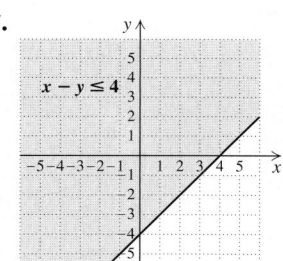

19. $2x + 3y < 6$

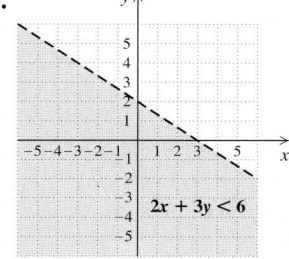

21. $2y - x \le 4$

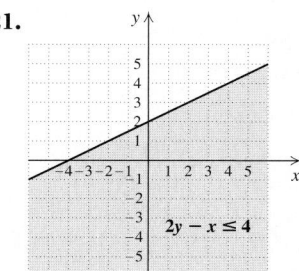

23.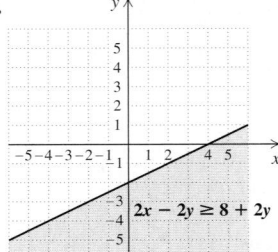

$2x - 2y \geq 8 + 2y$

25.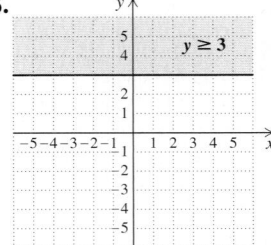

$y \geq 3$

27.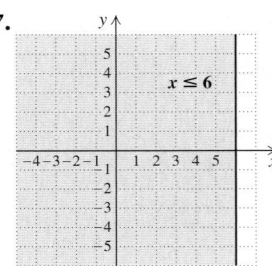

$x \leq 6$

29.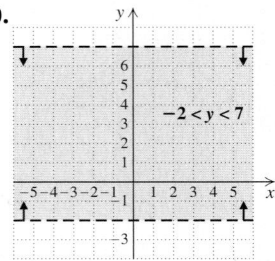

$-2 < y < 7$

31.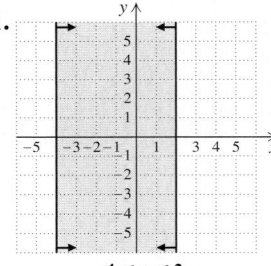

$-4 \leq x \leq 2$

33.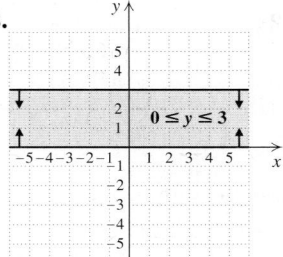

$0 \leq y \leq 3$

35.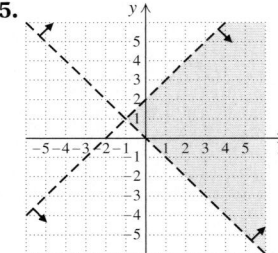

37.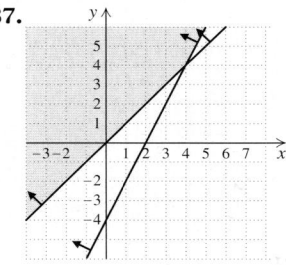

39.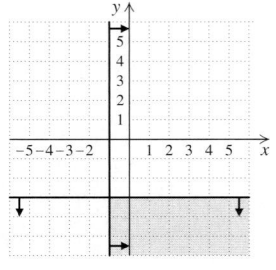

41.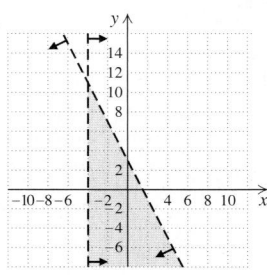

43.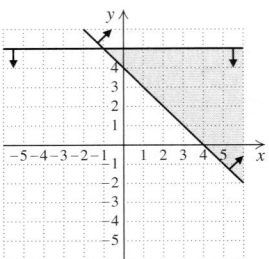

45.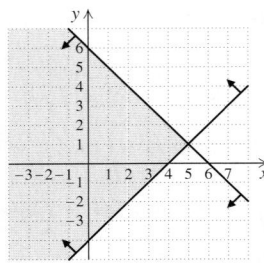

47.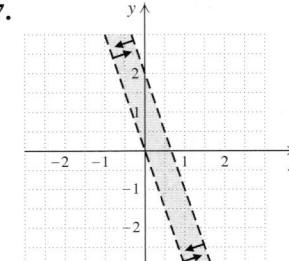

49.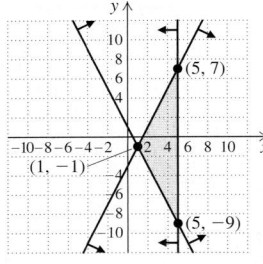

51.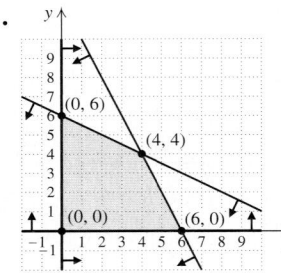

53.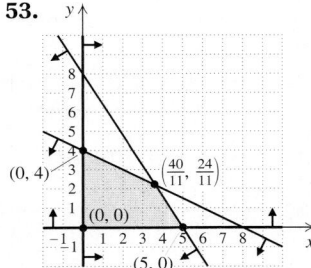

55.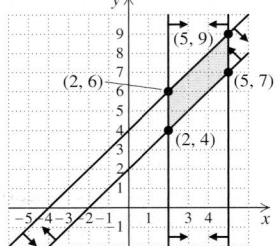

57. 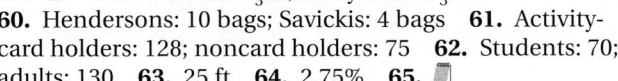   **59.** Peanuts: $6\frac{2}{3}$ lb; fancy nuts: $3\frac{1}{3}$ lb
60. Hendersons: 10 bags; Savickis: 4 bags   **61.** Activity-card holders: 128; noncard holders: 75   **62.** Students: 70; adults: 130   **63.** 25 ft   **64.** 2.75%   **65.**

67.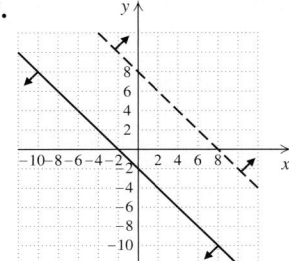

69.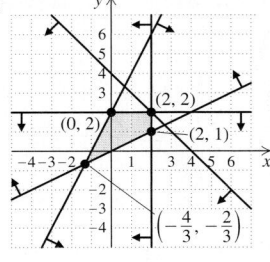

71. $w > 0,$
$\quad h > 0,$
$\quad w + h + 30 \le 62,$ or
$\quad w + h \le 32,$
$\quad 2w + 2h + 30 \le 130,$ or
$\quad w + h \le 50$

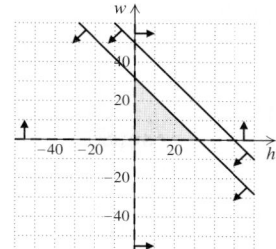

73. $35c + 75a > 1000,$
$\quad c \ge 0,$
$\quad a \ge 0$

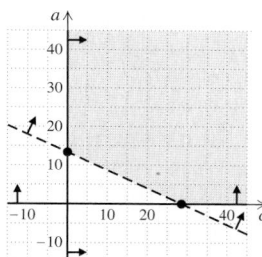

75. (a) $3x + 6y > 2$ **(b)** $x - 5y \le 10$

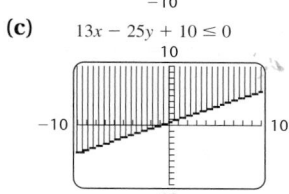   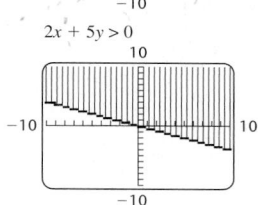

(c) $13x - 25y + 10 \le 0$ **(d)** $2x + 5y > 0$

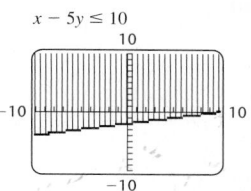

Exercise Set 9.5, pp. 623–625

1. Minimized   **3.** Constraints   **5.** Feasible
7. Maximum 84 when $x = 0, y = 6$; minimum 0 when
$x = 0, y = 0$ **9.** Maximum 76 when $x = 7, y = 0$;
minimum 16 when $x = 0, y = 4$ **11.** Maximum 5 when
$x = 3, y = 7$; minimum -15 when $x = 3, y = -3$
13. Car: 9 gal; moped: 3 gal; maximum: 480 mi
15. Lumber: 100 units; plywood: 300 units
17. Corporate bonds: \$22,000; municipal bonds: \$18,000;
maximum: \$3110
19. Matching: 5; essay: 15; maximum: 425   **21.** Merlot:
80 acres; Cabernet: 160 acres   **23.** Knit suits: 2; worsted
suits: 4   **25.** ▨   **27.** -40 **28.** 46   **29.** $10x + 5$
30. $26t + 20$ **31.** $3x - 6$ **32.** $2t + 2$ **33.** ▨   **35.** T3's:
30; S5's: 10   **37.** Chairs: 25; sofas: 9

Review Exercises: Chapter 9, pp. 627–628

1. True   **2.** False   **3.** True   **4.** False   **5.** True   **6.** True
7. True   **8.** False   **9.** False   **10.** False
11. $\{x \mid x \le -2\}$, or $(-\infty, -2]$;
12. $\{a \mid a \le -21\}$, or $(-\infty, -21]$;
13. $\{y \mid y \ge -7\}$, or $[-7, \infty)$;
14. $\{y \mid y > -\frac{15}{4}\}$, or $\left(-\frac{15}{4}, \infty\right)$;
15. $\{y \mid y > -30\}$, or $(-30, \infty)$;
16. $\{x \mid x > -\frac{3}{2}\}$, or $\left(-\frac{3}{2}, \infty\right)$;
17. $\{x \mid x < -3\}$, or $(-\infty, -3)$;
18. $\{y \mid y > -\frac{220}{23}\}$, or $\left(-\frac{220}{23}, \infty\right)$;
19. $\{x \mid x \le -\frac{5}{2}\}$, or $\left(-\infty, -\frac{5}{2}\right]$;
20. $\{x \mid x \le 4\}$, or $(-\infty, 4]$
21. More than 125 hr   **22.** \$3000   **23.** $\{1, 5, 9\}$
24. $\{1, 2, 3, 5, 6, 9\}$
25. ⟵○———•⟶ $(-5, 3]$
26. ⟵————⟶ $(-\infty, \infty)$
27. $\{x \mid -12 < x \le -3\}$, or $(-12, -3]$

28. $\{x \mid -\frac{5}{4} < x < \frac{5}{2}\}$, or $\left(-\frac{5}{4}, \frac{5}{2}\right)$

29. $\{x \mid x < -3 \text{ or } x > 1\}$, or $(-\infty, -3) \cup (1, \infty)$

30. $\{x \mid x < -11 \text{ or } x \ge -6\}$, or $(-\infty, -11) \cup [-6, \infty)$

31. $\{x \mid x \le -6 \text{ or } x \ge 8\}$, or $(-\infty, -6] \cup [8, \infty)$

32. $\{x \mid x < -\frac{2}{5} \text{ or } x > \frac{8}{5}\}$, or $\left(-\infty, -\frac{2}{5}\right) \cup \left(\frac{8}{5}, \infty\right)$

33. $(-\infty, 8) \cup (8, \infty)$ **34.** $[-5, \infty)$ **35.** $\left(-\infty, \frac{8}{3}\right]$
36. $\{-5, 5\}$ **37.** $\{t \mid t \le -3.5 \text{ or } t \ge 3.5\}$, or
$(-\infty, -3.5] \cup [3.5, \infty)$ **38.** $\{-4, 10\}$
39. $\{x \mid -\frac{17}{2} < x < \frac{7}{2}\}$, or $\left(-\frac{17}{2}, \frac{7}{2}\right)$
40. $\{x \mid x \le -\frac{11}{3} \text{ or } x \ge \frac{19}{3}\}$, or $\left(-\infty, -\frac{11}{3}\right] \cup \left[\frac{19}{3}, \infty\right)$
41. $\{-14, \frac{4}{3}\}$ **42.** $\varnothing$ **43.** $\{x \mid -16 \le x \le 8\}$, or $[-16, 8]$
44. $\{x \mid x < 0 \text{ or } x > 10\}$, or $(-\infty, 0) \cup (10, \infty)$ **45.** $\varnothing$

46.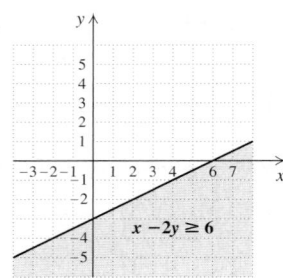

47.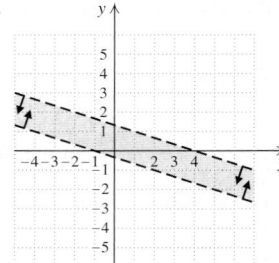

48.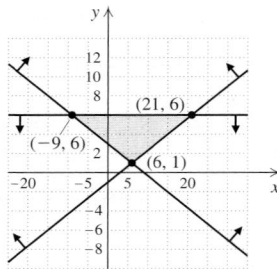

49. Maximum 40 when $x = 7$, $y = 15$; minimum 10 when $x = 1$, $y = 3$ **50.** Ohio plant: 120; Oregon plant: 40
51. ▨ The equation $|X| = p$ has two solutions when p is positive because X can be either p or $-p$. The same equation has no solution when p is negative because no number has a negative absolute value.   **52.** ▨ The solution set of a system of inequalities is all ordered pairs that make *all* the individual inequalities true. This consists of ordered pairs that are common to all the individual solution sets, or the intersection of the graphs.
53. $\left\{x \mid -\frac{8}{3} \le x \le -2\right\}$, or $\left[-\frac{8}{3}, -2\right]$ **54.** False; $-4 < 3$ is true, but $(-4)^2 < 9$ is false.   **55.** $|d - 1.1| \le 0.03$

Test: Chapter 9, p. 629

1. [9.1] $\{x \mid x < 12\}$, or $(-\infty, 12)$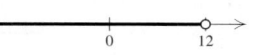

2. [9.1] $\{y \mid y > -50\}$, or $(-50, \infty)$

3. [9.1] $\{y \mid y \le -2\}$, or $(-\infty, -2]$

4. [9.1] $\left\{a \mid a \le \frac{11}{5}\right\}$, or $\left(-\infty, \frac{11}{5}\right]$

5. [9.1] $\left\{x \mid x > \frac{16}{5}\right\}$, or $\left(\frac{16}{5}, \infty\right)$

6. [9.1] $\left\{x \mid x \le \frac{7}{4}\right\}$, or $\left(-\infty, \frac{7}{4}\right]$

7. [9.1] $\{x \mid x > 1\}$, or $(1, \infty)$ **8.** [9.1] More than $166\frac{2}{3}$ mi
9. [9.1] Less than or equal to 2.5 hr   **10.** [9.2] $\{3, 5\}$
11. [9.2] $\{1, 3, 5, 7, 9, 11, 13\}$ **12.** [9.2] $(-\infty, 4]$
13. [9.2] $\{x \mid 1 < x < 8\}$, or $(1, 8)$

14. [9.2] $\left\{t \mid -\frac{2}{5} < t \le \frac{9}{5}\right\}$, or $\left(-\frac{2}{5}, \frac{9}{5}\right]$

15. [9.2] $\{x \mid x < 3 \text{ or } x > 6\}$, or $(-\infty, 3) \cup (6, \infty)$

16. [9.2] $\left\{x \mid x < -4 \text{ or } x > -\frac{5}{2}\right\}$, or $(-\infty, -4) \cup \left(-\frac{5}{2}, \infty\right)$

17. [9.2] $\left\{x \mid 4 \le x < \frac{15}{2}\right\}$, or $\left[4, \frac{15}{2}\right)$

18. [9.3] $\{-13, 13\}$

19. [9.3] $\{a \mid a < -7 \text{ or } a > 7\}$, or $(-\infty, -7) \cup (7, \infty)$

20. [9.3] $\left\{x \mid -2 < x < \frac{8}{3}\right\}$, or $\left(-2, \frac{8}{3}\right)$

21. [9.3] $\left\{t \mid t \le -\frac{13}{5} \text{ or } t \ge \frac{7}{5}\right\}$, or $\left(-\infty, -\frac{13}{5}\right] \cup \left[\frac{7}{5}, \infty\right)$

22. [9.3] $\varnothing$
23. [9.2] $\left\{x \mid x < \frac{1}{2} \text{ or } x > \frac{7}{2}\right\}$, or $\left(-\infty, \frac{1}{2}\right) \cup \left(\frac{7}{2}, \infty\right)$

24. [9.3] $\{1\}$
25. [9.4]

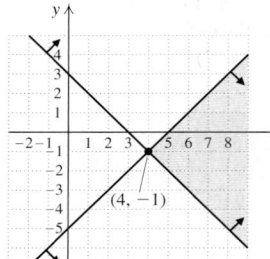

26. [9.4]

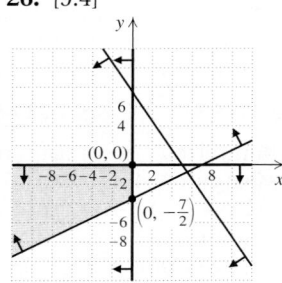

27. [9.5] Maximum 57 when $x = 6$, $y = 9$; minimum 5 when $x = 1$, $y = 0$ **28.** [9.5] Manicures: 35; haircuts: 15; maximum: $690   **29.** [9.3] $[-1, 0] \cup [4, 6]$ **30.** [9.2] $\left(\frac{1}{5}, \frac{4}{5}\right)$
31. [9.3] $|x + 3| \le 5$

Cumulative Review: Chapters 1–9, pp. 630–631

1. 10   **2.** 3.91×10^8 **3.** Slope: $\frac{7}{4}$; y-intercept: $(0, -3)$
4. $y = -\frac{10}{3}x + \frac{11}{3}$ **5.** $(-3, 4)$ **6.** $(-2, -3, 1)$
7. Small: 16; large: 29   **8.** $12, \frac{1}{2}, 7\frac{1}{2}$
9. $28.36 million per year   **10.** (a) $V(t) = \frac{5}{18}t + 15.4$;
(b) about 19.8 min   **11.** (a) $-\frac{1}{2}$;
(b) $\{x \mid x \text{ is a real number } and \ x \ne 5\}$ **12.** $\frac{1}{4}$

13. $-\frac{12}{7}, \frac{12}{7}$ **14.** $\{x \mid x > -3\}$, or $(-3, \infty)$ **15.** $\{x \mid x \geq -1\}$, or $[-1, \infty)$ **16.** $\{x \mid -10 < x < 13\}$, or $(-10, 13)$
17. $\{x \mid x < -\frac{4}{3} \text{ or } x > 6\}$, or $\left(-\infty, -\frac{4}{3}\right) \cup (6, \infty)$
18. $\{x \mid x < -6.4 \text{ or } x > 6.4\}$, or $(-\infty, -6.4) \cup (6.4, \infty)$
19. $\{x \mid -4 \leq x \leq \frac{16}{3}\}$, or $\left[-4, \frac{16}{3}\right]$ **20.** 2 **21.** -1
22. No solution **23.** $\frac{1}{3}$ **24.** $1, \frac{7}{3}$ **25.** $[9, \infty)$
26. $n = \dfrac{m - 12}{3}$ **27.** $a = \dfrac{Pb}{4 - P}$

28.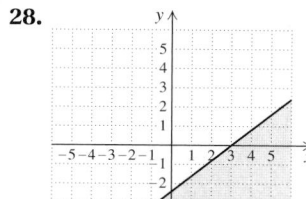

29.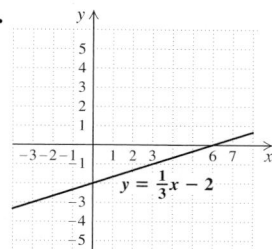
$y = \frac{1}{3}x - 2$

30. $-3x^3 + 11x^2 - 2x + 3$ **31.** $-24x^4y^4$
32. $5a + 5b - 5c$ **33.** $15x^4 - x^3 - 9x^2 + 5x - 2$
34. $9x^4 + 6x^2y + y^2$ **35.** $4x^4 - y^2$
36. $-m^3n^2 - m^2n^2 - 5mn^3$ **37.** $\dfrac{y - 6}{2}$ **38.** $x - 1$
39. $\dfrac{a^2 + 7ab + b^2}{(a - b)(a + b)}$ **40.** $\dfrac{-m^2 + 5m - 6}{(m + 1)(m - 5)}$ **41.** $\dfrac{3y^2 - 2}{3y}$
42. $\dfrac{y - x}{xy(x + y)}$ **43.** $9x^2 - 13x + 26 + \dfrac{-50}{x + 2}$
44. $2x^2(2x - 7)$ **45.** $(x - 6)(x + 14)$
46. $(4y - 9)(4y + 9)$ **47.** $8(2x + 1)(4x^2 - 2x + 1)$
48. $(t - 8)^2$ **49.** $x^2(x - 1)(x + 1)(x^2 + 1)$
50. $(0.3b - 0.2c)(0.09b^2 + 0.06bc + 0.04c^2)$
51. $(4x + 1)(5x - 3)$ **52.** $(3t - 4)(t + 7)$
53. $(x^2 - y)(x^3 + y)$
54. $\{x \mid x \text{ is a real number } and \ x \neq 2 \ and \ x \neq 5\}$
55. $3\frac{1}{3}$ sec **56.** 30 ft **57.** All such sets of even integers satisfy this condition. **58.** 25 ft
59. $x^3 - 12x^2 + 48x - 64$ **60.** $-3, 3, -5, 5$
61. $\{x \mid -3 \leq x \leq -1 \text{ or } 7 \leq x \leq 9\}$, or $[-3, -1] \cup [7, 9]$
62. All real numbers except 9 and -5 **63.** $-\frac{1}{4}, 0, \frac{1}{4}$

CHAPTER 10

Technology Connection, p. 637

1. False **2.** True **3.** False

Exercise Set 10.1, pp. 641–642

1. Two **3.** Positive **5.** Irrational **7.** Nonnegative
9. $7, -7$ **11.** $12, -12$ **13.** $20, -20$ **15.** $30, -30$
17. $-\frac{6}{7}$ **19.** 21 **21.** $-\frac{4}{9}$ **23.** 0.2 **25.** -0.05

27. $p^2 + 4$; 2 **29.** $\dfrac{x}{y + 4}$; 3 **31.** $\sqrt{20}$; 0; does not exist; does not exist **33.** -3; -1; does not exist; 0
35. 1; $\sqrt{2}$; $\sqrt{101}$ **37.** 1; does not exist; 6 **39.** $6|x|$
41. $6|b|$ **43.** $|8 - t|$ **45.** $|y + 8|$ **47.** $|2x + 7|$
49. 4 **51.** -1 **53.** $-\frac{2}{3}$ **55.** $|x|$ **57.** $6|a|$ **59.** 6
61. $|a + b|$ **63.** $|a^{11}|$ **65.** Cannot be simplified
67. $4x$ **69.** $3t$ **71.** $a + 1$ **73.** $2(x + 1)$, or $2x + 2$
75. $3t - 2$ **77.** 3 **79.** $2x$ **81.** -6 **83.** $5y$ **85.** t^9
87. $(x - 2)^4$ **89.** 2; 3; -2; -4 **91.** 2; does not exist; does not exist; 3 **93.** $\{x \mid x \geq 6\}$, or $[6, \infty)$ **95.** $\{t \mid t \geq -8\}$, or $[-8, \infty)$ **97.** $\{x \mid x \geq 5\}$, or $[5, \infty)$ **99.** $\mathbb{R}$ **101.** $\{z \mid z \geq -\frac{2}{5}\}$, or $\left[-\frac{2}{5}, \infty\right)$ **103.** $\mathbb{R}$ **105.** ▨ **107.** $a^9b^6c^{15}$ **108.** $10a^{10}b^9$
109. $\dfrac{a^6c^{12}}{8b^9}$ **110.** $\dfrac{x^6y^2}{25z^4}$ **111.** $2x^4y^5z^2$ **112.** $\dfrac{5c^3}{a^4b^7}$
113. ▨ **115.** (a) 34.6 lb; (b) 10.0 lb; (c) 24.9 lb; (d) 42.3 lb
117. $\{x \mid x \geq -5\}$, or $[-5, \infty)$ **119.** $\{x \mid x \geq 0\}$, or $[0, \infty)$

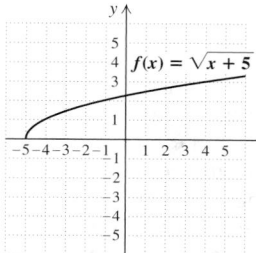
$f(x) = \sqrt{x + 5}$

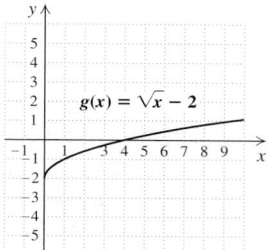
$g(x) = \sqrt{x} - 2$

121. $\{x \mid -3 \leq x < 2\}$, or $[-3, 2)$
123. $\{x \mid x < -1 \text{ or } x > 6\}$, or $(-\infty, -1) \cup (6, \infty)$ **125.** ▨

Technology Connection, p. 645

1. Without parentheses, the expression entered would be $\dfrac{7^2}{3}$. **2.** For $x = 0$ or $x = 1$, $y_1 = y_2 = y_3$; on $(0, 1)$, $y_1 > y_2 > y_3$; on $(1, \infty)$, $y_1 < y_2 < y_3$.

Technology Connection, p. 647

1. Most graphing calculators do not have keys for radicals of index 3 or higher. On those graphing calculators that offer $\sqrt[x]{\ }$ in a MATH menu, rational exponents still require fewer keystrokes.

Exercise Set 10.2, pp. 647–650

1. (g) **3.** (e) **5.** (a) **7.** (b) **9.** $\sqrt[6]{x}$ **11.** 4 **13.** 3
15. 3 **17.** $\sqrt[3]{xyz}$ **19.** $\sqrt[5]{a^2b^2}$ **21.** $\sqrt[5]{t^2}$ **23.** 8 **25.** 81
27. $27\sqrt[4]{x^3}$ **29.** $125x^6$ **31.** $20^{1/3}$ **33.** $17^{1/2}$ **35.** $x^{3/2}$
37. $m^{2/5}$ **39.** $(cd)^{1/4}$ **41.** $(xy^2z)^{1/5}$ **43.** $(3mn)^{3/2}$
45. $(8x^2y)^{5/7}$ **47.** $\dfrac{2x}{z^{2/3}}$ **49.** $\dfrac{1}{x^{1/3}}$ **51.** $\dfrac{1}{(2rs)^{3/4}}$ **53.** 8

55. $2a^{3/5}c$ **57.** $\dfrac{5y^{4/5}z}{x^{2/3}}$ **59.** $\dfrac{a^3}{3^{5/2}b^{7/3}}$ **61.** $\left(\dfrac{3c}{2ab}\right)^{5/6}$

63. $\dfrac{6a}{b^{1/4}}$ **65.** $7^{7/8}$ **67.** $3^{3/4}$ **69.** $5.2^{1/2}$ **71.** $10^{6/25}$

73. $a^{23/12}$ **75.** 64   **77.** $\dfrac{m^{1/3}}{n^{1/8}}$ **79.** $\sqrt[3]{x^2}$ **81.** a^3

83. a^2 **85.** x^2y^2 **87.** $\sqrt{7a}$ **89.** $\sqrt[4]{8x^3}$ **91.** $\sqrt[18]{a}$

93. x^3y^3 **95.** a^6b^{12} **97.** $\sqrt[12]{xy}$ **99.** ▨

101. $11x^4 + 14x^3$ **102.** $-3t^6 + 28t^5 - 20t^4$

103. $15a^2 - 11ab - 12b^2$ **104.** $49x^2 - 14xy + y^2$

105. \$93,500   **106.** 0, 1   **107.** ▨   **109.** $\sqrt[6]{x^5}$

111. $\sqrt[6]{p+q}$ **113.** $2^{7/12} \approx 1.498 \approx 1.5$ **115.** 53.0%

117. About 7.937×10^{-13} to 1   **119.** ▨

Technology Connection, p. 652

1. The graphs differ in appearance because the domain of y_1 is the intersection of $[-3, \infty)$ and $[3, \infty)$, or $[3, \infty)$. The domain of y_2 is $(-\infty, -3] \cup [3, \infty)$.

Exercise Set 10.3, pp. 655–657

1. True   **3.** False   **5.** True   **7.** $\sqrt{35}$ **9.** $\sqrt[3]{14}$ **11.** $\sqrt[4]{18}$

13. $\sqrt{26xy}$ **15.** $\sqrt[5]{80y^4}$ **17.** $\sqrt{y^2 - b^2}$ **19.** $\sqrt[3]{0.21y^2}$

21. $\sqrt[5]{(x-2)^3}$ **23.** $\sqrt{\dfrac{7s}{11t}}$ **25.** $\sqrt[7]{\dfrac{5x-15}{4x+8}}$ **27.** $3\sqrt{2}$

29. $3\sqrt{3}$ **31.** $2\sqrt{2}$ **33.** $3\sqrt{22}$ **35.** $6a^2\sqrt{b}$ **37.** $2x\sqrt[3]{y^2}$

39. $-2x^2\sqrt[3]{2}$ **41.** $f(x) = 5x\sqrt[3]{x^2}$ **43.** $f(x) = |7(x-3)|$, or $7|x-3|$ **45.** $f(x) = |x-1|\sqrt{5}$ **47.** $a^3b^3\sqrt{b}$

49. $xy^2z^3\sqrt[3]{x^2z}$ **51.** $-2ab^2\sqrt[5]{a^2b}$ **53.** $x^2yz^3\sqrt[5]{x^3y^3z^2}$

55. $-2a^4\sqrt[3]{10a^2}$ **57.** $3\sqrt{2}$ **59.** $3\sqrt{35}$ **61.** 3   **63.** $18a^3$

65. $a\sqrt[3]{10}$ **67.** $2x^3\sqrt{5x}$ **69.** $s^2t^3\sqrt[3]{t}$ **71.** $(x+5)^2$

73. $2ab^3\sqrt[4]{5a}$ **75.** $x(y+z)^2\sqrt[5]{x}$ **77.** ▨

79. $\dfrac{12x^2 + 5y^2}{64xy}$ **80.** $\dfrac{2a + 6b^3}{a^4b^4}$ **81.** $\dfrac{-7x - 13}{2(x-3)(x+3)}$

82. $\dfrac{-3x + 1}{2(x-5)(x+5)}$ **83.** $3a^2b^2$ **84.** $3ab^5$ **85.** ▨

87. 175.6 mi   **89. (a)** $-3.3°C$; **(b)** $-16.6°C$; **(c)** $-25.5°C$;
(d) $-54.0°C$ **91.** $25x^5\sqrt[3]{25x}$ **93.** $a^{10}b^{17}\sqrt{ab}$

95.

$f(x) = h(x); f(x) \neq g(x)$

97. $\{x \mid x \le 2 \ or \ x \ge 4\}$, or $(-\infty, 2] \cup [4, \infty)$ **99.** 6

101. ▨ , ▨

Exercise Set 10.4, pp. 663–665

1. (e)   **3.** (f)   **5.** (h)   **7.** (a)   **9.** $\dfrac{6}{5}$ **11.** $\dfrac{4}{3}$ **13.** $\dfrac{7}{y}$

15. $\dfrac{6y\sqrt{y}}{x^2}$ **17.** $\dfrac{3a\sqrt[3]{a}}{2b}$ **19.** $\dfrac{2a}{bc^2}$ **21.** $\dfrac{ab^2}{c^2}\sqrt[4]{\dfrac{a}{c^2}}$

23. $\dfrac{2x}{y^2}\sqrt[5]{\dfrac{x}{y}}$ **25.** $\dfrac{xy}{z^2}\sqrt[6]{\dfrac{y^2}{z^3}}$ **27.** $\sqrt{5}$ **29.** 3   **31.** $y\sqrt{5y}$

33. $2\sqrt[3]{a^2b}$ **35.** $\sqrt{2ab}$ **37.** $2x^2y^3\sqrt[4]{y^3}$

39. $\sqrt[3]{x^2 + xy + y^2}$ **41.** $\dfrac{\sqrt{6}}{2}$ **43.** $\dfrac{2\sqrt{15}}{21}$ **45.** $\dfrac{2\sqrt[3]{6}}{3}$

47. $\dfrac{\sqrt[3]{75ac^2}}{5c}$ **49.** $\dfrac{y\sqrt[3]{180x^2y}}{6x^2}$ **51.** $\dfrac{\sqrt[3]{2xy^2}}{xy}$ **53.** $\dfrac{\sqrt{14a}}{6}$

55. $\dfrac{3\sqrt{5y}}{10xy}$ **57.** $\dfrac{\sqrt{5b}}{6a}$ **59.** $\dfrac{5}{\sqrt{35x}}$ **61.** $\dfrac{2}{\sqrt{6}}$ **63.** $\dfrac{52}{3\sqrt{91}}$

65. $\dfrac{7}{\sqrt[3]{98}}$ **67.** $\dfrac{7x}{\sqrt{21xy}}$ **69.** $\dfrac{2a^2}{\sqrt[3]{20ab}}$ **71.** $\dfrac{x^2y}{\sqrt{2xy}}$ **73.** ▨

75. $\dfrac{3(x-1)}{(x-5)(x+5)}$ **76.** $\dfrac{7(x-2)}{(x+4)(x-4)}$ **77.** $\dfrac{a-1}{a+7}$

78. $\dfrac{t+11}{t+2}$ **79.** $125a^9b^{12}$ **80.** $225x^{10}y^6$ **81.** ▨

83. (a) 1.62 sec; **(b)** 1.99 sec; **(c)** 2.20 sec   **85.** $9\sqrt[3]{9n^2}$

87. $\dfrac{-3\sqrt{a^2 - 3}}{a^2 - 3}$, or $\dfrac{-3}{\sqrt{a^2 - 3}}$ **89.** Step 1: $\sqrt[n]{a} = a^{1/n}$, by

definition; Step 2: $\left(\dfrac{a}{b}\right)^n = \dfrac{a^n}{b^n}$, raising a quotient to a

power; Step 3: $a^{1/n} = \sqrt[n]{a}$, by definition

91. $(f/g)(x) = 3x$, where x is a real number and $x > 0$

93. $(f/g)(x) = \sqrt{x+3}$, where x is a real number and $x > 3$

Exercise Set 10.5, pp. 670–672

1. Radicands, indices   **3.** Bases   **5.** Numerator, conjugate

7. $9\sqrt{5}$ **9.** $2\sqrt[3]{4}$ **11.** $10\sqrt[3]{y}$ **13.** $7\sqrt{2}$

15. $13\sqrt[3]{7} + \sqrt{3}$ **17.** $9\sqrt{3}$ **19.** $23\sqrt{5}$ **21.** $9\sqrt[3]{2}$

23. $(1 + 6a)\sqrt{5a}$ **25.** $(x+2)\sqrt[3]{6x}$ **27.** $3\sqrt{a-1}$

29. $(x+3)\sqrt{x-1}$ **31.** $4\sqrt{3} + 3$ **33.** $15 - 3\sqrt{10}$

35. $6\sqrt{5} - 4$ **37.** $3 - 4\sqrt[3]{63}$ **39.** $a + 2a\sqrt[3]{3}$

41. $4 + 3\sqrt{6}$ **43.** $\sqrt{6} - \sqrt{14} + \sqrt{21} - 7$ **45.** 4

47. -2 **49.** $2 - 8\sqrt{35}$ **51.** $7 + 4\sqrt{3}$ **53.** $5 - 2\sqrt{6}$

55. $2t + 5 + 2\sqrt{10t}$ **57.** $14 + x - 6\sqrt{x+5}$

59. $6\sqrt[4]{63} + 4\sqrt[4]{35} - 3\sqrt[4]{54} - 2\sqrt[4]{30}$ **61.** $\dfrac{20 + 5\sqrt{3}}{13}$

63. $\dfrac{12 - 2\sqrt{3} + 6\sqrt{5} - \sqrt{15}}{33}$ **65.** $\dfrac{a - \sqrt{ab}}{a - b}$ **67.** -1

69. $\dfrac{12 - 3\sqrt{10} - 2\sqrt{14} + \sqrt{35}}{6}$ **71.** $\dfrac{3}{5\sqrt{7} - 10}$

73. $\dfrac{2}{14 + 2\sqrt{3} + 3\sqrt{2} + 7\sqrt{6}}$ **75.** $\dfrac{x - y}{x + 2\sqrt{xy} + y}$

77. $\dfrac{1}{\sqrt{a+h}+\sqrt{a}}$ **79.** $a\sqrt[4]{a}$ **81.** $b\sqrt[10]{b^9}$
83. $xy\sqrt[6]{xy^5}$ **85.** $3a^2b\sqrt[4]{ab}$ **87.** $a^2b^2c^2\sqrt[6]{a^2bc^2}$
89. $\sqrt[12]{a^5}$ **91.** $\sqrt[12]{x^2y^5}$ **93.** $\sqrt[10]{ab^9}$ **95.** $\sqrt[20]{(3x-1)^3}$
97. $\sqrt[15]{(2x+1)^4}$ **99.** $x\sqrt[6]{xy^5}-\sqrt[15]{x^{13}y^{14}}$
101. $2m^2+m\sqrt[4]{n}+2m\sqrt[3]{n^2}+\sqrt[12]{n^{11}}$ **103.** $\sqrt[4]{2x^2}-x^3$
105. x^2-7 **107.** $27+10\sqrt{2}$ **109.** $8-2\sqrt{15}$ **111.**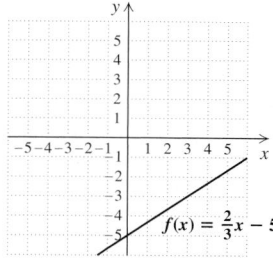
113. 8   **114.** $\frac{15}{2}$ **115.** $\frac{1}{5}$, 1   **116.** $\frac{1}{7}$, 1   **117.** -5, 4
118. Length: 20 units; width: 5 units   **119.**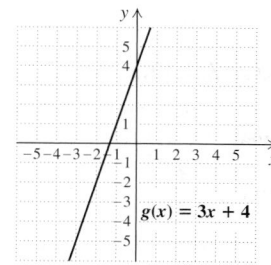
121. $f(x)=-6x\sqrt{5+x}$ **123.** $f(x)=(x+3x^2)\sqrt[4]{x-1}$
125. $ac^2\left[(3a+2c)\sqrt{ab}-2\sqrt[3]{ab}\right]$
127. $9a^2(b+1)\sqrt[6]{243a^5(b+1)^5}$ **129.** $1-\sqrt{w}$
131. $\left(\sqrt{x}+\sqrt{5}\right)\left(\sqrt{x}-\sqrt{5}\right)$
133. $\left(\sqrt{x}+\sqrt{a}\right)\left(\sqrt{x}-\sqrt{a}\right)$ **135.** $2x-2\sqrt{x^2-4}$

Technology Connection, p. 675

1. The x-coordinates of the points of intersection should approximate the solutions of the examples.

Exercise Set 10.6, pp. 677–679

1. False   **3.** True   **5.** True   **7.** $\frac{51}{5}$ **9.** $\frac{25}{3}$ **11.** 168
13. 56   **15.** 3   **17.** 82   **19.** 0, 9   **21.** 64   **23.** -27
25. 125   **27.** No solution   **29.** $\frac{80}{3}$ **31.** 57   **33.** $-\frac{5}{3}$
35. 1   **37.** $\frac{106}{27}$ **39.** 4   **41.** 3, 7   **43.** $\frac{80}{9}$ **45.** -1
47. No solution   **49.** 2, 6   **51.** 2   **53.** 4   **55.**
57. Height: 7 in.; base: 9 in.   **58.** 8 60-sec commercials
59.

60.

61.

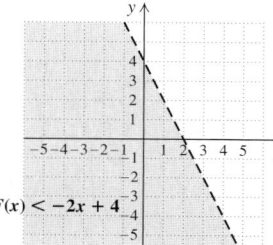

62.

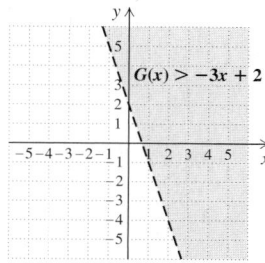

63. 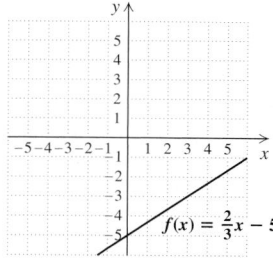   **65.** 524.8°C   **67.** $t=\dfrac{1}{9}\left(\dfrac{S^2\cdot 2457}{1087.7^2}-2617\right)$

69. 4480 rpm   **71.** $r=\dfrac{v^2h}{2gh-v^2}$ **73.** 72.25 ft   **75.** $-\frac{8}{9}$
77. -8, 8   **79.** 1, 8   **81.** $\left(\frac{1}{36},0\right)$, $(36,0)$ **83.**   **85.**

Exercise Set 10.7, pp. 685–688

1. Right, hypotenuse   **3.** Square roots
5. 30°, 60°, 90°, leg   **7.** $\sqrt{34}$; 5.831   **9.** $9\sqrt{2}$; 12.728
11. 5   **13.** 4 m   **15.** $\sqrt{19}$ in.; 4.359 in.   **17.** 1 m
19. 250 ft   **21.** $\sqrt{8450}$, or $65\sqrt{2}$ ft; 91.924 ft   **23.** 12 in.
25. $\left(\sqrt{340}+8\right)$ ft; 26.439 ft   **27.** $\left(110-\sqrt{6500}\right)$ paces;
29.377 paces   **29.** Leg = 5; hypotenuse = $5\sqrt{2}\approx 7.071$
31. Shorter leg = 7; longer leg = $7\sqrt{3}\approx 12.124$
33. Leg = $5\sqrt{3}\approx 8.660$; hypotenuse = $10\sqrt{3}\approx 17.321$
35. Both legs = $\dfrac{13\sqrt{2}}{2}\approx 9.192$
37. Leg = $14\sqrt{3}\approx 24.249$; hypotenuse = 28
39. $3\sqrt{3}\approx 5.196$ **41.** $13\sqrt{2}\approx 18.385$
43. $\dfrac{19\sqrt{2}}{2}\approx 13.435$ **45.** $\sqrt{10561}$ ft ≈ 102.767 ft
47. $h=2\sqrt{3}$ ft ≈ 3.464 ft   **49.** $(0,-4)$, $(0,4)$ **51.**
53. -47 **54.** 5   **55.** $x(x-3)(x+3)$
56. $7a(a-2)(a+2)$ **57.** $\left\{-\frac{2}{3},4\right\}$ **58.** $\left\{-\frac{4}{3},10\right\}$
59.   **61.** $36\sqrt{3}$ cm²; 62.354 cm²   **63.** $d=s+s\sqrt{2}$
65. 5 gal. The total area of the doors and windows is 134 ft² or more.   **67.** 60.28 ft by 60.28 ft

Exercise Set 10.8, pp. 695–696

1. False   **3.** True   **5.** True   **7.** False   **9.** $6i$ **11.** $i\sqrt{13}$,
or $\sqrt{13}i$ **13.** $3i\sqrt{2}$, or $3\sqrt{2}i$ **15.** $i\sqrt{3}$, or $\sqrt{3}i$ **17.** $9i$
19. $-10i\sqrt{3}$, or $-10\sqrt{3}i$ **21.** $6-2i\sqrt{21}$, or $6-2\sqrt{21}i$
23. $\left(-2\sqrt{19}+5\sqrt{5}\right)i$ **25.** $\left(3\sqrt{2}-10\right)i$ **27.** $11+10i$
29. $4+5i$ **31.** $2-i$ **33.** $-12-5i$ **35.** -42 **37.** -24
39. -18 **41.** $-\sqrt{10}$ **43.** $-3\sqrt{14}$ **45.** $-30+10i$
47. $-28-21i$ **49.** $1+5i$ **51.** $38+9i$ **53.** $2-46i$
55. $-11-16i$ **57.** $13-47i$ **59.** $12-16i$
61. $-5+12i$ **63.** $-5-12i$ **65.** $\frac{28}{17}-\frac{7}{17}i$ **67.** $\frac{6}{13}+\frac{4}{13}i$
69. $\frac{3}{17}+\frac{5}{17}i$ **71.** $-\frac{5}{6}i$ **73.** $-\frac{3}{4}-\frac{5}{4}i$ **75.** $1-2i$
77. $-\frac{23}{58}+\frac{43}{58}i$ **79.** $\frac{19}{29}-\frac{4}{29}i$ **81.** $\frac{6}{25}-\frac{17}{25}i$ **83.** $-i$ **85.** 1
87. -1 **89.** i **91.** -1 **93.** $-125i$ **95.** 0   **97.**
99.

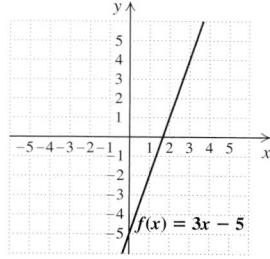

100.

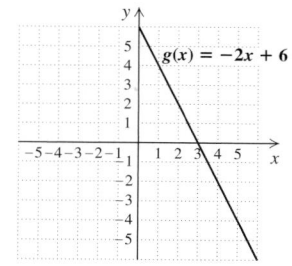

101.

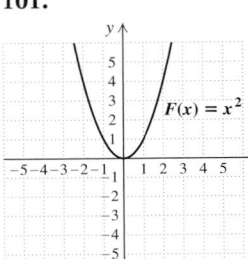

102.

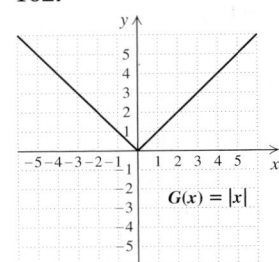

103. $-\frac{4}{3}, 7$ **104.** $\left\{x \mid -\frac{29}{3} < x < 5\right\}$, or $\left(-\frac{29}{3}, 5\right)$ **105.** ▨
107. $-9 - 27i$ **109.** $50 - 120i$ **111.** $\frac{250}{41} + \frac{200}{41}i$ **113.** 8
115. $\frac{3}{5} + \frac{9}{5}i$ **117.** 1

Review Exercises: Chapter 10, pp. 700–701

1. True   **2.** False   **3.** True   **4.** True   **5.** True   **6.** True
7. True   **8.** False   **9.** $\frac{7}{3}$ **10.** -0.5 **11.** 5

12. $\left\{x \mid x \geq \frac{7}{2}\right\}$, or $\left[\frac{7}{2}, \infty\right)$ **13.** $5|t|$ **14.** $|c + 8|$

15. $|x - 3|$ **16.** $|2x + 1|$ **17.** -2 **18.** $-\dfrac{4x^2}{3}$

19. $|x^3y^2|$, or $|x^3|y^2$ **20.** $2x^2$ **21.** $(5ab)^{4/3}$

22. $8a^4\sqrt{a}$ **23.** x^3y^5 **24.** $\sqrt[3]{x^2y}$ **25.** $\dfrac{1}{x^{2/5}}$ **26.** $7^{1/6}$

27. $f(x) = 5|x - 6|$ **28.** $\sqrt{6xy}$ **29.** $3a\sqrt[3]{a^2b^2}$

30. $-6x^5y^4\sqrt[3]{2x^2}$ **31.** $y\sqrt[3]{6}$ **32.** $\dfrac{5\sqrt{x}}{2}$ **33.** $\dfrac{2a^2\sqrt[4]{3a^3}}{c^2}$

34. $7\sqrt[3]{x}$ **35.** $\sqrt{3}$ **36.** $(2x + y^2)\sqrt[3]{x}$ **37.** $15\sqrt{2}$
38. $\sqrt{15} + 4\sqrt{6} - 6\sqrt{10} - 48$ **39.** $\sqrt[4]{x^3}$ **40.** $\sqrt[12]{x^5}$
41. $a^2 - 2a\sqrt{2} + 2$ **42.** $-4\sqrt{10} + 4\sqrt{15}$

43. $\dfrac{20}{\sqrt{10} + \sqrt{15}}$ **44.** 19   **45.** -126 **46.** 4   **47.** 14
48. $5\sqrt{2}$ cm; 7.071 cm   **49.** $\sqrt{32}$ ft; 5.657 ft
50. Short leg = 10; long leg = $10\sqrt{3} \approx 17.321$
51. $-2i\sqrt{2}$, or $-2\sqrt{2}i$ **52.** $-2 - 9i$ **53.** $6 + i$ **54.** 29
55. -1 **56.** 3   **57.** $-\frac{2}{5} + \frac{9}{10}i$ **58.** $9 - 12i$ **59.** $\frac{13}{25} - \frac{34}{25}i$
60. ▨ A complex number $a + bi$ is real when $b = 0$. It is imaginary when $b \neq 0$.
61. ▨ An absolute-value sign must be used to simplify $\sqrt[n]{x^n}$ when n is even, since x may be negative. If x is negative while n is even, the radical expression cannot be simplified to x, since $\sqrt[n]{x^n}$ represents the principal, or positive, root. When n is odd, there is only one root, and it will be positive or negative depending on the sign of x. Thus there is no absolute-value sign when n is odd.

62. $\dfrac{2i}{3i}$; answers may vary

Test: Chapter 10, p. 702

1. [10.3] $5\sqrt{2}$ **2.** [10.4] $-\dfrac{2}{x^2}$ **3.** [10.1] $9|a|$

4. [10.1] $|x - 4|$ **5.** [10.3] $x^2y\sqrt[5]{x^2y^3}$ **6.** [10.4] $\left|\dfrac{5x}{6y^2}\right|$, or
$\dfrac{5|x|}{6y^2}$ **7.** [10.3] $\sqrt[3]{15y^2z}$ **8.** [10.4] $\sqrt[5]{x^2y^2}$
9. [10.5] $xy\sqrt[4]{x}$ **10.** [10.5] $\sqrt[20]{a^3}$ **11.** [10.5] $6\sqrt{2}$
12. [10.5] $(x^2 + 3y)\sqrt{y}$ **13.** [10.5] $14 - 19\sqrt{x} - 3x$
14. [10.2] $(7xy)^{1/2}$ **15.** [10.2] $\sqrt[6]{(4a^3b)^5}$
16. [10.1] $\{x \mid x \geq 5\}$, or $[5, \infty)$ **17.** [10.5] $27 + 10\sqrt{2}$
18. [10.5] $\dfrac{5\sqrt{3} - \sqrt{6}}{23}$ **19.** [10.6] 7   **20.** [10.6] No solution
21. [10.7] $7\sqrt{3}$ cm or 12.124 cm (if the shorter leg is 7 cm);
$\dfrac{7}{\sqrt{3}}$ cm or 4.041 cm (if the longer leg is 7 cm)
22. [10.7] $\sqrt{10,600}$ ft; 102.956 ft   **23.** [10.8] $5i\sqrt{2}$, or $5\sqrt{2}i$
24. [10.8] $12 + 2i$ **25.** [10.8] -24 **26.** [10.8] $15 - 8i$
27. [10.8] $-\frac{11}{34} - \frac{7}{34}i$ **28.** [10.8] i **29.** [10.6] 3
30. [10.8] $-\frac{17}{4}i$ **31.** [10.7] The isosceles right triangle is larger by 1.206 ft^2.

CHAPTER 11

Technology Connection, p. 713

1. The right-hand x-intercept should be an approximation of $4 + \sqrt{23}$.   **2.** x-intercepts should be approximations of $-3 + \sqrt{5}$ and $-3 - \sqrt{5}$ for Example 7; approximations of $\left(-5 + \sqrt{37}\right)/2$ and $\left(-5 - \sqrt{37}\right)/2$ for Example 9(b)
3. Most graphing calculators can give only rational-number approximations of the two irrational solutions. An *exact* solution cannot be found with a graphing calculator.
4. The graph of $y = x^2 - 6x + 11$ has no x-intercepts.

Exercise Set 11.1, pp. 713–715

1. $\sqrt{k}$; $-\sqrt{k}$ **3.** $t + 3$; $t + 3$ **5.** 25; 5   **7.** $\pm\sqrt{5}$ **9.** $\pm\frac{4}{3}i$
11. $\pm\sqrt{\dfrac{7}{5}}$, or $\pm\dfrac{\sqrt{35}}{5}$ **13.** $-6, 8$ **15.** $13 \pm 3\sqrt{2}$
17. $-1 \pm 3i$ **19.** $-\dfrac{3}{4} \pm \dfrac{\sqrt{17}}{4}$, or $\dfrac{-3 \pm \sqrt{17}}{4}$
21. $-3, 13$ **23.** 1, 9   **25.** $-4 \pm \sqrt{13}$ **27.** $-14, 0$
29. $x^2 + 16x + 64 = (x + 8)^2$
31. $t^2 - 10t + 25 = (t - 5)^2$ **33.** $x^2 + 3x + \frac{9}{4} = \left(x + \frac{3}{2}\right)^2$
35. $t^2 - 9t + \frac{81}{4} = \left(t - \frac{9}{2}\right)^2$ **37.** $x^2 + \frac{2}{5}x + \frac{1}{25} = \left(x + \frac{1}{5}\right)^2$
39. $t^2 - \frac{5}{6}t + \frac{25}{144} = \left(t - \frac{5}{12}\right)^2$ **41.** $-7, 1$ **43.** 4, 6
45. $-9, -1$ **47.** $-4 \pm \sqrt{19}$
49. $\left(-3 - \sqrt{2}, 0\right), \left(-3 + \sqrt{2}, 0\right)$

51. $\left(-6 - \sqrt{11}, 0\right), \left(-6 + \sqrt{11}, 0\right)$
53. $\left(5 - \sqrt{47}, 0\right), \left(5 + \sqrt{47}, 0\right)$ **55.** $-\frac{4}{3}, -\frac{2}{3}$ **57.** $-\frac{1}{3}, 2$
59. $-\frac{2}{5} \pm \frac{\sqrt{19}}{5}, \text{ or } \frac{-2 \pm \sqrt{19}}{5}$
61. $\left(-\frac{1}{4} - \frac{\sqrt{13}}{4}, 0\right), \left(-\frac{1}{4} + \frac{\sqrt{13}}{4}, 0\right), \text{ or }$
$\left(\frac{-1 - \sqrt{13}}{4}, 0\right), \left(\frac{-1 + \sqrt{13}}{4}, 0\right)$
63. $\left(\frac{3}{4} - \frac{\sqrt{17}}{4}, 0\right), \left(\frac{3}{4} + \frac{\sqrt{17}}{4}, 0\right), \text{ or }$
$\left(\frac{3 - \sqrt{17}}{4}, 0\right), \left(\frac{3 + \sqrt{17}}{4}, 0\right)$ **65.** 10%   **67.** 18.75%
69. 4%   **71.** About 8.1 sec   **73.** About 9.5 sec   **75.**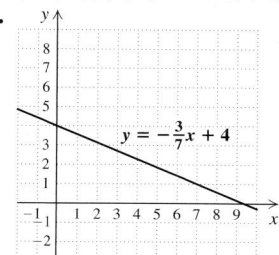
77. 28   **78.** −92   **79.** $3\sqrt[3]{10}$ **80.** $4\sqrt{5}$ **81.** 5   **82.** 7
83. 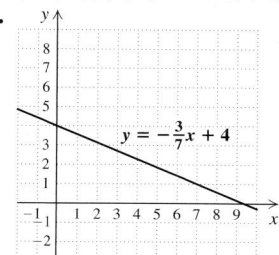   **85.** ±18   **87.** $-\frac{7}{2}, -\sqrt{5}, 0, \sqrt{5}, 8$
89. Barge: 8 km/h; fishing boat: 15 km/h
91.   **93.** ,

Exercise Set 11.2, pp. 720–722

1. True   **3.** False   **5.** False   **7.** $-\frac{7}{2} \pm \frac{\sqrt{61}}{2}$ **9.** $3 \pm \sqrt{7}$
11. $-\frac{1}{2} \pm \frac{\sqrt{3}}{2}i$ **13.** $2 \pm 3i$ **15.** $3 \pm \sqrt{5}$
17. $-\frac{4}{3} \pm \frac{\sqrt{19}}{3}$ **19.** $-\frac{1}{2} \pm \frac{\sqrt{17}}{2}$ **21.** $-\frac{3}{8} \pm \frac{\sqrt{129}}{24}$
23. $\frac{2}{5}$ **25.** $-\frac{11}{8} \pm \frac{\sqrt{41}}{8}$ **27.** 5, 10   **29.** $\frac{13}{10} \pm \frac{\sqrt{509}}{10}$
31. $2 \pm \sqrt{5}i$ **33.** $2, -1 \pm \sqrt{3}i$ **35.** $\frac{2}{3}, 1$ **37.** $5 \pm \sqrt{53}$
39. $\frac{7}{2} \pm \frac{\sqrt{85}}{2}$ **41.** $\frac{3}{2}, 6$ **43.** −5.31662479, 1.31662479
45. 0.7639320225, 5.236067978   **47.** −1.265564437,
2.765564437   **49.**   **51.** Kenyan: 30 lb; Kona: 20 lb
52. Cream-filled: 46; glazed: 44   **53.** $9a^2b^3\sqrt{2a}$
54. $4a^2b^3\sqrt{6}$ **55.** $\frac{3(x+1)}{3x+1}$ **56.** $\frac{4b}{3ab^2 - 4a^2}$ **57.**
59. (−2, 0), (1, 0)   **61.** $4 - 2\sqrt{2}, 4 + 2\sqrt{2}$
63. −1.1792101, 0.3392101   **65.** $\frac{-5\sqrt{2} \pm \sqrt{34}}{4}$ **67.** $\frac{1}{2}$
69.

Exercise Set 11.3, pp. 726–729

1. First part: 60 mph; second part: 50 mph   **3.** 40 mph
5. Cessna: 150 mph, Beechcraft: 200 mph; or
Cessna: 200 mph, Beechcraft: 250 mph
7. To Hillsboro: 10 mph; return trip: 4 mph
9. About 14 mph   **11.** 12 hr   **13.** About 3.24 mph
15. $r = \frac{1}{2}\sqrt{\frac{A}{\pi}}$ **17.** $r = \frac{-\pi h + \sqrt{\pi^2 h^2 + 2\pi A}}{2\pi}$

19. $s = \sqrt{\frac{kQ_1Q_2}{N}}$ **21.** $g = \frac{4\pi^2 l}{T^2}$
23. $c = \sqrt{d^2 - a^2 - b^2}$ **25.** $t = \frac{-v_0 + \sqrt{v_0^2 + 2gs}}{g}$
27. $n = \frac{1 + \sqrt{1 + 8N}}{2}$ **29.** $h = \frac{V^2}{12.25}$
31. $t = \frac{-b \pm \sqrt{b^2 - 4ac}}{2a}$ **33.** **(a)** 10.1 sec; **(b)** 7.49 sec;
(c) 272.5 m   **35.** 2.9 sec   **37.** 0.968 sec   **39.** 2.5 m/sec
41. 7%   **43.**   **45.** −104   **46.** $2i\sqrt{11}$ **47.** $\frac{x + y}{2}$
48. $\frac{a^2 - b^2}{b}$ **49.** $\frac{1 + \sqrt{5}}{2}$ **50.** $\frac{1 - \sqrt{7}}{5}$ **51.**
53. $t = \frac{-10.2 + 6\sqrt{-A^2 + 13A - 39.36}}{A - 6.5}$
55. $\pm\sqrt{2}$ **57.** $l = \frac{w + w\sqrt{5}}{2}$
59. $n = \pm\sqrt{\frac{r^2 \pm \sqrt{r^4 + 4m^4r^2p - 4mp}}{2m}}$
61. $A(S) = \frac{\pi S}{6}$

Technology Connection, p. 732

1. $(-0.4, 0)$ is the other x-intercept of $y = 5x^2 - 13x - 6$.
2. The x-intercepts of $y = x^2 - 175$ are $(-13.22875656, 0)$
and $(13.22875656, 0)$, or $(-5\sqrt{7}, 0)$ and $(5\sqrt{7}, 0)$.
3. The x-intercepts of $y = x^3 + 3x^2 - 4x$ are $(-4, 0)$,
$(0, 0)$, and $(1, 0)$.

Exercise Set 11.4, pp. 733–735

1. Discriminant   **3.** Two   **5.** Rational   **7.** Two irrational
9. Two imaginary   **11.** Two irrational   **13.** Two rational
15. Two imaginary   **17.** One rational   **19.** Two rational
21. Two rational   **23.** Two irrational   **25.** Two imaginary
27. Two irrational   **29.** $x^2 + 4x - 21 = 0$
31. $x^2 - 6x + 9 = 0$ **33.** $x^2 + 4x + 3 = 0$
35. $4x^2 - 23x + 15 = 0$ **37.** $8x^2 + 6x + 1 = 0$
39. $x^2 - 2x - 0.96 = 0$ **41.** $x^2 - 3 = 0$
43. $x^2 - 20 = 0$ **45.** $x^2 + 16 = 0$
47. $x^2 - 4x + 53 = 0$ **49.** $x^2 - 6x - 5 = 0$
51. $3x^2 - 6x - 4 = 0$ **53.** $x^3 - 4x^2 - 7x + 10 = 0$
55. $x^3 - 2x^2 - 3x = 0$ **57.**   **59.** $81a^8$ **60.** $16x^6$
61. $(-1, 0), (8, 0)$ **62.** $(2, 0), (4, 0)$ **63.** 6 commercials
64.

$y = -\frac{3}{7}x + 4$

65. 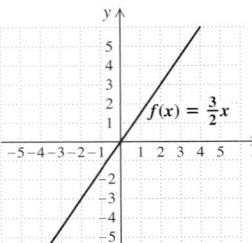 **67.** $a = 1, b = 2, c = -3$ **69. (a)** $-\frac{3}{5}$; **(b)** $-\frac{1}{3}$
71. (a) $9 + 9i$; **(b)** $3 + 3i$
73. The solutions of $ax^2 + bx + c = 0$ are

$x = \dfrac{-b \pm \sqrt{b^2 - 4ac}}{2a}$. When there is just one solution,

$b^2 - 4ac$ must be 0, so $x = \dfrac{-b \pm 0}{2a} = \dfrac{-b}{2a}$.

75. $a = 8, b = 20, c = -12$
77. $x^4 - 8x^3 + 21x^2 - 2x - 52 = 0$ **79.** ▢, ◿

Exercise Set 11.5, pp. 739–741

1. (f) **3.** (h) **5.** (g) **7.** (e) **9.** $\pm 1, \pm 2$ **11.** $\pm\sqrt{5}, \pm 2$

13. $\pm\dfrac{\sqrt{3}}{2}, \pm 2$ **15.** $8 + 2\sqrt{7}$ **17.** $\pm 2\sqrt{2}, \pm 3$

19. No solution **21.** $-\frac{1}{2}, \frac{1}{3}$ **23.** $-\frac{4}{5}, 1$ **25.** $-27, 8$

27. 729 **29.** 1 **31.** No solution **33.** $\frac{12}{5}$ **35.** $\left(\frac{4}{25}, 0\right)$

37. $\left(\dfrac{3}{2} + \dfrac{\sqrt{33}}{2}, 0\right), \left(\dfrac{3}{2} - \dfrac{\sqrt{33}}{2}, 0\right), (4, 0), (-1, 0)$

39. $(-243, 0), (32, 0)$ **41.** No x-intercepts **43.** ▢

45. 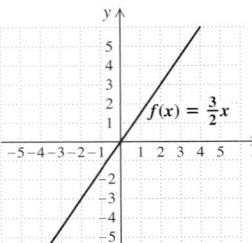   **46.**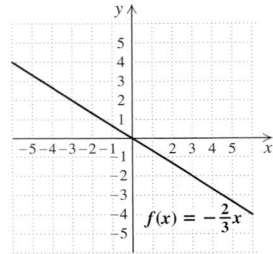

47. 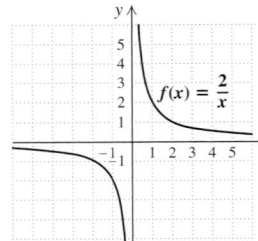   **48.**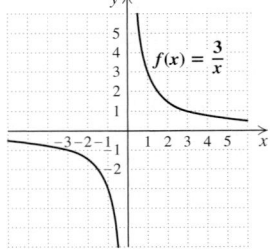

49. Hiker's Mix: 4 lb; Trail Snax: 8 lb **50.** $a^2 + a$ **51.** ▢

53. $\pm\sqrt{\dfrac{7 \pm \sqrt{29}}{10}}$ **55.** $-2, -1, 5, 6$ **57.** $\frac{100}{99}$

59. $-5, -3, -2, 0, 2, 3, 5$ **61.** $1, 3, -\dfrac{1}{2} + \dfrac{\sqrt{3}}{2}i,$

$-\dfrac{1}{2} - \dfrac{\sqrt{3}}{2}i, \dfrac{3}{2} + \dfrac{3\sqrt{3}}{2}i, -\dfrac{3}{2} - \dfrac{3\sqrt{3}}{2}i$

63. ◿ **65.** ▢, ◿

Technology Connection, p. 742

1. The graphs of y_1, y_2, and y_3 open upward. The graphs of y_4, y_5, and y_6 open downward. The graph of y_1 is wider

than the graph of y_2. The graph of y_3 is narrower than the graph of y_2. Similarly, the graph of y_4 is wider than the graph of y_5, and the graph of y_6 is narrower than the graph of y_5. **2.** If A is positive, the graph opens upward. If A is negative, the graph opens downward. Compared with the graph of $y = x^2$, the graph of $y = Ax^2$ is wider if $|A| < 1$ and narrower if $|A| > 1$.

Technology Connection, p. 744

1. Compared with the graph of $y = ax^2$, the graph of $y = a(x - h)^2$ is shifted left or right. It is shifted left if h is negative and right if h is positive. **2.** The value of A makes the graph wider or narrower, and makes the graph open downward if A is negative. The value of B shifts the graph left or right.

Technology Connection, p. 745

1. The graph of y_2 looks like the graph of y_1 shifted up 2 units, and the graph of y_3 looks like the graph of y_1 shifted down 4 units. **2.** Compared with the graph of $y = a(x - h)^2$, the graph of $y = a(x - h)^2 + k$ is shifted up or down. It is shifted down if k is negative and up if k is positive. **3.** The value of A makes the graph wider or narrower, and makes the graph open downward if A is negative. The value of B shifts the graph left or right. The value of C shifts the graph up or down.

Exercise Set 11.6, pp. 747–749

1. (h) **3.** (f) **5.** (b) **7.** (e)
9. 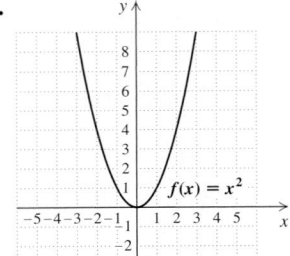   **11.**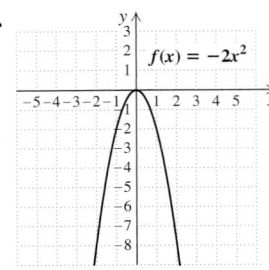

13. 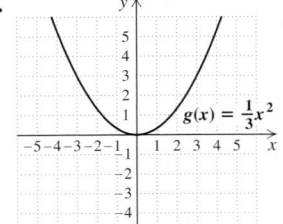   **15.**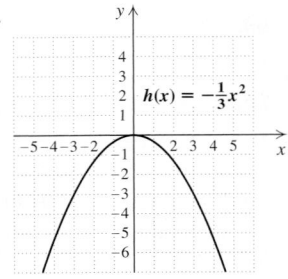

17.

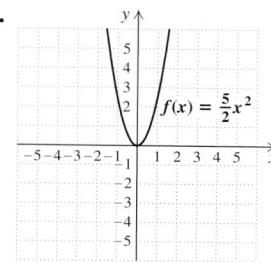

$f(x) = \frac{5}{2}x^2$

19. Vertex: $(-1, 0)$;
axis of symmetry: $x = -1$

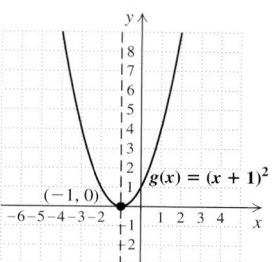

$g(x) = (x + 1)^2$

21. Vertex $(2, 0)$;
axis of symmetry: $x = 2$

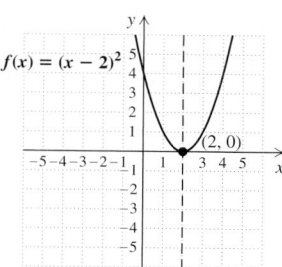

$f(x) = (x - 2)^2$

23. Vertex: $(3, 0)$;
axis of symmetry: $x = 3$

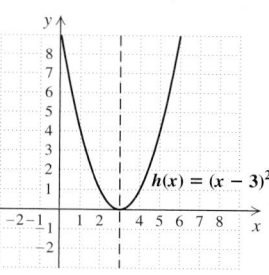

$h(x) = (x - 3)^2$

25. Vertex: $(-1, 0)$;
axis of symmetry: $x = -1$

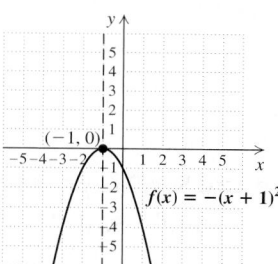

$f(x) = -(x + 1)^2$

27. Vertex: $(2, 0)$;
axis of symmetry: $x = 2$

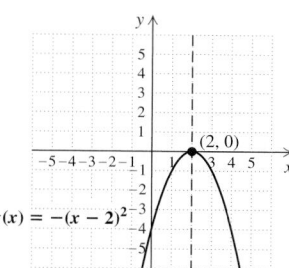

$g(x) = -(x - 2)^2$

29. Vertex: $(-1, 0)$;
axis of symmetry: $x = -1$

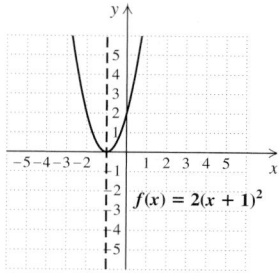

$f(x) = 2(x + 1)^2$

31. Vertex: $(4, 0)$;
axis of symmetry: $x = 4$

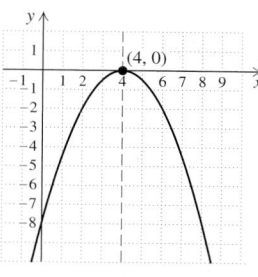

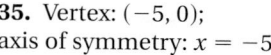

$h(x) = -\frac{1}{2}(x - 4)^2$

33. Vertex: $(1, 0)$;
axis of symmetry: $x = 1$

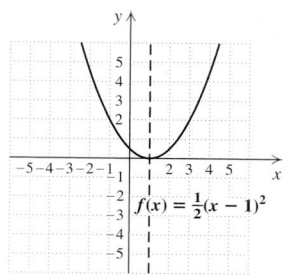

$f(x) = \frac{1}{2}(x - 1)^2$

35. Vertex: $(-5, 0)$;
axis of symmetry: $x = -5$

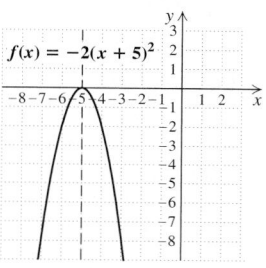

$f(x) = -2(x + 5)^2$

37. Vertex: $\left(\frac{1}{2}, 0\right)$;
axis of symmetry: $x = \frac{1}{2}$

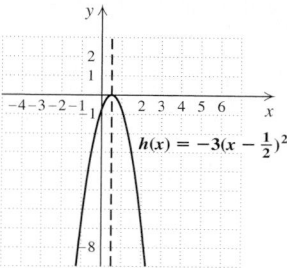

$h(x) = -3\left(x - \frac{1}{2}\right)^2$

39. Vertex: $(5, 2)$;
axis of symmetry: $x = 5$;
minimum: 2

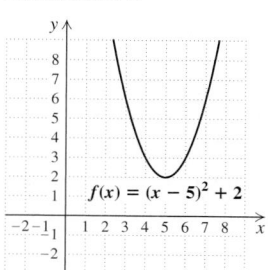

$f(x) = (x - 5)^2 + 2$

41. Vertex: $(-1, -3)$;
axis of symmetry: $x = -1$;
minimum: -3

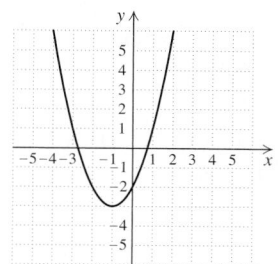

$f(x) = (x + 1)^2 - 3$

43. Vertex: $(-4, 1)$;
axis of symmetry: $x = -4$;
minimum: 1

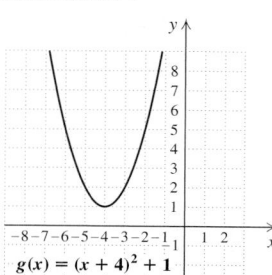

$g(x) = (x + 4)^2 + 1$

45. Vertex: $(1, -3)$;
axis of symmetry: $x = 1$;
maximum: -3

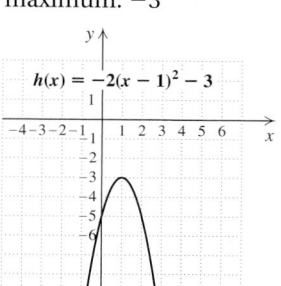

$h(x) = -2(x - 1)^2 - 3$

47. Vertex: $(-4, 1)$; axis of symmetry: $x = -4$; minimum: 1

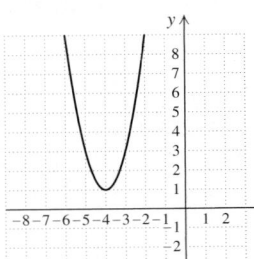

$$f(x) = 2(x + 4)^2 + 1$$

49. Vertex: $(1, 4)$; axis of symmetry: $x = 1$; maximum: 4

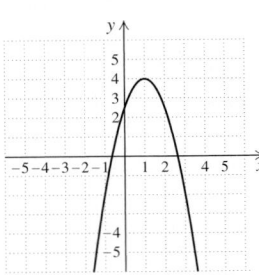

$$g(x) = -\frac{3}{2}(x - 1)^2 + 4$$

51. Vertex: $(8, 7)$; axis of symmetry: $x = 8$; minimum: 7
53. Vertex: $(-6, 11)$; axis of symmetry: $x = -6$; maximum: 11   **55.** Vertex: $\left(-\frac{1}{4}, -13\right)$; axis of symmetry: $x = -\frac{1}{4}$; minimum: -13
57. Vertex: $(-4.58, 65\pi)$; axis of symmetry: $x = -4.58$; minimum: 65π **59.**

61.

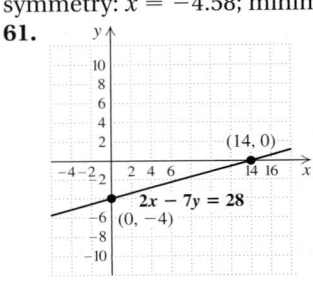

62.

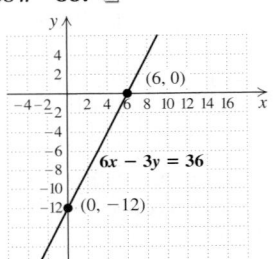

63. $(-5, -1)$ **64.** $(-1, 2)$
65. $x^2 + 5x + \frac{25}{4} = \left(x + \frac{5}{2}\right)^2$
66. $x^2 - 9x + \frac{81}{4} = \left(x - \frac{9}{2}\right)^2$ **67.**
69. $f(x) = \frac{3}{5}(x - 4)^2 + 1$ **71.** $f(x) = \frac{3}{5}(x - 3)^2 - 1$
73. $f(x) = \frac{3}{5}(x + 2)^2 - 5$ **75.** $f(x) = 2(x - 2)^2$
77. $g(x) = -2x^2 + 3$ **79.** $F(x) = 3(x - 5)^2 + 1$
81.

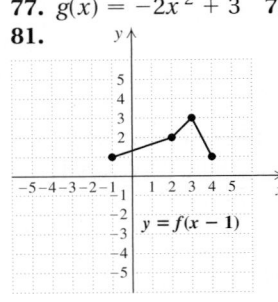

$$y = f(x - 1)$$

83.

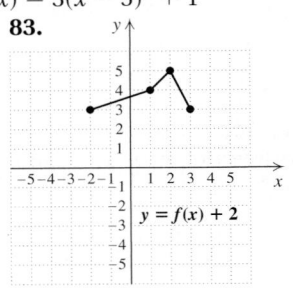

$$y = f(x) + 2$$

85.

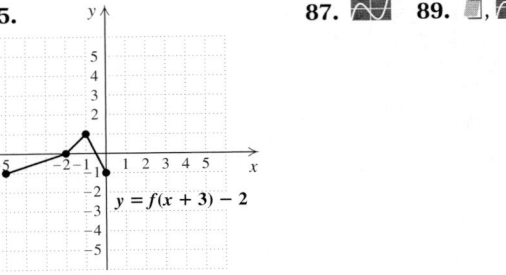

$$y = f(x + 3) - 2$$

87. 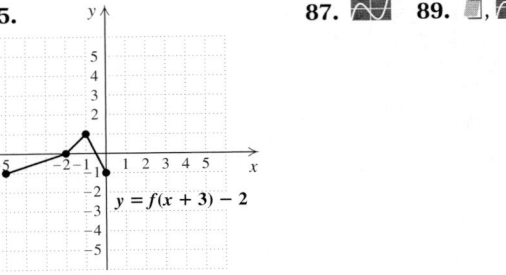   **89.**

Exercise Set 11.7, pp. 754–755

1. 9   **3.** 9   **5.** 3   **7.** $\frac{5}{2}, (-4)$
9. (a) Vertex: $(-2, 1)$; axis of symmetry: $x = -2$;
(b)

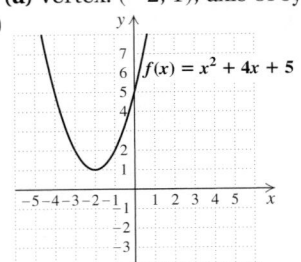

$$f(x) = x^2 + 4x + 5$$

11. (a) Vertex: $(3, 4)$; axis of symmetry: $x = 3$;
(b)

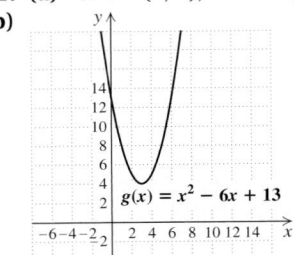

$$g(x) = x^2 - 6x + 13$$

13. (a) Vertex: $(-4, 4)$; axis of symmetry: $x = -4$;
(b)

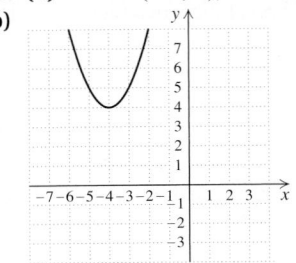

$$f(x) = x^2 + 8x + 20$$

15. (a) Vertex: $(4, -7)$; axis of symmetry: $x = 4$;
(b)

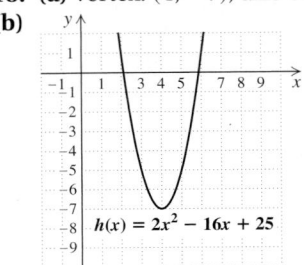

$$h(x) = 2x^2 - 16x + 25$$

17. (a) Vertex: $(1, 6)$; axis of symmetry: $x = 1$;
(b)

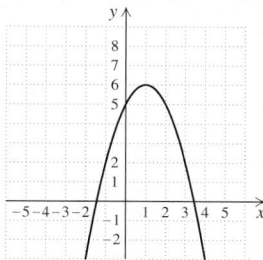

$f(x) = -x^2 + 2x + 5$

19. (a) Vertex: $\left(-\frac{3}{2}, -\frac{49}{4}\right)$; axis of symmetry: $x = -\frac{3}{2}$;
(b)

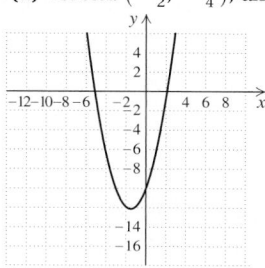

$g(x) = x^2 + 3x - 10$

21. (a) Vertex: $(4, 2)$; axis of symmetry: $x = 4$;
(b)

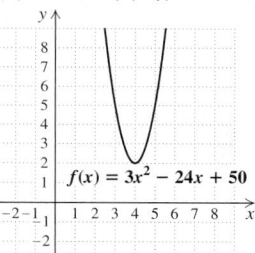

$f(x) = 3x^2 - 24x + 50$

23. (a) Vertex: $\left(-\frac{7}{2}, -\frac{49}{4}\right)$; axis of symmetry: $x = -\frac{7}{2}$;
(b)

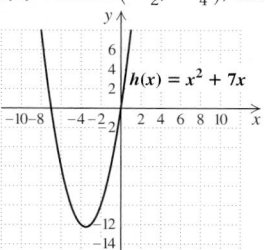

$h(x) = x^2 + 7x$

25. (a) Vertex: $(-1, -4)$; axis of symmetry: $x = -1$;
(b)

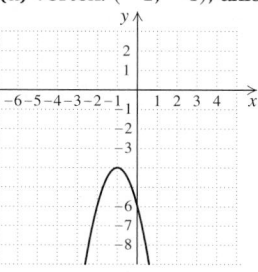

$f(x) = -2x^2 - 4x - 6$

27. (a) Vertex: $(2, -5)$; axis of symmetry: $x = 2$;
(b)

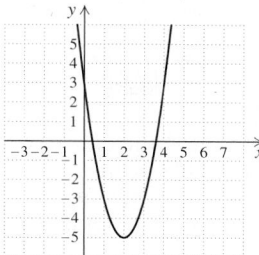

$g(x) = 2x^2 - 8x + 3$

29. (a) Vertex: $\left(\frac{5}{6}, \frac{1}{12}\right)$; axis of symmetry: $x = \frac{5}{6}$;
(b)

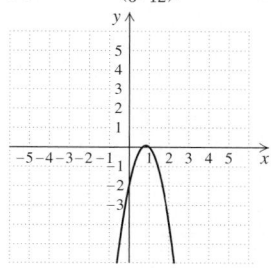

$f(x) = -3x^2 + 5x - 2$

31. (a) Vertex: $\left(-4, -\frac{5}{3}\right)$; axis of symmetry: $x = -4$;
(b)

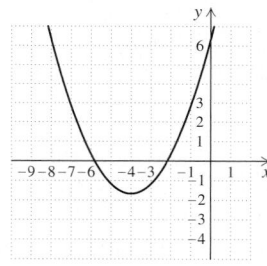

$h(x) = \frac{1}{2}x^2 + 4x + \frac{19}{3}$

33. $(3 - \sqrt{6}, 0), (3 + \sqrt{6}, 0); (0, 3)$ **35.** $(-1, 0), (3, 0);$
$(0, 3)$ **37.** $(0, 0), (9, 0); (0, 0)$ **39.** $(2, 0); (0, -4)$ **41.** No
x-intercept; $(0, 6)$ **43.**   **45.** $(2, -2)$ **46.** $(7, 1)$
47. $(3, 2, 1)$ **48.** $(1, -3, 2)$ **49.** 5   **50.** 4   **51.** ▥
53. (a) Minimum: -6.953660714; **(b)** $(-1.056433682, 0),$
$(2.413576539, 0); (0, -5.89)$ **55. (a)** $-2.4, 3.4$;
(b) $-1.3, 2.3$ **57.** $f(x) = m\left(x - \dfrac{n}{2m}\right)^2 + \dfrac{4mp - n^2}{4m}$
59. $f(x) = \frac{5}{16}x^2 - \frac{15}{8}x - \frac{35}{16}$, or $f(x) = \frac{5}{16}(x - 3)^2 - 5$
61.

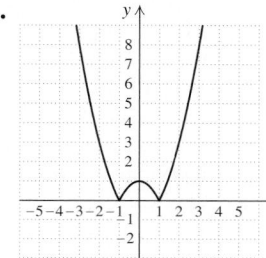

$f(x) = \left|x^2 - 1\right|$

63.

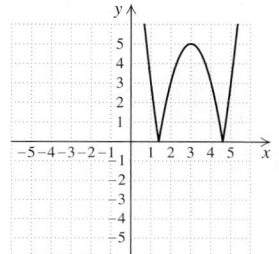

$f(x) = \left|2(x - 3)^2 - 5\right|$

Technology Connection, p. 760

1. 32.7%

Exercise Set 11.8, pp. 761–766

1. (e)   **3.** (c)   **5.** (d)   **7.** 11 days after the concert was announced; about 62 tickets   **9.** $120/Dobro; 350 Dobros   **11.** 32 in. by 32 in.   **13.** 450 ft^2; 15 ft by 30 ft (The house serves as a 30-ft side.)   **15.** 3.5 in.   **17.** 81; 9 and 9   **19.** -16; 4 and -4 **21.** 25; -5 and -5 **23.** $f(x) = mx + b$ **25.** $f(x) = ax^2 + bx + c, a < 0$ **27.** $f(x) = ax^2 + bx + c, a > 0$ **29.** $f(x) = ax^2 + bx + c, a < 0$ **31.** $f(x) = ax^2 + bx + c, a > 0$ **33.** $f(x) = mx + b$ **35.** $f(x) = 2x^2 + 3x - 1$ **37.** $f(x) = -\frac{1}{4}x^2 + 3x - 5$ **39.** (a) $A(s) = \frac{3}{16}s^2 - \frac{135}{4}s + 1750$; (b) about 531 accidents   **41.** $h(d) = -0.0068d^2 + 0.8571d$ **43.**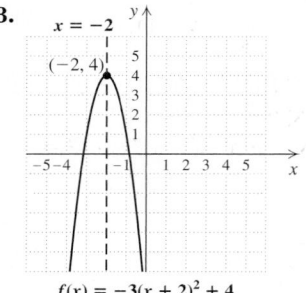

45. $\dfrac{x - 9}{(x + 9)(x + 7)}$ **46.** $\dfrac{(x - 3)(x + 1)}{(x - 7)(x + 3)}$ **47.** $\dfrac{(t + 2)(t + 8)}{(t - 3)(t + 1)}$ **48.** $\dfrac{2t(t + 2)}{(t - 7)(t - 3)(t + 7)}$ **49.** $\{x \mid x < 8\}$, or $(-\infty, 8)$ **50.** $\{x \mid x \geq 10\}$, or $[10, \infty)$ **51.** ▨   **53.** 158 ft   **55.** $15   **57.** The radius of the circular portion of the window and the height of the rectangular portion should each be $\dfrac{24}{\pi + 4}$ ft.

59. (a) $c(x) = 261.875x^2 - 882.5642857x + 2134.571429$; (b) 55,053 cars

Technology Connection, p. 771

1. $\{x \mid -0.78 \leq x \leq 1.59\}$, or $[-0.78, 1.59]$
2. $\{x \mid x \leq -0.21 \text{ or } x \geq 2.47\}$, or $(-\infty, -0.21] \cup [2.47, \infty)$
3. $\{x \mid x < -1.26 \text{ or } x > 2.33\}$, or $(-\infty, -1.26) \cup (2.33, \infty)$
4. $\{x \mid x > -1.37\}$, or $(-1.37, \infty)$

Exercise Set 11.9, pp. 774–775

1. True   **3.** True   **5.** True   **7.** False
9. $(-4, 3)$, or $\{x \mid -4 < x < 3\}$
11. $(-\infty, -7] \cup [2, \infty)$, or $\{x \mid x \leq -7 \text{ or } x \geq 2\}$
13. $(-\infty, -1) \cup (2, \infty)$, or $\{x \mid x < -1 \text{ or } x > 2\}$ **15.** $\varnothing$
17. $(-2, 6)$, or $\{x \mid -2 < x < 6\}$
19. $(-\infty, -2) \cup (0, 2)$, or $\{x \mid x < -2 \text{ or } 0 < x < 2\}$
21. $[-2, 1] \cup [4, \infty)$, or $\{x \mid -2 \leq x \leq 1 \text{ or } x \geq 4\}$
23. $[-2, 2]$, or $\{x \mid -2 \leq x \leq 2\}$ **25.** $(-1, 2) \cup (3, \infty)$, or $\{x \mid -1 < x < 2 \text{ or } x > 3\}$ **27.** $(-\infty, 0] \cup [2, 5]$, or $\{x \mid x \leq 0 \text{ or } 2 \leq x \leq 5\}$ **29.** $(-\infty, -5)$, or $\{x \mid x < -5\}$
31. $(-\infty, -1] \cup (3, \infty)$, or $\{x \mid x \leq -1 \text{ or } x > 3\}$
33. $(-\infty, -6)$, or $\{x \mid x < -6\}$ **35.** $(-\infty, -1] \cup [2, 5)$, or $\{x \mid x \leq -1 \text{ or } 2 \leq x < 5\}$ **37.** $(-\infty, -3) \cup [0, \infty)$, or $\{x \mid x < -3 \text{ or } x \geq 0\}$ **39.** $(0, \infty)$, or $\{x \mid x > 0\}$
41. $(-\infty, -4) \cup [1, 3)$, or $\{x \mid x < -4 \text{ or } 1 \leq x < 3\}$

43. $\left(-\frac{3}{4}, \frac{5}{2}\right]$, or $\left\{x \mid -\frac{3}{4} < x \leq \frac{5}{2}\right\}$ **45.** $(-\infty, 2) \cup [3, \infty)$, or $\{x \mid x < 2 \text{ or } x \geq 3\}$ **47.** ▨   **49.** $8a^9b^6c^{12}$ **50.** $25a^8b^{14}$ **51.** $\frac{1}{32}$ **52.** $\frac{1}{81}$ **53.** $3a^2 + 6a + 3$ **54.** $5a + 7$ **55.** ▨   **57.** $\left(-1 - \sqrt{6}, -1 + \sqrt{6}\right)$, or $\left\{x \mid -1 - \sqrt{6} < x < -1 + \sqrt{6}\right\}$ **59.** $\{0\}$ **61.** (a) $(10, 200)$, or $\{x \mid 10 < x < 200\}$; (b) $[0, 10) \cup (200, \infty)$, or $\{x \mid 0 \leq x < 10 \text{ or } x > 200\}$ **63.** $\{n \mid n \text{ is an integer } and\ 12 \leq n \leq 25\}$ **65.** $f(x) = 0$ for $x = -2, 1, 3$; $f(x) < 0$ for $(-\infty, -2) \cup (1, 3)$, or $\{x \mid x < -2 \text{ or } 1 < x < 3\}$; $f(x) > 0$ for $(-2, 1) \cup (3, \infty)$, or $\{x \mid -2 < x < 1 \text{ or } x > 3\}$ **67.** $f(x)$ has no zeros; $f(x) < 0$ for $(-\infty, 0)$, or $\{x \mid x < 0\}$; $f(x) > 0$ for $(0, \infty)$, or $\{x \mid x > 0\}$ **69.** $f(x) = 0$ for $x = -1, 0$; $f(x) < 0$ for $(-\infty, -3) \cup (-1, 0)$, or $\{x \mid x < -3 \text{ or } -1 < x < 0\}$; $f(x) > 0$ for $(-3, -1) \cup (0, 2) \cup (2, \infty)$, or $\{x \mid -3 < x < -1 \text{ or } 0 < x < 2 \text{ or } x > 2\}$ **71.** ▨

Review Exercises: Chapter 11, pp. 778–779

1. False   **2.** True   **3.** True   **4.** True   **5.** True   **6.** False   **7.** True   **8.** True   **9.** True   **10.** True   **11.** $\pm\frac{3}{2}$ **12.** $0, -\frac{3}{4}$ **13.** $3, 9$ **14.** $2 \pm 2i$ **15.** $3, 5$ **16.** $-\dfrac{9}{2} \pm \dfrac{\sqrt{85}}{2}$ **17.** $-0.3722813233, 5.3722813233$ **18.** $-\frac{1}{4}, 1$ **19.** $x^2 - 12x + 36 = (x - 6)^2$ **20.** $x^2 + \frac{3}{5}x + \frac{9}{100} = \left(x + \frac{3}{10}\right)^2$ **21.** $3 \pm 2\sqrt{2}$ **22.** 10%   **23.** 6.7 sec   **24.** About 153 mph   **25.** 6 hr   **26.** Two irrational   **27.** Two imaginary   **28.** $x^2 - 5 = 0$ **29.** $x^2 + 8x + 16 = 0$ **30.** $(-3, 0), (-2, 0), (2, 0), (3, 0)$ **31.** $-5, 3$ **32.** $\pm\sqrt{2}, \pm\sqrt{7}$ **33.**

$x = -2$

$(-2, 4)$

$f(x) = -3(x + 2)^2 + 4$
Maximum: 4

34. (a) Vertex: $(3, 5)$; axis of symmetry: $x = 3$; (b)

$f(x) = 2x^2 - 12x + 23$

35. $(2, 0), (7, 0); (0, 14)$ **36.** $p = \dfrac{9\pi^2}{N^2}$

37. $T = \dfrac{1 \pm \sqrt{1 + 24A}}{6}$ **38.** Quadratic   **39.** Linear

40. 225 ft^2; 15 ft by 15 ft

41. (a) $f(x) = -42x^2 + 167x + 281$;
(b) 277 million books

42. $(-1, 0) \cup (3, \infty)$, or $\{x \mid -1 < x < 0 \text{ or } x > 3\}$

43. $(-3, 5]$, or $\{x \mid -3 < x \leq 5\}$

44. 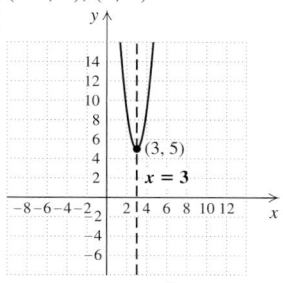 The x-coordinate of the maximum or minimum point lies halfway between the x-coordinates of the x-intercepts.

45. Yes; if the discriminant is a perfect square, then the solutions are rational numbers, p/q and r/s. (Note that if the discriminant is 0, then $p/q = r/s$.) Then the equation can be written in factored form, $(qx - p)(sx - r) = 0$.

46. Four; let $u = x^2$. Then $au^2 + bu + c = 0$ has at most two solutions, $u = m$ and $u = n$. Now substitute x^2 for u and obtain $x^2 = m$ or $x^2 = n$. These equations yield the solutions $x = \pm\sqrt{m}$ and $x = \pm\sqrt{n}$. When $m \neq n$, the maximum number of solutions, four, occurs.

47. Completing the square was used to solve quadratic equations and to graph quadratic functions by rewriting the function in the form $f(x) = a(x - h)^2 + k$.

48. $f(x) = \frac{7}{15}x^2 - \frac{14}{15}x - 7$ **49.** $h = 60, k = 60$

50. 18, 324

Test: Chapter 11, p. 780

1. [11.1] $\pm \dfrac{\sqrt{11}}{2}$ **2.** [11.2] 2, 9   **3.** [11.2] $-1 \pm \sqrt{2}i$

4. [11.2] $1 \pm \sqrt{6}$ **5.** [11.5] $-2, \frac{2}{3}$

6. [11.2] $-4.192582404, 1.192582404$ **7.** [11.2] $-\frac{3}{4}, \frac{7}{3}$

8. [11.1] $x^2 - 16x + 64 = (x - 8)^2$

9. [11.1] $x^2 + \frac{2}{7}x + \frac{1}{49} = \left(x + \frac{1}{7}\right)^2$ **10.** [11.1] $-5 \pm \sqrt{10}$

11. [11.3] 16 km/h   **12.** [11.3] 2 hr

13. [11.4] Two imaginary   **14.** [11.4] $3x^2 + 5x - 2 = 0$

15. [11.5] $(-4, 0), (4, 0)$

16. [11.6]

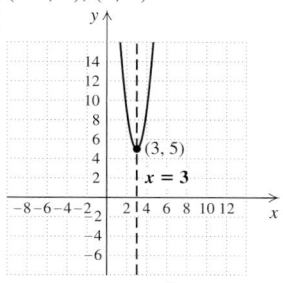

$f(x) = 4(x - 3)^2 + 5$
Minimum: 5

17. [11.7] **(a)** $(-1, -8), x = -1$;
(b)

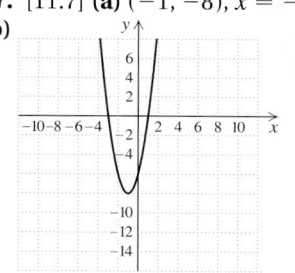

$f(x) = 2x^2 + 4x - 6$

18. [11.7] $(-2, 0), (3, 0); (0, -6)$

19. [11.3] $r = \sqrt{\dfrac{3V}{\pi} - R^2}$ **20.** [11.8] Neither

21. [11.8] Minimum: \$129/cap when 325 caps are built

22. [11.8] $f(x) = \frac{1}{5}x^2 - \frac{3}{5}x$ **23.** [11.9] $(-6, 1)$, or $\{x \mid -6 < x < 1\}$ **24.** [11.9] $[-1, 0) \cup [1, \infty)$, or $\{x \mid -1 \leq x < 0 \text{ or } x \geq 1\}$ **25.** [11.4] $\frac{1}{2}$

26. [11.4] $x^4 - 14x^3 + 67x^2 - 114x + 26 = 0$; answers may vary

27. [11.4] $x^6 - 10x^5 + 20x^4 + 50x^3 - 119x^2 - 60x + 150 = 0$; answers may vary

CHAPTER 12

Technology Connection, p. 784

1. To check $(f \circ g)(x)$, we use a table to show that $y_2 = y_3$.

$y_1 = x - 1$; $y_2 = \sqrt{y_1}$;
$y_3 = \sqrt{x - 1}$;

X	Y2	Y3
1	0	0
2	1	1
3	1.4142	1.4142
4	1.7321	1.7321
5	2	2
6	2.2361	2.2361
7	2.4495	2.4495

X =

A similar table shows that for $y_2 = \sqrt{x}$ and $y_4 = y_2(y_1)$, we have $y_3 = y_4$. The check for $(g \circ f)(x)$ is similar. A graph can also be used.

Technology Connection, p. 790

1. Graph each pair of functions in a square window along with the line $y = x$ and determine whether the first two functions are reflections of each other across $y = x$. For further verification, examine a table of values for each pair of functions.   **2.** Yes; most graphing calculators do not require that the inverse relation be a function.

Exercise Set 12.1, pp. 791–793

1. True **3.** False **5.** False **7.** True
9. $(f \circ g)(1) = 2; (g \circ f)(1) = 1;$
$(f \circ g)(x) = 4x^2 - 12x + 10;$
$(g \circ f)(x) = 2x^2 - 1$ **11.** $(f \circ g)(1) = -8; (g \circ f)(1) = 1;$
$(f \circ g)(x) = 2x^2 - 10; (g \circ f)(x) = 2x^2 - 12x + 11$
13. $(f \circ g)(1) = 8; (g \circ f)(1) = \frac{1}{64};$

$(f \circ g)(x) = \frac{1}{x^2} + 7; (g \circ f)(x) = \frac{1}{(x + 7)^2}$

15. $(f \circ g)(1) = 2; (g \circ f)(1) = 4;$
$(f \circ g)(x) = \sqrt{x + 3}; (g \circ f)(x) = \sqrt{x} + 3$

17. $(f \circ g)(1) = 2; (g \circ f)(1) = \frac{1}{2}; (f \circ g)(x) = \sqrt{\frac{4}{x}};$

$(g \circ f)(x) = \frac{1}{\sqrt{4x}}$ **19.** $(f \circ g)(1) = 4; (g \circ f)(1) = 2;$
$(f \circ g)(x) = x + 3; (g \circ f)(x) = \sqrt{x^2 + 3}$
21. $f(x) = x^2; g(x) = 7 + 5x$

23. $f(x) = \sqrt{x}; g(x) = 2x + 7$ **25.** $f(x) = \frac{2}{x}; g(x) = x - 3$

27. Yes **29.** No **31.** Yes **33.** No **35. (a)** Yes;

(b) $f^{-1}(x) = x - 4$ **37. (a)** Yes; **(b)** $f^{-1}(x) = \frac{x}{2}$

39. (a) Yes; **(b)** $g^{-1}(x) = \frac{x + 1}{3}$ **41. (a)** Yes;

(b) $f^{-1}(x) = 2x - 2$ **43. (a)** No **45. (a)** Yes;

(b) $h^{-1}(x) = \frac{x - 4}{-2}$ **47. (a)** Yes; **(b)** $f^{-1}(x) = \frac{1}{x}$

49. (a) No **51. (a)** Yes; **(b)** $f^{-1}(x) = \frac{3x - 1}{2}$ **53. (a)** Yes;

(b) $f^{-1}(x) = \sqrt[3]{x + 5}$ **55. (a)** Yes; **(b)** $g^{-1}(x) = \sqrt[3]{x} + 2$
57. (a) Yes; **(b)** $f^{-1}(x) = x^2, x \geq 0$
59.

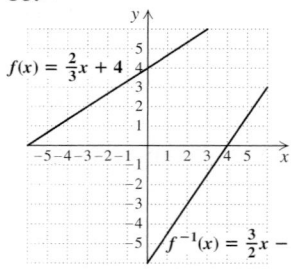

61.

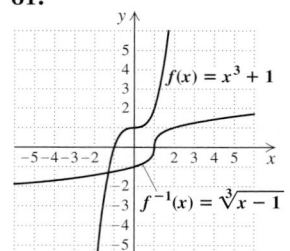

63.

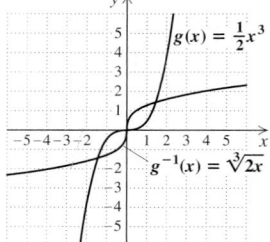

65.

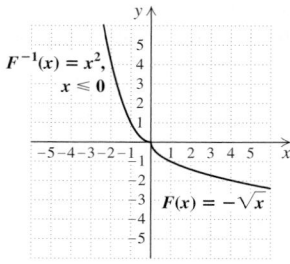

67.

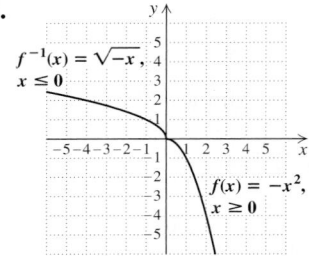

69. (1) $(f^{-1} \circ f)(x) = f^{-1}(f(x))$
$= f^{-1}(\sqrt[3]{x - 4}) = (\sqrt[3]{x - 4})^3 + 4$
$= x - 4 + 4 = x;$
(2) $(f \circ f^{-1})(x) = f(f^{-1}(x))$
$= f(x^3 + 4) = \sqrt[3]{x^3 + 4 - 4}$
$= \sqrt[3]{x^3} = x$

71. (1) $(f^{-1} \circ f)(x) = f^{-1}(f(x)) = f^{-1}\left(\frac{1 - x}{x}\right)$

$$= \frac{1}{\left(\dfrac{1 - x}{x}\right) + 1}$$

$$= \frac{1}{\dfrac{1 - x + x}{x}}$$

$$= x;$$

(2) $(f \circ f^{-1})(x) = f(f^{-1}(x)) = f\left(\dfrac{1}{x + 1}\right)$

$$= \frac{1 - \left(\dfrac{1}{x + 1}\right)}{\left(\dfrac{1}{x + 1}\right)}$$

$$= \frac{\dfrac{x + 1 - 1}{x + 1}}{\dfrac{1}{x + 1}} = x$$

73. (a) 40, 44, 52, 60; **(b)** $f^{-1}(x) = (x - 24)/2;$
(c) 8, 10, 14, 18 **75.** 🖩 **77.** $a^{13}b^{13}$ **78.** $x^{10}y^{12}$

79. 81 **80.** 125 **81.** $y = \frac{3}{2}(x + 7)$ **82.** $y = \frac{10 - x}{3}$
83. 🖩
85.

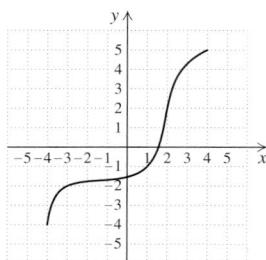

87. $g(x) = \frac{x}{2} + 20$ **89.** 🖩

91. Suppose that $h(x) = (f \circ g)(x)$. First, note that for $I(x) = x$, $(f \circ I)(x) = f(I(x)) = f(x)$ for any function f.
(i) $((g^{-1} \circ f^{-1}) \circ h)(x) = ((g^{-1} \circ f^{-1}) \circ (f \circ g))(x)$
$= ((g^{-1} \circ (f^{-1} \circ f)) \circ g)(x)$
$= ((g^{-1} \circ I) \circ g)(x)$
$= (g^{-1} \circ g)(x) = x$
(ii) $(h \circ (g^{-1} \circ f^{-1}))(x) = ((f \circ g) \circ (g^{-1} \circ f^{-1}))(x)$
$= ((f \circ (g \circ g^{-1})) \circ f^{-1})(x)$
$= ((f \circ I) \circ f^{-1})(x)$
$= (f \circ f^{-1})(x) = x.$
Therefore, $(g^{-1} \circ f^{-1})(x) = h^{-1}(x)$.   **93.** Yes   **95.** No
97. **(1)** C; **(2)** A; **(3)** B; **(4)** D   **99.**

Technology Connection, p. 796

1. $y_1 = \left(\frac{5}{2}\right)^x$; $y_2 = \left(\frac{2}{5}\right)^x$
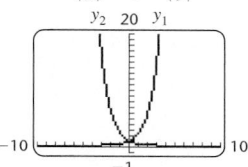

2. $y_1 = 3.2^x$; $y_2 = 3.2^{-x}$
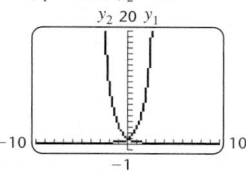

3. $y_1 = \left(\frac{3}{7}\right)^x$; $y_2 = \left(\frac{7}{3}\right)^x$
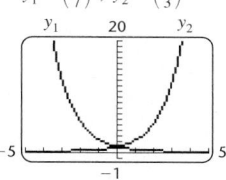

4. $y_1 = 5000(1.08)^x$; $y_2 = 5000(1.08)^{x-3}$
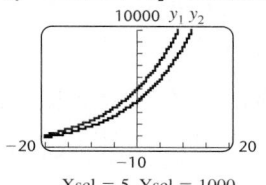
Xscl = 5, Yscl = 1000

Exercise Set 12.2, pp. 800–802

1. True   **3.** True   **5.** False
7.
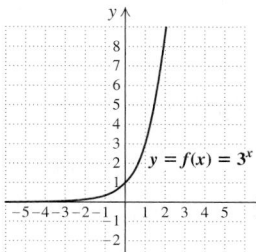
$y = f(x) = 3^x$

9.
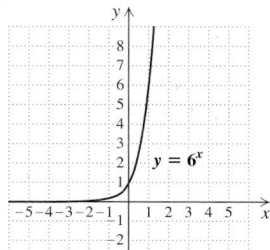
$y = 6^x$

11.
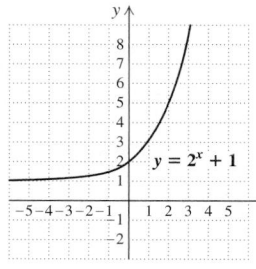
$y = 2^x + 1$

13.
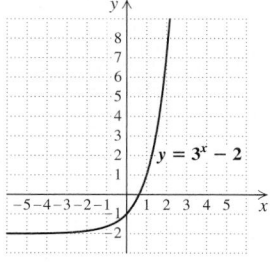
$y = 3^x - 2$

15.
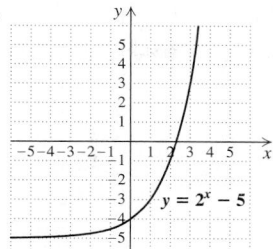
$y = 2^x - 5$

17.
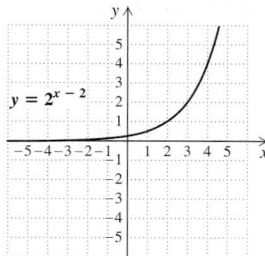
$y = 2^{x-2}$

19.
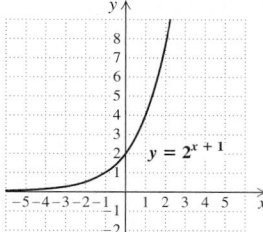
$y = 2^x + 1$

21.
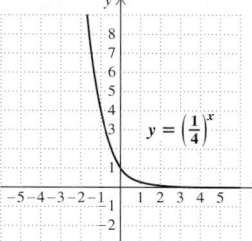
$y = \left(\frac{1}{4}\right)^x$

23.
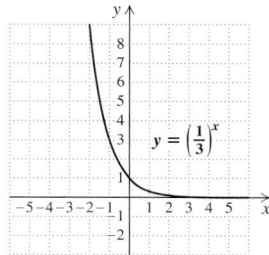
$y = \left(\frac{1}{3}\right)^x$

25.
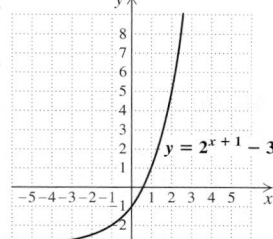
$y = 2^{x+1} - 3$

27.
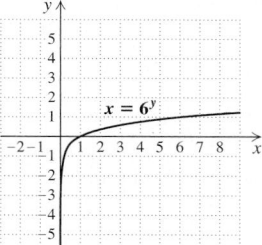
$x = 6^y$

29.
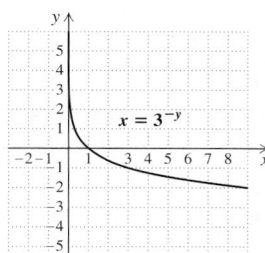
$x = 3^{-y}$

31.
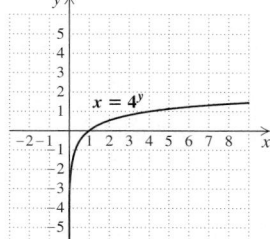
$x = 4^y$

33.
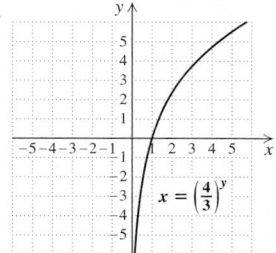
$x = \left(\frac{4}{3}\right)^y$

35.

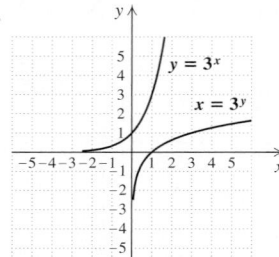

37.

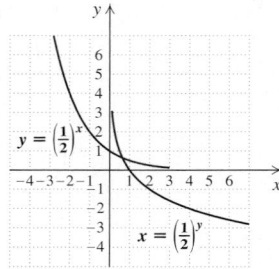

39. (a) About 6.8 billion; about 7.2 billion; about 7.7 billion;

(b)

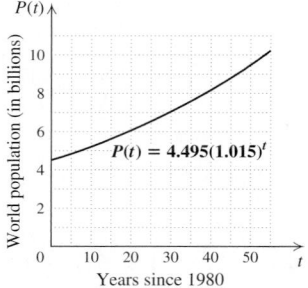

41. (a) 19.6%; 16.3%; 7.3%;

(b)

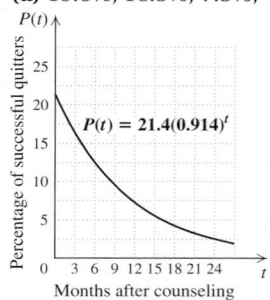

43. (a) About 44,079 whales; about 12,953 whales;

(b)

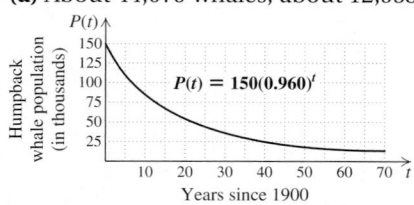

45. (a) About 8706 whales; about 15,107 whales;

(b)

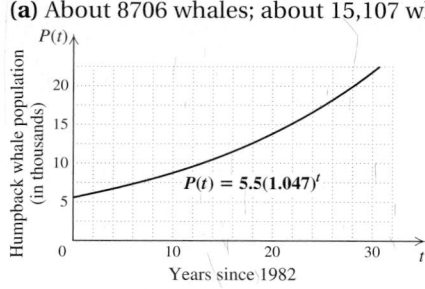

47. (a) 454,354,240 cm^2; 525,233,501,400 cm^2;

(b)

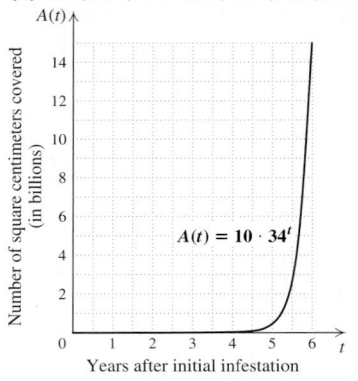

49.    **51.** $\frac{1}{25}$ **52.** $\frac{1}{32}$ **53.** 100   **54.** $\frac{1}{125}$ **55.** $5a^6b^3$

56. $6x^4y$ **57.**    **59.** $\pi^{2.4}$

61.

63

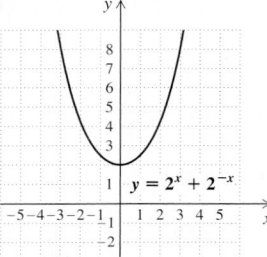

65.

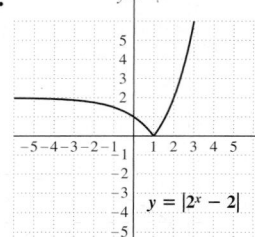

67.

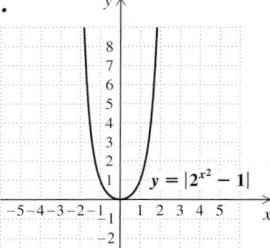

69.

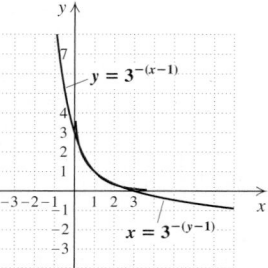

71. $A(t) = 200.7624553(1.992834389)^t$, where $A(t)$ is total sales, in millions of dollars, t years after 1997; $395,244,465,700

73. (a)

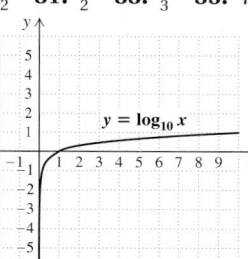

Years since 2000

(b) linear; the points appear to lie on a straight line.
75.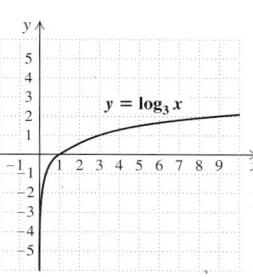

Exercise Set 12.3, pp. 808–809

1. (g)   **3.** (a)   **5.** (b)   **7.** (e)   **9.** 3   **11.** 4   **13.** 4
15. -2 **17.** -1 **19.** 4   **21.** 1   **23.** 0   **25.** 5   **27.** -2
29. $\frac{1}{2}$ **31.** $\frac{3}{2}$ **33.** $\frac{2}{3}$ **35.** 7

37. 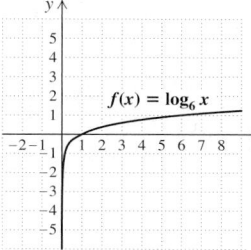   **39.**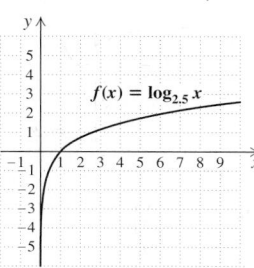

41. 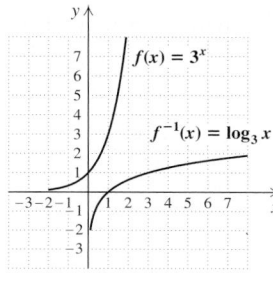   **43.**

45.

47. $5^t = 9$ **49.** $5^2 = 25$ **51.** $10^{-1} = 0.1$ **53.** $10^{0.845} = 7$
55. $c^8 = m$ **57.** $t^r = Q$ **59.** $e^{-1.3863} = 0.25$
61. $r^{-x} = T$ **63.** $2 = \log_{10} 100$ **65.** $-5 = \log_4 \frac{1}{1024}$
67. $\frac{3}{4} = \log_{16} 8$ **69.** $0.4771 = \log_{10} 3$ **71.** $m = \log_z 6$
73. $m = \log_p V$ **75.** $3 = \log_e 20.0855$
77. $-4 = \log_e 0.0183$ **79.** 9   **81.** 3   **83.** 4   **85.** 7
87. $\frac{1}{9}$ **89.** 4   **91.** 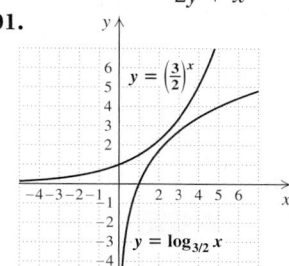   **93.** x^8 **94.** a^{12} **95.** $a^7 b^8$
96. $x^5 y^{12}$ **97.** $\dfrac{x(3y - 2)}{2y + x}$ **98.** $\dfrac{x + 2}{x + 1}$ **99.**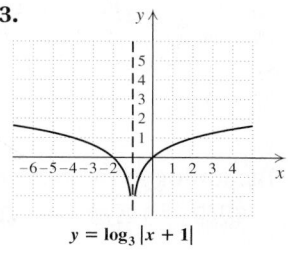
101.   **103.**

105. 6   **107.** $-25, 4$ **109.** -2 **111.** 0   **113.** Let $b = 0$,
and suppose that $x_1 = 1$ and $x_2 = 2$. Then $0^1 = 0^2$, but
$1 \neq 2$. Then let $b = 1$, and suppose that $x_1 = 1$ and $x_2 = 2$.
Then $1^1 = 1^2$, but $1 \neq 2$.

Exercise Set 12.4, pp. 815–817

1. (e)   **3.** (a)   **5.** (c)   **7.** $\log_3 81 + \log_3 27$
9. $\log_4 64 + \log_4 16$ **11.** $\log_c r + \log_c s + \log_c t$
13. $\log_a (5 \cdot 14)$, or $\log_a 70$ **15.** $\log_c (t \cdot y)$ **17.** $8 \log_a r$
19. $6 \log_c y$ **21.** $-3 \log_b C$ **23.** $\log_2 25 - \log_2 13$
25. $\log_b m - \log_b n$ **27.** $\log_a \frac{17}{6}$ **29.** $\log_b \frac{36}{4}$, or $\log_b 9$
31. $\log_a \frac{7}{18}$ **33.** $\log_a x + \log_a y + \log_a z$
35. $3 \log_a x + 4 \log_a z$ **37.** $2 \log_a x - 2 \log_a y + \log_a z$
39. $4 \log_a x - 3 \log_a y - \log_a z$
41. $\log_b x + 2 \log_b y - \log_b w - 3 \log_b z$
43. $\frac{1}{2}(7 \log_a x - 5 \log_a y - 8 \log_a z)$
45. $\frac{1}{3}(6 \log_a x + 3 \log_a y - 2 - 7 \log_a z)$ **47.** $\log_a (x^8 z^3)$
49. $\log_a x$ **51.** $\log_a \dfrac{y^5}{x^{3/2}}$ **53.** $\log_a (x - 2)$ **55.** 1.953
57. -0.369 **59.** -1.161 **61.** $\frac{3}{2}$ **63.** Cannot be found
65. 7   **67.** m **69.**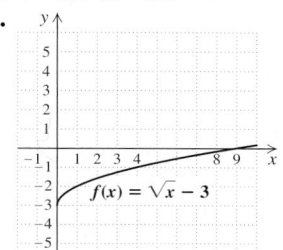
71.   **72.**

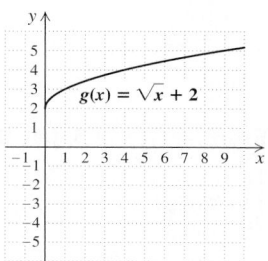

73. 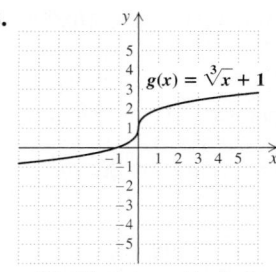 **74.**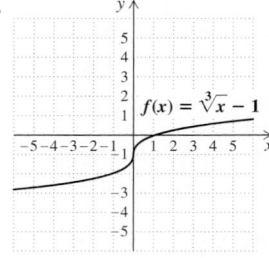

75. $a^{17}b^{17}$ **76.** $x^{11}y^6z^8$ **77.**
79. $\log_a (x^6 - x^4y^2 + x^2y^4 - y^6)$
81. $\frac{1}{2}\log_a (1 - s) + \frac{1}{2}\log_a (1 + s)$ **83.** $\frac{10}{3}$ **85.** -2
87. True

Technology Connection, p. 818

1.

Technology Connection, p. 819

1. As x gets larger, the value of y_1 approaches
2.7182818284.... **2.** For large values of x, the graphs of y_1
and y_2 will be very close or appear to be the same line,
depending on the window chosen. **3.** Using TRACE, no
y-value is given for $x = 0$. Using a table, an error message
appears for y_1 when $x = 0$. The domain does not include 0
because division by 0 is undefined.

Technology Connection, p. 822

1. $y = \log x/\log 7$ 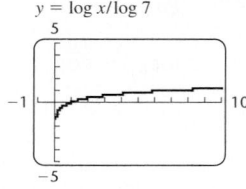 **2.** $y = \log (x+2)/\log 5$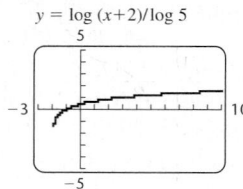

3. $y = \log x/\log 7 + 2$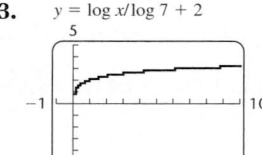

Exercise Set 12.5, pp. 822–824

1. True **3.** True **5.** True **7.** True **9.** 0.7782
11. 1.8621 **13.** 3 **15.** -0.2782 **17.** 1.7986
19. 199.5262 **21.** 1.4894 **23.** 0.0011 **25.** 1.6094
27. 4.0431 **29.** -5.0832 **31.** 96.7583 **33.** 15.0293

35. 0.0305 **37.** 109.9472 **39.** 2.5237 **41.** 6.6439
43. 2.1452 **45.** -2.3219 **47.** -2.3219 **49.** 3.5471
51. 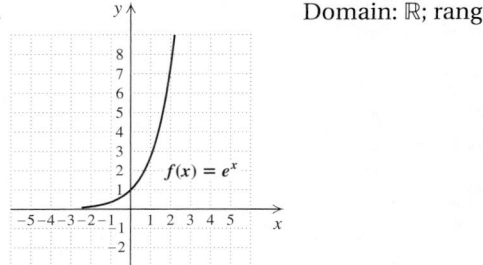 Domain: $\mathbb{R}$; range: $(0, \infty)$

53. 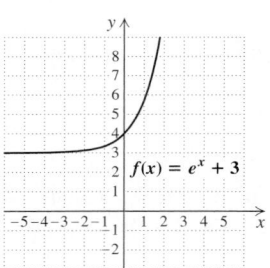 Domain: $\mathbb{R}$; range: $(3, \infty)$

55. 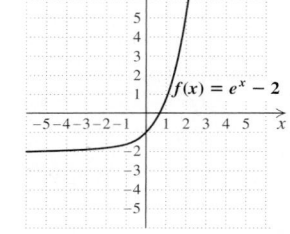 Domain: $\mathbb{R}$; range: $(-2, \infty)$

57. 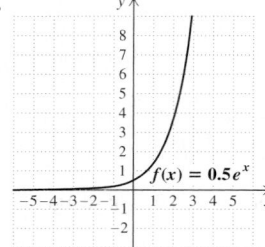 Domain: $\mathbb{R}$; range: $(0, \infty)$

59. 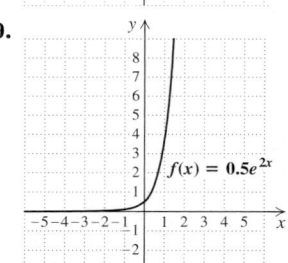 Domain: $\mathbb{R}$; range: $(0, \infty)$

61. 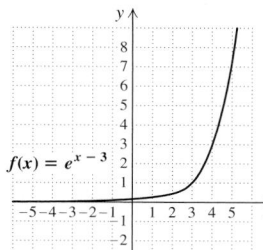   Domain: $\mathbb{R}$; range: $(0, \infty)$

63. 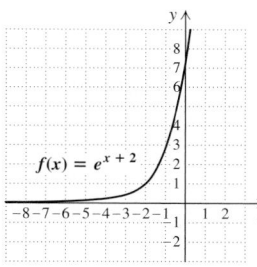   Domain: $\mathbb{R}$; range: $(0, \infty)$

65. 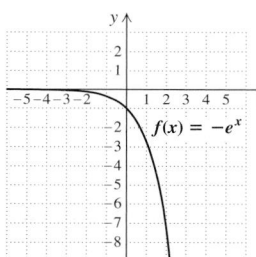   Domain: $\mathbb{R}$; range: $(-\infty, 0)$

67. 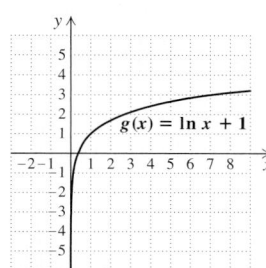   Domain: $(0, \infty)$; range: $\mathbb{R}$

69. 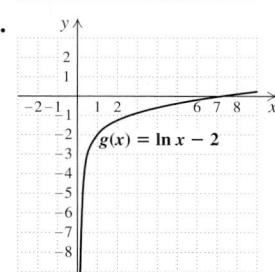   Domain: $(0, \infty)$; range: $\mathbb{R}$

71. 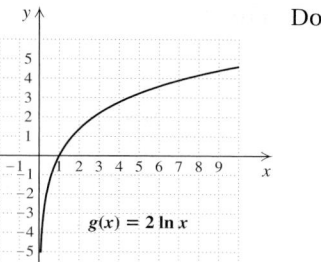   Domain: $(0, \infty)$; range: $\mathbb{R}$

73. 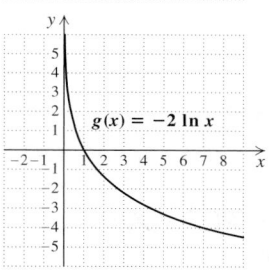   Domain: $(0, \infty)$; range: $\mathbb{R}$

75. 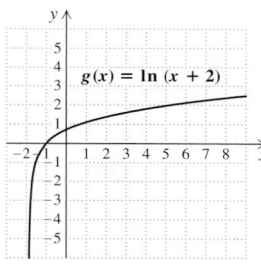   Domain: $(-2, \infty)$; range: $\mathbb{R}$

77. 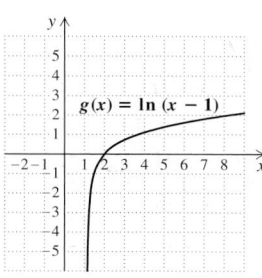   Domain: $(1, \infty)$; range: $\mathbb{R}$

79. ▨   **81.** $-\frac{5}{2}, \frac{5}{2}$ **82.** $0, \frac{7}{5}$ **83.** $\frac{15}{17}$ **84.** $\frac{9}{13}$ **85.** 16, 256
86. $\frac{1}{4}, 9$ **87.** ▨   **89.** 2.452   **91.** 1.442

93. $\log M = \dfrac{\ln M}{\ln 10}$ **95.** 1086.5129   **97.** 4.9855

99. (a) Domain: $\{x \mid x > 0\}$, or $(0, \infty)$;
range: $\{y \mid y < 0.5135\}$, or $(-\infty, 0.5135)$; **(b)** $[-1, 5, -10, 5]$;
(c) $y = 3.4 \ln x - 0.25 e^x$

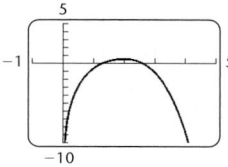

101. (a) Domain: $\{x \mid x > 0\}$, or $(0, \infty)$;
range: $\{y \mid y > -0.2453\}$, or $(-0.2453, \infty)$;
(b) $[-1, 5, -1, 10]$;
(c) $y = 2x^3 \ln x$

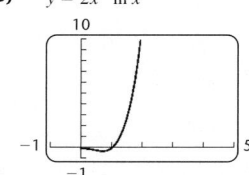

103.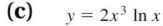

(c)

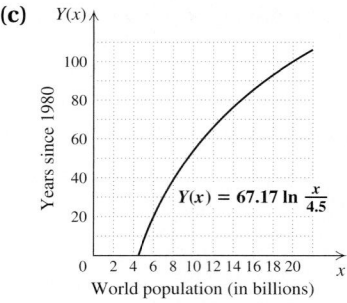

$Y(x) = 67.17 \ln \dfrac{x}{4.5}$

World population (in billions)

Technology Connection, p. 829

1. 0.38   **2.** -1.96 **3.** 0.90   **4.** -1.53 **5.** 0.13, 8.47
6. $-0.75, 0.75$

Exercise Set 12.6, pp. 830–831

1. (e)   3. (f)   5. (b)   7. (g)   9. $\dfrac{\log 19}{\log 2} \approx 4.248$

11. $\dfrac{\log 17}{\log 8} + 1 \approx 2.362$ **13.** $\ln 1000 \approx 6.908$

15. $\dfrac{\ln 5}{0.03} \approx 53.648$ **17.** $\dfrac{\log 5}{\log 3} - 1 \approx 0.465$ **19.** 1

21. $\dfrac{\log 87}{\log 4.9} \approx 2.810$ **23.** $\dfrac{\ln \left(\frac{19}{2}\right)}{4} \approx 0.563$ **25.** $\dfrac{\ln 2}{5} \approx 0.139$

27. 81   **29.** $\frac{1}{8}$ **31.** $e^5 \approx 148.413$ **33.** 2   **35.** $\dfrac{e^3}{4} \approx 5.021$

37. $10^{2.5} \approx 316.228$ **39.** $\dfrac{e^4 - 1}{2} \approx 26.799$ **41.** $e \approx 2.718$

43. $e^{-3} \approx 0.050$ **45.** -4 **47.** 10   **49.** No solution
51. $\frac{83}{15}$ **53.** 1   **55.** 6   **57.** 1   **59.** 5   **61.** $\frac{17}{2}$ **63.** 4

65.   **67.** $y = 9x$ **68.** $y = \dfrac{21.35}{x}$ **69.** $L = \dfrac{8T^2}{\pi^2}$

70. $c = \sqrt{\dfrac{E}{m}}$ **71.** $1\frac{1}{5}$ hr   **72.** $9\frac{3}{8}$ min   **73.**   **75.** $\frac{12}{5}$

77. $\sqrt[3]{3}$ **79.** -1 **81.** $-3, -1$ **83.** $-625, 625$
85. $\frac{1}{2}, 5000$ **87.** $-3, -1$ **89.** $\frac{1}{100,000}, 100,000$ **91.** $-\frac{1}{3}$
93. 38   **95.** 1

Exercise Set 12.7, pp. 839–844

1. (a) About 2001; **(b)** 1 yr   **3. (a)** 146,293; **(b)** 51
5. (a) 6.4 yr; **(b)** 23.4 yr   **7. (a)** 2005; **(b)** 2018
9. (a) 2019; **(b)** 15.1 yr   **11.** 4.9   **13.** 10^{-7} moles per liter
15. 65 dB   **17.** $10^{-1.5}$ W/m^2 **19. (a)** $P(t) = P_0 e^{0.025t}$;
(b) $5126.58; $5256.36; **(c)** 27.7 yr
21. (a) $P(t) = 292.80 e^{0.009t}$, where t is the number of years
after 2004 and $P(t)$ is in millions; **(b)** 295.45 million;
(c) 2016   **23.** 6.7 months
25. (a) About 2010; **(b)** about 2019;

27. (a) 68%; **(b)** 54%, 40%   **(d)** 6.9 months

(c)

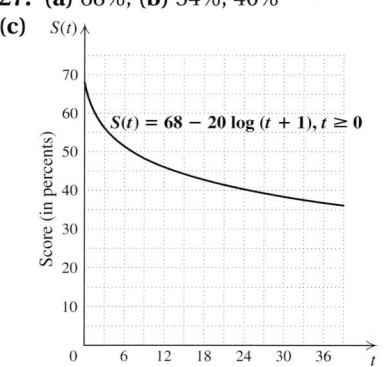

$S(t) = 68 - 20 \log (t + 1), t \geq 0$

Months

29. (a) $N(t) = 17e^{0.534t}$, where t is the number of years
since 2000; **(b)** about 419
31. (a) $k \approx 0.004$; $P(t) = 987 e^{-0.004t}$, where t is the
number of years after 1990 and $P(t)$ is in millions;
(b) 918 million acres; **(c)** about 2043   **33.** About 2103 yr
35. About 7.2 days   **37.** 69.3% per year
39. (a) $k \approx 0.099$; $V(t) = 451,000 e^{0.099t}$, where t is the
number of years after 1991; **(b)** about $1.99 million;
(c) 7.0 yr; **(d)** 2010   **41.**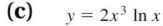

43.

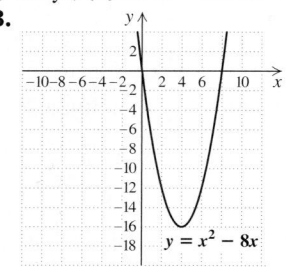

$y = x^2 - 8x$

44.

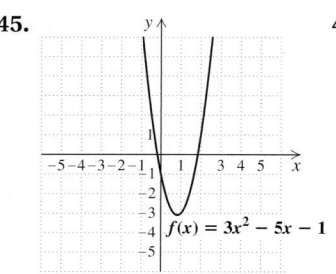

$y = x^2 - 5x - 6$

45.

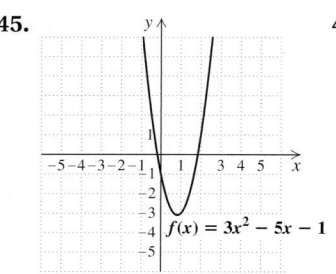

$f(x) = 3x^2 - 5x - 1$

46.

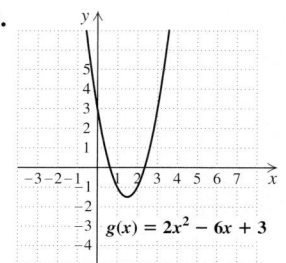

$g(x) = 2x^2 - 6x + 3$

47. $4 \pm \sqrt{23}$ **48.** $-5 \pm \sqrt{31}$ **49.** 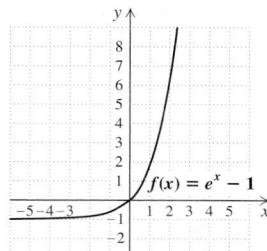   **51.** \$18.9 million
53. $P(t) = 100 - 63.03(0.95)^t$ **55.** About 80,922 yr, or with rounding of k, about 80,792 yr   **57.** Consider an exponential growth function $P(t) = P_0 e^{kt}$. At time T, $P(T) = 2P_0$.
Solve for T:
$$2P_0 = P_0 e^{kT}$$
$$2 = e^{kT}$$
$$\ln 2 = kT$$
$$\frac{\ln 2}{k} = T.$$

59. (a)

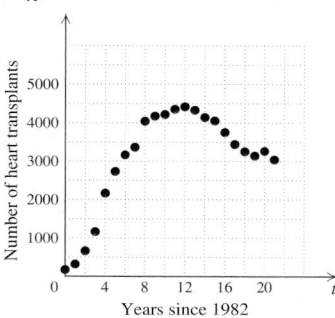

(b) ; **(c)**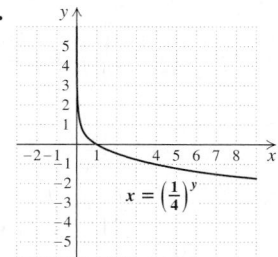

Review Exercises: Chapter 12, pp. 849–850

1. True   **2.** True   **3.** True   **4.** False   **5.** False   **6.** True
7. False   **8.** False   **9.** True   **10.** False
11. $(f \circ g)(x) = 4x^2 - 12x + 10$; $(g \circ f)(x) = 2x^2 - 1$
12. $f(x) = \sqrt{x}$; $g(x) = 3 - x$ **13.** No
14. $f^{-1}(x) = x + 8$ **15.** $g^{-1}(x) = \dfrac{2x - 1}{3}$
16. $f^{-1}(x) = \dfrac{\sqrt[3]{x}}{3}$

17.

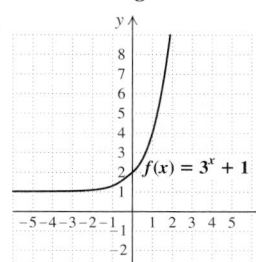

18.

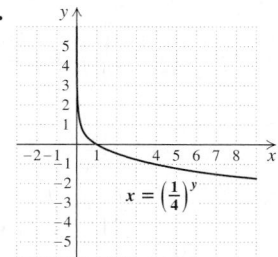

19.

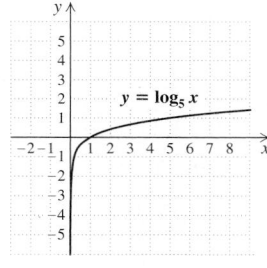

20. 2   **21.** -2 **22.** 7   **23.** $\frac{1}{2}$ **24.** $\log_{10} \frac{1}{100} = -2$
25. $\log_{25} 5 = \frac{1}{2}$ **26.** $16 = 4^x$ **27.** $1 = 8^0$
28. $4 \log_a x + 2 \log_a y + 3 \log_a z$
29. $5 \log_a x - (\log_a y + 2 \log_a z)$, or
$5 \log_a x - \log_a y - 2 \log_a z$ **30.** $\frac{1}{4}(2 \log z - 3 \log x - \log y)$
31. $\log_a (7 \cdot 8)$, or $\log_a 56$ **32.** $\log_a \frac{72}{12}$, or $\log_a 6$
33. $\log \dfrac{a^{1/2}}{bc^2}$ **34.** $\log_a \sqrt[3]{\dfrac{x}{y^2}}$ **35.** 1   **36.** 0
37. 17   **38.** 6.93   **39.** -3.2698 **40.** 8.7601   **41.** 3.2698
42. 2.54995   **43.** -3.6602 **44.** 1.8751   **45.** 61.5177
46. -2.9957 **47.** 0.3753   **48.** 0.4307   **49.** 1.7097
50.   Domain: $\mathbb{R}$; range: $(-1, \infty)$

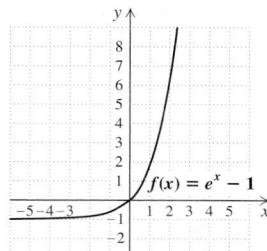

51.   Domain: $(0, \infty)$; range: $\mathbb{R}$

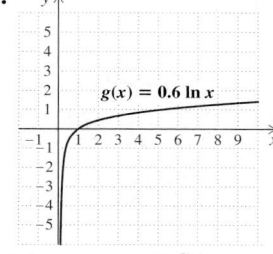

52. 5   **53.** -2 **54.** $\frac{1}{81}$ **55.** 2   **56.** $\frac{1}{1000}$
57. $e^{-2} \approx 0.1353$ **58.** $\dfrac{1}{2}\left(\dfrac{\log 19}{\log 4} + 5\right) \approx 3.5620$
59. $-5, 1$ **60.** $\dfrac{\log 8.3}{\log 4} \approx 1.5266$ **61.** $\dfrac{\ln 0.03}{-0.1} \approx 35.0656$
62. $e^{-3} \approx 0.0498$ **63.** 4   **64.** 8   **65.** 20   **66.** $\sqrt{43}$
67. (a) 82; **(b)** 66.8; **(c)** 35 months   **68. (a)** 6.6 yr; **(b)** 3.1 yr
69. (a) $k \approx 0.128$; $C(t) = 10e^{0.128t}$, where t is the number of years after 1999 and $C(t)$ is in thousands; **(b)** \$35,966; **(c)** 2009   **70.** 23.105% per year   **71.** 16.5 yr   **72.** 3463 yr
73. 6.6   **74.** 90 dB
75. 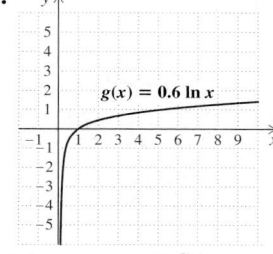 Negative numbers do not have logarithms because logarithm bases are positive, and there is no exponent to which a positive number can be raised to yield a negative number.   **76.** 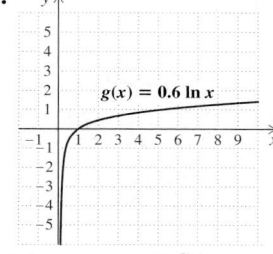 Taking the logarithm on each side of an equation produces an equivalent equation because the logarithm function is one-to-one. If two quantities are equal, their logarithms must be equal, and if the logarithms of two quantities are equal, the quantities must be the same.   **77.** e^{e^3} **78.** $-3, -1$ **79.** $\left(\frac{8}{3}, -\frac{2}{3}\right)$

Test: Chapter 12, pp. 851–852

1. [12.1] $(f \circ g)(x) = 2 + 6x + 4x^2$;

$(g \circ f)(x) = 2x^2 + 2x + 1$ **2.** [12.1] $f(x) = \dfrac{1}{x}$;

$g(x) = 2x^2 + 1$ **3.** [12.1] No   **4.** [12.1] $f^{-1}(x) = \dfrac{x-4}{3}$

5. [12.1] $g^{-1}(x) = \sqrt[3]{x} - 1$

6. [12.2]   **7.** [12.3]

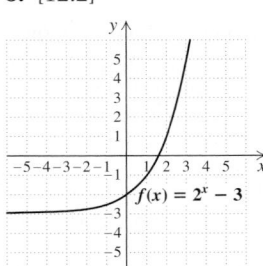

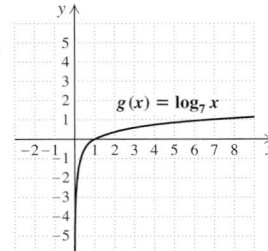

8. [12.3] 3   **9.** [12.3] $\frac{1}{2}$ **10.** [12.3] 18
11. [12.3] $\log_4 \frac{1}{64} = -3$ **12.** [12.3] $\log_{256} 16 = \frac{1}{2}$
13. [12.3] $49 = 7^m$ **14.** [12.3] $81 = 3^4$
15. [12.4] $3 \log a + \frac{1}{2} \log b - 2 \log c$
16. [12.4] $\log_a (z^2 \sqrt[3]{x})$ **17.** [12.4] 1   **18.** [12.4] 23
19. [12.4] 0   **20.** [12.4] 1.146   **21.** [12.4] 0.477
22. [12.4] 1.204   **23.** [12.5] 1.0899   **24.** [12.5] 0.1585
25. [12.5] -3.3524 **26.** [12.5] 121.5104   **27.** [12.5] 2.4022
28. [12.5]

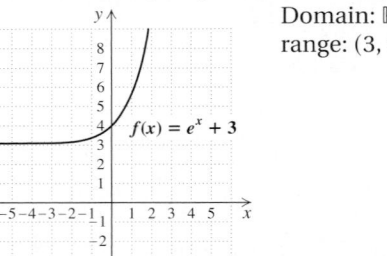

Domain: $\mathbb{R}$;
range: $(3, \infty)$

29. [12.5]

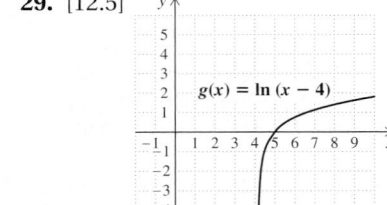

Domain: $(4, \infty)$;
range: $\mathbb{R}$

30. [12.6] -5 **31.** [12.6] 5   **32.** [12.6] 2

33. [12.6] 10,000   **34.** [12.6] $-\dfrac{1}{3}\left(\dfrac{\log 87}{\log 5} - 4\right) \approx 0.4084$

35. [12.6] $\dfrac{\log 1.2}{\log 7} \approx 0.0937$ **36.** [12.6] $e^{1/4} \approx 1.2840$

37. [12.6] 4   **38.** [12.7] **(a)** 2.56 ft/sec; **(b)** 711,637
39. [12.7] **(a)** $P(t) = 130.5e^{0.024t}$, where t is the number of years after 2002 and $P(t)$ is in millions; **(b)** 147.1 million; 165.9 million; **(c)** 2014; **(d)** 28.9 yr
40. [12.7] **(a)** $k \approx 0.031$; $C(t) = 19,070e^{0.031t}$, where t is the number of years after 1997; **(b)** \$28,535; **(c)** 2028
41. [12.7] 4.6%   **42.** [12.7] 4684 yr
43. [12.7] $10^{-4.5}$ W/m^2 **44.** [12.7] 7.0
45. [12.6] $-309,316$ **46.** [12.4] 2

Cumulative Review: Chapters 1–12, pp. 852–854

1. 2   **2.** 6   **3.** $\dfrac{y^{12}}{16x^8}$ **4.** $\dfrac{20x^6z^2}{y}$ **5.** $\dfrac{-y^4}{3z^5}$ **6.** $-2x - 1$

7. 25   **8.** $\frac{14}{5}$ **9.** $(3, -1)$ **10.** $(1, -2, 0)$ **11.** $-2, 5$

12. $\frac{9}{2}$ **13.** $\frac{5}{8}$ **14.** $\frac{3}{4}$ **15.** $\frac{1}{2}$ **16.** $\pm 5i$ **17.** 9, 25

18. $\pm 2, \pm 3$ **19.** 7   **20.** 6   **21.** $\frac{3}{2}$ **22.** $\dfrac{\log 7}{5 \log 3} \approx 0.3542$

23. $\dfrac{8e}{e-1} \approx 12.6558$ **24.** $(-\infty, -5) \cup (1, \infty)$, or $\{x \mid x < -5 \text{ or } x > 1\}$ **25.** $-3 \pm 2\sqrt{5}$
26. $\{x \mid x \le -2 \text{ or } x \ge 5\}$, or $(-\infty, -2] \cup [5, \infty)$

27. $a = \dfrac{Db}{b-D}$ **28.** $q = \dfrac{pf}{p-f}$ **29.** $B = \dfrac{3M - 2A}{2}$, or $B = \frac{3}{2}M - A$ **30.** 2   **31.** 3

32. $\{x \mid x \text{ is a real number } and \ x \ne -\frac{1}{3} \ and \ x \ne 2\}$

33. (a) $\dfrac{1.7 \text{ million barrels}}{13 \text{ yr}}$, or approximately

130,769 barrels/yr; **(b)** $g(t) = \dfrac{1.7}{13}t + 7.2$, where $g(t)$ is in millions of barrels; **(c)** $G(t) = 7.2e^{0.016t}$, where t is the number of years after 1990 and $G(t)$ is in millions of barrels
34. Length: 36 m; width: 20 m   **35.** A: 15°; B: 45°; C: 120°
36. $5\frac{5}{11}$ min   **37.** Thick and Tasty: 6 oz;
Light and Lean: 9 oz   **38.** $2\frac{7}{9}$ km/h   **39.** -49; -7 and 7
40. 78   **41.** 67.5   **42.** $P(t) = 33.7e^{0.01t}$
43. 35.8 million; 38.0 million   **44.** 69.3 yr   **45.** 18
46. $7p^2q^3 + pq + p - 9$ **47.** $8x^2 - 11x - 1$
48. $9x^4 - 12x^2y + 4y^2$ **49.** $10a^2 - 9ab - 9b^2$
50. $\dfrac{(x+4)(x-3)}{2(x-1)}$ **51.** $\dfrac{1}{x-4}$ **52.** $\dfrac{a+2}{6}$

53. $\dfrac{7x+4}{(x+6)(x-6)}$ **54.** $x(y + 2z - w)$
55. $(2 - 5x)(4 + 10x + 25x^2)$ **56.** $2(3x - 2y)(x + 2y)$
57. $(x^3 + 7)(x - 4)$ **58.** $2(m + 3n)^2$
59. $(x - 2y)(x + 2y)(x^2 + 4y^2)$ **60.** -12

61. $x^3 - 2x^2 - 4x - 12 + \dfrac{-42}{x-3}$ **62.** 1.8×10^{-1}

63. $2y^2\sqrt[3]{y}$ **64.** $14xy^2\sqrt{x}$ **65.** $81a^8b\sqrt[3]{b}$

66. $\dfrac{6 + \sqrt{y} - y}{4 - y}$ **67.** $\sqrt[10]{(x+5)^3}$ **68.** $18 - 2\sqrt{3}i$

69. $13 - i$ **70.** $f^{-1}(x) = \dfrac{x - 9}{-2}$, or $f^{-1}(x) = \dfrac{9 - x}{2}$

71. $y = -10x - 8$ **72.** $y = \frac{1}{2}x + 7$

73.

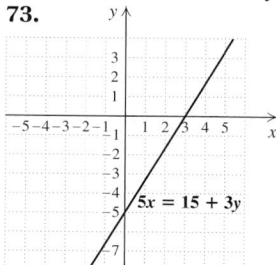

$5x = 15 + 3y$

74.

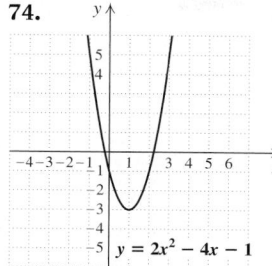

$y = 2x^2 - 4x - 1$

75.

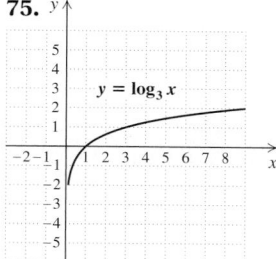

$y = \log_3 x$

76.

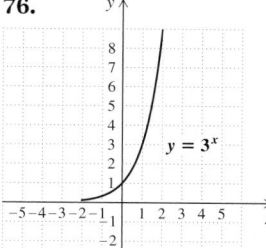

$y = 3^x$

77.

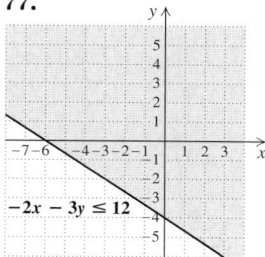

$-2x - 3y \le 12$

78.

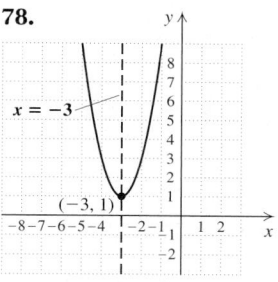

$x = -3$

$(-3, 1)$

$f(x) = 2(x + 3)^2 + 1$
Minimum: 1

79.

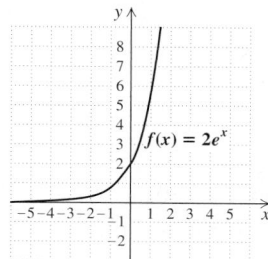

$f(x) = 2e^x$

Domain: $\mathbb{R}$; range: $(0, \infty)$

80. $2 \log a + 3 \log c - \log b$ **81.** $\log\left(\dfrac{x^3}{y^{1/2}z^2}\right)$

82. $a^x = 5$ **83.** $\log_x t = 3$ **84.** -1.2545 **85.** 776.2471
86. 2.5479 **87.** 0.2466 **88.** All real numbers except
1 and -2 **89.** $\frac{1}{3}, \frac{10{,}000}{3}$ **90.** 35 mph

CHAPTER 13

Technology Connection, p. 863

1. $x^2 + y^2 - 16 = 0$

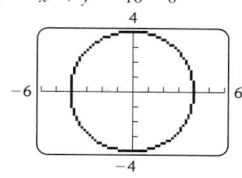

2. $(x - 1)^2 + (y - 2)^2 = 25$

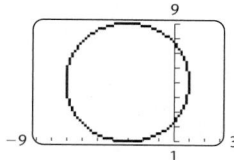

3. $(x + 3)^2 + (y - 5)^2 = 16$

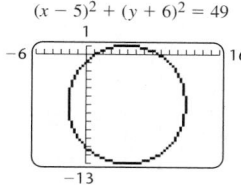

4. $(x - 5)^2 + (y + 6)^2 = 49$

Exercise Set 13.1, pp. 864–868

1. (f)   **3.** (g)   **5.** (c)   **7.** (d)

9.

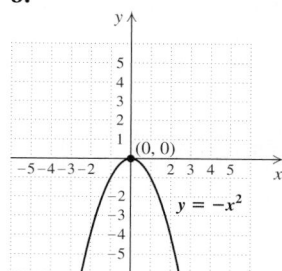

$(0, 0)$

$y = -x^2$

11.

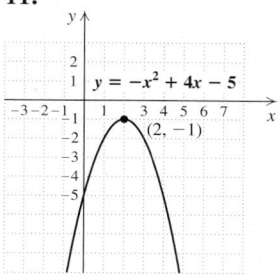

$y = -x^2 + 4x - 5$

$(2, -1)$

13.

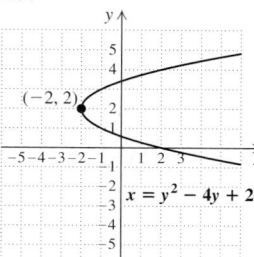

$(-2, 2)$

$x = y^2 - 4y + 2$

15.

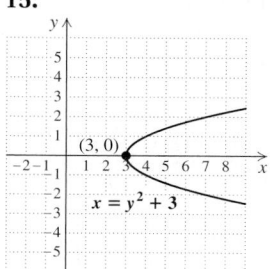

$(3, 0)$

$x = y^2 + 3$

17.

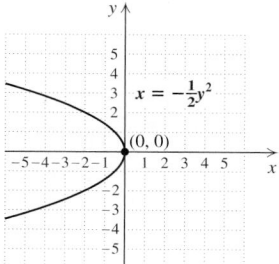

$x = -\frac{1}{2}y^2$

$(0, 0)$

19.

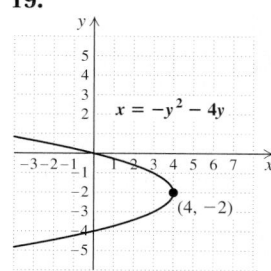

$x = -y^2 - 4y$

$(4, -2)$

21.

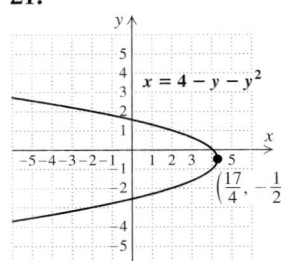

$x = 4 - y - y^2$

$\left(\frac{17}{4}, -\frac{1}{2}\right)$

23.

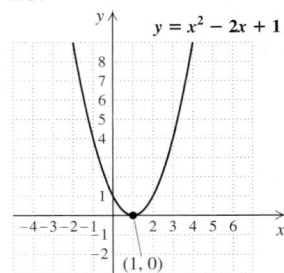

$y = x^2 - 2x + 1$

$(1, 0)$

25.

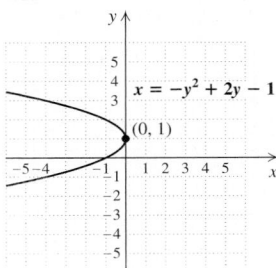

$x = -y^2 + 2y - 1$

$(0, 1)$

27.

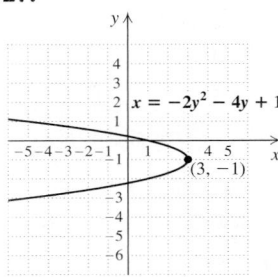

$x = -2y^2 - 4y + 1$

$(3, -1)$

29. 5   **31.** $\sqrt{18} \approx 4.243$ **33.** $\sqrt{200} \approx 14.142$ **35.** 17.8

37. $\dfrac{\sqrt{41}}{7} \approx 0.915$ **39.** $\sqrt{8} \approx 2.828$ **41.** $\sqrt{90} \approx 9.487$

43. $(1, 4)$ **45.** $\left(\frac{7}{2}, \frac{7}{2}\right)$ **47.** $(-1, -3)$

49. $(-0.25, -0.3)$ **51.** $\left(-\frac{1}{12}, \frac{1}{24}\right)$ **53.** $\left(\dfrac{\sqrt{2} + \sqrt{3}}{2}, \dfrac{3}{2}\right)$

55. $x^2 + y^2 = 36$ **57.** $(x - 7)^2 + (y - 3)^2 = 5$

59. $(x + 4)^2 + (y - 3)^2 = 48$

61. $(x + 7)^2 + (y + 2)^2 = 50$ **63.** $x^2 + y^2 = 25$

65. $(x + 4)^2 + (y - 1)^2 = 20$

67. $(0, 0)$; 8

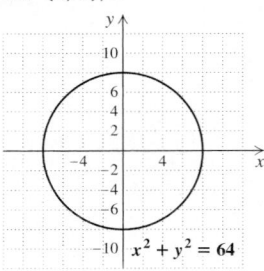

$x^2 + y^2 = 64$

69. $(-1, -3)$; 6

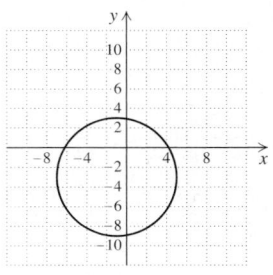

$(x + 1)^2 + (y + 3)^2 = 36$

71. $(4, -3)$; $\sqrt{10}$

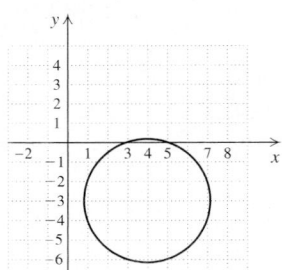

$(x - 4)^2 + (y + 3)^2 = 10$

73. $(0, 0)$; $\sqrt{10}$

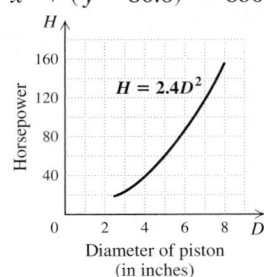

$x^2 + y^2 = 10$

75. $(5, 0)$; $\frac{1}{2}$

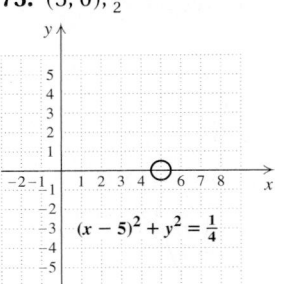

$(x - 5)^2 + y^2 = \frac{1}{4}$

77. $(-4, 3)$; $\sqrt{40}$, or $2\sqrt{10}$

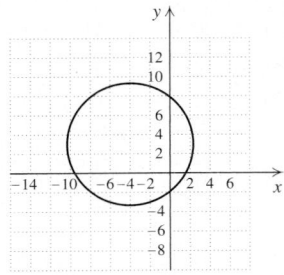

$x^2 + y^2 + 8x - 6y - 15 = 0$

79. $(4, -1)$; 2

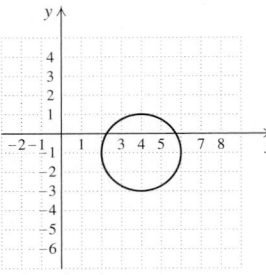

$x^2 + y^2 - 8x + 2y + 13 = 0$

81. $(0, -5)$; 10

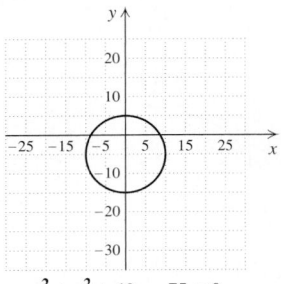

$x^2 + y^2 + 10y - 75 = 0$

83. $\left(-\dfrac{7}{2}, \dfrac{3}{2}\right)$; $\sqrt{\dfrac{98}{4}}$, or $\dfrac{7\sqrt{2}}{2}$ **85.** $(0, 0)$; $\frac{1}{6}$

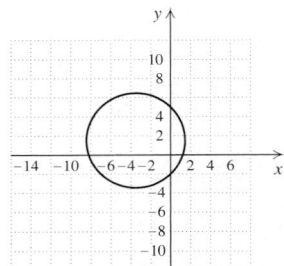

$x^2 + y^2 + 7x - 3y - 10 = 0$

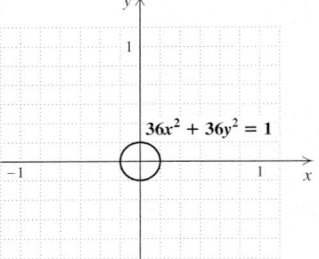

$36x^2 + 36y^2 = 1$

87. 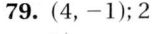   **89.** $-\frac{2}{3}$ **90.** $\frac{25}{6}$ **91.** 4 in.   **92.** 2640 mi

93. $\left(\frac{35}{17}, \frac{5}{34}\right)$ **94.** $\left(0, -\frac{9}{5}\right)$ **95.**

97. $(x - 3)^2 + (y + 5)^2 = 9$ **99.** $(x - 3)^2 + y^2 = 25$

101. $(0, 4)$ **103.** $\frac{17}{4}\pi$ m², or approximately 13.4 m²

105. 7451.2 mm   **107.** (a) $(0, -3)$; (b) 5 ft

109. $x^2 + (y - 30.6)^2 = 590.49$

111.

H = 2.4D²

Horsepower

Diameter of piston (in inches)

113. (a) $y = -1 \pm \sqrt{-x^2 + 6x + 7}$; **(b)** 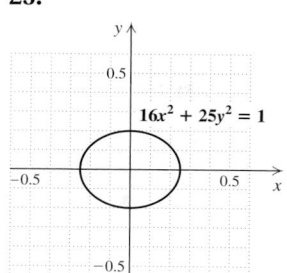 **115.**

Exercise Set 13.2, pp. 872–874

1. True   **3.** True   **5.** False   **7.** True
9.
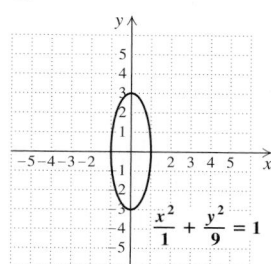
$$\frac{x^2}{1} + \frac{y^2}{9} = 1$$

11.
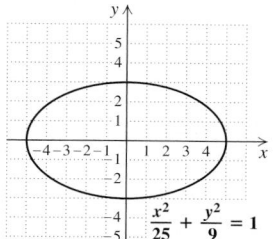
$$\frac{x^2}{25} + \frac{y^2}{9} = 1$$

13.
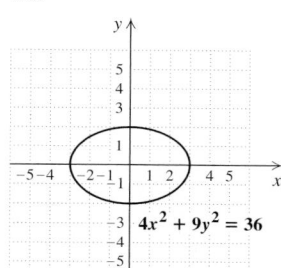
$$4x^2 + 9y^2 = 36$$

15.
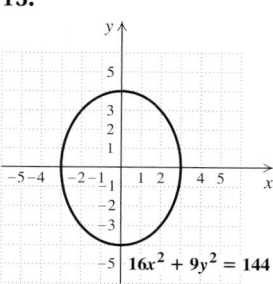
$$16x^2 + 9y^2 = 144$$

17.
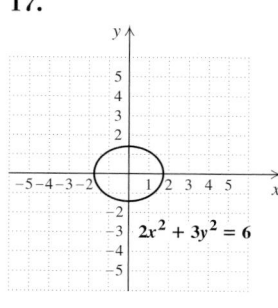
$$2x^2 + 3y^2 = 6$$

19.
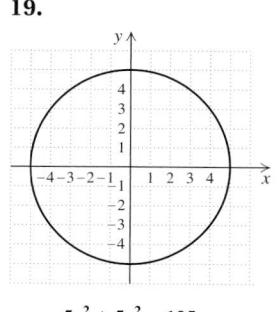
$$5x^2 + 5y^2 = 125$$

21.
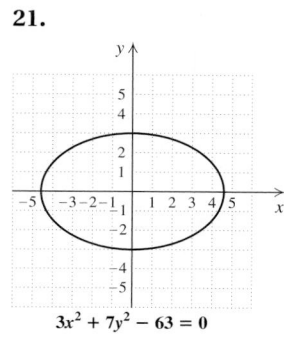
$$3x^2 + 7y^2 - 63 = 0$$

23.
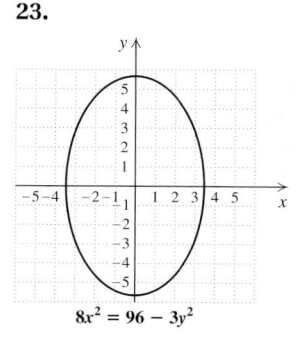
$$8x^2 = 96 - 3y^2$$

25.
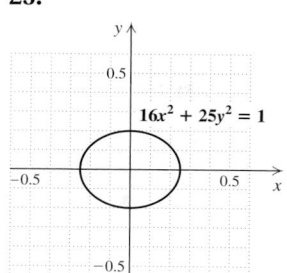
$$16x^2 + 25y^2 = 1$$

27.
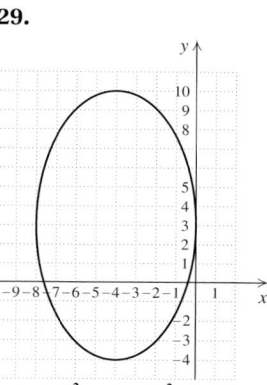
$$\frac{(x-3)^2}{9} + \frac{(y-2)^2}{25} = 1$$

29.
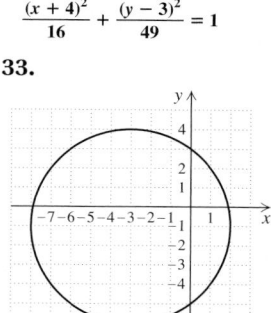
$$\frac{(x+4)^2}{16} + \frac{(y-3)^2}{49} = 1$$

31.
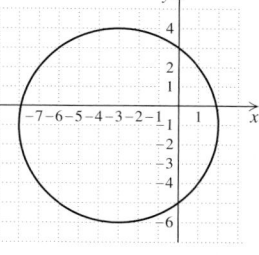
$$12(x-1)^2 + 3(y+4)^2 = 48$$

33.
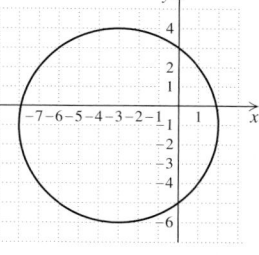
$$4(x+3)^2 + 4(y+1)^2 - 10 = 90$$

35. **37.** $\frac{16}{9}$ **38.** $-\frac{19}{8}$ **39.** $5 \pm \sqrt{3}$ **40.** $3 \pm \sqrt{7}$

41. $\frac{3}{2}$ **42.** No solution **43.** **45.** $\frac{x^2}{81} + \frac{y^2}{121} = 1$

47. $\frac{(x-2)^2}{16} + \frac{(y+1)^2}{9} = 1$ **49.** $\frac{x^2}{9} + \frac{y^2}{25} = 1$

51. (a) Let $F_1 = (-c, 0)$ and $F_2 = (c, 0)$. Then the sum of the distances from the foci to P is $2a$. By the distance formula,
$$\sqrt{(x+c)^2 + y^2} + \sqrt{(x-c)^2 + y^2} = 2a, \text{ or}$$
$$\sqrt{(x+c)^2 + y^2} = 2a - \sqrt{(x-c)^2 + y^2}.$$
Squaring, we get
$$(x+c)^2 + y^2 = 4a^2 - 4a\sqrt{(x-c)^2 + y^2} + (x-c)^2 + y^2,$$

or

$$x^2 + 2cx + c^2 + y^2 = 4a^2 - 4a\sqrt{(x-c)^2 + y^2}$$
$$+ x^2 - 2cx + c^2 + y^2.$$

Thus

$$-4a^2 + 4cx = -4a\sqrt{(x-c)^2 + y^2}$$
$$a^2 - cx = a\sqrt{(x-c)^2 + y^2}.$$

Squaring again, we get

$$a^4 - 2a^2cx + c^2x^2 = a^2(x^2 - 2cx + c^2 + y^2)$$
$$a^4 - 2a^2cx + c^2x^2 = a^2x^2 - 2a^2cx + a^2c^2 + a^2y^2,$$

or

$$x^2(a^2 - c^2) + a^2y^2 = a^2(a^2 - c^2)$$
$$\frac{x^2}{a^2} + \frac{y^2}{a^2 - c^2} = 1.$$

(b) When P is at $(0, b)$, it follows that $b^2 = a^2 - c^2$.
Substituting, we have

$$\frac{x^2}{a^2} + \frac{y^2}{b^2} = 1.$$

53. 5.66 ft
55.

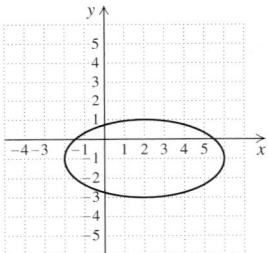

$$\frac{(x-2)^2}{16} + \frac{(y+1)^2}{4} = 1$$

57.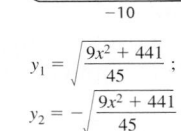

Technology Connection, p. 880

1.
$$y_1 = \frac{\sqrt{15x^2 - 240}}{2};$$
$$y_2 = -\frac{\sqrt{15x^2 - 240}}{2}$$

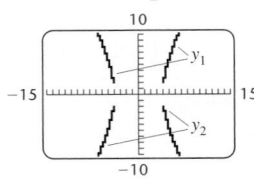

2.
$$y_1 = \sqrt{\frac{16x^2 - 64}{3}};$$
$$y_2 = -\sqrt{\frac{16x^2 - 64}{3}}$$

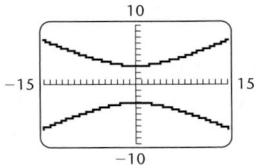

3.
$$y_1 = \frac{\sqrt{5x^2 + 320}}{4};$$
$$y_2 = -\frac{\sqrt{5x^2 + 320}}{4}$$

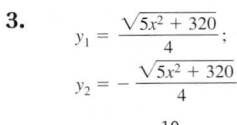

4.
$$y_1 = \sqrt{\frac{9x^2 + 441}{45}};$$
$$y_2 = -\sqrt{\frac{9x^2 + 441}{45}}$$

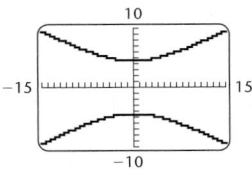

Exercise Set 13.3, pp. 885–886

1. (d) **3.** (h) **5.** (g) **7.** (c)
9.

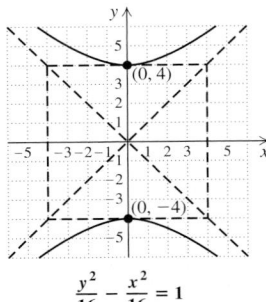

$$\frac{y^2}{16} - \frac{x^2}{16} = 1$$

11.

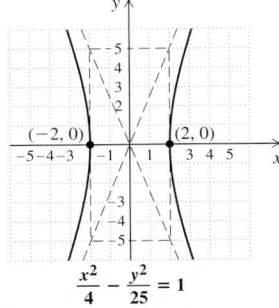

$$\frac{x^2}{4} - \frac{y^2}{25} = 1$$

13.

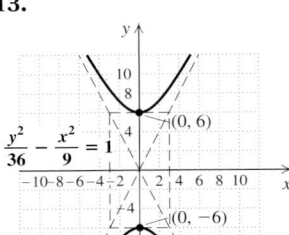

$$\frac{y^2}{36} - \frac{x^2}{9} = 1$$

15.

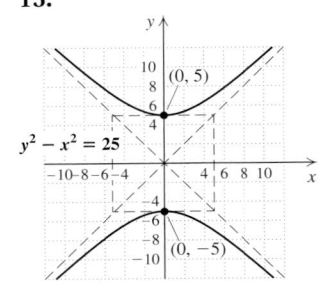

$$y^2 - x^2 = 25$$

17.

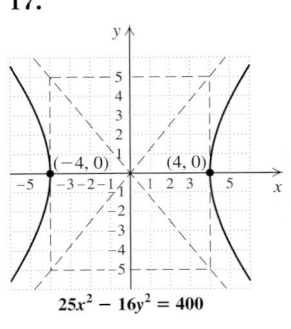

$$25x^2 - 16y^2 = 400$$

19.

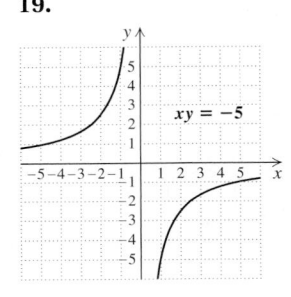

$$xy = -5$$

21.

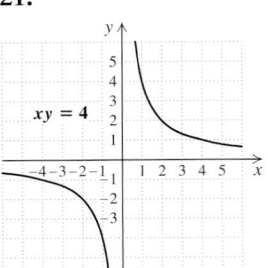

$$xy = 4$$

23.

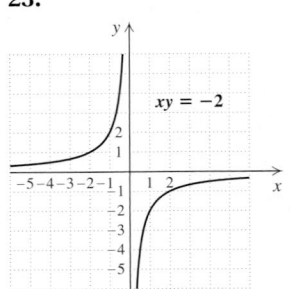

$$xy = -2$$

25.

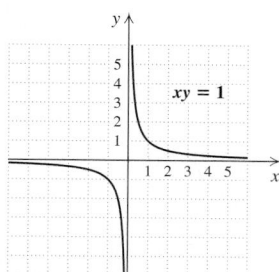

27. Circle **29.** Ellipse **31.** Hyperbola **33.** Circle
35. Ellipse **37.** Hyperbola **39.** Parabola
41. Hyperbola **43.** Circle **45.** Ellipse **47.**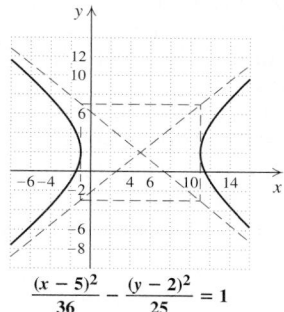
49. $\left(-\frac{22}{3}, \frac{37}{9}\right)$ **50.** $(9, -4)$ **51.** $-3, 3$ **52.** $-1, 1$

53. $35 **54.** 69 **55.** 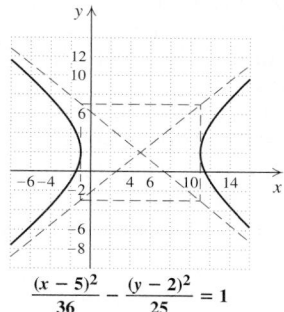 **57.** $\dfrac{y^2}{36} - \dfrac{x^2}{4} = 1$

59. C: $(5, 2)$; V: $(-1, 2)$, $(11, 2)$;
asymptotes: $y - 2 = \frac{5}{6}(x - 5)$, $y - 2 = -\frac{5}{6}(x - 5)$

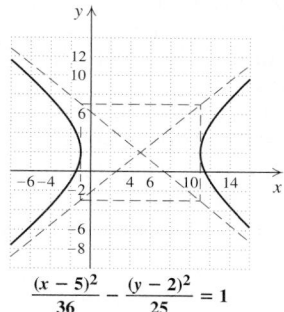

$$\frac{(x - 5)^2}{36} - \frac{(y - 2)^2}{25} = 1$$

61. $\dfrac{(y + 3)^2}{4} - \dfrac{(x - 4)^2}{16} = 1$; C: $(4, -3)$; V: $(4, -5)$, $(4, -1)$;
asymptotes: $y + 3 = \frac{1}{2}(x - 4)$, $y + 3 = -\frac{1}{2}(x - 4)$

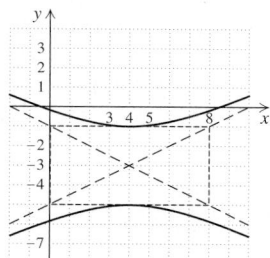

$$8(y + 3)^2 - 2(x - 4)^2 = 32$$

63. $\dfrac{(x + 3)^2}{1} - \dfrac{(y - 2)^2}{4} = 1$; C: $(-3, 2)$; V: $(-4, 2)$, $(-2, 2)$;
asymptotes: $y - 2 = 2(x + 3)$, $y - 2 = -2(x + 3)$

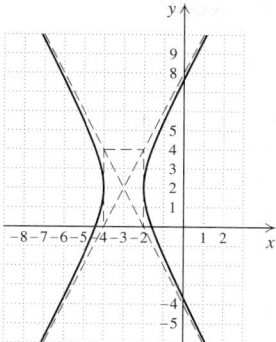

$$4x^2 - y^2 + 24x + 4y + 28 = 0$$

65.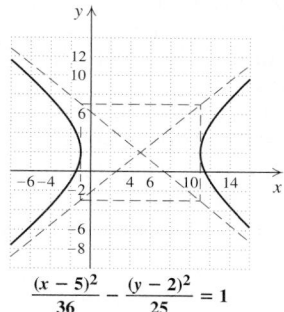

Technology Connection, p. 889

1. $(-1.50, -1.17)$; $(3.50, 0.50)$
2. $(-2.77, 2.52)$; $(-2.77, -2.52)$

Technology Connection, p. 891

1.

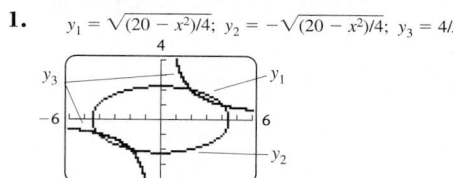

Exercise Set 13.4, pp. 894–896

1. True **3.** False **5.** True **7.** $(-4, -3)$, $(3, 4)$
9. $(0, 2)$, $(3, 0)$ **11.** $(-2, 1)$
13. $\left(\dfrac{5 + \sqrt{70}}{3}, \dfrac{-1 + \sqrt{70}}{3}\right)$, $\left(\dfrac{5 - \sqrt{70}}{3}, \dfrac{-1 - \sqrt{70}}{3}\right)$
15. $\left(4, \frac{3}{2}\right)$, $(3, 2)$ **17.** $\left(\frac{7}{3}, \frac{1}{3}\right)$, $(1, -1)$ **19.** $\left(\frac{11}{4}, -\frac{5}{4}\right)$, $(1, 4)$
21. $\left(\dfrac{7 - \sqrt{33}}{2}, \dfrac{7 + \sqrt{33}}{2}\right)$, $\left(\dfrac{7 + \sqrt{33}}{2}, \dfrac{7 - \sqrt{33}}{2}\right)$
23. $(3, -5)$, $(-1, 3)$ **25.** $(-5, -8)$, $(8, 5)$ **27.** $(0, 0)$, $(1, 1)$,
$\left(-\dfrac{1}{2} + \dfrac{\sqrt{3}}{2}i, -\dfrac{1}{2} - \dfrac{\sqrt{3}}{2}i\right)$, $\left(-\dfrac{1}{2} - \dfrac{\sqrt{3}}{2}i, -\dfrac{1}{2} + \dfrac{\sqrt{3}}{2}i\right)$
29. $(-3, 0)$, $(3, 0)$ **31.** $(-4, -3)$, $(-3, -4)$, $(3, 4)$, $(4, 3)$
33. $\left(\dfrac{16}{3}, \dfrac{5\sqrt{7}}{3}i\right)$, $\left(\dfrac{16}{3}, -\dfrac{5\sqrt{7}}{3}i\right)$, $\left(-\dfrac{16}{3}, \dfrac{5\sqrt{7}}{3}i\right)$,
$\left(-\dfrac{16}{3}, -\dfrac{5\sqrt{7}}{3}i\right)$ **35.** $(-3, -\sqrt{5})$, $(-3, \sqrt{5})$, $(3, -\sqrt{5})$,
$(3, \sqrt{5})$ **37.** $(4, 2)$, $(-4, -2)$, $(2, 4)$, $(-2, -4)$
39. $(4, 1)$, $(-4, -1)$, $(2, 2)$, $(-2, -2)$ **41.** $(2, 1)$, $(-2, -1)$
43. $\left(2, -\frac{4}{5}\right)$, $\left(-2, -\frac{4}{5}\right)$, $(5, 2)$, $(-5, 2)$

45. $\left(-\sqrt{2}, \sqrt{2}\right), \left(\sqrt{2}, -\sqrt{2}\right)$
47. Length: 8 cm; width: 6 cm
49. Length: 5 in.; width: 4 in.
51. Length: 12 ft; width: 5 ft   **53.** 6 and 10; −6 and −10
55. 24 ft, 16 ft   **57.** 13 and 12   **59.** 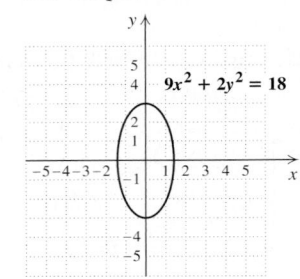   **61.** −16
62. −32   **63.** 1   **64.** $-\frac{1}{4}$ **65.** 44   **66.** 28   **67.**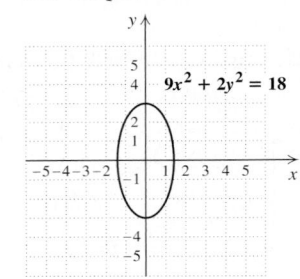
69. $(x + 2)^2 + (y - 1)^2 = 4$
71. $(-2, 3), (2, -3), (-3, 2), (3, -2)$ **73.** Length: 55 ft;
width: 45 ft   **75.** 10 in. by 7 in. by 5 in.
77. Length: 61.02 in.; height: 34.32 in.   **79.**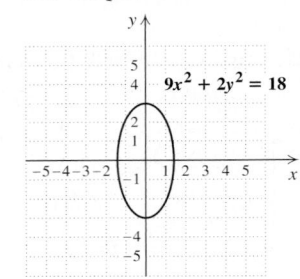

Review Exercises: Chapter 13, pp. 898–899

1. True   **2.** False   **3.** False   **4.** False   **5.** True   **6.** True
7. False   **8.** True   **9.** 4   **10.** 5   **11.** $\sqrt{90.1} \approx 9.492$
12. $\sqrt{9 + 4a^2}$ **13.** $\left(\frac{9}{2}, -1\right)$ **14.** $(-3, 7)$
15. $\left(\frac{3}{4}, \frac{\sqrt{3} - \sqrt{2}}{2}\right)$ **16.** $\left(\frac{1}{2}, 2a\right)$ **17.** $(-3, 2), \sqrt{7}$
18. $(5, 0), 7$ **19.** $(3, 1), 3$ **20.** $(-4, 3), \sqrt{35}$
21. $(x + 4)^2 + (y - 3)^2 = 48$
22. $(x - 7)^2 + (y + 2)^2 = 20$
23. Circle

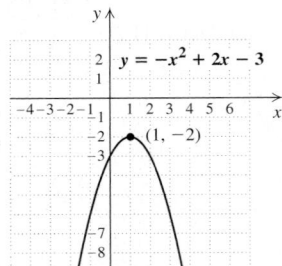

$$5x^2 + 5y^2 = 80$$

24. Ellipse

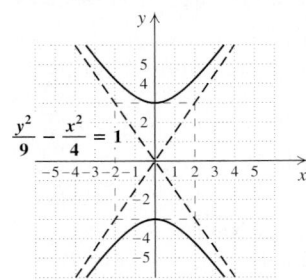

$9x^2 + 2y^2 = 18$

25. Parabola

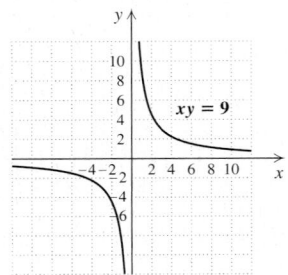

$y = -x^2 + 2x - 3$
$(1, -2)$

26. Hyperbola

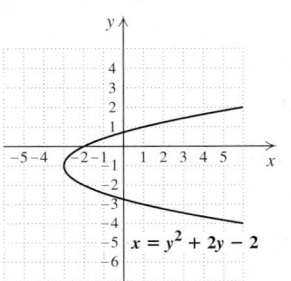

$$\frac{y^2}{9} - \frac{x^2}{4} = 1$$

27. Hyperbola

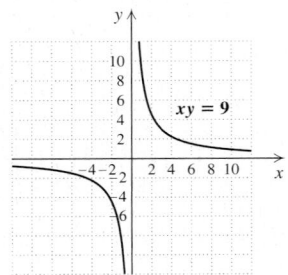

$xy = 9$

28. Parabola

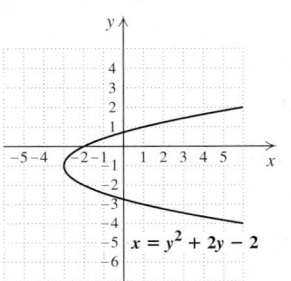

$x = y^2 + 2y - 2$

29. Ellipse

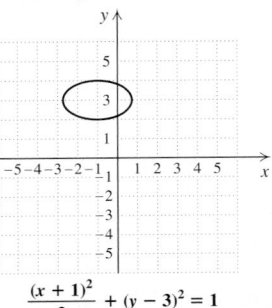

$$\frac{(x + 1)^2}{3} + (y - 3)^2 = 1$$

30. Circle

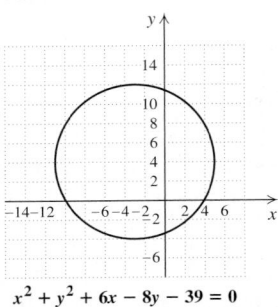

$$x^2 + y^2 + 6x - 8y - 39 = 0$$

31. $(7, 4)$ **32.** $(2, 2), \left(\frac{32}{9}, -\frac{10}{9}\right)$ **33.** $(0, -3), (2, 1)$
34. $(4, 3), (4, -3), (-4, 3), (-4, -3)$
35. $(2, 1), \left(\sqrt{3}, 0\right), (-2, 1), \left(-\sqrt{3}, 0\right)$ **36.** $(3, -3), \left(-\frac{3}{5}, \frac{21}{5}\right)$
37. $(6, 8), (6, -8), (-6, 8), (-6, -8)$
38. $(2, 2), (-2, -2), \left(2\sqrt{2}, \sqrt{2}\right), \left(-2\sqrt{2}, -\sqrt{2}\right)$
39. Length: 12 m; width: 7 m   **40.** Length: 12 in.;
width: 9 in.   **41.** 32 cm, 20 cm   **42.** 3 ft, 11 ft
43. 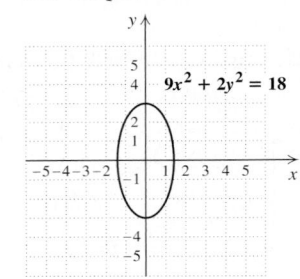 The graph of a parabola has one branch whereas
the graph of a hyperbola has two branches. A hyperbola
has asymptotes, but a parabola does not.
44. 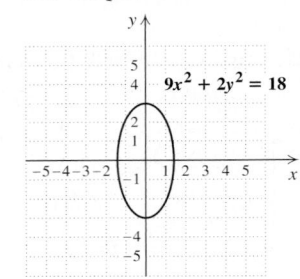 Function notation rarely appears in this chapter
because many of the relations are not functions. Function
notation could be used for vertical parabolas and for
hyperbolas that have the axes as asymptotes.
45. $\left(-5, -4\sqrt{2}\right), \left(-5, 4\sqrt{2}\right), \left(3, -2\sqrt{2}\right), \left(3, 2\sqrt{2}\right)$
46. $(0, 6), (0, -6)$ **47.** $(x - 2)^2 + (y + 1)^2 = 25$
48. $\dfrac{x^2}{81} + \dfrac{y^2}{25} = 1$ **49.** $\left(\frac{9}{4}, 0\right)$

Test: Chapter 13, p. 900

1. [13.1] $9\sqrt{2} \approx 12.728$ **2.** [13.1] $2\sqrt{9 + a^2}$
3. [13.1] $\left(-\frac{1}{2}, \frac{7}{2}\right)$ **4.** [13.1] $(0, 0)$ **5.** [13.1] $(-5, 1), 9$
6. [13.1] $(-2, 3), 3$
7. [13.1], [13.3] Parabola   **8.** [13.1], [13.3] Circle

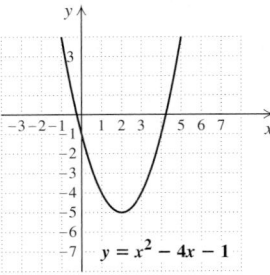

$y = x^2 - 4x - 1$

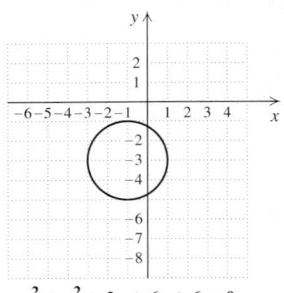

$$x^2 + y^2 + 2x + 6y + 6 = 0$$

9. [13.3] Hyperbola

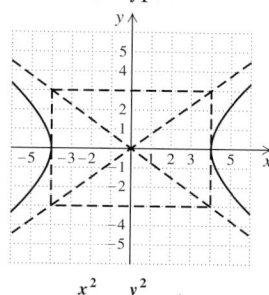

$$\frac{x^2}{16} - \frac{y^2}{9} = 1$$

10. [13.2], [13.3] Ellipse

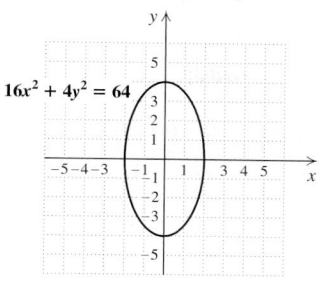

$16x^2 + 4y^2 = 64$

11. [13.3] Hyperbola

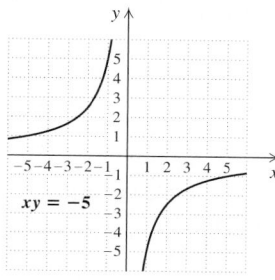

$xy = -5$

12. [13.1], [13.3] Parabola

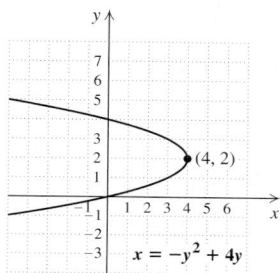

$x = -y^2 + 4y$

13. [13.4] $(0, 3)$, $\left(\frac{8}{5}, \frac{9}{5}\right)$ **14.** [13.4] $(4, 0)$, $(-4, 0)$
15. [13.4] $(3, 2)$, $(-3, -2)$
16. [13.4] $\left(\sqrt{6}, 2\right)$, $\left(\sqrt{6}, -2\right)$, $\left(-\sqrt{6}, 2\right)$, $\left(-\sqrt{6}, -2\right)$
17. [13.4] 2 by 11   **18.** [13.4] $\sqrt{5}$ m, $\sqrt{3}$ m
19. [13.4] Length: 32 ft; width: 24 ft

20. [13.4] $\$1200$, 6%   **21.** [13.2] $\dfrac{(x - 6)^2}{25} + \dfrac{(y - 3)^2}{9} = 1$

22. [13.1] $\left(0, -\frac{31}{4}\right)$ **23.** [13.4] 9   **24.** [13.2] $\dfrac{x^2}{16} + \dfrac{y^2}{49} = 1$

CHAPTER 14

Exercise Set 14.1, pp. 906–908

1. (f)   **3.** (d)   **5.** (c)   **7.** 5, 7, 9, 11; 23; 33
9. 3, 6, 11, 18; 102; 227   **11.** $0, \frac{3}{5}, \frac{4}{5}, \frac{15}{17}; \frac{99}{101}; \frac{112}{113}$
13. $1, -\frac{1}{2}, \frac{1}{4}, -\frac{1}{8}; -\frac{1}{512}; \frac{1}{16,384}$ **15.** $-4, 5, -6, 7; 13; -18$
17. $0, 7, -26, 63; 999; -3374$ **19.** 13   **21.** 364

23. -23.5 **25.** -363 **27.** $\dfrac{441}{400}$ **29.** $2n$ **31.** $(-1)^{n+1}$

33. $(-1)^n \cdot n$ **35.** $2n + 1$ **37.** $(-1)^n \cdot 2 \cdot (3)^{n-1}$

39. $\dfrac{n}{n + 1}$ **41.** 5^n **43.** $(-1)^n \cdot n^2$ **45.** 4   **47.** 30

49. $\dfrac{1}{2} + \dfrac{1}{4} + \dfrac{1}{6} + \dfrac{1}{8} + \dfrac{1}{10} = \dfrac{137}{120}$
51. $3^0 + 3^1 + 3^2 + 3^3 + 3^4 = 121$

53. $2 + \dfrac{3}{2} + \dfrac{4}{3} + \dfrac{5}{4} + \dfrac{6}{5} + \dfrac{7}{6} + \dfrac{8}{7} = \dfrac{1343}{140}$

55. $(-1)^2 2^1 + (-1)^3 2^2 + (-1)^4 2^3 + (-1)^5 2^4 + (-1)^6 2^5 + (-1)^7 2^6 + (-1)^8 2^7 + (-1)^9 2^8 = -170$
57. $(0^2 - 2 \cdot 0 + 3) + (1^2 - 2 \cdot 1 + 3) + (2^2 - 2 \cdot 2 + 3) + (3^2 - 2 \cdot 3 + 3) + (4^2 - 2 \cdot 4 + 3) + (5^2 - 2 \cdot 5 + 3) = 43$

59. $\dfrac{(-1)^3}{3 \cdot 4} + \dfrac{(-1)^4}{4 \cdot 5} + \dfrac{(-1)^5}{5 \cdot 6} = -\dfrac{1}{15}$ **61.** $\displaystyle\sum_{k=1}^{5} \dfrac{k + 1}{k + 2}$

63. $\displaystyle\sum_{k=1}^{6} k^2$ **65.** $\displaystyle\sum_{k=2}^{n} (-1)^k k^2$ **67.** $\displaystyle\sum_{k=1}^{\infty} 5k$

69. $\displaystyle\sum_{k=1}^{\infty} \dfrac{1}{k(k + 1)}$ **71.** 🗒   **73.** 77   **74.** 23
75. $x^3 + 3x^2y + 3xy^2 + y^3$ **76.** $a^3 - 3a^2b + 3ab^2 - b^3$
77. $8a^3 - 12a^2b + 6ab^2 - b^3$
78. $8x^3 + 12x^2y + 6xy^2 + y^3$ **79.** 🗒
81. 1, 3, 13, 63, 313, 1563   **83.** $\$5200$, $\$3900$, $\$2925$,
$\$2193.75$; $\$1645.31$, $\$1233.98$, $\$925.49$, $\$694.12$, $\$520.59$,
$\$390.44$ **85.** $S_{100} = 0$; $S_{101} = -1$ **87.** $i, -1, -i, 1, i; i$
89. 11th term

Exercise Set 14.2, pp. 915–917

1. True   **3.** False   **5.** True   **7.** False   **9.** $a_1 = 2$, $d = 4$
11. $a_1 = 7$, $d = -4$ **13.** $a_1 = \frac{3}{2}$, $d = \frac{3}{4}$
15. $a_1 = \$5.12$, $d = \$0.12$ **17.** 49   **19.** -94
21. $-\$1628.16$ **23.** 26th   **25.** 57th   **27.** 82   **29.** 5
31. 28   **33.** $a_1 = 8$; $d = -3$; 8, 5, 2, -1, -4
35. $a_1 = 1$; $d = 1$ **37.** 780   **39.** 31,375   **41.** 2550
43. 918   **45.** 1030   **47.** 35 marchers; 315 marchers
49. 180 stones   **51.** $\$49.60$ **53.** 722 seats   **55.** 🗒

57. $\dfrac{13}{30x}$ **58.** $\dfrac{23}{36t}$ **59.** $a^k = P$ **60.** $e^a = t$

61. $x^2 + y^2 = 81$ **62.** $(x + 2)^2 + (y - 5)^2 = 18$
63. 🗒   **65.** 33 jumps   **67.** $\$8760$, $\$7961.77$, $\$7163.54$,
$\$6365.31$, $\$5567.08$; $\$4768.85$, $\$3970.62$, $\$3172.39$, $\$2374.16$,
$\$1575.93$ **69.** Let $d = $ the common difference. Since p,
m, and q form an arithmetic sequence, $m = p + d$ and
$q = p + 2d$. Then $\dfrac{p + q}{2} = \dfrac{p + (p + 2d)}{2} = p + d = m$.

71. 156,375

Exercise Set 14.3, pp. 925–927

1. Geometric sequence   **3.** Arithmetic sequence
5. Geometric series   **7.** Geometric series   **9.** 2

11. -0.1 **13.** $-\frac{1}{2}$ **15.** $\frac{1}{5}$ **17.** $\dfrac{6}{m}$ **19.** 192

21. $112\sqrt{2}$ **23.** 52,488   **25.** $\$2331.64$ **27.** $a_n = 5^{n-1}$
29. $a_n = (-1)^{n-1}$, or $a_n = (-1)^{n+1}$

31. $a_n = \dfrac{1}{x^n}$, or $a_n = x^{-n}$ **33.** 3066   **35.** $\frac{547}{18}$ **37.** $\dfrac{1 - x^8}{1 - x}$,
or $(1 + x)(1 + x^2)(1 + x^4)$ **39.** $\$5134.51$ **41.** $\frac{64}{3}$ **43.** $\frac{49}{4}$
45. No   **47.** No   **49.** $\frac{43}{99}$ **51.** $\$25,000$ **53.** $\frac{7}{9}$ **55.** $\frac{830}{99}$
57. $\frac{5}{33}$ **59.** $\frac{5}{1024}$ ft   **61.** 155,797   **63.** 2710 flies

65. 10,723,491 apartments and houses   **67.** 3100.35 ft
69. 20.48 in.   **71.** ▧   **73.** $x^3 + 3x^2y + 3xy^2 + y^3$
74. $a^3 - 3a^2b + 3ab^2 - b^3$ **75.** $\left(-\frac{63}{29}, -\frac{114}{29}\right)$
76. $(-1, 2, 3)$ **77.** ▧   **79.** 54   **81.** $\dfrac{x^2[1 - (-x)^n]}{1 + x}$
83. $512\ \text{cm}^2$ **85.** ▧, ⌇

Technology Connection, p. 933

1. 479,001,600   **2.** 56; 792

Exercise Set 14.4, pp. 936–937

1. 2^5, or 32   **3.** 9   **5.** $\binom{8}{5}$ **7.** x^7y^2 **9.** 362,880
11. 39,916,800   **13.** 56   **15.** 3024   **17.** 35   **19.** 126
21. 4060   **23.** 780   **25.** $a^4 - 4a^3b + 6a^2b^2 - 4ab^3 + b^4$
27. $p^7 + 7p^6q + 21p^5q^2 + 35p^4q^3 + 35p^3q^4 + 21p^2q^5 + 7pq^6 + q^7$
29. $2187c^7 - 5103c^6d + 5103c^5d^2 - 2835c^4d^3 + 945c^3d^4 - 189c^2d^5 + 21cd^6 - d^7$
31. $t^{-12} + 12t^{-10} + 60t^{-8} + 160t^{-6} + 240t^{-4} + 192t^{-2} + 64$
33. $x^5 - 5x^4y + 10x^3y^2 - 10x^2y^3 + 5xy^4 - y^5$
35. $19{,}683s^9 + \dfrac{59{,}049s^8}{t} + \dfrac{78{,}732s^7}{t^2} + \dfrac{61{,}236s^6}{t^3} + \dfrac{30{,}618s^5}{t^4} + \dfrac{10{,}206s^4}{t^5} + \dfrac{2268s^3}{t^6} + \dfrac{324s^2}{t^7} + \dfrac{27s}{t^8} + \dfrac{1}{t^9}$
37. $x^{15} - 10x^{12}y + 40x^9y^2 - 80x^6y^3 + 80x^3y^4 - 32y^5$
39. $125 + 150\sqrt{5}t + 375t^2 + 100\sqrt{5}t^3 + 75t^4 + 6\sqrt{5}t^5 + t^6$
41. $x^{-3} - 6x^{-2} + 15x^{-1} - 20 + 15x - 6x^2 + x^3$
43. $15a^4b^2$ **45.** $-64{,}481{,}508a^3$ **47.** $1120x^{12}y^2$
49. $1{,}959{,}552u^5v^{10}$ **51.** y^8 **53.** ▧   **55.** 4   **56.** $\frac{5}{2}$
57. 5.6348   **58.** ± 5 **59.** ▧
61. List all the subsets of size 3: $\{a, b, c\}, \{a, b, d\}, \{a, b, e\}, \{a, c, d\}, \{a, c, e\}, \{a, d, e\}, \{b, c, d\}, \{b, c, e\}, \{b, d, e\}, \{c, d, e\}$.
There are exactly 10 subsets of size 3 and $\binom{5}{3} = 10$, so there are exactly $\binom{5}{3}$ ways of forming a subset of size 3 from $\{a, b, c, d, e\}$.
63. $\binom{8}{5}(0.15)^3(0.85)^5 \approx 0.084$
65. $\binom{8}{6}(0.15)^2(0.85)^6 + \binom{8}{7}(0.15)(0.85)^7 + \binom{8}{8}(0.85)^8 \approx 0.89$
67. $\binom{n}{n-r} = \dfrac{n!}{[n - (n - r)]!\,(n - r)!}$
$= \dfrac{n!}{r!\,(n - r)!} = \binom{n}{r}$

69. $\dfrac{-\sqrt[3]{q}}{2p}$ **71.** $x^7 + 7x^6y + 21x^5y^2 + 35x^4y^3 + 35x^3y^4 + 21x^2y^5 + 7xy^6 + y^7$

Review Exercises: Chapter 14, pp. 939–940

1. False   **2.** True   **3.** True   **4.** False   **5.** True   **6.** True
7. False   **8.** False   **9.** 1, 5, 9, 13; 29; 45
10. $0, \frac{1}{5}, \frac{1}{5}, \frac{3}{17}; \frac{7}{65}; \frac{11}{145}$ **11.** $a_n = 7n$
12. $a_n = (-1)^n(2n - 1)$
13. $-2 + 4 + (-8) + 16 + (-32) = -22$
14. $-3 + (-5) + (-7) + (-9) + (-11) + (-13) = -48$
15. $\sum_{k=1}^{5} 4k$ **16.** $\sum_{k=1}^{5} \dfrac{1}{(-2)^k}$ **17.** 85   **18.** $\frac{8}{3}$
19. $a_1 = \frac{45}{4}, d = \frac{5}{4}$ **20.** -544 **21.** 8580   **22.** $1024\sqrt{2}$
23. $\frac{2}{3}$ **24.** $a_n = 2(-1)^n$ **25.** $a_n = 3\left(\dfrac{x}{4}\right)^{n-1}$ **26.** 4095
27. $-4095x$ **28.** 12   **29.** $\frac{49}{11}$ **30.** No   **31.** No
32. \$40,000   **33.** $\frac{5}{9}$ **34.** $\frac{46}{33}$ **35.** \$24.30   **36.** 903 poles
37. \$15,791.18   **38.** 6 m   **39.** 5040   **40.** 56
41. $190a^{18}b^2$ **42.** $x^4 - 8x^3y + 24x^2y^2 - 32xy^3 + 16y^4$
43. ▧ For a geometric sequence with $|r| < 1$, as n gets larger, the absolute value of the terms gets smaller, since $|r^n|$ gets smaller.
44. ▧ The first form of the binomial theorem draws the coefficients from Pascal's triangle; the second form uses factorial notation. The second form avoids the need to compute all preceding rows of Pascal's triangle, and is generally easier to use when only one term of an expression is needed. When several terms of an expansion are needed and n is not large (say, $n \le 8$), it is often easier to use Pascal's triangle.
45. $\dfrac{1 - (-x)^n}{x + 1}$
46. $x^{-15} + 5x^{-9} + 10x^{-3} + 10x^3 + 5x^9 + x^{15}$

Test: Chapter 14, pp. 940–941

1. [14.1] 1, 7, 13, 19, 25; 67   **2.** [14.1] $a_n = 4\left(\frac{1}{3}\right)^n$
3. [14.1] $-1 + (-5) + (-13) + (-29) + (-61) = -109$
4. [14.1] $\sum_{k=1}^{5} (-1)^{k+1}k^3$ **5.** [14.2] -51 **6.** [14.2] $\frac{3}{8}$
7. [14.2] $a_1 = 31.2; d = -3.8$ **8.** [14.2] 2508   **9.** [14.3] $\frac{9}{128}$
10. [14.3] $\frac{2}{3}$ **11.** [14.3] 3^n **12.** [14.3] $511 + 511x$
13. [14.3] 1   **14.** [14.3] No   **15.** [14.3] $\frac{\$25{,}000}{23} \approx \$1086.96$
16. [14.3] $\frac{85}{99}$ **17.** [14.2] 63 seats   **18.** [14.2] \$17,100
19. [14.3] \$8981.05   **20.** [14.3] 36 m   **21.** [14.4] 220
22. [14.4] $x^{10} - 15x^8y + 90x^6y^2 - 270x^4y^3 + 405x^2y^4 - 243y^5$ **23.** [14.4] $220a^9x^3$ **24.** [14.2] $n(n + 1)$

25. [14.3] $\dfrac{1 - \left(\dfrac{1}{x}\right)^n}{1 - \dfrac{1}{x}}$, or $\dfrac{x^n - 1}{x^{n-1}(x - 1)}$

Cumulative Review: Chapters 1–14, pp. 941–943

1. $-35x^6y^{-4}$, or $\dfrac{-35x^6}{y^4}$ **2.** 6.3 **3.** $-4y + 17$ **4.** 280

5. $\dfrac{7}{6}$ **6.** $3a^2 - 8ab - 15b^2$ **7.** $13x^3 - 7x^2 - 6x + 6$

8. $6a^2 + 7a - 5$ **9.** $9a^4 - 30a^2y + 25y^2$ **10.** $\dfrac{4}{x + 2}$

11. $\dfrac{x - 4}{4(x + 2)}$ **12.** $\dfrac{(x + y)(x^2 + xy + y^2)}{x^2 + y^2}$ **13.** $x - a$

14. $(2x - 3)^2$ **15.** $(3a - 2)(9a^2 + 6a + 4)$

16. $(a + 3)(a^2 - b)$ **17.** $3(y^2 + 3)(5y^2 - 4)$ **18.** 20

19. $7x^3 + 9x^2 + 19x + 38 + \dfrac{72}{x - 2}$ **20.** $\dfrac{3}{5}$ **21.** $-\dfrac{6}{5}, 4$

22. $\mathbb{R}$, or $(-\infty, \infty)$ **23.** $(-1, 1)$ **24.** $(2, -1, 1)$ **25.** 2

26. $\pm 2, \pm 5$ **27.** $(\sqrt{5}, \sqrt{3}), (\sqrt{5}, -\sqrt{3}), (-\sqrt{5}, \sqrt{3}),$

$(-\sqrt{5}, -\sqrt{3})$ **28.** 1.4037 **29.** 1005 **30.** $\dfrac{1}{25}$ **31.** $-\dfrac{1}{2}$

32. $\{x \mid -2 \le x \le 3\}$, or $[-2, 3]$ **33.** $\pm i\sqrt{2}$

34. $-2 \pm \sqrt{7}$ **35.** $\{y \mid y < -5 \text{ or } y > 2\}$, or

$(-\infty, -5) \cup (2, \infty)$ **36.** $\{a \mid -6 \le a \le 8\}$, or $[-6, 8]$ **37.** 3

38. 5 ft by 12 ft **39.** More than 4 purchases

40. 65, 66, 67 **41.** $11\dfrac{3}{7}$ **42.** $2.68 herb: 10 oz;

$4.60 herb: 14 oz **43.** 350 mph **44.** $8\dfrac{2}{5}$ hr, or 8 hr 24 min

45. 20 **46.** 5000 ft^2

47.

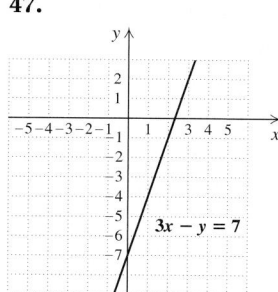

48.

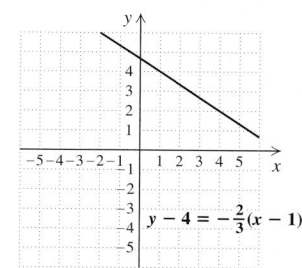

49.

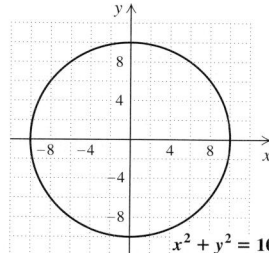

50.

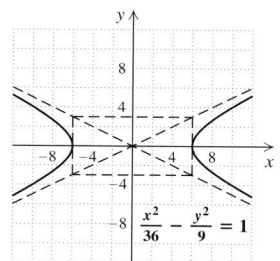

51.

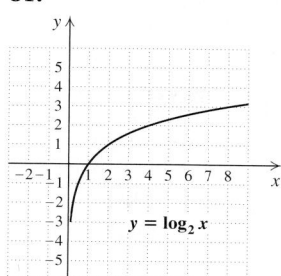

52.

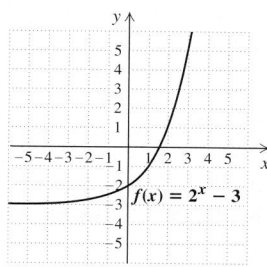

53.

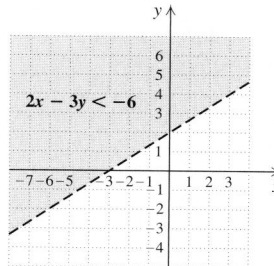

54.

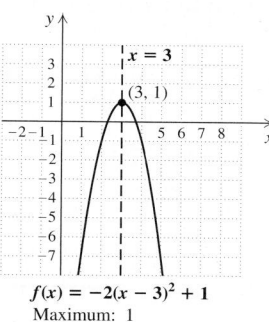

55. $r = \dfrac{V - P}{-Pt}$, or $\dfrac{P - V}{Pt}$ **56.** $R = \dfrac{Ir}{1 - I}$ **57.** $y = 3x - 8$

58. $\left\{x \mid x \le \dfrac{3}{4}\right\}$, or $\left(-\infty, \dfrac{3}{4}\right]$

59. $\{x \mid x \text{ is a real number } and \ x \neq 1\}$

60. 6.8×10^{-12} **61.** $8x^2\sqrt{y}$ **62.** $125x^2y^{3/4}$ **63.** $\dfrac{\sqrt[3]{5xy}}{y}$

64. $\dfrac{1 - 2\sqrt{x} + x}{1 - x}$ **65.** $26 - 13i$ **66.** $x^2 - 50 = 0$

67. $(2, -3)$; 6 **68.** $\log_a \dfrac{\sqrt[3]{x^2} \cdot z^5}{\sqrt{y}}$ **69.** $a^5 = c$

70. 3.7541 **71.** 0.0003 **72.** 8.6442 **73.** 0.0277

74. **(a)** $k \approx 0.219$; $C(t) = 0.12e^{0.219t}$; **(b)** about 29 million

computers **75.** 5 **76.** -121 **77.** 875

78. $16\left(\dfrac{1}{4}\right)^{n-1}$ **79.** $13{,}440a^4b^6$ **80.** $74.88671875x$

81. $652.39 **82.** All real numbers except 0 and -12

83. 81 **84.** y gets divided by 8 **85.** $-\dfrac{7}{13} + \dfrac{2\sqrt{30}}{13}i$

86. 84 yr

CHAPTER R

Exercise Set R.1, pp. 953–954

1. False **3.** True **5.** True **7.** 4 **9.** 1.3 **11.** -25

13. $-\dfrac{11}{15}$ **15.** -6.5 **17.** -15 **19.** 0 **21.** $-\dfrac{1}{2}$

23. 5.8 **25.** -3 **27.** 39 **29.** 175 **31.** -32

33. 16 **35.** -6 **37.** 9 **39.** -3 **41.** -16

43. 100 **45.** 2 **47.** -23 **49.** 36 **51.** 10 **53.** 10

55. 7　**57.** 32　**59.** 28 cm² **61.** $8x + 28$
63. $5x - 50$　**65.** $-30 + 6x$　**67.** $8a + 12b - 6c$
69. $2(4x + 3y)$　**71.** $3(1 + w)$　**73.** $10(x + 5y + 10)$
75. p　**77.** $-m + 22$　**79.** $3x + 7$　**81.** $6p - 7$
83. $-5x + 12y$　**85.** $36a - 48b$　**87.** $-10x + 104y + 9$
89. Yes　**91.** No　**93.** Yes　**95.** Let n represent the
number; $3n = 348$　**97.** Let c represent the number of
calories in a Taco Bell Beef Burrito; $c + 69 = 500$
99. Let l represent the amount of water used to produce
1 lb of lettuce; $42 = 2l$

Exercise Set R.2, pp. 961–963

1. 8　**3.** 12　**5.** $-\frac{1}{12}$　**7.** -0.8　**9.** $\frac{13}{3}$　**11.** $-\frac{5}{3}$
13. 42　**15.** -5　**17.** 2　**19.** $\frac{25}{3}$　**21.** $-\frac{4}{9}$
23. -4　**25.** $\frac{69}{5}$　**27.** $\frac{9}{32}$　**29.** -2　**31.** -15
33. $\frac{43}{2}$　**35.** $-\frac{61}{115}$　**37.** $l = \dfrac{A}{w}$　**39.** $q = \dfrac{p}{30}$
41. $P = IV$　**43.** $p = 2q - r$　**45.** $\pi = \dfrac{A}{r^2 + r^2 h}$
47. (a) No; (b) yes; (c) no; (d) yes
49. $\{x | x \le 12\}$
51. $\{m | m > 12\}$
53. $\{x | x \ge -\frac{3}{2}\}$
55. $\{t | t < -3\}$
57. $\{y | y > 10\}$　**59.** $\{a | a \ge 1\}$　**61.** $\{x | x \ge \frac{64}{17}\}$
63. $\{x | x > \frac{39}{11}\}$　**65.** $\{x | x \le -10.875\}$　**67.** 7
69. 16, 18　**71.** $166\frac{2}{3}$ pages　**73.** 4.5 cm, 9.5 cm
75. 900 cubic feet　**77.** 80¢

Exercise Set R.3, pp. 969–970

1.

3. I　**5.** IV

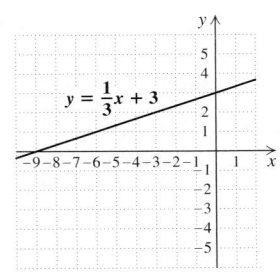

7. I, IV　**9.** No　**11.** Yes　**13.**

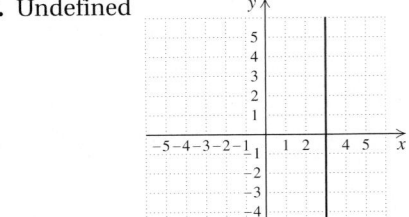

15.

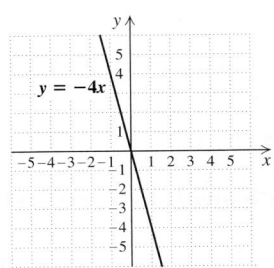

17. 1　**19.** 0　**21.** Slope: 2; y-intercept: $(0, -5)$
23. Slope: -2; y-intercept: $(0, 1)$

25.

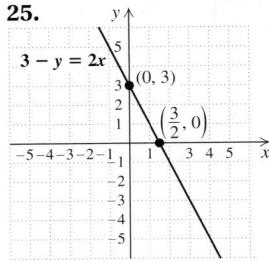

27.

29.

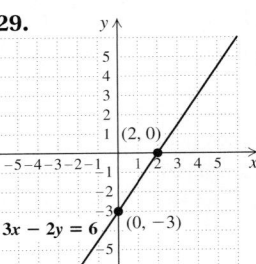

31.

33.

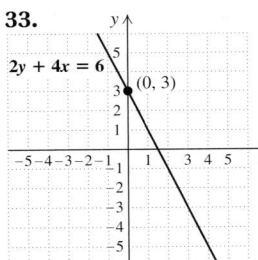

35. 0

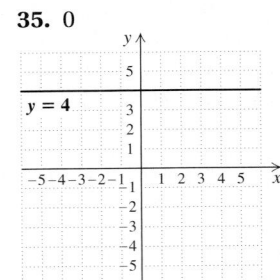

37. Undefined

39. $y = \frac{1}{3}x + 1$　**41.** $y = -x + 3$
43. Perpendicular　**45.** Neither

Exercise Set R.4, pp. 978–979

1. 1　**3.** -3　**5.** $\dfrac{1}{8^2} = \dfrac{1}{64}$　**7.** $\dfrac{1}{(-2)^3} = -\dfrac{1}{8}$

9. $\dfrac{1}{(ab)^2}$　**11.** y^{10}　**13.** y^{-4}　**15.** x^{-t}　**17.** x^{13}

19. a^6　**21.** $(4x)^8$　**23.** 7^{40}　**25.** $x^8 y^{12}$　**27.** $\dfrac{y^6}{64}$

29. $\dfrac{9q^8}{4p^6}$　**31.** $8x^3, -6x^2, x, -7$　**33.** $18, 36, -7, 3; 3, 9, 1,$
$0; 9$　**35.** $-1, 4, -2; 3, 3, 2; 3$　**37.** $8p^4; 8$
39. $x^4 + 3x^3$　**41.** $-7t^2 + 5t + 10$　**43.** 36
45. -14　**47.** 144 ft　**49.** $4x^3 - 3x^2 + 8x + 7$
51. $-y^2 + 5y - 2$　**53.** $-3x^2y - y^2 + 8y$
55. $12x^5 - 28x^3 + 28x^2$　**57.** $8a^2 + 2ab + 4ay + by$
59. $x^3 + 4x^2 - 20x + 7$　**61.** $x^2 - 49$
63. $x^2 + 2xy + y^2$　**65.** $6x^4 + 17x^2 - 14$
67. $a^2 - 6ab + 9b^2$　**69.** $42a^2 - 17ay - 15y^2$
71. $-t^4 - 3t^2 + 2t - 5$　**73.** $5x + 3$

75. $2x^2 - 3x + 3 + \dfrac{-2}{x+1}$　**77.** $5x + 3 + \dfrac{3}{x^2 - 1}$

Exercise Set R.5, pp. 989–990

1. $3x(x-1)(x+3)$　**3.** $(y-3)^2$
5. $(p+2)(2p^3 + 1)$　**7.** Prime
9. $(2t+3)(4t^2 - 6t + 9)$　**11.** $(m+6)(m+7)$
13. $(x^2 + 9)(x+3)(x-3)$　**15.** $(2x+3)(4x+5)$
17. $(x+2)(x+1)(x-1)$
19. $(0.1t^2 - 0.2)(0.01t^4 + 0.02t^2 + 0.04)$
21. $\left(x^2 + \frac{1}{4}\right)\left(x + \frac{1}{2}\right)\left(x - \frac{1}{2}\right)$　**23.** $(a+3+y)(a+3-y)$
25. $(m+15n)(m-10n)$　**27.** $2y(3x+1)(4x-3)$
29. $(y-11)^2$　**31.** $-7, 2$　**33.** $0, 4.7$　**35.** $-10, 10$
37. $-\frac{5}{2}, 7$　**39.** $0, 5$　**41.** $-2, 6$　**43.** $-11, 5$
45. -5　**47.** Base: 8 ft; height: 5 ft　**49.** 8 ft, 15 ft

Exercise Set R.6, pp. 1001–1002

1. $-\frac{2}{3}$　**3.** $-10, 10$　**5.** $\dfrac{8x}{9y}$　**7.** $\dfrac{t+2}{t+4}$　**9.** $\dfrac{-1}{x+2}$

11. $\dfrac{6}{x}$　**13.** $\dfrac{(a+1)^3(a-1)}{a^3}$　**15.** 1　**17.** $\dfrac{5x+6}{x^2}$

19. $\dfrac{6a+3b-4}{3a-3b}$　**21.** $\dfrac{-3}{5x(x+2)}$　**23.** $\dfrac{(x+1)^2}{x^2(x+2)}$

25. $\dfrac{x+3}{(x+1)^2}$　**27.** $\dfrac{-3x}{2}$　**29.** $\dfrac{-14x-7}{(x+5)(x-4)}$

31. $\dfrac{8x-4}{x^3}$　**33.** $\dfrac{3(x+1)}{(x-7)(4x+3)}$　**35.** $\dfrac{(x+1)^2}{(x-2)(x+4)}$

37. $\dfrac{-1}{x}$　**39.** $\dfrac{6}{5}$　**41.** 1　**43.** $\dfrac{31}{2}$　**45.** $-4, 4$
47. $22\frac{2}{9}$ hr　**49.** Jessica: 45 km/h; Josh: 25 km/h
51. 50

APPENDIXES

Exercise Set A, pp. 1006–1007

1. Mean: 21; median: 18.5; mode: 29　**3.** Mean: 21;
median: 20; mode: 5, 20　**5.** Mean: 5.2; median: 5.7;
mode: 7.4　**7.** Mean: 239.5; median: 234; mode: 234
9. Average: 23.8; median: 15; mode: 1　**11.** Mean: 897.2;
median: 798; mode: none　**13.** Average: \$8.19;
median: \$8.49; mode: \$6.99　**15.** 10 home runs
17. $a = 30, b = 58$

Exercise Set B, pp. 1010–1011

1. $\{3, 4, 5, 6, 7\}$　**3.** $\{41, 43, 45, 47, 49\}$　**5.** $\{-3, 3\}$
7. False　**9.** True　**11.** True　**13.** True　**15.** True
17. False　**19.** $\{c, d, e\}$　**21.** $\{1, 10\}$　**23.** $\varnothing$
25. $\{a, e, i, o, u, q, c, k\}$　**27.** $\{0, 1, 2, 5, 7, 10\}$
29. $\{a, e, i, o, u, m, n, f, g, h\}$　**31.** 🔲　**33.** The set of
integers　**35.** The set of real numbers　**37.** $\varnothing$　**39.** (a) A;
(b) A; (c) A; (d) $\varnothing$　**41.** True

Exercise Set C, pp. 1015–1016

1. $x^2 - 3x + 5 + \dfrac{-12}{x+1}$　**3.** $a + 5 + \dfrac{-4}{a+3}$

5. $x^2 - 5x - 23 + \dfrac{-43}{x-2}$　**7.** $3x^2 - 2x + 2 + \dfrac{-3}{x+3}$

9. $y^2 + 2y + 1 + \dfrac{12}{y-2}$　**11.** $x^4 + 2x^3 + 4x^2 + 8x + 16$

13. $3x^2 + 6x - 3 + \dfrac{2}{x+\frac{1}{3}}$　**15.** 6　**17.** 1　**19.** 54　**21.** 🔲
23. (a) The degree of R must be less than 1, the degree of
$x - r$; (b) Let $x = r$. Then
$$P(r) = (r - r) \cdot Q(r) + R$$
$$= 0 \cdot Q(r) + R$$
$$= R.$$
25. $0; -\frac{7}{2}, \frac{5}{3}, 4$　**27.** 0

Index

INDEX OF APPLICATIONS

Selected Keys of the Scientific Calculator

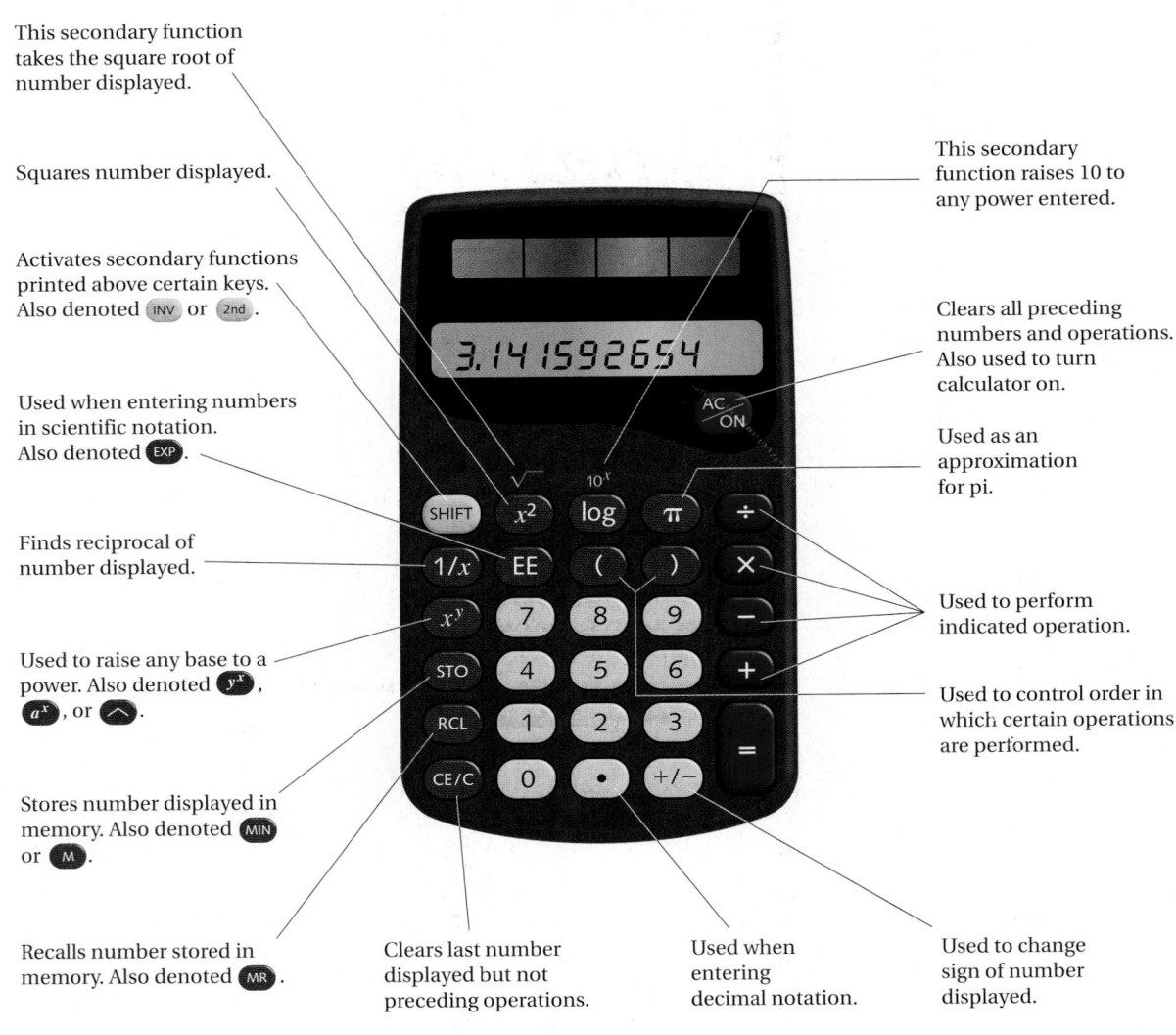

This secondary function takes the square root of number displayed.

Squares number displayed.

Activates secondary functions printed above certain keys. Also denoted INV or 2nd.

Used when entering numbers in scientific notation. Also denoted EXP.

Finds reciprocal of number displayed.

Used to raise any base to a power. Also denoted y^x, a^x, or ⌃.

Stores number displayed in memory. Also denoted MIN or M.

Recalls number stored in memory. Also denoted MR.

This secondary function raises 10 to any power entered.

Clears all preceding numbers and operations. Also used to turn calculator on.

Used as an approximation for pi.

Used to perform indicated operation.

Used to control order in which certain operations are performed.

Clears last number displayed but not preceding operations.

Used when entering decimal notation.

Used to change sign of number displayed.

Selected Keys of the Graphing Calculator*

Controls the values that are used when creating a table.

Determines the portion of the curve(s) shown and the scale of the graph.

Used to enter the equation(s) that is to be graphed.

Controls whether graphs are drawn sequentially or simultaneously and if the window is split.

Activates the secondary functions printed above many keys in blue or green.

Used to delete previously entered characters.

Accesses pre-programmed applications and tutorials.

These keys are similar to those found on a scientific calculator.

Magnifies or reduces a portion of the curve being viewed and can "square" the graph to reduce distortion.

Used to determine certain important values associated with a graph.

Used to display the coordinates of points on a curve.

Used to display x- and y-values in a table.

Used to graph equations that were entered using the 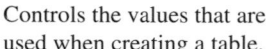 key.

Used to move the cursor and adjust contrast.

Used to fit curves to data.

Used to access a previously named function or equation.

Used to raise a base to a power.

Used to write the variable, x.

Used as a negative sign.

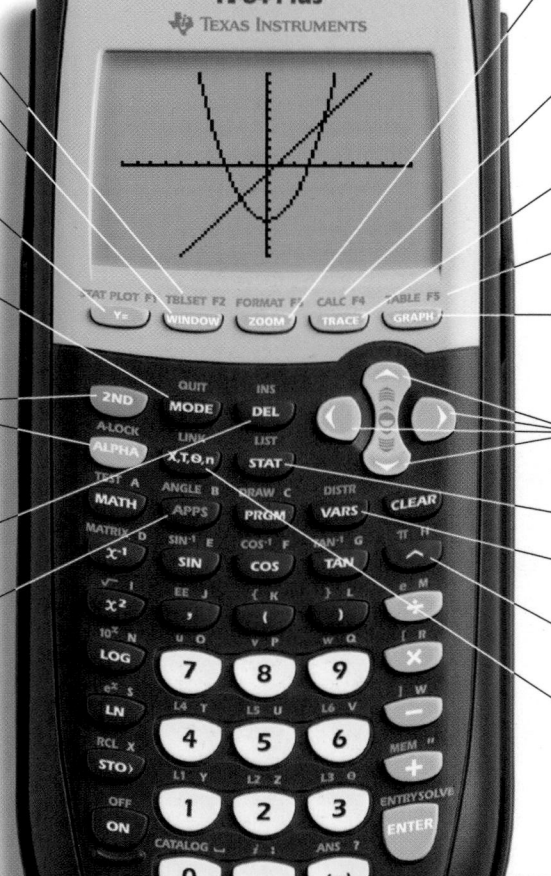

*Key functions and locations are the same for the TI-83 Plus.

Frequently Used Symbols and Formulas

SYMBOLS

$=$	Is equal to
$\approx$	Is approximately equal to
$>$	Is greater than
$<$	Is less than
$\geq$	Is greater than or equal to
$\leq$	Is less than or equal to
$\in$	Is an element of
$\subseteq$	Is a subset of
$|x|$	The absolute value of x
$\{x \mid x...\}$	The set of all x such that $x...$
$-x$	The opposite of x
$\sqrt{x}$	The square root of x
$\sqrt[n]{x}$	The nth root of x
LCM	Least Common Multiple
LCD	Least Common Denominator
π	Pi
i	$\sqrt{-1}$
$f(x)$	f of x, or f at x
$f^{-1}(x)$	f inverse of x
$(f \circ g)(x)$	$f(g(x))$
e	Approximately 2.7
Σ	Summation
$n!$	Factorial notation

FORMULAS

$m = \dfrac{y_2 - y_1}{x_2 - x_1}$	Slope of a line
$y = mx + b$	Slope–intercept form of a linear equation
$y - y_1 = m(x - x_1)$	Point–slope form of a linear equation
$(A + B)(A - B) = A^2 - B^2$	Product of the sum and difference of the same two terms
$\left.\begin{array}{l}(A + B)^2 = A^2 + 2AB + B^2, \\ (A - B)^2 = A^2 - 2AB + B^2\end{array}\right\}$	Square of a binomial
$d = rt$	Formula for distance traveled
$\dfrac{1}{a} \cdot t + \dfrac{1}{b} \cdot t = 1$	Work principle
$s = 16t^2$	Free-fall distance
$y = kx$	Direct variation
$y = \dfrac{k}{x}$	Inverse variation
$x = \dfrac{-b \pm \sqrt{b^2 - 4ac}}{2a}$	Quadratic formula
$P(t) = P_0 e^{kt},\ k > 0$	Exponential growth
$P(t) = P_0 e^{-kt},\ k > 0$	Exponential decay
$d = \sqrt{(x_2 - x_1)^2 + (y_2 - y_1)^2}$	Distance formula
$\dbinom{n}{r} = \dfrac{n!}{(n - r)!\,r!}$	$\dbinom{n}{r}$ notation